N_4-Macrocyclic Metal Complexes

N_4-Macrocyclic Metal Complexes

Edited by

JOSÉ H. ZAGAL

Departamento de Química de los Materiales
Facultad de Química y Biología
Universidad de Santiago de Chile
Santiago
Chile

FETHI BEDIOUI

Laboratoire de Pharmacologie Chimique et Génétique
UMR CNRS No. 8151/U 640 INSERM
Ecole Nationale Supérieure de Chimie de Paris
Paris
France

and

JEAN-POL DODELET

INRS-Énergie, Matériaux et Télécommunications
Varennes, Québec
Canada

Springer

Editors:

José H. Zagal
Departmento de Quimica de los Materiales
Facultad de Quimica y Biologia
Universidad de Santiago de Chile
Casilla 40
Correo 33, Santiago
Chile

Fethi Bedioui,
Laboratoire de Pharmacologie Chimique et Génétique
UMR CNRS no. 8151/U 640 INSERM
Ecole Nationale Supérieure de Chimie de Paris
11 rue Pierre et Marie Curie
75231 Paris cedex 05
France

Jean-Pol Dodelet,
INRS-Énergie Matériaux et Télécommunications
1650 boulevard Lionel Boulet
Varennes,
Québec, Canada J3X 1S2.

Library of Congress Control Number: 2005938667

ISBN 10: 0-387-28429-X
ISBN 13: 978-0387-28429-3

Printed on acid-free paper.

Printed in the United States of America.

9 8 7 6 5 4 3 2 1

springer.com

Contents

3. Oxygen Reduction in PEM Fuel Cell Conditions: Heat-Treated Non-Precious Metal-N_4 Macrocycles and Beyond

Jean-Pol Dodelet

4. Biomimetic NO_x Reductions by Heme Models and Proteins

Emek Blair, Filip Sulc, and Patrick J. Farmer

5. Electroreduction of CO_2 Catalyzed by Metallomacrocyclics

Juan A. Costamagna, Mauricio Isaacs, María J. Aguirre, Galo Ramírez, and Ignacio Azocar

6. Supramolecular Porphyrins as Electrocatalysts

Koiti Araki, Henrique Eisi Toma

7. Electrodes Modified with Monomeric M-N_4 Catalysts for the Detection of Environmentally Important Molecules

Tebello Nyokong

8. Electropolymerized Metalloporphyrin Metallophthalocyanine and Metal Schiff Base Complex Films: Applications to Biomimetic Electrocatalysis and Bioelectroanalysis

Alain Pailleret and Fethi Bedioui

11. Organisation and Photoelectrochemical Reactivity of Water-Soluble Metalloporphyrins at the Liquid/Liquid Interface

David Fermín and Nicolas Eugster

12. Theoretical Insights on the Chemical Reactivity of Metalloporphyrins Using Density Functional Theory

Ilaria Ciofini, Laurent Joubert, Michele Pavone, Vincenzo Barone and Carlo Adamo

13. Organized Multiporphyrinic Assemblies for Photoconduction and Electroconduction

Jean Weiss and Jennifer Wytko

Preface

Metal complexes of N_4-ligands, such as porphyrins and phthalocyanines, are widely studied due to their numerous physico-chemical properties and the great variety of their applications in many fields. For example, metalloporphyrins are used as biomimetic models for studying several biological redox processes, in particular for molecular oxygen transport and catalytic activation to mimic monooxygenase enzymes of the cytochrome P450. They are also well known as efficient catalysts for oxidative degradation of various types of pollutants (organohalides, for example) and residual wastes. The high stability of metallophthalocyanines makes them suitable for applications in various fields such as catalysis (for example, the MEROX process for the sweetening of oils), electrocatalysis (air batteries and fuel cells), dye stuffs, coloring for plastics and metal surfaces, sensor applications, chromatographic detectors, photoconducting agents, etc. Both these families of complexes are now also used for photobiology and photodynamic cancer therapy, electrochemical removal of organic wastes, display devices, electrochromism, electroluminescence, molecular metals, and nonlinear optical applications. Additionally, the intrinsic diversity and selectivity of axial ligation of these macrocycles confer to them formidable challenging potential uses as electrochemical-sensing devices for several fields of application in analytical, electro-, and spectrophotochemistry.

The rich and reversible redox chemistry of metalloporphyrins and metallophthalocyanines is the key factor that allows them to serve as mediators in many electron transfer reactions. Since the 1970s and more importantly in recent years, numerous reports have demonstrated that these complexes can be successfully used as electrocatalysts for a great variety of electrochemical reactions. One can cite some examples such as the oxidation of dopamine, thiols, H_2S, HS^-, reduced glutathione, L-cysteine, coenzyme A, penicillin, oxalic acid, NADH, hydroxylamine, hydrazine, nitrite, nitric oxide, cyanide, organic peroxides, hydrogen peroxide, propylgallate, ascorbic acid, hydroquinone, catechol, phenols, chlorophenols, sulfite, etc. and the reduction of molecular oxygen, hydrogen peroxide, carbon dioxide, L-cystine, disulfides, thionylchloride, etc. It is remarkable that a large number of the studied reactions involve significantly relevant biological compounds and the list keeps increasing as more publications appear in the literature reporting on new reactions and electrocatalytic processes. Concerning the fields of photoelectrochemistry and photocatalysis, although the list of the studied reactions and processes involving photoassistance is less abundant and fewer systems have been studied, this area of investigation is experiencing intense development due to the implication of these compounds in photobiology and nanosized semiconductor materials.

Nowadays, the double desire to mimic enzymatic or natural systems and to develop new complex structures that do combine a well-defined topology and a marked chemical flexibility allowing both the finetuning of the properties of the electron transfer reactions and the expansion of the supramolecular architectures, is incontestably leading to an active area of research devoted to the concept of "design of intelligent molecular material electrodes" with predetermined reactivity. To do so, highly elaborate synthesis routes are developed to design chemically modified metalloporphyrins and metallophthalocyanines that can then be simply strongly adsorbed on conventional materials, electropolymerized on conducting substrates, incorporated into hybrid organic/inorganic gel or solid matrix to produce catalytic electrodes with long-term stabilities, for expanded practical analytical applications.

Thus, it is clear that the numerous and varied possibilities of uses and of applications ensure that porphyrin and phthalocyanine compounds will remain of vital importance for many years to come and that the related fields of investigation are expected to have significant ramifications. The publication since 1997 of *The Journal of Porphyrins and Phthalocyanines*, an international journal of significant impact factor entirely devoted to these molecular materials, is a significant indicator. Also the Society of Porphyrins and Phthalocyanines provides a forum for interaction among researchers around the world.

The main objective of this monograph is to provide a general updated view of the vast applications of these materials in electrochemistry by focusing on a few significant topics and examples. It is also aimed at offering future projections and opening new fields of research and investigations.

J.H. Zagal
F. Bedioui
J.P. Dodelet

Contributors

Carlo Adamo, Ecole Nationale Supérieure de Chimie de Paris, Laboratoire d'Electrochimie et Chimie Analytique, UMR 7575, 11 rue P. et M. Curie, F-75231 Paris cedex 05, France

María J. Aguirre, Departamento de Química de los Materiales, Facultad de Química y Biología, Universidad de Santiago de Chile, Casilla 40, Correo 33, Santiago, Chile

Koiti Araki, Institute of Chemistry, University of São Paulo, Av. Prof. Lineu Prestes 748, Butantã, São Paulo 05508-900, Brazil

Ignacio Azocar, Departamento de Química de los Materiales, Facultad de Química y Biología, Universidad de Santiago de Chile, Casilla 40, Correo 33, Santiago, Chile

Vincenzo Barone, Laboratorio di Struttura e Dinamica Molecolare, Dipartimento di Chimica, Complesso Universitario Monte Sant'Angelo, Via Cintia, I-80126 Napoli, Italy

Fethi Bedioui, Laboratoire de Pharmacologie Chimique et Génétique, UMR CNRS no. 8151/U 640 INSERM, Ecole Nationale Supérieure de Chimie de Paris, 11 rue Pierre et Marie Curie, 75231 Paris cedex 05, France

Emek Blair, Department of Chemistry, University of California, Irvine, CA, USA

Roman Boulatov, Department of Chemistry, University of Illinois, 600 South Mathews Ave, Urbana, IL, USA

Marcelo M. Campos Vallette, Department of Chemistry, Faculty of Sciences, Universidad de Chile, P.O. Box 653, Santiago, Chile

Ilaria Ciofini, Ecole Nationale Supérieure de Chimie de Paris, Laboratoire d'Electrochimie et Chimie Analytique, UMR 7575, 11 rue P. et M. Curie, F-75231 Paris cedex 05, France

Juan A. Costamagna, Departamento de Química de los Materiales, Facultad de Química y Biología, Universidad de Santiago de Chile, Casilla 40, Correo 33, Santiago, Chile

Jean-Pol Dodelet, INRS-Énergie, Matériaux et Télécommunications, 1650 boulevard Lionel Boulet, Varennes, Québec, Canada J3X 1S2

Francis D'Souza, Department of Chemistry, Wichita State University, 1845 Fairmount, Wichita, KS 67260-0051, USA

Nicolas Eugster, Laboratoire d'Electrochimie Physique et Analytique, Institut de Sciences et Ingénierie Chimiques, Ecole Polytechnique Fédérale de Lausanne, Lausanne CH-1015, Switzerland

Patrick J. Farmer, Department of Chemistry, University of California, Irvine, CA, USA

David J. Fermín, Departement für Chemie und Biochemie, Universität Bern, Freiestrasse 3, Bern CH-3012, Switzerland

Mauricio A. Isaacs, Departamento de Química, Facultad de Ciencias, Universidad de Chile, Casilla 653, Santiago, Chile

Laurent Joubert, Ecole Nationale Supérieure de Chimie de Paris, Laboratoire d'Electrochimie et Chimie Analytique, UMR 7575, 11 rue P. et M. Curie, F-75231 Paris cedex 05, France

Karl M. Kadish, Department of Chemistry, University of Houston, Houston, TX 77005-2004, USA

Tebello Nyokong, Department of Chemistry, Rhodes University, P.O. Box 94, Grahamstown, South Africa

Maritza A. Páez, Departmento de Quimica de los Materiales, Facultad de Quimica y Biologia, Universidad de Santiago de Chile, Casilla 40, Correo 33, Santiago, Chile

Alain Pailleret, Laboratoire Interfaces et Systèmes Electrochimiques, UPR CNRS no. 15, Université Pierre et Marie Curie-Paris VI, 4 place Jussieu, 75252 Paris cedex 05, France

Michele Pavone, Laboratorio di Struttura e Dinamica Molecolare, Dipartimento di Chimica, Complesso Universitario Monte Sant'Angelo, Via Cintia, I-80126 Napoli, Italy

Galo Ramírez, Departamento de Química de los Materiales, Facultad de Química y Biología, Universidad de Santiago de Chile, Casilla 40, Correo 33, Santiago, Chile

Derck Schlettwein, Institute of Applied Physics, Justus-Liebig-University Gießen, Heinrich-Buff-Ring 16, D-35392 Gießen, Germany

J. Francisco Silva, Departmento de Quimica de los Materiales, Facultad de Quimica y Biologia, Universidad de Santiago de Chile, Casilla 40, Correo 33, Santiago, Chile

Filip Sulc, Department of Chemistry, University of California, Irvine, CA, USA

Henrique E. Toma, Institute of Chemistry, University of São Paulo, Av. Prof. Lineu Prestes 748, Butantã, São Paulo 05508-900, Brazil

Jean Weiss, Laboratoire de Chimie des Ligands à Architecture Contrôlée, Institut de Chimie, Université Louis Pasteur, 4 rue Blaise Pascal, 67070 Strasbourg, France

Jennifer Wytko, Laboratoire de Chimie des Ligands ‘a Architecture Contrôlée, Institut de Chimie, Universit'e Louis Pasteur, 4 rue Blaise Pascal, 67070 Strasbourg, France

José H. Zagal, Departmento de Quimica de los Materiales, Facultad de Quimica y Biologia, Universidad de Santiago de Chile, Casilla 40, Correo 33, Santiago, Chile

1

Billion-Year-Old Oxygen Cathode that Actually Works: Respiratory Oxygen Reduction and its Biomimetic Analogs

Roman Boulatov

1. The Basic Concepts of Energy Metabolism

Life exists only through a constant dissipation of energy. This energy is extracted from the environment, either as sunlight (photosynthesis) or as food (reduced organic matter, such as glucose or H_2). Two nonphotosynthetic forms of energy metabolism are fermentation and respiration[1]. In fermentation complex organic molecules, such as glucose, are broken down exergonically to simpler products (reactions 1.1), such as lactate (in muscle cells) or ethanol (in baker's yeast). Although fermentation proceeds by a series of electron transfer steps, it does not consume any external oxidants. The organic compounds of the food undergo disproportionation. In contrast, respiration involves the oxidation of food by an environmental oxidant (reaction 1.2). Mechanistically, food is converted into reduced respiratory electron carriers, such as NADH and $FADH_2$ (reaction 1.3), which enter the respiration cycle (reaction 1.4). Respiration proceeds by enzyme-catalyzed electron transfer between respiratory electron carriers, starting with the strongest reductants, NADH and $FADH_2$ and ending with the weakest reductants, quinols and cytochrome c[2]. Oxidation of the latter carriers in the final redox step of respiration requires an environmental (or terminal) oxidant.

$$\left.\begin{array}{lll} C_6H_{12}O_6 \rightleftharpoons 2CH_3CH(OH)CO_2H & \Delta G = -6\ \text{kcal mol}^{-1} & (1.1a) \\ C_6H_{12}O_6 \rightleftharpoons 2CO_2 + 2CH_3CH_2OH & \Delta G = -54\ \text{kcal mol}^{-1} & (1.1b) \end{array}\right\}\ \text{fermentation}$$

Roman Boulatov • Department of Chemistry, University of Illinois, 600 South Mathews Ave., Urbana, IL.

N_4-Macrocyclic Metal Complexes, edited by José H. Zagal, Fethi Bedioui and Jean-Pol Dodelet. Springer Science+Business Media, Inc., New York, 2006.

$$C_6H_{12}O_6 + 6O_2 \rightleftharpoons 6CO_2 + 6H_2O$$
$$\Delta G = -600\ \text{kcal mol}^{-1} \quad (1.2)\ \text{aerobic respiration}$$

$$C_6H_{12}O_6 + 6H_2O + 10NAD^{+}2FAD$$
$$\longrightarrow 6CO_2 + 10H^{+} + 10NADH + 2FADH_2 \quad (1.3)\ \text{glycolysis}$$

$$2NADH + 2H^{+} + O_2 \longrightarrow 2NAD^{+} + 2H_2O \quad (1.4)\ \text{respiration}$$

The free energy available from photosynthesis, fermentation, or respiration is captured in the form of transmembrane electrochemical gradients, generally of protons[3]. The gradient-generating system is embedded in a closed, relatively proton-impermeable membrane. The membrane-enclosed space (such as mitochondrial matrix, chloroplast's thylakoid space, or prokaryotic cytoplasm) is maintained at a pH and electrostatic potential different from that of the outside environment by membrane-embedded enzymes of the respiratory (or photosynthetic) cycle. Some of these enzymes use the free energy of exergonic chemical reactions they catalyze to actively translocate protons from the less acidic compartment to the more acidic one (proton pumps). Others create the transmembrane gradient by carrying out proton-consuming and proton-releasing reactions at the opposite sites of the membrane because of the appropriate location of the corresponding catalytic domains. The membrane-embedded ATP synthase converts the exergonic movement of protons down this electrochemical gradient into the free energy of the pyrophosphate bond in ATP (Figure 1.1)[2–5].

Depending on the environmental niche occupied by an organism, its respiration can be based on molecular oxygen, nitrate, fumarate, sulfate, carbon dioxide, etc. as the terminal oxidant[6]. Among these bioavailable oxidants, O_2 is by far the strongest, enabling an aerobe to extract the largest amount of energy from a given amount of food (Figure 1.2). Other advantages of O_2 as the terminal oxidant include its high abundance on the modern Earth, high permeation rate across biological membranes,* and the nontoxic reduction product, H_2O†. As a result, all multicellular organisms are obligatory aerobes, whereas fermentation and anaerobic respiration are limited to prokaryotes‡ and certain single-celled eukaryotes[1,6].

Although O_2 is a powerful four-electron/four-proton ($4e/4H^+$) oxidant, it is kinetically inert under ambient conditions. The inertness of O_2 arises from it

*A lipid membrane provides no barrier for O_2 diffusion, its permeability coefficient being ~40 cm s^{-1}, which is ~20% higher than the O_2 permeability coefficient of an H_2O layer of the same thickness[7]. Inside cells, most of O_2 flux occurs not by passive diffusion of O_2, which is insufficiently soluble in H_2O to provide required fluxes, but by diffusion of oxymyoglobin[8], which unloads O_2 at the point of O_2 consumption (e.g., at the outer mitochondrial membrane).

†However, partially reduced oxygen species, such as O_2^-, HO_2, H_2O_2, and $^\bullet OH$, are toxic.

‡Prokaryotes are single-celled organisms whose cell lacks a well-defined, membrane-enclosed nucleus and other organelles. Eukaryotes are cells with a distinct nucleus and cytoplasm.

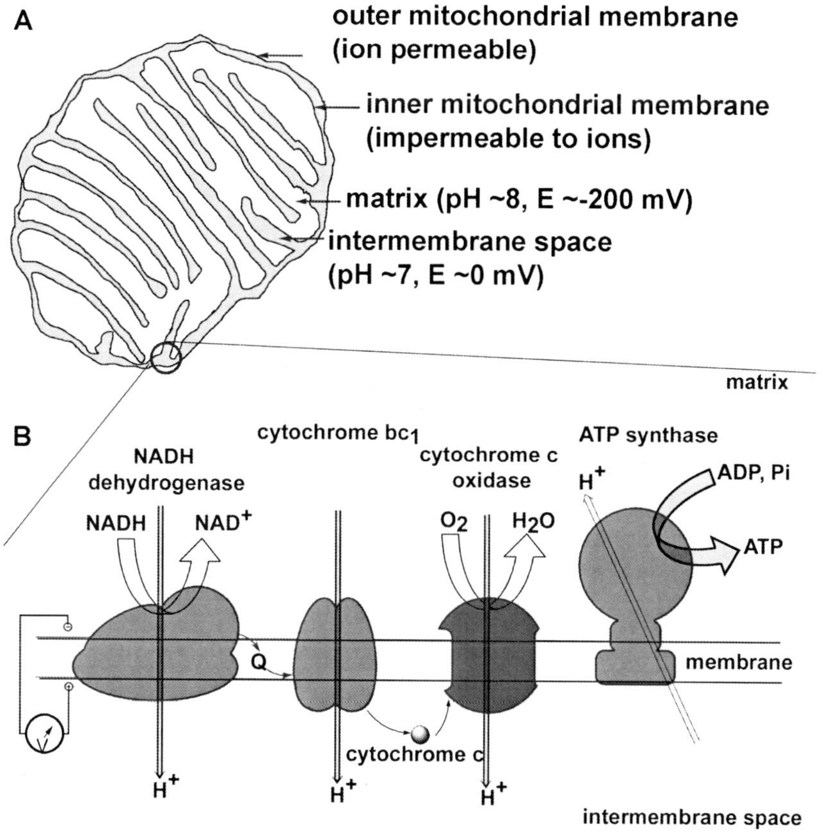

Figure 1.1. (A) A cross-sectional view of the mitochondrion indicating its four distinct components. (B) A schematic view of the mammalian respiratory electron transfer chain embedded in the inner mitochondrial membrane. Exergonic electron flow (narrow arrows) from NADH to quinone (Q), from quinol to ferricytochrome *c*, and finally, from ferrocytochrome *c* to O_2 are catalyzed by NADH dehydrogenase, cytochrome bc_1, and cytochrome *c* oxidase, respectively. Although thermodynamically favorable, electron transfer between the electron carriers bypassing the respiratory enzymes is too slow. The respiratory enzymes utilize a fraction of the free energy of these redox processes for endergonic translocation of H^+ from the basic, negatively charged matrix, to the more acidic, positively charged intermembrane space (gray block arrows)[2,3]. Protons spontaneously flow from the intermembrane space to the matrix, down the electrochemical H^+ gradient, through ATP synthase, powering the synthesis of ATP from ADP and inorganic phosphate (P_i). White and gray arrows depict exergonic and endergonic processes, respectively.

having a low affinity for H^+ (reaction 1.5), other electrophiles and H atoms[10] (reaction 1.6), being a poor 1e oxidant particularly in neutral aqueous media (reaction 1.7)[§], and having a triplet electronic ground state. The latter precludes

§ Superoxide is a moderately strong base (pK_a of the conjugate acid, HO_2, is ~4.2) so that at pH < 4.5 the formal potential of $1e^-$ reduction of O_2 to O_2^- depends on the concentration of H^+ as: $E^{o\prime} = E^o + RT/F \ln(1+[H^+]/K(HO_2))$.

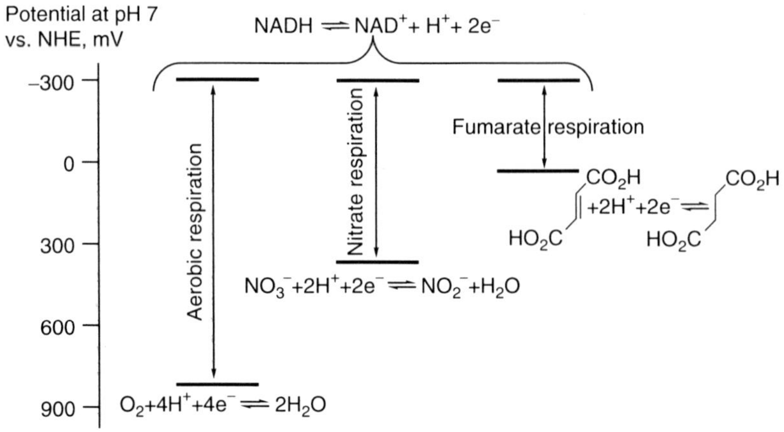

Figure 1.2. Comparison of the free energy available in three common forms of respiration: aerobic, based on nitrate, and on fumarate.

direct reaction between O_2 and most organic matter, which is in a singlet ground state. This inertness of O_2 allows for the existence of reduced organic matter in the highly oxidizing atmosphere of the modern Earth. However, this inertness also means that reduction of O_2 to H_2O at a rate and electrochemical potentials that are useful for energy metabolism requires catalysis.

$$O_2 + H^+ \rightleftharpoons HO_2^+ \quad \Delta G^0 \sim 95\,\text{kcal mol}^{-1}, \quad K \sim 10^{-70}(\text{est.})^{**} \tag{1.5}$$

$$O_2 + {}^1\!/\!{}_2H_2 \rightleftharpoons HO_2 \quad \Delta G^0 = 3\,\text{kcal mol}^{-1}, \quad K = 6 \times 10^{-3} \tag{1.6}$$

$$O_2 + e^- \rightleftharpoons O_2^- \quad \Delta G^0 = 8\,\text{kcal mol}^{-1}, \quad K = 3 \times 10^{-6}, \quad \Delta E^{0'} = -0.33\,\text{V(vs.NHE)}^{\dagger\dagger} \tag{1.7}$$

The appearance of photosynthesis and the subsequent increase in the concentration of O_2 in the Earth's atmosphere ~2 billion years ago[11] provided the environmental pressure for the evolution of enzymes to catalyze the $4e/4H^+$ reduction of O_2 to H_2O as an energy source. All known enzymes that perform this task, called terminal oxidases[12–22], are membrane-bound proteins. Two distinct classes of terminal oxidases are known. The better-studied superfamily of heme/Cu terminal oxidases[12–21] are distributed throughout all three domains of life (Bacteria, Archaea, and Eukarya) and contain a binuclear porphyrinatoiron(heme)–Cu site at which O_2 is reduced. The much less studied and smaller group of cytochrome *bd* terminal oxidases[22] is widely distributed in Gram-negative heterotrophs only‡‡. Cytochromes *bd* contain a binuclear heme–

** The thermodynamic values are referenced to the standard states of 1 atm partial pressure or 1 M aqueous solution at 298 K for gases, and H^+, HO_2^+, and HO_2, respectively.

†† The values are for neutral (pH 7) solution in equilibrium with 1 atm partial pressure of O_2.

‡‡ Gram-negative bacteria (e.g., *E. coli*) are microorganisms that have two cell membranes, separated by a periplasmic space; Gram-positive bacteria (e.g., *Streptococci*) have only one membrane. Heterotrophic organisms require complex organic compounds for metabolism as opposed to being able to utilize CO_2 as the only source of carbon.

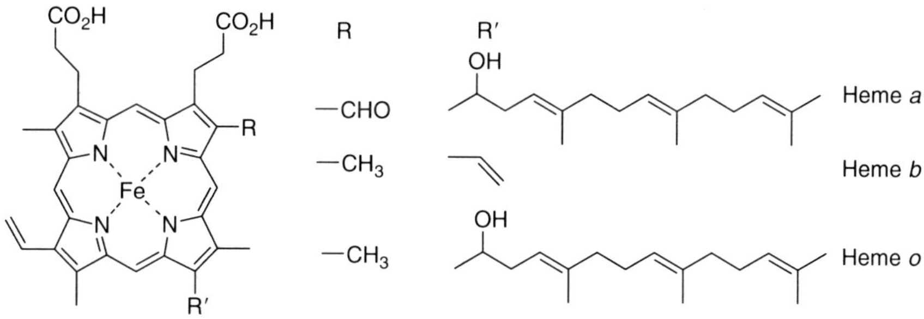

Figure 1.3. Chemical structures of some common hemes.

heme O_2 reduction site. Chemical structures of biologically significant hemes are illustrated in Figure 1.3.

2. Biological Catalysis of Respiratory Oxygen Reduction

2.1. Heme/Cu Terminal Oxidases (HCOs)

The superfamily comprises three classes: cytochrome *c* oxidases (C*c*Os)[12–20], quinol oxidases, and cytochrome *cbb*$_3$ oxidases[20,21]. C*c*Os are the only HCOs present in eukaryotes. Enzymes of all classes contain a binuclear heme–Cu catalytic site and all act as proton pumps. The water-soluble electron carrier, ferrocytochrome c, is the physiological electron donor for C*c*Os and cytochrome *cbb*$_3$ oxidases. Certain C*c*Os can also oxidize specific high-potential iron–sulfur proteins (HiPIPs)[23]. Quinol oxidases catalyze a $2e/2H^+$ oxidation of various quinols to quinones as the source of electrons for O_2 reduction.

There is substantial structural and functional homology among C*c*Os from different organisms and between C*c*Os and quinol oxidases (Figure 1.4)[24,25]. The vast majority of HCOs have at least two subunits: subunit I, containing the catalytic domain, and subunit II. These are the only subunits required for O_2 reduction and proton pumping. Mammalian C*c*Os contain as many as 11 additional subunits, of uncertain function[15,18]. HCOs reduce O_2 at a bimetallic heme/Cu site (Figure 1.5)[26], whose Fe–Cu distance varies around ~5 Å, depending on the exogenous ligation of the metal ions. The ~5-Å distance is appropriate for a bridging peroxide ligand, but whether such an intermediate ever forms at the heme/Cu site is uncertain. Most known C*c*Os and quinol oxidases appear to undergo a post-translational modification, linking one of the Cu-ligating imidazoles to a phenol residue of a tyrosine (Figure 1.5). The existence of this link is often taken to indicate that a phenoxyl radical is formed during initial enzyme turnovers[27].

All HCOs also have a six-coordinate heme in the same subunit as the catalytic site. In addition all cytochrome *c* oxidases have a binuclear Cu_A site and some also have a six-coordinate heme in the noncatalytic subunit (subunit II)[20]. These centers are absent in quinol oxidases. The six-coordinate heme(s) and Cu_A sites are responsible for electron-relay between the external electron

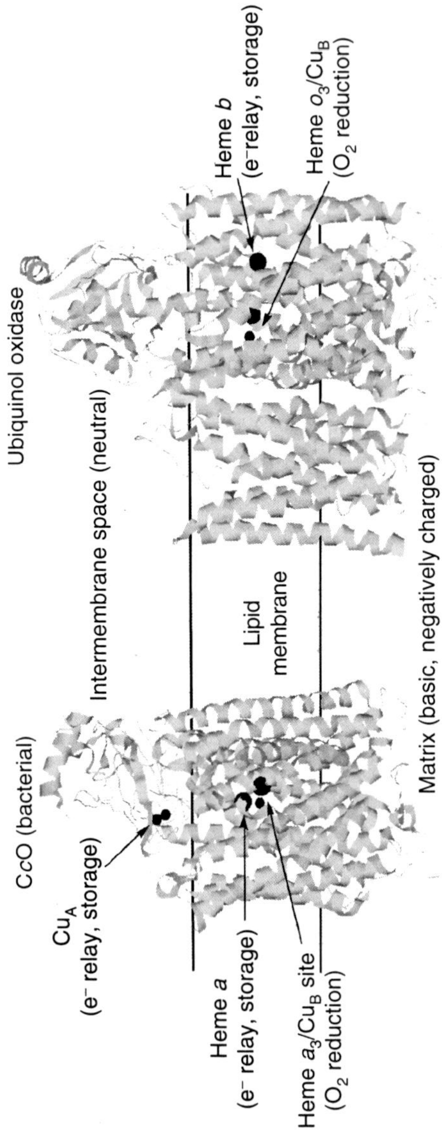

Figure 1.4. Position of the redox active metal sites relative to the lipid membrane in two-subunit bacterial C*c*O (*P. denitrificans*)[24] and ubiquinol oxidase (*E. coli*)[25] from solid-state X-ray structures. The subscript 3 in heme a_3 or heme o_3 signifies that O_2 binds to these sites.

donors (ferrocytochrome *c*, reduced HiPIP, or quinols) and the catalytic heme/Cu site (Figure 1.6)[26]. They may also be involved in the storage of external reducing equivalents during enzymatic turnover and/or controlling the redox potentials of the heme/Cu site (see below). The presence of additional electron-relay sites in C*c*Os relative to quinol oxidases is due to the different physicochemical

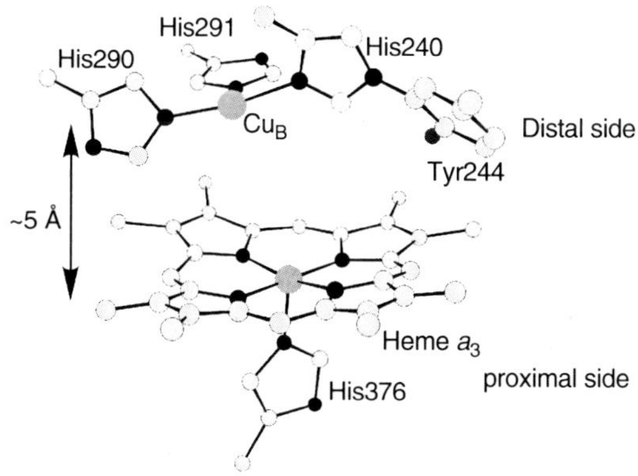

Figure 1.5. The catalytic O_2 reduction site of CcO[26]. The numbering of the residues is from the crystal structure of bovine heart CcO.

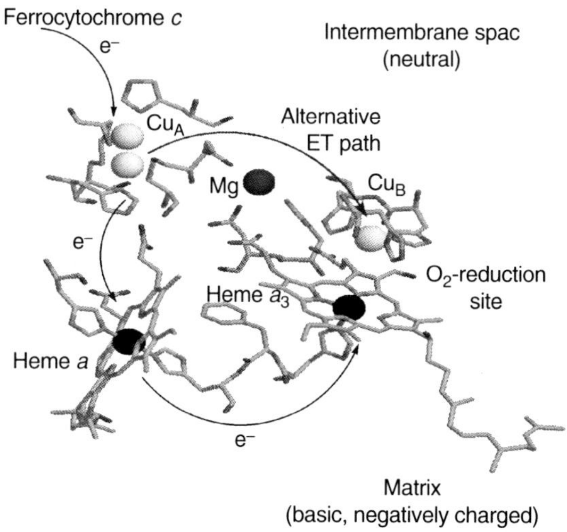

Figure 1.6. Electron transfer paths in mammalian CcO[26]. The six-coordinate heme *a* and the catalytic heme–Cu site are located at approximately the same depth within the dielectric.

properties of the electron donors utilized by these two classes of HCOs. CcOs accept electrons from water-soluble ferrocytochrome *c*. The corresponding docking site is located outside of the membrane and relatively far from the six-coordinate heme *a* (or the heme/Cu site). Hence, the need for an additional electron-relay site. In contrast, quinols are localized in the lipid bilayer and dock much closer to the six-coordinate heme *b* site[28].

With its four redox centers (Cu_A, heme *a*, heme a_3 and Cu_B) mammalian CcO can exist in one of four redox states: fully oxidized and singly reduced, which are aerobically stable; and mixed-valence (2e-reduced), $3e^-$, and fully reduced, all of which reduce O_2 to the redox level of water within $<200\ \mu s$ of O_2 binding[13,18,19]. This rate is substantially faster than the average rate of electron transfer within the mammalian respiratory chain under normal physiological conditions, which is 1 electron per 5–20 ms[1]. A comparison of these two numbers suggests that: (a) the major redox state of CcO that binds O_2 *in vivo* is mixed valence, whereas the $3e^-$ and fully reduced enzymes may be relevant only under hypoxic (low O_2) conditions; and (b) mixed-valence CcO reduces O_2 by four electrons without receiving any additional reducing equivalents from ferrocytochrome *c*. Thus, in the catalytic cycle of CcO, oxidation of four ferrocytochromes and the 4e reduction of O_2 are efficiently decoupled, allowing for complete reduction of O_2 to occur much faster than the rate of electron flow within the respiratory chain. Conceivably, this minimizes the chance of CcO "getting stuck" with a partially reduced dioxygen species (such as peroxide). This cytotoxic species could be released into the medium while the enzyme waits for an arrival of an electron from ferrocytochrome *c*.

The decoupling is accomplished by at least two routes. First, the redox potentials of the catalytic heme/Cu site are strongly dependent on the redox states of heme *a* and the Cu_A cofactor[12]. Whereas the heme/Cu site in 1e-reduced CcO (compound E[19], Figure 1.7) is in the aerobically stable Fe^{III}/Cu^{II} state, it becomes fully reduced (Fe^{II}/Cu^{I}) in mixed-valence (2e-reduced) CcO and thus binds O_2. Second, temporary oxidation of CcO by two equivalents above its resting, fully-oxidized, state, provides all four electrons for rapid reduction of O_2 to "oxides" ($1e^-$ each comes from $Cu^{I}_{B} \rightarrow Cu^{II}_{B}$ and TyrOH→TyrO$^{\bullet}$[19,27,29] and $2e^-$ come from $Fe^{II}_{a3} \rightarrow Fe^{IV}_{a3}$, Figure 1.7). The resulting P_M intermediate (Figure 1.7) is stable on the timescale of the electron flow between the components of the respiratory chain[19].

Thus, the heme/Cu site of HCOs appears to be optimized to reduce O_2 under kinetically limiting electron flux by converting bound O_2 to oxides 20–100 times faster than oxidizing a single molecule of ferrocytochrome *c* (Table 1.1). This rapid reduction minimizes the lifetime of intermediates containing partially reduced oxygen (such as peroxide), whose accidental release from the catalytic site would be cytotoxic. However, such rapid reduction of the O–O bond also results in the heme a_3/O_2 adduct (compound A) being the only spectroscopically observed intermediates of CcO containing an O–O bond.

The lack of spectroscopically observable intermediates between compounds A and P_M severely complicates attempts to understand the atomic mechanism of O–O bond reduction by studying the catalytic cycle spectroscopically. From computational studies the conversion of compound A to compound P_M was proposed to proceed via a ferric-hydroperoxo intermediate (Figure 1.7). This intermediate is generated by H atom transfer from Cu_B-bound H_2O to the distal atom of oxyheme[30,31]. The hydroperoxide, which may or may not interact strongly with Cu^{II}_{B}, is reduced to P_M by another H atom transfer, from Tyr244 (Figure 1.5), via a two-molecule H_2O bridge. The calculations indicate that the hydroperoxo intermediate is generated slower than it is consumed, making it spectroscopically unobservable.

The dual physiological role of HCOs determines the energetics of O_2 reduction by these enzymes (reactions 1.8–1.13, Figure 1.8). By clearing the respiratory electron transfer chain from low-potential (weakly reducing) electron CcOs

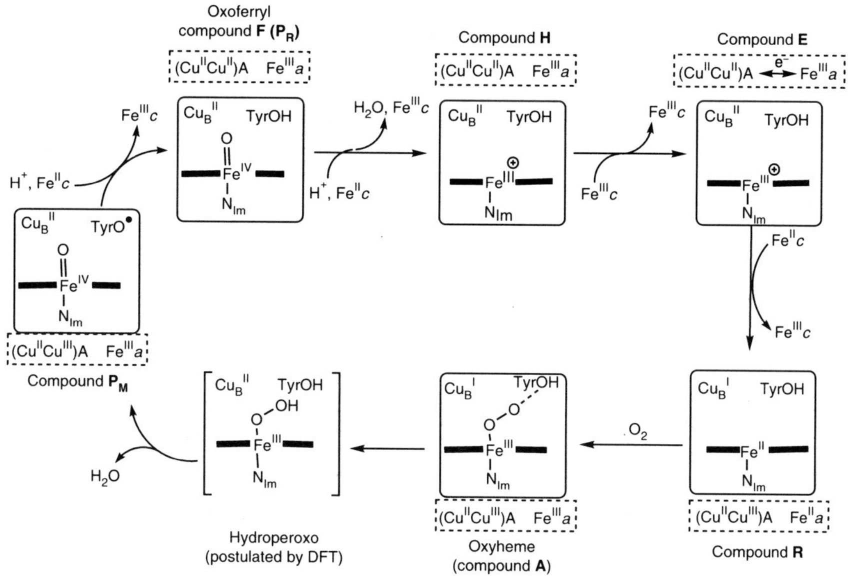

Figure 1.7. A plausible sequence of steps for O_2 reduction by mixed-valence C*c*O based on single-turnover spectroscopic studies of C*c*O and DFT calculations. Structures in the square frame depict the catalytic site (Figure 1.5); imidazole ligation of Cu_B is omitted for clarity; exogenous ligation of Cu_B is uncertain and is not shown. The rectangular dashed frame signifies the other redox-cofactors. Fe^{II}_c and Fe^{III}_c are ferro- and ferricytochrome *c*, respectively. Compound H is an "activated" analog of the resting-state C*c*O (compound O)[19]. The external reducing equivalent in singly reduced C*c*O (compound E) may be localized mainly on Cu_B[19] The heme a_3/O_2 adduct (oxyheme or compound A) is often considered as ferriheme/superoxide, $Fe^{III}_{a3}–(O_2^-)$, complex.

allow a continuous electron flow from NADH to quinones to ferricytochrome *c*. The redox potential of the last electron carrier, cytochrome *c* (~250 mV at pH 7), determines the overall potential drop (~550 mV) available for the NADH dehydrogenase and cytochrome *bc* components of the respiratory chain. The difference between the redox potentials of ferri-/ferrocytochrome *c* (~250 mV, reaction 1.8) and O_2/H_2O (~800 mV at pH 7, reaction 1.12, Figure 1.8) is utilized by C*c*O to increase the electrochemical potential gradient across the inner mitochondrial membrane (the protonmotive force, Figure 1.1).

This increase is brought about by two mechanisms. First, the enzyme draws four protons for the reduction of O_2 from the basic, negatively charged site of the membrane (N-side or matrix). Four electrons come from the opposite site (P-side or intermembrane space, Figure 1.1). The annihilation of these opposite charges at the O_2 reduction site (reactions 1.9 and 1.10) is equivalent to the translocation of four charges across the membrane against the electrostatic potential. This consumes ~220 mV (the value of the transmembrane gradient) of the free energy of reaction 1.12[19]. C*c*O expends another ~220 mV by physically moving four protons from the N-side (matrix) to the P-side (intermembrane space) of the

Table 1.1. Summary of Properties of Mammalian C*c*O

Catalyzed reaction[a]	$O_2 + 4(Fe^{II}c)_{IMS} + 8H^+_{matrix} \rightleftharpoons 2H_2O + 4(Fe^{III}c)_{IMS} + 4H^+_{IMS}$
Driving force	100 mV
Turnover frequency[b]	10–100 s^{-1} [19]
Power output	0.4–4 MW mol^{-1} or 2–20 W g^{-1} [c]
Energy-tranducing efficiency	80%
Lifetime	Days[13]
Selectivity[d]	>99%
Potential of the reductant	250 mV
Potential of the catalytic site in compound R[e]	> 350 mV[12,19]
Rates of intramolecular redox equilibration	$2 \times 10^4\ s^{-1}$ (Cu_A/heme *a*); > $3 \times 10^4\ s^{-1}$ (heme *a*/heme a_3)[f 12,18,70]
O_2 affinity, K_d	0.3 mM (ferroheme a_3); 8 mM (Cu^I_B)[18]
Inhibitors	Strong: CN^-, N_3^-; weak: CO[12,83]

[a] IMS: intermembrane space (Figure 1.1), Fe_c: cytochrome *c*.
[b] At physiologically relevant potentials; with $Ru(NH_3)_6^{2+}$ in air-saturated buffers, turnover frequency of up to 500 s^{-1} was observed[13].
[c] Molecular weight of C*c*Os is up to 200 kDa.
[d] Fraction of the O_2 flux converted to H_2O (vis-à-vis partially reduced oxygen species).
[e] The heme/Cu potential depends both on the site's protonation state[70] and on the redox states of heme a and Cu_A[12]; cited is "the best guess" potential in the mixed-valence intermediate ready to bind O_2.
[f] During turnover, the reduction rate of the heme/Cu site is limited by H^+ flow[70].

mitochondrion, by an unknown mechanism[14,16–19]. As a result, out of ~550 mV of the potential difference between the electron donor (ferrocytochrome *c*) and the electron acceptor (O_2), C*c*O captures ~450 mV (>80%) in a form that can be directly utilized by the cell to satisfy its energy-dissipating requirements. Quinol oxidases utilize a stronger reductant (e.g., ubiquinol, ~50 mV vs. ferrocytochrome *c*, ~250 mV)[13] and operate against a lower electrochemical gradient (~180 mV)[1] than C*c*Os do, so that their energy-transducing efficiency is correspondingly lower (~45%).

Cytochrome *cbb*$_3$ oxidases comprise the third class of enzymes belonging to the HCO superfamily[20,21]. Like other HCOs, cytochromes *cbb*$_3$ catalyze reduction of O_2 at a bimetallic heme/Cu site, contain an additional six-coordinate heme in the catalytic subunit, and act as proton pumps. Although both C*c*Os and cytochromes *cbb*$_3$ utilize cytochrome *c* as the physiological reductant, structurally and functionally they differ more than C*c*Os and quinol oxidases do. Unlike C*c*Os, which are distributed throughout all three domains of life, cytochromes *cbb*$_3$ are found almost exclusively in Proteobacteria (a group of Gram-negative

$$(Fe^{II}{}_{c})_{IMS} \rightleftharpoons (Fe^{III}{}_{c})_{IMS} + e^{-}{}_{IMS} \qquad E = -250 \text{ mV} \tag{1.8}$$

$$H^{+}{}_{matrix} \rightleftharpoons H^{+}{}_{heme/Cu} \qquad E = -60 \text{ mV} \tag{1.9}$$

$$e^{-}{}_{IMS} \rightleftharpoons e^{-}{}_{heme/Cu} \qquad E = -140 \text{ mV} \tag{1.10}$$

$$H^{+}{}_{matrix} \rightleftharpoons H^{+}{}_{IMS} \quad \text{Proton pump} \qquad E = -220 \text{ mV} \tag{1.11}$$

$$(O_2 + 4e^{-} + 4H^{+})_{heme/Cu} \rightleftharpoons 2H_2O \qquad E = 800 \text{ mV} \tag{1.12}$$

$$O_2 + 4(Fe^{II}{}_{c})^{IMS} + 8H^{+}{}_{matrix} \rightleftharpoons 2H_2O + 4(Fe^{III}{}_{c})^{IMS} + 4H^{+}{}_{IMS} \qquad E = 100 \text{ mV} \tag{1.13}$$

Figure 1.8. Energetics of O_2 reduction by mammalian C*c*O. Subscripts signify the location of charged species in the transmembrane electrochemical gradient: the intermembrane space (IMS); the heme/Cu site or the matrix (Figure 1.1). The potentials are approximate and referenced to the normal hydrogen electrode at pH 7.

bacteria, see footnote ‡‡ on page 4)[21]. Cytochromes *cbb*$_3$ appear to have evolved for respiration under microaerobic conditions. This is suggested by the very high O_2 affinity of their heme/Cu site ($K_M \sim 7$ nM[20] vs. 0.1–1 μM for C*c*Os and ubiquinol oxidases[18]; compare to the O_2 affinity of mammalian O_2 carrier, myoglobin: $K_d \sim 0.5$–2 μM[32]). In the absence of crystallographic data, the structural origin of this high affinity remains unknown. Analysis of the primary amino acid sequences suggests the absence of the histidine–tyrosine[21] linkage observed in catalytic sites of most C*c*Os and quinol oxidases (Figure 1.5). What moiety provides the fourth electron for O_2 reduction by the mixed-valence (2e-reduced) state of cytochromes *cbb*$_3$ is not known. Other data, such as gas binding and recombination studies, also suggest a unique organization of the heme/Cu site in cytochromes *cbb*$_3$. Cytochromes *cbb*$_3$ also lack the electron-relay Cu_A site, which is replaced by 3 six-coordinate hemes. The enzymes' energy-transducing efficiency is lower than that of C*c*Os. There is evidence suggesting that physiological roles of cytochromes *cbb*$_3$ may also include signal transduction and O_2 scavenging[21]. Cytochrome *cbb*$_3$ oxidases are the only type of terminal oxidases expressed in a number of pathogenic organisms, including *H. pylori*, *N. gonorrhoeae*, and *N. meningitidis*. Expression of cytochrome *cbb*$_3$ oxidases was suggested to be required for colonization of anoxic tissues and may be an important determinant of pathogenicity[21].

2.2. Cytochromes *bd*

Cytochrome *bd* oxidases are aerobic terminal oxidases unrelated to HCOs. They function as quinol oxidases and are widely distributed in Gram-negative bacteria (see footnote ‡‡ on page 4) and possibly some Archaea[§§ 22]. Like cytochromes *cbb*$_3$, cytochromes *bd* are also suggested to be essential for microaerobiosis and may protect anaerobic processes from O_2. Cytochromes *bd* are not known to operate as proton pumps. The enzymes generate a transmembrane electrochemical potential of ~180 mV relying solely on substrate protons, e.g., by

§§ Archaea comprise one of the two major divisions of prokaryotes.

releasing $4H^+$ in quinol oxidation at the acidic P-side of the membrane and uptaking $4H^+$ from the N-side of the membrane for O_2 reduction. It was reported, however, that cytochromes *cd* may translocate Na^+ [22].

Cytochromes *bd* are two-subunit enzymes, containing one heme *d* and two heme *b* groups, but not Cu or nonheme Fe. Whereas one heme *b* is six-coordinate and is thought to be the initial site of electron uptake from quinol, the other heme *b* and the heme *d* center are five-coordinate, although the nature of the axial ligands is not known. The two hemes share a binding pocket. In this pocket, however, only heme *d* is a high-affinity ligand binding site, which is also the O_2 reduction site. The $Fe_{\text{heme } b}$–$Fe_{\text{heme } d}$ distance within this pocket is too long for diatomic molecules to bridge the two metals and there is little if any bimetallic cooperativity in ligand binding (unlike that in C*c*Os, where O_2 first binds to Cu_B and is subsequently transferred to heme a_3 [18]). As expected for a terminal oxidase optimized for microaerobiosis, O_2 affinity of heme *d* is high ($K_d \sim 25$ nM).

The redox potentials of the hemes in cytochrome *bd* oxidases are ~100 mV more reducing than the potentials of the redox cofactors in HCOs. Up to three external reducing equivalents can be stored in these hemes. The primary O_2-binding redox form *in vivo* was hypothesized to be a complex between singly reduced cytochromes *cd* and the quinol, or the fully reduced enzyme, both of which would have a total of three external reducing equivalents. Only two oxygen-containing intermediates have been observed spectroscopically in the enzyme under turnover: a heme d/O_2 adduct, where O_2 does not appear to interact strongly with any other moieties in the binding pocket, and heme *d* oxoferryl. As in HCOs, no peroxo-level intermediates have convincingly been detected in cytochromes *bd*.

Cytochromes *cbb*$_3$ and particularly cytochromes *bd* catalyze O_2 reduction by mechanism(s) that are likely quite different from that observed in C*c*Os and quinol oxidases. Obtaining crystallographic structures of these enzymes would significantly contribute to our understanding of the diversity of the biochemical strategies of aerobic respiration.

3. Biomimetic Catalysis of O_2 Reduction

The significance of $4e/4H^+$ reduction of O_2 for the existence of the terrestrial biosphere and the importance of heme centers in catalyzing this reaction stimulated a significant effort aimed at replicating this reactivity in artificial systems containing a heme/Cu unit. No biomimetic studies of cytochromes *bd* have been carried out due to the narrower distribution of these enzymes in the biosphere and the lack of crystallographic data on their catalytic site. Biomimetic studies of HCOs currently develop along two major directions. First includes work on electrocatalytic O_2 reduction using electrode-confined Fe porphyrins (one so-called "functional" heme/Cu analogs). Second, O_2 reactivity of synthetic heme/Cu analogs is studied under stoichiometric conditions. A clever and promising approach, which is yet to be fully recognized, involves engineering the heme/Cu site into a simpler enzyme, such as myoglobin[33]. Synthetic systems designed to reproduce only the spectroscopic properties of the heme/Cu site, mainly of C*c*O, are of historic interest[28,34]. This review deals only with heme/Cu analogs whose electrocatalytic behavior has been characterized. Studies of the stoichiometric O_2 reactivity of biomimetic analogs and spectroscopic heme/Cu mimics have been extensively reviewed elsewhere[28,34,35]. The recent developments are described in refs [36–38].

Biomimetic studies usually have one of three objectives: (a) to reproduce in a synthetic system the reactivity pattern theretofore observed only in an enzyme, (b) to design useful synthetic catalysts based on the principles learned from studying the corresponding enzyme(s), and (c) to better understand the mechanism and structure/activity relationship of an enzymatic catalytic site by studying a properly designed biomimetic analog.

Initial work in biomimetic O_2 reduction was aimed almost exclusively at designing Fe porphyrin-based compounds that, when immobilized on an electrode, would catalyze electrochemical O_2 reduction mainly to H_2O (as opposed to H_2O_2) regardless of the electrochemical potential or the lifetime of the catalyst[28,32]. Such an objective is problematic in several aspects, as discussed below. It has evolved into studying the catalysts under conditions that would more closely resemble those under which HCOs operate *in vivo* (potentials, electron flow vs. O_2 flux, hydrophobicity of the microenvironment). More biologically relevant conclusions, particularly regarding the structure/activity relationship at the enzymatic catalytic site (objective (c) above) can be drawn from such experiments. It remains to be seen if a useful catalyst (e.g., for O_2 reduction in fuel cells) can be designed based on the knowledge acquired in biomimetic studies of HCOs.

Biomimetic electrocatalytic O_2 reduction is studied almost exclusively by rotating disk or rotating ring–disk voltammetry[39] (Figure 1.9). Usually, a water-insoluble catalyst is deposited on the graphite disk as a film of poorly defined morphology either by spontaneous adsorption from a solution of the catalyst in an organic solvent or by evaporation of an aliquot of such a solution onto the electrode. It is impossible to know the amount of catalyst immobilized on the electrode by spontaneous adsorption (although the amount of *electroactive* catalyst can be determined coulommetrically). Nor is it generally possible to vary the amount of the deposited catalyst. Spin casting is free of such limitations but it may produce more morphologically heterogeneous catalytic films due to nonuniform precipitation of the catalyst during the rapid evaporation of the solvent.

The modified electrode is immersed in a buffered air- (or O_2^-) saturated aqueous solution and the electrode potential is scanned. The selectivity of the catalysis, often represented as n_{av}, the average number of electrons delivered to an O_2 molecule (reaction 1.14), is determined either from the ratio of the ring and disk currents or from Koutecky–Levich plots (plots of inverse catalytic currents vs. inverse square–root of the rotational frequency of the electrode). The ring–disk ratio quantifies selectivity at both the potential-dependent and plateau parts of a linear-sweep voltammogram (Figure 1.9). However, for practical reasons only limiting currents can be acquired with adequate reproducibility to be used in Koutecky–Levich plots. Koutecky–Levich plots also afford information about catalytic turnover frequency at potentials corresponding to the plateau of the catalytic waves, provided that substrate and/or charge transfer within the catalytic film are not turnover limiting.

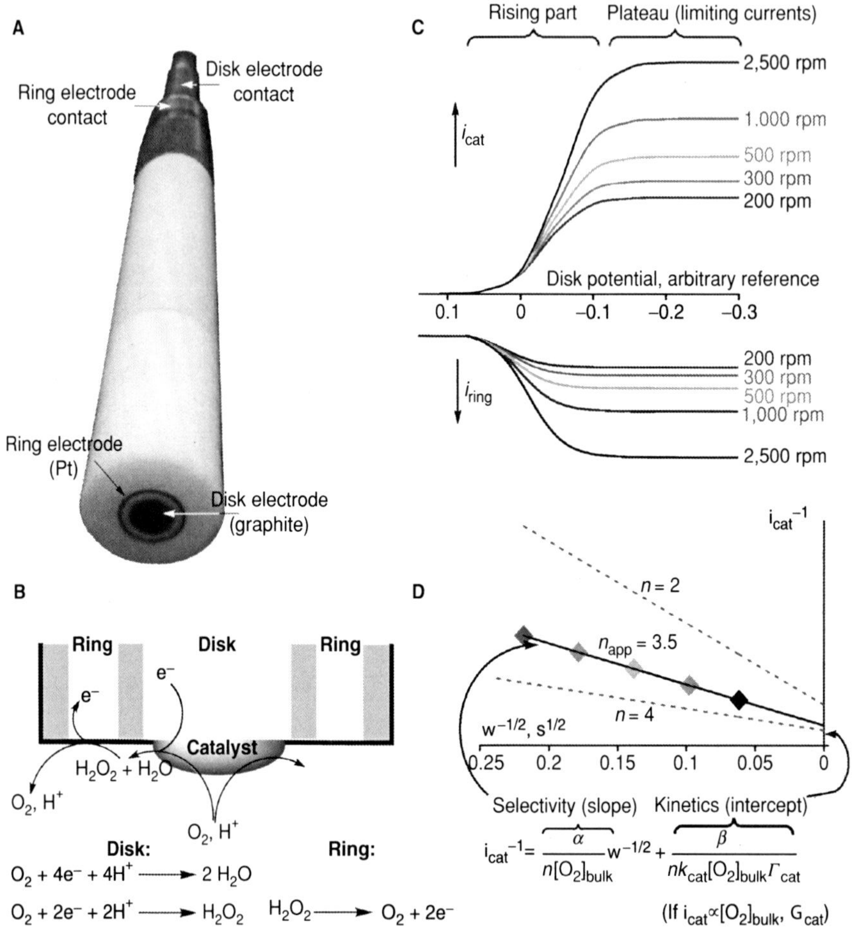

Figure 1.9. A. Rotating ring-disk electrode (RRDE). The RRDE employed in biomimetic O_2 reduction studies usually contains a graphite disk and Pt ring electrodes. A water-insoluble catalyst is deposited on the graphite disk as a film, the electrode is immersed in an aqueous medium and is rotated. This rotation causes a flow of the electrolyte to the electrode (B), bringing the reactants to the catalyst. The products of the catalysis (H_2O and, if the catalyst is not 100% selective, H_2O_2 and/or O_2^--HO_2) are transported to the ring electrode where H_2O_2 and/or $O_2^-HO_2$ are oxidized generating the ring current. A set of voltammograms at different rotational frequencies of the RRDE is obtained (C). From the ring-disk current ratio, the average redox stoichiometry of catalysis (n_{av}, reaction 1.14) can be determined if the collection efficiency of the ring towards H_2O_2 is known. When the catalytic currents are proportional to both the bulk concentration of O_2, $[O_2]_{bulk}$ and the surface coverage of the catalyst, Γ_{cat}, the value of n_{av} and the apparent second order rate constant can also be determined from a linear Koutecky-Levich plot[39] (D). Koutecky-Levich plots are plots of the inverse disk current vs. inverse square root of the rotational frequency.

$$O_2 + n_{av}H^+ + n_{av}e^- + \text{catalyst} \xrightarrow{k_{cat}} H_2O^{(n=4)} + H_2O_2^{(n=2)} + \text{catalyst} \quad (1.14)$$

The approach of characterizing catalytic properties of heme/Cu analogs using electrode-adsorbed films is quite useful but has several limitations that must be kept in mind when analyzing the literature data or designing new experiments. First, ring–disk voltammetry characterizes the catalytic properties of a film, which may differ significantly from those of individual, isolated molecules of the catalyst. For example, the fraction of O_2 reduced to H_2O_2 by films of Fe tetraphenylporphyrin, Fe(TPP) (Figure 1.10), decreases as the amount of the deposited catalyst increases[28]. One explanation for this is that individual molecules of Fe(TPP) reduce O_2 mainly to H_2O_2, but a significant residence time of H_2O_2 in the catalytic film increases the probability of its further reduction to H_2O by other molecules of the catalyst[40]. The residence time and hence the probability of H_2O_2 being reduced before escaping the film increases as more catalyst is deposited on the electrode. Second, neither RRDE nor Koutecky–Levich plots allow one to identify whether O–O bond homolysis is an important side reaction during catalytic turnover. Many simple Fe porphyrins induce O–O bond homolysis in H_2O_2[41] generating highly reactive $^\bullet$OH.[9] Hydroperoxyl radical reacts rapidly with components of the film so that it never escapes from the catalytic film and hence never reaches the ring electrode. A product of a reaction between $^\bullet$OH and a component of the catalytic film will become reduced by the electrode, thus contributing to the disk current. Third, as is the case of any chemically modified electrode[42], charge and/or substrate propagation in the catalytic film may affect the turnover frequency and this possibility must be taken into account when kinetic data are being sought. For a more detailed discussion of the limitations of RRDE as currently practiced in biomimetic O_2 reduction, see refs [28, 32]. Many of these limitations are addressed only in the latest studies of biomimetic O_2 reduction[43].

3.1. Simple Fe Porphyrins

The knowledge of the catalytic properties of simple Fe porphyrins toward electroreduction of O_2 and H_2O_2 is essential for understanding the structure/function relationship in electrocatalytic reduction of O_2 by structurally more elaborate synthetic analogs of the heme/Cu site. Electroreduction of O_2 and occasionally of H_2O_2 by simple Fe porphyrins (Figure 1.10) immobilized on a graphite electrode has been studied fairly extensively over the past 25 years[28]. Although there is substantial disagreement in the literature regarding the quantitative aspects of this reactivity a few generalizations can be made:

1. Electrocatalytic reduction of both O_2 and H_2O_2 starts at potentials close to the $Fe^{III/II}$ potential in the absence of a substrate (which for most porphyrins is ~0.2–0 V vs. NHE at pH $\leq$ 6; the exception being Fe(TMPyP), E ~ 0.5 V). Catalytic reduction of H_2O_2 by simple *ferric* porphyrins is too slow to be detectable in typical electrocatalytic experiments.
2. Depending on pH, increasing the acidity of the solution either makes the potential required to yield a fixed turnover frequency more oxidizing by

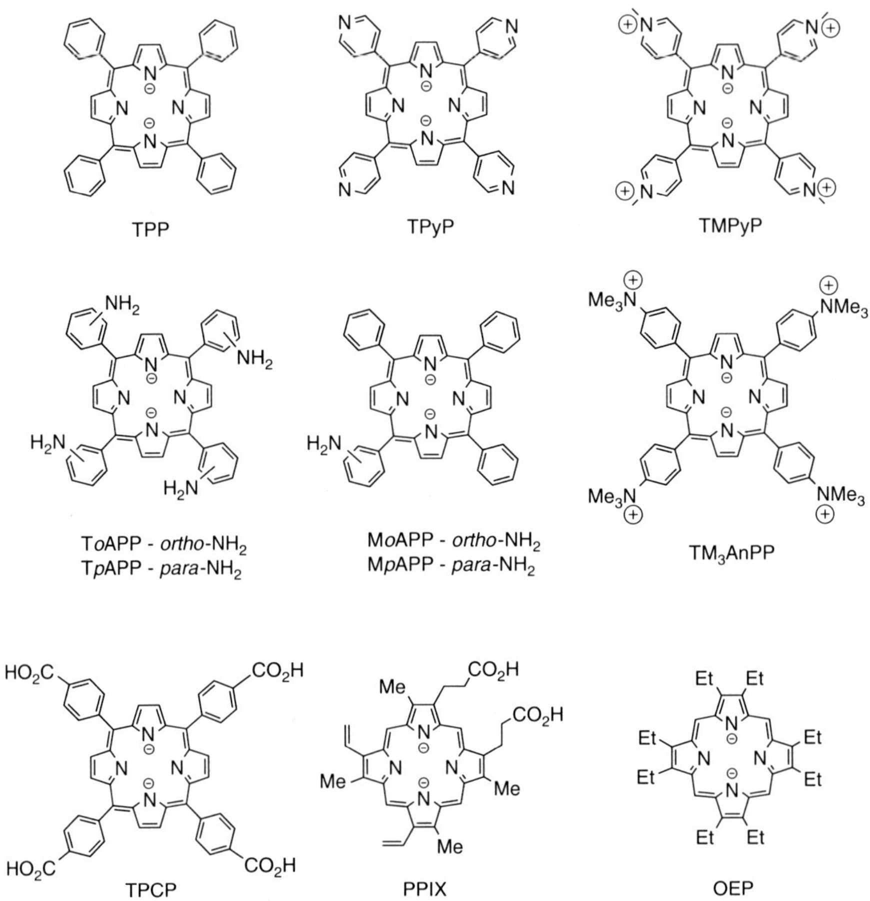

Figure 1.10. Chemical structures and abbreviated names of simple Fe porphyrins whose catalytic properties toward electrochemical reduction of O_2 and in some cases of H_2O_2 have been studied.

60 mV/pH or does not affect it. This pH dependence is in most cases the same as that of the $Fe^{III/II}$ couple in the absence of a substrate. These identical pH dependencies suggest a pre-equilibrium between the ferric and ferrous forms of the catalyst followed by a kinetically irreversible step that does not involve proton or electron transfer, e.g., O_2 binding.

3. The apparent redox stoichiometry of O_2 reduction catalysis (n_{av}, reaction 1.14) is pH independent but for many catalysts depends strongly on the applied potential (Figure 1.11)[40]. The apparent selectivity increases with the amount of deposited catalyst.
4. Usually simple Fe porphyrins degrade rapidly during catalytic O_2 reduction.

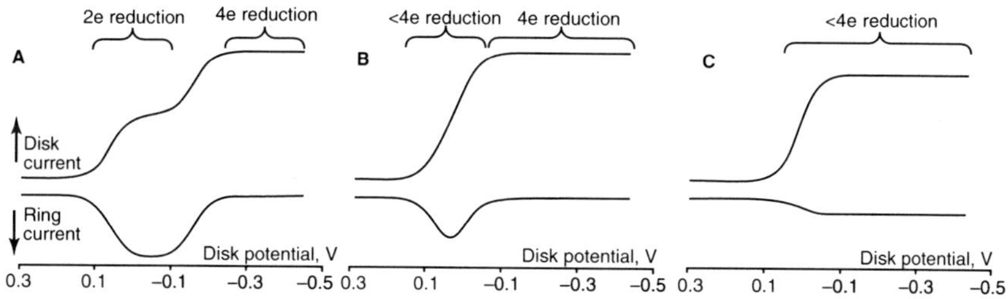

Figure 1.11. Typical current-potential curves for O_2 reduction by simple Fe porphyrins immobilized on a graphite electrode. The simulated traces are based on data in ref. [40] for (A) Fe(TPP); (B) Fe(PPIX); (C) Fe(TPyP) (see Figure 1.10 for chemical structures) in a pH 0 electrolyte. Qualitatively similar voltammograms are observed at other pH.

Most of these results can be understood within a generic mechanistic scheme (Figure 1.12)[44], based on numerous studies of reactions between O_2 or H_2O_2 and Fe porphyrins under stoichiometric conditions (see for example ref. [41]), and heme enzymes in single-turnover experiments[44]. Only five-coordinate ferrous centers have any measurable affinity for O_2. Ferric porphyrins are aerobically stable; four- and six-coordinate ferrous porphyrins do not bind O_2 but may reduce it to O_2^- by an outer-sphere mechanism, depending on the $Fe^{III/II}$ potential[45]. Anaerobic cyclic voltammetry suggests that simple Fe porphyrins deposited on an electrode in contact with an aqueous buffer are present mainly in the bis-aqua ligation state[28,40]. This coordinatively saturated complex must lose one water molecule to bind O_2. It is, however, also possible that a minority species, for example Fe(por) axially ligated by a carboxylate or quinol residue of the graphite electrode, is the catalytically active form.

Upon O_2 binding, the catalyst-O_2 adduct is probably reduced to the ferric-hydroperoxo complex, whose subsequent reactions determine the product distribution. In hemoproteins evolved for O–O bond reduction, such as peroxidase, this transient intermediate undergoes rapid O–O bond heterolysis yielding a high-valence oxoferryl-cation radical intermediate (Compound I)[44]. This reaction is facilitated both by the proximal ligand (such as imidazole, thiolate, or phenoxide) and the distal environment (see Figure 1.5 for the definition of proximal and distal sides). Since Fe porphyrins in Figure 1.10 lack appropriate distal/proximal structures it seems highly probable that O–O bond heterolysis in the ferric-hydroperoxo intermediate is inhibited relative to the competing reactions, which include O–O bond homolysis (reaction rxn10, Figure 1.12), hydrolysis of intact H_2O_2 (reaction rxn8) and reduction to a ferrous-hydroperoxo intermediate (reaction rxn4b). The propensity of simple Fe^{III} porphyrins to induce O–O bond homolysis in H_2O_2 is well established from studies of stoichiometric reactions between H_2O_2 and such porphyrins[41]. Predominance of this homolytic reduction pathway (reaction rxn10, Figure 1.12) during catalytic O_2 reduction by electrode-confined Fe porphyrins would explain the low stability of these catalysts. Indeed, simple Co porphyrins, which catalyze reduction of O_2 only to H_2O_2, retain their activity over a substantially larger number of turnovers than structurally analogous Fe porphyrins whose films reduce a substantial fraction of O_2 to H_2O. This

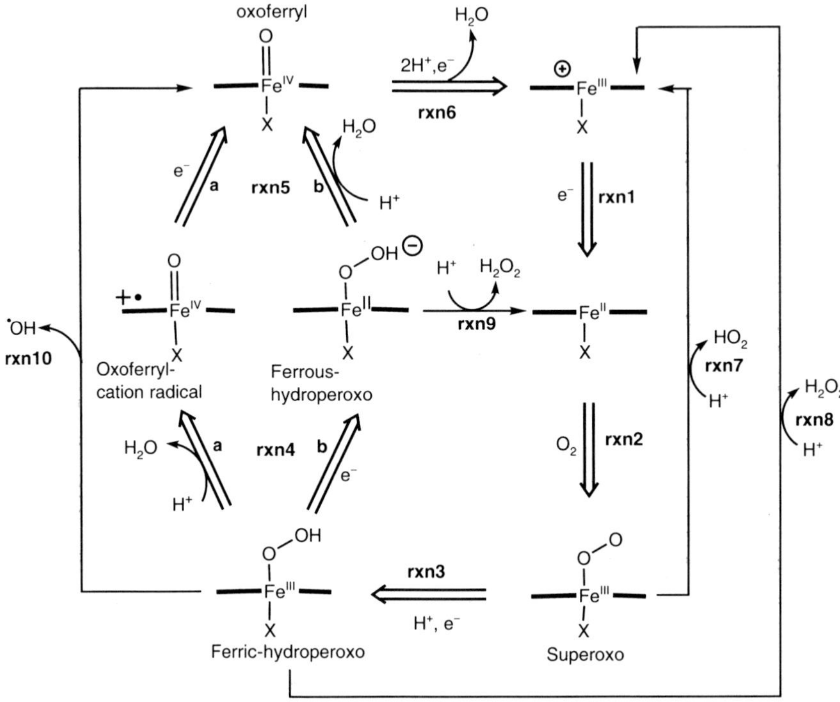

Figure 1.12. A plausible mechanism for electrochemical reduction of O_2 catalyzed by Fe porphyrins. For X = imidazole or a similar nitrogenous heterocycle the catalysis probably proceeds through steps rxn4a–rxn5a as it does in hemoproteins, such as peroxidase, catalase, and cytochromes P450[44], possibly with a minor contribution from side reaction rxn8. For simple Fe porphyrins lacking the suitable distal and proximal environments (X = H_2O or a residue of the graphite surface), the ferric-hydroperoxo intermediate decomposes via reactions rxn8 and rxn10, and at sufficiently reducing potentials it probably undergoes 1e reduction to the ferrous-hydroperoxo intermediate (reaction rxn4b), which is more susceptible to O–O bond heterolysis than the ferric analog.

is consistent with the primary role of hydroxyl radicals, not H_2O_2 as sometimes suggested[40], in rapid degradation of Fe porphyrin catalysts.

An increase in the fraction of the 4e reduction pathway at more reducing potentials (Figure 1.11a,b) is consistent with the kinetic competition between the release of H_2O_2 from the ferric-hydroperoxo intermediate and O–O bond heterolysis via a ferrous-hydroperoxo species. Because H_2O_2 and particularly HO_2^- are more basic ligands than H_2O, reduction of $[X(por)Fe^{III}(O_2H^-)]$ would require more reducing potentials than reduction of $[X(por)Fe^{III}(OH_2)]^+$ (the latter corresponds to the onset of catalytic O_2 reduction[28,32]). Since O–O bond heterolysis in the *ferrous*-hydroperoxo intermediate does not require oxidation of the porphyrin it is more facile than O–O bond heterolysis in the *ferric* analog. Hence, at potentials more oxidizing than that of the $[X(por)Fe^{III/II}(O_2H_x^{x-2})]^{x-1/x-2}$ couple, simple Fe porphyrins catalyze mainly 2e reduction of O_2 to H_2O_2, with possibly an important side reaction of O–O bond homolysis (reaction rxn10).

This reaction pathway dominates because the kinetics of O–O bond heterolysis is unfavorable relative to the other two competing pathways (reactions rxn8 and rxn10). At potentials comparable to, or more reducing than, that of $[X(por)Fe^{III/II}(O_2H_x^{x-2})]^{x-1/x-2}$ ($x = 1, 2$) the ferric-hydroperoxo intermediate is consumed in a 1e reduction, followed by irreversible O–O bond heterolysis to yield the oxoferryl (Fe^{IV}) intermediate. The flux of H_2O_2 decreases.

If this explanation is correct, Fe porphyrins in Figure 1.10 catalyze the 4e reduction of O_2 to H_2O by an abiological mechanism and the large overpotentials of the reduction (up to 1 V at pH 0) are due to the fairly reducing potential required for rapid 1e reduction of the ferric-hydroperoxo intermediate. One of the objectives of reproducing the stereoelectronics of the heme/Cu site of HCOs in a synthetic system is to design catalysts that would be free from these limitations.

3.2. Functional Analogs of the Heme/Cu Site

The stereoelectronic properties of the heme/Cu site of HCOs (Figure 1.5) are extraordinarily difficult to replicate in a small molecule. The difficult synthetic chemistry of imidazoles is compounded by the chelating nature of imidazole ligation to both Cu_B and Fe_{a3} (Figure 1.5). This chelating coordination, enforced by the conformation of the protein, rather than direct chemical bonds between metal-bound moieties, increases the dissociative stability of the metal–imidazole bonds. Synthetic analogs also need to be sufficiently flexible to accommodate the different stereoelectronic requirements of Cu^I (which prefers a tetrahedral coordination on steric grounds) and Cu^{II} (with its electronic preference for square–planar and square–pyramidal coordination) and the increasing steric bulk of O_2-derived intermediates during the catalytic cycle. The synthetic methodology for replicating the majority of the stereoelectronic features of the heme/Cu site (such as the proximal and distal imidazole environment and the imidazole–phenol moiety) has been developed only recently[38,46–48] (Figure 1.13, upper reaction scheme). Consequently, most of the work on biomimetic O_2 reduction was carried out on simplified heme/Cu analogs (Figure 1.14), which are also easier to synthesize (Figure 1.13).

All reported "functional" heme/Cu analogs (Figure 1.14) are designed on the tetraaminophenylporphyrin platform because convenient methods for selective derivatization of the two faces of the porphin exist (Figure 1.13). The steric repulsion between the *ortho-* substituents on the phenyl groups and β-pyrrolic hydrogens favors perpendicular orientation of the phenyl rings relative to the porphin plane in *meso*-aryl porphyrins. Free rotation around C_{meso}–C_{aryl} bonds occurs at any appreciable rate only at temperatures above ~80°C. Thus, heating is used to facilitate the enrichment of the sample in the desired atropomer***. Modern synthetic schemes utilize an $\alpha_3\beta$ atropomer and orthogonal protecting groups to attach different proximal and distal groups. Due to reversible metalation of the distal superstructure, but not of the porphin, Fe is introduced first (in the porphin), followed by metalation of the superstructure. The mono- and bimetallic

*** Atropomer, also called atropisomer, is a conformational isomer that converts to its other conformers slow on the NMR scale at room temperature; α and β refer to the top and bottom faces of porphin, with an arbitrarily chosen reference face.

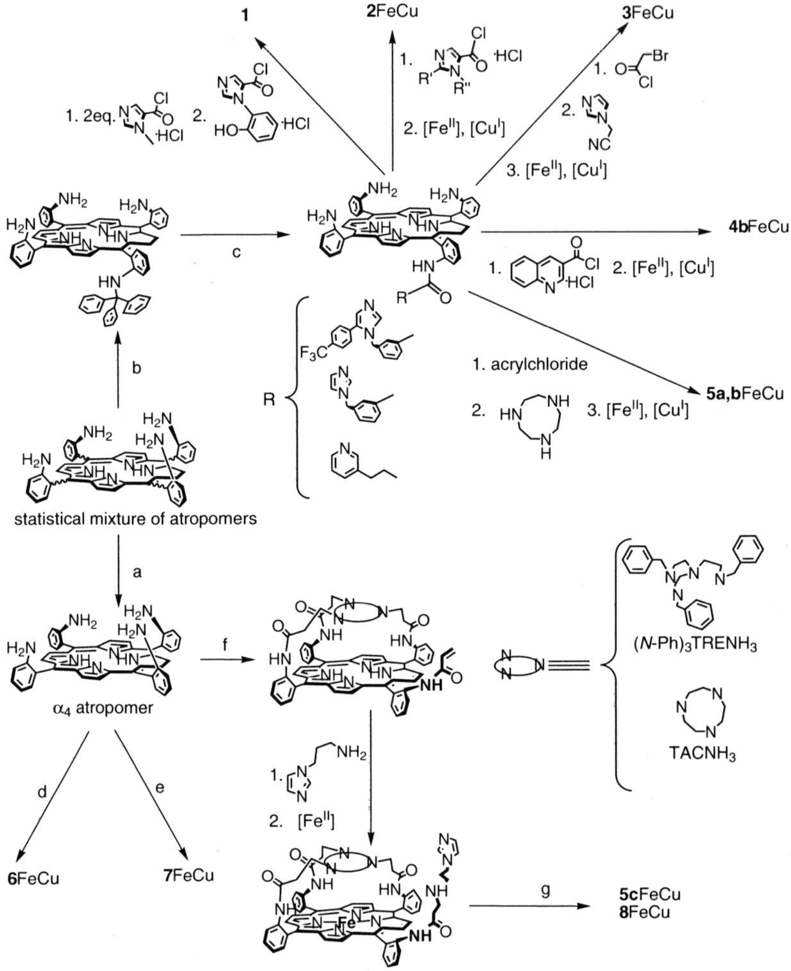

Figure 1.13. Synthetic schemes for all reported "functional" heme/Cu analogs. Fe metalation, [Fe^{II}], usually requires excess anhydrous $FeBr_2$ in glacial acetic acid in the presence of 2,6-lutidine as base under anaerobic conditions; Cu metalation, [Cu^I], is usually carried out with 1 equiv. $Cu(AcN)_4PF_6$ in acetonitrile at room temperature. Δ signifies heating. Experimental conditions: a) basic alumina, toluene, reflux (enrichment of the mixture in the α_4 atropomer). b) (i) trityl bromide, base; (ii) basic alumina, toluene, reflux (enrichment of the monotritylated porphyrin in the $\alpha_3\beta$ atropomer). c) (i) trifluoroacetic anhydride, base (protection of the α_3 amines); (ii) acid (deprotection of the β amine); (iii) appropriate acyl chloride; (iv) ammonia (deprotection of the α_3 amines). d) (i) 1eq acetyl chloride; (ii) excess of chloroacetyl chloride; (iii) Fe metalation; (iv) tris(2-aminoethyl)amine ($TRENH_6$), 50 °C; (v) CuBr. (e) (i) excess of acryloyl chloride, base; (ii) $TRENH_6$ or TACNH3 (f), 50 °C; (iii) metalation with Fe followed by metalation with Cu. f) (i) excess of acryloyl chloride, base; (ii) $(N\text{-Ph})_3$TREN or $TACNH_3$, 50 °C. g) (i) heating to facilitate the β-atropomerization of the imidazole-derivatized *o*-aniline driven by coordination of the imidazole to Fe; note that the imidazole cannot coordinate to Fe on the distal site because of the TREN or TACN superstructure (see Figure 1.5 for the definition of the distal site); (ii) Cu metalation with CuBr.

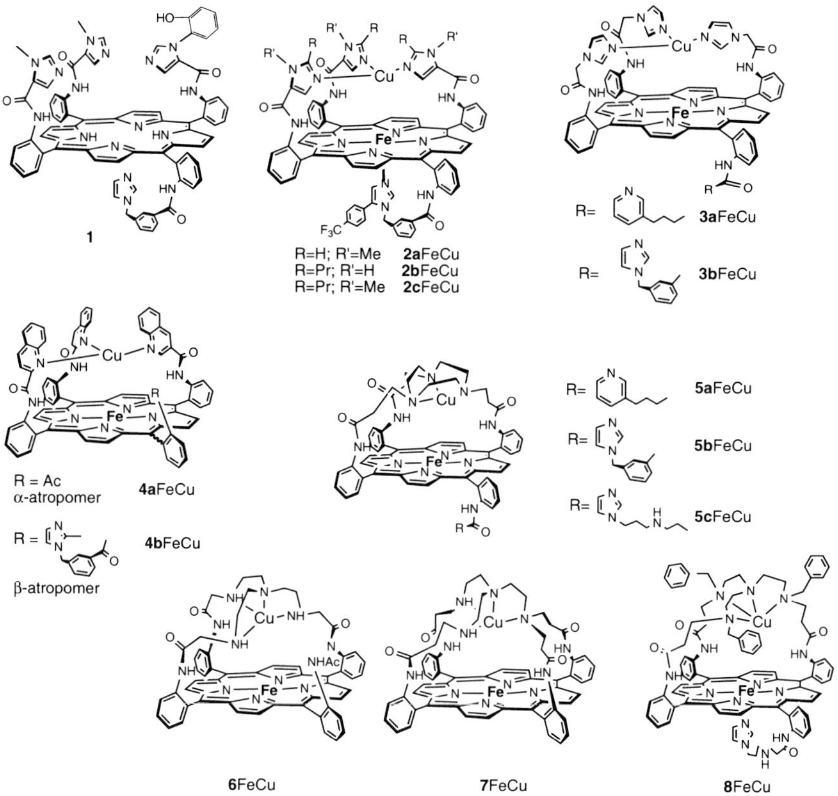

Figure 1.14. "Functional" analogs of the heme/Cu site reported to date.

products are analyzed as diamagnetic CO adducts. An important innovation involves the incorporation of a ^{19}F marker into porphyrins, which allows the use of ^{19}F NMR to monitor reactions and determine the ligand/Cu stoichiometry of FeCu derivatives using a fluorinated counterion, such as PF_6^- or BF_4^-. Methods to introduce F markers into distal imidazoles have also been developed[49]. For a detailed discussion of the synthetic methodologies, see reviews [28, 32, 50].

Complexes **2**FeCu are the best known biomimetic catalysts for electroreduction of O_2 and are the best studied[43,46,50–56]. They reproduce important kinetic and mechanistic aspects of O_2 and H_2O_2 reduction by HCOs and are discussed in detail in the subsequent section. Electrocatalytic properties of **3**FeCu have not been thoroughly studied in part due to the low dissociation stability of Cu[57].

Also poorly studied is the electrocatalytic behavior of complexes **5** and **8**, which both contain a proximal imidazole (pyridine) ligand and a distal chelate comprising aliphatic N donors. Complexes **5a–b**FeCu and **8**FeCu were reported to catalyze mainly 4e reduction of O_3 (although at unspecified and probably abiologically reducing potentials)[58,59]. In contrast, a structurally related complex **5c**FeCu releases a substantial fraction of reduced oxygen as H_2O_2 (Table 1.2) and is an inferior catalyst than most simple Fe porphyrins[59,60]. The origin of such a dramatic difference is unknown. Compounds **5a–b**FeCu adsorbed on the electrode demonstrate a single peak by cyclic voltammetry under N_2[58]. The presence of a single redox peak may result either from indistinguishable $Fe^{III/II}$ and $Cu^{II/I}$ potentials or loss of Cu from the electrode-confined catalysts. The latter possibility, that has not yet been addressed by the authors, may account for the observed differences in the electrocatalytic behavior of stereoelectronically similar **5a–b** and **5c**. Under anhydrous conditions, stoichiometric O_2 binding to the fully reduced Fe^{II}/Cu^{I} complexes **5**FeCu yields a bridging peroxo adduct, Fe^{III}–O–O–Cu^{II}, not the ferric-superoxide/Cu^{I} form of oxygenated heme/Cu site (compound A, Figure 1.7)[18] or of **2**FeCuO_2 adduct (Figure 1.15)[18,53,61]. Compounds **5**FeCu and **8**FeCu, which are relatively easy to synthesize, may play an important role in understanding the basic dioxygen chemistry of Fe^{II} and Cu^{I} ions, but they have not proved to be particularly suitable for understanding the structure/activity relationship at the heme/Cu site of HCO.

Related compounds with the tris-quinoline superstructure with and without Cu (**4b**FeCu and **4b**Fe-only) manifest electrocatalytic properties (Table 1.2) comparable to those of simple Fe porphyrins[62].

Although the elimination of the proximal nitrogenous axial ligand in **4a**Fe and **4a**FeCu has little apparent effect on the fraction of the 4e reduction pathway, the catalytic behavior of Fe and FeCu derivatives of **6** and **7** is quite unusual[63]. Whereas electrocatalytic properties of Fe-only complexes (Table 1.2) appear to be comparable to those of some simple Fe porphyrins, the introduction of Cu into the distal superstructure results in substantially lower selectivities toward the 4e pathway (Table 1.2) as well as decreased stability of the catalysts. No adequate explanation for this unexpected result was offered by the authors[63]. It may originate from Cu-mediated homolysis of the O-O bond in H_2O_2, which is generated during O_2 reduction at the Fe porphyrin moiety. Cu, being an efficient Fenton reagent, generates $^{\bullet}OH$ in the presence of H_2O_2. A substantial flux of hydroxyl

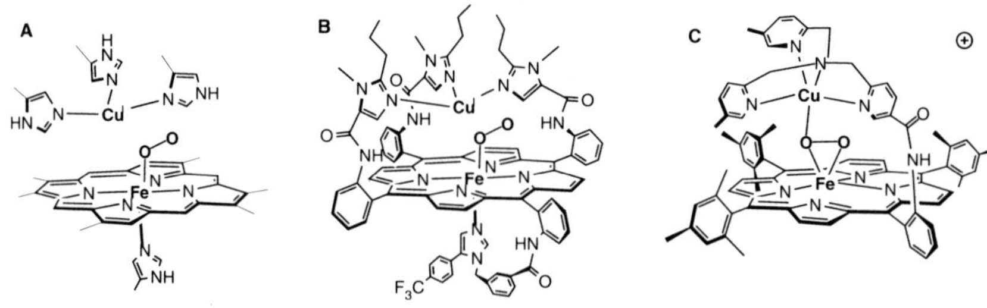

Figure 1.15. O_2 adducts of (A) the heme/Cu site of HCOs (compound A) and (B and C) two heme/Cu analogs[53,61]. Structure A is derived from resonance Raman spectroscopy[18]. structure B (derived from **2c**FeCu, Figure 1.14, Table 1.2) is based on resonance Raman and ^1H NMR spectroscopic data[53] and structure C is determined by single-crystal X-ray analysis; counterion (BPh_4^-) is omitted for clarity[61]. Replication of both the distal and proximal environment in **2c**FeCu was required to obtain the biologically relevant ferric-superoxide/Cu^{I} isomer of the O_2 adduct, not the more common ferric-peroxo-Cu^{II} isomer.

Table 1.2. Electrocatalytic Properties Toward O_2 Reduction of the Functional Heme/Cu Site Analogs. See Figure 1.14 for the Relevant Chemical Structures. All Potentials are vs. the Normal Hydrogen Electrode (NHE)

Catalust (Figure 1.14)	E_{cat}, V[a]	$E_{1/2}O_2$, V at pH 7	Selectivity (n_{av})		References
			At $E_{1/2}$	At plateau	
2FeCu[d]	0.1	0.2	~4	4	[43]
2Fe-only[d]	0.1	0.2	>3.7–3.8	4	[43]
3aFeCu	n.r.	0.2[e]	n.r.	4[f]	[57]
3bFeCu	~0.25	0.15	n.r.	4[f]	[57]
4aFeCu	n.r.	0	2.5–3.5	3.4 i_{lim} (3.8 i_r/i_d)	[62]
4aFe-only	n.r.	0	2.2–3.4	3.5 i_{lim} (3.9 i_r/i_d)	[62]
4bFeCu	n.r.	0.1	<3.1	3.5 i_{lim}	[62]
4bFe-only	n.r.	n.m.		3.6 i_{lim}	[62]
5aFeCu	n.r.	0.15	n.r.	3.9, KL	[58]
5bFeCu	n.r.	0.2	n.r.	3.9, KL	[58]
5cFeCu	~0.1 (Fe); ~0.5 (Cu)[g]	0.15	n.r.	2-3.1[h]	[59, 60]
8FeCu	~0.15 (Fe); ~ −0.2 (Cu)[g]	0.15[f]	n.r.	4-3.6[h], KL	[59, 60]
6FeCu	n.r.	0	<2.6	2.7 i_r/i_d	[63]
6Fe-only	n.r.	0	<3.8	4 i_r/i_d	[63]
7FeCu	n.r.	0.05	2.5–3	2.8, i_{lim} (3.2 i_r/i_d)	[63]
7Fe-only	n.r.	0	3.7–3.8	4 i_r/i_d	[63]

[a] The redox couple ($Fe^{III/II}/Cu^{II/I}$ or $Fe^{III}Cu^{II}/Fe^{II}Cu^{I}$) for the graphite-adsorbed catalyst in the absence of a substrate; n.r. – not reported.
[b] The half-wave potential of the catalytic wave for O_2 reduction by rotating disk voltammetry. This value depends on the electrode rotation rate, bulk O_2 concentration, pH, and in many cases on the scan rate, the amount of the adsorbed catalyst, and the nature of the supporting electrolyte; n.m. – not meaningful (no defined wave).
[c] The apparent redox stoichiometry of O_2 reduction. We have calculated the n_{av} values at $E_{1/2}$ from the linear-sweep voltammograms presented by the authors using the collection efficiencies reported by the authors or estimated from rotating ring–disk voltammogram of O_2 reduction on a bare graphite electrode. The method whereby n_{av} was estimated: KL—the Koutecky–Levich plot (Figure 1.9B); i_r/i_d—the ring-to-disk current ratio when the collection efficiency is reported or could be estimated; i_{lim}—comparison of the limiting current yielded by the catalyst with that given by the standard with a known n_{av} under the identical experimental conditions. The higher n_{av} estimated from the i_r/i_d ratio for several entries probably indicates that the reported collection efficiencies are significantly overestimated.
[d] The imidazole substitution pattern of the distal superstructure has only a minor effect on the electrochemical behavior.
[e] The numbers were extracted from the voltammograms given in the supplementary information.
[f] "No detectable peroxide leakage"; neither rotating ring–disk voltammograms nor i_r/i_d values are presented and the experimental collection efficiency is not specified.
[g] The peaks labeled as $Cu^{II/I}$ are poorly defined and only quasireversible, these peaks may not correspond to the $Cu^{II/I}$ couple.
[h] The first number refers to the originally reported selectivity; the second n value was reported later.

radicals would lead to a rapid degradation of the catalyst. Since partially degraded forms of **6**FeCu and **7**FeCu are likely less selective 4e catalysts than intact **6**FeCu and **7**FeCu, at the beginning of the degradation process the decrease in the catalytic currents may originate from lower average stoichiometry of O_2 reduction ($n_{av} < 4$). Only subsequently, the lower current may reflect the complete loss of the catalytic activity of a fraction of the catalytic film. Under anhydrous conditions, some Fe^{II}/Cu^{I} analogs of the heme/Cu site lacking the proximal imidazole bind O_2 in an unusual, biologically unprecedented mode (Figure 1.15). Other FeCu complexes form linear Fe^{III}–(O_2^{2-})–Cu^{II} units, based on their O–O bond frequencies[28,34]. With the exception of **2c**FeCu, no biomimetic analogs of the heme/Cu site reproduce the O_2 binding mode seen in compound A (Figure 1.7).

The literature data on synthetic Fe porphyrins convincingly suggest the importance of axial imidazole ligation in determining the reactivity of Fe porphyrins with O_2[43], ROOH (R = H, alkyl, aryl),[43,64] peroxysulfate and XO_n^- species ($X = Cl$, $n = 2$; $X = Br$, $n = 3$)[56]. Although addition of 1- or 1,2-substituted imidazoles to heme/Cu analogs lacking a chelating proximal imidazole in organic solvents yields five-coordinate Fe centers[65], there is no evidence that this strategy generates imidazole-ligated Fe porphyrins when they are adsorbed on an electrode[28]. Therefore, an efficient Fe-porphyrin based catalyst for electroreduction of O_2 would probably have to incorporate a proximal imidazole- or pyridine-derived ligand covalently attached to the porphin.

All reported synthetic heme/Cu analogs have Fe^{II} and Cu^{I} potentials ($\leq$150 mV at pH 7) that are substantially more reducing than the ferri-/ferroheme a_3 and $Cu_B^{II/I}$ potentials (~350 mV at pH 7)[12,19]. Increasing the $Fe^{III/II}$ potentials of heme/Cu analogs is important for at least two reasons. First, because ferric porphyrins do not bind O_2, the $Fe^{III/II}$ potential determines the onset potential of O_2 reduction catalysis and, thus, the overpotential of O_2 reduction. Second, O_2 catalysis at potentials more oxidizing than that of the O_2/H_2O_2 couple (~0.25 V at pH 7) is of particular interest, because thermodynamics limits the amount of H_2O_2 that can be generated as a byproduct under such conditions. The origin of the low $Fe^{III/II}$ potentials of synthetic heme/Cu analogs is not known. It may be related to the stabilization of the charged ferric state in the more polar environment experienced by an electrode-confined catalyst vis-à-vis the heme/Cu site buried in a nonpolar protein matrix[51].

3.3. Possible Physiological Role of Cu_B from Biomimetic Electrocatalytic Studies

Elucidating the function of distal Cu_B in O_2 reduction by HCOs has recently been the major focus of biomimetic studies of HCOs[32,34]. Cu_B has long been assumed to facilitate O–O bond heterolysis, or stabilize peroxo-level intermediates against dissociation, by forming a bridged Fe_{a_3}–(O_2^{2-})–Cu_B^{II} unit upon O_2 binding[66,67]. Such a situation would imply that a bimetallic catalytic site is required for efficient O_2 reduction. This would be comparable to the "dual site" mechanism of O_2 reduction at Pt[68], which is supported by some recent results[69]. However, no peroxo-level intermediates of any kind have ever been observed during O_2 reduction by HCOs and it remains uncertain whether an intermediate where

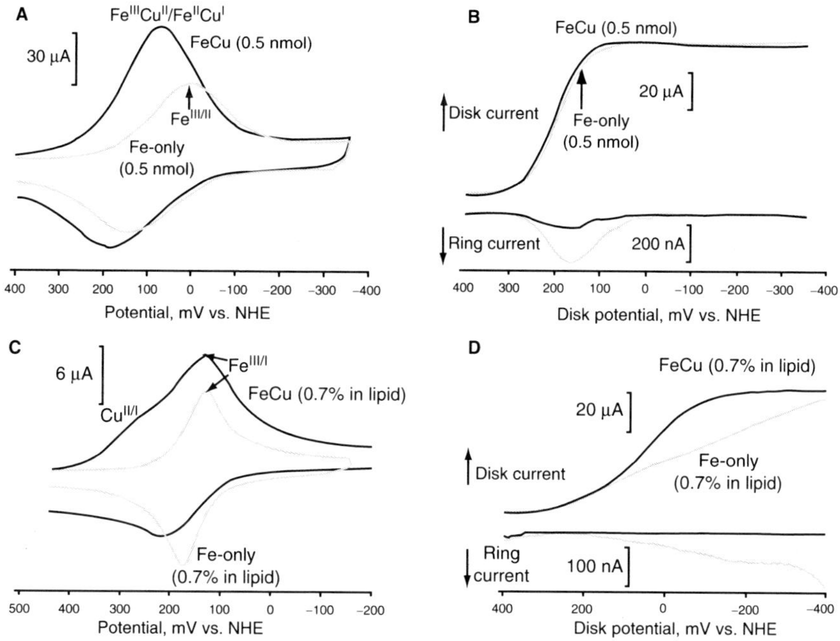

Figure 1.16. Cyclic voltammograms under N_2 (A,C) and rotating ring–disk current–potential curves in aqueous air-saturated pH 7 buffers (B,D) of **2**FeCu and **2**Fe-only directly adsorbed on a graphite electrode (A,B) and as a 0.7% (mol) suspension in a ~1-μm-thick phosphadytilcholine film on the electrode surface (C,D). The rapid charge transfer within the films of adsorbed catalysts is supported by the linear dependence of the peak currents on the scan rate. The non-ideal shape of the peaks is due to cooperative behavior of the catalytic films as a whole[51]. The $Fe^{III/II}$ and $Cu^{II/I}$ potentials are the same in the adsorbed catalysts (A) but separate when the catalysts are in the lipid film (C). Autooxidation of the catalyst–O_2 complex is the major source of ring-detectable byproducts (see below) and accounts for the potential-dependent selectivity of electrode-adsorbed catalysts (B). The measured collection efficiency of the ring electrode toward H_2O_2 in these experiments was 15%.

Fe and Cu are bridged forms even transiently[18]. Indeed, O_2 binding to the 2e reduced heme/Cu site yields a ferriheme-superoxide/Cu_B^I adduct (compound A, Figure 1.7) with little or no interaction between bound O_2 and Cu_B^I despite the suitable Fe–Cu distance (~5 Å) and $Cu^{II/I}$ potential.

To understand what role, if any, distal Cu plays in O_2 reduction at the heme/Cu site, electrochemical reduction of O_2 at physiologically relevant potentials (>50 mV) in pH 7 buffers was studied using a series of synthetic heme/Cu complexes **2** (Figure 1.14) in the bimetallic FeCu and monometallic Fe-only form[43,51,54,55].

The catalysts directly adsorbed on the electrode surface (up to an apparent surface coverage††† of ~2.5 nmol cm^{-2}) produced films within which both

††† Since the roughness of pyrolytic graphite is high, in the absence of the experimentally determined roughness factor, only geometric surface coverage can be calculated.

Table 1.3. Reactivity of Biomimetic Analogs 2FeCu Toward O_2 and H_2O_2 and the Relevant Data for Mammalian C*c*O

	FeCu[a]	Mammalian C*c*O[b]
E^o, mV (no substrate)	120	≥ 250[c]
% of the 4e O_2 reduction pathway	>96 – 98	>99 (?)
Turnover frequency of O_2 reduction at 250 mV, s^{-1}	>2	>50
Intermediate generated in the first kinetically irreversible step after O_2 binding	(por)Fe^{III}–OOH	(por)Fe^{III}–OOH (?) [d]
$(k_{cat})_{max}$ of O_2 reduction, $M^{-1}s^{-1}$ [e]	1×10^5	$1.5–2.5 \times 10^5$
k_{O_2}, $M^{-1}s^{-1}$ [e]	$>2 \times 10^7$	$1–3 \times 10^8$
Redox state of the O_2 adduct	$Fe^{III}(O_2^-)/Cu^I$ [53]	$Fe^{III}(O_2^-)/Cu^I$
Power output in O_2 reduction, W g^{-1}	100	<20
k_{app} of H_2O_2 reduction at 450 mV, $M^{-1}s^{-1}$ [e]	800	700 [14]
$(k_{app})_{max}$ of H_2O_2 reduction, $M^{-1}s^{-1}$ [e]	2×10^4	2×10^4 [f]

a Data from ref. 43 unless indicated otherwise.
b Data from ref. 13 unless indicated otherwise.
c The heme/Cu_B potential depends both on the site's protonation state[70] and on the redox states of heme *a* and Cu_A[12]. It may be higher or lower than indicated.
d Based on DFT calculations[30,31].
e k_{cat} is the apparent second-order catalytic rate constant for reduction of O_2 (reaction 1.14 page 14) or H_2O_2 at a specified potential; $(k_{cat})_{max}$ is the maximum (potential-independent) rate constant; k_{O_2}, second-order rate constant for binding of O_2 to the five-coordinate, ferrous catalyst (see below).
f Single-turnover experiment[14].

the charge transfer and the substrate transfer were more rapid than the catalytic reaction. By studying O_2 reduction under such conditions (Figure 1.16a,b)[51] it was established that these catalysts reproduce certain key kinetic (Table 1.3) and mechanistic aspects of O_2 reduction at the heme/Cu site. Using $^{\bullet}OH$ scavengers the catalysts were shown to generate undetectable amount of hydroxyl radicals (which results from homolysis of the O–O bond, Figure 1.12). Lack of O–O bond homolysis may explain why the catalysts retain their catalytic performance for over $>10^5$ turnovers. Based on the significantly greater turnover numbers of the catalysts in reduction of O_2 than of H_2O_2 (under N_2), the sequential mechanism of 4e O_2 reduction via the intermediacy of free H_2O_2 was estimated to contribute no more than 2–4% of the overall current consumption. The kinetic mechanism of O_2 electroreduction (Figure 1.17)[43,46] by these catalysts was derived from:

1. The pH dependence of catalytic currents under conditions of nonturnover-limiting substrate delivery.

2. The potentials for the (por)Fe^{IV}=O/(por)Fe^{III}OH and (por)$Fe^{III/II}(OH_2)$ couples and the pK_a of the coordinated H_2O determined by cyclic voltammetry in the absence of a substrate (Figures 1.16a and 1.17a).
3. The position of the equilibrium between aqua-ligated six-coordinate (por)$Fe^{II}(OH_2)$ and catalytically active five-coordinate ferrous catalysts (por)Fe^{II}, from inhibition of the O_2 reduction currents by CN^-[54].
4. The lower limit on the rate of the O–O bond heterolysis in the putative ferric-hydroperoxo intermediate from the rate of H_2O_2 reduction by the ferric catalyst (at potentials >400 mV at pH 7).

The mechanism most consistent with these data and the known dioxygen chemistry of Fe porphyrins (Figure 1.17) postulates that the product-determining intermediate, the ferric-hydroperoxo adduct, is generated slower than it is consumed by O–O bond heterolysis (i.e., the ferric-hydroperoxo intermediate is generated in the turnover-determining step). Such kinetics minimizes the steady-state concentration of the ferric-hydroperoxo intermediate and hence the amount of H_2O_2 that can be released during catalysis. Dioxygen reduction by CcO was also proposed to proceed by a slow generation of the ferric-hydroperoxo intermediate, followed by its rapid reduction[30,31].

To explain the potential-dependent redox stoichiometry of the catalysis (Figure 1.18) under rapid electron flow, the nature of the primary partially reduced oxygen species released by the catalyst was determined[43]. Because ring current itself does not allow the differentiation between H_2O_2 and O_2^-, selective superoxide scavengers (Figure 1.19) were incorporated into catalytic films. These compounds react rapidly with O_2^- (or its protonated form, hydroperoxyl radical, HO_2), reducing it to H_2O_2 without consuming current from the disk electrode. Subsequent oxidation of H_2O_2 to O_2 at the ring electrode generates twice as much current as oxidation of O_2^- would if superoxide were not reduced by the scavenger. As a result, if free O_2^- is generated within the film, with some probability, it will get reduced nonelectrochemically by a scavenger to H_2O_2 before escaping the film. An increase in the ring current due to oxidation of the product (H_2O_2) at the ring will be observed. The magnitude of the increase depends on the molar fraction of O_2^- in the primary flux of partially-reduced oxygen species released by the catalyst and the kinetic competition between O_2^- reduction by the scavenger and all other secondary sinks of O_2^-. The latter include recapture by the catalyst, escape from the film, electrochemical reduction or oxidation (depending on its protonation state), etc.

Catalytic films of **2a**Cu-free (Fe-only) catalysts containing 25% (mol) of one of the scavengers in Figure 1.19 were found to generate higher i_r/i_d ratios in O_2 reduction than similar films containing no additives, or additives structurally similar to the scavengers but inert to O_2^- (HO_2)[43]. The magnitude of the increase correlated with the reported rate constants for O_2^- reduction by these scavengers (Figure 1.19). No effect was observed with **2a**FeCu films. From these data it was concluded[43] that the primary side reaction at least for the **2a**Fe-only complex is superoxide-releasing hydrolysis of the catalyst-O_2 adduct (autooxidation— reaction 7 in Figure 1.12). Autooxidation is common for all Fe^{II}-porphyrin complexes

A

B

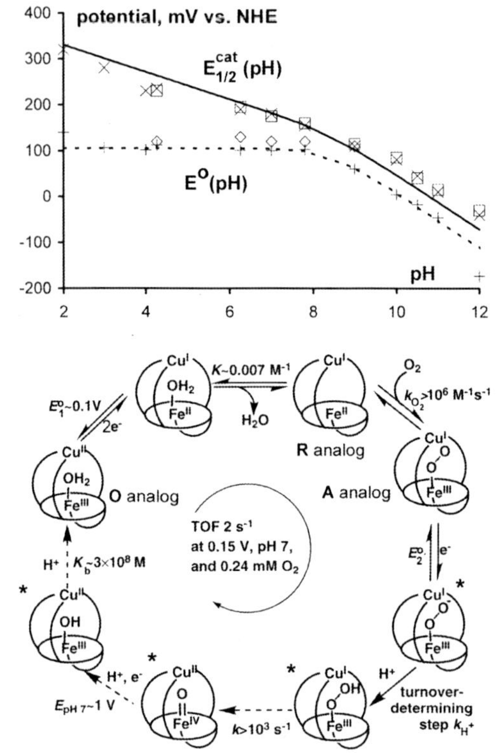

C

$$E_{\frac{1}{2}} = \frac{1}{f}\ln\left(\frac{\sqrt{\left(e^{-fE_1^o}\left(1+\frac{1}{K_b[H^+]}\right)+\frac{k_{-O_2}}{e^{fE_2^o}k_{H^+}[H^+]}\right)^2+172e^{-fE_1^o}\left(1+\frac{1}{K_b[H^+]}\right)\frac{k_{-O_2}}{e^{fE_2^o}k_{H^+}[H^+]}}-\left(e^{-fE_1^o}\left(1+\frac{1}{K_b[H^+]}\right)+\frac{k_{-O_2}}{e^{fE_2^o}k_{H^+}[H^+]}\right)}{e^{-fE_1^o}\left(1+\frac{1}{K_b[H^+]}\right)\frac{k_{-O_2}}{e^{fE_2^o}k_{H^+}[H^+]}}\right)$$

Figure 1.17. A plausible mechanism of O_2 reduction by catalysts **2**FeCu (Figure 1.14) in the physical contact with a graphite electrode. (A) The pH dependence of the formal potentials in the absence of O_2, $E^{o'}$ (◇, the $Fe^{III}Cu^{II}/Fe^{II}Cu^{I}$ potential of the FeCu complex; +, the $Fe^{III/II}$ potential of the Cu-free analog), and of the catalytic half-wave potentials, $E^{cat}_{1/2}$ (□, the FeCu catalyst; ×, the Cu-free catalyst). Note that the dependencies are identical for both the Cu-free and FeCu catalysts. The least-squares fit (LSF) of the $E^{o'}$ vs. pH data (A, broken line) indicates a water molecule (pK_a ~ 8.4) as an exogenous sixth ligand at the Fe^{III}. The $E^{cat}_{1/2}$ vs. pH data reveals two reversible electron transfer steps and a protonation as components of the turnover-determining part of the catalytic cycle. The mechanism in (B) is most consistent with these data (analogs of compounds O (H), R, and A of the C*c*O cycle, Figure 1.7, are indicated). Based on these data a rate law was derived (C) which fits well the experimental data (A, solid line), using a single adjustable parameter, $k_{-O_2}/e^{fE_2^o}k_{H^+}$ The reaction steps depicted with broken arrows are kinetically invisible at pH 7. Only the composition and the redox states (but not the structures) of intermediates in B indicated with asterisks are known. The other intermediates were prepared independently and characterized spectroscopically[43,46,53].

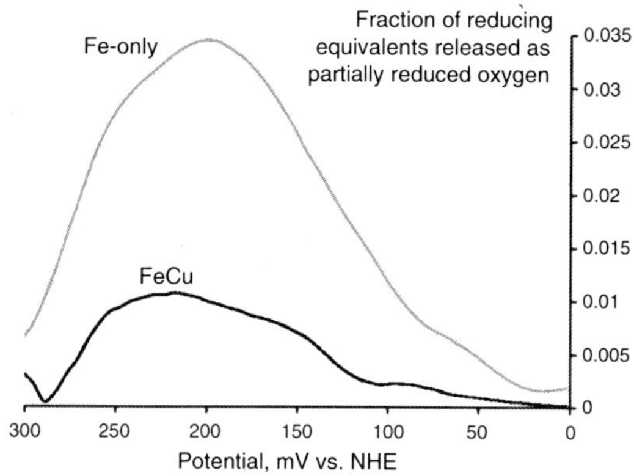

Figure 1.18. The potential-dependent selectivity of O_2 reduction by the **2b**FeCu and **2b**Fe-only catalysts under rapid electron flux represented as the ring-to-disk current ratio corrected for the collection efficiency of the ring.

AH

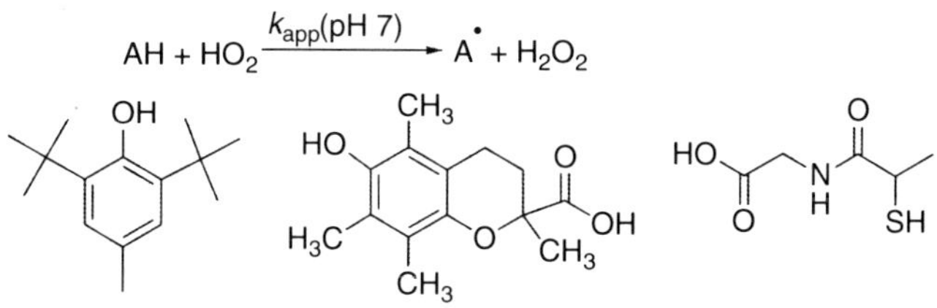

	BHT	trolox	MPG
k_{app}(pH 7), $M^{-1}s^{-1}$	2×10^4	10^5	$>2\times10^5$
i_r/i_d increase (±10 pcp)			
FeCu	<10%	<10%	<10%
Cu-free (Fe-only)	30%	80%	100%

Figure 1.19. Superoxide vs. H_2O_2 as the primary partially reduced oxygen species released within the catalytic film can be differentiated by observing the effect of superoxide scavengers incorporated in the catalytic film on the ring-to-disk current ratio (i_r/i_d). At a sufficient concentration in the film, the scavengers will reduce O_2^- (or HO_2) nonelectrochemically to H_2O_2, whose oxidation at the ring generates twice as much current as oxidation of O_2^- would. The apparent rate constants, k_{app}, are those reported in the literature for homogeneous reactions; pcp—percentage points.

(including hemoproteins) exposed to O_2, particularly in protic medium. Apparent autooxidation rate constants at pH 7 vary between 10^{-6} and 0.03 s^{-1} for hemoproteins and are even faster for certain synthetic macrocyclic Fe complexes[45]. Free O_2^- is also a possible intermediate in electrochemical O_2 reduction at surfaces[68].

The autooxidation rate constant for **2a**Fe-only was estimated to be ~0.03 s^{-1}, which is comparable to that of indoleamine dioxygenase[45] and the corresponding constant for the FeCu analog is probably ~5 times lower. A qualitatively similar difference in stabilities of the O_2 adducts of **2c**Fe-only and **2c**FeCu was observed in an anhydrous medium[53]. How distal Cu stabilizes the O_2 adduct is not known. Within this mechanism, the potential-dependent selectivity of the catalytic films (Figure 1.18) reflects the potential-dependent fraction of the catalyst-O_2 adduct (oxyheme or ferric-superoxide intermediate). The amount of the oxygenated catalyst increases as potentials decrease to ~200 mV, because a higher fraction of the catalytic film becomes reduced to the O_2-binding Fe^{II} state. As potentials become more reducing than ~200 mV, the fraction of the oxygenated catalyst decreases because: (a) the concentration of O_2 at the electrode/solution interface decreases due to its faster reduction by the modified electrode and (b) the redox equilibrium between the oxygenated catalyst and the ferric-peroxo intermediate (Figure 1.17) shifts toward the latter. At the plateau of catalytic waves, these two processes account for such a low steady-state concentration of the oxygenated catalyst that no detectable amount of O_2^- is released.

Although catalysts **2** successfully reproduce a number of stereoelectronic and catalytic properties of the heme/Cu site (Figure 1.5), their $Fe^{III}Cu^{II}/Fe^{II}Cu^{I}$ potential are lower than those of the heme a_3 and Cu_B (Figure 1.16a and Table 1.3) and their turnover frequency at potentials of cytochrome *c* are also likely lower (only the lower limit of the turnover frequency was established)[43]. It was argued that a very polar environment within the catalytic film efficiently stabilized the charged ferric state, thereby making the $Fe^{III/II}$ and $Cu^{II/I}$ potentials too reducing. In an environment whose dielectric constant is similar to that in the vicinity of the heme/Cu site, these synthetic catalysts were estimated to have potentials >250 mV[51]. The presumably lower turnover frequency of the catalysts at 250–50 mV was attributed[43] to: (a) the lower $Fe^{III/II}$ potentials as discussed above and (b) the very high population of the six-coordinate ferrous-aqua form of the catalyst (only <0.1% of the ferrous catalyst is present in the catalytically active five-coordinate state). The latter was also related to the high concentration of H_2O in the catalytic environment compared to that in C*c*O.

The major finding of this study was that despite the lower autooxidation rate of oxygenated **2a**FeCu, both the mechanism and the kinetics of O_2 reduction by electrode-adsorbed catalysts are comparable for the FeCu and Cu-free (Fe-only) forms. It was thus concluded that the distal Cu has no significant effect on catalysis when the electron flow to the catalytic sites is not kinetically limited (which is achieved by placing the catalysts in physical contact with an electrode).

However, as previously discussed, in vivo HCOs operate under kinetically limiting electron flux. To reproduce the physiologically relevant conditions of electron flux, the complexes were embedded at a low molar fraction in a lipid matrix on the electrode surface[55]. The kinetically limiting electron flux in such systems was established by: (a) the square root dependence of the $Fe^{III/II}$ peak currents (Figure 1.16c) under anaerobic conditions on the CV scan rate and (b) the independence of the catalytic currents (Figure 1.16d) on the electrode rotation frequency for air-saturated solutions. Under the conditions of kinetically limiting electron flow the Fe-only catalyst demonstrated much poorer catalytic activity than the FeCu analog (Figure 1.16d). This was explained by a substantially faster degradation of the Fe-only complex due to oxidation of the organic ligand, instead of Cu^I, during reductive O–O bond cleavage (Figure 1.20).

Although this degradation pathway may be important for a molecule in contact with an aqueous medium, many enzymatic redox reactions proceed by

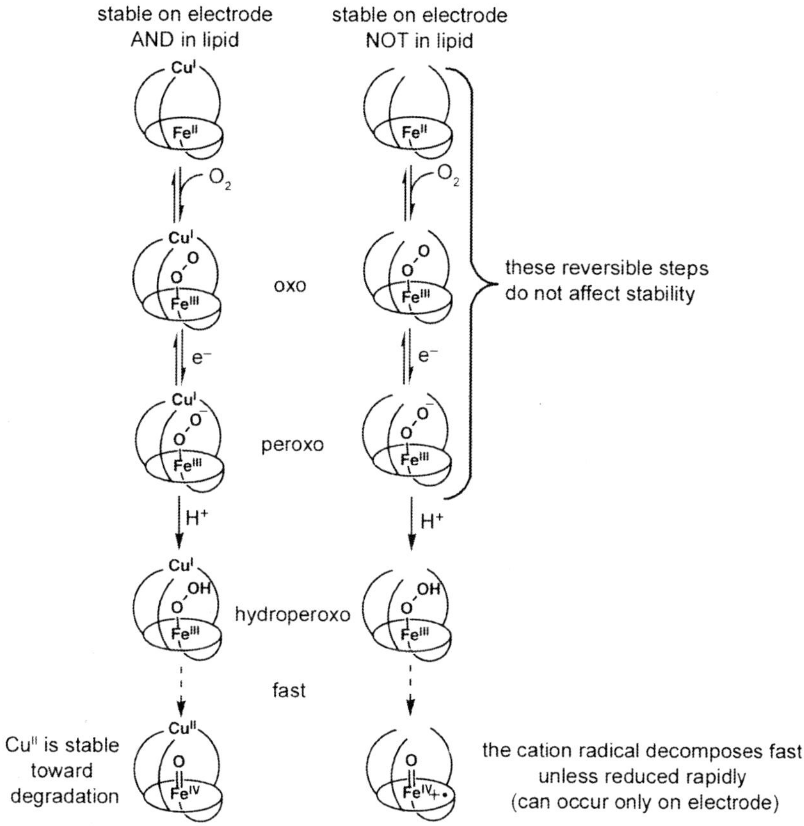

Figure 1.20. A plausible sequence of steps accounting for faster decomposition of **2**Fe vs. **2**FeCu in O_2 reduction under kinetically limiting electron flow, based on the kinetic mechanism in Figure 1.17.

oxidation of organic prosthetic groups and/or amino acid residues (see for example the postulated mechanism of C*c*O, Figure 1.7). In these enzymes the protein matrix protects these organic radicals from nucleophilic attack by H_2O and thus degradation. While it seems plausible that Cu_B (and Tyr244) in HCOs prevent oxidation of the heme to the cation radical, the function of such prevention is probably not to protect the heme against degradation. It would be very interesting to see how the replacement of Cu with phenol (the Fe-only derivative of **1**, Figure 1.14) would affect the catalytic behavior given that the phenol/phenoxyl-radical redox potential is more oxidizing that $Cu^{II/I}$ one.

The fact that only **2a**FeCu, but not **2a**Fe-only (Figure 1.14), has catalytic activity under kinetically limiting electron flux (whatever the origin of this might be) convincingly indicates that Cu in **2** can participate in O_2 reduction. On the other hand, Cu does not significantly affect the kinetics or the mechanism of O_2 reduction in the regime of rapid electron flow. This lack of an effect cannot be ascribed to the loss of Cu during electrocatalysis, because under identical conditions the FeCu catalyst manifests a significantly lower autooxidation rate and lower susceptibility to inhibition by CN^-, N_3^-, and CO (all biologically relevant poisons of HCOs, see Table 1.1)[54]. These observations suggest that distal Cu in **2** does not provide an O_2 reduction pathway that has a substantially lower activation barrier than that available to the Cu-free complex. However, under a biologically relevant kinetically limiting electron flux, when electrons for the reduction are not readily available from the environment (such as the electrode), Cu serves as an efficient intramolecular reductant. These heme/Cu analogs reproduce several major kinetic and mechanistic aspects of the O_2 reactivity of HCOs. Because of this functional fidelity, the effects of Cu on O_2 reduction seen in these biomimetic analogs may be reasonably expected to be present in C*c*Os as well. Consequently, Cu_B's function in the enzymatic O_2 reduction may be to store an electron in the immediate proximity of the catalytic center (heme a_3) rather than to provide a unique, low-activation reaction pathway (e.g., involving a bridged-peroxo intermediate) that would be inaccessible for the heme alone. Although HCOs have multiple electron storing sites besides Cu_B, the rate of electron transfer between heme *a* and heme a_3 is limited by the proton flow[70]. Hence, storing the fourth electron as ferroheme *a* instead of Cu_B^I may slow the 4e reduction of bound O_2 (compound A→P, Figure 1.7) enough for the release of H_2O_2 by hydrolysis of the ferric-hydroperoxo intermediate to become important. Certain biochemical experiments[33] independently suggest that the function of Cu_B in the O_2 reduction cycle of HCOs may be primarily electron storage. On the other hand, experiments with Cu_B-free mutants HCOs have not been conclusive[71–73]. Cu_B is well known to be the first site of O_2 binding to the heme/Cu center[18], although the physiological importance of this phenomenon is not known. Cu_B may also be involved in the proton pumping. This possibility presently cannot be addressed in biomimetic studies.

4. Summary and Conclusions: Lessons of Biomimetic O_2 Reduction for the Design of Fuel Cell Catalysts

Aerobic metabolism, based on molecular oxygen as the final acceptor of electrons of the respiratory electron transfer chain, is the most efficient form of energy metabolism known. Reduction of O_2 to H_2O is catalyzed by two distinct classes of enzymes: heme/Cu terminal oxidases (HCOs) and cytochrome *bd* oxidases, which span either mitochondrial or bacterial membranes. The major physiological functions of these enzymes are: (a) to clear the respiratory electron transfer chain from low-energy electrons and (b) to contribute to the build-up of transmembrane electrochemical proton gradients—the form of energy directly usable by cells. The terminal oxidases clear the respiratory electron transfer chain by oxidizing the least-reducing electron carriers, ferrocytochrome *c* (by cytochrome *c* oxidases (C*c*Os) and cytochromes *cbb*$_3$) or various quinols (by quinol oxidases and cytochromes *bd*). The lower the redox potentials of the least-reducing electron carriers are, the more energy is available for the build-up of transmembrane electrochemical gradients by the components of the respiratory electron transfer chain other than the terminal oxidase.

On the other hand, the overpotential of O_2 reduction (the difference in the standard redox potentials of the electron donor and the O_2/H_2O couple) provides the free energy that the HCO can transduce into the transmembrane potential gradient. All terminal oxidases contribute to the build-up of the transmembrane gradients by taking protons and electrons for $4e/4H^+$ reduction of O_2 from the opposite sites of the membrane, which expends 180–220 mV of the overpotential. In addition, all HCOs (but not cytochromes *bd* from prokaryotes) are proton pumps, i.e., they mechanically move protons (up to four per molecule of O_2) from the basic, negatively charged side of the membrane, to the acidic, positively charged side (another 180–220 mV, Figure 1.8). As a result, mammalian C*c*O reduces O_2 at an overpotential of $\sim$550 mV; it converts $>$80% of the free energy of this overpotential into protonmotive force at a power output of 4 MW mol^{-1}, while operating in a nearly neutral aqueous environment, at room temperature, and the O_2 pressure of $<$40 mmHg.

The catalytic O_2-reducing site of HCOs comprises a five-coordinate heme (Fe porphyrin) and a trisimidazole-ligated Cu. O_2 binds to and is reduced at the ferroheme of the 2e reduced (Fe^{II}/Cu^{I}) heme/Cu site. Reduction of O_2 to the redox level of H_2O is so rapid that no intermediates containing partially reduced oxygen species (e.g., various peroxo-type derivatives) have been observed during enzymatic turnover. Consequently, the exact mechanism of the O–O bond reduction is unknown. Likewise, Cu^{I} is oxidized during this step but whether Cu functions to form a transient, unobservable, bridged-peroxo intermediate, Fe^{III}–(O_2^{2-})–Cu^{II}, or serves simply as an electron storage site is also not known. Much less is established about the mechanism of O_2 reduction at the heme/heme catalytic site of cytochromes *bd*, whose crystal structure has not yet been determined. No peroxo-level intermediates have ever been observed during O_2 reduction by cytochromes *bd* either.

Studies of biomimetic respiratory O_2 reduction aim at: (a) designing small synthetic catalysts for electrochemical reduction of O_2 at biologically relevant electrochemical potentials ($>$50 mV vs. NHE) and in a neutral aqueous medium with the turnover frequency and the selectivity of HCOs and (b) using these catalysts to clarify the structure/activity relationship at the O_2-reducing heme/Cu site of HCOs. The best biomimetic catalysts reproduce key kinetic and mechanistic aspects of O_2 reduction by HCOs, but fall short of HCOs in their turnover frequencies at potentials $>$50 mV. Nor do any heme/Cu analogs operate as proton pumps. As a result the energy-tranducing efficiency and the power output of the best biomimetic electrocatalysts for O_2 reduction (using H_2 as the reductant) are $\sim$20% and $<$100 kW mol^{-1} ($\sim$100 W g^{-1}), respectively, at pH 7. The overpotentials are greater and the transducing efficiencies are lower in a more acidic medium. The degradation of the catalytic properties of these complexes during turnover is also fairly quick for industrial applications.

The main conclusion from the work on biomimetic O_2 reduction has been the sufficiency of a monometallic catalytic site for fairly selective reduction of the O–O or X–O bonds (be it in O_2, H_2O_2, peracids, or ClO_2^-) at potentials $>$50

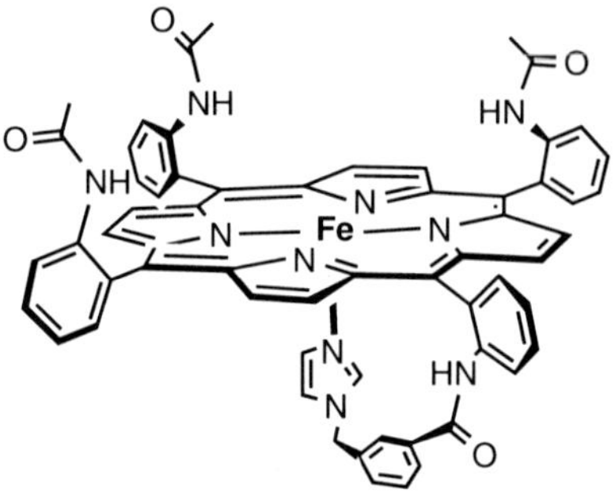

9

Scheme 1.1.

mV (at pH 7) using Fe porphyrin complexes in physical contact with an electrode. Arguably, a monometallic catalyst is easier to design and synthesize than a bimetallic one. The only structural feature of the heme/Cu site that appears to be critical is an axial ligation of Fe by imidazole or possibly pyridine. For this ligand to effectively compete with H_2O for Fe coordination, it must be covalently attached to the porphyrin in the appropriate (chelating) conformation. The nature of the distal environment has a smaller but still noticeable effect: replacing the "acetyl pickets" in **9** (Scheme 1.1) with imidazoles decreases the production of partially reduced oxygen species at 150 mV by a factor of ~3 (from ~7.5 to ~2.5%); an additional ~3-fold decrease (to ~0.5%, Figure 1.18) can be achieved by introducing the Cu ion. It is unclear if the improvements of this magnitude justify the synthetic effort required to create a tris-imidazole superstructure. We lack sufficient understanding of Fe-porphyrin-based reduction of O_2 to predict what, if any, groups can replace imidazoles to yield easier-to-synthesize and oxidatively more stable catalysts without diminishing their catalytic properties. Aryl substituents in the *meso* positions of the porphin increase the stability of the porphyrin toward oxidative degradation which probably makes *meso*-substituted porphyrins best suited for O_2-reduction electrocatalysts.

A strategic question is whether Fe-porphyrin-based systems represent the best candidates for the development of practical cathode catalysts for fuel cells. Although the catalytic sites for O_2 reduction in all known aerobic respiratory chains are exclusively hemes, the stereoelectronic properties of these enzymes have evolved in response to conditions that are irrelevant to those present at a fuel cell cathode, namely a turnover-limiting electron flux and the conversion of the free energy of O_2 reduction into a transmembrane electrochemical gradient of protons. For example, these enzymes utilize fairly reducing electron donors (no more than 250 mV vs. NHE at pH 7) and hence reduce O_2 at large overpotentials (at least 550 mV at pH 7). Indeed, it remains to be seen whether practically fast O_2 reduction at overpotentials <500 mV can be achieved with any metalloporphyrin site. As far as I am aware, knowledge to allow rational design of such catalysts is yet to be gathered.

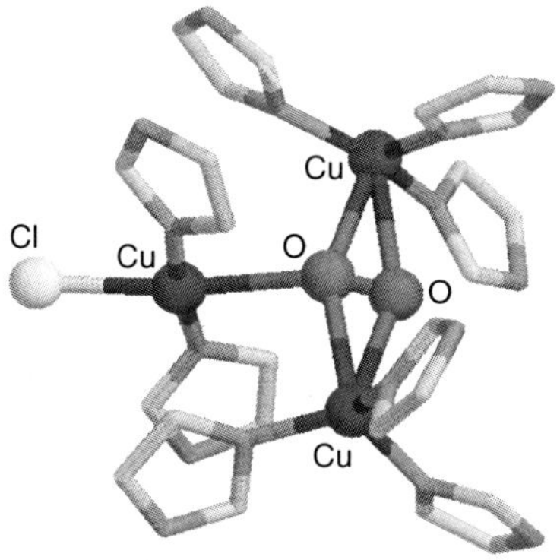

Figure 1.21. The catalytic O_2 reduction site of laccase (from *M. albomyces*)[76]. The unlabeled ligands are imidazoles of histidines. The amino acid chains are omitted for clarity.

A biomimetic catalyst based on the catalytic site of high-potential laccases may ultimately prove to be superior to any porphyrin-based complex. Laccases are widely distributed in fungi and plants and possibly in prokaryotic organisms, and are the simplest members of the family of multicopper oxidases[74,75]. The physiological function of laccases is 1e oxidation of various phenols, their derivatives (such as ethers), aromatic amines, and other nonphenolic substrates to highly reactive radicals using O_2 as the terminal oxidant. The latter is reduced to H_2O with small overpotentials at a tri-Cu catalytic site (Figure 1.21)[76]. The $Cu^{II/I}$ potentials of the pair of Cu ions that bind O_2 are ~800 mV at pH 5 (the optimal pH for these enzymes), compared to ~900 mV for the reversible potential of the O_2/H_2O couple at this pH. The catalytic mechanism is fairly well understood[77]. Electrode-immobilized laccases have been investigated for electrochemical oxygen reduction since at least 1984[78] and promising recent developments have been reported[79–82]. However, the narrow window of conditions (pH, ionic strength, temperature) under which the enzymes are active limits the suitability of such electrodes to niche applications. Arguably, a biomimetic analog might retain its activity over a wider spectrum of conditions. To the best of my knowledge, however, no biomimetic analogs of laccases exist. The trinuclear nature of the catalytic site undoubtedly makes it an exceptionally challenging synthetic target. Yet it appears that the challenge may be worth the effort not only for helping to clarify the enzymatic mechanism but also potentially for developing a practical catalyst for fuel cell cathode.

Acknowledgments

My interest in and appreciation of biological energy metabolism and catalytic O_2 reduction in particular originate from my biomimetic studies of HCOs carried out in the labs of Professor James P. Collman at Stanford. My own work on heme/Cu analogs, partially described in this review, would have been much more difficult without the help of Drs. Christopher J. Sunderland and Irina

M. Shiryaeva, and Professor Chris E.D. Chidsey. This review was written in the labs of Professor George M. Whitesides at Harvard.

References

1. Alberts, B., A. Johnson, J. Lewis, M. Raff, and P. Walter (2002). *Molecular Biology of the Cell*. Garland Science, New York.
2. Schults, B.E. and S.I. Chan (2001). Structures and proton-pumping strategies of mitochondrial respiratory enzymes. *Annu. Rev. Biophys. Biomol. Struct.* **30**, 23–65.
3. Nicholls, D.G. (1982). *Bioenergetics: An Introduction to the Chemiosmotic Theory*. Academic Press, London.
4. Dimroth, P., C.v. Ballmoos, T. Meier, and G. Kaim (2003). Electrical power fuels rotary ATP synthase. *Structure* **11**, 1469–1473.
5. Senios, A.E., S. Nadanaciva, and J. Weber (2002). The molecular mechanism of ATP synthesis by F_1F_0-ATP synthase. *Biochim. Biophys. Acta* **1553**, 188–211.
6. Moodie, A.D. and W.J. Ingledew (1990). Microbial anaerobic respiration. *Adv. Microb. Physiol.* **31**, 225–268.
7. See for example: Ligeza, A., A.N. Tikhonov, J.S. Hyde, and W.K. Subczynski (1998). Oxygen permeability of thylakoid membranes: Electron paramagnetic resonance spin labeling study. *Biochim. Biophys. Acta* **1365**, 453–463.
8. Wittenberg, J.B. and B.A. Wittenberg (2003). Myoglobin function reassessed. *J. Exper. Biol.* **206**, 2011–2020.
9. Huie, R.E. and P. Neta (1999). Chemistry of reactive oxygen species. In D.L. Gilbert and C.A. Colton, (eds). *Reactive Oxygen Species in Biological Systems*. Kluwer, New York, pp. 33–77.
10. Taube, H. (1986). Interaction of dioxygen species and metal ions—Equilibrium aspects. *Prog. Inorg. Chem.* **34**, 607–625.
11. Knoll, A.H. (2003). *Life on a Young Planet*, Princeton University Press, Princeton.
12. Rich, P.R. and A.J. Moody (1997). Cytochrome c oxidase. In P. Graber and G. Milazzo (eds), *Bioenergetics*, Vol. 4. Birkhauser, Basel, pp. 418–456.
13. Ferguson-Miller, S. and G.T. Babcock (1996). Heme/copper terminal oxidases. *Chem. Rev.* **96**, 2889–2907.
14. Zaslavsky, D. and R.B. Gennis (2000). Proton pumping by cytochrome oxidase: Progress, problems and postulates. *Biochim. Biophys. Acta* **1458**, 164–179.
15. Abramson, J., M. Svensson-Ek, B. Byrne, and S. Iwata (2001). Structure of cytochrome c oxidase: A comparison of the bacterial and mitochondrial enzymes. *Biochim. Biophys. Acta* **1544**, 1–9.
16. Michel, H. (1999). Cytochrome c oxidase: Catalytic cycle and mechanisms of proton pumping—A discussion. *Biochemistry* **38**, 15129–15140.
17. Mills, D.A., L. Florens, C. Hiser, J. Qian, and S. Ferguson-Miller (2000). Where is 'outside' in cytochrome c oxidase and how and when do protons get there? *Biochim. Biophys. Acta* **1458**, 180–187.
18. Michel, H., A. Harrenga, and A. Kannut (1998). Cytochrome c oxidase: Structure and spectroscopy. *Annu. Rev. Biophys. Biomol. Struct.* **27**, 329–356.
19. Wikstrom, M. (2004). Cytochrome c oxidase. *Biochim. Biophys. Acta* **1655**, 241–247.
20. Pereira, M.M., M. Santana, and M. Teixeira (2001). A novel scenario for the evolution of haem–copper oxygen reductases. *Biochim. Biophys. Acta* **1505**, 185–208 and references therein.
21. Pitcher, R.S. and N.J. Watmough (2004). The bacterial cytochrome cbb_3 oxidases. *Biochim. Biophys. Acta* **1655**, 388–399.

22. Junemann, S. (1997). Cytochrome bd terminal oxidase. *Biochim. Biophys. Acta* **1321**, 107–127.
23. Pereira, M.M., J.N. Carita, and M. Teixeira (1999). Membrane-bound electron transfer chain of the thermohalophilic bacterium *Rhodothermus marinus*: Characterization of the iron–sulfur centers from the dehydrogenases and investigation of the high-potential iron–sulfur protein function by in vitro reconstitution of the respiratory chain. *Biochemistry* **38**, 1276–1283.
24. Ostermeier, C., A. Harrenga, U. Ermler, and H. Michel (1997). Structure at 2.7 Å resolution of the *Paracoccus denitrificans* two-subunit cytochrome c oxidase complexed with an antibody Fv fragment. *Proc. Natl Acad. Sci. USA* **94**, 10547–10553.
25. Abramson, J., S. Riistama, G. Larsson, A. Jasaitis, M. Svensson-Ek, L. Laakkonen, A. Puustinen, and S. Iwata (2000). The structure of the ubiquinol oxidase from *Escherichia coli* and its ubiquinone binding site. *Nat. Struct. Biol.* **7**, 910.
26. Tsukihara, T., H. Aoyama, E. Yamashita, T. Tomizaki, H. Yamaguchi, K. Shinzawa-Itoh, R. Nakashima, R. Yaono, and S. Yoshikawa (1996). The whole structure of the 13-subunit oxidized cytochrome c oxidase at 2.8 Å, *Science* **272**, 1136–1141.
27. Rogers, M.S. and D.M. Dooley (2001). Posttranslationally modified tyrosines from galactose oxidase and cytochrome c oxidase. *Adv. Protein Chem.* **58**, 387–436.
28. Collman, J.P., R. Boulatov, and C.J. Sunderland (2003). Functional and structural analogs of the dioxygen reduction site in terminal oxidases. In K.M. Kadish, K.M. Smith and R. Guilard (eds), *The Porphyrin Handbook,* Vol. 11. Academic Press, Boston, pp. 1–49.
29. Proshlyakov, D.A., M.A. Pressler, C. DeMaso, J.F. Leykam, D.L. DeWitt, and G.T. Babcock (2000). Oxygen activation and reduction in respiration: Involvement of redox-active tyrosine 244. *Science* **290**, 1588–1591.
30. Blomberg, M.R.A., P.E.M. Siegbahn, and M. Wikstrom (2003). Metal-bridging mechanism for O–O bond cleavage in cytochrome c oxidase. *Inorg. Chem.* **42**, 5231–5243.
31. Blomberg, M.R.A., P.E.M. Siegbahn, G.T. Babcock, and M. Wikstrom (2000). Modeling cytochrome oxidase: A quantum chemical study of the O–O bond cleavage mechanism. *J. Am. Chem. Soc.* **122**, 12848–12858.
32. Collman, J.P., R. Boulatov, and C.J. Sunderland (2004). Functional analogs of cytochrome c oxidase, myoglobin and hemoglobin. *Chem. Rev.* **104**, 561–588.
33. Sigman, J.A., H.K. Kim, X. Zhao, J.R. Carey, and Y. Lu (2003). The role of copper and protons in heme-copper oxidases: Kinetic study of an engineered heme-copper center in myoglobin. *Proc. Natl Acad. Sci. USA* **100**, 3629–3634.
34. Kim, E., E.E. Chufan, K. Kamaraj, and K.D. Karlin (2004). Synthetic models for heme-copper oxidases. *Chem. Rev.* **104**, 1077–1134.
35. Naruta, Y., T. Sasaki, F. Tani, Y. Tachi, N. Kawato, and N. Nakamura (2001). Heme–Cu complexes as oxygen-activating functional models for the active site of cytochrome c oxidase. *J. Inorg. Biochem.* **83**, 239–246.
36. Kim, E., M.E. Helton, S. Lu, P. Moeenne-Loccoz, C.D. Incarvito, A.L. Rheingold, S. Kaderli, A.D. Zuberuehler, K.D. Karlin (2005). Tridentate copper ligand influences on heme-peroxo-copper formation and properties: Reduced, superoxo and m-peroxo iron/copper complexes. *Inorg. Chem.* **44**, 7014–7029.
37. Pratt, D.A., R.P. Pesavento, W.A. Van der Donk (2005). Model studies of the histidine-tyrosine cross-link in cytochrome c oxidase reveal the flexible substituent effect of the imidazole moiety. *Org. Letts.* **7**, 2735–2738.
38. Collman, J.P., R.A. Decreau, and C. Zhang (2004). Synthesis of cytochrome c oxidase models bearing a Tyr244 mimic. *J. Org. Chem.* **69**, 3546–3549.
39. Bard, A.J. and L.R. Faulkner (2001). *Electrochemical Methods*. Wiley, New York.

40. Shigehara, K. and F.C. Anson (1982). Electrocatalytic activity of three iron porphyrins in the reduction of dioxygen and hydrogen peroxide at graphite cathodes. *J. Phys. Chem.* **86**, 2776–2783.
41. Watanabe, Y. (2000). High-valent intermediates. In K.M. Kadish, K.M. Smith and R. Guilard (eds), *The Porphyrin Handbook*, Vol. 4. Academic Press, San Diego pp. 97–117.
42. Andrieux, C.P. and J.-M. Saveant (1992). Catalysis at redox polymer coated electrodes. In R.W. Murray (ed), *Molecular Design of Electrode Surfaces*. Wiley, New York pp. 207 - 270.
43. Boulatov, R., J.P. Collman, I.M. Shiryaeva, and C.J. Sunderland (2002). Functional analogs of the O_2 reduction site of cytochrome oxidase: Mechanistic aspects and possible effects of Cu_B. *J. Am. Chem. Soc.* **124**, 11923–11935.
44. See for example: Sono, M., M.P. Roach, E.D. Coulter, and J.H. Dawson (1996). Heme-containing oxygenases. *Chem. Rev.* **96**, 2841–2887.
45. Shikama, K. (1998). The molecular mechanism of autoxidation for myoglobin and hemoglobin: A venerable puzzle. *Chem. Rev.* **98**, 1357–1373.
46. Collman, J.P., C.J. Sunderland, and R. Boulatov (2002). Biomimetic studies of terminal oxidases: Trisimidazole picket metalloporphyrins. *Inorg. Chem.* **41**, 2282–2291.
47. Collman, J.P., M. Broring, L. Fu, M. Rapta, and R. Schwenninger (1998). Imidazole acid chlorides: Preparation and application in the syntheses of biomimetic heme models. *J. Org. Chem.* **63**, 8084–8085.
48. Collman, J.P., M. Broring, L. Fu, M. Rapta, R. Schwenninger, and A. Straumanis (1998). Novel protecting strategy for the synthesis of porphyrins with different distal and proximal superstructures. *J. Org. Chem.* **63**, 8085–8086.
49. Collman, J.P., R. Boulatov, C.J. Sunderland, and M. Zhong (2000). Fluorinated imidazoles as ^{19}F probes for biomimetic studies of Heme a_3–Cu_B site in cytochrome c oxidase. *J. Fluor. Chem.* **106**, 189–197.
50. Boulatov, R. (2004). Understanding the reaction that powers this world: Biomimetic studies of respiratory O_2 reduction by cytochrome oxidase. *Pure Appl. Chem.* **76**, 303–319.
51. Shiryaeva, I.M., J.P. Collman, R. Boulatov, and C.J. Sunderland (2003). Nonideal electrochemical behavior of biomimetic iron porphyrins: Interfacial potential distribution across multilayer films. *Anal. Chem.* **79**, 494–502.
52. Collman, J.P., I.M. Shiryaeva, and R. Boulatov (2003). Effect of electron availability on selectivity of O_2 reduction by synthetic monometallic Fe porphyrins. *Inorg. Chem.* **42**, 4807–4809.
53. Collman, J.P., C.J. Sunderland, K.E. Berg, M.A. Vance, and E.I. Solomon (2003). Spectroscopic evidence for a heme-superoxide/Cu(I) intermediate in a functional model of cytochrome c oxidase. *J. Am. Chem. Soc.* **125**, 6648–6649.
54. Collman, J.P., R. Boulatov, I.M. Shiryaeva, and C.J. Sunderland (2002). The distal Cu ion protects functional heme/Cu analogs of cytochrome oxidase from inhibition by cyanide and CO. *Angew. Chem., Int. Ed. Engl.* **41**, 4139–4142.
55. Collman, J.P. and R. Boulatov (2002). Electrocatalytic dioxygen reduction by synthetic analogs of the heme/Cu site of cytochrome oxidase incorporated in a lipid film. *Angew. Chem., Int. Ed. Engl.* **41**, 2487–2489.
56. Collman, J.P., R. Boulatov, C.J. Sunderland, I.M. Shiryaeva, and K.E. Berg (2002). Electrochemical metalloporphyrin-catalyzed reduction of chlorite. *J. Am. Chem. Soc.* **124**, 10670–10671.
57. Collman, J.P., M. Rapta, M. Broring, L. Raptova, R. Schwenninger, B. Boitrel, L. Fu, and M. L'Her (1999). Close structural analogues of the cytochrome c oxidase Fe_{a3}/Cu_B center show clean $4e^-$ electroreduction of O_2 to H_2O at physiological pH. *J. Am. Chem. Soc.* **121**, 1387–1388.

58. Collman, J.P., R. Schwenninger, M. Rapta, M. Broring, and L. Fu (1999). New 1,4,7-triazacyclononane-based functional analogues of the Fe/Cu active site of cytochrome c oxidase: Structure; spectroscopy and electrocatalytic reduction of oxygen. *Chem. Commun.*, 137–138.
59. Collman, J.P., L. Fu, P.C. Herrmann, Z. Wang, M. Rapta, M. Broring, R. Schwenninger, and B. Boitrel (1998). A functional model of cytochrome c oxidase: Thermodynamic implications. *Angew. Chem., Int. Ed. Engl.* **37**, 3397–3400.
60. Collman, J.P. and L. Fu (1999). Synthetic models for hemoglobin and myoglobin. *Acc. Chem. Res.* **32**, 455–463.
61. Chishiro, T., Y. Shimazaki, F. Tani, Y. Tachi, Y. Naruta, S. Karasawa, S. Hayami, and Y. Maeda (2003). Isolation and crystal structure of a peroxo-bridged heme–copper complex. *Angew. Chem., Int. Ed. Engl.* **42**, 2788–2791.
62. Ricard, D., A. Didier, M. L'Her, and B. Boitrel (2001). Application of 3-quinolinoyl picket porphyrins to the electroreduction of dioxygen to water: Mimicking the active site of cytochrome c oxidase. *Chembiochem* **2**, 144–148.
63. Ricard, D., B. Andrioletti, B. Boitrel, and M. L'Her (1999). Electrocatalytic reduction of dioxygen to water by tren-capped porphyrins, functional models of cytochrome c oxidase. *Chem. Commun.* 1523–1524.
64. Collman, J.P., M. Kaplun, C.J. Sunderland, and R. Boulatov (2004). Electrocatalytic reduction of ROOH. *J. Am. Chem. Soc.* **126**, 11166–11167.
65. Momenteau, M. and C.A. Reed (1994). Synthetic heme–dioxygen complexes. *Chem. Rev.* **94**, 659–698.
66. Babcock, G.T. and M. Wikstrom (1992). Oxygen activation and the conservation of energy in cell respiration. *Nature* **356**, 301–309.
67. Collman, J.P., L. Fu, P.C. Herrmann, and X. Zhang (1997). A functional model related to cytochrome c oxidase and its electrocatalytic four-electron reduction of O_2. *Science* **275**, 949–51.
68. Adzic, R. (1998). Recent advances in the kinetics of oxygen reduction. In J. Lipkowski and P.N. Ross (eds), *Electrocatalysis*. Wiley, New York, pp. 197–242.
69. Li, X. and A.A. Gewirth (2003). Electroreduction on bi-modified Au surfaces: Vibrational spectroscopy and density functional calculations. *J. Am. Chem. Soc.* **125**, 7086–7099.
70. Verkhovsky, M.I., J.E. Morhan, and M. Wikstrom (1995). Control of electron delivery to the oxygen reduction site of cytochrome c oxidase: A role for protons. *Biochemistry* **34**, 7483–7491.
71. Lemon, D.D., M.W. Calhoun, R.B. Gennis, and W.H. Woodruff (1993). The gateway to the active site of heme–copper oxidases. *Biochemistry* **32**, 11953–11956.
72. Mogi, T., T. Hirano, H. Nakamura, Y. Anraku, and Y. Orii (1995). Cu_B promotes both binding and reduction of dioxygen at the heme–copper binuclear center in the *Escherichia coli* bo-type ubiquinol oxidase. *FEBS Lett.* **370**, 259–263.
73. Moody, A.J., R. Mitchell, A.E. Jeal, and P.R. Rich (1997). Comparison of the ligand-binding properties of native and copper-less cytochromes *bo* from *Escherichia coli*. *Biochem. J.* **324**, 743–753.
74. Solomon, E.I., U.M. Sundaram, and T.E. Machonkin (1996). Multicopper oxidases and oxygenases. *Chem. Rev.* **96**, 2563–2605.
75. Claus, H. (2003). Laccases and their occurrence in prokaryotes. *Arch. Microb.* **179**, 145–150.
76. Hakulinen, N., L.-L. Kiiskinen, L. Kruus, M. Saloheimo, A. Paananen, A. Koivula, and J. Rouvinen (2002). Crystal structure of a laccase from *Melanocarpus albomyces* with an intact trinuclear copper site. *Nat. Struct. Biol.* **9**, 601–605.

77. Lee, S.-K., S.D. George, W.E. Antholine, B. Hedman, K.O. Hodgson, and E.I. Solomon (2002). Nature of the intermediate formed in the reduction of O_2 to H_2O at the trinuclear copper cluster active site in native laccase. *J. Am. Chem. Soc.* **124**, 6180–6193.
78. Lee, C.W., H.B. Gray, F.C. Anson, and B.G. Malmstroem (1984). Catalysis of the reduction of dioxygen at graphite electrodes coated with fungal laccase A. *J. Electroanal. Chem.* **172**, 289–300.
79. Barriere, F., Y. Ferry, D. Rochefort, and D. Leech (2004). Targeting redox polymers as mediators for laccase oxygen reduction in a membrane-less biofuel cell. *Electrochem. Comm.* **6**, 237–241.
80. Rowinski, P., R. Bilewicz, M.-J. Stebe, and E. Rogalska (2004). Electrodes modified with monoolein cubic phases hosting laccases for the catalytic reduction of dioxygen. *Anal. Chem.* **76**, 283–291.
81. Tarasevich, M.R., V.A. Bogdanovskaya, and A.V. Kapustin (2003). Nanocomposite material laccase/dispersed carbon carrier for oxygen electrode. *Electrochem. Comm.* **5**, 491–496 and references therein.
82. Barton, S.C. H.-H. Kim, G. Binyamin, Y. Zhang, and A. Heller (2001). Electroreduction of O_2 to water on the "wired" laccase cathode. *J. Phys. Chem. B* **105**, 11917–11921.
83. Rich, P.R. and J. Breton (2001). FTIR studies of the CO and cyanide adducts of fully reduced bovine cytochrome c oxidase. *Biochemistry* **40**, 6441–6449 and references therein.

2

Fundamental Aspects on the Catalytic Activity of Metallomacrocyclics for the Electrochemical Reduction of O_2

José H. Zagal, Maritza A. Páez, and J. Francisco Silva

1. Introduction

Molecular oxygen is certainly the most abundant oxidant on earth. Practically all of it is generated by living organisms, and without life it would not be present in the atmosphere. Molecular oxygen is generated by photosynthesis on earth and in the sea, so it can be considered a by-product of some living organisms and other living organisms profit from the free energy available from this molecule to sustain aerobic life. So molecular oxygen is a thermodynamically unstable molecule and will decay via reduction to form more stable species. So the reduction of molecular oxygen is favorable on thermodynamic grounds. However, at room temperature it reacts very slowly with reductants in the absence of catalysts. Indeed this is good, otherwise many things will burn rapidly at room temperature and life would not be possible as we know it.

In this chapter we discuss the catalytic properties of metallomacrocyclics for the electrochemical reduction of molecular oxygen in aqueous media. The different factors that affect their catalytic activity are discussed such as the nature of the central metal and the ligand, and the spatial conformation of some bimetallic molecules that possess special catalytic activity due to the binding of O_2 simultaneously to two metal centers. Catalytic activity of different metallomacrocyclics is discussed in terms of the different reduction pathways of O_2 that they promote, i.e., 2 vs. 4-electron reduction pathways.

José H. Zagal, Maritza A. Páez, and J. Francisco Silva • Departmento de Quimica de los Materiales, Facultad de Quimica y Biologia, Universidad de Santiago de Chile, Casilla 40, Correo 33, Santiago, Chile

N_4-Macrocyclic Metal Complexes, edited by José H. Zagal, Fethi Bedioui and Jean-Pol Dodelet. Springer Science+Business Media, Inc., New York, 2006.

2. Reaction Pathways for the Reduction of Molecular Oxygen

In spite of the considerable effort expended in trying to unravel the fundamental aspects of the O_2 electroreduction reaction, many details about the mechanism are not fully understood. The electrochemical reduction of oxygen is a multielectron reaction that occurs via two main pathways: one involving the transfer of two electrons to give peroxide, and the so-called direct four-electron pathway to give water. The latter involves the rupture of the O–O bond. The nature of the electrode strongly influences the preferred pathway. Most electrode materials catalyze the reaction via two electrons to give peroxide: Peroxide pathway

in acid

$$O_2 + 2H^+ + 2e^- \rightarrow 2H_2O_2 \qquad E_o = 0.67\text{V vs. NHE} \tag{2.1}$$

in basic media

$$O_2 + H_2O + 2e^- \rightarrow HO_2^- + OH^- \qquad E_o = -0.065\text{V} \tag{2.2}$$

Peroxide formation during O_2 reduction can be followed by its reduction

$$H_2O_2 + 2H^+ + 2e^- \rightarrow 2H_2O \qquad E_o = 1.77\text{V} \tag{2.3}$$

$$HO_2^- + H_2O + 2e^- \rightarrow 3OH^- \qquad E_o = 0.867\text{V} \tag{2.4}$$

or by its chemical decomposition

$$2H_2O_2 \rightarrow 2H_2O + O_2 \tag{2.5}$$

direct four-electron reduction pathway

in acid

$$O_2 + 4H^+ + 4e^- \rightarrow 2H_2O \qquad E_o = 1.229\text{V} \tag{2.6}$$

in alkali

$$O_2 + 2H_2O + 4e^- \rightarrow 4OH^- \qquad E_o = 0.401\text{V} \tag{2.7}$$

In strongly alkaline solutions or in organic solvents, $O_2$is reduced via the transfer of a single electron to give superoxide ion

$$O_2 + e^- \rightarrow O_2^- \qquad E_o = -0.33\text{V} \tag{2.8}$$

An important tool employed in the study of the reduction of dioxygen is the rotating ring–disk electrode technique[7]. The working electrode is the disk electrode, which is surrounded by a ring set to a constant potential acting as an amperometric sensor of peroxide (see Figure 1.9 of Chapter 1). The ring–disk set-up can be calibrated with well-known reactions in order to know the amount of reactant that is detected on the ring (this is called *collection efficiency*). When studying the reduction of O_2 on a ring–disk electrode it is possible to determine the amount of peroxide generated in the reaction. For example zero ring currents mean no peroxide is generated in the reaction and this is an indication of a four-electron reduction mechanism. The amount of peroxide is quantified as follows: $n = 4I_D/(I_D +$

$(I_R/N))$, where I_D, I_R and N are the faradaic current at the disk, the faradaic current at the ring, and the collection efficiency, respectively, and the percentage of hydrogen peroxide released in the reaction is then: $\%H_2O_2 = 100(4 - n)/2$.

In order to obtain the maximum free energy or the highest oxidant capacity of O_2 when this molecule reacts on the cathode of a fuel cell, it is necessary to reduce O_2 via four electrons as in Equations (2.6) and (2.7). So catalysts need to be developed that promote the four-electron reduction pathway. In practice, most electrode materials only promote the two-electron pathway, which releases almost one half of the free energy of the four-electron pathway. This can be attributed in part to the relatively high dissociation energy of the O–O bond (118 kcal/mol). Reduction of dioxygen to give water involves the rupture of the O–O bond and can involve the interaction of dioxygen with one site (single site) or with two active sites simultaneously (dual site) on the electrode surface. In this way, the energy of the O–O bond decreases, favoring bond rupture since electrons accepted by the oxygen molecule will occupy an antibonding π^* orbital (see Figure 2.1). Further, dioxygen is a triplet molecule in the fundamental state, with two unpaired electrons located in the double degenerate π^* antibonding orbital, so its interaction with singlet molecules is unfavorable. *A priori* one would expect that a catalyst whose fundamental state is not singlet could favor the reduction of O_2. On platinum, O_2 reduction occurs almost entirely via four electrons[7]. It is possible that on this metal a "bridge *cis*" type of interaction is favorable, which involves two metal active sites (see Figure 2.2) since the Pt–Pt separation in certain crystallographic orientations is optimal for this type of interaction. It is then very important to develop low-cost catalysts that decrease the overpotential of the reduction of oxygen and that can also promote the four-electron reduction.

Noble metals (e.g., platinum, gold, and silver), metal oxides, mixed oxides such as spinels, perovskites, and pyrochlores have been investigated, but by far the best catalytic material is highly dispersed platinum. The high cost of this metal stimulated research to look for less expensive alternative materials

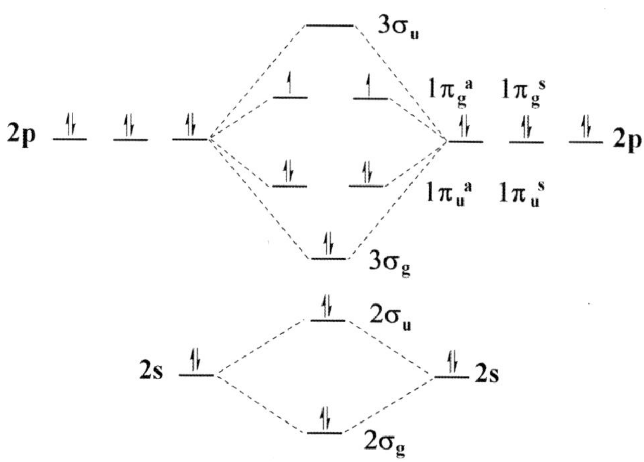

Figure 2.1. Molecular orbital diagram of molecular oxygen in the ground state. From ref. [6], reproduced with permission of Elsevier.

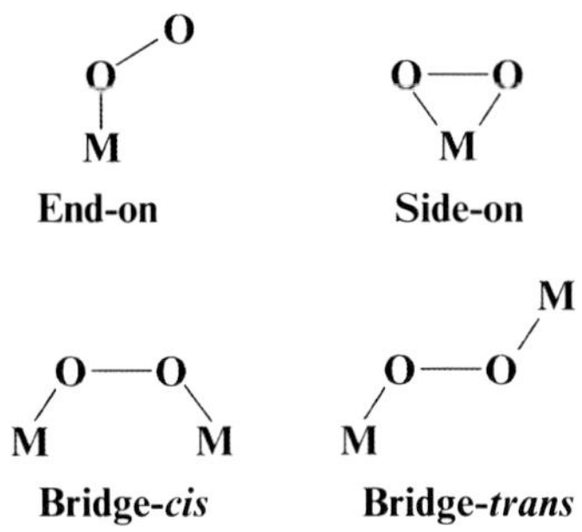

Figure 2.2. Different special configurations for molecular oxygen when it interacts with metal sites.

(non-noble metal catalysts). Inspired by biological catalysts like cytochrome *c* and hemoglobin, N_4 macrocyclics have been investigated very intensively in the last three decades, starting with the seminal work of Jasinski[1], who reported for the first time that cobalt phthalocyanine was catalytically active for the reduction of O_2, and several reviews have appeared in the literature in the last decades dealing with the electrocatalytic activity of MN_4 complexes for the reduction of dioxygen[2–9]. N_4 metal macrocyclic complexes are interesting because of the lower costs compared to noble metals, and for their high tolerance to methanol, for the case of methanol fuel-cell applications. They are also interesting because they provide models where active centers can be identified. The catalytic activity can also be modulated by changing the structure of the macrocyclic ligand. Figure 2.3 illustrates the chemical structures of some of the materials investigated.

Several authors have shown that the mechanism of the reaction is very sensitive to the nature of the metal center in the complex. For iron and manganese phthalocyanines, at low overpotentials a four-electron reduction is observed with rupture of the O–O bond[10]. In contrast Co, Ni, and Cu phthalocyanines only promote the reduction of O_2 via two-electrons to give peroxide[10]. Polymerized Co tetraamino phthalocyanines promote the four-electron[11,12] reduction whereas polymerized Fe tetraaminophthalocyanines only promote the two-electron reduction[13]. Vitamin B_{12}, which resembles the structure of cobalt porphyrin (see Figure 2.4) also promotes the four-electron reduction of O_2[14–16]. All this indicates that electronic factors around the metal center can dictate the course of the reaction. Two-electron catalysts such as cobalt porphyrins can be converted to four-electron catalysts by placing pendant groups on the periphery of the ligands that increase electron back-bonding[17]. Cofacial bimetallic complexes (see Figure 2.5) can be synthesized[18] to simultaneously bind both oxygen atoms in O_2, forming a bridge and promote the four-electron reduction. This is discussed in detail further on. The net catalytic activity of metal macrocyclics is linked to the redox potential of M(III)/(II) of the complexes, the more positive the redox potential, the higher the activity. This trend is opposite to what is expected from a simple redox catalysis mechanism, which is generally observed for the reduction of O_2 catalyzed by immobilized enzymes. The metal–N_4 chelates need to be

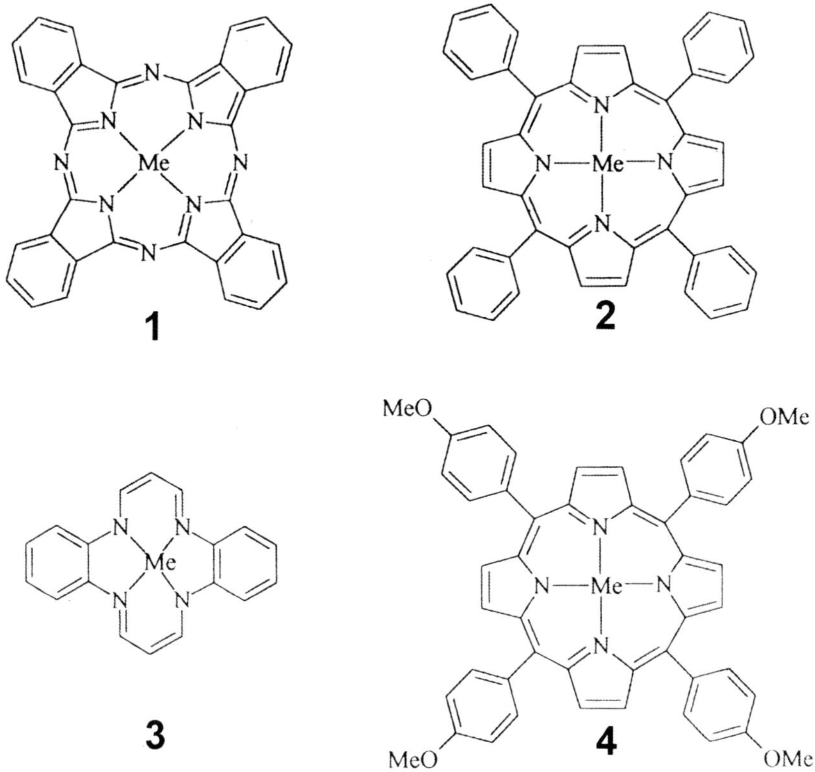

Figure 2.3. Structures of different MN_4 macrocycles. 1: metallophthalocyanine, 2: metal tetraphenylporphyrin, 3: metal tetraazaanuulene, and 4: metal tetramethoxyphenyl porphyrine.

supported on a conducting support, like carbon or graphite materials. Long-term stability is a problem with N_4-chelates. Heat treatment in an inert atmosphere increases both the stability and catalytic activity[19–23]. Heat-treated materials are discussed in detail in Chapter 3.

Even though a few studies have been carried out using the complexes in solution[24], most studies have been performed with the metal chelates confined on an electrode surface, generally graphite or carbon supports. Since the support can act as an axial ligand, the properties of the complexes in solution or on the adsorbed state could be different. So most studies discussed here have been carried out with the complexes immobilized on graphite or carbon supports. Smooth electrodes have been used to study mechanistic aspects of the reaction. In this chapter we discuss the catalytic activity of transition metal–N_4 chelates like those illustrated in Figures 2.3–2.5 and attention is paid to possible intermediates in the oxygen reduction reaction.

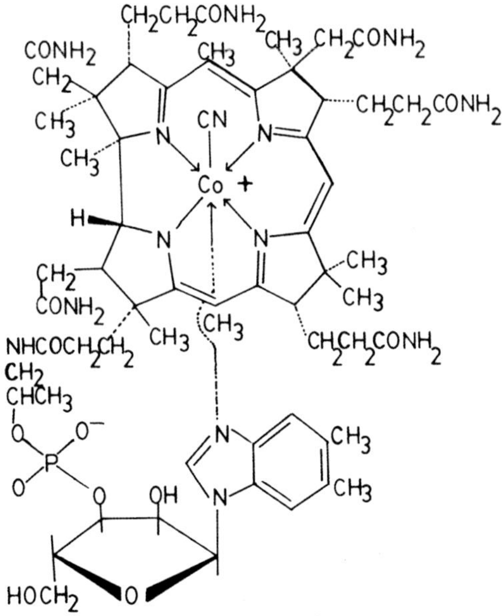

Figure 2.4. Molecular structure of vitamin B_{12} (cyanocobalamine).

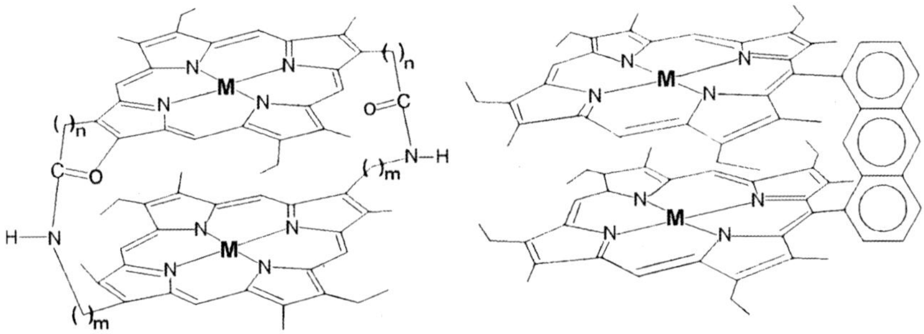

Figure 2.5. Molecular structure of some cofacial cobalt porphyrins.

3. Interaction of O_2 with Active Sites and the Redox Mechanism

O_2 reduction involving more than one electron is an inner-sphere reaction and involves a strong interaction of dioxygen with an active site. The outer sphere reduction of dioxygen involves the transfer of one single electron to give peroxide and probably occurs at the outer Helmholz plane. The kinetics of this outer-sphere reaction is independent of the nature of the catalyst and does not involve an interaction of dioxygen with the electrode surface. Since we are interested in studying the electrocatalysis of the reaction we focus our discussion on the inner-sphere reduction mechanisms.

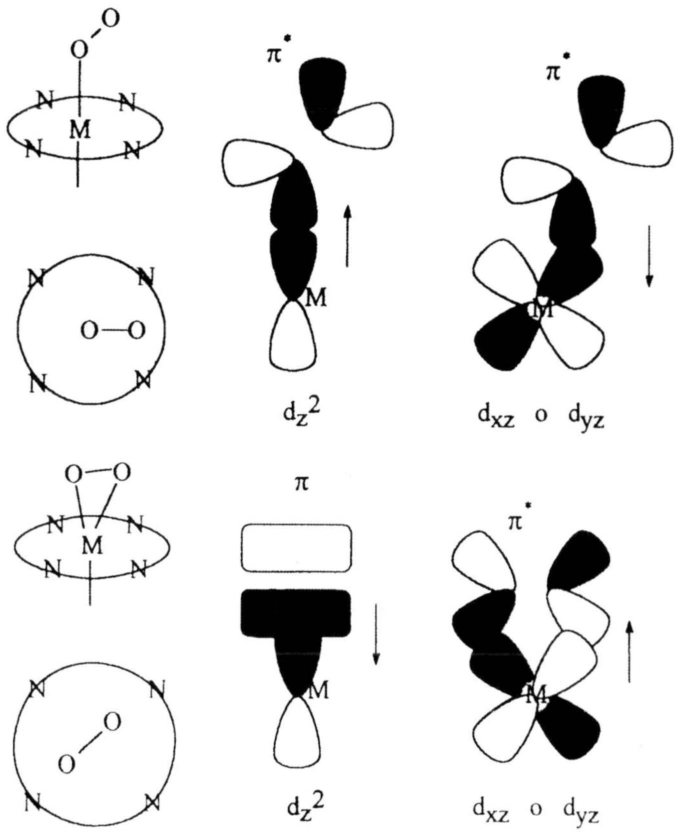

Figure 2.6. End-on and side-on interactions of frontier orbitals of molecular oxygen with orbitals of a metal site. From ref. [32], reproduced with permission of The Electrochemical Society Inc.

The binding of molecular oxygen to the N_4 catalyst involves binding to the d-orbitals of the central metal in the macrocyclic structure and will be influenced by the electronic density located on those orbitals. Figure 2.6 illustrates some of the different possible interactions (end-on and side-on) of the orbitals of the dioxygen molecule with the orbitals of the metal in the MN_4 molecule and Figures 2.7 and 2.8 illustrate the molecular orbitals diagram for an end-on and side interaction, respectively. For end-on M–O_2 complexes (see Figures 2.6 and 2.7), the major interaction for both σ and π bonding occurs with the π^* antibonding orbitals of the O_2. The σ interaction is primarily between the metal $3d_{z2}$ and the in-plane antibonding π_g^s orbital on the O_2, where the superscript "s" refers to whether or not the orbital is symmetric (or antisymmetric "a") with respect to the MO_2 plane. This can be viewed as an electron transfer (ET) to the O_2 from the metal ion. The π interaction is primarily between the metal $3d_{yz}$ and the $1\pi_g^a$ (π antibonding antisymmetric orbital) on the O_2 and can be viewed as a back-bonding interaction. In both of these interactions the π_u^s and π_u^a (bonding π orbitals symmetric and antisymmetric, respectively) play a lesser role

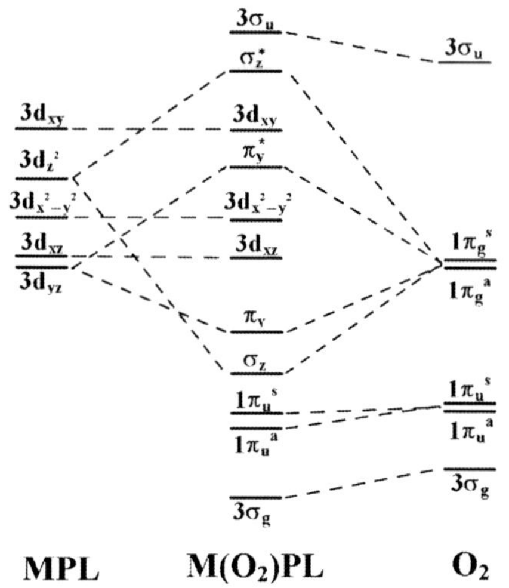

Figure 2.7. Qualitative molecular orbital diagram for the end-on $M(O_2)PL$ dioxygen adduct.(M = Fe,Co, P = porphyrin, L = NH_3 or imidazole). From ref. [25], reproduced with permission of the American Chemical Society.

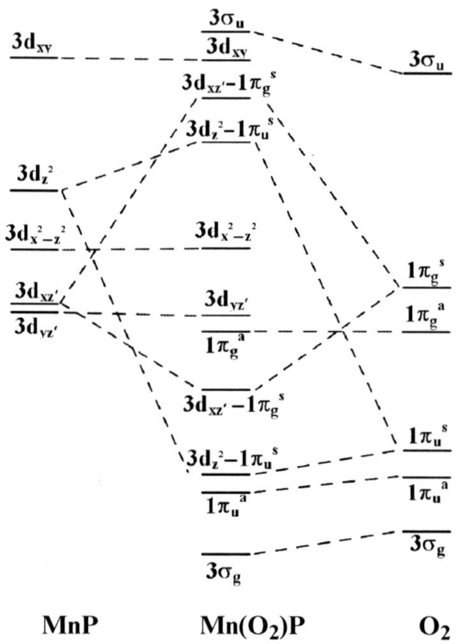

Figure 2.8. Qualitative molecular orbital diagram for a side-on $Mn(O_2)PL$ dioxygen adduct. From ref. [25], reproduced with permission of the American Chemical Society.

in the composition of the bonding orbitals[25]. For the side-on interaction a $3dz^2$ orbital of the metal will interact with the $1\pi_s^u$ bonding orbital (back-bonding interaction) and a 3 d_{xz} with a $1\pi_s^g$ antibonding orbital of dioxygen as illustrated in Figures 2.6 and 2.8.

The predominating interactions will weaken the O–O bond. The metal in the complex should be in the M(II) state, so for example in alkaline solution a step will require the reduction of M(III):

$$\text{M(III)–OH} + e \leftrightarrows \text{M(II)} + \text{OH}^- \tag{2.9}$$

An adduct will be formed according to:

$$\text{M(II)} + \text{O}_2 \rightarrow \text{M(III)–O}_2^- \quad \text{or} \quad \text{M(II)–O}_2 \tag{2.10}$$

This adduct must be short lived. Otherwise it will prevent further O_2 molecules from interacting with the active site. The adduct will undergo reduction as follows:

$$\text{M(III)–O}_2^- + e^- \rightarrow \text{M(II)} + \text{intermediates} \tag{2.11}$$

where M(II) is the active site. Equation (2.11) shows the process for basic media and could involve M(II)–O_2 instead of M(III)–O_2^- especially when Co is the metal center. In acid the process will involve a proton. The scheme above is applicable to Mn and Fe complexes. In the case of Co complexes, Co(III) is not formed upon its interaction with the dioxygen molecule but step 9 is still crucial since the catalytically active site is Co(II)[6].

When making comparisons between different catalysts it is important that the redox potentials or formal potentials are measured in the same conditions at which the oxygen reduction is investigated. This implies that the same electrolyte and the same pH is used since M(III)/(II) redox couples are pH dependent[10,26,27]. They should also be measured in the adsorbed state and not in solution since redox potentials of species in the homogeneous phase could differ from those measured with adsorbed species. For example for CoTMPyP and FeTMPyP (TMPyP = iron tetrakis(4-*N*-methylpyridyl)porphyrin) adsorbed on the edge plane of pyrolytic graphite (0.1 M H_2SO_4) differences of +0.41 and +0.3 V were found, respectively, for adsorbed and solution species[28,29].

According to the scheme above, the total driving force of the reaction will be given by the applied potential and also by the M(III)/M(II) formal potential of the catalyst, so the catalytic activity could be correlated with the formal potential of the catalyst. Many authors have discussed this issue and it is yet not clear what sort of correlation should be expected. Reduction should occur at the potential of reduction of the M(III)O_2^- adduct and not at the potential of the M(III)/(II) couple. The latter should only be observed if the reaction were outer sphere. For the particular case of iron phthalocyanines and other iron macrocyclics, O_2 reduction starts at potentials very close to the Fe(III)/(II) couple[2,6]. In contrast, for cobalt macrocyclics reduction of O_2 begins at potentials much more negative than those corresponding to the Co(III)/(II) couple[10]. Several authors have reported correlations between activity (measured as potential at constant current) and the M(III)/(II) formal potential and volcano-shaped curves have been obtained[5,10,30–32], see for

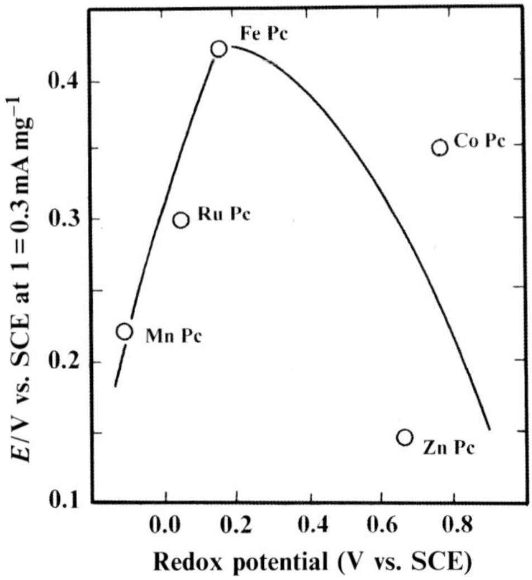

Figure 2.9. Volcano plot for the electroreduction of O_2 on different metal phthalocyanines. From ref. [6], reproduced with permission of Elsevier.

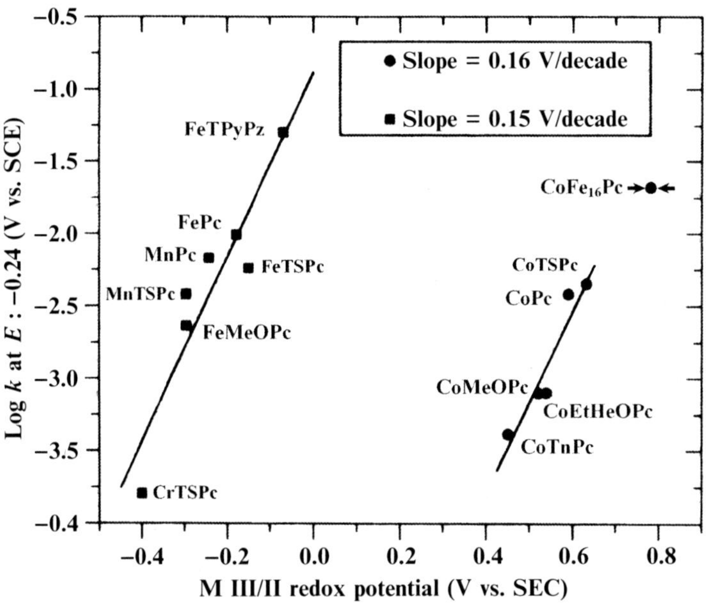

Figure 2.10. Plot of log k (at constant potential vs. the M(III)/(II) formal potential of the MN_4 macrocyclic for the reduction of oxygen in 0.2 M NaOH. From ref. [33], reproduced with permission of Elsevier.

example Figure 2.9. This could indicate that the redox potential needs to be located in an appropriate window to achieve maximum activity. In other words a M(III)/(II) formal potential that is too negative (easily oxidable metal center) or a M(III)/(II) formal potential that is too positive (metal center that is more difficult to oxidize) does not favor the catalysis. However more recent studies[33,34] have shown that when comparing families of metallophthalocyanines, linear correlations are obtained when plotting log k (rate constant at fixed potential) vs. the M(III)/(II) formal potential, as illustrated in Figure 2.10. (First-order rate constants were calculated as $k = I/nAFc$ where I is the current at a given potential; n is the total number of electrons transferred, which is 2 for the peroxide pathway and 4 for the reduction to H_2O, A is the area of the electrode in cm^2; F is the Faraday constant; and c is the oxygen concentration in moles per cm^3). What is interesting in the data of Figure 2.10 is that one linear correlation is obtained for Cr, Mn, and Fe complexes and these metals have configurations d^4(Cr), d^5(Mn), and d^6(Fe). Another linear correlation is obtained for Co complexes, which have a configuration d^7. The other interesting feature in the data of Figure 2.10 is that the lines are parallel with a slope close to 0.15 V/decade. It is not clear yet what is the meaning of this slope since there are no models available to explain these observations. However, one important conclusion for the data in Figure 2.10 is that the redox potential of the catalysts needs to be as positive as possible, and for a given complex this can be achieved by placing electron-withdrawing groups on the periphery of the macrocyclic ligand. The M(III)/(II) redox potential of some macrocyclics can be shifted in the positive direction with heat treatment. For example when iron tetraphenyl porphyrin[22], FeTPP, is heat treated the Fe(III)/(II) redox transition is shifted from 0.2 V vs. RHE for fresh FeTPP to 0.4 V for FeTPP heat treated at 700°C. Intermediate redox potentials are obtained for heat treatments at intermediate temperatures[22] and the catalytic activity increases with heat treatment, showing that a more positive redox potential of the catalyst favors the O_2 reduction reaction rate.

The increase in activity as the M(III)/(II) redox potential of the catalysts is more positive is in contrast to what was previously found in volcano correlations where a maximum activity is observed for intermediate redox potentials. It is also possible that the data in Figure 2.10 correspond to an incomplete volcano and further studies are necessary to clarify this point. A possible explanation for the results in Figure 2.10 (activity decreases as the driving force of the catalyst increases) is that the electronic coupling between the donor (MPc) and the acceptor (O_2) decreases as the electron-donating capacity of the substituents increases, due to a shift in the energy of the frontier orbitals of the metallophthalocyanine[34–36]. The shift in the energy of the frontier orbital with substituents on cobalt phthalocyanines has been estimated by Schlettwein[37] and Cárdenas-Jirón[38] using PM3 and ZINDO/S semiempirical theoretical calculations. There are several methods to estimate the electronic coupling matrix elements between the donor and the acceptor in ET reactions. One of them considers the energy difference between the LUMO (lowest unoccupied molecular orbital) of the electron acceptor and the HOMO (highest occupied molecular orbital) of the electron donor[39] but this requires a knowledge of the distance that separates the donor from the acceptor. This is not simple for an inner-sphere reaction where the $M\cdots O_2$ distance could vary from complex to complex. To avoid this, another reactivity index can be used to rationalize the data in Figure 2.10, and this is the concept of molecular hardness which is a commonly used criterion of reactivity in organic reactions and was proposed by Pearson[40,41]. The hardness η of a single molecule is approximately one-half of the energy gap of the HOMO–LUMO, so the larger the gap the greater the hardness of the molecule and the higher its stability (the harder the molecule the less its reactivity). The opposite would be a molecule with a narrow HOMO–LUMO gap (soft molecule) and this will make it very reactive. Now for a donor–acceptor pair it is more convenient to use the concept of donor-acceptor hardness η_{DA} which is one-half of the difference between the energy of the LUMO of the acceptor (O_2 molecule) and the energy of the HOMO of the donor (metal complex).

$$\eta_{DA} = \frac{1}{2}(\varepsilon LUMO_{Acceptor} - \varepsilon HOMO_{Donor}) \qquad (2.12)$$

Also, the donor–acceptor intermolecular hardness is one-half of the difference between the ionization potential of the donor and the electron affinity of the acceptor.

$$\eta_{DA} = -\frac{1}{2}(I_{Donor} - A_{Acceptor}) \qquad (2.13)$$

For a gas phase reaction involving the transfer of a single electron this will be equivalent to ΔG^o.

For the particular case of Co-phthalocyanine in its ground state, the highest occupied molecular orbital is occupied with a single electron (doublet state), so strictly speaking it corresponds to a single occupied molecular orbital (SOMO). The same is valid for molecular oxygen, which in its ground state has two unpaired electrons in two degenerate π^* antibonding orbitals. So the formation of an adduct CoPc···O_2 involves the interaction of two SOMOs and η_{DA} is given by:

$$\eta_{DA} = \frac{1}{2}(\varepsilon_{SOMO/Acceptor} - \varepsilon_{SOMO/Donor}) \qquad (2.14)$$

Figure 2.11 illustrates the calculated energy levels of the SOMOs of the different cobalt phthalocyanines with respect to the SOMO of dioxygen using PM3. Electron-withdrawing substituents (sulfonate, fluoro) on the phthalocyanine ring stabilize the SOMO and the opposite is true for electron-donating groups (methoxy and neopentoxy). So electron-withdrawing groups, even though they decrease the electron density on the cobalt (more positive redox potential), they also decrease the gap between the energy of the SOMO of the phthalocyanine and the energy of the SOMO of dioxygen. The bottom of Figure 2.11 shows that log k for O_2 reduction increases as the chemical hardness of the system decreases, or as the softness of the system increases (more reactivity). The trend in reactivity is exactly the same as that illustrated in Figure 2.10. So in qualitative terms it can be concluded that hardness could be used as a criterion for reactivity of these systems when comparing complexes that bear the same structure, and could explain why, for example, perfluorinated phthalocyanine, which has the most positive redox potential (the most oxidant), is the best catalyst for O_2 reduction in the series of cobalt phthalocyanines examined.

Quantum theories of elementary heterogeneous ET reactions in polar media have been extended to reactions which proceed through active intermediate electronic surface band states or bands. On the basis of this theoretical framework Ulstrup[42] has interpreted experimental data obtained for O_2 reduction catalyzed

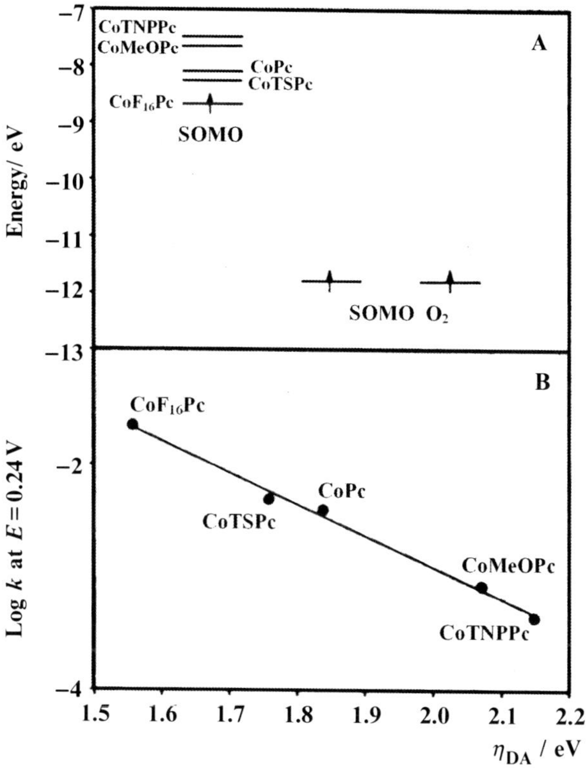

Figure 2.11. (A) Relative energies of frontier orbitals of dioxygen and substituted Co phthalocyanines. For simplicity, only one electron is shown on the SOMO of the CoPcs. (B) Plot of log k (at constant potential) vs. the donor–acceptor intermolecular hardness for the different O_2–CoPc pairs. From ref. [33], reproduced with permission of Elsevier.

by metal phthalocyanines. When comparing activities of complexes by plotting constant potential vs. redox potential of the catalyst, linear correlations are also obtained (see Figure 2.12) and this was predicted theoretically by the work of Ulstrup[42]. The slope of the lines in Figure 2.12 is less than 1 and this is also predicted by Ulstrup and is attributed to the excitation of intramolecular modes of relatively low frequencies in the cathodic range. The data shown in Figure 2.12 are essentially similar to those shown in Figure 2.10 but the comparison is made at constant current. So essentially, the graph of Figure 2.12 is a plot of driving force vs. driving force. This carries the assumption that the M(III)/(II) redox potential provides a measure of the driving force of the catalysts and this might not necessarily be true for inner-sphere reactions.

Not all metals of the first transition series exhibit the M(III)/(II) processes, so if one compares macrocyclics of different metals it is convenient to use another parameter, for example the number of d electrons in the metal as shown in Figure 2.13. In this figure since different Tafel slopes are obtained for the different catalysts it is not simple to compare activities as current at constant potential. So instead, as a criterion of activity, potential at constant currents is used. It can be clearly seen from this figure that Fe derivatives exhibit the highest

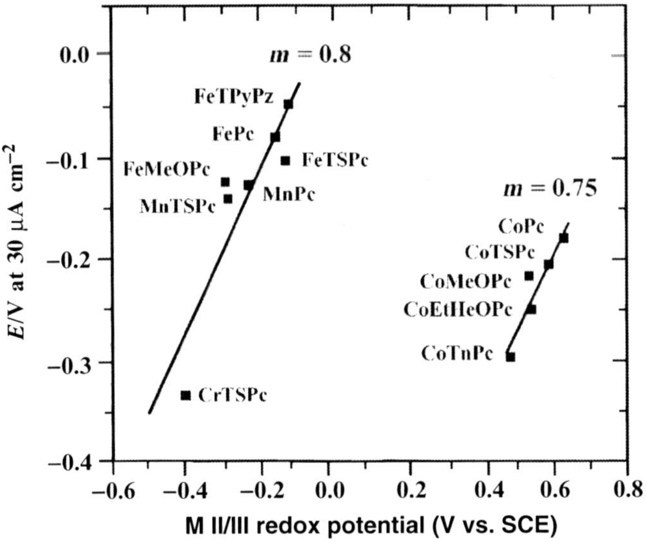

Figure 2.12. Plot of log E (at constant current) vs. the M(III)/(II) formal potential of the MN_4 macrocyclic for the reduction of oxygen in 0.2 M NaOH. From ref. [33], reproduced with permission of Elsevier.

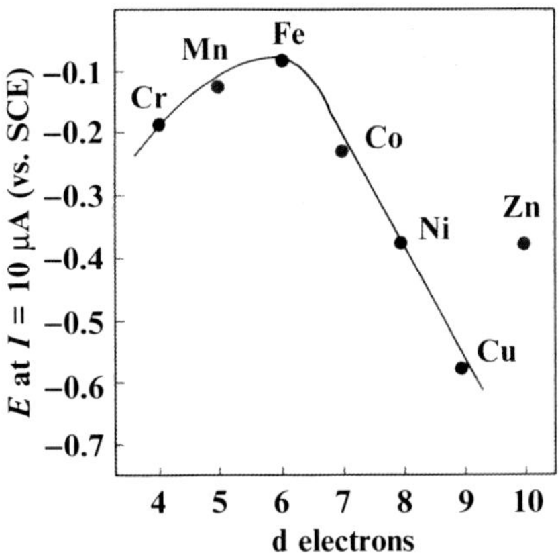

Figure 2.13. Volcano plot for the electrocatalytic activity of different M-tetrasulfonated phthalocyanines adsorbed on graphite for O_2 reduction in 0.1-M NaOH, as a function of the number of d-electrons in the metal. From ref. [6], reproduced with permission of Elsevier.

activity, followed by Mn and Co and also illustrates a common observation in catalysis that metals with nearly half-filled d-energy levels exhibit the highest activity. So a redox type of mechanism does not operate for metals that do not exhibit the M(III)/(II) transition. This is the case for Ni, Cu, and Zn phthalocyanines. It is important to comment at this point that in order to obtain catalytic activity, the frontier orbital of the M–N_4 needs to have some d character. The most active catalysts included in the data presented in Figure 2.13 (Cr, Mn, Fe, Co)

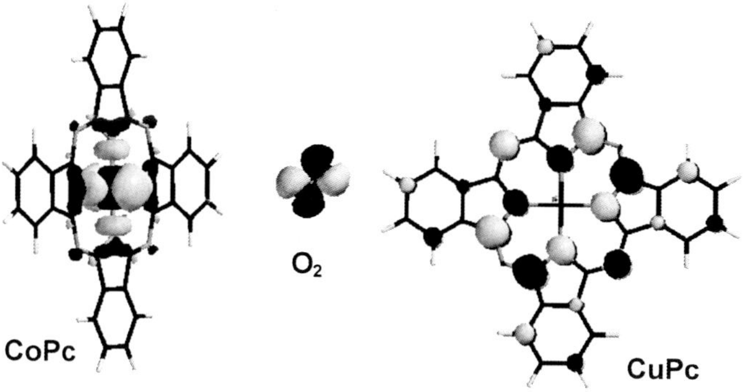

Figure 2.14. Illustration of the frontier orbitals involved in the interaction of cobalt phthalocyanine with O_2. From ref. [34], reproduced with permission of Elsevier.

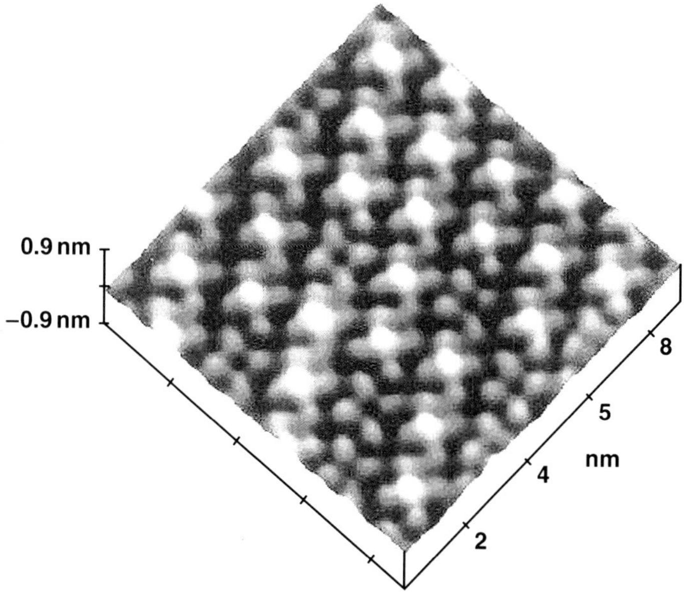

Figure 2.15. STM surface plot image of CoPc and CuPc coadsorbed on the (111) plane of Au. From ref. [44], reproduced with permission of the American Chemical Society.

have frontier orbitals with d character whereas Ni and Cu phthalocyanines, the frontier orbitals, have more ligand character[43]. This is graphically illustrated in Figure 2.14[34] that compares the frontier orbitals of CoPc and CuPc, and it is quite clear that CoPc shows a well, defined dz^2 orbital sticking out of the plane of the phthalocyanine whereas CuPc does not and the latter shows very low activity for O_2 reduction. There is experimental evidence to support this using tunneling electron microscopy. By this technique a strong d-orbital dependence on the images of metal phthalocyanines has been shown (see Figure 2.15) Unlike copper phthalocyanine where the metal appears as a hole in the molecular image, cobalt phthalocyanine shows the highest point in the molecular image[44]. The benzene regions of CoPc and CuPc show the same height. So essentially the data are in agreement with the theoretical calculations.

As discussed above, a redox type of mechanism would require that both the M(II) and M(III) oxidation states of the metal exist or are close to the potential region where oxygen reduction occurs. Fe and Co porphyrins are active as long as the M(III) state exists at a given potential[45]. FeTSPc (iron tetrasulfonated phthalocyanine) loses activity when the iron center is reduced from Fe(II) to Fe(I)[10] couple, and the complex goes from a four-electron reduction to a two-electron reduction catalyst, decreasing in this fashion the currents and increasing the amount of hydrogen peroxide formed.

Even though most authors agree that the M(II) state is the active site for O_2 reduction[2,4–6,10,26,30] for FePc (iron phthalocyanine) and FeNPc (iron naphthalocyanine)[46–48] it has been proposed that Fe(I) could also play a role in the electrocatalytic process. This was based on electroreflectance experiments that indicated that Fe(I) interacts with O_2 whereas Fe(II) does not. However, many authors have shown experimental evidence that O_2 reduction commences at potentials much more positive that those corresponding to the Fe(II)/(I) couple[10]. On the contrary, the reduction currents are observed at potentials close to the potential of the Fe(III)/(II) couple, so it seems unlikely that Fe(I) could be the active site. Worse, as shown from rotating ring–disk experiments, Fe(I) only favors the 2-electron reduction in contrast to Fe(II).[4,10]

Another factor that affects the catalytic activity is the amount of metal complex present on the electrode surface. In general, the amount of catalyst present on the surface is evaluated from cyclic voltammograms, measuring the electrical charge under reversible peaks. This carries the assumption that all adsorbed catalyst gives an electrochemical signal. This might not necessarily be true and there could be a fraction of complexes present on the surface that are electrochemically silent. It is assumed that the "electroactive" adsorbed species are also active for the reduction of O_2.

It has been found that the O_2 reduction currents are directly proportional to the amount of catalyst present [49,50], when the catalyst is adsorbed on the electrode surface indicating that the reaction is first order in the surface concentration of catalyst. This is not true for cases where the catalyst is incorporated to the surface by vapor deposition or when the catalyst is deposited from solutions and the solvent is completely evaporated[50]. An explanation for this observation is that when the catalyst is deposited by vapor deposition or from complete evaporation of solutions, multilayers are formed, and the metal active centers are not all completely accessible to O_2. This is also the case for polymerized multilayers of cobalt tetraamino phthalocyanines, where it has been shown that only the outermost layer was active for the reduction of O_2[11,12]. Scherson *et al.* have reported that when $(FeTMPP)_2O$ is deposited on a porous support, 30% of the amount deposited is found to be electrochemically active[51]. Anson *et al.*[52] have found that for the case of $CoPc(CN)_{16}$ and $CoPcF_{16}$ that were deposited from solutions where the solvent was completely evaporated, again it was found that only 30% of the amount deposited was electrochemically active. It was concluded that only those molecules that are attached to the surface of the electrode are active for the reduction of O_2. These results are not surprising since when multilayers are deposited, not all catalyst molecules are necessarily in electrical contact with the electrode, which is not the case for adsorbed layers, where molecules are probably lying flat on the electrode surface, interacting with the π system of the graphitic planes. Van der Putten *et al.*[53] have observed catalytic activity with vacuum-deposited layers in spite of the fact that these layers are electrochemically silent. Sometimes the redox response seems to be dependent on the electrolyte[54]. For example for adsorbed layers of CoTPyP on pyrolytic graphite in a solution of 0.5 M NH_4PF_6 in 0.5 M $HClO_4$no redox peaks were observed. However, when the same electrode is immersed in 0.5 M NH_4PF_6 in DMF, well-defined redox peaks are observed and this can be attributed to axial coordination of a DMF molecule, which improves the homogeneity of the adsorbed Co species, resulting in the well-defined redox transition.

With Co-facial cobalt porphyrins (Figure 2.5) adsorbed on pyrolytic graphite both cobalt centers are electrochemically active[18]. However if one of the cobalt centers is replaced by iron, this metal shows no redox activity . This cofacial porphyrin with two different metal centers shows catalytic activity for the reduction of O_2 and the foot of the reduction is observed at a potential where presumably the Fe(III)/(II) should appear. The fact that the Fe center is electrochemically silent is not clear[18].

As pointed out before when multilayers of metal phthalocyanines are deposited on an electrodic surface, only the outermost layer is active for the reduction of O_2[53] and this is also true for other electrochemical reactions. This shows that multilayers of phthalocyanines or polymerized multilayers of phthalocyanines are rather compact and the inner layers are not accessible to O_2 molecules[12].

4. Two-Electron Reduction Catalysts for the Reduction of Molecular Oxygen

The majority of mononuclear Co macrocyclics catalyze the reduction of dioxygen via two electrons to give peroxide[55]. The activity of iron phthalocyanines in general is higher than those of cobalt phthalocyanines and the opposite is true for porphyrins, which reveals the importance of the nature of the ligand in determining the catalytic activity [30,56]. The opposite is true for heat-treated materials[57] and this is discussed in detail in Chapter 3. Cobalt complexes are more stable than iron complexes and this trend is maintained after heat treatment[58]. However, iron complexes tend to promote the four-electron reduction of dioxygen and this is discussed further on. Cobaloxime complexes have catalytic activity for the two-electron reduction to give peroxide and modified electrodes using these complexes were obtained by preanodizing vitreous carbon in order to improve the adherence of these molecules to the electrode surface[59,60].

Surface groups present on the electrode seem to increase the catalytic activity[30,61–63]. These surface groups might axially coordinate to the cobalt center. The explanation given is that a basic group will enhance the electron density on the metal ion and lower the Co(III)/(II) redox potential. However, this statement disagrees with the fact that a more negative redox Co(III)/(II) redox potential does not favor the reaction as seen in the data of Figures 2.10 and 2.12[33] and by the results reported by Bouwkamp-Wijnoltz for heat-treated FeTPP and CoTPP[22]. However these surface groups seem to increase the stability of the electrode containing the Co complex[64,65]. O_2 reduction is not affected by the presence of cyanide ions which is surprising since electrochemical measurements in the absence of dioxygen have revealed that the Co(III)/(II) formal potential is shifted by almost 0.3 V in the negative direction in the presence of cyanide[66].

Even though most of the work reported has involved planar cobalt complexes nonplanar macrocylics also catalyze the reaction. For example, nonplanar cobalt tetrakis-(4-sulfonatophenyl)β-octabromoporphyrin catalyzed the reduction of O_2 via two electrons to give peroxide[67]. Vitamin B_{12}, which is also nonplanar, and its structure resembles that of a porphyrin (see Figure 2.4.), catalyzes the reduction of O_2 via two electrons to give peroxide at low potentials and via four electrons to give water at higher overpotentials[14–16].

Lamy *et al.*[68] and van der Putten *et al.*[69] conducted spectroscopic investigations of polymer-modified electrodes containing CoTSPc using UV–visible differential reflectance spectroscopy and were able to identify Co(III)→Co(II) transition when varying the electrode potential. They used electron spin resonance on the Ppy (polypyrrol) and Ppy-CoTSPc electrodes in deoxygenated and oxygen-saturated solutions. It was shown that the Co(III)TSPc species is effective in the electroreduction of oxygen and that this species is more stable in oxygen-saturated medium than in deoxygenated medium because of its stabilization under the following form: Co(III)..O_2. In the case of Ppy-CoTSPc film the polypyrrole matrix undergoes strong interactions with oxygen species, and most likely, with hydrogen peroxide.

Phosphoric acid is one of the electrolytes used in fuel cells and very few reports have focused their attention on the activity of metallomacrocyclics in this electrolyte (without heat treatment). Vasudevan *et al.*[70] investigated the electrocatalytic activity of cobalt phthalocyanine monomers and polymers with imido and carboxylic group ends. The complexes were mixed with carbon powder and polyethylene powder. The activity of the monomeric compounds was found to be higher than that of polymeric compounds.

With the aim of finding applications in polymer electrolyte fuel cells (PEFC) Co tetramethoxyphenylporphyrin (CoTMPP) and cobalt tetraazaannulene (CoTAA) were tested for O_2 reduction in acid media (1M H_2SO_4) with the catalysts bound with Nafion and deposited on a glassy carbon disk[71] and rotating-disk measurements revealed high peroxide yields, as expected for cobalt complexes.

The catalytic activity for O_2 reduction of CoTPP (cobalt 5, 10, 15, 20-tetraphenylporphyrin) was examined by Chung and Anson[72] by using the complex dissolved in a thin layer of acidified benzonitrile attached to a graphite electrode and equilibrated with an aqueous solution of $HClO_4$ saturated with air. It was found that the presence of benzonitrile enhanced the rate of the reduction of molecular oxygen compared to the case where CoTPP is simply adsorbed on graphite. With the aim of observing a Co(II)–O_2adduct directly Steiger and Anson[73] examined the catalytic activity of a "picket fence porphyrin," namely cobalt 5, 10, 15, 20-tetrakis($\alpha, \alpha, \alpha, \alpha$-2-pivalamido-phenyl)porphyrin (CoTpivPP) since it forms O_2

adducts in organic solvents at room temperature. They found that the O_2 adduct is formed in benzonitrile and axial coordination occurs with 1-methylimidazole (MeIm) present in the electrolyte. The coordination of molecular oxygen to (1-MeIm)Co^{II}TpivPP induces small changes in the electrochemical response of the complex. The coordination of O_2 seems to add to the delay in the rate of addition of a second 1-MeIm ligand to the oxidized Co(III) complex. In benzonitrile O_2 is reduced to O_2^-, and it is a non-catalytic diffusion-controlled process. CoTpivPP shows some catalytic activity for the reduction of O_2 in acid but it is less effective than CoTPP either when present in thin layers of benzonitrile attached to a graphite electrode or adsorbed on the electrode surface.

Murray *et al.*[74] prepared a tetraphenylporphyrin derivative with four thiol moieties, 5,10,15,20-tetrakis(*o*-(2-mercaptoethoxy)phenyl)porphyrin, Co(*o*-TMEPP) and was shown to form molecular films via irreversible chemisorption on a variety of gold substrates (see Figure 2.16). The presence of the adsorbed layers was verified by X-ray photoelectron spectroscopy which confirmed that, on average, about three of four thiolates per porphyrin become bound to the gold surface. Surface coverage determinations suggested that the molecules lie flat on the gold surface and that the adsorbed layers are far more stable than adsorbed layers of CoTPP (Co-tetraphenylporphyrin without thiols groups on the periphery of the ligand). This result is not surprising since molecules with thiol functionalities are known to adsorb strongly on gold substrates. Molecular films of Co(*o*-TMEPP) were found to reduce oxygen via two electrons to give hydrogen peroxide in both acid and alkaline media. The films retained their catalytic activity for more than 10^5 turnovers. The amount of peroxide formed during the O_2 reduction reaction was estimated using gold/platinum interdigitated array electrodes where the collection efficiency was as high as 90%. Murray *et al.*[75] also carried out visible spectroscopy together with electrochemical measurements of Co(*o*-TMEPP) monolayer films chemisorbed onto transparent gold electrodes and

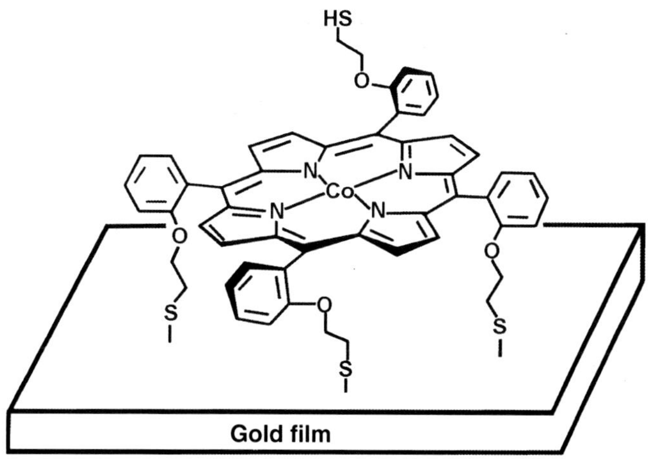

Figure 2.16. Co(*o*-TMEPP) adsorbed on a gold surface. From ref. [75], reproduced with permission of the American Chemical Society.

found that transmission spectra of these films indicate approximately monolayer coverage and edge–edge porphyrin–porphyrin interactions within the monolayer, as evidenced by red shifts and broadening in the Soret bands, compared to the spectra in solution. By the use of spectroelectrochemical measurements of Co(*o*-TMEPP) adsorbed on transparent gold it was possible to monitor the redox state of the porphyrin. It was found that the onset of O_2 reduction coincides with the Co(III) $\longrightarrow$ Co(II) transition in the adsorbed porphyrin, indicating that the active state is Co(II)(*o*-TMEPP).

If one moves in the periodic table or in the volcano plot of Figure 2.13 in the direction of metals that have more electrons in the d-orbitals, the activity decreases[5,10,30,31]. Ni, Cu, and Zn macrocyclics have low activity and catalyze the reduction of O_2 only to peroxide. The lowest activity is shown by CuPc and CuTSPc. Ni, Cu, and Zn complexes do not exhibit the M(III/(II) transition because between the energy levels of the HOMO and the LUMO of the ligand, they have no energy levels with d-character like for example Cr, Mn, Fe, and Co complexes[76,77]. So the low activity is associated to the fact that for these complexes the frontier orbitals have no d-character, so O_2 cannot bind to the metal center or the interaction is not favorable. This is clearly illustrated for CuPc in Figures 2.14 and 2.15. Their catalytic activity is then always lower than that found for iron and cobalt chelates[5,10,17,19]. Pt, Rh, and Os macrocyclics[56] show some activity but they give peroxide as the main product of the reaction[78].

5. Four-Electron Reduction Catalysts

As discussed before, a generally accepted thought in the electrocatalysis of the four-electron reduction, involving the rupture of the O–O bond is that the dioxygen molecule requires a simultaneous interaction with two active sites (dual-site mechanism). With this in mind, cobalt macrocyclic complexes having two Co centers have been designed, and having a Co–Co separation that would favor the simultaneous interaction of dioxygen with both active sites. Some typical complexes are shown in Figure 2.5. Anson and Collman were pioneers on this subject[17,79,80] and were the first to report a four-electron reduction process using cofacial cobalt porphyrins like the one illustrated in Figure 2.5 (left). It was found that the optimal Co–Co separation for achieving a four-electron reduction mechanism was 4 Å which seems to indicate that a Co–O–O–Co bridge is formed within the cavity of the cofacial complex. This was observed in acid media. In alkaline media only a two-electron reduction was observed[17]. Pillared dicobalt cofacial porphyrins complexes[81] (Figure 2.5, right), planar dinuclear[82], and polynuclear cobalt phthalocyanines[83] also promote the reduction of O_2 via 4 electrons[65]. However the four-electron pathway is favored in alkaline media whereas with cofacial porphyrins the four-electron pathway is favored in acid. This difference in reactivity was explained by considering several effects[64]. The formation of an O–O bridge in the cavity of the dinuclear complexes is a prerequisite for the splitting of the O–O bond and the electron density on the cobalt center determines the extent of weakening of the O–O bond. For planar dicobalt complexes[82] the opposite is observed since a two-electron reduction to peroxide is observed in acid media whereas in alkaline media a four-electron reduction to water is observed. A possible explanation for the differences in reactivities of cofacial and planar dicobalt complexes comes from the fact that O_2 could interact with the cobalt center in different fashions. For example for the cofacial complex it is likely that O_2 adopts a "bridge trans" configuration whereas for the planar dicobalt complexes, the configuration is "cis" (see Figure 2.2). A scheme for the reaction with cofacial porphyrins is shown in Figure 2.17. It can be seen in this scheme that protonation of the trans O_2 adduct leads to the rupture of the O–O bond.

Following the line of thought that cofacial structures would promote the four-electron reduction of O_2 some authors have suggested that it is possible that when depositing cobalt complexes, they could adopt configurations where two cobalt centers are separated by an optimal distance to promote the splitting of the O–O bond. An example of this is the work of Biloul *et al.*[84,85] using heat-treated CoNPc-deposited carbon black. These authors found evidence for a four-electron reduction process as they determined the number of electrons to be *ca*. 3.5.

Anson *et al.*[86] found evidence for a four-electron reduction mechanism on adsorbed multilayers of cobalt tetra-4-pyridylporphyrin on a graphite electrode but considerable amounts of peroxide were produced in the reaction. In a series of articles Anson and coworkers[17,54,65,87–93] have shown that by placing Ru(II)pentaamino complexes on the periphery of the ring in Co porphyrins

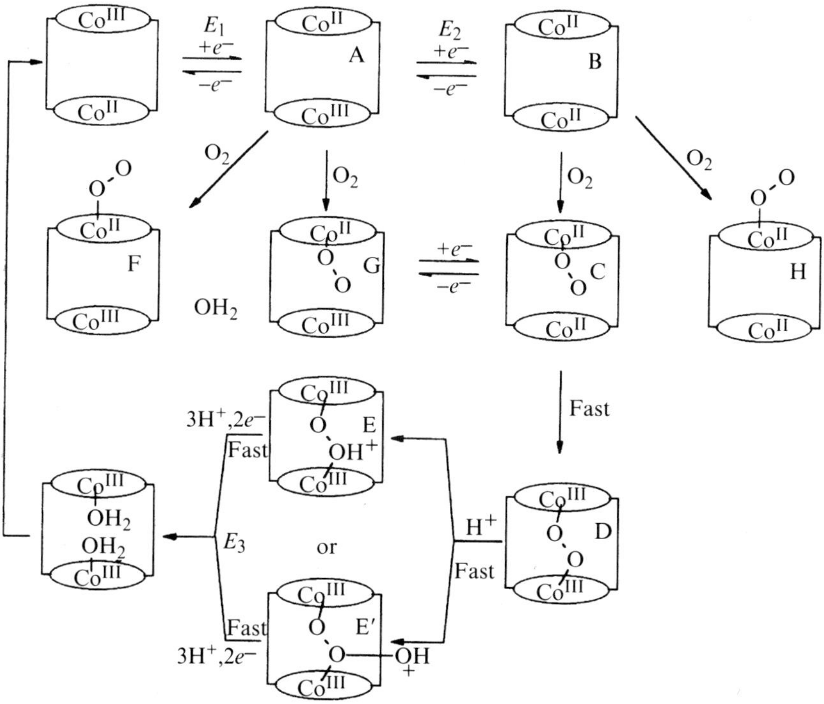

Figure 2.17. Reaction scheme for the reduction of O_2 on cofacial Co porphyrins. From ref. [18], reproduced with permission of the American Chemical Society.

it is possible to transform two-electron catalysts into four-electron catalysts. This is the case for Co(*meso*-tetrakis) (4-pyridyl) that contained 4 pentaamino Ru groups[87]. In this work it was found that the reaction between the Co porphyrin and the Ru complex, in Nafion films attached to graphite, generates a $CoP(pyRu(NH_3)_5)_4$complex. The $pyRu(NH_3)_4$ groups on each molecule interact amongst themselves very weakly and show practically the same formal potential. So the whole molecular structure catalyzes the four-electron reduction of O_2. It was found that the rate-determining step is the formation of the adduct and the transfer of electron from the Ru centers to the Co(II)-O_2 adduct is fast. It is concluded that the four Ru centers transfer their electrons simultaneously to the O_2 coordinated to Co(II). The ruthenated complex without Co is not catalytically active for the reduction of O_2. The catalytic reduction occurs at potentials at which the reduction of the Ru occurs, so this is consistent with the fact that electronic effects occur on the Co center, which are transmitted via π-back-bonding to the reduced Ru centers. This hypothesis is supported by studies of the catalytic activity of Co porphyrins which contain three $Ru(NH_3)_5^{+2}$ groups coordinated to cyanophenyl ligands[65] (see Figure 2.18). When a porphyrin bearing 3 cyanophenyl groups is tested for oxygen reduction it only promotes the two-electron reduction whereas if $RuNH_3)_5^{+2}$ groups are bound to the cyanophenyl groups, a four-electron reduction process is observed. The same authors[65] conclude that the back-bonding of d-electrons from Ru(II) centers to pyridine ligands onto the porphyrin ring, and ultimately to the Co(II)–O_2 adduct in the transition state, is believed to be the essential factor for achieving the four-electron reduction. None of the ruthenated cobalt porphyrins is active for peroxide reduction, which shows that a direct four-electron reduction is operative on this complex. In order to check even further the hypothesis that π-back-bonding interactions between Ru(II) centers and the porphyrin are responsible for the four-electron reduction mechanism, Steiger and Anson[17] prepared (5,10,15-tris(3-cyanophenyl)-20-(1-methylpyridinium-4-yl)porphyrinato Co(II) and its triply ruthenated derivative and compared its electrocatalytic activity with that of the corresponding triply ruthenated 4-cyanophenyl derivative since it is known[17] that back-bonding for the 4-cyano derivative is much greater than for the 3-cyano derivative. Indeed, they found much greater activity for the 4-cyanophenyl derivative, which reduced O_2 via 4 electrons compared to the 3-cyanophenyl derivative, which promoted the reduction of O_2 predominantly via 2 electrons. In order to check once more that the π-back-bonding interactions are responsible for the promotion of the four-electron reduction in tetraruthenated Co porphyrins, Anson *et al.*[91] prepared the (5,10,15, 20-tetrakis(4-((pentaamineruthenio)-cyano)phenyl)porphyrinato)cobalt(II) and its ruthenated derivative and compared the activity with the nonmethylated derivatives. The introduction of methyl groups at the 3 and 3′-positions of the 4,4′-bipyridine ligand prevented the pyridine rings from adopting a coplanar configuration, therefore interrupting π-conjugation between both rings. This decrease in coplanarity decreased the transmission of electronic effects between the coordinated acceptor and donor complexes. However, the ET rates changed very little by the presence of methyl groups. With this in mind Anson *et al.* thought that

1

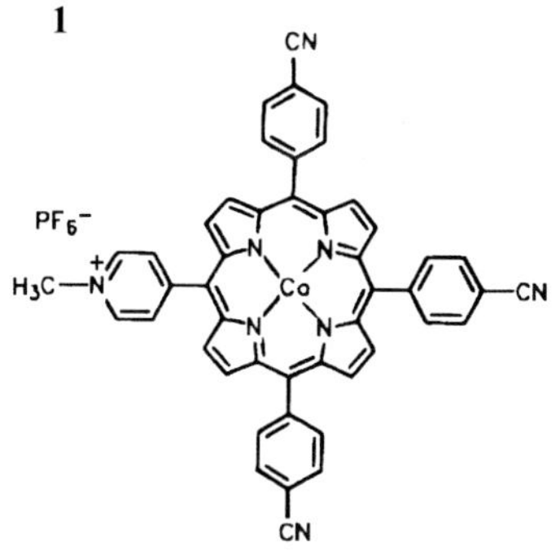

[CoP(PhCN)$_3$ (N–CH$_3$py)](PF$_6$)

2

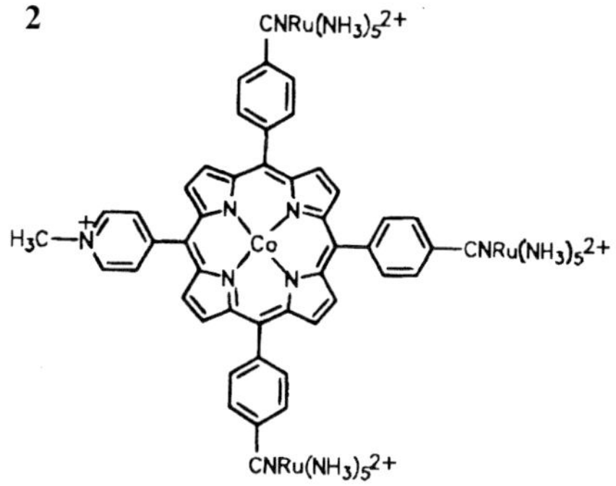

[CoP(PhCNRu(NH$_3$)$_5$)$_3$ (N–CH$_3$py)]$^{7+}$

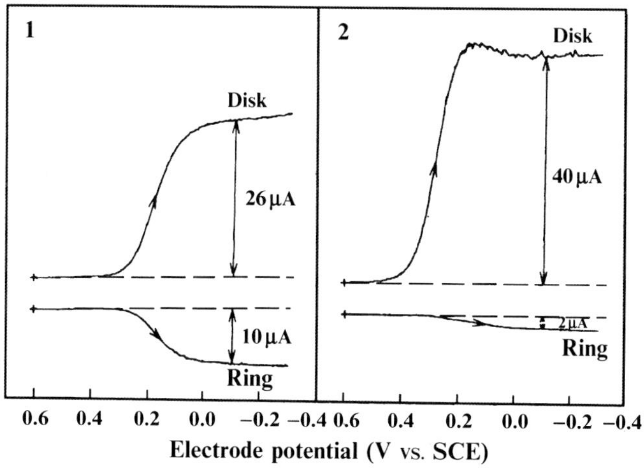

Figure 2.18. Structures of substituted and ruthenated tetraphenyl porphyrins and polarization curves rotating ring–disk curves for the reduction of O_2. From ref. [89], reproduced with permission of the American Chemical Society.

introducing methyl groups at the 2- and 6-positions of the 4-cyanophenyl ligands pendant to the porphyrin ring would lower π-conjugation between the porphyrin ring and the ligand and decrease the transmission to the porphyrin of the electron density that results from π-back-bonding of the Ru(II) complexes into the cyanophenyl rings, without an effect on the ET rates from the Ru(II) groups to reducible sites in the porphyrin. The authors concluded that back bonding is more important than intramolecular ET. Additional support to the idea that back bonding is a key factor in converting cobalt porphyrins into four-electron catalysts is the coordination of $Os(NH_3)_5^{3+/2+}$ to the ring of cobalt porphyrin[92]. This complex becomes a four-electron reduction catalyst when Os is converted to Os(II) on the periphery of the porphyrin ligand. The same authors[92] show that four (edta)Ru(II) groups coordinated to Co porphyrin are ineffective in promoting the four-electron reduction of O_2 despite their greater reducing strength, compared to $Os(NH_3)_5^{3+/2+}$. So when comparing porphyrins substituted with $Os(NH_3)_5^{3+/2+}$ and with (edta)Ru(II), the large differences in catalytic activity are attributed to differences in back-bonding strengths of the peripheral coordinated metal complexes and not to their relative reducing strengths. Anson *et al.*[93] conducted femtosecond dynamics of ruthenated and nonruthenated cobalt porphyrins and were able to elucidate the elementary steps involved in intramolecular relaxation and charge transfer in cobalt tetraphenylporphyrin from the porphyrin ligand $\pi(a_{2u})$ system to the $Co(d_z2)$ center. They found that the differences in the charge transfer state dynamics of the series of porphyrin molecules parallel the differences in their catalytic properties for the reduction of O_2. For example, when molecule I (a two-electron catalyst) is converted into molecule II by ruthenation (see Figure 2.19) it becomes a four-electron catalyst. In contrast, similar conversion of III into IV does change significantly the catalytic activity. The difference in activity of II and IV is attributed to the decrease of π-back-bonding by the Ru(II) center from the phenyl to the porphyrin ligand of IV compared to II due to the decrease in the ring-to-ring dihedral angle caused by the methyl groups in IV. The charge transfer from the porphyrin to Co(II) generates the porphyrin cation radical and Co(I). The charge transfer dynamics map the extent of the electronic coupling between the π system and the Co center that is the active site for O_2 reduction. From femtosecond transient absorbance of the different porphyrins depicted in Figure 2.19 it is observed that the recombination time of II almost doubles that of I whereas III and IV exhibit similar recombination times. All this indicates that back-transfer from Co(I) to the π-system is strongly affected by the presence of the $Ru(II)(NH_3)_5^{+2}$ groups only when the phenyl rings containing the Ru complex and the porphyrin ligand are in conjugation. All this agrees with the effect that the Ru complex has on the catalytic activity of the Co porphyrin since it back-bonds into the π-system, helping the Co center weaken the O–O bond via d–π charge transfer.

Anson *et al.*[65] have proposed the following mechanism for the four-electron reduction reaction catalyzed by ruthenated Co porphyrins:

$$[Co(III)P(pyRu(III)(NH_3)_5)]^{13+} + e^- \leftrightarrows [Co(II)P(pyRu(III)(NH_3)_5)]^{12+} \quad (2.15)$$

$$[Co(II)P(pyRu(III)(NH_3)_5)]^{12+} + O_2 \leftrightarrows [O_2CoP(pyRu(III)(NH_3)_5)]^{12+} \quad (2.16)$$

$$[O_2CoP(pyRu(III)(NH_3)_5)]^{12+} + 4e^- \leftrightarrows [O_2CoP(pyRu(II)(NH_3)_5)]^{8+} \quad (2.17)$$

$$\begin{aligned}&[O_2CoP(pyRu(II)(NH_3)_5)]^{8+} + 4e + 4H^+\\ &\longrightarrow [Co(II)P(pyRu(III)(NH_3)_5)]^{8+} + 4H_2O\end{aligned} \quad (2.18)$$

$$[Co(II)P(pyRu(III)(NH_3)_5)]^{8+} + O_2 \leftrightarrows [O_2CoP(pyRu(II)(NH_3)_5)]^{8+} \quad (2.19)$$

This reaction scheme involves the cobalt porphyrin confined on a graphite surface. The active Co(II) species is generated in step (2.15) at potentials more positive than those where Ru(III) and coordinated O_2 are reduced. The O_2 adduct is formed in step (2.16) that is reduced more easily than unbound O_2. The equilibrium constant of step (2.16) is small but the reaction rate for adduct formation

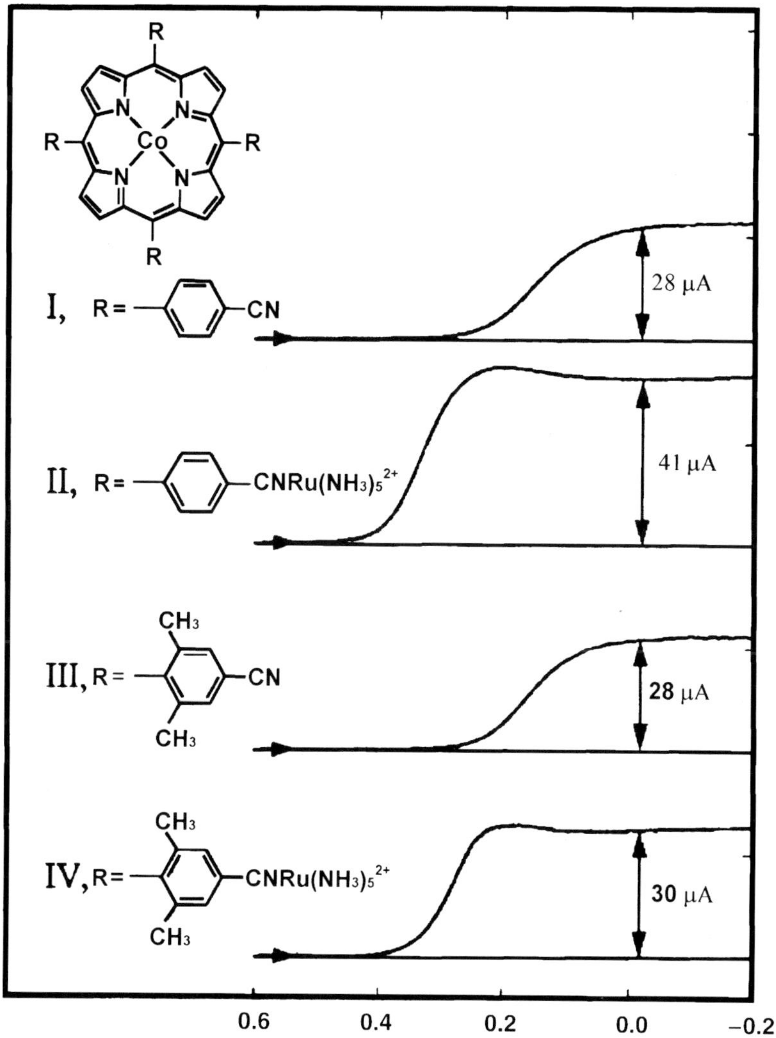

Figure 2.19. Structures of substituted cobalt(II) porphyrins and O_2 reduction waves at a rotating graphite disk electrode coated with the respective porphyrin ($\sim 2 \times 10^{-9}$ mol cm^{-2}). The four-electron reduction of O_2 produces a significantly larger plateau current for **II** than for **IV** where only a two-electron reduction proceeds. From ref. [93], reproduced with permission of the American Chemical Society.

is probably high enough to sustain substantial overall reduction rates. The back-bonding Ru(II) cocatalyst is formed in step (2.17) and this co-catalyst facilitates the four-electron reduction of the coordinated dioxygen molecule in step (2.18). Reaction (2.19) regenerates the Co(II)O_2 adduct. The overall catalytic cycle is then sustained by steps (2.18) and (2.19). According to these authors[65] reactions (2.17) and (2.18) occur at the same electrode potential indicating that the potential at which the back-bonding Ru(II) proceeds is near to, or more negative than, the potential required for the reduction of the Co(II)–O_2 adduct which is affected by the back-bonding interactions.

Imaoka and Yamamoto[94] reported that porphyrins possessing four ionic substituents of (*meso*-tetrakis(*N*-methyl-4-pyridiniumyl)porphyrinatoRu(II) (RuTMPyP) and (*meso*-tetrakis(4-sulfonatophenyl)porphyrinato)cobalt(II) (CoTPPS) associate spontaneously to form a dinuclear complex. Formation of CoTPPS-RuTMPyP was confirmed by UV–vis titration and TOF-mass spectra. The dinuclear complex exhibits an acceleration of intrinsic ET when it is present in a Nafion film deposited on glassy carbon. The CoTPPS–RuTMPyP–Nafion system catalyzes the four-electron reduction of O_2 with an efficiency of 95%.

Bettelheim *et al.*[95] developed bilayers of electropolymerized cobalt and manganese tetrakis(*o*-aminophenyl) porphyrins , i.e., polyCo(*o*-NH_2)TPP, and polyMn(*o*-NH_2)TPP, and have found that these polymeric bilayers are more efficient for the reduction of O_2 than polymeric Co(*o*-NH_2)TPP alone. These bilayers are not purely four-electron reduction catalysts but they substantially decrease the yields of peroxide (30%) compared to the polyCo(*o*-NH_2)TPP (75%). It is observed that the reduction of O_2 occurs at a potential close to the formal potential of Co(III)/(II) of Co(*o*-NH_2)TPP in both the polymeric bilayer or on the electrodes coated only with polyCo(*o*-NH_2)TPP. The lower yields in peroxide in the case of the polymeric bilayer films are attributed to the catalase activity of the manganese porphyrins that promotes the dismutation of H_2O_2. These authors also found[95] that the addition of 3×10^{-3} M imidazole decreases the H_2O_2 yield obtained at the polyCo(*o*-NH_2)TPP and polyMn(*o*-NH_2)TPP bilayers. This is attributed to a dismutation process of peroxide by the action of the imidazole-ligated Mn porphyrin. It is suggested that peroxide dismutation involves a μ-oxo-manganese (IV) or (V) adduct that is reduced by excess of H_2O_2 giving the original Mn(III)porphyrin. It was also found that the Co/Mn porphyrin bilayer exhibited much better long-term stability compared to polyCo(*o*-NH_2)TPP.

By using a Co-porphyrin–polyaniline complex deposited on glassy carbon Yamamoto and Taneichi[96] were able to observe a four-electron reduction of oxygen to water. It was proposed that the reduction reaction obeys a mechanism in which the four-electron reduction occurs through a multiple electron injection from the polyaniline to the cobalt complex after ET from Co to dioxygen. They propose that poly(2,3-dicarboxyaniline) acts as an excellent electron mediator from the electrode to the Co-porphyrin matrix, which shows stable redox activity in a wide pH range due to the self-dopable carboxy substituents in the polymeric chain. Yamamoto *et al.*[97] studied dinuclear Co porphyrin–polyaniline complexes. They found that poly(2,3-dicarboxyaniline) with CoTPPS–CoTMPyP show high catalytic for the four-electron reduction of oxygen. Several polymer complexes were prepared using synthesized electroresponsive polymers. It was found that the selectivity of the four-electron reduction decreases with an increase in the potential gap between the polymer and the porphyrin. This suggests that a sequential ET occurs from the polymer to the porphyrin with a potential similar to that of the cobalt porphyrin.

With the aim of modifying the catalytic properties of CoPcs and FePcs, in order to promote the four-electron reduction of O_2, Kobayashi, Sudo, and Osa[98] prepared ligand-substituted phthalocyanines with two dicarboximide groups on the four benzene groups of the indole unit (metal–1,2:3,4:8,9:10,11:15,16:22, 23:24,25-octakis(*N*-decyldicarboximide) phthalocyanine). Co and Fe phthalocyanines substituted with these groups exhibited visible spectra that are very different from those of nonsubstituted parent compounds, indicating that the substituents substantially change the electronic properties of the complexes and strongly affect the catalytic activity. Rotating ring–disk electrode studies indicated that dioxygen is reduced to water in both neutral and alkaline solutions for substituted FePc and in the whole pH range for substituted CoPc. The last result is remarkable since unsubstituted CoPc is a two-electron catalyst in the whole pH range.

An interesting strategy for achieving a four-electron reduction of O_2 and proposed by Tsuchida *et al.*[99–101] is by preparing μ-oxo metallomacrocyclics. For example μ-oxo-dimanganese(III)octaethylporphyrin[99] is a four-electron reduction catalyst in acid media. They also investigated the corresponding mononuclear complexes Mn(III)(OEP)Cl, Mn(III)(OEP)ClO_4, and Mn(III)(OEP)OH but they exhibited much less selectivity for the four-electron reduction. It is proposed that μ-oxo dimanganese complex has a cofacial configuration that allows the accommodation of molecular oxygen between the two Mn centers. A similar strategy was employed by Tsuchida *et al.*[100] using iron phthalocyanine (FePc). Mono nuclear iron phthalocyanine is a four-electron catalyst[4–8] but its selectivity for the reaction can be increased by using peroxo-bridged pyFe(III)Pc–O_2–PcFe(III)py where py is pyridine. When the dimer is reduced on the surface of a glassy carbon electrode, in acid media, a cofacially fixed Fe(II) phthalocyanine is produced, which can accommodate the dioxygen molecule between the two Fe(II) centers and then promote the rupture of the O–O bond.

Tsuchida *et al.*[101] used oxo-bridged (5,10,15,20)-tetraphenylporphyrinato iron(III) dimers adsorbed on carbon as electrocatalysts for the reduction of oxygen. They found that the electroreduction of the dimer adsorbed on carbon and immersed in an acid solution produced a cofacially fixed iron(II) porphyrin molecule, according to:

$$[P\mathrm{Fe(III)OFe(III)}P]_{\mathrm{ad}} + 2\mathrm{H}^+ + 2e^- \longrightarrow [P\mathrm{Fe(II)Fe(II)P}]_{\mathrm{ad}} + \mathrm{H_2O} \quad (2.20)$$

where P is porphyrin dianion.

Coordination of O_2 to the adjacent Fe(II) sites in acid, allows the formation of O_2-bridged Fe(II) porphyrin $[P\mathrm{Fe(III)O_2Fe(III)}P]_{\mathrm{ad}}$ on the electrode surface.

The following four-electron reduction catalytic cycle is proposed:

$$[P\mathrm{Fe(II)Fe(II)}P]_{\mathrm{ad}} + \mathrm{O_2} \longrightarrow [P\mathrm{Fe(III)O_2Fe(III)}P]_{\mathrm{ad}} \quad (2.21)$$

$$[P\mathrm{Fe(III)O_2Fe(III)}P]_{\mathrm{ad}} + 4\mathrm{H}^+ + 4e^- \longrightarrow [P\mathrm{Fe(II)OH_2\ H_2OFe(II)}P]_{\mathrm{ad}} \quad (2.22)$$

$$[P\mathrm{Fe(II)OH_2\ H_2OFe(II)}P]_{\mathrm{ad}} \longrightarrow [P\mathrm{Fe(II)Fe(II)}P]_{\mathrm{ad}} + 2\mathrm{H_2O} \quad (2.23)$$

From this work it is concluded that the electroreduction of the oxo-bridged dimer provides a cofacial geometry for Fe(II)P on the electrode surface which facilitates the binding and subsequent rupture of the dioxygen molecule upon reduction.

Anson *et al.* have reported[102] an interesting finding: the simplest cobalt porphyrin, cobalt porphine, when adsorbed at monolayer levels on a edge plane pyrolytic graphite, is capable of catalyzing the four-electron reduction of O_2. This is very unusual since most mononuclear porphyrins (except the ruthenated or osmium substituted) are two-electron catalysts. It also exhibited unusually positive Co(III)/(II) formal potential. Similar results were obtained by the same authors using (5, 10, 15, 20-tetramethylporphyrinato)cobalt(II)[103] These phenomena are probably the result of the high tendency of these molecules to form dimeric, cofacially oriented rings via van der Waall's interactions. The cofacial configuration could then accommodate the dioxygen molecule that would interact with two Co(II) centers.

Electrodes modified with films obtained from polymerization of CoTAPc (cobalt tetraamino phthalocyanine) show two waves for the reduction of O_2. At low polarizations two-electron reduction waves are observed with the formation of peroxide and a four-electron reduction to form water is observed at higher polarizations[11,12]. It is not clear why these polymeric films can promote the four-electron pathway. It can be argued that within the polymeric films Co atoms belonging to different MN_4 units become separated by the ideal distance for O_2 to bind to two cobalt centers forming a bridge. However there are no detailed studies of the structures of these polymeric films to clarify this point. A similar behavior has been observed with adsorbed layers of vitamin B_{12} where a four-electron reduction mechanism is observed at high polarizations[14,16]. In all these cases, it is likely that O_2 interacts with one single site at the time and vitamin B_{12} is still able to promote four-electron reduction process.

Iron macrocyclics catalyze the reduction of O_2 directly to water mainly in alkaline solution by providing only one site of adsorption[6–8,10,48,55,104–107,109,112]. It is not yet clear how these complexes promote the splitting of the O_2 bond in spite of providing one single site for O_2 adsorption. Many authors have suggested that the rupture of the O–O bond is possible when a peroxo dimer is formed which involves two active sites[49,53,69] (dual-site mechanism). Other authors have suggested that with a single site type of mechanism the rupture of the O–O bond is also possible[55,58]. It is also possible that Fe complexes catalyzed the reduction of peroxide, giving them an apparent direct four-electron reduction mechanism but this does not seem to be a good explanation because experiments conducted with only peroxide in the solution have shown that Fe phthalocyanines do not catalyze the reduction of peroxide[55]. Yeager *et al.* have suggested a dual-site mechanism where O_2 would coordinate to the Fe center and to a N atom on the macrocyclic ligand[105].

Uchida *et al.*[106] have prepared electrostatic aggregates of FeT(p-SO_3)PP (anionic) and FeTMPyP (cationic) porphyrins deposited on glassy carbon electrodes, where FeT(p-SO_3)PP = iron *meso*-tetrakis(p-sulfophenyl)porphyrin and FeTMPyP = iron *meso*-tetrakis(N-methyl-4-pyridyl)porphyrin. The authors conclude that a face-to-face configuration is obtained with these two complexes of opposite electrical charges. They promote the two-electron reduction of O_2 at low overpotentials but at high overpotentials a four-electron pathway is observed. They claim that the reduced form of the iron porphyrin ion complex is capable of reducing hydrogen peroxide to water at high polarizations.

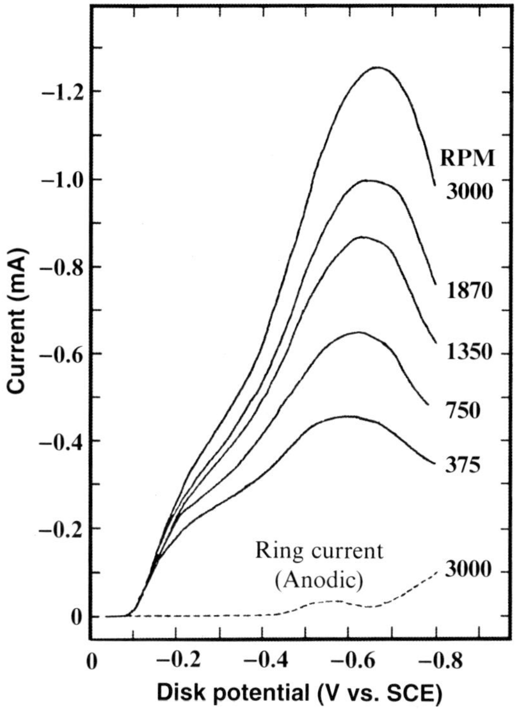

Figure 2.20. Polarization curves for the rotating ring–disk electrode for the reduction of O_2 on FeTPc adsorbed on graphite. From ref. [55], reproduced with permission of the Electrochemical Society Inc.

Elzing *et al.*[107] have observed that when FeTSPc (iron tetrasulfonatephthalocyanine) is incorporated into a polypyrrole film, the 4-electron reduction pathway is enhanced, especially when the films are thick. It was suggested that peroxo dimeric species could be formed within the film that could not only increase the catalytic activity but also shift the on-set potential for the reduction of O_2.

Interesting results have been obtained with monolayers of FeTSPc adsorbed on pyrolytic graphite and basal plane of graphite. The reduction of O_2 on these electrodes in alkaline solution gives two waves as illustrated in Figure 2.20 which shows rotating ring–disk data for several rotation rates. For the Fe-TSPc modified surface the first wave or prewave corresponds to the four-electron reduction of O_2 entirely to water without detectable amounts of peroxide since the ring currents, which are proportional to the peroxide generated on the rotating-disk electrode, are practically 0 (potentials more positive than −0.4 V vs. SCE). The ring electrode exhibits a small wave which tracks the shape of the second O_2 reduction wave, showing a maximum at *ca.* −0.6 vs. SCE. So the second wave corresponds to the reduction of O_2 via a mixed mechanism, *i.e.*, four- and two-electron reduction pathways and some peroxide is formed. At more negative potentials the two-electron reduction predominates (potentials more negative than

−0.7 V vs. SCE, (see Figure 2.20) and this is concomitant with the reduction of Fe(II) to Fe(I). So in general this is opposite to what is observed with some Co complexes where four-electron reduction mechanisms are observed at high polarizations[11,12]. It could be possible that the four-electron reduction mechanism could be attributed to the formation of dimeric Fe species, but this can be ruled out on the light *in situ* spectroreflectance measurements conducted with FeTSPc adsorbed on the basal plane of graphite[69]. Elzing *et al.*[49] studied the effect of FeTSPc surface concentration on the O_2 reduction currents. They found that both waves for the reduction of O_2 are dependent on the FeTSPc surface concentration. However the dependence of the second wave is less pronounced. A mechanism has been proposed for the reaction based on these observations. The first step would involve the reduction of Fe(III) to Fe(II):

$$\text{TSPcFe(III)–OH} + e^- \rightarrow \text{TSPcFe(II)} + \text{OH}^- \tag{2.24}$$

The Fe(II) interacts with oxygen and is oxidized to Fe(III)

$$\text{TSPcFe(II)Pc} + \text{O}_2 \rightarrow \text{TSPcFe(III) Pc–O}_2^- \tag{2.25}$$

With porphyrins an adduct involving Fe(IV) has also been proposed[25].

Elzing *et al.* have proposed[108] that the presence of a basic axial ligand will lessen the charge transfer from the Fe(II) center to dioxygen:

$$\text{L–FePc} + \text{O}_2 \rightarrow \text{L–Fe(II)Pc–O}_2 \tag{2.26}$$

This was supported by Mulliken population analysis but it showed that some net negative is transferred to dioxygen.

In order to check if FeTSPc could be active in the electroreduction of peroxide a study was conducted using a mixture of CoTSPc and FeTSPc in different proportions and adsorbed on graphite. As CoTSPc catalyzes the reduction of O_2 only to peroxide, if FeTSPs had any catalytic activity it would reduce the peroxide generated on the CoTSPc sites. However this was not observed. Both catalysts behave independently. Peroxide generated on Co sites was not reduced or decomposed on Fe sites[109]. These authors found that the amount of peroxide generated was inversely proportional to the fraction of FeTSPc present on the surface (see Figure 2.21). In spite of these results, the possibility for the Fe centers to form hydrogen peroxide and promote its decomposition cannot be ruled out completely since Fe(II) sites are known for their catalase activity[110]. Indeed for some metal chelates it has been found that their catalytic activity for peroxide decomposition is directly proportional to its activity for O_2 reduction[111].

It has been shown for complexes of the same metal (Fe), that the nature of the ligand can also influence the pathways for O_2 reduction. For example when studying FeTPP (iron tetraphenyl porphyrin) and FeT-PyP (iron tetrapyridino porphyrin)[112] peroxide reduction did not occur with Fe-T-PyP.

Tanaka *et al.*[105] when studying adsorbed layers of FeTPyPz (Fe-tetrapyridinoporphyrazine), at low polarizations, observed no peroxide formation. Peroxide was only formed at high polarizations. Again, as observed with FeTSPc and FePc[10,107,113] a prewave is observed for the reduction currents and corresponding to the direct four-electron reduction. In contrast to FeTSPc or FePc, production

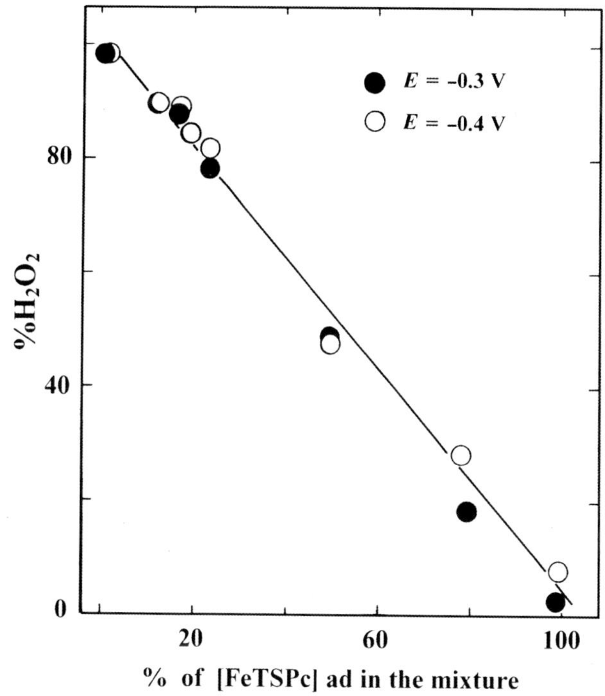

Figure 2.21. Peroxide yield as a function of surface composition of electrode with mixtures of Co-TSPc and FeTSPc coadsorbed on graphite. From ref. [6], reproduced with permission of Elsevier.

of peroxide was attributed to reduction of the ligand and not to reduction of the metal (Fe(I) formation). For this particular catalyst it was suggested that dioxygen can bind to the Fe center and to a highly electronegative nitrogen in the ring, which will avoid the desorption of peroxide before it is reduced. This dual-site mechanism would aid charge injection via backbonding from the macrocycle into antibonding orbitals of O_2 or bound peroxide causing the destabilization and further rupture of the O–O bond[105]. At more negative potentials at which reduction of the ligand takes place this mechanism becomes inoperative, the O_2 molecule only binds to the Fe center and peroxide can desorb into the solution. In a study involving heat-treated FeTPP[56] and deposited by thick layers on glassy carbon it was found that the amount of hydrogen peroxide decomposed, compared to the amount of oxygen and hydrogen peroxide reduced, was so small that chemical decomposition was ruled out.

Van den Brink *et al.*[113] when using vacuum-deposited layers of FePc have found that when examining the catalytic activity of these layers the first reduction wave scan was very different from subsequent scans, indicating that some reorganization of the deposited layers took place since this phenomenon is not observed on adsorbed layers of FePc. The effect of irreversible changes of FePc when treated under potential load with oxygen is only observed using vacuum-deposited multilayers[46].

Recent theoretical studies conducted by Anderson *et al.*[115] using spin-unrestricted hybrid gradient-corrected density functional calculations have predicted that Fe(II) is the active site for four-electron reduction of oxygen by iron in the N_4Fe systems employed in the calculation and it may be suggested that the same should be expected for heat-treated iron macrocyclics. The calculations have shown that Fe(II) is favored over Fe(III) because H_2O bonds strongly to the Fe(III) site, preventing O_2 adsorption and water does not bond strongly to Fe(II). In the first step, –OOH bonds more strongly to Fe(II) than to Fe(III), which results in a calculated more reversible potential for its formation over Fe(II). Calculations show that subsequent reduction steps have very reversible potentials over both centers (Fe(II) and Fe(III)). Calculations also show a hydrogen bonding interaction between –(OHOH) bonded to Fe(II) and to a nitrogen lone-pair orbital in the N_4 chelate. This interaction prevents peroxide from desorbing as a two-electron reduction product. So essentially these studies show that adsorbed hydrogen peroxide in an intermediate formed from N_4Fe–OOH on Fe(II) sites and can be released into the solution at more negative potentials as found experimentally[10,55,94,113].

As illustrated in Figures 2.7 and 2.11, by far the best catalysts are those macrocyclic containing Co and Fe as metal centers. However, complexes of other metals have been investigated. For example CrTSPc and MnTSPc were studied[10] and they resemble somehow the behavior of Fe complexes, mainly MnTSPc in the sense that it shows a prewave where O_2 reduction proceeds entirely via four electrons to give water. Peroxide is produced at higher polarizations. Even though it is still not clear whether linear or volcano correlations are valid when comparing activities, the low activity of Cr and Mn can be attributed to their low redox potential, i.e., they are easily oxidized[5,31]. Most macrocyclic metal complexes increase their activity after heat treatment[57]. However, for manganese complexes heat treatment decreases their activity probably because the metal is lost from the N_4 structure[32,116].

Collman *et al.*[117] investigated IrOEP adsorbed on a pyrolytic graphite electrode and found that it catalyzed the direct four-electron reduction to water in acid media. This is the only single site catalyst known that promotes the four-electron pathway in acid media. IsTPP was also investigated but showed no activity. The direct four-electron reduction is likely to occur since this complex does not catalyze the reduction of peroxide. The authors interpret these observations on the basis of a dual-site mechanism where dioxygen interacts with two Ir centers with an intermediate N_4Ir–O–O–IrN_4 and N_4Ir–IrN_4 being the configuration that accepts the oxygen molecule. N_4Ir–IrN_4 is supposed to be formed by oxidation of monomeric IrN_4–H. Since IrOEP is flat, these dimeric species can be formed. In contrast, IrTPP is sterically hindered by the peripheral phenyl groups so it is more difficult for these species to form dimers.

Complexes of Mo can only be used in alkaline solution since they are not stable in acid. MoNPc shows less catalytic activity than FeNPc[118]. Ru complexes are active in both acid and alkaline solutions[30,31], but in acid the direct four-electron reduction was observed whereas in basic solution hydrogen peroxide was the main product of the reaction[119].

6. Conclusions

As a final conclusion of the work performed on this subject it can be said that in spite of the considerable amount of work published in the literature, there are still many questions that remain unsolved about the electrocatalytic reduction of O_2 mediated by metallomacrocyclics confined on electrode surfaces. However some clear trends do exist. It is now well established that Fe and Co macrocyclic complexes are by far the best catalysts for oxygen reduction even though recent studies demonstrate that cofacial Ir complexes are also very active. These complexes are characterized by exhibiting a reversible redox transition involving the M(III)/(II) couple. Some authors have found volcano-shaped correlations between activity (measured as current at constant potential or potential at constant current) vs. M(III)/(II) formal potential of the catalyst suggesting that an optimal M(III)/(II) redox potential does exist for maximum activity. However other authors have found only linear correlations between activity and M(III)/(II) redox potential, which indicates that the more positive the redox potential the higher the activity. In the latter correlations the activity decreases with increasing the driving force of the catalyst. This finding is important since *a priori* one would expect that the more negative the M(III)/(II) formal potential the higher the activity, since this could favor the partial reduction of O_2 upon interacting with the metal center, i.e., M(III)–O_2^-. It is also possible that the linear correlations found are part of an incomplete volcano. However, in these correlations it was found that Cr, Mn, Fe, and Co complexes, which exhibit a M(III)/(II) transition give rise to two separate correlations or families of compounds. This is a reflection of the fact that the reduction wave for O_2 reduction on for example Mn and Fe complexes starts at a potential very close to the M(III)/(II) formal potential of the catalyst. In contrast, for Co macrocyclics, the reduction wave starts at potentials far more negative than the Co(III)/(II) formal potentials. The proximity of the O_2 reduction wave to the M(III)/(II) for some complexes is also reflected in the observation that a direct four-electron reduction process operates, as observed for Fe and Mn complexes. For most monomeric or monolayers of Co complexes the onset for O_2 reduction is far removed from the Co(III/(II) transition and only the peroxide pathway is observed. Dual-site mechanisms for O_2 reduction are observed for cofacial Cu complexes or dimetallic planar Co complexes, so two Co sites are required for a four-electron reduction mechanism. Polymeric Co complexes exhibit four-electron reduction waves at high polarizations only. Vitamin B_{12} seems to be very special in the sense that in spite of providing one single site for the reaction, it can catalyze the four-electron reduction at high polarizations. Anson *et al.*in a series of papers have demonstrated that two-electron reduction catalysts can be turned into four-electron reduction catalysts, using cobalt porphyrins by placing pyRu(II)$(NH_3)_5$groups on the periphery of the ligand. Placing Os(II)$(NH_3)_5$has similar effects on promoting four-electron reduction mechanisms and the effect of this groups is attributed to back-bonding from the Ru or Os centers to the bound O_2 molecule, so substitution of the periphery of the ligand with the appropriate redox groups can have dramatic effects on the selectivity of the macrocyclic catalysts.

In spite of the good results obtained with iridium complexes since they catalyze the direct four-electron reduction of O_2 to water rather few studies have focused their attention on these promising catalysts. Most of the catalysts discussed in this chapter provide good models for O_2 reduction in aqueous media. However many of them lack the necessary long-term stability for fuel cell applications. As discussed in great detail in a separate chapter, heat treatment of these catalysts mixed with or without mixtures with carbon support increases both activity and stability. Nonheat-treated materials like the ones discussed in this chapter are still very interesting due to their potentialities for making O_2 sensors and sensors of a great variety of other molecules. This is discussed in great detail in other chapters of the book.

Finally it is worth mentioning that one of the advantages of MN_4 macrocyclics over Pt catalysts is their tolerance to methanol cross over, which is a serious problem in methanol–air fuel cells[7]. When methanol crossover from the anode through the electrolytic membrane to the cathode occurs, electroreduction of dioxygen and electrooxidation of methanol occur simultaneously. If this happens, it is detrimental to the overall performance of the fuel cell since the fuel efficiency decreases as does the output power. In general MN_4 macrocyclics are poor catalysts for the oxidation of methanol, so this reaction should not occur and this would avoid the problems mentioned above. As an example, Lamy *et al.* have studied the catalytic activity of Co-tetraazaannulene (CoTAA) deposited on graphite powder embedded in a cast Nafion film[120] and found that the activity of this catalyst for the reduction of O_2 is not affected by the presence of methanol in the electrolyte. Jiang and Chu[121] found excellent methanol tolerance for heat-treated CoTPP/FeTPP mixtures and high catalytic activity for O_2 reduction, with an onset potential of 0.9 V vs. RHE, comparable with platinum black (onset potential of 1.0 V vs. RHE). Savinell *et al.*[122, 123] studied heat-treated μ-oxo-iron(III) tetramethoxyphenyl porphyrin $(FeTMPP)_2$), iron(III) tetramethoxyphenyl porphyrin (FeTMPP–Cl) and iron(III) octamethoxyphenylporphyrin (FeFeOEP–Cl) adsorbed on high-area carbons finding high stability in their catalytic activity for the reduction of $O_2$and insensitivity to the presence of methanol. Bettelheim *et al.*[124] have found that tetra(*ortho*-aminophenyl)porphyrin (*o*-NH_2)TPP can be used as a methanol barrier when electropolymerized on a direct methanol fuel cell cathode. The polymeric films were found to be nonporous, 0.55 μm thick, of high density (0.94 g/cm^3), and of high porphyrin site concentration (1.4 M). Experiments in a two-electrode configuration showed that it is possible to operate a direct methanol fuel cell with low current output using an electrolytic membrane separating the anode from the cathode, when the cathodic electrode is covered with a bilayer of polyaniline sulfonic acid/poly(*o*-NH_2)TPP. Besides these properties of metallomacrocyclics, in particular, tetraphenylporphyrinCo(II), CoTPP, has been shown to possess O_2-binding capabilities when complexed with poly(3,4-azopyridylene) (PAP)[125, 126]. In this fashion CoTPP can act as an oxygen carrier at the surface of the electrode and can enhance the rate of O_2 reduction. The complexed CoTPP combined with a conventional Pt/carbon catalyst increases substantially the steady-state four-electron reduction currents, especially at low overpotentials, which are common for fuel-cell operation. So CoTPP facilitates

O_2 transport from the atmosphere to the surface of the catalyst present on the electrode surface. All these findings indicate that metallomacrocyclics are still interesting for O_2 electroreduction and for possible applications in fuel cells. Long-term stability, which is one of the limiting factors, can be improved dramatically by heat treatment and this is discussed in detail in Chapter 3.

Ackowledgments

The authors acknowledge financial support of Fondecyt, Líneas Complementarias 8010006. F.S. acknowledges support from doctoral fellowship from Conicyt. J.Z. acknowledges support from Vicerrectoria de Investigación Científica and Dicyt from Usach.

References

1. Jasinski, R. (1964). A new fuel-cell cathode catalyst. *Nature* **201**, 1212–1213.
2. van den Brink, F., E. Barendrecht, and W. Visscher (1980). The cathodic reduction of oxygen—A review with emphasis on macrocyclic organic metal-complexes as electrocatalysts. *Recl. Trav. Chim. Pays-Bas*, **99**, 253–262.
3. Rayushkina, K.A. and M.R. Tarasevich (1986). Electrocatalytic properties of pyrolytic polymers produced from N4-complexes (review). Translated from *Elektrokhimiya* **22**, 1155–1170.
4. Yeager, E. (1984). Electrocatalysis for O_2 reduction. *Electrochim. Acta* **29**, 1527–1537.
5. Vasudevan, P., S.N. Mann, and S. Tyagi (1990). Transition-metal complexes of porphyrins and phthalocyanines as electrocatalysts for dioxygen reduction. *Transition Met. Chem.* **15**, 81–90.
6. Zagal, J.H. (1992). Metallophthalocyanines as catalysts in electrochemical reactions. *Coord. Chem. Rev.* **119**, 89–136.
7. Adzic, R. (1998). Recent advances in the kinetics of oxygen reduction. In J. Lipkowski and P.N. Ross (ed), *Electrocatalysis, Frontiers in Science*. Wiley-VCH, New York, 197–242.
8. Zagal, J.H. (2003). Macrocycles. In W. Vielstich, A. Lamm and H.A. Gasteiger (eds.), *Handbook of Fuel Cells—Fundamentals, Technology and Applications*, Vol. 2, Part 5. John Wiley & Sons Ltd, Chichester, pp. 544–554.
9. Collman, J.P., R. Boulatov, and C.J. Sunderland (2003). Functional and structural analogs of the dioxygen reduction site in terminal oxidases. In K.M. Kadish, K.M. Smith, and R. Guilard (eds.), *The Porphyrin Handbook*, Chapter 23. Academic Press, Boston, pp. 1–49.
10. Zagal, J.H., M. Páez, A.A. Tanaka, J.R. dos Santos, and C. Linkous (1992). Electrocatalytic activity of metal phthalocyanines for oxygen reduction. *J. Electroanal. Chem.* **339**, 13–30.
11. Tse, Y.H., P. Janda, H. Lam, J. Zhang, W.P. Pietro, and A.B.P. Lever (1997). Monomeric and polymeric tetraaminophthalocyanatocobalt (II) modified electrodes: Electrocatalytic reduction of O_2. *J. Porphyrins Phthalocyanines* **1**, 3–16.
12. Pavez, J., M. Páez, A. Ringuedé, F. Bedioui, and J.H. Zagal (2005). Effect of film thickness on the electro-reduction of molecular oxygen on electropolymerized cobalt tetra-aminophthalocyanine films. *Solid State Electrochem.* **9**, 21–29.

13. Ramirez, G., E. Trollund, M. Isaacs, J. Zagal, J.C. Costamagna, and M.J. Aguirre (2002). Electroreduction of dioxygen on poly-fetetraaminophthalocyanine modified electrodes. *Electroanalysis* **14**, 540–545.
14. Zagal, J.H., M. Páez, and C. Páez (1987). Electroreduction of O_2 catalyzed by vitamin B_{12} adsorbed on a graphite electrode. *J .Electroanal. Chem.* **237**, 145–148.
15. Qiu, Q. and S. Dong (1993). Rapid determination of kinetic parameters for the electrocatalytic reduction of dioxygen by vitamin B_{12} adsorbed on glassy carbon. *Electrochim. Acta* **38**, 2297–2302.
16. Zagal, J.H., M.J. Aguirre, and M.A. Páez (1997). O_2 reduction kinetics on a graphite electrode modified with adsorbed vitamin B_{12}. *J. Electroanal. Chem.* **437**, 45–52.
17. Steiger, B. and F.C. Anson (1995). Evidence of the importance of back-bonding in determining the behavior of ruthenated cyanophenyl cobalt porphyrins as electrocatalysts for the reduction of dioxygen. *Inorg. Chem.* **34**, 3355–3357.
18. Durand, R.R., C.S. Bencosme, J.P. Collman, and F.C. Anson (1983). Mechanistic aspects of the catalytic reduction of dioxygen by cofacial metalloporphyrins. *J. Am. Chem. Soc.* **105**, 2710–2718.
19. Lalande, G., R. Cote, D. Guay, J.P. Dodelet, L.T. Weng, and P. Bertrand (1997). Is nitrogen important in the formulation of Fe-based catalysts for oxygen reduction in solid polymer fuel-cells? *Electrochim. Acta* **42**, 1378–1388.
20. Bouwkamp-Wijnoltz, A.L., W. Visscher, and J.A.R. Van Veen (1998). The selectivity of oxygen reduction by pyrolysed iron porphyrin supported on carbon. *Electrochim. Acta* **43**, 3141–3152.
21. Lefevre, M., J.P. Dodelet, and P. Bertrand (2002). Molecular oxygen reduction in PEM fuel cells: Evidence for the simultaneous presence of two active sites in Fe-based catalysts. *J. Phys. Chem., B* **106**, 8705–8713.
22. Bouwkamp-Wijnoltz, A.L., W. Visscher, J.A.R. Van Veen, J.A.R. Boellaard, A.M. Van der Kraan, and S.C. Tang (2002). On active-site heterogeneity in pyrolyzed carbon supported iron porphyrin catalysts for the electrochemical reduction of oxygen: An in situ Mössbauer study. *J. Phys. Chem., B* **106**, 12993–13001.
23. Lefevre, M. and J.P. Dodelet (2003). Fe-based catalysts for the reduction of oxygen in polymer electrolyte membrane fuel cell conditions: Determination of the amount of peroxide released during electroreduction and its influence on the stability of the catalysts. *Electrochim. Acta* **48**, 2749–2760.
24. Kobayashi, N. and W.A. Nevin (1996). Electrocatalytic reduction of oxygen using water-soluble iron and cobalt phthalocyanines and porphyrins. *Appl. Organomet. Chem.* **10**, 579–590.
25. Bytheway, I. and M.B. Hall (1994). Theoretical calculations of metal–dioxygen complexes. *Chem. Rev.* **94**, 639–658.
26. van der Putten, A., A. Elzing, W. Visscher, and E. Barendrecht (1987). Redox potential and electrocatalysis of O_2 reduction on transition metal chelates. *J. Electroanal. Chem.* **221**, 95–104.
27. Zecevic, S., B. Simic-Glavaski, E. Yeager, A.B.P. Lever, and P.C. Minor (1985). Spectroscopic and electrochemical studies of transition metal tetrasulfonated phthalocyanines: Part V. Voltammetric studies of adsorbed tetrasulfonated phthalocyanines (MTsPc) in aqueous solutions. *J. Electroanal. Chem.* **196**, 339–358.
28. Ni, C.L. and F.C. Anson (1984). Relation between the potentials where adsorbed and unadsorbed cobalt(III) tetrakis(*N*-methylpyridinium-4-yl)porphyrin is reduced and those where it catalyzes the electroreduction of dioxygen. *Inorg. Chem.* **24**, 4754–4756.
29. Shi, C. and F. C. Anson (1990). Catalytic pathways for the electroreduction of oxygen by iron tetrakis(4-*N*-methylpyridyl)porphyrin or iron tetraphenylporphyrin adsorbed on edge plane pyrolytic graphite electrodes. *Inorg. Chem.* **29**, 4298–4305.

30. van Veen, J.A.R. and C. Visser (1979). Oxygen reduction on monomeric transition metal phthalocyanines in acid electrolyte. *Electrochim. Acta* **24**, 921–928.
31. van Veen, J.A.R., J.F. van Baar, C.J. Croese, J.G.F. Coolegem, N. de Wit, and H.A. Colijin (1981). Oxygen reduction on transition metal porphyrins in acid electrolyte: I activity. *Ber. Bunsenges. Phys. Chem.* **85**, 693–700.
32. Zagal, J.H., M.J. Aguirre, L. Basaez, J. Pavez, L. Padilla, and A. Toro-Labbé (1995). Possible explanations for the volcano-shaped plots for the electrocatalytic reduction of O_2 on electrodes modified with N-4 macrocyclics. In F.C. Anson, R.R. Adzic, and K. Kinoshita, *Oxygen Electrochemistry. The Electrochemical Society Symposium Series* 95–26, pp. 89–100.
33. Zagal, J.H., M. Gulppi, M. Isaacs, G. Cárdenas-Jirón, and M.J. Aguirre (1998). Linear vs. volcano correlations between electrocatalytic activity and redox properties of metallophthalocyanines. *Electrochim. Acta* **44**, 1349–1357.
34. Cárdenas-Jirón, G., M.A. Gulppi, C.A. Caro, R. del Rio, M. Paez, and J.H. Zagal (2001). Reactivity of electrodes modified with substituted metallophthalocyanines. Correlations with redox potentials, Hammett parameters and donor–acceptor intermolecular hardness. *Electrochim. Acta* **46**, 3227–3235.
35. Zagal, J.H. and G. Cárdenas-Jiron (2000). Reactivity of immobilized cobalt phthalocyanines for the electroreduction of molecular oxygen in terms of molecular hardness. *J. Electroanal. Chem.* **489**, 96–100.
36. Cárdenas-Jirón, G.I. and J.H. Zagal (2001). Donor–acceptor intermolecular hardness on charge transfer reactions of substituted cobalt phthalocyanines. *J. Electroanal. Chem.* **497**, 55–60.
37. Schlettwein, D. and T.Yoshida (1998). Electrochemical reduction of substituted cobalt phthalocyanines adsorbed on graphite. *J. Electroanal. Chem.* **441**, 148.
38. Zagal, J.H., M.A. Gulppi, and G. Cárdenas-Jirón (2000). Metal-centered redox chemistry of substituted cobalt phthalocyanines adsorbed on graphite and correlations with MO calculations and Hammett parameters. Electrocatalytic reduction of disulfide. *Polyhedron* **19**, 2255–2260.
39. Newton, M.D. (1991). Quantum chemical probes of electron-transfer kinetics: The nature of donor–acceptor interactions. *Chem. Rev.* **91**, 767–792.
40. Pearson, R.G. (1986). Absolute electronegativity and hardness correlated with molecular orbital theory. *Proc. Natl Acad. Sci. USA* **83**, 8440–8441.
41. Parr, R.G. and R.G. Pearson (1983). Absolute hardness: Companion parameter to absolute electronegativity. *J. Am. Chem. Soc*, **105**, 7512–7516.
42. Ulstrup, J. (1977). Catalysis of the electrochemical reduction of molecular dioxygen by metal phthalocyanines. *J. Electroanal. Chem.* **79**, 191–197.
43. Rosa, A. and E.J. Baerends (1994). Metal macrocycle interaction in phthalocyanines—density functional calculations of ground and excited states. *Inorg. Chem.* **33**, 584–595.
44. Hipps, K.W., X. Lung, X.D. Wang, and U. Manzur (1996). Metal d-orbital occupation-dependent images in the scanning tunnelling microscopy of metal phthalocyanines. *J. Phys. Chem., B* **100**, 11207–11210.
45. Theodoridou, E., A.D. Jannakoudakis, P.D. Jannakoudakis, and J.O. Besenhard (1992). Carbon fibre based conductive ion exchangers as support for metal and ligand charged metalloporphyrin electrocatalysts. *J. Appl. Electrochem.* **22**, 733–737.
46. Hinnen, C., F. Coowar, and M. Savy (1989). Oxygen reduction in acid media: Investigations by electroreflectance on adsorbed iron phthalocyanine and naphthalocyanine layers. *J. Electroanal. Chem.* **264**, 167–180.
47. van den Ham, D., C. Hinnen, G. Magner, and M. Savy (1987). Electrocatalytic oxygen reduction: The role of oxygen bridges as a structural factor in the activity of transition-metal phthalocyanines. *J. Phys. Chem.* **91**, 4743–4748.

48. Coowar, F., O. Contamin, M. Savy, and G. Scarbeck (1998). Electrocatalysis of O_2 reduction to water in different acid media by iron naphthalocyanines. *J. Electroanal. Chem.* **246**, 119–138.
49. Elzing, A., A. van der Putten, W. Visscher, and E. Barendrecht (1987). The cathodic reduction of oxygen at metal tetrasulfonato-phthalocyanines: Influence of adsorption conditions on electrocatalysis. *J. Electroanal. Chem.* **233**, 99–112.
50. Elzing, A., A. van der Putten, W. Visscher, and E. Barendrecht (1986). The cathodic reduction of oxygen at cobalt phthalocyanine: Influence of electrode preparation on electrocatalysis. *J. Electroanal. Chem.* **200**, 313–322.
51. Fierro, C.A., M. Mohan, and D.A. Scherson (1990). In situ Mossbauer spectroscopy of a species irreversibly adsorbed on an electrode surface. *Langmuir* **6**, 1338–1342
52. Ouyang, J., K. Shigehara, A. Yamada, and F.C. Anson (1991). Hexadecafluoro- and octacyano phthalocyanines as electrocatalysts for the reduction of dioxygen. *J. Electroanal. Chem.*, **297**, 489–498.
53. van der Putten, A., A. Elzing, W. Visscher, and E. Barendrecht (1986). Oxygen reduction on vacuum-deposited and absorbed transition-metal phthalocyanine films. *J. Electroanal. Chem.* **214**, 523–533.
54. Shi, C. and F. C. Anson (1992). Electrocatalysis of the reduction of molecular oxygen to water by tetraruthenated cobalt meso-tetrakis(4-pyridyl)porphyrin adsorbed on graphite electrodes. *Inorg. Chem.* **31**, 5078–5083.
55. Zagal, J., P. Bindra, and E. Yeager (1980). A mechanistic study of O_2 reduction on water soluble phthalocyanines adsorbed on graphite electrodes. *J. Electrochem. Soc.* **127**, 1506–1517.
56. Ikeda, O., H. Fukuda, and H. Tamura (1986). The effect of heat treatment on group VIII B porphyrins as electrocatalysts in the cathodic reduction of oxygen. *J. Chem. Soc. Farad. Trans. I* **82**, 1561–1573.
57. van Veen, J.A.R., J.F. van Baar, and K.J. Krose (1981). Effect of heat treatment on the performance of carbon-supported transition-metal chelates in the electrochemical reduction of oxygen. *J. Chem. Soc. Farad. Trans.* **77**, 2827–2843.
58. van Veen, J.A.R. and H.A. Colijn (1981). Oxygen reduction on transition-metal porphyrins in acid electrolyte. Stability. *Ber Bunsenges. Phys. Chem.* **85**, 700–704.
59. Shamsipur, M., A. Salimi, H. Haddadzadeh, and M.F. Mousavi (2001). Electrocatalytic activity of cobaloxime complexes adsorbed on glassy carbon electrodes toward the reduction of dioxygen. *J. Electroanal. Chem.* **517**, 37–44.
60. Salimi, A. and M. Ghadermazi (2001). Electrocatalytic reduction of dioxygen on a glassy carbon electrode modified with adsorbed cobaloxime complex. *Anal. Sci.* **17**, 1165–1170.
61. Chan, R.J., Y.O. Su, and T. Kuwana (1985). Electrocatalysis of oxygen reduction. 5. Oxygen to Hydrogen peroxide conversion by cobalt (II) tetrakis (*N*-methyl-4pyridyl) porphyrin. *Inorg. Chem.* **24**, 3777–3784.
62. Atoguchi, T., A. Aramata, A. Kazusaka, and M. Enyo (1991). Cobalt (II)–tetraphenylporphyrin–pyridine complex fixed on a glassy carbon electrode and its prominent catalytic activity for reduction of carbon dioxide. *J. Chem. Soc. Chem. Commun.* 156–157.
63. Zeng, Z.Y., S.L. Gupta, H. Huang, and E.B Yeager (1991). Oxygen reduction on poly (4-vinylpyridine)-modified ordinary pyrolytic graphite electrodes with adsorbed cobalt tetra-sulphonated phthalocyanine in acid solution. *J. Appl. Electrochem.* **21**, 973–981.
64. van der Putten, A., A. Elzing, W. Visscher, E. Barendrecht, and R.D. Harcourt (1988). Increased valence theory and the 4 electron reduction of O_2 to H_2O. *J. Mol. Struct.* (Theochem) **180**, 309–318.

65. Anson, F.C., C. Shi, and B. Steiger (1997). Novel multinuclear catalysts for the electroreduction of dioxygen directly to water. *Acc. Chem. Res.* **30**, 437–449.
66. Gupta, S., C. Fierro and E.Yeager (1991). The effects of cyanide on the electrochemical properties of transition metal macrocycles for oxygen reduction in alkaline solutions. *J. Electroanal. Chem.* **306**, 239–250.
67. D'Souza, F., Y.Y. Hsieh, and G.R. Deviprasad (1997). Electrocatalytic reduction of molecular oxygen using non-planar cobalt tetrakis-(4-sulfonatophenyl)β-octabromoporphyrin. *J. Electroanal. Chem.* **426**, 17–21.
68. Coutanceau, C., A. Rakotondrainibe, P. Crouigneau, J.M. Léger and C. Lamy (1995). Spectroscopic investigations of polymer-modified electrodes containing cobalt phthalocyanine: Application to the study of oxygen reduction at such electrodes. *J. Electroanal. Chem.* **386**, 173–182.
69. Elzing, A., A. van der Putten, W. Visscher, E. Barendrecht, and C. Hinnen (1990). Spectroscopic measurements on metal tetrasulphonato-phthalocyanines. *J. Electroanal. Chem.* **279**, 137–156.
70. Phougat, N. and P. Vasudevan (1997). Electrocatalytic activity of some metal phthalocyanine compounds for oxygen reduction in phosphoric acid. *J. Power Sources* **69**, 161–163.
71. Lin, A.S. and J.C. Huang (2003). Oxygen reduction on Nafion-bound unpyrolyzed metal macrocyclic complexes. *J. Electroanal. Chem.* **541**, 147–151.
72. Chung, T.D. and F.C. Anson (2001). Catalysis of the electroreduction of $O_2$by cobalt 5,10,15,20-tetraphenylporphyrin dissolved in thin layers of benzonitrile on graphite electrodes. *J. Electroanal. Chem.* **508**, 115–122.
73. Steiger, B. and F.C. Anson (2000). Examination of cobalt "picket fence" porphyrin and its complex with 1-methylimidazole as catalysts for the reduction of dioxygen. *Inorg. Chem.* **39**, 4579–4585.
74. Hutchison, J.E., T.A. Postlethweite, and R.W. Murray (1993). Molecular films of thiol-derivatized tetraphenylporphyrins on gold: Film formation and electrocatalytic dioxygen reduction. *Langmuir* **9**, 3277–3283.
75. Postlethweite, T.A., J.E. Hutchison, K.W. Hathcock, and R.W. Murray (1995). Optical electrochemical, and electrocatalytic properties of self-assembled thiol-derivatized porphyrins on transparent gold electrodes. *Langmuir* **11**, 4109–4116.
76. Lever, A.B.P., E.R. Milaeva, and G. Speir (1993). The redox chemistry of metallophthalocyanines in solution. In C.C. Leznoff, A.B.P. Lever (eds), *Phthalocyanines, Properties and Applications*, Vol. 3, VCH Publishers, New York, pp. 1–70.
77. Lever, A.B.P. (1999). The phthalocyanines-molecules of enduring value, a two dimensional analysis of redox potentials. *J. Porphyrins Phthalocyanines,* **3**, 488–499.
78. Paliteiro, A., A. Hamnett, and J.B. Goodenough (1984). Cathodic reduction of oxygen by platinum phthalocyanine thin films. *J. Electroanal. Chem.* **160**, 359–367.
79. Collman, J.P., M. Marrocco, P. Denisevich, C. Koval, and F.C. Anson (1979). Potent catalysis of the electroreduction of oxygen to water by dicobalt porphyrin dimers adsorbed on graphite electrodes. *J. Electroanal. Chem.* **101**, 117–122.
80. Collman, J.P., P. Denisevich, Y. Konai, M. Marrocco, C. Koval, and F.C. Anson (1980). Electrocatalysis of the four-electron reduction of oxygen to water by dicobalt face-to-face porphyrins. *J. Am. Chem. Soc.* **102**, 6027–6036.
81. Chang, C.K., H.Y. Liu, and I. Abdalmuhdi (1984). Electroreduction of oxygen by pillared cobalt(II) cofacial diporphyrin catalysts. *J. Am. Chem. Soc.* **106**, 2725–2726.
82. van der Putten, A., A. Elzing, W. Visscher, and E. Barendrecht (1986). The four-electron reduction of oxygen to water on a planar cobalt chelate. *J. Chem. Soc. Chem. Commun.* 477–479.

83. Janda, P., N. Kobayashi, P.R. Auburn, H. Lam, C.C. Leznoff, and A.B.P. Lever (1989). Dioxgen reduction at a graphite electrode modified by mononuclear tetraneopentoxyphthalocyaninatocobalt (II) and Related Polynuclear Species. *Can. J. Chem.* **67**, 1109–1119.
84. Biloul, A., F. Coower, O. Contamin, G. Scarbeck, M. Savy, D. van den Ham, J. Riga, and J.J. Verbist (1993). Oxygen reduction in an acid medium: Electrocatalysis by CoNPc(1,2) impregnated on a carbon black support; effect of loading and heat treatment. *J. Electroanal. Chem.* **350**, 189–204.
85. Biloul, A., O. Contamin, G. Scarbeck, M. Savy, B. Palys. J. Riga, and J. Verbist (1994). Oxygen reduction in acid media: influence of the activity of CoNPc(1,2) bilayer deposits in relation to their attachment to the carbon black support and role of surface groups as a function of heat treatment. *J. Electroanal. Chem.* **365**, 239–246.
86. Anson, F.C., C.L. Ni, and J.M. Saveant (1985). Electrocatalysis at redox polymer electrodes with separation of the catalytic and charge propagation roles. Reduction of dioxygen to hydrogen peroxide as catalyzed by cobalt(II) tetrakis(4-*N*-methylpyridyl)porphyrin. *J. Am. Chem. Soc.* **107**, 3442–3450.
87. Shi and F.C. Anson (1991). Multiple intramolecular electron transfer in the catalysis of the reduction of dioxygen by cobalt *meso*-tetratrakis(4pyridyl)porphyrin to which four $Ru(NH_3)_5$ groups are coordinated. *J. Am Chem. Soc.* **113**, 9564–9570.
88. Steiger, B., C. Shi, and F.C. Anson (1993). Electrocatalysis of the reduction of dioxygen by adsorbed cobalt 5,10,15,20-tetraarylporphyrins to which one, two, or three pentaammineruthenium(2+) centers are coordinated. *Inorg. Chem.* **32**, 2107–2113.
89. Steiger, B. and F.C. Anson (1994). New electrocatalysts for the four-electron reduction of dioxygen based on (5,10,15-tris(pentaammineruthenium(II)-4-cyanophenyl)-20-(1methylpyridinium-4-yl)porphyrinato)cobalt(II) immobilized on graphite electrodes. *Inorg. Chem.* **33**, 5767–5779.
90. Shi, C. and F.C. Anson (1995). Coordination chemistry in electrocatalysis. *Pure Appl. Chem.* **67**, 319–322.
91. Steiger, B. and F.C. Anson (1997). (5,10,15,20-Tetrakis(4-((pentaammineruthenio)-cyano)phenyl)porphyrinato)cobalt(II) immobilized on graphite electrodes catalyzes the electroreduction of O_2 to H_2O, but the corresponding 4-cyano-2,6-dimethylphenyl derivative catalyzes the reduction only to H_2O_2, *Inorg. Chem.* **36**, 4138–4140.
92. Shi, C. and F.C. Anson (1996). Cobalt *meso*-Tetrakis(*N*-methyl-4 pyridinium)-porphyrin becomes a catalyst for the electroreduction of O_2 by four electrons when $((NH_3)_5Os)^{n+}$ $(n = 2, 3)$ groups are coordinated to the porphyrin ring. *Inorg. Chem.* **35**, 7928–7931.
93. Zhong, H.-Z., J.S. Baskin, B. Steiger, F.C. Anson, and A.H. Zewail (1999). Femtosecond dynamics and electrocatalysis of the reduction of O_2: Tetraruthenated cobalt porphyrins. *J. Am. Chem. Soc.* **121**, 484–485.
94. Imaoka, T. and K. Yamamoto (2001). Electrochemistry of Co–Ru hetero-dinuclear porphyrin complex in a Nafion matrix. *Phys. Chem. Chem. Phys.* **3**, 4462–4468.
95. Bettelheim, A., D. Ozer, R. Harth, and R. Ydgar (1990). Dioxygen reduction and hydrogen peroxide dismutation using electropolymerized bilayers of cobalt + manganese tetrakis(*o*-aminophenyl)porphyrins. *J. Electroanal. Chem.* **281**, 147–161.
96. Yamamoto, K. and D. Taneichi (1999). Electrochemical catalytic reduction of oxygen by self-doped polyaniline–Co porphyrin complex-modified glassy carbon electrode. *J. Inorg. Organomet. Polymers.* **9**, 231–243.
97. Matsufuji, A., S. Nakazawa, and K. Yamamoto (2001). Electrocatalysis of the reduction of dioxygen by π-conjugated polymer complexes with dinuclear cobalt porphyrin. *J. Inorg. Organomet. Polym.* **11**, 47–61.

98. Kobayashi, N., K. Sudo, and T. Osa (1990). Preparation of iron and cobalt phthalocyanine-1,2:3,4:8,9;10,11:15,16:22,23:24,25-octakis(*N*-decyldicarboximide) and their catalytic behavior for electroreduction of oxygen, *Bull. Chem. Soc. Jpn* **63**, 571–575.
99. Oyaizu, K., A. Haryono, H. Yonemaru, and E. Tsuchida (1998). Catalytic behavior of a μ-oxo dimanganese(III)octaethylporphyrin in O_2 reduction. *J. Chem. Soc. Farad. Trans.* **94**, 3393–3399.
100. Oyaizu, K., A. Haryono, J. Natori, and E. Tsuchida (1998). (μ-peroxo)bis [pyridine(phthalocyanato)iron(III)] as a convenient catalyst for the four-electron reduction of dioxygen. *J. Chem. Soc. Farad. Trans.* **94**, 3737–3742.
101. Oyaizu, K., A. Haryono, J. Natori, H. Shinoda, and E. Tsuchida (2000). Electroreduction of μ-oxo iron(III) porphyrins adsorbed on an electrode leading to a cofacial geometry for the iron(II) complex: Unexpected active site for the catalytic reduction of O_2 toH_2O. *Bull. Che. Soc. Jpn* **73**, 1153–1163.
102. Shi, C., B. Steiger, M. Yuasa, and F.C. Anson (1997). Electroreduction of O_2 to H_2O at unusually positive potentials catalyzed by the simplest of the cobalt porphyrins. *Inorg. Chem.* **36**, 4294–4295.
103. Shi, C. and F.C. Anson (1998). (5,10,15,20-Tetramethylporphyrinato)Co(II): A remarkably active catalyst for the electroreduction of O_2 to H_2O. *Inorg. Chem.* **37**, 1037–1043.
104. Chaves, J.A.P., M.F.A. Araujo, J. de J.G. Varela Júnior, and A.A. Tanaka (2003). Electrocatálise da reaçao de reduçao de oxigenio sobre eletrodos de grafite modificados com ftalocianina tetracarboxilada de ferro. *Ecl. Quím., Sao Paulo* **28**, 9–20.
105. Tanaka, A.A., C. Fierro, D.A. Scherson, and E. Yeager (1989). Oxygen reduction on adsorbed iron tetrapyridinoporphyrazine. *Mat. Chem. Phys.* **22**, 431–456.
106. Sawaguchi, T., T. Matsue, K. Itaya, and I. Uchida (1991). Electrochemical catalytic reduction of molecular oxygen by iron porphyrin ion-complex modified electrode. *Electrochim. Acta* **36**, 703–708.
107. Elzing, A., A. van der Putten, W. Visscher, and E. Barendrecht (1987). The mechanism of oxygen reduction at iron tetrasulfonato-phthalocyanine incorporated in polypyrrole. *J. Electroanal. Chem.* **233**, 113–123.
108. Elzing, A., A van der Putten, W. Visscher, and E. Barendrecht (1990). Models for adsorption of dioxygen on metal-chelates. *Recl. Trav. Chim. Pays-Bas* **109**, 31–39.
109. Zagal, J.H., M. Páez, J. Sturm, and S. Ureta-Zañartu (1984). Electroreduction of oxygen on mixtures of phthalocyanines co-adsorbed on a graphite electrode. *J. Electroanal. Chem.* **181**, 295–300.
110. Sheldon, R.A. and J.K. Kochi (1981). *Metal-Catalysed Oxidations of Organic Compounds*. Academic Press, New York.
111. van Veen, J.A.R. and J.F. van Baar (1982). Electrocatalysis on transition metal chelates. *Rev. Inorg. Chem.* **4**, 293–327.
112. Shigehara, K. and F.C. Anson (1982). Electrocatalytic activity of three iron porphyrins in the reduction of dioxygen and hydrogen peroxide at graphite cathodes. *J. Phys. Chem.* **86**, 2776–2783.
113. van den Brink, F., W. Visscher, and E. Barendrecht (1984). Electrocatalysis of cathodic oxygen reduction by metal phthalocyanines: Part III. Iron phthalocyanine as electrocatalyst: Experimental part. *J. Electroanal. Chem.* **172**, 301–325.
114. Liu, H.Y., I. Abdalmuhdi, C.K. Chang, and F.C. Anson (1986). Catalysis of the electroreduction of dioxygen and hydrogen peroxide by an anthracene-linked dimeric cobalt porphyrin. *J. Phys. Chem.* **89**, 665–670.
115. Anderson, A.B. and R.A. Sidik (2004). Oxygen electroreduction on Fe^{II} and Fe^{III} coordinated to N4 chelates. Reversible potentials for the intermediate steps from quantum theory. *J. Phys. Chem. B* **108**, 5031–5035.

116. Abraham, K.M., M. Alamgir, E.B. Willstaedt, and W.P. Kilroy (1992). Metal phthalocyaninc-catalysed Li/$SOCl_2$ cells. *Electrochim. Acta* **37**, 531–543.
117. Collman, J.P. and K. Kim (1986). Electrocatalytic four-electron reduction of dioxygen by iridium porphyrins adsorbed on graphite. *J. Am. Chem. Soc.* **108**, 7847–7849.
118. Magner, G., M. Savy, G. Scarbeck, J. Riga, and J.J. Verbist (1981). Effects of substitution of iron by molybden in the naphthalocyanine structures upon their electrocatalytic properties for O_2 reduction and evolution in alkaline media. *J. Electrochem. Soc.*, **128**, 1674–1680.
119. Bettelheim, A., D. Ozer, R. Harth, and R.W. Murray (1988). Redox and electrocatalytic properties towards dioxygen reduction of ruthenium tetra(*ortho*-aminophenyl)porphyrin complexes with various axial ligands. *J. Electroanal. Chem.* **246**, 139–154.
120. Convert, P., C. Coutanceau, P. Crouigneau, F. Gloagen, and C. Lamy (2001). Electrodes modified by electrodeposition of CoTAA complexes as selective oxygen cathodes in a direct methanol fuel cell. *J. Appl. Electrochem.* **31**, 945–952.
121. Jiang, R. and D. Chu (2000). Remarkably active catalysts for the electroreduction of O_2 to H_2O for use in an acidic electrolyte containing concentrated methanol. *J. Electrochem. Soc.* **147**, 4605–4609.
122. Gupta, S., D. Tryk, S.K. Zecevic, W. Aldred, D. Guo, and R.F. Savinell (1998). Methanol-tolerant electrocatalysts for oxygen reduction in a polymer electrolyte membrane fuel cell. *J. Appl. Electrochem.* **28**, 673–682.
123. Sun, G.Q., J.T. Wang, S. Gupta, and R.F. Savinell (2001). Iron(III) tetramethoxyphenylporphyrin (FeTMPP–Cl) as electrocatalyst for oxygen reduction in direct methanol fuel cells. *J. Appl. Electrochem.* **31**, 1025–1031.
124. Bettelheim, A., L. Soifer, E. Korin (2004). Electropolymerized porphyrin films as methanol barriers in direct methanol fuel cells. *J. Electroanal. Chem.* **571**, 265–272.
125. Shentu, B, K. Oyaizu, and H. Nishide (2004). Electroreduction of oxygen at a Pt/C-modified electrode with a cobalporphyrin complex. *Bull. Soc. Chem. Jpn* **77**, 401–405.
126. Shentu, B., K. Oyaizu, and H. Nishide (2004), Preparation of a tetraphenylporphyrinato-cobalt(II)-poly(3,4-azopyridylene)complex and its oxygen enrichment effect at an oxygen electrode. *J. Mater. Chem.* **14**, 3308–3311.

3

Oxygen Reduction in PEM Fuel Cell Conditions: Heat-Treated Non-Precious Metal-N_4 Macrocycles and Beyond

Jean-Pol Dodelet

1. Why Search for a Non-Pt Based Catalyst for the Reduction of O_2 in PEM Fuel Cells?

Traditionally, fossil fuels (coal, oil, gas) have been used to satisfy the world's energy needs. However, these resources are not endless. For instance, at the present 2% growth in demand, the peak of world petroleum production is estimated to occur between 2004 and 2025[1,2]. This will be followed a few decades later by the peak of natural gas production. Furthermore, fossil fuels account for most of the 23–25 Gt of CO_2 that people around the world vent into the atmosphere every year[3,4].

CO_2, along with nitrous oxide, methane, chlorofluorocarbons, tropospheric ozone, and water vapor is a greenhouse gas[5]. With the exception of water vapor, the concentration of all these gases is controlled more or less directly by human activities. CO_2 is the most important greenhouse gas because it has an atmospheric life of about 100 years and therefore builds up in the atmosphere over long periods[5]. While the preindustrial CO_2 concentration was about 280 ppmv, its present concentration has increased 32% since preindustrial times to reach 370 ppmv today. Half of the increase has been since 1965[6].

The energy produced by human activity cannot alter global climate directly because it represents only 1/10,000 of the total energy (120×10^{15} J)[6] absorbed in 1 s by the earth and its atmosphere. On the other hand, human activity is able to alter global climate indirectly by interfering with natural flows of energy through

Jean-Pol Dodelet • INRS-Énergie, Matériaux et Télécommunications, 1650 boulevard Lionel Boulet, Varennes, Québec, Canada J3X. 1S2.

N_4-Macrocyclic Metal Complexes, edited by José H. Zagal, Fethi Bedioui and Jean-Pol Dodelet. Springer Science+Business Media, Inc., New York, 2006.

changes in the atmospheric composition. Indeed, the greenhouse gases trap outgoing radiation from earth to space, creating a warming of the planet. Even small changes in the concentration of a gas like CO_2, which is a minor component of the atmosphere, may have a very large effect when the total energy (175×10^{15} J)[6] received per second at the top of the earth atmosphere is considered. Unless huge reductions are made in fossil fuel use, the atmospheric CO_2 concentration is expected to reach 560–1,000 ppmv by the end of this century, increasing the global temperature a further 1.4–5.8°C[5]. Consequences of global heating on the troposphere are accelerated land-surface drying and more atmospheric water vapor. Accelerated drying increases the frequency and severity of droughts, whereas additional water vapor increases the risk of heavy precipitation events[6]. A perfect demonstration of the correlation between world temperature change and CO_2 concentrations over 420,000 years has been given by measuring the content of these gases in the Vostok Antarctica ice core[7]. Accordingly, an energy economy based on hydrogen seems like the perfect alternative[8]. When burned or oxidized in a fuel cell, it emits no pollution, including no greenhouse gases, except of course for water vapor. Furthermore, gram for gram, hydrogen releases more energy than any other fuel.

In the developed world, the most important sector of energy consumption is transportation. It represents 56.5 GJ per capita and per year, which is 36% of the total energy consumption[9]. Replacing fossil fuel by H_2 in that sector alone would already have a big impact on CO_2 release, provided that H_2 is not produced from fossil fuels. Following that line, General Motors has announced plans to be the first company to sell one million hydrogen fuel cell cars by the middle of the next decade. The development of fuel cell cars is a worldwide priority. Indeed, in 2003, the Bush Administration made a commitment to the development by 2020 of hydrogen economy, pledging US $1.7 billion over 5 years for its Hydrogen Fuel Initiative and Freedom CAR partnership to develop hydrogen fuel cell-powered vehicles. European Commission followed by launching the first phase of an expected 10-year, €2.8 billion public–private partnership to develop hydrogen fuel cells. In 2003, the Japanese government nearly doubled its fuel cell R&D budget to US $268 million. Other countries, including Canada, also have fuel cell programs[8].

Fuel cells have a long history. Since their invention in 1839 by William R. Grove, many fuel cell types have appeared. However, in technical terms, polymer electrolyte membrane (PEM) fuel cells dominate in transport. In 2002, 97% of the total fuel cell development for transport was focused on PEM fuel cells[10]. They have the high power density required to meet the space constraints in light-duty vehicles. Their low operating temperature should also lead to quick start up and shut down. The tolerance of PEM fuel cells to the presence of CO_2 is attractive in comparison with alkaline fuel cells, which can also exhibit high power density and fast start up. PEM fuel cells may also become important in other sectors of energy consumption, such as residential as well as commercial and public services, which together represent another 31% of the total energy consumption per capita in developed countries[9].

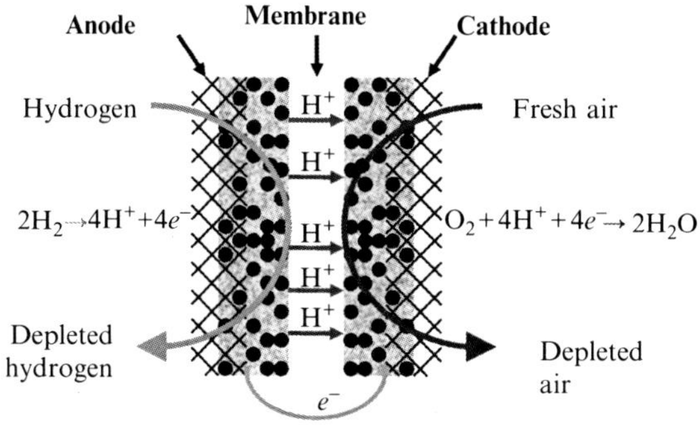

Figure 3.1. Schematic representation of a PEM fuel cell single membrane electrode assembly (MEA) with electrochemical reactions occurring at catalyst particles (dark circles).

PEM fuel cells are electrical power generators that involve the oxidation of hydrogen at the anode and the reduction of oxygen at the cathode. The electrodes are separated by an ionomer material, an H^+- (or proton-) conducting polymer electrolyte. Different companies produce PEMs . The dominant membranes have, however, a common theme, which is the use of sulfonated fluoropolymers. The best known of these is Nafion®, an ionomer from Dupont. Besides being chemically highly resistant and rather mechanically strong, even in very thin films (down to 50μm), well-hydrated Nafion® is a good proton conductor. In the H^+ form, it is an acid medium characterized by a pH between 0 and 1[11].

The simplest PEM fuel cell is made of a single membrane electrode assembly (MEA) presented schematically in Figure 3.1. Work is extracted from this fuel cell by adding a load on the external circuit of electrons. PEM fuel cells are usually run at about 80°C. Higher temperatures in the neighborhood of 100°C and above dry the membrane, which becomes much more resistive, thereby impeding the flow of protons from the anode to the cathode. Since PEM fuel cells are low-temperature fuel cells, electrocatalysts are necessary to perform both H_2 oxidation and O_2 reduction at a useful rate. Today, the only catalyst used in PEM fuel cell commercial prototypes is essentially Pt, a noble metal that is chemically stable in the acid environment of the cell, at 80°C and under the operating potentials of the anode and cathode. In the early days[12], Pt was used unsupported at the loading of up to 28 mg/cm^2, but its usage has been drastically reduced to around 0.6–0.8 mg/cm^2 today[13]. Pt catalysts for fuel cells are now made of Pt nanoparticles (from 2 to 5 nm) loaded on high surface area carbon blacks with a high mesoporous area and a degree of graphitic character[14]. A common support is, for instance, aggregates of the ~30 nm diameter spheres of Vulcan XC72R carbon black from Cabot Corporation.

Pt is a precious metal with low abundance. Two problems may be associated with the use of this metal in PEM fuel cells. They are: (i) aspects related to Pt cost and reserves and (ii) aspects related to Pt supply. Figure 3.2 presents a chart of Pt cost since 1992. The source of this information is the web site of Platinum Today from Johnson Matthey[15]. Figure 3.2 shows that the price of one troy ounce (31.1 g) of Pt remained quite stable at around US $400 for several years but has

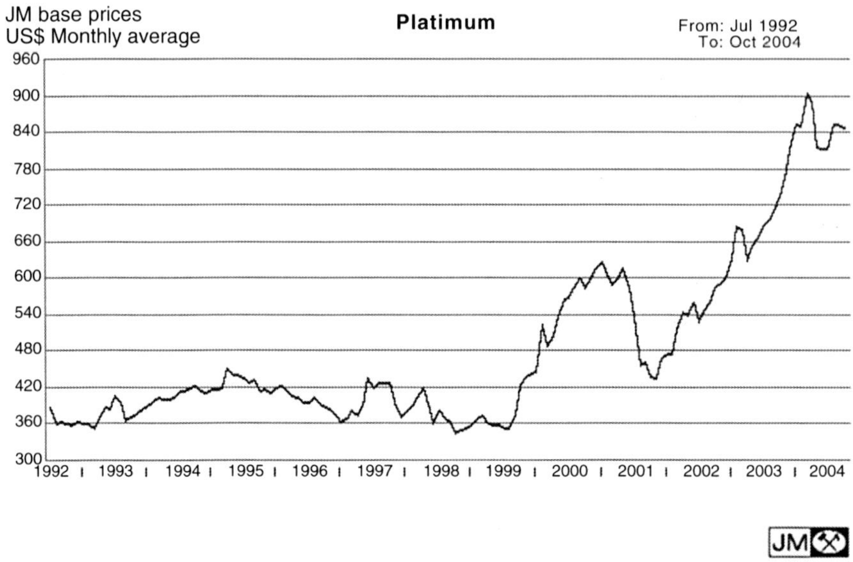

Figure 3.2. Chart of Pt cost since 1992 (according to ref. [15]).

been rising since 2000. At the end of 2004, one troy ounce of Pt costs more than US \$800. A first solution to Pt cost in electrocatalysts is to use as little of it as possible. In 2003, DOE released technical targets on precious metal loadings in PEM fuel cells for transportation[16]. On a basis of 50 kWe fuel cell systems, the targets were < 2.0 g_{Pt}/kW in 2003 0.6 g_{Pt}/kW in 2005, and will be 0.2 g_{Pt}/kW in 2010.

The total Pt mined in 2002 and 2003 was 185.7 and 194.1 metric tons, respectively, while the total Pt demand during the same years was 201.2 and 202.8 metric tons, respectively. The production of 10 millions vehicles per year (about 1/5 of the global auto market[17]), using 75 kWe PEM fuel cells with a loading of 0.2 g_{Pt}/kW would create a demand of 150 metric tons of Pt per year. This would practically double the present demand for Pt. However, even against the most optimistic of fuel cell vehicle projections, there are substantial Pt reserves at current extraction depths (the South Africa's Bushveld Complex, which represents ~75% of the world's platinum resources, alone has 30,000 metric tons reserve[17]). To avoid price volatility, an important issue will be to closely match mining and demand. Operations in South Africa typically require 3–4 years to make substantial mining expansions. Furthermore, it is also anticipated that Pt recovery, which is about 20% today for autocatalysts[18], will drastically improve for PEM fuel cell stacks. At US \$800 per troy ounce and at a loading of 0.2 gPt/kW, the cost of Pt in a 75 kW fuel cell is US \$386. If neither the cost of Pt (considering its recovery) nor its reserves are restricting the use of Pt in PEM fuel cells, are there other aspects that would restrict its availability?

In 2003, only two countries supplied most of the Pt used in the world. They are South Africa (74.8%) and Russia (16.8%). The amount of Pt mined in North America represents only 4.7%, while 3.7% are supplied by the rest of the world[19].

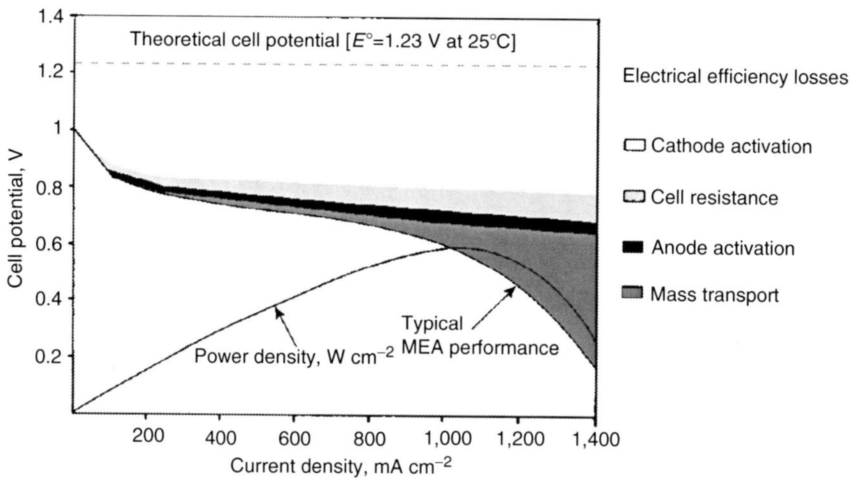

Figure 3.3. Typical polarization curve for a H_2/O_2 PEM fuel cell single membrane electrode assembly with individual electrical efficiency losses (according to Figure 3 in ref. [14]; reproduced with permission of JOhnson Matthey PLC).

It is clear that, in a hydrogen economy, if PEM fuel cells are to generate electricity for the complete transport and the other sectors previously mentioned, the dependence of Pt availability on a limited number of countries will be worse than the present dependence on oil, which is better distributed worldwide than the precious metal. An alternative to Pt as catalyst in PEM fuel cells therefore becomes attractive. This novel catalyst needs, however, to be of comparable activity with Pt, besides being cheap and available everywhere.

Pt is used as electrocatalyst at the anode and the cathode of PEM fuel cells. The electrical efficiency losses in a typical PEM fuel cell are shown in Figure 3.3. Among them are the efficiency losses related to the kinetics of electron transfer at both electrodes. It is clear that cathode activation is much larger than its anode counterpart. On the one hand, at the anode, the small efficiency loss reflects the large activity of Pt toward H_2 oxidation[20] and, on the other hand, the much larger efficiency loss at the cathode reflects its poor activity for the electrochemical reduction of oxygen[21]. The reduction of Pt anode loading from 0.2 to 0.4 mg_{Pt}/cm^2 down to 0.05mg_{Pt}/cm^2 is straightforward[22], while lowering today's cathode loading of $\sim$0.4 mg_{Pt}/cm^2 is limited by the poor activity of Pt for the oxygen reduction reaction (ORR)[21]. It is therefore more productive to seek to replace Pt at the cathode than at the anode.

In the acid medium of PEM fuel cells that may also contain some levels of fluoride derived from membrane degradation, Pt cannot be replaced with a non-noble metal or a metallic oxide, as both will corrode in such environment. This chapter describes some of the efforts that have been made over 40 years to obtain other non-precious metal electrocatalysts for ORR in acidic medium. They all started with the discovery in 1964 that CoN_4 phthalocyanine was capable of oxygen reduction in an alkaline solution[23].

2. Activity of Electrocatalysts based on Fe–N_4 and Co–N_4 Macrocycles and Beyond

2.1. The Early Days: 1964–1989

The discovery of N_4-metal chelate capacity to reduce oxygen electrochemically was reported in 1964 by Jasinsky[23]. He showed that cobalt phthalocyanine (CoPc; see Figure 3.4) mixed with Ni powder could be used as a cathode electrocatalyst for O_2 reduction at room temperature and in 25% KOH electrolyte. The same author also used CoPc impregnated in a Ni plaque from a CoPc-pyridine slurry. Cu-, Ni-, and PtPc's were also measured, but their performance was inferior to that of CoPc. This discovery created a large interest in molecular-based catalysts for ORR, and the work performed during the following 10 years was reviewed by Jahnke *et al.*[24]. During these years, it was demonstrated that several unsupported metallic phthalocyanines, in which the metal is surrounded by four nitrogen ligands, are also active in sulfuric acid-containing media with an activity increasing in the order Cu $\sim$ Ni $<$ Co $<$ Fe. The interaction essential for the catalysis takes place between the oxygen and the central metal ion.

At the beginning of the 1970s, Beck and coworkers discovered the electrocatalytic activity of metal-dibenzo-tetra-aza-annulene (MeTAA) and Sandstede and coworkers that of metal tetraphenyl porphyrin (MeTPP) and its tetramethoxy derivative (MeTMPP)[24]. The molecular structures of CoTAA and CoTMPP are also presented in Figure 3.4. All Me–N_4 chelates were also active on a carbon substrate. For the Pcs, which are insoluble in conventional solvents and water, the catalysts were prepared by dissolving the chelate in concentrated sulfuric acid, then precipitating it onto a carbon substrate by addition of water. TPPs and TAAs on carbon were prepared by impregnation from an appropriate organic solvent. It was shown[24] that the nature of the oxygen-containing groups on the carbon surface is important; alkaline surface groups must be present if the Me–N_4 chelate is to have a catalytic effect. The carbons with alkaline surface groups that were used were acetylene black, Norit BRX, or Vulcan 3H. It was demonstrated[24] by Mössbauer spectroscopy that there is an intimate interaction between FePc and alkaline surface groups at the surface of the carbon support.

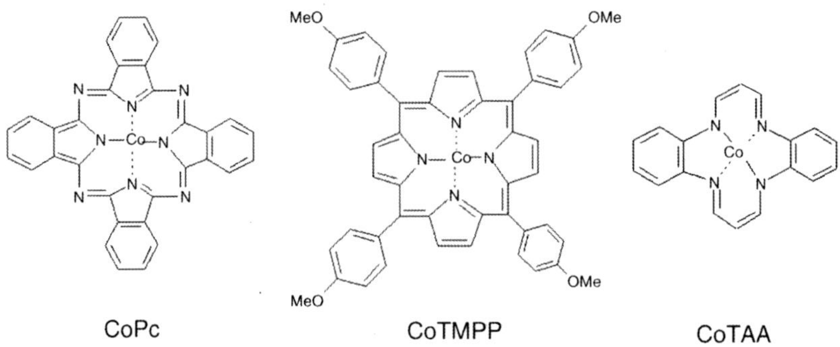

Figure 3.4. Molecular structure of cobalt phthalocyanine (CoPc), cobalt tetramethoxy phenylporphyrin (CoTMPP), and cobalt tetraazaannulene (CoTTAA).

There was, however, a rapid decline in the activity of both unsupported and carbon supported Me–N_4 chelate electrocatalysts. Their lifetime was restricted by direct chemical attack of the catalyst molecule by H_2O_2 released during oxygen reduction. *An important discovery during the first 10 years was to show that the catalyst stability as well as its activity toward ORR could be improved by subjecting the Me–N_4 chelate/carbon sample to thermal pretreatment in an inert gas (N_2 or Ar).* The best heat-treating temperatures depend on the particular Me–N_4 chelate, but 500–600°C was recommended. At higher temperatures, the catalyst activity decreased but its stability increased up to 900°C. A good example of improved stability after heat treatment was reported in 1977 by Bagotzky *et al.*[25]. They loaded by impregnation 5 wt% of CoTMPP (0.37 wt% Co) on Norit BRX and heated this material in inert atmosphere at various temperatures from 400 to 1,000°C. After a heat treatment at 400°C, the time of operation without loss of activity at 10 mA/cm^2 in 4.5N H_2SO_4 and 20°C was 150 h, while the catalyst functioned for more than 10,000 h (417 days) after a heat treatment at 800°C. They also reported that the heterogeneous decomposition of H_2O_2 was much higher on a heat-treated powder catalyst than on untreated ones.

If the heat treatment of Co- or Fe–N_4 chelates adsorbed on carbon improves not only the stability, but also the activity, of the resulting catalyst, the question is: what happens to the chelate molecule during that heat treatment and what is the structure of the resulting catalytic site? This topic has been strongly debated during the 1980s and still remains incompletely resolved. Seven reviews[26–32], which appeared between 1986 and 1990, summarize the structures of the catalytic site that have been proposed during that decade. Essentially, there were three schools of thoughts led by van Veen, Yeager, Wiesener, and their respective collaborators.

1. Based on X-ray photoelectron spectroscopy (XPS), Fourier transform infrared (FTIR), electron spin resonance (ESR), Mössbauer, and extended X-ray absorption fine structure spectroscopy (EXAFS)[26,28,33,34], van Veen and collaborators concluded that the thermal treatment at temperatures where catalytic activity is maximum (~500–600°C) does not lead to complete destruction of the macrocycles, but rather to a ligand modification which preserves the Me–N_4 moiety intact. Furthermore, the stability of this catalytic site is improved because the reactive parts of the ligands are bound to the carbon support and thus are no longer susceptible to an oxidative attack. Thermal treatments at higher temperatures (up to 850°C) led to some decomposition of the Me–N_4 moiety, and thus to a decrease of the catalytic activity, and to the reduction of some of the ions to their metallic state.
2. On the other hand, based on pyrolysis mass spectrometry, pyrolysis gas chromatography, microanalysis and mainly Mössbauer spectroscopy[35–37], Yeager and his collaborators who were mainly working at high pyrolysis temperatures (800–850°C) disagreed with the previous school. They concluded that the decomposition of Me–N_4 macrocycles starts at about 400–500°C. At 800°C, despite the fact that heat-treated samples retain more than 80% of the nitrogen atom content measured for the untreated

samples, a significant fraction of the iron or cobalt ions does not seem to remain coordinated to the original ligating atoms. Most of the metal appears as a mixture of oxides and metal, the latter spontaneously oxidizing when exposed to air at room temperature. Upon contact with an electrolyte solution, oxide species may undergo partial or total dissolution. Metallic ions subsequently adsorb or coordinate to thermally formed sites on the carbon surface, most likely involving nitrogen. The resulting structure is believed to be the true catalytic site for oxygen reduction. Here, it is important to notice that, according to Yeager and his collaborators, the catalytic site is obtained not during, but after, the heat treatment at high temperature of $Me–N_4$ chelates adsorbed on the carbon support.

3. A third school, led by Wiesener[29], proposed that the Co or Fe ions of the adsorbed N_4 chelates promote the decomposition of the chelate upon thermal treatment followed by the formation, at high temperature, of a special form of carbon that would be the true catalyst. In this scenario the metal is only an intermediate and has no active role in the electroreduction of oxygen. In a later publication[38], they concluded that nitrogen is involved in the electrocatalytic active group on carbon.

An important discovery for non-noble metal PEM fuel cell catalysts for oxygen reduction was reported in 1989. During that year, *Yeager and collaborators published a paper about the preparation and characterization of a novel catalyst active for oxygen reduction that was not made by the adsorption on a carbon support and subsequent heat treatment of a* $Me–N_4$ *macrocycleprecursor, but rather was prepared by the adsorption of polyacrylonitrile (PAN) mixed with* Co^{II} *or* Fe^{II} *salts dissolved in dimethylformamide*[39]. After pyrolysis of that material in argon at high temperature (800°C), these catalysts were studied in 85% H_3PO_4 at 100°C in porous gas-fed electrodes. The activity of one of the novel catalysts containing 1.18 wt% Co was shown to be equivalent to that of a catalyst prepared at the same heat-treatment temperature with 5% CoTMPP (0.37 wt% Co). The retention of cobalt ions in these catalysts was explained in terms of a possible coordination interaction of cobalt with the nitrogen of the pyrido-rings resulting from the pyrolysis of adsorbed PAN.

This report followed a seminal work from the same group using heat-treated pyrrole-black polymer mixed with Co^{II} or Fe^{II} salts[40]. The catalytic activity of those materials was also interpreted by the retention during pyrolysis of nitrogens of the pyrrole groups as binding sites for the transition metal ions, which then provided the catalytic activity. Both reports from Yeager's group clearly showed that catalytic activity for O_2 reduction in acid medium requires the presence of three components: (i) a nitrogen precursor, (ii) a transition metal (Fe or Co) precursor, and (iii) a carbon support, plus a heat treatment at relatively high temperature of the material resulting from the adsorption of the nitrogen and metal precursors on the carbon support. According to Yeager, the last step is necessary to obtain nitrogen functionalities on the carbon support and to reduce the metal ions. The catalytic site is not, however, produced during that heat treatment but is self-assembled afterward during contact with the electrolyte that dissolves metal particles, which had been generated during pyrolysis. This novel preparation procedure represented a crossing point from which two different roads were followed to prepare nonprecious metal-based catalysts for the electroreduction of oxygen: the first one pursued the work on Me–N_4 chelates for which metal and nitrogen are part of the same molecular precursor. This work is reviewed in Section 2.2.1. The second road explored another research direction in which metal and nitrogen belonged to two different molecular precursors. Besides giving more flexibility in the preparation of the catalysts, this latter avenue also led to the use of much cheaper molecular precursors than Me–N_4 chelates. This work is reviewed in Section 2.2.2.

2.2. Molecular Structure of the Catalytic Site(s)

2.2.1. Fe- and Co Macrocycles After the 1980s

In Sect. 2.1 it was shown that three molecular structures were proposed during the 1980s by van Veen, Yeager, Wiesener, and their respective collaborators, as potential catalytic sites for the reduction of oxygen in acid medium. In this section, how these three molecular models for the catalytic site evolved since their first appearances in the literature is described.

During the 1990s, van Veen and collaborators mainly studied the electrochemical kinetics of oxygen reduction. Their results are presented in Sect. 3. These mechanistic studies were, however, always based on the model in which the Co–N_4 or Fe–N_4 moieties of the respective macrocycles were retained intact at *all* pyrolysis temperatures. Their last contribution to the molecular structure of the catalytic site was a study in 2002 of catalysts obtained by adsorption of iron tetramethoxyphenyl porphyrin chloride (ClFeTMPP) on Vulcan XC-72, heat treated between 325 and 800°C in inert atmosphere, and characterized by EXAFS and Mössbauer spectroscopy, as well as by cyclic voltammetry[41]. The loading of these catalysts was 7 wt% chelate (~0.5 wt% Fe).

EXAFS confirmed that unpyrolyzed ClFeTMPP adsorbed intact on Vulcan and was still intact up to 325°C. After pyrolysis at 700°C (the optimum temperature for catalytic activity) the first shell of atoms around the Fe ion was still Fe–N_4, but the interatomic distance contracted and the Fe ion was now in the N_4 plane (Fe ion was above that plane for the intact molecule). The Mössbauer spectrum of samples heat treated at 800°C revealed the formation of magnetic iron oxides that could be chemically leached out. The leached material yielded a spectrum similar to that recorded at 700°C and attributed to the Fe–N_4 moiety. Van Veen and collaborators proposed a reaction scheme of what they believed happened to the ClFeTMPP chelates upon pyrolysis. This scheme is illustrated in Figure 3.5. However, they warned the reader that, in Figure 3.5, while the overall structure of the iron catalytic site remained close to the original Fe–N_4 moiety, there must be substantial local variation in how this unit is attached to the subjacent carbon support. This study is in agreement with an earlier study by McBreen and collaborators[42], who analyzed 9 wt% CoTMPP and FeTMPP chelates (about 0.7 wt% metal) adsorbed on Vulcan XC-72 and heat treated in inert atmosphere between 600 and 1,000°C. They found that the heat treatment of the macrocycles

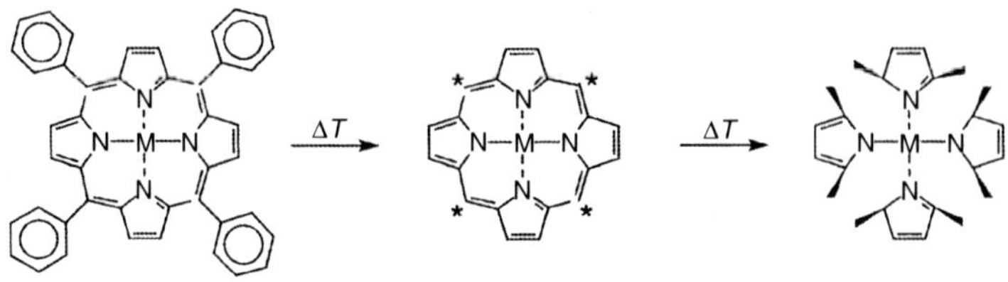

Figure 3.5. Visualization of the reaction of a porphyrin with the carbon support during heat treatment (according to Figure 9 in ref. [41]; reproduced with permission of the American Chemical Society).

also provided a mixture of several compounds, including metal and metal oxides. Chemical leaching removed metal and oxides revealing the electrocatalyst, which was an isolated ion in both cases. Its first coordination shell was apparently four nitrogen atoms at a distance similar to that in the original macrocycle. With a further increase of the temperature, formation of the catalyst and its decomposition to metal occurred simultaneously. In the case of FeTMPP, there were some Fe–N_4 moieties even after pyrolysis temperatures up to 1,000°C, while the Co chelates were reduced to metal at 900°C.

The last two publications of Yeager and his collaborators appeared in 1992 and 1994. In the first one[43], FeTMPP or CoTMPP (4.8 wt% chelate, ~0.4 wt% Co or Fe) was adsorbed on Vulcan XC-72 and heat treated in inert atmosphere at 800°C. The Mössbauer spectra indicated that FeTMPP was adsorbed on carbon black as PPTMFe–O–FeTMPP after facile and irreversible oxidation. At 800°C, the doublet characteristic of the Fe–N_4 coordination was transformed into a singlet indicative of a pronounced change in the coordination, which was no longer consistent with the Fe–N_4 structure of the catalytic site. This new moiety corresponded either to metallic iron or oxide in highly divided form. Furthermore, it was shown by cyclic voltammetry in de-oxygenated 0.05 M H_2 SO_4 that the initial material displayed oxidation and reduction peaks at 0.50 V vs. SCE assigned to the Co^{II}/Co^{III} couple or peaks at –0.1 V vs. SCE assigned to the Fe^{II}/Fe^{III} couple. Cyclic voltammograms after heat treatment at 800°C were featureless, providing evidence that the initial materials underwent profound changes in their structure — an observation in favor of Yeager's model for the catalytic site.

In their second publication[44], yeager and his collaborators used the same CoTMPP and FeTMPP chelates adsorbed on de-ashed RB carbon (20 wt% chelate; ~1.5 wt% Fe or Co) and heat treated these materials at 800°C in inert atmosphere to show that the resulting catalysts were inactive for the oxidation of methanol. This demonstration was performed in a mini-fuel cell using a Nafion 117 membrane equilibrated with H_3PO_4 and running at 105°C. The methanol tolerance of these catalysts as well as the use of H_3PO_4 equilibration of the membranes to enable higher operating temperatures were investigated further by other researchers, in particular by Savinell and his collaborators. Methanol tolerance is an important factor for direct methanol fuel cells, which are plagued with the problem of methanol cross-over through the polymeric membrane from the anodic compartment to the cathodic one. Mixed potentials and decreased net current densities result from the simultaneous reduction of O_2 and the oxidation of methanol at the cathode, greatly affecting the power output of the cell.

It is clear from the analysis of the recent results presented by van Veen and Yeager that neither van Veen, Yeager, nor their respective collaborators, changed their opinion about the molecular structure of their respective catalytic site. *According to van Veen the catalytic site is the Me–N_4 moiety at all pyrolysis temperatures, while Yeager maintains that it is a C–N_x–Me complex of unknown structure. The complex is formed not during, but after, pyrolysis at high (800°C and above) temperatures by the dissolution of metal oxide when the catalyst is brought in contact with an electrolyte, and the subsequent readsorption of metal ions on the C–N_x functionalities of the carbon support.*

An important element of support for Yeager's model of the catalytic site for O_2 reduction came from Scherson and his collaborators[45], who ran *in situ* EXAFS and XANES (X-ray absorption near edge structure) as well as cyclic voltammetry experiments in 1.0 M H_3PO_4 on ClFeTMPP adsorbed on Black Pearls carbon (22 wt% chelate; ~1.7 wt% Fe; 1/4 monolayer Black Pearls coverage based on a 4 nm^2 coverage per ClFeTMPP) and heat treated at 800°C in an inert atmosphere. Figure 3.6 presents their voltammetry results in a deaerated solution. They showed that the intact adsorbed macrocycle (A) was characterized by the Fe^{II}/Fe^{III} couple in the Fe–N_4 complex with $E° = -0.1$ V vs. DHE, while after heat treatment at 800°C, the same Fe^{II}/Fe^{III} couple was characterized by an ill-defined broad structure shifted toward more positive potentials (B).

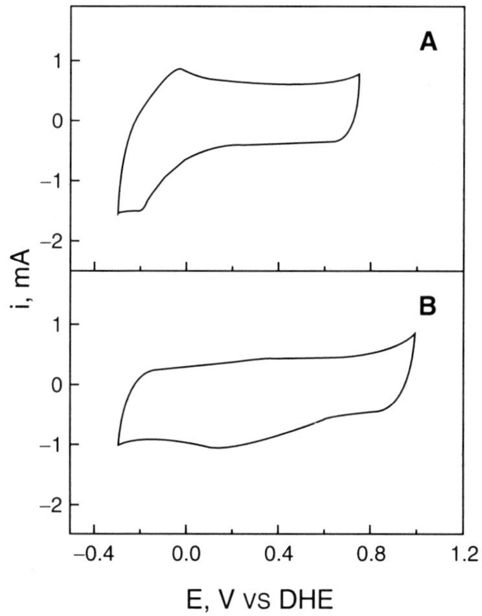

Figure 3.6. Cyclic voltammograms of Teflon-bonded electrodes (2 mg active material, 0.15 cm^2 projected area) containing intact (panel A) and heat treated ClFeTMPP/BP (panel B) in 1 M H_3PO_4 (according to Figure 1 in ref. [45]; reproduced with permission of the American Chemical Society).

Fe^{II} began to appear at about 0.8V vs. DHE, a potential consistent with the relatively positive potentials at which O_2 is reduced at significant current density in H_3PO_4-equilibrated Nafion at 125°C. XANES experiments indicated that the dry heat-treated ClFeTMPP/Black Pearls catalyst contained elemental iron and possibly also iron carbide. The same spectroscopy also ascertained a sharp energy transition between Fe^{II} and Fe^{III} oxidation states for the intact macrocycle, while after heat treatment at 800°C, the same energy transition between Fe^{II} and Fe^{III} still occurred but was no longer so sharp. This indicated the presence of many different catalytic sites that were no longer in the original $Fe–N_4$ environment of the chelate, which was consistent with the cyclic voltammetry data and the ill-defined molecular structure of the catalytic site ($C–N_x–Fe$). The latter is not obtained during the heat treatment at high temperature, but results, as proposed by Yeager, from coordination, in acid solution, of Fe ions to the $C–N_x$ groups formed during pyrolysis. Furthermore, CO had no effect on the pyrolyzed catalyst, while a strong effect of CO was observed for the intact chelate.

Publications from several new groups, including ours, appeared during the 1990s and the idea of two catalytic sites, one at low pyrolysis temperatures and another one at high pyrolysis temperatures, was put forward. The most important players during these years (at least by the number of their publications) were Savy, Savinell, and their respective collaborators, as well as our group.

Savy and his collaborators published 14 papers on heat-treated metal chelates during the period spanning from 1990 to 2000. The earliest papers from this group[46–51] reported results obtained with catalysts made by the adsorption on Norit BrX, Printex, and Black Pearls of Fe or Co naphtophthalocyanines (Fe or CoNPc) at various loadings (between 10 and 40 wt% chelate; ~0.7–3.0 wt% metal). These materials were pyrolyzed at 500°C in inert atmosphere. The catalysts (mainly those based on CoNPc) were studied by IR spectroscopy, XPS, and rotating disk electrode in 0.25M H_2SO_4. It was concluded that the heat treatment at 500°C preserved the $Me–N_4$ unit. A face-to-face, stacked phthalocyanine dimer configuration was proposed as an active site on the basis of: (i) an optimal activity observed for the loading of two monolayers of phthalocyanines, the molecular structure being retained on the carbon support as evidenced by carboxylic and quinonic groups detected in FTIR; (ii) an apparent number of 3.8 electrons exchanged during electroreduction of oxygen (unpyrolyzed covalent face-to-face dimeric cobalt porphyrin structures have been synthesized by Anson and his group; they are known to reduce O_2 to H_2O by 4 electrons[52]); and (iii) mixed valences of Co^{II}/Co^{III} (50%; 50%) detected by XPS indicating that one Co ion of the dimer may be found as Co^{II}, while the other one was found as Co^{III}. A schematic representation of this dimer is given in Figure 3.7. Figure 3.8 presents the mechanism proposed by Savy for the 4 electron reduction of oxygen catalyzed by the face to face $Co–N_4$ dimer proposed as a catalytic site obtained after pyrolysis at 500°C.

Further studies by the same group involved mainly CoTAA and also CoTMPP. These molecules were adsorbed on Norit SX Ultra and were usually pyrolyzed between 600 and 800°C. The optimum CoTAA loading was 13 wt% chelate (~2.3 wt% Co). At the lowest heat-treatment temperature (600°C), the

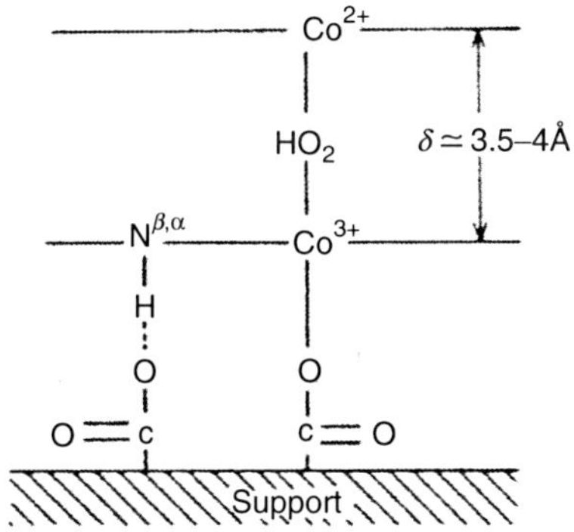

Figure 3.7. Schematic representation of an active Co-based dimer proposed by Savy and collaborators and obtained after heat treatment at 500°C, in inert atmosphere, of Co naphthophthalocyanine adsorbed on a carbon support. The remaining portions of the macrocycle structures lie in the planes represented by the two lines parallel to the support. (according to Figure 8b in ref. [50]; reproduced with permission of Elsevier).

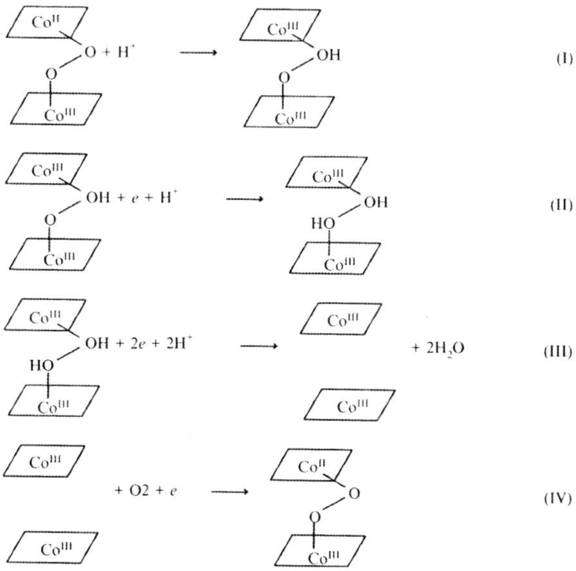

[I] Initial chemical step [proton fixation]
[II] First electron transfer
[III] 2- electron transfer [rds]
[IV] Electron transfer and μ-preoxo formation

Figure 3.8. Mechanism of the 4-electron reduction of oxygen catalyzed by the Co-dimer whose detailed sructure is illustrated in Figure 3.7 (according to Figure 8 in ref. [49]; reproduced with permission of Elsevier).

catalytic activity was attributed to Co in the Co–N_4 moiety preserved from destruction (a face-to-face Co–N_4 dimeric structure was again mentioned here[53–55]). The ESR signal disappeared after heat treatment at 800°C[56]. In other papers[56–59] no Co was detected by XPS for the 800°C heat-treated catalyst after a longevity test in 0.25M H_2SO_4 solution, while before the test Co was detected as CoO. A metal-less site based on nitrogen–oxygen, and carbon atoms and on various oxidation states of the nitrogen atom, was invoked as the origin of the catalytic activity in samples pyrolyzed at high temperatures. This metal-less catalytic site would be involved in the following sequence of reactions during the reduction of oxygen:[57]

1. Lower nitrogen oxidation states such as NH_3OH^+ or $N_2H_5{}^+$ are highly favorable to a protonation reaction such as:

$$\text{R–NH}_2\ \text{OH}^+ + \text{H}^+ + e^- \Rightarrow \text{R–NH}_2^+ + \text{H}_2\text{O (fast)}$$

2. To yield a catalytic cycle, this protonation should be followed by reoxidation of RNH_2^+ by oxygen via a chemical step which should not be rate limiting :

$$2\text{R–NH}_2^+ + (x + 1/2)\text{O}_2 \Rightarrow 2\text{R–NO}_x + 2\text{H}^+ + \text{H}_2\text{O}$$

3. The rate-limiting step would consist of an electrochemical reduction of R–NO_x (stable) into R–NH_2OH^+ (unstable) according to:

$$\text{R–NO}_x + (2x + 1)\ \text{H}^+ + 2xe^- \Rightarrow \text{R–NH}_2\text{OH}^+ + (x - 1)\text{H}_2\ \text{O}$$

It is clear that, according to Savy and his collaborators, there are two types of catalytic sites: (i) the low temperature (LT) one occurring at 500–600°C and the high temperature (HT) one occurring at 800°C. The structure of the LT catalytic site proposed by Savy was inspired by the structure proposed by van Veen (a face-to-face Me–N_4 *dimeric* structure has, however, never been proposed by van Veen). The HT catalytic site proposed by Savy is metal-less. This type of molecular structure is therefore related to the catalytic structure proposed by Wiesener.

Face-to-face dimeric structures were expected when Co or Fe cationic and anionic porphyrins were adsorbed together on a carbon support[60,61]. It was suggested by Sawagushi, Okada, and their respective collaborators that the Fe–N_4 or Co–N_4 dimeric complexes remained associated at low pyrolysis temperatures; but the face-to-face structures would not be preserved after a heat treatment at 800°C. The idea of an unexpected formation of a face-to-face Me–N_4 dimeric structure obtained by pyrolysis of nonionic porphyrins in the 600°C range reappeared later on in 2000 and 2002 in publications of Chu and Jiang[62,63], who reported results obtained after the pyrolysis of unsupported CoTPP and FeTPP. Again, on the basis of an apparent 4-electron transfer measured for the mixed Co-Fe-based catalyst, a face-to-face dimeric structure containing one Co–N_4 and one Fe–N_4 complex was proposed as the catalytic site. The latter nanostructure was thought to have the typical 0.4 nm[52] distance between the two metal centers to accommodate O_2 in a bridged configuration. These catalysts showed methanol tolerance.

The group that did most work on methanol tolerance of non-precious metal cathode catalysts in PEM fuel cells was certainly that of Savinell. Out of the six papers[64–69] that were published by that group on non-precious metal cathode catalysts between 1998 and 2001, four[64–66,68] mentioned that these catalysts prepared with various Fe porphyrins were not active for methanol oxidation. Furthermore, some of these studies reported results obtained with gas diffusion electrodes (GDEs) in fuel cells at high temperatures, using either H_3PO_4-equilibrated Nafion 117[64] or H_3PO_4-doped polybenzimidazole membrane at 150°C[65,66]. The highest fuel cell performances reported to date with non-precious metal catalysts were obtained under these conditions. For the sake of comparison with the results obtained in fuel cells by other groups, the performance with hydrated Nafion was also reported by Savinell *et al.* for two cathode catalysts made by adsorption of ClFeTMPP on RB carbon (12.3 and 24.2 wt% chelate; ~0.8 and 1.6 wt% Fe), followed by heat treating at 800°C. Their polarization curves are presented in Figure 3.9 and compared to that of Pt.

Savinell and collaborators also characterized their catalysts. They found that: (i) the resistivity of the catalysts decreased as the heat-treatment temperature increased; (ii) most of the typical chelate structure of their FTIR spectra disappeared at pyrolysis temperatures > 400°C; (iii) XPS indicated the presence of the Fe–N_4 structure below 800°C, but metallic and/or iron carbide clusters surrounded with a carbon envelope were detected above 800°C. As they attributed no catalytic activity to those particles, they concluded[68] that a plausible explanation for the remaining catalytic activity after pyrolysis at these high temperatures might be active sites containing nitrogen and oxygen. Therefore, for Savinell's group, there are also two catalytic sites, an Fe–N_4 structure at a low pyrolysis temperature (~600°C), which follows van Veen's suggestion, and possibly a metal-less catalytic site of the type discussed earlier by Wiesener and Savy.

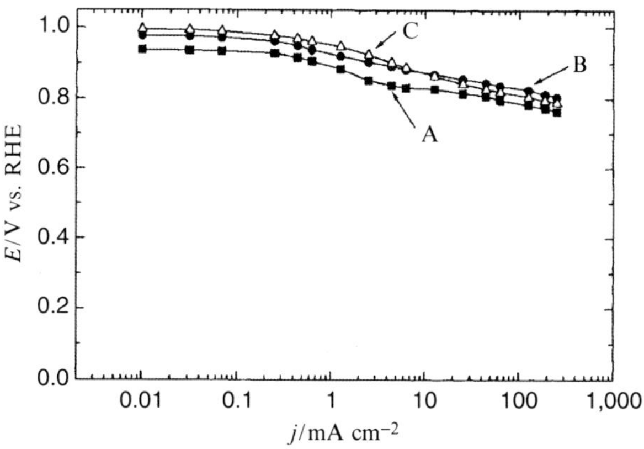

Figure 3.9. Steady-state polarization curve for oxygen reduction (1 atm.) on ClFeTMPP/RB electrode of different loadings with a Pt/C (E-Tek) electrode for comparison in an H_2/O_2 minifuel cell with hydrated Nafion 117 at 60°C: (A, dark squares) 12.3% ClFeTMPP (0.17 mgF/cm^2) on RB (800°C H.T.); (B, dark circles) 24.2% ClFeTMPP (0.34 mgFe/cm^2) on RB (800°C H.T.); and (C, open triangles) Pt/Xc-72 (E-Tek, 2 mg Pt/cm^2) (according to Figure 9 in ref. [64]; reproduced with permission of Springer Science and Busines Media).

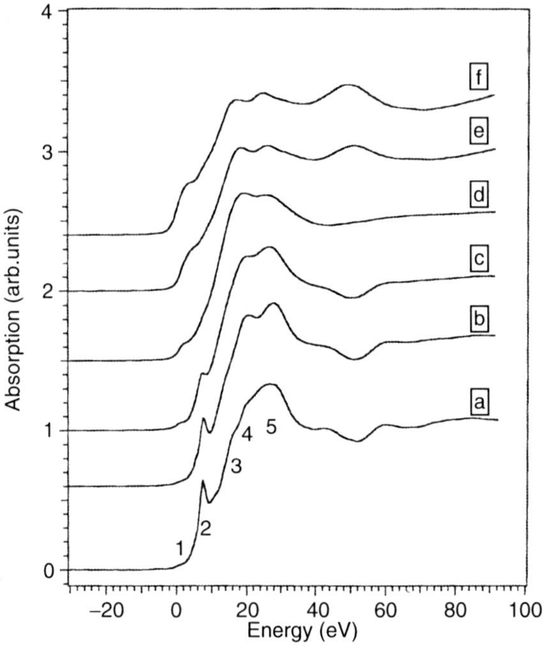

Figure 3.10. Co K edge XANES for CoPc on Vulcan XC-72: (a) Pure CoPc; (b) CoPc adsorbed on Vulcan XC-72; heat treated CoPc adsorbed on Vulcan XC-72 at (c) 700°C; (d) 800°C; (e) 1,000°C; (f) Co metal. The zero energy reference corresponds to 7,709 eV (according to Figure 3 in ref. [70]; reproduced with permission of the Amercian Chemical Society).

Our group began to publish on non-precious metal catalysts for oxygen reduction in acid medium in 1992. Our first article[70] reported mainly XANES and EXAFS characterizations of catalysts obtained by the adsorption of CoPc (10.7 wt% chelate; 1.1 wt% Co) on Vulcan XC-72. That material was heat treated in inert atmosphere at temperatures ranging from 250 to 1,100°C. A XANES characterization of heat treated CoPc/Vulcan clearly revealed that the Co–N_4 structure was retained up to a pyrolysis temperature of 700°C. This is shown in Figure 3.10. Cobalt phthalocyanine has a D_{4h} symmetry where the metallic center atom is in a square planar environment. The CoPc spectrum exhibited several transitions labeled 1, 2, 3, 4, and 5 in Figure 3.10. The first peak was assigned to a dipole forbidden 1s $\rightarrow$ 3d transition. The second transition was assigned to a shake down satellite involving the 1s $\rightarrow$ $4p_z$ transition with simultaneous ligand to metal charge transfer. The third transition was pure 1s $\rightarrow$ $4p_z$. The fourth and fifth features were due to 1s $\rightarrow$ $4p_{x,y}$ transitions. The relative energies of the p_x, p_y, p_z orbitals depend on the coordination of the metallic center. In a D_{4h} symmetry, the absence of axial ligands stabilize the p_z orbital relative to the p_y or p_z ones and the 1s $\rightarrow$ $4p_z$ transition appears at lower energy than the 1s $\rightarrow$ $4p_{x,y}$ ones. The transition labeled 2, which was observed for all compounds in a square-planar environment, is a fingerprint of the Co–N_4 structure. Any modification of the coordination greatly affected this transition.

For temperatures above 700°C, important changes in the XANES pre-edge region occurred. The Co–N_4 configuration was destroyed (loss of transition 2) and the formation of Co particles became more and more important as pyrolysis temperature increased. The evolution of the coordination number of metallic cobalt suggested that very small (about 2 nm diameter) Co clusters were obtained at 900°C. Etching the samples did not change the XANES spectra. High-resolution TEM studies on similar samples (2.0 wt% Co)[71] indicated that the characterization by X-ray absorption spectroscopy underestimated the size (4 ± 1 nm) of the βCo metallic particles which were found in the catalytic material pyrolyzed at 700°C, while an average diameter of 13 nm was observed for material pyrolyzed at 900°C. Most particles at 900°C were also surrounded by an outer shell of graphite, as shown in Figure 3.11.

Following this first study, we prepared a series of catalysts with the following chelates: CoPc[72–75], FePc[76], tetracarboxylic CoPc (CoPcTc)[76,77], tetracarboxylic FePc (FePcTc)[76], CoTMPP[78,79], and FeTMPP[79]. Their nitrogen and metal concentrations were determined in the bulk and at the surface of these materials. The same catalysts were also characterized by XRD analysis, time-of-flight secondary ion mass spectrometry (ToF SIMS), and TEM microscopy, as well as by cyclic voltammetry in H_2SO_4 solution at pH 0.5 and GDE tests in fuel cells. The results obtained with various chelates followed always the same pattern: (i) The bulk metal concentration remained practically constant in all these catalysts over the entire pyrolysis temperature range, while the bulk N concentration decreased with increasing temperature to reach the detection limit around 1,000°C. (ii) At the same time, the metal and N surface concentrations measured by XPS first increased for CoPc and FePc to reach a maximum at about 500–600°C, then decreased at higher temperatures. This is shown in the left part of Figure 3.12 for FePc/C. It was interpreted as a sublimation and subsequent readsorption on the carbon support of the chelate molecules or molecular fragments still containing the intact metal–N_4 complex up to ∼500–600°C, followed by decomposition of the metal–N_4 site at higher temperatures and the formation of metal aggregates,

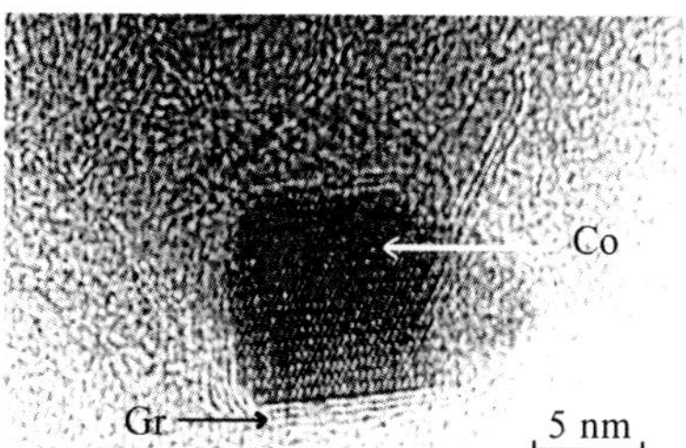

Figure 3.11. High-resolution transmission electron microscopy image of a Co particle after pyrolysis of CoPc/C at 900°C. A graphitic layered structure (Gr) is present on the surface of the particle (according to Figure 5 ref. [71]; reproduced with permission of the Materials Research Society).

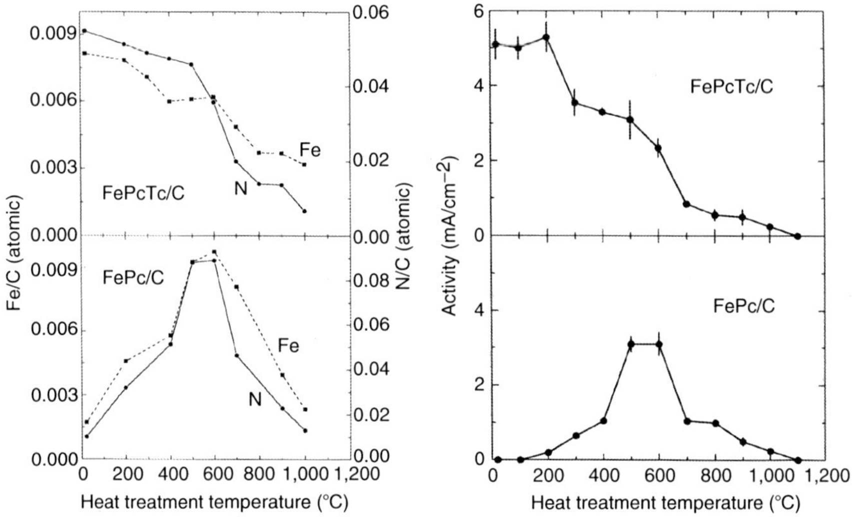

Figure 3.12. Left panel: Change of the Fe and N surface concentrations (relative to C) for FePc/C and FePcTc/C with the heat treatment temperature. Right panel: Change of the catalytic activity for O_2 reduction as a function of the heat treatment temperature for FePc/C and FePcTc/C. The catalytic activity was determined by taking the difference between the current measured at 0.7 V vs. NHE when the electrode was rotating at 1,500 rpm and when it was stationary (according to Figures 1 and 7 in ref. [76]; reproduced with permission of Elsevier).

whose diameter increased with the pyrolysis temperature. Tetracarboxylic FePcTc and CoPcTc do not sublime. They polymerize on the surface of the carbon support. The N and metal surface concentrations do not go through a maximum but simply decline when the pyrolysis temperature increases. The change with heat-treatment temperature of N and Fe surface concentration measured by XPS for all these catalysts evolves in parallel with the catalytic activity measured by cyclic voltammetry in an H_2SO_4 solution. Catalyst activity is shown in the right part of Figure 3.12. (iii) Although the polarization curves in fuel cell tests initially follow the same order of performance than the catalyst activity measured by cyclic voltammetry, short stability tests (10 h) in fuel cell drastically perturb that initial order as shown in Figure 3.13 for tetracarboxylic FePc/C. Catalysts containing the highest concentration of $Fe–N_4$ sites at the surface of the support are much less stable than catalysts obtained at high pyrolysis temperatures. (iv) For the same catalysts, the relative abundance of metal–N_x–C_y ions detected by ToF SIMS, which were believed to be the typical ions released by the $Me–N_4$ complex assigned to the low-temperature catalytic site, decreased upon increasing the pyrolysis temperature to become undetectable at and above 800°C.

The fuel cell behavior, which is illustrated in Figure 3.13, combined with the disappearance of metal–N_x–C_y ions at and above 800°C convinced us of the existence of LT and HT catalytic sites, the LT one, on the one hand, being characterized by a structure in agreement with van Veen's model. On the other hand, we assigned the HT catalytic site to either metallic Co or metallic Fe/iron carbide

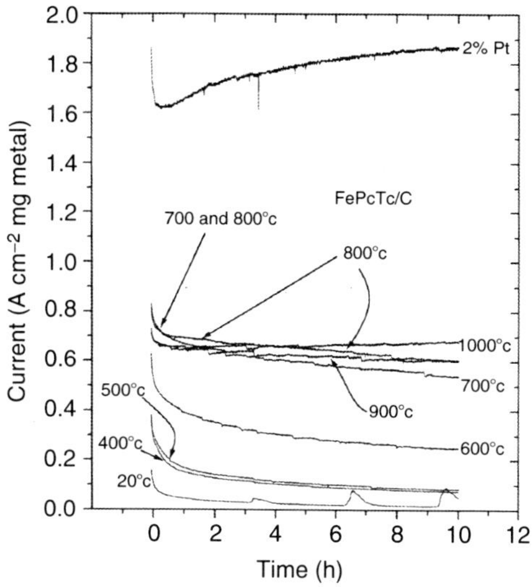

Figure 3.13. H_2/O_2 PEM fuel cell short-term stability test performed at 0.5 V on various MEAs using non-noble metal cathode catalysts obtained by heat treating FePc/C at various temperatures. The curve for 2 wt% Pt is given for comparsion (according to Figure 4 in ref. [76]; reproduced with permission of Elsevier).

particles surrounded by a graphitic envelope. Being more stable than the LT catalytic site, the HT catalytic site was the one most interesting for fuel cell applications. The graphitic envelope around the metallic particles was due to the strong graphitization capabilities of Co and/or Fe (these metals are indeed among the best catalysts to prepare carbon nanotubes[80]). In identifying the HT catalytic site as a metal aggregate surrounded by a graphite envelope, we were, however, confronted by a serious problem! How does this site interact with oxygen? The unprotected metal was not catalytic since 25% of the total metal content of these catalysts could be leached in acid solution without changing the catalytic activity[79]. The metal and its carbide found in the aggregates by XRD and by Mössbauer[81] would not survive unprotected in acid solution[82]. If these particles were at the origin of the catalytic activity, the catalytic sites had to be at the surface of the graphitic envelopes. Were these envelopes a special carbon like the one proposed by Wiesener or some kind of graphite-intercalated metal that could not be seen by TEM?

To verify this hypothesis, we first synthesized Fe– and Co-intercalated graphites[83]. These intercalates had very little catalytic activity for O_2 reduction (the results of the electrochemical reduction of oxygen by Fe and Co graphite intercalates were never published). After those experiments, it occurred to us that the entire metal particle surrounded with a graphitic envelope might not be the HT catalytic site that we were trying to identify. In pursuing this possibility, we used two different precursors to obtain the catalysts, one for the metal and another one for nitrogen[84]. In doing so, we were able to test the role and importance of each component: the metal, the nitrogen atoms, and the carbon support, in producing the HT catalytic site. Catalysts were, therefore, prepared with ferrocene as Fe precursor (a vinyl ferrocene was polymerized on the carbon black (Vulcan XC-72)). The N atom precursor was acetonitrile, which was introduced in the pyrolysis oven by an inert carrier gas (Ar). The synthesis was performed at 1,000°C in order to be unambiguously in the HT catalytic site regime. It was found that active catalysts were obtained only when both Fe and N precursors were in the reactor at the same time. No catalytic activity was found by pyrolyzing just the N precursor adsorbed on carbon and no catalytic activity was found from pyrolyzing just the adsorbed Fe precursor under the same conditions. *In the latter case, only metallic Fe and iron carbide particles surrounded by carbon were obtained, thereby demonstrating clearly that these species were not the HT catalytic site.*

At this point a question remained unanswered. Could the model proposed by Yeager and collaborators be the HT catalytic site? To test this hypothesis we first loaded H_2TPP on Vulcan XC-72 and then pyrolyzed that material at 1,000°C[79]. The nitrogen atom concentration at the surface of the resulting powder was 1.29%. Then the powder was placed in contact with a solution containing an excess of Fe^{II} ions (as $FeSO_4$). The resulting material, which corresponded exactly to the catalytic site described by Yeager, was tested by voltammetry. Its activity was found to be very poor as may be seen in Figure 3.14B. A strong improvement of this activity, as shown by the shift to positive potentials of the oxygen reduction peak in Figure 3.14C could, however, be obtained by heat treating this poor catalyst at high temperature (1,000°C). *This experiment revealed the importance of the temperature step in obtaining catalytic sites. In other words, the C–N_x–Me structure proposed as catalytic site by Yeager and collaborators is only the precursor of the catalytic site.* This C–N_x–Me complex has to be thermally activated at high temperature to become an actual catalytic site that will be labeled Me–N_x/C to differentiate it from Me–N_4/C, the catalytic site proposed by van Veen and his collaborators. Section 2.2.2 reviews the catalysts prepared that way, i.e., without the adsorption on the carbon support of either Co– or Fe–N_4 macrocycles.

2.2.2. Beyond Fe– and Co–N_4 Macrocycles After 1989

At the end of the previous section, we came to the conclusion that, except for van Veen, several authors proposed the existence of two catalytic sites (a low temperature one, and a high temperature one) that are obtained after pyrolysis of Co– or Fe–N_4 chelates adsorbed on carbon. The LT catalytic site would be Me–N_4/C, a Me–N_4 structure bound to the carbon support, such as the one proposed by van Veen (Figure 3.5). However, contrary to the ideas defended by van Veen that the same structure is also responsible for catalysis after high-temperature pyrolysis, there have been several propositions to define the nature of the HT catalytic site. The only proposition that appears consistent with all the data discussed to this point is a modified Yeager structure, Me–N_x/C, i.e., the molecular C–N_x–Me structure proposed by Yeager and obtained after adsorption of metal ions on nitrogen functionalities of the carbon, that is rendered catalytically active after a heat treatment at high temperature. Me–N_x/C is exactly what Yeager and collaborators obtained in 1989 when they heat treated at 800°C, in an inert atmosphere, PAN and iron acetate adsorbed on Vulcan.

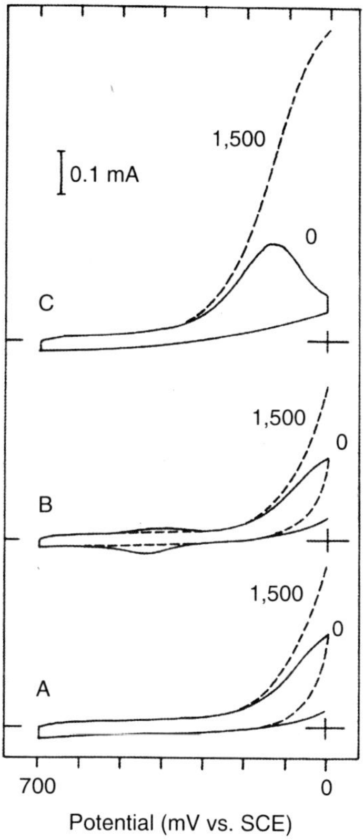

Figure 3.14. Cyclic voltammograms at 0 and 1,500 rpm, in O_2 saturated H_2SO_4 of H_3TPP/XC-72 after various successive treatments: (A) H_2TPP/XC-72 heat treated at 1,000°C; (B) after soaking sample (A) in an aqueous solution containing Fe^{2+} ions; (C) after heat treatment of sample (B) at 1,000°C in inert atmosphere. Cathodic currents are above the (+) sign indicating the position of 0 mV and 0 mA for figures A, B, and C (according to Figure 5 in ref. [79]; reproduced with permission of Elsevier).

Let us go back now to 1989 and see how the idea to use two precursors, one for nitrogen and the other one for the metal, was utilized in the preparation of non-precious metal catalysts for oxygen reduction. The first ones to follow up on the work of Yeager and his team with PAN were Wiesener and his collaborators[85], who used PAN as nitrogen precursor with a variety of transition metal salts of the first row (Mn, Fe, Co, Ni, Cu, Zn). The best catalytic activities were obtained by the adsorption of $CoSO_4$ or $FeSO_4$ (12.3 wt% metal) as metal precursors on carbon P33. Monitoring the metal, they showed that a large fraction of this metal dissolved from the catalyst: after 2,000 h, only 11% of the initial Co loading and 20% of the initial Fe loading were still present in the catalysts. This result indicated that only a small fraction of the metal was activated by PAN to become a catalyst, the activity of which remained, however, smaller than that of a catalyst made with Co–N_4 or Fe–N_4 chelate precursors. Catalysts obtained with PAN as nitrogen precursor were also studied later on by Ye and Vijh[86], who used a PAN aerogel as carbon support for the adsorption of Fe (10.4 wt%) and Co (6.3 wt%) from iron nitrate, cobalt chloride, or cobalt acetylacetonate as metal precursors. A catalytic activity for ORR was reported for both PAN–Fe and PAN–Co catalysts obtained after a pyrolysis step at 900°C in inert atmosphere.

Besides PAN, various pyrrole derivatives were also adsorbed as nitrogen precursors on carbon. Fabjan *et al.*[87] reported the use of polypyrrole impregnated on graphite with Co and Fe salts. The resulting material was then heat treated in inert atmosphere at 820°C, the optimum temperature, to obtain a catalyst. These catalysts showed methanol tolerance. Their catalytic activity was attributed to the Me–N_4 structure that would be formed during pyrolysis. These authors were followed by Seeliger and Hamnett[88], who used pyrrole oxidatively polymerized in the presence of anionic complexes like $Co(CN)_6^{3-}$ and anionic CoTPP. The resulting material was then heat treated at 820°C to obtain a catalyst. The best documented work using several pyrrole derivatives was that of van Veen and collaborators[89]. These nitrogen precursors were used in conjunction with cobalt acetate loaded on Vulcan (1.5 wt% Co) and on Norit BRX (4.5 wt% Co), then heat treated at 700°C in inert atmosphere. Vulcan was used as received or was preoxidized first with HNO_3 to add carboxyl and lactone functionalities. The best catalysts were obtained for a Co:N ratio of 1:10 with 2,5-dimethylpyrrole adsorbed on Vulcan. This catalyst was found to be equivalent to that obtained with CoTMPP loaded at 7 wt% chelate (0.6 wt% Co) on Vulcan and also heat treated at 700°C. The presence of surface groups on Vulcan lowered the catalytic activity. The catalysts prepared with the pyrrole precursors and those prepared with CoTMPP were analyzed by EXAFS. Heat-treated CoTMPP/Vulcan displayed a first shell of four nitrogen atoms at 0.190 nm around the Co ion and a second shell of carbon atoms at 0.284 nm, while heat-treated 2,5-dimethylpyrrole and cobalt acetate/Vulcan displayed a first shell of four nitrogen atoms at 0.190 nm around the Co ion and a second shell of carbon atoms at 0.280 nm. Residual Co metal that could not be removed and could not be seen by TEM was also detected by EXAFS. These authors concluded that Co–N_4 was the catalytic site in their two catalysts and noticed that the nitrogen precursor (2,5-dimethylpyrrole) has a structure corresponding to one-fourth of that of their proposed catalytic site illustrated in Figure 3.5.

Sawaï and Suzuki[90] used complexes related to Prussian blue as nitrogen and metal precursors. These were adsorbed on acetylene black from a $M'SO_4$ and a $K_3M''(CN)_6$ solution, where M' and M'' are either Co and/or Fe ions. This material was heat treated at 800°C in inert atmosphere. Its metal loading was not specified but it was very high (~50 wt%, based on the loading of the initial Prussian blue derivative). The catalysts prepared with the Fe/Fe derivative displayed the lowest activity, while that prepared with the Co/Fe derivative was the most active. From these results, the importance of having Co catalytic sites in the neighborhood of Fe catalytic sites was stressed. No activity was found for heat treatments $<$ 500°C, but the authors claimed that their catalysts, after pyrolysis at 800°C, displayed an activity similar to that of Pt measured in the same conditions (floating electrode in 0.25M H_2SO_4 at room temperature). These catalysts showed methanol tolerance.

Various Co complexes with 3, 4, and 6 coordinating atoms, including Co salen (N, N'-bis(salicylidene)ethylenediamino cobalt) and Co anten (*o*-amino-benzaldehydeethylenediamino cobalt, for which the two complexing oxygen atoms found in Co salen are replaced with two NH groups) were also used by Okada and his collaborators as Co and nitrogen precursors[91]. They were mixed with graphite and heat treated at 600°C in inert atmosphere to obtain catalysts having a nominal Co loading of 12 wt%. These authors concluded that the best catalysts were obtained with complexes with the initial number of four coordinating atoms as, for instance, salen and anten complexes of Co. They suggested that the catalytic site was always the Co–N_4 unit (even for the salen Co precursor for which Co is initially coordinated by two nitrogen and two oxygen atoms).

Our group published many studies on the use of two different precursors for nitrogen and metal to obtain active catalysts for ORR in acid medium. The first paper[84] has already been introduced at the end of Section 2.2.1. It reported the use of ferrocene and acetonitrile that allowed us to demonstrate: (i) that aggregates containing iron metal and carbide, surrounded by a graphitic envelope, were not catalytic and (ii) that a metal precursor adsorbed on carbon had to be heat treated in the presence of a nitrogen precursor (acetonitrile) to obtain a catalyst, while (iii) a material containing only C and N was found to be inert toward ORR. The inactivity of similar aggregates containing Co, which were also surrounded with a carbon envelope, was also confirmed[92]. In this case, metal particles were obtained by the Kratschmer–Hoffman carbon arc discharge. A fraction of the cobalt could again be activated toward oxygen reduction by a heat treatment at 1,000°C in acetonitrile.

In four other publications[93–96], catalysts were prepared by first adsorbing iron hydroxide onto Vulcan. This iron precursor was then reduced to Fe metal in a stream of H_2 at 600°C before being activated at 1,000°C in a stream of Ar + acetonitrile. The iron loading was as high as 10 wt%. Other procedures[93], such as the chemical reduction in non-aqueous solution of $FeCl_2$ adsorbed on Vulcan or the reduction of an $FeCl_3$ Vulcan intercalate, were also used to obtain Fe metal that was then activated for oxygen reduction by a heat treatment in acetonitrile at 1, 000°C. The best catalyst was obtained by activation of the iron intercalate. In this case, iron was progressively released, during the heat treatment, from the intercalate and reacted with the nitrogen precursor to produce the catalytic sites. The Fe loading of this catalyst was 1.5 wt%.

The ability of several nitrogen precursors to activate $Fe(OH)_2$ adsorbed on Vulcan was also evaluated[94]. Solid nitrogen precursors such as PAN, tetracyanoquinodimethane (TCNQ), or H_2Pc were coadsorbed with $Fe(OH)_2$. The resulting materials were then heat treated in Ar at 1,000°C, while $Fe(OH)_2$ alone adsorbed on Vulcan was also heat treated at the same temperature but, this time, in a stream of a gaseous nitrogen precursors such as CH_3–CN or NH_3. All these catalysts obtained using the same Fe loading (10 wt%) were not equivalent. For the solid N precursors, the activity increased as PAN $<$ TCNQ $<$ H_2Pc, while for the gaseous precursors, the activity of catalysts obtained by activation of $Fe(OH)_2$ with acetonitrile or ammonia was similar, but lower than when solid H_2Pc was used as nitrogen precursor.

The catalyst preparation procedure starting with the adsorption of a metal hydroxide followed by its reduction in a stream of H_2 at 600°C and a further activation step in acetonitrile at 1,000°C was expanded to all transition metals of the first row[95]. Catalysts for O_2 reduction were obtained only with Cr, Fe, and Co. Cr_2O_3/C, Fe/C, and Co/C were detected after the reduction step in H_2, while Co, Fe_3C, and a chromium carbide–nitride ($Cr_{6.2}$ $C_{3.5}$ $N_{0.3}/C$) were detected in the catalyst. The nominal loading of all metals was 10 wt%. Catalytic activity decreased as Cr > Fe > Co. Tests in fuel cells indicated that the Cr-based catalyst was not stable, while Fe and Co-based catalysts were stable (see Section 4 for details).

After completing these experiments, we came to one important conclusion: although it was possible, when two different precursors were used for metal and nitrogen, to increase the metal content well above that of the catalysts obtained by the pyrolysis of Me–N_4 chelates adsorbed on carbon, the resulting activity was not proportional to the metal content. On the contrary, *using high metal contents generated a lot of inactive material that dominated catalyst characterization and did not allow conclusions to be drawn about the identification of the catalytic site.* In a first attempt to remedy this situation, a catalyst starting from the adsorption of $Fe(OH)_2$ on Vulcan (10 wt% Fe) was prepared. The metal precursor was first reduced to metallic Fe at 600°C in a stream of H_2, then activated to 1,000°C in a stream of acetonitrile[96]. It was then leached in H_2SO_4 (pH 0.5) for 12 h. The leached catalyst, which still contained 6.2 wt% Fe, was then exposed to a stream of Cl_2 at 650°C for 2 h in order to remove all extractable Fe as volatile $FeCl_3$. The Fe content after the second leaching step could not be reduced below 3.3 wt%. The fully extracted catalyst was stable for 200 h in fuel cell tests.

When it became clear that it was impossible to leach all inactive metal from the catalysts, we completely changed our approach to their synthesis and tried to find out what minimum metal loading would produce a maximum catalytic activity. Following the findings that a C–N_x–Me structure was the precursor of the HT catalytic site, Vulcan XC-72 was first refluxed in concentrated HNO_3, then reacted in NH_3 in order to maximize the number of C–N_x functionalities[97]. A nitrogen atom surface concentration of 7.3 at% could be reached under these conditions. Various amounts of Fe were then loaded as $FeSO_4$ on the nitrogen enriched carbon support. These materials were heat treated under Ar at 900°C to obtain catalysts. It was found that the catalytic activity increased steadily with Fe loading up to an Fe/C ratio of about 0.1 wt%, which is much less than the 3.3 wt% Fe loading previously reached, even after leaching in hot Cl_2! After a heat treatment of the C–N_x–Me precursor at 900°C, the nitrogen surface concentration of the catalyst was 3.6 at%. Another increase in the Fe loading by one order of magnitude did not further improve the catalytic activity. This result was interpreted as a saturation of Fe ion coordination to the nitrogen existing on the carbon surface. Once all coordination sites were used up, a further increase in the Fe loading produced only inactive material upon pyrolysis, yielding iron-containing aggregates that became covered with a graphitic layer if the heat treatment was done at high enough temperature. A similar experiment was also performed by Savinell and his group[98] with cobalt sulfate and acetonitrile as metal and nitrogen precursors, respectively. The best catalytic activities were obtained when the carbon support (Vulcan) was pyrolyzed first with CH_3CN at 1,000°C before loading the cobalt. This material was then pyrolyzed a second time in CH_3CN at 1,000°C. It was also found that the surface concentration of Co was directly related to the concentration of nitrogen atoms on the carbon. These authors also recognized that the C–N_x–Me structure that was proposed as the HT catalytic site by Yeager was not the HT site itself, but rather its precursor. A high-temperature pyrolysis step is required after adsorption of Co ions to obtain a catalyst with good performance.

At the end of the 1990s, our group therefore concluded that, in order to obtain more information about the catalytic sites, it was imperative to keep the Fe loading below the saturation level (full occupation of coordinating C–N functionalities). Doing so, XPS spectra became meaningful, while XRD diffractograms of the catalysts became quite similar to that of the carbon support alone[99]. ToF SIMS spectra of catalysts could not yet be fully exploited because the background signal of Vulcan, the usual carbon support, was quite high and would hide important peaks coming from ions released by the catalytic sites. This is the reason we sought a substitute for Vulcan and found pyrolyzed perylene tetracarboxylic dianhydride (PTCDA) to be suitable[100]. PTCDA is a dye that loses its carboxylic moiety upon heat treatment above 520°C, a temperature at which it begins to polymerize into carbon fibers[101]. Information about the duality of catalytic sites, obtained by ToF SIMS measurements using PTCDA as carbon precursor, is presented in the next section.

2.2.3. Simultaneous Presence of Two Catalytic Sites in all Fe-Based Catalysts

At the end of the 1990s, our group believed, on the one hand, that pyrolyzing Me–N_4 chelates adsorbed on carbon produced, depending on the pyrolysis temperature, either an LT active site of the type Me–N_4/C or an HT active site of the type Me–N_x/C. On the other hand, we also believed that pyrolyzing metal and nitrogen precursors simultaneously adsorbed on carbon produced only the HT active site Me–N_x/C. The nitrogen precursor could be either an N-bearing molecule adsorbed on the carbon support or a gas that needed to be present with the metal precursor in the reactor during the pyrolysis step. The dividing line between low and high temperatures was believed to be around 800°C.

PTCDA was introduced as a precursor of the carbon support at the end of the previous section. Besides its low background in ToF SIMS spectra, the use of pyrolyzed PTCDA as carbon support had another advantage: the possibility to be practically devoid of metallic impurities[102]. This was obtained by purifying PTCDA in an acid medium before pyrolyzing this precursor at 900°C in a stream of H_2:Ar:NH_3. H_2 was added to the pyrolysis gas in order to help remove oxygen as H_2O while PTCDA was losing its tetracarboxylic moiety above 520°C; NH_3 was added to the gas stream to dope the resulting carbon with N-bearing functionalities that would help the adsorption of metallic ions on the new carbon support.

Figure 3.15 shows that a saturation of the catalytic activity for ORR with Fe loading already happened above ~0.5 wt% Fe when iron acetate was used as

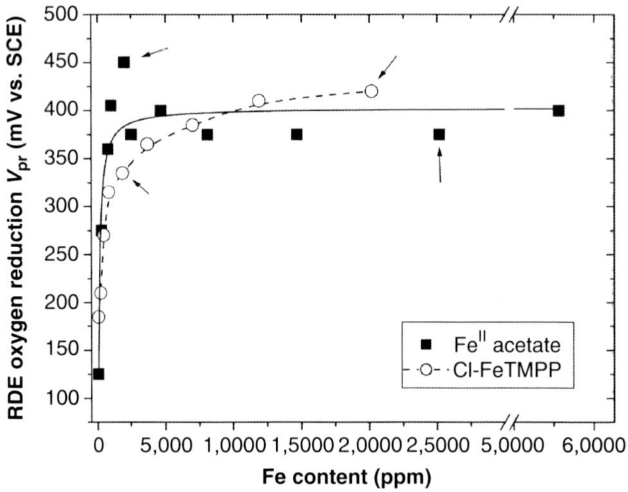

Figure 3.15. Changes in the catalytic activity for O_2 reduction, expressed as V_{pr} (potential at peak reduction for 0 rpm) with the Fe content for materials prepared by adsorbing either iron acetate or ClFeTMPP on PTCDA and pyrolyzing the material at 900°C in H_2:Ar:NH_3(1:1:2) (according to Figure 2 in ref. [100]; reproduced with permission of the American Chemical Society).

Fe precursor, while larger Fe loadings (up to about 2 wt%) could still increase the ORR activity when the Fe precursor was a porphyrin (ClFeTMPP)[100]. All catalysts in Figure 3.15 were prepared by the adsorption of the Fe precursor on PTCDA. The resulting material was then pyrolyzed at 900°C in H_2:Ar:NH_3. The difference between iron acetate and porphyrin precursors was interpreted in terms of a better interaction of the PTCDA with the Fe-porphyrin than with iron acetate, leading to a better dispersion of the iron ions on the carbon precursor.

The XPS spectra of the catalysts prepared with iron acetate as Fe precursor were quite instructive[99]. Narrow-scan spectra of Fe $2p_{3/2}$ and $2p_{1/2}$ (Figure 3.16) revealed that Fe was always found in the ionic form for all Fe loadings but that a reduced form of Fe (either Fe^0 or an iron carbide) was also detected above 8,100 pm (0.81 wt%) Fe, in agreement with the saturation effect of the catalytic activity observed in Figure 3.15. Narrow-scan spectra of N1s of the same catalysts (Figure 3.17) revealed that only the low binding energy peak was shifting toward higher binding energy upon increasing the Fe loading. A saturation of this effect also appeared above 4,660 ppm Fe, again in agreement with saturation of the catalytic activity observed in Figure 3.15. the N1s XPS narrow-scan spectra were deconvoluted into four components, which were assigned to pyridinic, nitrile, pyrrolic, and graphitic types of nitrogen atoms[103,104] (Figure 3.18). The shift in energy with increasing Fe loading of the peak assigned to pyridinic nitrogen was interpreted in terms of a preferential interaction of Fe ion with pyridinic nitrogen atoms in the HT catalytic site. These nitrogen atoms are located at the edge of a graphene layer and contribute one electron to the π band (Figure 3.18).

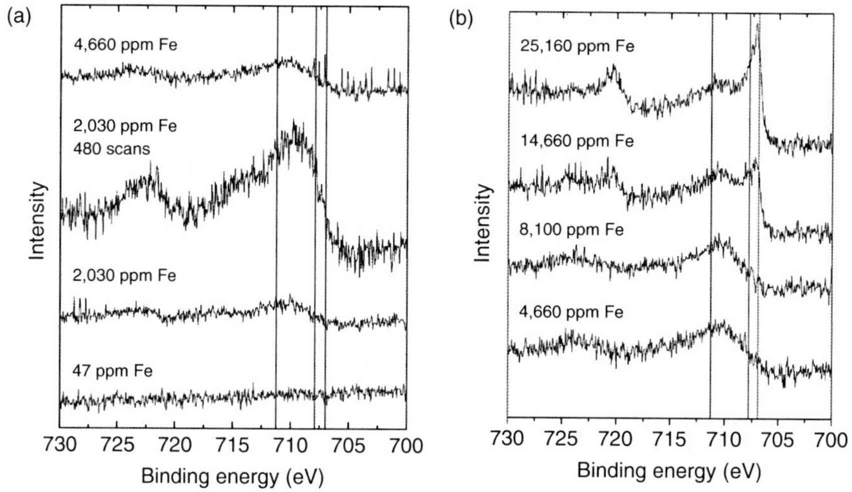

Figure 3.16. XPS Fe $2p_{3/2}$ and $2p_{1/2}$ narrow scan spectra of catalysts having various Fe contents: Panel (a) from 47 to 4,660 ppm Fe; panel (b) from 4,660 to 25,160 ppm Fe. The catalyst were prepared by adsorbing iron acetate on PTCDA and pyrolyzing the material at 900°C in H_2:Ar:NH_3 (1:1:2) (according to Figure 9 in ref. [99]; reproduced with permission of Elsevier).

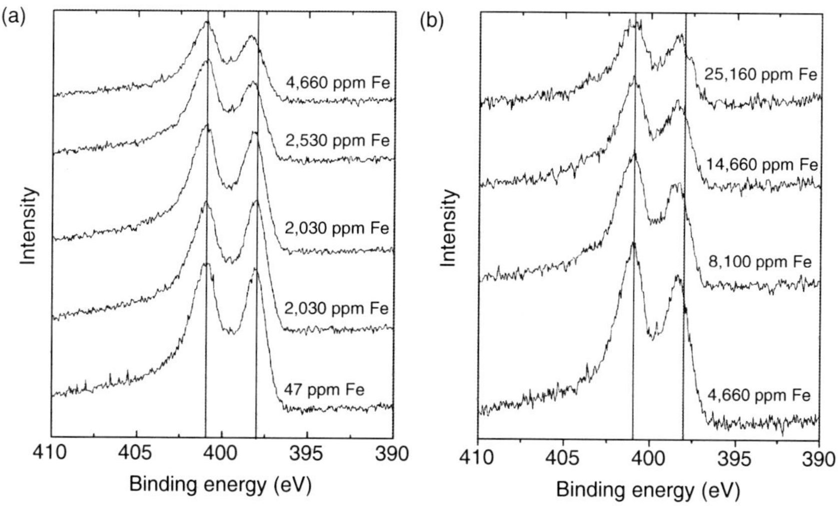

Figure 3.17. XPS N 1s narrow scan spectra of catalysts having various Fe contents: Panel (a) from 47 to 4,660 ppm Fe; panel (b) from 4,660 to 25,160 ppm Fe. The catalysts were prepared by adsorbing iron acetate on PTCDA and pyrolyzing the material at 900°C in H_2:Ar:NH_3 (1:1:2). Right panel: graphene sheet with pyridinic, pyrrolic, nitrile, and graphitic types of nitrogen atoms (according to Figure 10 in ref. [99]; reproduced with permission of Elsevier).

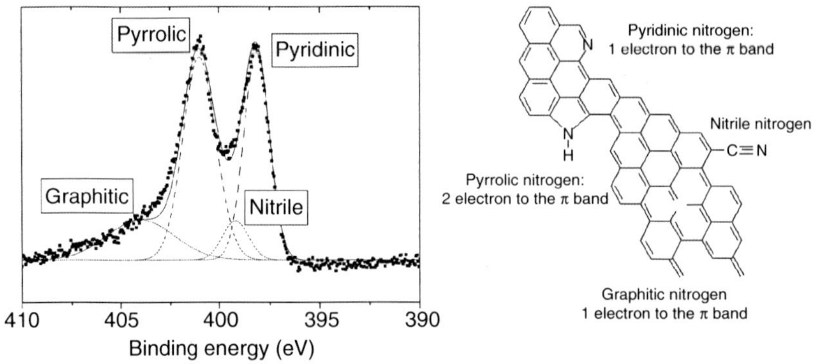

Figure 3.18. Left panel: Deconvolution of a typical XPS N1s narrow scan spectrum of a catalyst containing 2,030 ppm Fe. The catalyst was prepared by adsorbing iron acetate on PTCDA and pyrolyzing the material at 900°C in H_2:Ar:NH_3 (1:1:2). Right panel: graphene sheet with pyridinic, pyrrolic, nitrile, and graphitic types of nitrogen atoms (according to Figure 11 in ref. [99]; reproduced with permission of Elsevier).

They have also an orbital in the plane of the graphene layer containing two non-bonding electrons, which are available to coordinate a metal ion such as iron. Similar experiments were also performed with metal precursors other than Fe. Cr, Mn, Co, Ni, or Cu were loaded on PTCDA at various contents up to about 2 wt% metal. Among the materials obtained after pyrolysis at 900°C in H_2:Ar:NH_3, only Mn < Cr < Co displayed catalytic activity[102]. XPS 2p narrow scan spectra of the metal for all the catalysts loaded below 5,000 ppm metal indicated that metals in the catalysts were in the ionic form. Similarly, N1s XPS narrow scan spectra of all the catalysts showed a shift of the binding energy of the pyridinic-type nitrogen peak similar to the shift already observed for various Fe metal loadings as depicted in Figire 3.17. A smaller but definite shift of the binding energy was also observed for the nitrile type of nitrogen when the metal content was increased. No such shift was observed for Cu and Ni, which both gave quite inactive materials for ORR.

If the N1s XPS spectra of Fe-based catalysts revealed a specific interaction in the high-temperature catalytic site between pyridinic-type nitrogen atoms and the Fe ion, ToF SIMS experiments performed on similar catalysts revealed that:

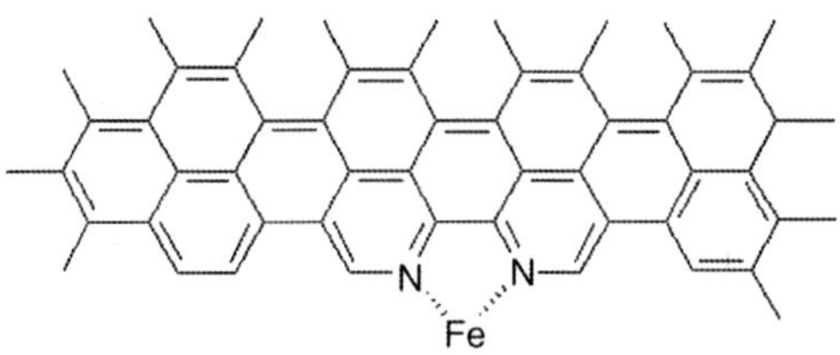

Figure 3.19. Proposed moiety of the high-temperature catalytic site labeled Fe–N_2/C. The complete structure of this site is still undetermined (according to Figure 1 in ref. [105]; reproduced with permission of the American Chemical Society).

(i) the high-temperature catalytic site contained at least two pyridinic nitrogens and (ii) these nitrogen atoms were in a 1,10-phenanthrolinic type structure like the one reproduced in Figure 3.19[105]. In order to reach this conclusion, a series of catalysts were prepared with iron acetate. The Fe precursor was loaded at 2,000 ppm (0.2 wt%) Fe on pre-pyrolyzed PTCDA (900°C in H_2:Ar:NH_3). The resulting material was again heat-treated, but this time, at various temperatures ranging from 400 to 1,000°C in an inert atmosphere in order to obtain catalysts of various activities for ORR. By doing so, we hoped to find one or several ToF SIMS ions whose pattern of relative abundance would follow the pattern of catalytic activities. One ion came to the fore out of all ions detected by ToF SIMS for these materials. It was ion $FeN_2C_4^+$. The change of the relative intensity (abundance) of that ion is compared in Figure 3.20 with the evolution with the pyrolysis temperature of the catalytic activity for the same Fe-based catalysts with the pyrolysis temperature. The coincidence of the two patterns indicated that $FeN_2C_4^+$ was a ToF SIMS signature of the HT catalytic site. This ion was assigned to a structure where an Fe ion is coordinated to two nitrogen atoms, each nitrogen atom being itself bound to two carbon atoms like in a 1,10-phenanthroline. The proposed structure is presented in Figure 3.19. *The full coordination of the catalytic site is not complete and is still unknown. Other coordinating atoms are expected to be either nitrogen or oxygen* because, except for carbon and iron, nitrogen and oxygen are the only other atoms detected by XPS in these catalysts (XPS is unable to detect hydrogen).

Besides $FeN_2C_4^+$, other ions of the type $FeN_xC_y^+$ were also detected in the ToF SIMS spectra of these catalysts. They were of much lower relative abundance than $FeN_2C_4^+$. A refined analysis of the ToF SIMS spectra took them into

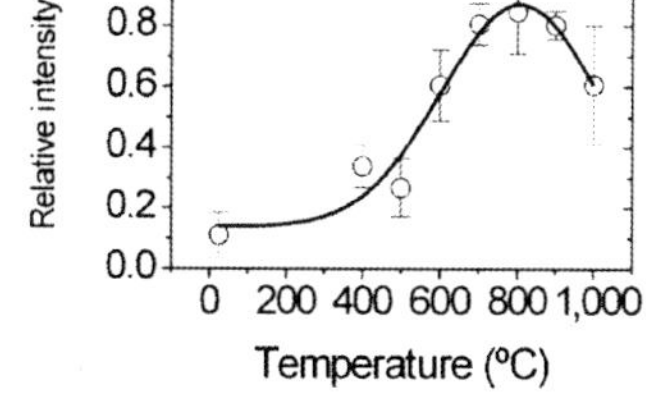

Figure 3.20. Comparison between changes with the pyrolysis temperature, of the catalyst activity (top panel) and the relative intensity of $FeN_2C_y^+$ (bottom panel), measured in cataysts obtained by adsorbing iron acetate (0.2 wt% Fe) on prepyrolyzed PTCDA and heat treating that material in inert atmosphere at various temperatures, ranging from 400 to 1,000°C. Prepyrolyzed PTCDA is obtained by heat treating PTCDA at 900°C in H_2:Ar:NH_3 (1:1:2) (according to Figure 8 in ref. [100]; reproduced with permission of the American Chemical Society).

Table 3.1. Relative Abundance in % of $FeN_xC_y^+$ Ions as a Function of the Pyrolysis Temperature for Catalysts Using Iron Acetate (0.2 wt% Fe) as Fe precursor (Reproduced from ref. [105] with Permission of the American Chemical Society)

Ions	400°C	500°C	600°C	700°C	800°C	900°C	1,000°C
$FeNC^+$	28.27	23.82	17.08	0.76	5.76	1.77	2.50
$FeNC_2^+$	0.78	0.00	1.61	2.63	0.00	0.40	2.50
$FeNC_3^+$	0.78	1.41	2.06	1.82	0.63	1.33	7.50
FeN_2C^+	14.94	7.78	4.80	2.38	1.56	2.56	3.75
$FeN_2C_2^+$	1.78	4.70	1.44	0.53	0.63	0.83	0.00
$FeN_2C_3^+$	0.40	1.26	0.53	1.85	0.00	0.93	12.50
$FeN_2C_4^+$	**23.92**	**26.93**	**46.95**	**64.89**	**78.38**	**76.38**	**49.86**
$FeN_2C_5^+$	1.10	0.00	0.35	2.08	0.00	0.93	0.00
$FeN_2C_6^+$	1.47	0.00	0.64	0.53	1.32	0.40	0.00
FeN_3C^+	1.75	4.97	5.68	1.32	0.00	0.00	2.78
$FeN_3C_2^+$	0.00	1.36	0.00	0.00	1.32	0.42	0.00
$FeN_3C_3^+$	1.78	2.98	1.91	1.82	0.00	0.81	2.78
$FeN_3C_4^+$	0.00	1.10	0.00	1.06	0.00	0.00	0.00
$FeN_3C_5^+$	2.33	0.37	0.29	0.53	1.56	0.85	0.00
$FeN_3C_6^+$	2.64	0.47	0.35	0.76	1.56	0.00	2.78
$FeN_3C_7^+$	0.00	0.00	0.00	0.00	0.00	0.42	0.00
$FeN_3C_8^+$	0.81	3.98	3.36	2.89	0.00	1.67	1.25
$FeN_3C_9^+$	7.33	3.29	1.16	0.76	0.63	3.06	4.03
FeN_4C^+	3.01	0.68	1.12	1.32	0.00	0.42	0.00
$FeN_4C_2^+$	0.34	1.31	1.67	2.38	0.00	0.93	0.00
$FeN_4C_3^+$	0.34	1.89	0.00	0.76	1.56	0.42	0.00
$FeN_4C_4^+$	0.72	0.52	1.08	0.00	0.00	0.42	2.50
$FeN_4C_5^+$	0.40	0.37	2.18	1.32	1.56	0.00	0.00
$FeN_4C_6^+$	0.00	0.00	0.00	0.53	0.00	0.40	1.25
$FeN_4C_7^+$	0.00	0.00	0.00	0.00	0.00	0.00	0.00
$FeN_4C_8^+$	**5.08**	**8.05**	**5.45**	**4.17**	**1.88**	**4.64**	**4.03**
$FeN_4C_9^+$	0.00	1.72	0.29	1.60	0.00	0.00	0.00
$FeN_4C_{10}^+$	0.00	0.68	0.00	0.00	0.00	0.00	0.00
$FeN_4C_{11}^+$	0.00	0.00	0.00	1.32	0.00	0.00	0.00
$FeN_4C_{12}^+$	0.00	0.37	0.00	0.00	1.67	0.00	0.00

account as well. Their relative abundances as a function of pyrolysis temperature are reported in Table 3.1[105]. In the refined analysis, it was assumed that all these ions originated from the catalytic site(s), either being extracted as such from the surface of the catalyst or being the result of fragmentation of a larger ion on its way to the detector. A glance at Table 3.1 shows that these ions may be divided into four families, each one having a different number of nitrogen atoms bound to a variable number of carbon atoms. Surprisingly, some of the ions contained up to four nitrogen atoms. In particular, $FeN_4C_8^+$ was of relatively high abundance. It was assigned to an Fe ion coordinated to four nitrogen atoms, each nitrogen atom being itself bound to two carbon atoms. This ion is typical of the fragmentation of $ClFeTMPP^+$ and is characteristic of the Fe–N_4 moiety of the porphyrins[105]. $FeN_4C_8^+$ is also reminiscent of the structure of the catalytic site proposed by van Veen and collaborators, which is illustrated in Figure 3.5, and would appear with a maximum probability at 500°C in Fe-based catalysts obtained from iron acetate as the Fe precursor.

In order to extract more information from all the ions of Table 3.1, we made the following assumption: the ions with the same number of nitrogen atoms should all originate from the same type of catalytic site. By doing so, one is able to reduce each column of the table into four numbers, one number per nitrogen ion families. Figure 3.21 is obtained when those combined relative abundances are plotted against the pyrolysis temperature. It is obvious that Figure 3.21 displays two different behaviors. On the one hand, the family of ions of the type $FeN_2C_y^+$ increases in relative abundance with the pyrolysis temperature, goes through a maximum at 800–900°C, then decreases at 1,000°C. On the other hand, the relative intensity of the three other families behave similarly, but with a maximum abundance at a low pyrolysis temperature. From this behavior, one may conclude that the ions belonging to the N_1, N_3, and N_4 families have the same origin, a catalytic site of the Fe–N_4 type bound to the carbon support (N_1 and N_3 ions would all originate from the fragmentation of N_4-containing ions). This site was labeled Fe–N_4/C. On the other hand, the origin of the ions of the N_2 family is a different catalytic site of the type Fe–N_2 bound to the carbon

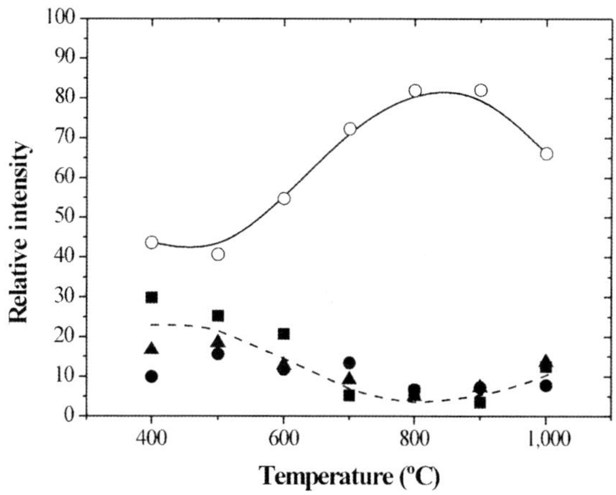

Figure 3.21. Relative intensity of $\Sigma FeN_2C_y^+$ (open circles), $\Sigma FeN_1C_y^+$ (dark squares), $\Sigma FeN_3C_y^+$ (dark triangles), and $\Sigma FeN_4C_y^+$ (dark circles) as a function of the pyrolysis temperature for catalysts analysed by ToF SIMs. The catalysts were obtained by adsorbing iron acetate (0.2 wt% Fe) on prepyrolyzed PTCDA and heat treating this material in inert atmosphere at various temperatures, ranging from 400 to 1,000°C. Prepyrolyzed PTCDA was obtained by heat treating PTCDA at 900°C in H_2:Ar:NH_3 (1:1:2) (according to Figure 4 in ref. [105]; reproduced with permission of the American Chemical Society).

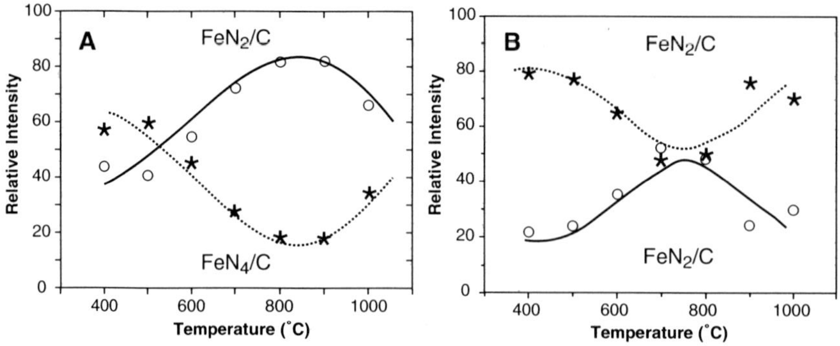

Figure 3.22. Relative abundance, according to ToF SIMS results, of the two catalytic sites (Fe–N_2/C and Fe–N_4/C) as a function of the heat treatment temperature of the catalysts obtained with iron acetate (panel A) and with ClFeTMPP (panel B) as Fe precursors. The catalysts were obtained by adsorbing either iron acetate (0.2 wt% Fe) or ClFeTMPP (0.2 wt% Fe) on prepyrolyzed PTCDA and heat treating that material in inert atmosphere at various temperatures, ranging from 400 to 1,000°C. Prepyrolyzed PTCDA was obtained by heat treating PTCDA at 900°C in H_2:Ar:NH_3 (1:1:2) (according to Figure 5 in ref. [105]; reproduced with permission of the American Chemical Society).

support. It was labeled Fe–N_2/C. By grouping all relative abundances related to the N_1, N_3, and N_4 families of ions one ends up with Figure 3.22A *that shows the evolution of the two catalytic sites that were found simultaneously, but in various proportions according to the pyrolysis temperature, in catalysts obtained with iron acetate as Fe precursor!* One of the catalytic sites is the expected Fe–N_2/C identified previously as the HT catalytic site (Figure 3.19). The other one is Fe–N_4/C identified previously as the LT catalytic site (Figure 3.5). Similar conclusions were obtained when ToF SIMS experiments were performed on catalysts made by adsorbing an Fe porphyrin (ClFeTMPP) at the same Fe loading than iron acetate (0.2 wt% Fe) on prepyrolyzed PTCDA (900°C; H_2:Ar:NH_3) followed by the heat treatment of this material at various temperatures between 400 and 1,000°C. Four families of N-containing ions were again found, the N_2 family again behaving differently from the N_1, N_3, and N_4 families of ions. Two catalytic sites emerged again from this analysis and their relative abundances for the porphyrin precursor are reported in Figure 3.22B. One of the catalytic sites is the expected Fe–N_4/C site, identified previously as the LT catalytic site (Figure 3.5). The other one is Fe–N_2/C identified previously as the HT catalytic site (Figure 3.19). *As a general conclusion from these experiments, it looks like, irrespective of the iron or nitrogen precursor used to obtain a catalyst for ORR, this material always contains two kinds of catalytic sites identified as Fe–N_2/C and Fe–N_4/C, the relative proportions of which depend upon the pyrolysis temperature and also upon the particular iron and nitrogen precursors used.*

These results influenced the choice of the iron precursor and also the structures for the catalytic site proposed later by other groups. Indeed, Bron and his colleagues published three articles reporting activity for ORR from catalysts prepared by adsorption of an iron 1,10-phenanthroline complex (Fe^{III} $Phen_3$) on Vulcan, Printex, and Black Pearls[106–108], and heat treated at 800°C in inert or NH_3 atmosphere. Their current density for O_2 reduction improved by one order of magnitude when a Black Pearls' carbon support was used instead of Vulcan. This effect was attributed to the much larger specific area of the former (1,500 m^2/g for Black Pearls) compared to the latter (250 m^2/g for Vulcan). Under these conditions, the Fe phenanthroline-based catalyst became as active as ClFeTMPP on Black Pearls pyrolyzed at the same temperature and for similar Fe loadings (1.14 wt% Fe for $Fe^{III}Phen_3$ and 1.5 wt% Fe for ClFeTMPP). They proposed an active center of the $Fe–N_{2+2}$ type that would be made by bringing together, around the Fe ion, two graphene platelets, each one bearing the 1,10- phenanthroline type structure like the one shown in Figure 3.19 for $Fe–N_2/C$. Such an $Fe–N_{2+2}$ structure was suggested earlier by our group to complete the coordination of the $Fe–N_2/C$ catalytic site[100]. The same authors also performed EXAFS experiments on their catalysts and found that approximately four nitrogen atoms coordinate around the central iron atom, at an Fe-N distance of 0.196 nm, in agreement with their model. A very similar model of $Fe–N_{2+2}/C$ was also proposed by Schulenburg and collaborators[109] on the basis of Mössbauer measurements of catalysts made by adsorption of ClFeTMPP on Black Pearls and heat treated at 900°C in inert atmosphere. The initial Fe loading of these catalysts was 2.05 wt%, but it was lowered to 0.47 wt% after leaching the excess Fe. Three Mössbauer subspectra were reported for the catalyst before acid leaching. They were assigned to small (< 1.5 nm) Fe particles, Fe_3C, and α-Fe_2O_3. After acid leaching, the Fe_3C disappeared as well as the Fe metal particles. α-Fe_2O_3, which is poorly soluble in acid solution, was still observed.

Two new subspectra, labeled A and B, were also detected in the leached material. Subspectrum A was assigned to tetracoordinate Fe^{III}, while subspectrum B was assigned to hexacoordinate Fe^{III}. Longevity tests and H_2O_2 attack experiments were also performed on these leached catalysts. Mössbauer measurements indicated that α-Fe_2O_3 still survived but subspectrum A was now absent, while the catalyst remained still active. The catalytic activity was therefore assigned to sites giving rise to subspectrum B. A model of the catalytic site responsible for subspectrum B was given by these authors. This model is illustrated in Figure 3.23. At first glance, the FeN_{2+2}/C structure proposed by Schulenburgh and collaborators as model for the catalytic site looks like van Veen's model reproduced in Figure 3.5, but it is not. Here the two upper nitrogen atoms belong to a 1,10-phenanthrolinic type structure, and the two lower nitrogen atoms to another such structure (a total of four *pyridinic* type nitrogens), while in Figure 3.5, the coordination around the metal atom is of *pyrrolic* type. Since, in Figure 3.23, the four nitrogen atoms coordinate to Fe by lone pair donation, it implies that Fe is being oxidized by other ligands. Figure 3.23 might be a possible fully coordinated representation of what was previously labeled as the $Fe–N_2/C$ catalytic site (presented in Figure 3.19), while Figure 3.5 is a representation of what was previously labeled as the $Fe–N_4/C$ catalytic site. Further work is still needed to ascertain the exact structure of fully coordinated FeN_{2+2}/C.

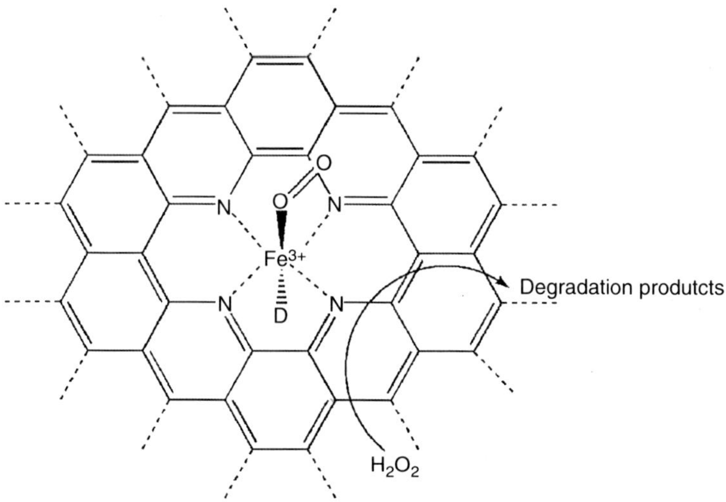

Figure 3.23. Proposed complete molecular structure for the catalytic active site Fe–N_2/C presented in Figure 3.19. This structure is labeled Fe–N_{2+2}/C in the text (according to Figure 12 in ref. [109]; reproduced with permission of the American Chemical Society).

2.2.4. Influence of the Carbon Support on the Surface Density of Catalytic Sites

Research in non-noble metal electrocatalysts for oxygen reduction in the acidic environment of PEM fuel cell has mainly dealt with improving the activity of these catalysts by exploring a large range of metal precursors of the catalytic sites. It was also recognized that nitrogen atoms were an essential part of the catalytic sites, allowing the coordination of the metal ion to the carbon support. Very few studies, however, focused on the role of the carbon support and its influence on the activity of the electrocatalysts. Among these studies, the following publications are particularly worth mentioning.

In 2001, Tarasevich and his collaborators reported a comparison between electrocatalysts for oxygen reduction prepared using a disperse synthetic diamond powder promoted with CoTMPP and its pyropolymers[110]. Two types of diamond powders with specific area of 5.8 and 170 m^2/g were used as catalyst supports and the activity of the catalysts obtained with the diamond supports was compared to that obtained with the same CoTMPP precursor loaded on acetylene black. In all cases, the loading was one monolayer of CoTMPP. These authors found a much lower activity for the electrocatalysts prepared on synthetic diamonds than for that catalyst prepared on acetylene black. The kinetic mechanisms of ORR was, however, the same for both supports.

In 2002, Bron and his colleagues compared the activity of catalysts made by the adsorption of Fe phenanthroline on three different supports: Vulcan (254 m^2/g), Printex (900 m^2/g), and Black Pearls (1,500 m^2/g)[107]. The Fe loading was 1.14 wt% for the three carbon supports, and all catalysts were obtained after

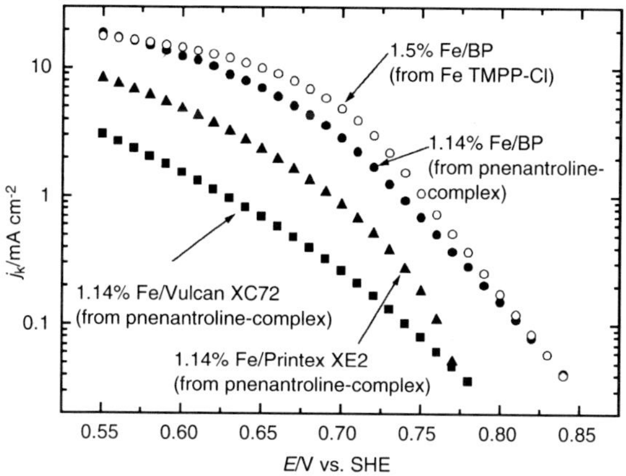

Figure 3.24. Tafel plots for the oxygen reduction at various iron-based catalysts, as obtained from RDE measurements. All catalysts were heat treated in Ar at 900°C (according to Figure 2 in ref. [107]; reproduced with permission of Elsevier).

pyrolysis at 900°C in Ar. As depicted in Figure 3.24, there is a difference of about one order of magnitude in the current densities measured at the same potentials for these catalysts on different carbon supports, with the catalyst on Black Pearls providing the highest activity for ORR. This difference was attributed to the larger specific area of Black Pearls compared with that of Vulcan, which provided the possibility of a larger density of catalytic sites on the former support than on the latter.

Our group also performed some studies on catalysts made with various carbon supports[111, 112]. In the first study published in 2003, we used: (i) six commercial carbon supports (Printex XE-2, Norit SX Ultra, Ketjenblack EC-600JD, acetylene black, Vulcan XC-72R, and Black Pearls 2000); (ii) three developmental carbon supports (HS 300 from Lonza , RC1 and RC2 from Sid Richardson Carbon Corporation; RC1 was enriched in nitrogen, while RC2 was not); (iii) the same nine previous supports prepyrolyzed at 900°C in an atmosphere containing NH_3 to increase their N content; and (iv) a synthetic carbon made by pyrolyzing PTCDA at 900°C in an atmosphere containing NH_3. The same Fe precursor, iron acetate, with a loading of 0.2 wt% Fe, was adsorbed on all these carbon supports, and the resulting materials were heat treated at 900°C in an atmosphere containing NH_3.

The variation of activity for these catalysts, all made under the same experimental conditions, except for the carbon support, was stunning and is presented as fuel cell polarization curves in Figure 3.25. On the one hand, the specific area of all these catalysts was measured and no correlation with the catalytic activity was found. On the other hand, there was a definite correlation between the catalytic activity and the surface nitrogen concentration measured by XPS. This correlation is illustrated in Figure 3.26. It is obvious that the catalytic activity for ORR increases when the surface of the catalyst is richer in nitrogen atoms. This is a logical result, since the precursor of the Fe–N_2/C catalytic site (the most

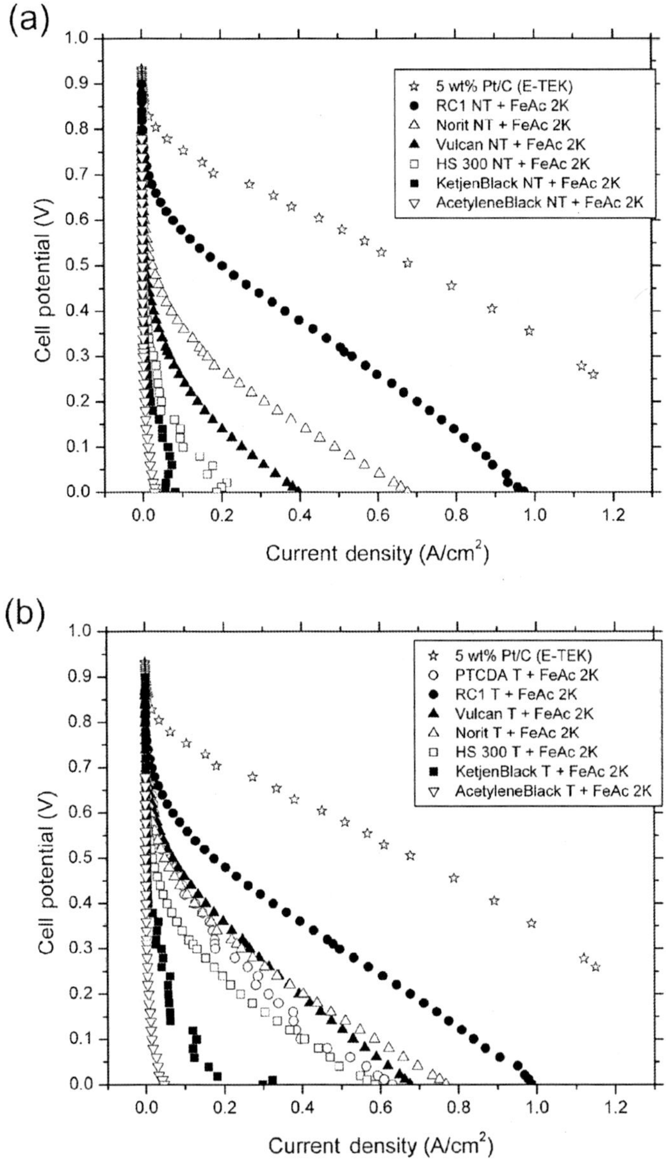

Figure 3.25. PEM fuel cell polarization curves at 80°C of several MEAs made with non-noble metal cathode catalysts. The cathode catalysts were prepared by adsorption of iron acetate (0.2 wt% Fe) on various untreated (NT) (panel a) or heat treated (T) carbon supports (panel b). The (T) carbon supports were obtained by heat treatment of (NT) carbon supports at 900°C in H_2:Ar:NH_3 (1:1:2). The polarization curve for 5 wt% Pt/C at the cathode is given for comparison (according to Figure 3 in ref. [111]; reproduced with permission of the American Chemical Society).

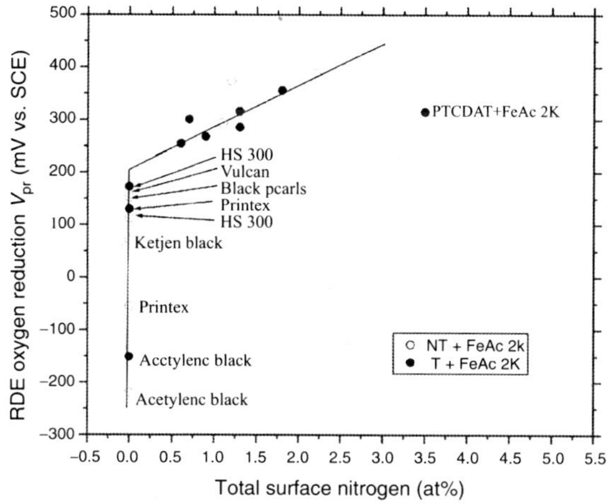

Figure 3.26. Change in the catalytic activity (expressed as V_{pr}: potential at peak reduction at 0 rpm) with te total N surface concentration of the catalysts derived from XPS results. The catalysts are the same as the ones used in Figure 3.25 (according to Figure 8 in ref. [111]; reproduced with permission of the American Chemical Society).

favorable site in the present experimental conditions) is C–N_x–Me. Therefore, it is expected that the carbon support that is able to accumulate the largest density of C–N_x functionalities will also be able to accommodate the largest density of the catalytic site's precursor: C–N_x–Me, and finally of catalytic sites after pyrolysis. It also explains why better catalysts were obtained when the carbon support was prepyrolyzed first in NH_3 atmosphere before loading iron acetate and heat treated again at 900°C in the same NH_3 atmosphere (Figure 3.25B). For the catalysts that contain N at concentrations below the detection limit of XPS ($\leq$ 0.1 at%), a correlation was found between their activity and the surface oxygen content. It is possible that an iron oxide or hydroxide is responsible for the low, but non-zero, catalytic activity of these materials.

After demonstrating that the density of nitrogen-containing functionalities on the carbon support seemed to be the key to catalytic activity, the next logical step was to enhance this nitrogen atom concentration of the carbon support. Based on the premise that ammonia reacts with oxygen-containing functionalities existing at the surface of some carbons leading to a carbon surface with an improved nitrogen content[112], four selected carbons were oxidized in air, oxygen, or HNO_3 before being treated in NH_3[113]. Catalysts were obtained by pyrolyzing (at 900°C in ammonia) iron acetate (loaded at 0.2 wt% Fe) adsorbed on these modified carbon supports. The selected carbons were Vulcan XC-72R, Black Pearls 2000, Norit SX Ultra, and a developmental carbon, RC2, from Sid Richardson Carbon Corporation. The best results were obtained by refluxing Norit in HNO_3, then loading the iron acetate and heat treating that material in an NH_3 atmosphere at 900°C. Figure 3.27 presents the Tafel plot of the polarization curve obtained at 80°C with that catalyst when the GDE current is expressed in A/mg metal. It is compared with the polarization curve obtained for 10 wt% Pt/C under the same experimental conditions, and with the state-of-the-art activity obtained at

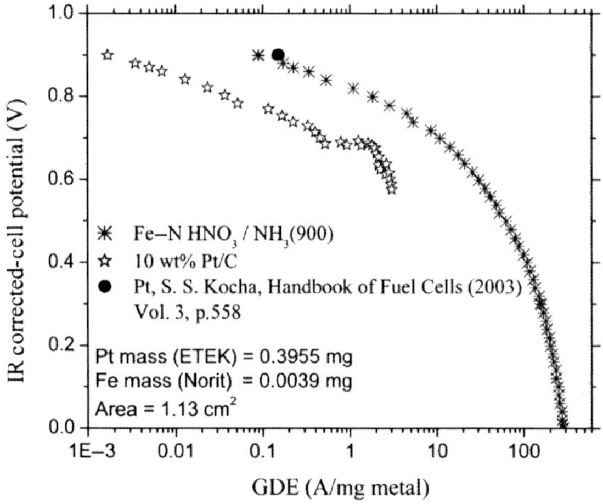

Figure 3.27. Tafel plots of the polarization curve for the best non-noble metal catalyst obtained on Norit compared to a catalyst from E-Tek loaded with Pt at 10 wt%. The GDE currents are expressed in A/mg metal. Dark circle: reported state-of-the-art activity at 900 mV for H_2/saturated O_2 at 65°C, 100 kPa, for a cathode with 0.4 mg Pt/cm^2. The non-noble metal catalyst was made by adsorbing iron acetate (0.2 wt% Fe) on pretreated Norit. This material was heat treated at 900°C in H_2:Ar:NH_3 (1:1:2). The pretreated Norit was obtained by refluxing Norit in HNO_3 (according to Figure 10 in ref. [113]; reproduced with permission of The Electrochemical Society).

65°C for Pt. The latter result, which has been measured by Kocha, is shown as a black circle in Figure 3.27. It should also be mentioned that the Tafel slopes at low current densities for the Fe-based and Pt catalysts are practically identical: ~70 mV/decade. These results indicate that the mass activity of these non-noble metal catalysts is quite good and practically at the level of the mass activity of Pt. *Better activity for non-noble metal catalysts would be expected if it were possible to increase the density of C–N_x functionalities on the carbon support.*

Before ending this section, it is worth mentioning two particular cases for which the carbon support was made from a carbon precursor during the pyrolysis step that also generated the catalyst. The first case was that of PTCDA, which was introduced in Section 2.2.3. Active catalysts were produced by adsorbing a metal precursor on that carbon precursor, then heat treating the resulting material, usually at 900°C in NH_3 atmosphere. Our best results in fuel cells were obtained with catalysts made by this procedure, either with iron acetate (0.2 wt% Fe)[99], or with ClFeTMPP (2 wt% Fe) as an Fe precursor[114].

Recently, another procedure had appeared in the literature using a mixture of CoTMPP/iron-oxalate (< 3 wt% Co/Fe loading) as carbon, nitrogen, and metal precursors[115]. The oxalate was used as a foaming agent to improve the specific area of the pyrolyzed catalyst. The latter was prepared by first mixing CoTMPP and iron oxalate. The mixture was then heat treated, initially at 450°C, then at 750°C. The resulting material was quenched at room temperature before being exposed to air. Finally, HCl was conditioned in order to remove inactive byproducts. It was assumed that molten CoTMPP encapsulated the oxalate crystals at about

400°C before pyrolysis started. During decomposition of the oxalate at higher temperatures, cavities were formed by the release of CO_2. These cavities were then filled up with molten CoTMPP. Due to the successive pyrolysis of CoTMPP within the cavities, carbon was formed and partially covered the reaction products of the oxalate decomposition. The surface area of leached samples was determined. It ranged from 400 to 700 m^2/g. This new preparation method led to an enhanced activity in terms of current densities. The catalyst was more active than that made at the same pyrolysis temperature by impregnation of CoTMPP on Vulcan. Pyrolysis of pure CoTMPP led only to dense particles with low catalytic activity.

3. Kinetic and Mechanistic Aspects of Electrochemical Oxygen Reduction

3.1. Kinetic Aspects of Electrochemical Oxygen Reduction

Oxygen reduction in an acid medium may occur according to the following reactions:

$$O_2 + 4\,H^+ + 4e^- \leftrightarrows 2H_2O \qquad E_1{}^{o} = 1.23\ V \qquad (3.1)$$

$$O_2 + 2H^+ + 2e^- \leftrightarrows H_2O_2 \qquad E_2{}^{o} = 0.68\ V \qquad (3.2)$$

$$H_2O_2 + 2\,H^+ + 2e^- \leftrightarrows 2\,H_2O \qquad E_3{}^{o} = 1.77\ V \qquad (3.3)$$

The standard potentials of reactions (3.1) – (3.3) are given for a temperature of 25°C.

H_2O_2 may also decompose into O_2 and H_2O.

The apparent number of electrons transferred during ORR, n, has been measured for several Fe-based catalysts obtained by the pyrolysis of various Fe precursors, including Fe–N_4 chelates, iron phenanthroline or salts such as iron acetate. Values of n and the associated yields of hydrogen peroxide, $\%H_2O_2$, are reported in Table 3.2 for Fe precursors and in Table 3.3 for Co precursors. $\%H_2O_2$ is obtained from the following equation[114]: $\%H_2O_2 = 100\,(4 - n)/2$.

According to these tables, reduction of O_2 on Fe-based catalysts occurs mainly to water ($n = 4$), except for cationic and anionic FeTMPP, while the lower value of n obtained for Co-based precursors indicates a larger release of peroxide ($n = 2$) by these catalysts. The reduction of O_2 to water may either occur directly with the transfer of 4 electrons ($4e^-$) or O_2 may be first reduced to H_2O_2 with the transfer of two electrons, followed by the transfer of another two electrons during a subsequent reduction of H_2O_2 ($2 \times 2e^-$). The rate constant for reaction (3.1) in the $4e^-$ transfer mechanism is k_1 in the following scheme, while two rate constants k_2 and k_3 for the consecutive reactions (3.2) and (3.3) are involved in the $2 \times 2e^-$ transfer mechanism for ORR to water.

Wroblowa proposed an analysis to discriminate between k_1 and k_2 using the ring and disk currents, I_R and I_D , respectively measured at various angular rates, ω, and at a fixed potential value with a rotating-ring disk electrode (RRDE)[117]. The following relation has been demonstrated:

$$I_D N/I_R = 1 + 2k_1/k_2 + (1/Z_{H_2O_2})[k_3 + (k_3 + k_4)(1 + 2k_1/k_2)]\omega^{-0.5} \qquad (3.1)$$

Table 3.2. Apparent Number of Electrons Transferred, n, and Percentage Yield of Peroxide, $\%H_2O_2$, for Fe-Based Catalysts

Precursor	Carbon support	Heat-treatment temperature (°C)	n	$\%H_2O_2$	References
1,2 FeNPc	Printex XE2	500	3.5	25	[48]
ClFeTMPP	Black Pearls	200–1,000	3.45–4.0[a]	28–0	[69]
FeTMPP	RB Carbon	800	3.5	25	[44]
c- and a-FeTMPP[b]	Vulcan XC-72	800	2.7	65	[60]
c-FeTMPP	Vulcan XC-72	800	2.2	90	[60]
a-FeTMPP	Vulcan XC-72	800	2.0	100	[60]
FeTMPP	RB Carbon	800	3.96	2	[40]
ClFeTMPP	p-PTCDA[c]	900	> 3.9	< 5	[114]
Fe-acetate	Pyrrole Black	800	3.90	5	[40]
Fe-phenanthroline	Vulcan XC-72	800	3.7	15	[106]
Fe-acetate	p-PTCDA[c]	900	> 3.9	< 5	114
CoTMPP+ FeTMPP	No carbon support	600	4.0	0	[62]

[a] n changes with applied potential but not with heat-treatment temperature.
[b] c- and a-FeTMPP indicate cationic and anionic FeTMPP.
[c] p-PTCDA stands for prepyrolyzed PTCDA at 900°C in ammonia atmosphere.

Table 3.3. Apparent Number of Electrons Transferred, n, and Percentage Yield of Peroxide, $\%H_2O_2$, for Co-Based Catalysts

Precursor	Carbon support	Heat-treatment temperature (°C)	n	$\%H_2O_2$	References
1,2 CoNPc	Printex XE2	500	2.33–3.55	83.5–22.5	[49]
1,2 CoNPc	Printex XE2/ Black Pearls 2000	500	3.8/3.8	10/10	[50]
1,2 CoNPc	Printex XE2/ Black Pearls 2000	500	3.8/3.8	10/10	[51]
2,3 CoNPc	Printex XE2/ Black Pearls 2000	500	3.63/3.63	18.5/18.5	[51]
CoTMPP	Norit SX Ultra	600–800	3.8	10	[56]
CoTMFPP	Norit SX Ultra	600–800	3.7–3.0	15–50	[56]
CoTFPP	Norit SX Ultra	600–800	3.3–3.0	35–50	[56]
CoTAA	Norit SX Ultra	600	3.3–2.8	35–60	[57]
CoTAA	Norit SX Ultra	600/800	3.3/3.0	35/50	[116]
CoTAA	Norit SX Ultra	600/800	2.8–3.8/ 2.5–3.6	60–10/ 75–20	[59]
CoTMPP	Printex XE2	600	> 3.8	< 10	[55]
c- and a-CoTPP[a]	Graphite powder	600/800	3.72/3.50	14/25	[61]
a-CoTPP	Graphite powder	600/800	3.08/3.08	46/46	[61]
c-CoTPP	Graphite powder	600/800	2.74/2.88	63/56	[61]
Co Salen	Graphite powder	600	2.96	52	[91]
Co-Anten	Graphite powder	600	3.04	48	[91]

[a] c- and a-CoTPP stand for cationic and anionic CoTPP.

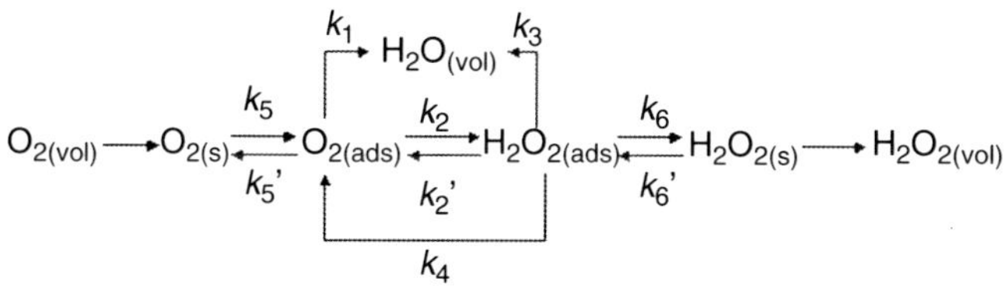

Scheme 3.1. Overall scheme for parallel and successive reactions involved in ORR (according to ref. [117]; reproduced with permission of Springer Science and Business Media).

where N is the collection efficiency of the RRDE, $Z_{H_2O_2} = 0.62\ D^{2/3}{}_{H_2O_2}\ \nu^{-1/6}$, with D being the diffusion coefficient of oxygen in the solution of viscosity ν. Using $I_D N/I_R$ and $\omega^{-0.5}$ as y and x coordinates, respectively, Equation (3.1) is a straight line with an intercept A described by $A = 1 + 2\ k_1/k_2$ and with a slope B described by $B = (1/Z_{H_2O_2})\ [k_3 + (k_3 + k_4)\ (1 + 2k_1/k_2)]$. The magnitude of intercept A characterizes the ratio k_1/k_2, or the relative occurrence of the $4e^-$ transfer vs. $2 \times 2e^-$ transfers for ORR, while slope B reflects the subsequent behavior of H_2O_2 formed during reaction 3.2. Measurements are performed at various applied potentials yielding a straight line and the possibility to evaluate k_1/k_2 for each potential.

This Wroblowa analysis has been performed on non-noble metal electrocatalysts by three different research groups. The first study was that of van Veen and his colleagues[118] on a 7 wt% ClFeTMPP (~0.5 wt% Fe)/Vulcan heat treated at 700°C in inert atmosphere and analyzed in 0.5M H_2SO_4 as a function of rotation frequency, f, between 2 and 64 Hz. A correction was applied to N that decreased with increasing f, due to the destruction of laminar flow by the rough surface of the electrode. It was concluded that the desired $4e^-$ reduction plays a dominant role only at high potentials, where the current densities are low. At lower potentials, the production of hydrogen peroxide dominates. The direct formation of water was related to the possibility of forming an Fe(IV)=O intermediate adduct (this point will be discussed in more detail later on). In their work on 24 wt% ClFeTMPP (1.6 wt% Fe)/Black Pearls heat treated at 800°C in inert atmosphere and analyzed in 0.5M H_2SO_4 as a function of rotation frequency between 250 and 1,500 rpm (4.2–25 Hz), Savinell and his colleagues reported that the rate constant for the direct ORR path was about 2.5 times higher than the rate constant for the series path[69]. Finally, a recent work has also been performed by Tarasevich and his colleagues[117] on 25 wt% CoTMPP (1.9 wt% Co)/acetylene black heat treated at 800°C in inert atmosphere and analyzed in 0.5M H_2SO_4 between 610 and 1,960 rpm (10.2 to 32.7 Hz). Their dependence of $I_D N/I_R$ on $\omega^{-0.5}$ is presented in Figure 3.28, and the potential dependence of the ratio k_1/k_2 for ORR as a function of applied potential is presented in Figure 3.29. From these three analyses, it is concluded that the direct $4e^-$ reduction of O_2 to H_2O is, indeed, occurring on these electrodes; and it may represent an important fraction of the two possible paths for the ORR.

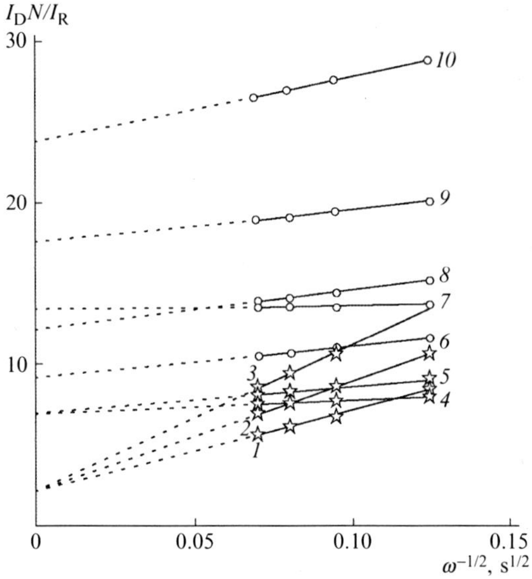

Figure 3.28. Dependence of $I_D N/I_R$ on $\omega^{-0.5}$ for (4–10) AD-100 promoted with a pyroporphyrin (CoTMPP adsorbed on activated AD-1000 and heat treated at 800°C in inert atmosphere). I_D and I_R were obtained from polarization curves in H_2SO_4 at : (4) 0.40, (5) 0.50, (6) 0.65, (7) 0.75, (8) 0.10, (9) 0.0, and (10) −0.10V (RHE) (according to Figure 9 in ref. [117]; reproduced with permission of Springer Science and Business Media).

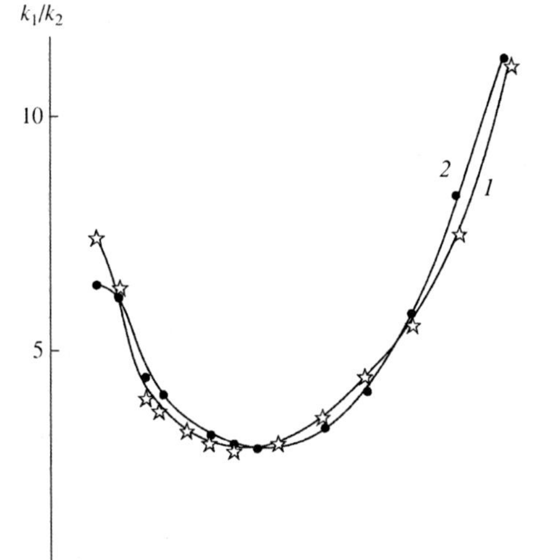

Figure 3.29. Potential dependence of the ratio k_1/k_2 for ORR on polarization curves with pure activated AD-100 carbon support promoted with pyroporphyrin (CoTMPP adsorbed on activated AD-1000 and head treated at 800°C in inert atmosphere) for the following amounts of active mass deposited onto the electrode (1) 20 and (2) 40 μg (according to Figure 10 in ref. [117]; reproduced with permission of Springer Science and Business Media.)

In the previous section, we came to the conclusion that two catalytic sites were always obtained simultaneously, but not in the same proportions, when an Fe precursor and an N precursor were present at the same time in the pyrolysis reactor. This demonstration was performed with the help of ToF SIMS analysis of either heat-treated iron acetate or ClFeTMPP adsorbed on N-enriched prepyrolyzed PTCDA[105]. The two catalytic sites were labeled $Fe–N_4/C$ and $Fe–N_2/C$, according to the relative abundance of their typical ions detected by ToF SIMS. While $Fe–N_4/C$ corresponds to the catalytic site proposed by van Veen and illustrated in Figure 3.5, the full coordination of $Fe–N_2/C$, illustrated in Figure 3.19, is not completely known. Possible $Fe–N_{2+2}/C$ catalytic structures have been proposed by various authors[100, 107, 109] but have not yet been confirmed. In the following discussion we will continue to use the $Fe–N_4/C$ and $Fe–N_2/C$ labels to identify these catalytic sites.

The same catalysts as those used in ToF SIMS experiments were analyzed for their capacity to reduce oxygen electrochemically. Their apparent number of electrons transferred, as well as the percentage of H_2O_2 released by these catalysts during ORR, was determined. The results are presented in Figures 3.30 and 3.31 for iron acetate (0.2 wt%Fe) and for ClFeTMPP (0.2 wt% Fe), respectively. Panel A of both figures represents the relative abundance vs. heat-treatment temperature of either $Fe–N_2/C$ (full line) or $Fe–N_4/C$ (broken line), as deduced by ToF SIMS

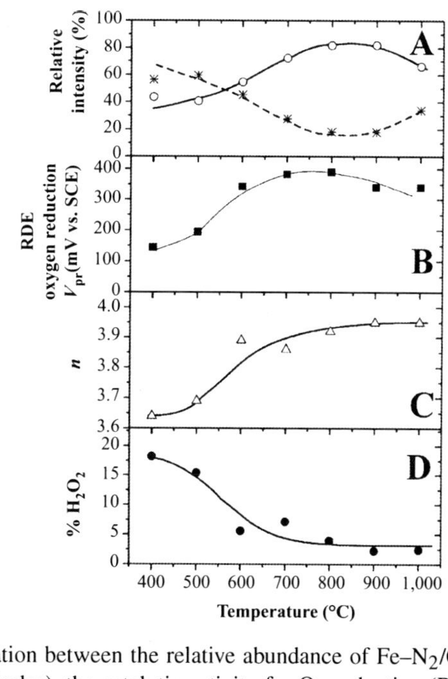

Figure 3.30. Correlation between the relative abundance of $Fe–N_2/C$ in non-noble metal catalysts (A: open circles), the catalytic activity for O_2 reduction (B), the value of n (C), and the value of $\%H_2O_2$ (D) for catalysts made with iron acetate (0.2 wt% Fe) as Fe precursor. The catalysts were the same as those used in Figures 3.20, 3.21, and 3.22A. The relative abundance of $Fe–N_4/C$ in the catalysts is given by stars in panel A (according to Figure 6 in ref. [114]; reproduced with permission of Elsevier).

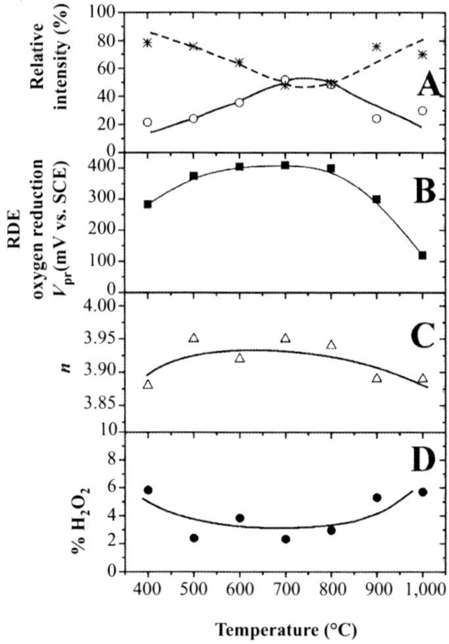

Figure 3.31. Correlation between the relative abundance of Fe–N_2/C in non-noble metal catalysts (A: open circles), the catalytic activity for O_2 reduction (B), the value of n (C), and the value of %H_2O_2 (D) for catalysts made with ClFeTMPP (0.2 wt% Fe) as Fe precursor. The catalysts were the same as those used in Figures 3.22B. The relative abundance of Fe–N_4/C in the catalysts is given by stars in panel A (according to Figure 7 in ref. [114]; reproduced with permission of Elsevier).

in these materials. Panel B represents the ORR activity of these catalysts as a function of heat-treatment temperature as well. In both Figures 3.30 and 3.31, we see that the shape of the activity curve of panel B follows that of the relative abundance of the Fe–N_2/C catalytic sites of panel A. A corollary of that observation is that the Fe–N_2/C catalytic sites are therefore more electrochemically active than the Fe–N_4/C ones for both Fe precursors. Panels C and D of Figures 3.30 and 3.31 present the change with heat-treatment temperature of the apparent number of electrons transferred during ORR for both catalysts and the percentage of H_2O_2 released by these catalysts. Again, we see that the shape of the curve for n follows that of Fe–N_2/C and not that of Fe–N_4/C for both Fe precursors. Of course, the value of %H_2O_2 and n have the opposite trend. A corollary of that observation is that the Fe–N_2/C catalytic sites have the property to reduce O_2 mainly to water. It has not been determined so far whether ORR on Fe–N_2/C involves a direct 4-electron transfer to H_2O or a series of 2 × 2-electron transfers with H_2O_2 as an intermediate.

Another observation worth mentioning is that the carbon support also has an effect on the apparent number of electrons transferred during ORR. These results have been reported for cobalt acetate (0.2 wt% Co) as Co precursor, adsorbed on various carbon supports and heat treated at 900°C in an atmosphere containing

NH_3[119]. In that work, it was shown that n increased toward 4 as the nitrogen concentration detected by XPS at the surface of the catalyst increased.

3.2. Mechanistic Aspects of Electrochemical Oxygen Reduction

The dioxygen molecule, in its ground state, has two unpaired electrons located in a doubly degenerate $\pi*$ antibonding orbital. This corresponds to a triplet state. When O_2 undergoes reduction, the added electrons occupy antibonding orbitals, thereby increasing the O–O distance. The plausible 1:1 and 2:1 metal–dioxygen complex structures are illustrated in Figure 3.32[120]. Geometries (1) and (2) are labeled Griffiths and Pauling models, respectively. The Griffiths model (1) involves a lateral interaction of the π orbitals of O_2 (+ and − rectangle-shaped orbitals in Figure 3.32) with the empty d_{z^2} orbitals of a transition element, ion or metal atom with back bonding from at least partially filled d_{xz} or d_{yz} orbitals of the transition element (four leaf clover-shaped orbitals centered on M in Figure 3.32) to the $\pi*$ orbital of O_2[121]. This type of interaction leads to a weakening of the O–O bond with a corresponding lengthening of this bond. With most transition metal catalysts, the most probable structure for O_2 adsorption is the Pauling model, i.e., adsorption through single O and metal atoms[121]. The bridge interaction (3) was proposed by Yeager and it is likely to occur on noble metal such as Pt, where O_2 is reduced to water with little or no peroxide formed. A *trans* configuration has been proposed to occur with dimeric cobalt dioxygen complexes[120].

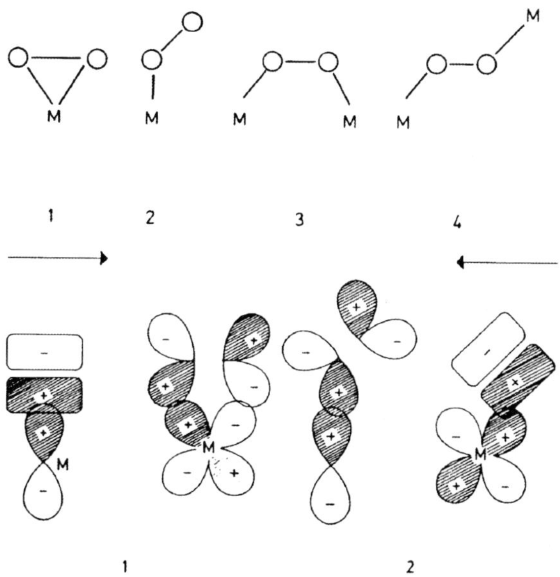

Figure 3.32. Top panel: Possible configurations of dioxygen interaction with a metal in a catalytic complex; bottom panel: Molecular orbitals involved in different interactions of O_2 with a metal center (according to Figures 3 and 4 in ref. [120]; reproduced with permission of Elsevier.

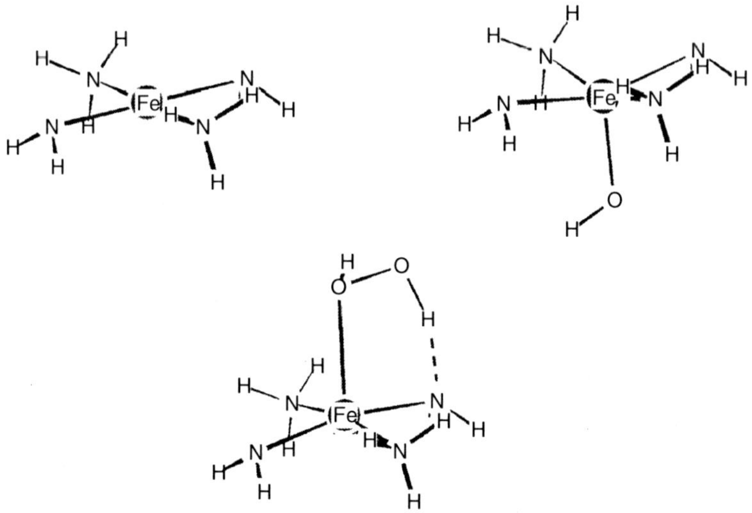

Figure 3.33. Top panel: Fe^{II} catalytic site modeled by $Fe(NH_2)_2(NH_3)_2$ on the left and Fe^{III} site modeled by $Fe(NH_2)_2(NH_3)_2OH$ on the right. Bottom panel: strucutre of L_4 Fe–(OH–OH), i.e., H_2O_2 bound to the Fe^{II} catalytic site (according to Figures 1 and 3 in ref. [122]; reproduced with permission of the American Chemical Society).

A quantum theoretical approach predicting the activation energies and reversible potentials for forming surface intermediates during the course of electrocatalytic reactions was developed by Anderson and Sidik[122]. This approach was applied to O_2 reduction on transition metal macrocycles that had undergone pyrolysis. In that article, quantum theory was used to study O_2 reduction on Fe^{II} and Fe^{III} centers. The 2- and 4-electron reductions in acid electrolyte were applied to the iron centers. In their analysis, Fe^{II} is coordinated to two NH_2^- and two NH_3 ligands placed in a square, diagonally from each other, making it an approximate model for Fe in the center of a porphyrin. The actual porphyrin site has all four Fe–N bonds essentially identical. However, if the result of heat treatment is to break off some of the conjugated ring systems, then the active Fe^{II} site might well have different types of coordinated nitrogen groups, similar to this model. The theme of that study was the dependence of the electrochemistry on the oxidation state of the metal site.

A representation of the two possible starting sites, one for Fe^{II} (L_4Fe^{II}) and the other one for Fe^{III} ($HO–L_4Fe^{III}$) is depicted in the top of Figure 3.33. Each NH_2^- ligand carries a lone pair of electrons. In the absence of oxygen, a water molecule is coordinated to these starting sites. The complexes are represented either by $L_4Fe^{II}–OH_2$ or by $HO–L_4Fe^{III}–OH_2$. A water molecule would, according to the calculations, form relatively strong hydrogen bonding interactions with the nitrogen lone pairs of these complexes. The calculations show that O_2 is capable of displacing H_2O from $L_4Fe^{II}–OH_2$, but not from $HO–L_4Fe^{III}–OH_2$. Therefore, $L_4Fe^{II}–OH_2$ is the only possible starting point of the electrocatalytic reduction of O_2 on these catalytic sites. The electrocatalytic reduction of O_2

involves four successive reaction steps for which reversible potentials (U°) were calculated. These reaction steps are:

$$\mathbf{L_4Fe^{II}-OH_2}+O_2+H^+(aq)+e^- \rightarrow L_4Fe^{III}-O-OH+H_2O \quad (3.4) \quad U^\circ = 0.91\text{ V}$$

$$L_4Fe^{III}-O-OH+H^+(aq)+e^- \rightarrow L_4Fe^{II}-(OH-OH) \quad (3.5) \quad U^\circ = 0.96\text{ V}$$

$$L_4Fe^{II}-(OH-OH)+H^+(aq)+e^- \rightarrow L_4Fe^{III}-OH+H_2O \quad (3.6) \quad U^\circ = 2.43\text{ V}$$

$$L_4Fe^{III}-OH+H^+(aq)+e^- \rightarrow \mathbf{L_4Fe^{II}-OH_2} \quad (3.7) \quad U^\circ = 0.64\text{ V}$$

An alternative reaction path involving Fe^{IV} would be possible after (3.4). This path is described by the following reaction steps:

$$L_4Fe^{III}-O-OH+H^+(aq)+e^- \rightarrow L_4Fe^{IV}-O+H_2O \quad (3.8) \quad U^\circ = 0.44\text{ V}$$

$$L_4Fe^{IV}-O+H^+(aq)+e^- \rightarrow L_4Fe^{III}-OH \quad (3.9) \quad U^\circ = 2.96\text{ V}$$

The succession of reaction steps (3.4), (3.8), (3.9), and (3.7), involving Fe^{IV} was found to not be favored. Therefore this path is not followed and ORR on such pyrolyzed materials involves only the Fe^{II}/Fe^{III} couple. An important difference compared to O_2 reduction on Pt is the H-bonding interaction between –(OH–OH) bonded to Fe^{II} after reaction (3.5) and the nitrogen lone pair orbitals in the N_4 structure. The structure of H_2O_2 bound to Fe^{II} is illustrated at the bottom of Figure 3.33. This particular structure prevents H_2O_2 from leaving the catalytic site and provides a path for direct reduction to water on Fe-based catalysts.

A reaction mechanism for oxygen reduction in acid media was also proposed by van Veen and his collaborators[118]. However, these authors proposed the alternative reaction path (involving reactions (3.8) and (3.9)) as the main path followed during ORR at high applied potential. For them, the ability for the transition metal center to stabilize the $Fe^{IV} = O$ adduct, gives these catalysts the possibility, at a high applied potential, to reduce oxygen to water by a direct 4-electron transfer. This could explain why, for most Co–N_4 chelates, no direct water formation was observed, other than the apparent formation caused by H_2O_2 decomposition. Anderson and Sidik[122] mentioned this contradiction in their work, where they found that adsorbed H_2O_2 was the favored intermediate formed from L_4Fe^{III}–O–OH in the 0.44–0.96 V range. This H_2O_2 can be released into solution only at potentials less than the reversible potential for $H_2O_2(aq)$ formation, which has the standard value of 0.695 V. At potential higher than 0.695 V, i.e., in the potential region where van Veen proposes the involvement of $Fe^{IV} = O$, Anderson and Sidik argue that the (H_2O_2) adsorbed intermediate can only be reduced to L_4Fe^{III}–OH without the participation of Fe^{IV}.

Although there is some disagreement between these two groups about the exact reaction path to be followed during ORR, they agree on the starting oxidation state of iron, i.e., Fe^{II}, in the catalytic site, and on the fact that the first electron transfer is the rate-limiting step[98,122,123]. The activation energies for steps in oxygen reduction over Fe^{II} and Fe^{III} sites might be anticipated to behave similarly to those on Pt[122]. If so, the first reduction step on Fe^{II} sites will be responsible for the observed effective activation energy, just as it is generally believed to be on Pt. This would explain the 60 mV/decade (at 25°C) and 70 mV/decade (at 80°C)

Table 3.4. Tafel Slope, *b*, Reported for Fe- and Co-Based Electrocatalysts for the Reduction of Oxygen in Acid Medium at Room Temperature

Precursor	Carbon support	Heat-treatment temperature (°C)	*b* (mV/ decade)	Electrolyte	References
1,2 CoNPc	Printex XE2	500	53	0.25 MH_2SO_4	[49]
1,2 CoNPc	Printex XE2 and Black Pearls	500	42	0.25 MH_2SO_4	[50]
1,2 CoNPc and 2,3 CoNPc	Printex XE2 and Black Pearls	500	42 and 40	0.25 MH_2SO_4	[51]
CoTAA	Norit SX Ultra	600 and 800	60	0.25 MH_2SO_4	[57]
CoTAA	Norit SX Ultra	600	50–70	0.25 MH_2SO_4	[59]
CoTAA	Norit SX Ultra	600	59	0.25 MH_2SO_4 0.5 M$HClO_4$	[54]
CoTMPP	Acetylene black	800	60	0.5 MH_2SO_4	[124]
CoTMPP	Acetylene black and diamond	800	60	0.5 MH_2SO_4	[110]
ClFeTMPP	Black Pearls	800	60–68	0.1 MH_2SO_4, $HClO_4$, H_3PO_4	[68]
Various Co and Fe porphyrins	Norit BRX	500–900	60	8 NH_2SO_4	[123]
ClFeTMPP	Black Pearls	900	61	0.5 MH_2SO_4	[109]
Co and Fe salts and polypyrrole	Graphite powder	820	75	1 N$HClO_4$	[87]
Fe phenanthroline	Vulcan XC-72	900	64–74	0.5 MH_2SO_4	[106]
Fe acetate	RC2	900	67[a]	Nafion/80°C	[111]
Fe acetate	RC2 and Norit SX Ultra	900	67[a]	Nafion/80°C	[113]

[a] Experiments done in fuel cells at 80°C (theoretical *b* value at 80°C; 70 mV/decade).

Tafel slopes which are usually found for O_2 reduction on Fe- and Co-based catalysts and on Pt. A list of Tafel slopes experimentally found for these catalysts is given in Table 3.4.

4. Important Factors for the Use of Non-Noble Metal Electrocatalysts in PEM Fuel Cells for Automotive Applications

In their article on activity benchmarks and requirements for Pt, Pt alloys, and non-Pt oxygen reduction catalysts for PEM fuel cells, Gasteiger, Kocha, Sompalli, and Wagner[13] came to the conclusion that, while the technology development shows that the DOE target of $< 0.2\ g_{Pt}/kWe$ (at ≥ 0.65 V) can be accomplished using Pt-alloy catalysts, there is always a strong interest to search for Pt-free cathode catalysts as they would potentially reduce the cost of membrane–electrode-assemblies to even lower levels. Two essential points remain, however, to be considered when a low-cost or, even a costless catalyst, is considered. The first one is the performance of the catalyst, the second one is its stability. These two points are now considered.

If a costless non-Pt catalyst were available, a significant larger amount of such catalyst might be used in MEAs. This would compensate for an eventual lower catalytic activity compared with Pt. However, a rigorous comparison between precious metal and nonprecious metal-based catalysts implies a comparison between the respective turn-over frequencies (TOF, defined as the number of electrons transferred per active site and per second at a given potential and experimental conditions) and the volumetric site densities (SD, defined as the number of active site per cm^3 of supported catalyst) of the different catalysts. The comparison of these materials is, therefore, made in terms of current per unit volume of supported catalyst (A/cm^3), which equals ($1.6 \times 10^{-19}[C/e^-] \times TOF[e^-/site.s] \times SD[sites/cm^3]$) with a limit of a 100-$\mu$m thick catalytic layer, a thickness above which mass transport limitation is expected to become too large. The current per unit volume of supported non-precious metal catalyst was compared to Pt at the same cell resistance-corrected potential of 800 mV RHE, where mass transport effects should be small. Considering all these factors, the activity of a costless cathode catalyst for automotive applications needs to be no less than $\sim 1/10_{th}$ of the current industrial Pt activity at equivalent conditions. This value is referred to in Table 3.5 as costless catalyst target.

Table 3.5 compares TOF, SD, and A/cm^3 for selected Pt and non-noble metal catalysts reported in the literature. All these data have been corrected by Gasteiger and his collaborators[13] to allow comparison at 80°C with Pt catalysts at the same temperature and experimental conditions as in Figure 4 of their article (fully humidified [80°C dew point] reactants at 150 kPa_{abs} [H_2/O_2 stoichiometric flow of $s = 2/9.5$]). From Table 3.5, it can be seen, on the one hand, that the TOF of actual non-noble metal catalysts is of the same magnitude as that of the costless catalyst target, at least for the best performing catalysts. *On the other hand, the*

Table 3.5. Turnover frequencies (TOFs) for O_2 Reduction [e^-/site.s], Site Densities (SD; in [sites/cm^3]), and Their Product $\times$ $1.6 \times 10^{-19}[C/e^-] = A/cm^3$ of Supported Catalyst, at $800mV_{iR-free}$ for 47 wt% Pt Compared to Activity Requirements for a Costless Catalyst and the Experimental Absolute Activities of Non-Pt Catalysts from Selected References. TOFs are Shown Both at the Conditions of Measurement and Corrected for the Pt Reference Industrial Conditions, Assuming pO_2 to the First Order and $E_a = 54.7$ kJ/mol O_2, i.e., Corrected Using Parameters for Pt (Reproduced from Ref.[13], with the Permission of Elsevier)

	T (°C)	pO_2 (kPa_{abs})	TOF(e^-/sites) at reported conditions	TOF(e^-/sites) at Pt reference conditions	SD ($\times 10^{20}$ (sites/cm^3)	A/cm^3 at Pt reference conditions
47 wt% Pt/C reference	80	100	25	25	3.2	1,300
Costless catalyst targets	80	100	2.5	2.5	3.2	130
Fe/PTCDA[99]	50	498	1.45	1.7	0.09	2.4
Fe-phenanthroline[108]	21	100	0.023	1.0	0.49	7.8
Fe/PTCDA[111]	80	498	0.11	0.023	0.09	0.03

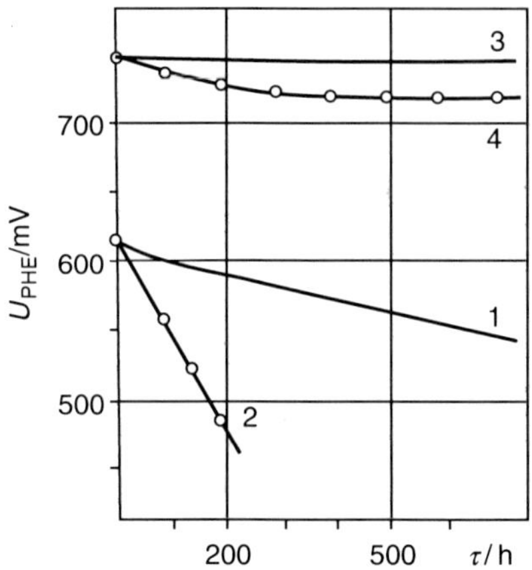

Figure 3.34. Galvanostatic potential–time curves ($i = 5$ mA/cm^2) of oxygen electrodes with an active material from active carbon P33 + 15% CoTAA. Temperature 25°C, electrolyte 4.5 N H_2SO_4 . (1) Without thermal treatment of the active material, (2) without thermal treatment of the active material adding H_2O_2 up to 1% in concentration in the electrolyte every 100 h, (3) with thermal treatment of the active material in Ar at 630°C adding H_2O_2 up to 1% in concentration in the electrolyte every 100 h (according to Figure 6 in ref. [29]; reproduced with permission of Elsevier).

site density of actual catalysts needs to be significantly increased in order to reach the costless catalyst target. In Sections 2.2, and, in particular, 2.2.4, we argued that *an increase in the site density of these catalysts is possible if the amount of nitrogen atoms at the surface of the carbon support is increased*, since these atoms are used in the coordination of the non-precious metal ions in the two types of catalytic sites: Me–N_4/C and Me–N_2/C.

The second factor to consider for a costless catalyst is its stability under PEM fuel cell conditions. This is an important point that is not often addressed by authors working on non-noble metal-based catalysts. The origin of catalyst instability in acid medium has been attributed to the release of peroxide during oxygen reduction when ORR does not occur directly to water in a 4-electron transfer. Figure 3.34 illustrates this point for a 15 wt% CoTAA (~ 2.7 wt% Co) on P33. Curves 1 and 3 display the unstable and stable catalytic activity behavior for the adsorbed complex before and after heat treatment at 630°C, respectively, while curves 2 and 4 display the catalytic activity of the same catalysts, but this time, with addition of H_2O_2 to the 4.5NH_2SO_4 electrolyte every 100 h. Although H_2O_2 may be directly responsible for some of the observed instabilities among non-noble metal catalysts, there may be also another source of instability: the HO• radical, a much more ferociously oxidizing species than H_2O_2! The occurrence of HO• release is especially expected when Fe is the active ion in the catalytic site. Any trace of free Fe^{2+} in the presence of H_2O_2 will, indeed, generate HO radicals according to:

$$H_2O_2 + Fe^{2+} \rightarrow Fe^{3+} + HO^- + HO\bullet \tag{3.10}$$

The combination of H_2O_2 and Fe^{2+} is known as Fenton's reagent[125]. Reaction (3.10) will be followed by:

$$HO\bullet + RH \rightarrow H_2O + R\bullet \tag{3.11}$$

where RH is either the catalytic site or the carbon support.

R• will then yield reaction degradation products or regenerate Fe^{2+} according to:

$$R\bullet + Fe^{3+} \rightarrow Fe^{2+} + \text{degradation reaction products} \tag{3.12}$$

Unprotected Fe aggregates may be a source of Fe ions in Fe-based catalysts, especially when they are made with high Fe loadings. Fe ions are also quite common in the corrosion products of steel, a metal frequently used in PEM fuel cells. The only way to avoid the occurrence of Fenton's reaction is either by completely avoiding the release of H_2O_2 or by drastically reducing the free Fe^{2+} concentration. The latter is easily obtained by increasing pH and precipitating Fe^{2+} as $Fe(OH)_2$. This is probably the main reason why the same non-precious metal catalysts are stable in alkaline media. In PEM fuel cells, where Fenton's reagent is also a troubling issue for the membrane, the only alternative left is to design catalysts that will electrochemically reduce oxygen to water by a direct $4e^-$ transfer.

Details on the stability of non-noble catalysts have been collected in the literature and are presented in Table 3.6 for catalysts in thin films, mainly from RDE-type electrodes in solution; in Table 3.7 for catalysts in single layer of the GDE type; and in Table 3.8 for catalysts used in single membrane electrode assemblies. In these tables stability means that current density and applied potential (after stabilization, if any) remained constant during the test. Longevity tests as short as 40 h and as long as 10,000 h were reported for catalysts prepared by several groups with a large variety of metal and nitrogen precursors as well as with a rather large range of heat-treatment temperatures. It is not easy to rationalize these results, but some conclusions may be drawn from the longevity tests that lasted at least 1 week (168 h) and during which at least 5,000 C were delivered by the electrode. The catalysts that were stable under those conditions are highlighted in bold in Tables 3.7 and 3.8. It is clear that a single GDE-type electrode in contact with an electrolyte was found to be the most stable structure. Five out of seven catalysts tested under these conditions were stable for extended periods of time (Table 3.7). The activities of the two catalysts that were found unstable did not decrease dramatically either. As seen in Ref. [126], the potential at a constant current of 20 mA/cm^2 changed from 0.73 to only 0.71 V after 2000 h, while in Ref. [127], the potential at 10 mA/cm^2 changed from 0.75 to 0.64 V, after 800 h. Any peroxide released by the ORR probably found its way to the acid solution before reacting with the catalyst. The same was also true for unprotected metal which simply dissolved out of the catalyst without causing any harm. Wiesener and collaborators[85] even used leaching of Fe and Co into the electrolyte *during the longevity test* to measure what percentage of the initial metal loading of the catalyst remained in the electrode. This occurred without affecting the electrode performance or stability. A common factor among all catalysts performing in single layer GDE was also the low imposed current densities (low rates of H_2O_2 and

Table 3.6. Long-Term Experiments for Catalysts in Thin Films

Precursor on carbon	Heat-treatment temperature (°C)	Electrolyte and its temperature (°C)	Constant current mA/cm^2	Constant potential (V vs. RHE)	Experiment duration (h)	Coulombs	Stability[a] (yes/no)	References
CoTAA	600	0.25 MH_2SO_4 (stirred) / 22°C	—	0.5	100	380	No	[57]
	800					310	Yes	
CoTAA	600	0.25 MH_2SO_4 (stirred) / 22°C	—	0.5 (0–100 h)	300	710	No	[59]
	800			0.4 (100–300 h)		950	No	
80 % Co (dqph) + 20 % $Pt(NH_3)_4$ Cl_2 (in wt%)	600	0.05 MH_2SO_4 (stirred) / 80°C	—	0.182	50	135	Yes	[128]
Co salt / polypyrrole	820	1 N$HClO_4$ (stirred?) / 60°C	0.05	—	400	72	No	[87]
Fe phenanthroline	900	0.5 MH_2SO_4 (stirred) / 21°C	—	?	40	0.6	No	[108]

[a] Is stability reached during longevity test?

Table 3.7. Long-Term Experiments for Catalysts in Gas Diffusion Electrode in solution

Precursor on carbon	Heat-treatment temperature (°C)	Electrolyte and its temperature (°C)	Constant current mA/cm^2	Constant potential (V vs. RHE)	Experiment duration (h)	Coulombs	Stability[a] (yes/no)	References
CoTMPP	800	4.5 NH_2SO_4/20°C	10	0.77	10,000	360,000	**Yes**	[25]
			30	0.75	3,000	324,000	**Yes**	
CoTMPP	830	4.5 NH_2SO_4/50°C	20	—	2000	158,400	No	[126]
CoTMPP	830	4.5 NH_2SO_4/20°C	10	—	800	28,000	No	[127]
CoTAA[b]	630	4.5NH_2SO_4/25°C	5	0.75	700	12,600	**Yes**	[29]
Co salt/	850	2.25 MH_2SO_4/25°C	5	0.70	500	36,000	**Yes**	[85]
PAN[c] Fe salt/PAN[c]	850		5	—	500	36,000	No	
ClFeTMPP[d]	900	2.25 MH_2SO_4/25°C	18	0.7	100	5,000	**Yes**	[109]

[a] Is stability reached during longevity test?
[b] Co loading = 2.7 wt% — no acid digestion.
[c] Leaching performed during stability test; at the end of the test: Fe loading = 2.5 wt% and Co loading = 1.4 wt%.
[d] Fe loading = 0.47 wt% after H_2SO_4 digestion.

Table 3.8. Long-Term Experiments for Catalysts in Fuel Cell Single Membrane Electrode Assemblies

Precursor on carbon	Heat-treatment temperature (°C)	Electrolyte and its temperature (°C)	Constant current mA/cm^2	Constant potential (V vs. RHE)	Experiment duration (h)	Coulombs	Stability[a] (yes/no)	References
CoTAA	800	Nafion, 80°C	—	100	200	99,000	No	[58]
Ferrocene/ AN[b,c]	1,000	Nafion, 50°C	45	0.5	200	32,400	**Yes**	[84]
Co aggregates/ AN	1,000	Nafion, 50°C	—	0.5	350	24,520	No	[92]
$Cr(OH)_3/AN$[d]	1000	Nafion, 50°C	—	0.5	300	135,000	No	[95]
$Fe(OH)_2/AN$[d]	1,000		80	0.5	300	86,400	**Yes**	
$Co(OH)_2/AN$[d]	1,000		55	0.5	300	59,400	**Yes**	
$Fe(OH)_2/AN$[e]	1,000	Nafion, 50°C	100	0.5	180	64,800	**Yes**	[96]
FeTPP	1,000	Nafion, 50°C	—	0.5	225	96,700	No	[79]
FeTMPP	800	H_3PO_4-equilibrated Nafion, 125°C	100	Succession of polarization curves	24	No graph	(**Yes**)	[64]
ClFeTMPP/ PTCDA/NH_3	900	Nafion, 80°C	—	0.5	625	226,500	No	[114]

[a] Is stability reached during longevity test?
[b] AN = acetonitrile.
[c] Fe loading = 2 wt% — no acid digestion.
[d] After HNO_3 digestion, Fe, Co, Cr loadings = 8.5, 6.2, and 7.5 wt%, respectively.
[e] After H_2SO_4 digestion (at RT), and Cl_2 digestion (at 650°C), Fe loading = 3.3 wt%.

metal ion release, if any) that probably favors catalyst stability. The story is, however, different for catalysts tested in fuel cells (Table 3.8). Here current densities were larger (at least the starting current densities were larger: up to $0.25\text{A}/\text{cm}^2$ for Ref. [114]) and there is no solution to carry away the released metal ions or peroxide. Curiously, the catalysts that were not made with metal–N_4 chelates and for which the metal loading (even after chemical leaching of the catalyst, if any) was quite high (2 wt% Fe for ferrocene, 3.3 and 8.5 wt% Fe for $Fe(OH)_2$ and 6.2 wt% Co for $Co(OH)_2$) were found to be the most stable. However, FeTMPP-based catalysts were also found to be stable at 125°C in fuel cell tests using H_3PO_4-equilibrated Nafion[64] but, in that case, their stable behavior was estimated by running several consecutive polarization curves and comparing the potential of these curves at $100\ \text{mA}/\text{cm}^2$.

5. Conclusions

The electrochemical reduction of oxygen to water is a difficult reaction which requires a large overpotential, especially in acid media. The reaction pathway that involves a direct transfer of 4 electrons to oxygen (as opposed to a sequence of 2×2-electron transfer steps with the release of an undesirable intermediate: H_2O_2) is the preferred cathodic reaction for PEM fuel cells, which are low temperature and low-pH fuel cells using an acid polymeric ionomer as electrolyte. In order to deliver the practical current densities required in automotive applications, electrocatalysts are needed at the cathode (as well as at the anode) of these fuel cells. Today, only Pt and its alloys are used in PEM fuel cells, but the replacement of Pt and Pt-based catalysts would be highly desirable due to the high cost and potential availability problems of that metal. A possible solution is the development of Fe and/or Co-based electrocatalysts for ORR, especially catalysts prepared with low-cost metal and nitrogen precursors.

It has been reported in the course of this review that a recent study of the targets of a costless electrocatalyst to replace Pt in automotive applications requires that such non-noble metal catalysts have an activity no less than 1/10th of the current industrial Pt activity under equivalent conditions[13]. This requires mainly a sizeable increase in the site density (defined as catalytic sites/cm^3 in the electrocatalytic layer) of the non-noble metal catalysts. A knowledge of the molecular structure of the catalytic site for the electrochemical reduction of oxygen in acid medium is, therefore, essential in order to increase the site density on the carbon support for those catalysts. The long-term stabilities of the same catalysts under current industrial conditions are yet to be demonstrated, as well.

Today, we know that it is possible to produce these Fe- and/or Co-based electrocatalysts by adsorbing related metal–N_4 macrocycles on a carbon support and heat-treating this material at about 600°C, the optimum temperature in terms of activity. More stable, but less active catalysts are, however, obtained for heat-treatment temperatures $\geq$ 800°C. Similar catalysts may also be obtained with cheaper metal and nitrogen precursors (like metal salts and ammonia, for instance). For all these catalysts, it is generally now believed that two types of catalytic sites are obtained simultaneously, but not in the same proportions, irrespective of the preparation procedure used. These catalytic sites have been labeled $Fe–N_4/C$ and $Fe–N_2/C$ when Fe is used as metal precursor. The structure of the $Fe–N_4/C$ catalytic site, which is favored when $Fe–N_4$ chelates are used as Fe and N precursors, was shown by van Veen and his collaborators to be an iron ion coordinated to four nitrogen atoms of the pyrrolic type, which are themselves bound to the carbon support (Figure 3.5). The complete molecular structure of the $Fe-N_2/C$ catalytic site is still undetermined, but it is known that it is partially made of an iron ion coordinated to at least two pyridinic nitrogen atoms in a 1,10-phenanthrolinic structure bound to the carbon support (Figure 3.19). Two complete molecular structures have been proposed so far for the $Fe–N_2/C$ catalytic site. They have been labeled $Fe–N_{2+2}/C$ and are not yet fully confirmed. In one of these structures, the iron ion would be coordinated to four pyridinic nitrogen atoms in the same graphene plane, with each pair of nitrogen atoms being organized in a 1,10-phenanthroline structure (Figure 3.23). A juxtaposition of two graphene sheets around the Fe ion, with each graphene sheet carrying two pyridinic nitrogens in a 1,10-phenanthrolinic structure has been proposed in the second $Fe–N_{2+2}/C$ structure (not shown). Other complete molecular structures for $Fe–N_2/C$ are also plausible. They all have the common point of requiring high temperature (optimally $\geq$ 800°C) to be assembled and bound to the structure of the carbon support. So does also $Fe–N_4/C$, but in that case, bonding to the carbon support occurs at a lower temperature (optimally $\leq$ 800°C).

Whatever preparation procedure is used to produce these non-noble metal catalysts, any excess metal ions, which cannot be coordinated to the carbon support, will aggregate and form metal and/or metal carbide particles that will eventually be surrounded with a carbonaceous envelope during the heat treatment at high temperature. These surrounded particles have no catalytic activity for ORR and do not seem to contaminate with metal ions the membrane and the ionomer in the catalyst layer during fuel cell stability tests.

Since N are involved as coordinating atoms in both catalytic sites, it is clear that an increase of the catalytic site density, as required by the analysis previously done for a costless electrocatalyst to replace Pt in automotive applications, will be obtained by increasing the density of nitrogen atoms at the surface of the carbon supports prior to the adsorption of the metal precursor.

As far as the kinetics and mechanistic aspects of oxygen reduction on these non-noble metal electrocatalysts are concerned, it has been shown that these catalysts may reduce O_2 to water with an apparent number of electrons transferred, n, that may reach values very close to 4. This is especially true for Fe-based electrocatalysts made either from $Fe–N_4$ chelates or from cheaper Fe salt precursors. It seems also that the $Fe–N_2/C$ catalytic site, which is the most active site in catalysts obtained after a pyrolysis temperature $\geq$ 800°C, is characterized by a low release of peroxide. Co-based catalysts release, on average, more peroxide than the corresponding Fe-based materials. Studies that were undertaken to decouple the direct 4-electron reduction of oxygen to water from the successive 2×2-electron reduction indicate that the direct 4-electron reduction path may be important for these catalysts. This result is in agreement with the quantum theoretical approach of Anderson and Sidik[122] about a model of the pyrolyzed

Fe-based catalytic site, which concluded that, in the absence of oxygen, a water molecule forms a complex with the catalytic site with relatively strong binding interactions to the nitrogen lone pairs of that site. As O_2 is capable of displacing H_2O from Fe^{II}, but not from Fe^{III}, L_4Fe^{II}–OH_2 is therefore the starting point of the electrocatalytic reduction of oxygen on these catalytic sites. An important point to notice is that the peroxide in the L_4Fe^{II}–(OH–OH) intermediate remains bonded to one nitrogen lone pair of the catalytic site, preventing H_2O_2 from leaving and thereby providing a path for direct reaction to water on these Fe-based catalysts.

The stability of the non-noble metal-based catalysts remains the most elusive point, especially when Fe precursors are used. It was shown that the release of H_2O_2 is detrimental to the longevity of the catalytic sites, but there is a strong possibility in acid media that H_2O_2 would only be an intermediate to another much more oxidizing species: the HO• radical. This species may result from the reaction of free Fe^{2+} with H_2O_2 (Fenton's reagent). It is possible that long-term stability will remain the last challenge after the site density of these catalysts has been dramatically increased.

Acknowledgments

The author of this review is thankful to all research groups that he contacted in obtaining their permission to reproduce some of the figures of their work. He is also thankful to Dr. Frederick T. Wagner from General Motors Corporation Fuel Cell Activities, Prof. Alfred Anderson, Prof. Robert F. Savinell, and Dr. Reyimjan Sidik, all from Case Western Reserve University, as well as Prof. Roman Boulatov from University of Illinois for reading and commenting on this review. Help from Dr. Michel Lefèvre from INRS with the preparation of the figures of this review is also acknowledged.

References

1. Barlett, A.A. (July 2004). Thoughts on long-term energy supplies: Scientists and the silent lie. *Phys. Today*. 53–55.
2. Weisz, P.B. (July 2004). Basic choices and constraints on long-term energy supplies. *Phy. Today*. 47–52.
3. Petit, M. Qu'est ce que l'effet de serre? Ses conséquences sur l'avenir du climat, Vuibert, Paris, France, 2003.
4. Service, R.F. (2004). The carbon corrundum. *Science* **305**, 962–963.
5. Hileman, B. (December 15, 2003). Climate change. *Chem. Eng. News* 27–37.
6. Karl, T.R. and K.E. Trenberth. (2003). *Science* **302**, 1719–1723.
7. Petit, J.R., J. Jouzel, D. Raynaud, N.I. Barkov, J.M. Barnola, I. Basile, M. Benders, J. Chapellaz, M. Davis, G. Delaygue, M. Delmotte, V.M. Kotlyakov, M. Legrand, V.Y. Lipenkov, C. Lorius, L. Pépin, C. Ritz, E. Saltzman, and M. Stievenard. (1999). Climate and atmospheric history of the past 420,000 years from the Vostok ice core, Antarctica. *Nature* **399**, 429–436.
8. Service, R.F. (2004). The hydrogen backlash. *Science* **305**, 958–961.

9. Chow, J., R.J. Kopp, and P.R. Portney (2003). Energy resources and global development. *Science* **302**, 1528–1531.
10. Cropper, M.A.J., S. Geiger, D.M. Jollie (2004). Fuel cells: a survey of current development. *J. Power Sources* **131**, 57–61.
11. Gottesfeld, S. and T. Zawodzinski (1997). Polymer Electrolyte Fuel Cells. *Adv. Electrochem. Sci. Eng.* **5**, 195–301.
12. Larminie, J. and A. Dicks (2003). *Fuel Cell System Explained*, Second Edition, John Wiley & Sons, Inc., Hoboken, NJ, USA p. 72.
13. Gasteiger, H.A., S.S. Kocha, B. Sompalli, and F. Wagner (2005). Activity benchmark and requirements for Pt, Pt-alloy, and non-Pt oxygen reduction catalysts for PEMFCs. *Appl. Catal. B* **56**, 9–35.
14. Ralph, T.R. and M.P. Hogarth (2002). Catalysis for low temperature fuel cells, Part I: The cathode challenges. *Platinum Metals Rev.* **46**, 3–14.
15. Matthey, J. Platinumtoday; http://www.platinum.matthey.com.
16. Garland, N. DOE Non-Platinum Electrocatalyst Workshop Overview (28 Aug. 2003); http://www.eere.energy.gov/hydrogenandfuelcells/pdfs/garland_nancy.pdf
17. Jaffray, C. and G. Hards (2003). Precious metal supply requirements, *Handbook of Fuel Cells—Fundamentals, Technology and Applications*, W. Vielstich, H. Gasteiger, A. Lamm, eds., John Wiley & Sons, Chichester, UK Volume 3, Chapter 41, pp. 509–513.
18. Bassett, G.H. *Platinum* 2004, Johnson Matthey North America, New York 10016, USA, pp. 25–27.
19. Bassett, G.H. *Platinum* 2004, Johnson Matthey North America, New York 10016, USA, p. 3.
20. Markovic, N. (2003). The hydrogen electrode reaction and the electrooxidation of CO and H_2/CO mixtures on well characterized Pt and Pt-bimetallic surfaces, *Handbook of Fuel Cells – Fundamentals, Technology and Applications*, W. Vielstich, H. Gasteiger, A. Lamm, eds., John Wiley & Sons, Chichester, UK Volume 2, Chapter 26, pp. 368–393.
21. Gasteige, H.A., W. Gu, R. Makharia, M.F. Mathias, and B. Sompalli, Beginning-of-life MEA performance: Efficiency loss contributions, *Handbook of Fuel Cells – Fundamentals, Technology and Applications*, W. Vielstich, H. Gasteiger, A. Lamm, eds., John Wiley & Sons, Chichester, UK (2003) Volume 3, Chapter 46, 2003, pp. 593–610.
22. Gasteiger, H.A., J.E. Panels, and S.G. Yan (2004). Dependence of PEM fuel cell performance on catalyst loading. *J. Power Sources* **127**, 162–171.
23. Jasinski, R. (1964). A new fuel cell cathode catalyst. *Nature* **201**, 1212–1213.
24. Jahnke, H., M. Schönborn, G. Zimmermann (1976). Organic dyestuffs as catalysts for fuel cells. *Top. Cur. Chem.* **61**, 133–181.
25. Bagotzky, V.S., M.R. Tarasevich, K.A. Radyushkina, O.E. Levina, and S.I. Andrusyova (1977). Electrocatalysis of the oxygen reduction process on metal chelates in acid electrolyte. *J. Power Sources* **2**, 233–240.
26. Van der Putten, A., A. Elzing, W. Visscher, E. Barendrecht (1986). Oxygen reduction on pyrolyzed carbon supported transition metal chelates. *J. Electroanal. Chem.* **205**, 233–244.
27. Radyushkina, K.A. and M.R. Tarasevich (1986). Electrocatalytic properties of pyrolytic polymers produced from N_4-complexes (Review), Translated from *Elektrokhimiyia* **22**, 1155–1170.
28. van Veen, J.A.R., H.A. Colijn, and J.F. Van Baar (1988). On the effect of heat treatment on the structure of carbon-supported metalloporphyrins and phthalocyanines. *Electrochim. Acta* **33**, 801–804.

29. Wiesener, K. N_4-chelates as electrocatalysts for cathodic oxygen reduction (1986). *Electrochim. Acta* **31**, 1073–1078.
30. Tarasevich, M.R. and K.A. Radyuskina, Pyropolymers of N_4-complexes: structure and electrocatalytic properties (1989). *Mater. Chem. Phys.* **22**, 477–502.
31. Wiesener, K., D. Ohms, V. Neumann, and R. Franke (1989). N_4-Macrocycles as electrocatalysts for the reduction of oxygen, *Mater. Chem. Phys.* **22**, 457–475.
32. Vasudevan, P., S.N. Mann, and S. Tyagy (1990). Transition metal complexes of porphyrins and phthalocyanines as electrocatalysts for dioxygen reduction. *Transit. Met. Chem.* **15**, 81–90.
33. van Wingerden, B., J.A.R. Van Veen, and C.T.J. Mensch (1984). An extended X-ray absorption fine structure study of heat-treated cobalt porphyrin catalysts supported on active carbon. *J. Chem. Soc. Faraday Trans.* **1**, 84, 65–74.
34. van Veen, J.A.R., J.F. van Baar, and K.J. Kroese (1981). Effect of heat treatment on the performance of carbon-supported transition metal chelates in the electrochemical reduction of oxygen. *J. Chem. Soc. Faraday Trans.* **1**, 77, 2827–2843.
35. Scherson, D.A., S.L. Gupta, C. Fierro, E.B. Yeager, M.E. Kordesch, J. Eldridge, R.W. Hoffman, and J. Blue (1983). Cobalt tetramethoxyphenyl porphyrin emission Mössbauer spectroscopy and O_2 reduction electrochemical studies. *Electrochim. Acta* **28**, 1205–1209.
36. Yeager, E. (1984). Electrocatalysis for O_2 reduction. *Electrochim. Acta* **29**, 1527-1537.
37. Scherson, D. A.A. Tanaka, S.L. Gupta, D. Tryk, C. Fierro, R. Holze, and E.B. Yeager (1986). Transition metal macrocycles supported on high area carbon : Pyrolysis-mass spectrometry studies. *Electrochim. Acta* **31**, 1247–1258.
38. Franke, R., D. Ohms, and K. Wiesener (1989). Investigation of the influence of thermal treatment on the properties of carbon materials modified by N_4-chelates for the reduction of oxygen in acidic media. *J. Electroanal. Chem.* **260**, 63–73.
39. Gupta, S., D. Tryk, I. Bae, W. Aldred, and E. Yeager (1989). Heat-treated polyacrylonitrile-based catalysts for oxygen electroreduction. *J. Appl. Electrochem.* **19**, 19–27.
40. Gupta, S.L., D. Tryk, M. Daroux, W. Aldred, and E. Yeager (1986). Heat-treated pyrrole black-based electrocatalysts for oxygen reduction. In: D. Chin (ed), *Proceedings of the Symposium on "Load levelling and energy conservation in industrial processes"*. The Electrochemical Society, Pennington, N.J. 207–218 (1986).
41. Bouwkamp-Wijnoltz, A.L., W. Visscher, J.A.R. Van Veen, E. Boellaard, A.M. Van der Kraan, and S.C. Tang (2002). On active-site heterogeneity in pyrolyzed carbon supported iron porphyrin catalysts for the electrochemical reduction of oxygen: An in situ Mössbauer study. *J. Phys. Chem. B* **106**, 12993–13001.
42. McBreen, J., W.E. O'Grady, D.E. Sayers, C.Y. Yang, and K.I. Pandya (1987). An EXAFS study of pyrolyzed metal macrocyclic electrocatalysts. *Proc. Electrochem. Soc. 1987, 87–12 (Proc Symp. Electrode Mater. Processes Energy Convers. Storage; 2nd edition*, 182–197.
43. Tanaka, A., S.L. Gupta, D. Tryk, C. Fierro, E.B. Yeager, and D.A. Scherson (1992). Electrochemical and spectroscopic aspects of heat-treated transition metal macrocycles as electrocatalysts for oxygen reduction. In: D. Scherson, D. Tryk, M. Daroux, and X. Xing (eds), *Proceedings of the Symposium on "On structural effects in electrocatalysis and oxygen electrochemistry"*. The Electrochemical Society, Inc., Pennington, NJ, pp. 555–572.
44. Tryk, D.A., S.L. Gupta, W.H. Aldred, and E.B. Yeager, Approaches to methanol tolerant air cathodes for methanol-air fuel cells, *Proceedings of the Symposium on "On electrode materials and processes for energy conversion and storage"*, S. Srinivasan,

D.D. Macdonald, A.C. Khandkar, ed., The Electrochemical Society, Inc. Pennington, NJ (1994) 94–23, 294–301.
45. Bae, I.T., D.A. Tryk, and D.A. Scherson (1998). Effect of heat treatment on the redox properties of iron porphyrins adsorbed on high area carbon in acid electrolytes: An in situ Fe K-edge X-ray absorption near-edge structure study. *J. Phys. Chem. B* **102**, 4114–4117.
46. Biloul, A., F. Coowar, O. Contamin, G. Scarbeck, M. Savy, D. van de Ham, J. Riga, and J.J. Verbist (1990). Oxygen reduction in acid media on supported iron naphthalocyanine. *J. Electroanal. Chem.* **289**, 189–201.
47. Biloul, A., F. Coowar, O. Contamin, G. Scarbeck, M. Savy, D. van den Ham, J. Riga, and J.J. Verbist (1992). Oxygen reduction in acid media: Influence of the heat treatment on the FeNPc(1–2) isomer mixture impregnated on carbon blacks and active charcoals. *J. Electroanal. Chem.* **328**, 219–232.
48. Biloul, A., O. Contamin, G. Scarbeck, M. Savy, D. van den Ham, J. Riga, and J.J. Verbist (1992). Oxygen reduction in acid media: Effect of iron substitution by cobalt on heat-treated naphthalocyanine impregnations supported on preselected carbon blacks. *J. Electroanal. Chem.* **335**, 163–186.
49. Biloul, A., F. Coowar, O. Contamin, G. Scarbeck, M. Savy, D. Van den Ham, J. Riga, and J.J. Verbist (1993). Oxygen reduction in acid medium: Electrocatalysis by CoNPc(1,2) impregnated on carbon black support; effect of loading and heat treatment. *J. Electroanal. Chem.* **350**, 189–204.
50. Biloul, A., O. Contamin, G. Scarbeck, M. Savy, B. Palys, J. Riga, and J.J. Verbist (1994). Oxygen reduction in acid media: Influence of the activity of CoNPc(1,2) bilayer deposits in relation to their attachment to the carbon black support and role of surface groups as function of heat treatment. *J. Electroanal. Chem.* **365**, 239–246.
51. Biloul, A., O. Contamin, M. Savy, G. Scarbeck, J. Riga, and J.J. Verbist (1994). Oxygen reduction in acid media: Effect of heat-treatment on the activity of Co naphthalocyanine (2,3) isomer impregnations on carbon black and life test comparison with the (1,2) analogue isomer mixture. *J. Electroanal. Chem.* **379**, 321–328.
52. Collman, J.P., P. Denisevich, Y. Konai, M. Marrocco, C. Koval, and F.C. Anson (1980). Electrode catalysis of the four-electron reduction of oxygen to water by dicobalt face-to-face porphyrins. *J. Am. Chem. Soc.* **102**, 6027–6036.
53. Contamine, O., C. Debiemme-Chouvy, M. Savy, and G. Scarbeck (1999). Oxygen reduction catalysis : Effect of sulfur addition on cobalt tetraannulene precursors. *Electrochim. Acta* **45**, 721–729.
54. Contamin, O., C. Debiemme-Chouvy, M. Savy, and G. Scarbeck (2000). O_2 electroreduction catalysis: Effect of sulfur addition on some cobalt macrocycles. *J. New Mat. Electrochem. Systems* **3**, 67–74.
55. Claude, E., T. Adou, J.M. Latour, and P. Aldebert (1998). A new method for electrochemical screening based on the rotating disc electrode and its application to oxygen reduction catalysts. *J. Appl. Electrochem.* **28**, 57–64.
56. Gouerec, P., A. Biloul, O. Contamin, G. Scarbeck, M. Savy, J. M. Barbe, and R. Guilard (1995). Dioxygen reduction electrocatalysis in acid media: Effect of peripheral ligand substitution on cobalt tetraphenylporphyrin, *J. Electroanal. Chem.* **398**, 67–75.
57. Gouérec, P., A. Biloul, O. Contamin, G. Scarbeck, M. Savy, J. Riga, L.T. Weng, and P. Bertrand (1997). Oxygen reduction in acid media catalyzed by heat treated cobalt tetraazaannulene supported on an active charcoal: Correlations between the performance after longevity tests and the active site configuration as seen by XPS and ToF–SIMS. *J. Electroanal. Chem.* **422**, 61–75.

58. Biloul, A., P. Gouérec, M. Savy, G. Scarbeck, S. Besse, and J. Riga (1996). Oxygen electrocatalysis under fuel cell conditions: Behaviour of cobalt porphyrins and tetraazaannulene analogues. *J. Appl. Electrochem.* **26**, 1139–1146.
59. Gouérec, P. and M. Savy (1999). Oxygen reduction electrocatalysis: Ageing of pyrolyzed cobalt macrocycles dispersed on an active carbon. *Electrochim. Acta* **44**, 2653–2661.
60. Sawaguchi, T., T. Itabashi, T. Matsue, and I. Uchida (1990). Electrochemical reduction of oxygen by metalloporphyrin ion-complexes with heat-treatment. *J. Electroanal. Chem.* **279**, 219–230.
61. Okada, T., M. Gokita, M. Yusada, and I. Sekine (1999). Oxygen reduction characteristics of heat-treated catalysts based on cobalt–porphyrin ion complexes. *J. Electrochem. Soc.* **145**, 815–822.
62. Jiang, R. and D. Chu (2000). Remarkably active catalysts for the electroreduction of O_2 to H_2O for use in an acidic electrolyte containing concentrated methanol. *J. Electrochem. Soc.* **147**, 4605–4609.
63. Chu, D. and R. Jiang (2002). Novel electrocatalysts for direct methanol fuel cells. *Solid State Ionics* **148**, 591–599.
64. Gupta, S., D. Tryk, S.K. Zecevic, W. Aldred, D. Guo, and R.F. Savinell (1998). Methanol tolerant electrocatalysts for oxygen reduction in a polymer electrolyte membrane fuel cell. *J. Appl. Electrochem.* **28**, 673–682.
65. Sun, G.Q., J.T. Wang, S. Gupta, and R.F. Savinell (2001). Iron(III) tetramethoxyphenylporphyrin (FeTMPP-Cl) as electrocatalyst for oxygen reduction in direct methanol fuel cells. *J. Appl. Electrochem.* **31**, 1025–1031.
66. Sun, G.Q., J.T. Wang, and R.F. Savinell (1998). Iron(III) tetramethoxyphenylporphyrin (FeTMPP) as methanol tolerant electrocatalyst for oxygen reduction in direct methanol fuel cells. *J. Appl. Electrochem.* **28**, 1087–1093.
67. Gojkovic, S. Lj., S. Gupta, and R.F. Savinell (1998). Heat-treated iron(III) tetramethoxyphenyl porphyrin supported on high area carbon as an electrocatalyst for oxygen reduction, I. Characterization of the electrocatalyst. *J. Electrochem. Soc.* **145**, 3493–3499.
68. Gojkovic, S. Lj., S. Gupta, and R.F. Savinell (1999). Heat-treated iron(III) tetramethoxyphenyl porphyrin chloride supported on high area carbon as an electrocatalyst for oxygen reduction, Part II. Kinetics of oxygen reduction. *J. Electroanal. Chem* **462**, 63–72.
69. Gojkovic, S. Lj., S. Gupta, and R.F. Savinell (1999). Heat-treated iron(III) tetramethoxy porphyrin chloride supported on high area carbon as an electrocatalyst for oxygen reduction: Part III. Detection of hydrogen-peroxide during oxygen reduction. *Electrochim. Acta* **45**, 889–897.
70. Martins Alves, M.C., J.P. Dodelet, D. Guay, M. Ladouceur, and G. Tourillon (1992). Origin of the electrocatalytic properties for O_2 reduction of some heat-treated polyacrylonitrile and phthalocyanine cobalt compounds adsorbed on carbon black as probed by electrochemistry and X-ray absorption spectroscopy. *J. Phys. Chem.* **96**, 10898–10905.
71. Dignard-Bailey, L., M.L. Trudeau, A. Joly, R. Schulz, G. Lalande, D. Guay, and J.P. Dodelet (1994). Graphitization and particle size analysis of pyrolyzed cobalt phthalocyanine/carbon catalysts for oxygen reduction in fuel cells. *J. Mater. Res.* **9**, 3203–3209.
72. Ladouceur, M., G. Lalande, D. Guay, J.P. Dodelet, L. Dignard-Bailey, M. L. Trudeau, and R. Schulz (1993). Pyrolyzed cobalt phthalocyanine as electrocatalyst for oxygen reduction. *J. Electrochem. Soc.* **140**, 1974–1981.

73. Tamizhmani, G., J.P. Dodelet, D. Guay, G. Lalande, and G.A. Capuano (1994). Electrocatalytic activity of Nafion-impregnated pyrolyzed cobalt phthalocyanine. *J. Electrochem. Soc.* **141**, 41–45.
74. Weng, L.T., P. Bertrand, G. Lalande, D. Guay, and J.P. Dodelet (1995). Surface characterization of time of flight SIMS of a catalyst for oxygen electroreduction: pyrolyzed cobalt phthalocyanine-on-carbon black. *Appl. Surf. Sci* **84**, 9–21.
75. Lalande, G., G. Tamizmani, R. Côté, L. Dignard-Bailey, M.L. Trudeau, R. Schulz, D. Guay, and J.P. Dodelet (1995). Influence of loading on the activity and stability of heat-treated carbon-supported cobalt phthalocyanine electrocatalysts in solid polymer electrolyte fuel cells. *J. Electrochem. Soc.* **142**, 1162–1168.
76. Lalande, G., G. Faubert, R. Côté, D. Guay, J.P. Dodelet, L.T. Weng, and P. Bertrand (1996). Catalytic activity and stability of heat-treated iron phthalocyanines for the electroreduction of oxygen in polymer electrolyte fuel cells. *J. Power Sources* **61**, 227–237.
77. Lalande, G., R. Côté, G. Tamizhmani, D. Guay, J.P. Dodelet, L. Dignard-Bailey, L.T. Weng, and P. Bertrand (1995). Physical, chemical and electrochemical characterization of heat-treated tetracarboxylic cobalt phthalocyanine adsorbed on carbon black as electrocatalyst for oxygen reduction in polymer electrolyte fuel cells. *Electrochim. Acta* **40**, 2635–2646.
78. Faubert, G., G. Lalande, R. Côté, D. Guay, J.P. Dodelet, L.T. Weng, P. Bertrand, and G. Dénès (1996). Heat-treated iron and cobalt tetraphenylporphyrins adsorbed on carbon black: Physical characterization and catalytic properties of these materials for the reduction of oxygen in polymer electrolyte fuel cells. *Electrochim. Acta* **41**, 1689–1701.
79. Faubert, G., R. Côté, D. Guay, J.P. Dodelet, G. Dénès, and P. Bertrand (1998). Iron catalysts prepared by high-temperature pyrolysis of tetraphenylporphyrins adsorbed on carbon black for oxygen reduction in polymer electrolyte fuel cells. *Electrochim. Acta* **43**, 341–353.
80. Moisala, A., A.G. Nasibulin, and E.I. Kauppinen (2003). The role of metal nanoparticles in the catalytic production of single-walled carbon nanotubes—a review. *J. Phys.: Condens. Matter* **15**, S3011–S3035.
81. Blomquist, J., H. Lang, R. Larson, and A. Widelöv (1992). Pyrolysis behaviour of metalloporphyrins, Part 2—A Mössbauer study of pyrolyzed Fe^{III} tetraphenylporphyrin chloride. *J. Chem. Soc. Faraday Trans.* **88**, 2007–2011.
82. Pierson, H.O. (1996) *Handbook of Refractory Carbides and Nitrides, Properties, Characteristics, Processing and Applications.* Noyes Publications, Westwood, New Jersey, U.S.A. p. 15.
83. Tilquin, J.Y., J. Fournier, D. Guay, J.P. Dodelet, and G. Dénès (1997). Intercalation of $CoCl_2$ into graphite: Mixing method vs. molten salt method. *Carbon* **35**, 299–306.
84. Lalande, G., R. Côté, D. Guay, J.P. Dodelet, L.T. Wang, and P. Bertrand (1997). Is nitrogen important in the formulation of Fe-based catalysts for oxygen reduction in solid polymer fuel cells? *Electrochim. Acta* **42**, 1379–1388.
85. Ohms, D., S. Herzog, R. Franke, V. Neumann, K. Wiesener, S. Gamburcev, A. Kaisheva, and I. Iliev (1992). Influence of metal ions on the electrocatalytic oxygen reduction of carbon materials prepared from pyrolyzed polyacrylonitrile. *J. Power Sources* **38**, 327–334.
86. Ye, S. and A.K. Vijh (2003). Non-noble metal-carbonized aerogel composites as electrocatalysts for the oxygen reduction reaction, *Electrochem. Comm.* **5**, 272–275.
87. Fabjan, Ch. G. Frithum, and H. Hartl (1990). On the electrocatalysis of the oxygen reduction in acid electrolyte. *Ber. Bunsenges. Phys. Chem.* **94**, 937–941.
88. Seeliger, W. and A. Hamnett (1992). Novel electrocatalysts for oxygen reduction. *Electrochim. Acta* **37**, 763–765.

89. Bouwkamp-Wijnoltz, A.L., W. Visscher, J.A.R. Van Veen, and S.C. Tang (1999). Electrochemical reduction of oxygen: An alternative method to prepare active CoN_4 catalysts. *Electrochim. Acta* **45**, 379–386.
90. Sawai, K. and N. Suzuki (2004). Heat-treated transition metal hexacyanometallates as electrocatalysts for oxygen reduction insensitive to methanol. *J. Electrochem. Soc.* **151**, A682–A688.
91. Okada, T., S. Gotou, M. Yoshida, M. Yuasa, T. Hirose, and I. Sekine (1999). A comparative study of organic cobalt complex catalyst for oxygen reduction in polymer electrolyte fuel cells. *J. Inorg. Organometallic Polymers* **9**, 199–219 (1999).
92. Lalande, G., D. Guay, J.P. Dodelet, S.A. Majetich, and M.E. McHenry (1997). Electroreduction of oxygen in polymer electrolyte fuel cells by activated carbon coated cobalt nanocrystallites produced by electric arc discharge. *Chem. Mater.* **9**, 784–790.
93. Fournier, J., G. Lalande, R. Côté, D. Guay, and J.P. Dodelet (1997). Activation of various Fe-based precursors on carbon black and graphite supports to obtain catalysts for the reduction of oxygen in fuel cells. *J. Electrochem. Soc* **144**, 218–226.
94. Côté, R., G. Lalande, D. Guay, J.P. Dodelet, and G. Dénès (1998). Influence of nitrogen-containing precursors on the electrocatalytic activity of heat-treated $Fe(OH)_2$ on carbon black for O_2 reduction. *Electrochem. Solid State Lett.* **145**, 2411–2418.
95. Côté, R., G. Lalande, G. Faubert, D. Guay, J.P. Dodelet, and G. Dénès (1998). Non-noble metal-based catalysts for the reduction of oxygen in polymer electrolyte fuel cells. *J. New Mat. Electrochem. Systems* **1**, 7–16.
96. Faubert, G., R., Côté, D. Guay, J.P. Dodelet, G. Dénès, C. Poleunis, and P. Bertrand (1998). Activation and characterization of Fe-based catalysts for the reduction of oxygen in polymer electrolyte fuel cells. *Electrochim. Acta* **43**, 1969–1984.
97. Wang, H., R. Côté, G. Faubert, D. Guay, and J.P. Dodelet (1999). Effect of the pretreatment of carbon black supports on the activity of Fe-based electrocatalysts for the reduction of oxygen. *J. Phys. Chem. B* **103**, 2042–2049.
98. Wei, G., J.S. Wainright, and R.F. Savinell (2000). Catalytic activity for oxygen reduction reaction of catalysts consisting of carbon, nitrogen and cobalt. *J. New Mat. Electrochem. Systems* **3**, 121–129.
99. Faubert, G., R. Côté, J.P. Dodelet, M. Lefèvre, and P. Bertrand (1999). Oxygen reduction catalysts for polymer electrolyte fuel cells from the pyrolysis of Fe^{II} acetate adsorbed on 3,4,9,10-perylenetetracarboxylic dianhydride. *Electrochim. Acta* **44**, 2589–2603.
100. Lefèvre, M., J.P. Dodelet, and P. Bertrand (2000). O_2 reduction in PEM fuel cells: Activity and active site structural information for catalysts obtained by the pyrolysis of high temperature of Fe precursors. *J. Phys. Chem. B* **104**, 11238–11247.
101. Murakami, M., S. Iijima, and S. Yoshimura (1986). Morphology and structure of a one dimensional graphite polymer, poly-per-naphthalene. *J. Appl. Phys.* **60**, 3856–3863.
102. He, P., M. Lefèvre, G. Faubert, and J.P. Dodelet (1999). Oxygen reduction catalysts for polymer electrolyte fuel cells from the pyrolysis of various transition metal acetates adsorbed on 3,4,9,10-perylenetetracarboxylic dianhydride. *J. New Mat. Electrochem. Systems* **2**, 243–251.
103. Casanovas, J., J. Manel Ricart, J. Rubio, F. Illas, and J.M. Jiménez-Mateos (1996). Origin of the large N1s binding energy in X-ray photoelectron spectra of calcined carbonaceous materials. *J. Am. Chem. Soc.* **118**, 8071–8076.
104. Jiménez-Mateos, J.M. and J.L.G. Fierro (1996). X-ray photoelectron spectroscopic study of petroleum fuel cokes. *Surf. Interface Anal.* **24**, 223–236.

105. Lefèvre, M., J.P. Dodelet, and P. Bertrand (2002). Molecular oxygen reduction in PEM fuel cells: Evidence for the simultaneous presence of two active sites in Fe-based catalysts. *J. Phys. Chem. B* **106**, 8705–8713.
106. Bron, M., S. Fiechter, M. Hilgendorff, and P. Bogdanoff (2002). Catalysts for oxygen reduction from heat-treated carbon-supported iron phenanthroline complexes. *J. Appl. Electrochem.* **32**, 211–216.
107. Bron, M., J. Radnik, M. Fieber-Erdmann, P. Bogdanoff, and S. Fiechter (2002). EXAFS, XPS and electrochemical studies on oxygen reduction catalysts obtained by heat-treatment of iron phenanthroline complexes supported on high surface area carbon black. *J. Electroanal. Chem.* **535**, 113–119.
108. Bron, M., S. Fiechter, P. Bogdanoff, and H. Tributsch. (2002). Thermogravimetry/mass spectrometry investigations on the formation of oxygen reduction catalysts for PEM fuel cells on the basis of heat-treated iron phenanthroline complexes. *Fuel Cells* **2**, 127–142.
109. Schulenburg, H., S. Stankov, V. Schünemann, J. Radnik, I. Dorbandt, S. Fiechter, P. Bogdanoff, and H. Tributsch (2003). Catalysts for the oxygen reduction from heat-treated iron (III) tetramethoxyphenylporphyrin chloride: Structure and stability of active sites. *J. Phys. Chem. B* **107**, 9034–9041.
110. Zhutaeva, G.V., K.A. Radyushkina, M.A. Marinich, G.P. Bogatyreva, and M.R. Tarasevich (2001). Electrocatalysis of the oxygen reaction on electrodes prepared using disperse synthetic diamond promoted with cobalt porphyrin and its pyropolymer. *Russ. J. Electrochem.* **37**, 1059–1064.
111. Jaouen, F., S. Marcotte, J.P. Dodelet, and G. Lindbergh (2003). Oxygen reduction catalysts for polymer electrolyte fuel cells from the pyrolysis of iron acetate adsorbed on various carbon supports. *J. Phys. Chem. B* **107**, 1376–1386.
112. Stöhr, B., H.P. Boehm, and R. Schlögl (1991). Enhancement of the catalytic activity of activated carbons in oxidation reactions by thermal treatment with ammonia or hydrogen cyanide and observation of a superoxide species as a possible intermediate. *Carbon* **29**, 707–720.
113. Villers, D., X. Jacques-Bédard, and J.P. Dodelet (2004). Fe-based catalysts for oxygen reduction in PEM fuel cells: Pretreatment of the carbon support. *J. Electrochem. Soc.* **151**, A1507–A1515.
114. Lefèvre, M., and J.P. Dodelet (2003). Fe-based catalysts for the reduction of oxygen in polymer electrolyte membrane fuel cell conditions: Determination of the amount of peroxide released during electroreduction and its influence on the stability of the catalysts. *Electrochim. Acta* **48**, 2749–2760.
115. Bogdanoff, P., I. Herrmann, M. Hilgendorff, I. Dorbrandt, S. Fiechter, and H. Tributsch (2004). Probing structural effects of pyrolyzed CoTMPP-based electrocatalysts for oxygen reduction via new preparation strategies. *J. New. Mat. Electrochem. Syst.* **7**, 85–92.
116. Gouérec, P., M. Savy, and J. Riga (1998). Oxygen reduction in acidic media catalyzed by pyrolyzed cobalt macrocycles dispersed on an active carbon: The importance of the content of oxygen surface groups on the evolution of the chelate structure during the heat treatment. *Electrochim. Acta* **43**, 743–753.
117. Tarasevich, M.R., L.A. Beketaeva, B.N. Efremov, N. M. Zagudaeva, L.N. Kuznetsova, K.V. Rybalka, and V.E. Sosenkin (2004). Electrochemical properties of carbon black AD-100 and AD-100 promoted with pyropolymer of cobalt tetra (*p*-methoxyphenyl)porphyrin. *Russ. J. Electrochem.* **40**, 542–551.
118. Bouwkamp-Wijnoltz, A.L., W. Visscher, and J.A.R. Van Veen. (1998). The selectivity of oxygen reduction by pyrolysed iron porphyrin supported on carbon. *Electrochim. Acta* **43**, 3141–3152.

119. Marcotte, S., D. Villers, N. Guillet, L. Roué, and J.P. Dodelet (2004). Electroreduction of oxygen on Co-based catalysts: Determination of the parameters affecting the two-electron transfer reaction in an acid medium. *Electrochim. Acta* **50**, 179–188.
120. Zagal, J.H. (1992). Metallophthalocyanines as catalysts in electrochemical reactions. *Coord. Chem. Revs.* **119**, 89–136.
121. Yeager, E. (1976). Mechanism of electrochemical reactions on non-metallic surfaces. *Electrocatalysis of Non Metallic Surfaces*, NBS Publications, Vol. 455, pp. 203–219.
122. Anderson, A.B., and R.A. Sidik (2004). Oxygen electroreduction on $Fe^{II}$and Fe^{III} coordinated to N_4 chelates. Reversible potentials for the intermediate steps from quantum theory. *J. Phys. Chem. B* **108**, 5031–5035.
123. Van Veen, J.A.R., J.F. Van Baar, C.J. Kroese, J.G.F. Coolegem, N. De Wit, and H.A. Colijn (1981). Oxygen reduction on transition metal porphyrins in acid electrolyte: I. Activity. *Ber. Bunsenges. Phys. Chem.* **85**, 693–700.
124. Tyurin, V.S., K.A. Radyushkina, O.A. Levina, and M.R. Tarasevich (2001). Electrocatalytic properties of a composite based on cobalt porphyrin pyropolymer and Nafion. *Russ. J. Electrochem.* **37**, 843–847.
125. Walling, C. (1975). Fenton's reagent revisited, *Acc. Chem. Res.* **8**, 125–131.
126. Wiesener, K. (1982). Electrocatalysts based on N_4 organic chelates for the cathodic reduction of oxygen in acidic solutions. Translated from *Élektrokhimiyia* **18**, 758–765.
127. Gambutrtsev, S., I. Iliev, and A. Kaisheva (1983). Influence of the pyrolysis products of cobalt tetramethoxyphenylporphyrin in activated charcoal on the behavior of hydrophobic carbon air electrodes during prolonged operation. Translated from *Élektrokhimiya* **19**, 1261–1264.
128. Okada, T., M. Yoshida, T. Hirose, K. Kasuga, T. Yu, M. Yuasa, and I. Sekine (2000). Oxygen reduction characteristics of graphite electrodes modified with cobalt diquinolyldiamine derivatives. *Electrochim. Acta* **45**, 4419–4428.

4

Biomimetic NO_x Reductions by Heme Models and Proteins

Emek Blair, Filip Sulc, and Patrick J. Farmer

1. Introduction

Heme cofactors are widely used in nature to catalyze multielectron reductions. As illustrated in Scheme 4.1, these include the two-electron activation of O_2 by the cytochrome P450s[1], the four-electron reduction of O_2 to H_2O by cytochrome oxidase[2], the six-electron reductions of NO_2^- and SO_3^{2-} by the nitrite and sulfite reductases[3]. All of these processes involve reductive cleavage of substrate-oxygen bonds from Fe-bound intermediates, a reactivity that is inherent to a heme or metalloporphyrin. The protein matrix controls access of substrates and electrons to the active site, and influences reaction pathways by altering the stability of transient intermediates. Most often, in these enzymatic systems, the required electrons are delivered to the active site from single-electron donors or associated redox cofactors. In the last two decades, electrochemical research has provided significant insights into these enzymatic transformations using simple N4-chelates such as porphyrins, phthalocyanines, heme protein substitutes as well as the enzymes themselves. Electrochemistry allows a way to directly test

$O_2 + 4e^- + 4H^+ \longrightarrow 2H_2O$	Cytochrome oxidase
$RH + O_2 + 2e^- + 2H^+ \longrightarrow ROH + H_2O$	Cytochrome P450
$NO_2^- + 6e^- + 8H^+ \longrightarrow NH_4^+ + 2H_2O$	Nitrite reductase
$SO_3^{2-} + 6e^- + 8H^+ \longrightarrow H_2S + 3H_2O$	Sulfite reductase
$NO_2^- \xrightarrow[\textbf{\textit{NiR}}]{e^-/2H^+} NO \xrightarrow[\textbf{\textit{NoR}}]{2e^-/2H^+} N_2O \xrightarrow[\textbf{\textit{No}}_2\textbf{\textit{R}}]{2e^-/2H^+} N_2$	Denitrification

Scheme 4.1.

Emek Blair, Filip Sulc, and Patrick J. Farmer • Department of Chemistry, University of California, Irvine, CA.

N_4-Macrocyclic Metal Complexes, edited by José H. Zagal, Fethi Bedioui and Jean-Pol Dodelet. Springer Science+Business Media, Inc., New York, 2006.

the stepwise sequences of these transformations, by linking current and potential to the generation of intermediates or products, or by coupling structural changes with catalytic abilities.

Our focus in this review is on the reactivity associated with the NO_x reductases, which perform several steps in the biochemical cycling of nitrogen, Figure 4.1[4–9]. These enzymes play key roles in the uptake and loss of agricultural fertilizers as well as the generation of atmospheric N_2O, a potent greenhouse gas[10–13]. Two distinct reactivities are observed: assimilatory nitrite reductases (aNiR) convert nitrite to ammonia for incorporation into biomass; while dissimilatory denitrifying nitrite reductases (dNiR) take part in multienzyme denitrification processes yielding the N–N coupled gases N_2O and N_2, functioning as important terminal electron acceptors during anaerobic respiration[3,14]. Three different porphyrinoid structures are known in NiRs: isobacteriochlorin (siroheme), dioxoisobacteriochlorin (heme d_1), and the more common protoporphyrin (hemin), Scheme 4.2[7]. All these reductases have associated redox sites such as Fe_4S_4 clusters or other Fe–heme sites which aid in the reduction of NO_x substrates.

The complicated redox chemistry of NO_x species is both a blessing and a curse in this endeavor, as is illustrated in Figure 4.1. Unlike SO_x chemistry, many NO_x species are stable and may be used directly as substrates or analyzed for as products; these include nitrate, nitrite, hydroxylamine, and ammonia and the gases nitric oxide, nitrous oxide, and dinitrogen. Still, there remains much confusion about NO_x chemistry, for example the potential of the simple one-electron reduction of nitric oxide (NO) to the nitroxyl (NO^-) state has been widely misrepresented, a recent re-evaluation has set it at *ca.* −800 mV NHE, a revision of more than one volt from previously cited values[15]. Likewise, the pK_a of the

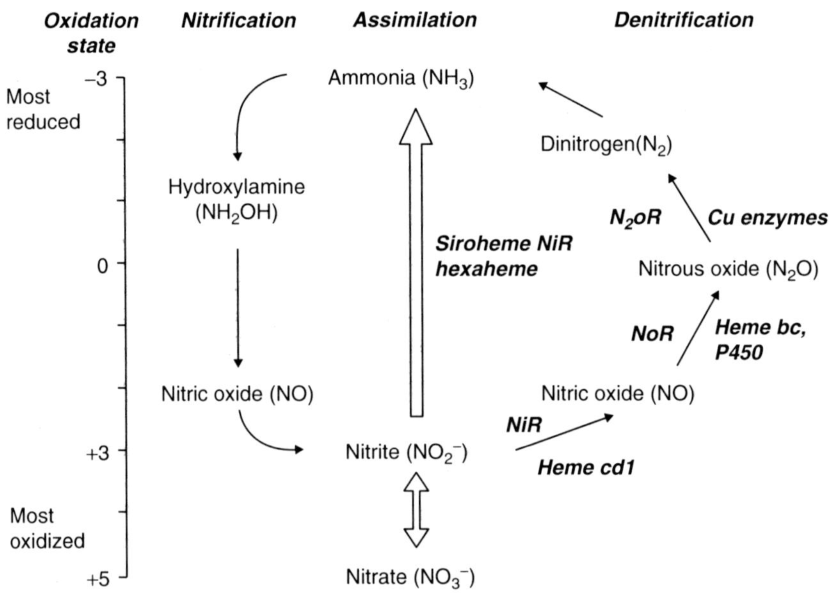

Figure 4.1. Formal nitrogen oxidation states of NO_x species, and the biological pathways for their formations and conversions.

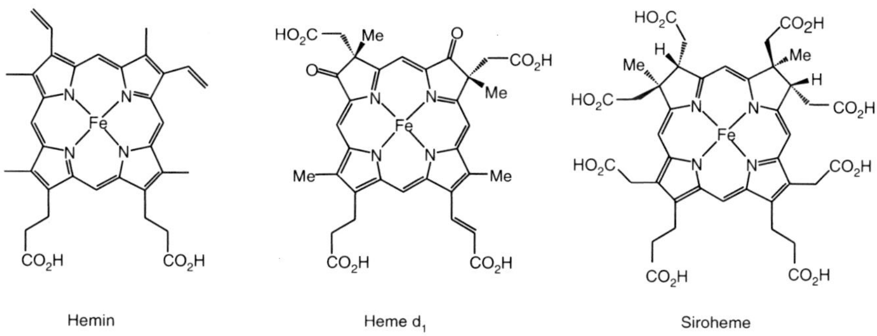

Scheme 4.2.

reduced product, HNO, was previously reported as 4.7, but has now been revised to above 11.5.

Several different approaches to protein voltammetry have been developed, and there are a number of reviews on the subject[16–23]. One widely applied method is protein film voltammetry (PFV), as developed by Armstrong, in which the protein is adsorbed as a monolayer at low temperatures, allowing direct electrochemical measurements[17]. This is a powerful method, with which the kinetics of intraprotein electron transfers can be analyzed, but the signal response is low and protein concentrations are often limiting. An alternative methodology developed by Rusling allows the direct electrochemistry of proteins contained in fluid surfactant films[23]. An advantage is that the current response is greatly enhanced over that of a monolayer and a minimum of protein may be used to greatest effect. As these methods are sometimes incompatible with the enzymes of interest, a more traditional methodology using redox mediators to drive the catalytic activity in bulk solutions can be applied; these mediators may be small molecules (e.g. methyl viologen) or small redox proteins (e.g. ferredoxins) that undergo facile electron exchange with both the electrode surface and the enzyme itself[24].

As our intent in this review is to address questions of NO_x reductions at heme active sites from a mechanistic inorganic perspective, the use of simple heme model compounds is quite useful and will be thoroughly described. Here similar issues of solubility and reactivity come into play, as biomimetic NO_x catalysis requires either water-soluble porphyrins or surface-adsorbed catalysts. As might be expected, questions of intersite reactivity come into play both in the reactivity of transient intermediate species such as nitroxyl-adducts, and in multi-electron reactivity required for assimilatory reductase activity. We will begin with brief descriptions of the known families of heme-based NO_x reductases and the all-too-few electrochemical investigations of these native enzymes.

2. Native Enzymes for NO_x Reductions

2.1. Assimilatory Nitrite Reductases

The siroheme aNiR, a crucial component of plant metabolism, contains an isobacteriochlorin cofactor[3,25,26], Scheme 4.3, which is tightly coupled to a

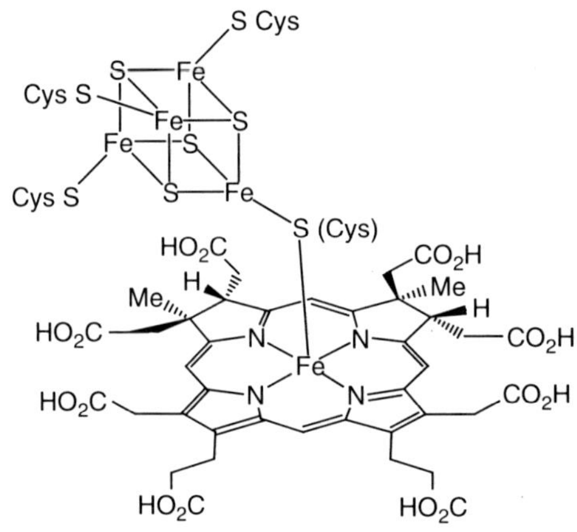

Scheme 4.3.

4Fe–4S cluster. An identical active site is found in the siroheme sulfite reductases (SiR) that catalyze the six-electron reduction of SO_3^{2-} to H_2S[27]. The similarity of the enzymes extends to their activity, SiR will catalyze nitrite reduction and NiR will catalyze sulfite reduction; the main difference being a greater affinity (by several orders of magnitude) for each enzyme's specific substrate[28].

It has been shown that NO_x substrates bind to the Fe(siroheme), and suggested that the 4Fe–4S cluster serves as a reservoir of reducing equivalents[29], but the pathway of the overall six electron reduction remains unclear. The required electrons are delivered from the single-electron donor ferredoxin, leading to the necessity of long-lived, nonlabile intermediates during catalysis. Both the siroheme and the 4Fe–4S cluster are redox active, and the electronic coupling of these two redox active cofactors in both NiR and SiR is well established by EPR and Mossbauer studies[28,30–32]. Cowan has reported voltammetric measurements on ferredoxidin:siroheme enzymes, but none as of yet has followed the electrochemistry during turnover[33,34]; nevertheless, steady-state measurements have shown that the siroheme cofactor itself is as competent and fast a catalyst as the holoenzyme[35].

In a study of the reduction of nitrite by the SiR from *Desulfovibrio vulgaris*, Cowan and coworkers proposed sequential two electron cleavages of the N–O bonds, Figure 4.2[34,36,37]. One reducing equivalent is envisioned to come from the $Fe^{III/II}$ couple of the Fe–heme itself, and one from the $Fe_4S_4^{II/I}$ cluster, Scheme 4.3. Thus the active site is "primed" in the reduced state to perform two-electron reductions of substrates[37]. As described, the ferric-nitrosyl intermediate (Fe^{III}–NO) obtained from dehydration of the ferrous nitrite adduct is further reduced by the $Fe_4S_4^{+}$ cluster to a ferric-nitroxyl, Fe^{III}–NO^{-}, which may more reasonably be formulated as a ferrous-nitrosyl, Fe^{II}–NO. A second proton-coupled two-electron reduction generates a nitrosyl hydride intermediate, Fe^{II}–HNO, with an oxidized $Fe_4S_4^{2+}$ cluster. Subsequent protonation of this species engenders an electronic reorganization, yielding a ferric complex of a hydroxylamine anion, formally the product of a two-electron reduction of HNO.

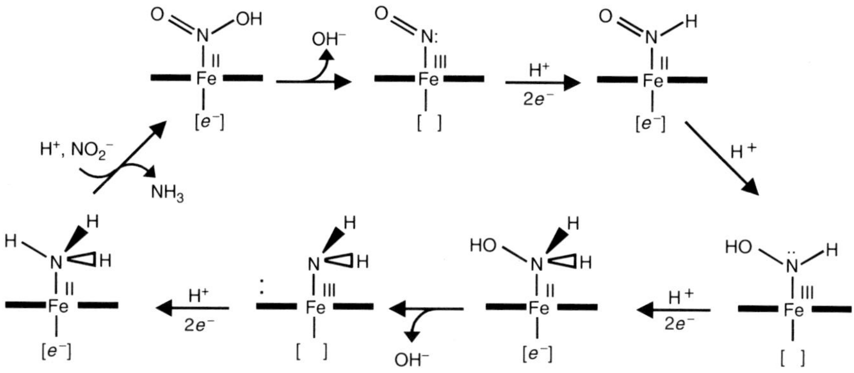

Figure 4.2. Proposed mechanism for siroheme NiR catalytic cycle[37].

A very similar catalytic cycle has been proposed for the cytochrome *c* nitrite reductases, ccNiR, which serve in some anaerobes as terminal electron acceptors[38]. The ccNiR has five type *c* hemes per monomer and a lysine-coordinated heme at the active site. Potentiometric titrations showed enzyme's unique lysine-coordinated heme, labeled heme 1, has an $Fe^{III/II}$ reduction potential at −107 mV. The other hemes with bis-histidine coordination, termed hemes 2 through 5, have reduction potentials between −37 and −323 mV. Tentative assignments of each heme's potentials have been proposed using EPR titration data correlated with a recently determined structure[39].

Butt and coworkers have reported electroanalytical experiments on the cytochrome *c* nitrite reductase (ccNiR) from *Escherichia coli*, NrfA[39]. The catalysis of nitrite and hydroxylamine reduction was investigated using PFV, in which the enzyme is directly deposited as a monolayer on freshly cleaned pyrolytic graphite edge (PGE) electrode. This technique allows for observation of the enzymatic catalytic currents, but could not observe the noncatalytic reduction couples of individual hemes. Therefore, the absolute turnover could only be estimated by assuming partial monolayer enzyme coverage. Assigning specific sequence and roles for the different heme sites is problematic because of their overlapping potentials, as well as the likelihood that these will change upon substrate binding, but some strong correlations were observed. Rotating disc voltammetry showed substrate-limited current for catalytic nitrite reduction; during Nerstian titrations the two substrates, nitrite and hydroxylamine, displayed different voltammetric fingerprints, seen in Figure 4.3. The first derivative graphs of catalytic current vs. potential were used to analyze catalytic potentials and number of electrons involved. The onset of catalytic currents and the slope of the current peaks were then interpreted in terms of rate-limiting steps under specific conditions.

Under nitrite-limited conditions, an increased catalytic activity occurred at −105 mV, close to the reduction potentials of heme 1 and 3 (*ca.* −107 mV) and decreased *ca.* −320 mV, possibly due to a conformational change caused by reduction of hemes 4 and/or 5 (*ca.* −323 mV), or inhibition due to ferrous nitrosyl formation[40]. As the concentration of nitrite increased, the catalytic peak broadened and moved toward a lower potential; under these enzyme-limited conditions,

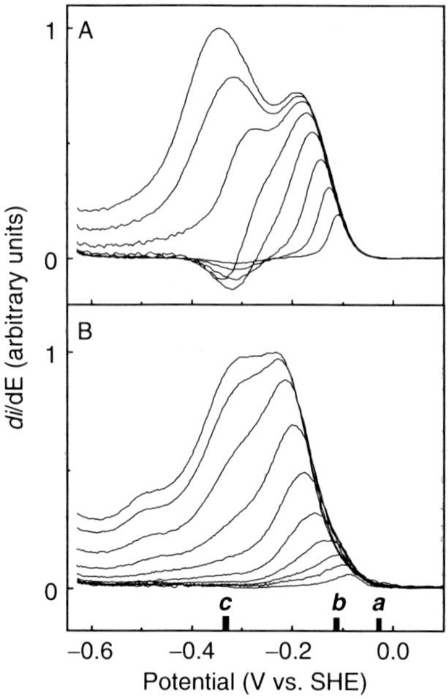

Figure 4.3. First-order derivatives of RDE voltammograms, demonstrating catalytic reduction of nitrite and hydroxylamine by the five-heme ccNiR, NrfA.[39] (A) Catalytic current derivatives for nitrite concentrations from 1.7 to 265 μM. (B) Same for hydroxylamine concentrations from 1.1 to 374 μM. Experiments were performed in 2 mM CaC12, 50 mM Hepes, pH 7.0, at 20°C. Approximate potentials of the five hemes are marked on the *x*-axis: (a) heme 2(b) heme 1 and 3; (c) heme 4 and 5.

activity increased at *ca.* −320 mV with the appearance of a second catalytic peak. These voltammetric currents describe an initial rate-limiting two electron reduction, generating the ferrous nitrosyl, and a subsequent potential-limited single electron reduction, which would then generate a nitroxyl substrate state. Hydroxylamine reductions showed much lower activity, and were not substrate limited. Voltammetric current analysis suggests that upon reduction of heme 1 and 3, the hydroxylamine is bound and reduced by two electrons. This two-electron reduction also likely occurs during nitrite reduction, but only after four initial electron equivalents. As the concentration is increased, there is no observed decrease in activity and no loss of activity is seen if the film is kept at negative potentials, which was seen for nitrite reductions. The kinetics of nitrite and hydroxylamine reduction by enzyme film were also compared to parallel solution studies. The overall rate of electrochemical nitrite reduction showed good agreement with rates found in spectrophotometric assays, but hydroxylamine reduction proved to be faster electrochemically than in solution assays[39].

Recently, several substrate-bound intermediates for ccNiR from *Wolinella succinogenes*, including the nitrite and hydroxylamine adducts, have been structurally characterized by Kronek and coworkers[41–43]. Their proposed mechanism points out an inherent problem associated with generating a ferrous nitroxyl intermediate, Fe^{II}–NO^-, by sequential electron transfers, Scheme 4.4. If transiently

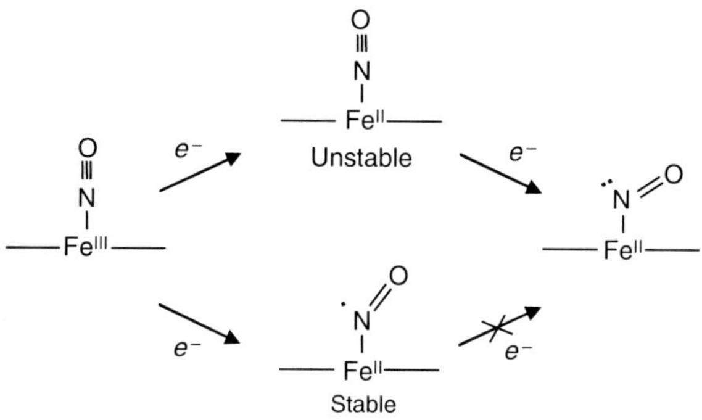

Scheme 4.4.

formed, the thermostable ferrous nitrosyl, Fe^{II}–NO, may likely act as a kinetic dead end, inhibiting catalytic turnover. The authors suggest that the Fe^{II}–NO^- state may be attained by two rapid single electron reductions of the linear Fe^{III}–NO, such that both occur before the nuclear rearrangement to the stable bent form of Fe^{II}–NO. A more reasonable alternative was suggested in a computational study which found that the amino ligation in ccNiR destabilizes the nitric oxide adduct relative to the nitroxyl, making the second reduction more kinetically feasiblc[44].

2.2. Dissimilatory Nitrite Reductases

Similar questions of the thermostability of the ferrous nitrosyl have arisen in mechanistic studies of the cytochrome cd_1 NiR from denitrifying bacteria, which catalyze the single-electron reduction of nitrite to nitric oxide[3,45]. These bacterial proteins contain a Fe-dioxoisobacteriochlorin, heme d_1, as shown in Figure 4.1, as well as a covalently bound heme *c* group that acts as an electron reservoir[46]. The electrons that are shuttled to NiR come from physiological electron donor *Pseudomonas aeruginosa*, with a heme *c* site, or from azurin during cellular stress, with type I copper active site. Recent structural studies of the enzyme in different oxidation and ligation states have divulged apparent ligation switching during the catalytic cycle at both the heme *c* and d_1 centers[47,48].

In the 1990s, Averill described the putative intermediate of the initial steps in nitrite reduction, a ferric nitrosyl, as characterized by FTIR[49]. Recent work has identified this ferric species, using time-resolved stop-flow experiments, that is in apparent equilibrium with a ferrous nitrosyl, previously thought to be a catalytic "dead end"[50]. Both the ferric and ferrous nitrosyl adducts are relatively long-lived, and no study has thus far demonstrated catalytically competent NO release from cytochrome cd_1. Therefore, certain mechanistic aspects remain obscure.

Lojou and coworkers have investigated catalytic nitrite reduction by cytochrome cd_1 NiR using cyclic voltammetry[51]. For this enzyme, direct electrochemistry was not possible and superimposition of absorption spectra made spectroelectrochemical analysis difficult; but catalysis could be driven

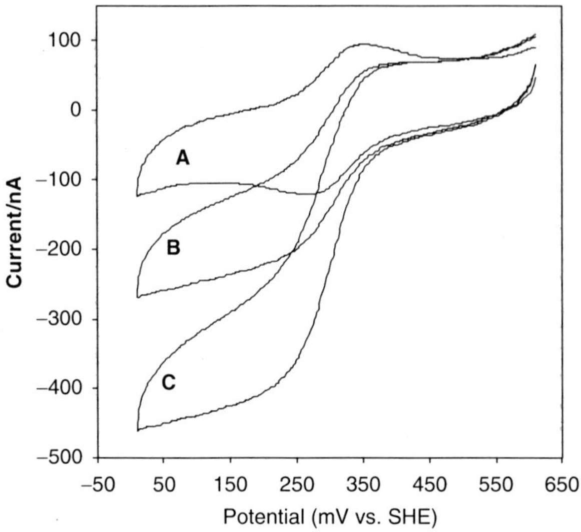

Figure 4.4. Electrochemical response of 38 μM cytochrome *c*551 in 5 mM $NaNO_2$pH 6.5, with (A) no, (B) 0.25 μM, and (C) 1.5μ M of cytochrome cd_1 NiR. Data adapted from ref. [51].

using electrochemical mediators such as cytochrome *c*551 or azurin, Figure 4.4. Direct electrochemistry of both cytochrome *c*551 and azurin, at pH 6.3 in 50 mM HEPES buffer, showed quasireversible behavior with midpoint potentials of 305 and 343 mV (SHE), respectively. When nitrite was added to a solution of cd_1 NiR with cytochrome *c*551, a catalytic current appeared.

$$NO_2^- + H^+ + e^- \rightarrow NO + H_2O \quad (4.1)$$

$$\text{Cyt } c_{551\text{ox}} + e^- \leftrightharpoons \text{Cyt } c_{551\text{red}} \quad (4.2)$$

$$\text{Cyt}_{c551\text{red}} + \text{NiR}_{\text{ox}} \leftrightharpoons \text{Cyt}_{c551\text{ox}} + \text{NiR}_{\text{red}} \quad (4.3)$$

$$\text{NiR}_{\text{red}} + NO_2^- + 2H^+ \rightarrow \text{NiR}_{\text{ox}} + NO + H_2O \quad (4.4)$$

The obtained catalytic currents under varied conditions of substrate, enzyme and mediator were then modeled using the reaction sequence described by Equation (4.1)–(4.4). Based on the Nicholson and Shain analysis, a second order rate constant of 2×10^6 M^{-1} s^{-1} was determined for reduction of cd_1 NiR by its physiological electron donor cytochrome *c*551, and a smaller constant, 1.4×10^5 M^{-1} s^{-1}, for azurin which is its electron donor under stress. Nitrite reductase activity was independent of ionic strength but highly pH dependent, with maximum activity occurred at pH 6.5. The lowered activity at more acidic pH was attributed to an acid form of cytochrome *c*551, which had a less efficient intermolecular electron transfer with cd_1 NiR. At higher pH, it was suggested that the formation of the highly stable NO-ferrous heme adduct inhibits catalysis, as was previously suggested for the ccNiR[51].

2.3. Nitric Oxide Reductases

A number of conflicting mechanisms have been proposed for heme-based nitric oxide reduction, the main argument being as to how the N–N coupling between substrates occurs to form N_2O and N_2 gases. From initial studies on the cytochrome cd_1 NiR, Averill and Teidje proposed that the dehydration of the ferrous nitrite complex produces an electrophilic Fe^{II}–(NO^+) intermediate, Figure 4.5, which can be attacked by a incoming nitrite to form a trioxodinitrate, $N_2O_3^{2-}$, as a precursor to N_2O[49,52,53,53,55]. Only subsequently, it has been established that free NO is an obligate intermediate in many denitrifying organisms, i.e., there are separate enzymes that reduce NO_2^- to NO and NO to N_2O.

In the heme *bc* nitric oxide reductase (NoR), Hollocher proposed that N_2O is formed after reduction of a nitrosyl to nitroxyl (NO^-), and its subsequent free coupling in solution, Figure 4.5[56–59]. The bacterial NorBC are integral membrane proteins that catalyze the reduction of nitric oxide to nitrous oxide and water, within a larger denitrification complex containing both nitrite and nitrous oxide reductases. The possibility of a nitroxyl intermediate came as a result of N labeling experiments in denitrifying bacterial cell cultures[59,60]; a nitroxyl adduct was suggested as a branching point in these N-oxide reductases between pathways leading to N_2O by N–N coupling and NH_3 by further reduction and protonation[61,62]. Another alternative was proposed by Caughey, who found that cytochrome *c* oxidase is capable of NoR activity, reducing NO to N_2O[63]. He proposed that the reductive dehydration of Fe-bound NO to an Fe-nitrido intermediate, analogous to the Fe-oxo intermediates important to peroxidase and P450 chemistry, as the precursor to N_2O production, Figure 4.5.

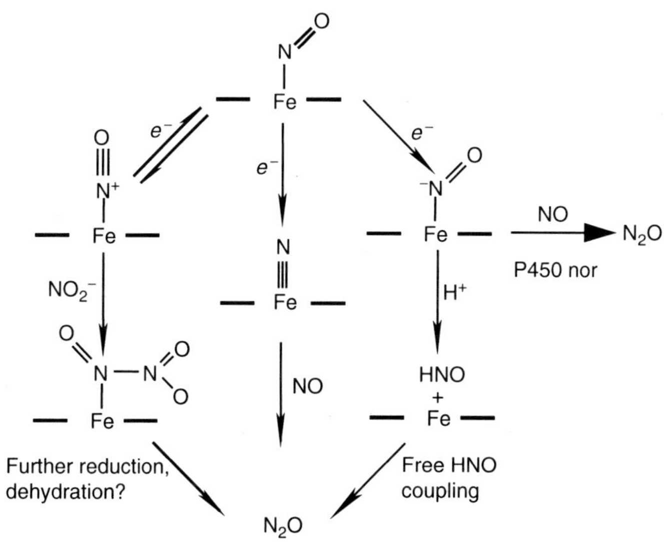

Figure 4.5. Proposed pathways for N–N couplings as proposed by Averill and Teijde (left)[50], Caughey (center)[49], Hollocher(right)[51].

Recent work has shown NorBC to contain an Fe heme *b*/nonheme Fe binuclear center, termed NorB[64], which is both structurally and genomically quite similar to the Fe–Cu center of cytochrome *c* oxidases[65]. A number of spectroscopic experiments suggest that the diferrous form is required for NO reduction[64,66,67]. However, the reactivity of a model binuclear heme/nonheme Fe complex suggests that a simple coupling of Fe^{II}–NO species at the two active site metals can yield N_2O and an Fe^{III}–O–Fe^{III} bridging species[68], and a similar bridging oxo species has been observed in the inactive enzyme[69].

The best studied N–N coupling system is cytochrome P450nor, a fungal enzyme which does not exhibit the usual mono-oxygenation activity of other P450 enzymes. In its catalytic cycle, the ferric-nitrosyl complex of P450nor is reduced by NADH to generate an identifiable intermediate (λ_{max} bat 444 nm) that reacts rapidly with nitric oxide to give nitrous oxide, Scheme 4.5[70–72]. Resonance Raman of isotopically labeled samples identified the Fe–NO stretch for the intermediate at 596 cm^{-1}, significantly higher than that for the Fe^{III}–NO at 550 cm^{-1}. It was reasoned that the addition of two electrons to π^* orbitals of Fe^{III}–NO increases metal–ligand back-donation, thus strengthening the Fe–NO bond and shifting the ν_{Fe-NO} to a higher frequency. It was also demonstrated that the intermediate does not undergo deuterium exchange with D_2O, which was used as evidence against an Fe^{II}–HNO adduct (vide supra).

Subsequent kinetic and spectroscopic experiments show that the 444 nm intermediate may be formed from reaction of $NaBH_4$ with the ferric nitrosyl adduct of P450nor at $-10°C$ and also by reaction of the ferric P450nor with hydroxylamine radicals generated by pulse radiolysis oxidation of hydroxylamine[73]. Thus the intermediate was formulated as a ferric hydroxylamine radical adduct, Fe^{III}–NHOH, which would then react with NO directly to generate N_2O and the Fe^{III}–OH_2 resting state. The sequential mechanism proposed, as adapted in Figure 4.6, includes rather unusual formulations of Fe-based oxidation states and it's coupling with the proximal cysteine ligand. The direct addition of a hydride to the Fe^{III}–NO avoids the thermodynamically stable Fe^{II}–NO and its potentially unfavorable reduction. The authors failed to note that the proposed Fe^{III}–NHOH intermediate is inconsistent with the lack of deuterium exchange. But the forma-

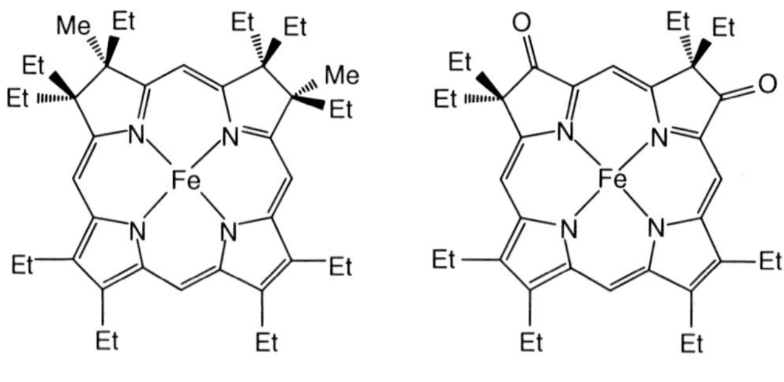

Fe^{II}–2,4-dimethyloctaethyl-isobacteriochlorin

Fe^{II}–2,4-octaethyl porphinedione

Scheme 4.5.

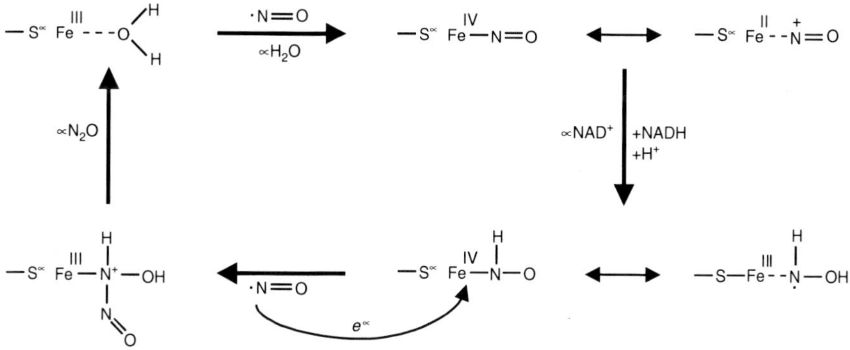

Figure 4.6. Proposed mechanism for P450nor[73].

tion of long-lived nitroxyl adducts and their coupling reactions with free NO have been recently confirmed in both heme protein[74] and small molecule[75] models.

3. Electrochemical Investigations of NO_x Reduction

3.1. Nitrite

The capability of nitrite to oxidize hemeproteins has been recognized for more than 100 years[76]. Nitrite is known to be causitive agent in the development of methemoglobinemia, a serious medical condition which results from ferric hemoglobin's inability to bind oxygen[77,78]. As a result, significant effort has been made in understanding the interactions of hemoglobin, Mb, P450s, and Cyt *c* with nitrite and alkyl nitrites[79,80]. These reactivity are related to those involved in the electrocatalytic reduction of nitrite by hemes and heme model compounds, and will be used to help interpret mechanistic aspects of the observed electrochemical behavior.

3.1.1. Small Molecule Models

Early investigations of electrocatalytic nitrite reductions by heme models were reported by the group of Meyer[81,82], using water-soluble Fe-porphyrins (e.g., Fe^{II}-tetra(4-sulfonatophenyl)-porphyrin (TPPS) in acidic solutions, Eq. (4.5). An important observation was the direct formation of Fe^{II}–NO from dehydration of nitrite by the ferrous catalyst Eq. (4.6). Cyclic voltammograms of the Fe-porphyrins in nitrite solutions showed a greatly diminished anodic current at the $Fe^{III/II}$ couple, a result of the depletion of Fe^{II}(TPPS) by Eq. (4.6), and also a growth of waves corresponding to the nitrosyl. They estimated the equilibrium constant K at $> 2 \times 10^{14}\,M^{-1}$ for Eq. (4.6), corresponding to a potential of 350 mV SCE for the NO–$Fe^{III/II}$ couple. Differential pulse polarography showed a large current at 0.0 V that suggested that at open circuit the ferric nitrosyl is indeed reduced. Hence, the first reduction is assigned to a one-electron reduction of nitrosyl to form the nitrosyl radical followed by another one-electron reduction to form a negatively charged nitroxyl, Eq. (4.7). At more negative potentials, a greater catalytic response is observed which was suggested to go through an iron(I) intermediate, Eq. (4.8)

$$Fe^{II}(NO_2^-) + 2H^+ \rightarrow Fe^{III}(NO) + H_2O \quad (4.5)$$

$$Fe^{II}(P)(NO^+) + e^- \leftrightarrows Fe^{II}(P)(NO\bullet) \quad (4.6)$$

$$Fe^{II}(P)(NO\bullet) + e^- \leftrightarrows Fe^{II}(P)(NO^-) \quad (4.7)$$

$$Fe^{II}(P)(NO^-) + e^- \leftrightarrows Fe^{I}(P)(NO^-) \quad (4.8)$$

The products of the aqueous catalytic reduction, as determined by tandem bulk electrolysis and mass spectrometry, were mainly NH_2OH, NH_3, and N_2O with very minor amounts of N_2. Meyer attributed the N_2O formation as evidence of a bimolecular dimerization between reduced nitroxyl adducts(Fe^{II}–NO^-) as shown in Eqs. (4.9)–(4.10) and Figure 4.7. Su and coworkers utilized Co tetra(*N*-methyl-4-pyridyl)porphyrin (TMPyP) as an electrocatalyst to examine the pH dependence for nitrite reduction in aqueous solution[83]. At low pH, both hydroxylamine and ammonia are formed; at high pH only hydroxylamine is detected with no observable N–N coupled products.

$$2Fe^{II}(P)(NO^-) \rightarrow (P)Fe^{II}(N_2O_2)Fe^{II}(P) \quad (4.9)$$

$$(P)Fe^{II}(N_2O_2)Fe^{II}(P) + 2H^+ + H_2O \rightarrow N_2O + 2Fe^{II}(P)(H_2O) \quad (4.10)$$

A subsequent report by Meyer's group, using films of polymerized Fe protoporphyrin IX dimethyl ester to reduce aqueous solutions of nitrite, confirmed the importance of bimolecular reactivity, Scheme 4.6[84]. The same complex mixture of products was found, but N_2 became the second most dominant

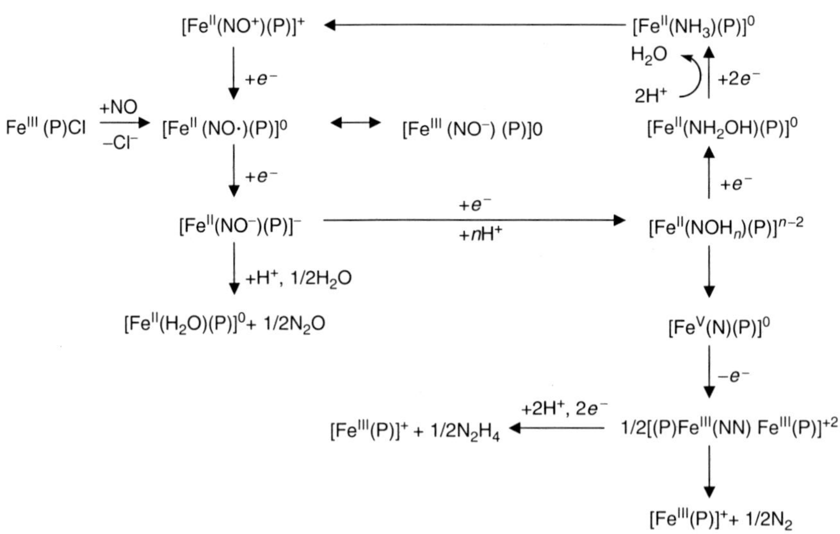

Figure 4.7. Proposed mechanism for electrocatalytic reduction of nitrite, adapted from Meyer[84].

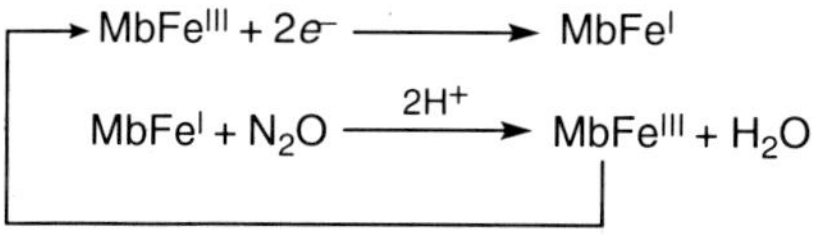

Scheme 4.6.

product and a sharp increase in N_2O was also seen. The increase in N–N coupled products apparently resulted from the close proximity of the catalytic sites. Bettelheim's group have also investigated polymeric films of iron *ortho* aminophenyltetraphenylporphyrin[85]. In the enzymatic NoR systems, such coupling should be precluded by protein-isolated active sites.

The issue of intersite interactions during catalysis has also been seen for phthalocyanin catalysts. Lever showed that monomeric or binuclear species Co-phthalocyanine derivatives, when adsorbed onto PG electrodes, were unable to reduce nitrite as efficiently as tetranuclear species[86]. They hypothesized that the tetra-nuclear complex was required to carry out a concerted multielectron reduction. Using a variety of electrochemical techniques and product analysis, they analyzed the mechanism of nitrite reduction by a dinuclear Ru-phthalocyanine adsorbed onto a PG electrode $(RuPc)_2$[87]. At low nitrite concentrations, the number of electrons in the Ru-dimer catalyzed reduction of nitrite was 6, corresponding to the production of ammonia, as determined by Koutecky–Levich (K–L) plots of rotating disc voltammetric data; at higher nitrite concentrations (> 2 mM) the number decreased, approaching two electrons which suggests formation of N_2O as the main pathway. Catalytic voltammetric waves for the reduction of nitric oxide and hydroxylamine occurred at more positive reduction potentials than nitrite for this system, therefore it was reasoned that they could not be the final product of nitrite reduction. By the same method, the reductions of NH_2OH and N_2O were also found to be two electron processes, suggesting that they are reduced to NH_3 and N_2, respectively. Bulk electrolysis and product analysis confirmed the production of N_2O and NH_3. Rotating ring disc experiments (RRDE) demonstrated production of NH_2OH, but no N_2H_2 was observed.

Characterization of nitrite adducts of Fe^{II} porphyrins has proven difficult, as isolation often engenders O-atom transfer reactivity[88]. Only recently have the structures of sterically protected adducts been crystallographically determined[89]. However, nitrite–heme complexes have been electrochemically characterized in both the ferric and ferrous forms in aprotic solvents[90,91]. Ryan *et al.* have shown that ferric and ferrous tetraphenylporphyrin (TPP) and octaethylporphyrin (OEP) readily form apparent bis complexes in the presence of excess nitrite; in DMF the TPP and OEP complexes had $Fe^{III/II}$ couples at -450 and -830 mV vs. SCE, respectively[90]. The structure of these complexes is still unknown, as the nitrite may bind through N or O atoms, or bridge two metal centers. However, once formed the observed nitro complexes decompose to new products, with mono reacting more rapidly than the bis-nitro complex[92].

Ryan's group analyzed this reactivity by electrochemical and solution-based reaction studies. Cyclic voltammetry of Fe(OEP) showed that upon the addition of nitrite, the $Fe^{III/II}$ couple shifts negatively; at higher nitrite concentrations, the potential vs. log [NO_2^-] changes greater than 59 mV, consistent with more than one nitrite ion lost upon reduction. Overall, three distinct reduction currents were observed; the first wave correlates to the $Fe^{III/II}$ couple and the last wave to the $Fe^{II/I}$ couple, but at high scan rates the last reduction wave splits, indicative of the generation of two products. This may be accounted for by the interconversion of bis and mono complexes, which also correlates with the observed changes in UV–Vis spectra. The addition of Fe(TPP)NO to the Fe(TPP) in nitrite solution induced the second wave to become more predominant and reversible, suggesting that nitrosyl adduct is indeed formed during the reduction.

Spectroelectrochemical absorbance and IR measurements were used to study the reduction of Fe(OEP)Cl in the presence of stoichiometric nitrite. Spectral analysis of bulk reduction at the observed potential of the nitrosyl adduct showed that the dimeric μ-oxo and nitrosyl complexes are produced in a 1:1 ratio with respect to Fe(OEP), while a significant amount of ferrous Fe(OEP) is unreacted. Thus, both ferrous and ferric oxidation states are required for the generation of the nitrosyl adduct, hypothesized to occur during a mixed oxidation state bridged intermediate, Eqs. (4.11)–(4.13).

$$\mathrm{Fe^{II}(OEP) + Fe^{III}(OEP)(NO_2^-)_2 \rightarrow (OEP)Fe^{II}\text{–}NO_2\text{–}Fe^{III}(OEP) + NO_2^-} \quad (4.11)$$

$$\mathrm{(OEP)Fe^{II}\text{–}NO_2\text{–}Fe^{III}(OEP) \rightarrow Fe^{II}(OEP)(NO) + O{=}Fe^{IV}(OEP)} \quad (4.12)$$

$$\mathrm{O{=}Fe^{IV}(OEP) + Fe^{II}(OEP) \rightarrow [Fe^{III}(OEP)]_2O} \quad (4.13)$$

When the potential of the bulk reduction was shifted to that of the third reduction wave, Fe^{II}(OEP) generation increased with a corresponding loss of the μ-oxo complex, in addition to a new $Fe^{II}(OEP)NO^-$ band at 538 nm. The formation of the nitrosyl adduct by reduction of Fe(OEP)Cl in the presence of nitrite was also characterized by spectroelectrochemical IR measurements. The intermediacy of μ-oxo dimer reduction was also observed by Ikeda *et al.* during the reduction of NO by Fe(TMPyP)[93].

Other researchers have reported electrochemical nitrite reduction using aquocobalamin[94] and water-insoluble metallo-phthalocyanines absorbed onto electrodes[95]. Nyokong and coworkers found that the potential for catalytic nitrite reduction varied with metal in a series of metallo-phthalocyanine complexes adsorbed onto a glassy carbon electrode (GCE) in basic solutions, Cu(Pc) > Fe(Pc) > Ni(Pc) > Co(Pc) > Mn(Pc) > Zn(Pc)[96].

3.1.2. Heme Protein Models

Several important model systems for the reactivity of NO_x have been developed using simple heme proteins such as myoglobin and hemoglobin in surfactant film modified electrodes. Rusling first reported that myoglobin contained within a dimethyldidodecylammonium bromide (DDAB) films on an electrode showed greatly enhanced electrochemical response, indicative of diffusion controlled currents within the thin film[97]. Such a prepared electrode, termed Mb/DDAB, is stable to repeated cycling, can be moved from solution to solution, and will remain active for several weeks with proper storage. Coulometry of Mb/DDAB modified electrodes consistently shows that a large percentage of the deposited Mb remain electrochemically active. The temperature dependence for the $Fe^{III/II}$ current-response follows that for the solid-to-liquid-crystal phase transition for DDAB films[98]. Importantly, these Mb/DDAB films allow access to a wide potential range, with both the $Fe^{III/II}$ (−0.22 V vs. SCE) and $Fe^{II/I}$ (−1.09 V vs. SCE) couples observable at pH 7, in Figure 4.8 and Eqs. (4.14)–(4.15).

$$\text{Mb–Fe}^{III} + e^- \leftrightarrows \text{Mb–Fe}^{II} \quad (4.14)$$

$$\text{Mb–Fe}^{II} + e^- \leftrightarrows \text{Mb–Fe}^{I} \quad (4.15)$$

$$NO_2^- + ne^- + mH^+ \rightarrow NH_4^+ + NH_3OH^+ + N_2O + NO + N_2 \quad (4.16)$$

Lin first showed that when nitrite is introduced to the solution above, two new catalytic reduction waves appeared close to the $Fe^{II/I}$ couple, Figure 4.8, both of which displayed complex pH dependence indicative of multiple reaction pathways[99]. A small negative shift is also observed in the $Fe^{III/II}$ wave, attributable to the effect of nitrite binding at the Fe center. The pH dependence of the second wave (*ca.* −800 mV NHE) suggested a multielectron (*ca.* $3e^-/H^+$) limiting step associated with the production of NH_3. Bulk electrolysis at this wave yielded both NH_3 and NH_2OH as aqueous products, but their yield accounted for less than 25% of the consumed nitrite. On-line mass spectral analysis of the head gas over electrolyzed solutions of $^{15}NO_2^-$ demonstrated production of $^{15}N_2O$, ^{15}NO, and $^{15}N_2$. By varying the potential during electrolysis, it was shown that the production of N_2O initiated at the first catalytic wave, *ca.* −600 mV NHE, which was tentatively assigned to the reduction of the ferrous NO adduct[62]. The catalysis followed saturation kinetics, with an apparent K_m of *ca.* 3.3 mM, close to that of nitrite binding to the ferric protein.

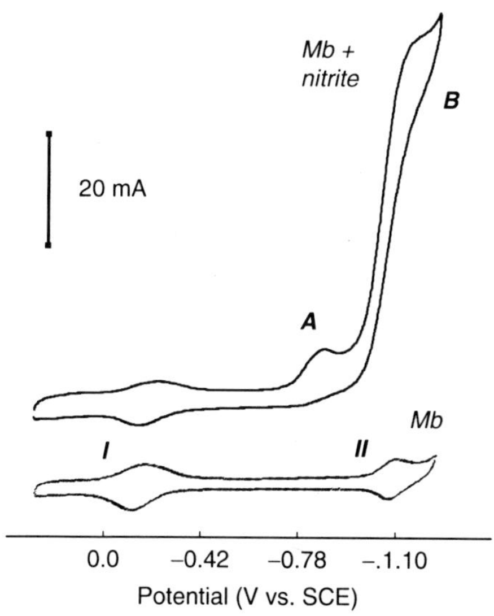

Figure 4.8. Voltammograms of Mb/DDAB at 100 mV s^{-1} demonstrating catalytic activity (bottom) pH 7.5 buffer, $Fe^{III/II}$ couple and $Fe^{II/I}$ labeled I and II, respectively (top) in presence of 5 mM $NaNO_2$, catalytic current waves labeled A and B[99].

A subsequent comparison of the electrocatalytic activity of Mb/DDAB with similar film-modified electrodes using a thermophilic cytochrome P450, CYP119/DDAB, was performed in order to investigate the effect of the axial ligand on biomimetic NO_x reductions[100]. The enzyme CYP119 has the characteristic heme of P450s with axial thiolate ligation, while the Fe heme in Mb is ligated by histidine. Voltammograms of nitrite reduction by CYP119/DDAB electrodes were quite similar to that of Mb/DDAB: two catalytic waves at analogous potentials are observed, the first corresponding to the formation of N_2O, the second to that of NH_3. CYP119 is the much more selective catalyst, producing NH_3 almost exclusively during the initial half hour of reductive electrolysis of nitrite, as compared with less than 20% conversion by Mb under the same conditions. This was attributed to rigidity of the thermophilic protein, which traps the substrate within the active site and facilitates multielectron reductions of single substrate. As cytochrome *P*450 CYP119 is a thermophilic protein, its activity when immobilized in dimethyldidodecylammonium poly(*p*-styrene sulfonate) on pyrolytic graphite was examined for temperature ranging between 25 and 75°C. Increasing temperature dramatically enhances the nitrite reduction rate over twofold while decreasing the reduction potential over 150 mV.[101]

Further insight was provided by comparing the effect of nitrite concentration on the catalytic current for CYP119. Steady state current analysis of the nitrite concentration dependence gave an apparent K_m of 34 mM, which should correlate somewhat to the binding affinity of the substrate for the enzyme. However, no binding of nitrite to the ferric CYP119 was observable in solution studies at even much higher concentrations. Likewise, no shift of the $Fe^{III/II}$ wave of CYP119 was observed with increasing nitrite up to 100 mM, suggesting that nitrite did not bind to the enzyme. As above, the formation of the Fe^{II}–NO could be observed indirectly by the loss of the $Fe^{III/II}$ redox couple during successive voltammetric scans in nitrite solution. Plotting the loss of the $Fe^{III/II}$ couple with time gave a rate of dehydration for CYP119 ($9.0 \times 10^{-3}\ s^{-1}$) virtually identical to that of Mb under identical conditions. This implies both systems have similar rate-limiting step that is independent of the distal ligation, as modeled by Eqs. (4.17)–(4.22).

$$Fe^{III} + NO_2^- \leftrightarrows Fe^{III}\text{–}NO_2^- \tag{4.17}$$

$$Fe^{III}\text{–}NO_2^- + e^- \leftrightarrows Fe^{II}\text{–}NO_2^- \tag{4.18}$$

$$Fe^{II}\text{–}NO_2^- + 2H^+ \rightarrow Fe^{III}\text{–}NO + H_2O \tag{4.19}$$

$$Fe^{II} + NO_2^- + H^+ \rightarrow Fe^{III}\text{–}NO + OH^- \tag{4.20}$$

$$Fe^{III}\text{–}NO + e^- \leftrightarrows Fe^{II}\text{–}NO \tag{4.21}$$

All native NiR enzymes are believed to undergo a ferric nitrite-to-nitrosyl transformation, as in Eqs. (4.17)–(4.19)[49,50], but this may not be the case with other heme proteins. Doyle first reported that the reaction of ferrous hemoglobin (Hb) with nitrite was zero order in protein[79,80], but first order in nitrite and protons; a similar dependence has been reported for the reaction of ferrous Mb with nitrite[102]. This implies that the reductive dehydration is dependent on the reaction of Fe^{II} with HONO, or its equivalent as NO^+, as in Equation (4.22)–(4.24), and that this reaction may generate free NO by dissociation of the ferric nitrosyl, as suggested by the mechanism of Doyle.

$$NO_2^- + H^+ \leftrightarrows HONO \quad (4.22)$$

$$Fe^{II} + HONO \leftrightarrows Fe^{II}\text{–}HONO \quad (4.23)$$

$$Fe^{II}\text{–}HONO \rightarrow Fe^{III}\text{–}NO + OH^- \quad (4.24)$$

$$Fe^{III}\text{–}NO \leftrightarrows Fe^{III} + NO \quad (4.25)$$

Equations (4.22)–(4.25) suggest that nitrite reduction by Mb at the $Fe^{III/II}$ potential may be catalytic, as in the heme cd_1 NiR enzymes. Bedioui and coworkers were the first to demonstrate such catalysis at the $Fe^{III/II}$ couple in a comparison of nitrite reduction by Mb to that by hemoglobin (Hb) and hemin (Hm) itself in DDAB-modified electrodes[103]. They first examined the effect of scan rate on catalytic current of nitrite reduction, which was relatively unchanged between the various DDAB-modified catalysts. At slow scan rates two irreversible cathodic peaks are present near $Fe^{II/I}$ couple while the $Fe^{III/II}$ couple shifts more negatively, indicative of the binding of nitrite. At faster scan rates (1 $V^{-1}s$) the slow kinetics of nitrite binding diminishes the reduction of nitrite and both of the original couples are largely conserved, except for a new cathodic peak at −0.85 V.

Rotating disc electrode (RDE) voltammetry for Hm/DDAB showed two steady state waves at −0.1 and −1.1 V, with larger limiting current at the second wave due to apparent multielectron reduction; in agreement with this analysis, strong nitrite dependence was found only for the second wave. Analogous experiments with Mb/DDAB and Hb/DDAB show similar behavior, but much greater nitrite dependence was observed for the first wave, suggestive of catalytic reduction at the $Fe^{III/II}$ couple, as in Eqs. (4.19)–(4.21) or Eqs. (4.22)–(4.25). The initial catalytic wave was assigned to both the reductive dehydration of the Fe^{III}-bound nitrite and also the subsequent reduction of the resulting Fe^{III}–NO, the latter not observable by cyclic voltammetry because of slow charge-transfer kinetics. This interpretation was supported by analysis of the nitrite reaction order, as determined by the slopes of linear plots of log I_k vs. log[NO_2^-]; RDE voltammograms of Mb/DDAB at varying nitrite concentrations is shown in Figure 4.9. At the $Fe^{III/II}$ wave, for Hm/DDAB the reaction order is low, *ca.* 0.19, but for Mb/DDAB the order is close to unity. Tafel plots of the $Fe^{III/II}$ wave were indicative of a one-electron process, implying that reduction of the Fe^{III}–NO (or equivalently, the Fe^{II}–NO^+) is kinetically slow step and mass transport limited. Thus, for the reduction of nitrite by Mb or Hb in DDAB, a reaction sequence was proposed invoking first the binding of nitrite, reductive dehydration forming Fe^{III}–NO, rate-limiting reduction to Fe^{II}–NO, and subsequent multielectron reductions at potentials close to the $Fe^{II/I}$ couple. For this second reduction, which forms ammonia, the steady state substrate nitrite concentration limits were identical for Hm, Mb and Hb, at 6 mM; this would correspond to a K_m ca. 3 mM, as had previously been reported for Mb.

Liu and coworkers utilized Hb adsorbed on colloidal gold nanoparticles in carbon paste electrode (CPE) to investigate the reduction of nitrite at pH 4[104]. (One criticism of this work is that, at this low pH, aqueous Hb is denatured,

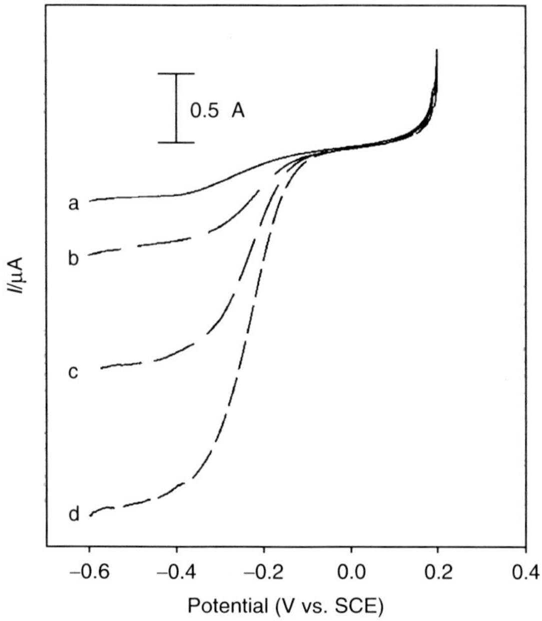

Figure 4.9. RDE voltammograms response of Mb/DDAB film modified GCE around the $Fe^{III/II}$ redox couple in the presence of various nitrite concentrations. Curve (a) 2.0 mM; (b) 4.0 mM; (c) 6.0 mM; (d) 8.0 mM. Electrolyte:phosphate buffer solution, 50 mM NaBr (pH = 7.4). Rotation rate 500 rpm;scan rate: 0.02 Vs^{-1}[103].

therefore raising the question of whether holoprotein (Hb) or the free cofactor (Hm) is the electroactive species investigated.) A plot of the cathodic peak potentials vs. the logarithm of the scan rates, obtained a slope of 35.4, close to the expected value for the formation of Fe^{II}–NO as in Eqs. (4.26)–(4.28). When the scan range is limited to −0.6 V in the presence of nitrite the Fe^{III}/Fe^{II} couple starts losing reversibility, again indicative of the formation of the stable Fe^{II}–NO during the Fe^{III}/Fe^{II} reduction process.

$$\text{Hb–Fe}^{III} + H^+ + e^- \leftrightarrows \text{HbH–Fe}^{II} \quad (4.26)$$

$$NO_2^- + 2H^+ + e^- \rightarrow NO + H_2O \quad (4.27)$$

$$\text{HbH–Fe}^{II} + NO \leftrightarrows \text{HbH–Fe}^{II}NO \quad (4.28)$$

Similar results were reported by Fan *et al.* in their voltammetric studies of nitrite reduction by Hb contained in DNA film electrodes at pH 4[105]. These films showed well-defined Fe^{III}/Fe^{II} couple at −0.36 V SCE, Eq. (4.26), and upon addition of nitrite two new cathodic peaks were observed. Also, when the lower scan potential was limited to −0.5 V, the $Fe^{III/II}$ peak current decreased linearly due to formation of stable Hb–Fe^{II} NO complex, as previously seen for Mb. Pre-electrolysis followed by linear scan voltammetric sweeps showed that at +0.3 V the predominant species is met-Hb, at −0.5 V the Hb–NO complex is the predominates species, and at −0.9 V the Hb–Fe^{II} predominates, apparently due to NO depletion in the electroactive layer.

Many additional reports have mentioned nitrite reduction by Mb[99, 103, 106–111], Hb[103, 104, 106–109, 111–115], and horse radish peroxidase (HRP)[106, 108, 109, 111, 116, 117], modified electrodes, but their primarily focus is on film methodology or biosensor application. The NO_x reactivity of heme proteins is apparently widely applicable, and the reported potentials for both the $Fe^{III/II}$ couples and catalytic nitrite reductions are quite similar regardless of the mode of deposition or protein. The results are summarized in Table 4.1, which include deposited films made of water-insoluble surfactants (DDAB)[99, 103, 115], amphiphilic polyacrylamide (PAM)[116], carboxymethyl cellulose (CMC)[107], chitosan (CS)[108], ionomer poly(estersulfonic acid) or Eastman AQ29 (AQ)[116]; proteins trapped within sol–gel film derived from tetraethylorthosilicate (TEOS)[110, 114] or layer by layer

Table 4.1. Film Methodologies Used in Examinations of Nitrite Reduction by Heme Proteins

Methodology	Protein	pH	Fe^{II}/Fe^{III} (V)	pH	NO_2^- red. (V)	Detection range (mM)	Detection limit (mM)	References
HRP-PAM on PG	HRP	7	−0.33[a]		−0.84[b]			[116]
Clay on PG	Mb	5.5	−0.28[a]		−0.8[b]	0.05–0.82	0.03	[106]
	Hb		−0.28[a]			0.05–0.82	0.03	
	HRP		−0.28[a]				5	
PS on PG	Hb	7	−0.36[a]	5.5	−0.75[c]	6–16	1	[112]
CMC on PG	Mb	7	−0.34[a]	5.5	−0.8[d]	0.6–8	0.32	[107]
	Hb	7	−0.34[a]			0.6–9	1.32	
CS on PG	Mb	7	−0.33[a]	5.5	−0.77[d]	0.1–3	0.017	[108]
	Hb	7	−0.33[a]				0.037	
	HRP	7	−0.33[a]				0.87	
AQ on PG	HRP	7	−0.33[a]	7	−0.84[d]			[116]
Fe_3O_4 on PG	Mb	7	−0.34[a]			2.6–16.3	0.2	[109]
	Hb	7	−0.34[a]	5.5	−0.8[c]	0.4–11.9	0.18	
	HRP	7	−0.32[a]			3.1–18.2	0.44	
PVS on PG	Hb	5.5	−0.28[a]	5.5	−0.8[d]			[113]
Sol–gel on CPE	Mb	7	−0.298[e]	7	−0.774[c]			[110]
TEOS on CPE	Hb	7	−0.312[e]	5.5	−0.6[d]			[114]
PAMAM on PG	Mb	7	−0.34[a]					[111]
	Hb	7	−0.34[a]	5.5	−0.7[c]			
	HRP	7	−0.34[a]					
Au on CPE	Hb	5.5	−0.42[f]	4	−0.72[d]			[104]
DDAB on PG	Mb	7[e]	−0.221[a]	7.0	−0.895[g]			[99]
	Mb	7.4	−0.24[a]	7.4	−0.85[b]			[103]
	Hb	7.4	−0.25[a]					
	Hm	7.4	−0.22[a]					

[a] SCE.
[b] 200 mV^{-1} s.
[c] 100 mV^{-1} s.
[d] 150 mV^{-1} s.
[e] Ag/AgCl.
[f] NHE.
[g] 500 mV^{-1} s.
[h] pH dependence was analyzed.

in polyanionic poly(vinyl sulfonate) (Hb-PVS)[113]; and protein films applied to nanosized particles of polystyrene latex beads (PS)[112], Fe_3O_4[109], clay films[106], polyamidoamine dendrimers (PAMAM)[111], or colloidal gold (Au)[104], and then subsequently deposited on an electrode.

3.2. Nitric Oxide, NO

The growing understanding of the ubiquitous involvement of hemes in NO physiology has resulted in widespread interest in the possible chemical reactivity of NO. The formation of NO in our bodies is directly attributable to the heme cofactor in nitric oxide synthase[118]; furthermore, NO binding to other heme proteins has been linked to important physiological processes such as vasodilation[119], neurotransmission[120], and the immune response[121]. The reduced form of nitric oxide, termed a nitroxyl or nitrosyl hydride (NO^- or HNO), has attracted much interest in recent medical studies, for HNO precursor compounds have unusual cardiovascular effects, and may be useful as protective drugs for heart disease[122–124]. Although we restrict ourselves here to reduction of NO by hemes and heme model compounds as applicable towards an understanding of bacterial and plant enzymes, these discussions may likewise apply to the physiological biochemistry of humans as well.

3.2.1. Small Molecule Catalysts

Important early electrochemical characterizations of nitrosyl adducts of porphyrin complexes were performed by the Kadish group[125]. Voltammetric studies of the nitrosyl adduct of Fe^{II}(TPP) demonstrated an electrochemically reversible reduction which was substantially positive to that of the parent Fe^{II}(TPP). The product of the reduction, a nitroxyl adduct formulated as $Fe^{II}(TPP)(NO)^-$, could be generated in situ and its absorbance spectra observed; reduction of the dinitrosyl, $Fe^{II}(TPP)(NO)_2$, who's formation is characterized by a positive $E_{1/2}$ shift, Table 4.2[126], also forms the same product. Nevertheless, attempts to prepare the reduced nitrosyl product via bulk electrolysis were unsuccessful. Up to ten reducing equivalents were passed through a sample solution electrolytically, but the major species in solution remained the ferrous nitrosyl, i.e., the nitroxyl apparently decomposed via some unspecified reaction to regenerate the more stable nitrosyl.

Kadish and others have examined the effects of a variety of parameters on the $E_{1/2}$ of nitrosyl metalloporphyrins. Because of the high covalency of metal–nitrosyl bonding, it is often difficult to distinguish between metal-, nitrosyl-, or even porphyrin-based reductions; we here try to limit discussion to systems where nitrosyl-reductions are invoked. For an analogous series of metal–nitrosyl compounds, the solvent effect on the $E_{1/2}$ potential is most closely correlated to the solvent property as measured by the Dimroth–Reichardt parameter, a measure of the ionizing power or polarity of the solvent[127]. The porphyrin structure also plays a significant role in the redox potential of the nitrosyl complex, Table 4.2. As may be expected, electron-withdrawing substituents shift the $E_{1/2}$ positively while electron-donating groups do the opposite[128], as has been demonstrated for substituted-TPP complexes in Fe^{II}–NO[129] and Co^{II}–NO reductions[130], Tables 4.3

Table 4.2. Reduction Potentials for Fe(por)NO[a]

Macrocycle	Solvent	$E_{1/2}$ R1	References
TPPS	H_2O	−0.63	[81]
OEC	PhCN	−0.41	[131]
OEP	PhCN	−1.08	[135, 128]
OEP[b]	PhCN	−0.88	[128]
TPP	PhCN	−0.88	[135]
OEP	CH_2Cl_2	−1.02	[129]
MOEC	CH_2Cl_2	−1.02	[129]
OEiBC	BuCN	−1.11	[129]
2,4-DMOEiBC	CH_2Cl_2	−1.17	[129]
OEPone	CH_2Cl_2	−0.81	[129]
2,4-OEPdione	CH_2Cl_2	−0.69	[129]
2,3-OEPdione	THF	−0.66	[129]

[a] vs. SCE.
[b] bis–NO complex.

and 4.4. Kadish also compared Fe porphyrins to corroles[131]. Other substituent effects have also been examined by various groups: nitration of the porphyrin ring causes a *ca.* + 0.4 V shift[132]; fluorine to chlorine substitution in polyhalogenated corroles causes a *ca.* + 0.08 V shift[133]. Increasing the bonding strength of the counter ion in nonpolar solvents is found to shift the potential negatively ($ClO_4^- < Br^- < Cl^- < N_3^- < F^-$); however, the opposite is true for coordinating solvents. But the potential shift due to metal-substitution is much less than might be expected: between Cr[134], Fe[135], Mn[126], Co and Ru[136] TPP nitrosyl adducts, the potentials vary less than 400 mV, Table 4.5. Of the various properties including solvent, metal ion, counter ion, and axial ligand, the macrocyclic structure itself appears the main determinant of the potential for this apparent ligand-based reduction.

Electrocatalytic nitric oxide reductions were noted in the previously described studies by Meyer's group on nitrite reduction by water-soluble porphyrins[81]. As NO is a proposed intermediate, it was used independently as a substrate to verify the assigned potential for the ferrous nitrosyl to nitroxyl transformation. Bettelheim *et al.* also investigated NO reduction by Fe(TMPyP) immobilized onto a GCE with a Nafion coating that repelled anions like nitrite from the electrode surface[137]. Differential pulse voltammetry of this system obtained three distinct reduction waves at potentials *ca.* 0.5, −0.6, and −0.8 V, assigned using the mechanistic scheme of Meyer[84] to sequential reductions of $Fe^{II}(NO^+)$, $Fe^{II}(NO)$, and Fe^{II} (NO^-).

Similarly, Bedioui *et al.* used Fe(TPPS) immobilized in poly(pyrrole-alkylammonium) films to investigate the reversible reduction of $[Fe^{II}(NO)]$ to $[Fe^{II}\ (NO^-)]^-$[138]. They also examined similar reactivity with electropolymerized hemin, hematin, Fe(TMPyP), and Fe(TSPP). They found that a polymeric

Table 4.3. Reduction Potentials for Substituted Fe(x-TPP)NO[a]

x-TPP[b]	4σ[c]	$E_{1/2}$
p-OMe	−1.08	−1.45
m-Me	−0.28	−1.44
H	0.00	−1.44
p-F	0.24	−1.34
p-Cl	0.92	−1.35
m-Cl	1.48	−1.33
p-CF_3	2.20	−1.32

[a] In THF, as $E_{1/2}$ vs. SCE.
[b] *x* is the *meta*- or *para*-substituent on each phenyl.
[c] The Hammet parameter for the phenyl substituent[129].

Table 4.4. Reduction Potentials for Substituted Co(*x*-TPP)NO Reduction[a]

x[b]	4σ[c]	$E_{1/2}$
p-OMe	−1.08	−1.20
p-Me	−0.68	−1.20
m-Me	−0.28	−1.18
H	0.00	−1.16
m-OMe	0.48	−1.15
p-OCF_3	1.40	−1.10
p-F_3	2.20	−1.06
p-CN	2.52	−1.05

[a] In CH_2Cl_2, as $E_{1/2}$ vs. SCE.
[b] *x* is the *meta*- or *para*-substituent on each phenyl.
[c] The Hammet parameter for the phenyl substituent[129].

Table 4.5. Reduction potentials of M^{II}(TPP)NO[a]

Metal	$E_{1/2}$	References
Ru[b]	−1.46	[136]
Ru	−1.06	[136]
Co	−1.18	[126]
Mn	−0.92	[126]
Cr	−1.24	[134]
Fe	−0.93	[135]

[a] In CH_2Cl_2, $E_{1/2}$ vs. SCE.
[b] Nitrite as axial ligand.

film of hemin, coated with Nafion, loses its $Fe^{III/II}$ couple during voltammetric cycling in the presence of NO (formed by low pH disproportion of HONO), and a new cathodic peak appears at *ca.* −600 mV, Figure 4.10[139]. If such a treated electrode is subsequently transferred to a solution without nitrite, the same features are seen during the initial cycle, but regain its $Fe^{III/II}$ couple during the second cycle through apparent loss of NO. The peak at −600 mV (−800 mV at neutral pH) was attributed to the reduction of Fe^{III}(NO) was suggested to occur at the irreversible current at −600 mV, although this is far from the potential assigned by Meyer for the (NO)Fe^{III}(TPPS) couple at *ca.* + 300 mV[81].

In an important series of papers, Ryan and coworkers performed extensive investigations into the effect of the porphyrin structure on the electrochemical reduction of the iron–nitrosyl complex[129, 140–143] Several different types of porphyrins were examined, including tetrahydroporphins that model the siroheme NiR (e.g., 2,4-dimethyloctaethylisobacteriochlorin (2,4-DMOEiBC) and oxochlorins which model heme cd_1 NiR (e.g., 2,4-octaethylporphinedione (2,4-OEPdione), Scheme 4.5.

Further investigation concentrated upon the reduction mechanism of the iron nitrosyl. Ryan *et al.* found that the electrochemical reduction of Fe(TPP)(NO) was highly reversible and, in the absence of excess NO, leads to a diamagnetic product (Fe(TPP)$(NO)^-$) that they characterized by NMR, UV–Vis, and resonance Raman spectroscopy[140]. The porphyrin vibrations for both Fe(TPP)(NO) and Fe(TPP)$(NO)^-$ were consistent with low-spin ferrous complexes, thus the reduction is suggested to be ligand-based. Once generated, Fe(TPP)$(NO)^-$ reacts slowly with proton sources such as phenols to ultimately regenerate Fe(TPP)(NO); in contrast, the two electron reduced product, Fe(TPP)$(NO)^{2-}$, reacts rapidly with

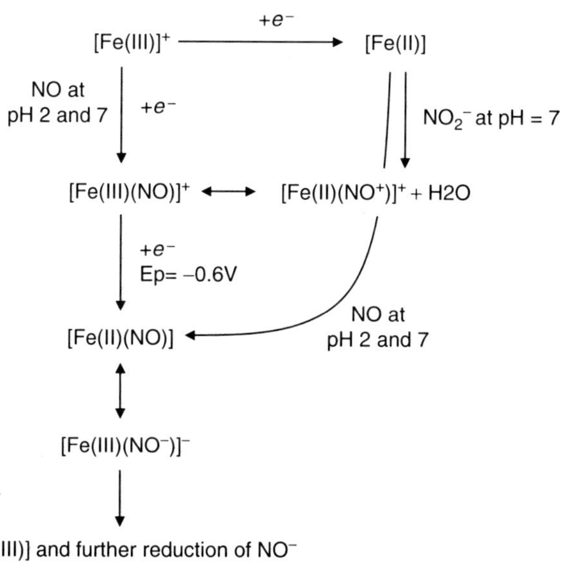

Figure 4.10. Proposed mechanism for electrocatalytic reduction of NO_2^- or NO, as per Bedioui[139].

phenols to yield ammonia[143]. The proton-dependences of these reductions were studied using *para*-substituted phenols as a proton source in nonaqueous solvents[143]. As shown in Figure 4.11, both the potential and limiting current of the second current wave (IIa and IIb), corresponding to the proton-limited formation of hydroxylamine, were dramatically affected by the presence of proton donors. In contrast, the first current wave (*I*), which corresponds to the one-electron reduction of Fe(TPP)(NO), shifted positively only at much higher phenol concentrations, and its limiting current was unaffected. The resulting mechanistic interpretation included the initial reversible reduction of Fe(TPP)(NO), followed by two reversible protonations before an irreversible three-electron reduction and protonation forms hydroxylamine, Eqs. (4.29)–(4.32).

$$\mathrm{Fe(TPP)(NO)} + \mathrm{e}^- \leftrightarrows \mathrm{Fe(TPP)(NO)}^- \tag{4.29}$$

$$\mathrm{Fe(TPP)(NO)}^- + \mathrm{H}^+ \leftrightarrows \mathrm{Fe(TPP)(HNO)} \tag{4.30}$$

$$\mathrm{Fe(TPP)(HNO)} + \mathrm{H}^+ \leftrightarrows \mathrm{Fe(TPP)(H_2NO)}^+ \tag{4.31}$$

$$\mathrm{Fe(TPP)(H_2NO)}^+ + \mathrm{H}^+ + 3\mathrm{e}^- \rightarrow \mathrm{Fe(TPP)}^- + \mathrm{NH_2OH} \tag{4.32}$$

This reactivity was used to directly compare porphyrin derivatives that model siroheme and heme d_1, Scheme 4.5. The activities of the porphyrin derivatives were correlated to the apparent pK_a of their reduced Fe-nitroxyl, Eq. (4.30). Iron-isobacteriochlorins, which model the assimilatory siroheme enzymes, were the most basic while the oxo-derivatives, which model the cytochrome d_1 of the denitrifying enzymes, were the least. These results mirror their enzymatic activity,

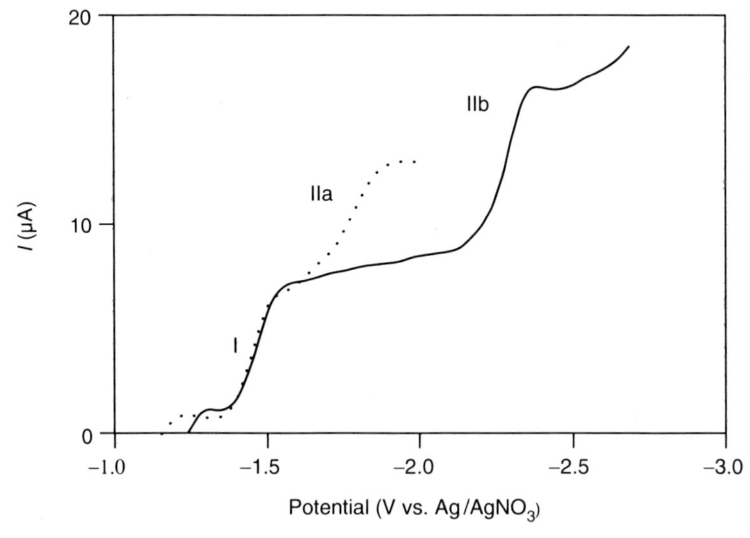

Figure 4.11. Normal pulse polarographs of the reduction of Fe(TPP)(NO) in the presence (dotted) and absence (solid) of 4.0 mM2,3-dicholrophenol. Current wave I corresponds to one reversible one-electron reduction, wave IIa and IIb correspond to a proton-limited multielectron eduction forming hydroxylamine. Data adapted from reference (Ryan EAC1994).

in that the former produces NH_3 and the latter NO and N_2O. Thus the macrocyclic structure plays an important role in tuning the reactivity and ultimately the reduction products.

Su and coworkers examined the effect of different metals like Mn[144] and Co[83] as well as Fe[145] on this reactivity. For Fe porphyrins, e.g., Fe(TMPyP), an EC mechanism, encompassing Fe reduction preceding NO coordination, is proposed[145]. Water soluble Mn(TMPyP), Mn(2-TMPyP), Mn(TSS), and Mn(TSPP) illustrate that porphyrins with electron withdrawing groups exhibit higher rates in electrocatalytic NO reduction activity. While the reduction products are similar, the mechanism for Mn is proposed to be via a ECE mechanism, i.e., initial reduction of the Mn^{III} to Mn^{II} precedes NO coordination and reduction. The Co(2-TMPyP) illustrated that a two electron couple, $Co^{III/I}$, is more efficient than a one electron couple, $Fe^{III/II}$, in yielding the N–N coupled product N_2O.

Bettelheim *et al.* investigated nitrosyl adducts of cobalt tetrasulfonated phthalocyanine, $Co(TSPc)^{4-}$, dissolved in aqueous solution or incorporated into a DDAB surfactant film where $Co(TSPc)^{4-}$ acts as the counterion for four DDA^+ forming $Co(TSPc)(DDA)_4$ (this was also done with dodecyltrimethylammonium)[146]. Using FTIR and electrochemical experiments the conformational orientation of the NO adduct was determined to be bent, Eq. (4.33). When exposing the phthalocyanin to NO in solution, a sharp band at 1,646 cm^{-1} appears but is quickly replaced by a broader band at about 1,700 cm^{-1} implicit of the N–Co bond shortening, this also occurs for the solid form of Co(TSPc) and $Co(TSPc)(DDA)_4$, Eq. (4.34). A plot of E v. 1/[NO], using CV, yielded a slope of 65 mV, ruling out the possibility of a $Co(TSPc)(NO)_2^{4-}$, complex.

$$Co^{II}(TSPc) + NO \leftrightarrows Co^{II}(TSPc)(NO)_{bent} \tag{4.33}$$

$$Co^{II}(TSPc)(NO)_{bent} \leftrightarrows Co^{II}(TSPc)(NO)_{linear} \leftrightarrows Co^{I}(TSPc)(NO^+) \tag{4.34}$$

Nykong *et al.* have also studied a variety of cobalt containing macrocycles. The electrocatalytic properties of cyanocobalamin adsorbed onto GCEs in aqueous solution was examined[147]. Along with adsorbed cobalt(II) tetrasulfonated phthalocyanine (Co(TSPc))[148], adsorbed cobalt phthalocyanine is examined (CoPc) in DMSO[149]. The proposed mechanism invokes both a ring- and metal-centered reduction during NO catalysis, Eqs. (4.35)–(4.39).

$$Co^{II}Pc(-2) + NO \rightarrow (NO)Co^{III}Pc(-2) \tag{4.35}$$

$$(NO)Co^{III}Pc(-2) + e^- \rightarrow (NO)Co^{II}Pc(-2) \tag{4.36}$$

$$(NO)Co^{II}Pc(-2) + e^- \rightarrow (NO)Co^{I}Pc(-2) \tag{4.37}$$

$$(NO)Co^{I}Pc(-2) + e^- \rightarrow (NO)Co^{I}Pc(-3) \tag{4.38}$$

$$(NO)Co^{I}Pc(-3) + ne^- \rightarrow NH_3 \text{ and other products.} \tag{4.39}$$

Cyanocobalamin adsorbed onto GCEs also readily reduces NO (nitrite as well), giving sequential reduction peaks associated with the effect of NO ligation onto the $Co^{III/II}$ and NO reduction $Co^{II/I}$ couples[147].

3.2.2. Heme Protein Models

The ability of Mb to catalyze the reduction of NO was first demonstrated by Bayachou in our group[150]. Voltammograms of Mb/DDAB under an NO solution, C in Figure 4.12, show the loss of the $Fe^{III/II}$ couple and an introduction of a large cathodic current at *ca.* −600 mV NHE due to the catalytic reduction of NO. The reduction of NO by Mb/DDAB is distinct from the reduction of NO_2^-, the catalytic current occurred at more positive potentials and remained catalytic at all pH tested. Bulk electrolysis of saturated ^{15}NO solutions produced $^{15}N_2O$, as identified by mass spectroscopy, with no $^{15}NH_3$ formation observed. Thus, the N–N coupling reaction, analogous to that of P450nor, dominates in the presence of excess NO.

Significantly, voltammetry at fast scan rates (>1 V^{-1} s) under NO gas yielded anodic currents attributed to the re-oxidation of the Fe^{II} nitroxyl complex, which implied that a nitroxyl adduct had a measurable lifetime during the catalytic reaction. A similar voltammetric fingerprint was apparent after succes-

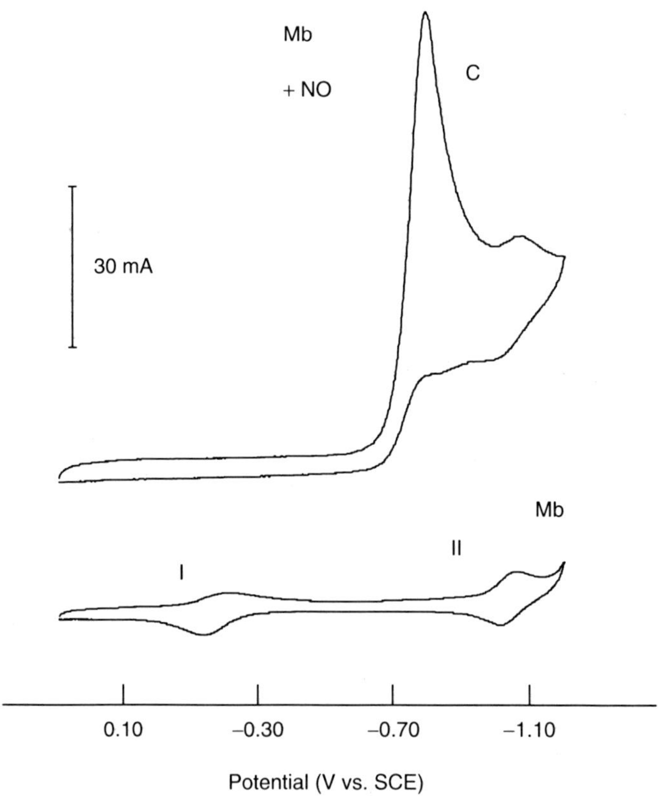

Figure 4.12. Voltammograms of Mb/DDAB at 100 mV^{-1} s demonstrating catalytic NoR activity (bottom) pH 7.5 buffer, $Fe^{III/II}$ couple and $Fe^{II/I}$ labeled I and II, respectively (top) in presence of sat. NO, catalytic current wave labeled **C**[150].

sive reductive scans under nitrite solution, consistent with a common intermediate in both catalytic reductions. The apparent reversibility decreased with increasing concentrations of H^+ or NO, implying that both these species react with the Fe-bound nitroxyl, as in Eqs. (4.40)–(4.43).

$$\text{Mb–Fe}^{\text{II}}\text{–NO} + e^- \leftrightarrows \text{Mb–Fe}^{\text{II}}\text{–NO}^- \quad (4.40)$$

$$\text{Mb–Fe}^{\text{II}}\text{–NO}^- + H^+ \rightarrow \text{Mb–Fe}^{\text{II}}\text{–HNO} \quad (4.41)$$

$$\text{Mb–Fe}^{\text{II}}\text{–HNO} \rightarrow \text{Mb–Fe}^{\text{II}} + \text{HNO} \quad (4.42)$$

$$\text{Mb–Fe}^{\text{II}}\text{–NO}^- + \text{NO} + H^+ \rightarrow \text{Mb–Fe}^{\text{III}} + N_2O + HO^- \quad (4.43)$$

In order to determine the lifetime of the nitroxyl intermediate, modified electrodes were fabricated directly from preformed Mb–NO with hopes to observe a single-turnover reduction in the absence of excess NO. A reversible single-electron reduction was in fact observed; the reduction potential of this couple, Eq. (4.40), was pH independent but its reversibility highly pH dependent. The singly reduced product was long-lived at pH 10, as demonstrated by the reversibility at slow scan rate in Figure 4.13, but the lifetime diminished at lower pH. At pH 7 the lifetime was determined by digital simulation as on the order of a tenth of a second.

These results were interpreted in terms of two possible mechanisms for N_2O formation[62]; by coupling of free HNO as was proposed by Hollocher[61] and

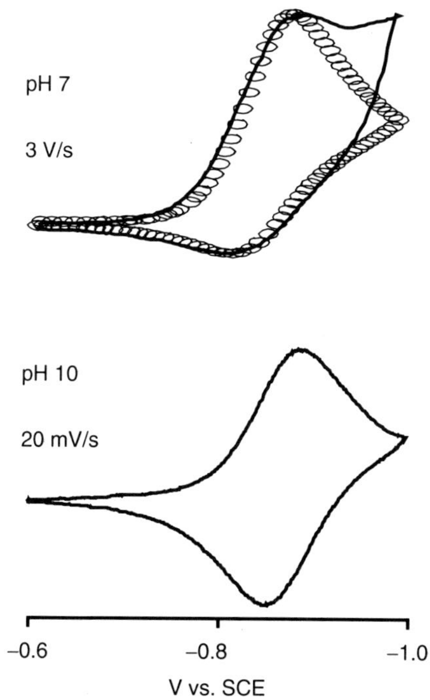

Figure 4.13. Reversibility of Mb–NO reduction in pH 10 and 7 solutions. Circles in top represent simulation used to derive lifetime of Mb–NO^-[150].

others[140], or by the coupling of heme-bound nitroxyl with exogenous NO has been proposed for the P450nor mechanism[70]. Digital simulations of the catalytic voltammograms gave estimates for the N–N coupling reaction, Eq. (4.43), as 10^8 M^{-1} s^{-1}[150]. Subsequently, it was found that the protonation of $Mb–NO^-$ does not induce the release of free HNO but gives the stable nitrosyl hydride adduct, Fe^{II}–HNO in Eq. (4.42)[74].

As a better model for the native P450nor, the ability of CYP119 as an electrocatalyst for NO was investigated and compared to that of Mb[100]. In contrast to Mb, CYP119 does not form a thermostable nitrosyl adduct that would allow for its formation and purification before film deposition. A simple method was described to generate the NO adduct *in situ* via the reductive dehydration of nitrite. Holding the electrode at negative potentials for 20 s prior to rapidly scanning to negative potentials resulted in a reversible redox couple at −850 mV vs. Ag/AgCl, corresponding to the reduction of the ferrous nitrosyl to nitroxyl, Eq. (4.40). Under similar conditions and scan rates, Mb/DDAB also exhibits a redox couple corresponding to nitrosyl/nitroxyl reduction, at potential identical to that previously reported, Figure 4.14.

The generation of the ferrous nitrosyl adduct in the presence of nitrite was first observed using small molecule porphyrins[81] and in polymerized films by Bedioui[138, 139]; here it shown to be a widely applicable method to characterization the nitrosyl/nitroxyl redox couple. As was seen for the preformed Mb–NO adduct[17], the reversibility of this redox couple for CYP119 is strongly dependent on the scan rate, becoming more reversible as the scan rate is increased. A fully reversible signal, where $i_{pa}/i_{pc} \sim 1$, is only observed at scan rates greater than

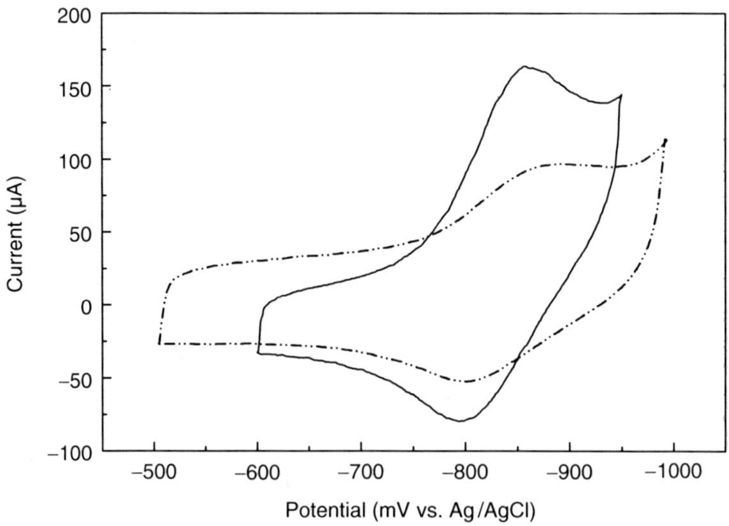

Figure 4.14. Comparison the reversibility of the single electron reduction of NO adducts generated *in situ* by prereduction of protein films nitrite solutions at pH 7. Solid line: Mb/DDAB; dashed line; CYP/DDAB[100].

1 V^{-1} s, Figure 4.14. At slower scan rates, the reduction is irreversible and the $Fe^{III/II}$ redox couple reappears. Although almost identical potentials and lifetimes for ferrous nitroxyl intermediate (Fe^{II}–NO^-) were found for both CYP119 and Mb, the catalytic efficiency the N_2O formation is much reduced for CYP119. This was attributed to both a lower affinity of the protein for NO, as well as a decreased rate of N–N coupling as determined by digital simulations. Thus, the thiolate ligation of CYP119 does not increase its effectiveness in N–N coupling reaction central to the $P450_{NoR}$ reactivity.

3.3. Hydroxylamine, NH_2OH

The next isolable intermediate in NO_x reduction is hydroxylamine, and in this area Ryan's group has provided insight[151,152]. Cyclic voltammograms of Fe(TPP) in a hydroxylamine solution show a one-electron reduction wave similar to the nitrosyl adduct, with complete conversion to a new species observable by both electrochemistry and visible absorbance measurements. Solution studies show that at low temperatures (−4°C), a stable $[Fe(TPP)(NH_2OH)_2]^+$ complex is formed, while at room temperature ($E_{1/2}$ of -1.22 V vs. $Ag/AgNO_3$ in methylene chloride) hydroxylamine reacts with ferric and ferrous Fe(TPP) to form Fe(TPP)NO and ammonia, Eq. (4.44).

$$[Fe(TPP)(NH_2OH)_2]^+ \rightarrow Fe(TPP)(NO) + NH_4{}^+ + H_2O \qquad (4.44)$$

Such a metal-assisted disproportionation of hydroxylamine may be initiated by a bimolecular reaction with uncoordinated hydroxylamine or by intramolecular electron transfer[152]. The presence of a reducible metal allows a concurrent three electron oxidation of hydroxylamine to nitric oxide with the two electron reduction of hydroxylamine to ammonia, Eqs. (4.45) and (4.46). Ultimately, the reaction proved to be an efficient synthetic route to the ferrous nitrosyl adduct, which rapidly precipitates from solution in high yield.

Similar reactions of hydroxylamine with Mn(TPP)Cl, Cr(TPP)Cl, and Co(TPP)Cl were also attempted: Mn(TPP)Cl reacts readily to form the nitrosyl complex, but Cr(TPP)Cl remained unreactive. Anaerobic Co^{II}(TPP) or Co^{III}(TPP) Cl showed no reactivity with hydroxylamine, while aerobic addition to Co^{II}(TPP) yielded a stable Co^{III} adduct, $[Co(TPP)(NH_2OH)_2]^+$; attempts to electrochemically reduce this species only led to degradation[152]. Thus reaction with hydroxylamine is dependent on both the binding affinity and reduction potential of the metal; it was concluded that the metal-based reduction potentials are most predictive of this reductive nitrosylation.

$$NH_2OH \rightarrow NO + 3H^+ + 3e^- \qquad (4.45)$$

$$NH_2OH + 2e^- + 2H^+ \rightarrow NH_3 + H_2O \qquad (4.46)$$

3.4. Co-Denitrification

Certain denitrifying bacteria (*Pseudomoras stutzeri*) and fungi (*Fusarium oxysporum*) can incorporate nitrogen atoms into N_2O from sources besides nitrite or nitric oxide, a process that has been termed "co-denitrification"[153,154]. Most typically, these utilize other nitrogen sources, like ammonia or azide, which cannot by themselves serve as N_2O precursors, but their incorporation into the gas during nitrite turnover is proven by ^{15}N labeling studies[155]. Immoos and coworkers from our lab have shown that co-dentrification of ammonia occurred during nitrite reduction by Mb/DDAB and CYP119/DDAB[100], and tested possible N–N coupled species as potential precursors for solution-based uptake of ^{15}N-labeled ammonia into N_2O. Still, the mechanism and chemical species responsible for these phenomena are not readily identifiable.

3.5. Nonbiomimetic substrates

3.5.1. Nitrate, NO_3^-

The biological reduction of nitrate is not heme-related; a well-known family of Mo-oxo pterin enzymes facilitate O-atom transfers between nitrite and nitrate, sulfite and sulfate, and other substrates[156,157]. Still, the electrochemical reduction of nitrate on bare metal electrodes is well known, and can also be facilitated by addition of N4 chelates. For instance, Hobbs *et al.* have examined the electrochemical reduction of nitrate in basic solutions using iron and nickel electrodes coated with phthalocyanin yielding nitrite, hydroxylamine, and ammonia[158]. The iron electrodes efficiency increased when coated with phthalocyanin while the nickel's activity decreased. A related study has been done by Shibata *et al.* to determine the synthetic applications of simultaneous reduction of nitrate and carbon dioxide using phthalocyanin complexes of most of the first-row transition metals and other metals with mixed success[159].

In a similar study, Cu, Ni, and unmetallated tetrakis (*ortho* or *para*-aminophenyl) porphyrin were polymerized onto SnO_2:F on glass electrodes to study the reduction of nitrate[160]. After 6 hr of bulk electrolysis the Cu and Ni *ortho* polymer produced nitrite and nitrous oxide while the other four polymers only produced nitrite. Nyokong and coworkers also studied nitrate reduction on Cu, Fe, Ni, Co, Mn, and Zn phthalocyanin adsorbed onto a GCE in basic solutions[96]. The main product was ammonia without any detectable nitrite; however,the current efficiency is lower than the GCE alone with only the Ni and Cophthalocyanin curbing the overpotential required to reduce nitrate.

3.5.2. Nitrous Oxide, N_2O; Azide, N_3^-, and Nitromethane, CH_3NO_2

The reduction of N_2O to N_2, carried by nitrous oxide reductases, N_2ORs, is the last step of denitrification, and is also used as an independent respiratory process by certain anaerobic bacteria[65,161]. But the catalytic centers of native N_2ORs known to date are almost exclusively copper-based. The Cu-containing N_2OR from *Wolinella* contains an Fe–heme suggested to be a part of a dinuclear Cu-heme catalytic center[162,163]. Though its role in the catalysis is controversial, the heme does seem to be obligatory for the activity of this enzyme[164].

The ability of simple N4 metal complexes and heme proteins to electrocatalytically reduce N_2O was first described for (tetraamino-phthalocyaninato)cobalt (Co(TAP)) adsorbed on graphite electrodes[165]. The reduction current plateaus observed during voltammetric N_2O reduction were essentially pH independent, suggesting that protons may not be involved in the rate-determining step. However, RDE measurements show that the catalytic halfwave potential shifts −56 mV/pH unit, indicative of a 1:1 electron to proton ratio. The k_f of the cobalt system was determined to be $7.5 \times 10^4\ M^{-1}\ s^{-1}$, and analysis of the L–K plots show that two electrons are involved, as expected for reduction of N_2O to N_2, with two protons included to form water in proposed reaction, Eq. (4.47). Other work by Lever, using a dinuclear ruthenium phthalocyanine, demonstrated that N_2O has the most negative reduction potential among NO, NH_2OH, and NO_2^-[87].

$$N_2O + 2H^+ + 2e^- \rightarrow N_2 + H_2O \tag{4.47}$$

The reduction of N_2O to N_2 was also examined using Mb/DDAB electrodes[166]. The addition of N_2O to electrochemical cell shows no change in the Fe^{III}/Fe^{II} redox couple but a catalytic current appears at potentials close to the Fe^{II}/Fe^{I} couple. The catalytic current diminishes to the noncatalytic, reversible Fe^{II}/Fe^{I} couple as the scan rate is increased above $1\ V^{-1}$ s. The head gas after bulk electrolysis at 1.1 V/SCE of doubly-labeled $^{15}N_2O$ showed $^{15}N_2$ as a product, without detected $^{15}NH_3$. The reactivity was attributed to the reaction of Fe^{I} with N_2O, and a catalytic cycle proposed as in Scheme 4.5.

When the isoelectronic anion azide (N_3^-) was used instead of N_2O, a similar catalytic current was also observed at the formal $Fe^{II/I}$ couple. But unlike voltammograms under N_2O, the presence of azide also produces a negative shift of the $Fe^{III/II}$ couple, *ca.* 100 mV, due to its binding at the $Fe^{III/II}$ centers. Bulk electrolysis confirmed production of ammonia, as expected for the overall two electron reduction in Eq. (4.48).

$$N_3^- + 2e^- + 3H^+ \rightarrow N_2 + NH_3 \tag{4.48}$$

The ability of the Fe^{I} state as a catalyst was also demonstrated in the reduction of nitromethane, CH_3NO_2[167]. Voltammetry of Mb/DDAB under solutions of nitromethane at pH 5.5, obtained a catalytic current at the formal $Fe^{II/I}$ couple, without an observable shift at the $Fe^{II}I/^{II}$ couple. Mass spectrometry analysis of aqueous products from bulk electrolysis confirmed production of methylhydroxylamine, CH_3NHOH. The catalytic efficiency, I_c/I_d current ratio, decreased significantly with increased pH or scan rate, indicative of a strongly proton dependant process. The slope of the catalytic potential was *ca.* 30 mV/pH unit, in agreement with the proposed $1e^-/2\ H^+$ limiting step, Eqs. (4.49)–(4.51).

$$\text{Mb–Fe}^{III} + 2e^- \rightarrow \text{Mb–Fe}^{I} \tag{4.49}$$

$$\text{Mb–Fe}^{I} + CH_3NO_2 + 1e^- + 2H^+ \rightarrow [CH_3N(O)\text{–Mb–Fe}^{II}] + H_2O \tag{4.50}$$

$$[CH_3N(O)\text{–Mb–Fe}^{II}] + 2e^- + 2H^+ \rightarrow \text{Mb–Fe}^{II} + CH_3NHOH \tag{4.51}$$

In this reaction sequence a ferrous-bound nitrosomethane intermediate ($MbFe^{II}$–CH_3NO) is suggested, and an absorbance change characteristic of such a species was observed in Mb/DDAB films on ITO after addition methylhydroxylamine. When the voltammetric window was expanded to -1.5 V an additional reduction wave, at a potential ~ -1.39 V SCE was observed with nitrosomethane substrate, the only catalytic wave with methylhydroxylamine substrate. This observation served as indirect proof of methylhydroxylamine as the catalytic product[167].

4. Conclusions

The recent advances in protein voltammetry hold much promise for the future as they now allow direct electrochemical investigations of the native enzymes. The NO_x enzymes, as multielectron reductases with sequential products isolable, can provide a clear target for demonstration of these methods. Still, there are many complicating factors with multicofactor enzymes and there remains a significant role for investigations using simpler small molecule catalysts. Below, we have listed several remaining questions which may be addressed by electrochemical investigations in both the native system and electrochemical models.

Is Fe^{II}–NO a kinetic trap? A common mechanistic quandary in both NiR and NoR chemistries is the possible inhibition of catalysis due to formation of a stable ferrous nitrosyl. Electrochemical studies of heme model systems suggest that the potential needed to reduce these species are well out of the normal range of biological reductants. Is the Fe^{II}–NO state truly a kinetic trap, and if so, how is it avoided during biological catalysis? There is most obviously a correlation between macrocyclic structure and the potential of the ferrous NO adduct, and therefore structural even for the isobacteriochlorin model complex would require a powerful reductant to generate a nitroxyl adduct.

What is the reactivity of the nitroxyladduct? As suggested above, the nitroxyl adduct plays a central role in the several NO_x reduction pathways. Recent results have shown that the N–N coupling reaction in P450nor enzymes proceeds through a nitroxyl intermediate, and that this species is formed by a hydride transfer to the coordinated ferric nitrosyl. Is this species protonated, as is the stable adduct of Mb? And, following the work of Ryan, how does the pK_a of this species control its subsequent reactions, N–N coupling or further reduction?

What is the sequence of multielectron reductions in the NiR? There is as yet no consensus on the sequence of multielectron transformations between NO and NH_3 in the aNiR, and likewise no identifiable Fe-bound species past the nitroxyl adduct. The two electron priming of the siroheme/FeS site as proposed by Cowan remains difficult to rationalize with the observed one and three electron reduction steps seen by Ryan for the simple model Fe(TPP)(NO).

Which are the roles played by the distal pocket residues? What are the requirements for specific active-site residues that may facilitate binding and protonation of nitrite to initiate the dehydration reaction, stabilize certain reduced intermediates, and perhaps serve as redox-active participants (e.g. hydride sources)?

Acknowledgments

The authors wish to thank several researchers for access to their insight and data in the compilation of this review, including Dr. Elisabeth Lojou at BIP-CNRS, Professor Michael Ryan at Marquette University, Professor Fethi Bedioui at Ecole Nationale Supérieure de Chimie de Paris, and Professor Julea Butt at the University of East Anglia. We also acknowledge the financial support by the National Science Foundation (PJF CHE-0100774) and the UC Toxic Substances Research and Teaching Program (EB). We also wish to thank the editors for their patience and assistance during the writing of this review.

References

1. Dawson, J.H. (1988). Probing structure–function relations in heme-containing oxygenases and peroxidases. *Science* **240**, 433–439.
2. Babcock, G.T., R. Floris, T. Nilsson, M.A. Pressler, C. Varotsis, and E. Vollenbroek (1996). Dioxygen activation in enzymic systems and in inorganic models. *Inorg. Chim. Acta* **243**, 345–353.
3. Averill, B.A. (1996). Dissimilatory nitrite and nitric-oxide Reductases. *Chem. Rev.* **96**, 2951–2964.
4. Guerrero, M.G., J.M. Vega, and M. Losada (1981). The assimilatory nitrate-reducing system and its regulation. *Annu. Rev. Plant Physiol.* **32**, 169–204.
5. Brittain, T., R. Blackmore, C. Greenwood, and A.J. Thomson (1992). Bacterial nitrite-reducing enzymes. *Eur. J. Biochem.* **209**, 793–802.
6. Silvestrini, M.C., S. Falcinelli, I. Ciabatti, F. Cutruzzola, and M. Brunori (1994). *Pseudomonas aeruginosa* nitrite reductase (or cytochrome oxidase): An overview. *Biochimie* **76**, 641–654.
7. Zumft, W.G., (1994). The Denitrifying Prokaryotes. In A. Balows, H.G. Truper, M. Dworkin, W. Harder, and K.H. Schleifer (Eds), *The Prokaryotes: A Handbook on the biology of bacteria: Ecophysiology; isolation; identification; applications.* Springer-Verlag, New York, pp. 554–582.
8. Ferguson, S.J. (1994). Denitrification and its control. *Anton. Leeuw.* **66**, 89–110.
9. Henry, Y. and P. Bessieres (1984). Denitrification and nitrite Reduction. *Biochimie* **66**, 259–289.
10. Rhee, S.K., J.J. Lee, and S.T. Lee (1997). Nitrite accumulation in a sequencing batch reactor during the aerobic phase of biological nitrogen removal. *Biotechnol. Lett.* **19**, 195–198.
11. Wicht, H. (1996). A model for predicting nitrous-oxide production during denitrification in activated-sludge. *Water Sci. Technol.* **34**, 99–106.
12. Apel, W.A. and C.E. Turick (1993). The use of denitrifying bacteria for the removal of nitrogen oxides from combustion gases. *Fuel* **72**, 1715–1718.
13. Kleifges, K.H., G. Kreysa, and K. Juttner (1997). An indirect electrochemical process for the removal of NO_x from industrial waste gases. *J. Appl. Electrochem.* **27**, 1012–1020.

14. Hollocher, T.C. (1996). The enzymology and occurrence of nitric oxide in the biological nitrogen cycle. In J. Lancaster, Jr. (ed), *Nitric oxide: Principles and actions*. Academic Press, San Diego, p. 289.
15. M.D., Bartberger, W. Liu, E. Ford, K.M. Miranda, C. Switzer, J.M. Fukuto, P.J. Farmer, D.A. Wink, and K.N. Houk (2002). The reduction potential of nitric oxide (NO) and its importance to NO biochemistry. *Proc. Natl Acad. Sci.* **99**, 10958–10963.
16. Leger, C., S.J. Elliott, K.R. Hoke, L.J.C. Jeuken, A.K. Jones, and F.A. Armstrong (2003). Enzyme electrokinetics: Using protein film voltammetry to investigate redox enzymes and their mechanisms. Biochemistry **42**, 8653–8662.
17. Armstrong, F.A., H.A. Heering, and J. Hirst (1997). Reactions of complex metalloproteins studied by protein-film voltammetry. *Chem. Soc. Rev.* **26**, 169–179.
18. Rusling, J.F. and Z. Zhang (2002). Polyion and surfactant films on electrodes for protein electrochemistry. *Electroanal. Meth. Bio. Mat.* 195–231.
19. Rusling, J.F. and R.J. Forster (2003). Electrochemical catalysis with redox polymer and polyion-protein films. *J. Colloid Interf. Sci.* **262**, 1–15.
20. Davis, J.J., H.A.O. Hill, and A.M. Bond (2000). The application of electrochemical scanning probe microscopy to the interpretation of metalloprotein voltammetry. *Coord. Chem. Rev.* **200–202**, 411–442.
21. Armstrong, F.A. (2002). Insights from protein film voltammetry into mechanisms of complex biological electron-transfer reactions. *Dalton Trans.* **5**, 661–671.
22. Bond, A.M. (1993). New understandings gained from chemical and electrochemical investigations of redox reactions of electron transfer metalloproteins and enzymes: Relevance to voltammetric (amperometric) biosensors. *Anal. Proc.* **30**, 218–226.
23. Rusling, J.F. (1998). Enzyme bioelectrochemistry in cast biomembrane-like films. *Acc. Chem. Res.* **31**, 363–369.
24. Guo, M., F. Sulc, M.W. Ribbe, P.J. Farmer, and B.K. Burgess (2002). Direct assessment of the reduction potential of the $[4Fe\text{-}4S]^{1+/0}$ couple of the Fe protein from *Azotobacter vinelandii*. *J. Am. Chem. Soc. Commun.* **124**, 12100–12101.
25. Vega, J.M., J. Cardenas, and M. Losada (1980). Ferrodoxin-nitrite Reductase. *Methods Enzymol.* **69**, 255–270.
26. Fujita, E. and J. Fajer (1983). Models for nitrite reductases. Redox chemistry of iron-nitrosyl porphyrins, chlorins, and isobacteriochlorins and p cation radicals of cobalt-nitrosyl isobacteriochlorins. *J. Am. Chem. Soc.* **105**, 6743–6745.
27. Krueger, R.J. and L.M. Siegel (1982). Spinach siroheme enzymes: Isolation and characterization of ferredoxin-sulfite reductase and comparison of properties with ferredoxin-nitrite reductase. *Biochemistry* **21**, 2892–2904.
28. Lancaster, J.R., J.M. Vega, H. Kamin, N.R. Orme-Johnson, W.H. Orme-Johnson, R.J. Krueger, and L.M. Siegel (1979). Identification of the Iron–sulfur center of spinach ferredoxin-nitrite reductase as a tetranuclear center, and preliminary EPR studies of mechanism. *J. Biol. Chem.* **254**, 1268–1272.
29. Mikami, B. and S. Ida (1989). Spinach ferredoxin-nitrite reductase: Characterization of catalytic activity and interaction of the enzyme with substrates. *J. Biochem.* **105**, 47–50.
30. Christner, J.A., E. Munck, T.A. Kent, P.A. Janick, J.C. Salerno, and L.M. Siegel (1984). Exchange coupling between siroheme and [4Fe–4S] cluster in *E. Coli* sulfite reductase. Mossbauer studies and coupling models for a 2-electron reduced enzyme state and complexes with sulfide. *J. Am. Chem. Soc.* **106**, 6786–6794.
31. Christner, J.A., E. Munck, P.A. Janick, and L.M. Siegel (1981). Mossbauer spectroscopic studies of *Escherichia coli* sulfite reductase. *J. Biol. Chem.* **256**, 2098–2101.

32. Wilkerson, J.O., P.A. Janick, and L.M. Siegel (1983). Electron paramagnetic resonance and optical spectroscopic evidence for interaction between siroheme and tetranuclear iron–sulfur center prosthetic groups in spinach ferredoxin-nitrite reductase. *Biochemistry* **22**, 5048–5054.
33. Hirasawa, M., G. Tollin, Z. Salamon, and D.B. Knaff (1994). Transient kinetic and Oxidation–reduction studies of spinach ferredoxin:nitrite Oxidoreductase. *Biochim. Biophys. Acta* **1185**, 336–345.
34. Lui, S.M. and J.A. Cowan (1994). Direct reversible protein electrochemistry at a pyrolytic graphite electrode. Characterization of the redox thermodynamics of the Fe_4S_4-siroheme prosthetic center in the hexameric dissimilatory sulfite reductase and the monomeric assimilatory sulfite reductase from desulfovibrio vulgaris (Hildenbourgh). Systematic pH titration experiments and implications for active site chemistry. *J. Am. Chem. Soc.* **116**, 11538–11549.
35. Soriano, A. and J.A. Cowan, Sulfite reductase: Active site residues are "noncatalytic." Comparison of reaction energetics for enzyme- and siroheme-catalyzed reduction of inorganic substrates. *J. Am. Chem. Soc.* **117**, 4724–4725.
36. Lui, S.M., A. Soriano, and J.A. Cowan (1993). Enzymatic reduction of inorganic anions. Pre-steady-state kinetic analysis of the dissimilatory sulfite reductase (desulfoviridin) from desulfovibrio vulgaris (Hildenborough). Mechanistic implications. *J. Am. Chem. Soc.* **115**, 10483–10486.
37. Lui, S.M., W. Liang, A. Soriano, and J.A. Cowan (1994). Enzymatic reduction of inorganic anions-variable-temperature steady-state and pre-steady-state kinetics experiments to map the energy profile of an enzymatic multielectron redox reaction — application to the dissimilatory sulfite reductase from desulfovibrio-vulgaris (Hildenborough). *J. Am. Chem. Soc.* **116**, 4531–4536.
38. Einsle, O., A. Messerschmidt, R. Huber, P.M.H. Kroneck, and F. Neese (2002). Mechanism of the six-electron reduction of nitrite to ammonia by cytochrome *c* nitrite reductase. *J. Am. Chem. Soc.* **124**, 11737–11745.
39. Angove, H.C., J.A. Cole, D.J. Richardson, and J.N. Butt (2002). Protein film voltammetry reveals distinctive fingerprints of nitrite and hydroxylamine reduction by a cytochrome c nitrite reductase. *J. Biol. Chem.* **277**, 23374–23381.
40. Gwyer, J.D., H.C. Angove, D.J. Richardson, and J.N. Butt (2004). Redox-triggered events in cytochrome c nitrite reductase. *Bioelectrochemistry* **63**, 43–47.
41. Einsle, O., A. Messerschmidt, P. Stach, G.P. Bourenkov, H.D. Bartunik, R. Huber, and P.M.H. Kroneck (1999). Structure of cytochrome *c* nitrite reductase. *Nature (London)* **400**, 476–480.
42. Rudolf, M., O. Einsle, F. Neese, and P.M.H. Kroneck (2002). Pentahaem cytochrome *c* nitrite reductase: Reaction with hydroxylamine, a potential reaction intermediate and substrate. *Biochem. Soc. Transact.* **30**, 649–653.
43. Einsle, O., A. Messerschmidt, R. Huber, P.M.H. Kroneck, and F. Neese (2002). Mechanism of the six-electron reduction of nitrite to ammonia by cytochrome *c* nitrite reductase. *J. Am. Chem. Soc.* **124**, 11737–11745.
44. Silaghi-Dumitrescu, R. (2003). Nitric oxide reduction by heme-thiolate enzymes (P450nor): A reevaluation of the mechanism. *Eur. J. Inorg. Chem.* **6**, 1048–1052.
45. Fülöp, V., J.W.B. Moir, S.J. Ferguson, and J. Hajdu (1995). The anatomy of a bifunctional enzyme: Structural basis for reduction of oxygen to water and synthesis of nitric oxide by cytochrome cd_1, *Cell* **81**, 369–377.
46. Chang, C.K., R. Timkovich, and W. Wu (1986). Evidence that heme d_1 is a 1,3porphyrindione. *Biochem.* **25**, 8447–8453.
47. Allen, J.W.A., N.J. Watmough, and S.J. Ferguson (2000). A switch in heme axial ligation prepares *Paracoccus pantotrophus* cytochrome cd_1 for catalysis. *Nat. Struct. Biol.* **7**, 885–888.

48. Steensma, E., E. Gordon, L.M. Oster, S.J. Ferguson, and J. Hajdu (2001). Heme ligation and conformational plasticity in the isolated c domain of cytochrome cd_1 nitrite reductase. *J. Biol. Chem.* **276**, 5846–5855.
49. Wang, Y. and B.A. Averill (1996). Direct observation by FTIR spectroscopy of the ferrous heme–NO+ intermediate in reduction of nitrite by a dissimilatory heme cd_1 nitrite reductase. *J. Am. Chem. Soc.* **118**, 3972–3973.
50. George, S.J., J.W.A. Allen, S.J. Ferguson, and R.N.F. Thorneley (2000). Time-resolved infrared spectroscopy reveals a stable ferric heme–NO intermediate in the reaction of *Paracoccus pantotrophus* cytochrome cd_1 nitrite reductase with nitrite. *J. Biol. Chem.* **275**, 33231–33237.
51. Lojou, E., F. Cutruzzola, M. Tegoni, and P. Bianco (2003). Electrochemical study of the intermolecular electron transfer to *Pseudomonas aeruginosa* cytochrome cd_1 nitrite reductase. *Electrochim. Acta* **48**, 1055–1064.
52. Averill, B.A. and J.M. Tiedje (1982). The chemical mechanism of microbial denitrification. *FEBS Lett.* **138**, 8–12.
53. Aerssens, E., J.M. Tiedje, and B.A. Averill (1986). Isotope labeling studies on the mechanism of N–N bond formation in denitrification. *J. Biol. Chem.* **261**, 9652–9656.
54. Shearer, G. and D.H. Kohl (1988). Nitrogen isotopic fractionation and ^{18}O exchange in relation to the mechanism of denitrification of nitrite by *Pseudomonas stutzeri*. *J. Biol. Chem.* **263**, 13231–13245.
55. Ye, R.W., I. Toro-Suarez, J.M. Tiedje, and B.A. Averill (1991). $H_2{}^{18}O$ isotope exchange studies on the mechanism of reduction of nitric oxide and nitrite to nitrous oxide by denitrifying bacteria. Evidence for an electrophilic nitrosyl during reduction of nitric oxide. *J. Biol Chem.* **266**, 12848–12851.
56. Dermastia, M., T. Turk, and T.C. Hollocher (1991). Nitric oxide reductase: Purification from *Paracoccus denitrificans* with the use of a single column and some characteristics. *J. Biol. Chem.* **266**, 10899–10905.
57. Goretski, J. and T.C. Hollocher (1990). The kinetic and isotopic competence of nitric oxide as an intermediate in denitrification. *J. Biol.Chem.* **265**, 889–895.
58. Goretski, J., O.C. Zafiriou, and T.C. Hollocher (1990). Steady-state nitric oxide concentrations during denitrification. *J. Biol. Chem.* **265**, 11535–11538.
59. Garber, E.A.E. S. Wehrli, and T.C. Hollocher (1983). Nitrogen-15 tracer and NMR studies on the pathway of denitrification. Evidence against trioxodinitrate but for nitroxyl as an Intermediate. *J. Biol. Chem.* **258**, 3587–3591.
60. Garber, E.A.E. and T.C. Hollocher (1982). Positional isotopic equivalence of nitrogen in nitrous oxide produced by the denitrifying bacterium *Pseudomonas stutzeri*. Indirect evidence for a nitroxyl pathway. *J. Biol. Chem.* **257**, 4705–4708.
61. Kim, C.H. and T.C. Hollocher (1983). ^{15}N tracer studies on the reduction of nitrite by the purified dissimilatory nitrite reductase of *Pseudomonas Aeruginosa*. *J. Biol. Chem.* **258**, 4861–4863.
62. Farmer, P.J., R. Lin, and M. Bayachou (1998). Electrochemistry and catalysis by myoglobin in surfactant films. *Commun. Inorg. Chem.* **20**, 101–120.
63. Zhao, X.J., V. Sampath, and W.S. Caughey (1995). Cytochrome *c* oxidase catalysis of the reduction of nitric oxide to nitrous oxide. *Biochem. Biophys. Res. Commun.* **212**, 1054–1060.
64. Hendriks, J., A. Warne, U. Gohlke, T. Haltia, C. Ludovici, M. Lübben, and M. Saraste (1998). The active site of the bacterial nitric oxide reductase is a dinuclear iron center. *Biochemistry* **37**, 13102–13109.
65. Zumft, W.G. (1997). Cell biology and molecular basis of denitrification. Microbiology and molecular biology reviews. *Mol. Biol. Rev.* **61**, 533–616.

66. Moënne-Loccoz, P. and S. de Vries, (1998). Structural characterization of the catalytic high-spin heme *b*of nitric oxide reductase: A resonance raman study. *J. Am. Chem. Soc.* **120**, 5147–5152.
67. Cheesman, M.R., W.G. Zumft, and A.J. Thomson (1998). The MCD and EPR of the heme centers of nitric oxide reductase from *Pseudomonas stutzeri*: Evidence that the enzyme is structurally related to the heme–copper oxidases. *Biochemistry* **37**, 3994–4000.
68. Ju, T.D., A.S. Woods, R.J. Cotter, P. Moenne-Loccoz, and K.D. Karlin (2000). Dioxygen and nitric oxide reactivity of a reduced heme/non-heme diiron(II) complex [(5L)FeII—FeII-Cl]+. Using a tethered tetraarylporphyrin for the development of an active site reactivity model for bacterial nitric oxide reductase. Nitric oxide reductase from *Paracoccus denitrificans* contains an oxo-bridged heme/non-heme diiron Center. *Inorganica Chim. Acta* **297**, 362–372.
69. Moeenne-Loccoz, P., O.M.H. Richter, H.W. Huang, I.M. Wasser, R.A. Ghiladi, and K.D. Karlin, and S. de Vries (2000). Nitric oxide reductase from *Paracoccus denitrificans* contains an oxo-bridged heme-non-heme diiron center. *J. Am. Chem. Soc.* **122**, 9344–9345.
70. Shiro, Y., M. Fujii, T. Iizuka, S.I. Adachi, K. Tsukamoto, K. Nakahara, and H. Shoun (1995). Spectroscopic and kinetic studies on reaction of cytochrome P450nor with nitric oxide. Implication for its nitric oxide reduction mechanism. *J. Biol. Chem.* **270**, 1617–1623.
71. Shiro, Y., M. Fujii, Y. Isogai, S.I. Adachi, T. Iizuka, E. Obayashi, R. Makino, K. Nakahara, and H. Shoun (1995). Iron-ligand structure and iron redox property of nitric oxide reductase cytochrome P450nor from *Fusarium oxysporum*: Relevance to its NO reduction activity. *Biochem.* **34**, 9052–9058.
72. K. Nakahara, T. Tanimoto, K. Hatano, K. Usuda, and H. Shoun (1993). Cytochrome P-450 55A1 (P-450dNIR) acts as nitric oxide reductase employing NADH as the direct electron donor. *J. Biol. Chem.* **268**, 8350–8355.
73. Daiber, A., T. Nauser, N. Takaya, T. Kudo, P. Weber, C. Hultschig, H. Shoun, and V. Ullrich (2002). Isotope effects and intermediates in the reduction of NO by P450nor. *J. Inorg. Biol. Chem.* **88**, 343–352.
74. Lin, R. and P.J. Farmer (2000). The HNO adduct of myoglobin: Synthesis and characterization. *J. Am. Chem. Soc.* **122**, 2393–2394.
75. Marchenko, A.V., A.N. Vedernikov, D.F. Dye, M.Pink, J.M. Zaleski, and K.G. Caulton (2004). Reactivity of the hydrido/nitrosyl radical $MHCl(NO)(CO)(P^iPr_3)_2$, M = Ru, Os. *Inorg. Chem.* **43**, 351–360.
76. Gamgee, A. (1868). Researches on the blood. On the action of nitrites on blood. *Phil. Trans. R. Soc. Lond.*, **158**, 589–625.
77. Kiese, M. (1966). Biochemical production of ferrihemoglobin-forming derivatives from aromatic amines and mechanisms of ferrihemoglobin formation. *Pharmacol. Rev.* **18**, 1091–1161.
78. Swann, P.F. (1975). Toxicology of nitrate, nitrite, and N-nitroso Compounds. *J. Sci. Food Agric.* **26**, 1761–1770.
79. Doyle, M.P., D.M. LePoire, and R.A. Pickering (1985). Oxidation of hemoglobin and myoglobin by alkyl nitrites inhibition by oxygen. *J. Biol. Chem.* **256**, 12399–12404.
80. Doyle, M.P., J.G. Herman, and R.L. Dykstra (1985). Autocatalytic oxidation of hemoglobin induced by nitrite: Activation and chemical inhibition. *J. Free Radicals Biol. Med.* **1**, 145–153.
81. Barley, M.H., K.J. Takeuchi, and T.J. Meyer (1986). Electrocatalytic reduction of nitrite to ammonia based on a water-soluble iron porphyrin. *J. Am. Chem. Soc.* **108**, 5876–5885.

82. Barley, M.H., M.R. Rhodes, and T.J. Meyer (1987). Electrocatalytic reduction of nitrite to nitrous oxide and ammonia based on the N-methylated, cationic iron porphyrin complex $[Fe^{III}(H_2O)(TMPyR)]^{5+}$. *Inorg. Chem.* **26**, 1746–1750.
83. Cheng, S.-H. and Y.O. Su (1994). Electrocatalysis of nitric oxide reduction by water-soluble cobalt porphyrin. Spectral and electrochemical studies. *Inorg. Chem.* **33**, 5847–5854.
84. Younathan, J.N., K.S. Wood, and T.J. Meyer (1992). Electrocatalytic reduction of nitrite and nitrosyl by iron(III) protoporphyrin IX dimethyl ester immobilized in an electropolymerized film. *Inorg. Chem.* **31**, 3280–3285.
85. Hayon, J., A. Raveh, and A. Bettelheim (1993). Electrocatalytic properties of chemically polymerized films of cobalt, iron and manganese tetrakis(*o*-aminophenyl)porphyrins. *J. Electroanal. Chem.* **359**, 209–221.
86. Nevin, W.A., W. Liu, S. Greenberg, M.R. Hempstead, S.M. Marcuccio, M. Melnik, C.C. Leznoff, and A.B.P. Lever (1987). Synthesis, aggregation, electrocatalytic activity, and redox properties of a tetranuclear cobalt phthalocyanine. *Inorg. Chem.* **26**, 891–899.
87. Ebadi, M. and A.B.P. Lever (2003). Electroreduction of nitrite catalyzed by a dinuclear ruthenium phthalocyanine modified graphite electrode. *J. Porphyrins Phthalocyanines,* **7**, 529–539.
88. Finnegan, M.G., A.G. Lappin, and W.R. Scheidt (1990). Instability of the nitrite/iron(III) porphyrinate system. *Inorg. Chem.* **29**,181–185.
89. Nasri, H., M.K. Ellison, C. Krebs, B.H. Huynh, and W.R. Scheidt (2000). Highly variable π-bonding in the interaction of iron(II) porphyrinates with nitrite. *J. Am. Chem. Soc.* **122**, 10795–10804.
90. Fernandes, J.B., D.W. Feng, A. Chang, A. Keyser, and M.D. Ryan (1986). Electrochemistry of nitrite redcutase model compounds. 2. Formation of an iron bis-nitro porphyrin complex. *Inorg. Chem.* **25**, 2606–2610.
91. Wei, Z. and M.D. Ryan (2001). Electrochemistry and spectroelectrochemistry of iron porphyrins in the presence of nitrite. *Inorg. Chim. Acta* **314**, 49–57.
92. Munro, O.Q. and W.R. Scheidt (1998). (Nitro)iron(III) porphyrins. EPR detection of a transient low-spin iron(III) complex and structural characterization of an O atom transfer product. *Inorg. Chem.* **37**, 2308–2316.
93. Trofimova, N.S., A.Y. Safronov, and O. Ikeda (2003). Electrochemical and spectral studies on the reductive nitrosylation of water-soluble iron porphyrin. *Inorg. Chem.* **42**, 1945–1951.
94. Zheng, D., L. Yan, and R.L. Burke (2002). Electrochemical and spectral studies of the reactions of aquocobalamin with nitric oxide and nitrite ion. *Inorg. Chem.* **41**, 2548–2555.
95. Choi, H.J., G. Kwag, and S. Kim (2001). Electrochemical and XAFS investigation of nitrite reduction by heat-treated μ-oxo derivative if iron phthalocyanin supported on high area carbon. *J. Electroanal. Chem.* **508**, 105–114.
96. Chebotareva, N. and T. Nyokong (1997). Metallophthalocyanine catalysed electroreduction of nitrate and nitrite ions in alkaline media. *J. Appl. Electrochem.* **27**, 975–981.
97. Rusling, J.F. and A.E.F. Nassar (1993). Enhanced electron transfer for myoglobin in surfactant films on electrodes. *J. Am. Chem. Soc.* **115**, 11891–11897.
98. Nassar, A.-E., Z. Zhang, V. Chynwat, H.A. Frank, J.F. Rusling, and K. Suga (1995). Orientation of myoglobin in cast multibilayer membranes of amphiphilic molecules. *J. Phys. Chem.* **99**, 11013–11017.
99. Lin, R., M. Bayachou, and P.J. Farmer (1997). Nitrite Reduction by myoglobin in surfactant films. *J. Am. Chem. Soc.* **119**, 12689–12690.

100. Immoos, C.E., J. Chou, M. Bayachou, E. Blair, and P.J. Farmer (2004). Electrocatalytic reductions of nitrite, nitric oxide and nitrous oxide by Cytochrome P450 CYP 119. *J. Am. Chem. Soc.* **126**, 4934–4942.
101. E. Blair, J. Greaves, P.J. Farmer, High-temperature electrocatalysis using thermophilic P450 CYP119: Dehalogenation of CCl_4 to CH_4, *J. Am. Chem. Soc.* **126**, 8632–8633 (2004).
102. Sulc, F., E. Fleischer, P.J. Farmer, D. Ma, and G. La Mar (2003). ^{1}H NMR structure of the heme pocket of HNO-myoglobin. *J. Biol. Inorg. Chem.* **8**, 348–352.
103. Mimica, D., J.H. Zagal, and F. Bedioui (2001). Electroreduction of nitrite by hemin, myoglobin and hemoglobin in surfactant films. *J. Electroanal. Chem.* **497**, 106–113.
104. Liu, S. and H. Ju (2003). Nitrite reduction and detection at a carbon paste electrode containing hemoglobin and colloidal gold. *Analyst* **128**, 1420–1424.
105. Fan, C., X. Chen, G. Li, J. Zhu, D. Zhua, and H. Scheer (2000). Direct electrochemical characterization of the interaction between haemoglobin and nitric oxide. *Phys. Chem. Chem. Phys.* **2**, 4409–4413.
106. Zhou, Y., N. Hu, Y. Zeng, and J.F. Rusling (2002). Heme protein–clay films: Direct electrochemistry and electrochemical catalysis. *Langmuir* **18**, 211–219.
107. Huang, H., P. He, N. Hu, and Y. Zeng (2003). Electrochemical and electrocatalytic properties of myoglobin and hemoglobin incorporated in carboxymethyl cellulose films. *Bioelectrochemistry* **61**, 29–38.
108. Huang, H., N. Hu, Y. Zeng, and G. Zhou (2002). Electrochemistry and electrocatalysis with heme proteins in chitosan biopolymer films. *Anal. Biochem.* **308**, 141–151.
109. Cao, D., P. He, and N. Hu (2003). Electrochemical biosensors utilizing electron transfer in heme proteins immobilised on Fe_3O_4 nanoparticles. *Analyst* **128**, 1268–1274).
110. Wang, Q., G. Lu, and B.Yang (2004). Myoglobin/sol–gel film modified electrode: Direct electrochemistry and electrochemical catalysis. *Langmuir* **20**, 1342–1347.
111. Shen, L. and N. Hu (2004). Heme protein films with polyamidoamine dendrimer: Direct electrochemistry and electrocatalysis. *Biochim. Biophys. Acta* **1608**, 23–33.
112. Sun, H. and N. Hu (2004). Voltammetric studies of hemoglobin-coated polystyrene latex bead films on pyrolytic graphite electrodes. *Biophys. Chem.* **110**, 297–308.
113. Wang, L. and N. Hu (2001). Direct electrochemistry of hemoglobin in layer-by-layer films with poly(vinyl sulfonate) grown on pyrolytic graphite electrodes. *Bioelectrochemistry* **53**, 205–212.
114. Wang, Q., G. Lu, and B. Yang (2004). Direct electrochemistry and electrocatalysis of hemoglobin immobilized on carbon paste electrode by silica sol–gel film. *Biosensors Bioelectron* **19**, 1269–1275.
115. Chen, S. and C. Tseng (2004). The characterization and bioelectrocatalytic properties of hemoglobin by direct electrochemistry of DDAB film modified electrodes. *Electrochimica Acta* **49**, 1903–1914.
116. Huang, R. and N. Hu (2003). Direct voltammetry and electrochemical catalysis with horseradish peroxidase in polyacrylamide hydrogel films. *Biophys. Chem.* **104**, 199–208.
117. Huang, R., and N. Hu (2003). Direct electrochemistry and electrocatalysis with horseradish peroxidase in Eastman AQ films. *Bioelectrochemistry* **54**, 75–81.
118. Thomas, D.D., K.M. Miranda, C.A. Colton, D. Citrin, M.G. Espey, and D.A. Wink (2003). Heme proteins and nitric oxide (NO): The neglected, eloquent chemistry in NO redox signaling and regulation. *Antioxid. Redox Signal.* **5**, 307–317.
119. Clancy, R.M. and S.B. Abramson (1995). Nitric oxide: A novel mediator of Inflammation. *Proc. Soc. Exp. Biol. Med.* **210**, 93–101.
120. Tsai, A.L. (1994). How does NO activate hemeproteins? *FEBS Lett.* **341**, 141–145.

121. Kurata, S., M. Matsumoto, and U. Yamashita (1996). Concomitant transcriptional activation of nitric oxide synthase and heme oxygenase genes during nitric oxide-mediated macrophage cytostasis. *J. Biochem.* **120**, 49–52.
122. Paolocci, N., W.F. Saavedra, K.M. Miranda, C. Martignani, T. Isoda, J.M. Hare, M.G. Espey, J.M. Fukuto, M. Feelisch, D.A. Wink, and D.A. Kass (2001). Nitroxyl anion exerts redox-sensitive positive cardiac inotropy in vivo by calcitonin gene-related peptide signaling. *Proc. Natl Acad. Sci. USA* **98**, 10463–10468.
123. Paolocci, N., T. Katori, H.C. Champion, M.E. St. John, K.M. Miranda, J.M. Fukuto, D.A. Wink, and D.A. Kass (2003). Positive inotropic and lusitropic effects of HNO/NO^- in failing hearts: independence from β-adrenergic signaling. *Proc. Natl Acad. Sci. USA* **100**, 5537–5542.
124. Feelisch, M. (2003). Nitroxyl gets to the heart of the matter. *Proc. Natl Acad. Sci. USA* **100**, 4978–4980.
125. Olson, L.W., D. Schaeper, D. Lancon, and K.M. Kadish (1982). Characterization of several novel iron nitrosyl porphyrins. *J. Am. Chem. Soc.* **104**, 2042–2044.
126. Kelly, S., D. Lancon, and K.M. Kadish (1984). Electron-transfer and ligand-addition reactions of (TPP)Mn(NO) and (TPP)Co(NO) in nonaqueous media. *Inorg. Chem.* **23**, 1451–1458.
127. Kadish, K.M., J.-L. Cornillon, and C.-L. Yao (1987). Solvent effects on electrode potentials of metalloporphyrins. Reduction of 5, 10, 15, 20-tetraphenylporphinato complexes in non-aqueous media. *J. Electroanal. Chem.* **235**, 189–207.
128. Kadish, K.M. (1984). Redox tuning of metalloporphyrin reactivity. *J. Electroanal. Chem.* **108**, 261–274.
129. Liu, Y., C. DeSilva, and M.D. Ryan (1997). Electrochemistry of nitrite reductase model compounds. 6. Voltammetric and spectroelectrochemical studies of iron(II) nitrosyl complexes with porphyrins, hydroporphyrins and porphinones. *Inorg. Chim. Acta* **258**, 247–255.
130. Richter-Addo, G.B., S.J. Hodge, G.-B. Yi, M.A. Khan, T.Ma, E.V. Caemelbecke, N. Guo, and K.M. Kadish 1996. Synthesis, characterization, and spectroelectrochemistry of cobalt porphyrins containing axially bound nitric oxide. *Inorg. Chem.* **35**, 6530–6538.
131. Autret, M., S. Will, E.V. Caemelbecke, J. Lex, J.-P. Gisselbrecht, M. Gross, E. Vogel, and K.M. Kadish (1994). Synthesis and electrochemistry of iron(III) corroles containing a nitrosyl axial ligand. Spectral charecterization of $[(OEC)Fe^{III}(NO)]^n$ where $n = 0, 1, 2$, or -1 and OEC is the trianion of 2, 3, 7, 8, 12, 13, 17, 18-octaethylcorrole. *J. Am. Chem. Soc.* **116**, 9141–9149.
132. Kadish, K.M., Z. Ou, X. Tan, T. Boschi, D. Monti, V. Fares, and P. Tagliatesta (1999). Synthesis and electrochemistry of cobalt β-halogenated meso-tetraphenylporphyrins containing a nitrosyl axial ligand. Crystal structure of $(TPPBr_4NO_2)Co(NO)$. *J. Chem. Soc Dalton Trans.* 1595–1601.
133. Simkhovich, L., I. Luobeznova, I. Goldberg, and Z. Gross (2003). Mono- and binuclear ruthenium corroles: Synthesis, spectroscopy, electrochemistry, and structural characterization. *Chem. Eur. J.* **9**(1), 201–208.
134. Kelly, S., D. Lancon, and K.M. Kadish (1984). Electron-transfer and ligand-addition reactions of (TPP)Mn(NO) and (TPP)Co(NO) in nonaqueous media. *Inorg. Chem.* **23**, 1451–1458.
135. Lancon, D. and K.M. Kadish (1983). Electrochemical and spectral characterization of iron mono- and dinitrosyl porphyrins. *J. Am. Chem. Soc.* **105**, 5610–5617.
136. Kadish, K.M., V. A. Adamian, E.V. Caemaelbecke, Z. Tan, P. Tagliatesta, P. Bianco, T. Boschi, G.-B. Yi, M.A. Khan, and G.B. Richter-Addo (1996). Synthesis, characterization, and elecectrochemistry of ruthenium porphyrins containing a nitrosyl axial ligand. *Inorg. Chem.* **35**, 1343–1348.

137. Hayon, J., D. Ozer, J. Rishpon, and A. Bettelheim (1994). Spectroscopic and electrochemical response to nitrogen monoxide of a cationic iron porphyrin immobilized in nafion-coated electrodes or membranes. *J. Chem. Soc., Chem. Commun.* 619–620.
138. Bedioui, F., Y. Bouhier, C. Sorel, J. Devynck, L. Coche-Guerente, A. Deronzier, and J.C. Moutet (1993). Incorporation of anionic metalloporphyrins into poly(pyrrole-alkylammonium) films— Part 2. Characterization of the reactivity of the iron(III) porphyrinic-based polymer. *Electrochim. Acta* **38**, 2485–2491.
139. Bedioui, F., S. Trevin, V. Albin, M.G.G. Villegas, and J. Devynck (1997). Design and characterization of chemically modified electrodes with iron(III) porphyrinic-based polymers: Study of their reactivity toward nitrites and nitric oxide in aqueous solution. *Anal. Chim. Acta* **341**, 177–185.
140. Choi, I.-K., Y. Liu, D. Feng, K.-J. Paeng, and M.D. Ryan (1991). Electrochemical and spectroscopic studies of iron porphyrin nitrosyls and their reduction products. *Inorg. Chem.* **30**, 1832–1839.
141. Choi, I.-K. and M.D. Ryan (1992). Electrochemistry of nitrite reductase model compounds. 4. Electrochemistry and spectroelectrochemistry of iron chlorin and isobacteriochlorin complexes. *New J. Chem.* **16**, 591–597.
142. Liu, Y. and M.D. Ryan (1994). Electrochemistry of nitrite reductase model compounds. 5. Electrochemistry and spectrolelectrochemistry of iron porphinone, porphindione and isobacteriochlorin complexes. *Inorg. Chim. Acta* **225**, 57–66.
143. Liu, Y. and M.D. Ryan (1994). The electrochemical reduction of iron porphyrin nitrosyls in the presence of weak acids. *J. Electroanal. Chem.* **368**, 209–219.
144. Yu, C.H. and Y. O. Su (1994). Electrocatalytic reduction of nitric oxide by water-soluble manganese porphyrins. *J. Electroanal. Chem.* **368**, 323–327.
145. Chen, S.-M. and Y.O. Su (1990). Electrocatalytic reduction of nitric oxide to ammonia by water-soluble iron porphyrin. *J. Electroanal. Chem.* **280**, 189–194.
146. Zilbermann, I., J. Hayon, T. Katchalski, R. Ydgar, J. Rishpon, A.I. Shames, E. Korin, and A. Bettelheim (2000). Spectroscopic and electrochemical characterization of the interaction of nitrogen monoxide and cobalt tetrasulfonated phthalocyanine in aqueous solutions and surfactant films. *Inorg. Chim. Acta*, **305**, 53–60.
147. Vilakazi, S.L. and T. Nyokong (2000). Electrocatalytic properties of vitamin B_{12} towards oxidation and reduction of nitric oxide. *Electrochim. Acta* **46**, 453–461.
148. Vilakazi, S. and T. Nyokong (2000). Interaction of nitric oxide with cobalt(II) tetrasulfophthalocyanine. *Polyhedron* **19**, 229–234.
149. Vilakazi, S.L. and T. Nyokong (1998). Interaction of nitric oxide with cobalt(II) phthalocyanine: Kinetics, equilibria and electrocatalytic studies. *Polyhedron* **17**, 4415–4423.
150. Bayachou, M., R. Lin, W. Cho, and P.J. Farmer (1998). Electrochemical reduction of NO by myoglobin in surfactant film: characterization and reactivity of the nitroxyl (NO^-) adduct. *J. Am. Chem. Soc.* **120**, 9888–9893.
151. Feng, D.W. and M.D. Ryan (1987). Electrochemistry of nitrite reductase model compounds. 3. Formation and characterization of a bis(hydroxylamine) (tetraphenylporphyrinato)iron(III) complex. *Inorg. Chem.* **26**, 2480–2483.
152. Choi, I.-K., Y. Liu, Z. Wei, and M.D. Ryan (1997). Reactions of hydroxylamine with metal porphyrins. *Inorg. Chem* **36**, 3113–3118.
153. Tanimoto, T., K. Hatano, D. Kim, H. Uchiyama, and H. Shoun (1992). Co-denitrification by the denitrifying system of the fungus Fusarium oxysporum. *FEMS Microbiol. Lett.* **93**, 177–180.
154. Kumon, Y., Y. Sasaki, I. Kato, N. Takaya, H. Shoun, and T. Beppu (2002). Codenitrification and denitrification are dual metabolic pathways through which dinitrogen evolves from nitrate in Streptomyces antibioticus. *J. Bacteriol.* **184**, 2963–2968.

155. Laughlin, R.J. and R.J. Stevens (2002). Evidence for fungal dominance of denitrification and codenitrification in a grassland soil. *Soil Sci. Soc. Am.* **66**, 1540–1548.
156. Kroneck, P.M.H. and D.J. Abt (2002). Molybdenum in nitrate reductase and nitrite oxidoreductase. *Metal Ions Bio. Syst.* **39**, 369–403.
157. Blasco, F., B. Guigliarelli, A. Magalon, M. Asso, G. Giordano, and R.A. Rothery (2001). The coordination and function of the redox centers of the membrane-bound nitrate reductases. *Cell. Mol. Life Sci.* **58**(2), 179–193.
158. Li, H.L., J.Q. Chambers, and D.T. Hobbs (1988). Electroreduction of nitrate ions in concentrated sodium hydroxide solutions at lead, zinc, nickel and phthalocyanine-modified electrodes. *J. Appl. Electrochem.* **18**(3), 454–458.
159. Shibata, M. and N. Furuya (2003). Simultaneous reduction of carbon dioxide and nitrate ions at gas-diffusion electrodes with various metallophthalocyanin catalysts. *Electrochim. Acta* **48**, 3953–3958.
160. Armijo, F., M. Isaacs, G. Ramírez, E. Trollund, J. Canales, and M. Aguirre (2004). Electrocatalytic reduction of nitrate ion on Cu and Ni poly-tetraaminophenylporphyrin-modified electrodes. *J. Electroanal. Chem.* **566**, 315–322.
161. McEwan, A., A.J. Greenfield, H.G. Wetzstein, J.B. Jackson, and S.J. Ferguson (1985). Nitrous oxide reduction by members of the family Rhodospirillaceae and the nitrous oxide reductase of Rhodopseudomonas capsulate. *J. Bacteriol.* **164**(2), 823–830.
162. Teraguchi, S. and T.C. Hollocher (1989). Purification and some characteristics of a cytochrome c-containing nitrous oxide reductase from *Wolinella succinogenes*. *J. Biol. Chem.* **264**, 1972–1979.
163. Farrar, J.A., W.G. Zumft, and A.J. Thomson (1989). CuA and CuZ are variants of the electron transfer center in nitrous oxide reductase. *Proc. US Acad. Sci.* **95**, 9891–9896.
164. Zhang, C.S., T.C. Hollocher, A.F. Kolodziej, and W.H. Orme-Johnson (1991). Electron paramagnetic resonance observations on the cytochrome c-containing nitrous oxide reductase from *Wolinella succinogenes*. *J. Biol. Chem.* **266**, 2199.
165. Zhang, J., Y.-H. Tse, A.B.P. Lever, and W.J. Pietro (1997). Electrochemical reduction of nitrous oxide (N_2O) catalysed by tetraaminophtalocyanatocobalt(II) adsorbed on a graphite electrode in aqueous solution. *J. Porph. Phthal.* **1**, 323–331.
166. Bayachou, M., L. Elkbir, and P.J. Farmer (2000). Catalytic two-electron reductions of N_2O and N_3^- by myoglobin in surfactant films. *Inorg. Chem.* **39**, 289–293.
167. Boutros, J. and M. Bayachou (2004). Myoglobin as an efficient electrocatalyst for nitromethane reduction. *Inorg. Chem.* **43**(13), 3847–3853.

5

Electroreduction of CO_2 Catalyzed By Metallomacrocycles

Juan A. Costamagna, Mauricio A. Isaacs, María J. Aguirre, Galo Ramírez and Ignacio Azocar

1. Introduction

The use of fossil fuels in the past century has generated a continuous increase in the carbon dioxide atmospheric concentration, creating the "greenhouse effect", which produces undesirable changes in the global climate[1]. Carbon dioxide is a small molecule of utmost importance for the production and/or energy storage in nonconventional resources. Its reduction is a process of applicability in environmental chemistry related issues, such as the synthesis of ecologic fuels or the control of the greenhouse effect, responsible for the natural environment disruption. Carbon dioxide fixation is an important area of research, oriented to overcome problems related to the global environment and to the depletion of fossil fuels[2]. The conversion of CO_2 to organic compounds, principally formic acid, for using as fuel and for industrial applications has been studied since the beginning of this century[3,4]. During the last decades several methods of reducing CO_2 to CO, formaldehyde, formate, methanol, methane, oxalate and others, using electrochemical methods which consume less energy, have been developed[5–11]. However, a large scale method of reducing CO_2, equivalent to photosynthesis in green plants, has not yet been developed. If this is to be achieved, technologies

Juan A. Costamagna, María J. Aguirre, Galo Ramírez and Ignacio Azocar • Universidad de Santiago de Chile, Facultad de Química y Biología, Departamento de Química de los Materiales, Casilla 40- Correo 33, Santiago Chile. maguirre@lauca.usach.cl

Mauricio A. Isaacs • Universidad de Chile, Facultad de Ciencias, Departamento de Química, Casilla 653 Santiago, Chile. misaacs@uchile.cl

Juan A. Costamagna • To whom correspondence should be addressed. Juan A. Costamagna, Universidad de Santiago de Chile, Facultad de Química y Biología, Departamento de Química de los Materiales, Casilla 40- Correo 33, Santiago. Chile. jcostama@lauca.usach.cl; Fax: 56-2-6812108

N_4-Macrocyclic Metal Complexes, edited by José H. Zagal, Fethi Bedioui and Jean-Pol Dodelet. Springer Science+Business Media, Inc., New York, 2006.

for the removal and reduction of CO_2 will need to become more sophisticated. Since atmospheric CO_2 is one the gases contributing to the green house effect, and is produced by technologically oriented societies, this aim is interesting also in other areas of science[12]. The chemical fixation involves conversion of recovered or separated CO_2 into useful chemicals and fuels. Economical fixation of CO_2 is only feasible when solar or hydraulic energy are converted to chemical energy.

Several procedures have been reported that allow the use of solar energy for CO_2 fixation[13]. During the last years many studies of matters such as reduction at metallic electrodes, solvent effects on mechanisms, reduction in the presence of coordination compounds with open and closed macrocycles ligands, identification and characterization of products and mechanisms, photoelectroactivation and catalytic activity in homogeneous and heterogeneous media and modified electrodes have been reported[14–25].

However, the reduction (chemical, electrochemical or photochemical) of CO_2 is difficult. Carbon dioxide is the ultimate product of oxidation of carbon and its compounds and is thermodynamically stable. To convert CO_2 in carbon monoxide, high energy ($\Delta G^0 = -394\,\text{kJ mol}^{-1}$., cf $= -137\,\text{kJ mol}^{-1}$) is required[18,26,27]. The bond strength of C–O for carbon monoxide is the largest known ($1{,}076\,\text{kJ mol}^{-1}$)[18]. For carbon dioxide, that bond strength is lesser ($532\,\text{kJ mol}^{-1}$).

The redox potentials for the reduction of carbon dioxide have been determined:

$$CO_2 + e^- \longrightarrow CO_2^{\cdot -} \qquad E^o = -1.9\text{V vs. NHE} \tag{5.1}$$

$$CO_2^{\cdot -} + e^- \longrightarrow CO_2^{2-} \qquad E^o = -1.2\text{V vs. NHE} \tag{5.2}$$

For the first reduction, there is a change in geometry from the linear CO_2 to a bent $CO_2^{\cdot -}$. This structural change gives rise to a very slow self-exchange rate for the $CO_2/CO_2^{\cdot -}$ couple[28] and to a significant overpotential in the reduction of CO_2[28]. As noted previously the nature of $CO_2^{\cdot -}$ may be significant in the consideration of aspects of the activation of CO_2[28,29].

Coordination compounds, in particular, have shown catalytic activity toward the above-mentioned reactions, and have become very promising because they allow their rational optimization by means of properly designed modifications of either the ligand or the central metal atom. Coordination compounds, according to electrochemical reduction of CO_2 studies, can be separate in several structural groups. In general these groups include:

I. Phosphine complexes.
II. Complexes containing 2,2′-bypiridine and other related.
III. Metallic clusters and other polymetallic complexes.
IV. Cyclams, porphyrins, phthalocyanines and other N-macrocyles ligands.

In the group IV, macrocyclic complexes containing transition metals have been tested for the electrocatalytic reduction of carbon dioxide. $[Ni(cyclam)]^{2+}$

being one of the complexes most extensively studied[16]. Moreover, other macrocycles like porphyrins, phthalocyanines and other poly-aza-macrocycles also show electrocatalytic activity toward this reaction[21–24]. The aim of this chapter is to summarize recent research studies on electroinduced activation of CO_2 by using azamacrocyclic ligands. Detailed reviews of these matters appeared some years ago[15–25]. Additionally, some emphasis will be placed on the use of macrocyclic chemistry in the formation of supramolecular structures and modified electrodes as electrocatalysts for this reaction.

2. Electroreduction of CO_2 on Metallic Cathodes

The reduction of CO_2 at metallic cathodes has been studied with almost every element in the periodic table[30]. This reaction can be driven electrochemically or photochemically[5,30–46] and semiconductors have been used as cathodic materials in electrochemical or photoelectrochemical cells[30,44,45]. The aim of these studies has been to find cathodes that discriminate against the reduction of H_2O to H_2 and favor the reduction of CO_2 and also to find a cathode selective for one product in the reduction of CO_2. A fundamental requirement is that the latter process occurs at a lower overpotential on such electrodes. However the purposes mentioned before in metallic cathodes depends on a series of factors such a solvent, support electrolyte, temperature, pressure, applied overpotential, current density, etc. (we will see the same factors again in macrocyclic electrocatalysis). For instance when protons are not readily available from the solvent (e.g., *N,N'*-dimethylformamide), the electrochemical reduction involves three competing pathways–oxalate association through self-coupling of $CO_2^{\cdot-}$ anion radicals, production of CO via O–C coupling between and $CO_2^{\cdot-}$ and CO_2, and formate generation by interaction of $CO_2^{\cdot-}$ with residual or added water[46].

Copper electrodes have shown interesting electrocatalytic properties towards the reduction of CO_2 because they produce hydrocarbons such as CH_4 and C-2 compounds and considering the high accessibility of this metal comparing with noble metals its use as a electrode material is an important topic of research. For this reason we include some of the current literature in this chapter[5,31,32].

The low temperature (0–60°C) reduction of CO_2 to CH_4[30] on Cu foils near room temperature was first reported by Hory and coworkers[31]. Hori and coworkers [31] used neutral bicarbonate, and low temperature down 0°C but, for the reduction of CO_2, high overpotentials (>1 V) were needed. Methane formation at Cu was investigated[30] under much different conditions with observed low CH_4 production. High efficiencies for the generation of other hydrocarbon products were found, when copper foil electrodes were used during short electrolysis times. At longer times, the reaction efficiency decreases with a simultaneous increase in hydrogen evolution and the electrodes become almost completely inactive[36]. Electrode poisoning has been attributed to the formation of an adsorbed intermediate which accelerates hydrogen evolution[47]. This is caused by the deposition of compounds such as graphitic carbon originated from the adsorption of CO. In the last years many efforts have been focused to settle this problem.

Yano and coworkers[48] have developed an electrolysis system in which CO_2 is reduced at three phases (gas/liquid/solid) interface on a Cu-mesh electrode. In these studies a Cu-mesh electrode is modified beforehand by copper(I) halides, and CO_2 is reduced at constant potential with the modified electrode in an acidic solution of potassium halide. The faradaic efficiency for C_2H_4 is increased and for H_2 is decreased by the presence of the copper(I) halide. In the case of CuBr, the conversion of CO_2 to C_2H_4 and the hydrogen evolution are observed with faradaic efficiencies of about 80 and 90%, respectively. Such a contribution of copper(I) halide to the CO_2 reduction is related to its reversible combination with CO and C_2H_4. At the three-phase interface, CO_2(g) is first reduced to CO (g). This gas is readily adsorbed on the copper(I) halide with its π-bond perpendicular to the surface, and the CO is subjected to electron injection from the electrode to be reduced to the methylene radical. The coupling of methylene radicals results in the formation of C_2H_4 and this product is stabilized by adsorbing on copper(I) halide[48]. However one of the drawbacks of this method is the high overpotential needed to perform the reaction.

Also high selective reduction of CO_2 to C-2 compounds such as CH_3CHO, C_2H_5OH, and C_2H_4 has been achieved using CuAg alloy electrodes by pulsed electroreduction[49]. The faradaic efficiencies of C-2 compounds produced on CuAg alloy electrodes were varied with the atomic ratio of Cu to Ag. The total value of faradaic efficiencies for these compounds was 53.2% for the pulsed electroreduction on CuAg alloy electrode (Cu/Ag= 28/72) with an anodic bias of V–a = −0.4 V and cathodic bias of V–c = −2.0 V vs. Ag/AgCl. In this work it was found that the formation of an oxide layer on Cu and the desorption of intermediates on the alloy electrode under anodic bias were the key factors for the selective reduction of CO_2 to C-2 compounds[49].

Another example using alloys was presented by Schrebler and coworkers[50] where the electrocatalytic reduction of CO_2 was studied in CuRe alloy highly dispersed in polypirrole films at Au electrodes (Au/PpyCuRe). These electrodes were stable over 30 hr of electrolysis and the applied overpotential was less than −1.5 V. The faradaic efficiency for methane formation was 31% on Au/PpyCuRe and the intermediate CH_2 was detected[50].

It has also been suggested that the loss of catalytic activity of copper electrodes depends on the crystallographic properties of the electrode, the surface characteristics and the morphology[36,39,51]. The rate of methanol synthesis from a 1:1 mixture of CO_2 and H_2 at a Cu(100) single crystal has been measured and a kinetic model has been proposed[52,53]. This model correctly predicts the rates of methanol production in catalysts under industrial conditions.

When the electrochemical reduction of CO_2 was studied using single-crystal Cu electrode, Cu(111), Cu(100), Cu(s)-[n(100) × (111)] and Cu(s)-[n(100) × (110)] at constant current density of 5 mA cm^{-2} in bicarbonate aqueous solutions, the Cu(111) electrode yielded mainly CH_4 and Cu(100) gave C_2H_4. Introduction of (111) steps to Cu(100) basal plane leading to Cu(s)-[n(100) × (111)] orientations promoted C_2H_4 formation and inhibited CH_4 formation[20,52,53].

On the other hand, efforts to obtain more complicated products have been reported. The simultaneous reduction of CO_2 and NO_2^- anion was examined at gas-diffusion electrodes with various catalysts (Cr, Mo, Mn, Ru, Co, Ir, Pd, Pt, Cu,

Ag, Au, Zn, Cd, In, Tl, Sn and Pb)[54]. The formation of urea, CO, formic acid, and ammonia at gas diffusion electrodes with catalysts from groups 11–14 was found on the simultaneous reduction. The maximum current efficiency (c.e.) of urea formation at Cd catalyst is approximately 55% at −1.0 V. The formation of urea at the gas diffusion electrodes with catalysts from groups 6–10, except Pd, was not found for the simultaneous reduction of CO_2 and nitrite. Relationship of the ability of urea formation to the ability of CO and NH_3 formation was investigated at various catalysts. The c.e. of urea increases when there is an increasing on the c.e. of CO and NH_3 on the reduction of CO_2 alone and NO_2^- alone, respectively. The authors[54] conclude that the ability of urea formation depends on the ability of CO and NH_3 formation. The catalyst with the high ability for CO and NH_3 formation could form large amounts of CO-like and ammonia-like precursors. Urea would be formed from both the ammonia-like precursor formed from nitrite ions and the CO-like precursor formed from CO_2 catalyzed by elements from the 11–14 groups[54].

When using nitrate, the formation of urea, CO, formic acid, nitrite ions and ammonia[55,56] at the gas diffusion electrodes with elements from groups 11–14 as catalysts was found in the simultaneous reduction, except for Au. The maximum faradaic efficiency of urea formation on Zn catalyst is approximately 35% at −1.75 V. The formation of urea at gas diffusion electrodes with catalysts from groups 6–10 was not found in the simultaneous reduction of CO_2 and nitrate.

Electroreduction mechanism has been studied by adsorption/desorption behavior of reacting species by using an *in situ* electrochemical quartz crystal microbalance[57]. The surface changes were measured by *ex situ* SEM, AES, and XRD analysis. During the cathodic reduction of CO_2 on Cu, the adsorption of amorphous carbon was observed. After electrolysis time of 1 hr at constant potential, the poisoning of amorphous carbon resulted in the decrease of faradaic efficiency for the formation of hydrocarbons such as CH_4 and C_2H_4. On the other hand, the potential modulation method caused the change of the surface structure of copper, i.e., the formation of oxide (Cu_2O). This structural change prevented the adsorption of amorphous graphite and the constant production rate of methane was obtained in long terms electrolysis[57].

The brief description mentioned above shows the advance in the development of new kinds of metallic cathodes modified as new materials for the reduction of CO_2. There is an important improvement in the amount of obtained hydrocarbons and better selectivity in the highly sophisticated new electrodes. Among the series of metallic cathodes shown here, the copper electrodes were the most studied although some problems still remains associated with the potential where the reaction takes place. However, the use of copper electrodes seems to be the cheapest and feasible catalysts for this reaction.

3. Biphenantroline and Bypiridine Hexaazacyclophane Systems

There are few studies about other systems containing extended π-electron orbitals differing from phthalocyanines and porphyrins. Since hexaazacyclophanes

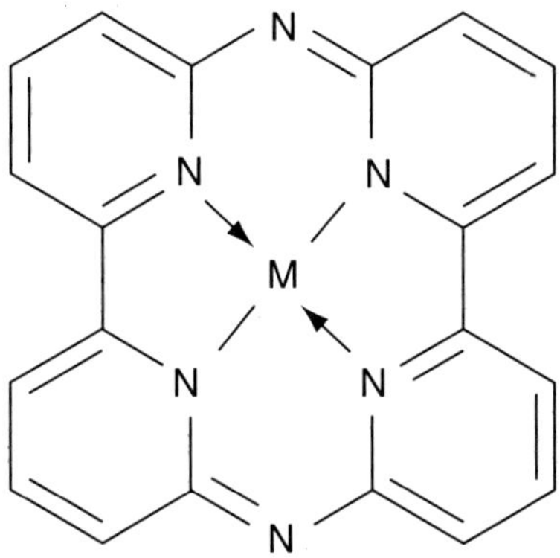

(1)

Structure 5.1. Molecular structures of cobalt(II), nickel(II) and copper(II) complexes with hexaaza-macrocyclic ligands derived from the condensation of bipyridines. Reprinted from Figure 1A: M. Isaacs, J.C. Canales, M.J. Aguirre, G. Estiú, F. Caruso, G. Ferraudi and J. Costamagna, Electrocatalytic reduction of CO_2 by aza-macrocyclic complexes of Ni(II), Co(II) and Cu(II). Theoretical contribution to probable mechanism, Inorganica Chimica Acta, 339 (2002) 224–232. Copyright 2002, with permission of Elsevier.

can be considered as relatives of the phthalocyanines and porphyrins, they may also be catalysts in several reactions. Preparative procedures for some azacyclophanes have been reported[58–64]. Recent studies of the chemical properties of azacyclophanes (**5.1**) and (**5.2**), where (**5.2**) is a hexaazacyclophane macrocyclic complex derived from 1,10-phenantroline, demonstrated that they are promising catalysts for CO_2 reduction. Large yields in the preparations of this hexaaza-macrocyclic ligands and its Cu(II), Ni(II), Co(II) complexes, have been reported[65]. Pulse radiolysis studies[65,66] suggest that CO_2 adducts could be formed during the reduction observed in the Cu(II) complex. Semi-empirical self-consistent field and CI calculations of INDO type have been recently applied on the analysis of the electronic transitions of azacyclophanes[67–69]. The obtained results confirm the activity of the structures toward electrochemical reduction processes involving the metal center.

Also a related family of complexes bis-bipyrine hexaaza-macrocycles (**5.1**) azacyclophanes[69] present electrocatalytic activity toward the electroreduction of CO_2[70] (see Figure 5.1). Canales and coworkers[70] compare the activity of the Ni(II) complex with the Zn(II) complex the role of the metal center is evident.

Electrochemical results of the bis-aza bpy ligand and its Ni(II) complex in N_2 and CO_2 atmosphere show the presence of a low-intensity redox process in the free ligand (in CO_2) suggesting that the CO_2 reduction process in the Ni(II) complex is probably occurring on the metal center[70]. The Zn azabipy complex shows a redox behavior similar to the free ligand, i.e., a significant catalytic process was not observed. Also depicted in Figure 5.1[70] it is clear the modification of the cyclic voltammogram for the Ni(II) complex, measured in CO_2 atmosphere compared to its behavior in N_2 atmosphere. The cathodic peak at -1.5 V is three times larger

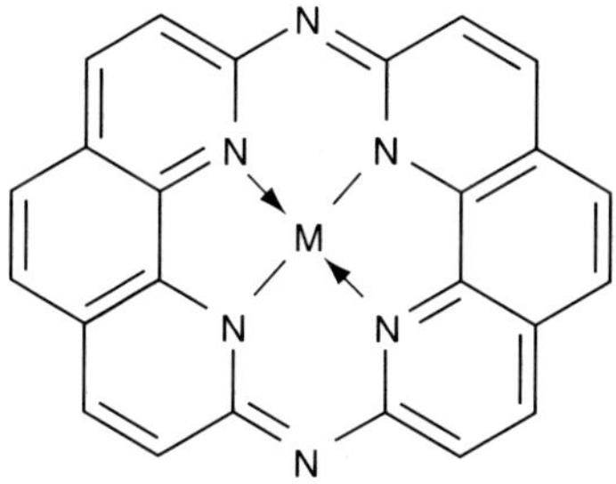

(2)

M = Co, Ni, Cu

Structure 5.2. Molecular structures of cobalt(II), nickel(II) and copper(II) complexes with hexaaza-macrocyclic ligands derived from the condensation of phenanthrolines. Reprinted from Figure 1B: M. Isaacs, J.C. Canales, M.J. Aguirre, G. Estiú, F. Caruso, G. Ferraudi and J. Costamagna, Electrocatalytic reduction of CO_2 by aza-macrocyclic complexes of Ni(II), Co(II) and Cu(II). Theoretical contribution to probable mechanism, Inorganica Chimica Acta, 339 (2002) 224–232. Copyright 2002, with permission of Elsevier.

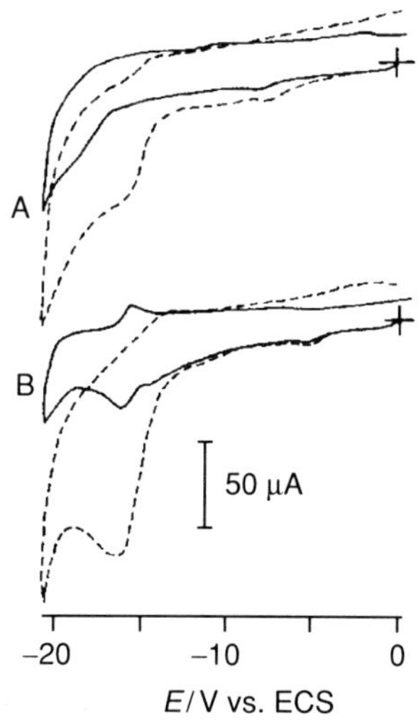

Figure 5.1. Cyclic voltammograms, in *N,N'*-dimethylformamide, under N_2 (—) and CO_2 (- - -) for : (A) bisazabipy; (B) Ni-bis-azabipy; approximately 0.1 mM; scan rate 0.2 $V\,s^{-1}$. Reprinted from Figure 3: J. Canales, J. Ramírez, G. Estiú and J. Costamagna, Bis-bipyridine hezaaza-macrocyle complexes of Zinc(II) and Nickel(II) and the catalytic reduction of carbon dioxide, Polyhedron 19(22–23), 2373–2381 (2000). copyright 2000, with permission of Elsevier.

than in N_2. The electrocatalytic character of the Ni(II) complex was demonstrated by coulometric measurements (controlled potential electrolysis at -1.6 V) where, in CO_2 atmosphere shows a constant current, but in N_2 atmosphere the current was diminishing and shows one electron reduction process[70].

Quantum chemical calculations[70] were used to understand the origin of the properties of both the ligands and the metal complexes. They show in this case the importance of delocalized π interactions, which are responsible for their different structural and electronic characteristics in protic and nonprotic media. As an example, the π interactions favored molecular stacking for the complexes in aprotic solvents. This behavior in solution is important in the catalytic process but it is not yet well understood.

On another hand, several techniques have been applied in the elucidation of the mechanism for the electroreduction of CO_2 mediated by biphenantrolines[69–71]. Voltammetric studies for the Co complex are shown in the Figure 5.2[69]. The voltammogram recorded in N_2 presents two peaks at −0.94 and −1.48 V, respectively (see the insert in Figure 5.2). The peak observed at −0.94 V is quasireversible and could be assigned to the [Co(II)L]/[Co(I)L] redox process because of its reversibility and because related macrocycles complexes present the Co(II)/Co(I) couple close to the potential observed in this case[72].

The peak located at −1.48 V is completely irreversible and could be assigned to a redox process of the ligand. Drastic changes are observed when

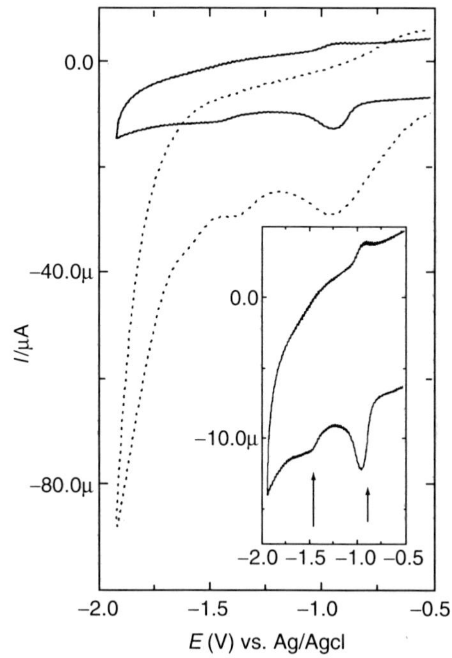

Figure 5.2. Cyclic voltammogram for the complex of structure 1(b) for M = cobalt. [CoL] $\approx 5 \times 10^{-4}$ M, in N,N′-dimethylformamide. Scan rate: 200 mV s^{-1}. (—) in nitrogen; (- - -) in CO_2. Insert is for the voltammogram registered in nitrogen atmosphere. Reprinted from Figure 3: M. Isaacs, J.C. Canales, M.J. Aguirre, G. Estiú, F. Caruso, G. Ferraudi and J. Costamagana, "Electrocatalytic reduction of CO_2 by aza-macrocylic complexes of Ni(II), Co(II) and Cu(II). Theoretical contribution to probable mechanism, Inorganica Chimica Acta, 339 (2002) 224–232. Copyright 2002, with permission of Elsevier.

N_2 is replaced by CO_2 as can be seen in Figure 5.2. Around −0.94 V, a broad signal appears which could be assigned to a first reduction that involves the metal center and the CO_2 molecule. At this potential, only a peak appears and there is not a discharge of current[69]. This behavior does not correspond to a "true" reduction of CO_2 but probably to an adduct or intermediate formation. After the redox signal attributed to the ligand (not simultaneous), a discharge of current is observed. This phenomena could indicate that in the first electronic transfer, a species $[M(II)L–CO_2]^{-1}$ is formed. When the ligand is reduced, a second electronic transfer takes place from the ligand to the adduct species promoting the reduction of CO_2[69]. This explains the apparition of the current discharge after the second reduction peak at approximately −1.50 V. These facts indicate that not only the reduced cobalt is needed to promote the electroreduction of CO_2, but also a second electron transfer process is required. This second electron transfer process must involve the ligand because there is no metal redox couple at those negative potentials[69].

Although the structure of the macrocycle used in this study is similar to that of porphyrins or phthalocyanines, it is not possible to apply the Gouterman's assignments of the bands that appear in the UV–Vis spectrum[69–71]. The assignments have been discussed previously in the literature[67,68]. Figure 5.3[69] shows the UV–Vis. spectra registered in N_2 or CO_2 atmosphere for the complex (**5.2**) (where M = Co). These spectra correspond to the species generated at open circuit and at high cathodic potentials (ca. −1.65 V). At can be seen in Figure 5.3, when cathodic potentials are applied, drastic differences appear compared to the spectrum at open circuit potentials under N_2 atmosphere[69]. Under N_2 (at cathodic potentials) a broad band appears between 350 and 400 nm with a maximum at 380 nm and the band with a maximum at 423 nm decreases its intensity.

A similar behavior has been observed when cobalt phthalocyanines are electrolyzed at potentials close to the Co(II)/Co(I) redox couple[72,73]. At those potentials, the metal center Co(II) is reduced to Co(I); then the appearance of bands in the zone of 400 nm can be attributed to a metal-to-ligand charge transfer bands[74,75], i.e., Co(I)·L where L = macrocyclic ligand. Figure 5.3[69] depicts the spectrum of (**5.2**) at −1.65 V in the presence of CO_2. It is observed that the charge transfer band located at 385 nm (that appears under N_2 at the same potential) does not appear. The applied potential corresponds to the 2-electron reduction of the complex (**5.2**), and the disappearance of this band should be explained by the regeneration of the catalyst with the metal center in a state of oxidation (II) after the reduction of CO_2[69–76]. When spectroelectrochemical experiments are performed with the free ligand, there are not significant changes (see Figure 5.4[71]) when the potential is changed to negative values or if the spectrum is recorded in N_2 or CO_2 atmosphere.

Controlled potential electrolysis at approximately −1.4 V for these metalled complexes (**5.2**) produces CO as the main reaction product[69–71]. A very small amount of water needs to be added because the process without water is very slow. Minor amounts of formic acid were also detected for this complex[69–71]. For the free ligand H_2 was the only product detected.

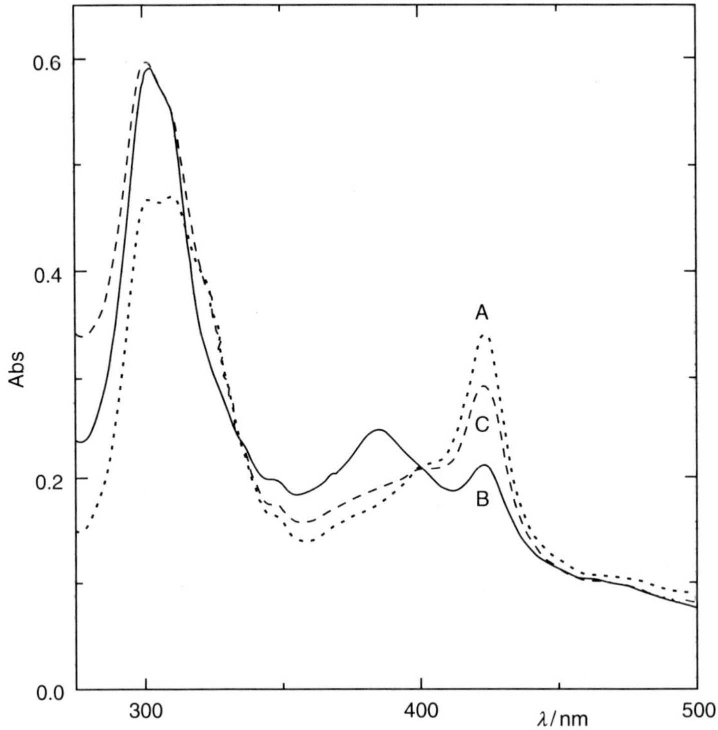

Figure 5.3. UV–Vis spectra for the complex of structure 1(b) for M = cobalt. [CoL]: 5×10^{-4} M, in N,N′-dimethylformamide, [PTEA] = 0.01 M. (a) (·········): under N_2 at open circuit; (b) (——): under N_2 at *ca.* −1.65 V; (c) (- - - - - -): under CO_2 at *ca.* −1.65 V. Spectrum at open circuit in CO_2 remains unchanged compared to the spectrum in N_2. Reprinted from Figure 4 : M. Isaacs, J.C. Canales, M.J. Aguirre, G. Estiú, F. Caruso, G. Ferraudi and J. Costamagana, "Electrocatalytic reduction of CO_2 by aza-macrocylic complexes of Ni(II), Co(II) and Cu(III). Theoretical contribution to probable mechanism, Inorganic Chimica Acta, 339 (2002) 224–232. Copyright 2002, with permission of Elsevier.

The interaction of CO_2 with the metal complexes has been analyzed by means of semiempirical PM3 calculations, for Ni(II)-bis-aza-phenanthroline as model complex[69]. According to the previous discussion, coordination to $[M(II)L]^{2-}$ has been evaluated. Optimization of the geometry gives planar structures for both the neutral and the negatively charged species, with Ni–N distances of 1.852 and 1.847Å, respectively. The optimized structural parameters have been compared with those derived from density functional calculations, which renders Ni–N bond lengths of 1.815Å. PM3 calculations do not show structural differences for the Cu(II) complex[69]. On the other hand, structural characteristics are not modified, in the −2 charged species[69]. The theoretical study of CO_2 coordination has considered its interaction as either monodentate or bidentate ligand. As a monodentate ligand, CO_2 can interact through the C atom, leading to a η^1-C complex, or through an oxygen atom, resulting in a η^1-O species. As a bidentate ligand, CO_2 can form η^2-C,O complexes, when the metal interacts with the CO

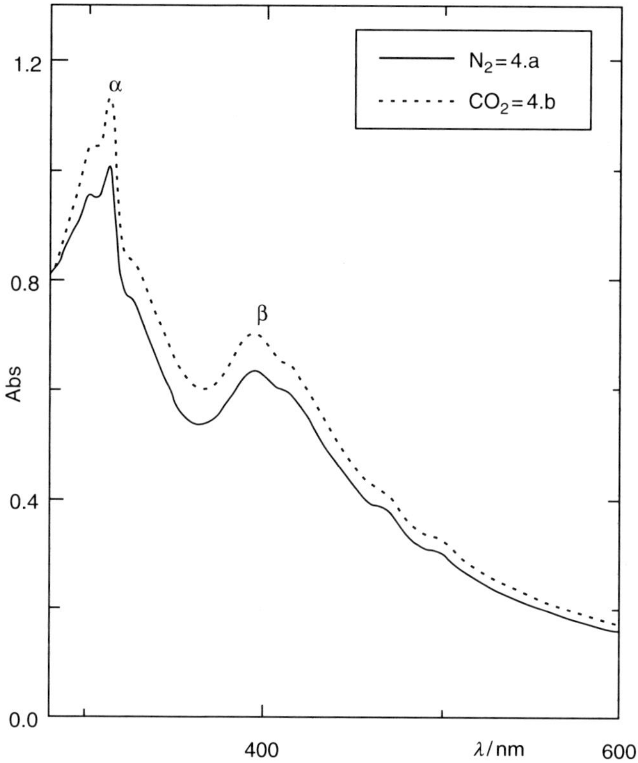

Figure 5.4. Electronic spectra of the hexaaza-macrocyle-H_2 (free ligand) in DMF, (a) under N_2. (b) under CO_2. [HAM-H2] $\approx 5 \times 10^{-5}$ M. [PTBA] = 0.01 M. registered at OPC. Reprinted from Figure 4: M. Isaacs, J.C. Canales, A. Riquelme, M. Lucero, M.J. Aguirre and J. Costamagna, Contribution of the ligand to the electroreduction of CO_2 catalyzed by a cobalt(II) macrocylic complex, Journal of Coordination Chemistry 56(14), 1193–1201 (2003). With permission of Taylor and Francis (http://www.tandf.co.uk/journals)

bond, or η^2-O,O complexes, when the bond occurs with the two oxygen atoms[69]. On the negatively charged $[Ni(II)L]^{2-}$ structures, η^2 coordinations are not stabilized, but evolve to η^1 adducts. Monodentate coordination results from σ and π CO–Ni contributions, and stabilizes the system in 31.3 and 23.4 kcal mol^{-1} for the η^1-O and the η^1-C geometries, respectively (see Figures 5.5 and 5.6[69]). Due to the small energy difference, both coordination geometries have been considered as capable of occurring. η^1-C coordination slightly shifts the Ni atom out of the macrocycle plane, defining a torsion angle NiNCN of 2.6°, for a C–Ni distance of 2.005Å. Adsorbed CO_2 is also bent, with an angle of 149°. Coordination polarizes the CO_2 molecule, which develops negative charges of −0.452 on the O atoms according to a Mulliken population partitioning scheme. η^1-O coordination, with an O–Ni distance of 1.996Å, also distorts the macrocycle structure, leading to a NiNCN torsion angle of 2.3°[69]. A calculated negative charge (−0.36) characterizes the terminal O atom[69].

In the experimental conditions previously described, protonation of the negatively charged centers will certainly occur. The process is energetically favored, according to the calculations, by more than 900 kcal mol^{-1}. In the η^1-O structure, the sp hybridization of CO_2 goes to sp^2 after protonation of the terminal O (see Figure 5.5[69]). Both the C and O atoms, not directly involved in Ni interaction, bear negative density charges, a fact that justifies further protonation. The η^1-C structure evolves to a η^1-O one after O-proton coordination (see Figure 5.6), with negative charges on the remaining C and O atoms. Distances and angles are depicted in the above-mentioned figure.

A calculated energy difference of 2.5 kcal mol^{-1} favors the structure belonging to the η^1-C coordination. On the η^1-O adduct, a second proton can coordinate to either the C atom or the terminal oxygen. Coordination to the C atom results in an energy lowering of 330 kcal mol^{-1}, and formic acid bonded to the Ni atom through the carbonyl oxygen. Coordination to the terminal oxygen results in free water, and a residual CO bonded to the Ni atom. The first process is

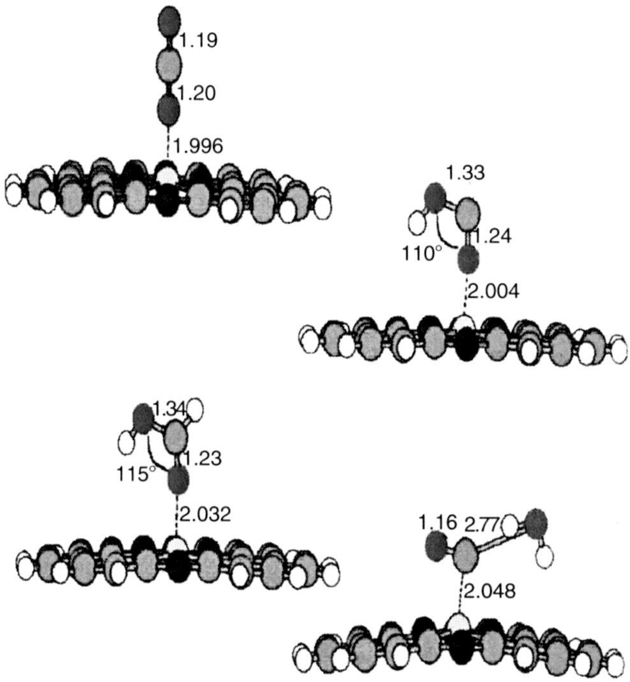

Figure 5.5. $[NiL]^{-2}$ (bypiridine complex) + CO_2+2H + system. Optimized stable species originally oxygen coordinated. Reprinted from Figure 7.6: M. Isaacs, J.C. Canales, M.J. Aguirre, G. Estiú, F. Caruso, G. Ferraudi and J. Costamagana, "Electrocatalytic reduction of CO_2 by aza-macrocyclic complexes of Ni(II), Co(II) and Cu(II). Theoretical contribution to probable mechanism, Inorganica Chimica Acta, 339 (2002) 224–232. Copyright 2002, with permission of Elsevier.

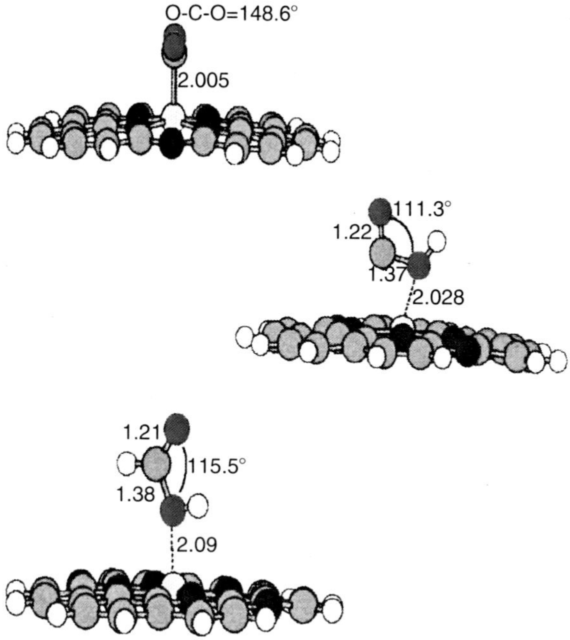

Figure 5.6. $[NiL]^{-2}$ (bypiridine complex) + CO_2+2H + system. Optimized stable species originally oxygen coordinated. Reprinted from Figure 7: M. Isaacs, J.C. Canales, M.J. Aguirre, G. Estiú, F. Caruso, G. Ferraudi and J. Costamagana, "Electrocatalytic reduction of CO_2 by aza-macrocyclic complexes of Ni(II), Co(II) and Cu(II). Theoretical contribution to probable mechanism, Inorganica Chimica Acta, 339 (2002) 224–232. Copyright 2002, with permission of Elsevier.

favored by 10 kcal mol^{-1}. Proton coordination to the C-atom of the structure that originally belongs to the η^1-C adduct also results in formic acid, this time bonded by the hydroxy O. This coordination geometry is isoenergetic to the previously described for formic acid coordination[69].

To the date, the studies reviewed here about hexaaza-cyclophane systems have been focused in mechanistic aspects. These studies demonstrate that the nature of the interaction between the CO_2 molecule and the complexes depends on the metal and on the ligand contrarily to that observed for cyclam systems (see the next section). However, all this information is still not enough to fully understand the relevant steps in the formation or in the kinds of intermediates in order to design a more catalytic and selective complex based on these systems. Also, it is necessary to improve the synthetic approach of these systems with different substituents in order to gain systematic knowledge about the new structures and the understanding of its chemical nature and kind of products obtained in each case. In this way, a feedback between the experimental and theoretical data will be possible and therefore a more efficient system for the target molecule.

4. Cyclam and Derivative Systems

4.1. Studies in Solution for Ni Cyclam Systems and Electrochemical Reduction of CO_2

The complex of Ni(II) and Co(II) with macrocyclic ligand cyclam (1,4,8,11-tetraazacyclotetradecane) (**5.3**)[20] has been shown to be particularly effective and selective catalysts for the electrochemical reduction of CO_2 to CO at mercury electrodes in water at potentials much less negative than those required for the uncatalyzed reduction[77–84]. The ability of $[Ni(cyclam)]^{2+}$ to serve as a catalyst precursor for the electroreduction of CO_2 has prompted many studies. These studies have established the following[16,20]:

The active catalyst is a form of $Ni(cyclam)^+$ adsorbed on the surface of mercury electrodes[77–80,82,83].

$Ni(cyclam)^+$ is adsorbed on mercury over an unusually wide range, including potentials that are much more positive than those where $Ni(cyclam)^{+2}$ is reduced to unadsorbed $Ni(cyclam)^+$[83,84].

CO is the product of the catalyzed reduction of CO_2[77,78,82–84].

A reasonable explanation for the fact that the adsorbed Ni(I) complex $Ni(cyclam)^+_{ads}$, but not the same complex in solution, is catalytically active towards the reduction of CO_2 has been provided on the basis of theoretical calculations[85]. The catalytic activity is severely diminished in the presence of CO on unstirred mercury electrodes and as CO is the primary product of the reduction. This behavior limits the long term effectiveness of the catalyst. It is proposed that the decrease in activity is due to Ni(cyclam)CO, an insoluble complex of Ni(0), which is formed during the reduction of CO_2[20]. Unfavorable shifts in the

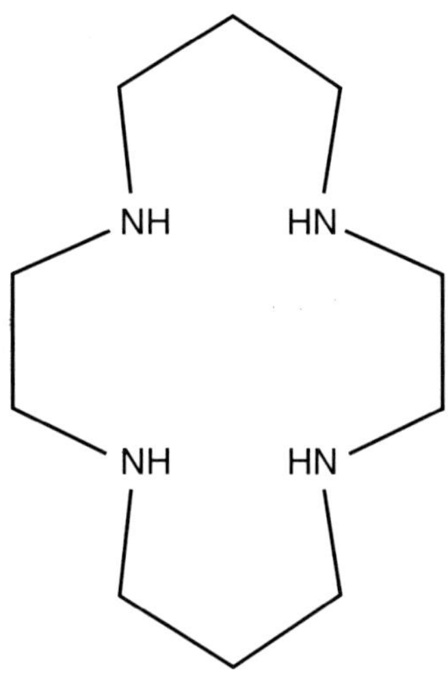

(3)

Structure 5.3. Macrocyclic ligand: (1,4,8,11-tetraazacyclotetradecane = cyclam). Reprinted from Structure 4: J. Costamagna, G. Ferraudi, J. Canales and J. Vargas, Carbon dioxide activation by aza-macrocyclic complexes, Coordination Chemistry Reviews 148 (1996) 221–248.Copyright 1996, with permission of Elsevier.

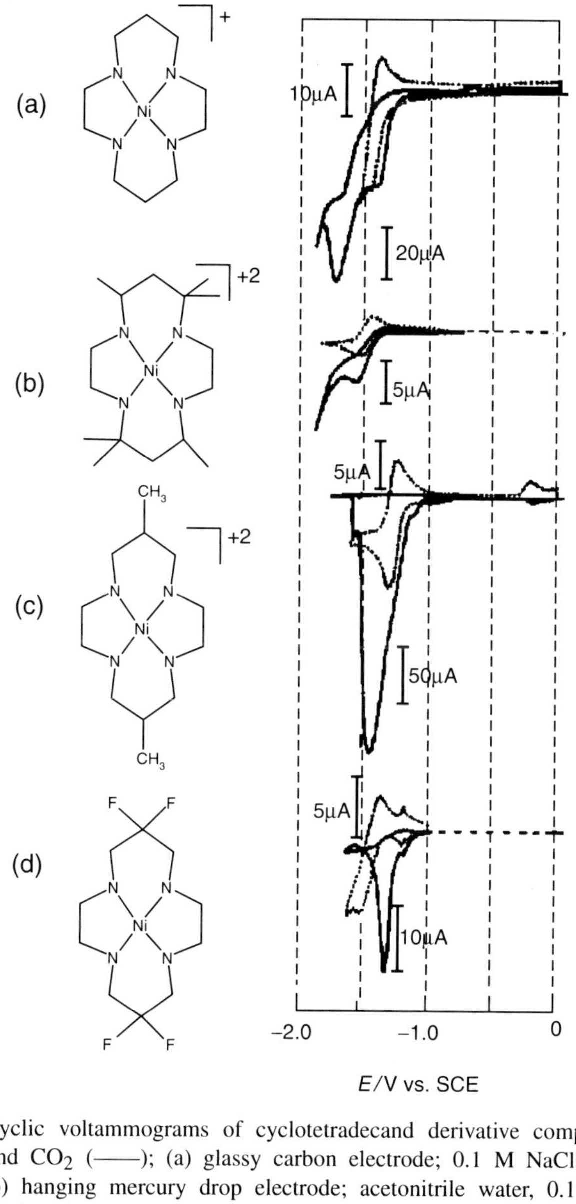

Figure 5.7. Cyclic voltammograms of cyclotetradecand derivative complexes in N_2 (·········) and CO_2 (——); (a) glassy carbon electrode; 0.1 M $NaClO_4$ (pH 6.2), 0.1 V s^{-1}, (b) hanging mercury drop electrode; acetonitrile water, 0.1 M $NaClO_4$, 0.02 V s^{-1}; (c) hanging mercyry drop electrode; acetonitrile water, 0.1 M $NaClO_4$, 0.1 V s^{-1}; (d) glassy carbon electrode; 0.5 M Na_2SO_4, 0.2 V s^{-1}. Reprinted from Figure 1: J. Costamagna, G. Ferraudi, J. Canales and J. Vargas, Carbon dioxide activation by aza-macrocylic complexes, Coordination Chemistry Reviews 148 (1996) 221–248. Copyright 1996, with permission of Elsevier.

potential and decreased catalytic activity were observed when carbon rather than mercury was used as working electrode with Ni(cyclam)$^{2+}$ as catalysts[20]. This electrochemical behavior is shown in the cyclic voltammograms (see Figure 5.7) for various Ni(II) complexes[20,80].

Mechanisms for the electrochemical processes at mercury electrodes in solutions of $[Ni(cyclam)]^{2+}$ and CO_2 have been proposed (see Seheme 5.1[20]). Scheme 5.1 shows the formation of a carbon-bonded Ni(II) complex by reaction of CO_2 with $Ni(cyclam)^+$. The formation of such a complex is considered to be a fundamental step in the mechanism of the $[Ni(cyclam)]^{2+}$-catalyzed electrochemical reaction[16]. The overall process for the transformation of CO_2 into CO also involves inner-sphere reorganization[20]. Scheme 5.1 includes the formation of sparingly soluble complex containing Ni(0), cyclam and CO which is a product of the reduction of $[Ni(cyclam)]^{2+}$ under CO. Depositation of a precipitate of the Ni(0) complex on the mercury electrodes inhibits catalysis and removes the catalyst from the cycle. The potential at which the $[Ni^{I}L–CO_2H]^{2+}$ intermediate (see lower left hand of Scheme 5.1) accepts electrons from the electrode. This potential is not affected by substitution on the cyclam ring, as shown by comparison of $[Ni(cyclam)]^{2+}$ and $[Ni(TMC)]^{2+}$ (TMC = tetra-*N*-methylcyclam)

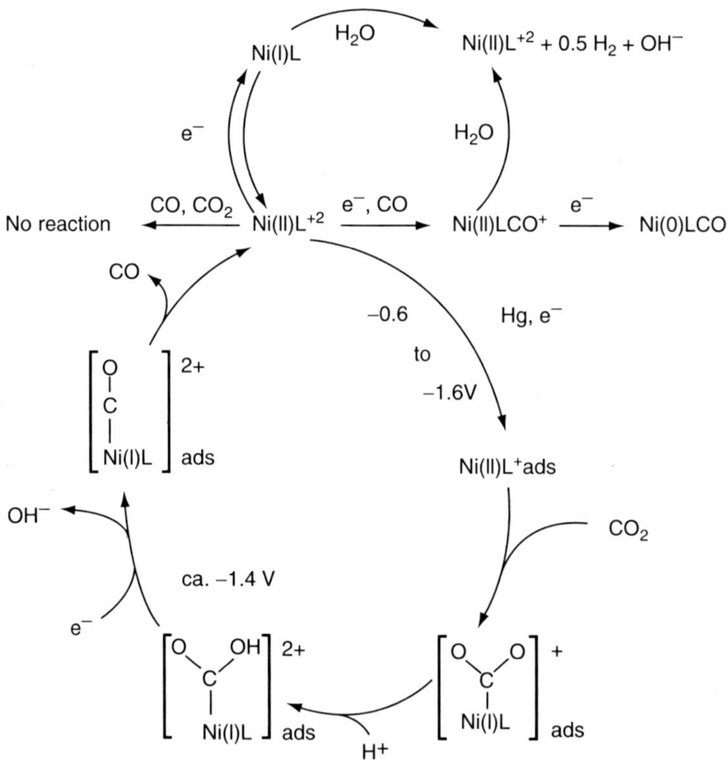

Scheme 5.1. This scheme shows the formation of a carbon-bonded Ni(II) complex by reaction of CO_2 with $[Ni(cyclam)]^+$. Reprinted from Scheme 3: J. Costamagna, G. Ferraudi, J. Canales and J. Vargas, Carbon dioxide activation by aza-macrocyclic complexes, Coordination Chemistry Reviews 148 (1996) 221–248. Copyright 1996, with permission of Elsevier.

(**5.4**) as catalysts[20,83]. However, in some cases the catalytic activity of [NiL] (L = 5, 7, 7, 12, 14, 14 hexamethyl-1,4,8,11-tetraaza-cyclotetradecane) (**5.5**) can be strongly decreased (see Figure 5.7b)[20].

Rotating-copper-disk electrode techniques have been used to evaluate the efficiency of the nickel macrocycle catalyst for the reduction of CO_2 to CO[16,86]. Studies have been performed using Ni(diazacyclam)$^{2+}$ (diazacyclam = 3,10-dimethyl-1,3,5,8,10,12-hexaazacyclophane) (**5.6**), a complex derived from cyclam, which appears to be more active than [Ni(cyclam)$_2$]$^{2+}$ under the same conditions (see Figure 5.7c)[16,20,86]. These results are consistent with a mechanism proposed by other authors[16,77,78].

Also, interesting *ab initio* MO-SDCI studies of model complexes of the intermediates in the electrochemical reduction of CO_2 catalyzed by $NiCl_2$(cyclam) have been performed[87]. The calculations of several Ni^I– and Ni^I–CO complexes indicate that CO_2 can coordinate to $Ni^IF(NH_3)_4^+$, yielding a stable Ni–CO_2 complex (**5.7**), but not to $[Ni^{II}F(NH_3)_4]^+$, $[Ni^{II}F(NH_3)_4]^+$ or $[Ni^I(NH_3)_5]^+$[85]. The HOMO of $Ni^IF(NH_3)_4(\eta^1\text{-}CO_2)$ is largely the oxygen pπ orbital at a higher energies than the HOMO (nonbonding π orbital) of the uncomplexed CO_2. In addition, the electron density increases around the oxygen atom upon CO_2 coordination. As result, the coordinated CO_2 in $Ni^IF(NH_3)_4(\eta^1\text{-}CO_2)$ is activated with regard to electrophilic attack and is expected to undergo a facile protonation[20,87]. MO calculations also show that the second one-electron reduction can easily occur in the protonated species $[NiF(NH_3)_4(CO_2H)]^+$ yielding the triplet state $[NiF(NH_3)_4(CO_2H)]$, but cannot occur in the unprotonated species $[NiF(NH_3)_4(CO_2)]$ (**5.7**). The second reduction significantly weakens the C–OH

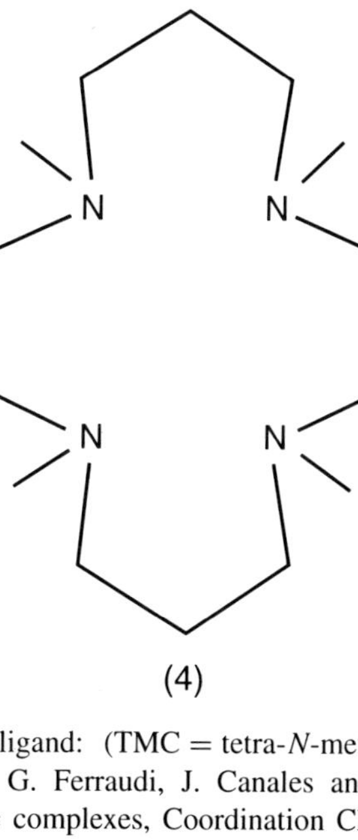

(4)

Structure 5.4. Macrocyclic ligand: (TMC = tetra-*N*-methylcyclam). Reprinted from Structure 5: J. Costamagna, G. Ferraudi, J. Canales and J. Vargas, Carbon dioxide activation by aza-macrocyclic complexes, Coordination Chemistry Reviews 148 (1996) 221–248. Copyright 1996, with permission of Elsevier.

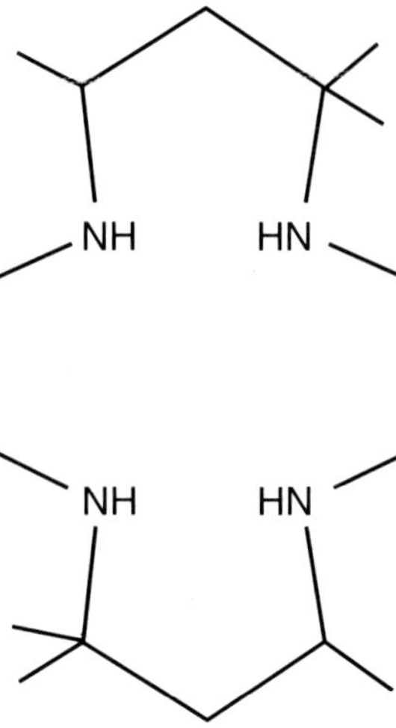

(5)

Structure 5.5. Macrocyclic ligand: (L = 5, 7, 7, 12, 14, 14 hexamethyl-1,4,8,11-tetraaza-cyclotetradecane). Reprinted from Structure 6: J. Costamagna, G. Ferraudi, J. Canales and J. Vargas, Carbon dioxide activation by aza-macrocyclic complexes, Coordination Chemistry Reviews 148 (1996) 221–248. Copyright 1996, with permission of Elsevier.

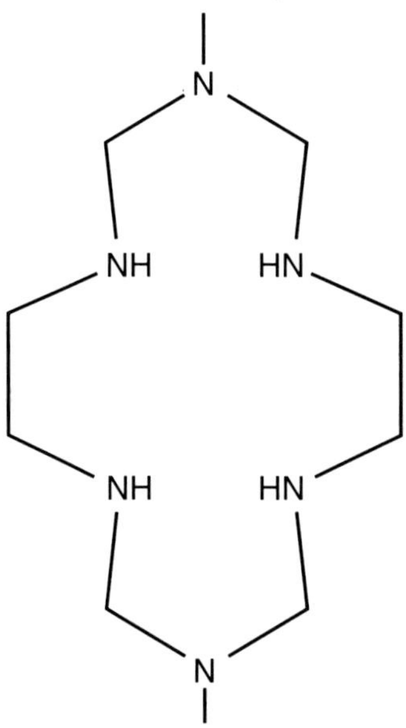

(6)

Structure 5.6. Macrocyclic ligand: (diazacyclam = 3,10-dimethyl-1,3,5,8,10,12-hexaazacyclophane). Reprinted from Structure 5.7: J. Costamagna, G. Ferraudi, J. Canales and J. Vargas, Carbon dioxide activation by aza-macrocyclic complexes, Coordination Chemistry Reviews 148 (1996) 221–248. Copyright 1996, with permission of Elsevier.

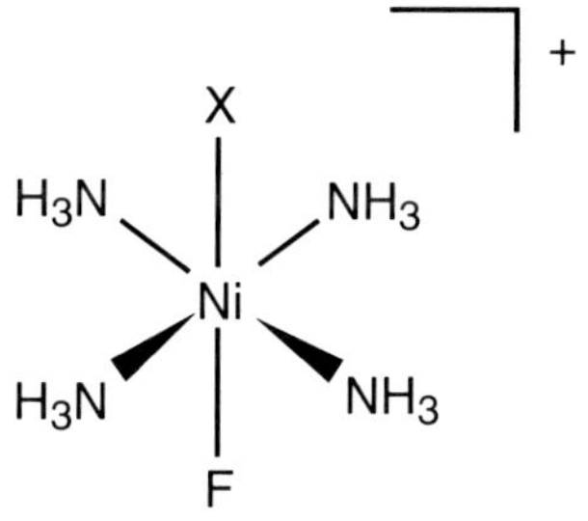

(7)

Structure 5.7. Complex: $[NiF(NH_3)_4(X)^+$. Reprinted from Structure 5.8: J. Costamagna, G. Ferraudi, J. Canales and J. Vargas, Carbon dioxide activation by aza-macrocyclic complexes, Coordination Chemistry Reviews 148 (1996) 221–248. Copyright 1996, with permission of Elsevier.

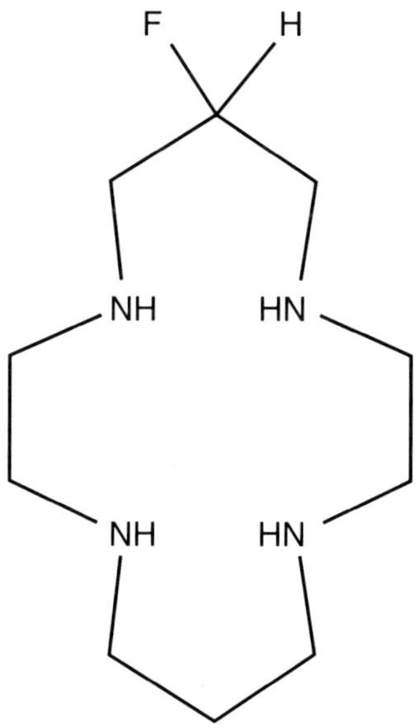

(8)

Structure 5.8. Macrocyclic ligand: (monofluorinated cyclam). Reprinted from Structure 5.9: J. Costamagna, G. Ferraudi, J. Canales and J. Vargas, Carbon dioxide activation by aza-macrocyclic complexes, Coordination Chemistry Reviews 148 (1996) 221–248. Copyright 1996, with permission of Elsevier.

bond, which suggests that OH^- easily dissociates from $[NiF(NH_3)_4(CO_2H)]$, yielding the triplet state of $[NiF(NH_3)_4(CO]^+$. The CO bond to Ni(II) is calculated to be weak, which suggests that CO easily dissociates from Ni(II). All these results support a reaction mechanism proposed previously[16] for the reduction of CO electrocatalyzed by $NiCl_2$(cyclam).

On the other hand, a series of mono-, di- and tetrafluorinated cyclams (**5.8**), (**5.9**) and (**5.10**) have been prepared. The electrocatalytic reduction of CO_2

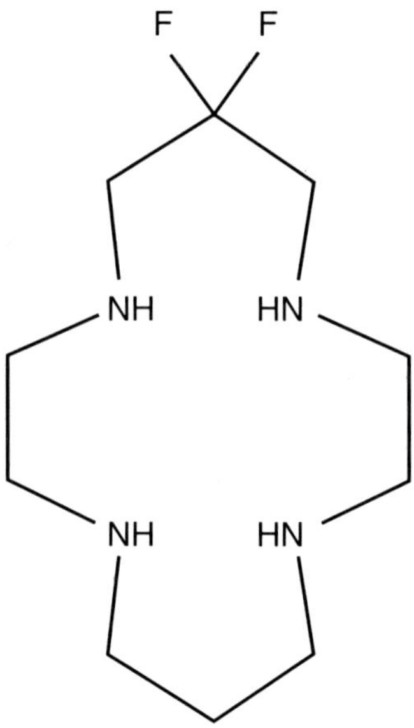

(9)

Structure 5.9. **N** Macrocyclic ligand: (difluorinated cyclam). Reprinted from Structure 5.9: J. Costamagna, G. Ferraudi, J. Canales and J. Vargas, Carbon dioxide activation by aza-macrocyclic complexes, Coordination Chemistry Reviews 148 (1996) 221–248. Copyright 1996, with permission of Elsevier.

showed that the efficiency of ($CO + H_2$) product yields and the H_2/CO selectivity depending on the number of fluorine atoms available[20]. The tetrafluorinated complexes shows more efficiency and selectivity at potentials lower than those measured for some nonfluorinated cyclam complexes[20,88]. Figure 5.7d shows the results obtained for the 3,3,10,10,-tetrafluoro-cyclam complex. The CO yield decreases as the number of fluorine atoms increases[20,88]. However, at a potential of −1.10 V, $[Ni^{II}$(3,3,10,10-tetrafluoro-cyclam)]$(ClO_4)_2$ (**5.10**) showed more efficient selective catalytic activity than the nonfluorinated [Ni(cyclam)]Cl_2 complex. On the other hand, perchlorate complexes have higher catalytic activity than chloride complexes at higher overpotentials.

Recent studies[89] of the of the stereochemical effect of substituents on the electrocatalytic reduction of CO_2 have shown that the complexes RSSR-Ni(2,3,9,10-tetramethyl-cyclam)$^{2+}$ (**5.11**), RRSS-Ni(2,3,9,10–tetramethyl-cyclam)$^{2+}$ (**5.12**) in aqueous KCl solution show increase in the catalytic current by a factor of 50–100, with the current densities of the RRSS-Ni(2,3,9,10-tetra-methyl-cyclam)$^{2+}$ and Ni(5,12-dimethyl-cyclam)$^{2+}$ complexes being higher than that observed for Ni(cyclam)$^{2+}$. Thus, these complexes appear to be better catalysts than Ni(cyclam)$^{2+}$ in terms of their larger catalytic currents and more positive potentials. Hydrogen and formate production is less than 1.5% of CO for each mole of complex[20]. However, the geometric isomer RSSR-Ni(2,3,9,10-tetramethyl-cyclam)$^{2+}$ (**5.11**) shows a lower catalytic activity because the axial methyl

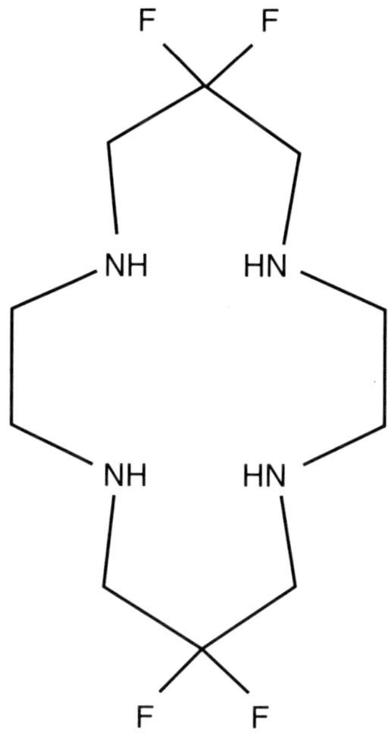

(10)

Structure 5.10. Macrocyclic ligand: (tetrafluorinated cyclam). Reprinted from Structure 5.11: J. Costamagna, G. Ferraudi, J. Canales and J. Vargas, Carbon dioxide activation by aza-macrocyclic complexes, Coordination Chemistry Reviews 148 (1996) 221–248. Copyright 1996, with permission of Elsevier.

group may sterically hinder CO_2 coordination[20]. Catalytic activity has not been observed in aprotic solvents because the mechanism (Scheme 5.1) requires a proton source. Experimental observations have been reported for Ni(cyclam)$^{2+}$ (**5.3**) in DMSO and for RSSR-Ni(2,3,9,10-tetra-methyl-cyclam)$^{2+}$ (**5.11**), RRSS-Ni(2,3,9,10-tetra-methyl-cyclam)$^{2+}$ (**5.12**) and Ni(5,12-dimethyl-cyclam)$^{2+}$ (**5.13**) in ACN[20]. Other examples with substituents effect in the Ni(cyclam)$^{2+}$ type complexes are illustrated in Scheme 5.2[90].

Studies performed with $[NiL]^{+2}$ and $CO_2^{\cdot-}$ as a reducing agent allow to predict the behavior of the complexes in the electrochemical reduction of CO_2. The rate constant obtained are close to rate of diffusion, except in the cases where the complex present steric effect to the inner sphere reaction[22,90]. The following proposed mechanism that adjusts with the rate constants and the potential (pH range where the potential of the redox couple corresponds to an exothermic reaction) is presented:[22,90]

$$NiL^{+} \cdot CO_2 + 2H_3O^{+} + e- \longrightarrow NiL^{2+} + CO + 3H_2O \qquad (5.3)$$

Alternatively

$$NiL^{+} \cdot CO_2 + NiL^{i+} + 2H_3O^{+} \longrightarrow 2NiL^{+2} + CO + 3H_2O \qquad (5.4)$$

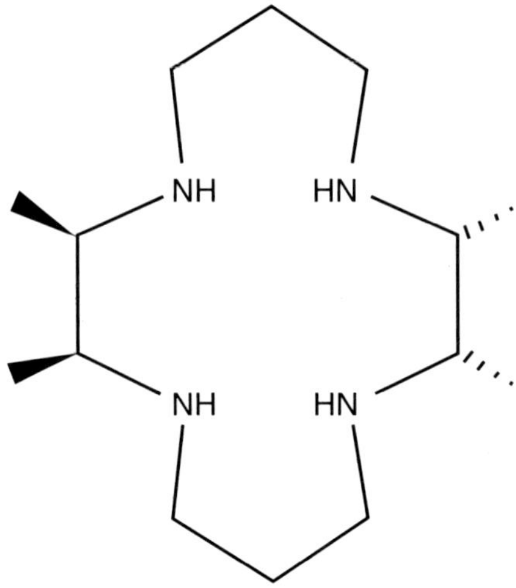

(11)

Structure 5.11. Macrocyclic ligand: RSSR-(2,3,9,10-tetramethyl-cyclam). Reprinted from Structure 5.12: J. Costamagna, G. Ferraudi, J. Canales and J. Vargas, Carbon dioxide activation by aza-macrocyclic complexes, Coordination Chemistry Reviews 148 (1996) 221–248. Copyright 1996, with permission of Elsevier.

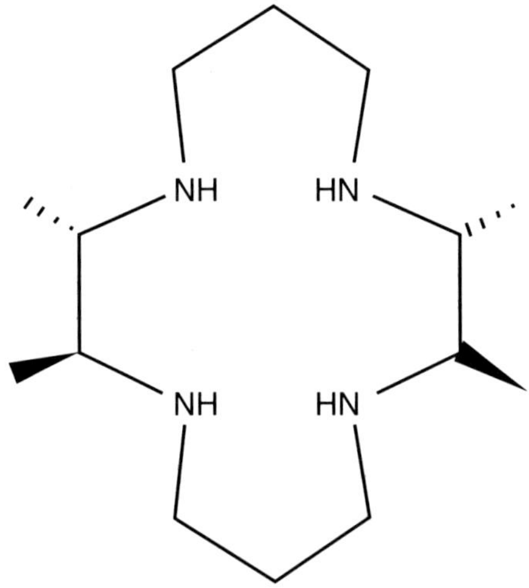

(12)

Structure 5.12. Macrocyclic ligand: RRSS-(2,3,9,10-tetramethyl-cyclam). Reprinted from Structure 5.13: J. Costamagna, G. Ferraudi, J. Canales and J. Vargas, Carbon dioxide activation by aza-macrocyclic complexes, Coordination Chemistry Reviews 148 (1996) 221–248. Copyright 1996, with permission of Elsevier.

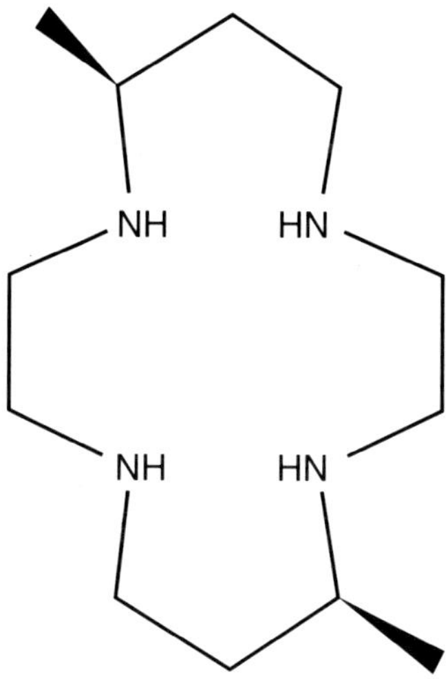

(13)

Structure 5.13. Macrocyclic ligand: (5,12-dimethyl-cyclam). Reprinted from Structure 5.14: J. Costamagna, G. Ferraudi, J. Canales and J. Vargas, Carbon dioxide activation by aza-macrocyclic complexes, Coordination Chemistry Reviews 148 (1996) 221–248. Copyright 1996, with permission of Elsevier.

On the other hand, clear differences are expected between the electrochemical reaction and those obtained by pulse radiolysis techniques[22, 91]. In the last case, the complex is reduced without regard of its more stable conformation. On the other hand, in an electrochemical activation, the conformation of the complex, either in a monovalent or divalent oxidation state, should be similar to allow the charge transfer and this is precisely that it happens[22,91].

Electrochemical experiments show that the redox couples Ni(III)/Ni(II) and Ni(II)/Ni(I) of the complexes showed in Scheme 5.2 are strongly affected by the nature of the substituents[90]. This effect could be interpreted as an interaction between the metal center and the fifth N in the azacyclam macrocycle. The soluble complexes in water are catalytic towards the electrochemical reduction of CO_2, and their efficiencies are comparable to the Ni(cyclam)$^{2+}$ complexes. The efficiency is related with the structural characteristics of the macrocycle cyclam and the azacyclam: a 14-member ring forming a sequence 5,6,5,6 of quelate rings. Any variation of this geometrical order changes or decreases the electrocatalytic activity. The macrocyclic effect in the Ni^{2+} metal center (14 ring members tetraamino cyclam), allows for stabilizing oxidation states Ni(I) and Ni(III) that will not be stable in another type of coordination. The azacyclam, with a ring of five N atoms, kept a cyclam structure but the fifth N atom does not coordinate to the Ni^{2+} metal center. The complexes show catalytic activity comparable to the Ni(cyclam)$^{2+}$ complex, suggesting that the replacement of a CH_2 for an amide group does not produce a decrease in the electrocatalytic activity[22]. The results show that electron withdrawing effect shifts the potential of the redox couple Ni(II)/Ni(I) and the reduction of CO_2 to CO takes place at a more positive

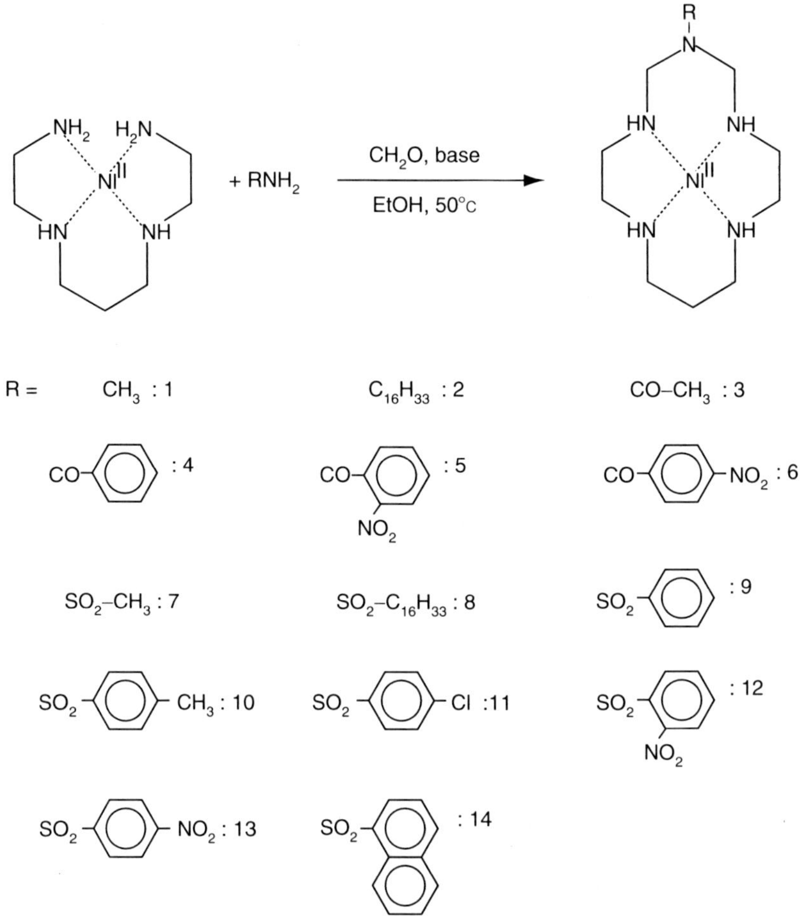

Scheme 5.2. Structures of Ni-azacyclams complexes obtained by template synthesis. Reprinted from Scheme 5.1: F. Abba, G. De Santis, L. Fabrizzi, M. Licchelli, A.M. Manotti Lanfredi, P. Pallavicini, A. Poggi and F. Ugozzoli, Nickel (II) complexes of azacyclams: Oxidation and reduction behavior and catalytic effects in the elctroreduction of carbon dioxide, Inorganic Chemistry 33 (1994) 1366–1375. Copyright 1994, with permission of American Chemical Society.

potential compared with the Ni(cyclam)$^{2+}$ complex[22]. The substituents effect in the potential is not a fact related with the coordination to the metallic center but to a variation in the electronic density in the central ion[22]. The half wave potentials data demonstrate that the replacement of a CH_2 group in the middle of the cyclam ring (six member quelate ring) for a R fragment notoriously affect the half wave potential of the Ni(III)/Ni(II) redox couple in the azacyclam systems. In particular, the replacement for an N–CH_3 does not change the potential of the redox couple, but the introduction of carboxamide group increases the potential from 100 to 130 mV[22]. A bigger increment is observed with a sulfamide group. This effect could be explained by π interaction between the metallic center and the amide group in the azacyclam ligand[22]. In the other hand, the potential is related with the stability of the oxidation states involved. It has been shown that Ni(II) (d8) is able to transfer its charge by a d–π mechanism to a π^* molecular orbital of the amide group, which is especially favorable for short distances between Ni(II)–N[22]. In the oxidation, the electronic density in the metallic center decreases and the magnitude of the donation π to R also decreases. This fact makes that the average stability of the trivalent complex decreases and the potential becomes more positive[22]. The destabilization of the trivalent state and the shift of the potential in the Ni(III)/Ni(II) redox–couple are greater when the substituent is a sulfonamide group, where the donation π g M–N is more important[22]. The steric effects could also be important in the shift of the potential. For example, in derivatives of cyclam C-substituted the Ni(III)/Ni(II) redox couple[22,92] is shifted to more positive potentials than the cyclam original complex. This fact has been attributed to nonbonding interactions between CH_3 groups in the C skeleton and axial ligands (i.e., solvent molecules[22,92]). A similar effect also has been described for 14 members-Ni(II) complexes where one or two cyclohexanes have been fused in a cyclam ring[93]. For the studied complexes, in this case, the steric effect is difficult to predict. Whereas the substituents in the amide group, R′ o R″ can be large, the long distance to the metallic center and the relative rigidity of the amide group avoids interactions with coordinated axial groups (solvent molecules), and apparently the influence is neglected. The Ni(II)/Ni(I) redox couple also is influenced for the insertion of the amide group, but in less magnitude that Ni(III)/(II) redox couple[22]. The half wave potential values are shifted 50–90 mV to less negative values compared with the cyclam parent. In this case there is a variation in the strength of the π interaction M–L associated with the change in the oxidation state. The π contribution should stabilize the oxidation state I more than the oxidation state II and thus the reduction requires a less negative potential[22].

4.2. Studies in Supramolecular Systems

Advances in the design of structures which show combinations of simple complexes and complicated free ligands have appeared in the literature[94,95]. These assemblies, in which several subunits are linked together through noncovalent interactions, have been defined as supramolecular systems. The behavior observed for these superstructures can be described in terms of basic concepts of supramolecular chemistry[94–96], i.e., they act in molecular recognition, have supramolecular reactivity and transport, and show self-organizing tendencies with a positive cooperative display[20,94–97]. In this sense, coordinatively linked multicenter systems can be classified as supramolecular coordination compounds. During the last 100 years coordination chemists have produced hundreds of metal-centered systems, many with redox activity, whose electrode potential can be conveniently modulated by synthetic modification of the coordination framework. More sophisticated systems, which are able to exchange a specified number of electrons, according to a predetermined sequence, at desired potential values, can now be prepared according to the principles of supramolecular chemistry. Moreover, the use of metal centers as structural elements allows the topology of the

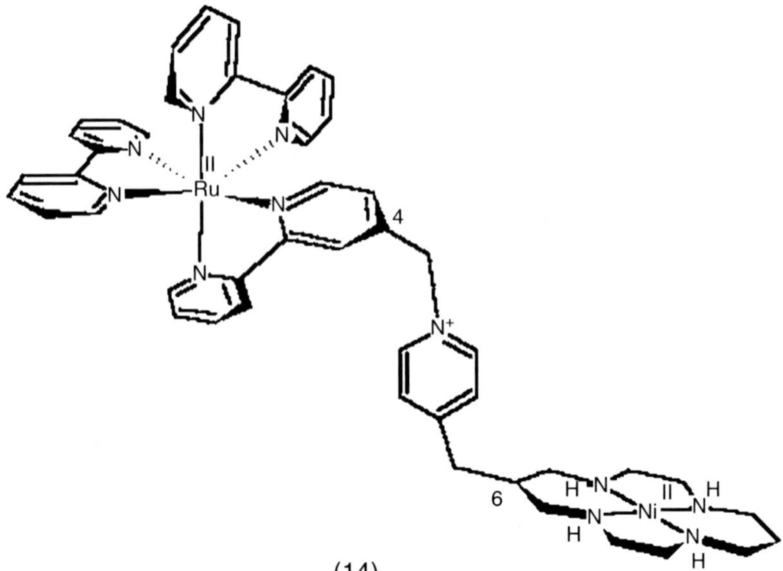

(14)

Structure 5.14. Bifunctional supermolecule. Reprinted from Structure 5.18: J. Costamagna, G. Ferraudi, J. Canales and J. Vargas, Carbon dioxide activation by azamacrocyclic complexes, Coordination Chemistry Reviews 148 (1996) 221–248. Copyright 1996, with permission of Elsevier.

multielectron redox systems to be controlled easily so that redox sites can be placed in chosen positions at predetermined distances[20, 98].

A bifunctional supramolecule (**5.14**[20]) has been synthesized and its redox and CO_2 catalytic properties have been investigated[99]. Although this process induced reductive photocleavage of the supramolecule, catalytic studies show twice the amount of CO is produced compared with the multimolecular system composed of $Ru(bpy)_3$, pyridinium salt and $Ni(Cyclam)^{2+}$[20]. This example shows how the Ru(II) unit can be used as an antenna and allows the charge transfer reaction and to obtain the cooperative effects.

On the other hand multimetallic cyclam systems have been prepared in the last 10 years[100–104]. Its preparation includes different numbers of cyclam units and also different bridge molecules affecting in multiple ways the electroreduction of CO_2. The electrocatalytic abilities of $Ni(cyclam)^{2+}$ and $Ni(biscyclam)^{4+}$ were compared for CO_2 and H_2 reduction[98]. The bimetallic complex was a better catalyst than its monomer analogue for the H_2 evolution from water, however for the electroreduction of CO_2 both complexes present similar properties leading only to C1 products. Also the electrochemical characteristics did not change from the monomer to the dimer, the Ni(II)/Ni(I) redox process only change by 30 mV in ACN/HME (HME, hanging mercury electrode) although it was clear for the authors[98] that in each voltammetric process two electron were involved.

Another interesting work[103] shows a series of binuclear cyclam systems with varying chain linking the two macrocyclic rings. These dimers were characterized by cyclic voltammetry under Ar or CO_2. The first series consisted of binuclear complexes $[NiL^{2-6}]^{4+}$ with $(CH_2)_n$ bridges ($n = 2, 3, 4, 6$) or p-xylyl linkage (L^6). In general the two nickel sites in the binuclear complexes behave independently with the currents corresponding to the simultaneous transfer of two electrons. The redox potentials are remarkably constant along this series, but the peak separations increase, reflecting slower electron transfer due to more effective adsorption on the electrode HME[103].

Electrochemical data for the electrochemical reduction of CO_2 in ACN/10% H_2O revealed catalytic waves for CO_2 reduction with E_p^c close to -1.7 V and catalytic currents (i_p^c) which are about half those of the mononuclear complex, proposed to be due to steric constraints allowing strong interaction of only one nickel center of the binuclear one the surface[103]. The catalytic currents increased slightly as the linking chain length increased as the stereochemical constrains were relaxed somewhat. There was also a splitting in the catalytic peaks of bis-macrocyclic complexes which could reflect two types of adsorbed catalyst sites. In the more sterically crowded series of complex, $[Ni_2L^7]^{4+}$ along with the series of linked heptaaza macrocyclic complexes $[Ni_2L^{9-11}]^{4+}$ much more positive redox potentials were observed due to both alkylation of the coordinated nitrogen atoms, which decreases the ligand field, and the introduction of steric barriers to axial coordination[103]. These steric barriers prevented strong electrode interaction and led to a lower catalytic activity. Indeed, the complex $[Ni_2L^7]^{4+}$ did not show any interaction with CO_2 in ACN. The authors[103] proposed a less effective absorption on the electrode arising from ligand steric interactions places far fewer stereochemical constraints on the absorption of both nickel centers to the same extent as the binuclear complex, and hence the catalytic currents for binuclear complex and mononuclear complex are comparable.

Trimetallic nickel compounds also have been prepared[104]. The complex $[Ni_3(X)](ClO_4)_6$ where X = 8, 8′, 8″-{2, 2′, 2″(-nitrilotriethyl) -tris(1,3,6,8,10, 13,15-heptaazatricyclo(11.3.1.1) octadecane} prepared by template condensation has showed to have a poor axial ability. The cyclic voltammetry experiments revealed that the complex suffer three electron reversible processes at negative potential and one electron reversible process at positive potential pulse technique (ousteryang square wave voltammetry and differential pulse voltammetry) demonstrated that these processes are electrochemically independent and when $[Ni_3(X)](ClO_4)_6$ was tested in CO_2 atmosphere the complex showed electrocatalytic activity for the electrochemical reduction of CO_2 to CO as a main product[104]. However the catalytic efficiency of $[Ni_3(X)](ClO_4)_6$ is lower than that of $[Ni(cyclam)]^{2+}$. The relationship Ip (CO_2)/Ip (N_2) is greater in $[Ni_3(X)](ClO_4)_6$ than $[Ni(cyclam)]^{2+}$ but the first one produces less CO per nickel center compared with the monomer relative. From the results showed above, it is clear that if there are interactions between the metal centers, the catalytic behavior could be enhanced, but the degree of interactions is strongly dependent on the characteristics of the bridge between metals.

4.3. Modified Electrodes with [Ni (cyclam)]$^{2+}$

Many applications of modified electrodes with [Ni (cyclam)]$^{2+}$ or derivatives have been reported in the past years either on mercury or glassy carbon electrodes[105–107]. Modifications include cubic phases[105], Langmuir-blodget films[106], Nafion inclusion[106], supramolecular assemblies[106] and adsorption in thin mercury films[107]. Several works have focused in the determination of SCN^- or electrocatalytic reduction of aldehydes and ketones[108, 109], but the most important goal is towards the determination of CO_2 in aqueous or gas phase.

Jacquinot and Hauser[107] reported preparation of an amalgamated Pb, Tl, Au-poly(tetrafluoroethylene) gas diffusion electrode and an internal electrolyte containing [Ni (cyclam)]$^{2+}$. This preparation involves the formation of various metal films by electrolytic deposition or in the case of Hg by dipping the Au disk electrode into elemental Hg. For concentrations between 0.1 and 1% the electrochemical cell showed a sensitivity of 3.58 mA%$^{-1}$ and the detection limit is 500 ppm. The optimum pH range was determined between 3.5 and 6 and a selectivity ratio of the catalyst for CO_2/H^+ of 5:1 was found. The electrochemical process is limited by diffusion of CO_2 demonstrated by rotating disk experiments; it means that the relationship between reduction current and the square root of the angular speed was linear[107]. Figure 5.8[107] presents the response of a thin mercury film electrode (TMFE) when exposed to solutions equilibrated with CO_2. As can be seen a well established reduction wave for CO_2 is only obtained in presence of the [Ni (cyclam)]$^{2+}$ catalysts.

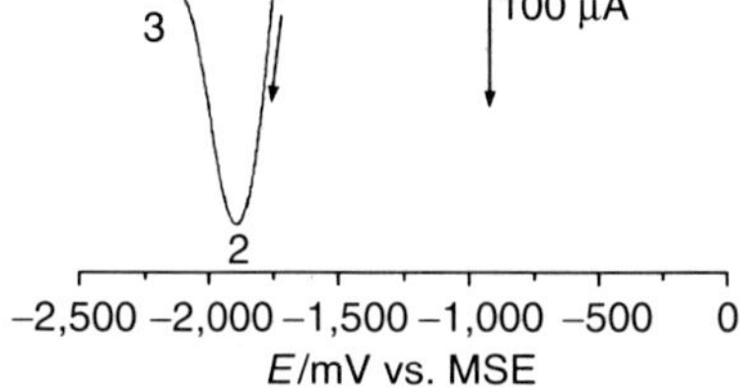

Figure 5.8. Cyclic voltammogram of a Hg-film electrode in contact with 100 mM $NaClO_4$ (pH 4) equilibrated with 20% CO_2 (A, C) or N_2 (B). B and C were recorded with the electrolyte solution also containing 0.2 mM Ni(cyclam)$^{2+}$. (1) Reductive adsorption of Ni(cyclam)$^{2+}$; (2) Electrocatalytic reduction of CO_2; (3) H_2-evolution; (4) electrocatalytic reduction of H^+; (5) anodic dissolution of precipitated Ni(cyclam)-CO. Hg-film on rotating Au-disk electrode (diameter 1.5 mm), $f = 3000\ \text{min}^{-1}$, $\upsilon = 100\ \text{mV s}^{-1}$. Reprinted from Figure 1: P. Jacquinot and P.C. Hauser, Ni(II) cyclam Catalyzed Reduction of CO_2 Towards a Voltammetric Sensor for in Gas Phase, Electronalysis 15(8), 1437–1444 (2003). With permission of Wiley-VCH, Germany.

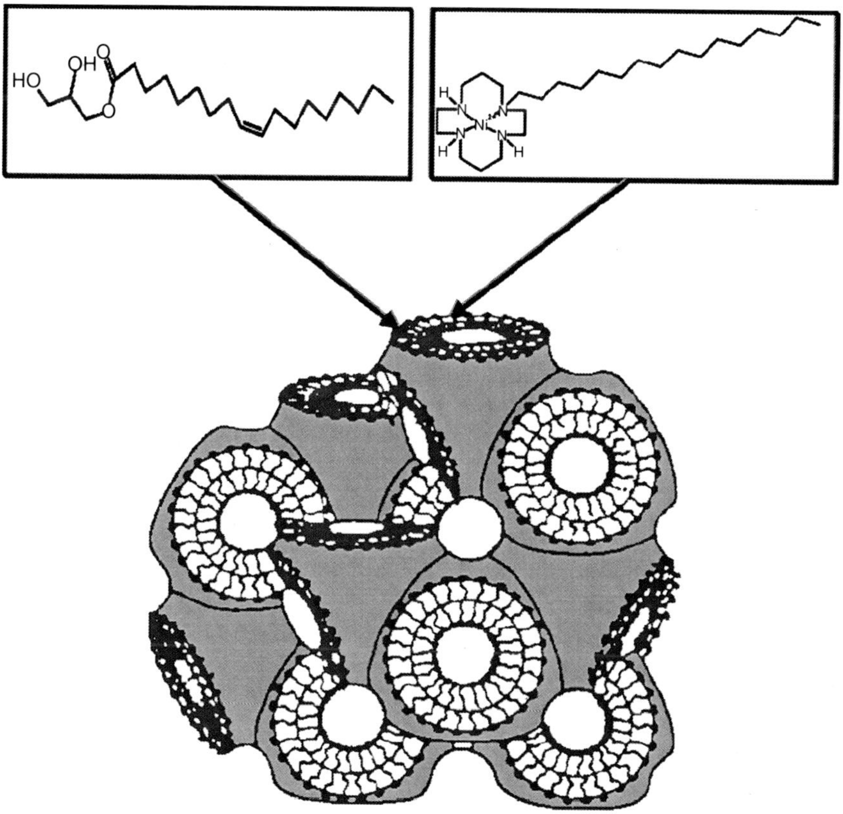

Figure 5.9. *Pn* 3 m cubic phase matrix containing the catalyst. Reprinted from Scheme 5.1: P. Rowinski, R. Bilewicz, M.J. Stebe and E. Rogalska, A concept for immobilizing catalytic complexes on electrodes: cubic phase layers for carbon dioxide sensing, Analytical Chemistry, 74(7), 1554–1559 (2002). Copyright 2002, with permission of American Chemical Society.

Another novel system is presented in Figure 5.9[110]. This is a crystalline cubic phase formed with monoolein and Myverol[105, 110]. This structure was used to host [Ni (cyclam)]$^{2+}$ or the derivative 1-hexadecyl-1,4,8,111 tetraazacyclotetradecane. After obtaining a mixture of the lipids and the catalyst, the mixture was spread over a glassy carbon electrode or a thin mercury silver film. The best catalyst for the reaction was the substituted complex because the hydrophilic [Ni(cyclam)]$^{2+}$ is easily removed from the cubic phase. The voltammetric results[110] indicate that the best response is obtained on the thin mercury film with formation of CO and regeneration of the catalyst. A further reaction of [Ni(cyclam)]$^{2+}$ in the presence of CO generates [Ni(cyclam)CO].

Figure 5.10[110] shows the response of the TMFE electrode for increasing concentration of CO_2, and the catalytic peak current–concentration plot is shown in the inset. The detection limit for CO_2 determination was 0.0001 mol dm^3. The

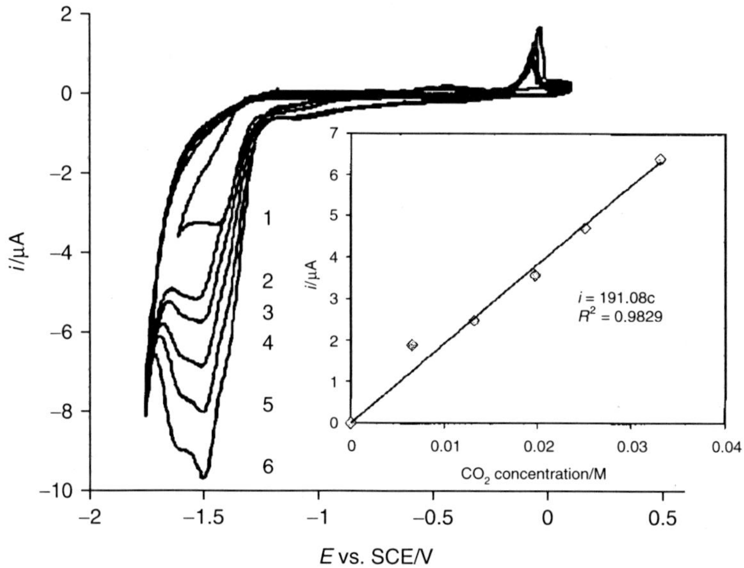

Figure 5.10. Cyclic voltammograms recorded on TMFE modified with a monoolein cubic phase layer containing $NiLC_{16}$, in 0.1 M $NaClO_4$, pH 4.8, for increasing concentrations of CO_2: (1) 0, (2) 0.0066, (3) 0.0132, (4) 0.0198, (5) 0.025, and (6) 0.033 M. Scan rate, 0.05 V s^{-1}. Inset: catalytic peak current vs. concentration of CO_2 in the solution. Reprinted from: Figure 3, P. Rowinski, R. Bilowicz, M.J. Stebe and E. Rogalska. A concept for immobilizing catalytic complexes on electrodes: cubic phase layers for carbon dioxide sensing, Analytical Chemistry, 74(7), 1544–1559 (2002). Copyright 2002, with permission of American Chemical Soceity.

results are sufficiently reproducible to assume that changes of the cubic phase structure do not affect the measured electrochemical signal.

In this section the electrocatalytic properties of the Ni(cyclam)$^{2+}$ complex and its derivatives toward the electrochemical reduction of CO_2 have been revised. This complex shows high selectivity and high electrocatalytic activity at Hg electrodes and in the last years the complex and supramolecular derivatives structures have been used as modified electrodes on the road to the design of sensor electrodes for CO_2. These new applications are in direct proportion with the knowledge generated in the past two decades. Probably many new applications will be found for the Ni(cyclam)$^{2+}$ family of complexes in the future, all of them closely related with the carbon dioxide reduction.

5. Phthalocyanines and Porphyrins Complexes

5.1. Phthalocyanines Complexes in Solution

Metallophthalocyanines have been reported to be active toward the electroreduction of CO_2[111–114]. Most of them modified electrodes[115], and not many examples have been presented for the electrocatalytic activity in solution in contrast with other tetrazamacrocycles (cyclams, etc.). In consequence this part of the chapter will have emphasis on the modified electrodes.

Nevertheless the voltammetry of mixed solutions of CO_2 and sulfophthalocyanines was informed in 1998[116] 1998[116] and the results indicate that an observed at potentials more negative than the second reduction process for both, Ni and Cu tetrasulfophthalocyanine under CO_2 atmosphere, and this wave depends on the concentration of carbon dioxide. This behavior can be connected with the formation of an adduct since Gangi and Durand[117] demonstrated by pulse differential voltammetry the formation of adducts with similar macrocycles. On the other hand Co tetrasulphophthalocyanine did not present any other additional wave indicating different mechanism for different metals[116]. The reduction products of mixed solutions of CO_2 and sulfonatedphthalocyanines were mostly oxalate and CO[116].

For Co and Fe phthalocyanines, Grodowski and coworkers[118] studied the electrochemical and photochemical reduction of CO_2. The combined results of flash photolysis, chemical reduction, cyclic voltammetry and IR and UV–Vis spectroscopy allowed the authors to conclude that the cation radical $[Co(I)Pc^{\cdot -}]^{2-}$ is the species responsible of the catalysis. However for Fe phthalocyanine[118] the results are not very clear. Other contributions to the reduction mechanism with Co phthalocyanine have been report recently[119,122]. The authors found a dependence of the substituents in the periphery of the ring with the mechanism of the reaction. The reaction studied by *in situ* potential-step chronoamperospectroscopy (PSCAS) inferred the formation of different radical species[119–122]. For Co-phthalocyanine and Co-octacyano-phthalocyanine the anion radical species is $[Co(I)Pc(-3)X]^{2-}$ (where X is the peripheric substituents of the phthalocyanine ring)[122], for Co-octabutoxy-phthalocyanine the responsible species is $[Co(I)Pc(-3)X(H)]^{-}$[121]. The formation of the hydride intermediate is explained in terms of the electron donor properties of the butoxy substituents. This fact could affect the stability of the dianion species and for stabilization purposes the dianion formed needs atomic H before its coordinates to the CO_2 molecule[121,122].

From the results shown here, it is clear that depending on the metal, the formation of an adduct is feasible. However, in the majority of the cases, the species responsible for the catalysis is an anion radical complex instead the double reduced complex. For that reason in some cases oxalate is obtained as product.

5.2. Porphyrins Complexes in Solution

Cobalt porphyrins have been used for several studies and exhibit catalytic activity[123,124]. For example when Co-tetraphenylporphyrin (**5.15**) and Co-octamethylporphyrin complexes are used, CO is obtained catalytically whereas palladium and silver porphyrins in DCL produced oxalate. However, demetallation of the complexes rapidly deactivates the catalyst[20,123,124]. In these cases, the catalytic process was interpreted in terms of the anion radical species of the reduced state of Pd (II) and Ag (II) porphyrins rather than the Pd(I) and Ag(I) states[125].

Also Co porphyrins have been studied by photophysical[126,127], photochemical[126,127] and electrochemical techniques[128,129]. The authors[126,127] found that

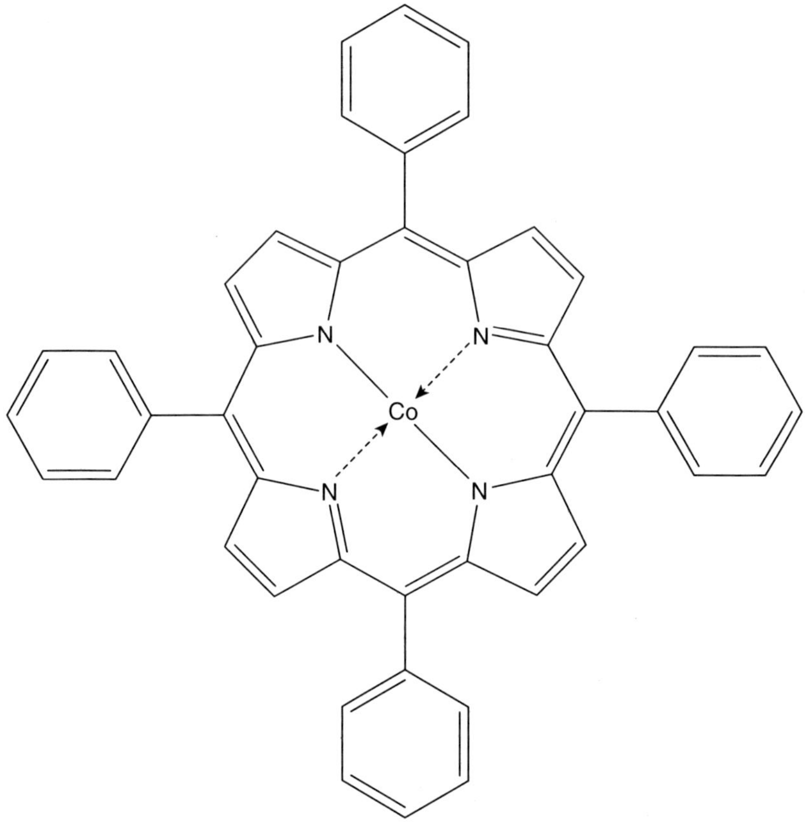

(15)

Structure 5.15. Co-tetraphenylporphyrin. Reprinted from Structure 5.15: J. Costamagna, G. Ferraudi, J. Canales and J. Vargas, Carbon dioxide activation by aza-macrocyclic complexes, Coordination Chemistry Reviews 148 (1996) 221–248. Copyright 1996, with permission of Elsevier.

a $Co^{I}P$ species is highly stable at basic pH but is short-lived in neutral or acidic solutions. Also, the species $Co^{I}P$ is unreactive towards CO_2 at high pH, but this is probably due to the formation of stable CO_3^{2-} that does not coordinate with the Co center[126, 127]. On the other hand, the existence of an intermediate $Co^{0}P$ more stable than the π-radical anion that reacts with CO_2 and the catalytic formation of CO and HCO_2^- was demonstrated in an organic solvent[126, 127]. The results of cyclic voltammetry also confirmed that the catalytic specie is $Co^{0}P$. The fluorinated derivatives were more active since they require less potential to carry out the catalysis[126–129].

It has been reported that the macrocycle known as "Corrole" is an effective catalyst toward CO_2 reduction[130, 132], (**5.16**)[132]. The Co (II) corrole $[Ph_3PCo^{III}$ (tpfc)] (tpfc = 5, 10, 15-tris(pentafluorophenyl) corrole) exhibit catalytic activity in ACN. Stepwise reduction to the $[Co^{II}(tpfc)]^-$ and $[Co^{I}(tpfc)]^{2-}$ states was

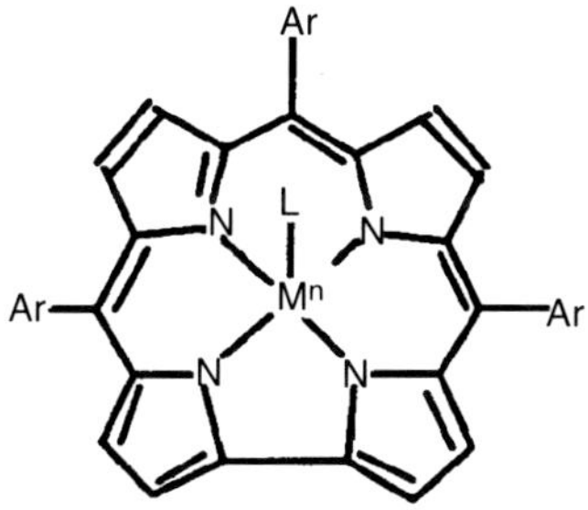

M	n	Ar	L	Abbreviation
Co	III	C_6F_5	Ph_3P	$Ph_3PCo(tpfc)$
Fe	IV	C_6F_5	Cl	ClFe(tpfc)
Fe	IV	2,6,-$C_6H_3HI_2$	Cl	ClFe(tdcc)

(16)

Structure 5.16. Corrole structures of different metals. Reprinted from Figure in the Experimental Section: J. Grodkowski, P. Neta, E. Fujita, A. Mahammed, L. Simkhovich and Z. Gross, Reduction of cobalt and iron corroles and catalyzed reduction of CO_2, Journal of Physical Chemistry A 106(18), 4772–4778 (2002). Copyright 2002, with permission of American Chemical Society.

observed in all cases[132]. Cyclic voltammetry in CO_2 saturated solutions indicated that Co^I complex react with CO_2[132]. Photochemical reduction in CO_2-saturated acetonitrile solutions containing *p*-terphenyl (TP) as a sensitizer and triethylamine (TEA) as the reductant led to the production of CO and H_2[132]. This finding is very interesting from a macrocycle-coordination chemistry point of view because there is a contrast with the other macrocyclic relatives (phthalocyanines and porphyrins) where the macrocycle should be reduced beyond the M^I sate in order to react with CO_2.

Porphyrin Iron(III) chloride, in general catalyzes electrochemical reduction of CO_2 to CO. Tetraphenylporphyrin Iron(III) chloride, with basket-handle of the picket-fence type with secondary amide groups in close vicinity to the porphyrin ring (**5.17**) belongs to the group mentioned before[20, 125]. The activity of this complex decays after a few cycles of work. Degradation of the porphyrin appears to be the result of progressive saturation of the ring through carboxylation and/or hydrogenation. However the addition of Mg^{2+} ion dramatically improves the rate of the reaction and the stability of the catalyst[125]. The reaction mechanism, studied by UV–Vis spectroelectrochemistry, preparative electrolysis and one of the finest studies in cyclic voltammetry published until now, involves the introduction of one molecule of CO_2 into the iron coordination sphere[20, 125]. The addition of a second molecule of CO_2 acts as a Lewis acid and then allows the breaking of one C–O bond of the first CO_2 molecule thus leading to CO. This process is accelerated by Mg^{2+} ions in a way that depends upon the temperature. At low temperatures (−40°C), the Mg^{2+} ions facilitate the decomposition of the complex containing two molecules of CO_2, whereas, at room temperature, Mg^{2+} ions

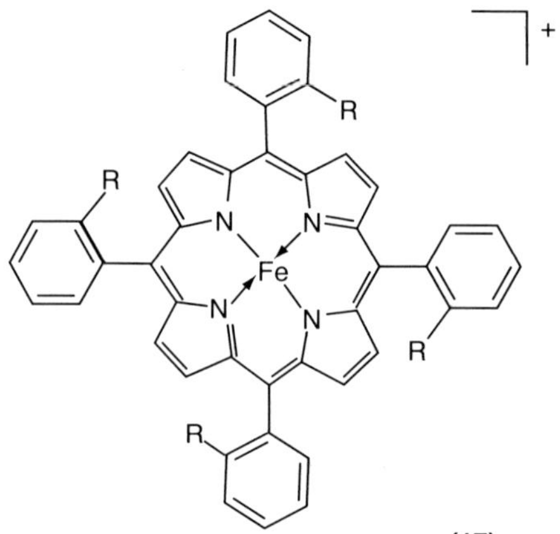

(17)

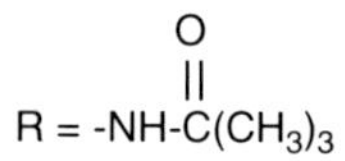

Structure 5.17. Tetraphenylporphyrin Iron(III) chloride, with basket-handle of the picket-fence type with secondary amide groups in close vicinity to the porphyrin ring. Reprinted from Structure 16: J. Costamagna, G. Ferraudi, J. Canales and J. Vargas, Carbon dioxide activation by aza-macrocyclic complexes, Coordination Chemistry Reviews 148 (1996) 221–248. Copyright 1996, with permission of Elsevier.

triggers the breaking of the bond at the level of the complex containing a single molecule of CO_2 in its coordination sphere. The combined action of iron("0") porphyrins and of Mg^{2+} ions offers a remarkable example of a bimetallic catalysis where an electron-rich center starts the reduction process and an electron-deficient center assists the transformation of the bond system[20, 125].

The same stabilization effect was corroborated with another cations (Brønsted and Lewis acids)[20, 133, 134]. The authors demonstrated that the addition of weak Brønsted acids such as 1-propanol, 2-pyrolidone, and CF_3CH_2OH or Lewis acid such as Mg^{2+}, Ca^{2+}, Ba^{2+}, Li^+, and Na^+ resulted in a considerable improvement of the catalysis of CO_2 by Fe(0) tetraphenylporphyrins[20]. Both the catalytic currents and the lifetime of the catalysts increase without significant formation of hydrogen. The authors[133, 134] have also shown that typical cyclic voltammogram for Fe(III)(tetraphenylporphyrin)Cl in the absence of CO_2exhibits three successive waves corresponding to the reversible formation of the Fe(II), Fe(I) and Fe(0) species. The authors[133, 134] find typical enhancements and changes of the Fe(I):Fe(0) wave in nitrogen and CO_2 in the presence of the Brønsted acids. The voltammograms and results from the preparative electrolysis suggest the mechanism depicted in Scheme 5.3[21].

Other added cations also change the catalytic activity as shown in the mechanism corresponding to Scheme 5.4[21]. The order of reactivity $Mg^{2+} = Ca^{2+}$ and $Ba^+ > Li^+ > Na$ was deduced from the enhancement of the Fe(I):Fe(0) wave. These cations accelerate the catalytic process by creating ion-pairing with the negatively charged oxygen atoms of CO_2 after it coordinates to the iron

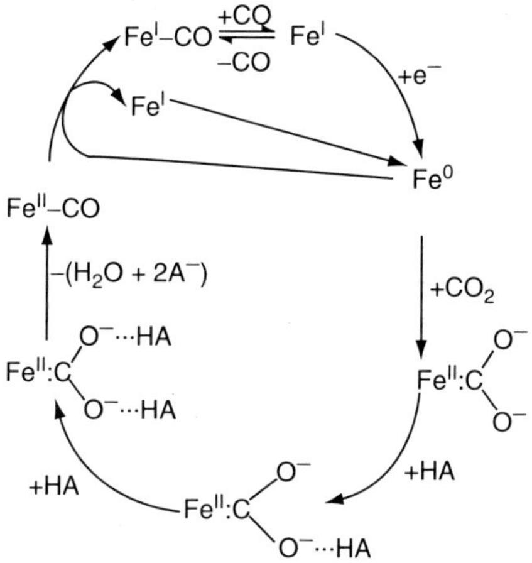

Scheme 5.3. Suggested mechanism for the electrocatalytic behavior of $Fe^{(0)}$ tetraphenylporphyrins in the presence of Brönsted acids. Reprinted from Scheme 5.6: J. Costamagna, G. Ferraudi, B. Matsuhiro, M. Campos-Vallette, J. Canales, M. Villagrán, J. Vargas and M.J. Aguirre, Complexes of macrocycles with pendant arms as models for biological molecules, Coordination Chemistry Reviews 196 (2000) 125–164. Copyright 2000, with permission of Elsevier.

atom. This mechanism, proposed by Saveant and coworkers, is shown in Scheme 5.4[21, 133, 134]. The nature of the synergetic effect is, therefore, similar to the action of a weak Brønsted acid, i.e., where hydrogen binding in Scheme 5.3 has been replaced by ion-paring in Scheme 5.4.

From the results showed here, it is clear that in the case of porphyrins, a wide variety of mechanisms are operating depending on the metal. As examples, Co-porphyrins produce Co while Ag and Pd porphyrins give oxalate. It is noticeable that Co^0P in organic media is the reactive species and not, the double reduced complex (metal and ligand). In the case of corroles (related species), Co(I) reacts with CO_2. Also, the corrole complexes show a different behavior compared with other azamacrocyclic complexes whereas the ligand must be reduced in order to catalyze the reaction. In the case of Fe-porphyrins, it is necessary to add Mg^{2+} to improve the reaction. It was postulated that after one CO_2 molecule is coordinated to the ion; a second is required to act as a Lewis acid allowing the breaking of the C–O bond of the first CO_2. This kind of stabilization was demonstrated for Brønsted and Lewis acids. This few examples depicts the complexity of the reduction of CO_2 proceess and the great importance of the kind of coordination of the carbon dioxide on the metal in order to transfer the electrons.

5.3. Studies in Supramolecular Systems

In the last years[135–140] there has been an increase interest for the studies of supramolecular properties of tetra-ruthenated porphyrins (M-TRP), especially after the discovery of the exceptional electrocatalytic efficiency of the penta-aminoruthenium-cobalt porphyrin species in the four electron reduction of

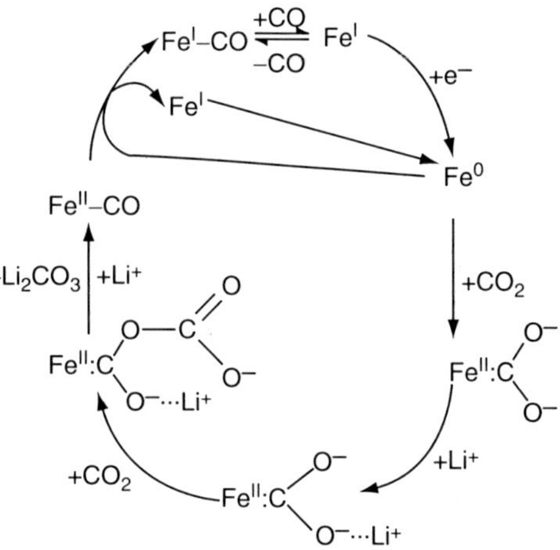

Scheme 5.4. Suggested mechanism for the electrocatalytic behavior of $Fe^{(0)}$ tetraphenylporphyrins in the presence of metal cations. Reprinted from Scheme 7: J. Costamagna, G. Ferraudi, B. Matsuhiro, M. Campos-Vallette, J. Canales, M. Villagrán, J. Vargas and M.J. Aguirre, Complexes of macrocycles with pendant arms as models for biological molecules, Coordination Chemistry Reviews 196 (2000) 125–164. Copyright 2000, with permission of Elsevier.

molecular oxygen[135]. Also this kind of compounds could present particular applications in catalysis, artificial photosynthesis and molecular devices[136]. However, few examples of electrochemical reduction of CO_2 have been published.

Toma and coworkers[136] studied the electrochemistry in solution of [Ni-TRP] (**5.18**)[136]. The cyclic voltammograms corresponding to the structure (**5.18**) are shown in Figure 5.11[136].

Contrasting behavior has been observed for the Ni-TRP first and second reduction waves at −0.8V and −0.97, which differ appreciably from those of the typical species e.g., $[Ni\text{-}TMPyP]^{4+}$ at −0.37 and −0.73, reflecting the influence of the π-back bonding interactions from the attached Ru^{II} groups. The existence of Ni(I) character in the reduced complex generated at −0.8 V has been investigated by adding methyl iodide[136]. Ni(I) species, exhibiting d^9 configuration, and CH_3 are isolobal species and should react readily, in contrast to the Ni(II) species. Methyl iodide enhanced by a factor of 2, the cathodic current at −0.8 V. No such effect has been observed when the porphyrin anion presents essentially a π-radical character[136,141,142]. Therefore, the results indicate a significant Ni(I) character for the reduced tetraruthenated nickel porphyrin species. In the presence of CO_2, there is a small increase in the reduction wave at −0.8 V, but a great intensification at −1.3 V. In this case the authors[136] assigned this increment in the cathodic current to the CO formation. The peripheral $[Ru(bpy)_2Cl]^+$ groups enhance the catalytic activity of the Ni porphyrin center, by increasing the Ni(I) character of the reduced species[136,141]. The results showed above indicate that the electronic communication between two metallic centers coud enhance the electrocatalytic activity of a system stabilizing the reactive oxidation state.

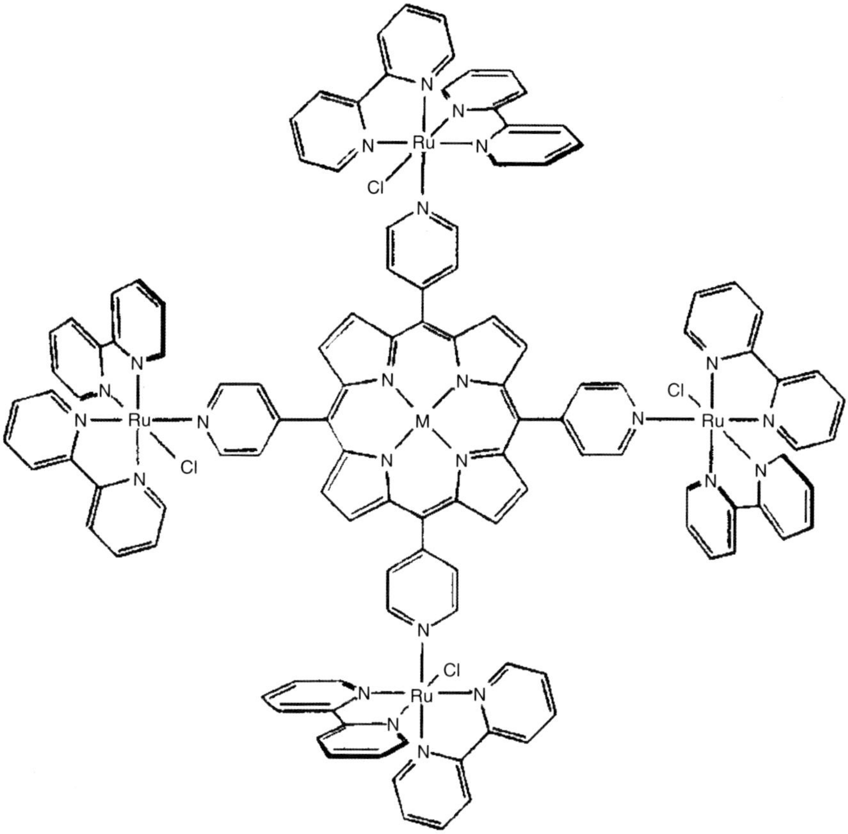

(18)

Structure 5.18. Structural representation of $[M(II) - TPyP\{Ru(bipy)_2Cl\}_4]^{4+}$ complex, TRP. Reprinted from Figure 3: H.E. Toma and K. Araki, Supramolecular assemblies of ruthenium complexes and porphyrins, Coordination Chemical Reviews, 196 (2000) 307–329. Copyright 2000, with permission of Elsevier.

5.4. Studies in Modified Electrodes

In the case of metallophthalocyanines and metalloporphyrins different ways to modified electrodes have been reported: deposited onto electrodic surfaces[112–114, 122, 143], complexes immobilized or incorporated in polymeric membranes as Nafion®[119–121, 144, 145], electropolymerized complexes[146, 147], gas diffusion electrodes[111, 148–152] and complexes chemically attached to carbon electrodes[153–157].

The intermittent plasma-assisted vacuum deposition technique has been found to introduce the effective electrocatalytic activity and stability for CO_2 reduction into metal phthalocyanine thin films formed on a glassy carbon[114]. The films properties are significantly influenced by the chemical state of the film. It has been suggested that the electrode process is determined by the surface chemical reaction involving adsorbed $H^{\cdot}$ and/or H^+ and a carbon containing intermediate[114].

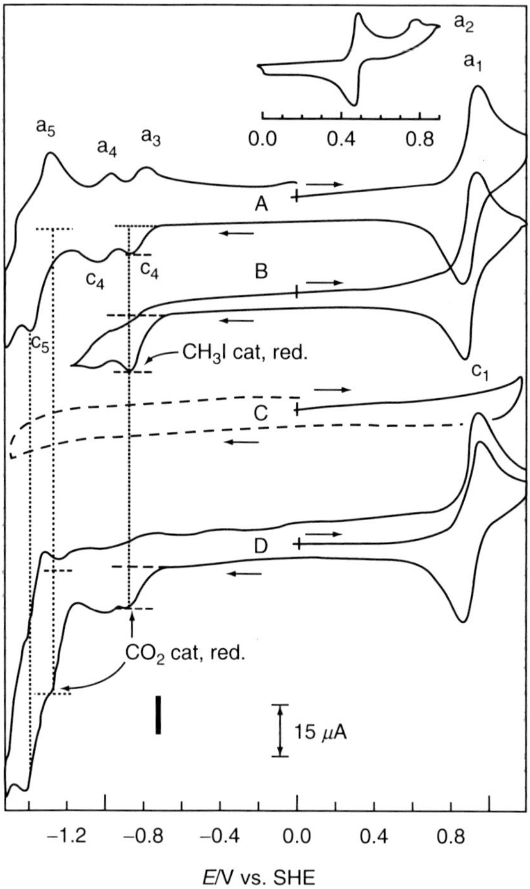

Figure 5.11. Cyclic voltammograms (50 $mV\,s^{-1}$) of: (A) [Ni-TRP](TFMS)$_4$ (1.1 $mmol\,dm^{-3}$) in DMF, TEA ClO_4 0.10 $mol\,dm^{-3}$, under an argon atmosphere; (B) in the presence of 16.2 $mmol\,dm^{-3}$ of CH_3I; (C) of an electroltye solution saturated with CO_2; and (D) of the complex solution saturated with CO_2. Insert: Voltammogram of a 1.1 $mmol\,dm^{-3}$ solution of the complex in acetonitrile, TEA ClO_4 0.10 $mmol\,dm^{-3}$, Reprinted from Figure 10: H.E. Toma and K. Araki, Supramolecular assemblies of ruthenium complexes and porphyrins, Coordination Chemical Reviews, 196 (2000) 307–329. Copyright 2000, with permission o Elsevier.

In 1974, Meshituka and coworkers[143] described the electrochemical reduction of CO_2 with modified electrodes containing transition metal phthalocyanine on graphite electrodes where the method for modification was the adsorption of a monolayer. The authors used as supporting electrolytes organic salts such Net_4CLO_4, NetCl and Nbu_4ClO_4. In the Co case the height of the reduction peak was linear with the square root of the sweep rate suggesting that the diffusion of CO_2 is the rate-determining step. Also the peak height was proportional with the concentration of CO_2[143]. CoPc and NiPc were very active in these media when quaternary ammonium salts were used for the supporting electrolyte, oxalic acid and glycolic acid were detected instead of CO or $HCOO^-$[143]. Then, the hydrophobic nature of the quaternary salts played a very important role and probably produces a suitable environment to obtain products with more than one carbon atom. MnPc, PdPc, were not active and CuPc, FePc were just slightly active in these conditions[143]. But when the same method of modification is used, with similar complexes but using inorganic salts such KCl, as supporting electrolytes the results of the reaction are very different.

Kapusta and Hackerman in 1984[113], reported that working at pH (3–7) the main reaction product was formic acid when the catalysts were CoPc and NiPc. Formate ion was also produced at pH > 5. Methanol is also produced at lower pH values. And the phthalocyanines become inactive when the medium is less acid. The order in activity was found to be $Co^{2+} > Ni^{2+} >> Fe^{2+} = Cu^{2+} > Cr^{3+}$, Sn[113].

In the cases described[113, 143] the order of activity was similar and also a very important dependence with negative limit of the potential and the stability of the macrocycles was found. Indeed, if the modified electrode is swept at more negative potentials, they become less active with the time. This fact is probably associated with the deactivation of the macrocycle. Also the poor selectivity of these modified electrodes has been noted.[113, 143]. With these factors in mind another kind of modification has been developed, the incorporation of the macrocycles either phthalocyanines or porphyrins in polymeric membranes.

In 1995 Zhang and coworkers[144] studied the electrochemical detection of CO_2 in rotating ring (Pt)-disk graphite electrode coated with the macrocyle Co N, N′, N″, N″″-tetramethyl-3,4-pyridoporphyrazine and protected by a Nafion film. In this work it was informed that the CO detection on the Pt ring is proportional to the CO_2 concentration. CO_2 reduction was analyzed in the concentration range 1.0×10^{-4}–3.2×10^{-2} by the rotating ring disk method. The method provided a selective way of identifying and quantifying CO_2 in a mixture which contained other gases which could only interfere if they or their redox products reacted on the ring at potentials near that of the CO oxidation. The role played by the Nafion film was to improve the selectivity to one product (CO) and an enhancement of the stability of the catalyst since it was proved that if the protection is omitted, there is a gradual reduction in the current, with a scanning from 0 to −1.0 V vs. SCE, due to gradual loss of surface catalyst[144].

Another selective electrocatalyst Co phthalocyanines-poly-4-vinylpiridine (PVP) has been used to modified electrodes[119–121]; in this case the method of modification involves the casting on the surface of basal plane graphite electrode (BPG) of a mixture of CoPc/PVP. The mixture composed by 1 wt% PVP + DMF solution containing 10 μM of CoPc or another related macrocycles and letting the solvent (2 μL) evaporate under air at room temperature. Enhanced catalysis and selectivity was observed with the CoPc/PVP/BPG electrode comparing with the CoPc/ BPG but both factors were very dependent on the pH of the solution and the potential where the reaction takes place. For instance the selectivity ratio at the same potential (−1.3 V) CO/ H_2 at pH 2.3 was 0.16 for CoPc/ BPG and 0.54 for CoPc/PVP/BPG while at pH 6.8 CO/H_2 was 2.28 for CoPc/BPG and 4.65 for CoPc/PVP/BPG. The selectivity toward CO was explained in terms of electron donation of the pyridine group to the central cobalt by an axial coordination, a proton-exchanging property of the PVP and the increased local concentration of CO_2 in the PVP layer owing to their hydrophobic and/ or acid–base interaction[119–121]. The first factor is explained by the well established, axial coordination of an electron-donating ligand such as pyridine or imidazole to the cobalt ion of macrocyclic cobalt complexes greatly stabilizes the *trans* coordination of O_2[120, 158–160] as well as CO_2[120, 161–163]. It has also been reported that the presence of such ligands enhances the catalytic activity of macrocyclic cobalt complexes for the reduction of O_2[164] or CO_2[120, 161]. The coordination of the pyridine residue of PVP probably stabilizes the electronegativity of the oxygen atom of the coordinated CO_2, thus favoring coordination of the CO_2 molecule to the central cobalt of CoPc, (see Scheme 5.5[120]). When Co octabutoxy-phthalocyanine is used with the same method of modification the catalytic activity is lower compared with the parent complex (CoPc), however the activity of Co octabutoxy-phthalocyanine is higher than CoPc if the modification is by the "casting" method[119–121]. The low catalytic activity is explained by the authors[119–121] in terms of to the electrochemical inertness due to some specific interactions between the hydrophobic substituents and the PVP macromolecule.

Another attempt to obtain stable and selective modified electrodes with transition metal phthalocyanines or porphyrins is the design of carbon gas diffusion

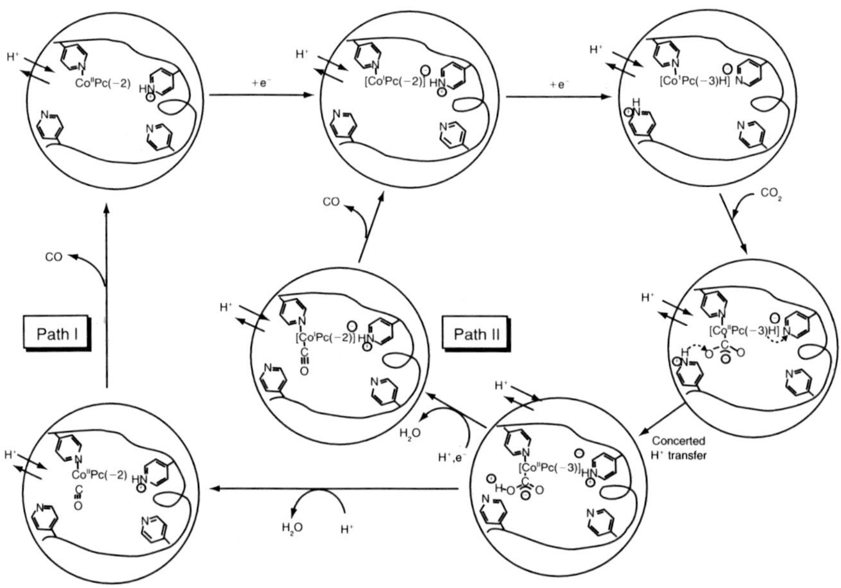

Scheme 5.5. Possible mechanism for electrochemical CO_2 reduction catalyzed by CoPc-PVC. Reprinted from Figure 10: T. Abe, T. Yoshida, S. Tokita, F. Taguchi, H. Imaya and M. Kaneko, Factors affecting selective electrocatalytic CO_2 reduction with cobalt phthalocyanine incorporated in a polyvinylpyridine membrane coated on a graphite electrode, Journal of Electroanalytical Chemistry 412 (1996) 125–132. Copyright 1996, with permission of Elsevier.

electrodes. Mahmood and coworkers[111] reported the use of polytetrafluorethylene-bonded, carbon gas diffusion impregnated with metal phthalocyanines. The authors[111] designed a method to obtain this kind of electrodes and it is briefly described here: Carbon black is added to a stirred solution of the appropriate metal phthalocyanine dissolved in concentrated sulfuric acid. The resulting mixture is quickly poured into water, to precipitate the metal phthalocyanine onto the carbon substrate. The impregnated carbon is filtered, washed and dried. A sample of the impregnated carbon is micronized for 10 min in water to form uniform slurry. Then PTFE dispersion was added with gentle stirring to the slurry to give a thick paste. A small quantity of this paste was spread onto a lead-plated nickel gauze, dried and compacted onto the gauze until the carbon layer was firmly embedded onto the current collector. This procedure was repeated until the required loading had been achieved. The electrodes were then cured in a tube furnace in a nitrogen atmosphere at 300°C for 3 hr before being given a final compression at a pressure approximately 10^5 kPa.

The results obtained in this work[111] are related mainly with the high rates for the reduction process achieved with this kind of electrodes. With Co phthalocyanine electrodes the only significant products were CO and H_2 and the results are highly dependent on the potential used for the electrolysis experiments. Other important factor described by the authors was the amounts of carbon monoxide and hydrogen produced from sample to sample, reflecting probably variations in the exact composition of different Co phthalocyanines electrodes. For Mn, Cu, and Zn impregnated electrodes both formic acid and hydrogen were produced.

Also Furuya and Matsui[148] reported electrocatalytic activity with gas-diffusion carbon electrodes with 17 kinds of different phthalocyanines. When cobalt and nickel phthalocyanines are used as electrocatalysts high yields of CO are obtained, in these cases a current efficiency of *ca.* 100% was observed. When tin, lead and indium phthalocyanines are used, the main products are formic acid and H_2, while for copper, gallium and titanium phthalocyanines the main products are CO and H_2. Methane is also produced with a good yield (about 30%). Iron, zinc and palladium also give CO as the leading reduction product, but in a lower yield compared with those obtained with cobalt and nickel phthalocyanines. Hydrogen is exclusively obtained with magnesium, vanadium, manganese, and platinum derivatives and with the free base.

Then, it is evident from this work[148] and the others mentioned above that the selectivity in this case is driven by the metal in the carbon hydrophobic environment. These distributions of products and current efficiencies depend strongly on the chemical properties of the metal more than the ligand[148] and suggest that the mechanism for producing a particular product may be determined by the interaction between the phthalocyanine ring, the central ion and CO_2 or CO molecules.

On the other hand, phthalocyanines and porphyrins transition metal complexes have been adsorbed on activated carbon fiber nanopouros support in the form of gas diffusion electrodes[152]. The loading of the complexes in the support involve the dissolution of the macrocycles in the proper solvent and then the addition of the (activated carbon fibers ACF). The mixtures were stirred on a magnetic stirrer for 3–4 days at room temperature and then vacuum filtered. The ACF with the catalyst adsorbed was dried in a vacuum oven at 100°C for 1 hr. Gas diffusion electrodes were prepared according to the following procedure: for each electrode, a 25 mg portion of ACF loaded with the catalyst, 25 mg of acetylene black and 100 μL of polytetrafluorethylene (PTFE) aqueous solution were mixed on a PTFE sheet with a spatula to obtain a homogeneous paste, suitable for the preparation of the active catalyst layers. This paste and gas diffusion layer, which was cut in the shape of a disk with 13 mm diameter, were pressed together under vacuum, with steadily increasing pressure up to 4.5 tons for 15 min. The gas diffusion electrode thus obtained was then heat treated in Ar at 350°C for 1 hr. All the complexes used in this work, (Co(II) 5,10,15,20-tetrakis-(4-methoxyphenyl)-porphyrin, CoPc, CuPc, and Cu(II)2,9,16,23-tetra-tertbutylphthalocyanine, Co(II) 2,9,16,23-tetra-tertbutylphthalocyanine) demonstrated catalytic activity for CO_2 electroreduction yielding CO and H_2 as the main products[152]. In this case it is surprising that the selectivity of the modified electrode is not dependent on the metal as was mentioned above. It is noticeable because CuPc gives some CH_4 production[148]. For the complex with bulky substituents[152], the performance for CO production is poor compared with the unsubstitued complexes, due to size restrictions of the narrow ACF pores and for the complexes that are less bulky, the CO production appears to be very similar for all the compounds, irrespective of the metal or ring type. This result shows a lack of sensitivity for the detailed characteristics of the macrocycle complexes. In all cases[152] a current increased was observed when the loaded catalyst in the ACF support was higher, but the increase is limited by the adsorption limit of the catalyst on the ACF in a flat orientation. In the case of bulky compounds, the opposite situation was observed. Smaller amounts of the complex can be absorbed on ACF, and the increase in the catalyst amount results in a small decrease of catalytic efficiency[152]. This is probably due to the molecular dimensions, which are the largest among all of the investigated compounds and, hence, the adsorption limits are expected to be significantly lower, as already suggested. In addition the authors[152] mentioned that when the pores are filled with these bulky molecules, it may be more difficult for CO_2 molecules to penetrate, thus resulting in a decreased current efficiency for CO production.

Another very interesting approach with diffusion gas electrodes modified with transition metal phthalocyanines deals with the simultaneous reduction of CO_2 and NO_3^- or NO_2^- ions[150,151] toward the production of urea or other C–N bonded compounds. The simultaneous reduction was investigated at gas diffusion electrodes with phthalocyanines containing the following metal centers: Cr, Mo, Mn, Ru, Co, Rh, Ir, Ni, Pd, Pt, Cu, Ag, Au, Zn, Cd, In, Tl, Sn and Pb[150,151]. The formation of urea, CO, formic acid, and ammonia was confirmed for all the metals except for Al and Ge. The best catalyst in this case was NiPc with a maximum current efficiency of 40% at -1.5 V. Again the selectivity is dependent on the metal center and the electrochemical potential used in the experiments, but the formation of urea specifically depends on the ability of the metal center to form CO alone and ammonia alone[150,151]. The mechanistic study[150,151] showed that CO obtained from the electroreduction of CO_2 cannot be combined with ammonia in the electrolyte to form urea and the ammonia formed in the reduction of nitrite ion cannot be combined neither CO nor CO_2 dissolved in the solution to form urea. Also the urea formation cannot be obtained with in the simultaneous reduction of the following pairs of reagents: CO–ammonia, formic acid–nitrite, formic acid–ammonia. Thus the authors[151] established that formation of urea only is possible when precursor like CO and ammonia are formed at the same electrodic surface. Although this kind of precursor was not found, authors propose the following mechanism for the different reactions:

(I) CO_2 reduction:

$$CO_2(g) \longrightarrow CO_2(ad) \quad (5.5)$$

$$CO_2(ad) + e^- \longrightarrow CO_2^{\cdot-} \quad (5.6)$$

$$CO_2^{\cdot-} + 2H^+ + e^- \longrightarrow CO(pre) + H_2O \quad (5.7)$$

$$CO(pre) \longrightarrow CO\ (g) \quad (5.8)$$

where CO(pre) = CO like precursor.

(II) NO_2^- reduction:

$$NO_2^- + 6H^+ + 5e^- \longrightarrow NH_2(pre) + 2H_2O \quad (5.9)$$

where NH_2(pre) = ammonia-like precursor

$$NH_2(pre) + H^+ + e^- \longrightarrow NH_3 \quad (5.10)$$

(III) Urea formation:

$$CO(pre) + 2NH_2(pre) \longrightarrow (NH_2)_2CO \quad (5.11)$$

From kinetic considerations, the authors[150, 151] find a linear relationship between the current efficiency (CE) of the urea formation and the square root of the current efficiency of the product concentrations: CO and NH_3. It means:

$CE_{\text{urea formation}}$ vs. $(CE_{CO} \cdot CE_{NH3})^{1/2}$. In other words the relationship suggests that ammonia-like precursor and a CO-like precursor would coexist and react with each other. Thus the authors present the following new mechanism[151]:

$$NH_2(pre) + CO(pre) \longrightarrow (NH_2)CO\ (ad) \quad (5.12)$$

$$(NH_2)CO\ (ad) + NO_2^- + 6H^+ + 5e^- \longrightarrow (NH_2)_2CO + 2H_2O. \quad (5.13)$$

Here a nitrite ion would be reduced on an adsorbed $(NH_2)CO_{(ad)}$ and urea would be formed. However, this mechanism it still very speculative and it needs to be strictly confirmed. Also the same reaction was studied but with NO_3^- instead of NO_2^-.

The complexes used in the case of the nitrate[151] nitrate[151] were the same described before but that phthalocyanines are unable to reduce NO_3^- to ammonia and the reaction just produces NO_2^-. Thus the production of urea was not possible in this kind of electrodes[151].

Another method for the modification of carbon electrodes with macrocycles towards the electroreduction of CO_2 is to form chemical bonds between the electrode and the macrocycle. Several studies have shown that it is possible to attach small organic molecules, to the polished gassy carbon surface and then this attached molecule could be an anchor for another molecule, in this case Co-porphyrins or Co-phthalocyanine[153–157]

The general procedure to obtain this kind of modified electrode is described as follows:[155]

After being polished to a mirror finish, the GC electrode was subjected to potential sweeps from 0 to 1.4 V (SCE) in 0.1 M $LiClO_4$ ethanol solution. At 1.4 V, the amine containing molecule was added to make a 0.1 mM solution, and the potential of the GC was kept there for 5 s and then swept to 0 V, where the GC was taken from the solution. The amine containing molecules should posses a group which can coordinate as a fifth ligand of porphyrin. After the above bridging species were bonded on GC's the modified electrode was rinsed with ethanol, water and sonicated for 15 min in pH 6.8 phosphate buffer for removal of any physically adsorbed materials. To attach the porphyrin ligand, the modified electrode (GC-amino molecule) was subjected to 2 hr reflux in a proper concentration of CoTTP in benzene + dichloromethane mixture (4:1) and bonded to Co(II)TPP as a fifth ligand through the nitrogen of pyridyl group to GC. The electrodes obtained can be characterized by several spectroscopic techniques, the analysis involve diffuse reflectance FT-IR, diffuse reflectance UV–Vis, XPS, AFM, among others[155].

Then when CoTTP-py-NHCO-GC modified electrode is obtained[155], the catalytic activity towards the electrochemical reduction of CO_2 showed a current efficiency of 92% for CO production and also an overall turnover number of 10^7 at -1.1 V vs. SCE. As mentioned before the CoTTP in CoTTP-py-NHCO-GC electrode is in the form of a five coordinate complex with a pyridine ligand, with a vacant site at the *trans* position to pyridine. It is expected that CO_2 reduction occurs at vacant site to which pyridine gives specific reactivity through the called *trans* effect. The axial pyridine ligand could favor an electron transfer from Co^I of the complex to CO_2, since it increases the electron transfer ability of Co, as in the case of an O_2 carrier, and facilitates the reduction of CO_2[155].

The modified electrode with the attached CoTPP can reduce CO_2 to CO[155] whereas the electrode modified with adsorbed CoTPP cannot. The cyclic voltammogram of the modified porphyrin-amine-GC electrode in pH 6.8 phosphate buffer shows the appearance of the anodic hump after the electrode is polarized to potentials where the evolution of hydrogen takes place; under N_2 atmosphere see Figure 5.12[155]. The ligand bonded to the glassy carbon does not show this hump, and then the authors associate it with the cobalt center. The hump would correspond to the oxidation of the formed hydride. Over the first-bonded porphyrin there are more than 100 layers of stacked porphyrins. Stacked Co(II)TPP could accept a hydrogen atom from CoTPP bonded GC to form HCo(II)TAPP[155].

However, when the stacked electrode is cycled under CO_2 the hump disappears indicating that here is not a formation of hydride in the stacked porphyrins.

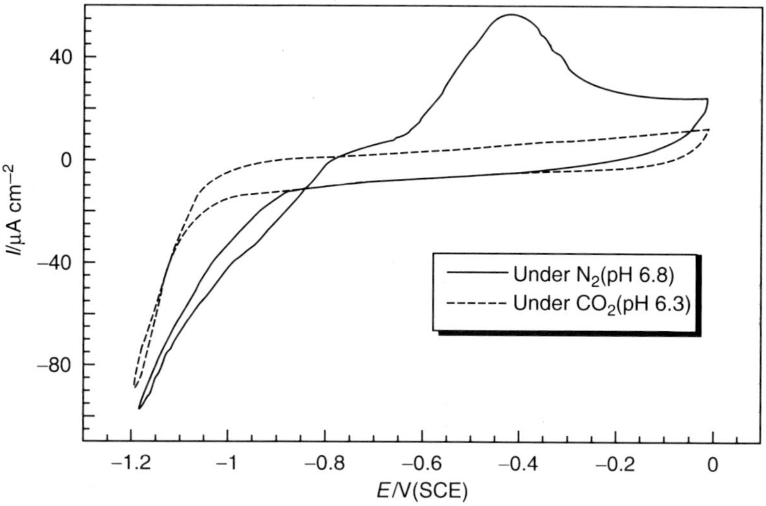

Figure 5.12. Cyclic voltammograms at 50 mV s^{-1} on "CoTPP bonded GC" in N_2 and Co_2 atmospheres. Bridging compound is 4-aminopyridine. The solution was pH 6.8 phosphate buffer solution. Reprinted from Figure 5: H. Tanaka, A. Aramata, Aminopyridyl cation radical method for bridging between metal complex and glassy carbon: Cobalt(II) tetraphenylporphyrin bonded on glassy carbon for enhancement of CO_2 electroreduction, Journal of Electroanalytical Chemistry, 437 (1997) 29–35. Copyright 1997, with permission of Elsevier.

Only, the porphyrin directly bonded can undergo the reaction forming the hydride according the following mechanism[155]:

$$\text{Co(II)TPP} + e^- \longrightarrow \text{Co(I)TPP} \tag{5.14}$$

$$\text{Co(I)TPP} + H^+ \longrightarrow \text{HCo(II)TPP} \tag{5.15}$$

The HCo(II)TPP bonded on GC reacts with CO_2 giving CO. On the other hand if the electrode is heavily rinsed, the packed structure disappears. In this case, the electrolysis measurements show higher current efficiencies for the non-stacked structure. In fact, according these results the stacked porphyrins promote the evolution of hydrogen instead the reduction of CO_2[155].

Contrarily to the above results[157], using a similar method to modified the electrode but in this case with Co(II) benzoporphyrin shows a very interesting activity of the stacked system in the reduction of CO_2 . The obtained current for the evolution of hydrogen in the case of the packed system compared to the aminopyridine-glassy carbon or the physically adsorbed layer of porphyrin is more than one order of magnitude higher. This fact indicates the difference in the amount of porphyrin (or active sites, in the case of the amine-GC) in each case (see Figure 5.13[157]). On the other hand, the current obtained for the reduction of CO_2 in the case of the packed system is very similar to that obtained in the case of the reduction of hydrogen for the same system (see Figure 5.14[157]). These results indicate that in this case, the entire packed system reacts with the CO_2 in the reduction process. In this study the appearance of the anodic hump was not reported[157]). On the other hand, as in the case of Tanaka and Aramata[155], the

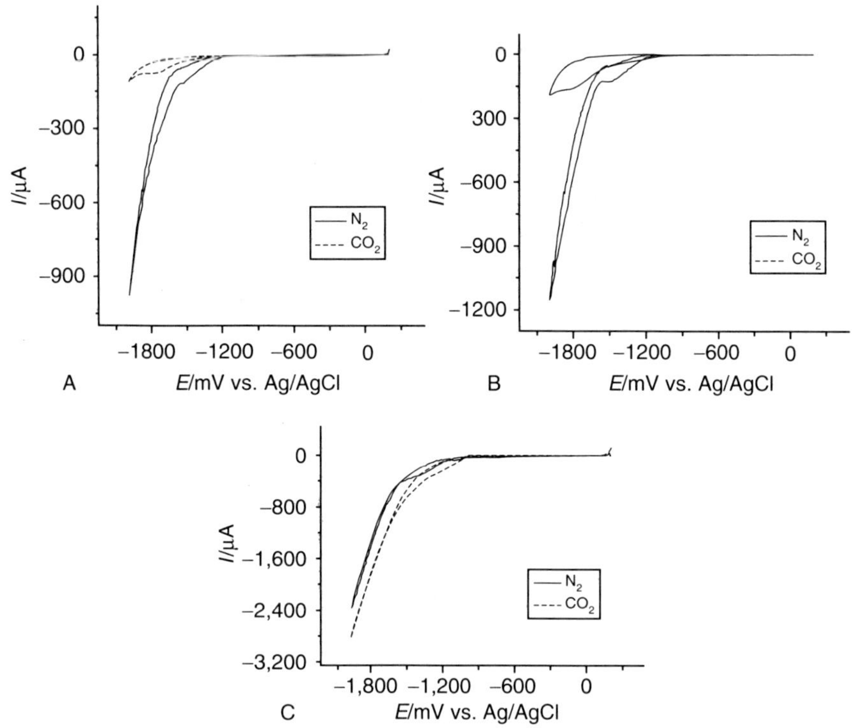

Figure 5.13. Voltammetric response of a glassy carbon (A), 4-aminopyridine-electrode (B), and the supramolecular electrode (formed by packed Co(II) benzoporphyrins) (C) under N_2 (continuous line) and CO_2 (dashed line). Scan rate 0.05 V s^{-1}. Electrolyte: phosphate/biphosphate (pH 6.8 under N_2, pH 5.2 under Co_2) buffered aqueous solution. Reprinted from Figure 4A (A), Figure 4B (B) and Figure 5(C): G. Ramírez, M. Lucero, A. Riquelme, M. Villagrán, J. Costamagna, E. Trollund and M.J. Aguirre, A supramolecular Cobaltporphyrin-modified electrode toward the electroreduction of CO_2, Journal Coordination Chemistry, 57 (2004) 249–255. With permission of Taylor and Francis (http://www.tandf.co.uk/journal).

Co(II) benzoporphyrin physically adsorbed on the glassy carbon does not reduce CO_2[157].

In the same way, various Co macrocycles were chemically bonded to the glassy carbon electrodes by the same method[156]. A 4-aminopyridine was the bridging molecule, the attached macrocycles were: Co(II) naphthalocyanine, Co(II) phthalocyanine,dibromo (11-hydroxyimino-4,10-dipropyl-5,9-diazatrideca-5,9-dien-3one oximate) Co(III), (denoted as CoDO), two kinds of hydrophobic B_{12} vitamins (heptamethyl cobyrinate perchlorate and heptapropyl cobyrinate perchlorate), Co(II) tetraphenyl porphyrin and Co(II) 5,10,15,20-tetrakis-(4-methoxyphenyl) porphyrin. Not all of the macrocycles were active toward the electroreduction of CO_2. In fact, Co naphthalocyanine and Co phthalocyanine (CoPc) produce larger amounts of H_2; even CoPc was the macrocycle with the most positive potential to give H_2. CoDO was unable to produce CO at

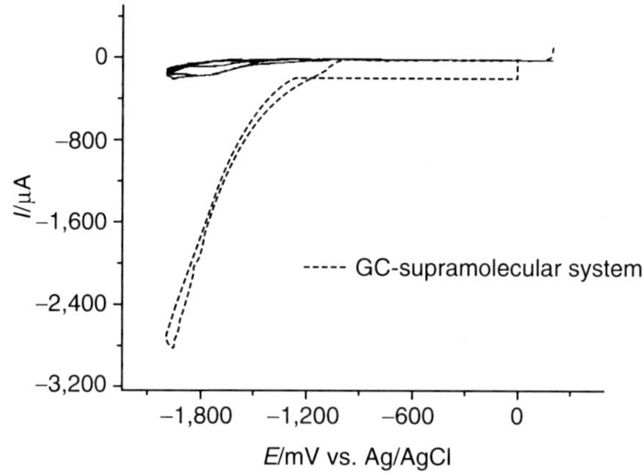

Figure 5.14. Voltammetric response of all the systems toward the reduction of CO_2. The dashed line shows the response of the supramolecular electrode. In the continuous line, the response of other systems (GC, GC-4-aminopyridine, GC-absorbed layer and GC-4-aminopyridine-adsorbed layer) (adsorbed layer and supramolecular assembly formed by Co(II) benzoporphyrin). Scan rate: 0.05 V s^{-}. Electrolyte: phosphate/biphosphate (pH 6.8 under N_2, pH 5.2 under CO_2) buffered aqueous solution. Reprinted from Figure 6: G. Ramírez, M. Lucero, A. Riquelme, M. Villagrán, J. Costamagna, E. Trollund and M.J. Aguirre, A supramolecular Cobalt-porphyrin-modified electrode toward the electroreduction of CO_2, Journal of Coordination Chemistry, 57 (2004) 249–255. With permission of Taylor and Francis (http://www.tandf.co.uk/journal.).

any potential studied. The highest current efficiency was observed in the case of Co-porphyrins. In all the cases, during a voltammetry where the lower potential is less than the required to the evolution of hydrogen, during the reverse scan an anodic hump current appears. This hump disappears when CO_2 is introduced into the solution. The authors[155,156] propose that the appearing of this hump is due to the formation of a cobalt hydride (see Figure 5.12). This species would be the responsible for the catalysis. However, the authors do not explain why in the case of CoDO, where the hump appears, there is not reduction of CO_2 . However, the results are very interesting and the best catalyst is tetraphenyl-porphyrinato cobalt(II) chemically modified.

On the other hand, there are few studies about polymerized systems used as catalysts in the reduction of CO_2 . For example, there is a study from Riquelme and coworkers[147] where a glassy carbon electrode was modified with an electropolymerized Co-tetra(3-aminophenyl)porphyrin. In this work, authors made a comparison between the electrode modified with a physically adsorbed layer of porphyrin by deposition of a drop of solution (DMF containing the porphyrin) and an electrode modified with the same porphyrin electropolymerized by cyclic voltammetry. The response of the polymeric-modified electrode in NaOH under N_2 atmosphere shows a voltammetric reversible couple at *ca.* −1.0V vs. Ag/AgCl corresponding to the couple Co(II)/Co(I) and at potentials more negative than −1.2 V (*ca.* close to −1.6 V) a current discharge corresponding to the evolution of hydrogen reaction. In the presence of CO_2 , the redox signal corresponding to the couple diminishes in intensity but is broader and the current discharge appears at potential slightly negative than in the case of N_2. It is possible that under CO_2 atmosphere two reduction processes take place simultaneously: evolution of hydrogen and reduction of CO_2[147]. On the other hand, the open circuit potential (OPC) of the modified electrode changes when the system is submerged in a solution saturated with N_2 or CO_2. In the first case, the value is -0.13 V and in the second, -0.278 V. The change indicates the possible possible formation of an adduct between CO_2 and the metal center because, if the system is maintained under N_2 and pH varies in the same sense of the variation produced by CO_2, there is a practically no change in the OCP[147]. The formation of a kind of adduct could explain the change in the voltammetric couple of the Co(II)/Co(I) when the solution is saturated with N_2 or CO_2. The monomeric-modified electrode does not catalyze the reduction of CO_2, probably due to its low stability. Electrolysis of the polymeric system at -1.25 V gives CO, formic acid and H_2. The main products are CO and H_2. From the polarization curves it is possible to obtain kinetic data. The Tafel plots have a slope of -0.12 V/decade indicating that the rate-determining step of the reaction is the first electron transfer:

$$[\mathrm{Co(II)} \cdots \mathrm{CO_2}] + e^- \longrightarrow [\mathrm{Co(I)} \cdots\cdots \mathrm{CO_2}]. \tag{5.16}$$

The polymeric electrode is very stable and products were detected 4 hr of electrolysis[147].

In an interesting study, Magdesieva and coworkers[146] compared the catalytic behavior of mono, di, and electropolymerized phthalocyanine complexes in methanol, toward the electroreduction of CO_2. In the case of mono- and diphthalocyanines, they were used in impregnating the electrode (graphite) from a solution of CH_2Cl_2 or DMF. The electropolymerized complexes were obtained by cyclic voltammetry. The reason for using methanol instead of water is the enhanced solubility of CO_2 in methanol and the diminishing of the hydrogen evolution reaction. In this work Cu 2,9,16,23-tetra-tert-butylphthalocyanine (Bu^tPcCu), Cu 2,9,16,23-tetraaminophthalocyanine ($^{NH}_{2}PcCu$), Lu monophthalocyanine (PcLuOAc), Lu (Pc_2Lu) and Dy diphthalocyanine (Pc_2Dy), Lu 2,3,9,10, 16, 17, 23, 24, 2′, 3′, 9′, 10′, 16′, 17′, 23′, 24′-hexadecapentoxydiphthalocyanine ($^{CH\ O}_{5\ 11}Pc_2Lu$), and electropolymerized tetraamino Cu monophthalocyanine (poly $-^{NH}_{2}$ PcCu) as electrocatalysts were studied. All the complexes catalyze the electroreduction of CO_2. The products depend on the complexes and the applied potential. For example, the quantity of methane changes from 20 to 30% when the potential was varied from -1.8 to -2.5 V vs. Ag/AgCl in the case of $^{NH}_{2}PcCu$. In the case of Pc_2Lu, there are a decrease in the amount of CO and an increase in the production of methylformate. The polymer is destroyed at potentials beyond -1.5 V. In all the cases, the current efficiencies for the products vary depending on the time of electrolysis. During the first 30 min there is no methane. CO appears between 60 and 90 min of electrolysis. However, this variation can be an artifact of the higher solubility of these products in the solvent (products were measured in the aqueous fraction). The average of the current efficiencies measured in the gas phase show noticeable variation depending on the different kinds of complexes. The higher current efficiencies for methane were obtained for Bu[1]PcCu at two different potentials[146]. All complexes showed more than one product indicating that the selectivity in these cases is poor. Based on these results and published data, authors[146] proposed the mechanism in Scheme 5.6[146] for the CuPc.

According to the mechanism, the extent of CO_2 reduction (and the consequent kind of products) is determined by the reduction ability of the anionic form of the complex and the stability of the intermediate products[146]. On the other hand, the difference in the ability to produce methane between the amino and the butyl Cu phthalocyanine indicates the strong dependence of the reduction on the substituents in the phthalocyanine ring. Then the mechanism is based on the strong reducing behavior of the reduced form of CuPc. Then in the first stage, the reduced complex can transform CO_2 in the radical cation and some amount of this radical becomes HCOOH whereas another coordinates with the Cu center (paramagnetic). After various steps, the coordinated radical anion terms into CO. If the adduct is sufficiently stable, the consecutive reduction steps permits the transfor-

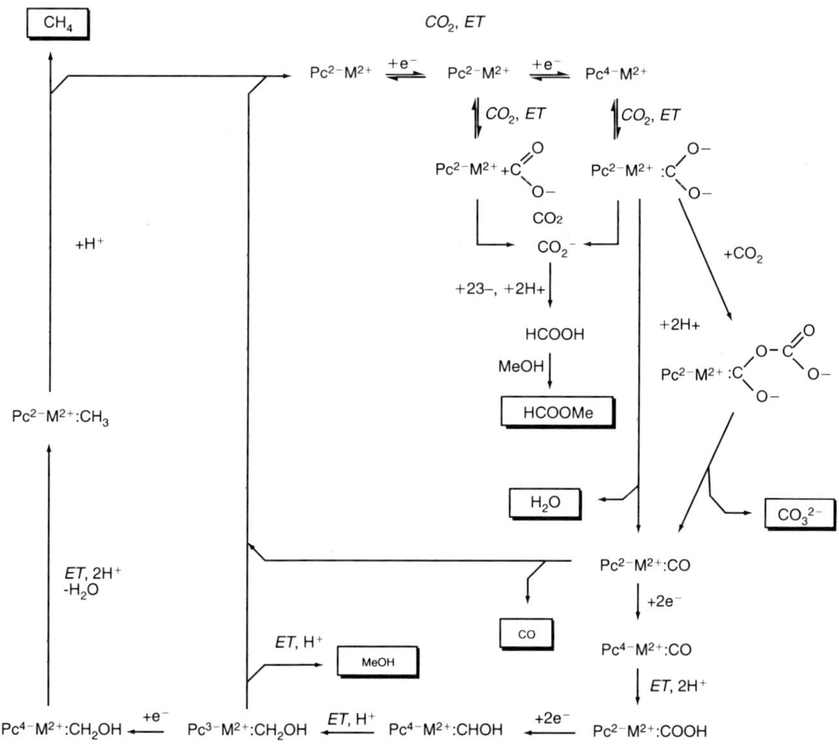

ET is electron transfer

Scheme 5.6. Catalytic cycle for cooper phthalocyanine. Reprinted from Scheme 1: T.V. Magdesieva, I.V. Zhukov, D.N. Kravchuk, O.A. Semenikhin, L.G. Tomilova, and K.P. Butina, Electrocatalytic CO_2 reduction in methanol catalyzed by mono-, di- and electropolymerized phthalocyanine complexes, Russian Chemical Bulletin 41 (2002) 805–812. With permission of Springer, Kluwer Academic/Plenum Publishers.

mation of the coordinated radical anion into methane[146]. In the case of the amino substituents, its electron donating character enhances the electronic density of the metal center and the ring. The energy of the d orbital that interacts with CO_2 increases diminishing the overlapping with the CO orbitals. The formation of CO as the main product in the case of the electropolymerized Cu phthalocyanine was not explained. In order to understand the obtained results, a theoretical approach can be used, based in literature[146]. In fact, the catalytic activity of a neutral phthalocyanine can be related to the energy of the frontier orbitals of the complex. Also the catalytic activity of monophthalocyanines correlates with their ability to coordinate a fifth axial ligand, and hence with the differences in energy of the d orbitals of the metal and the frontier orbitals of the coordinating ligand[146,166]. The reduced phthalocyanine complex has a high-energy HOMO formed from of the neutral complex[165] mainly located at the N atoms of the ring. Then, the complex is a strong reducing agent and the main product of the reaction will be CO, in the case of a weak bound between the CO and the metal. This situation appears when the metal is bond to the CO by $d_{xz}(d_{yz})$ orbitals, but not d_z^2. It can occur if it is doubly occupied. In this case, the back donation with the π*-orbital of CO should involve the d_z^2-orbital and will be weak. But, if the CO is bound to the metal by a σ M–C bond (involving the d_{z2} and $(n+1)p_z$ and $(n+1)s$ orbitals), methane can be produced. In the case of low-lying d-levels (Sn, Pb, In, Zn, and others), they are not involved in the ligand binding. The binding occurs only through the s and p orbitals. The LUMO of the neutral molecule is less localized and then, higher in energy compared to the first case. The complex in this case will be a weakly reducing agent. It is possible that in this case the formation of the radical anion does not generate CO but only formic acid[146,165]. In the case of rare-earth porphyrins, d and f-orbitals are very low in energy and do not participate in bonding. The reaction only involves the orbitals of the ring and will produce formic acid as the main product. In the case of diphthalocyanines of rare-earths, the orbitals of the metal and the ring are highly separated in energy, and there is a negligible interaction between them[146,167]. Then the reduced form of the complex is a weaker reducing agent compared to monophthalocyanines.

Semi-empirical calculation[168] shows that in the case of diphthalocyanines, the coordination of CO_2 or CO molecule to the metal atom is sterically impeded. Probably, the peripheral N atom is involved enabling the formation of methyl formate and CO. When the diphthalocyanine has bulky substituents, only H_2 is the product of the reaction due to steric factors. In order to confirm those conclusions authors performed quantum-chemical calculations using porphyrins as models: Co, Ni, Cu and Zn tetraaza porphyrins. In all the cases except the Ni, the monoreduced or double reduced complex, in the LM–CO molecule, where L represents the ligand, the axial C–O group is perpendicular to the plane of the ligand, and the metal center is shifted from the plane toward C–O. In the case of the Ni complexes, the angle is 94° for the mono and dianion (see Figure 5.15)[146]. On the other hand, when the charge of the complex varies from -1 to -2, the M–C bond shortens in the Co and Ni complexes but in the case of Cu and Zn, elongates. The reduction of CO elongates the bond C–O much more that in the case of M–CO complexes. These results can be explained because in the case of Co and Ni complexes, the back donation $\pi^*(\text{CO})\text{–d}_{xz(yz)}(\text{M})$ interaction (see Figure 5.16) plays the main role, whereas in the case of the Cu and Zn complexes, the normal binding of CO as a Lewis acid, $n(\text{CO})\text{-d}_{z2(pz-s)}(\text{M})$ interaction is the main interaction. This assumption was confirmed by the data of adsorption on metal surfaces from UPS, EEIS and IRAS[169].

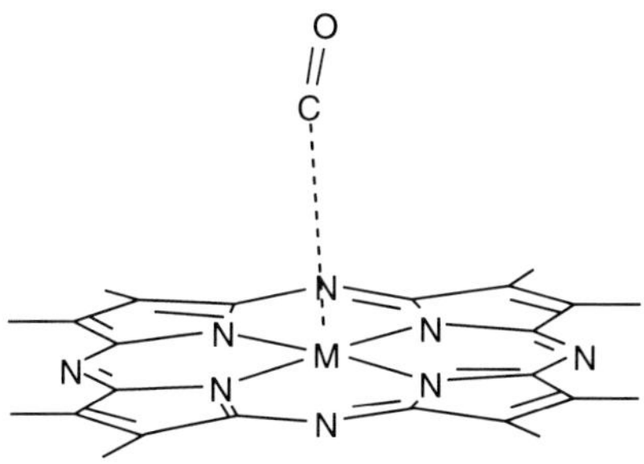

M = Fe, Co, Cu, Zn

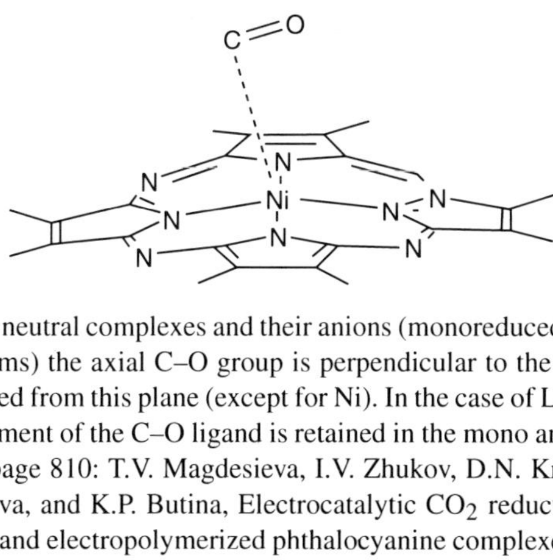

Figure 5.15. For the neutral complexes and their anions (monoreduced forms) and dianions (doubly reduced forms) the axial C–O group is perpendicular to the ligand plane and the atom is slightly shifted from this plane (except for Ni). In the case of L–Ni–CO, the angle is 94° and this arrangement of the C–O ligand is retained in the mono and dianion. Reprinted from Figure in the page 810: T.V. Magdesieva, I.V. Zhukov, D.N. Kravchuk, O.A. Semenikhin, L.G. Tomilova, and K.P. Butina, Electrocatalytic CO_2 reduction in methanol catalyzed by mono, di- and electropolymerized phthalocyanine complexes, Russian Chemical Bulletin, 41 (2002) 805–812. With permission of Springer, Kluwer Academic/Plenum Publishers.

Ab initio calculations of the coordination of the CO_2 molecule to several metals have recently been reported[165]. The relative stability of different modes of CO_2 coordination was calculated. One factor in the bonding between metal atoms and CO_2 is that the ground state has the maximum spin electron configuration[20, 165]. Other high-spin states, which only differ from the ground state by electronic permutations within the nonbonding d subshell, must be very close to the ground state. The spin populations show a transfer of one metal valence electron to the CO_2 group, i.e., into a MO largely localized at the sp-hybridized atomic orbital of the carbon atom. Differences in the products of electrocatalysis (CO or CH_4 with some metallic phthalocyanines and HCOOH with others) are rationalized on the basis of the electronic configuration of the metal atom. The reduction of CO_2 to CO is attributed to the strongly electron-donating HOMO

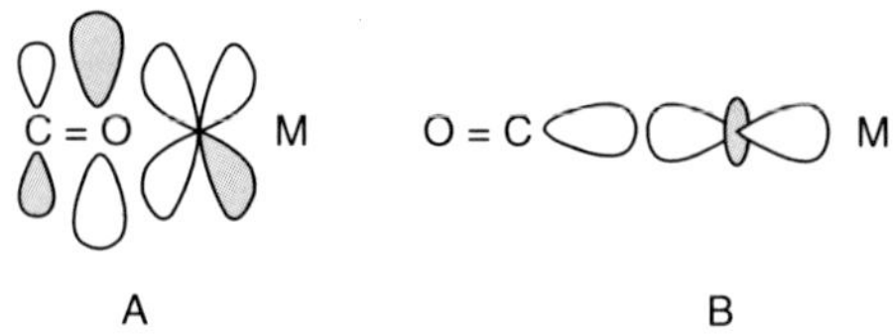

Figure 5.16. In the case of the Co and Ni complexes the back donation (π^*(CO))–$d_{xz(yz)}$(M) interaction (A) plays the main role. In the case of the Cu and Zn complexes, the normal binding of CO an a Lewis acid; n(CO)–$d_{z2(pzs)}$(M) interaction (B) is the most important. Reprinted from Figure in the page 810: T.V. Magdesieva, I.V. Zhukov, D.N. Kravchuk, O.A. Semenikhin, L.G. Tomilova, and K.P. Butina, Electrocatalytic CO_2 reduction in methanol catalyzed by mono, di- and electropolymerized phthalocyanine complexes, Russian Chemical Bulletin, 41 (2002) 805–812. With permission of Springer, Kluwer Academic/Plenum Publishers.

of $[MPc]^{n-}$. This orbital, which derives from the electron-accepting LUMO of neutral MPc, spreads over the nitrogen atoms surrounding the metal atom M. The final step in the generation of CO takes place when it is rapidly separated from the metal if this metal has a doubly occupied $2a_{1g}(dz^2)$ orbital[20, 165].

However if CO molecule is bound to the metallic atom by σ-bonding, the reduction may proceed up to the CH_4 product[20, 165]. Electron occupation of the LUMO by ionization does not take place in metals with outermost s or p electrons and extensive reduction to CO may not be possible. Electron transfer from MPc to CO_2 will end in CO_2^-, followed by separation of HCOOH instead of CO. In the "carbon dioxide complex" the anion is bent with an angle O–C–O equal to 134° and C is probably σ-bonded using the dz^2 or pz orbitals of M[165]. Figure 5.17 depicts the interaction between orbitals of the CO_2 molecule and the metallophthalocyanine.

The studies shown here depict a wide variety of modified electrodes. In all the cases, the aim was to enhance the selectivity and the catalytic activity of differents systems toward the electroreduction of CO_2. In fact, the modification of electrodic surfaces with metalled phthalocyanines or porphyrins, as adsorbed layers, electropolymerized films or supramolecular assemblies show very different behavior depending on the kind of modification. In the case of adsorption of phthalocyanines, the products are strongly dependent on the electrolyte, pH and negative limit of potential used. When the same complexes are protected with membranes as Nafion or PVP, there is an improvement in the selectivity and stability of the electrocatalyst. However, as in the other cases, each complex has a different behavior that depends on the experimental conditions. Also, the behavior of carbon gas diffusion electrodes where the electrode is modified with phthalocyanines was investigated. For this kind of systems, an enhancement in the rate of the process is obtained. The selectivity depends on the metal in the carbon (electrode) hydrophobic environment. It was also studied, with diffusion gas electrodes, the simultaneous reduction of CO_2 and nitrates or nitrites, using metallophthalocyanines, obtaining urea as the main product. On the other hand, supramolecular assemblies of porphyrins anchored to the electrodic surface through a covalently bonded molecule that acts as a fifth ligand of the first porphyrin show an

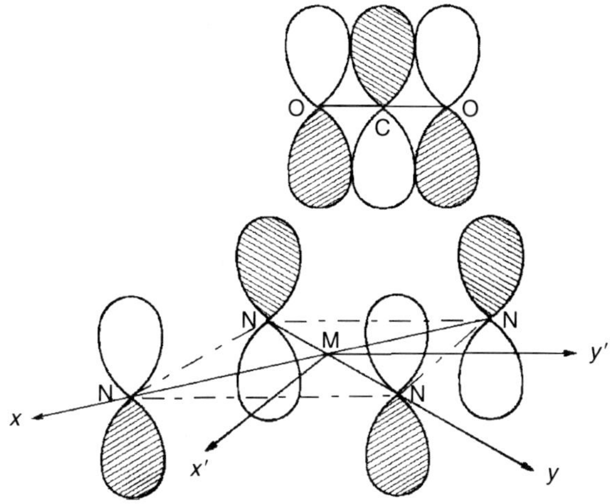

Figure 5.17. Relevant part of Pc's and the antibonding MO and CO_2. Reprinted from Figure 5.2: N. Furuya and S. Koide, Electroreduction of carbon dioxide by metal phthalocyanines, Electrochimica Acta, 36(8), 1309–1313 (1991). Copyright 1991, with permission of Elsevier.

interesting catalytic behavior toward the reduction of CO_2 in spite of the noncatalytic behavior of the same porphyrin physically adsorbed on the same surface. In the case of electropolymerized azamacrocycles, the stability and catalytic activity increases compared to the adsorbed layers of monomers.

6. Conclusions

The global warming is probably one of the most important environmental issues for this century. Every day there is new evidence of the climate change, but perhaps all this evidence is still not totally verified. However it is evident that the problem is real and the electrochemical reduction of carbon dioxide is one of the ways to solve the problem. In this chapter we discuss three families of compounds that catalyze the reduction of carbon dioxide. The general property of these macrocyclic complexes is that they reduce the overpotential required for this reaction to a more positive potential than metallic cathodes.

On the other hand the selectivity is not very high with few exceptions, but the general drawback is that always the most frequent product and in larger amounts is carbon monoxide. Therefore, the design of new macrocycles that allow multielectronic transfer reactions that lead to more reduced products is one of the possibilities in the next steps of this field. All these examples depict the great interest on the reduction of CO_2 catalyzed by azamacrocyclic complexes of transition metals. Progress in the obtention of better systems is an interesting goal. However, there is a long way to walk in order to reach economical systems that permit the conversion of CO_2 in combustible using the solar light as the energy required for the reaction. However any contribution is valuable, because the topic involve the future of the next generations.

Acknowledgments

M.J.A acknowledges Fondecyt project 1010695, G.R. and I.A acknowledge Conicyt Doctoral Fellowship. G.R. acknowledges Conicyt thesis research project AT 403139.

J.C. acknowledges Fondecyt project 8010006.

M.I. acknowledges Universidad de Chile I2 04/09-2 project.

References

1. Bockris, J.O.M. (1989). Modern aspects electrochemistry. Electrochemical and Photoelectrochemical Reduction of Carbon Dioxide. Plenum Press, New York.
2. Vasudevan, P., N. Phougat, and A.K. Shukla (1996). Metal phthalocyanines as electrocatalysts for redox reactions. *Appl. Organometal. Chem.* **10**(8), 591–604.
3. Udupa, K.S., G.S. Subramanian, and H.V.K. Udupa (1971). The electrolytic reduction of carbon dioxide to formic acid. *Electrochim. Acta*, **16**(9) 1593–1598 and references therein.
4. Russell, P.G., N. Kovac, S. Srinivasan, and M. Steinberg (1977). The electrochemical reduction of carbon dioxide, formic acid, and formaldehyde. *J. Electrochem. Soc.* **124**(9), 1329–1338 and references therein.
5. Kyriacouand, G. and A. Anagnostopoulos (1992). Electroreduction of CO_2 on differently prepared copper electrodes: The influence of electrode treatment on the current efficiencies. *J. Electroanal. Chem.* **322**(1–2), 233–246.
6. Arana, C., S. Yan, M. Keshavaraz, K.K.T. Pots, and H.D. Abruña (1992). Electrocatalytic reduction of carbon dioxide with iron, cobalt, and nickel complexes of terdentate ligands. *Inorg Chem.* **31**, 3680–3682.
7. Ogura, K. and H. Uchida (1987). Electrocatalytic reduction of carbon dioxide to methanol. Part 5. Relationship between the ability of metal complexes to engage in homogeneous catalysis and their coordination chemistry. *J. Chem. Soc. Dalton Trans.* **6**, 1377–1380.
8. Frese, K.W. and S. Leach (1985). Electrochemical reduction of carbon dioxide to methane, methanol, and CO on Ru electrodes. *J. Electrochem. Soc.* **132**, 259–261.
9. O'Toole, T.R., L.D. Margerum, T.D. Westmoreland, W.J. Vining, R.W. Murray, and T.J. Meyer (1985). Electrocatalytic reduction of CO_2 at chemically modified electrodes. *J. Chem. Soc. Chem. Commun.* **20**, 1416–1417.
10. Ogura, K. and K. Takamagari (1986). Electrocatalytic reduction of carbon dioxide to methanol. Part 2. Effects of metal complex and primary alcohol. *J. Chem. Soc. Dalton Trans.* **10**, 1519–1523.
11. Slater, S. and J.W. Wagenknecht (1984). Electrochemical reduction of CO_2 catalyzed by $Rh(diphos)_2Cl$. *J. Am. Chem. Soc.* **106**, 5367–5368.
12. Naitoh, A., K. Otha, T. Mizuno, H. Yoshida, M. Sakai, and H. Noda (1993). Electrochemical reduction of carbon dioxide in methanol at low temperature. *Electrochim. Acta* **38**(15) 2177–2179.
13. Fujita, E. (1996). DOE Report, DE 96007, 869, 66.
14. Behr, A. (1988). Carbon Dioxide Activation by Metal Complexes. VCH, New York.

15. Abruña, H.D. (1988). Coordination chemistry in two dimensions: Chemically modified electrodes. *Coord. Chem. Rev.* **86**, 135–189.
16. Collin, J.P. and J.P. Sauvage (1989). Electrochemical reduction of carbon dioxide mediated by molecular catalysts. *Coord. Chem. Rev.* **93**(2), 245–268.
17. Vigato, P.A., S. Tamburrini, and D.E. Fenton (1990). The activation of small molecules by dinuclear complexes of copper and other metals. *Coord. Chem. Rev.* **106**, 25–170.
18. Sullivan, B.P. (ed) (1993). Electrochemical and Electrocatalytic Reduction of Carbon Dioxide. Elsevier, Amsterdam.
19. Fujita, E., B.S. Brunschwig, T. Ogata, and S. Yanagida (1994). Toward photochemical carbon dioxide activation by transition metal complexes. *Coord. Chem. Rev.* **132**, 195–200.
20. Costamagna, J., G. Ferraudi, J. Canales, and J. Vargas (1996). Carbon dioxide activation by aza-macrocyclic complexes. *Coord. Chem. Rev.* **148**, 221–248.
21. Costamagna, J., G. Ferraudi, B. Matsuhiro, M. Campos-Vallette, J. Canales, M. Villagrán, J. Vargas, and M.J. Aguirre (2000). Complexes of macrocycles with pendant arms as models for biological molecules. *Coord. Chem. Rev.* **196**(1), 125–164.
22. Zagal, J.H., M.J. Aguirre, M. Isaacs, and J. Pavez (2003). Electrocatálisis sobre electrodos modificados con complejos macrocíclicos. In N. Alonso-Vante (ed), *Electroquímica y Electrocatálisis. Nuevos Materiales: Aspectos Fundamentales y Aplicaciones*. e-libro.net, Buenos Aires.
23. Scibioh, M.A. and V.R. Vijayaraghavan (1998). Electrocatalytic reduction of carbon dioxide: Its relevance and importance. *J. Sci. Ind. Res.* **57**(3), 111–123.
24. Liu, J., H.Q. Yan, Y.F. Yang, K.L. Wang, and E.F. Wang (2004). Electroreduction–oxidation and quantitative determination of CO_2 on a new SPE-based system. *J. Appl. Electrochem.* **34**(7), 757–762.
25. Yin, X. and J.R. Moss (1999). Recent developments in the activation of carbon dioxide by metal complexes. *Coord. Chem. Rev.* **181**(1), 27–59.
26. Latimer, W.L. (1952). The Oxidation States of the Elements and Their Potentials in Aqueous Solutions, 2nd edition. Prentice Hall, USA.
27. Bard, A.J., R. Parsons, and J. Jordan (eds) (1985). Standard Potentials in Aqueous Solutions. IUPAC, Physical and Analytical Chemistry Divisions. Marcel Dekker, New York.
28. Schwarz, H.A., C. Creuz, and N. Suttin (1985). Cobalt(I) polypyridine complexes. Redox and substitutional kinetics and thermodynamics in the aqueous 2, 2′-bipyridine and 4, 4′-dimethyl-2, 2′-bipyridine series studied by the pulse-radiolysis technique. *Inorg. Chem.* **24**, 433–439.
29. Amatore, C. and J.M. Saveant (1981). Mechanism and kinetic characteristics of the electrochemical reduction of carbon dioxide in media of low proton availability. *J. Am. Chem. Soc.* **103**, 5021–5023.
30. Frese, Jr., K.W. (1993). in B.P. Sullivan, K. Krist, H.E. Guard, (eds), *Electrochemical and Electrocatalytic Reduction of Carbon Dioxide*. Elsevier, Amsterdam, Chapter 6, and references therein.
31. Hori, Y., K. Kikuchi, and S. Suzuki (1985). Production of CO and CH_4 in electrochemical reduction of CO_2 at metal-electrodes in aqueous hydrogen carbonate solution. *Chem. Lett.* **11**, 1695.
32. Azuma, M., K. Hashimoto, M. Miramoto, M. Watanabe, and T. Sakata (1990). Electrochemical reduction of carbon dioxide in various metal electrodes in low-temperature aqueous $KHCO_3$ media. *J. Electrochem. Soc.* **137**(6), 1772–1778.
33. Hori, Y., K. Kikuchi, A. Murata, and S. Suzuki (1986). Production of methane and ethylene in electrochemical reduction of carbon-dioxide at copper electrode in aqueous hydrogen carbonate solution. *Chem. Lett.* **6**, 897.

34. Hori, Y., A. Murata, R. Takahashi, and S. Suzuki (1988). Enhanced formation of ethylene and alcohols at ambient temperature and pressure in electrochemical reduction of carbon dioxide at a copper electrode. *J. Chem. Soc. Chem. Commun.* **1**, 17–19.
35. Hori, Y., A. Murata, and R. Takahashi (1989). Formation of hydrocarbons in the electrochemical reduction of carbon dioxide at a copper electrode in aqueous solution, *J. Chem. Soc. Faraday Trans. 1*, **5**(8), 2309–2326.
36. DeWulf, D., T. Jin, and A. Bard (1989). Electrochemical and surface studies of carbon dioxide reduction to methane and ethylene at copper electrodes in aqueous solutions. *J. Electrochem. Soc.* **13**, 1686–1691.
37. Kim, J., D. Summers, and K. Frese (1988). Reduction of CO_2 and CO to methane on Cu foil electrodes. *J. Electroanal. Chem.* **245**(1–2), 223–244.
38. Cook, R., R. MacDuff, and A. Sammels (1988). On the electrochemical reduction of carbon dioxide at *in situ* electrodeposited copper. *J. Electrochem. Soc.* **135**(6), 1320–1326.
39. Wasmus, S., E. Cattaneo, and W. Vielstich (1990). Reduction of carbon dioxide to methane and ethane: An on-line MS study with rotating electrodes. *Electrochim. Acta* **35**(4), 771–775.
40. Boffa, A., C. Lin, A.T. Bell, and G.A. Somorjal (1994). Promotion of CO and CO_2 hydrogenation over Rh by metal oxides: The influence of oxide Lewis acidity and reducibility. *J. Catal.* **149**(1), 149–158.
41. Kudo, A., S. Nakagawa, A. Tsuneto, and T. Sakata (1993). Electrochemical reduction of high pressure CO_2 on Ni electrodes. *J. Electrochem. Soc.* **140**(6), 1541–1545.
42. Koga, O. and Y. Hori (1993). Reduction of adsorbed Co on a Ni electrode in connection with the electrochemical reduction of CO_2. *Electrochim. Acta* **38**(10), 1391–1394.
43. Schwartz, M., R.L. Cook, V.M. Kehoe, R.C. MacDuff, J. Patel, and A.F. Sammels (1993). Carbon dioxide reduction to alcohols using Perouskite-type electrocatalysis. *J. Electrochem. Soc.* **140**(3), 614–618.
44. Frese, K.W. and D. Canfield (1984). Reduction of CO_2 on n-GaAs electrodes and selective methanol synthesis. *J. Electrochem. Soc.* **131**(12), 2518–2522.
45. Kondelka, M., A. Monnier, and J. Augustyniski (1984). Electrocatalysis of the cathodic reduction of carbon dioxide on platinized titanium dioxide film electrode. *J. Electrochem. Soc.* **131**(4), 745–750.
46. Chardon-Noblat, S., A. Deronzier, R. Ziessel, and D.Z. Zsoldos (1998). Electroreduction of CO_2 catalyzed by polymeric $[Ru(bpy)(CO)_2]_n$ films in aqueous media: Parameters influencing the reaction selectivity. *J. Electroanal. Chem.* **444**, 253–260.
47. De Jesus-Cardona, H., C. del Moral, and C.R. Cabrera (2001). Voltammetric study of CO_2 reduction at Cu electrodes under different $KHCO_3$ concentrations, temperatures and CO_2 pressures. *J. Electroanal. Chem.* **513**(1), 45–51.
48. Yano, H., T. Tanaka, M. Nakamaya, and K. Ogura (2004). Selective electrochemical reduction of CO_2 to ethylene at a three-phase interface on copper(I) halide-confined Cu-mesh electrodes in acidic solutions of potassium halides. *J. Electroanal. Chem.* **565**(2), 287–293.
49. Ishimaru, S., R. Shiratsuchi, and G. Nogami (2000). Pulsed electroreduction of CO_2 on Cu–Ag alloy electrodes. *J. Electrochem. Soc.* **147**(5), 1864–1867.
50. Screbler, R., P. Cury, C. Suarez, E. Muñoz, H. Gomez, and R. Cordova, Study of the electrochemical reduction of CO_2 on a polypyrrole electrode modified by rhenium and copper–rhenium microalloy in methanol media. *J. Electroanal. Chem.* **533**(1–2), 167–175.
51. Rassmussen, P.B., P.M. Holmblad, T. Askgaard, C.V. Ovesen, P. Stolze, J.K. Norskov, and I. Chrokendorff (1994). Methanol synthesis on Cu(100) from a binary gas-mixture of CO_2 and H_2, *Catal. Lett.* **26**(3–4), 373–381.

52. Hori, Y., I. Takahashi, O. Koga, and N. Hoshi (2002). Selective formation of C2 compounds from electrochemical reduction of CO_2 at a series of copper single crystal electrodes. *J. Phys. Chem. B* **106**, 15–17.
53. Takahashi, I., O. Koga, N. Hoshi, and Y. Hori (2002). Electrochemical reduction of CO_2 at copper single crystal $Cu(S) - [n(111) \times (111)]$ and $Cu(S) - [n(110) \times (100)]$ electrodes. *J. Electroanal. Chem.* **533**(1–2), 135–143.
54. Shibata, M., K. Yoshida, and N. Furuya (1998). Electrochemical synthesis of urea at gas-diffusion electrodes III. Simultaneous reduction of carbon dioxide and nitrite ions with various metal catalysts. J. El*ectrochem. Soc.* **145**(2), 595–600.
55. a) Shibata, M., K. Yoshida, and N. Furuya (1998). Electrochemical synthesis of urea at gas-diffusion electrodes—IV. Simultaneous reduction of carbon dioxide and nitrate ions with various metal catalysts. *J. Electrochem. Soc.* **145**(7), 2348–2353.
b) Shibata, M., K. Yoshida, and N. Furuya (1995). Electrochemical synthesis of urea on reduction of carbon dioxide with nitrate and nitrite ions using Cu-loaded gas-diffusion electrode. *J. Electroanal. Chem.* **387**(1–2), 143–145 .
56. Shibata, M., K. Yoshida, and N. Furuya (1998). Electrochemical synthesis of urea at gas-diffusion electrodes—Part II. Simultaneous reduction of carbon dioxide and nitrite ions at Cu, Ag and Au catalysts. *J. Electroanal. Chem.* **442**(1–2), 67–72.
57. Lee, J. and Y. Tak (2001). Electrocatalytic activity of Cu electrode in electroreduction of CO_2, *Electrochim. Acta* **46**(19), 3015–3022.
58. Ogawa, S., T. Yamaguchi, and N. Gotoh (1974). Synthesis of a novel macrocyclic compound and of its copper complex. *J. Chem. Soc. Chem. Commun.* **10**, 577–579.
59. Ogawa, S., T. Yamaguchi, and N. Gotoh (1974). Preparation of a conjugated tautomer of 1,14:7,8-diethenotetrapyrido-[2,1,6-*de*:2,1,6-*gh*:2,1,6-*kl*:2,1,6*na*][1,3,5,8,10,12] hexa-azacyclotetradecine and its metal derivatives]. *J. Chem. Soc. Perkin Trans. 1*, 976–978.
60. Ogawa, S., (1977). Preparation of macrocyclic compounds by thermal dimerization of 1,10-phenanthroline derivatives. *J. Chem. Soc. Perkin Trans. 1* **2**, 214–216.
61. Seno, M., S. Tsuchida, and S. Ogawa (1977). Electronic structure of macrocyclic compounds revealed by X-ray photoelectron spectroscopy. *J. Am. Chem. Soc.* **99**(9), 3014–3018.
62. Cheng, C.-C., Y.-N. Kuo, K.-S. Chuang, C.-F. Luo, and W.J. Wang (1999). A new Co-II complex as a bulge-specific probe for DNA. *Angew. Chem. Int. Ed.* **38**(9), 1255–1257.
63. Costamagna, J., J. Canales, J. Vargas, M. Camalli, F. Caruso, and E. Rivarola (1993). Precursors of aza-macrocycles—characterization of substituted phenanthrolines and related bases—crystal and molecular-structure of dichloro-di-*n*-butyl(2, 2′, 6, 6′-bipyrimidine)–tin(IV). *Pure Appl. Chem.* **65**(7), 1521–1526.
64. Lewis, J. and T. O'Donoghue (1980). Chemistry of polydentate ligands. Part 5. Complexes of 2,9-dihydrazino-derivates of 1,10-phenanthroline. Dependence of co-ordination number of a ligand on the anion present. *J. Chem. Soc. Dalton.Trans.* **5**, 736–742.
65. Costamagna, J., J. Canales, J. Vargas, A. Alvarado, and G. Ferraudi (1993). Spectroscopic and electrochemical characterization of some complexes with azamacrocyclic ligands. *J. Indian Chem. Soc.* **70**(11–12), 877–883.
66. Costamagna, J., J. Canales, J. Vargas, and G. Ferraudi (1995). Electrochemical reduction of carbon-dioxide by hexa-azamacrocyclic complexes. *Pure Appl. Chem.* **67**(7), 1045–1052.
67. Estiu, G., A. Jubert, J. Molina, J. Costamagna, J. Canales, and J. Vargas (1995). Semiempirical SCF/CI interpretation of the origin of the catalytic activity of transition metal-macrocycle complexes. *J. Mol. Struct: Teochem.* **330**(1–3), 201–210.

68. Estiú, G. and A. Jubert (1995). A theoretical investigation of the structure and electronic spectra of porphyrin homologue macrocycles. Hexaazacyclophane and its nickel and copper complexes. *Inorg. Chem.* **34**, 1212–1220.
69. Isaacs, M., J.C. Canales, M.J. Aguirre, G. Estiú, F. Caruso, G. Ferraudi, and J. Costamagna (2002). Electrocatalytic reduction of CO_2 by aza-macrocyclic complexes of Ni(II), Co(II), and Cu(II). Theoretical contribution to probable mechanisms. *Inorg. Chim. Acta* **339**, 224–232.
70. Canales, J., J. Ramírez, G. Estiú, and J. Costamagna (2000). Bis-bipyridine hexa-aza-macrocycle complexes of zinc(II) and nickel(II) and the catalytic reduction of carbon dioxide. *Polyhedron* **19**(22–23), 2373–2381.
71. Isaacs, M., J.C. Canales, A. Riquelme, M. Lucero, M.J. Aguirre, and J. Costamagna (2003). Contribution of the ligand to the electroreduction of CO_2 catalyzed by a cobalt(II) macrocyclic complex. *J. Coord. Chem.* **56**(14), 1193–1201.
72. Kadish, K., M. Royal, E. Van Caemelbecke, and L. Gueletti (2000). Metalloporphyrins in nonaqueous Media: Database of Redox Potentials. In K.M. Kadish, K.M. Smith, and R. Guilard (eds), *The Porphyrin Handbook*, Vol. 9. Chapter 1, Academic Press, New York.
73. Lever, A.B.P., E.R. Milaneva, and G. Speir (1993). The redox chemistry of metallophthalocyanines in solution. in C.C. Leznoff and A.B.P.Lever, (eds), *Phthalocyanines Properties and Applications*, Vol. 3, Chapter 1, VCH, Weinheim.
74. Day, P., H.A.O. Hill, and M.G. Price (1968). Some reaction of cobalt phthalocyanines. *J. Chem. Soc. A* **1**, 90–93.
75. Nevin, W.A., W. Liu, M. Melnick, and A.B.P. Lever (1986). Spectro-electrochemistry of cobalt and iron tetrasulphonated phthalocyanines. *J. Electroanal. Chem.* **213**(2), 217–234.
76. Ardiles, P., E. Trollund, M. Isaacs, F. Armijo, and M.J. Aguirre (2001). Evidence of a stable charge-transfer adduct between 2-mercaptoethanol and cobalt-poly-tetra-aminophthalocyanine. *J. Coord. Chem.* **54**(3–4), 183–191.
77. Beley, M., J. Collin, R. Rupert, and J. P. Sauvage (1986). Electrocatalytic reduction of CO_2 by Ni cyclam^{2+} in water: Study of the factors affecting the efficiency and the selectivity of the process. *J. Am. Chem. Soc.* **108**, 7461–7467.
78. Beley, M., J. Collin, R. Rupert, and J.P. Sauvage (1984). Nickel (II)-cyclam: An extremely selective electrocatalyst for reduction of CO_2 in water. *J. Chem. Soc. Chem. Commun.* **19**, 1315–1316.
79. Pearce, D.J. and D. Pletcher (1986). A study of the mechanism for the electrocatalysis of carbon dioxide reduction by nickel and cobalt square planar complexes in solution, *J. Electroanal. Chem.* **197**(1–2), 317–330.
80. Fujihira, M., Y. Hirata, and K. Kuga (1990). Electrocatalytic reduction of CO_2 by nickel(II) cyclam: Study of the reduction mechanism on mercury by cyclic voltammetry, polarography and electrocapillarity. *J. Electroanal. Chem.* **292**(1–2), 199–215.
81. Grant, J.L., K. Goswami, L.O. Spreer, J.W. Otvos, and M. Calvin (1987). Photochemical reduction of carbon dioxide to carbon monoxide in water using nickel(II) tetra-azamacrocycle complex as catalyst, *J. Chem. Soc. Dalton Trans.* **9**, 2105–2109.
82. Collin, J.P., A. Jouaiti, and J.P. Sauvage (1988). Electrocatalytic properties of Ni(cyclam)$^{2+}$ and Ni_2(biscyclam)$^{4+}$ with respect to CO_2 and H_2O reduction. *Inorg. Chem.* **27**, 1986–1990.
83. Balazs, G.B., and F.C. Anson (1992). The adsorption of Ni(cyclam)$^+$ at mercury electrodes and its relation to the electrocatalytic reduction of CO_2, *J. Electroanal. Chem.* **322**(1–2), 325–345 .
84. Balazs, G.B., and F.C. Anson (1993). Effects of CO on the electrocatalytic activity of Ni(cyclam)$^{2+}$ toward the reduction of CO_2. *J. Electroanal. Chem.* **361**(1–2), 149–157.

85. Sasaki, S. (1990). Can CO_2 Coordinate to a Ni(1) complex? An ab-initio MO/SD-CI study. *J. Am. Chem. Soc.* **11**(2), 7813–7814.
86. Smith, C.I., J.A. Crayston, and R.W. Hay (1993). Reduction of carbon dioxide by nickel macrocyclic catalysts adsorbed on a mercury electrode or a copper rotating disc electrode. *J. Chem. Soc. Dalton Trans.* **21**, 3267–3269.
87. Sasaki, S. (1992). An ab-initio MO/SD-CI study of model complexes of intermediates in electrochemical reduction of CO_2 catalyzed by $NiC1_2$(cyclam). *J. Am. Chem. Soc.* **114**, 2055–2062.
88. Shionoya, M., E. Kimura, and Y. Iitaka (1990). Mono-, di-, and tetrafluorinated cyclams. *J. Am. Chem. Soc.* **112**, 9231–9245.
89. Fujita, E., J. Haff, R. Sanzenbacher, and H. Elias (1994). High electrocatalytic activity of RRSS-$[Ni^{II}HTIM](C10_4)_2$ and $[Ni^{II}DMC](C10_4)_2$ for carbon dioxide reduction (HTIM = 2, 3, 9, 10-tetramethyl-l,4,8,11-tetraazacyclotetradecane, DMC = C-meso-5,12-dimethyl-l,4,8,11-tetraazacyclotetradecane). *Inorg. Chem.* **33**, 4627–4628.
90. Abbi, F., G. de Santis, L. Fabbrizzi, M. Licchelli, A.M.P. Manotti-Lanfredi, P. Pallavicini, A. Poggi, F. Ugozzoliz (1994). Nickel(II) complexes of azacyclams: oxidation and reduction behavior and catalytic effects in the electroreduction of carbon dioxide. *Inorg. Chem.* **33**, 1366–1375.
91. Zilbermann, I., M. Winnik, D. Sagiv, A. Rotman, A.H. Cohen, and D. Meyerstein (1995). Properties of monovalent nickel complexes with tetraaza-macrocyclic ligands in aqueous solutions. *Inorg. Chim. Acta* **240**(1–2), 503–514.
92. Lovecchio, F.V., E.S. Gore, and D.H. Busch (1974). The oxidation and reduction behavior of macrocyclic complexes of nickel. Electron spin resonance studies. *J. Am. Chem. Chem.* **96**(10), 3109–3118.
93. Kobiro, K., A. Nakayama, T. Hiro, M. Suwa, and Y. Tobe (1992). Synthesis and molecular structures of nickel(II) alkyl-substituted cyclam complexes. *Inorg. Chem.* **31**, 676–685.
94. Beer, P. D., P.A. Gale, and D.K. Smith (1999). *Supramolecular Chemistry*. Oxford University Press, New York.
95. Kaifer, A. and M. Gomez-Kaifer (1999). *Supramolecular Electrochemistry*. Wiley-VCH, Weinheim, Germany.
96. Lehn, J.-M. (1992). In A.F. Williams, C. Floriani, and A.E. Merbach (eds), *Perspectives in Coordination Chemistry*. VHCA, Basel p. 447.
97. Krämer, R., J.-M. Lehn, A. De Cian, and J. Fisher (1993). Self-assembly, structure, and spontaneous resolution of a trinuclear triple helix from an oligobipyridine ligand and Ni^{II} Ions. *Angew. Chem. In. Ed. Engl.* **32**(5), 703–706.
98. De Blas, A., G. De Santis, L. Fabrizzi, M. Licchelli, A.M. Manotti, L.P. Pallavicini, A. Poggi, and F. Ugozzoli (1993). Pyridines with an appended metallocyclam subunit. Versatile building blocks to supramolecular multielectron redox systems. *Inorg. Chem.* **32**, 106–113.
99. Kimura, E., S. Wada, M. Shionoya, and Y. Okazaki (1994). New series of multifunctionalized nickel(II)-cyclam (cyclam = 1,4,8,11 tetraazacyclotetradecane) complexes. Application to the photoreduction of carbon dioxide. *Inorg. Chem.* **33**, 770–778.
100. Taraszewska, J., G. Roslonek, Y.D. Lampeka, and I.M. Maloshtan (1998). Oxalate-bridged dinuclear nickel azacyclam macrocyclic complexes: Synthesis and electrochemical study. *J. Electroanal. Chem.* **452**(1), 49–56.
101. Taraszewska, J., L. Tsymbal, and Y.D. Lampeka (1999). Polynuclear oxalate-bridged nickel complexes based on bis(macrocyclic) building blocks: Synthesis, spectral and electrochemical studies. *Supramol. Chem.* **10**(4), 287–295.

102. Taraszewska, J., G. Roslonek, and B. Korybut-Daszkiewicz (2000). Redox properties of novel dinuclear Ni(II) bis-tetraazamacrocyclic complex. *Supramol. Chem.* **12**(1), 115–121.
103. de Alwis, C., J. Crayston, T. Cromie, T. Eisenblätter, R. Hay, Y.D. Lampeka, and L. Tsymbal (2000). Cyclic voltammetry study of the electrocatalysis of carbon dioxide reduction by bis(polyazamacrocyclic) nickel complexes. *Electrochim. Acta* **45**(13), 2061–2074.
104. Lee, E., D. Hong, H. Park, and M. Suh (2003), Synthesis, properties, and reactions of trinuclear macrocyclic nickel(II) and nickel(I) complexes: Electrocatalytic reduction of CO_2 by nickel(II) complex. *Eur. J. Inorg. Chem.* **17**, 3242–3249.
105. Rowinski, P. and R. Bilewicz (2001). Carbon dioxide electrochemical sensor based on lipid cubic phase containing tetraazamacrocyclic complexes of Ni(II). *Materials Science and Engineering C* **18**(1–2), 177–183.
106. Jarzebinska, A., P. Rowinski, I. Zawisza, R. Bilewicz, L. Siegfried, and T. Kaden (1999). Modified electrode surfaces for catalytic reduction of carbon dioxide. *Anal. Chim. Acta* **396**(1), 1–12.
107. Jacquinot, P. and P.C. Hauser (2003). Ni(II)cyclam catalyzed reduction of CO_2—towards a voltammetric sensor for the gas phase. *Electroanal.* **15**(18), 1437–1444.
108. Ganjali, M., M. Yousefi, M. Javanbakht, T. Poursaberi, M. Salavati-Niasari, L. Hajiagha-Babael, E. Latifi, and M. Shamsipur (2002). Determination of SCN^- in urine and saliva of smokers and non-smokers by SCN^--selective polymeric membrane containing a nickel(II)–azamacrocycle complex coated on a graphite electrode. *Anal. Sci.* **18**(8), 887–892.
109. Kashiwagi, Y., C. Kikuchi, F. Kurashima, and J. Anzai (2002). Electrocatalytic reduction of aldehydes and ketones on nickel(II) tetraazamacrocyclic complex-modified graphite felt electrode. *J. Organomet. Chem.* **662**(1–2), 9–13.
110. Rowinski, P., R. Bilewicz, M.J. Stébé, and E. Rogalska (2002). A concept for immobilizing catalytic complexes on electrodes: Cubic phase layers for carbon dioxide sensing. *Anal. Chem.* **74**(7), 1154–1559.
111. Mahmood, M.N., D. Masheder, and C.J. Harty (1987). Use of gas-diffusion electrodes for high-rate electrochemical reduction of carbon dioxide. II. Reduction at metal phthalocyanine-impregnated electrodes. *J. Appl. Electrochem.* **17**(6), 1223–1227.
112. Christensen, P.A., A. Hammnett, and A.V.G. Muir (1988). An in-situ FTIR study of the electroreduction of CO_2 by CoPc-coated edge graphite electrodes. *J. Electroanal. Chem.* **241**(1–2), 361–371.
113. Kapusta, S. and N. Hackerman (1984). Carbon dioxide reduction at a metal phthalocyanine catalyzed carbon electrode. *J. Electrochem. Soc.* **131**(7), 1511–1514.
114. Tanabe, H. and K. Ohno (1987). Electrocatalysis of metal phthalocyanine thin film prepared by the plasma-assisted deposition on a glassy carbon in the reduction of carbon dioxide. *Electrochim. Acta*, **32**(7), 1121–1124.
115. Zagal, J.H. (1992). Metallophthalocyanines as catalysts in electrochemical reactions. *Coord. Chem. Rev.* **119**, 89–136.
116. Eggins, B.R., J.T.S. Irvine, and J. Grimshaw (1998). The voltammetry of mixed solutions of carbon dioxide and metal phthalocyanines in DMSO. *J. Electroanal. Chem.* **266**, 125–131.
117. Gangi, D.A., R.R. Durand, Jr., (1986). Binding of carbon dioxide to cobalt and nickel tetra-aza macrocycles. *J. Chem. Soc. Chem. Commun.* **9**, 697–699.
118. Grodowski, J., T. Dhanasekaran, P. Neta, P. Hambright, B.S. Brunschwig, K. Shinozaki, and E. Fujita (2000). Reduction of cobalt and iron phthalocyanines and the role of the reduced species in catalyzed photoreduction of CO_2. *J. Phys. Chem. A* **104**(48), 11332–11339.

119. Yoshida, T., K. Kamato, M. Tsukamoto, T. Iida, D. Schletwein, D. Wöhrle, and M. Kaneko (1995). Selective electroacatalysis for CO_2 reduction in the aqueous phase using cobalt phthalocyanine/poly-4-vinylpyridine modified electrodes. *J. Electroanal. Chem.* **385**(2), 209–225.
120. Abe, T., T. Yoshida, S. Tokita, F. Taguchi, H. Imaya, and M. Kaneko (1996). Factors affecting selective electrocatalytic CO_2 reduction with cobalt phthalocyanine incorporated in a polyvinylpyridine membrane coated on a graphite electrode. *J. Electroanal. Chem.* **412**(1–2), 125–132.
121. Abe, T., F. Taguchi, T. Yoshida, S. Tokita, G. Schnupfeil, D. Wöhrle, and M. Kaneko (1996). Electrocatalytic CO_2 reduction by cobalt octabutoxyphthalocyanine coated on graphite electrode. *J. Mol. Catal. A* **112**(1), 55–61.
122. Abe, T., H. Imaya, T. Yoshida, S. Tokita, D. Schletwein, D. Wöhrle, and M. Kaneko (1997). Electrochemical CO_2 reduction catalysed by cobalt octacyanophthalocyanine and its mechanism. *J. Porhyrins Phthalocyanines* **1**, 315–321.
123. Becker, J.Y., B. Vainas, R. Eger, and L. Kaufman (1985). Electrocatalytic reduction of CO_2 to oxalate by Ag^{II} and Pd^{II} porphyrins. *J. Chem. Soc. Commun.* **21**, 1471–1472.
124. Prasad, R., V.K. Gupta, and A. Kumar (2004). Metallo-tetraazaporphyrin based anion sensors: Regulation of sensor characteristics through central metal ion coordination. *Anal. Chim. Acta* **508**, 61–70.
125. Hammouche, M., D. Lexa, M. Momentau, and J.-M. Saveant (1991). Chemical catalysis of electrochemical reactions. Homogeneous catalysis of the electrochemical reduction of carbon dioxide by iron ("O") porphyrins. Role of the addition of magnesium cations. *J. Am. Chem. Soc.* **113**(22), 8455–8466.
126. Behar, D., T. Dhanasekaran, P. Neta, C.M. Hosten, D. Ejeh, P. Hambright, and E. Fujita (1998). Cobalt porphyrin catalyzed reduction of CO_2. Radiation chemical, photochemical, and electrochemical studies. *J. Phys. Chem. A* **102**(17), 2870–2877.
127. Dhanasekaran, T., J. Grodowski, P. Neta, P. Hambright, and E. Fujita (1999). P-terphenyl-sensitized photoreduction of CO_2 with cobalt and iron porphyrins. Interaction between CO and reduced metalloporphyrins. *J. Phys. Chem. A* **103**(38), 7742–7748.
128. Swistak, C. and K.M. Kadish (1987). Electrochemistry of iron porphyrins under a CO atmosphere. Interactions between CO and pyridine. *Inorg. Chem.* **26**, 405–412.
129. Qiu, A. and D.T. Sawyer (1997). The electrochemical evaluation of the metal–carbon bond energies ($-\Delta G_{BF}$) of alkylated iron and cobalt porphyrins [(por)-M-R]. *J. Porphyrins Phthalocyanines* **1**, 125–134.
130. Guilard, R., F. Jérôme, J.M. Barbe, C. Gross, Z. Ou, J. Shao, J. Fischer, R. Weiss, and K.M. Kadish (2001). Alkyl and aryl substituted corroles. 2. Synthesis and characterization of linked "face-to-face" biscorroles. X-ray structure of (BCA)Co-2(py)(3), where BCA represents a biscorrole with an anthracenyl bridge. *Inorg. Chem.* **40**(19), 4845–4855.
131. Guilard, R., D.T. Gryko, G. Canard, J.M. Barbe, B. Koszarna, S. Brandes, and M. Tasior (2002). Synthesis of corroles bearing up to three different meso substituents. *Org. Lett.* **4**(25), 4491–4494.
132. Grodkowski, J., P. Neta, E. Fujita, A. Mahammed, L. Simkhovic, and Z. Gross (2002). Reduction of cobalt and iron corroles and catalyzed reduction of CO_2. *J. Phys. Chem. A* **106**(18), 4772–4778.
133. Bhugun, I., D. Lexa, and J.M. Saveant (1996). Catalysis of the electrochemical reduction of carbon dioxide by iron(0) porphyrins: Synergystic effect of weak Brönsted acids. *J. Am. Chem. Soc.* **118**, 1769–1776.

134. Bhugun, I., D. Lexa, and J.M. Saveant (1996). Catalysis of the electrochemical reduction of carbon dioxide by iron(0) porphyrins. Synergistic effect of Lewis acid cations. *J. Phys. Chem.* **100**, 19981–19985.
135. Anson, F.C., C. Shi, and B. Steiger (1997). Novel multinuclear catalysts for the electroreduction of dioxygen directly to water. *Acc. Chem. Res.* **30**, 437–444.
136. Toma, H. and K. Araki (2000). Supramolecular assemblies of ruthenium complexes and porphyrins. *Coord. Chem. Rev.* **196**(1), 307–329.
137. Souza, F.D. (2002). Recent advances in the electrochemistry of porphyrins and phthalocyanines. *J. Porphyrins Phthalocyanines* **6**(4), 285–288.
138. Prodi, A., M.T. Indelli, C.J. Kleverlaan, E. Alessio, and F. Scandola (2002). Energy transfer pathways in pyridylporphyrin metal adducts and side-to-face arrays. *Coord. Chem. Rev.* **229**(1–2), 51–58.
139. Baerends, E.J., G. Ricciardi, A. Rosa, and S.J.A. van Gisbergen (2002). A DFT/TDDFT interpretation of the ground and excited states of porphyrin and porphyrazine complexes. *Coord. Chem. Rev.* **230**(1–2), 5–27.
140. Iengo, E., E. Zangrando, and E. Alessio (2003). Discrete supramolecular assemblies of porphyrins mediated by coordination compounds. *Eur. J. Inorg. Chem.* **13**, 2371–2384.
141. Araki, K. and H.E. Toma (1994). Spectroelectrochemistry and electrocatalytic properties of a tetraruthenated nickel porphyrin. *J. Chem. Res. (S)* **7**, 290–290.
142. Araki, K. and H.E. Toma (1994). Synthesis and properties of a new polymetallated iron porphyrin. *J. Chem. Res. (M)*. **3**, 82–83.
143. Meshituka, S., M. Ichikawa, and K. Tamaru (1974). Electrocatalysis by metal phthalocyanines in the reduction of carbon dioxide. *J. Chem. Soc. Chem. Commun.* **5**, 158–159.
144. Zhang, T., W.J. Pietro, and A.B.P. Lever (1996). Rotating ring-disk electrode analysis of CO_2 reduction electrocatalyzed by a cobalt tetramethylpyridoporphyrazine on the disk and detected as CO on a platinum ring. *J. Electroanal. Chem.* **403**(1–2), 93–100.
145. Yoshida, T., K. Tsutsumida, S. Teratani, K. Yasufuku, and M. Kaneko (1993). Electrocatalytic reduction of CO_2 in water by [Re(bpy)$(CO)_3$Br] and [Re(bpy)$(CO)_3$Br] complexes incorporated into coated nafion membrane (bpy = 2,2-bipyridine; terpy = 2,2; 6,2-terpyridine). *J. Chem. Soc. Chem. Commun.* **7**, 631–633.
146. Magdesieva, T.V., I.V. Zhukov, D.N. Kravchuk, O.A. Semeninkhin, L.G. Tomilova, and K.P. Butin (2002). Electrocatalytic CO_2 reduction in methanol catalyzed by mono-, di-, and electropolymerized phthalocyanine complexes. *Russ. Chem. Bull. Int. Ed.* **51**(5), 805–812.
147. Riquelme, A., M. Isaacs, M. Lucero, E. Trollund, and M. J. Aguirre (2003). Electrocatalytic reduction of carbon dioxide at polymeric cobalt tetra (3-amino (phenyl) porphyrin glassy carbon-modified electrodes. *J. Chil. Chem. Soc.* **48**(2), 89–92.
148. Furuya, N. and K. Matsui (1989). Electroreduction of carbon dioxide on gas-diffusion electrodes modified by metal phthalocyanines. *J. Electroanal. Chem.* **271**(1–2), 181–191.
149. Sonoyama, N., M. Kirii, and T. Sakata (1999). Electrochemical reduction of CO_2 at metal-porphyrin supported gas diffusion electrodes under high pressure CO_2. *Electrochem. Commun.* **1**(6), 213–216.
150. Shibata, M. and N. Furuya (2001). Electrochemical synthesis of urea at gas-diffusion electrodes: Part VI. Simultaneous reduction of carbon dioxide and nitrite ions with various metallophthalocyanine catalysts. *J. Electroanal. Chem.* **507**(1–2), 177–184.
151. Shibata, M. and N. Furuya (2003). Simultaneous reduction of carbon dioxide and nitrate ions at gas-diffusion electrodes with various metallophthalocyanine catalysts. *Electrochim. Acta* **48**(25–26), 3953–3958.

152. Magdesieva, T.V., T. Yamamoto, D.A. Tryk, and A. Fujishima (2002). Electrochemical reduction of CO_2 with transition metal phthalocyanine and porphyrin complexes supported on activated carbon fibers. *J. Electrochem. Soc.* **149**(6), D89–D95.
153. Atoguchi, T., A. Aramata, A. Kazusaka, and M. Enyo (1991). Cobalt (II)–tetraphenylporphyrin–pyridine complex fixed on a glassy carbon electrode and its prominent catalytic activity for reduction of carbon dioxide. *J. Chem. Soc. Chem. Commun.* **13**, 156–157.
154. Atoguchi, T., A. Aramata, A. Kazusaka, and M. Enyo (1991). Electrocatalytic activity of CO^{II} TPP–pyridine complex modified carbon electrode for CO_2 reduction. *J. Electroanal. Chem.* **318**(1–2), 309–320.
155. Tanaka, H. and A. Aramata (1997). Aminopyridyl cation radical method for bridging between metal complex and glassy carbon: Cobalt(II) tetraphenylporphyrin bonded on glassy carbon for enhancement of CO_2 electroreduction. *J. Electroanal.Chem.* **437**(1–2) 29–35.
156. Aga, H., A. Aramata, and Y. Hisaeda (1997). The electroreduction of carbon dioxide by macrocyclic cobalt complexes chemically modified on a glassy carbon electrode. *J. Electronal. Chem.* **437**(1–2), 11–118.
157. Ramírez, G., M. Lucero, A. Riquelme, M. Villagrán, J. Costamagna, E. Trollund, and M.J. Aguirre (2004). A supramolecular cobalt–porphyrin-modified electrode, toward the electroreduction of CO_2. *J. Coord. Chem.* **57**(3), 249–255.
158. Floriani, C., and F. Calderazzo (1969). Oxygen adducts of Schiffs base complexes of cobalt prepared in solution. *J. Chem. Soc. A* **6**, 946–953.
159. Cariati, F., D. Galizzioli, F. Morazzoni, and C. Busetto (1975). New adducts of phthalocyaninatocobalt(II) with pyridine and 4-methylpyridine and their vibrational, magnetic, and electronic properties. Part 1. Reactivity towards oxygen. *J. Chem. Soc. Dalton Trans.* **7**, 556–561.
160. Collman, J.P., R.R. Gagne, J. Kouba, and H. Ljusberg-Wahren (1974). Reversible oxygen adduct formation in cobalt(II) picket fence porphyrins. *J. Am. Chem. Soc.* **96**(21), 6800–6802.
161. Floriani, C. and G. Fachinetti (1974). Sodium [N, N'-ethylenebis (salicylideneiminato) cobaltate(I)], a reversible carbon dioxide carrier. *J. Chem. Soc. Chem. Commun.* **15**, 615–616.
162. Gambarotta, S., F. Arena, C. Floriani, and P. Zanazzi (1982). Carbon dioxide fixation. Bifunctional complexes containing acidic and basic sites working as reversible carriers. *J. Am. Chem. Soc.* **104**(19), 5082–5092.
163. Schmidt, M.H., G.M. Miskelley, and N.S. Lewis (1990). Effects of redox potential, steric configuration, solvent, and alkali-metal cations on the binding of carbon-dioxide to cobalt(I) and nickel(I) macrocycles. *J. Am. Chem. Soc.* **112**(9), 3420–3426.
164. Zeng, Z.Y., S.L. Gupta, H. Huang, and E.B. Yeager (1991). Oxygen reduction on poly(4-vinyl pyridine)-modified ordinary pyrolytic graphite electrodes with adsorbed cobalt tetra-sulphonated phthalocyanine in acid solutions. *J. Appl. Electrochem.* **21**(11), 973–981.
165. Furuya, N. and S. Koide (1991). Electroreduction of carbon dioxide by metal phthalocyanines. *Electrochim. Acta* **36**(8), 1309–1313.
166. Zagal, J., M. Páez, and C. Fierro (1997). In: S. Srinivasan, S. Wagner, and H. Wrobloba (eds), *Electrode Materials and Processes for Energy Conversion and Storage*. Electrochemical Society, Pennington, New York.
167. Rodríguez-Méndez, M.L., R. Aroca, and A. De Saja (1993). Electrochromic and gas adsorption properties of Langmuir–Blodgett films of lutetium bisphthalocyanine complexes. *Chem. Mater.* **4**, 933–937.

168. Magdeviesa, T.V., S.V. Milanov, B.V. Lokshin, Z.S. Klemenkova, L.G. Tomilova, K.P. Butin, and N.S. Zefirov (1998). Study of the catalytic activity of electrochemically reduced forms of phthalocyanines in the reaction of CO_2 with epoxides. *Russ. Chem. Bull.* **47**(11), 2137–2145.
169. Heskett, A., B. Plummer, and C. Messmer (1984). A correlation between anomalous electronic and vibrational properties of Chemisorbed molecules. *Surf. Sci.* **139**(2–3), 558–568.

6

Supramolecular Porphyrins as Electrocatalysts

Koiti Araki and Henrique E. Toma

1. Build-up of Supramolecular Porphyrins Based on Metal–Ligand Coordination

Diverse approaches are being employed to build up supramolecular systems, including covalent and coordination bonds, electrostatic assembly, hydrogen bond, and even weaker interactions like π-stacking and dipole–dipole interactions. The use of weak bonds has been majoritary in supramolecular chemistry, but the metal–ligand coordination approach is becoming popular because well-defined tridimensional structures can be assembled using this strategy. In addition, the properties of the inserted metal complexes can be varied almost at will, thanks to the large number of metallic elements associated with a plethora of available oxidation states and ligands. Consequently, the electronic, catalytic and electrocatalytic, photochemical, magnetic, electron- and energy-transfer properties can be conveniently controlled, in order to assemble functional systems and materials. However, this is not an easy task because it implies a fine tuning on the structural, electronic, kinetic, thermodynamic, and even entropic factors. Nature performs such a complicated task to perfection in biological systems using only a small number of molecules from an infinitely large library of possible molecular systems.

Among the various systems that have been subject of research, we will focus on the catalytic, electrocatalytic, and photoelectrochemical properties of supramolecular systems containing porphyrins, more specifically pyridylporphyrins linked to transition metal complexes. The advances on the field of supramolecular poly- and oligometalloporphyrin arrays were reviewed until the year 2000 by the excellent articles of Wojaczynski *et al.*[1], Imamura *et al.*[2], Sanders[3], and Chambron *et al.*[4]; while we have reviewed the properties of TPyP

Koiti Araki and Henrique E. Toma • Institute of Chemistry, University of São Paulo, Av. Prof. Lineu Prestes 748, Butantã, São Paulo 05508–900, Brazil.

N_4-Macrocyclic Metal Complexes, edited by José H. Zagal, Fethi Bedioui and Jean-Pol Dodelet. Springer Science+Business Media, Inc., New York, 2006.

coordinated to $[Ru(bpy)_2Cl]$ and $[\mu_3 - ORu_3(OAc)_6(py)_2]$ in the same period[5,6]. More recently, the chemistry of porphyrin arrays obtained by axial coordination and the energy transfer in some side-to-face pyridylporphyrin adducts were reviewed, respectively, by Baldini and Hunter[7] and Prodi *et al.*[8] In this article we intend to update and extend the existing view, covering also other systems containing pyridylporphyrins directly linked to transition metal ions.

Porphyrins are aromatic tetraazamacrocyles capable to bind most of the transition metal ions, generating complexes with four fold symmetry. The porphyrin ring is somewhat flexible, as attested by the saddle conformation generally assumed by the free-base and the corresponding di-acid. The porphyrins possess three possible sites for metal–ligand coordination: (a) the axial positions at the central metal ion in metalloporphyrins, (b) the ligands attached to the *meso*- or (c) pyrrole β-carbon positions. The coordination geometry at the central metal ion is generally fixed, giving rise to a pentacoordinate square pyramidal geometry, like in zinc porphyrins, or hexacoordinate complexes with more or less tetragonal distortion. The structures of the most commonly used pyridylporphyrins MPyP (**1**), *trans*-DPyP (**2**), *cis*-DPyP (**3**), TriPyP (**4**), and TPyP (**5**) building blocks are shown in Scheme 6.1. The spheres indicate the peripheral coordination points and M represents $2H^+$ or a transition metal ion bond to the porphyrin macrocycle. In general, only the most relevant structural characteristics will be shown in the schemes, such that the readers are invited to look at the original papers for the details.

1.1. Supramolecular Porphyrin Assemblies by Axial Coordination

Many molecular architectures can be designed by axial coordination approach, which is suitable for obtaining photochemically active supramolecular systems, exhibiting intramolecular energy- and electron-transfer (ET) processes. Fleischer *et al.*[9] pioneered the assembly of side-to-face arrays of pyridylporphyrins exploring the coordination properties of *meso*-(phenylpyridyl)porphynato zinc to get dimers, trimers, and polymers, depending on the number of pyridyl-substituents and the relative position of the N-atoms in the ring. A dimer was obtained simply by mixing ZnTPP and ZnMPyP, and a trimer resulted from the mixture of two equivalents of ZnTPP with DPyP in $CHCl_3$, by axial coordination of the pyridyl N-atom to the zinc ion. Spectroscopic studies of more concentrated ZnMPyP solutions ($> 10^{-4}$ M) revealed the presence of oligomers and polymers (Scheme 6.2). This approach is certainly the most widely used one to obtain discrete porphyrin oligomers and polymers (**6–8**), by self-assembly. Also, a strategy for the preparation of supramolecular systems containing cavities with molecular recognition properties (**9**), was demonstrated by Sanders *et al.*[3,10–15]

Interesting results were also obtained by self-assembly of phenyl-bridged Zn-diporphyrins with pyridylporphyrins, in which energy transfer was observed from the ZnP to the free-base moiety (**10**, **11** without the quinone, *Q* and **12**, in Scheme 6.3)[16–19]. A series of triads were obtained by linking *para*-benzoquinone, pyromellitimide, or anthraquinone, as electron-acceptor (*Q*), to the 5-position of a Zn-diporphyrin and self-assembly with *cis*- or *trans*-dipyridylporphyrins, *trans*-dipyridylchlorin, and tetrahydroporphyrin (**10**, **11**)[20–25]. A strong quenching of the dimer fluorescence was observed in toluene due to energy and sequential ET

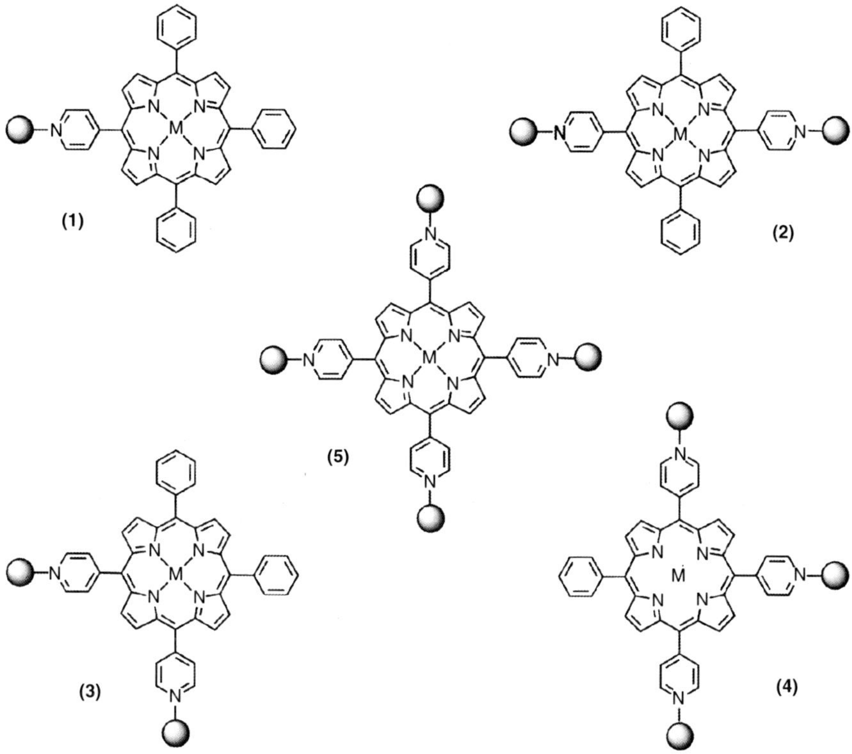

Scheme 6.1.

to the free-base moiety (0.9 and 1.7 ps, respectively; 293 K), which are faster than the ET (34 and 135 ps) from the dimer to the covalently linked electron-acceptor *Q*. The photophysical processes are strongly influenced by the spacer in between the dimer and the quinone, as well as the position of the pyridyl N-atoms in the free-base.

One of the most impressive recent examples using this approach was given by Michelsen *et al.*[26] and extended by Kobuke *et al.*[27–29], using *meso*-bis-(*N*-methyl-3-imidazolyl)-bis-porphyrinato zinc and analogous species. The linear bis-imidazolylporphyrins afforded the giant staircase arrays **13**, **14** shown in Scheme 6.4, by successive complementary coordination of the imidazolyl moiety to the axial positions of Zn^{2+}, Mg^{2+}, or Co^{3+} ions of two adjacent porphyrin rings, in a slipped conformation[29–35]. Highly stabilized structures were obtained when the central metal ion was hexacoordinating and allowed the formation of an extra bond, such that very long arrays could be obtained. For example, chains with an average length of 110 nm and MW $= 10^5$ were characterized for **13** and up to 500 nm long nanostructures were found by AFM for **14**. In contrast, a well-defined macroring **15** (Scheme 6.4) was spontaneously assembled when a canted bis-porphyrin, forming an angle of 120°, was used instead[27,36]. Furthermore, it was demonstrated that the nanostructures formed with zinc porphyrins can be reversibly dismounted and mounted, simply by controlling the amount of

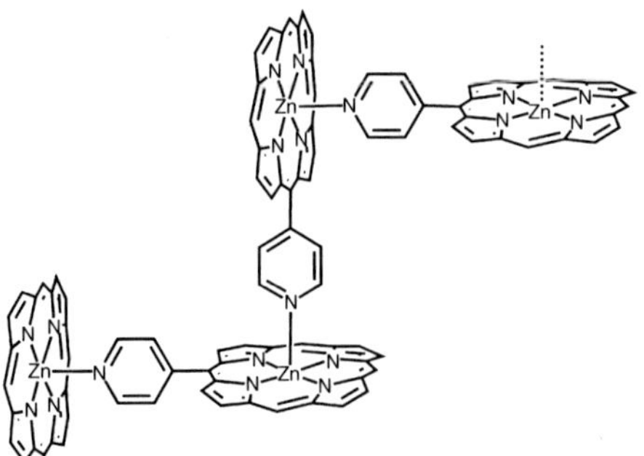

(6)

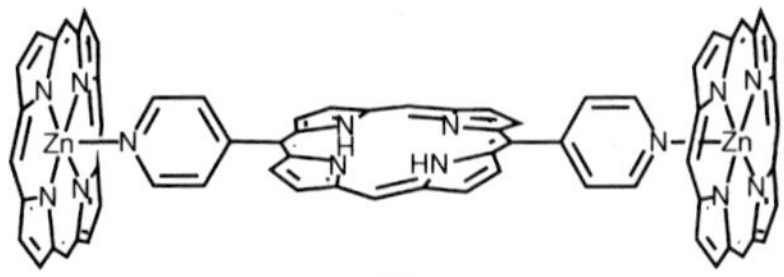

(8)

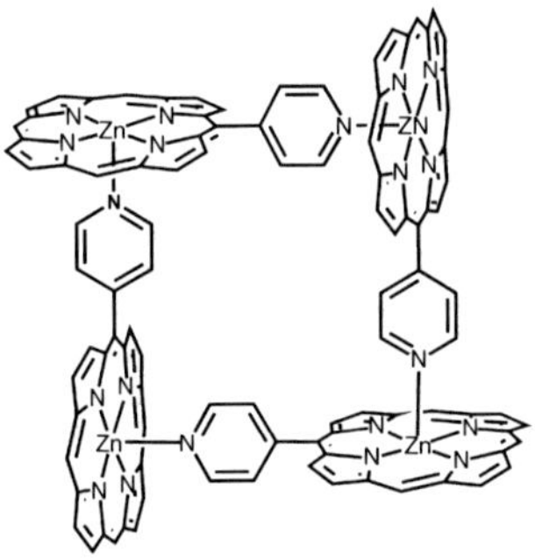

(7)

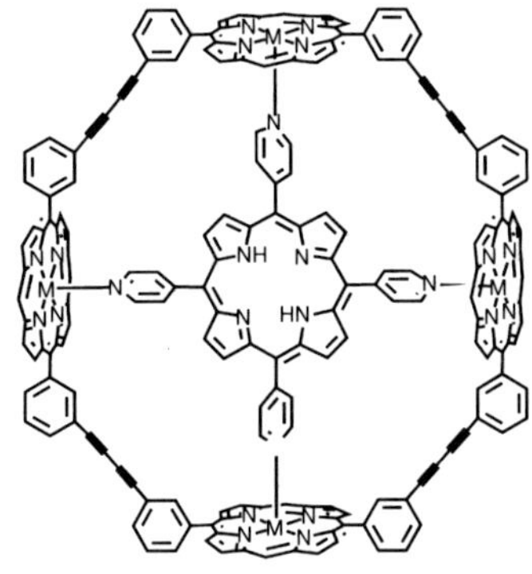

(9)

Scheme 6.2.

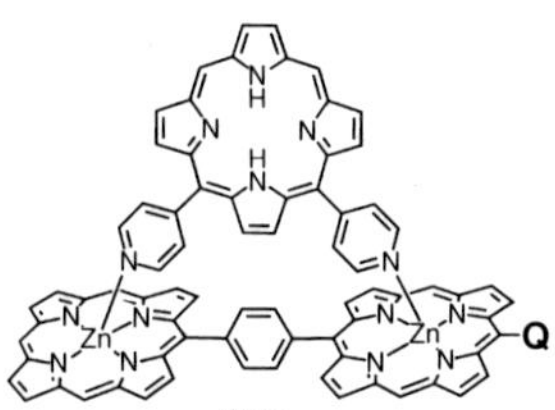

(10)

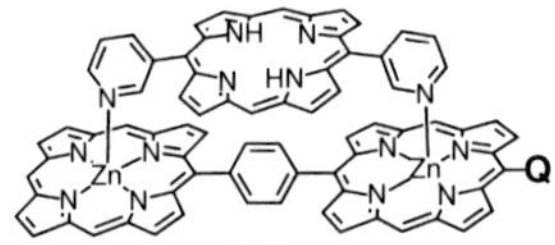

(11)

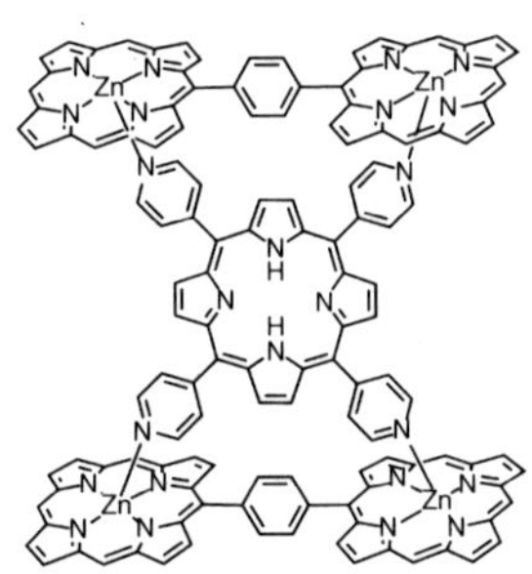

(12)

Scheme 6.3.

CH_3OH added to the $CHCl_3$ solution, thanks to the relatively high formation constant ($K \sim 10^{10}$ in $CHCl_3$) and labile methylimidazole–ZnP bonds. Accordingly, molecular wires and circular structures, that mimic the light-harvesting center of the photosynthetic bacteria, could be obtained in a controlled way. Self-assembled monolayers (SAM) of porphyrin arrays were also obtained on Au surfaces, using a three-step procedure. First, an imidazolyl-appended zincporphyrin bearing a ω-mercaptohexylphenyl substituent in *trans*-position was attached forming a monolayer. A solution of monomeric **13** in a 1:1 mixture of nitrobenzene and pyridine was then transferred onto it and the subsequent evaporation of the pyridine ligand (50°C for 10 min) lead to films of standing porphyrin arrays[30,37].

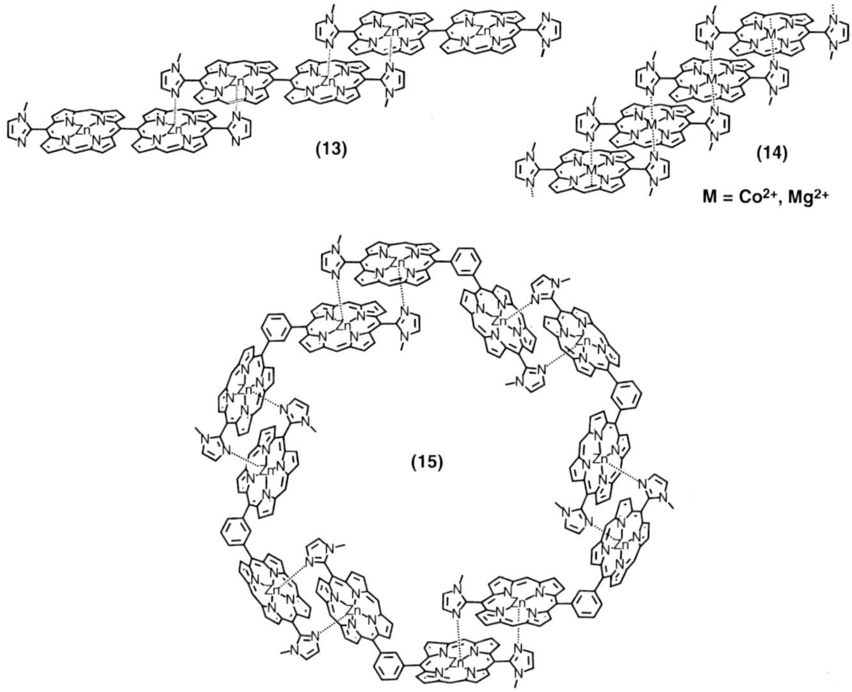

Scheme 6.4.

Photocurrent efficiencies of about 0.1% and consistent photoaction spectra were measured, when the modified Au plates were immersed in a 5 mM MV^{2+} solution and irradiated with visible light.

The variety of this face-to-edge assemblies can be extended further by using more inert hexacoordinating transition metal ions, such as Ru(II), Os(II), Rh(III), etc., in association with Zn(II) and free-base porphyrins. In this way, supramolecular species with structures analogous to **6–8** (Scheme 6.2) and **16–18** (Scheme 6.5) or even more complex arrangements were obtained, by exploring the available axial coordination sites. Supramolecular porphyrins like **16** and its analogues were obtained by using **1–5** and M-OEP or M-TPP (M = Ru(II) or Os(II)), in which the sixth axial position was blocked by CO[8,38–49]. The singlet state of these complexes is strongly perturbed by the heavy-atom effect, which enhanced the intersystem crossing and quenching processes. The singlet and triplet states of the RuP were generally at higher energies than the free-base and ZnP excited states, such that efficient energy-transfer processes from the ruthenium porphyrin moieties were observed. Interestingly, the photophysical properties of the *para*- and *meta*-pyridyl derivatives were similar[8,45,48]. Zig-zag and cyclic structures like **6** and **7** were also prepared using Ru(II) and Os(II) porphyrins. On the other hand, slipped dimers **17** (Scheme 6.5) were obtained by assembling Ru-2MPyP having one of the axial positions protected by pyridine or CO. Then, more complex supramolecular architectures **19**, **20** (Scheme 6.5) were obtained by extending those structures through such coordination sites. Wires and

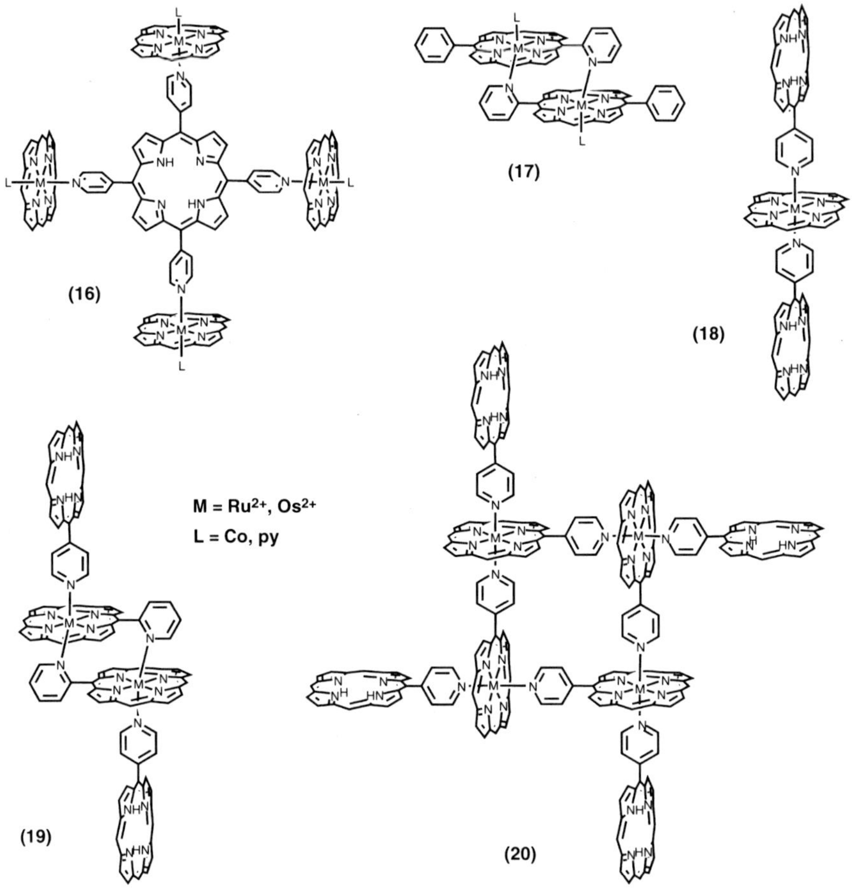

Scheme 6.5.

macrorings[50–52] with phosphine-substituted porphyrins were also obtained, as well as porphyrin arrays containing ruthenium and osmium polypyridine complexes[53].

1.2. Supramolecular Assembly Using Peripheral Pyridyl-Substituents

Coordinating substituents can be used to assemble supramolecular systems without blocking the metalloporphyrin center. This is the most interesting approach when the catalytic and electrocatalytic properties are considered, since these processes generally involve the coordination of the substrate to the central metal ion.

The assembly of supramolecular structures by peripheral coordination of transition metal complexes is a very versatile approach, as will be shown below using essentially the same pyridylporphyrins presented in the previous section. Many intrincate new architectures can be built-up generating exquisite nanosized structures and even polymers, by controlling the geometry and bonding characteristics of the molecularly engineered components. The research on the use of Pd(II), Pt(II), Re(I), Ru(II), etc. complexes as linear and angular molecular joints was developed by many groups. From those studies it becomes clear that cyclic structures, generated by coordinative self-assembly, can be thermodynamically stable and exhibit molecular recognition properties[54–67]. This same approach is useful for building-up many supramolecular species[1,2,5,6,8,64,68–70], that can be hierarchically self-assembled further to generate highly organized molecular materials, by exploring increasingly weaker interactions[66,71,72].

Many supramolecular systems can be devised according to the number, position, and nature of the coordinating groups. For example Pt(II) and Pd(II) form well-defined square planar complexes that can be used to build-up supramolecular architectures in a controlled way, by judicious selection of the ligands and geometry of the starting reagents. One, two, and four **1** can be coordinated to such complexes forming stable compounds, where the bis-porphyrins can be selectively obtained in the *cis*- (**21**, Scheme 6.6) or *trans*-conformations. Asymmetric substitution was also demonstrated in Pt(II) and Ru(II) derivatives. For example, Yuan *et al.* showed the controlled bonding of one, two (**21**), and four **1** to $[M(dmso)_2Cl_2]$, where M=Pd or Pt; and the preparation of the asymmetric species **22** (Scheme 6.6) containing **1** and 4′-methylpyridinium-4-pyridine coordinated to MCl_2[73]. However, the formation of the metal–ligand bond can be used to assemble more sophisticated supramolecular systems by the coordinative self-assembly strategy. In this case, the structure of the porphyrin ligands and the metal complexes used as building blocks should be complementary. For example, the reaction of *cis*-DPyP (**3**) with *cis*-metal complexes leads to cyclic dimers **23** (Scheme 6.6)[8,48,65,74]. A more complex structure **24** (Scheme 6.6) was obtained by using TPyP, with two vicinal pyridyl substituents capped with $[RuCl_2CO(dmso)_2]$ units, to assemble a dimer with a Pd(II) complex[75,76]. However, the reaction of *cis*-DPyP with *trans*-Pt(II) or Pd(II) complexes generate tetrameric cyclic species (**25**, Scheme 6.7) where the corners are the porphyrins. A similar approach can be used to obtain cyclic tetramers (**26**, Scheme 6.7), whose corners are the transition metal complexes[62,65,66,72,77–82].

Drain *et al.* have shown that this strategy can be used to self-assemble a spectacular π-stacking grid-like nonameric porphyrin **27** (Scheme 6.8), by employing porphyrins and transition metal components with adequate structures and stoichiometry[8,83]. The molecular-grid was obtained by reacting TPyP, TriPyP, *cis*-DPyP and *trans*-$[PdCl_2(benzonitrile)_2]$ in 1:4:4:12 ratio, in $CHCl_3$. The yield was as high as 90%, demonstrating the specificity of this multicomponent self-assembly process. Interestingly, tape or linear porphyrin arrays were obtained by reacting TPyP and *cis*-DPyP with *cis*-Pd(II) or Pt(II) complexes, or *trans*-DPyP and MPyP with *trans*-complexes, respectively. Grid and square porphyrins can be used to obtain interesting nanomaterials, such as films with very well-defined channels useful in molecular sieving or recognition[72,80,85]. More interesting porphyrin materials and properties will be found as the diversity of building-blocks and systems are augmented. The ruthenium dimethylsulphoxide complexes studied by Alessio *et al.* are one of the most versatile building-blocks

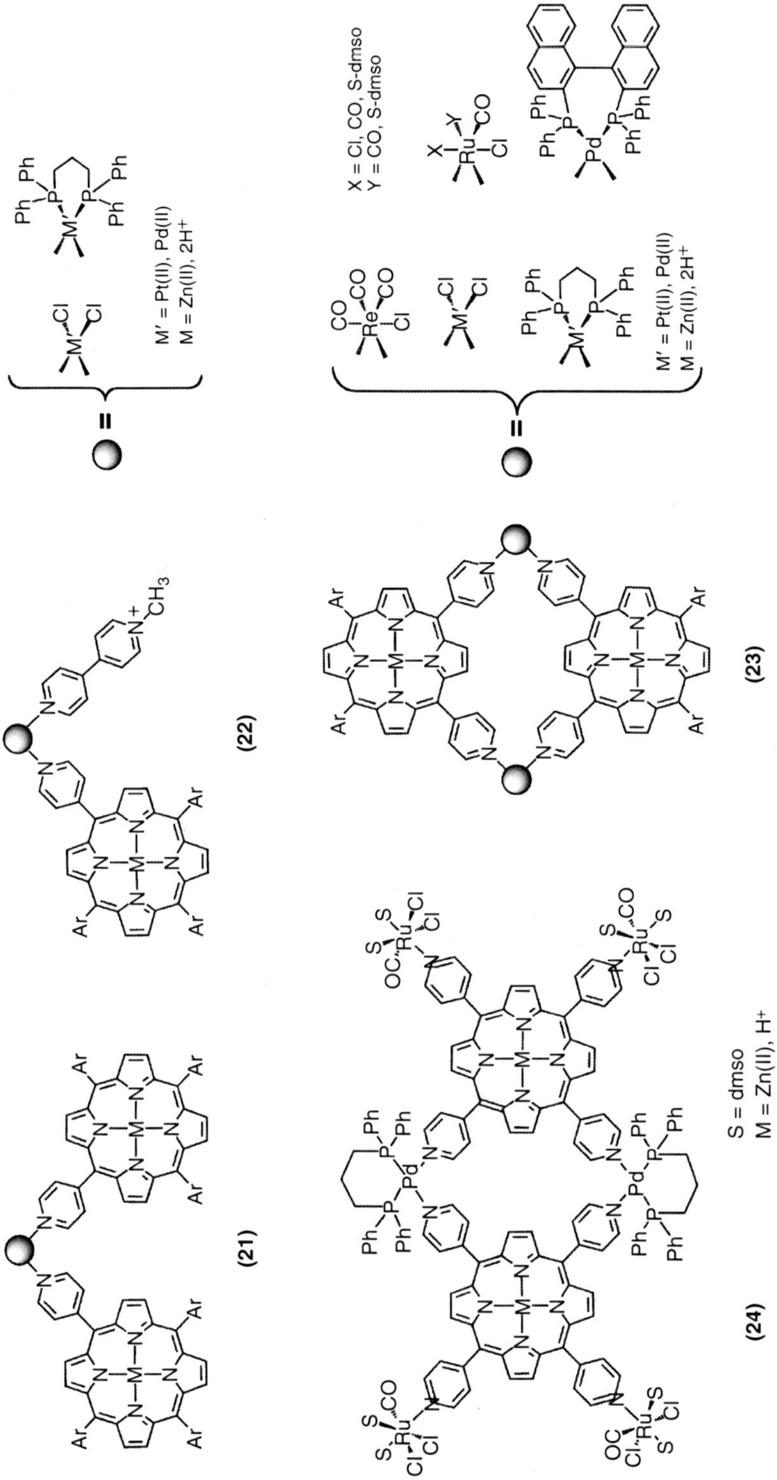
(21)
(22)
(23)
(24)
M′ = Pt(II), Pd(II)
M = Zn(II), 2H+
X = Cl, CO, S-dmso
Y = CO, S-dmso
S = dmso
M = Zn(II), H+

Scheme 6.6.

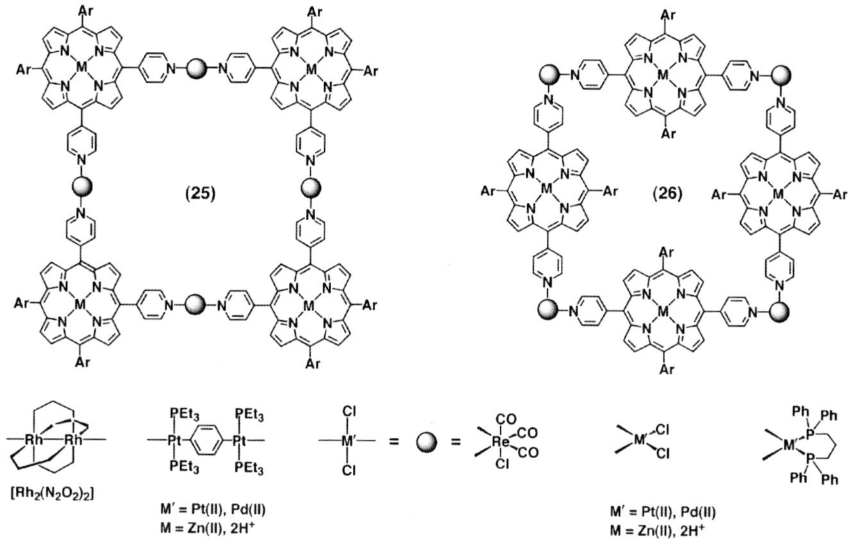

Scheme 6.7.

found to date, allowing the assembly of elaborate supramolecular systems in a stepwise manner[44,69,70,74–76].

A related class of compounds has been derived by using porphyrins connected to polypyridines such as 2, 2′-bipyridine, 1,10-phenantholine and 2, 2′:6′, 2″-terpyridine. These chelating ligands form thermodynamically more stable and inert transition metal complexes than the pyridine ligand, and such porphyrins have been extensively explored for the assembly of photochemically active supramolecular systems, especially by Sauvage *et al.*[86–97] The oblique bis-porphyrins (Scheme 6.9) obtained by covalent bonding to the 2,9- (**28**) or 4,7-phen, or 3, 3′- or 5, 5′-bpy (**29**) positions can be used to form catenanes and rotaxanes or dimers and trimers by coordination to appropriate metal ions. $[Ru(bpy)_3]^{2+}$ like complexes can be generated as redox and photoactive components of the supramolecular systems[98–100]. Linear bis-porphyrins connected through 4, 4′-bpy[100] and a cross-shaped species have been synthesized by the reaction of Cu(I) with a cofacial dimer[101].

A series of linear supramolecular porphyrins have been assembled by using terpy complexes as connectors or end-groups and varying the bridge and the ring substituents (Scheme 6.10)[102–111]. The photochemical properties of those supramolecular porphyrins, as well as of oblique bis-porphyrins and related compounds, have been extensively explored particularly by Flamigni, Hambright, Sauvage, and Balzani. Multistep energy and electron-transfer processes have been observed in the diads **30** and triads **31**, **32**, assembled by combining Ru(II), Rh(II), Os(II) and more recently Ir(III) terpy complexes with free base, zinc and Au(III) porphyrins. When $[Ru(terpy)_2]$ is used in combination with porphyrins, the energy transfer to that complex is the predominant decay pathway[103,105–107,109]. Nevertheless, relatively long lasting (450 ns) charge-separated species has been observed for a ZnP–Ir–AuP triad[108] in contrast with the analogous H_2P–Ir–AuP

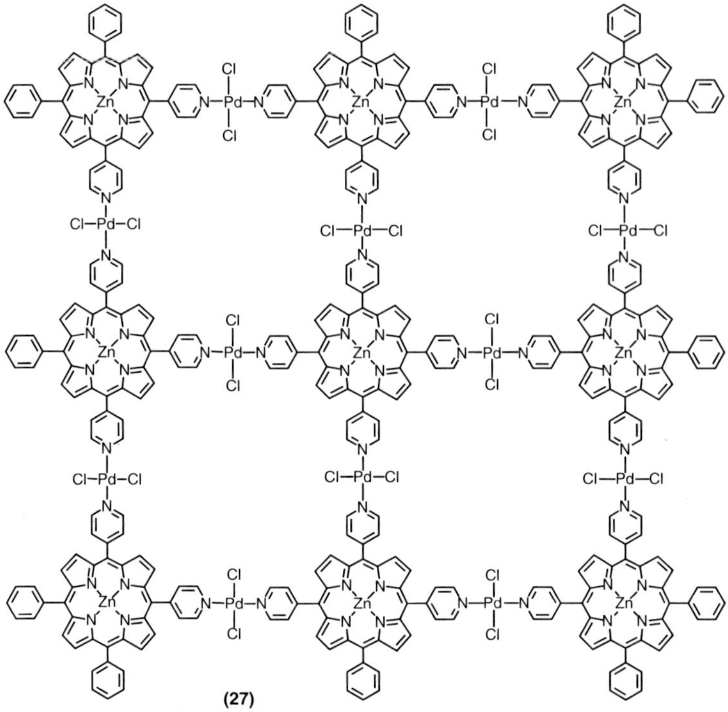

(27)

Scheme 6.8.

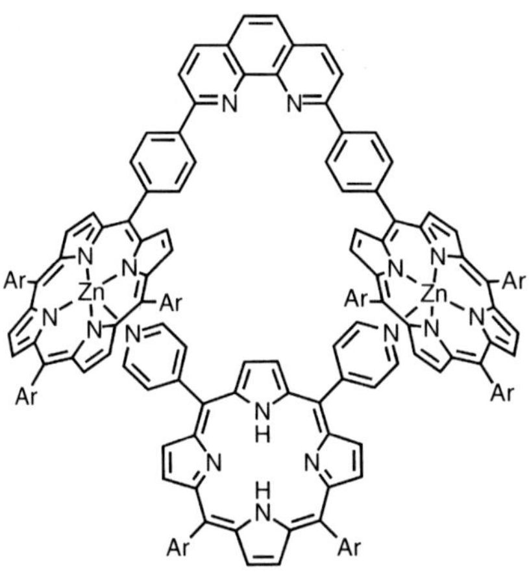

(28)

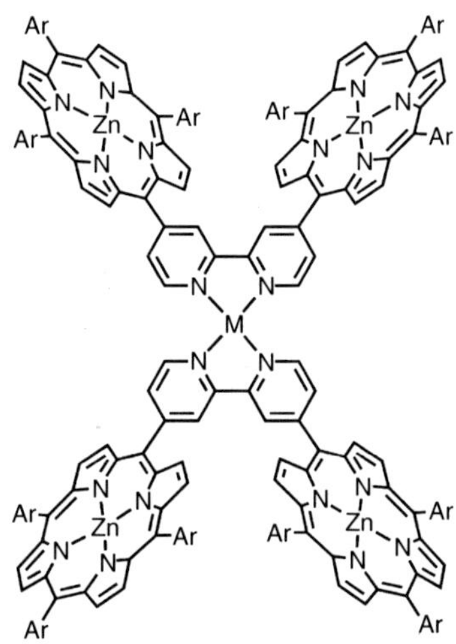

(29)

Scheme 6.9.

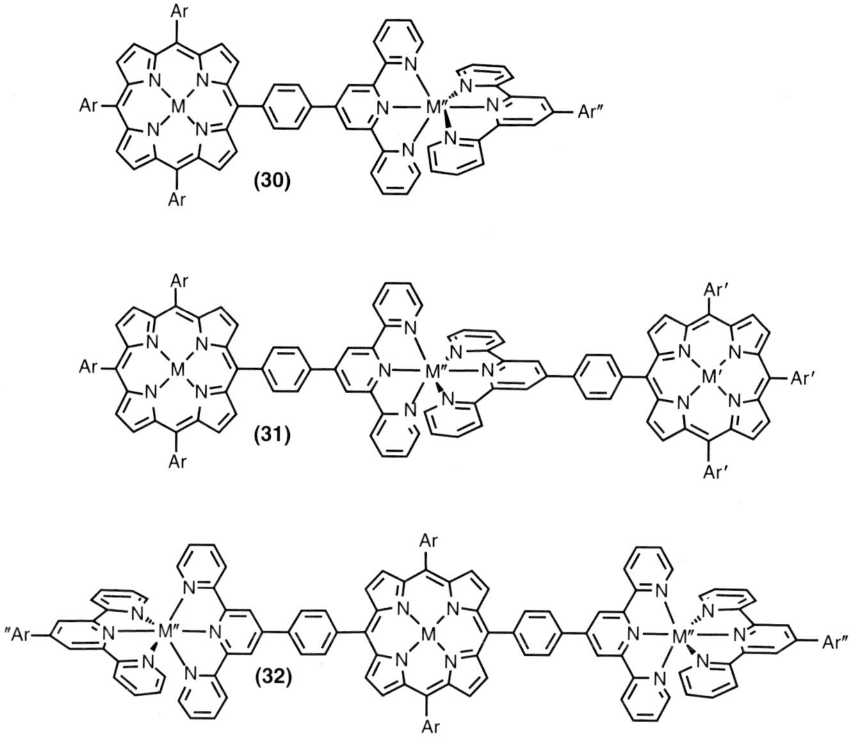

Scheme 6.10.

complex, in which a fast decay to the H_2P^* triplet state was found. Large nonlinear optical properties for this kind of donor–acceptor complexes have also been reported[111].

2. Synthesis and Characterization of Tetrametallated Pyridyl Porphyrins

A variety of supramolecular species containing porphyrins and transition metal complexes (many have been omitted), with increasingly higher complexity, have been synthesized and characterized, promoting an extraordinary advancement of the chemistry of such compounds. Consistent efforts directed to the study of the molecular sieving and photoinduced charge separation, have been carried out by many groups. These points were rapidly reviewed in the previous paragraphs, in order to show the current status and the yet unexplored potentiality of the supramolecular chemistry based on the metal–ligand coordination approach. Now we will focus on a more specific series of porphyrins bond to transition metal complexes, with particular emphasis on their catalytic and electrocatalytic properties which have been subject of systematic studies by ourselves and other researchers. The prototype porphyrin is M-TPyP coordinated to a variety of terminal complexes as shown in Figure 6.1, where M = first series transition metal ions.

The study of this series of compounds was begun with the intention of preparing multielectronic redox catalysts based on the coordination of redox active transition metal complexes such as $[Ru(NH_3)_5]^{2+}$ and $[Ru(edta)]^-$ to the pyridyl N-atoms of M-TPyP (**33, 34**)[112–115]. The basic idea was that the binding of peripheral ruthenium complexes could help delivering electrons to the catalytic active metalloporphyrin center, activating it for multielectronic processes; for example, the tetraelectronic reduction of O_2 to water. This was an alternative mechanism to the bis-coordination insight proposed by Collman and Anson, based on the study of a face-to-face cobalt diporphyrin[116–118]. In fact such hypothesis was demonstrated by Anson *et al.*, who started an exhaustive work on the electrocatalytic, properties of the Co-**33** complex in 1991[119]. Soon after, we started a systematic study on the catalytic, electrocatalytic, and photochemical properties of the M-TRP series (**35**)[120]. The coordination of $[Ru(bipy)_2Cl]^+$ groups to M-TPyPs showed to be very convenient and opened a wide range of interesting possibilities, particularly due to the tendency of **35** to stack, forming homogeneous films[121–123]. This was followed by the work of Alessio *et al.*[69] that showed

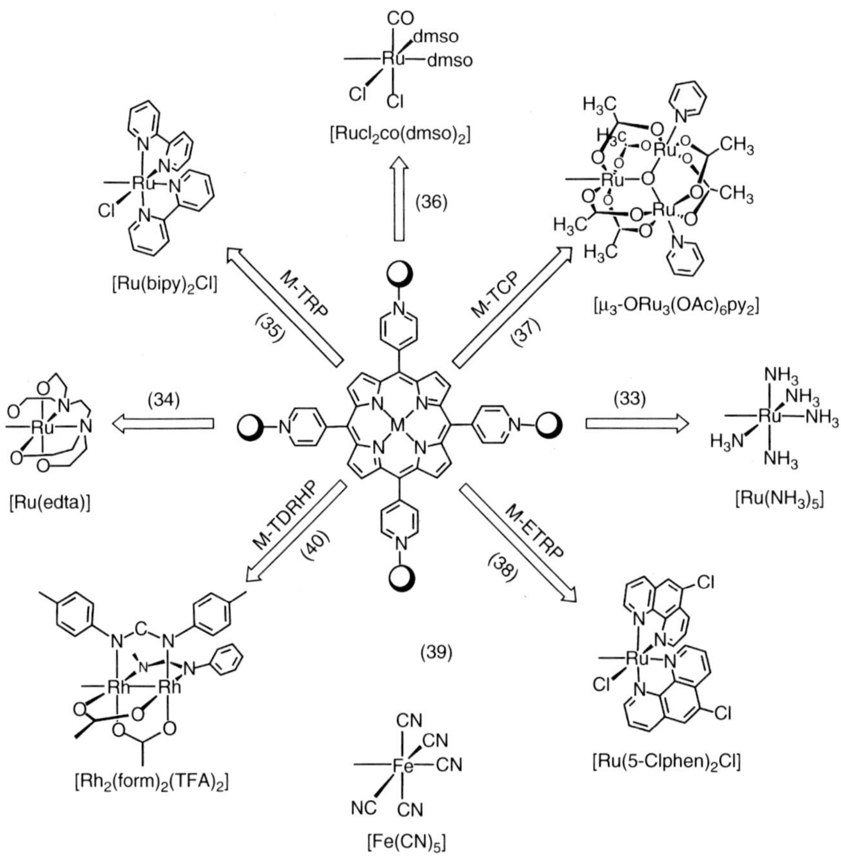

Figure 6.1. Scheme showing the structures of the peripherally metallated tetrapyridylporphyrins.

the potentiality of $[RuCl_2(CO)(dmso)_2]$ (**36**) and similar complexes for coordinative assembly of supramolecular porphyrins. After that we have extended the series including the $[Ru(phen)_2Cl]$[124], $[\mu\text{-}ORu_3(OAc)_6(py)_2]^+$ (**37**)[125], [Ru(5-Clphen)$_2$Cl] (**38**)[126], $[Ru(bpy)_3]^{2+}$ (**41**)[127], $[Fe(CN)_5]^{3-}$ (**39**)[128], and $[Ru(bpy)_2(OH_2)]^{2+}$ derivatives[129]. A 3,4-pyridylporphyrazine (TPyPz) derivative (**42**) has been also presented[130]. Recently, dirhodium $[Rh_2(form)_2(TFA)_2]$ porphyrin derivatives (**40**) were explored as photoactive multiredox systems by Lo Schiavo *et al.*[131] where form = *N*, *N*-9-di-p-tolylformamidinate and TFA = trifluoroacetate. The porphyrin fluorescence was significantly quenched by the electron-transfer from the dirhodium moieties. Hereafter, the synthesis, spectroscopic, and spectroelectrochemical properties of such supramolecular porphyrins in solution will be described, followed by a section about the conduction, catalytic, electrocatalytic, and photochemical properties of the molecular films in a more integrated way.

2.1. Syntheses

The peripherally polymetallated porphyrins have been obtained by the reaction of the corresponding pyridylporphyrins with $[Ru(edta)(OH_2)]^-$, $[Ru(NH_3)_5(OH_2)]^{2+}$, $[Ru(bpy)_2Cl_2]$, $[Ru(bpy)_2Cl(OH_2)]^+$, $[Ru(phen)_2Cl_2]$, [Ru(5-Clphen)$_2$Cl$_2$], $[Ru(bpy)_2(OH_2)_2]^{2+}$, $[\mu_3\text{-}ORu_3(OAC)_6(py)_2(MeOH)]^+$, $[RuCl_2(CO)(dmso)_3]$, $[Fe^{II}(CNH)_3(CN)_2(NH_3)]$ or $[Rh_2(form)_2(TFA)_2(OH_2)]$ complexes. Typically, the M-**35** complexes have been obtained by reacting the corresponding metallated porphyrin with $[Ru(bpy)_2Cl_2]$ in 1:4.1 molar stoichiometry, in refluxing acetic acid. The yields are generally high (about 90%), but the corresponding peripherally ruthenated derivatives of MPyP, *cis*- and *trans*-DPyP and TriPyP cannot be obtained by this method. In these cases, the more reactive $[Ru^{II}(bpy)_2Cl(OH_2)]^+$ species should be used instead[132]. The insertion of Zn(II), Cu(II), Ni(II) into H_2-TRP can be readily performed in CH_3OH, and used as an alternative preparation method of the corresponding metallated derivatives.

The preparation of $[Ru(NH_3)_5]$, [Ru(edta)], $[\mu_3 - ORu_3(OAc)_6(py)_2]$ and $[Fe(CN)_5]$ derivatives (**33, 34, 37, 39**) is usually complicated by the low solubility of the porphyrin or the transition metal complexes in a same solvent. For example, the $[Ru(edta)(OH_2)]^-$ and $[Fe(CN)_5(NH_3)]^{3-}$ complexes are soluble only in water, while the M-TPyPs are only soluble in organic solvents such as $CHCl_3$. Protonation can help dissolving the M-TPyPs in water, however, H^+ competes with the metal complexes, hindering the reaction. An exception is $[Ru(edta)(OH_2)]^-$ which reacts rapidly enough with the pyridyl N-atoms to avoid the precipitation of the porphyrin during the neutralization reaction, leading to the formation of **34**. This strategy was also used to obtain its Co and Fe derivatives in high yield[11,113,133], but was not successful for the remaining complexes. In the case of the $[Fe(CN)_5(NH_3)]^{3-}$ complex three of the five CN^- ligands can be protonated in concentrated HCl solution, generating a neutral species which is soluble enough in trifluoroethanol to allow the formation of **39**[128]. Only spectroscopic evidence of **39** was obtained when a large excess of a concentrated $[Fe(CN)_5(NH_3)]^{3-}$ solution was added into an acidified MeOH solution of **5**.

All attempts to prepare the M-**33** derivatives, where M= Fe(III) and Co(III), starting from the protonated porphyrins and $[Ru(NH_3)_5(OH_2)]^{2+}$ have failed, always leading to an insoluble material. Anson *et al.* suggested a possible explanation based on a polymerization process, involving the dissociation of an ammine ligand and formation of bridging ruthenium sites[134]. For this reason, the majority of their work has been carried out with samples generated "*in situ*" inside Nafion films, previously prepared on carbon electrodes and loaded with Co-TPyP[109, 134–138]. It should be noticed that, the reaction of *meso*-pyridyltritolylporphyrin with $[Ru(NH_3)_5(OH_2)](PF_6)_2$, yielding a binuclear complex, was successfully carried out by McLendon *et al.* in acetone[139], at room temperature. In this case, a 40% decrease in the fluorescence was reported for the Ru(III) derivative and assigned to a redox quenching mechanism.

Similar solubility problems hindered the preparation of the $[\mu_3 - ORu_3(OAc)_6(py)_2]$ derivatives in our laboratory for quite a long time. Acidified solvents could not be used because H^+ competes for the pyridyl sites, the coordination reaction is slow and the product **37** has relatively low solubility in the majority of the solvents. Accordingly, only a solvent that truly dissolves the reagents and intermediates would enable the preparation of this series of supermolecular porphyrins. An important finding was that 2,2,2-trifluoroethanol satisfy such requirements. In this way it was used for the preparation of the free-base, Zn(II), Co(III), and Mn(III) porphyrin derivatives[125, 140–142]. That solvent also dissolves tetrapyridylporphyrazine, allowing the coordination of $[Ru(bpy)_2Cl]$ complexes to the ring fused pyridyl N-atoms (**42**, Scheme 6.11)[130]. Zn-**41** (Scheme 6.11) was also synthesized in this way from a *meso*-tetra(2, 2′-bipyridyl)porphyrin, allowing the investigation of vectorial energy transfer under the influence of the axial ligand coordinated to the zinc porphyrin[127].

Cis-$[RuCl_2(dmso)_4]$ complex possesses S-bonded dmso ligands in *trans*-position to the Cl^-, being stabilized by strong donor–acceptor *trans*-influence. One of the remaining dmso is O-bonded and labile, being easily exchanged by ligands such as CO to generate [*cis*, *fac*-$RuCl_2(dmso)_3CO$]. This species

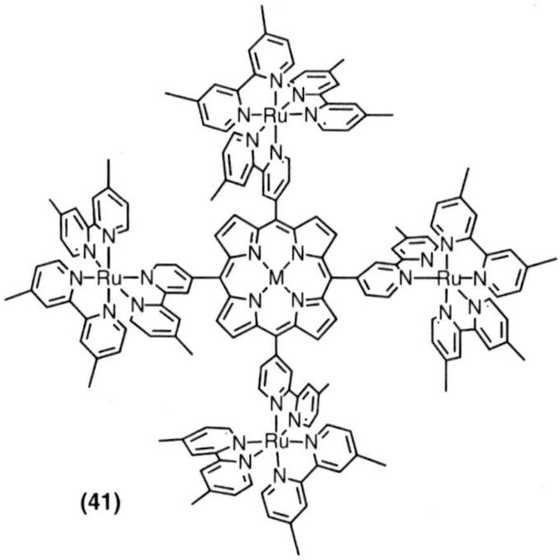

(41)

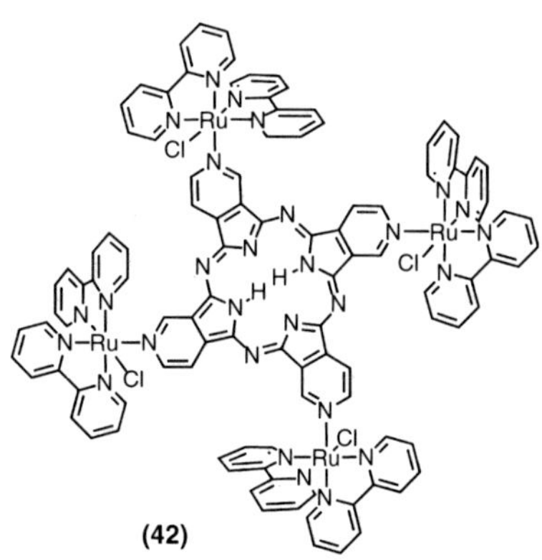

(42)

Scheme 6.11.

was combined with TPyP to give **36** in $CHCl_3$. The Zn and Co porphyrin derivatives were obtained by reacting further with the corresponding acetates. Analogous tetrarutenated porphyrins were obtained by coordinating four [cis,fac-$RuCl_2(CO)_3$], [trans-$RuCl_4$(dmso)] or [trans-$RuCl_4$(CO)] to TPyP[69,70].

The dirhodium carboxylate formamidinate complexes can be used as linear bridging and ending groups, by exploring the classical axial coordination chemistry of this type of dinuclear complexes. Only one of the axial positions is used for coordination to the *meso*-pyridylporphyrin when it is employed as ending group (reaction with an excess of the dirhodium complex), but both axial sites can be used for the assembly of linear structures and squares[131], in the same way as the *trans*-$PtCl_2$ complex. However, functionalized carboxylates such as *meso*-tetra(*p*-carboxyphenyl)porphyrins are bonded to the equatorial positions of that dirhodium complex forming a canted diporphyrin and a square[131,143]. The combination of such chemistry is expected to give rise to more complex multiporphyrinic architectures.

2.2. Characterization

All tetrametallated porphyrins dealt with in this section have been extensively characterized by electrochemical and various spectroscopic techniques such as NMR, IR, luminescence, and UV-Vis, among others. But, the following discussion will be focused mainly on the electrochemistry, spectroelectrochemistry, and electronic spectroscopy, and the evaluation of the electronic coupling between the components of the peripherally metallated porphyrins. This is an important parameter for the discussion of the catalytic and electrocatalytic properties as shown below.

2.2.1. Spectroscopic Characterization

The peripherally metallated porphyrins are soluble in water and conventional organic solvents such as dmf, $CHCl_3$, acetonitrile, and ethanol, in contrast with M-TPyP which are usually soluble only in acidic solvents and 2,2, 2-trifluoroethanol. The coordination of the peripheral complexes perturbs the porphyrin π-system and this effect was shown to be strong enough to influence significantly the electronic spectra. However, the general pattern is given by the sum of the spectra of the chromophores constituting those supramolecular porphyrins (Figure 6.2A). Also, when the peripheral complexes are asymmetric all the diasteroisomers seem to have the same spectroscopic, electrochemical and photophysical properties. Accordingly, the nonresolved samples actually behave as pure compounds. Spectral characterization of [H_2TRP (**35**) will be described in the following paragraphs, but the data for the remaining derivatives were summarized in Table 6.1. The other peripherally metallated porphyrins exhibit similar behavior, as can be inferred from spectral data of the M-TCP derivatives listed in Table 6.2. However, the absorption bands can be slightly shifted depending on the solvent.

H_2-TRP exhibits the porphyrin Soret and the four *Q* bands at 414, 518, 560, 584, and 642 nm, respectively; and the bands of peripheral complexes at 292 (bipy $\pi \rightarrow \pi^*$), 356 (MLCT2 $Ru^{II}(d\pi) \rightarrow bipy(\pi_2^*)$) and ~ 470 nm (MLCT1

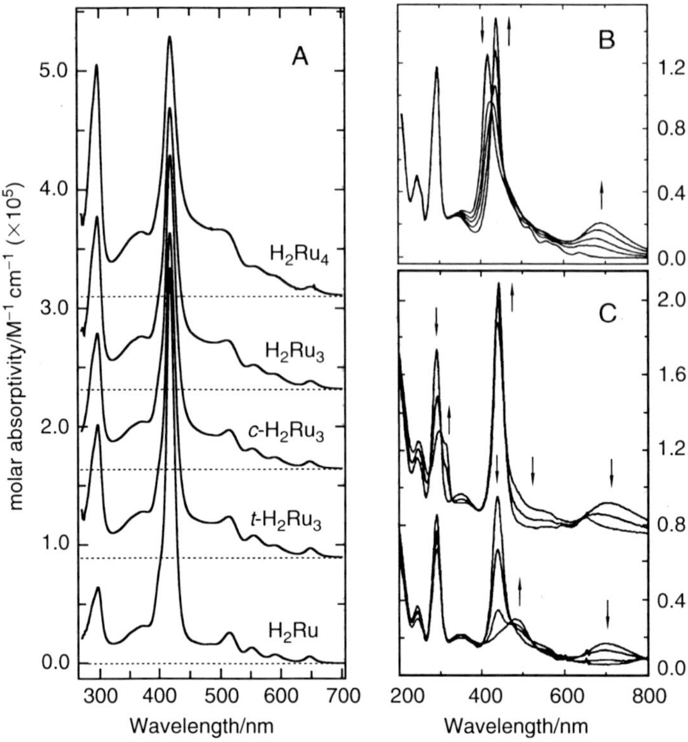

Figure 6.2. (A) Spectra of phenylpyridylporphyrins, with $(4-n)$ *meso*-pnenyl substituents ($n = 0$ to 3), coordinated to $n[Ru(bpy)_2Cl]^+$ complexes in dmf; (B) acid–base titration of a 5.8×10^{-6} M solution of $[H_2TRP]Cl_4$, and (C) spectral changes as a 6.4×10^{-4} M solution.

$Ru^{II}(d\pi) \rightarrow bipy(\pi_1^*)$). The protonation of the porphyrin ring ($pK_a \cong 2.0$), by addition of HCl, shifts the Soret band to 440 nm. Nevertheless, the rise of a new broad and strong band at 690 nm has also been observed (Figure 6.2B)[120], contrasting with the conventional two-peaked Q bands at 590 and 640 nm for the 3TPyP derivative[128]. The fact that it disappeared when the ruthenium complexes were oxidized or the porphyrin ring reduced (Figure 6.2C), supported a $Ru^{II}(d\pi) \rightarrow H_4\text{-}4TPyP(\pi_1^*)$ charge-transfer transition, providing a direct evidence of the existence of electronic coupling between those components. Such interaction is also reflected in the observed shift (by 6 nm) and enhancement of the Soret band (~30%) when the peripheral complexes are oxidized to Ru(III). The presence of a nodal plane in the π-system of the pyridyl bridges of the 3TPyP derivative would lead to much smaller interactions, as confirmed by the absence of that MLCT band.

The vibronic coupling between the chromophores of ZnTRP has been studied by resonance Raman spectroscopy using six laser lines in the 450–515 nm range[144], in which there is a superposition of the Soret, $Ru^{II}(d\pi) \rightarrow (p\pi^*)$bipy and $Ru^{II}(d\pi) \rightarrow (p\pi^*)$pyP charge-transfer transitions. Intense peaks

Table 6.1. Absorption Bands of M-35 in dmf Solution, Where $M = 2H^+, Zn^{2+}, Cu^{2+}, Ni^{2+}, Co^{3+}$ or Fe^{3+}. The Logarithm of the Molar Absorptivity is Shown in Between Parenthesis

Complex	bpy($\pi - \pi*$)	MLCT2	MLCT	Soret	Q_{1-0}	Q_{0-0}
Cu-**35**	298 (5.3)	356 (3.9)	490 (4.7)	415 (5.3)	544 (4.2)	582 (3.7)
Fe-**35**	298 (5.0)	362 (3.9)	490 (4.5)	416 (4.9)	568 (3.9)	616 (2.3)
Co-**35**	298 (5.2)	356 (4.1)	490 (4.6)	434 (5.2)	546 (4.0)	594 (3.0)
Ni-**35**	298 (5.2)	354 (4.2)	490 (4.7)	414 (5.2)	530 (4.1)	580 (3.7)
Zn-**35**	298 (5.3)	358 (4.1)	490 (4.7)	430 (5.5)	562 (4.4)	606 (3.9)
H_2-**35**	298 (5.2)	368 (4.7)	490 (4.7)	418 (5.3)	$Q_y = 514(4.7)$ $Q_x = 594(4.2)$	$Q_y = 558(4.3)$ $Q_x = 650(3.8)$
H_2Ru_3	296 (5.1)	368 (4.6)	480 (4.6)	418 (5.3)	$Q_y = 513(4.6)$ $Q_x = 591(4.1)$	$Q_y = 556(4.3)$ $Q_x = 649(3.8)$
t-H_2Ru_2	297 (5.0)	368 (4.6)	480 (4.5)	418 (5.4)	$Q_y = 513(4.5)$ $Q_x = 647(3.8)$	$Q_y = 555(4.2)$ $Q_x = 589(3.9)$
c-H_2Ru_2	297 (5.0)	369 (4.6)	480 (4.4)	418 (5.4)	$Q_y = 514(4.5)$ $Q_x = 590(3.9)$	$Q_y = 554(4.2)$ $Q_x = 647(3.7)$
H_2Ru	297 (4.8)	369 (4.5)	480 (4.2)	418 (5.5)	$Q_y = 514(4.4)$ $Q_x = 591(3.9)$	$Q_y = 551(4.1)$ $Q_x = 647(3.7)$

H_2Ru = [MPyP{Ru(bpy)$_2$Cl}], H_2Ru_2 = [DPyP{Ru(bpy)$_2$Cl}$_2$], H_2Ru_3 = [TriPyP {Ru(bpy)$_2$Cl}$_3$]. Lowering of the symmetry from D_{4h} (metallated) to D_{2d} (free-base) splits the $Q_{0–0}$ and $Q_{1–0}$ bands into their x and y polarized components, Q_x and Q_y.

Table 6.2. UV-Vis Absorption Bands of M-TCP in Acetonitrile Solution

Porphyrin	Soret	Q_{1-0}	Q_{0-0}	ORu_3 (MLCT)	Intracluster
H_2TCP	414 (5.42)	518, 579	552, 645	314, 351	707
ZnTCP	436 (5.59)	572 (4.55)	619 (4.56)	324 (4.81)	685 (4.52)
CoTCP	435 (5.01)	557 (4.22)	611 (4.23)	318 (4.61)	693 (4.28)
MnTCP	475 (5.01), 370 (4.86)[a]	584 (4.30)	621 (4.36)	~ 320	694 (4.34)

[a] Band VI.

characteristic of the totally symmetric modes of ZnTPyP and bpy-ligands have been found and classified in four groups according to the excitation pattern. The characteristic ZnTPyP modes are intensified by excitation at the Soret band while the bpy modes at 1491, 1321, and 669 cm^{-1} followed the contour of the MLCT band with a maximum at 495 nm. However, some of them were intensified both at the Soret and a band at 466 nm. The last one has been assigned to a $Ru^{II}(d\pi) \rightarrow (p\pi^*)$pyP charge-transfer transition and constitutes direct evidence of the vibronic coupling between the ZnTPyP and [Ru(bpy)$_2$Cl] moieties[102, 121]. Accidentally degenerate porphyrin and bpy modes have also been identified. A similar behavior has been observed for other analogous compounds.

2.2.2. Electrochemical Characterization

The peripherally metallated porphyrins exhibit two electrochemically active sites: the central metalloporphyrin and the peripheral transition metal complexes. Because of the presence of multiple (up to four) peripheral redox sites, the voltammetric behavior is dependent on the electronic coupling proportionated by the pyridylporphyrin bridge. Each site is oxidized or reduced at a different potential in the strong coupling limit, as observed in the [μ_3-$ORu_3(OAc)_6(py)_3$] and analogous complexes[6], in which the monoelectronic processes involving the ruthenium ions are separated by ~1V. As the electronic coupling decreases, ΔE becomes smaller and finally the sites behave independently of each other in the weak coupling limit. In this case, only a single couple of waves exhibiting intensity proportional to the number of equivalent active sites, is observed in the CVs. However, the change of oxidation state dramatically alters their donor–acceptor character, influencing the redox potential of the site to which they are bond, in the same way as the effect induced by the ligand substituents. This means that one can get a system equivalent to a "whole series of compounds" from a single supermolecule, just by modifying the oxidation state of its components. In fact, more or less strong donor or acceptor groups can be generated by adding or withdrawing electrons from any molecular species, including organic compounds, but the transition metal complexes exhibit a comparativelly rich electrochemistry. The question that remains is, how good are the pyridylporphyrins as electronic connectors (for a discussion on the current status on charge-transfer see ref[145]. and refs. therein). The electrochemistry data for all the supramolecular porphyrins shown in Figure 6.1, including the [μ_3-$ORu_3(OAc)_6(py)_2$] (**37**) and the dirhodium derivatives (**40**), indicate that the pyridylporphyrins are moderate or weakly coupling bridges, in contrast to pyrazine and analogous ligands. In fact, the peripheral redox sites behave essentially as a single species, giving rise to only one voltammetric wave with intensity proportional to the number of equivalent sites.

The electrochemistry of [Ru(edta)], [$Ru(NH_3)_5$] and [$Fe(CN)_5$] derivatives (**33, 34, 39**) has been performed exclusively in aqueous solution (−1 to + 1 V range), including the M(III/II) process of the peripheral complexes, as well as the Fe(III/II)P, Co(III/II)P and Mn(III/II)P processes. The potential of the [$Ru^{III/II}$(edta)(pyP)] process in Fe-**34** and Co-**34** differs by 30 mV, reflecting the change in the redox state of the central metal ion, i.e., the peripheral sites are reduced before the iron porphyrin center in the Fe(III)P derivative[113,133], while Co(II)P is formed before the reduction of those sites[114]. The same effect was expected for the [$Ru(NH_3)_5$] derivatives (Table 6.3), but the corresponding Co(III/II)P wave has not been observed by CV[138]. The [$Fe^{III/II}(CN)_5$pyP] potential is similar to that of analogous complexes, however, the oxidation of the porphyrin ring occurs at comparatively low potentials (~ 0.7 V). This suggests a significant stabilization of the oxidized porphyrin probably by electrostatic effects, since the complex is present as a Fe(III) complex and is unable to induce such kind of effect by electronic means[128].

The electrochemistry of M-TRP ([$Ru(bpy)_2Cl$] derivatives) has been carried out in dmf or acetonitrile solutions and a typical set of voltammograms for the Cu(II) derivative is shown in Figure 6.3[146]. The Ru(III/II) redox processes

Table 6.3. Redox Potentials (E vs. SHE) of Some Peripherally Metallated Porphyrins

Porphyrin	Peripheral M(III/II)	M′(III/II)P
Fe-**34**	−0.02	0.11
Co-**34**	0.30	0.08
H_2-**33**	0.33	—
Co-**33**	0.38*	not obs.
[CoTriPyP{Ru(NH_3)$_5$}$_3$]	0.38*	not obs.
[CoDPyP{Ru(NH_3)$_5$}$_2$]	0.38*	not obs.
[CoMPyP{Ru(NH_3)$_5$}]	0.36*	not obs.
[MPyTTP{Ru(NH_3)$_5$}]	0.4[139]	—
[3TPyP{Fe(CN_5)}$_4$]	0.45	—

KCl 0.5 M or *in Nafion film, 0.5 M $HClO_4$ + 0.5 M NH_4PF_6 aqueous solution[138].

associated with peripheral ruthenium complexes are responsible for a reversible wave around 0.9 V, invariably exhibiting an intensity four times bigger than that centered at the metalloporphyrin moiety[120, 122, 124, 132, 147]. This process is followed by an EC process (> 1.2 V) assigned to the oxidation of the metalloporphyrin (Table 6.4). At negative potentials, two successive monoelectronic ring reduction waves to the radical anion (−0.7 V) and the dianion (−1.1 V) can be observed, followed by the reduction of the bpy-ligands. This process occurs in steps, such that one of the rings is reduced around −1.4 V and the second at −1.6 V. Also, the Co(III), Fe(III), and Mn(III) porphyrins can be reduced to the respective M(II)P species. Normally, the Co(III/II)P wave is not observed in the CVs because of its slow kinetics, but is expected around 0.1 V, by analogy with other *meso*-substituted CoPs. On the other hand, the Fe(III)P derivative can form the highly stabilized μ-oxo dimer, μ-O(Fe^{III}P)$_2$, which rapidly dissociates after the reduction at −0.76 V[148]. In addition, the porphyrin ring reduction potential is sensitive to the coordinated metal ion and its oxidation state. For example, the reduction of the [Co^{II}TPyP{Ru^{II}(bpy)$_2$Cl}$_4$] species is cathodically shifted due to the prior formation of Co(I)P, while the ring oxidation potentials remain almost invariant at 1.5 V. Although M(I)P state has been reported for Ni, Co, and Fe porphyrins, the only example in the case of supramolecular porphyrins is Co^{I}TRP. This species was detected around −1.1 V by spectroelectrochemistry[142], and is characterized by a double Soret band. Its formation usually shifts the porphyrin and bpy reduction to more negative potentials (Table 6.4). In all cases, the spectroelectrochemistry technique has proven essential for the proper assignment of the redox processes (a typical set is shown in Figure 6.3)[5, 114, 122, 133]. Considering the currently available information, the changes at −0.5 V, previously assigned to the formation of a Co^{0}P species, would better be ascribed to a change in the axial coordination, as found out for the ZnTCP (**37**) and ZnTRP complexes[140]. This revision was considered in the redox potentials listed in Table 6.4.

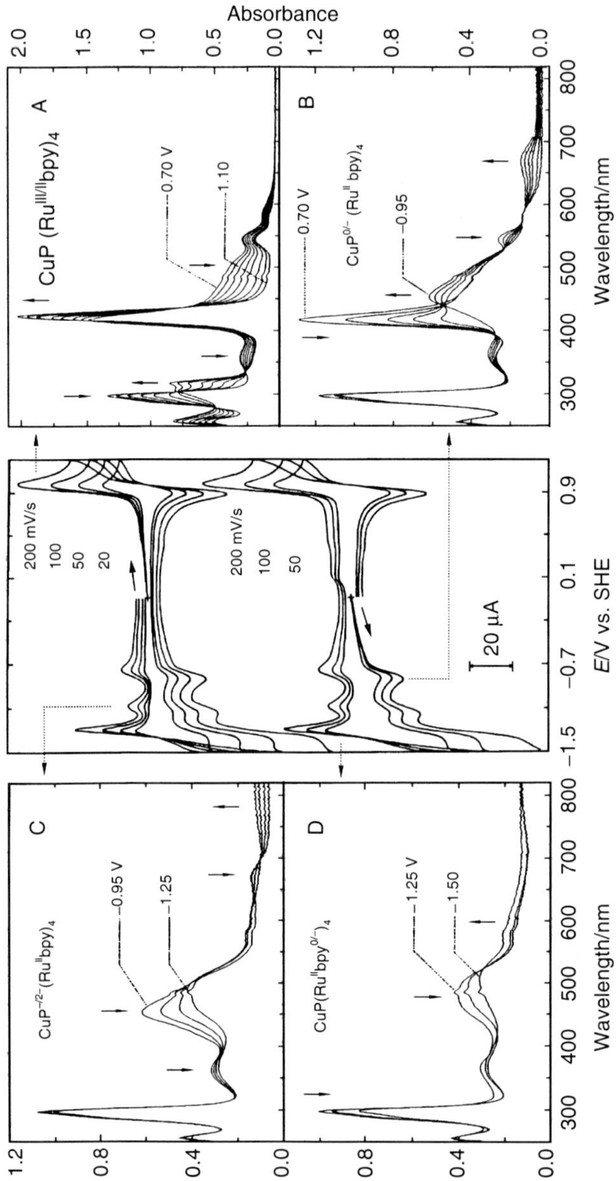

Figure 6.3. CVs of a 2.5 mM [CuTRP] $(CF_3SO_3)_4$ (center) and spectroelectrochemical changes in the (A) 0.70 to 1.10 V, (B) 0.70 to −0.95 V, (C) −0.95 to −1.25 V and (D) −1.25 to −1.50 V vs. SHE, in 0.1 M $TEAClO_4$ dmf solution (ref [146]).

The replacement of bpy by phen in the peripheral complexes does not seem to influence significantly the Ru(III/II) process[124], however, in the case of 5-Clphen that wave is shifted to 1.0 V[126]. Although those changes were not expected to modify the porphyrin ring localized processes, the reduction potentials were significantly shifted, as shown in Table 6.4. Analogously, the substitution of

Table 6.4. Redox Potentials (E vs. SHE) of Supramolecular Porphyrins Obtained by Coordination of $[Ru(bpy)_2Cl]^+$ and Analogous Complexes in dmf Solution

Porphyrin	$bpy^{-/2-}$	$bpy^{0/-}$	$P^{-/2-}$	$P^{0/-}$	$M^{III/II}P$	$Ru^{3+/2+}$	$P^{+/0}$
Cu-**35**	—	−1.34	−1.11	−0.79	—	0.93	1.5*,a
Co-**35**	∼ −1.9c	∼ −1.6c	—	∼ −1.8c	∼ 0.1	0.93	—
Mn-**35**	—	−1.35c	−1.16c	−0.90	−0.04	0.94	1.8*,a
Fe-**35**	—	−1.34c	not obs.	∼ −1.15c	0.16	0.91	1.5*,a
μ-O(Fe-**35**)$_2$	—	−1.34c	−1.23c	−1.05c	∼ −0.8c	0.91	—
Ni-**35**	−1.67c	−1.37c	−0.97	−0.80	—	0.92	1.5*,a
Zn-**35**	−1.67c	−1.39c	−1.18c	−0.97c	—	0.92	1.5*,a
H_2-**35**	−1.52	−1.42c	−0.93	−0.68	—	0.92	1.6,a
$H_2Ru_4^{**}$	−1.5c	−1.3	−1.01	−0.69	—	0.91	—
H_2-**38**	—	−1.11c	−1.04c	−0.65	—	1.00	> 1.5
H_2Ru_3	—	−1.39c	−1.02c	−0.76c	—	0.86	1.4a
c-H_2Ru_2	—	−1.39	−1.19	−0.86	—	0.86	1.43a
t-H_2Ru_2	—	−1.43	−1.14	−0.84	—	0.85	1.42a
H_2Ru	—	−1.56c	−1.22	−0.90	—	0.85	1.26a

c = cathodic peak, a = anodic peak, * = in CH_3CN solution. **[TPyP{Ru(phen)$_2$Cl}$_4$] complex.

the pyridyl by phenyl-substituents affected the porphyrin localized redox potentials as well as the ruthenium complexes[132]. However, the effect on the Ru(III/II) potential was pronounced only for the insertion of the first phenyl ring; and a general tendency for a cathodic shift was observed for the other reduction processes, as expected for a successive substitution by a weaker electron withdrawing group. Unfortunately, the electrochemistry of the $[RuCl_2(dmso)_2CO]$ and other similar derivatives has not been published yet.

The voltammetric behavior of the $[\mu_3\text{-}ORu_3(OAc)_6(py)_2]$ and $[Rh_2(form)_2(COO)_2]$ derivatives is similar to that described above, but with some peculiarities. Those complexes are different from the previous ones not just because of the higher nuclearity, but also due to the strong metal–metal interactions, responsible for the splitting of the redox potentials involving those sites. Accordingly, the two waves at 0.60 and 1.50 V vs. SCE found for **40** were assigned to the $Rh^{III}Rh^{II}/Rh^{II}Rh^{II}$ and $Rh^{III}Rh^{III}/Rh^{III}Rh^{II}$ processes, while an irreversible wave at −1.02 V was attributed to the monoelectronic reduction of the porphyrin ring, in 1,2-dichloroethane[131]. In the case of M-TCP, the $Ru^{III}Ru^{III}Ru^{II}/Ru^{III}Ru^{II}Ru^{II}$, $Ru^{III}Ru^{III}Ru^{III}/Ru^{III}Ru^{III}Ru^{II}$, $Ru^{III}Ru^{III}Ru^{IV}/Ru^{III}Ru^{III}Ru^{III}$ and $Ru^{III}Ru^{IV}Ru^{IV}/Ru^{III}Ru^{III}Ru^{IV}$ redox processes were easily discernible in the CVs, being assigned accordingly[125, 140–142]. In contrast, the porphyrin waves are generally hidden, but can be found by spectroelectrochemistry (Table 6.5, Figure 6.4). In fact, four regions can be distinguished in the voltammograms and subtle changes in the intensity or shape are indicative of additional redox processes involving the metalloporphyrin moiety. In the case of ZnTCP, the reversible wave around 0.2 V was attributed to the $Ru_3^{III,III,III}/Ru_3^{III,III,II}$ process. Nevertheless,

Table 6.5. Redox Potentials (E vs. SHE) of M-TCP

Porphyrin	$Ru_3^{IV,IV,III}$/ $Ru_3^{IV,III,III}$	$Ru_3^{IV,III,III}$/ $Ru_3^{III,III,III}$	$Ru_3^{III,III,III}$/ $Ru_3^{III,III,II}$	$Ru_3^{III,III,II}$/ $Ru_3^{III,II,II}$	$P^{0/+}$	$P^{0/-}$	$P^{-/2-}$	$M^{3+/2+}P$
H_2TCP	2.3[a]	1.23	0.16	−1.14	–	−0.72	−1.05[c]	–
ZnTCP	2.18[a]	1.19	0.15	−1.13	~ 1.35	~ −1.1	~ −1.27	–
CoTCP	2.33[a]	1.25	0.16	−1.05	> 2.0	–	–	31, −0.65*
MnTCP	~ 2	1.17	0.16, 0.01#	−1.17	–	−0.94	–	05, 0.20 #

0.1 M $TEAClO_4$ CH_3CN solution. c = cathodic peak; a = anodic peak; *$Co^{II/I}$ TCP process; #in 1,2-dichloroethane.

the spectroelectrochemistry at the distorted wave around −1.1 V revealed the occurrence of both, the $ZnP^{0/-}$ and $Ru_3^{III,III,II}/Ru_3^{III,II,II}$ reactions in a very narrow range, followed by the second reduction of the porphyrin ring to the dianion species. In fact, all three processes were hidden under a single voltammetric wave. The almost reversible wave at 1.2 V was also found to be composed by the $Ru_3^{IV,III,III}/Ru_3^{III,III,III}$ and $ZnP^{0/+}$ processes, and was followed by an irreversible wave at 2.1 V ($Ru_3^{IV,IV,III}/Ru_3^{IV,III,III}$) exhibiting a characteristic EC profile.

The electrochemistry of CoTCP is apparently similar (Figure 6.4A), but the hidden CoP behavior can be misleading. In this case, the waves at 1.2 and 2.1 V were assigned to the consecutive monoelectronic oxidations of the peripheral ruthenium clusters (Figure 6.4B), and no evidence of $Co^{III}P$ moiety oxidation could be found up to 2.0 V. Although the wave at 0.2 V looks like a conventional cluster wave, the spectroelectrochemistry in the 0.7 to 0.0 V revealed the presence of the $Co^{III/II}P$ process. As the potential was shifted from 0.7 to 0.20 V, the Soret and Q bands shifted from 435 and 537 nm to 412 and 533 nm, respectively, while the ruthenium cluster bands remained almost unchanged (Figure 6.4C). In contrast, the broad Ru-cluster band at 700 nm were shifted to 913 nm in the 0.25 to 0.0 V range, but the porphyrin bands were only slightly affected (Figure 6.4D). It is practically impossible to analyze those processes by CV, however, the redox potential of the $Ru_3^{III,III,III}/Ru_3^{III,III,II}$ couple (0.16 V) can be determined straightforwardly by the linear coefficient of the Nernst plot of the spectroelectrochemistry data at 913 nm. The $Co^{III/II}P$ potential (0.31 V) was determined in a similar way, after some considerations about the effect of the reduction of the peripheral complexes on the Soret band at 412 nm[142].

Going to a more negative region, weak waves can be seen at 0.65 V. The spectroelectrochemistry in this region showed an enhancement of the 913 nm band and of the absorbance below 550 nm, as the potential was stepped from −0.5 to −0.8 V. In addition, new bands appeared at 366 and 300 nm and the Soret band was broadened and shifted from 412 to 421 nm (Figure 6.4E). This behavior is completely different from that expected for the reduction of the porphyrin ring or the peripheral ruthenium clusters and was assigned to the $Co^{II/I}P$

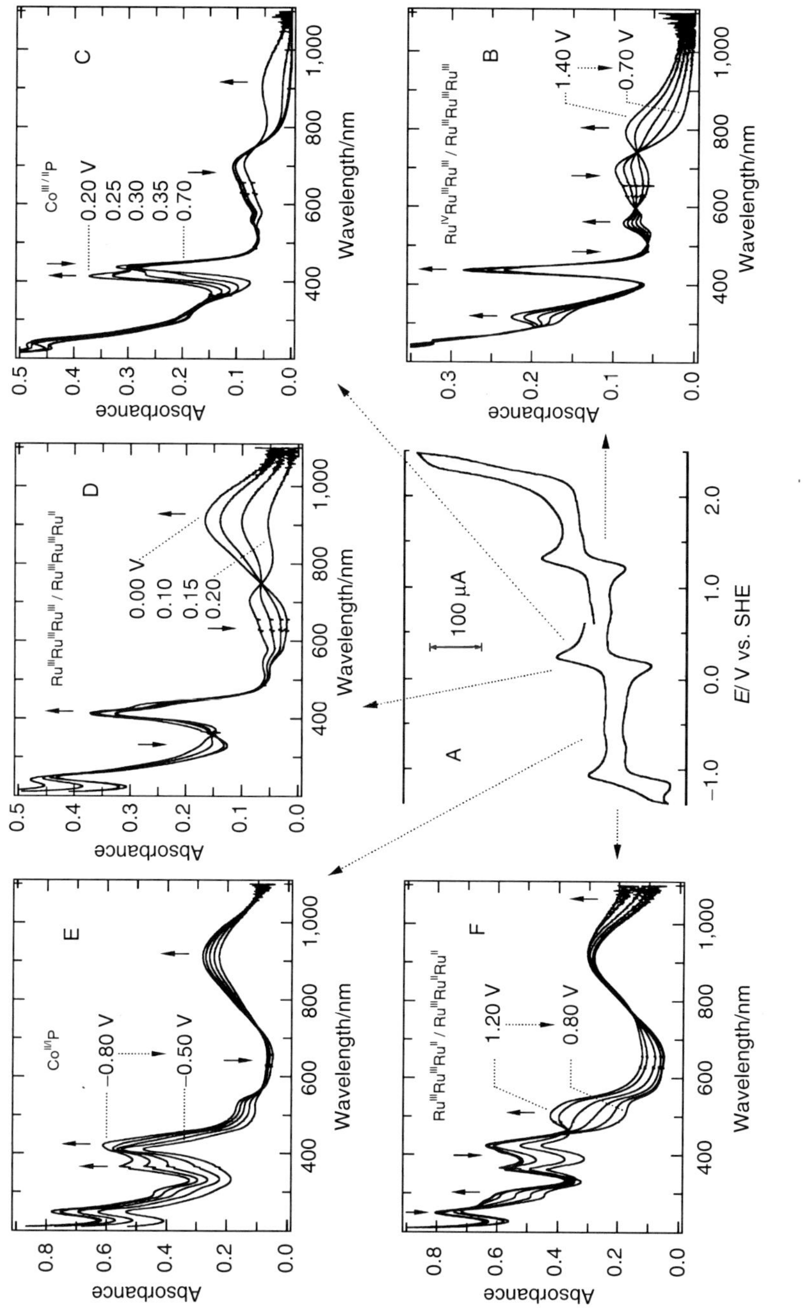

Figure 6.4. (A) CV of a 3.0×10^{-4} M CoTCP in the -1.5 to 2.5 V vs. SHE range and spectroelectrochemical changes in the (B) 0.7 to 1.4, (C) 0.2 to 0.7, (D) 0.0 to 0.2, (E) -0.5 to -0.8, and (F) -0.8 to -1.2 V range, in 0.1 M $TEAClO_4$ CH_3CN solution.

process. The double Soret band of the $Co^{I}P$ species becomes more evident when the ruthenium clusters are further reduced to the $Ru_3^{III,II,II}$ species (−0.8 to −1.2 V range), shifting the MLCT band from 400 to 502 nm and the intracluster band from 913 to 938 nm (Figure 6.4F).

Changes in the oxidation state at the metalloporphyrin center can modify the absorption spectra of the peripheral complexes and vice versa, as shown in the spectroelectrochemical behavior of CoTCP. Significant shifts of the redox potentials can be observed, also. For example, the ring oxidation of the $Co^{III}P$ species was shifted to potentials above 2.0 V (generally found at 1.5 V), as a consequence of the four electron withdrawing $Ru_3^{IV,III,III}$ species bond to the peripheral pyridyl sites. Similar behavior has been observed for MnTCP and the dirhodium derivative **40**, in which the porphyrin oxidation is anodically shifted from ~1.2 to >1.6 V due to the presence of four $Rh^{III}Rh^{II}$ complexes coordinated to TPyP[131,143]. Furthermore, the reduction of the porphyrin ring to the radical anion seems to perturb more intensely the $Ru_3^{III,III,II}/Ru_3^{III,II,II}$ than a reduction centered on the porphyrin metal ion, as can be inferred from the 100 mV difference between CoTCP and the other metalloporphyrin derivatives (Table 6.5). In summary, the electrochemically active components of a supramolecular system are not simply electron-transfer relays, electron sources or drains, but active components that can be used to alter the electronic properties of the components to which they are bond. Further evidences will be shown in the following sections.

3. Catalytic and Electrocatalytic Properties

The catalytic and electrocatalytic properties of the metalloporphyrins are strongly influenced by the metal-ion, because the activation of a catalytic process generally depends on the coordination of the substrate to an active metal site. Consequently, its electronic and redox properties control the activity. On the other hand, these characteristics are more or less influenced by the nature of the transition metal complex bond to the peripheral pyridyl-substituents and its oxidation state. In this section, the catalytic properties of the peripherally metallated porphyrins will be presented starting with their behavior in solution and then as thin films.

Among the metalloporphyrins, the nickel derivative exhibited electrocatalytic activity for the reduction of CO_2 and should have some Ni(I)P character in the active form. Such species constains a d^9 metal ion which is isolobal with the methyl-radical, being able to interact with substrates with similar orbital symmetry and configuration. Accordingly, the ability to reduce CH_3I can be used as an activity test. NiTRP presents two reversible waves assigned to the successive monoelectronic reductions of the NiP at –0.87 and –0.97 V, followed by the $bpy^{0/-}$ process at –1.37 V, in dmf solution (Figure 6.5). The first cathodic wave is intensified when CH_3I is added, but only minor changes could be observed in the presence of CO_2. In fact, a stronger effect has been noted only at the onset of the $bpy^{0/-}$ process and assigned to two effects: (1) the increase of the number of electrons available for multielectronic processes and (2) to electronic effects induced

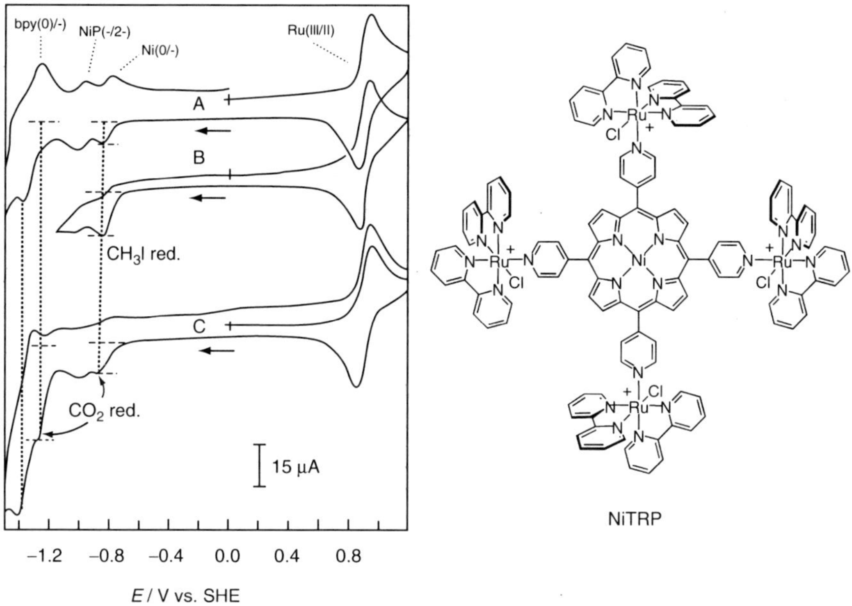

Figure 6.5. Cyclic Voltammograms of a 1.1×10^{-3} M [NiTRP]$(CF_3SO_3)_4$ dmf solution, $TEAClO_4$ 0.10 M in (A) Ar atmosphere, (B) 1.6×10^{-2} M of CH_3I and (C) CO_2 saturated solution. $\upsilon = 50\,mV\,s^{-1}$.

by the formation of the bpy^- radicals, which possibly is increasing the Ni(I)P character of the metalloporphyrin. This last effect probably is the most important one for the activation of the catalytic site. The electrocatalytic properties of the other M-TRPs will be discussed in the section devoted to the porphyrin films.

3.1. Hydrocarbon Oxidation

The M-TCP series (where, M $= H_2, Zn^{2+}, Mn^{2+}$ and Co^{3+}) have been prepared aiming their use as multielectronic redox catalysts. The axial ligand is known to influence the spectroscopic and electrochemical properties, such that the association equilibria of CH_3CN($K = 4.5$), py (4.8×10^4) and imidazole (1.7×10^7) were investigated by monitoring the Soret band[140]. The constants found in CH_2Cl_2 were significantly higher than for ZnTPP (py $= 6.7 \times 10^3$ and im $= 1.9 \times 10^5$) and tended to increase as a function of the basicity of the ligand L, suggesting that the bond is predominantly electrostatic. Hence, the comparatively higher equilibrium constants found for ZnTCP were assigned to the more pronounced acidity of ZnP, induced by the electron withdrawing character of the four coordinated $[Ru_3O(OAc)_6(py_2)]^+$ groups.

Manganese and iron porphyrins have been extensively investigated as cytochrome P450 model, because they catalyze the oxidation of organic substrates by oxygen donors like iodosylbenzene, persulphate, and H_2O_2, among others[151–156]. This subject has been extensively discussed by many authors[151, 157–162] including a series of articles in The Porphyrin Handbook[163–168], and is out of our scope.

Hence, only the aspects strictly necessary for our discussion will be presented here.

Porphyrins such as MnTPP and FeTPP are relatively susceptible to ring oxidation and are rapidly decomposed during the catalytic reaction. Consequently, electron-withdrawing substituents such as halogens and nitro groups were introduced in the phenyl rings (second generation catalysts) and subsequently at the β-pyrrole positions, in order to shift that reaction to more positive potentials and increase their durability and reactivity. In fact, some of them were significantly more stable and active than the first generation porphyrins for the epoxidation and hydroxylation of organic substrates such as cyclohexane. In addition, bulky substituents can preclude the catalyst autodegradation by steric effects, because it involves the collision of activated metalloporphyrin molecules. Accordingly, the peripherally metallated porphyrins are potentially interesting as cytochrome oxidase models, particularly MnTCP[141]. In this series of compounds, the peripheral μ_3-ORu$_3$ complexes are electron withdrawing groups even in the $Ru^{III}Ru^{III}$, Ru^{III} oxidation state[140]. Hence, $Ru^{IV}Ru^{III}Ru^{III}$ is a strong and the $Ru^{IV}Ru^{IV}Ru^{III}$ species an even stronger electron-acceptor. Consequently, the electronic density on the porphyrin ring should decrease significantly as the peripheral complexes are oxidized, in the same way as the covalent binding of chloro or nitro groups, shifting the $Mn^{III}P$ ring oxidation to more positive potentials ($> +1.5$ V). Interestingly, MnTCP and MnTPyP presented similar catalytic activity for the epoxidation of cyclooctene by iodosylbenzene, in acetonitrile solution. However, MnTCP is a much more active and selective catalyst for the less reactive cyclohexane, producing cyclohexanol as the sole oxidation product in 45% yield, while MnTPyP gave a mixture of cyclohexanol (27%) and cyclohexanone (13%).

From this reaction pattern we can suppose that the high valent $O = Mn^{V}P$ species is the catalytic active species, but it would be difficult to detect this intermediate due to its high reactivity and short lifetime. Nevertheless, Groves *et al.*[169,170] succeeded in monitoring the formation of $O = Mn^{V}$TMPyP 30 ms after the reaction of Mn^{III}TMPyP with chlorobenzoic acid in aqueous solution. That intermediate decayed slowly ($k = 5.7\,s^{-1}$) to the less reactive $O = Mn^{IV}$TMPyP species, such that both were characterized spectroscopically. Accordingly, the reaction of Mn^{III}TCP with a suspension of iodosylbenzene in CH_2Cl_2 was carried out in a flow-cell and monitored by UV-Vis spectroscopy. The shift of the Soret band (band V) from 473 to 427 nm is consistent with the rapid formation of the $O = Mn^{IV}$TCP species (Figure 6.6)[141]. However, no spectral change could be observed when cyclohexane was added to the reaction mixture. Those apparently contradictory results do make sense if one conceives that a very reactive intermediate $O = Mn^{V}$TCP is being initially formed by the reaction of MnTCP and iodosylbenzene. Subsequently, it is rapidly reacting with cyclohexane regenerating the starting Mn^{III}TCP or being converted to $O = Mn^{IV}$TCP. Considering that $O = Mn^{IV}$TCP was found to be a very poor catalyst, the oxygen transfer from that active species should be highly efficiency and quantitative, otherwise $O = Mn^{IV}$TCP should build-up in solution. The high reactivity also explains why $O = Mn^{V}$TCP does not accumulate, remaining below the detection limit of the spectrophotometric technique ($< 0.1\,\mu$M). A more detailed analysis

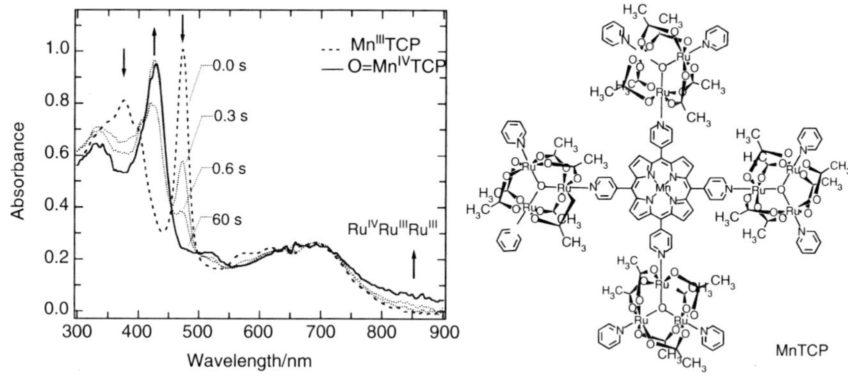

Figure 6.6. UV-Vis spectra of a dichloroethane solution obtained by mixing equivalents volumes of $Mn^{III}TCP(2 \times 10^{-5}$ M) and iodosylbenzene (1.4×10^{-3} M) solutions in a flow-cell, as a function of the time.

of Figure 6.6, suggests the presence of some oxidized $Ru^{IV}Ru^{III}Ru^{III}$ in equilibrium with $O = Mn^{IV}P$. Hence, a large amount of oxidized peripheral complexes should be in equilibrium with $O = Mn^{V}TCP$, enhancing the stability and catalytic activity in a synergistic way.

Interesting results have been obtained by Hupp *et al.*[171] when Mn-**2** was used as catalyst for the epoxidation of styrene by iodosylbenzene in $CHCl_3$, in the presence and absence of Zn- or MgTPP. The presence of these Lewis acids enhanced significantly the durability and the turnover number of the catalyst and such effects were assigned to the formation of supramolecular assemblies like **8**. Hence, the M-TPP rings should be providing steric protection against bimolecular decomposition and inducing some supramolecular effects. In fact, the addition of those metalloporphyrins led to the full activity recovery of some deactivated catalysts. On the other hand, Belvedere and Breslow[172] showed that Cu(II) ions bond to a *meso*-tetra(bipyridyl)porphyrin can be used to orient steroids containing an adequate auxiliary coordinating group, leading to selective hydroxylation of a particular carbon atom.

3.2. Tetraelectronic O_2 Reduction

The efficient tetraelectronic reduction of O_2 to H_2O remains as a challenge for the development of efficient and durable fuel cells operating at room temperature. This reaction is routinely performed with almost 100% yield during the respiration of cells and is responsible for the conversion of food stuffs chemical energy into the cellular currency ATP, from ADP and phosphate. This does not mean that the reaction is trivial, as can be inferred from the involvement of four electrons and four protons, evidencing the necessity to control a multielectronic process coupled with mass transport. Furthermore, O_2 can be reduced by one and two electron processes leading to the superoxide and peroxide (Figure 6.7A), respectively; the last one being the most prevalent pathway in reactions catalyzed by metalloporphyrins. These are not desirable because they decrease the efficiency

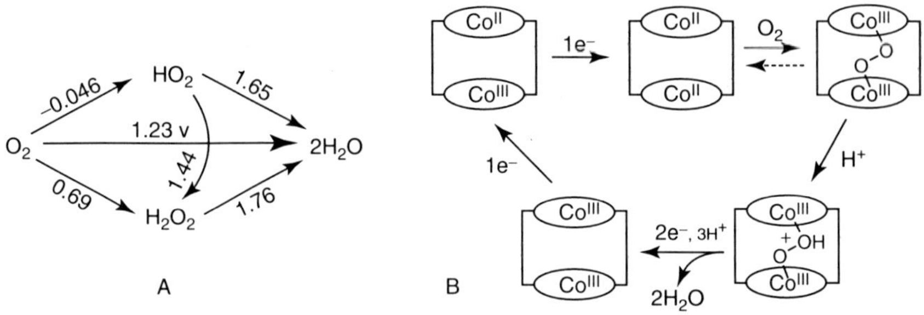

Figure 6.7. (A) Redox potentials for the reactions involving dioxygen in acidic media; (B) scheme of the mechanism proposed for the tetraelectronic reduction of O_2 catalyzed cofacial cobalt prophyrins.

and generate highly reactive species that may damage the devices. Many cells are operating at high temperatures and pH's in order to overcome such drawbacks.

The mitochondrial power plants work at mild conditions and the energy is released in a stepwise manner ($\Delta E \sim 0.2\,\mathrm{V}$) as the electron hops between the redox sites in the respiratory chain, finally reaching an enzyme known as cytochrome oxidase. This is responsible for the last process, the transfer of four electrons to O_2 generating water and more energy. It contains heme-*a* and bis-copper centers as electron donors, and a couple of heme/Cu active sites (Figure 6.8A). Dioxygen was previously believed to be activated by simultaneous coordination to a heme-a_3 and a Cu(I) complex (Cu_B), generating an activated state with peroxide character. Then, two more electrons and protons would be pumped in to produce water before dissociation takes place (Figure 6.8C). Any HO_2 and H_2O_2 generated are rapidly converted to H_2O and O_2 by the action of superoxide dismutase and catalase, respectively, to avoid damaging the cell.

Collman *et al.* prepared many FeP/Cu model compounds modeling the enzyme active site. The most recently obtained supramolecular systems are remarkably similar to the natural reaction center, reproducing the Fe to Cu distance and their first coordination sphere (Figure 6.8B)[173–177]. A series of studies were carefully performed tailoring the model compound in order to give insights on the mechanism and the role of the Cu_B center, which may provide an unique reaction pathway. The compound was immobilized on an edge-plane graphite electrode surface and investigated by rotating ring-disk electrode (RRDE) voltammetry. The potentials of both Fe(III/II)P and Cu(II/I) redox processes were about 100 mV shifted anodically compared to cytochrome-*c* oxidase, probably because of the more hydrophobic environment provided by the apoprotein. H_2O and OH^- are expected to be the axial and fourth ligands of the $Fe^{III}P$ and Cu^{II} sites, because the redox potentials remained unchanged in the pH 4–9 range. The Cu(II) ion dissociates below pH $= 4$. The studies by RRDE voltammetry indicated that the catalyst is highly efficient, as attested by the linear Levich and Koutecky–Levich plots with slopes similar to that found for the well-known tetraelectronic reaction at Pt electrode. Furthermore, the ring current was negligible and the limiting current independent of the pH in that range.

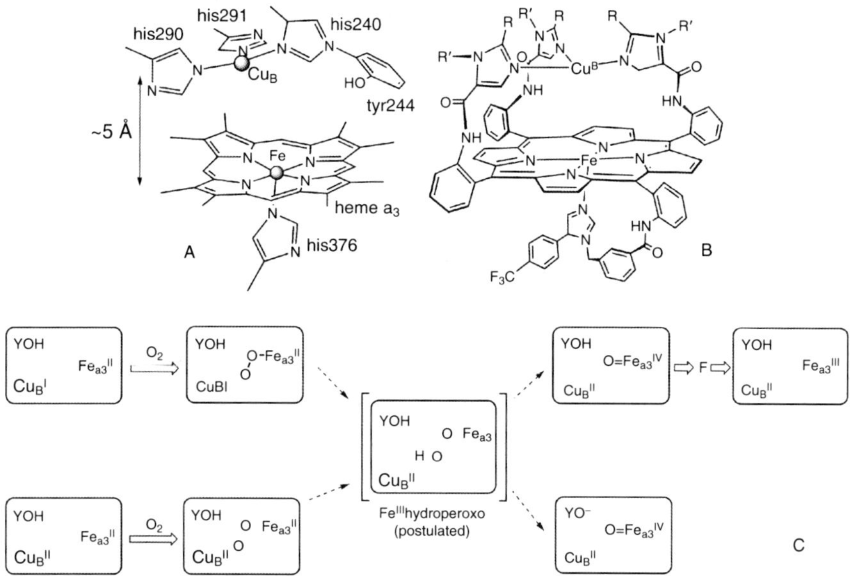

Figure 6.8. (A) Structure of cytochrome-*c* oxidase and (B) synthetic model compound; (C) mechanism proposed for O_2 reduction by the fully reduced and mixed valence cytochrome-*c* oxidase. The rectangle represents the heme-*a*3/CuB active site; YOH is a tyrosine residuc; F is an intermediate.

The amount of peroxide and superoxide produced in that reaction was estimated using scavengers. The comparison with the Cu-free complexes suggested that the copper site does not influence the activity, mechanism, or stability at mass transport limited conditions, but is essential under diffusion-limited electron flux. In conclusion, the Cu_B site seems to behave simply as an electron storage site in the first case and may not provide any special O_2 reduction pathway. However, it seems to influence the activity by nonredox mechanisms suppressing the superoxide-release at steady-state conditions, accelerating the substrate binding and minimizing homolytic O–O bond cleavage in the reduction of H_2O_2[173–175, 177]. Studies were also carried out with the model compound incorporated in lipid membranes and as multilayer films[176, 178].

At the beginning of the 1980s, Collman *et al.* also reported that a cofacial bis-porphyrinato cobalt complex (CoFTF4) adsorbed in pyrolitic graphite electrodes was able to promote the reduction of O_2 to H_2O[116–118, 179]. This finding was followed by a series of investigations that led to the conclusion that the bis-coordination to the cobalt porphyrin sites is essential for the activation of CoPs to the tetraelectronic process (Figure 6.7B). Note that the catalytic active species is the reduced $Co^{II}Co^{II}$ porphyrin, which transfers two electrons to O_2 generating a μ-peroxo intermediate, which is subsequently protonated. Then, two more electrons and three protons are transferred generating two molecules of water and regenerating the starting $Co^{III}Co^{III}$ species.

However, electronic factors and/or the lack of sufficient number of electrons could be limiting the tetraelectronic process, and the study of some peripherally metallated pyridylporphyrins was started. The first goals were the Co-**33** and Fe-**33** complexes. Unfortunately the formation of an insoluble polymeric by-product hindered their preparation, but the properties of the free-base was reported[112,121]. Subsequently, the [Ru(edta)]derivatives (**34**) were obtained and the catalytic properties evaluated in acetate buffer[113,114,133]. The Fe and Co derivatives were shown to be more efficient than the corresponding M-TMPyP species, because the four ruthenium complexes act as reservoirs and charge-transfer relays for electron exchange with the electrode surface[114]. This was confirmed by reacting the totally reduced $[Co^{II}TPyP\{Ru^{II}(edta)\}_4]$ complex and air or O_2 saturated solutions in a stopped-flow apparatus. Both metal sites were shown to be oxidized to M(III) within the mixing time-scale, while the reaction of $[Ru^{II}(edta)py]$ complex was much slower. The rapid ET from the peripheral complexes did not change the reduction mechanism which remained essentially bi-electronic, in solution or adsorbed on pyrolitic graphite electrodes[119]. This clearly indicated that factors other than electron availability are responsible for the activation of CoPs toward the tetraelectronic process.

New insights came from the work of Shi and Anson, who succeeded in the preparation of Co-**33** directly in Nafion films adsorbed on graphite electrodes and explored its electrocatalytic properties[119]. The CoTPyP was incorporated in the Nafion film by ion-exchange in acidic solution and then reacted with electrochemically generated $[Ru^{II}(NH_3)_5(H_2O)]$. The formation of the desired complex was followed by the shift of the $[Ru^{III/II}(NH_3)_5L]$ redox potential from about –0.25 to 0.1 V, due to the substitution of the water ligand for the pyridyl group of CoTPyP. In contrast with Co-**34**, a very strong catalytic peak was observed at 0.0 V vs. SCE and assigned to the reduction of O_2 to H_2O. The mechanism was confirmed by RRDE voltammetry and the slope of the Koutecky–Levich plot. The influence of the number of coordinated ruthenium complexes (1–4) on the catalytic properties was then investigated, using Nafion films loaded with the *meso*-phenylpyridylporphyrin derivatives instead of Co-**33**. As a conclusion, they found out that at least three peripheral complexes were necessary to activate the CoP center[138]. Analogous results were obtained when a series of cobalt *meso*-(4-cyanophenyl)porphyrins were used instead of the pyridylporphyrins (Figure 6.9A and B)[180]. Interestingly, when the phenyl rings were forced to be almost orthogonal to the porphyrin ring, due to steric effects induced by two methyl groups at the 2,6-positions, the O_2 reduction changed back to the bi-electronic mechanism (Figure 6.9C and D)[181]. A similar result was obtained for the cobalt *meso*-(tetra-3-cyanophenyl)porphyrin derivative. Consequently, electronic effects induced by the peripheral complexes are playing a major role and a strong contribution of back-bonding interactions to the activation of CoTPyP was proposed[181,182].

The electrocatalytic activity of CoTCP films was evaluated by RDE and RRDE voltammetry and eqs. 6.1 and 6.2,

$$I_L = 0.620\, n\ F\ A\ D_0^{2/3} \omega^{1/2} \nu^{-1/6} C_0 \tag{6.1}$$

$$I_{KL}^{-1} = (nFA\, k_f \Gamma C_0)^{-1} + (0.620 nFAD_0^{2/3} \omega^{1/2} \nu^{-1/6} C_0)^{-1} = I_k^{-1} + I_L^{-1} \tag{6.2}$$

where $n = 4$, $F = 96,484.6\,C\ mol^{-1}$, $A = 0.196\,cm^2$, $D_0 = 2.0 \times 10^{-5}\,cm^2\,s^{-1}$, $\omega =$ angular frequency ($2\pi \times$ rotation rate in Hz), $\nu = 0.01\,cm^2\,s^{-1}$, $\Gamma =$ surface

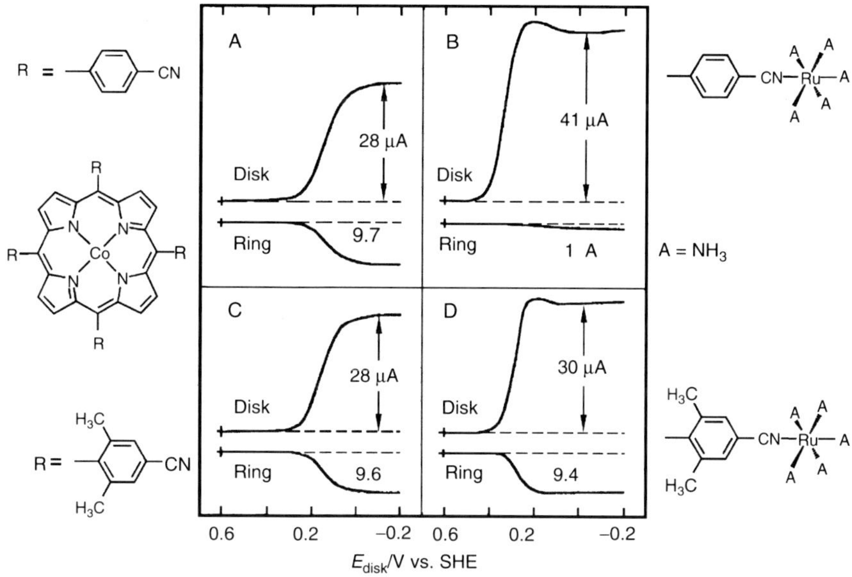

Figure 6.9. RRDE voltammograms for the (A) *meso*-tetra (4-cyanophenyl) porphynato cobalt (II) and (B) the same species coordinated to $[Ru(NH_3)_5]^{2+}$ complexes in air saturated 0.5 M $HClO_4$/0.5 M NH_4PF_6 solution (ω = 100 rpm, υ = 5 mV s^{-}, $\Gamma \sim 2.0 \times 10^{-9}$ mol cm^{-2}). The comparative effect of the insertion of methyl groups at 2,6-positions keeping the phenyl rings orthogonal to the prophyrin ring is shown in (C) and (D), respectively.

concentration of active sites, k_f = heterogeneous charge-transfer rate constant and $C_0 = 2.4 \times 10^{-4}$ M for an air saturated aqueous solution; I_k and I_L are the kinetic and mass-transfer limited currents, respectively. The linear plots virtually passing through the origin (Figure 6.10D, 100–4,000 rpm) indicated that there are no charge or mass-transport limitations for the reaction, and that the heterogeneous rate constant should be quite high. Furthermore, the slopes of the Levich and Koutecky–Levich plots were consistent with a tetraelectronic process, while the peripheral ruthenium clusters were found to be inactive, since ZnTCP modified electrodes showed no catalytic current at all[142]. Hence, the high catalytic activity of CoTCP was ascribed to the electronically induced activation of the cobalt porphyrin center by the μ_3-ORu_3 complexes. However, there was no clear evidence to rule out the bis-coordination mechanism (Figure 6.7B), because porphyrins tend to aggregate forming π-stacks. In order to shed some more light on this question, the properties of electrostatic assembled films of CoTCP and M-TPPS (where M = $2H^+$ or Zn^{2+}) were evaluated varying the pH and the electrode material[183].

The electrostatic assembled porphyrin films were obtained by piling-up alternating layers of ion-paired CoTCP and M-TPPS (scheme in Figure 6.12), where M = $2H^+$ or Zn^{2+}, as described in the following section. The M-TPPS species is tetra-anionic and has a suitable structure to stack face-to-face with CoTCP, sterically hindering the bis-coordination mechanism. Sigmoidal RRDE curves, with well-defined limiting currents and CVs with intense reduction waves

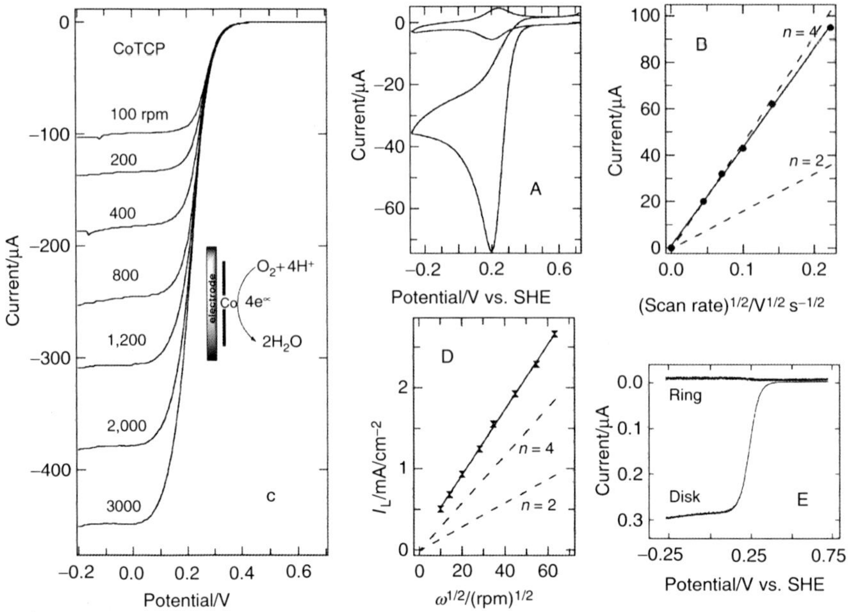

Figure 6.10. (A) CVs at $\upsilon = 20\,\mathrm{mV\ s^{-1}}$ in Ar and air saturated solutions; (B) i_{cat} vs. $\upsilon^{1/2}$ plot and linear fitting for the experimental data (circles and solid line) compared with the calculated curves considering for $ne^- = 2$ and 4 (dashed lines); (C) RDE voltammograms of a CoTCP modified glassy carbon electrode in air saturated solution; (D) Levich plot of the calculated and experimental data; (E) RRDE curves using a glassy carbon disk and Pt ring electrode at $\omega = 1,000\,\mathrm{rpm}$ ($E_{ring} = 1.22\,\mathrm{V}$, $\upsilon = 10\,\mathrm{mV\ s^{-1}}$). Electrolyte: 0.50 M KNO_3 solution, acetate buffer 0.05 M, pH = 4.7, T = 25°C.

at 0.2 V, were obtained for the CoTCP/ZnTPPS and CoTCP/H_2TPPS electrostatic assembled films, in O_2(sat.) and air(sat.) solutions. Once again, linear Levich and Koutecky–Levich plots were obtained and the slopes almost matched the one expected for a four-electron process. However, the curvature at higher rotation rates (above 1,000 rpm) suggests that the CoTCP/H_2TPPS catalyzed process may be limited by charge diffusion. The four-electron reduction mechanism was observed in the $3 < \mathrm{pH} < 5.5$ range, but the two-electron process became increasingly important up to pH = 7. At this pH, the limiting catalytic current became independent of the rotation rate, indicating that the catalytic reaction is not controlled by mass transfer anymore. Rate constants in the $10^6\,\mathrm{M^{-1}\,s^{-1}}$ and turnover frequencies in the $9 \times 10^3\,\mathrm{s^{-1}}$ range, higher than for cytochrome-*c* oxidase, were determined. The tetraelectronic pathway remained as the predominant process even when a gold electrode was used, in contrast with the face-to-face cobalt porphyrin[184,185]. A change in the reduction mechanism to bi-electronic was also observed for Co-**33** when $[Ru^{II}(NH_3)_6]$ was used as reductant and assigned to the relatively slow delivery of multiple electrons by a diffusion-controlled process[182]. Similar tendency was reported by Collman et at for the biomimetic FeP/Cu catalyst[173,177].

In conclusion, FeP/Cu and peripherally metallated CoTPyPs such as CoTCP were found to be excellent catalysts for the four-electron reduction of O_2 to H_2O in slightly acidic or neutral aqueous solutions. The first one was conceived as a truly biomimetic compound while the second are functional models. It is impressive how closely Collman *et al.*[177,186] were able to reproduce the structure and activity of the natural enzyme, making fundamental contributions for the understanding of the reduction mechanism and role of Cu_B site. On the other hand, the activation of Co porphyrins by coordination of peripheral transition-metal complexes depends on induced electronic effects in addition to their ability to function as electron reservoirs and relays, rapidly delivering multiple electrons to the catalytic active sites.

4. Electrochemical and Photoelectrochemical Properties of Porphyrin Films

The properties of molecular species can be transferred to electrodes by immobilizing them on the surface using chemical or physical interactions, as shown in the previous section[187–190]. Such modified electrodes can be used as convenient photoelectrochemical interfaces for the preparation of molecular devices such as chemical sensors, photoelectrochemical cells, light emitting, and electrochromic devices, etc. Accordingly, a detailed knowledge of their catalytic, conduction, and photochemical properties is demanded.

Many techniques can be employed to obtain films, but only those using low-energy conditions are adequate for molecular materials. Among them, the self-assembly[191,192] and the Langmuir–Blodget[193] methods are commonly employed to obtain molecular films constituted by mono- or multiple layers with excellent homogeneity, thickness, and composition control. The self-assembly method is generally used for the preparation of monolayers, for example by covalently bonding thiols on gold surface. But, other interactions such as electrostatic, hydrogen bonding, and transition metal coordination oriented assembly have been developed, for the preparation of multilayered films. Other broadly used techniques are spin-coating, dip-coating, and electropolymerization[157,194–202], all allowing good control on the thickness and the amount of deposited material but at expense of a lower homogeneity. Nevertheless, they are less time demanding and simpler, being suitable for the preparation of modified electrodes.

The above discussion reminds us of the importance of molecular interactions on the preparation and properties of molecular materials. In fact, this process should be addressed in a much broader way considering the principles of hierarchical self-assembly[71], which states that the matter is organized according to the rules defined by the interaction strengths. Strong covalent bonds hold the atoms together generating molecular building-blocks, which can be assembled into discrete supramolecular entities by using relatively strong interactions, such as covalent, coordination or electrostatic, hydrogen bond, etc. Of course, such species can be assembled further generating higher order aggregates, since such assembly processes involve noncovalent interactions and take place at multiple distinct levels with specific interaction strengths.

4.1. Electrochemical Properties of M-TRP Films

4.1.1. Dip-Coated Films

Homogeneous films of [M-TRP]($CF_3SO_3^-$) can be easily deposited on glassy carbon or ITO substrates, by dip-coating from a methanolic solution. The transmission UV-Vis spectra generally exhibit a broadened and less intense Soret band, probably due to exciton coupling with neighboring porphyrin rings. This hypothesis is supported by the fact that such effect is virtually absent in the less intense Q bands.

The modified electrodes are electrochemically active and present a pair of sine-shaped waves at $E_{pc} = 0.92$ and $E_{pa} = 0.99\,V$ ($20\,mV\,s^{-1}$), whose intensities are directly proportional to the scan rate, as expected for redox species immobilized on the electrode surface[203]. A linear Cottrell plot has been obtained for the chronoamperometry data but with a negative deviation at longer times, where the charge propagation is not diffusion controlled. The estimated charge-diffusion constant $D_{CT}^{1/2}C = 1.0 \times 10^{-8}\,mol\,cm^{-2}\,s^{-1/2}$ is similar to that found for poly-[Ru(vbipy)$_3$] films in acetonitrile, confirming the high conductivity of the porphyrin material. The conduction mechanism has been probed registering the CVs of the modified electrodes in the presence of ferrocyanide[203]. In both cases, no wave associated with the dissolved species has been found at its normal potential; but a sharp oxidation peak has appeared at the onset of the Ru(III/II) wave (Figure 6.11A), indicating that the conduction is taking place by electron-hopping mechanism. In addition, those results are consistent with the absence of holes, cracks, and other types of imperfections that would allow the diffusion of ferrocyanide and the direct electron exchange with the electrode surface. Similar behavior was also observed for [Ru^{II}(edta)(H_2O)], ascorbic acid and NADH indicating that the size of the channels is relatively small[122, 204].

Considering the electrochemical properties mentioned above, it was proposed the application of those modified electrodes as amperometric detectors. Two situations are possible: (a) the metalloporphyrin is the electrocatalytic active site or (b) the peripheral complexes act as redox catalyst by outer sphere mechanism. The possibility of having two independently accessible active sites was explored using O_2 (previous section) and reducing substrates such as ascorbic acid, NADH, dopamine, acetaminophen[205], phenol, nitrite, and sulfite. However, only in the first case the metalloporphyrin center was the active site, because of the good matching between the redox processes involving the peripheral and CoP sites, in addition to synergistic induced electronic effects.

All species mentioned above are substrates of biological and environmental interest. For example, nitrite is usually employed for color enhancement and preservation of industrialized meat, but can promote the production of nitrosamines, which are carcinogenic substances. Sulfite is also commonly used as food preservative to prevent browning. However about 1% of the population is sensitive and its concentration should be controlled at few ppm level. It is also used in large amounts in the pulp and cellulose industry as delignifying agent which has cause serious environmental concerns. Various analytical methods can be employed but the electrochemical detection can be faster, sensitive, and convenient. Nevertheless, the direct electrochemical detection using bare electrodes is hampered by adsorption problems and slow electrode reactions. The [M-TRP]$(CF_3SO_3)_4$ modified electrodes were shown to be much more resistant to surface poisoning and enhanced electron-transfer rates to those substrates. Such was the case for NADH, which presents a broad oxidation wave at 0.74 V in a bare GC electrode, but a sharp and catodically shifted (100 mV) wave at the modified one (Figure 6.11C and D)[204]. A more striking evidence of the catalytic activity of M-TRP films was obtained for sulfite, a species which shows almost no signal in a bare GC electrode but yields a relatively sharp peak around 1.0 V in a modified one[122]. CV has a low sensibility ($\sim$ 0.1 mM) and flow-injection analysis (FIA) was considered for analytical applications. In this case, the limit of detection is $\sim$ 0.1 μM; however, mechanically stable and less soluble materials are required in this case, because of the more stringent hydrodynamic conditions.

4.1.2. Electrostatic Assembled Films (EAF)

The tetra-cationic species $[M\text{-TRP}]^{4+}$ interacts strongly with the tetra-anionic $[M'\text{-TPPS}]^{4-}$ molecules by electrostatic and π-interactions. This ion-pairing process is directed by molecular recognition as can be inferred from the structural and electric charge complementarity of those species (Figure 6.12)[123,125,130]. A face-to-face dimer is initially formed when they are mixed in dilute methanol solution, according to the titration and spectral data shown in Figure 6.12A. Note that the Soret band is blue-shifted and remain very sharp, as expected for a strong exciton coupling between parallel transition dipole moments. The ion-pairs aggregate further generating an amorphous precipitate, which is insoluble in water and many organic solvents. This material can be grown layer-by-layer in a controlled way, generating molecularly organized materials on substrates such as silicon wafers and ITO plates[124,148,205,206]. The procedure is very simple and based on the deposition of successive monolayers of the tetra-cationic and tetra-anionic porphyrins, taking advantage of the strong interactions between them, the low solubility of the ion-paired material and the solubility of the starting components. Initially, a layer of [M-TRP]$(CF_3SO_3)_4$ is deposited by dipping the substrate in a 0.1 mM methanol solution and allowing it to dry in air. Next, it is dipped in an aqueous Na_4[M′-TPPS] solution and washed with water to obtain a monolayer of the ion-paired porphyrins. This process can be repeated several times to get multilayered films in a reproducible way, as can be inferred from the linear increase of the absorbance during the deposition process (Figure 6.12B). Surprisingly, the thickness of the film on a silicon wafer, measured by AFM, also showed to be a linear function of the number of bilayers, and the thickness of a single electrostatic assembled ZnTRP/TPPS bilayer was determined to be 12.7 Å from the slope. This implies that the porphyrin molecules are laying flat on the surface and the TPPS is ion-pairing face-to-face like in solution, such that the thickness is being essentially defined by ZnTRP peripheral ruthenium

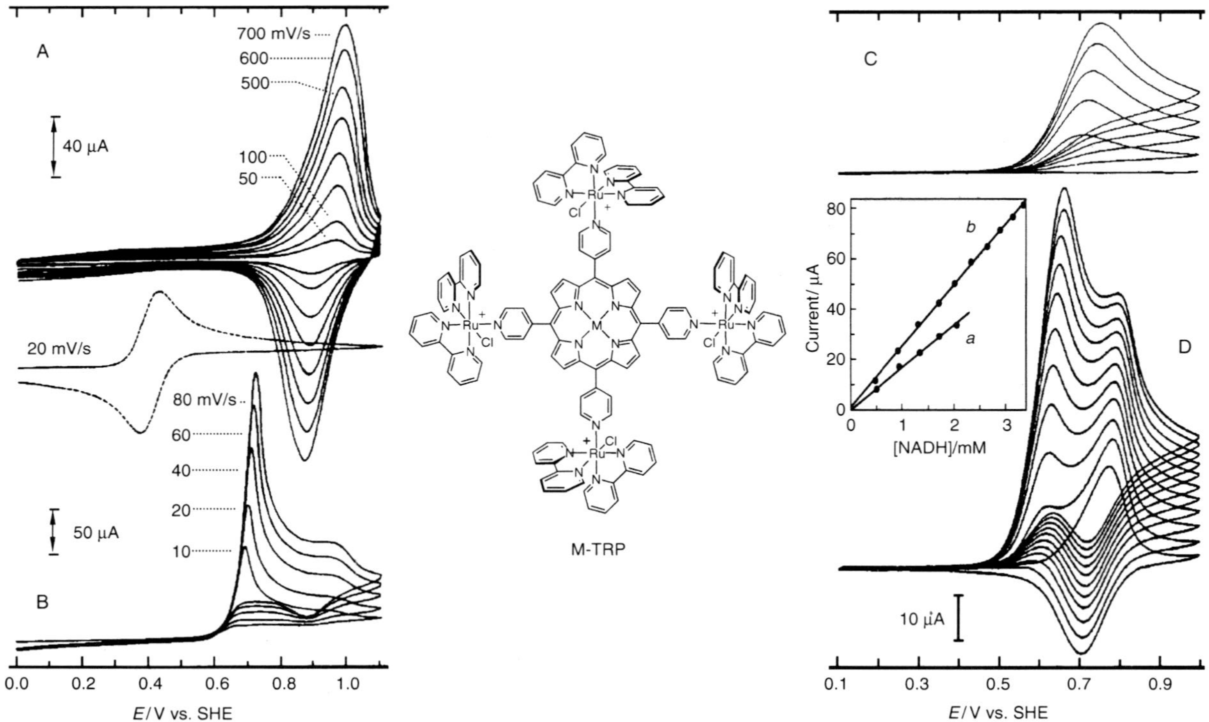

Figure 6.11. CVs of a glassy carbon electrode modified with $[NiTRP](CF_3SO_3)_4$ by dip-coating, $\Gamma \sim 1 \times 10^{-9}$ mol cm^{-2}, in (A) neat electrolyte solution and (B) 5.37 mM $K_4[Fe(CN)_6]$ solution. The CV of ferrocyanide with a bare glassy carbon electrode is also shown (dashed line). CVs of a NADH aqueous solution using a (C) bare GC electrode (0.5–2.0 mM range) and (D) $[CoTRP](CF_3SO_3)_5$ modified electrode, $\Gamma = 1.1 \times 10^{-9}$ mol cm^{-2} (0.5–3.5 mM range), $\nu = 100$ mV s^{-1} (inset: i_{peak} vs. [NADH]). Electrolyte 0.25 M $LiCF_3SO_3$, acetate buffer 0.05 M, pH = 4.7. The structure of M-TRP is shown at the center.

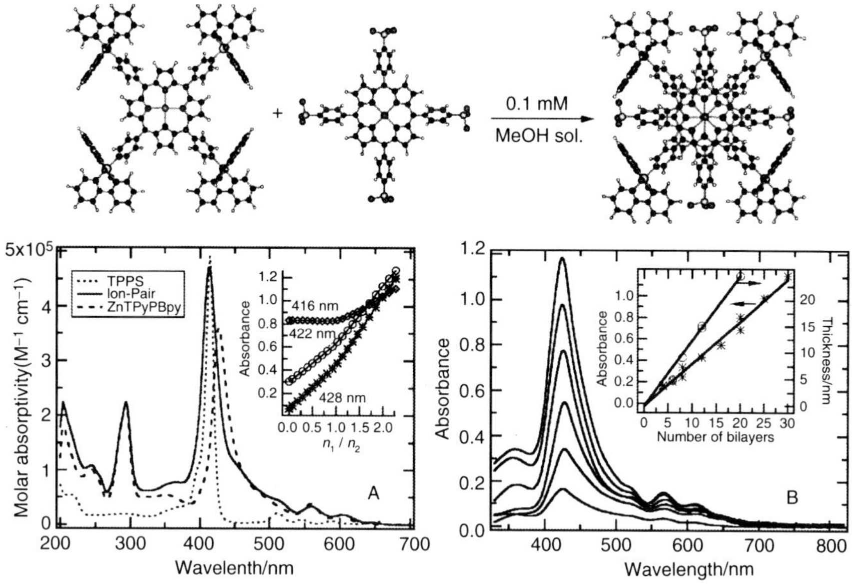

Figure 6.12. (A) Spectra of Na_4TPPS, $[ZnTRP](CF_3SO_3^-)_4$ and the respective ion-pair in methanol; inset: absorbance vs. molar ratio. (B) Growth of an electrostatic assembled Zn-TRP/TPPS film on ITO(4, 8, 16, 20, 25, and 30 bilayers) followed by UV-Vis spectroscopy; inset: plots of absorbance at 424 nm and thickness vs. number of bilayers. The thickness of films deposited on silicon wafer was measured by AFM. The computer simulated structures of TRP, TPPS, and corresponding ion-pair is shown at the top.

complexes. The formation of a 1:1 material was confirmed by XPS spectroscopy of ZnTRP/CuTPPS samples[123].

The electrochemical behavior of the electrostatic assembled films is essentially identical to that of pure M-TRP, exhibiting a pair of sine-shaped waves at $E_{pc} = 0.90$ and $E_{pa} = 0.98$ V. The surface concentration per bilayer was estimated as 7.5×10^{-11} mol cm^{-2}, being slightly superior to the value found for TMPyP on Au(111) surface[208], reflecting the roughness of the ITO surface. The charge-transfer properties of a multibilayer EAF was studied by RDE voltammetry and electrochemical impedance spectroscopy[207]. The RDE experiments were carried out using CoTRP/ZnTPPS films and solutions of $K_4[Fe(CN)_6]$. Linear Levich plots with increasingly higher linear coefficients were obtained in the 0.1–10 mM range, as expected from Equation (6.1). The charge diffusion through the film was found to be fast by chronoamperometry ($D_{ct} = 3 \times 10^{-11}$ cm^2 s^{-1}). In fact, no Ru(III) complex reduction wave was superimposed to the typical sigmoidal RDE voltammogram in 10 mM ferrocyanide solution, indicating that they are being rapidly reduced. The oxidation of NO_2^- in the same modified electrode is slower, showing a typical saturation behavior in the Levich plots. However, the Koutecky–Levich plots were straight lines and the linear and angular coefficients increased proportionally as a function of the substrate concentration in solution. The limiting step was found to be the heterogeneous charge transfer in

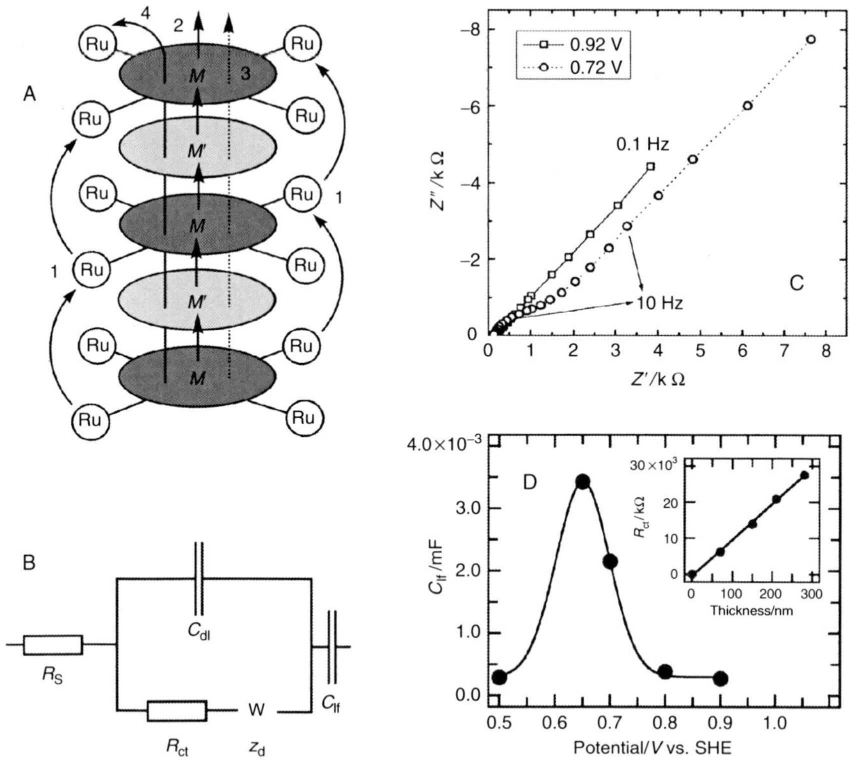

Figure 6.13. (A) Conduction mechanisms in electrostatic assembled π-stacked materials: (1) electron hopping involving the ruthenium and (2) M and M′ sites, (3) electronic conduction through the porphyrin π-stacks, and (4) hole transfer to a Ru(II) complex followed by electronic conduction through the porphyrin π-stack; (B) Randles type equivalent circuit proposed for the films; (C) Nyquist diagrams of a modified electtrode ($\Gamma = 3.5 \times 10^{-8}$ mol cm^{-2}) at 0.72 and 0.92 V; and (D) low-frequency capacitance vs. potential and charge-transfer resistance vs. thickness (inset), for electrostatic assembled CoTRP/ZnTPPS films.

the film/solution interface ($k = 2.7 \times 10^3\ \mathrm{M^{-1}\ s^{-1}}$), since the conductivity of the film itself is quite high as shown below.

The impedance spectra were analyzed considering that the material is constituted by stacks of alternating ruthenated and sulfonated porphyrins[207]. In this case, the following conduction mechanisms are possible: (1) hopping through the peripheral ruthenium complexes; (2) or involving the transition metal ions coordinated to the porphyrin ring; (3) electronic conduction through the porphyrin π-stacks; and (4) hole transfer from the electrode to the ruthenium complex and propagation through the porphyrin π-stacks (Figure 6.13A). The impedance experiments carried out in the 0.72–1.12 V vs. SHE range indicated that the conductivity of the film is relatively high even at the lower limit, increasing as the potential approaches the $E_{1/2}$ of the Ru(III/II) process (Figure 6.13C). In fact, the shape of the voltammetric curve was reproduced by the low-frequency capacitance vs.

potential plot (Figure 6.13D), confirming the predominance of a redox conduction mechanism. In addition, the charge-transfer resistance of the film increased linearly as a function of the thickness, like in conventional conductors.

An interesting case is provided by the peripherally metallated tetra-3,4-pyridylporphyrazine (TRPyPz) derivative (Figure 6.14), which exhibits an enhanced electronic coupling between the central and the peripherally bond $[Ru(bpy)_2Cl]$ complexes. That species formed a face-to-face ion-pair but with tetrasulfonated Cu(II)-phthalocyanine (CuTSPc), which formed a more stable complex than M-TPPS[209]. Adherent films of TRPyPz/CuTSPc were obtained by electrostatic assembly, showing a pair of bell-shaped waves at 0.95 V. However, the Ru(III/II) process was preceded by the oxidation of the porphyrazine ring around 0.8 V. This means that a second conduction pathway is available for charge transfer through this material. In fact, the CVs of $[Fe^{II}(CN)_5(Mepz)]^{2-}$ ($E_{1/2} = 0.78$ V, Mepz = *N*-methylpyrazinium) with the modified electrode was very similar to that obtained with a bare electrode. Hence the film behaves as a metallic conductor when the material is doped (Figure 6.14A).

The spectrum at 0.60 V exhibited the beginning of a semicircle in the high-frequency region, followed by a diffusion controlled and saturation regions (inset Fig. 6.14B). This behavior was interpreted based on a Randles type circuit modified with a low-frequency capacitor (Figure 6.13B). The conduction at that potential is governed by the π-stacks and the moderate resistivity ($\sim 500\Omega$) reflects the low doping level of the material. But, two semicircles assigned to a capacitive and an inductive contribution, followed by saturation at low frequencies, were observed at 0.95 V (Figure 6.14B). The inductive effect was represented by a component in series and explained based on mechanism 4 (Figure 6.13A). This is feasible since the porphyrazine π-stacks can be oxidized by the ruthenium complexes, which is in contact with the electrode surface mediating the ET process. An enhanced photoelectrochemical response was observed for the EAFs in O_2 saturated solution, in contrast with the pure TRPyPz films (Figure 6.14C)[209].

The electrocatalytic behavior of electrodes modified with porphyrin EAFs was probed using NO_2^-, SO_3^{2-}, phenol, etc. as substrates by CV, RDE, and FIA. The charge transport is very fast as shown above, hence the limiting step is the heterogeneous ET to the substrate in solution, except in the case of thick films. Generally, the electrocatalytic behavior was similar to that exhibited by pure M-TRP modified electrodes, but the electrostatic assembled films were more stable and durable, giving reproducible FIA peaks for more than a hundred of successive analysis (Figure 6.15A and B)[206]. In addition, they were less prone to surface passivation even in the case of phenol that has a strong tendency to form a polymeric insulating material when oxidized. No significant change was observed in the response for successive measurements with CoTRP/ZnTPPS modified electrodes using a 1×10^{-5} M solution, while a steady decrease was evident for the bare one. A higher current intensity would be expected for a 1×10^{-4} M solution but a decrease of about 40% was observed instead, implying that the surface of both electrodes was partially covered and inactivated when the first sample reached the electrode surface. Although the passivation rate remained fast for the bare electrode, the subsequent poisoning of the modified electrode

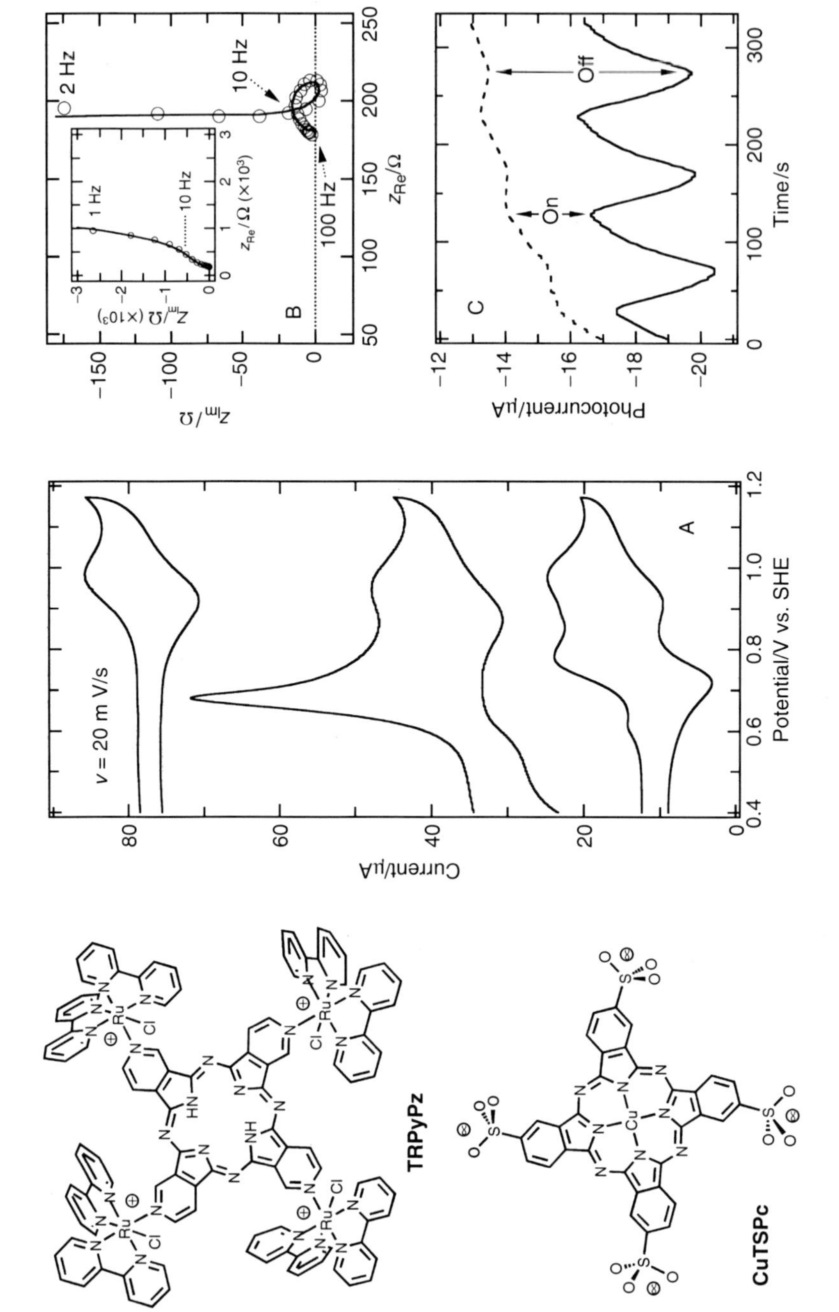

Figure 6.14. Structures of CuTSPc and TRPyPz; (A) CV of a TRPyPz/CuTSPc modified electrode ($\Gamma = 3.8 \times 10^{-10}$ mol cm^{-2}) in pure electrolyte (top), 5 mM $[Fe(CN)_5(NH_3)]$(center) and $[Fe(CN)_5(Mepz)]$ solution; Nyquist diagram for a $\Gamma = 2.3 \times 10^{-10}$ mol cm^{-2} film on ITO at 0.95 and 0.60 V (inset); and photoelectrochemical response of TRPyPz(----) and TRPyPz/CuTSPc(—,$\Gamma = 7.5 \times 10^{-10}$ mol cm^{-2}) in O_2 saturated solution. Electrolyte: 0.10 M $LiCF_3SO_3$ aqueous solution.

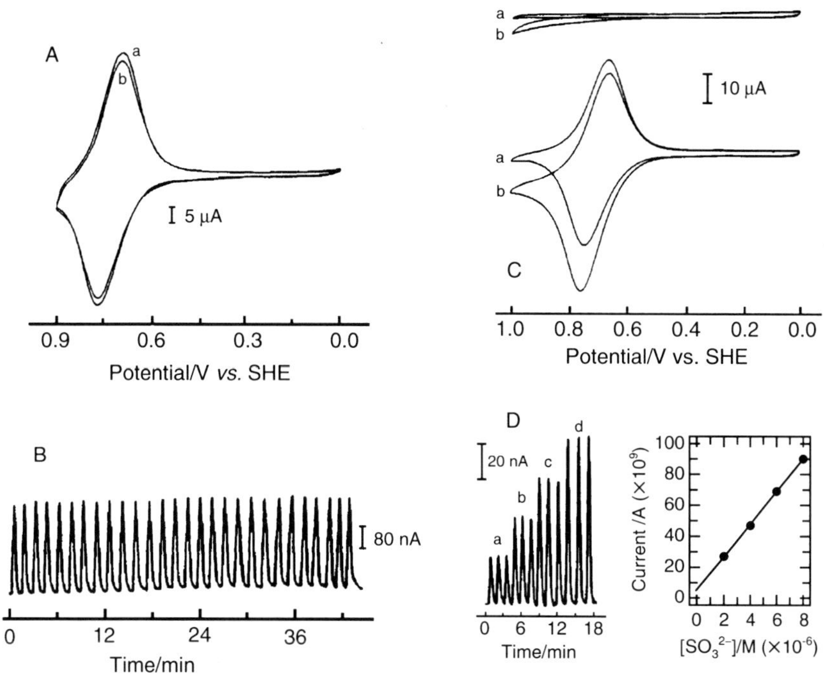

Figure 6.15. (A) CVs at $100\,mV\,s^{-1}$ of a CoTRP/ZnTPPS modified electrode before (a) and after (b) a series of successive analysis of a $10\,\mu M$ NO_2^- solution at 0.90 V, shown in (B); (C) CVs of a bare (top) and modified electrodes in the absence (a) and presence of SO_3^{2-} (b), and (D) a typical FIA response and respective i vs. $[SO_3^{2-}]$ plot.

was much slower. The above-mentioned characteristics are interesting for analytical purposes, such that FIA detectors were devised for analysis of sulfite and nitrite/nitrate and tested for real samples.

For example, small amounts of sulfite are normally added to wine as preservative. The Monier–Williams is the standard analytical method but was substituted by the iodometric method in the industry, because it is too laborious and time consuming for practical purposes. However, colored samples impose serious problems in the visualization of the endpoint and the measurements become highly dependent on the operator acuity. In order to minimize the effect of interferants keeping the high sensitivity of FIA, the samples were injected into a 2 M H_2SO_4 solution and the SO_2 gas released in the reaction separated and quantified in a cell based on CoTRP/ZnTPPS EAFs (Figure 6.15C and D). The separation was carried out by diffusing the gas through a teflon membrane and collecting with a 0.2 M $NaClO_4$ solution (acetate buffer, pH $= 4.7$) flowing in counter-current[210]. Reproducible and consistent results were obtained using this gas-diffusion system in a fraction of the time necessary for the iodometric analysis ($\sim$ 30 samples/h, Table 6.6). The detector showed to be sensitive (μM range) and reasonably durable, showing only $\sim$ 14% decrease in the signal after 5 days of continuous use. A similar arrangement was used in the analysis of nitrite and

Table 6.6. Comparitive Free and Total SO_2 Content in Wine Determined by Iodometric and FIA Method. Adapted with Permission from Ref [210].

Sample	Free SO_2 (mg L^{-1})		Total SO_2 (mg L^{-1})	
	i	Iodometric	*i*	Iodometric
Dry white wine	16.8	18.8	62.6	60.0
Sweet white wine	44.7	46.4	70.9	69.9
Dry read wine	12.0	12.9	83.7	82.1

Table 6.7. Comparative Nitrite and Nitrate Concentrations in Saliva, Mineral water, and Cured Meat Determined by Spectrophotometric and FIA Methods. Adapted with Permission from Ref [211].

Sample	Nitrite (μmol L^{-1})		Nitrate (μmol L^{-1})	
	i	$h\upsilon$	*i*	$h\upsilon$
Saliva	18.9	19.4	45.5	46.6
Min.water 1	–	–	51.5	53.6
Min.water 2	–	–	139.1	138.3
Min.water 3	–	–	253.5	250.4
Min.water 4	–	–	117.0	113.1
Min.water 5	–	–	318.9	316.8
Cured meat	1.3*	1.4*	12.8*	12.2*

* mmol kg^{-1}.

nitrate in mineral water, saliva, and cured meat[211]. The nitrite was directly analyzed but nitrate had to be reduced in a copperized cadmium column, inserted before the gas-diffusion cell. Results in good agreement with the spectrophotometric Griess method were obtained (Table 6.7).

Recently, a MnTPyP species with four $[Ru^{II}(bpy)_2(OH_2)]$ complexes coordinated to the pyridyl-substituents were synthesized and characterized[129]. The oxidation of the peripheral aqua-complexes led to formation of $[O=Ru^{IV}(bpy)_2]$, a known oxygen-atom transfer catalyst. The films of this peripherally metallated porphyrin are too soluble, but can be stabilized by anion-exchange with $[Fe(CN)_6]$ or M-TPPS. The oxidation of nitrite occurred at 1.0 V and was catalyzed by that high-valent species.

4.1.3. Polymeric Films

The peripherally polymetallated porphyrins are interesting materials for the preparation of amperometric detectors and an excellent control on the deposition process is possible by the electrostatic assembly technique, as shown above. A further improvement in stability and deposition control can be obtained by making them electropolymerizable. This can be done by introducing groups such as pyrrol, thiophene, aniline, phenol, aldehyde, etc., in the molecular structure. In particular, 5-chlorophenanthroline has been chosen because it is a ligand similar to bpy and is polymerizable by a reductive mechanism, as reported by Fussa-Rydel *et al.*[212] The reduction of the 5-Clphen ligand is followed by the elimination of a Cl^- ion generating a phenanthroline radical, which couples forming a dimer through the 5-position (Figure 6.16). Accordingly, ETRP (**38**) becomes electropolymerizable because each of the four peripheral [Ru(5-Clphen)$_2$Cl] complexes has two coupling sites, making possible the formation of an extended tridimensional network.

A typical voltammogram with a Ru(III/II) and two monoelectronic reduction waves of the porphyrin ring has been observed in the −1.1 to 1.2 V range, in dmf. However, a continuous increase of the voltammetric waves as a function of the number of scans has been observed when the potential was scanned until −1.35 V and the 5-Clphen ligand reduced, confirming the build-up of an electrochemically active film on the electrode surface[126]. The polymerization process is effective but the modified electrodes exhibit high resistances, leading to CVs with diffusion-controlled characteristics. This indicates that the charge transfer is limited by the slow movement of the electrolyte in the material. Hence, a new method has been employed for the preparation of ETRP films in order to improve the conductivity by controlling the packing of the ruthenated porphyrins. The monomer has been dip-coated from a methanolic solution, dried under vacuum, and electropolymerized in an acetonitrile solution. The monomer film dissolves slowly in this solvent and a couple of scans until −1.35 V is enough to perform the polymerization reaction. This method is much more effective than the direct deposition of the polymer from a solution, because it assures a very high local concentration of the reactive monomer facilitating the coupling of the phenanthroline radicals. Furthermore, the molecules should form the polymer keeping more or less the same structure of the dip-coated material, since their mobility is hindered by steric and intermolecular interactions. In fact, a highly reversible voltammetric behavior has been observed for this modified electrode (Figure 6.16)[126], exhibiting sine-shaped waves with average FWHM $= 130$ mV and $\Delta E = 50$ mV for the Ru(III/II) process (FWHM = Full Width at Half of the Maximum).

The electrochemistry of poly-ETRP in aqueous solution shows a reversible process at 1.0 V, but a broadening and shift of this Ru(III/II) waves have been observed when the surface concentration is higher than about 1×10^{-8} mol cm^{-2}, probably due to slow electrolyte diffusion in the film. The impedance spectroscopy is consistent with a redox conduction mechanism and the charge-transfer impedance reached a minimum ($\sim 30\Omega$) at 0.97 V, increasing sharply at lower and higher potentials (Figure 6.17A). An intense anodic wave is observed in the CVs, in the presence of nitrite, and the kinetics of this catalytic reaction was probed by RDE voltammetry, as a function of the substrate concentration (Figure 6.17B). Sigmoidal shaped voltammograms typical of a reversible process have been obtained, giving straight Levich plots with a small curvature at the highest

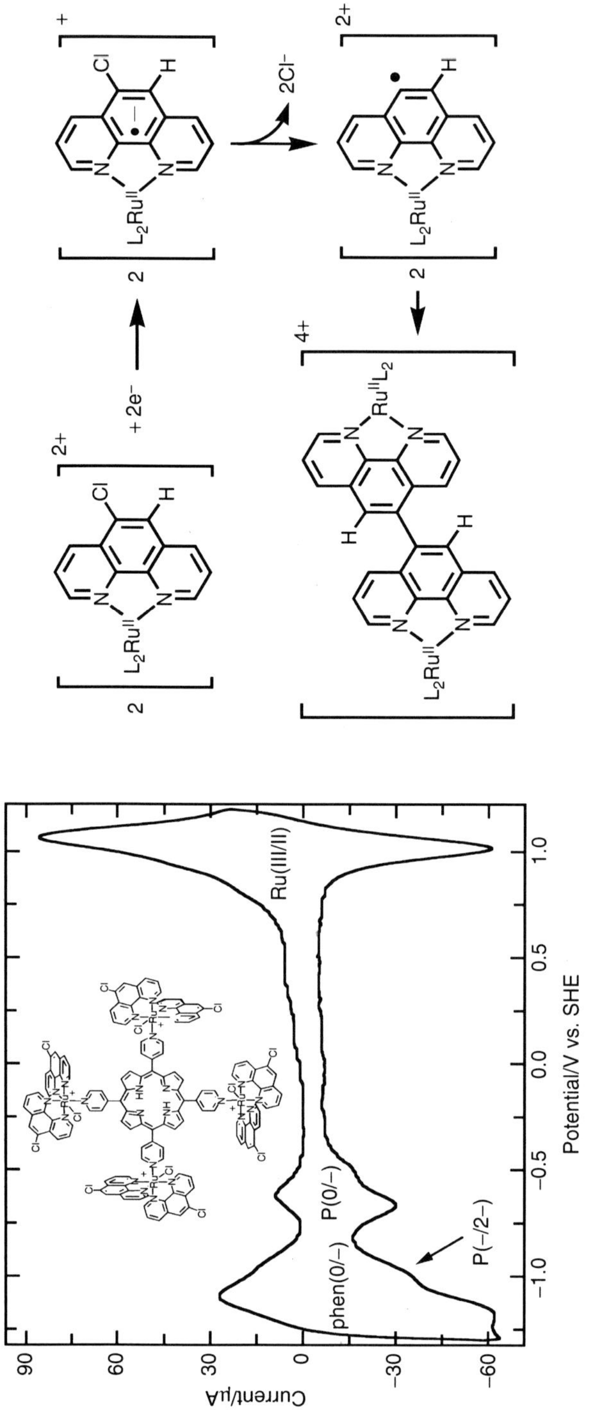

Figure 6.16. CV of an electrode modified with a polymeric ETRP film CH_3CN, 0.1 M $TEAClO_4$ solution (100 mV s^{-1}, inset: structure of the monomer). The electropolymerization reaction mechanism is shown on the right.

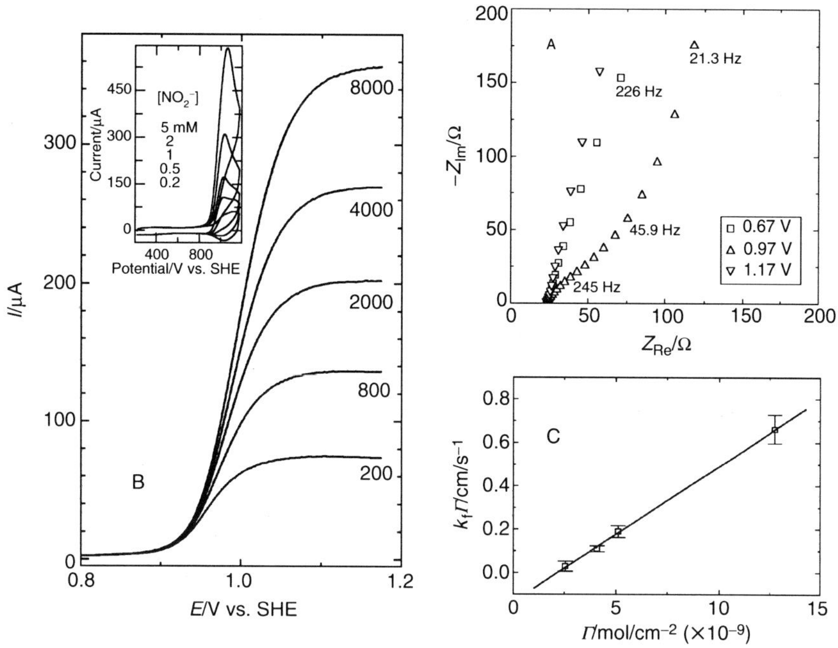

Figure 6.17. (A) Nyquist plots for a GC modified with poly-ETRP film ($\Gamma = 4 \times 10^{-9}$ mol cm^{-2}) in 5 mM KNO_3 solution; (B) RDE voltammograms in 3.85×10^{-4} M $NANO_2$ soltuion (inset CVs at increasing [NO_2^-]); and C) $k_f\Gamma$ vs. Γ, where $k_f\Gamma$ was obtained from the slope of the Koutecky–Levich plots as function of the substrate concentration.

concentrations (0.01–1 mM) and rotation rates (200–8,0000 rpm). The Koutecky–Levich plots are linear and $k_f\Gamma = 0.11$ cm s^{-1} can be obtained from the slope of the linear coefficient vs. [NO_2–] graphic. The heterogeneous rate constant can be estimated using the Γ values evaluated from the integrated charge under the voltammetric wave. Nevertheless, this gives an estimate of the lower limiting value since it considers that all the electrochemically active sites are effectively participating in the catalytic reaction. To overcome this problem, similar RDE experiments have been carried out varying the surface concentration of poly-ETRP. The slope of $k_f\Gamma$ vs. Γ plot gave $k_f = 6.2 \times 10^4$ M^{-1}s^{-1} (Figure 6.17C), a value 50% higher than that determined by the previous procedure. This is 30 times superior to that found for CoTRP/ZnTPPS modified electrodes, suggesting that the detection limit by FIA may possibly be lowered to the nanomolar range.

4.2. Photocatalytic Properties

The incidence of light on polymetallated porphyrin films, in the presence of O_2, leads to the flow of photocurrent (Figure 6.18). This behavior was observed for the pure M-TRP and the electrostatic assembled M-TRP/M′-TPPS films deposited on ITO[5,122,123,147,203,206,210]. In both cases the photoaction spectrum was

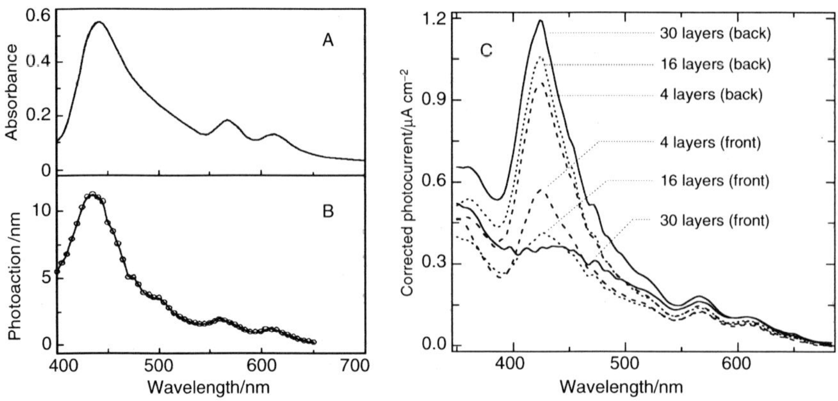

Figure 6.18. (A) UV-Vis spectrum of a ZnTRP film on Pt and (B) corresponding photoaction spectrum spectrum at 0.22 V; (C) photoaction spectra of an electrostatic assembled ZnTRP/TPPS film on ITO as a function of the thickness and illumination (front or back). Electrolyte: O_2 saturated $LiCF_3SO_3$ aqueous acetate buffer (pH = 4.7) solution.

very similar to the corresponding absorbance spectrum, suggesting that the porphyrin is the active species. The $Ru^{II}(d\pi) \rightarrow bpy(p\pi^*)$ MLCT band was also present in the photoaction spectrum, indicating that both sites are photocatalytic active. The mechanism involves the photoinduced reduction of dissolved O_2 and formation of the oxidized sensitizer, which is regenerated by the electrode. However, the photocurrent intensity is an inverse function of the thickness of the film, indicating that only few layers next to the electrode surface is contributing to the process. In fact, the remaining layers are absorbing the incoming light and diminishing the photoresponse.

Dip-coated TRPyPz films show a very low efficiency in comparison with TRPyPz/CuTSPc EAFs (Figure 6.14C). In this case, the photoresponse is more intense and a broad band at 700 nm and a shoulder at 500 nm, consistent with the Q and MLCT transitions, were observed in the action spectrum[209]. The difference has been assigned to the presence of an efficient conduction pathway involving the porphyrazine π-stacks, which makes possible a fast hole transfer to the electrode surface. A similar behavior is expected for the Prussian blue type films prepared with 3-TPyP coordinated to four $[Fe(CN)_5]^{3-}$ complexes[128].

The polymetallated porphyrins have been explored as sensitizers in Grätzel-type cells also, using I_3^-/I^- couple as redox electron-transport material to close the circuit[213,214]. In those compounds the peripheral ruthenium complexes act as antenna and can contribute to improve their photostability. The porphyrin radical-cation generated after injection of an electron to the TiO_2 conduction band ($\Delta G \sim 0.7$ V) is rapidly reduced back by ET from a ruthenium complex, which retains the hole until being regenerated by the supersensitizer. Considering that the electron injection from the excited sensitizer is much faster than the decay rate to the ground state, the efficiency of the device should be limited basically by the competition with the recombination reactions. However, many other factors seems to be contributing to the overall efficiency, since the ZnTPyP coordinated to

[Ru(phen)$_2$Cl] is more efficient than Zn-**35** and a 13% IPCE has been measured at the Soret band. Also, much lower efficiencies have been obtained for the free-base derivatives and assigned to the lack of a coordination point to the nanocrystalline TiO_2 surface. Even the solvent used for the adsorption of the sensitizer has been found to influence the IPCE (Incident Photon to Current Conversion Efficiency $= 1,250(J_{sc}/\lambda\phi)$, where J_{sc} is the short-circuit photocurrent density, λ the wavelength, and ϕ the photon flux).

5. Final Remarks

Both supramolecular chemistry and porphyrin chemistry have undergone a dramatic evolution in the last two decades, making possible the preparation of very sophisticated structures by exploiting covalent and noncovalent interactions. The chemistry of supramolecular porphyrins based on coordinative-assembly has also evolved rapidly, making possible the preparation of incredibly complex structures like the porphyrin macrorings, mimicking the light harvesting center of photosynthetic systems, and the porphyrin grids. All this has been accomplished by exploiting the metalloporphyrin axial sites and the coordinating groups attached to the periphery of that macrocycle. The bonding geometry, strengths, and lability/inertness of the transition metal complexes can be controlled to a much wider extent than in conventional organic compounds. In addition, they have characteristic redox, catalytic, electrocatalytic, and photochemical properties that can be used to introduce new functionalities or to tune the electronic properties of the materials. These points have been illustrated by the enhanced activity of MnTCP and CoTCP for the oxidation of organic compounds and reduction of O_2, respectively. In these cases, the peripheral μ_3-ORu_3 complexes activate the metalloporphyrin center in several ways: (a) by inducing electronic effects, (b) by acting as electron sources or drains, and (c) by acting as ET relays. The peripherally bound complexes seem to adjust automatically their redox state to the electrochemical potential of the environment, acting in a synergistic way. For example, in reducing conditions, the peripheral complexes are reduced enhancing the activity of the porphyrin and metalloporphyrin for reduction reactions and vice versa, tuning the electronic properties of the active site and concomitantly acting as electron sources or drains for multielectronic processes. Such unique characteristics are transferred to the molecular films, where they also act as electron-transfer relays for the rapid diffusion of electrons from the electrode surface to the electrocatalytic active sites. Nevertheless, the number of transition metal complexes that has been used for the assembly of well-characterized supramolecular systems is still insignificant, considering the variety and number of known compounds. The number of systems whose electrochemical, catalytic, photochemical, and conduction properties have been characterized is even meager, inspite of the significant potential uses of these materials. Hence, we foresee a surge of activity in a near future, exploiting the recent advances in many fields of chemistry, supramolecular chemistry, and nanoscience.

References

1. Wojaczynski, J. and L. Latos-Grazynski (2000). Poly- and oligometalloporphyrins associated through coordination. *Coord. Chem. Rev.* **204**, 113–171.
2. Imamura, T. and K. Fukushima (2000). Self-assembly of metallopyridylporphyrin oligomers. *Coord. Chem. Rev.* **198**, 133–156.
3. Sanders, J.K.M. (2000). Coordination chemistry of oligoporphyhrins. In K.M. Kadish, K.M. Smith, and R. Guilard (eds.), *The Porphyrin Handbook*. Academic Press, New York, pp. 347–368.
4. Chambron, J.C., V. Heitz, and J.P. Sauvage (2000). Non-covalent multiporphyrin assemblies. In K.M. Kadish, K.M. Smith, and R. Guilard, (eds.), *The Porphyrin Handbook*. Academic Press, New York, pp. 1–42.
5. Toma, H.E. and K. Araki (2000). Supramolecular assemblies of ruthenium complexes and porphyrins. *Coord. Chem. Rev.* **196**, 307–329.
6. Toma, H.E., K. Araki, A.D.P. Alexiou, S. Nikolaou, and S. Dovidauskas (2001). Monomeric and extended oxo-centered triruthenium clusters. *Coord. Chem. Rev.* **219**, 187–234.
7. Baldini, L. and C.A. Hunter (2000). Self-assembly of porphyrin arrays. *Adv. Inorg. Chem.* **53**, 213–259.
8. Prodi, A., M.T. Indelli, C.J. Kleverlaan, E. Alessio, and F. Scandola (2002). Energy transfer pathways in pyridylporphyrin metal adducts and side-to-face arrays. *Coord. Chem. Rev.* **229**, 51–58.
9. Fleischer, E.B. and A.M. Shachter (1991). Coordination oligomers and a coordination polymer of zinc tetraarylporphyrins. *Inorg. Chem.* **30**, 3763–3769.
10. Anderson, H.L., A. Bashall, K. Henrick, M. McPartlin, and J.K.M. Sanders (1994). Crystal-structure of a supramolecular dimer formed by π–π interactions between 2 interlocked cyclic zinc porphyrin trimers. *Angew. Chem. Int. Ed. Engl.* **33**, 429–431.
11. Anderson, S., H.L. Anderson, and J.K.M. Sanders (1995). The roles of templates in the syntheses of porphyrin oligomers. *J. Chem. Soc. Perkin Trans. 1* 2255–2267.
12. Anderson, H.L., S. Anderson, and J.K.M. Sanders (1995). Ligand-binding by butadiyne-linked porphyrin dimers, trimers and tetramers, *J. Chem. Soc. Perkin Trans. 1* 2231–2245.
13. Anderson, S., H.L. Anderson, and J.K.M. Sanders (1995). Template-directed synthesis of linear and cyclic butadiyne-linked porphyrin oligomers up to a linear octamer. *J. Chem. Soc. Perkin Trans. 1* 2247–2254.
14. Anderson, H.L., C.J. Walter, A. Vidalferran, R.A. Hay, P.A. Lowden, and J.K.M. Sanders (1995). Octatetrayne-linked porphyrins—stretched cyclic dimers and trimers with very spacious cavities. *J. Chem. Soc. Perkin Trans. 1* 2275–2279.
15. Anderson, H.L. and J.K.M. Sanders (1995). Enzyme mimics based on cyclic porphyrin oligomers—strategy, design and exploratory synthesis. *J. Chem. Soc. Perkin Trans. 1* 2223–2229.
16. Rempel, U., B. Vonmaltzan, and C. von Borczyskowski (1991). Picosecond time-resolved investigation of singlet energy-transfer in porphyrin hetero dimers. *J. Lumin.* **48–49**, 415–418.
17. Rempel, U., B. Vonmaltzan, and C. von Borczyskowski (1992). Electron-transfer in self-organized porphyrin–quinone compounds on a picosecond time scale. *J. Lumin.* **53**, 175–178.
18. Rempel, U., B. Vonmaltzan, and C. von Borczyskowski (1993). Temperature and solvent dependent charge-transfer in self-organized porphyrin–quinone compounds. *Pure Appl. Chem.* **65**, 1681–1685.

19. Chernook, A.V., U. Rempel, C. von Borczyskowski, A.M. Shulga, and E.I. Zenkevich (1996). Formation and optical properties of self-organized pentameric porphyrin arrays. *Chem. Phys. Lett.* **254**, 229–241.
20. Rempel, U., B. Vonmaltzan, and C. von Borczyskowski (1995). Competition between charge-transfer via superexchange and thermally activated energy-transfer in porphyrinheterodimer-quinone systems. *Chem. Phys. Lett.* **245**. 253–261.
21. Chernook, A.V., A.M. Shulga, E.I. Zenkevich, U. Rempel, and C. von Borczyskowski (1996). Complexation and interchromophoric interactions in self-organized porphyrin and chlorin triads. *J. Phys. Chem.* **100**, 1918–1926.
22. Zenkevich, E.I., C. von Borczyskowski, A.M. Shulga, S. Bachilo, U. Rempel, and A. Willert (2002). Self-assembled nanoscale photomimetic models: Structure and related dynamics. *Chem. Phys.* **275**, 185–209.
23. Zenkevich, E.I., A.M. Shulga, S.M. Bachilo, U. Rempel, J. von Richthofen, and C. von Borczyskowski (1998). Energy and charge-transfer dynamics in self-organized multimolecular arrays. *J. Lumin.* **76–77**, 354–358.
24. Willert, A., S. Bachilo, U. Rempel, A. Shulga, E. Zankevich, and C. von Borczyskowski (1999). Efficient low temperature charge transfer in a self-assembled porphyrin aggregate. *J. Photochem. Photobiol. A* **126**, 99–109.
25. Rempel, U., S. Meyer, B. von Maltzan, and C. von Borczyskowski (1998). Energy transfer and distance independent charge separation in self-organized porphyrin–quinone aggregates. *J. Lumin.* **78**, 97–110.
26. Michelsen, U. and C.A. Hunter (2000). Self-assembled porphyrin polymers. *Angew. Chem. Int. Ed. Engl.* **39**, 764–767.
27. Takahashi, R. and Y. Kobuke (2003). Hexameric macroring of gable-porphyrins as a light-harvesting antenna mimic. *J. Am. Chem. Soc.* **125**, 2372–2373.
28. Inaba, Y. and Y. Kobuke (2004). Synthesis of a complementary dimer from mono(imidazolyl)-substituted cobalt(II) porphyrin as a new artificial T-form hemoglobin. *Tetrahedron* **60**, 3097–3107.
29. Ogawa, K. and Y. Kobuke (2000). Formation of a giant supramolecular porphyrin array by self-coordination. *Angew. Chem. Int. Ed. Engl.* **39**, 4070–4073.
30. Kobuke, Y. and K. Ogawa (2003). Porphyrin supramolecules for artificial photosynthesis and molecular photonic/electronic materials. *Bull. Chem. Soc. Jpn* **76**, 689–708.
31. Nagata, N., S. Kugimiya, and Y. Kobuke (2000). Antenna functions of 5,15-bis(imidazol-4-yl)-10,20-bis(4-dodecyloxyphenyl)-porphyrin supramolecular assembly through imidazole–imidazole hydrogen bonding. *Chem. Commun.* 1389–1390.
32. Nagata, N., S. Kugimiya, E. Fujiwara, and Y. Kobuke (2003). Staircase-form assembly with 5,15-bis(imidazol-4-yl)porphinatogallium steps. *New J. Chem.* **27**, 743–747.
33. Ikeda, C., E. Fujiwara, A. Satake, and Y. Kobuke (2003). Long rod-like array of bis(imidazolyl)porphyrinatocobalt(III) by successive complementary coordination. *Chem. Commun.* 616–617.
34. Kobuke, Y. and H. Miyaji (1996). Supramolecular stacks of bis(imidazolyl)porphyrin through metal coordination. *Bull. Chem. Soc. Jpn* **69**, 3563–3569.
35. Rubtsov, I.V., Y. Kobuke, H. Miyaji, and K. Yoshihara (1999). Energy transfer in a porphyrin chelate assembly. *Chem. Phys. Lett.* **308**, 323–328.
36. Ikeda, C., A. Satake, and Y. Kobuke (2003). Proofs of macrocyclization of gable porphyrins as mimics of photosynthetic light-harvesting complexes. *Org. Lett.* **5**, 4935–4938.
37. Nomoto, A. and Y. Kobuke (2002). Photocurrent generation system incorporated with antenna function. *Chem. Commun.* 1104–1105.
38. Kimura, A., K. Funatsu, T. Imamura, H. Kido, and Y. Sasaki (1995). Vertically Linked Ruthenium(II) Porphyrin Oligomers. *Chem. Lett.* 207–208.

39. Funatsu, K., A. Kimura, T. Imamura, A. Ichimura, and Y. Sasaki (1997). Perpendicularly arranged ruthenium porphyrin dimers and trimers. *Inorg. Chem.* **36**, 1625–1635.
40. Funatsu, K., T. Imamura, A. Ichimura, and Y. Sasaki (1998). Novel cofacial ruthenium(II) porphyrin dimers and tetramers. *Inorg. Chem.* **37**, 4986–4995.
41. Funatsu, K., T. Imamura, A. Ichimura, and Y. Sasaki (1998). Synthesis and properties of cyclic ruthenium(II) porphyrin tetramers. *Inorg. Chem.* **37**, 1798–1804.
42. Kariya, N., T. Imamura, and Y. Sasaki (1998). Synthesis and properties of osmium(II) porphyrin oligomers linked by 3-pyridylporphyrins. *Inorg. Chem.* **37**, 1658–1660.
43. Kariya, N., T. Imamura, and Y. Sasaki (1997). Synthesis, characterization, and spectral properties of new perpendicularly linked osmium(II) porphyrin oligomers. *Inorg. Chem.* **36**, 833–839.
44. Alessio, E., M. Macchi, S. Heath, and L.G. Marzilli (1996). A novel open-box shaped pentamer of vertically linked porphyrins that selectively recognizes S-bonded Me_2SO complexes. *Chem. Commun.* **12**, 1411–1412.
45. Prodi, A., M.T. Indelli, C.J. Kleverlaan, F. Scandola, E. Alessio, T. Gianferrara, and L.G. Marzilli (1999). Side-to-face ruthenium porphyrin arrays: Photophysical behavior of dimeric and pentameric systems. *Chem. Eur. J.* **5**, 2668–2679.
46. Alessio, E., S. Geremia, S. Mestroni, E. Iengo, I. Srnova, and M. Slouf (1999). Solution and solid state structure of a canted, side-to-face, bis(porphyrin) adduct. *Inorg. Chem.* **38**, 869–875.
47. Alessio, E., S. Geremia, S. Mestroni, I. Srnova, M. Slouf, T. Gianferrara, and A. Prodi (1999). Porphyrin "flying-saucers": Solid state and solution structure of a novel pentameric array of axially-ligated canted porphyrins. *Inorg. Chem.* **38**, 2527–2529.
48. Prodi, A., C.J. Kleverlaan, M.T. Indelli, F. Scnadola, E. Alessio, and E. Iengo (2001). Photophysics of pyridylporphyrin Ru(II) adducts: Heavy-atom effects and intramolecular decay pathways, *Inorg. Chem.* **40**, 3498–3504.
49. Mak, C.C., N. Bampos, and J.K.M. Sanders (1999). Ru(II)-centred porphyrin pentamers as coordination building blocks for large porphyrin arrays. *Chem. Commun.* 1085–1086.
50. Stulz, E., S.M. Scott, A.D. Bond, S.J. Teat, and J.K.M. Sanders (2003). Selection and amplification of mixed-metal porphyrin cages from dynamic combinatorial libraries. *Chem. Eur. J.* **9**, 6039–6048.
51. Darling, S.L., E. Stulz, N. Feeder, N. Bampos, and J.K.M. Sanders (2000). Phosphine-substituted porphyrins as supramolecular building blocks. *New J. Chem.* **24**, 261–264.
52. Stulz, E., S.M. Scott, Y.F. Ng, A.D. Bond, S.J. Teat, S.L. Darling, N. Feeder, and J.K.M. Sanders (2003). Construction of multiporphyrin arrays using ruthenium and rhodium coordination to phosphines. *Inorg. Chem.* **42**, 6564–6574.
53. Chichak, K. and N.R. Branda (1999). Self-assembly of a linear multicomponent porphyrin array through axial coordination. *Chem. Commun.* 523–524.
54. Fujita, M., J. Yazaki, and K. Ogura (1990). Preparation of a macrocyclic polynuclear complex, $[(en)Pd(4,4'\text{-bpy})]_4(NO_3)_8$, which recognizes an organic-molecule in aqueous-media. *J. Am. Chem. Soc.* **112**, 5645–5647.
55. Fujita, M., J. Yazaki, and K. Ogura (1991). Macrocyclic polynuclear complexes $[(en)M(4,4'\text{-bpy})]_4(NO_3)_8$ (M = Pd or Pt) as inorganic cyclophane—their ability for molecular recognition. *Tetrahedron Lett.* **32**, 5589–5592.
56. Fujita, M., J. Yazaki, and K. Ogura (1991). Spectroscopic observation of self-assembly of a macrocyclic tetranuclear complex composed of Pt^{2+} and $4,4'$-bipyridine. *Chem. Lett.* 1031–1032.
57. Fujita, M., S. Nagao, M. Iida, K. Ogata, and K. Ogura (1993). Palladium(II)-directed assembly of macrocyclic dinuclear complexes composed of $(en)Pd^{2+}$ and

bis(4-pyridyl)-substituted bidentate ligands—remarkable ability for molecular recognition of electron-rich aromatic guests. *J. Am. Chem. Soc.* **115**, 1574–1576.
58. Fujita, M., Y.J. Kwon, M. Miyazawa, and K. Ogura (1994). One-dimensional coordinate polymer involving heptacoordinate cadmium(II) ions. *J. Chem. Soc. Chem. Commun.* 1977–1978.
59. Fujita, M. and G. Ogura (1996). Supramolecular self-assembly of macrocycles, catenanes, and cages through coordination of pyridine-based ligands to transition metals. *Bull. Chem. Soc. Jpn* **69**, 1471–1482.
60. Fujita, M. and K. Ogura (1996). Transition-metal-directed assembly of well-defined organic architectures possessing large voids: From macrocycles to [2]catenanes. *Coord. Chem. Rev.* **148**, 249–264.
61. Olenyuk, B., J.A. Whiteford, A. Fechtenkotter, and P.J. Stang (1999). Self-assembly of nanoscale cuboctahedra by coordination chemistry. *Nature* **398**, 796–799.
62. Olenyuk, B., A. Fechtenkotter, and P.J. Stang (1998). Molecular architecture of cyclic nanostructures: Use of coordination chemistry in the building of supermolecules with predefined geometric shapes. *J. Chem. Soc. Dalton Trans.* 1707–1728.
63. Manna, J., C.J. Kuehl, J.A. Whiteford, P.J. Stang, D.C. Muddiman, S.A. Hofstadler, and R.D. Smith (1997). Nanoscale tectonics: Self-assembly, characterization, and chemistry of a novel class of organoplatinum square macrocycles. *J. Am. Chem. Soc.* **119**, 11611–11619.
64. Stang, P.J. and B. Olenyuk (1997). Self-assembly, symmetry, and molecular architecture: Coordination as the motif in the rational design of supramolecular metallacyclic polygons and polyhedra. *Acc. Chem. Res.* **30**, 502–518.
65. Fan, J., J.A. Whiteford, B. Olenyuk, M.D. Levin, P.J. Stang, and E.B. Fleischer (1999). Self-assembly of porphyrin arrays via coordination to transition metal bisphosphine complexes and the unique spectral properties of the product metallacyclic ensembles. *J. Am. Chem. Soc.* **121**, 2741–2752.
66. Slone R.V., K.D. Benkstein, S. Belanger, J.T. Hupp, I.A. Guzei, and A.L. Rheingold (1998). Luminescent transition-metal-containing cyclophanes ("molecular squares"): Covalent self-assembly, host–guest studies and preliminary nanoporous materials applications. *Coord. Chem. Rev.* **171**, 221–243.
67. Leininger, S., B. Olenyuk, and P.J. Stang (2000). Self-assembly of discrete cyclic nanostructures mediated by transition metals. *Chem. Rev.* **100**, 853–907.
68. Stang, P.J. (1998). Molecular architecture: Coordination as the motif in the rational design and assembly of discrete supramolecular species—Self-assembly of metallacyclic polygons and polyhedra. *Chem. Eur. J.* **4**, 19–27.
69. Alessio, E., M. Macchi, S.L. Heath, and L.G. Marzilli (1997). Ordered supramolecular porphyrin arrays from a building block approach utilizing pyridylporphyrins and peripheral ruthenium complexes and identification of a new type of mixed-metal building block. *Inorg. Chem.* **36**, 5614–5623.
70. Alessio, E., E. Ciani, E. Iengo, V.Y. Kukushkin, and L.G. Marzilli (2000). Stepwise assembly of unsymmetrical supramolecular arrays containing porphyrins and coordination compounds. *Inorg. Chem.* **39**, 1434–1443.
71. Elemans, J., A.E. Rowan, and R.J.M. Nolte (2003). Mastering molecular matter. Supramolecular architectures by hierarchical self-assembly. *J. Mater. Chem.* **13**, 2661–2670.
72. Belanger, S., M.H. Keefe, J.L. Welch, and J.T. Hupp (1999). Rapid derivatization of mesoporous thin-film materials based on Re(I) zinc–porphyrin 'molecular squares': selective modification of mesopore size and shape by binding of aromatic nitrogen donor ligands. *Coord. Chem. Rev.* **192**, 29–45.

73. Yuan, H.P., L. Thomas, and L.K. Woo (1996). Synthesis and characterization of mono-, bis-, and tetrakis-pyridyltriarylporphyrin Pd(II) and Pt(II) supramolecular assemblies. Molecular structure of a Pd-linked bisporphyrin complex. *Inorg. Chem.* **35**, 2808–2817.
74. Iengo, E., B. Milani, E. Zangrando, S. Geremia, and E. Alessio (2000). Novel ruthenium building blocks for the efficient modular construction of heterobimetallic molecular squares of porphyrins. *Angew. Chem. Int. Ed. Engl.* **39**, 1096–1099.
75. Iengo, E., R. Minatel, B. Milani, L.G. Marzilli, and E. Alessio (2001). Metal-mediated self-assembly of molecular squares of porphyrins rimmed with coordination compounds. *Eur. J. Inorg. Chem.* 609–612.
76. Iengo, E., E. Zangrando, S. Geremia, R. Graff, B. Kieffer, and E. Alessio (2002). Two-point self-coordination of a dizinc(II) bispyridylporphyrin ruthenium complex leading selectively to a discrete molecular assembly: Solution and solid-state characterization. *Chem. Eur. J.* **8**, 4670–4674.
77. C.M. Drain, R. Fischer, E.G. Nolen, and J.M. Lehn (1993). Self-assembly of a bisporphyrin supramolecular cage induced by molecular recognition between complementary hydrogen-bonding sites. *J. Chem. Soc. Chem. Commun.* 243–245.
78. Drain, C.M. and J.M. Lehn (1994). Self-assembly of square multiporphyrin arrays by metal-ion coordination. *J. Chem. Soc. Chem. Commun.* 2313–2315.
79. Stang, P.J., J. Fan, and B. Olenyuk (1997). Molecular architecture via coordination: Self-assembly of cyclic cationic porphyrin aggregates via transition-metal bisphosphane auxiliaries. *Chem. Commun.* 1453–1454.
80. Belanger, S. and J.T. Hupp (1999). Porphyrin-based thin-film molecular materials with highly adjustable nanoscale porosity and permeability characteristics. *Angew. Chem. Int. Ed. Engl.* **38**, 2222–2224.
81. Slone, R.V. and J.T. Hupp (1997). Synthesis, characterization, and preliminary host–guest binding studies of porphyrinic molecular squares featuring fac-tricarbonylrhenium(I) chloro corners. *Inorg. Chem.* **36**, 5422.
82. Aspley, C.J., J.R.L. Smith, R.N. Perutz, and D. Pursche (2002). Synthesis and photochemistry of free base and zinc tetraaryl porphyrins mono-substituted with tungsten pentacarbonyl via a pyridine linker. *J. Chem. Soc. Dalton Trans.* 170–180.
83. Drain, C.M., F. Nifiatis, A. Vasenko, and J.D. Batteas (1998). Porphyrin tessellation by design: Metal-mediated self-assembly of large arrays and tapes. *Angew. Chem. Int. Ed. Engl.* **37**, 2344–2347.
84. Drain, C.M., J.D. Batteas, G.W. Flynn, T. Milic, N. Chi, D.G. Yablon, and H. Sommers (2002). Designing supramolecular porphyrin arrays that self-organize into nanoscale optical and magnetic materials. *Proc. Natl. Acad. Sci. USA*, **99**, 6498–6502.
85. Czaplewski, K.F., J.T. Hupp, and R.Q. Snurr (2001). Molecular squares as molecular sieves: Size-selective transport through porous-membrane-supported thin-film materials. *Adv. Mater.* **13**, 1895–1897.
86. Sauvage, J.P., J.P. Collin, J.C. Chambron, S. Guillerez, C. Coudret, V. Balzani, F. Barigelletti, L. Decola, and L. Flamigni (1994). Ruthenium(II) and osmium(II) bis(terpyridine) complexes in covalently-linked multicomponent systems—synthesis, electrochemical-behavior, absorption-spectra, and photochemical and photophysical properties. *Chem. Revs.* **94**, 993–1019.
87. Iengo, E., E. Zangrando, E. Alessio, J.C. Chambron, V. Heitz, L. Flamigni, and J.P. Sauvage (2003). A functionalized noncovalent macrocyclic multiporphyrin assembly from a dizinc(II) bis-porphyrin receptor and a free-base dipyridylporphyrin. *Chem. Eur. J.* **9**, 5879–5887.

88. Andersson, M., M. Linke, J.C. Chambron, J. Davidsson, V. Heitz, L. Hammarstrom, and J.P. Sauvage (2002). Long-range electron transfer in porphyrin-containing 2-rotaxanes: Tuning the rate by metal cation coordination. *J. Am. Chem. Soc.* **124**, 4347–4362.
89. Paul, D., J.A. Wytko, M. Koepf, and J. Weiss (2002). Design and synthesis of a self-assembled photochemical dyad based on selective imidazole recognition. *Inorg. Chem.* **41**, 3699–3704.
90. Ochsenbein, P., M. Bonin, K. Schenk, J. Froidevaux, J. Wytko, E. Graf, and J. Weiss (1999). X-ray structures of a phenanthroline-strapped porphyrin and its dihydrated zinc(II) complex: Convergent distal hydrogen bonds and "CH center dot center dot center dot O" interactions. *Eur. J. Inorg. Chem.* 1175–1179.
91. Hungerford, G., M. van der Auweraer, and D.B. Amabilino (2001). Intramolecular fluorescence quenching in porphyrin-bearing (2)catenates. *J. Porph. Phthalocyan.* **5**, 633–644.
92. Chambron, J.C., A. Harriman, V. Heitz, and J.P. Sauvage (1993). Effect of the spacer moiety on the rates of electron-transfer within bis-porphyrin-stoppered rotaxanes. *J. Am. Chem. Soc.* **115**, 7419–7425.
93. Chambron, J.C., A. Harriman, V. Heitz, and J.P. Sauvage (1993). Ultrafast photoinduced electron-transfer between porphyrinic subunits within a bis(porphyrin)-stoppered rotaxane. *J. Am. Chem. Soc.* **115**, 6109–6114.
94. Chambron, J.C., V. Heitz, J.P. Sauvage, J.L. Pierre, and D. Zurita (1995). Bis-porphyrins containing diimine chelates of variable geometry as spacer. *Tetrahedron Lett.* **36**, 9321–9324.
95. Brun, A.M., S.J. Atherton, A. Harriman, V. Heitz, and J.P. Sauvage (1992). Photophysics of entwined porphyrin conjugates—competitive exciton annihilation, energy-transfer, electron-transfer, and superexchange processes. *J. Am. Chem. Soc.* **114**, 4632–4639.
96. Hungerford, G., M. Van der Auweraer, J.C. Chambron, V. Heitz, J.P. Sauvage, J.L. Pierre, and D. Zurita (1999). Intramolecular energy transfer in bis-porphyrins containing diimine chelates of variable geometry as spacers. *Chem. Eur. J.* **5**, 2089–2100.
97. Tomohiro, Y., A. Satake, and Y. Kobuke (2001). Synthesis of bipyridylene-bridged bisporphyrin by nickel-mediated coupling reaction: ON–OFF control of cofacial porphyrin unit by reversible complexation. *J. Org. Chem.* **66**, 8442–8446.
98. Sessler, J.L., V.L. Capuano, and A.K. Burrell (1993). Orthogonal Ru(bpy)3 complexes in mesosubstituted porphyrins. *Inorg. Chim. Acta* **204**, 93–101.
99. Kus, P., G. Knerr, and L. Czuchajowski (1991). Diporphyrinyl derivatives of 1,10-phenanthroline and 2, 2′-bipyridyl. *J. Heterocycl. Chem.* **28**, 7–11.
100. Bruce, J.I., J.C. Chambron, P. Kolle, and J.P. Sauvage (2002). Synthesis of a linear bis-porphyrin with a u(phen)$_2^{2+}$-complexed 2, 2′-bipyridine spacer. *J. Chem. Soc. Perkin Trans. 1* 1226–1231.
101. Cheng, K.F., C.M. Drain, and K. Grohmann (2003). Porphyrins linked directly to the 5, 5′ positions of 2, 2′-bipyridine: A new supramolecular building block and switch. *Inorg. Chem.* **42**, 2075–2083.
102. Collin, J.P., A. Harriman, V. Heitz, F. Odobel, and J.P. Sauvage (1994). Photoinduced electron-transfer and energy-transfer processes occurring within porphyrin–metal–bisterpyridyl conjugates. *J. Am. Chem. Soc.* **116**, 5679–5690.
103. Harriman, A., F. Odobel, and J.-P. Sauvage (1995). Multistep electron transfer between porphyrin modules assembled around a ruthenium center. *J. Am. Chem. Soc.* **117**, 9461–9472.
104. Collin, J.P., A. Harriman, V. Heitz, F. Odobel, and J.P. Sauvage (1996). Transition metal-assembled multiporphyrinic systems as models of photosynthetic reaction centre. *Coord. Chem. Rev.* **148**, 63–69.

105. Flamigni, L., N. Armaroli, F. Barigelletti, V. Balzani, J.P. Collin, J.O. Dalbavie, V. Heitz, and J.P. Sauvage (1997). Photoinduced processes in dyads made of a porphyrin unit and a ruthenium complex. *J. Phys. Chem. B* **101**, 5936–5943.
106. Flamigni, L., F. Barigelletti, N. Armaroli, J.P. Collin, J.P. Sauvage, and J.A.G. Williams (1998). Photoinduced processes in highly coupled multicomponent arrays based on a ruthenium(II)bis(terpyridine) complex and porphyrins. *Chem. Eur. J.* **4**, 1744–1754.
107. Flamigni, L., F. Barigelletti, N. Armaroli, B. Ventura, J.P. Collin, J.P. Sauvage, and J.A.G. Williams (1999). Triplet–triplet energy transfer between porphyrins linked via a ruthenium(II) bisterpyridine complex. *Inorg. Chem.* **38**, 661–667.
108. Flamigni, L., I.M. Dixon, J.P. Collin, and J.P. Sauvage (2000). A Zn(II) porphyrin–Ir(III) bis-terpyridine–Au(III) porphyrin triad with a charge-separated state in the microsecond range. *Chem. Commun.* 2479–2480.
109. Dixon, I.M., and J.P. Collin (2001). Synthesis and properties of diads based on tetra-aryl porphyrins and ruthenium bis-terpyridine-type complexes. *J. Porph. Phthalocyan.* **5**, 600–607.
110. Flamigni, L., G. Marconi, I.M. Dixon, J.P. Collin, and J.P. Sauvage (2002). Switching of electron- to energy-transfer by selective excitation of different chromophores in arrays based on porphyrins and a polypyridyl iridium complex. *J. Phys. Chem. B* **106**, 6663–6671.
111. Uyeda, H.T., Y.X. Zhao, K. Wostyn, I. Asselberghs, K. Clays, A. Persoons, and M.J. Therien (2002). Unusual frequency dispersion effects of the nonlinear optical response in highly conjugated (polypyridyl)metal–(porphinato)zinc(II) chromophores. *J. Am. Chem. Soc.* **124**, 13806–13813.
112. Toma, H.E. and K. Araki (1988). Spectroelectrochemical behavior of a polymetallated porphyrin. *Proc. VI Braz. Symp. Electrochem. Electroanal.* 183–190.
113. Toma, H.E., and K. Araki (1990). Synthesis and properties of a new polymetallated iron porphyrin. *J. Chem. Res. S* 82–83.
114. Araki, K., and H.E. Toma (1991). Synthesis and electrochemical-behavior of a tetrametallated cobalt porphyrin. *Inorg. Chim. Acta* **179**, 293–296.
115. Araki, K., P.S. Santos, and H.E. Toma (1993). Electronic and resonance raman-spectra of a multibridged iron porphyrin. *Spectrosc. Lett.* **26**, 1417–1426.
116. Collman, J.P., P. Denisevich, Y. Konai, M. Marrocco, C. Koval, and F.C. Anson (1980). Electrode catalysis of the four-electron reduction of oxygen to water by dicobalt face-to-face porphyrins. *J. Am. Chem. Soc.* **102**, 6027–6036.
117. Collman, J.P., C.S. Bencosme, R.R. Durand, R.P. Kreh, and F.C.Anson (1983). Mixed-metal face-to-face porphyrin dimers. *J. Am. Chem. Soc.* **105**, 2699–2703.
118. Collman, J.P., F.C. Anson, C.E. Barnes, C.S. Bencosme, T. Geiger, E.R. Evitt, R.P. Kreh, K. Meier, and R.B. Pettman (1983). Further-studies of the dimeric beta-linked face-to-face 4 porphyrin—FTF4. *J. Am. Chem. Soc.* **105**, 2694–2699.
119. Shi, C., and F.C. Anson (1991). Multiple intramolecular electron-transfer in the catalysis of the reduction of dioxygen by cobalt meso-tetrakis(4-pyridyl)porphyrin to which 4 $Ru(NH_3)_5$ groups are coordinated. *J. Am. Chem. Soc.* **113**, 9564–9570.
120. Araki, K., and H.E. Toma (1993). Synthesis and characterization of a multibridged porphyrin complex containing peripheral bis(bipyridine)-ruthenium(II) groups. *J. Coord. Chem.* **30**, 9–17.
121. Araki, K. (1989). *Preparation and Properties of New Polymetallated Porphyrins.* Master. Institute of Chemistry, University of Sao Paulo.
122. Araki, K., L. Angnes, C.M.N. Azevedo, and H.E. Toma (1995). Electrochemistry of a tetraruthenated cobalt porphyrin and its use in modified electrodes as sensors of reducing analytes. *J. Electroanal. Chem.* **397**, 205–210.

123. Araki, K., M.J. Wagner, and M.S. Wrighton (1996). Layer-by-layer growth of electrostatically assembled multilayer porphyrin films. *Langmuir* **12**, 5393–5398.
124. Araki, K., A.L. Araujo, M.M. Toyama, M. Franco, C.M.N. Azevedo, L. Angnes, and H.E. Toma (1998). Spectroscopic and electrochemical study of a tetrapyridylporphyrin modified with four bis-(1,10-phenanthroline)chlororuthenium(II) complexes. *J. Porph. Phthalocyan.* **2**, 467–472.
125. Toma, H.E., K. Araki, and E.O. Silva (1998). Synthesis and characterization of a novel dodecanuclear porphyrin ruthenium cluster. *Monats. Chem.* **129**, 975–984.
126. Winnischofer, H., S.D. Lima, K. Araki, and H.E. Toma (2003). Electrocatalytic activity of a new nanostructured polymeric tetraruthenated porphyrin film for nitrite detection. *Anal. Chim. Acta* **480**, 97–107.
127. Araki, K., P. Losco, F.M. Engelmann, H. Winnischofer, and H.E. Toma (2001). Modulation of vectorial energy transfer in the tetrakis[tris(bipyridine) ruthenium(II)]porphyrinate zinc complex. *J. Photochem. Photobiol. A* **142**, 25–30.
128. Winnischofer, H., F.M. Engelmann, H.E. Toma, K. Araki, and H.R. Rechenberg (2002). Acid–base and spectroscopic properties of a novel supramolecular porphyrin bonded to four pentacyanoferrate(II) groups. *Inorg. Chim. Acta* **338**, 27–35.
129. Araki, K., H. Winnischofer, H.E.B. Viana, M.M. Toyama, F.M. Engelmann, I. Mayer, A.L.B. Formiga, and H.E. Toma (2004). Enhanced electrochemical and electrocatalytic activity of a new supramolecular manganese-porphyrin species containing four bis(bipyridine)(aqua)ruthenium(II) complexes. *J. Electroanal. Chem.* **562**, 145–152.
130. Toyama, M.M., M. Franco, L.H. Catalani, K. Araki, and H.E. Toma (1998). Spectroelectrochemical and photophysical properties of a (3,4-pyridyl)porphyrazine supermolecule containing four $[Ru(bipy)_2Cl]^+$ groups. *J. Photochem. Photobiol. A* **118**, 11–17.
131. Lo Schiavo, S., S. Serroni, F. Puntoriero, G. Tresoldi, and P. Piraino (2002). Synthesis and characterization of dirhodium(II,II)–porphyrin-based multiredox systems. *Eur. J. Inorg. Chem.* 79–86.
132. Engelmann, F.M., P. Losco, H. Winnischofer, K. Araki, and H.E. Toma (2002). Synthesis, electrochemistry, spectroscopy and photophysical properties of a series of meso-phenylpyridylporphyrins with one to four pyridyl rings coordinated to $[Ru(bipy)_2Cl]^+$ groups. *J. Porph. Phthalocyan.* **6**, 33–42.
133. Araki, K. and H.E. Toma (1991). Spectroelectrochemical analysis of overlapping redox processes in a tetrametallated iron porphyrin. *J. Electroanal. Chem.* **297**, 301–307.
134. Shi, C.N. and F.C. Anson (1994). Electrocatalysts for the 4-electron reduction of dioxygen based on adsorbed cobalt tetrapyridylporphyrin molecules linked by aquaamine complexes of ruthenium(II). *Inorg. Chim. Acta* **225**, 215–227.
135. Shi, C.N. and F.C. Anson (1992). Electrocatalysis of the reduction of O_2 to H_2O by tetraruthenated cobalt meso-tetrakis(4-pyridyl)porphyrin adsorbed on graphite-electrodes. *Inorg. Chem.* **31**, 5078–5083.
136. Shi, C.N. and F.C. Anson (1994). Potential-dependence of the reduction of dioxygen as catalyzed by tetraruthenated cobalt tetrapyridylporphyrin. *Electrochim. Acta* **39**, 1613–1619.
137. Shi, C.N. and F.C. Anson (1995). Comparison of the catalytic reduction of dioxygen by 5,10,15,20-tetrakis((pentaammineruthenino(II))-4-pyridyl)porphyrinato cobalt(II) in solution and on graphite electrode surfaces. *Inorg. Chem.* **34**, 4554–4561.
138. Steiger, B., C. Shi, and F.C. Anson (1993). Electrocatalysis of the reduction of dioxygen by adsorbed cobalt 5,10,15,20-tetraarylporphyrins to which one-Ru, 2-Ru, or 3-$Ru(NH_3)_5^{2+}$ centers are coordinated. *Inorg. Chem.* **32**, 2107–2113.

139. Franco, C. and G. McLendon (1984). Bimetallic porphyrins: Synthesis and rapid intramolecular electron transfer of meso-tritolyl[*N*-(pentaammineruthenium)-pyridyl]porphyrin. *Inorg. Chem.* **23**, 2370–2372.
140. Dovidauskas, S., K. Araki, and H.E. Toma (2000). Electrochemical and binding properties of a meso-tetra(4-pyridyl)porphyrinatozinc supermolecule containing four μ_3-oxo-triruthenium acetate clusters. *J. Porph. Phthalocyan.* **4**, 727–735.
141. Dovidauskas, S., H.E. Toma, K. Araki, H.C. Sacco, and Y. Iamamoto (2000). (5,10,15,20-Tetra(4-pyridil)porphinato)manganes(III) acetate modified by four μ_3-oxo-triruthenium acetate clusters: Synthesis, characterization, electrochemical behavior and catalytic activity. *Inorg. Chim. Acta* **305**, 206–213.
142. Araki, K., S. Dovidauskas, H. Winnischofer, A.D.P. Alexiou, and H.E. Toma (2001). A new highly efficient tetra-electronic catalyst based on a cobalt porphyrin bound to four mu_3-oxo-ruthenium acetate clusters. *J. Electroanal. Chem.* **498**, 152–160.
143. Lo Schiavo, S., G. Pocsfalvi, S. Serroni, P. Cardiano, and P. Piraino (2000). Self-assembly of square molecular boxes containing dirhodium(II,II) units. *Eur. J. Inorg. Chem.* **6**, 1371–1375.
144. Araki, K., P.S. Santos, L.F.C. de Oliveira, and H.E. Toma (1995). Resonance-raman spectra of a supramolecular species containing 4 ruthenium(II) bipyridine complexes attached to zinc tetrapyridyl porphynate. *Spectrosc. Lett.* **28**, 118–126.
145. Adams, D.M., L. Brus, C.E.D. Chidsey, S. Creager, C. Creutz, C.R. Kagan, P.V. Kamat, M. Lieberman, S. Lindsay, R.A. Marcus, R.M. Metzger, M.E. Michel-Beyerle, J.R. Miller, M.D. Newton, D.R. Rolison, O. Sankey, K.S. Schanze, J. Yardley, and X. Zhu (2003). Charge transfer on the nanoscale: Current status. *J. Phys. Chem. B* **107**, 6668–6697.
146. Araki, K. (1994).*Supramolecular Properties of Polynuclear Porphyrins*. PhD. Institute of Chemistry, University of Sao Paulo.
147. Araki, K. and H.E. Toma (1994). Luminescence, spectroelectrochemistry and photoelectrochemical properties of a tetraruthenated zinc porphyrin. *J. Photochem. Photobiol. A Chem.* **83**, 245–250.
148. Araki, K. and H.E. Toma (1999). Electrochemistry of a tetraruthenated iron porphyrin and its electrostatically assembled bilayered films. *Electrochim. Acta* **44**, 1577–1583.
149. Araki, K. and H.E. Toma (1994). Spectroelectrochemistry and electrocatalytic properties of a tetraruthenated nickel porhyrin. *J. Chem. Res. (M)* 1501–1515.
150. Araki, K. and H.E. Toma (1994). Spectroelectrochemistry and electrocatalytic properties of a tetraruthenated nickel porphyrin. *J. Chem. Res. S* 290–290.
151. Gunter, M.J. and P. Turner (1991). Metalloporphyrins as models for the cytochromes P-450. *Coord. Chem. Rev.* **108**, 115–161.
152. Meunier, B. (1992). Metalloporphyrins as versatile catalysts for oxidation reactions and oxidative DNA cleavage. *Chem. Rev.* **92**, 1411–1456.
153. Iamamoto, Y. M.D. Assis, K.J. Ciuffi, C.M.C. Prado, B.Z. Prellwitz, M. Moraes, O.R. Nascimento, and H.C. Sacco (1997). Manganese(III) porphyrins: Catalytic activity and intermediate studies in homogeneous systems. *J. Mol. Catal. A: Chem.* **116**, 365–374.
154. Iamamoto, Y., Y.M. Idemori, and S. Nakagaki (1995). Cationic ironporphyrins as catalyst in comparative oxidation of hydrocarbons—homogeneous and supported on inorganic matrices systems. *J. Mol. Cat. A* **99**, 187–193.
155. Groves, J.T., J.B. Lee, and S.S. Marla (1997). Detection and characterization of an oxomanganese(V) porphyrin complex by rapid-mixing stopped-flow spectrophotometry. *J. Am. Chem. Soc.* **119**, 6269–6273.
156. Groves, J.T., S.J. Crowley, and K.V. Shalyaev (1998). Paramagnetic H-1-NMR relaxation probes of stereoselectivity in metalloporphyrin catalyzed olefin epoxidation. *Chirality* **10**, 106–119.

157. Bedioui, F., J. Devynck, and C. Bied-Charreton (1995). Immobilization of metalloporphyrins in electropolymerized films: Design and applications. *Acc. Chem. Res.* **28**, 30–36.
158. Groves, J.T. (2000). Reactivity and mechanisms of metalloporphyrin-catalyzed oxidations. *J. Porph. Phthalocyan.* **4**, 350–352.
159. Mansuy, D. (1993). Activation of alkanes—the biomimetic approach. *Coord. Chem. Rev.* **125**, 129–141.
160. Bonnet, M., L. Schmid, A. Baiker, and F. Diederich (2002). A new mesoporous hybrid material: Porphyrin-doped aerogel as a catalyst for the epoxidation of olefins. *Adv. Funct. Mater.* **12**, 39–42.
161. Guo, C.C., J.X. Song, X.B. Chen, and G.F. Jiang (2000). A new evidence of the high-valent oxo-metal radical cation intermediate and hydrogen radical abstract mechanism in hydrocarbon hydroxylation catalyzed by metalloporphyrins. *J. Mol. Cat. A* **157**, 31–40.
162. Schenning, A., J.H.L. Spelberg, D.H.W. Hubert, M.C. Feiters, and R.J.M. Nolte (1998). A supramolecular cytochrome P450 mimic. *Chem. Eur. J.* **4**, 871–880.
163. Mansuy, D. and P. Battioni (2000). Diversity of reactions catalyzed by heme–thiolate proteins. In K.M. Kadish, K.M. Smith, and R. Guilard (eds), *The Porphyrin Handbook*. Academic Press, New York, pp. 1–15.
164. Groves, J.T., K. Shalyaev, and J. Lee (2000). Oxometalloporphyrins in oxidative catalysis. In K.M. Kadish, K.M. Smith, and R. Guilard (eds), *The Porphyrin Handbook*. Academic Press, New York, pp. 17–40.
165. Suslick, K.S., (2000). Shape-selective oxidation by metalloporphyrins. In K.M. Kadish, K.M. Smith, and R. Guilard (eds), *The Porphyrin Handbook*. Academic Press, New York, pp. 41–63.
166. Weiss, R., A. Gold, A.X. Trautwein, and J. Terner (2000). High-valent iron and manganese complexes of porphyrins and related macrocycles. In K.M. Kadish, K.M. Smith, and R. Guilard (eds), *The Porphyrin Handbook*. Academic Press, New York, pp. 65–96.
167. Watanabe, Y. (2000). High-valent intermediates. In K.M. Kadish, K.M. Smith, and R. Guilard (eds), *The Porphyrin Handbook*. Academic Press, New York, pp. 97–117.
168. Meunier, B., A. Robert, G. Pratirel, and J. Bernadou (2000). Metalloporphyrins in catalytic oxidations and oxidative DNA cleavage. In K.M. Kadish, K.M. Smith, and R. Guilard (eds), *The Porphyrin Handbook*. Academic Press, New York, pp. 119–187.
169. Jin, N. and J.T. Groves (1999). Unusual kinetic stability of a ground-state singlet oxomanganese(V) porphyrin. Evidence for a spin state crossing effect. *J. Am. Chem. Soc.* **121**, 2923–2924.
170. Sorokin, A., A. Robert, and B. Meunier (1993). Intramolecular kinetic isotope effects in alkane hydroxilations catalyzed by manganese and iron porphyrin complexes. *J. Am. Chem. Soc.* **115**, 7293–7299.
171. Merlau, M.L., W.J. Grande, S.T. Nguyen, and J.T. Hupp (2000). Enhanced activity of manganese(III) porphyrin epoxidation catalysts through supramolecular complexation. *J. Mol. Cat. A* **156**, 79–84.
172. Belvedere, S. and R. Breslow (2001). Regioselective oxidation of steroids by a manganese porphyrin carrying metal coordinating groups. *Bioorg. Chem.* **29**, 321–331.
173. Collman, J.P., C.J. Sunderland, and R. Boulatov (2002). Biomimetic studies of terminal oxidases: Trisimidazole picket metalloporphyrins. *Inorg. Chem.* **41**, 2282–2291.
174. Collman, J.P., I.M. Shiryaeva, and R. Boulatov (2003). Effect of electron availability on selectivity of O_2 reduction by synthetic monometallic Fe porphyrins. *Inorg. Chem.* **42**, 4807–4809.

175. Collman, J.P., R. Boulatov, I.M. Shiryaeva, and C.J. Sunderland (2002). Distal Cu ion protects synthetic heme/Cu analogues of cytochrome oxidase against inhibition by CO and cyanide. *Angew. Chem. Int. Ed. Engl.* **41**, 4139–4142.
176. Collman, J.P. and R. Boulatov (2002). Electrocatalytic O_2 reduction by synthetic analogues of the heme/Cu site of cytochrome oxidase incorporated in a lipid film. *Angew. Chem. Int. Ed. Engl.* **41**, 3487–3489.
177. Boulatov, R., J.P. Collman, I.M. Shiryaeva, and C.J. Sunderland (2002). Functional analogues of the dioxygen reduction site in cytochrome oxidase: Mechanistic aspects and possible effects of Cu–B. *J. Am. Chem. Soc.* **124**, 11923–11935.
178. Shiryaeva, I.M., J.P. Collman, R. Boulatov, and C.J. Sunderland (2003). Nonideal electrochemical behavior of biomimetic iron porphyrins: Interfacial potential distribution across multilayer films. *Anal. Chem.* **75**, 494–502.
179. Collman, J.P., M. Marrocco, P. Denisevich, C. Koval, and F.C. Anson (1979). Potent catalysis of the electroreduction of oxygen to water by dicobalt porphyrin dimers adsorbed on graphite-electrodes. *J. Electroanal. Chem.* **101**, 117–122.
180. Steiger, B. and F.C. Anson (1994). New electrocatalysts for the 4-electron reduction of dioxygen based on (5,10,15-tris(pentaamineruthenium(II)-4-cyanophenyl)-20-(1-methylpyridinium-4-yl) porphyrinato)cobalt(II) immobilized on graphite-electrodes. *Inorg. Chem.* **33**, 5767–5779.
181. Steiger, B. and F.C. Anson (1997). [5,10,15,20-tetrakis(4-((pentaammineruthenio)-cyano)phenyl)porphyrinato]cobalt(II) immobilized on graphite electrodes catalyzes the electroreduction of O_2 to H_2O, but the corresponding 4-cyano-2, 6-dimethyl-phenyl derivative catalyzes the reduction only to H_2O_2. *Inorg. Chem.* **36**, 4138–4140.
182. Anson, F.C., C.N. Shi, and B. Steiger (1997). Novel multinuclear catalysts for the electroreduction of dioxygen directly to water. *Acc. Chem. Res.* **30**, 437–444.
183. Winnischofer, H., V.Y. Otake, S. Dovidauskas, M. Nakamura, H.E. Toma, and K. Araki (2004). Supramolecular tetracluster-cobalt porphyrin: A four electron-transfer catalyst for dioxygen reduction. *Electrochim. Acta* **49**, 3711–3718.
184. Hutchison, J.E., T.A. Postlethwaite, C.H. Chen, K.W. Hathcock, R.S. Ingram, W. Ou, R.W. Linton, R.W. Murray, D.A. Tyvoll, L.L. Chng, and J.P. Collman (1997). Electrocatalytic activity of an immobilized cofacial diporphyrin depends on the electrode material. *Langmuir* **13**, 2143–2148.
185. Collman, J.P., M.S. Ennis, D.A. Offord, L.L. Chng, and J.H. Griffin (1996). Electrocatalytic reduction of dioxygen by diruthenium cofacial diporphyrins axially-bound to a gold-supported, self-assembled monolayer. *Inorg. Chem.* **35**, 1751–1752.
186. Collman, J.P., C.J. Sunderland, K.E. Berg, M.A. Vance, and E.I. Solomon (2003). Spectroscopic evidence for a heme-superoxide/Cu(I) intermediate in a functional model of cytochrome C oxidase. *J. Am. Chem. Soc.* **125**, 6648–6649.
187. Arana, C., M. Keshavarz, K.T. Potts, and H.D. Abruna (1994). Electrocatalytic reduction of CO_2 and O_2 with electropolymerized films of vinyl-terpyridine complexes of Fe, Ni and Co. *Inorg. Chim. Acta* **225**, 285–295.
188. Bard, A.J. (1994). *Integrated Chemical Systems. A Chemical Approach to Nanotechnology*. John Wiley & Sons, Inc., New York.
189. Murray, R.W.(1984). In A.J. Bard (ed), *Electroanalytical Chemistry*. Marcel Dekker, New York, p. 191.
190. Murray, R.W., A.G. Ewing, and R.A. Durst (1987). Chemically modified electrodes: Molecular design for electroanalysis. *Anal. Chem.* **59**, 379A–390A.
191. Arrigan, D.W.M. and L. Le Bihan (1999). A study of L-cysteine adsorption on gold via electrochemical desorption and copper(II) ion complexation. *Analyst* **124**, 1645–1649.
192. Mandler, D. and I. Turyan (1996). Applications of self-assembled monolayers in electroanalytical chemistry. *Electroanalysis* **8**, 207–213.

193. Abatti, D., M.E.D. Zaniquelli, Y. Iamamoto, and Y.M. Idemori (1997). Porphyrin LB film as a catalyst for alkene epoxidation. *Thin Solid Films* **310**, 296–302.
194. Blair, T.L., J.R. Allen, S. Daunert, and L.G. Bachas (1993). Potentiometric and fiber optic sensors for pH based on an electropolymerized cobalt porphyrin. *Anal. Chem.* **65**, 2155–2158.
195. Bettelheim, A., B.A. White, and R.W. Murray (1987). Electrocatalysis of dioxygen reduction in aqueous acid and base by multimolecular layer films of electropolymerized cobalt tetra(ortho-aminophenyl)porphyrin. *J. Electroanal. Chem.* **217**, 271–286.
196. Macor, K.A., Y.O. Su, L.A. Miller, and T.G. Spiro (1987). Electrochemical and resonance raman spectroscopic characterization of polyaniline and polyaniline–metalloporphyrin electrode films. *Inorg. Chem.* **26**, 2594–2598.
197. Daunert, S., S. Wallace, A. Florido, and L.G. Bachas (1991). Anion-selective electrodes based on electropolymerized porphyrin films. *Anal. Chem.* **63**, 1676–1679.
198. Kliza, D.M. and M.E. Meyerhoff (1992). Potentiometric anion response of poly(tetrakis (*p*-aminophenyl)porphyrin film modified electrodes. *Electroanalysis* **4**, 841–849.
199. Hayon, J., A. Raveh, and A. Bettelheim (1993). Electrocatalytic properties of chemically polymerized films of cobalt, iron and manganese tetrakis(*o*-aminophenyl)porphyrins. *J. Electroanal. Chem.* **359**, 209–221.
200. Younathan, J., K.S. Wood, and T.J. Meyer (1992). Electrocatalytic reduction of nitrite and nitrosyl by iron(III) protoporphyrin-IX dimethyl ester immobilized in an electropolymerized film. *Inorg. Chem.* **31**, 3280–3285.
201. Macor, K.A. and T.G. Spiro (1983). Porphyrin electrode films prepared by electrooxidation of metalloprotoporphyrins. *J. Am. Chem. Soc.* **105**, 5601–5607.
202. Macor, K.A. and T.G. Spiro (1984). Oxidative electrochemistry of electropolymerized metalloporphyrin films. *J. Electroanal. Chem.* **163**, 223–236.
203. Araki, K., L. Angnes, and H.E. Toma (1995). Rectifying properties and photoconductivity of tetraruthenated nickel porphyrin films. *Adv. Materials* **7**, 554–559.
204. Angnes, L., C.M.N. Azevedo, K. Araki, and H.E. Toma (1996). Electrochemical detection of NADH and dopamine in flow analysis based on tetraruthenated porphyrin modified electrodes. *Anal. Chim. Acta* **329**, 91–95.
205. Quintino, M.S.M., K. Araki, H.E. Toma, and L. Angnes (2002). Batch injection analysis utilizing modified electrodes with tetraruthenated porphyrin films for acetaminophen quantification. *Electroanalysis* **14**, 1629–1634.
206. Azevedo, C.M.N., K. Araki, L. Angnes, and H.E. Toma (1998). Electrostatically assembled films for improving the properties of tetraruthenated porphyrin modified electrodes. *Electroanalysis* **10**, 467–471.
207. da Rocha, J.R.C., G.J.F. Demets, M. Bertotti, K. Araki, and H.E. Toma (2002). Charge transfer at electrostatically assembled tetraruthenated porphyrin modified electrodes. *J. Electroanal. Chem.* **526**, 69–76.
208. Kunitake, M., N. Batina, and K. Itaya (1995). Monolayers of 4-TMPyP on Au-111) surface by STM. *Langmuir* **11**, 2337–2340.
209. Toyama, M.M., G.J.F. Demets, K. Araki, and H.E. Toma (2000). Highly conductive electrostatically assembled porphyrazine films. *Electrochem. Commun.* **2**, 749–753.
210. Azevedo, C.M.N., K. Araki, H.E. Toma, and L. Angnes (1999). Determination of sulfur dioxide in wines by gas-diffusion flow injection analysis utilizing modified electrodes with electrostatically assembled films of tetraruthenated porphyrin. *Anal. Chim. Acta* **387**, 175–180.
211. da Rocha, J.R.C., L. Angnes, M. Bertotti, K. Araki, and H.E. Toma (2002). Amperometric detection of nitrite and nitrate at tetraruthenated porphyrin-modified electrodes in a continuous-flow assembly. *Anal. Chim. Acta* **452**, 23–28.

212. Fussa-Rydel, O., H.-T. Zhang, J.T. Hupp, and C.R. Leidner (1989). Electrochemical assembly of metallopolymeric films via reaction of coordinated 5-chlorophenanthroline. *Inorg. Chem.* **28**, 1533–1537.
213. Nogueira, A.F., L.F.O. Furtado, A.L.B. Formiga, M. Nakamura, K. Araki, and H.E. Toma (2004). Sensitization of TiO_2 by supramolecules containing zinc porphyrins and ruthenium-polypyridyl complexes. *Inorg. Chem.* **43**, 396–398.
214. Nogueira, A.F., A.L.B. Formiga, H. Winnischofer, M. Nakamura, F.M. Engelmann, K. Araki, and H.E. Toma (2004). Photoelectrochemical properties of supramolecular species containing porphyrin and ruthenium complexes on TiO_2 films. *Photochem. Photobiol. Sci.* **3**, 56–62.

7

Electrodes Modified with Monomeric M–N$_4$ Catalysts for the Detection of Environmentally Important Molecules

Tebello Nyokong

1. Introduction

Macrocyclic M–N$_4$ complexes such as metallophthalocyanines (MPc, Figure 7.1a) and metalloporphyrins (MP, Figure 7.1b), have been extensively studied as electrocatalysts for many reactions with the aim of developing sensitive, selective, and stable sensors. A good electrocatalyst must lower the redox potential, increase sensitivity and selectivity, and should show stability. MPc and MP complexes readily form thin film coatings for transforming nonselective unmodified surfaces into sensitive and selective sensors. The choice of these complexes as suitable electrocatalysts arises from their ability to change their oxidation states, but retain their stability and molecular structure during electrocatalysis. Modification of the ring by attaching different substituents and varying the nature of the central metal ion, can result in dramatic changes in the redox properties of the macrocyclic M–N$_4$ complexes. The central metal coordinated to the N$_4$ ring plays the main role in the electrocatalytic process. The potential at which catalytic currents are observed being closely related to the redox potentials of the central metals in these complexes[1–3]. Catalytic activity has in the past been observed mainly for those MP or MPc complexes containing electrochemically active central metals such as Fe, Co, Mn, Cr[1,4]. The generally accepted[1,3–7] mechanism for catalytic oxidation using M–N$_4$ complexes containing an electroactive central metal may be represented by Equations (7.1) and (7.2) as follows:

$$M^{n+}\text{–}N_4 \longrightarrow M^{(n-1)+}\text{–}N_4 + e^- \tag{7.1}$$

$$M^{(n-1)+}\text{–}N_4 + R \longrightarrow M^{n+}\text{–}N_4 + O \tag{7.2}$$

Tebello Nyokong • Department of Chemistry, Rhodes University, P.O. Box 94, Grahamstown, South Africa.

N$_4$-Macrocyclic Metal Complexes, edited by José H. Zagal, Fethi Bedioui and Jean-Pol Dodelet. Springer Science+Business Media, Inc., New York, 2006.

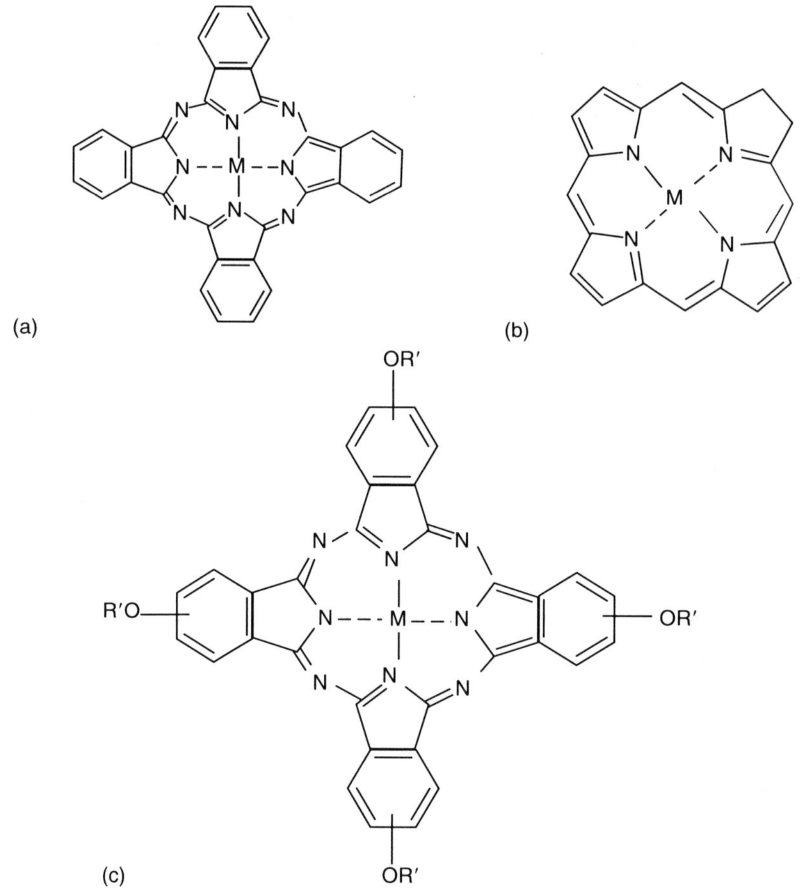

Figure 7.1. Molecular structures of (a) metallophthalocyanine (MPC), (b) metalloporphyrin (MP), and (c) ring substituted MPc where $R = (CH_2)_2N^+(R)_xI^-$.

where R and O are the reduced and oxidized forms of the species catalyzed. The catalyst is regenerated following the electron transfer. For reduction, catalysis would be mediated by the reduced form of the catalyst (e.g., $M^{(n+1)+}$–N_4).

M–N_4 complexes containing Ni(II) and Cu(II) also show catalytic activity[8–10]. In CuPc, ZnPc, and NiPc complexes the central metals are electrochemically inactive. The involvement of Ni^{III}/Ni^{II} couple in NiPc catalyzed reactions has been suggested[11], but solution electrochemistry of NiPc does not show the presence of this couple[12]. However, in porphyrins, electropolymerized $Ni^{II}P$ film is readily oxidized to the Ni^{III} species[7]. The oxidation of $Ni^{II}P$ to $Ni^{III}P$ occurs with difficulty in nonaqueous solutions. For NiP modified electrodes, the role played by the Ni porphyrin catalyst was described as being that of a simple organic layer, modifying the electrode surface without any specific chemical interaction with the analyte[8,9] since both the NiP and the unmetallated form H_2P showed the same catalytic behavior. This may suggest that catalytic activity in some cases is mediated by ring-based processes. In fact electrocatalytic reactions mediated by ring-based MPc redox processes have been reported[13,14]. The catalytic activities of the $M–N_4$ complexes containing a central metal which is not electroactive are sometimes much larger than for complexes containing electroactive central metal ion.

Monomeric $M–N_4$ complexes show catalytic activity either in solution or when adsorbed onto electrodes. For heterogeneous catalysis, the complexes are used to modify electrodes, resulting in chemically modified electrodes. Chemically modified electrodes (CMEs) are very useful in lowering the overpotential, increasing the rate of electrochemical reactions, hence increasing sensitivity and selectivity.

The $M–N_4$ monomers are widely employed for electrocatalysis[1,7]. The monomers readily adsorb onto electrodes to form electroactive surfaces. The methods employed for the adsorption are often fast. Carbon electrodes are popular for the fabrication of CME using $M–N_4$ complexes. The following are some of the methods which have been employed in electrode modification using $M–N_4$ monomers:

- (i) Dip-dry method[15–22]: where the electrode is immersed in a solution of the monomer. The thin film of the monomer occurs by spontaneous adsorption.
- (ii) Drop-dry method[23–27]: where a few drops of the monomer solution are applied to the electrode surface and the solvent allowed to dry.
- (iii) Spin-coating[28]: where a droplet of a solution of the monomer is applied to the surface of a rotating electrode. Multiple layers may be applied in this way until the desired thickness is obtained.
- (iv) Electrodeposition[29,30]: the monomer is adsorbed onto the electrode by cyclic voltammetry without polymer formation. And the first scan is generally similar to the second and subsequent cyclic voltammetry scans, in terms of potential values, but the current increases with scan number. The amount of catalyst on the surface may be controlled by the number of cyclic voltammetry scans.
- (v) Modified carbon paste electrodes[6,31–38]: These are prepared by thoroughly mixing paraffin oil, the solid monomer and graphite powder to form a paste, which is then packed into Teflon or glass tubes.
- (vi) Vacuum deposition[39], and screen printing[40].
- (vii) $M–N_4$ complexes have also been used to modify electrodes by incorporation into other polymers[41–50], such as plasticized poly(vinyl chloride) and polypyrrole.
- (viii) Self-assembled monolayers (SAMs): $M-N_4$ complexes substituted with sulfur donors are self-assembled onto substrates such as gold[51–59].
- (ix) Langmuir–Blodgett films[60,61].

Most electrocatalytic reactions are based on a signal generated by heterogeneous electron exchange between the dissolved analyte and the electrode which has been modified as described above. However, the use of $M–N_4$ complexes as homogeneous electrocatalysts where both the catalyst and the analyte are in solution has also received attention[23].

This review focuses on the use of monomers based on M–N_4 complexes for analysis of molecules which are of environmental importance. The inorganic pollutants discussed include sulfur dioxide, carbon dioxide, nitrites, cyanides, etc., and organic pollutants include phenols and chlorinated organics. Where necessary a comparison with analysis using polymeric M–N_4 complexes will be included.

2. Phenols, Organohalides, and Pesticides

Phenols are widespread in nature. Phenol is also produced by normal human metabolism from tyrosines. The bactericidal properties of phenol make it a useful substance for natural defense mechanisms in biological systems. Phenolic compounds are also released into the environment in a number of ways including during the manufacture of industrial products such as plastics, pharmaceuticals, dyes, and pesticides (e.g., pentachlorophenol). Insecticides containing toxic nitrophenols are used in agriculture. Many phenols are toxic and tend to persist in the environment. Organohalides are recognized as important environmental pollutants arising from industrial as well as biological sources. Halogenated methanes and ethanes are often used as solvents. Trichloroacetic acid is for example used as plaguicide in some countries, but it is known to be toxic. Organohalides have been a subject of government regulations in many countries.

Electrochemical methods are preferred in analysis of phenols and halogenated organics since often there is no need for extensive separation. However direct determination on noble metal electrodes is not favored due to high overpotentials. Electrochemical oxidation of phenols readily occurs on unmodified electrodes, but oxidation results in the formation of dimers which poison the electrodes, decreasing the oxidation currents[62]. In order to improve sensitivity and selectivity, chemically modified electrodes are employed. In this regard M–N_4 complexes have shown remarkable catalytic activity towards the detection of phenols and other species when either employed as homogeneous catalysts or when adsorbed to electrodes.

The term pesticides includes insecticides, fungicides, molluscicides, bactericides, rodentcides, fumigants, and herbicides. Pesticides are highly toxic. A number of methods have been employed in the determination of pesticides in the environment, these include spectrophotometry and GC coupled with MS. The latter being the most common[63]. However, electrochemical methods are increasingly been employed. The use M–N_4 modified electrodes for the detection of organo pollutants, including organochlorinated plaguicides has been reviewed[64].

2.1. Porphyrin and Salen Complexes

2.1.1. Organohalides

In general, phthalocyanines, porphyrins and corrins catalyze the reduction of organohalides and the overall reaction is shown by Equation (7.3)[65].

$$RX + H^+ + 2e^- \rightarrow RH + X^- \qquad (7.3)$$

Table 7.1 lists redox potentials for the catalysis of organohalides and phenols using MPc and MP catalysts. The reduction of *trans*-1,2-dibromocyclohexanes was catalyzed by Co(II) porphrin monomers at potentials $\geq$ 1 V positive of the uncatalyzed case[66]. High catalytic currents were obtained when using siloxane films on carbon electrodes containing specially designed diethylene glycol spaced Co porphyrin[66]. Fe^{III}TAPP monomers adsorbed onto carbon electrodes and following reduction of Fe^{III} to Fe^{I}, showed catalytic activity towards the reduction of alky halides such as benzyl bromide, Table 7.1[67]. A decrease in overvoltage of up to 1.0 V was obtained compared to unmodified carbon electrode. Catalytic activity was also observed for the Fe^{II} species[67].

Zheng *et al.* reported on the electrochemical reduction of the intermediates generated from the reaction of MTPP with benzyl chloride or butyl chloride and showed that CoTPP (Table 7.1) and FeTPP have higher catalytic activities than MnTPP, NiTPP, CuTPP, or ZnTPP[68,69]. The latter two complexes are not expected to show good catalytic activity since the central metals are electrochemically inactive. Electron-donating substituents on the porphyrin ring of these molecules enhanced the catalytic activity while the electron-withdrawing substituents led to a decrease in activity. Dobson and Saini reported on the catalytic activity of CoTPP towards the reduction of organohalides under both homogeneous and heterogeneous (drop dry method on graphite) catalysis[65]. The Co^{II}/Co^{I} couple catalyzed the reduction. The ease of detection of the organochloride depended on the chemical nature as follows: haloalkane > haloalkene > haloarene.

The reduction of benzyl bromide mediated by vitamin B12r, CoTSPc, and Co(salem) in a microemulsion under homogeneous conditions, resulted in products derived from the resulting radicals and anions[70]. The reactions were mediated by Co^{I} species[70]. Priyantha and Tambalo employed GCE coated with Co^{II}TPP or Fe^{III}TPP as electrocatalytic amperometric sensors for the detection of herbicides including 2,3-dichlorophenoxyacetic acid[71].

Salem derivatives have been employed (mainly under homogeneous conditions) for the catalytic reduction of organohalides. Salen complexes containing Co and Ni metals have received considerable attention compared to other metal(salen) derivatives. Co(salen) incapsulated into zeolyte was found to show less deactivation than when in homogeneous solutions[72]. Catalytic activity of Co^{II}salen towards reduction of organohalides was studied in organic solvents and ionic liquids[73,74], and was found to be simpler in the latter medium[73]. Electrogenerated Co^{I} salen catalyzed the reduction of 2,6-bis(chloromethyl)pyridine[75], and 1,1,2-trichloro-1,2,2-trifluoroethane[76]. Electrocatalytic reduction of dihalocyclopropanes (using various metal(salen) complexes) resulted mainly in the formation of monohalocyclopropanes[77]. Reductive coupling of benzyl bromide catalyzed by a cobalt complex containing two salen units resulted in the formation bibenzyl as the main product[78]. However when Co^{II} contained only one salen unit, the main product was toluene[78,79]. Benzyl chloride is converted to a mixture of *cis*- and *trans*- stilbene in the presence of electrogenerated Co^{I} salen[80]. When Co(salen) films were employed for the detection of organohalides (e.g., trichlophenol, TCA, dichloroacetic acid, etc.), a 1.0 V decrease in reduction potential was achieved accompanied by a large increase in the reduction current[81].

Table 7.1. Electrochemical Data for the Electrocatalytic Detection of Organohalides and Phenols Using M–N_4 complexes[a]

M–N_4 complex	Electrode	Method of modification	Analyte	E (V) (SCE)	Medium	O/R	References
Organohalides and pesticides							
FeTAPP	GCE	Drop dry	Benzyl bromide	−0.9	DMF/$TBAPF_6$	R	[67]
FeTAPP	GCE	Drop dry	Ph_2MeBr	−0.2	DMF/$TBAPF_6$	R	[67]
FeTAPP	GCE	Drop dry	C_2Cl_6	−1.0	DMF/$TBAPF_6$	R	[67]
CoTPP	Pt	Solution	CCl_4	−0.75	DMF/TBAP	R	[67]
CoTPP	Pt	Solution	DCPA	−0.9	DMF/TBAP	R	[67]
CoTPP	Graphite	Drop dry	CCl_4 or DDT	−0.4	Aq. CH_3CN	R	[67]
CoTPP	Graphite	Drop dry	PCP	−0.42	Aq. CH_3CN	R	[67]
CoTPP	Pt	Solution	Benzyl chloride	−1.38	DMSO/TBAP	R	[69]
CoPc	CPE		Amitrole	+0.4	0.1 M NaOH	O	[103]
FePc/DDAB/Clay	PGE	Membrane	TCA	−1.3	0.1 M KBr	R	[91]
NiTSPc/DDBA	GCE	Dip dry	TCA /DBB/t-DBCH	−1.26	Microemulsion	R	[92]
NiPc/DDBA	GCE	Dip dry	TCA or DBB or t-DBCH	−1.38	Microemulsion	R	[92]
CuTSPc/DDBA	GCE	Dip dry	TCA or DBB or t-DBCH	−1.39	Microemulsion	R	[92]
CuPc/DDBA	GCE	Dip dry	TCA or DBBor t-DBCH	−1.41	Microemulsion	R	[92]
Phenols							
NiPPIX	GCE	Polymer	Aminophenol	+0.42	pH 6–8	O	[88]
NiPPIX	GCE	Polymer	*p*-chlorophenol	+0.76	pH 6–8	O	[88]
NiPPIX	GCE	Polymer	*p*-nitrophenol	+1.04	pH 6–8	O	[88]
FeTMPyP/DNA	GCE	Solution	*p*-nitrophenol	−0.85	pH9	R	[87]
CoPc	GCE	Drop dry	phenol	−1.08	0.05 M H_2SO_4	O	[108]
CoPc	GCE	Drop dry	2-chlorophenol	−1.05	0.05 M H_2SO_4	O	[108]
CoPc	GCE	Drop dry	4-chlorophenol	−1.06	0.05 M H_2SO_4	O	[108]

O: oxidation; R: Reduction;

[a] To convert potential vs. Ag|AgCl to SCE, a correction factor of −0.045 V (ref. [12]) has been applied.

Reduction of bromo hexene was achieved in the presence of Ni salen and cyclam complexes[82]. Electropolymerized Ni(II) salen catalyzed the reduction of iodoethane and 2-iodopropane[83], with efficient regeneration of the Ni(II)salen catalyst. Under homogeneous conditions, electrogenerated Ni^{I} catalyses the reduction of dihaloalkanes[84] and cyclohexanecarbonyl chloride[85]. Ni(salen) has a high catalytic activity towards the electrochemical reduction of perfluoroalkyl chlorides[86].

2.1.2. Phenols

Reports on the use of monomeric or polymeric porphyrins as electrocatalysts for the detection of phenols are limited. Chen and Chen reported on the use of FeTMPyP in solution, in the presence of DNA or electrodeposited on a GCE previously modified with DNA, for the catalysis of the reduction of *p*-nitrophenol[87].

MPPIX (M = Ni^{II}, Co^{II}, and Cu^{II}) complexes polymerized onto a GCE showed good catalytic activity towards the detection of nitrophenol, aminophenol, and chlorophenol, Table 7.1[88]. For the NiPPIX complex, the potential for the oxidation of aminophenol was much lower than for chlorophenol and nitrophenol, with nitrophenol being oxidized at a much larger potential, Table 7.1. The catalytic response depended on the nature of the central metal ion since the catalytic activity was mediated by an interaction of the phenol (as an axial ligand) with the central metal[88]. A review by Biesaga *et al.*[89] summarized the results obtained on using porphyrins as potentiometric and amperometric catalysts for analysis of many species. The analysis of phenols and organohalides were included in the review.

2.2. Phthalocyanine Complexes

2.2.1. Organohalides and Pesticides

MPc complexes have been employed as catalytic components in microemulsions and as composite films for the analysis of phenols and organohalides. Rusling and coworkers reported on FePc/DDAB, CuPc/DDAB, CuTSPc/DDAB, NiPc/DDAB, NiTSPc-DDAB, ZnPc/DDAB or ZnPc/CTAB surfactant films[90–97], and Jiang *et al.*[98] on CoTSPc/DDAB for use in catalyzed reduction of trichloroacetic acid (TCA) and other organohalides. The catalytic reduction of TCA was more efficient in the acetonitrile/water solvent mixture than in the microemulsions[97]. Table 7.1 shows that lower potentials were observed for catalysis of TCA on surfactant films containing NiPc complexes than for the corresponding CuPc species.

Rusling and coworkers also studied debromination and dechlorination of organohalidess by MPc in microemulsions[96, 99–101]. Reductive dechlorination on clay surfactant films containing MPc complexes was better on CoPc compared to FePc, since the former remained intact in the film while the latter decomposed[100]. The order of the catalytic activity was CoPc>FePc>ZnPc. $Co^{I}Pc$ species was implicated as an intermediate in the catalytic process[99]. The high activity of CoPc was correlated with an inner sphere mechanism, while the low activity of ZnPc is consistent with outer sphere pathways for complexes which accept electrons on the ring[99]. de la Fuente *et al.*[102] reported that polypyrrole electrodes modified with NiPc electrocatalyzed the determination of pollutants and drugs.

CoPc modified carbon paste electrodes were reported by Chicharo *et al.* to show good catalytic activity towards the measurement of triazolic herbicides such as amitrole at low oxidation potential (+0.4 V, Table 7.1) in basic media, a detection limit of 0.04 μg mL^{-1} was obtained using a injection system[103]. A screen-printed carbon electrode which was impregnated with CoPc electrocatalyst, was employed in conjunction with acetylcholinesterase by Hartley and Hart for the reduction of organophosphate pesticides[104]. The detection limits were of the order of 10^{-8} and 10^{-7} M[104].

Using CuTSPc as component of the microemulsion lowered the overpotential for the reduction of *o*-chlorophenol by 1 V, hence showed promise for catalytic decomposition of DDT[105].

2.2.2. Phenols

Reports on the use of monomeric MPc complexes for the electrocatalysis of phenols are rare, hence the use of some polymeric MPc complexes will be included in this section. Biosensors based on enzymes have been developed for many electrochemical analyses. The use of MPc complexes as part of the enzyme improves the sensitivities of the biosensors. Ozsoz *et al.* used CoPc monomer dispersed in mushroom tissue electrode as a catalyst for the analysis of phenolic compounds[106]. The enzymatic reaction between mushroom and the phenolic compounds was coupled with the catalytic activity of CoPc. The CoPc dispersed electrodes gave shorter response times and lower potentials compared to conventional tissue biosensors[106].

The use of MPc complexes for the oxidation of phenols was explored by Zagal and coworkers[107] by employing Ni as the central metal in the catalytic oxidation of phenols. Electro-oxidation of 2-chlorophenol occurred on polymerized NiTSPc[107]. The use of CoPc monomer to modify GCE resulted in less fouling of the GCE and improved sensitivity[108].

3. Thiols

Thiols are impurities distributed among petroleum products. They cause foul odor and deterioration of additives in finished products. Some thiol compounds such as 6-mercaptopurine and 6-thioguanine are used in medicine (e.g., treatment of leukemia) while others form essentials components of biological systems (e.g., amino acids). Some thiols (e.g., cysteine) are not readily detected on bare electrodes, hence the need for chemical modification of electrodes with electroactive catalysts. The use of CME also enhances sensitivity for the detection of thiols even for those that can be determined directly on bare electrodes. Phthalocyanines[1] and porphyrins have been studied extensively as electrocatalysts for the detection of thiols. Electrocatalytic oxidations of cysteine and 2-mercaptoethanol have received considerable attention over many years. The aim is to lower the oxidation potentials and to improve the detection limits to the values obtained in biological systems.

3.1. Porphyrin Complexes

As expected, the potentials vary with the nature of the catalyst and with the media. Chen reported on a series of porphyrins for the detection of thiols[109,110]. In aqueous solutions, electrochemically generated Fe^{I}TSPP (from the reduction of Fe^{III}TSPP or the $Fe^{III}\mu$ oxo dimer) or Co^{I}TMPyP (from the reduction of Co^{II}TMPyP) electrocatalytically reduced L-cystine and the oxidized form of glutathionine; whereas Fe^{III}TSPP, Co^{III}TMPyP and Mn^{III}TMPyP, electrocatalyzed the oxidation of L-cysteine or the reduced form of glutathionine, Table 7.2[109,110]. Electrochemically generated Mn^{IV}TMPyP could also oxidize L-cysteine and the reduced form of glutathionine. Table 7.2 shows that as expected, the potentials for the oxidation or reduction of thiols depends on the catalyst and the medium (e.g., pH). Bedioui and coworkers employed polypyrrole films (electrodeposited on ITO or Pt) containing alkylammonium groups with tectracationic Co porphyrin incorporated into the film to reach total saturation of the alkyammonium sites, for the catalytic oxidation of *N*-acetyl-L-cysteine in aqueous solutions[111].

Wang, Pang, and coworkers showed that water soluble porphyrins (H_2TTMAPP, CoTTMAPP, and CoTSPP, Table 7.2) catalyze the oxidation of cysteine, 3-mercaptopropionic acid, mercaptoethanoic acid, thioacetamide, ascorbic acid, hydrazine, and hydroxylamine[112,113]. There was spectroscopic evidence for the formation of an adduct between the porphyrins and the analyte[112]. It was demonstrated that the N_4 internal ring of the porphyrin plays an important role in the electrocatalytic reduction of cystine[114]. Water soluble CoTTMAPP was found to be a good catalyst for the homogeneous reduction of dithiodipropionic acid[115]. Rea *et al.*[116] and Azevedo *et al.*[117] reported on the use of mono-, di-, tri- and tetra-ruthenated MTMPyP and *trans* di-ruthenated MOEP coated on carbon electrodes as effective sensors for sulfide using the oxidation wave of Ru^{III}/Ru^{II} couples[116,117]. At least two peripheral Ru atoms are necessary to trigger the catalysis, and no major influence of the central metal ion in the porphyrin was detected[116].

3.2. Phthalocyanine Complexes

A large amount of data exists for the oxidation of thiols using monomeric MPc catalysts compared to porphyrins, Table 7.2. As Table 7.2 shows the potential for the oxidation of cysteine is highly dependent on pH being lower in basic media. The study of the oxidation of 2-ME has mainly been conducted in basic media, and this species is readily oxidized on MPc modified electrodes, as judged by the low oxidation potential, Table 7.2. In general the potentials for the oxidation of thiols is dependent on the MPc catalysts and the electrolyte as was the case with porphyrin catalysts.

Halbert and Baldwin[6,118] as well as Zagal and coworkers[25,119] were among the first groups to report on the use of MPc complexes for the electrocatalytic oxidation of thiols (especially cysteine). Halbert and Baldwin employed a CPE modified with CoPc for the electrocatalytic oxidation of sulfhydryl containing

Table 7.2. Electrochemical Data for the Electrocatalytic Detection of Thiols Using M–N_4 Complexes[a]

$M–N_4$ complex	Electrode	Method of modification	Analyte	E(V) (SCE)	Medium	O/R	References
H_2TTMAPP	GCE	Solution	Cysteine	+0.60	0.05 M H_2SO_4	O	[112]
Co(2-TMPyP)	GCE	Solution	Cysteine	+0.54	pH 4.5	O	[109]
FeTSPP	GCE	Solution	Cysteine	+0.96	pH 2.0	O	[109]
CoTCPP	Pt/poly Pyrrole	poly pyrrole	Acetyl-L-Cysteine	+0.2	0.01 M KNO_3	O	[111]
FeTSPP	GCE	Solution	Cysteine	+0.88	pH 2.0	O	[110]
MnTMPyP	GCE	Solution	Cysteine	+0.21	pH 13.0	O	[110]
Co(2-TMPyP)	GCE	Solution	Cystine	−0.64	pH 4.5	R	[109]
FeTSPP	GCE	Solution	Cystine	−1.08	pH 5.5,11.5	R	[110]
CoTTMAPP	GCE	Solution	Cystine	−0.77	0.2 M H_2SO_4	R	[114]
CoTSPP	GCE	Solution	Cystine	−0.74	0.2 M H_2SO_4	R	[114]
Co(2-TMPyP)	GCE	Solution	GSSG	−0.65	pH 4.5	R	[109]
FeTSPP	GCE	Solution	GSSH	−1.08	pH 5.5,11.5	R	[110]
FeTSPP	GCE	Solution	GSH	+0.88	pH 2.2	O	[110]
FeTSPP	GCE	Solution	GSH	+0.96	pH 2.0	O	[109]
Co(2-TMPyP)	GCE	Solution	GSH	+0.51	pH 4.5	O	[109]
MnTMPyP	GCE	Solution	GSH	+0.30	pH 11.5	O	[110]
H_2TTMAPP	GCE	Solution	Thioacetamide	+0.90	0.05 M H_2SO_4	O	[112]
CoTCPc	GCE	Electrodep	Cysteine	+0.78	pH 3.5	O	[3]
CoTSPc	GCE	Electrodep	Cysteine	+0.77	pH 3.5	O	[3]
CoPc	GCE	Electrodep	Cysteine	+0.63	pH 3.5	O	[3]
CoTBPc	GCE	Electrodep	Cysteine	+0.69	pH 3.5	O	[3]
CoTNPc	GCE	Electrodep	Cysteine	+0.70	pH 3.5	O	[3]
CoTCPc	GCE	Electrodep	Cysteine	+0.16	pH 9	O	[3]
CoTSPc	GCE	Electrodep	Cysteine	+0.10	pH 9	O	[3]
CoPc	GCE	Electrodep	Cysteine	−0.03	pH 9	O	[3]
CoTBPc	GCE	Electrodep	Cysteine	+0.08	pH 9	O	[3]
CoTNPc	GCE	Electrodep	Cysteine	−0.03	pH 9	O	[3]
CoPc	CPE		Cysteine	+0.32	pH 5.5	O	[140]
CoTSPc	OPGE	Dip dry	Cysteine	+0.2	0.2 M NaOH	O	[25]
CoOBTPc	Au	SAM	Cysteine	+0.38	pH 4	O	[51]
CoOHETPc	Au	SAM	Cysteine	+0.46	pH 4	O	[150]
FeOBTPc	Au	SAM	Cysteine	+0.29	pH 4	O	[150]
FeOHETPc	Au	SAM	Cysteine	+0.44	pH 4	O	[150]
RhPc(Cl)(py)	GCE	Drop dry	Cysteine	+1.05	pH 7.2	O	[13]
RhPc(Cl)(DMSO)	GCE	Drop dry	Cysteine	+0.90	pH 7.2	O	[13]
$[RhPc(CN)_2]^-$	GCE	Drop dry	Cycteine	+0.81	pH 7.2	O	[13]
$OsPc(py)_2$	GCE	Drop dry	Cysteine	+0.99	pH 7.2	O	[13]
$[OsPc(CN)_2]^{2-}$	GCE	Drop dry	Cycteine	+1.05	pH 7.2	O	[13]
$OsPc(DMSO)_2$	GCE	Drop dry	Cysteine	+1.01	pH 7.2	O	[13]

Table 7.2. Continued.

$RuPc(py)_2$	GCE	Drop dry	Cysteine	+1.15	pH 7.2	O	[13]
$[RuPc(CN)_2]^{2-}$	GCE	Drop dry	Cycteine	+0.87	pH 7.2	O	[13]
$RuPc(DMSO)_2$	GCE	Drop dry	Cysteine	+0.94	pH 7.2	O	[13]
MnTSPc	OPGE	Dip dry	Cysteine	+0.15	0.2 M NaOH	O	[25]
OMoPc	CPE		Cysteine	+0.29	0.05 M H_2SO_4	O	[139]
OMoTSPc	CPE		Cysteine	+0.25	0.05 M H_2SO_4	O	[139]
FeTSPc	OPGE	Dip dry	Cysteine	+0.2	0.2 M NaOH	O	[25]
CoOBTPc	Au	SAM	Homocysteine	+0.48	pH 4	O	[150]
CoOHETPc	Au	SAM	Homocysteine	+0.50	pH 4	O	[150]
FeOBTPc	Au	SAM	Homocysteine	+0.40	pH 4	O	[150]
FeOHETPc	Au	SAM	Homocysteine	+0.41	pH 4	O	[150]
CoOBTPc	Au	SAM	Penicillamine	+0.54	pH 4	O	[150]
CoOHETPc	Au	SAM	Penicillamine	+0.51	pH 4	O	[150]
FeOBTPc	Au	SAM	Penicillamine	+0.41	pH 4	O	[150]
FeOHETPc	Au	SAM	Penicillamine	+0.40	pH 4	O	[150]
FePc	OPGE	Drop dry	2-ME	−0.28	pH 10.5	O	[121]
FeOMePc	OPGE	Drop dry	2-ME	−0.33	pH 10.5	O	[121]
CoTAPc	VCE	Polymer	2-ME	−0.29	0.5 M NaOH	O	[120]
CoTAPc	VCE	Drop dry	2-ME	−0.32	0.5 M NaOH	O	[120]
CoTSPc/Ppy	GCE	Addition to Ppy	2-ME	−0.2	0.1 M NaOH	O	[151]
CoPc	CPE		6-MP	+0.23	pH 7	O	[118]
CoPc	CPE		6-MPR	+0.29	pH 7	O	[118]
CoPc	CPE		6-TG	+0.29	pH 7	O	[118]
CoPc	CPE		6-TGR	+0.28	pH 7	O	[118]
CoPc	CPE		6-TX	+0.41	pH 7	O	[118]
CoPc	CPE		6-TU	+0.34	pH 7	O	[118]
CoPc	CPE		6-MTG	+0.69	pH 7	O	[118]
CoPc	CPE		6-MMP	+0.59	pH 7	O	[118]
CoPc	CPE		glutathionine	+0.78	pH 2.4	O	[119]
CoPc	CPE		MBT	+0.23	pH 10	O	[35]
CoPc	CPE		MBI	+0.41	pH 10	O	[35]
CoTTMAPP	GCE	Solution	MEA	+0.7	0.05 M H_2SO_4	O	[112]

O: oxidation; R: Reduction;
[a] To convert potential vs. Ag/AgCl to SCE, a correction factor of −0.045 V (ref. [12]) has been applied. Electrodep: electrodeposition.

compounds (cysteine, homocysteine, *N*-acetylcysteine, glutathionine, 6-mercaptopurine, 6-thioguanine and glutathionine)[6, 118]. The catalytic peaks were obtained in a narrow range of 0.75–0.85 V.

Zagal, Bediuoi, and coworkers have continued to study monomeric MPc complexes as electrocatalysts for the oxidation of thiols in recent years[120–126]. From their study on the catalytic behavior of adsorbed or polymeric CoTAPc towards the oxidation of 2-mercaptoethanol (2-ME), Griveau *et al.*[120] derived the mechanism given by Equations (7.4)–(7.7) for the electrocatalytic process.

$$Co^{II}TAPc \leftrightarrow [Co^{II}TAPc]^{+\bullet} + e^- \quad (7.4)$$

$$[Co^{II}TAPc]^{+\bullet} + RS^- \rightarrow [RS\text{–}Co^{II}TAPc]^{-\bullet} \quad (7.5)$$

$$[\text{RS–Co}^{\text{II}}\text{TAPc}]^{-\bullet} \rightarrow [\text{RS–Co}^{\text{II}}\text{TAPc}]^{+\bullet} + \text{RS}^{\bullet} + e^{-} \quad (7.6)$$

$$\text{RS}^{\bullet} + \text{RS}^{\bullet} \rightarrow \text{RSSR} \quad (7.7)$$

The potential for the oxidation of 2-ME was almost similar for both the monomer and polymer, Table 7.2.

3.2.1. Adsorbed Monomer: Effects of Ring Substituents

Zagal and coworkers studied the effect of ring substituents on the electrocatalytic behavior of MPc complexes (FeTSPc, FeTCPc, FeOMePc, $FePc(Cl)_{16}$ and FePc) adsorbed on OPGE, for the oxidation of 2-mercaptoethanol[121].

Complexes containing electron-withdrawing groups ($-CO_2^-$, $-SO_3^-$, Cl^-) shifted the catalytic peak to more positive potentials, and the ones containing electron-donating methoxy groups to more negative potentials[121]. The catalytic activity was measured by the difference in peak potentials between the oxidation peak of the RSH and the reduction peak of the product (RSSR). That is, the higher the reversibility of the RSH/RSSR couple, the higher the catalytic activity. Based on this definition, the catalytic activities of the MPc complexes towards the oxidation of 2-mercaptoethanol was found to decrease as follows: FeOMePc > FePc > FeTCPc > FeTSPc > $FePc(Cl)_{16}$[121]. The last species contains electron-withdrawing substituents hence will be difficult to oxidize and will be a poor catalyst for oxidation, whereas the electron-donating Me group will make the complex more readily oxidized and hence a good electrocatalyst for oxidation. CoOEHPc monomer adsorbed onto graphite electrodes shows a higher catalytic activity towards oxidation of 2-ME than adsorbed CoTSPc or CoPc[125]. This was attributed to the electron-donating ability of the ethylhexyloxy substituents in the former[125]. Ring substituted $CoPc(X)_4$ (where X = NH_2, NO_2, $C(CH_3)_3$, SO_3H and COOH, giving CoTAPc, CoTNPc, CoTBPc, $CoTSPc(H)_4$, and CoTCPc, respectively) showed catalytic activity towards the oxidation of cysteine, with the catalytic activity increasing with the ring substituent as follows: COOH > $C(CH_3)_3$ > SO_3H > NH_2 > NO_2. The potential for the catalytic oxidation of cysteine was closely related to the Co^{III}/Co^{II} couple of the CoPc species in acid media and to Co^{II}/Co^{I} couple in basic media, Table 7.2[3]. The redox couple Co^{II}/Co^{I} was also implicated for the oxidation of glutathionine[1]. The Fe^{II}/Fe^{I} couple has also been implicated for electrocatalyzed oxidation of cysteine by FeTSPc[126]. Kimura *et al.* showed that the catalytic activity of the cationic complexes shown in Figure 7.1c towards the oxidation of 2-mercaptoethanol depended on the length of the alkyl chain, with those catalysts having a long chain showing lower catalytic activity[127]. Perchlorinated FePc has also been employed in another study for the catalytic oxidation of 2-ME[128].

3.2.2. Adsorbed Monomer: Effects of Central Metals

Electrocatalytic oxidation of cysteine on graphite electrodes modified with MTSPc complexes showed that the catalytic activity decreased with M as follows: Co > Fe > Mn > Ni > Cu, whereas the electrocatalytic reduction of cystine showed the following trend Mn > Fe > Co > Cu > Ni[25,129]. In both cases, the MPc complexes containing electroactive central metal shows better catalytic activity. The highest catalytic activity for the oxidation of 2-ME was observed for CoTSPc[1, 130] followed by FeTSPc[130]. FePc complexes were also found to be the most active (compared to MnPc, NiPc, and CuPc) for the electrocatalytic oxidation of cysteine[131] and differential oxidation of thiols occurred on CPE modified with CoPc[132].

Oxidation of 2-mercaptoethanol and other thiols on CoTAPc electropolymerized on carbon electrodes, showed that the activity of the electrode remained similar to that of the adsorbed monomer, with the polymer showing higher stability than the monomer[124]. The mechanisms for the catalytic oxidation of 2-ME by CoTAPc monomer or polymer were similar.

3.2.3. Adsorbed Monomer: Effects of Axial Ligands

Platinum group metal phthalocyanine monomers adsorbed on GCE autocatalyze the oxidation of cysteine depending on the nature of axial ligands[13], Table 7.2. When DMSO or cyanide were employed as axial ligands, autocatalytic behavior was observed, Figure 7.2. These complexes also catalyzed the oxidation of methionine, hydroxylamin, and hydrazine[13]. Ring based redox processes were implicated in the catalytic process shown by the following mechanism (Equations (7.8)–(7.10)):

$$\mathrm{MPc}(-2) \longrightarrow [\mathrm{MPc}(-1)]^+ + e^- \tag{7.8}$$

$$2[\mathrm{MPc}(-1)]^+ \longrightarrow ([\mathrm{MPc}(-1)]^+)_2 \tag{7.9}$$

$$\mathrm{MPc}(-1)]^+ + \mathrm{RSH} \longrightarrow \mathrm{MPc}(-2) + \mathrm{RSSR} \tag{7.10}$$

where RSH and RSSR are cysteine and cystine, respectively.

The catalytic process was also dependent on the nature of the central metal, with lower potentials being observed on RuPc and RhPc complexes compared to OsPc complexes[13], Table 7.2.

3.2.4. Other Methods of Electrode Modification

In addition to studies on adsorbed monomers, other methods of electrode modifications for the detection of thiols have appeared in the literature. Zagal and

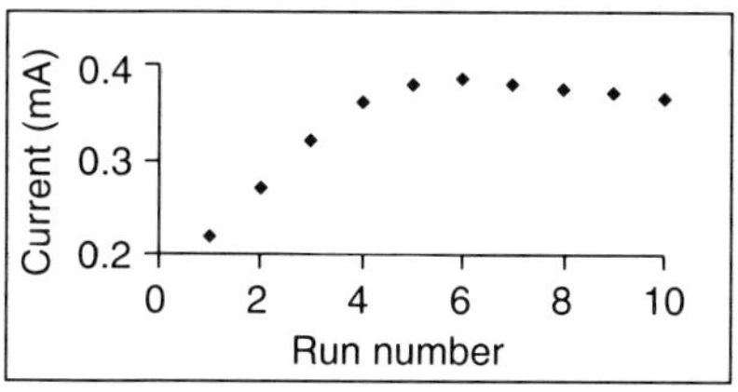

Figure 7.2. The variation of peak currents with scan number for the electrocatalytic oxidation of cysteine on a GCE modified with $[(CN)_2RhPc]^-$. Scan rate = 100 mV s^{-1}. Cysteine concentration = 0.03 mol dm^{-3}. (Reproduced with permission from Ref. [13] Figure 7.2).

coworkers studied the supported CoPc (on poly(2-chloroaniline)) as catalysts for the electrooxidation of 2-mercaptoethanol, glutathione, and hydrazine[133]. Wring *et al.* reported on a graphite–epoxy resin composite electrode modified with CoPc for the detection of glutathionine in whole blood[134] using HPLC connected to electrochemical detection system. The use of CoPc resulted in the lowering of overpotential for glutathionine oxidation by 750 mV[134]. Wang *et al.* reported on the use of graphite epoxy electrodes modified with CoPc, PVP and cation exchange resin for the detection of hydrazine, L-cysteine, and penicillamine[135]. Retamal *et al.* showed that the diffusion rate of the electroactive species (e.g., 2-ME) was improved by using an electrode modified with CoPc together with PANCl[136].

Ballarin *et al.* used sonogel–carbon composite electrodes containing CoPc for the oxidation of cysteine[137, 138]. The catalytic process is mediated by the generation of the $Co^{III}Pc$ species. Oxidation of sulfide ion was catalyzed by microcrystals of CoPc immobilized at a surface of paraffin impregnated graphite electrode[2]. Filanovsky reported the use of a carbon electrode prepared by hand pressing carbon powder mixed with CoPc and a polymer solution (poly(methyl methacrylate)) in dichloromethane, followed by evaporating the dichloromethane[32], for the analysis of mercaptoethanol in petroleum and cysteine in human urine with detection limits of 8×10^{-7} M for the former and 2×10^{-7} M for the latter[32].

MPc modified CPEs have been employed by several researchers for the analysis of thiols. Carbon paste electrodes incorporating $OMo^{V}(OH)Pc$ catalyzed the oxidation of cysteine with a considerable reduction in overpotential[139], Table 7.2. The oxidation of cysteine was mediated by $Mo^{VI}Pc$ species[139]. CoPc modified CPME was used as a sensor for analysis of cysteine and glutathionine in urine using electrophoresis[140], with detection limits of 3.1×10^{-8} and 3.0×10^{-7} M, respectively. CoPc modified CPE was employed for the potentiometric oxidation of 2-mercaptobenzimidazole and 2-mercaptobenzothiazole[35]. Low peak potentials of 0.45 V for the former and 0.27 V for the latter were obtained[35]. A detection limit of 5×10^{-7} M was obtained. CPEs modified with CoPc showed catalytic activity for the determination of thiocarbonyl compounds (thiourea, thioacetamide, thiobenzamide, and dithiooxamide)[141] and the oxidation of thioglyconic acid[142]. The latter occurred via a two step process, which leads to the formation of the dimer of the thiol[142]. A carbon paste electrode constructed from NiTSPc immobilized on silica gel modified with TiO_2 catalyzed the oxidation of cysteine[143]. The immobilization caused the increase in the monomeric form of the catalysts.

The effect of light on the catalytic activities of MPc complexes was examined by Karmann *et al.*[144] Thin films of ZnOCPc, CoOCPc, CuOCPc, TpaTAPz or TpiTAPz on ITO substrates show higher catalytic activity towards the oxidation 2-mercaptoethanol, in the presence of a light beam compared to when the electrodes were not illuminated[144]. The cyclic voltammetry in the presence of light showed larger currents than in the dark[144]. Hart and coworkers employed CoPc modified screen printed electrode for the electrocatalyzed oxidation of propanethiol and other sulfur containing compounds[145–148].

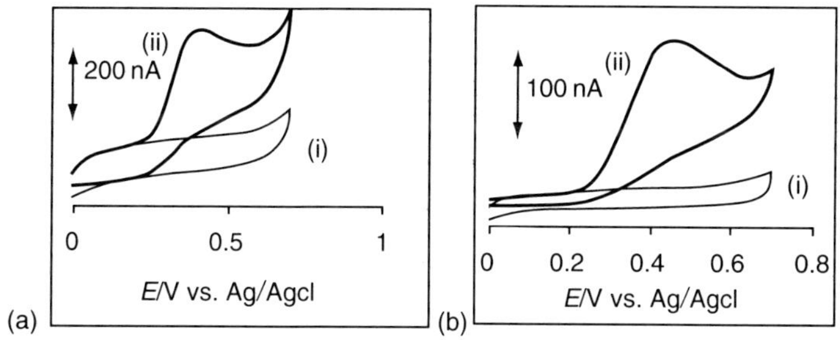

Figure 7.3. Cyclic voltammgrams for the oxidation of (a) cysteine (1×10^{-5} mol dm^{-3}) and (b) penicilamine (1×10^{-5} mol dm^{-3}) on SAM of FeOHTPc. FeOHTPc-SAM in (i) buffer (pH 4) alone and (ii) in the presence of the analyte. Scan rate = 25 mV^{-1} s. (Reproduced with Permission from Ref. [150], Figures 5(A) and (C)).

The use of self–assembled monolayers (SAMs) for the analysis of species has not been explored sufficiently. MPc complexes substituted with thiols may be used to form SAMs. SAMs of CoOBTPc or FeOBTPc on gold were used to detect L-cysteine in acid media[51,149]. The electrodes were very stable and reproducible. Detection limits for cysteine as low as 3.1×10^{-7} M were obtained[51]. SAMs of FeOBTPc, FeOHETPc, CoOHETPc were used as electrocatalysts for the detection of the thiols: L-cysteine (Figure 7.3a), homocysteine and penicillamine (Figure 7.3b) with detection limits in the 10^{-7}–10^{-6} M range[150]. Lower oxidation potentials for these thiols were observed for the FePc derivatives when compared to CoPc derivatives, Table 7.2.

Since pyrrole and its derivatives form very stable conducting surfaces, Zagal, Bedioui, and coworkers used ultramicro-carbon-fiber electrode modified with polypyrrole-dopped cobalt-tetrasulfonated phthalocyanine for the electrocatalytic oxidation of 2-ME[151]. The resulting electrode was stable and showed a detection limit of $8 \sim 10^{-5}$ M. Pyrolle substituted phthalocyanines have been used for the analysis of cysteine[152]. Electrocatalytic behavior of MPc complexes has also been examined under homogeneous conditions. Water soluble OMo^{V}(OH)TSPc catalyzed the oxidation of cysteine under homogeneous conditions[139].

Sensors based on CoTM-3,4pyPz or CoTAPc coated HOPG electrode and further coated with Nafion ion exchange membranes[30,153] or FePc modified carbon paste electrode[154] were employed for potentiometric analysis of sulfide ion and 2-mercaptoethanol[30,153,154], and 2-pyridinethiol[154]. The sensors showed a faster response time than commercial potentiometric sulfide indicators[30,153]. The overpotential of the analytes were lowered by more than 400 mV in the case of FePc modified carbon paste electrodes[154].

4. Sulfur Dioxide and Sulfur Oxoanions

The emission of SO_2 by burning fuel is responsible for air pollution, hence its monitoring is of great concern[155,156]. Sulfides are used as preservatives in the food industry due to their antioxidant and antiseptic properties. However high levels of sulfite salts can produce harmful effects towards hypersensitive people along with distortion of the flavor of food where these have been added.

4.1. Porphyrin Complexes

4.1.1. Sulfur Oxoanions

Several researchers have reported on the use of porphyrin complexes as electrocatalysts for the detection of sulfites. Table 7.3 lists potentials for the oxidation and reduction of sulfites on porphyrin catalysts. Under homogeneous conditions, Kline *et al.* employed Fe^{III}TSPP to catalyze the reduction of sulfite[157]. The Fe^{III}/Fe^{II} couple mediated the reduction process. Chen and coworkers studied the electrocatalytic transformation of HS^-, $S_2O_3^{2-}$, $S_4O_6^{2-}$, and SO_3^{2-} to SO_4^{2-} by water-soluble Fe, Mn, and Co porphyrin complexes[109, 158–160], with lower reduction potentials (SO_3^{2-} reduction) being observed for FeTSPP, Table 7.3. Fe(2-TMPyP) catalyzes the oxidation of $S_2O_3^{2-}$ to SO_4^{2-} through the Fe^{IV} species of the porphyrin. Fe^{IV} species also oxidizes SO_3^{2-} to SO_4^{2-} and

Table 7.3. Electrochemical Data for the Electrocatalytic Detection of Sulfur Dioxide and Sulfur Oxoanions Using M–N_4 complexes[a]

M–N_4 complex	Electrode	Method of modification	Analyte	*E* (V) (SCE)	Medium	O/R	Reference
Co(2-TMPyP)	GCE	Solution	$S_2O_3^{2-}$	+0.55	pH 5.5	O	[109]
Co(2-TMPyP)	GCE	Solution	SO_3^{2-}	+0.58	pH 2.2	O	[109]
CoTPyP-[Ru(bipy)$_2$Cl]$_4$	GCE	Dip dry	SO_3^{2-}	+0.91	pH 4.7, 0.25 LiTFMS	O	[200]
CoTPyP-[Ru(bipy)$_2$Cl]$^+$/ ZnTSTPP	GCE	Drop dry	SO_3^{2-}	+0.76	pH 4.7	O	[117]
CoDPP[Rubipy)$_2$]		Dip dry	SO_3^{2-}	+0.65	0.25 M LiTFMS	O	[116]
Fe(2-TMPyP)	GCE	Solution	SO_3^{2-}	+0.82	pH 9.2	O	[109]
FeTSPP	GCE	Solution	SO_3^{2-}	+0.94	pH 2.0	O	[109]
Mn(2-TMPyP)	GCE	Solution	SO_3^{2-}	+0.30	pH 11.5	O	[109]
FeTSPP	GCE	Solution	SO_3^{2-}	−0.40	pH 2.2	R	[109]
Co(2-TMPyP)	GCE	Solution	SO_3^{2-}	−0.56	pH 2.2	R	[109]
Fe(2-TMPyP)	GCE	Solution	SO_3^{2-}	−0.58	pH 2.0	R	[109]
Fe(3-TMPyP)	GCE	Solution	$S_4O_6^{2-}$	−0.78	pH 7.6	R	[109]
Co(2-TMPyP)	GCE	Solution	$S_4O_6^{2-}$	−0.58	pH 5.5	R	[109]
Fe(2-TMPyP)	GCE	Solution	$S_4O_6^{2-}$	−0.73	pH 7.6	R	[109]
FeTSPP	GCE	Solution	HSO_3^-	−0.64	pH 3.48	R	[157]
CoTNPPc	OPGE	Drop dry	HEDS	−0.9 V	0.2 M NaOH	R	[169]
CoPc	PIGE	Abrasive	HS^-	+0.13	pH[a]	O	[2]
FePc(Cl)$_{16}$	BPG	Dip dry	HS^-	−0.33	0.1 M NaOH	O	[17]
FePc	OPGE	Dip dry	SO_2	−0.12	pH 1	R	[168]

O: oxidation; R: reduction.

[a]To convert potential vs. Ag/AgCl to SCE, a correction factor of −0.045 V (ref. [12]) has been applied.

HS^- can be oxidized to $S_2O_3^{2-}$ in the presence of oxygen[158]. Co(2-TMPyP), Mn(2-TMPyP), and FeTSPP catalyze the oxidation of SO_3^{2-}[109], the potential being dependent on pH and the nature of the catalyst, Table 7.3. Biesaga *et al.* included the use of carbon electrodes modified with CoTPyP for the detection of SO_3^{2-} in a review[89]. Fe^I porphyrin species reduces $S_4O_6^{2-}$ to $S_2O_3^{2-}$[158]. Co(2-TMPyP) and Fe(2-TMPyP) catalyzed the reduction of $S_4O_6^{2-}$ to $S_2O_3^{2-}$, lower potentials are observed for the former, Table 7.3[109, 159, 160]. Chen and coworkers also compared the catalytic activities of Fe^{II}(2-TMPyP) and Fe^{II}(3-TMPyP) and showed that electrocatalytic reduction of $S_4O_6^{2-}$ by the latter occurred at more negative potentials and that $S_4O_6^{2-}$ coordinates to porphyrin molecule.

4.1.2. Sulfur Dioxide

The use of porphyrin complexes for the electrocatalytic detection of SO_2 is limited. Iron and titanium porphyrin peroxo complexes (TPPFe-O-FeTPP or TiP(O_2)) interact with SO_2 to give sulfato complexes of the porphyrins or to give free sulfate[161, 162]. Carballo *et al.* showed that NiPPIX polymerized on GCE catalyzed the oxidation of SO_2, and the modified electrode was stable under hydrodynamic conditions of a flow cell[163].

4.2. Phthalocyanine complexes

The use of phthalocyanine complexes towards the electrocatalytic detection of sulfites is limited. The use of these complexes for the electrocatalytic oxidation of sulfur dioxide has however received considerable attention. When adsorbed on GCE, FeTSPc catalyzes the oxidation of SO_2 and depending on pH, $SO_2 \cdot xH_2O$, HSO_3^- and/or SO_3^{2-} are the main compounds in solution[164]. It was found out that these products behaved differently at the electrode surface and that the modified electrode surface was stable over the entire pH-range[164]. Nikolov and coworkers have extensively studied the use of MPc complexes as catalysts for the oxidation of sulfur dioxide[165–167]. In the presence of a combination catalyst "active carbon plus CoPc", which has been subjected to pyrolysis, electrochemical oxidation of SO_2 was affected by the active centers produced following pyrolysis of the ring[165]. This combination catalyst resulted in significant increase in the catalytic activity for SO_2 oxidation[165]. Electrocatalytic oxidation of SO_2 by pyrolytic CoPc on activated carbon and in the presence of nitrogen oxides, showed that the nitrogen oxides block the catalyst surface without themselves undergoing any electrochemical changes[166]. This led to a decrease in SO_2 catalysis[166]. SO_2 can be oxidized at room temperature, in the presence of air on active carbon modified with CoPc without use of external current[167].

Yu *et al.* showed that when FePc adsorbed on graphite electrodes using the drip dry method was employed as a catalyst for the electroreduction of SO_2 in aqueous solution[168], the highest activity is observed in strongly acidic media and decreases with increase in pH due to limitation in H^+. The reducible forms are SO_2 and HSO_3^- only. SO_2 reduction current was independent of surface coverage and the detection limit was in the parts per billion range[168]. The electrocatalytic activities of MPc complexes adsorbed onto PGE towards the oxidation of SO_2 increased as follows[129]:

FePc, MnPc > CoPc > VoPc > MoOPc ≫ CuPc, CrPc.

This trend is common for electrocatalytic reactions involving MPc complexes where CoPc, FePc, and MnPc show better catalytic activities than the other MPc complexes.

Zhang *et al.* showed that a graphite electrode modified with $FePc(Cl)_{16}$ catalyzes the oxidation of sulfide in basic media[17]. Zagal *et al.*[169] reported on the electrocatalytic activity for the reduction of 2-hydroxy-ethyldisulfide on CoPc derivatives (CoOEHPc, CoTMePc, CoTNPPc, CoTAPc, CoTSPc, $CoPcF_{16}$) adsorbed onto OPGE[169]. The catalytic activity increased with the driving force of the catalyst.

5. Carbon Dioxide/Carbon Monoxide

The increase of carbon dioxide in the atmosphere is of global environmental interest due to the greenhouse effect. Thus accurate analysis of carbon dioxide is of environmental importance. The distribution of products obtained on electrocatalyzed reduction of CO_2 on M–N_4 complexes, depend of factors such as applied potential, pH, nature of macrocycle and the nature of the central metal.

5.1. Porphyrin Complexes

CO_2 is reduced electrochemically in the presence of porphyrins and phthalocyanines[170,171]. The products of reduction of CO_2 on electrodes modified with porphyrins are mainly CO and hydrogen, see Table 7.4. The potentials for CO_2 reduction are affected drastically by changes on the ligands coordinated to the porphyrin ring, Table 7.5.

5.1.1. CO_2 Reduction: M“0” Porphyrin Catalysts

The use of gas diffusion electrodes modified with porphyrins or phthalocyanines for CO_2 catalysis has been reported by several workers. The involvement of M^0 state of the metal porphyrin complex has been reported[14,172,173]. For CoTPP and FeTPP, the CO_2 reduction occurs at the potentials where M^{I}TPP is reduced to M^0TPP. For CuTPP and ZnTPP, reduction of CO_2 proceeds via the electron transfer from the reduced porphyrin ring since the metal centers are electrochemically inactive in these complexes. In all cases, the increase in CO_2 pressure enhanced the electrocatalytic activity of the MTPP catalysts[14]. The products formed using the MTPP complexes were mainly H_2 or CO, and the efficiency of product formation for CoTPP or FeTPP were comparable to those of CuTPP or ZnTPP, Table 7.4[14]. Table 7.4 shows that larger current efficiencies for CO_2 reduction were observed for MgTPP and MnTPP, compared to NiTPP, CuTPP, and under the same electrolyte conditions.

Table 7.4. Bulk Electrolysis Products Obtained Using M–N4 Complexes[a]

M–N4 complex	Electrode	Method of modification	Reactant	Main product	E.(V)[b] vs. (SCE)	Electrolyte	Current efficiency (%)	References
Carbon dioxide/carbon monoxide								
CoTMPP	ACF	Dip dry	CO_2	CO	−1.3	0.5 M $KHCO_3$	70	[20]
CoTPP	Pt	Drop dry	CO_2	CH_3OH	−0.5	pH 3.5	15.1	[131]
CoOEP	PGE	Dip dry	CO	CO_2				
FeTPP /Mg^{2+}	Hg pool	Solution	CO_2	CO	−1.7	DMF/0.1 M TEAP	60–70	[173, 176]
Fe(III)TPP	Pt	Drop dry	CO_2	CH_3OH	−0.5	pH 3.5	10.3	[181]
FeTPP /Mg^{2+}	GCE	Solution	CO_2	CO	−1.6	DMF	70	[172]
FeTPP	Pt	Drop dry	CO_2	CH_3OH	−0.5	pH 3.5	12.5	[181]
Cr(III)TPP	Pt	Drop dry	CO_2	CH_3OH	−0.5	pH 3.5	11.6	[181]
NiTPP	Pt	Drop dry	CO_2	CH_3OH	−0.5	pH 3.5	14.4	[181]
CoTPP	GDE	Dip dry	CO_2	CO		0.5 M $KHCO_3$	75	[14][c]
FeTPP	GDE	Dip dry	CO_2	H_2		0.5 M $KHCO_3$	53	[14][c]
NiTPP	GDE	Dip dry	CO_2	H_2		0.5 M $KHCO_3$	63	[14][c]
ZnTPP	GDE	Dip dry	CO_2	H_2		0.5 M $KHCO_3$	66	[14][c]
CuTPP	GDE	Dip dry	CO_2	H_2		0.5 M $KHCO_3$	63	[14][c]
MgTPP	GDE	Dip dry	CO_2	H_2		0.5 M $KHCO_3$	91	[14][c]
MnTPP	GDE	Dip dry	CO_2	H_2		0.5 M $KHCO_3$	88	[14][c]
AgOEP	GCE or Pt	Solution	CO_2	H_2	−1.50	CH_2Cl_2/TBAF		[183]
PdOEP	GCE or Pt	Solution	CO_2	H_2	−1.65	CH_2Cl_2/TBAF		[183]
H_2TPP	GDE	Dip dry	CO_2	H_2		0.5 M $KHCO_3$	90	[14][c]
GaPc	GDE	Dip dry	CO_2	Methane	−2.6	0.5 M $KHCO_3$	40	[191, 192]
TiPc	GDE	Dip dry	CO_2	Methane	−2.0	0.5 M $KHCO_3$	30	[191, 192]
FePc	GDE	Dip dry	CO_2	CO	−1.2	0.5 M $KHCO_3$	50	[191, 192]
PdPc	GDE	Dip dry	CO_2	CO	−1.5	0.5 M $KHCO_3$	80	[191, 192]
ZnPc	GDE	Dip dry	CO_2	CO	−1.9	0.5 M $KHCO_3$	15	[191, 192]
AlPc	GDE	Dip dry	CO_2	Formic acid	−1.8	0.5 M $KHCO_3$	15	[191, 192]
H_2Pc	GDE	Dip dry	CO_2	H_2	−1.6	0.5 M $KHCO_3$	100	[191, 192]
MgPc	GDE	Dip dry	CO_2	H_2	−1.6	0.5 M $KHCO_3$	100	[191, 192]
VPc	GDE	Dip dry	CO_2	H_2	−1.6	0.5 M $KHCO_3$	100	[191, 192]
PtPc	GDE	Dip dry	CO_2	H_2	−1.6	0.5 M $KHCO_3$	100	[191, 192]
MnPc	GDE	Dip dry	CO_2	H_2	−1.6	0.5 M $KHCO_3$	100	[191, 192]
CoPc	GCE	Drop dry	CO_2	Formade ion	−1.6	pH > 5	60	[27]
CoPc	PGE or carbon cloth	Drop dry/dip dry	CO_2	CO	−1.15	pH 5	87	[15]
CoPc	BPG/PVP	Drop dry	CO_2	CO/H_2	−1.25	pH 5		[41]
CoPc	Graphite	Membrane	CO_2	CO/H_2	−1.25	pH 5		[41]
CoPc	GCE	Chemical bonding (py)	CO_2	CO	−1.0		20	[179]

Table 7.4. *Continued*

$M–N_4$ complex	Electrode	Method of modification	Reactant	Main product	$E.(V)^b$ vs. (SCE)	Electrolyte	Current efficiency (%)	References
Nitrite								
FeTSPP	Hg pool	Solution	NO_2^-	NH_3	−0.9	pH 6.7	97	[203]
FeTSPP	GCE	Solution	NO_2^-	NH_3	−0.9	pH 6.5–7.4	46–72	[204]
FePc	GDE	Dip dry	NO_2^-	NH_3	−1.5	0.2 M $KHCO_3$	6	[192]
TiPc	GDE	Dip dry	NO_2^-	NH_3	−1.5	0.2 M $KHCO_3$	3	[192]
VOPc	GDE	Dip dry	NO_2^-	NH_3	−1.5	0.2 M $KHCO_3$	7	[192]
CrPc	GDE	Dip dry	NO_2^-	NH_3	−1.5	0.2 M $KHCO_3$	10	[192]
MoPc	GDE	Dip dry	NO_2^-	NH_3	−1.5	0.2 M $KHCO_3$	15	[192]
RuPc	GDE	Dip dry	NO_2^-	NH_3	−1.5	0.2 M $KHCO_3$	10	[192]
NiPc	GDE	Dip dry	NO_2^-	NH_3	−1.5	0.2 M $KHCO_3$	17	[192]
PdPc	GDE	Dip dry	NO_2^-	NH_3	−1.5	0.2 M $KHCO_3$	11	[192]
CuPc	GDE	Dip dry	NO_2^-	NH_3	−1.5	0.2 M $KHCO_3$	14	[192]
ZnPc	GDE	Dip dry	NO_2^-	NH_3	−1.5	0.2 M $KHCO_3$	16	[192]
CdPc	GDE	Dip dry	NO_2^-	NH_3	−1.5	0.2 M $KHCO_3$	16	[192]
GaPc	GDE	Dip dry	NO_2^-	NH_3	−1.5	0.2 M $KHCO_3$	4	[192]
InPc	GDE	Dip dry	NO_2^-	NH_3	−1.5	0.2 M $KHCO_3$	14	[192]
GePc	GDE	Dip dry	NO_2^-	NH_3	−1.5	0.2 M $KHCO_3$	3	[192]
SnPc	GDE	Dip dry	NO_2^-	NH_3	−1.5	0.2 M $KHCO_3$	14	[192]
PdPc	GDE	Dip dry	NO_2^-	NH_3	−1.5	0.2 M $KHCO_3$	12	[192]
$[CoTM-2,3pyPz]^{4+}$	C rod	Drop dry	NO_2^-	NH_3	−1.65	0.5 M NaOH	97	[213]
$[CoTM-3,4pyPz]^{4+}$	C rod	Drop dry	NO_2^-	NH_3	−1.65	0.5 M NaOH	72	[213]
CoTpyPz	C rod	Drop dry	NO_2^-	NH_3	−1.65	0.5 M NaOH	70	[213]
CoPc	C rod	Drop dry	NO_2^-	NH_3	−1.65	0.5 M NaOH	63	[213]
CoPc	GDE	Dip dry	NO_2^-	NH_3	−1.5	0.2 M $KHCO_3$	16	[192]
Sulfur oxides								
Fe(2–TMPyP)	GCE	Solution	$S_2O_3^{2-}$	$S_4O_6^{2-}$	+0.81	pH 9	56	[158]
Fe(3–TMPyP)	GCE	Solution	$S_2O_3^{2-}$	SO_4^{2-}	+1.06	pH 9	70	[158]
Fe(4–TMPyP)	GCE	Solution	$S_2O_3^{2-}$	SO_4^{2-}	+1.06	pH 9	90	[158]

O: oxidation; R: Reduction.

[a] To convert potential vs. Ag/AgCl to SCE, a correction factor of −0.045 V (ref. [12]) has been applied.

[b] Electrolysis potential.

[c] Pressure 1 atm., current density 100 mA cm^2.

Table 7.5. Electrochemical Data for the Electrocatalytic Detection of Carbon Dioxide and Carbon Monoxide Using M–N$_4$ Complexes[a]

MN_4 complex	Electrode	Method of modification	Analyte	E (V) (SCE)	Medium	O/R	References
FeTPP/Mg^{2+}	HMDE	Solution	CO_2	−1.7	0.1 M TEAP/DMF	R	[173, 177]
CoTPP	GCE	Solution	CO_2	−1.96	0.1 M TEAP/DMF	R	[184]
CoTPP bridged by AP	GCE	Axial ligation of bonded AP	CO_2	−1.1	pH 6.3	R	[180]
CoOEP	PGE	Dip dry	CO	+0.5	pH 0, 7, 8	O	[188]
CoTSPc	Pt	Solution	CO_2	−1.2	pH 7	R	[198]
NiTSPc	Pt	Solution	CO_2	−1.5	pH 7	R	[198]
CoPc	Graphite rod	Dip dry	CO_2	−1.7	TEAP (aq.)	R	[19]
CoPc	GCE	IPAVD	CO_2	−1.6	pH 6.65	R	[39]
CoPc	GCE	Drop dry	CO_2	−1.55	pH 3–7	R	[27]
CoPc	Graphite	Membrane	CO_2	−1.20	pH 4.4	R	[41]
CoPc	Graphite	Membrane	CO_2	−1.20	pH 4.4	R	[41]
CoPc	PGE or carbon cloth	Dip dry/ drop dry	CO_2	−0.95	pH 5	R	[15]
CoPc/PVP	membrane	Dip dry	CO_2	−1.1	pH 4.4	R	[194]

O: oxidation; R: reduction.
[a]To convert potential vs. AG/AgCl to sCE, a correction factor of −0.045 V (ref. [12]) has been applied.

The involvement of the M^0P species in the mechanism was also confirmed by van Behar *et al.*[174]. Co^IP species formed from reduction of Co porphyrin by electrochemical, photochemical, and radiation methods was found to be unreactive towards reduction of CO_2[174]. But one electron reduction of the Co^IP resulted in a species which bound CO_2 and reduced it with the formation of CO and formic acid as products[174]. However, Riqelme *et al.* reported that the Co^{II}/Co^I couple (rather than the M^0/M^I couple) was responsible for the catalytic reduction of CO_2 to CO and formic acid by polymeric CoTAPP (on GCE)[175]. The monomeric form of the catalyst did not catalyze the reduction[175]. Grodkowski *et al.* found the Co^I and Fe^I complexes to be the active species when Co and Fe corroles electrocatalyzed the reduction of CO_2[176]. The Fe corroles showed better catalytic activity than the Co corroles. The catalytic behavior of the corroles towards the reduction of CO_2 is different from that of MP complexes in that the latter do not react with CO_2 until they are reduced to beyond M^I state[176].

5.1.2. CO_2 Reduction: Effects of Lewis and Brønsted Acids

Lexa and coworkers studied the effects of Lewis and Brønsted acids on the catalytic efficiencies of Fe^0TPP complexes towards the reduction of CO_2[172, 173, 177].

The potentials observed for the reduction of CO_2 under these conditions were still relatively high (e.g., −1.7 V, Table 7.5). The efficiency of the electrocatalytic reduction of CO_2 by Fe^0TPP improves in the presence of Lewis acid cations such as Mg^{2+}, Ca^{2+}, Ba^{2+}, Li^+, and Na^+[172] or Bronsted acid co-catalyst such as 2-trifluoroethanol[173] 1-propanol or 2-pyrrolidone[177,178]. The lifetime of the catalyst also improves in the presence of these acids. The main product of reduction is CO, with formate being formed in lesser amounts[173]. The order of reactivity of the Lewis acids is as follows: $Mg^{2+} \sim Ca^{2+} > Ba^{2+} > Li^+ > Na^+$. The study confirmed the push–pull mechanism, whereby electrons are pushed into the CO_2 molecule by the electron-rich catalyst, and the cleavage of one of the C–O bonds is helped by the presence of the electron deficient Lewis acid cation[172,178]. It was proposed that Mg^{2+} or the other cations, break the CO bond following the formation of a complex between the Fe^0TPP and CO_2[178].

5.1.3. CO_2 Reduction: Effect of Changes in the Ring and Central Metal

Aga *et al.*[179] showed that CO_2 can be catalytically electroreduced to form CO as the main product, on N_4 cobalt complexes chemically bonded to GCE through –CONH–pyridine, with the N of the pyridine forming a coordinate bond with the Co center of the N_4 complexes. Tanaka *et al.* employed aminopyridine as a bridge between the metal complex and the GCE, and the potential for the reduction of CO_2 of −1.1 V was obtained[180]. The N_4 complexes used as catalysts were CoNPc, CoTMPP, CoPc, vitamin B_{12}, CoTPP, and CoDO (Figure 7.4a) and the modified electrode may be denoted as GC-py-Co. CoDO was however not able to reduce CO_2. The mechanism for the reduction of CO_2 by the rest of the N_4 complexes was proposed as follows[179] (Equation 7.11–7.14):

$$\mathrm{GC-py-Co^{II}+e^- \rightarrow GC-py-Co^{I}} \quad (7.11)$$

$$\mathrm{GC-py-Co^{I}+H^+ \rightarrow GC-py-Co^{II}H} \quad (7.12)$$

$$\mathrm{GC-py-Co^{II}H+CO_2 \rightarrow GC-py-Co^{II}COOH} \quad (7.13)$$

$$\mathrm{GC-py-Co^{II}COOH+H^+ +e^- \rightarrow GC-py-Co^{II}+CO+H_2O} \quad (7.14)$$

Ogura and Yoshida examined the electrocatalytic CO_2 reduction at electrodes modified with a number of MTPP complexes[181]. Co^{II}TPP showed better catalytic activity followed by Ni^{II}TPP, then by Fe^{II}TPP, with Cr^{III}TPP showing the least catalytic activity[181]. Low current efficiencies were obtained[181], Table 7.4. Tatahashi *et al.* reported on the use of Co porphyrin complexes for the catalytic reduction of CO_2 on Hg(Pt) electrodes[182], HCO_2H was found to be the reduction product. Metal fee, Cu, and Fe porphyrin complexes showed no catalytic activity.[182].

When Ag^{II} and Pd^{II} porphyrins were employed as catalysts for the electrochemical reduction of CO_2 in dichloromethane, H_2, and oxalic acid were the detected products[183], Table 7.4, the latter being a rare product. Qiu and Sawyer reported that alkalyated iron and cobalt porphrins dianioins electrocatalytically reduce carbon dioxide to give CO and CO_3^{2-}[184].

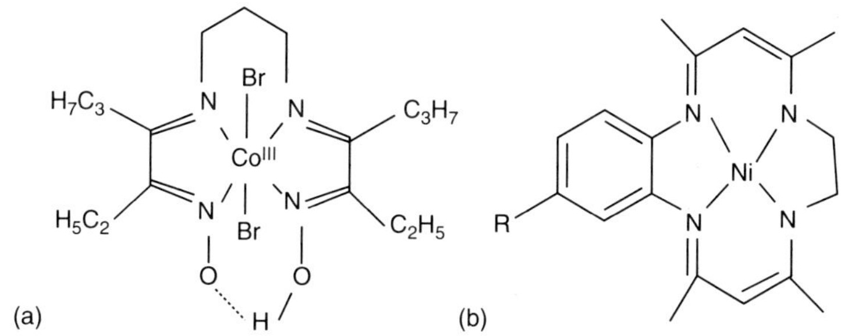

Figure 7.4. Molecular structures of (a) dibromo (11-hydroxyimino-4,10-dipropyl-5,9-diazatrideca-4,9-dien-3-one oximato) cobalt (III) (CoDO) and (b) an unsymmetrical N$_4$Ni complex.

Cofacial dinuclear CoTMPyP:MTSTPP was employed by Enoki et al. as a homogeneous catalyst for the electrochemical reduction of CO_2[185]. The catalytic activity was depended on the central metal[185]. Bernard et al. reported that iron porphycene complexes show electrocatalytic behavior towards the reduction of carbon dioxide in the same manner as porphyrins[186].

5.1.4. Carbon Monoxide Oxidation

Venkataramab *et al.* reported that MP, MPc, and cyclam complexes, employed as co-catalysts with Pt electrodes, acted as redox mediators which generated a species that catalyzes CO oxidation[187]. Shi and Anson showed that adsorbed Co^{II}OEP catalyzed the oxidation of CO to CO_2 in aqueous media[188]. The first step was the oxidation of Co^{II}OEP to Co^{III}OEP, followed by coordination of CO. Van Baar et al. reported that electrocatalytic oxidation of CO occurs in the presence of Rh and Ir porphyrins in aqueous acid solutions[189]. In strongly alkaline media Co and Fe porphyrin counterparts showed excellent catalysis for CO oxidation[189]. The proposed catalytic mechanism was as follows: (i) CO adsorption to metal center, (ii) nucleophilic attack by H_2O (acid media) or OH^- (basic media) on the adsorbed CO, and (iii) decarboxylation[189, 190]. The differences in the behavior of the metalloporphyrin complexes was explained in terms of CO affinity for the central metal in the different oxidation states[189].

5.2. Phthalocyanine Complexes

There are several reports on the use of MPc complexes as electrocatalysts for the reduction of CO_2, but reports on the use of these complexes for the oxidation of CO are scarce. Diffusion electrodes have been employed extensively for the use of MPc complexes as electrocatalysts for CO_2 reduction. Using porous gas diffusion electrodes, Furuya and coworkers studied the activity of a series of MPc complexes towards the reduction of CO_2 and found the activity to be depended on the nature of the central metal, Table 7.4[191, 192]. On FePc and PdPc modified electrodes both hydrogen and CO were obtained, on ZnPc and AlPc, the main products were hydrogen, carbon dioxide and formic acid and on H_2Pc, MgPc, MnPc, and PtPc, the main product was hydrogen only[191]. The main product of reduction was CO on NiPc and CoPc modified electrodes[191]. On SnPc, PbPc, and InPc modified electrodes, the main product was either formic acid or hydrogen. On CuPc, GaPc, and TiPc modified electrodes, methane was the main product of CO_2 reduction[191]. The current efficiencies of the main products are listed on Table 7.4. CoPc, CuPc, CoTMPP, and CoTBPc adsorbed on activated carbon fibers (ACF) supports, in the form of gas diffusion electrodes were found to be efficient electrocatalysts for the reduction of CO_2[1,193]. The product of the reduction was CO.

Meshitsuka *et al.* also found NiPc and CoPc complexes to be effective catalysts for the reduction of CO_2[19], with the reduction potential depending on the sweep rate[19]. Lieber and Lewis[15], Tanabe and Ohno[39], and Kaputsa and Hackerman[27] (Table 7.5) also found CoPc to be a good catalyst for the reduction of CO_2. CoPc modified pyrolytic graphite (PGE) or carbon cloth electrodes were found to catalyze the reduction of CO_2[15], with CO being the main reduction product, Table 7.4. Using IPAVD technique for modifying GCE, CoPc was found to be active for the reduction of CO_2, and the efficiency of the catalysis varied with MPc as follows: CoPc $>>$ NiPc $>$ FePc $\sim$ MgPc $\sim$ MnPc $\sim$ ZnPc[39]. On a GCE modified using the drop dry method, the order of activity was found to be CoPc $>$ NiPc $>>$ FePc $=$ CuPc $>$ CrPc, SnPc[27]. In all cases CoPc seems to be a much better catalyst than other MPc complexes for the reduction of CO_2.

Kaneko and coworkers reported that a BPG electrode modified with PVP membrane containing CoPc was selective to (and enhanced) CO_2 reduction (with the formation of CO and H_2) than if CoPc only was employed[41,194]. The enhanced catalytic activity in the presence of PVP was discussed in terms of (i) electron donation of the pyridine group (of PVP) to the central Co metal in CoPc by axial coordination, (ii) proton exchange property of PVP, and (iii) the increased local concentration of CO_2 in the PVP layer[194].

Magdesieva and coworkers employed mono- and di-phthalocyanine films for the electrocatalytic reduction of CO_2[20,195,196] in methanol. Methyl formate was found to be the main reaction product for the catalytic reduction using mono and di-phthalocyanine complexes of the rare earth metals. The lutetium mono- and di-Pc complexes contain an electroinactive central metal, but were found to catalyze the reduction of CO_2[196]. It was proposed that a key step in the catalytic behavior of these complexes was the coordination of CO_2 molecule to the external N-atom of the NPc or Pc[196]. CuTBPc transformed CO_2 to methane (with $\sim$30% yield), and CuTAPc resulted in the formation of CO and methyl formate as the main products[195]. Abe *et al.*[197] reported on CoOBPc coated graphite electrodes (dip dry method) for the electrocatalytic reduction of CO_2. CO was produced at a higher selectivity on CoOBPc compared to CoPc[197]. The electron-donating octa butoxy substituents facilitated the coordination of CO_2 as well as electron transfer in CoOBPc[197]. Magdesieva *et al.* also reported CoTBPc and CuTBPc complexes containing bulky tetra-tert-butyl substituents to be more effective catalysts for CO reduction[20].

Using spectroelectrochemistry (reflectance spectra), Christensen *et al.* probed the mechanism of CoPc assisted reduction of CO_2, and the involvement of $Co^{I}Pc$ intermediate during the reduction process was confirmed[20].

Catalytic behavior of MTSPc (M = Co, Ni, Cu, and Fe) towards the reduction of CO_2 under homogeneous conditions was reported by Hiratsuka *et al.*[198]. NiTSPc and CoTSPc were found to be effective catalysts[198], while CuTSPc and FeTSPc were ineffective. The reduction process consisted first of the formation of CO_2 complex with the MTSPc, followed by electron transfer from the electrode to CO_2 through MTSPc[198].

6. Nitrites and Nitrates

Nitrates and nitrites are spread within environmental, food, industrial, and physiological systems. Nitrate may be reduced by bacteria to nitrite, and nitrite produces carcinogenic nitrosamines. Nitrite is also commonly employed in cured meat products to provide antimicrobial action, color fixation, and preservation. Nitrite can oxidize hemoglobin iron(II) to iron(III) resulting in methemoglobin, leading to a condition known as methemoglobinemia in children mainly. In excess, nitrates and nitrites may be toxic.

6.1. Porphyrin Complexes

Table 7.6 lists the potentials for the catalytic oxidation of nitrites on electrodes modified with MP or MPc complexes. Chen and Chen reported on the use of FeTMPyP film adsorbed on DNA modified carbon, gold, platinum, or transparent semiconductor tin oxide electrodes for the catalytic reduction of nitrite[87], through the Fe(I) species. Oxidation (at 0.81 V, Table 7.6) of nitrite occurred through the Fe(IV) species as follows[87] (Equation 7.15 and 7.16):

$$Fe^{III}TMPyP/DNA \rightarrow Fe^{IV}TMPyP/DNA + e^- \quad (7.15)$$

$$Fe^{IV}TMPyP/DNA + NO_2^- \rightarrow Fe^{III}TMPyP/DNA + NO_2 \quad (7.16)$$

Araki, Toma, and coworkers, reported that thin films of the polymeric tetraruthenated porphyrin material were one of the most active electrocatalysts for the oxidation of nitrite to nitrate[199,200], with activities 30 times higher than that of electrostatically assembled porphyrin films[200]. Electrostatically assembled porphyrin films of tetraruthenated cobalt porphyrin/(*meso*-tetra(4-sulfonatephenyl) porphyrinate zinc(II) could be used to detect nitrite with a detection limit of 0.1 μM[117,201]. The bilayered film allowed for the determination of nitrite at less positive potentials than the individual components and with enhanced sensitivity[201]. The coating also prevented surface poisoning. The detector allowed for the determination of nitrate after reduction to nitrite in mineral water, saliva, and cured meats. The use of alternating layers of porphyrins for the reduction of nitrite and nitrate was also described in a review by Moorcroft *et al.*[202].

Barley, Meyer and coworkers showed that Fe^{III}TMPyP and Fe^{III}TSPP were effective electrocatalysts for the reduction of NO_2^- to N_2O, NH_3, and NH_2OH[203–205]. Current efficiencies as high as 97% were obtained for the formation of ammonia in some cases, Table 7.4. Under acidic conditions where nitrite disproportionates into mainly NO, a nitrosyl intermediate was implicated during

Table 7.6. Electrochemical Data for the Electrocatalytic Detection of nitrites, Cyanides and Thiocyanades using $M - N_4$ Complexes[a]

$M - N_4$ complex	Electrode	Method of modification	Analyte	E(V) (SCE)	Medium	O/R	References
Nitrites							
FeTMPyP/DNA	GCE	Electrodep	Nitrite	+0.81	pH 9	O	[87]
$H_2TPyP - [Ru(5 - ClPhen_2Cl)]$	GCE or Pt	Polymer	Nitrite	+1.0	pH 4.5	O	[199]
$CoTPyP - [Ru(bipy)_2Cl]_4$	GCE	Dip dry	Nitrite	+0.91	pH4.7, 0.25 LiTFMS	O	[200]
$CoTPyP - [Ru(bipy)_2Cl]_4$: ZnTSPP	GCE	Drop dry	Nitrite	+0.71	0.2 M $BaClO_4$	O	[201]
$CoTPyP - [Ru(bipy)_2Cl]_4^+$/ZnTSTPP	GCE	Drop dry	Nitrite	+0.76	pH 4.7	O	[117]
Hm/DDAB	GCE	Drop dry	Nitrite	−1.1	pH 7.4	R	[208]
Mb/DDAB	PGE	Drop dry	Nitrite	−1.2	pH 7.4	R	[209]
CoPc	VCE	Drop dry	Nitrite	+0.87	pH 7.3	O	[210]
CoTpyPz	GCE	Electrodep	Nitrite	−1.4	0.5M NaOH	R	[213]
$[CoTMT - 2, 3pyPz]^{4+}$	GCE	Electrodep	Nitrite	−1.44	0.5M NaOH	R	[213]
$[CoTMT - 3, 4pyPz]^{4+}$	GCE	Electrodep	Nitrite	−1.40	0.5M NaOH	R	[213]
Thiocyanade/cyanide							
CoOBTPc	Au	SAM	Thiocyanate	+0.74	pH 4	O	[150]
CoOHETPc	Au	SAM	Thiocyanate	+0.71	pH 4	O	[150]
FeOBTPc	Au	SAM	Thiocyanate	+0.72	pH 4	O	[150]
FeOHETPc	Au	SAM	Thiocyanate	+0.70	pH 4	O	[150]
CoPc/PAA	Pt	Drop dry	CN^-	+1.51	0.1 MTEAP/CH_3CN	O	[44]

O: oxidation; R: reduction:

[a]To convert potential vs. Ag/AgCl to SCE, a correction factor of −0.045 V (ref. [12]) has been applied. Electrodep: electrodeposition.

the catalytic reduction[203]. Bedioui *et al.* reported that electropolymerized iron porphyrin or iron porphyrin incorporated in pre-electropolymerized pyrrole derivatives showed catalytic activity towards the reduction of nitrite[206].

Gao *et al.* reported on nitrite sensitive liquid membrane electrodes based on metalloporphyrin derivatives; Ir(TPP)Cl, Ir(TMPP)Cl and chloro(2-nitro-5,10, 15,20-tetraphenylporphyrinato)indium[207]. The electrodes were found to have a high selectivity for nitrite in the presence of a number of interfering ions with the following selectivity pattern:

$$NO_2^- > SCN^- > I^- > F^- > Br^- > Cl^- > ClO_4^- > NO_3^-$$

A detection limit of 5.0×10^{-6} mol dm^{-3} was obtained[207].

Mimica *et al.* and Lin *et al.* employed haemin[208], myoglobin[208,209] and hemoglobin[208] to electrocatalytically reduce of nitrite. These iron porphrin complexes were embedded in DDAB surfactant films[208] and were found to be active catalysts. The catalytic currents were observed at potentials close to the Fe^{II}/Fe^{I} couple[209].

6.2. Phthalocyanine Complexes

Table 7.6 shows that the potentials for the reduction of nitrite on MPc complexes are relatively high compared to when Hb, Mb, or FeTMPyP/DNA are employed. Zagal, Bedioui, and coworkers employed CoPc adsorbed (as monomer) or electropolymerized onto a vitreous carbon electrode (VCE), to catalyze the oxidation of nitrite[210]. The activity of the electrode was found not to increase with the amount of CoPc adsorbed onto the electrode[210]. At low potentials (at the foot of the electrocatalytic wave), the proposed mechanism is similar to that as proposed for thiols Equations (7.4)–(7.7).

Kim and Kwag showed that heat treatment affected the catalytic activity of supported FePc for the electrocatalytic reduction of nitrite[211]. The highest catalytic activity was obtained at 800°C, at this temperature the FePc complex existed as either Fe–N complex or in the form of small iron clusters, both of which exhibited catalytic activities[211]. The decrease in activity at temperatures greater than 800°C could be attributed to formation of larger iron clusters which may be catalytically inactive. The study showed that the Fe–N$_4$ structure was not important to maintain catalytic activity[211]. Electrocatalytic reduction of nitrate and nitrite in alkaline media by a series of MPc complexes (adsorbed on GCE) showed that the overpotential for the reduction was lowered by the MPc complex as follows: CuPc > FePc > NiPc > CoPc > MnPc > ZnPc > unmodified GCE[10]. The reduction of nitrite occurred at slightly less negative potentials compared to nitrate reduction[10]. The high catalytic activity of CuPc is again suggesting the involvement of ring-based processes. In fact Hwang *et al.* showed that FePc adsorbed on the edge of PGE was an efficient catalyst for the reduction of nitrite in acid solutions and the $Fe^{II}Pc^{2-}/Fe^{II}Pc^{3-}$ ring-based couple was found to be responsible for the high catalytic activity[212]. Carbon electrodes modified with $[CoTMTpyPz]^{4+}$, CoTpyPz, $[CoTSPc]^{4-}$ and by the 1:1 electrostatically assembled dimer $[CoTMTpyPz]^{4+}$: $[CoTSPc]^{4-}$ showed the latter to have the best catalytic activity[213]. Bulk electrolysis of nitrite gave ammonia and hydroxylamine as the main products[213], Table 7.4. Nevin *et al.* showed that tetranuclear CoPc complexes reduced nitrite in basic media[214]. Iron electrodes modified with FePc catalyzed the reduction of nitrite[1].

Simultaneous reduction of nitrite and carbon dioxide was studied on gas-diffusion electrodes modified with MPc complexes (where M = Cr, Mo, Mn, Ru, Co, Rh, Ir, Ni, Pd, Pt, Cu, Ag, Au, Zn, Cd, In, Tl, Sn, and Pb) by Shibata and Furuya[192]. Urea, CO, formic acid, and NH_3 were formed (with low current efficiencies for NH_3 formation, Table 7.4) at the gas diffusion electrodes with Group 8–14 MPc catalysts (except for Al and Ge). No urea was formed for MPc complexes of Group 4–7[192].

7. Cyanides/Thiocyanades

Cyanide is a well known toxic molecule which gets into the environment through its industrial use in mainly mining. Biological sources of cyanide are also known. Cyanogenic glucosides found in plants release HCN following ingestion of these plants by humans. Thiocyanade is a detoxification product of hydrogen cyanide. Its determination in body fluids has been used to monitor exposure to hydrogen cyanide from tobacco smoke. Thiocyanade possesses many industrial and biological applications and its toxicity is not comparable to that of cyanide, but it is harmful to aquatic life.

The use of porphyrin and phthalocyanine complexes for the electrocatalysis of cyanide and thiocyanade is limited. Most studies for the detection of thiocyanade using porphyrin or phthalocyanine catalysts have been based mainly on potentiometic methods, Table 7.7. Abbaspour *et al.* reported on the use of unsymmetrical benzo N_4 nickel(II) macrocyclic complexes (Figure 7.4b) as ionophores for preparing PVC-based membranes for selective detection of thiocyanade[215]. A membrane electrode based on MnOBTPP (reported by Shamsipur *et al.*) was found to be selective for thiocyanade[216]. The electrode was employed for direct analysis of thiocyanade in urine[216]. Amini and coworkers developed thiocyanade selective electrodes based on NiPc, MnPc, CoPc, and FePc ionophores incorporated into plastisized PVC on graphite electrodes[47,48]. These electrodes gave good detection limits in the pH range 3–10[47,48]. MnPc was found to show better selectivity for anions than CoPc[47]. Using thiol substituted MPc complexes of Co and Fe, thiocyanade could be electrocatalysed on SAMs of these complexes[150], Figure 7.5, with the potentials shown in Table 7.6. A platinum disc electrode modified with PAA which has been modified with CoPc was reported by Tatsuka *et al.* for the selective detection of cyanide ions[44]. The catalysis was mediated by cyanide coordination to the CoPc complex.

8. Hydrazine/hydroxylamine

Hydrazine is volatile and toxic, and is readily absorbed by oral, dermal, or inhalation routes or exposure. Hydroxylamine and its derivatives are moderately

Table 7.7. Potentiometric Determination of Sulfur Oxides and Thiocyanade Using $M–N_4$ as Electrocatalysts

$M–N_4$ complex	Electrode matrix	Analyte	Medium	LOD ($\mu g\ ml^{-1}$)	References
CoTM-3,4pyPz	HOPG	Sulfide	pH 7	$\mu g\ ml^{-1}$	[30]
NiPPIX	GCE	SO_3^{2-}	$0.05MKH_2\ PO_4$	0.15	[163]
NiTACTD	PVC	SCN^-	pH 3.5–10.5	1×10^{-7}	[215]
MnOBTPP	PVC	SCN^-	pH 3	3.2×10^{-7}	[216]
NiPc	PVC/DOP	SCN^-	pH 3–10	5×10^{-7}	[48]
FePc	PVC/DOP	SCN^-	pH 3–10	1×10^{-5}	[48]
MnPc	PVC	SCN^-	pH 2–10	5×10^{-7}	[47]
CoPc	PVC	SCN^-	pH 2–10	5×10^{-7}	[47]

LOD: limit of detection.

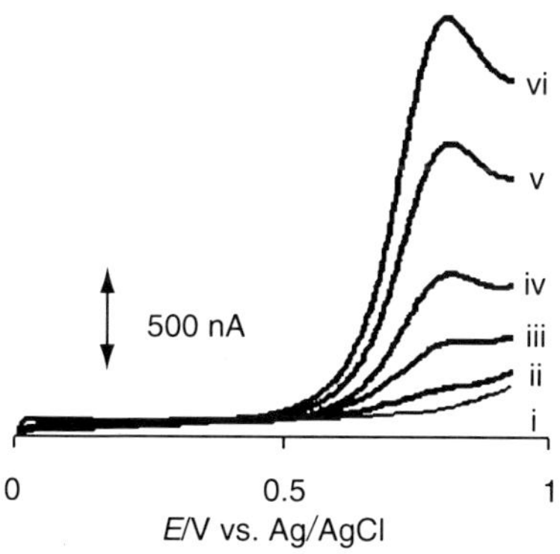

Figure 7.5. Typical Osteryoung square wave voltammograms of CoOBTPc-SAM modified gold electrodes in pH 4 buffer solutions before (i) and after the addition of (ii) 9.30×10^{-6} mol dm^{-3}, (iii), 1.98×10^{-6} mol dm^{-3} (iv) 40.5×10^{-6} mol dm^{-3} (v) 80.0×10^{-6} mol dm^{-1} and (vi) 11.90×10^{-6} mol dm^{-3} thiocyanate. (Reproduced with Permission frm Ref.[150] Figure 7.5D.

toxic. Hydroxylamine is presumed to be an intermediate in the reduction of nitrosyl to ammonia.

8.1. Porphyrin Complexes

Most reports are on the catalytic oxidation of hydrazine using polymerized porphyrins. Hence polymerized porphyrins are included here. There has been less attention paid to catalysis of hydroxylamine. Table 7.8 shows that the data reported for the oxidation of hydrazine is mainly in basic media. Unmetalated porphyrin complexes show catalytic activity, again suggesting the involvement of the ring-based processes in the catalytic processes, Table 7.8[112,113] Under the same

Table 7.8. Electrochemical Data for the Electrocatalytic Detection of Hydrazine and Hydroxylamine Using M–N_4 Complexes[a]

M–N_4 complex	Electrode	Method of modification	Analyte	*E* (V) (SCE)	Medium	O/R	References
Hydrazine							
H_2TTMAPP	GCE	Solution	Hydrazine	0.65	pH 7	O	[112,113]
CoTTMAPP	GCE	Solution	Hydrazine	0.18	pH 7	O	[112]
CoTSPP	GCE	Solution	Hydrazine	0.18	pH 7	O	[112]
CoPCPP	Pt or GCE	Polymer	Hydrazine	+0.45	0.1M NaOH	O	[218]
CoTPCPP	Pt or GCE	Polymer	Hydrazine	+0.56	0.1M NaOH	O	[218]
H_2TPCPP	Pt or GCE	Polymer	Hydrazine	+0.68	0.1M NaOH	O	[218]
H_2PCPP	Pt or GCE	Polymer	Hydrazine	+0.40	0.1M NaOH	O	[218]
MnTPCPP	Pt or GCE	Polymer	Hydrazine	+0.65	0.1M NaOH	O	[218]
NiTPCPP	Pt or GCE	Polymer	Hydrazine	+0.40	0.1M NaOH	O	[218]
NiTSPP	VCE	Polymer	Hydrazine	+0.60	0.1M NaOH	O	[219]
CoTAPc	GCE	Polymer	Hydrazine	−0.30	pH 13	O	[234]
CoTAPc	GCE	Polymer	Hydrazine	−0.42	pH 13	O	[235]
$(RuPc)_2$	RDE	Dip dry	Hydrazine	−0.34	pH 13	O	[236]
CoTAPc	GCE	Polymer	Hydrazine	−0.30	pH 13	O	[221]
CoPc	CPE		Hydrazine	+0.1	0.2 M NaOH	O	[36]
CoTSPc	PGE	Dip dry	Hydrazine	−0.53	0.2 M NaOH	O	[21]
FeTSPc	PGE	Dip dry	Hydrazine	−0.53	0.2 M NaOH	O	[21]
Hydroxylamine							
H_2TTMAPP	GCE	Solution	NH_2OH	+0.75	pH 7.0	O	[112,113]
$(RuPc)_2$	BPG	Dip dry	NH_2OH	−0.16	pH 13	O	[237]

O; oxidation; R; Reduction;

[a]To convert potential vs. Ag/AgCl to SCE, a correction factor of –0.045 V (ref.[12]) has been applied.

conditions, unmetalated porphyrin complexes give lower oxidation potentials for hydrazine than Co or Mn porphyrin complexes.

Malinski and coworkers employed polymerized metal porphyrins for the catalytic oxidation of hydrazine in basic media[217,218]. NiTMTHPP, CoPCPP, CoTPCPP, MnTPCPP, NiTPCPP, CuTDEAPP, CoTDEAPP, MnTDEAPP, NiTDEAPP, and PdTDEAPP complexes were found to catalyze the electrochemical oxidation of hydrazine[217,218]. Table 7.8 shows that low oxidation potentials for the oxidation of hydrazine were observed on NiTPCPP compared to the corresponding Mn, Co, and unmetalated derivatives. Trevin *et al.*[219] examined the role of the Ni central metal in M–N$_4$ complexes on the catalytic activity towards the electrooxidation of hydrazine. The role of the central metal could be explained by the fact that when nickel-based modified electrodes, prepared by oxidative polymerization of nickel porphyrin, nickel salen, and nickel cyclam complexes from 0.1M NaOH solutions, and used for electrocatalytic oxidation of hydrazine[219], the behavior of the polymerized films was similar to that shown by nickel hydroxide modified electrodes[219]. A carbon paste electrode modified with CuTPP was discussed in a review by Biesaga *et al.* for the analysis of hydrazine[89]. Pessoa *et al.* reported that CoTCPP immobilized onto Nb oxide grafted-silica gel surface, catalyzes the oxidation of hydrazine with low response times (<1 s)[220].

8.2. Phthalocyanine Complexes

8.2.1. Hydrazine

There has been relatively more research activity on the use of MPc complexes for electrocatalysis of hydrazine compared to porphyrins. Isaacs *et al.*[221] reported that when CoPc and CoNPc are adsorbed on a GCE, they show different catalytic behavior towards the oxidation of hydrazine. The differences were explained by the fact than in CoNPc the redox couples are centered only on the ring and not on the metal as is the case with CoPc[221].

MPc and MTSPc catalyzed the oxidation of hydrazine[1,21], with the catalytic order depending on the metal as follows: MnPc > MnTSPc > CrTSPc > NiTSPc > VOPc > NiPc ~ CuPc ~ VOTSPc ~ ZnTSPc[222]. The mechanism proposed for the oxidation of hydrazine suggests the involvement of the M^{II}/M^{I} couple of the MPc complex[1]. This may explain the low catalytic activities of CuPc and ZnTSPc since these MPc complexes do not contain the M^{II}/M^{I} couple. Also CoTSPc and FeTSPc showed good catalytic activities towards the electrocatalytic oxidation of hydrazine compared to NiTSPc and CuTSPc[223]. FeTSPc (adsorbed on platinum) was found to be a better electrocatalyst than CoTSPc for the oxidation of hydrazine in basic media[21]. Zagal and coworkers studied a series of unsubstituted MPc complexes for their effectiveness as electrocatalysts for the oxidation of hydrazine[131,224]. FePc complexes were found to be the most active (compared to MnPc, CoPc, NiPc, and CuPc) for the electrocatalytic oxidation of hydrazine and reduction of cystine[131,224].

Different methods have been employed for the modification of electrodes for use in the electrocatalytic oxidation of hydrazine in addition to the ones listed above. For example, Li *et al.* showed that self-assembled multilayer films containing CoTSPc and diazo-resin have excellent response to the oxidation of hydrazine[225]. Vinod *et al.* showed intrazeolitically prepared FePc to have a good electrocatalytic behavior towards the oxidation of hydrazine[226]. Parez *et al.* immobilized NiTSPc onto titanized silica gel and used it to modify carbon paste electrode[227], the Ni^{III}/Ni^{II} couple catalyzed the oxidation of hydrazine[227]. Wang and coworkers prepared an electrode by dispersing CoPc and graphite in a sol gel derived network, which showed improvement over CPE containing CoPc in terms of stability and electrocatalytic behavior towards hydrazine and some thiols[228]. They also prepared screen-printed electrodes containing CoPc[229], a multifunctional electrode coating based on a mixture of CoPc and cellulose acetate[230], and a CoPc modified carbon wax composite electrodes[231], as amperometric sensors for hydrazine. The latter electrodes showed the same attractive features of CPE (e.g., low background current, flexible modifier loading, surface renewal, and low cost)[231]. A multilayer film containing CoPc and a bipolar pyridine salt on a surface of a gold electrode was reported by Sun *et al.* to show excellent electrocatalytic response for the oxidation of hydrazine[232]. FeTSPc immobilized on silica gel with titanium oxide also showed catalytic activity towards the oxidation of hydrazine[233].

Peng and Guarr[234] and Trollund *et al.*[235] reported on the electrocatalytic oxidation of hydrazine at a GCE modified with CoTAPc polymer, with a well-defined catalytic peak at –0.30 V[234], Table 7.8. The central Co(II) metal was found to act as the electronic receptor for hydrazine[234]. H_2TAPc on GCE or conducting glass electrodes has also shown catalytic activity towards the oxidation of hydrazine[234]. Ebadi reported on the use of a dimeric $(RuPc)_2$ complex for the electrocatalytic reduction of hydrazine to nitrogen in basic media[236]. The mechanism for this process may be similar to that reported below for hydroxylamine. The potentials for the oxidation of hydrazine on FeTSPc and CoTSPc are similar, Table 7.8.

8.2.2. Hydroxylamine

Ebadi reported that hydroxylamine is electrocatalytically oxidized to N_2O via a two electron process by $(RuPc)_2$ adsorbed onto a graphite electrode (by dip dry method) in the pH range 9 –13, Table 7.8[237]. The catalytic process is mediated by the Ru^{III}/Ru^{II} redox couple and the coordination of hydroxylamine as follows[237] (Equations (7.17) and (7.18)):

$$(Ru^{II}Pc)_2 \leftrightarrow [(Ru^{III,II}Pc)_2]^+ + e^- \quad (7.17)$$

$$[(Ru^{III,II}Pc)_2]^+ + NH_2OH \leftrightarrow [NH_2OH(Ru^{III,II}Pc)_2]^+ \quad (7.18)$$

MnPc, FePc, CoPc, NiPc, and CuPc catalyze the oxidation of hydroxylamine, with the NiPc showing the highest catalytic activity[129]. On CoPc modified electrodes, the electrogenerated Co^{III}Pc species was involved in the oxidation process. Whereas when FeTSPc was employed, fouling of the electrode was observed during the catalytic process[129]. Qi and Baldwin reported on CoPc modified electrodes for the electroxidation of hydroxylamine and its derivatives[236]. Co^{III}Pc species mediated the catalysis[238]. Zagal *et al.* found that NiPc complexes were the most active (compared to MnPc, CoPc, FePc, and CuPc) for the electrocatalytic oxidation of hydroxylamine[131].

9. Conclusions

It is clear from this review that it is not only M–N_4 complexes containing electroactive central metal which show catalytic activity. Unmetalated complexes as well as Zn and Cu complexes, which show ring-based processes only, often show electrocatalytic behavior towards the detection of some of the pollutants discussed in this work. There has been controversy surrounding the electrocatalytic activity of NiP or NiPc complexes. It seems the Ni porphyrin complexes do exhibit the Ni^{III}/Ni^{II} couple (at high potential in solution and more readily in the polymeric state) which may be involved in the electrocatalytic reactions involving these complexes. The Ni^{III}/Ni^{II} couple has not been identified electrochemically for the NiPc complexes in solution, but has been implicated in catalysis as an adsorbed polymer. It would be of interest to determine the values of the couple on polymeric NiPc complexes.

The potentials for the oxidation of the pollutants discussed in this chapter vary with the nature of the catalyst, the medium and the nature of the electrode. At this stage, there is not enough work done on different types on M–N_4 complexes under the same conditions, so that a trend on the best ring system for the catalytic process could be determined. Also it would be useful to study one catalyst system (for a particular analyte) using different pHs, electrolytes and electrodes, and also under homogenous and heterogeneous conditions in order to determine the best conditions for the catalytic activity. Such conditions once determined may be used for other catalysts.

10. Acknowledgment

Financial support by the National Research Foundation and by Rhodes University in South Africa is acknowledged.

References

1. Zagal, J.H. (1992). Metallophthalocyanines as catalysts in electrochemical reactions. *Coord. Chem. Rev.* **119**, 89–136.
2. Komorsky-Lovrić, S., M. Lovrić, and F. Scholz (1997). Sulfide ion electrooxidation catalysed by cobalt phthalocyanine microcystals. *Microchim. Acta* **127**(1–2), 95–99.
3. Maree, S. and T. Nyokong (2000). Electrocatalytic behaviour of substituted cobalt phthalocyanines towards the oxidation of cysteine. *J. Electroanal. Chem.* **492**(2), 120–127.
4. Shi, C. and F.C. Anson (1998). (5,10,15,20-tetramethylporphyrinato)cobalt(II): A remarkably active catalyst for the reduction of O_2 to H_2O. *Inorg. Chem.* **37**(5), 1037–1043.

5. Vilakazi, S.L. and T. Nyokong (2001). Voltammetric determination of nitric oxide on cobalt phthalocyanine modified microelectrodes. *J. Electroanal. Chem.* **512**, 56–63.
6. Halbert, M.K. and R.P. Baldwin (1985). Electrocatalytic and analytic response of cobalt phthalocyanine containing carbon paste electrodes towards sulfhydryl compounds. *Anal. Chem.* **57**(3), 591–595.
7. Malinski, T. (2000). Porphyrin-based electrochemical sensors. In K.M. Kadish, K.M. Smith, and R. Guilard (eds), *Porphyrin Handbook. Applications: Past, Present and Future*. Vol. 6, Chapter 44. Academic Press, New York, pp. 231–255.
8. Ciszewski, A., E. Kubaszewski, and M. Łożyński (1996). The role of nickel as central metal in conductive polymeric porphyrin film for electrocatalytic oxidation of nitric oxide. *Electroanalysis* **8**(3), 293–295.
9. Trevin, S., F. Bedioui, and J. Devynck (1996). Electrochemical and spectrophotometric study of the behaviour of electropolymerized nickel porphyrin films in the determination of nitric oxide. *Talanta* **43**, 303–311.
10. Chebotareva, N. and T. Nyokong (1997). Metallophthalocyanine catalysed electroreduction of nitrate and nitrite ions in alkaline media. *J. Appl. Electrochem.* **27**(8), 975–981.
11. Zilbermann, I., J. Hayon, T. Katchalski, O. Raveh, J. Rishpon, A.I. Shames, and A. Bettelheim (1997). Ligation and mediated oxidation of nitrogen monoxide by nickel(II) tetrasulfonated phthalocyanine. *J. Electrochem Soc.* **144**(8), L228–L230.
12. Lever, A.B.P., E.R. Milaeva, and G. Speier (1993). The redox chemistry of metallophthalocyanines in solution. In C.C. Leznoff and A.B.P. Lever, (eds), *Phthalocyanines: Properties and Applications*, Vol. 3. pp. VCH, New York, 1–69.
13. Sekota, M. and T. Nyokong (1999). Catalytic behaviour of osmium(II), rhodium(II) and ruthenium(II) phthalocyanines towards the electrooxidation of cysteine on glassy carbon electrodes. *Electoanalysis* **9**(16), 1257–1261.
14. Sonoyama, N., M. Kirii, and T. Sakata (1999). Electrochemical reduction of CO_2 at metal–porphyrin supported gas diffusion electrodes under high pressure. *Electrochem. Commun.* **1**(6), 213–216.
15. Lieber, C.M. and N.S. Lewis (1984). Catalytic reduction of CO_2 at carbon electrodes modified with cobalt phthalocyanine. *J. Am. Chem. Soc.* **106**(17), 5033–5034.
16. Zagal, J.H. and G.I. Cárdenas-Jirón (2000). Reactivity of immobilized cobalt phthalocyanines for the electroreduction of molecular oxygen in terms of molecular hardness. *J. Electroanal. Chem.* **489**, 96–100.
17. Zhang, J., A.B.P. Lever, and W.J. Pietro (1995). Electrocatalytic activity of a graphite electrode coated with hexadecachlorophthalocyanato iron(II) toward sulfide oxidation, and its possible application in electroanalysis. *Can. J. Chem.* **73**, 1072–1077.
18. Christensen, P.A., A. Hamnett, and A.V.G. Muir (1988). An in-situ FTIR study of the electroreduction of CO_2 by CoPc-coated edge graphite electrodes. *J. Electroanal. Chem.* **241**(1–2), 361–371.
19. Meshitsuka, S., M. Ichikawa, and K.Tamaru (1974). Electrocatalysis by metal phthalocyanines in the reduction of carbon dioxide. *J. Chem. Soc. Chem. Commun.* **5**, 158–159.
20. Magdesieva, T.V., T. Yamamoto, D.A. Tryk, and A. Fujishima (2002). Electrochemical reduction of CO_2 with transition metal phthalocyanine and porphyrin complexes supported on activated carbon fibers. *J. Electrochem. Soc.* **149**(6), D89–D95.
21. Zagal, J. (1980). Electrocatalysis of hydrazine electrooxidation by phthalocyanines adsorbed on graphite. *J. Electroanal. Chem.* **109**(1–3), 389–393.
22. Liu, S.-Q., J.-Q. Xu, H.-R. Sun, and D.-M. Li (2000). *meso*-Tetrakis(4-*N*-benzylpyridyl)porphyrin and its supramolecular complexes formed with anionic metal-oxo cluster: Spectroscopy and electrocatalytic reduction of dioxygen. *Inorg. Chim. Acta* **306**, 87–93.

23. Zecevic, S., B. Simic-Glavaski, E. Yeager, A.B.P. Lever, and P.C. Minor (1985). Spectroscopic and electrochemical studies of transition metal tetrasulfonated phthalocyanines—part V: Voltammetric studies of adsorbed tetrasulfonated phthalocyanines (MTSPc) in aqueous solutions. *J. Electroanal. Chem.* **196**, 339–358.
24. Oyaizu, K., A. Haryono, J. Natori, and E. Tsuchida (1998). (μ-Peroxo)bis[pyridine (phthalocyaninato)iron(III) as a convenient catalyst for the four-electron reduction of dioxygen. *J. Chem. Soc. Faraday Trans.* **94**, 3737–3742.
25. Zagal, J.H. and P. Herrera (1985). Electrochemistry of cysteine and cystine on metal–phthalocyanines adsorbed on a graphite electrodes. *Electrochim. Acta* **30**(4), 449–454.
26. Aquirre, M.J., M. Isaacs, F. Armijo, L. Basáez, and J. Zagal (2002). Effect of the substituents on the ligand of iron phthalocyanines adsorbed on graphite electrodes on their activity for the electrooxidation of 2-mercaptoethanol. *Electroanalysis* **14**(5), 356–362.
27. Kapusta, S. and N. Hackerman (1984). Carbon dioxide reduction at a metal phthalocyanine catalysed carbon electrode. *J. Electrochem. Soc.* **131**(7), 1511–1514.
28. Janda, P., N. Kobayashi, P.R. Auburn, H. Lam, C.C. Leznoff, and A.B.P. Lever (1989). Dioxygen reduction at a graphite electrode modified by mononuclear tetraneopentoxyphthalocyaninatocobalt(II) and related polynuclear species. *Can. J. Chem.* **67**, 1109–1019.
29. Janda, P., J. Weber, L. Dunsch, and A.B.P. Lever (1996). Detection of ascorbic acid using a carbon fiber microelectrode coated with cobalt tetramethylpyridoporphyrazine. *Anal. Chem.* **68**(6), 960–965.
30. Tse, Y.-H., P. Janda, and A.B.P. Lever (1994). Electrode with electrochemically deposited N, N', N'', N'''-tetramethyltetra-3,4-pyridinoporphyrazino-cobalt(I) for detection of sulfide ion. *Anal. Chem.* **66**(3), 384–390.
31. Huang, X. and W.Th. Kok (1993). Conductive carbon cement as electrode matrix for cobalt phthalocyanine modified electrodes for detection in flowing solutions. *Anal. Chim. Acta* **273**(1–2), 245–253.
32. Filanovsky, B. (1999). Electrochemical response of new carbon electrodes bulk modified with cobalt phthalocyanine to some thiols in the presence of heptane or human urine. *Anal. Chim. Acta* **394**, 91–100.
33. Kuwano, J., T. Eguchi, and Y. Sait (1997). Ambient temperature oxygen sensors based on fluorine solid electrolyte: The roles of constituents in the sensing electrode mixtures containing phthalocyanines. *Talanta* **44**, 705–712.
34. Kilinc, E., A. Erdem, L. Gokgunnec, T. Dalbasti, M. Karaoglan, and M. Ozsoz (1998). Buttermilk based cobalt phthalocyanine dispersed ferricyanide mediated amperometric biosensor for the determination of xanthine. *Electroanalysis* **10**(4), 273–275.
35. Shahrokhian, S., M.K. Amini, I. Mohammadpoor-Baltork, and S. Tangestaninejad (2000). Potentiometric detection of 2-mercaptobenzimidazople and 2-mercaptobenzothiazole at cobalt phthalocyanine modified carbon-paste electrode. *Electroanalysis* **12**(11), 863–867.
36. Korfhage, K.M., K. Ravichandran, and R.P. Baldwin (1984). Phthalocyanine-containing chemically modified electrodes for electrochemical detection in liquid chromatography/flow injection systems. *Anal. Chem.* **56**(8), 1517.
37. Mizutani, F., S. Yabuki, and S. Iijima (1995). Amperometric glugose-sensing electrode based on carbon paste containing poly(ethylene glycol)-modified glucose oxidase and cobalt octaethoxyphthalocyanine. *Anal. Chim. Acta* **300**, 59–64.
38. Mizutani, F., S. Yabuki, and S. Iijima (1995). Carbon paste electrode incorporated with cobalt(II) octaethoxyphthalocyanine for the amperometric detection of hydrogen peroxide. *Electroanalysis* **7**(8), 706–709.

39. Tanabe, H. and K. Ohno (1987). Electrocatalysis of metal phthalocyanine thin films prepared by the plasma-assisted deposition on a glassy carbon in the reduction of carbon dioxide. *Electrochim. Acta* **32**(7), 1121–1124.
40. Napier, A. and J.P. Hart (1996). Voltammetric and amperometric studies of selected thiols and dimethyldisulfide using a screen-printed carbon electrode modified with cobalt phthalocyanine: Studies towards a gas sensor. *Electroanalysis* **8**(11), 1006–1013.
41. Abe, T., T. Yoshida, S. Tokita, F. Taguchi, H. Imaya, and M. Kaneko (1996). Factors affecting selective electrocatalytic CO_2 reduction with cobalt phthalocyanine incorporated in a polyvinylpyridine membrane coated on a graphite electrode. *J. Electroanal. Chem.* **412**(1–2), 125–132.
42. Coutanceau, C., P. Crouigneau, J.M. Léger, and C. Lamy (1994). Mechanism of oxygen electroreduction at polypyrrole electrodes modified by cobalt phthalocyanine. *J. Electroanal. Chem.* **379**, 389–397.
43. Zhao, F., J. Zhang, T. Abe, D. Wöhrle, and M. Kaneko (1999). Electrocatalytic proton reduction by phthalocyanine cobalt derivatives incorporated in poly(4-vinylpyridine-co-styrene) film. *J. Mol. Cat. A: Chem.* **145**, 245–256.
44. Tatsuta, H., T. Nakamura, and T. Hinoue (2001). Thermal modulation voltammetric observation of cyanide ion in the membrane part of an ion-selective electrode based on a polymer modified with cobalt phthalocyanine in acetonitrile. *Anal. Sci.* **17**, 991–994.
45. de la Fuente, C., J.A. Acuña, M.D. Vázquez, M.L. Tascón, and M.I. Gómez (1997). Preparation of a polypyrrole electrode modified with a nickel phthalocyanine complex: Application to the determination of antioxidant (propylgallate) in foods. *Talanta* **44**(4), 685–695.
46. Leyzerovich, N.N., N.V. Shvedene, Y.N. Blikova, L.G. Tomilova, and I.V. Pletnev (2001). Comparative study of the metal phthalocyanates as active components in salicylate-selective electrodes. *Electroanalysis* **13**(13), 245–252.
47. Amini, M.K., S. Shahrokhian, and S. Tangestaninejad (1999). PVC-based cobalt and manganese phthalocyanine coated graphite electrodes for determination of thiocyanate. *Anal. Lett.* **32**(14), 2737–2750.
48. Amini, M.K., S. Shahrokhian, and S. Tangestaninejad (1999). Thiocyanate-selective electrodes based on nickel and iron phthalocyanines. *Anal. Chim. Acta* **402**, 137–143.
49. Nwachukwu, F.A. and M.G. Baron (2003). Polymeric matrices for immobilizing zinc tetraphenylporphyrin in absorbance based gas sensors. *Sensors Actuators B* **90**, 276–285.
50. Devenney, M., J. Grimshaw, and J. Trocha-Grimshaw (1998). Preparation and characterization of electrodes modified with iron porphyrins immobilized in a poly(amino acid) matrix. *J. Chem. Soc. Perkin Trans. 2*, 917–924.
51. Ozoemena, K., P. Westbroek, and T. Nyokong (2001). Long-term stability of a gold electrode modified with a self-assembled monolayer of octabutylthiophthalocyaninato-cobalt(II) towards L-cysteine detection. *Electrochem. Commun.* **3**, 529–534.
52. Yamada, H., H. Imohari, Y. Nishimura, I. Yamazaki, T.K. Ahn, S.K. Kim, D. Kim, and S. Fukuzumi (2003). Photovoltaic properties of self-assembled monolayers of porphyrin and porphyrin–fullerene dyads on ITO and gold surfaces. *J. Am. Chem. Soc.* **125**(30), 9129–9139.
53. Somashekarappa, M.P. and S. Sampath (2002). Orientaion dependent electrocatalysis using self-assembled molecular films. *Chem. Commun.* 1262–1263.
54. Somashekarappa, M.P., J. Keshavayya, and S. Sampath (2002). Self-assembled molecular films of tetraamino metal (Co,Cu,Fe) phthalocyanines on gold and silver. Electrochemical and spectroscopic characterization. *Pure Appl. Chem.* **74**(9), 1609–1620.

55. Revell, D.J., I. Chambrier, M.J. Cook, and D.A. Russell (2000). Formation and spectroscopic characterization of self-assembled phthalocyanine monolayers. *J. Mater. Chem.* **10**, 31–37.
56. Chabrier, I., M.J. Cook, D.A. Russell (1995). Synthesis and characterization of functionalised phthalocyanine compounds for fabrication of self-assembled monolayers. *Synthesis* 1283–1284.
57. Li, Z., M. Lieberman, and W. Hill (2001). XPS and SERS study of silicon phthalocyanine monolayers: Umbrella versus octopus design strategies for formation of oriented SAMs. *Langmuir* **17**(16), 4887–4894.
58. Ozoemena, K., P. Westbroek, and T. Nyokong (2002). Cyclic voltammetric studies of octabutylthiophthalocyaninatocobalt(II) and its self-assembled monolayers (SAMs) on gold electrode. *J. Porphyrins Phthalocyanines* **6**, 98–106.
59. Cook, M.J. (1999). Phthalocyanine thin films. *Pure Appl. Chem.* **71**(11), 2145–2151.
60. Xie, D., Y. Jiang, J. Jiang, Z. Wu, and Y. Li (2001). Gas sensitive Langmuir–Blodjectt films based on erbium bis[octakis(octyloxy)phthalocyaninato) complex. *Sensor Actuator B: Chem.* **77**, 260–263.
61. Del Caño, T. and R. Aroca (2003). Langmuir–Blodgett mixed films of titanyl(IV) phthalocyanine and arachidic acid: Molecular orientation and film structure. *Langmuir* **19**(9), 3747–3751.
62. Sharma, L.R., G. Singh, and A. Sharma (1988). Cyclic voltammetric study of the redox mechanism of cresols in aqueous solutions. *B. Electrochem.* **4**(7), 679–682.
63. Oubina, A., D. Puig, J. Gascón, and D. Barceló (1997). Determination of pentachlorophenol in certified waste waters, soil samples and industrial effluents using ELISA and liquid solid extraction followed by liquid chromatography. *Anal. Chim. Acta* **346**, 49–59.
64. Bedioui, F., S.G. Granados, and C. Bied-Charreton (1999). Metalloporphyrin, metallophthalocyanine and related macrocycle complex-based film modified electrodes: Review of selected significant designs and applications to the electrochemical detection of pollutants. *Recent Res. Dev. Electrochem.* **2**(1), 91–108.
65. Dobson, D.J. and S. Saini (1997). Porphrin-modified electrodes as biomimetic sensors for the determination of organohalide pollutants in aqueous samples. *Anal. Chem.* **69**(17), 3532–3538.
66. Demel, R. and A.M. Eleonore (1999). Catalytic electrochemical reduction of 1,2-dibromocyclohexane derivative at carbon electrodes modified with cobalt porphyrin siloxane films. *Acta Chemica Scandinavica* **53**(11), 1038–1042.
67. Elliott, C.M. and C.A. Marrese (1981). Catalytic reduction of some alkyl halides by iron porphyrin modified carbon electrodes. *J. Electroanal.* Chem. **119**(2), 395–401.
68. Zheng, G., J.D. Guodong, and L. Li (2000). Metalloporphyrin catalysed electrocarboxylation of alkyl halides: Elucidate mechanism by electrochemical, in situ UV–Vis, and FT–IR spectrochemical methods. *Recent Dev. Electroanal. Chem.* **2**, 89–115.
69. Zheng, G., M. Stradiotto, and L. Li (1998). An electrochemical and spectroelectrochemical (IR) investigation of the reduction of Rco(II)TPP (R = benzyl or butyl; TPP = tetraphenylporphyrin): Mechanistic implications in the CoTPP catalysed electrocarboxylation of alkyl halides. *J. Electroanal. Chem.* **453**(1–2), 79–88.
70. Zhou, D.-L., H. Carrero, and J.F. Rusling (1996). Radical vs anionic pathway in mediated electrochemical reduction of benzyl bromide in a bicontinuous microemulsion. *Langmuir* **12**(12), 3067–3074.
71. Priyantha, N. and M. Tambalo (1992). Metalloporphyrin-coated electrodes for detection of 2,4-D. *ACS Symp. Ser.* **511**, 41–47.
72. Bessel, C.A. and D.R. Rolison (1997). Electrocatalytic reactivity of zeolite-encapsulated Co(salen) with benzyl chloride. *J. Am. Chem. Soc.* **119**(51), 12673–12674.

73. Shen, Y., T. Tajima, M. Atobe, and T. Fuchigami (2004). Electrocatalytic debromination of organic bromides using a cobalt(II) salem complex in ionic liquids. *Electrochemistry* **72**(12), 849–851.
74. Gaillon, L. and F. Bedioui (2004). Voltammetric analysis of the catalytic reactivity of electrogenerated CoI-salen with organohalogenated derivatives in an ionic liquid at room temperature. *J. Mol. Cat. A: Chem.* **214**(1), 91–94.
75. Ji, C., D.G. Peters, J.A. Karty, J.P. Reilly, and S.M. Mubarak (2001). Direct and cobalt(I) salen-catalysed reduction of 2,6-bis(chloromethyl)pyridine at carbon electrodes in acetonitrile. *J. Electroanal. Chem.* **516**(1–2), 50–58.
76. Persinger, J.D., J.L. Hayes, J.L. Klein, D.G. Peters, J.A. Karty, and J.P. Reilly (2004). Catalytic reduction of 1,1,2-trichloro-1,2,2-trifluoroethane (CFC-113) by cobalt(I) salen electrogenerated at vitreous carbon electrode. *J. Electroanal. Chem* **568**(1–2), 157–165.
77. Yanikin, V.V., E.I. Strunskaya, N.V. Nastapova, N.I. Iakskimyuk, Z.A. Bredikhina, D.R. Sharafutdinova, and A.A. Bredikhin (2003). Metal complex catalytic reduction of 1,1-dihalocyclopropanes. *Russ. Chem. Bull.* **52**(4), 923–928.
78. Shimakoshi, H., W. Ninomiya, and Y. Hisaeda (2001). Reductive coupling of benzyl bromide catalysed by a novel dicobalt complex having two salen units. *J. Chem. Soc. Dalton Trans.* **13**, 1971–1974.
79. Fry, A.J. and A.H. Singh (1994). Cobalt(salen)-electrocatalyzed reduction of benzal chloride. Dependence of products upon electrocatalysis potential. *J. Org. Chem.* **59**(26), 8172–8177.
80. Fry, A.J. and U.N. Sirisoma (1993). Electrocatalytic reduction of benzal bromide by cobalt(I)salen. A mechanistic investigation. *J. Org. Chem.* **58**18), 4919–4124.
81. Ordaz, A.A., J.M. Rocha, F.J.A. Aguilar, S.G. Granados, and F. Bedioui (2000). Electrocatalysis of the reduction of organic halide derivatives at modified electrodes coated by cobalt and iron macrocyclic complex-based films: Application to the electrochemical determination of pollutants. *Analusis* **28**(3), 238–244.
82. Ataide, A.L.R., A.P. Esteves, A.M. Freitas, M.J. Medeiros. V. Mota, D. Pletrcher, and P. Rodrigues (1999). Catalytic reduction of unsaturated halides by Ni(II) complexes. *Portugaliae. Electrochi. Acta* **17**(2/3), 215–219.
83. Dahm, C.E. and D.G. Peters (1994). Catalytic reduction of iodoethane and 2-iodopropane at carbon electrodes coated with anodically polymerized films of nickel(II) salen. *Anal. Chem.* **66**(19), 3117–3123.
84. Mubarak, M.S. and D.G. Peters (1995). Homogeneous catalytic reduction of α,ω-dihaloalkanes with electrogenerated nickel(I) salen. *J. Electroanal. Chem.* **388**(1–2), 195–198.
85. Bhattacharya, D., M.J. Samide, and D.G. Peters (1998). Catalytic reduction of cyclohexanecarbonyl chloride with electrogenerated nickel(I) salen in acetonitrile. *J. Electroanal. Chem.* **441**(1–2), 103–107.
86. Stepanov, A.A., M.G. Peterleitner, S.M. Peregudova, and L.I. Denisovich (2002). Electrocatalytic reduction of perfluoroalkyl chlorides in the presence of a nickel complex. *Russ. J. Electrochem.* **38**(5), 565–566.
87. Chen, S.-M. and S.-V. Chen (2003). The interaction of water-soluble iron porphyrins with DNA films and the electrocatalytic properties for inorganic and organic nitro compounds. *Electrochim. Acta* **48**(27), 4049–4060.
88. Dall'Orto, V.C., C. Danilowicz, J. Hurst, A. Lo Balbo, and I. Rezzano (1998). Studies of the interaction between metalloporphyrin films and ohenols in a preconcentration type sensor. *Electroanalysis* **10**(2), 127–131.
89. Biesaga, M., K. Pyrzyska, and M, Trojanowicz (2000). Porphyrins in analytical chemistry: A review. *Talanta* **51**, 209–224.

90. Zhang, S. and J.F. Rusling (1993). Dechlorination of polychlorinated biphenyls by electrochemical catalysis in a bicontinuous microemulsion. *Env. Sci. Technol.* **27**(7), 1375–1380.
91. Hu, N. and J.F. Rusling (1991). Surfactant-intercalated clay films for electrochemical catalysis: Reduction of trichloroacetic acid. *Anal. Chem.* **63**(19), 2163–2168.
92. Kamau, N.G. and J.F. Rusling (1996). Enhanced rates of organic dehalogenations in a microemulsion using adsorbed metal phthalocyanines on electrodes. *Langmuir* **12**(11), 2645–2649.
93. Zhang, S. and J.F. Rusling (1996). Evaluation of microemulsions of cationic surfactants and a polyoxyethylene cosurfactant for electrolytic dechlorinations of chlorobiphenyls. *J. Colloid Interf. Sci.* **182**(2), 558–563.
94. Zelina, J.P. and J.F. Rusling (1995). Molecular orientation in self-assembled films of copper phthalocyanine tetrasulfonate and cationic surfactant. *Micropor. Mat.* **5**(4), 203–210.
95. Zhang, H. and J.F. Rusling (1993). Preconcentration and catalysis in reduction of aliphatic organohalides using surfactant-coated electrodes. *Talanta* **40**(5), 741–747.
96. Rusling, J.F., N. Hu, H. Zhang, D.J. Howe, C.I. Miaw, and E.C. Couture (1992). Electrochemical catalysis in surfactant media. *Electrochem. Colloids Dispers.* 303–318.
97. Kamau, G.N., N. Hu, and J.F. Rusling (1992). Rate enhancement and control in electrochemical catalysis using a bicontinuous microemulsion. *Langmuir* **8**(4), 1042–1044.
98. Jiang, X.-E., L-P. Guo, and X.-G. Du (2003). Electrochemistry and electrocatalysis of binuclear cobalt phthalocyaninehexasulfonate-surfactant film modified electrode. *Talanta* **61**(3), 247–256.
99. Iwunze, M.O., N. Hu, and J.F. Rusling (1992). Electrocatalysis in organized assemblies—part X: Debromination of alkyl vicinal dibromides with neutral metal phthalocyanines in bicontinuous microemulsion. *J. Electroanal. Chem.* **333**(1–2), 353–361.
100. Rusling, J.F., M.F. Ahmadi, and N. Hu (1992). Surfactant-intercalated clay films containing metal phthalocyanines. *Langmuir* **8**(10), 2455–2460.
101. Iwunze, M.O. and J.F. Rusling (1989). Aqueous lamellar surfactant system for mediated electrolytic dechlorination of polychlorinated biphenyls. *J. Electroanal. Chem.* **266**(1), 197–201.
102. de la Fuente, C., F. Bedioui, J.A. Acuna, M.L. Vazquez, M.L. Tascon, M.I. Gomez, Y. Castrillejo, and P. Sanchez-Batanero (1994). Modification of polypyrrole electrodes by using phthalocyanine complexes: Study of their stability and morphology. *Quimica Analitica* **13**(3), 152–1577.
103. Chicharro, M., A. Zapardiel, E. Bermejo, M. Moren, and E. Madrid (2002). Electrocatalytic amperometric determination of amitrole using a cobalt-phthalocyanine-modified carbon paste electrode. *Anal. Bioanal. Chem.* **373**(4–5), 277–283.
104. Hartley, I.C. and J.P. Hart (1994). Amperometric measurement of organophosphate pesticides using screen-printed disposable sensor and biosensor based on cobalt phthalocyanine. *Anal. Proc.* **31**(11), 333–337.
105. Kiflom, W.G., O.S. Wandiga, and N.G. Kamau (2001). Status of organochlorine (DDT) pollutants and steps toward electrocatalytic reductions. *Pure Appl. Chem.* **73**(12), 1907–1916.
106. Ozsoz, M., A. Erdem, E. Kilinc, and L. Gokgunnec (1996). Mushroom-based cobalt phthalocyanine dispersed amperometric biosensor for the determinination of phenolic compounds. *Electroanalysis* **8**(2), 147–150.

107. Ureta-Zunartu, M.S., C. Berrios, J. Pavez, J. Zagal, C. Gutierrez, and J.F. Marco (2003). Electrooxidation of 2-chlorophenol on polyNiTSPc-modified glassy carbon electrodes. *J. Electroanal. Chem.* **553**, 147–156.
108. Mafatle, T. and T. Nyokong (1997). The use of cobalt(II) phthalocyanine to improve the sensitivity and stability of glassy carbon electrodes for the detectuon of cresols, chlorophenols and phenol. *Anal. Chim. Acta* **354**, 307–314.
109. Chen, S.-M. (1997). Reversible electrocatalytic reaction of sulfur oxoanions and sulfides by one catalyst of water-soluble cobalt porphyrin. *J. Electroanal. Chem.* **432**(1–2), 101–109.
110. Chen, S.-M. (1997). The electrocatalytic reactions of cysteine and cystine by water-soluble iron porphyrin, manganese porphyrin and iron(II) phenanthrolines. *Electrochim. Acta* **42**(11), 1663–1673.
111. De Gregori, I., M. Carrier, A. Deronzier, J.-C. Moutet, F. Bedioui, and J. Devynck (1992). Incorporation of anionic cobalt porphyrin by anion exchange into polypyrrole films containing alkylammonium groups. *J. Chem. Soc. Faraday Trans.* **88**(11), 1567–1572.
112. Pang, D.-W. and Z.-L. Wang (1993). Electrocatalysis of metalloporphyrins-part 13: electrocatalysis of several water-soluble porphyrins for the oxidation of some small molecules. *J. Electroanal. Chem.* **358**(1–2), 235–246.
113. Pang, D.-W., B.-H. Deng, and Z.-L. Wang (1994). Electrocatalysis of metalloporphyrins—part 14: Electro-oxidation of hydrazine catalysed by water-soluble tetrakis(4-trimethylammoniumphenyl) porphyrin and its cobalt complex. *Electrochim. Acta* **39**(6), 847–851.
114. Wang, Z. and D. Pang (1990). Electrocatalysis of metalloporphyrins—part 9: Catalytic electroreduction of cysteine using water-soluble cobalt porphyrins. *J. Electroanal. Chem.* **283**(1–2), 349–358.
115. Pang, D., Z. Wang, and C.C. Zongli (1992). Electrocatalysis of metalloporphyrins—Part 11: Catalytic electroreduction of dithiodipropionic acid with water-soluble cobalt tetrakis(4-trimethylammoniophenyl)porphyrin catalyst. *Electrochim. Acta* **37**(14), 2591–2594.
116. Rea, N., B. Loock, and D. Lexa (2001). Porphyrins bound to $Ru(bpy)_2$ clusters: Electrocatalysis of sulfite. *Inorg. Chim Acta* **312**(1–2), 53–66.
117. Azevedo, C.M.N., K. Araki, L. Angnes, and H.E. Toma (1998). Electrostatically assembled films for improving the properties of tetraruthenated porphyrin modified electrodes. *Electroanalysis* **10**(7), 467–471.
118. Halbert, M.K. and R.P. Baldwin (1986). Amperometric detection of thiopurines in blood plasma with a cobalt-phthalocyanine chemically modified electrode after liquid chromatography. *Anal. Chim. Acta* **187**, 89–97.
119. Zagal, J.H., C. Fierro, and R. Rozas (1981). Electrocatalytic effects of adsorbed cobalt phthalocyanine tetrasulfonate in the anodic oxidation of cysteine. *J. Electroanal. Chem.* **119**(2), 403–408.
120. Griveau, S., J. Pavez, J.H. Zagal, and F. Bedioui (2001). Electro-oxidation of 2-mercaptoethanol on adsorbed monomeric and electropolymerized cobalt tetra-aminophthalocyanine films: Effect of film thickness. *J. Electroanal. Chem.* **497**, 75–83.
121. Aguirre, M.J., M. Isaacs, F. Armijo, L. Basáez, and J.H. Zagal (2002). Effect of the substituents on the ligand of iron phthalocyanines adsorbed on graphite electrodes on their activity for the electrooxidation of 2-mercaptoethanol. *Electroanalysis* **14**(5), 356–362.
122. Gulppi, M., S. Griveau, F. Bedioui, and J.H. Zagal (2001). Electrocatalysis of 2-mercaptoethanesulfonic acid oxidation on cobalt phthalocyanine modified elec-

trodes: Effect of surface concentration on the catalyst. *Electrochim. Acta* **46**(22), 3397–3404.
123. Zagal, J.H., M.A. Gulppi, C.A. Caro, and I.G. Cardenas-Jiron (1999). Paradoxical effect of the redox potential of adsorbed metallophthalocyanines on their activity for the oxidation of 2-mercaptoethanol: Inner versus outer sphere electrocatalysis. *Electrochem. Comm.* **1**(9), 389–393.
124. Greveau, S., M. Gulppi, J. Pavez, J.H. Zagal, and F. Bedioui (2003). Cobalt phthalocyanine-based molecular materials for the electrocatalysis and electroanalysis of 2-mercaptoethanol, 2-mercaptoethanesulfonic acid, reduced glutathione and L-cysteine. *Electoanalysis* **15**(9), 779–785.
125. Zagal, J.H., M.A. Gulppi, C. Depretz, and D. Lelievre (1999). Synthesis and electrocatalytic properties of octaalkoxycobalt phthalocyanine for the oxidation of 2-mercaptocthanol. *J. Porphyrins Phthalocyanines* **3**(5), 355–363.
126. Lezna, R.O., S. Juanto, and J.H. Zagal (1998). Spectroelectrochemical studies of tetrasulfonated metallophthalocyanine adsorbed on the basal plane of graphite in the presence of cysteine. *J. Electroanal. Chem.* **452**(2), 221–228.
127. Kimura, M., Y. Yamaguchi, T. Koyama, K. Hanabusa, and H. Shirai (1997). Catalytic oxidation of 2-mercaptoethanol by cationic water-soluble phthalocyaninatocobalt(II) complexes. *J. Porphyrins Phthalocyanines* **1**, 309–313.
128. Aguirre, J.M., M. Isaacs, F. Armijo, N. Bocchi, and J.H. Zagal (1998). Catalytic electrooxidation of 2-mercaptoethanol on perchlorinated iron phthalocyanine adsorbed on a graphite electrode. *Electroanalysis* **10**(8), 571–575.
129. Vasudevan, P., N. Phouget, and A.K. Shukla (1996). Metal phthalocyanines as electrocatalysts for redox reactions. *Appl. Organomet. Chem.* **10**(8), 591–604.
130. Zagal, J.H. and C. Paez (1989). Catalytic electrooxidation of 2-mercaptoethanol on a graphite electrode modified with metal–phthalocyanines. *Electrochim. Acta* **34**(2), 243–247.
131. Zagal, J.H., P. Herrera, K. Brinck, and S. Ureta-Zanartu (1984). Electrocatalytic properties of phthalocyanines. *Proc. Electrochem. Soc.* **84**(12), 602–617.
132. Linders, C.R., G.J. Patriarche, J.M. Kauffmann, and G.G. Guilbault (1986). Electrochemical study of thiols and disulfides using modified electrodes. *Anal. Lett.* **19**(1–2), 193–203.
133. Zagal, J.H., M.E. Vaschetto, and B.A. Retamal (1999). Electrocatalytic response of poly(2-chloroaniline)/cobalt phthalocyanine modified electrodes. *Int. J. Polym. Mater.* **44**(3–4), 225–239.
134. Wring, S.A., J.P. Hart, and B.J. Birch (1991). Determination of reduced glutathione in human whole blood by high-performance lquid chromatography with electrochemical detection by a graphite-epoxy resin electrode chemically modified with cobalt phthalocyanine. *Talanta* **38**(11), 1257–1260.
135. Wang, J., T. Golden, K. Varughese, and I. El-Rayes (1989). Polishable robust modified graphite epoxy electrodes. *Anal. Chem.* **61**(5), 508–512.
136. Retamal, B.A. and M.A. Vaschetto (1997). Catalytic electro-oxidation of 2-mercaptoethanol using cobalt phthalocyanine+poly(2-chloroaniline) modified electrodes. *J. Electroanal. Chem.* **431**(1), 1–5.
137. Ballarin, B., M. Gazzano, J.L. Hidalgo-Hidalgo de Cisneros, D. Tonelli, and R. Seeber (2002). Electrocatalytic activity of cobalt phthalocyanine stabilized by different matrixes. *Anal. Bioanal. Chem.* **374**(4), 891–897.
138. Ballarin, B. and M. Del Mar Cordero-Rando (2003). New rigid conducting composites for electrochemical sensors. *Collect. Czech. Chem Commun.* **68**(8), 1420–1436.
139. Mafatle, T.J. and T. Nyokong (1996). Electrocatalytic oxidation of cysteine by molybdenum(V) phthalocyanine complexes. *J. Electroanal. Chem.* **408**(1–2), 213–218.

140. O'Shea, T.J. and S.M. Lunte (1994). Chemically modified microelectrodes for capillary electrophoresis/electrochemistry. *Anal. Chem.* **66**(2), 307–311.
141. Popesku, L.G., L.G. Shaidarova, and G.K. Budnikov (1999). Electrocatalytic determination of organic thiocarbonyl compounds using a carbon paste electrode modified with cobalt phthalocyanine. *Ind. Lab. (Diag. Mater.)* **65**(7), 423–425.
142. Shahrokhian, S. and J. Yazdani (2003). Electrocatalytic oxidation of thioglycolic acid at carbon paste electrode modified with cobalt phthalocyanine: Application as a potentiometric sensor. *Electrochim. Acta* **48**(28), 4143–4148.
143. Parez, E.F., L.T. Kubota, A.A. Tanaka, and G. De Oliveira Neto (1998). Anodic oxidation of cysteine catalysed by nickel tetrasulfonated phthalocyanine immorbilized on silica gel modified with titanium(IV) oxide. *Electrochim. Acta* **43**(12–13), 1665–1673.
144. Karmann, E., D. Schlettwein, and N.I. Jaeger (1996). Photoelectrochemical oxidation of 2-mercaptoethanol at the surface of octacyanophthalocyanine thin films. *J. Electroanal. Chem.* **405**, 149–158.
145. Abaas, A.K. and J.P. Hart (1997). Electrocatalytic, diffusional and analytic characteristics of a cobalt phthalocyanine modified, screen-printed, amperometric gas sensor for propanethiol. *Sensor Actuator B: Chem.* **B42**(1–3), 169–175.
146. Hart, J.P. and A.K. Abass (1997). A disporsable amperometric gas sensor for sulfur-containing compounds based on a chemically modified screen printed carbon electrode coated with hydrogel. *Anal. Chim. Acta* **342**(2–3), 199–206.
147. Hart, J.P. and I.C. Hartley (1994). Development of amperometric sensors for uric acid based on chemically modified graphite-epoxy resin and screen-printed electrodes containing cobalt phthalocyanine. *Analyst* **119**(2), 259–263.
148. Wring, S.A., J.P. Hart, L. Bracely, and B.J. Birch (1990). Development of screen-printed carbon electrodes chemically modified with cobalt phthalocyanine, for electrochemical sensor applications. *Anal. Chim. Acta* **231**(2), 203–212.
149. Ozoemena, K. and T. Nyokong (2002). Voltammetric characterization of the self-assembled monolayer (SAM) of octabutylthiophthalocyaninatoiron(II): A potential electrochemical sensor. *Electrochim. Acta* **47**(25), 4035–4043.
150. Ozoemena, K.I., T. Nyokong, and P. Westbrook (2003). Self-assembled monolayers of cobalt and iron phthalocyanine complexes on gold electrodes: Comparative surface electrochemistry and electrocatalytic interaction with thiols and thiocyanate. *Electroanalysis* **15**(22), 1762–1770.
151. Gulppi, M., F. Bedioui, and J.H. Zagal (2001). Overoxidized polypyrrole/cobalt tertasulfonated phthalocyanine modified ultramicro-carbon-fiber electrodes for the electrooxidation of 2-mercaptoethanol. *Electroanalysis* **13**(13), 1136–1139.
152. Obirai, J., N.P Rodrigues, F. Bedioui, and T. Nyokong (2003). Synthesis, spectral and electrochemical properties of a new family of pyrrole substituted cobalt, iron, manganese, nickel and zinc phthalocyanine complexes. *J. Porphyrins Phthalocyanines* 7(7), 508-520.
153. Tse, Y.-H., P. Janda, H. Lam, and A.B.P. Lever (1995). Electrode with electropolymerized tetraaminophthalocyanatocobalt(II) for detection of sulfide ion. *Anal. Chem.* **67**(5), 981–985.
154. Amini, K.M., S. Shahrokhian, S. Tangestaninejad, and I. Mahammadpoor-Baltork (2001). Voltammetric and potentiometric behaviour of 2-pyridinethiol, 2-mercaptoethanol and sulfide at iron(II) phthalocyanine modified carbon-paste electrode. *Iran J. Chem. Chem. Eng.* **20**(1), 29–36.
155. Cape, J.N., D. Fowler, and A. Davison (2003). Ecological effects of sulfur dioxide, fluorides, and minor air pollutants: Recent trends and research needs. *Environ. Int.* **29**(2–3), 201–211.

156. Crabbe, H., R. Beaumont, and D. Norton (2004). Assessment of air quality, emissions and management in a local urban environment. *Environ. Monitor. Assess.* **65**(1–2), 435–442.
157. Kline, M.A., M.H. Barley, and T.J. Meyer (1987). Electrocatalytic reduction of HSO_3^- to H_2S based on a water-soluble iron porphyrin. *Inorg. Chem.* **26**(14), 2196–2197.
158. Chen, S.-M. and S.-W. Chiu (2000). The electrocatalytic transformation of HS^-, $S_2O_3^{2-}$, $S_4O_6^{2-}$ and SO_3^{2-} to SO_4^{2-} by water-soluble iron porphyrins. *Electrochim. Acta* **45**(27), 4399–4408.
159. Chen, S.-M. (1996). Bicatalyst electrocatalytic reaction of sulfur oxoanions by water-soluble iron porphyrins and iron(II) tris(1,10-penanthroline). *Inorg. Chim. Acta* **244**(2), 155–164.
160. Chen, S.-M. (1996). Electrocatalytic oxidation of thiosulfate by metalhexacyanoferrate film modified electrodes. *J. Electroanal. Chem.* **417**(1–2), 145–153.
161. Mikisztal, A.R. and J.S. Valentine (1984). Reactivity of the peroxo ligand in metalloporphyrin complxes: Reaction of sulfur dioxide with iron and titanium porphyrin peroxo complxes to give sulfato complexes or sulfate. *Inorg. Chem.* **23**(22), 3548–3552.
162. Shin, K. and H.M. Goff (1990). Iron(III) porphyrin promoted aerobic oxidation of sulfur dioxide. *J. Chem. Soc. Chem. Commun.* 461–462.
163. Carballo, R., V.C. Dall'Orto, A. Lo Balbo, and I. Rezzano (2003). Determination of sulfite by flow injection analysis using poly[Ni-(protoporphyrin IX)] chemically modified electrode. *Sensor Actuators B: Chem.* **88**, 155–161.
164. Thamae, M., P. Westbroek, and T. Nyokong (2002). pH study of the SO_2 detection at a glassy carbon electrode modified with iron(II)tetrasulfophthalocyanine. *Acta Microkim. Acta* **140**, 233–239.
165. Nikolov, I., I. Vitanova, V. Najdenov, T. Milusheva, and T. Vitanov (1997). Effect of pyrolysis temperature of the catalytic activity of active carbon + cobalt phthalocyanines in sulfur dioxide oxidation by oxygen. *J. Appl. Electrochem.* **27**, 77–82.
166. Iliev, P., I. Nikolov, T. Vitanov, and E. Budevski (1992). Influence of nitrogen oxides on the electrocatalytic oxidation of sulfur dioxide. *J. Appl. ElectroChem.* **22**(5), 425–428.
167. Nikolov, I., K. Petrov, and T. Vitanov (1996). Low temperature electrochemical oxidation of sulfur dioxide. *J. Appl. ElectroChem.* **26**(7), 703–709.
168. Yu, C.S., H. Choi, and S. Kim (2002). Electrocatalytic reduction of sulfur dioxide by iron phthalocyanine monolayer in acid conditions. *Chem Lett.* **7**, 648–649.
169. Zagal, J.H., M.A. Gulppi, and G. Cárdenas-Jirón (2000). Metal-centered redox chemistry of substituted cobalt phthalocyanines adsorbed on graphite and correlations with MO calculations and Hammett parameters: electrocatalytic reduction of a disulfide. *Polyhedron* **19**(22–23), 2255–2260.
170. Collin, J.P. and J.P. Sauvage (1989). Electrochemical reduction of carbon dioxide mediated by molecular catalysts. *Coord. Chem. Rev.* **93**(2), 245–268.
171. Matsuoka, S., K. Kiichi, T. Ogata, M. Kasuba, N. Nakashima, E. Fujita, and S. Yanagida (1993). Efficient and selective electron mediation of cobalt complexes with cyclam and related macrocycles in the *p*-terphenyl-catalyzed photoreduction of carbon dioxide. *J. Am. Chem Soc* **115**(2), 601–609.
172. Bhugun, I., D. Lexa, and J.-M. Saveant (1996). Catalysis of the electrochemical reduction of carbon dioxide by iron(0) porphyrins: Synergistic effect of Lewis acid cations. *J. Phys. Chem.* **100**(51), 19981–19985.
173. Bhugun, I., D. Lexa, and J.-M. Savéant (1994). Ultraefficient selective homogeneous catalysis of the electrochemical reduction of carbon dixode by an iron(0) por-

phyrin associated with a weak Broensted acid cocatalyst. *J. Am. Chem. Soc.* **116**(11), 5015–5016.

174. Behar, D., T. Dhanasekaran, P. Neta, C.M. Hosten, D. Ejeh, P. Hambright, and E. Fujita (1998). Cobalt porphyrin catalysed reduction of CO_2: Radiation chemical, photochemical, and electrochemical studies. *J. Phys. Chem. A* **102**(17), 2870–2877.
175. Riqelme, M.A., M. Isaacs, M. Lucero, E. Trullund, M.J. Aguirre, and J. Canales (2003). Electrocatalytic reduction of carbon dioxide at polymeric cobalt tetra(3-aminophenyl) porphyrin glassy carbon-modified electrodes. *J. Chilean Chem. Soc.* **48**(2), 89–92.
176. Grodkowski, J., P. Neta, E. Fujita, A. Mahammed, L. Simkhovich, and Z. Gross (2002). Reduction of cobalt and iron corroles and catalysed reduction of CO_2. *J. Phys. Chem. A.* **106**(1), 4772–4778.
177. Bhugun, I., D. Lexa, and J.-M. Saveant (1996). Catalysis of the electrochemical reduction of carbon dioxide by iron(0) porphyrin: Synergistic effect of weak Broensted acids. *J. Am. Chem. Soc.* **118**(7), 1769–1776.
178. Hammouche, M., D. Lexa, M. Momenteau, and J.M. Savéant (1991). Chemical catalysis of electrochemical reactions: Homogeneous catalysis of the electrochemical reduction of carbon dioxide by iron("0") porphyrins—role of the addition of magnesium cations. *J. Am. Chem Soc.* **113**(22), 8455–8466.
179. Aga, H., A. Aramata, and Y. Hisaeda (1997). The electroreduction of carbon dioxide by macrocyclic cobalt complexes chemically modified on a glassy carbon electrodes. *J. Electroanal. Chem.* **437**, 111–118.
180. Tanaka, H. and A. Aramata (1997). Aminopyridyl cation radical method for bridging between metal complex and glassy carbon: Cobalt(II) tetraphenylporphyrin bonded on glassy carbon for enhancement of CO_2 electroreduction. *J. Electroanal. Chem.* **437**, 29–35.
181. Ogura, K. and I. Yoshida (1988). Electrocatalytic reduction of CO_2 to methanol—part 9: Mediation with metal porphyrins. *J. Mol. Catal.* **47**(1), 51–57.
182. Takahashi, K., K. Hiritsuka, H. Sasaki, and S. Toshima (1979). Electrocatalytic behaviour of metal porphyrins in the reduction of carbon dioxide. *Chem. Lett.* **4**, 305–308.
183. Becker, J.Y., B. Vainas, R. Eger, and L. Kaufman (1985). Electrocatalytic reduction of CO_2 to oxalate by Ag^{II} and Pd^{II} porphyrins. *J. Chem. Soc. Chem. Commun.* **21**, 1471–1472.
184. Qiu, A. and D.T. Sawyer (1997). The electrochemical evaluation of the metal–carbon bond energies ($-\Delta G_{BF}$) of alkylated iron and cobalt porphyrins[(por)M-R]. *J. Porphyrins Phthalocyanines* **1**(2), 125–134.
185. Enoki, O., T. Imaoka, and K. Yamamoto (2003). Electrochemical reduction of carbon dioxide catalysed by cofacial dinuclear metalloporphyrin. *Electrochemitry* **71**(4), 254–256.
186. Bernard, C., Y. Le Mest, and J.P. Gisselbrecht (1998). Coordination chemistry of iron porphycenes in the presence of CO, CO_2 and N-methylimidazole: Electrochemical, ESR, and UV–Vis study. *Inorg. Chem.* **37**(2), 181–190.
187. Venkataraman, R., H.R. Kunz, and J.M. Fenton (2001). Systematic approach to CO-tolerant catalyst development using metal macrocycle complexes with platinum. *Proc. Electrochem. Soc.* **22**, 1–13.
188. Shi, C. and F.C. Anson (2001). Catalysis of the electro-oxidation of carbon monoxide by cobalt octaethylporphyrin. *Inorg. Chem.* **40**(23), 5829–5833.
189. Van Baar, J.F., J.A.R. Van Veen, J.M. Van der Eijk, T.J. Peters, and N. De Wit (1982). Electrochemical oxidation of carbon monoxide with carbon-supported Group VIII metal chelates: Mechanistic aspects. *Electrochim. Acta* **27**(9), 1315–1319.

190. Van Baar, J.F., J.A.R. Van Veen, and N. De Wit (1982). Selective electrooxidation of carbon monoxide with carbon-supported rhodium and iridium porphyrins at low potentials in acid electrolyte. *Electrochim. Acta* **27**(1), 57–59.
191. Furuya, N. and K. Matsui (1989). Electroreduction of carbon dioxide on gas-diffusion electrodes modified by metal phthalocyanines. *J. Electroanal. Chem.* **271**(1–2), 181–191.
192. Shibata, M. and N. Furuya (2001). Electrochemical synthesis of urea at gas-diffusion electrodes part IV: Simultaneous reduction of carbon dioxide and nitrite ions with various metallophthalocyanine catalysts. *J. Electroanal. Chem.* **507**(1–2), 177–184.
193. Savinova, E.R., S.A. Yashnik, E.N. Savinov, and V.N. Parmon (1992). Gas-phase electrocatalytic reduction of carbon dioxide to carbon monoxide on carbon gas-diffusion electrode promoted by cobalt phthalocyaninre. *React. Kinet. Catal. Lett.* **46**(2), 249–254.
194. Yoshida, T., K. Tsukasa, M. Tsukamoto, T. Iida, D. Schlettwein, D. Woehrle, and M. Kaneko (1995). Selective electrocatalysis for CO_2 reduction in the aqueous phase using cobalt phthalocyanine/poly-4-vinylpyridine modified electrodes. *J. Electroanal. Chem.* **385**(2), 209–225.
195. Magdesieva, T.V., I.V. Zhukov, D.N. Kravchuk, O.A. Semenikhin, L.G. Tomilova, and K.P. Butin (2002). Electrocatalytic CO_2 reduction in methanol catalysed by mono-, di-, and electropolymerized phthalocyanine. *Russ. Chem. Bull.* **51**(5), 805–812.
196. Magdesieva, T.V., K.P. Butin, T. Yamamoto, D.A. Tryk, and A. Fujishima (2003). Lithium monophthalocyanine and diphthalocyanine complexes and lithium naphthalocyanine as catalysts for electrochemical CO_2 reduction. *J. Electrochem. Soc.* **150**(12), E608–E613.
197. Abe, T., F. Tagushi, T. Yoshida, S. Tokida,G. Schnurpfell, D. Woehrle, and M. Kaneko (1996). Electrocatalytic CO_2 reduction by cobalt octabutoxyphthalocyanine coated on graphite electrode. *J. Mol. Catal. A: Chem.* **112**(1), 55–61.
198. Hiratsuka, K., K.Takahashi, H. Sasaki, and S. Toshima (1977). Electrocatalytic behavior of tetrasulfonated metal phthalocyanines in the reduction of carbon dioxide. *Chem. Lett.* **1977**(10), 1137–1138.
199. Winnischofer, H., L. de Souza, K. Araki, and H.E. Toma (2003). Electrocatalytic activity of a new nanostructured polymeric tetraruthenated porphyrin film for nitrite detection. *Anal. Chim. Acta* **480**(1), 97–107.
200. Araki, K., L. Angnes, C.M.N. Azevedo, and H.E. Toma (1995). Electrochemistry of a tetraruthenated cobalt porphyrin and its use in modified electrodes as sensors of reducing agents. *J. Electroanal. Chem.* **397**, 205–210.
201. Caetano da Rocha, J.R., L. Angnes, M. Bertotti, K. Araki, and H.E.Toma (2002). Amperometric detection of nitrite and nitrate at tetraruthenated porphyrin-modified electrodes in a continuous-flow assembly. *Anal. Chim. Acta* **452**, 23–28.
202. Moorcroft, M.J., J. Davis, and R.G. Compton (2001). Detection and determination of nitrate and nitrite: A review. *Talanta* **54**, 785–803.
203. Barley, M.H., K.J. Takeuchi, and T.J. Meyer (1986). Electrocatalytic reduction of nitrite to ammonia based on a water-soluble iron porphyrin. *J. Am. Chem. Soc.* **108**(19), 5876–5885.
204. Barley, M.H., K. Takeuchi, W.R. Murphy, Jr., and T.J. Meyer (1985). Iron porphyrin-based electrocatalytic reduction of nitrite to ammonia. *J. Chem. Soc. Chem Commun.* **8**, 507–508.
205. Barley, M.H., M.R. Matthew, and T.J. Meyer (1987). Electrocatalytic reduction of nitrite to nitrous oxide and ammonia based on the N-methylated, cationic iron porphyrin complex $[Fe^{III}(H_2O)(TMPyP)]^{5+}$. *Inorg. Chem.* **26**(11), 1746–1750.

206. Bedioui, F., S. Trevin, V. Albin, M. Guadalupe, G. Villegas, and J. Devynck (1997). Design and characterization of chemically modified electrodes with iron(III) porphyrinic-based polymers: Study of their reactivity toward nitrites and nitric oxide in aqueous solution. *Anal. Chim. Acta* **341**, 177–185.
207. Gao, D., J. Gu, R.-Q. Yu, and G.-D. Zheng (1995). Nitrite-sensitive liquid membrane electrodes based on metalloporphyrin derivatives. *Analyst* **120**, 499–502.
208. Mimica, D., J.H. Zagal, and F. Bedioui (2001). Electroreduction of nitrite by hemin, mygoglobin and hemoglobin in surfactant films. *J. Electroanal. Chem.* **497**, 106–113.
209. Lin, R., M. Bayachou, J. Greaves, and P.J. Farmer (1997). Nitrite reduction by myoglobin in surfactant films. *J. Am. Chem. Soc.* **119**(51) 12698–12690.
210. Caro, C.A., F. Bedioui, and J.H. Zagal (2002). Electrocatalytic oxidation of nitrite on a vitreous carbon electrode modified with cobalt phthalocyanine. *Electrochim. Acta* **47**(9), 1489–1494.
211. Kim, S. and G. Kwag (2002). Increased catalytic activity of iron phthalocyanine on the electrochemical nitrite reduction upon heat-treatment probed by x-ray absorption fine structure. *Bull. Korean Chem. Soc.* **23**(1), 25–26.
212. Hwang, D.Y., J. Park, and S. Kim (1998). Electrocatalytic reduction of nitrite by iron phthalocyanine monolayer irreversibly adsorbed on the edge plane of pyrolytic graphite. *Bull. Korean Chem. Soc.* **19**(7), 795–797.
213. Thamae, M. and T. Nyokong (1999). Cobalt(II) porphyrazine catalysed reduction of nitrite. *J. Electroanal. Chem.* **470**, 126–135.
214. Nevin, W.A., W. Liu, S. Greenberg, M.R. Hempstead, S.M. Marcuccio, M. Melnik, C.C. Leznoff, and A.B.P. Lever (1987). Synthesis, aggregation, electrocatalytic activity, and redox properties of a tetranuclear cobalt phthalocyanine. *Inorg. Chem.* **26**(6), 891–899.
215. Abbaspour, A., M.A. Kamyabi, A.R. Esmaeiling, and K. Kia (2002). Thiocyanate-selective electrodes based on unsymmetrical benzoN_4 nickel(II) macrocyclic complexes. *Talanta* **57**, 859–867.
216. Shamsipur, M., G. Khayatian, and S. Tangestaninejad (1999). Thiocyanate-selective membrane electrode based on (octabromotetraphenylporphyrinato)manganese(III) chloride. *Electroanalysis* **11**(18), 1340–1344.
217. Malinski, T., A. Ciszewski, J. Bennett, and J.R. Fish (1991). Characterization of conductive polymeric nickel(II) tetrakis(3-methoxy-4-hydroxy-phenyl)porphyrin as an anodic material for electrocatalysis. *J. Electrochem. Soc.* **138**(7), 2008–2015.
218. Bennett, J.E. and T. Malinski (1991). Conductive polymetric porphyrin films: Application in the electrocatalytic oxidation of hydrazine. *Chem. Mater.* **3**(3), 490–495.
219. Trevin, S., F. Bedioui, G.M.G. Villegas, and C. Bied-Charreton (1997). Electropolymerized nickel macrocyclic complex-based flms: Design and electrocatalytic application. *J. Mat. Chem.* **7**(6), 923–928.
220. Pessoa, C.A., Y. Gushikem, and S. Nakagaki (2002). Cobalt porphyrin immobilized on a niobium(V) oxide grafted-silica gel surface: Study of the catalytic oxidation of hydrazine. *Electroanalysis* **14**(15–16), 1072–1076.
221. Isaacs, M., M.J. Aguirre, A. Totor-Labbe, J. Costamagna, M. Paez, and J.H. Zagal (1998). Comparative study of the electrocatalytic activity of cobalt phthalocyanine and cobalt naphthalocyanine for the reduction of oxygen and the oxidation of hydrazine. *Electrochim. Acta* **43**(12–13), 1821–1827.
222. Zagal, J.H., S. Lira, and S. Ureta-Zanartu (1986). A mechanistic study of the electro-oxidation of hydrazine on phthalocyanines of vanadyl, chromium, manganese, nickel, copper and zinc attached to graphite electrodes. *J. Electroanal. Chem.* **210**(1), 95–110.

223. Zagal, J. and S. Ureta-Zanaru (1982). Electrooxidation of hydrazine catalysed by sulfonated phthalocyanines adsorbed on a graphite electrode. *J. Electrochem. Soc.* **129**(10), 2242–2247.
224. Zagal, J. (1982). Catalytic electrooxidation of hydrazine on phthalocyanines deposited on graphite electrodes. *Electrochim. Acta* **27**(10), 1373–1377.
225. Li, X., S. Zhang, and C. Sun (2003). Frabrication of a covalently attached multiplayer film electrode containing cobalt phthalocyanine and its electrocatalytic oxidation of hydrazine. *J. Electroanal. Chem.* **553**, 139–145.
226. Vinod, M.P., K.T. Das, A.J. Chandawadkar, K. Vijayamohanan, and J.G. Chandwadkar (1999). Catalytic and electrocatalytic properties of intrazeolitically prepared iron phthalocyanine. *Mater. Chem. Phys.* **58**(1), 37–43.
227. Parez, E.F., G. De Oliveira Neto, A.A. Tanaka, and L.T. Kubota (1998). Electrochemical sensor for hydrazine based on silica modified with nickel tetrasulfonated phthalocyanine. *Electroanalysis* **10**(2), 111–115.
228. Wang, J., P.V.A. Pamidi, C. Parrado, S.D. Park, and J. Pingarron (1997). Sol–gel-derived cobalt phthalocyanine-dispersed carbon composite electrodes for electrocatalysis and amerometric flow detection. *Electroanalysis* **9**(12), 908–911.
229. Wang, J. and P.V.A. Pamidi (1995). Disporsable screen-printed electrodes for monitoring hydrazines. *Talanta* **42**(3), 463–467.
230. Wang, J., T. Golden, and R. Li (1988). Cobalt phthalocyanine/cellulose acetate chemically modified electrodes for electrochemical detection in flowing streams: Multifunctional operation based upon the coupling of electrocatalysts and permselectivity. *Anal. Chem.* **60**(15), 1642–1645.
231. Wang, J. and N. Naser (1995). Modified carbon–wax composite electrodes. *Anal. Chim.* **316**(2), 253–259.
232. Sun, C., Y. Sun, Z. Zhang, D. Jiang, Q. Gao, H. Xu, and J. Shen (1996). Fabrication of a multiplayer film containing cobalt phthalocyanine on the surface of a gold electrode based on electrostatic interaction and its application as an amperometric sensor of hydrazine. *Thin Solid Films* **288**(1–2), 291–295.
233. Kobuta, L.T., Y. Gushikem, J. Perez, and A.A. Tanaka (1995). Electrochemical properties of iron phthalocyanine immorbilized on titanium(IV) oxide coated on silica gel surface. *Langmuir* **11**(3), 1009–1013.
234. Peng, Q.-Y. and T.F. Guarr (1994). Electro-oxidation of hydrazine at electrodes modified with polymeric cobalt phthalocyanine. *Electrochim. Acta* **39**(17), 2629–2632.
235. Trollund, E., P. Ardiles, M.J. Aguire, S.R. Biaggo, and R.C. Rocha-Filho (2000). Spectroelectrochemical and electrical characterization of poly(cobalt-tetraaminophthalocyanine)-modified electrodes: electrocatalytic oxidation of hydrazine. *Polyhedron* **19**, 2303–2312.
236. Ebadi, M. (2003). Electrocatalytic oxidation and flow amperometric detection of hydrazine on a dinuclear ruthenium phthalocyanine-modified electrode. *Can. J. Chem.* **81**(2), 161–168.
237. Ebadi, M. (2003). Electrocatalytic oxidation of hydroxylamine by $(RuPc)_2$. *Electrochim. Acta* **48**(28), 4233–4238.
238. Qi, X. and R.P. Baldwin (1994). Liquid chromatography/electrochemical detection of hydroxylamines by oxidation at a cobalt phthalocyanine chemically modified electrode. *Electroanalysis* **6**(5–6), 353–360.

8

Electropolymerized Metalloporphyrin, Metallophthalocyanine and Metal Schiff Base Complex Films: Applications to Biomimetic Electrocatalysis and Bioelectroanalysis

Alain Pailleret and Fethi Bedioui

1. Introduction

Molecular engineering and design of new controlled spatial assemblies and architectures is a field undergoing wide growth[1]. A large variety of disciplines, especially bioinorganic chemistry, are now making profit from the design of molecular systems. Indeed, bioinorganic structural motifs can potentially model metalloenzyme structures and functions. The aim of such modelling is to mimic natural properties in order to elucidate fundamental aspects of reactivity and mechanism.

The desire to mimic enzymatic systems has led to an active area of research involving synthetic porphyrin models of enzyme active sites, especially for monooxygenase enzymes of the cytochrome P-450[2–4]. An analysis of comparative studies dealing with the selectivity, efficiency and stability for both synthetic porphyrin models and natural systems has shown that efficiency arises from the control of the environment of the enzyme active site[4–7]. Thus, in a first approach, the design of synthetic supramolecular architecture for porphyrin models

---

Alain Pailleret • Laboratoire Interfaces et Systèmes Electrochimiques, UPR CNRS no. 15, Université Pierre et Marie Curie-Paris VI, 4 place Jussieu, 75252 Paris cedex 05, France.

Fethi Bedioui • Laboratoire de Pharmacologie Chimique et Génétique, UMR CNRS no. 8151/ U 640 INSERM, Ecole Nationale Supérieure de Chimie de Paris, 11 rue Pierre et Marie Curie, 75231 Paris cedex 05, France.

N_4-Macrocyclic Metal Complexes, edited by José H. Zagal, Fethi Bedioui and Jean-Pol Dodelet. Springer Science+Business Media, Inc., New York, 2006.

with an elaborated steric environment (such as picnic basket porphyrins, strapped porphyrins, etc.) has been the subject of intense effort[8]. A similar approach in concept which has been advanced during the last two decades involves replacing the protein envelop of natural enzymes by a mineral or organic polymeric matrix[9–18]. This strategy is based on the fact that a polymeric matrix may provide the best arrangement for a catalytically active centre as well as prevent auto-destruction of the enzyme model during catalysis. Additionally, the intrinsic diversity and selectivity of axial ligation of the porphyrinic macrocycles confer to this strategy a wide field of applications in analytical chemistry.

Using the same method, related complexes such as metallophthalocyanines and Schiff bases are known to be efficient catalysts, especially for the electrochemical activation of molecular oxygen and the oxidation of thiols. For this purpose, it has been reported that modified electrodes coated with adsorbed cobalt phthalocyanines show a substantial electrocatalytic activity for the electro-oxidation of thiols such as L-cysteine, 2-mercaptoethanol and reduced glutathione, by lowering particularly the overpotential of the electrochemical processes. Only few attempts have been reported to develop the use of multilayer metallophthalocyanine-based films, aimed at offering a better alternative to the design of efficient and stable electrocatalysts. In addition, the intrinsic diversity and selectivity of the redox activities and the axial ligation properties of metalloporphyrins, metallophthalocyanines and metal Schiff base complexes confer on them formidable challenging potential uses as materials for the design of electrochemical sensing devices and catalytic electrodes.

For this purpose and in the special case of modelling the enzymatic systems, a fruitful strategy has been developed based on the use of polymeric porphyrins, as molecular devices combined with electrochemistry. Therefore, the electroassisted design of modified electrode surfaces by organized porphyrinic structures has been the subject of an intense effort. In fact, one of the challenges in developing this concept is the necessity to achieve rapid electron transfer to the enzyme model active site at the electrode surface[19–23]. Conducting polymers such as polypyrrole, polythiophene or polyaniline are attractive as possible materials for this approach[19–22, 24–27]. Meanwhile, the charge transport in the porphyrin films may act like a conducting polymer or as a hopping process between complex sites. Such an approach has been successfully extended afterward to metallophthalocyanine and Schiff base complexes.

This report is aimed first at showing that the use of electropolymerization strategies provides an elegant way to build up metalloporphyrin-and metallophthalocyanine-based multilayered structures and architectures. Besides, the idea of building such molecular devices having designated catalytic and analytical properties is shown to be realistic. In the case of biomimetic electrocatalysis, the polymer matrix prevents supported complexes from degradation, controls the access of substrate molecules to the reactive centres and makes the reactivity of the supported catalysts largely enhanced. The electropolymerized metalloporphyrin and metallophthalocyanine films also offer powerful materials for analytical tools, especially for sensing biologically relevant species. The significant developments in sensor science and technology, such as the design of ultramicroelectrodes and arrays, are now intensively stimulating and expanding the exploration of this field of research.

2. The Electrochemical Polymerization Strategy

Electrochemical polymerization is an elegant, attractive and easy strategy for the immobilization of metal complexes[19–21, 24–26] on the surface of electrodes. The principle is based on the electrochemical oxidation (or reduction) of a suitably designed monomer to form a polymeric film incorporating the metal complex. The obtained polymeric films should be electronic conductors to ensure electron transfer within the matrix (and then the continuous growth of the polymers). Pyrrole-, thiophene- and aniline-based monomers have been the most commonly used materials[24–27]. Such chemically substituted monomers have many interesting features, including a high flexibility in their molecular design. Additionally, such materials offer the possibility of using either aqueous or organic solutions to carry out the electropolymerization.

The electrochemical deposition process is controlled by the electrode potential. This can be achieved either by controlled potential, current electrolysis or cyclic voltammetry within a well-defined potential range. Growth of the polymeric films or more precisely, control of the amount of deposited materials (or the polymer film thickness), can be easily achieved by monitoring the total charge passed during the electrooxidative (or reductive) polymerization process. In the general case, the mechanism of electropolymerization of such substituted monomers has still not been completely explained, despite the large quantity of data on this subject. However, it is now well accepted that in the case of pyrrole and aniline derivatives, the first step in the electropolymerization process is the electrooxidative formation of a radical cation from the chosen monomer[24–27]. This oxidation reaction is followed by a dimerization process, followed by further oxidation and coupling reactions. This leads to the formation of oligomers and polymers on the electrode surface. It also appears that the morphology and the physical properties of the polymer films depend largely upon the electrochemical polymerization conditions. Thus, one can induce a supplementary design parameter during the polymerization step by adjusting the solution composition or the electrode potential. In this way, electrochemical polymerization can be carefully controlled, resulting in multilayered structures and copolymers from multicomponent solutions.

3. Design and Characterization of Electropolymerized Metalloporphyrin and Metallophthalocyanine Film

3.1. Doped Electropolymerized Polypyrrole and Polythiophene Films

One of the first examples that involved the incorporation of metalloporphyrin and metallophthalocyanine complexes into polypyrrole films was based on the ion-exchange properties of the oxidized polymer. Tetrasulfonated and tetracarboxy-substituted metalloporphyrins[28–31] and metallophthalocyanines[32–40] have been introduced into polypyrrole films as counter-ions (or "doping" ions). In this purpose, the polymers were electrochemically grown in solutions containing pyrrole and anionic complexes as supporting electrolyte (see significant examples in Figure 8.1). Both the electrochemical polymerization of pyrrole and the doping of polypyrrole by anionic complexes were obtained in one step from aqueous solutions[28–40]. The electrochemical doping level of polypyrrole by the complexes was usually estimated from chemical and spectrophotometric analysis[31]. Thus, it has been suggested that the number of pyrrole units interacting with one anionic group of a metalloporphyrin is 3 or 4 as mentioned by Diaz *et al.*[41,42] for simple anions such as ClO_4^-.

The electrochemical characterization of the polypyrrole films doped by anionic complexes was generally achieved by cyclic voltammetry. Unfortunately, no well-defined waves for the porphyrin or phthalocyanine redox couples were observed which may be because they are superimposed on the large polypyrrole background. Therefore, the reported voltammograms[29,31] could either be those for the redox process of the incorporated complexes or those of the

Porphyrins	Metal	R_1, R_2, R_3, R_4	Identification number	References
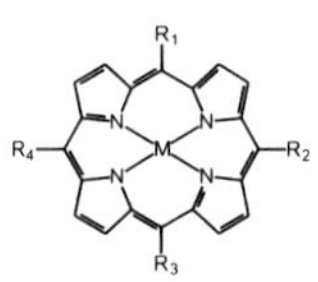		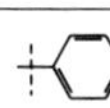	1	
	Co(II), Fe(III), Co(III), Mn(III)	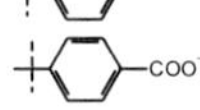	2	[29, 31, 90, 91]
	Fe(III)	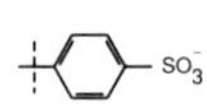	3	[29, 31, 46, 90, 91]
	Fe(III), Co(II), 2H		3	[47]
	Fe(III)	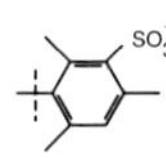	4	[48]
	Co(III), Fe(III)		3	[49]
	Co(II), 2H		3	[50]

Phthalocyanines	Metal	R_1, R_2, R_3, R_4	Identification number	References
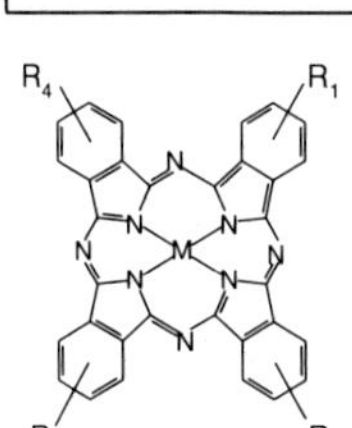	Fe(II), Co(II)	SO_3^-	5	[51]
	Fe(III), Co(II)		5	[52]
	Cu(II)		5	[50, 53]
	Cu(II)	$R_1, R_2, R_3 = SO_3^-$	6	[54]
	Cu(II),Ni(II), Fe(II), Co(II)		5	[55]
	Cu(II), Ni(II), Co(II), Pb(II)		5	[56]

Figure 8.1. Examples of metalloporphyrin or metallophthalocyanine derivatives used as doping anions in electronically conducting polymers.

polypyrrole redox system. That is why it appears essential to couple the electrochemical characterization with spectrophotometric measurements.

However, this method of incorporating metalloporphyrins or metallophthalocyanines into a polypyrrole matrix has a severe limitation, namely, the electrostatic binding capacity of the polymer film depends on the potential and disappears when the polypyrrole is reduced[24,41–45]. Moreover, this ion-exchange capacity is restricted to one negative charge for 3 or 4 pyrrole units.

In the following Section we report on some significant examples described in the literature in the last decade in order to update the previously published overview on this matter[15].

3.1.1. Significant Recent Examples of Electropolymerized Films Doped with Porphyrins.

The major significant examples of doped electropolymerized polypyrrole films with porphyrins were developed to design sensing materials or biomimetically active catalytic membranes for targeted applications[15]. For example, Bedioui *et al.* investigated in the last decade the electrodeposition of polypyrrole (PPy) films doped with tetrasulphonated iron porphyrin Fe(III)-**3** in view of a study of the catalytic behaviour of the corresponding film towards the electrochemical oxidation of nitric oxide in acidic and neutral aqueous solution, confirming thus the formation of iron-nitrosyl intermediate $[Fe(III)(NO)]^{+}$[46]. Later on, it was shown that this doping process could be achieved either by electropolymerizing pyrrole in the presence of M-**3** (M = Fe or Co) or by metallizing the free base ligand H_2-**3** doped into polypyrrole through ion exchange. The resulting film was then used for achieving the electrochemical reduction of oxygen in acidic and basic solutions[47]. Also, the trapping of a similar anionic porphyrinic derivative, Fe(III) *meso*-tetrakis(3-sulphonatomesityl)porphyrin, during the electropolymerization of pyrrole was investigated by Liu and Su and confirmed using electrochemical quartz crystal microbalance (EQCM). The resulting film was applied to the study of oxygen reduction[48]. Paik *et al.* developed one of the rare examples of conducting polythiophene film, namely poly(3,4-ethylenedioxythiophene) (PEDOT) that was used as the polymeric backbone instead of polypyrrole to immobilize a sulphonated Co(III) porphyrin (Co(III)-**3**)[49]. Interestingly, it was suggested by the authors that, in contrast to the polypyrrole-based electrodes, this kind of modified electrode was stable with wider potential windows, including the oxygen reduction potential.

In a more recent work, Paik *et al.* described transport properties and stability of sulphonated porphyrin derivatives (**3**) in films of conducting polymers PPy and PEDOT films using EQCM[50]. As expected, it was evidenced that such anions were not expelled from these polymer films by extensive potential cycles. The authors also analyzed the effects of potential cycling on PEDOT-H_2-**3** films obtained by two different modes of electropolymerization: (i) at a constant current and (ii) with repeated cyclic potential sweeps. Figure 8.2 shows the results of potential cycles in 0.1 M NaCl solution for the two types of films (noted A and B, respectively). In both cases, the general feature of mass increases of the negative

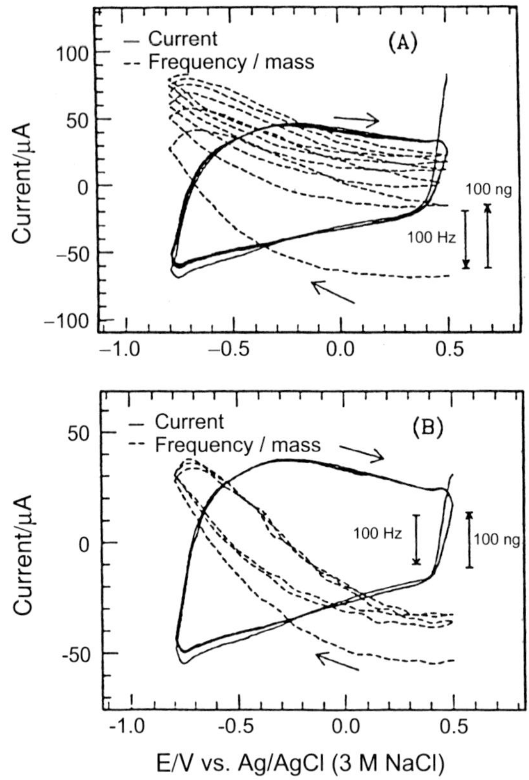

Figure 8.2. Cyclic voltammetry and mass change of PEDOT-[H_2-**3**] flims in 0.1 M NaCl solution. (A) Film from galvanostatic polymerization; (B) film from polymerization with cyclic potential. Potential sweep rate: 50 mVs^{-1}. Reprinted from ref.[50]. Copyright (2000), with permission from Elsevier.

side of the potential range, which is accounted for by cation insertion. However, there is a subtle difference between the two films (Figure 8.2A,B). Larger increases in the mass with the number of successive potential cycles were observed in the film obtained by polymerization at a constant current (film A) than in the film prepared with cycling potential (film B). Thus film A seems to permit more salt accumulation inside the backbone compared with film B. The accumulation of salt is the result of transport of ions across the film/solution interface, entering into the film in larger amount than leaving the film. With such a study, the authors showed that with EQCM experiments, it was possible to demonstrate that transport properties and stability of ionic species in films of conducting polypyrrole and substituted thiophene can be drastically influenced by the mode of polymerization by which the polymer film was prepared as well as by the structure of the involved ions.

3.1.2. Significant Recent Examples of Electropolymerized Films Doped with Phthalocyanines

The elaboration of electropolymerized polypyrrole, polyaniline and polythiophene doped by metallophthalocyanines was mainly aimed at developing catalytic materials for oxygen reduction. Indeed, in 1987, iron tetrasulphonato–phthalocyanine Fe(II)-**5** was used as a doping anion in electrochemically synthesized polypyrrole films for the electrochemical reduction of oxygen[51]. In this study, thickness of the film was shown to strongly alter the reduction mechanism of oxygen. Few years later, polypyrrole and polyaniline films doped with iron or cobalt tetrasulphonated phthalocyanines inserted in the polymeric backbone during the electrodeposition process were characterized in view of the same application, namely oxygen reduction[52]. Also, polypyrrole films doped with copper tetrasulphonated phthalocyanine were the target of UV–visible spectroscopic investigation immediately after electrochemical synthesis and during heat treatment[53]. Interestingly, when conducting polymers such as PPy or PEDOT are doped with phthalocyanines derivatized with sulphonate anionic groups, these complexes were not expelled from the polymeric structure by extensive cycling, proving thus the efficiency and the stability of the doping process[50]. Copper phthalocyanine trisulphonate was also used as doping anion in polypyrrole films. The resulting films were then covered with evaporated silver so as to investigate the electronic structure of the resulting systems using photoemission electron spectroscopy, photoemission spectroscopy and near edge X-ray absorption spectroscopy[54]. Tetrasulphonated metallophthalocyanine were also used to dope polypyrrole in a polypyrrole–polyoxyphenylene composite material whose electrical conductivity was the main target of the reported work[55]. Finally, one can mention the case of metallophthalocyanine, toluenesulphonate-doped polypyrrole films and silicon that were assembled so as to make heterojunctions[56].

3.2. Electropolymerization of Pyrrole or Thiophene-Substituted N_4-Complexes

The most elegant approach to design polypyrrole, polyaniline or polythiophene-based porphyrin, phthalocyanine or Schiff base matrices involves the electrochemical polymerization of suitably designed substituted N_4-macrocyclic monomers. We and others have shown that the electro-oxidative polymerization of such species (see significant examples in Figure 8.3) leads to the formation of films having the electrochemical properties of the monomeric complex[46–52].

In the case of the pyrrole-substituted N_4-macrocyclic complexes, in a typical experiment, the electrochemical polymerization is achieved by cyclic voltammetry of acetonitrile, dimethylformamide, dimethylsulfoxide or dichloromethane solutions containing the complex monomer and the supporting electrolyte, through a well-defined potential range. The evolution of the cyclic voltammograms during repeated potential scans in the well-designated potential range thus shows a continuous increase in the amplitude of the cyclic voltammetric peaks of the parent

Porphyrins	Metal	R_1, R_2, R_3, R_4	Identification number	References
	2H, Fe(II), Co(II), Mn(II)	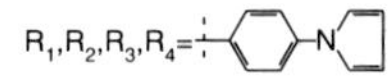	**7**	[58]
	Mn(III), Zn(II)	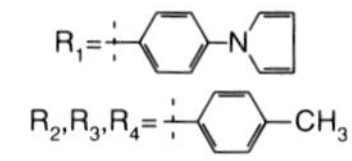	**8**	[59, 62, 65]
	Co(II), Mn(III)	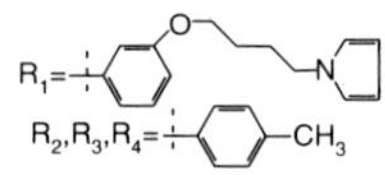	**9**	[60, 65]
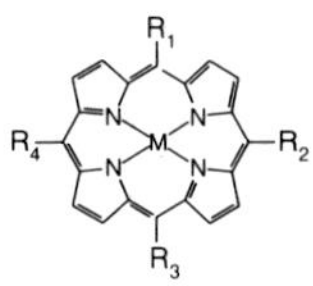	Ni(II), Zn(II), Co(II), Cu(II), Mn(III)	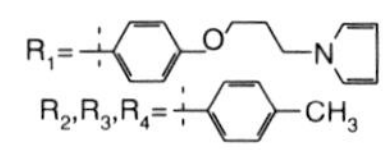	**10**	[63, 65]
	Co(II)	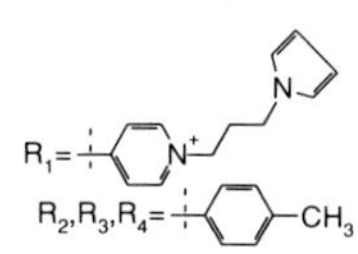	**11**	[90]
	Mn(III)		**7**	[64]
	Mn(III)	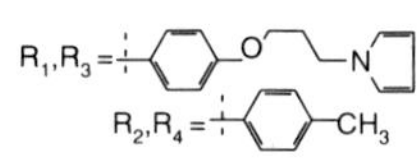	**12**	[65]
	Mn(III)	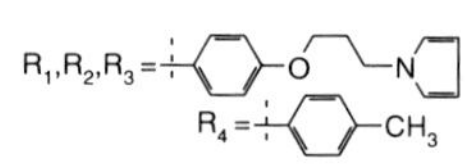	**13**	[65]
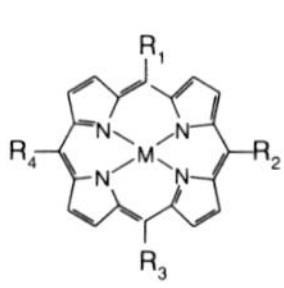	Ni(II)	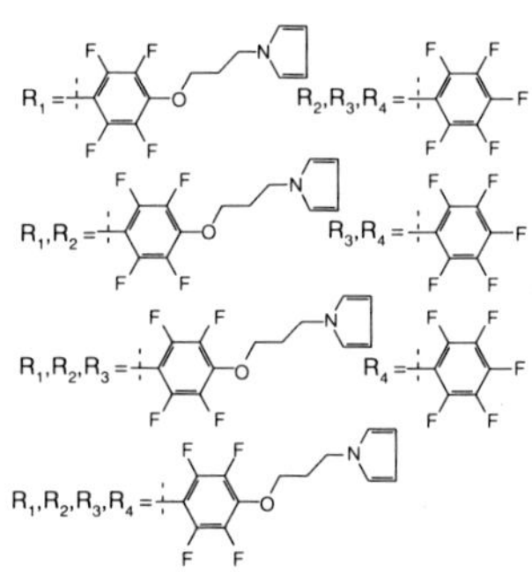	**14, 15, 16, 17**	[66]
	Cu(II), Mn(III), Fe(III)	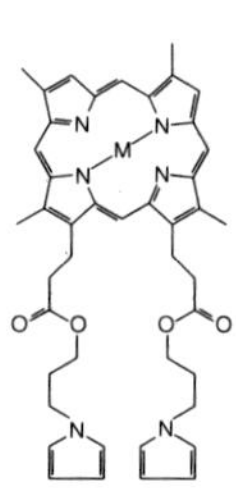	**18**	[67]
	Co(II)		**18**	[68]
	Mn(III)	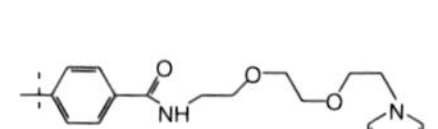	**19**	[69]

Figure 8.3. Examples of metalloporphyrin, metallophthalocyanine and salen derivatives bearing electropolymerizable pyrrole or thiophene units.

Porphyrins	Metal	R_1, R_2, R_3, R_4	Identification number	References
	Fe(III)–Cl, Mn(III)–Cl, Co(II)	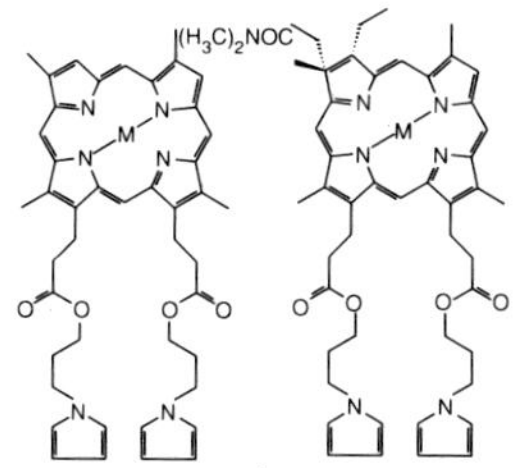	**20, 21**	[70]
	Ni(II)	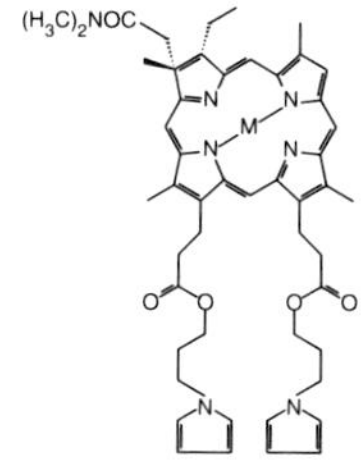	**22**	[71]
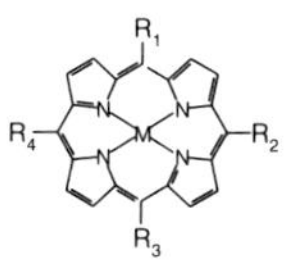	P(OR')$_2$	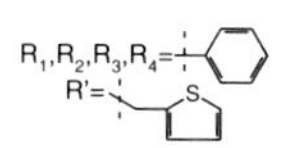	**23**	[72–75]
	P(OR')$_2$	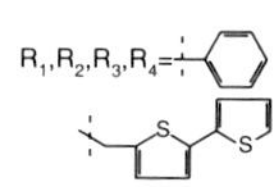	**24**	[72–75]
	Zn(II)	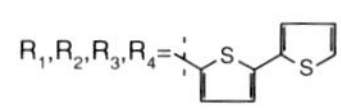	**25**	[72–75]
	P(OR')$_2$	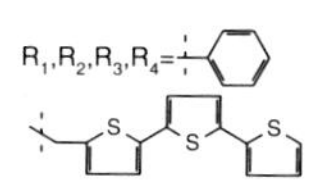	**26**	[72–75]
	2H, Co(II), Ni(II), Zn(II), Fe(III)–Cl, Mn(III)–Cl	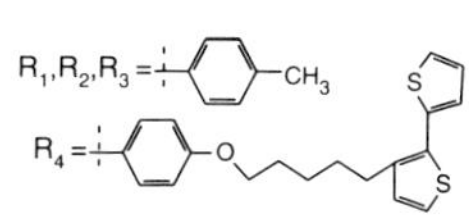	**27**	[76]
	2H, Fe(III)-Cl, Mn(III)-Cl	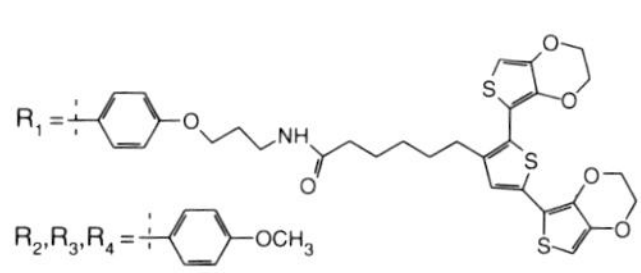	**28**	[77]
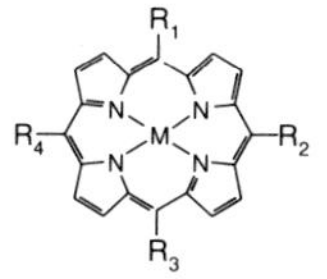		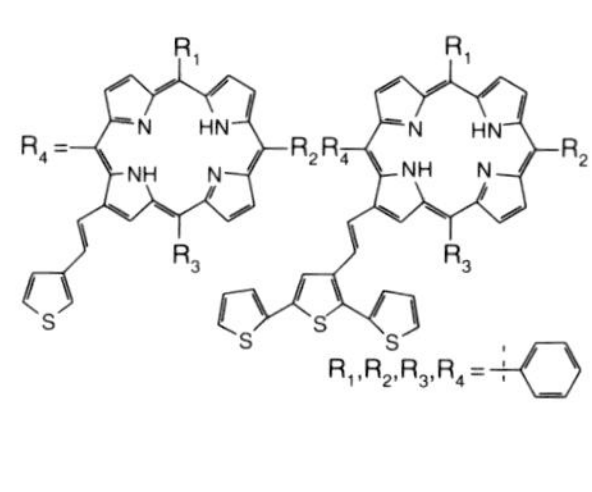	**29**	[78]
	1/2 Ce(III)	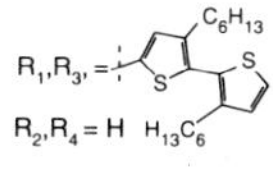	**30**	[79]

Figure 8.3. *Continued*

Phthalocyanines	Metal	R_1, R_2, R_3, R_4	Identification number	References
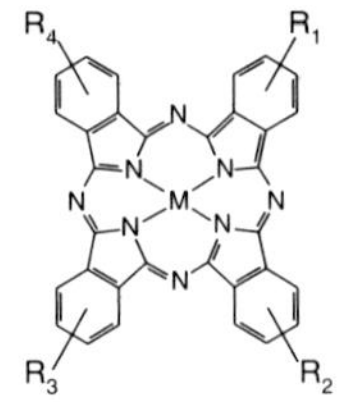	2H, Co(II), Ni(II), Zn(II),	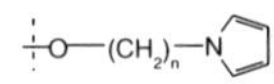	**31**	[80, 81]
	Co(II), Fe(II), Mn(III), Ni(II), Zn(II)	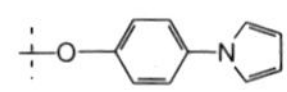	**32**	[82, 241]

Salen derivatives	Metal	Identification number	References
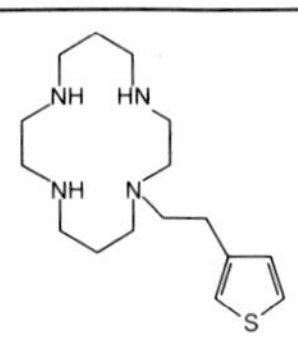	Ni(II)	**33**	[83, 84]
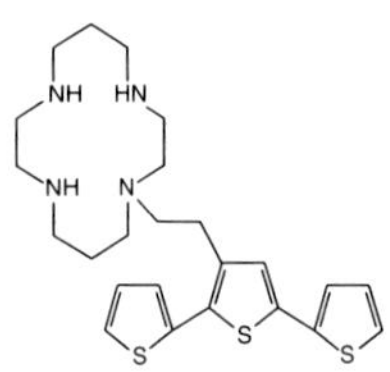	Ni(II)	**34**	[84]
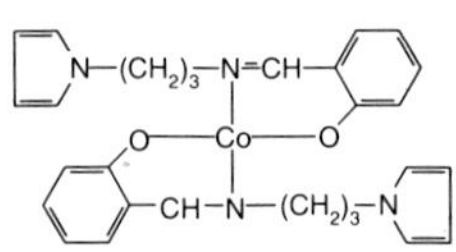	Co(II)	**35**	[85]
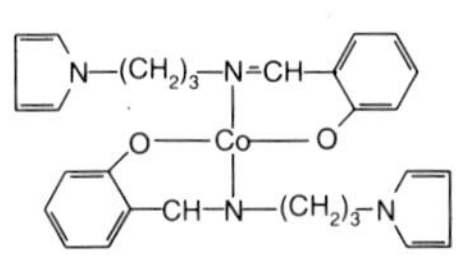	Ni(II)	**36**	[86]
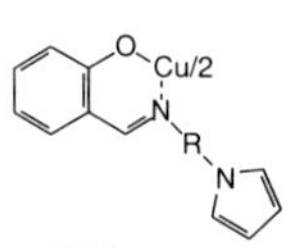 R = —$(CH_2)_3$—, — or —pC_6H_4—	Cu(II)	**37**	[87]
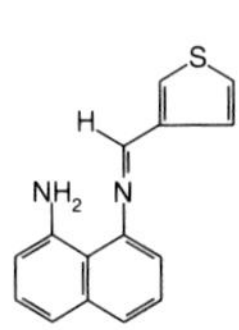		**38**	[88]
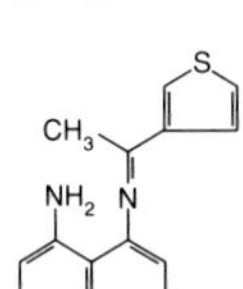		**39**	[89]

Figure 8.3. *Continued*

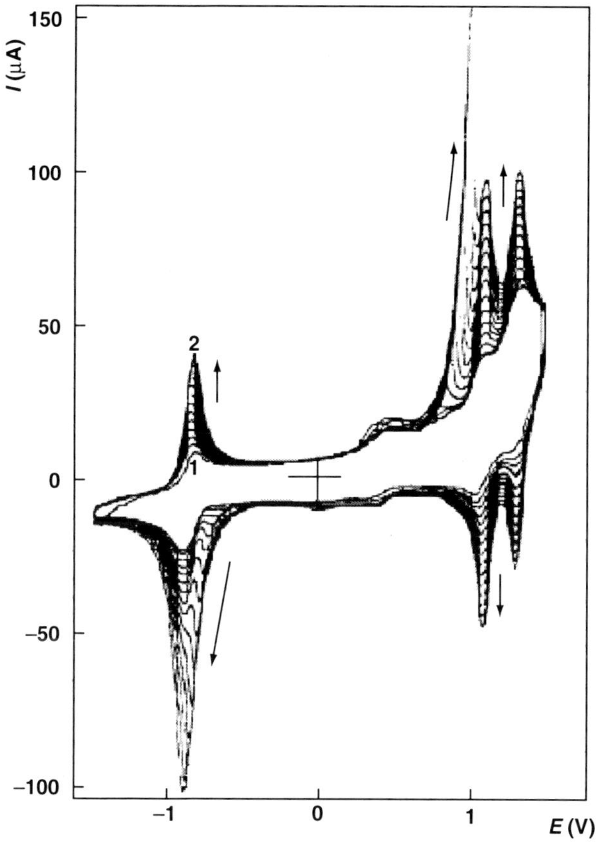

Figure 8.4. Cyclic voltammetry at a vitreous carbon disk electrode ($A = 0.071$ cm^2) of monomer Co(II)-**9** (2 mM) in acetonitrile + 0.1 M TBFA. $V = 0.15$ V s^{-1}. (1) 1 cycle; (2) 20th cycle. Reprinted from ref.[61]. Copyright (1991), with permission from Elsevier.

complex. This indicates that a film is formed on the electrode as a consequence of electrochemical polymerization of the attached pyrrole groups. Figure 8.4 shows the case of a pyrrole-substituted cobalt porphyrin, Co(II)-**9** during its electropolymerizing scans from −1.5 to 1.5 V/SCE in acetonitrile solution[61].

The electropolymerized films are usually characterized by cyclic voltammetry and UV–visible spectrophotometry on optically transparent electrodes[62]. These techniques reveal the formation of supported N_4-complexes on the electrode surface as well as provide evidence for the similarity of the electrochemical and spectrophotometric behaviour of the polymer and the starting monomers. Indeed, as it is shown in Figure 8.5, the absorption spectrum of a pyrrole-substituted Zn(II)-**8** monomer complex in dichloromethane solution (Figure 8.5a) is compared with that of the electropolymerized film on ITO transparent electrode prepared by cyclic voltammetry (Figure 8.5b) and by controlled-potential electrolysis at 1.7 V (Figure 8.5c)[62]. On film formation, all the bands (Soret band and Q band of the π/π^* transitions) are red shifted by about 10–15 nm. For the Soret band, a

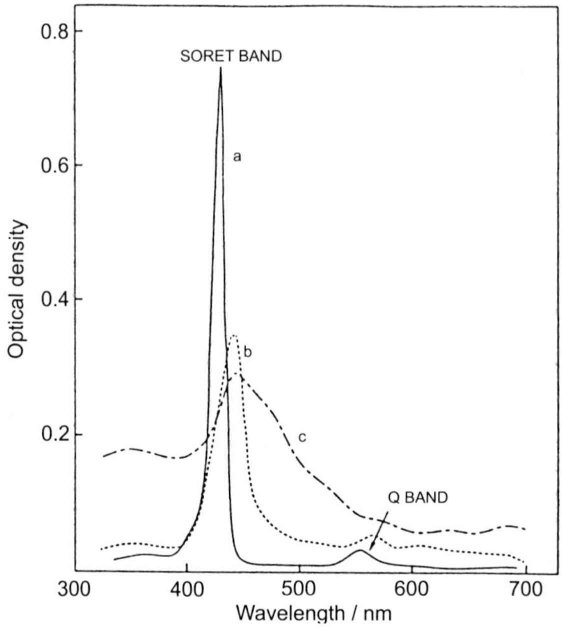

Figure 8.5. Absorption spectrum of (a) Zn(II)-**8** monomer (1 μM) in DCM solution, (b) poly[Zn(II)-**8**] film on transparent ITO electrode ($A = 0.8$ cm^2) prepared by 55 electropolymerizing scans and (c) poly[Zn(II)-**8**] on transparent ITO electrode ($A = 0.8$ cm^2) prepared by controlled-potential electrolysis at 1.7 V for 1 hr. Reprinted from ref.[62]. Copyright (1991), with permission from Elsevier.

red shift of 15 nm is observed, as well as considerable broadening. These are primarily due to aggregation and stacking of porphyrin molecules on the electrode surface. Similar features are observed for the Q band also. Comparison between Figures 8.5b,c shows that the broadening of the Soret band is more considerable for the film prepared by controlled-potential electrolysis. This indicates that during the preparation of the porphyrin film by electropolymerizing scans between –1.5 and 1.7 V, the negative potential field exercises a neat influence on the polymer structuration and chain formation.

Figures 8.6a,b corroborate this observation. Indeed, Figure 8.6a shows the spectrum of the polypyrrole-Zn(II)-**8** film obtained by controlled-potential electrolysis, and without any further treatment. Figure 8.6b shows the spectrum of the same film after electrochemical reduction of the polymer by scanning the electrode potential from 0 to -1.7 V in dichloromethane $+ 0.1$ M TBFA. It appears that after reduction of the polymer film, the broadening of the Soret band is lessened. One can conclude that the stacking of porphyrin molecules on the electrode surface is strongly influenced by the scanning of the electrode potential to negative values.

Cyclic voltammetry also provides useful information on the apparent surface coverage of the electropolymerized film-coated electrodes. In the case

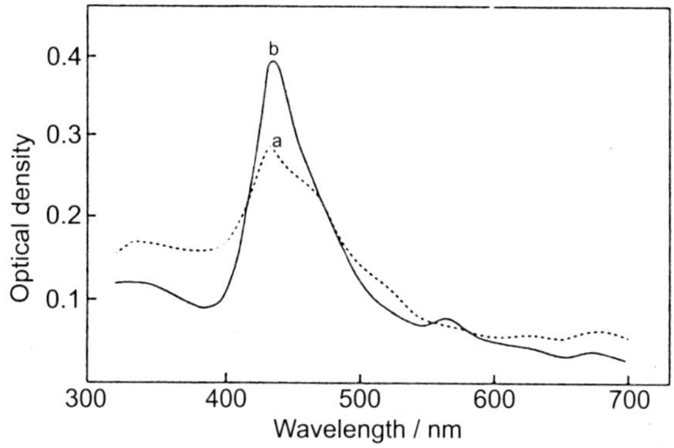

Figure 8.6. Absorption spectrum of a poly[Zn(II)-**8**] film on transparent ITO electrode ($A = 0.8$ cm^2) prepared by controlled-potential electrolysis at 1.7 V for 1 hr. (a) Polymer without any further treatment; (b) polymer after electrochemical reduction by potential scans between 0 and −1.7 V. Reprinted from ref.[62]. Copyright (1991), with permission from Elsevier.

of pyrrole-substituted porphyrins, it was shown that the configuration of the starting pyrrole, mono- or tetra-substituted monomers (with the pyrrole group directly linked on the porphyrin macrocycle or through a flexible tether) induces a cross-linking effect due to the steric hindrance of the macrocomplex. This dramatically affects the efficiency of the electropolymerization reactions and limits the electrode surface coverages by the supported porphyrin[58–61].

Also, the available data show that the conductivity of the obtained films decreases at low potential values ($E < -1$ V/SCE) as the film thickness increases[60,61]. In contrast, the polymer conductivity seems to be unchanged at positive potential values. This evokes a voltage-dependent conductivity similar to that described for "simple" polypyrrole films. Anyhow, the incorporation of the complexes into the polymer films imparts a typical "redox conductivity" to these materials that allows their use over a large potential range. One can even assert that at the potentials where many of the effects are expected, the electron hopping process between macrocyclic complex sites dominates the global charge transport mechanism.

In the following Section we report on some significant examples described in the literature in the last decade in order to update the previously published overview on this matter[15].

3.2.1. Significant Recent Examples of Electropolymerized Pyrrole- and Thiophene-Substituted Porphyrins

In the last decade, Mn(III) porphyrin-modified polypyrrole films were reported and used to mimic electrochemically and advantageously the functioning of cytochrome P450, an enzyme of the monooxygenase type[64,65]. Indeed, various families of compounds, namely Mn(III) porphyrin derivatives such as **8, 9, 10, 12** and **13** were the subject of intense research devoted to the electrocatalytic activity of these compounds towards the oxidation of hydrocarbons, olefins and a thioacetamide derivative using molecular oxygen. Also, the synthesis and characterization of a series of electropolymerized films based on innovative nickel(II) *meso*-tetrakis(polyfluorophenyl) porphyrins bearing pyrrole groups **14, 15, 16** and **17** have been successfully reported[66], as well as pyrrole moieties modified with Cu(II), Mn(III) and Fe(III) deuteroporphyrins **18** which were also electropolymerized, leading to interesting electroactive films[67]. The latter two compounds even showed electrocatalytic properties in the presence of molecular oxygen and benzoic anhydride and recognition properties towards cyanide in organic and aqueous electrolyte, respectively. Lately, has been reported a new Co(II) deuteroporphyrin substituted by two pyrrole units which was investigated in organic solvents and electropolymerized to the corresponding conducting film without any substantial modification of the initial monomer characteristics, allowing then its application to the recognition of nitrite ions in solution[68]. Meanwhile, a polymeric film obtained from a Mn(III)-**19** linked via a spacer chain to a pyrrole unit was used for the making of a porphyrin-based NO electrochemical sensor[69]. One can also cite the work reported by Monforts *et al.* who proposed new metalloporphyrinoid complexes functionalized with two pyrrole groups and bearing two phosphonate residues, these latter being aimed at the immobilization of these compounds on polycrystalline titanium dioxide electrodes and showed them to be useful in electrocatalysis purposes or for the design of electrochemical sensors[70]. Finally, in a preliminary study, the electrochemical behaviour of a Ni(II) hydroporphyrin bearing two pyrrole groups was shown first to be highly dependent on the nature of the solvent and second to lead to the first example of an electrogenerated polypyrrole–metachlorin film[71].

In the last decade, a significant amount of work had also been carried out concerning the electrochemical polymerization of thiophene monomers bearing covalently linked porphyrinic moieties, as reviewed for example by Shimidzu[72–75]. The most striking work reported therein deals with the electrochemical polymerization of thienyl phosphorus (V) porphyrin derivatives, in which the thiophene units were attached to their axial or lateral position. As a result of their polymerization, one-dimensional (1D) or two-dimensional (2D) porphyrin polymers with ordered oligothienyl molecular wires can be obtained. Various other types of P(V) porphyrin arrays were synthesized. After connection with conducting and/or insulating molecular wires, such arrays were considered to be converted into elements of molecular photo-electronic devices. Later on, bithiophene was shown to be a usable monomer for obtaining electronically conducting polymers bearing porphyrins. Indeed, **27**, a *meso*-tetraphenylporphyrin derivative linked to an ω-bromoalkyl-substituted bithiophene via an ether bond and then metallized, was shown to lead to the corresponding metalloporphyrin-functionalized poly(bithiophenes)[76]. More recently, the same authors had shown that conducting films, obtained by cyclic voltammetry of the precursor *N*-hydroxysuccinimidester-functionalized terthiophene as indicated in Figure 8.7 allow the post-grafting of iron and manganese amino-substituted porphyrins[77]. The resulting hybrid molecular materials were tested as model compounds for redox enzymes.

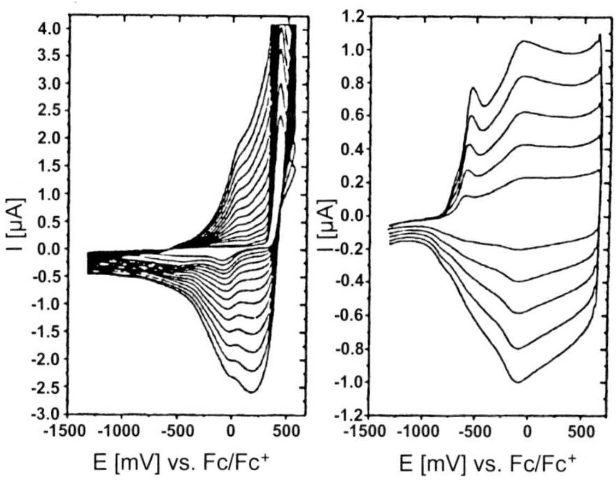

Figure 8.7. Potentiodynamic polymerization of *N*-hydroxysuccinimidester-functionalized terthiophene precursor (left) and polymer characterization (right) at different scan rates (50, 100, 150, 200, 250 mV s^{-1}) in CH_2Cl_2/Bu_4NPF_6. Reprinted from ref.[77]. Copyright (2001), with permission from Elsevier.

More recently, polythiophenes, some of which were substituted with porphyrin moieties as light harvesters, had been investigated for their performances in photovoltaic devices[78]. In an original work, Shinkai et al. succeeded in the electropolymerization of a Ce(III) triple decker porphyrin complex bearing bithienyl substituents at an ITO electrode surface[79].

3.2.2. Significant Examples of Electropolymerized Pyrrole-Substituted Phthalocyanines

Only two examples had been recently reported in the literature describing the electropolymerization of pyrrole-substituted metallophthalocyanine complexes[80,81]. In the first one, the electropolymerizable pyrrole group was separated from the phthalocyanine macrocycle by an alkylene spacer. The authors reported on the formation of thick films by oxidative electropolymerization under potentiodynamic conditions or constant potential, but no electrochemical and electrocatalytic behaviours were thoroughly investigated. Later on, the synthesis of a new series of pyrrole-substituted phthalocyanines in which the pyrrole is separated from the phthalocyanine by a phenoxy group (namely tetrakis-4-(pyrrol-1-yl)phenoxy metal phthalocyanines noted as M-**32**, where M is the metallic cation) has been reported[82]. Also, the electrochemical polymerization of these newly synthesized pyrrole-substituted phthalocyanines has been demonstrated in the case of the cobalt complex and the electrocatalytic activity of the obtained film has been tested towards the oxidation of L-cysteine. Figure 8.8A shows the evolution of the cyclic voltammograms of Co(II)-**32** during repeated potential scans from −0.5 to 1.2 V/Ag–AgCl in dichloromethane. The continuous increase in the amplitude of the cyclic voltammetric peaks for the systems related to Co(II)-**32**/Co(III)-**32** and Co(II)-**32**/Co(I)-**32** processes indicates that a film was formed

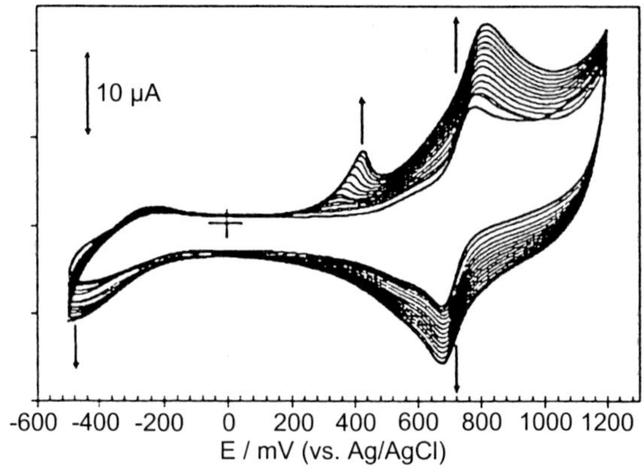

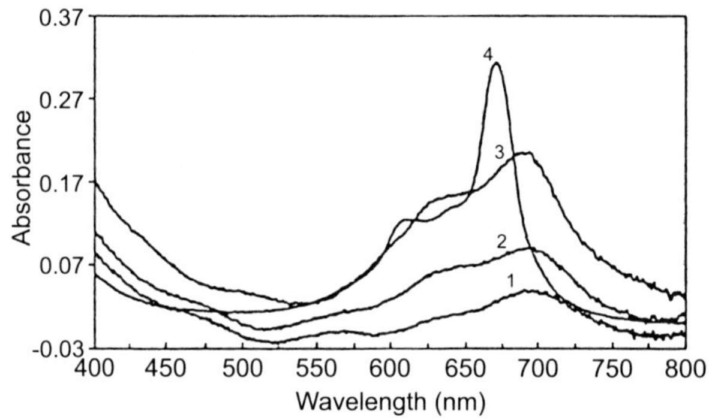

Figure 8.8. (A) Evolution of the cyclic voltammograms of complex Co(II)-**32** in DCM containing 0.1 M $TBABF_4$ during repeated successive scans (only 15 scans are shown). Scan rate = 200 mV s^{-1}, concentration ≈ 3 mM, electrode = glassy carbon. (B) Absorption spectra of poly[Co(II)-**32**] films on ITO electrodes prepared with 30 (curve 1), 60 (curve 2) and 120 (curve 3) electropolymerizing scans. Curve 4 corresponds to Co(II)-**32** monomer in DCM solution. Reprinted from ref.[82]. Copyright (2003) Society of Porphyrins & Phthalocyanines.

on the electrode as a consequence of electrochemical polymerization of the attached pyrrole groups due to their irreversible electrochemical oxidation. In addition, after five potential scans, a new peak appears around 0.4 V/Ag–AgCl. Its intensity increases with the number of scans and its potential shifts gradually to higher values. This new anodic process depicts the electroactivity of the polymer matrix formed during the electropolymerizing scans, as it has been previously reported with pyrrole-substituted porphyrins. Note that the number of the electropolymerizing scans controls the amount of the deposited complex and thus the thickness of the obtained film. Also, several attempts to electropolymerize the complex by scanning the potential electrode through wider ranges, by varying either the anodic or the cathodic limits, have not allowed to obtain the expected typical evolution for the electropolymerization of the complex.

After a period of scanning and transfer of the electrode, with thorough rinsing to a dichloromethane +0.1 M $TBABF_4$ solution (without monomeric phthalocyanine in solution), the obtained cyclic voltammogram exhibited the electrochemical responses previously described for the complex in solution. This indicates that a polymeric phthalocyanine film, noted as poly(Co(II)-**32**), was formed on the electrode. The two well-defined pairs of peak, which can be attributed to the Co(II)-**32**/Co(III)-**32** and Co(II)-**32**/Co(I)-**32** redox reactions, are observed at nearly the same potential as for the monomer redox systems in solution. The electropolymerization of the Co(II)-**32** complex has been further studied on an optically transparent electrode ITO to allow the monitoring of their growth by the changes observed in the UV–visible spectra. Typical spectra for three films of poly(Co(II)-**32**) films prepared by 30, 60 and 120 electropolymerizing scans are shown in Figure 8.8B. Curve 4 corresponds to the absorption spectra of Co(II)-**32** monomer in solution and shows bands at 600 and 680 nm. The UV–visible spectra obtained for the polymer films are similar to monomer species, but exhibit a notorious broadening of the bands along with a red shift. Such a considerable broadening of UV–visible bands have been previously reported in the case of electropolymerized metalloporphyrin and phthalocyanine films and analyzed in terms of stacking and high aggregation of the complexes on the electrode surface. The intensity of the absorption bands does increase with the number of electropolymerizing scans, indicating that the electroformation process of the film progresses during the cycling of the potential. The similarity of the polymer and monomer spectra indicates that the phthalocyanine conjugated-π system has remained intact upon electropolymerization.

3.2.3. Significant Examples of Electropolymerized Pyrrole- or Thiophene-Substituted Schiff Bases and Related Derivatives

Among the rare examples of conducting polymers covalently functionalized with salen and related derivatives, one can cite the case of Ni and Co tetraazamacrocycles functionalized at nitrogen with a pendant 3-thiophene moiety which were shown to be electrodeposited on electrode surfaces via an electrochemical copolymerization process in the presence of 3-methylthiophene[83,84]. Also, **35**, a Schiff base (namely 2-(3-pyrrole-1-yl-propylimino-methyl)-phenol was successfully electropolymerized, leading to conducting polymeric films that are catalytically active towards the electroreduction of both oxygen and carbon dioxide[85] (in the case where M = Co). The oxidative electropolymerization of the cobalt complex is shown in Figure 8.9 (curve a)[85]. The continuous increase in the size of the cyclic voltammogram peaks during repeated potential scans from 0.0 to 1.0 V indicates the growth of a polymer film on the electrode. The thin polymeric film formed adheres strongly to glassy carbon, platinum and gold electrodes. When the modified electrode is rinsed and transferred to a fresh acetonitrile solution, two pairs of peaks, which can be ascribed to the Co(II)/Co(I) and Co(II)/Co(III) processes, are observed at *ca.* −1.01 and 0.6 V, respectively (see Figure 8.9, curve b), almost the same potential values as the corresponding

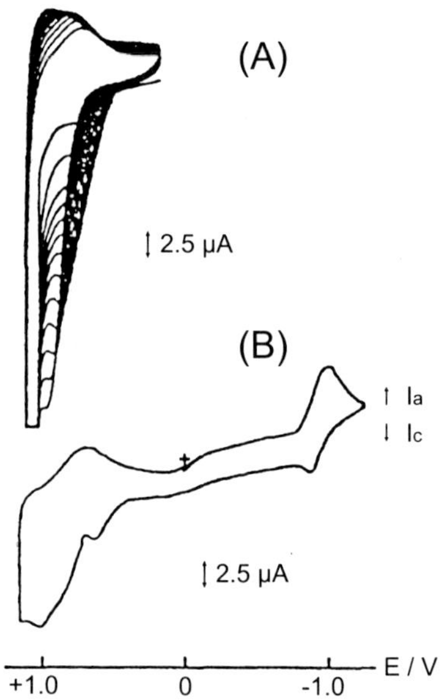

Figure 8.9. (A) Consecutive cyclic voltammograms of Co(II)-**35** in acetonitrile + 0.1 M Bu_4NClO_4 at a gold electrode (υ = 100 m V s^{-1}; (B) Cyclic voltammogram of a gold electrode coated with a film of poly[Co(II)-**35**] ($\Gamma = 1.5 \times 10^{-9}$ mol cm^{-2}) in acetonitrile + 0.1 M Bu_4NClO_4 (υ = 100 mV s^{-1}). Reprinted from ref.[85]. Copyright (1995), with permission from Elsevier.

monomer in solution. The electrochemical polymerization and electrochemistry of the resulting film were also investigated for the Ni complex of this same Schiff base[86].

Other examples of Schiff-base ligands able to bind copper and bearing pyrrole moieties were also successfully electropolymerized in acetonitrile to give conducting polymer films at the electrode surface[87]. Finally, one can cite the case of a Schiff base resulting from the reaction of 3-thiophene carboxaldehyde with 1,8-diaminonaphhalene that was electropolymerized in acetonitrile solution, leading to an electroactive film displaying electrocatalytic activity towards hydroquinone[88,89].

3.3. Immobilization of Multicharged Porphyrins into Pre-Electropolymerized Polypyrrole Films Bearing Functional Groups

Attempts to incorporate water-soluble tetracarboxylated and tetrasulphonated porphyrins into polypyrrole films containing alkylammonium groups have been reported[90,91]. Polypyrrole-alkylammonium polymers, prepared from 3-(pyrrol-1-yl)propyltriethylammonium (noted **40**) and 12-(pyrrol-1-yl)dodecyltriethylammonium (noted **41**) monomers (see examples in Figure 8.10) exhibit

Monomeric unit	Identification number	References
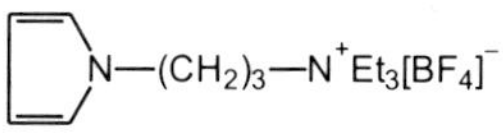	**40**	[90-96]
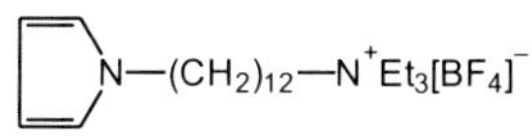	**41**	[90-96]
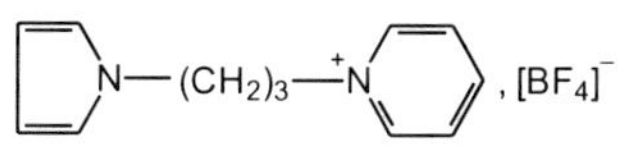	**42**	[96]
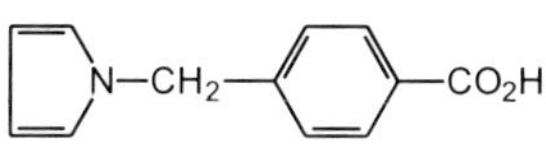	**43**	[46, 97]

Figure 8.10. Example of N-functionalized electropolymerizable pyrrole units leading to pre-doped electronically conducting polymers used for post-electropolymerization incroporation of pophyrin derivatives.

potential-independent anion-exchange behaviour as well as extraction and binding of organic and inorganic anions from organic and aqueous solutions[92–96]. Furthermore, the incorporation of anionic porphyrins in the polypyrrole films was easily achieved without any limitation due to the electropolymerization reaction of the starting monomers[90, 91].

The electrochemical polymerization of the pyrrole-substituted alkylammonium monomer **41** was achieved according to the following procedure: a drop of a 3-mM solution of the monomer in ultra-pure water was deposited on the electrode surface and then air dried. Next, the electrode recovered by the monomer was placed in $H_2O + 0.1$ M $LiClO_4$ and the adsorbed layers were electropolymerized by scanning the electrode potential repeatedly from -0.2 V to 0.8 V. Figure 8.11 shows the evolution of the repeated cyclic voltammograms of 3×10^{-7} mole of **41** deposited on an ITO electrode ($A = 1$ cm^2) in $H_2O + 0.1$ M $LiClO_4$ solution. The appearance of a pair of peaks centered on 0.5 V and their continuous increase in amplitude indicate that a film is formed on the electrode as a consequence of electrooxidative polymerization of the pyrrole groups of the adsorbed monomer. These new peaks are attributed to the reversible redox process of the polymer film. In addition, it is important to note that no increase in intensity was observed after 15 scans due to the complete oxidation of the adsorbed monomers. This simple and monomer-saving procedure makes this approach extremely useful.

The incorporation of anionic metalloporphyrins Mn(III)-**2**, Zn(II)-**2** and Fe(III)-**3** into poly(**41**) was achieved by soaking the polymer deposited on electrode in an aqueous solution of these anionic complexes ($\sim$ 0.2 mM) over several hours. The absorption spectra recorded after rinsing the electrodes with water (ITO transparent electrodes) (see Figure 8.12, curves a–c) are compared to those of the porphyrin complexes dissolved in aqueous solutions[91]. These data clearly show that the metalloporphyrins are retained by poly(**41**) films, owing to electrostatic binding. On incorporation into the polymer films, a red shift of 12–25 nm

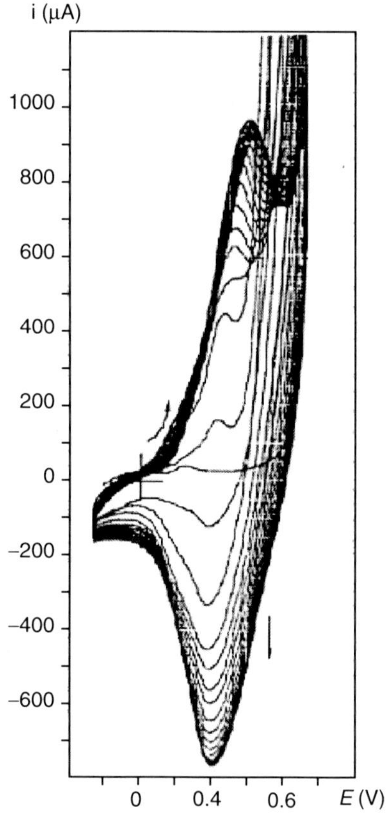

Figure 8.11. Electropolymerization of **41** (3×10^{-7} mole) adsorbed on an ITO electrode ($A = 1\ cm^2$) by repeated potential scans in $H_2O + 0.1\ M\ LiClO_4$ solution ($\upsilon = 50\ mV\ s^{-1}$). Reprinted from ref.[91]. Copyright (1993), with permission from Elsevier.

is observed for the Soret band. This red shift can be attributed to the aggregation and stacking of porphyrin molecules in the polymer.

It is important to note that the anionic porphyrins are strongly held by these polymer films after soaking the films in saturated KNO_3 solutions for more than 18 hours. The reported results[90,91] show clearly that the metalloporphyrins are retained by the poly-(pyrrole–alkylammonium) polymer, due to electrostatic binding. It is noteworthy that the ability of these polymers to retain anionic porphyrins is strong. The available data also showed that, besides their large size, the bulky porphyrins can be efficiently incorporated into the poly-(pyrrole-alkylammonium) films until total saturation of all the cationic sites is complete. Finally, the very high affinity of poly(**41**) amphiphilic films for anionic porphyrins allows the use of such molecular materials in concentrated ionic media without a significant loss of complexes.

In an innovative study, polypyrrole films exhibiting cation-exchange ability were functionalized via the incorporation of 5,10,15,20-tetrakis(1-methyl-4-

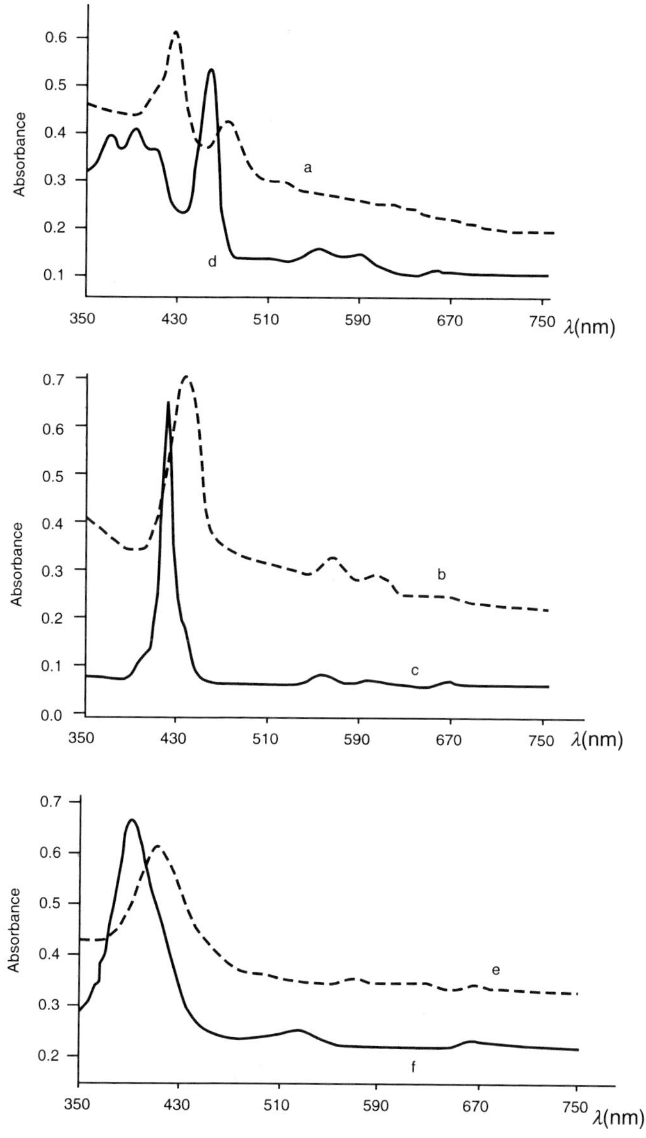

Figure 8.12. Absorption spectra of poly[**41**] films on transparent electrodes ($\Gamma = 1.6 \times 10^{-8}$ mole cm^{-2}) containing (A) Mn(III)-**2**, (B) Zn(II)-**2** and (C) Fe(III)-**3** compared to those of the dissolved complexes in aqueous solutions ((D) Mn(III)-**2**, 10^{-5} M at pH 8, (E) Zn(II)-**2**, 2×10^{-6} M at pH 8 and (F) Fe(III)-**3**, 6×10^{-6} M at pH 2.8). Reprinted from ref.[91]. Copyright (1993), with permission from Elsevier.

pyridyl)-21*H*,23*H*-porphyrin cations, allowing the pK_a of the resulting polymeric material and exchange capacity to be determined[97]. In a comparative study, a multicharged iron porphyrin, cationic Fe(III) tetrakis(*N*-methyl-4-pyridyl) porphyrin, was trapped into pre-electropolymerized polypyrrole films bearing carboxylic groups (obtained from the electropolymerization of carboxyphenyl-substituted pyrrole) by ion exchange[46]. Such a pre-electropolymerization strategy was also used to elaborate new oxygen sensors based on gold electrodes coated with thin polymer film matrices of either neutral β-cyclodextrin polymer or cation-exchange carboxymethylated β-cyclodextrin polymer hosting electrocatalytically active centres of [Co(II)-**1**], cobalt tetrakis-(*N*-methylpyridyl) porphyrin or [Co(II)-**3**][98].

3.4. Electropolymerization of Amino-, Hydroxy- and Vinyl-Substituted N_4-Complexes

Several types of porphyrin and phthalocyanine ligand modification have been performed to create a new family of electrochemically polymerizable complexes. The most commonly used porphyrins are the amino[58,99–104], hydroxy[58,105], methoxy[106–108], vinyl-[109–113] and other[114] substituted macrocyclic complexes (see examples in Figure 8.13). In the case of phthalocyanine, tetra-amino-substituted macrocycle was exclusively and intensively developed. Studies of these complexes have focused on the electrochemical synthesis and characterization of conductive polymeric or copolymeric materials.

In all cases, the films were obtained by oxidative electropolymerization of the cited substituted complexes from organic or aqueous solutions. The mechanism of metalloporphyrin film formation was suggested to be a radical-cation induced polymerization of the substituents on the periphery of the macrocycle. As it was reported for the case of polypyrrole-based materials[57–63], cyclic voltammetry and UV–visible spectroscopy with optically transparent electrodes were extensively used to provide information on the polymeric films (electroactivity, photometric properties, chemical stability, conductivity, etc.). Based on the available data, it appears that the electrochemical polymerization of the substituted complexes leads to well-structured multilayer films. It also appears that the low conductivity of the formed films, combined with the cross-linking effects due to the steric hindrance induced by the macrocyclic ligand, confers to these materials a certain number of limitations such as the limited continuous growth of the polymers due to the absence of electronic conductivity of the films. Indeed, the charge transport in many of these films acts only by electron-hopping process between porphyrin sites.

3.4.1. Significant Examples of Electropolymerized Films from Aminophenyl-, Hydroxyphenyl- and Vinyl-Substituted Porphyrins

In 1988, a pioneering work dealing with electrode characteristics of an immobilized enzyme chemically modified electrode based on bilayer-film coating for amperometric determination of glucose was reported[115]. In this work, a substrate was coated with two kinds of polymeric films in a bilayer state, that is, first with the cobalt tetrakis(*o*-aminophenyl)porphyrin polymer film and then with the enzyme film consisting of bovine serum albumin and glucose oxidase that were

Porphyrins	Metal	R_1, R_2, R_3, R_4	Identification number	References
	Fe(II), Co(II), Mn(II)	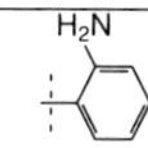	**44**	[58, 99, 104]
	Fe(III)		**45**	[109, 113]
	Ni(II)	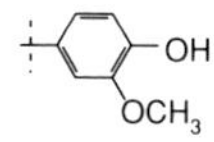	**46**	[114]
	Co(II)		**44**	[115–117, 131–133]
	2H, Co(II)	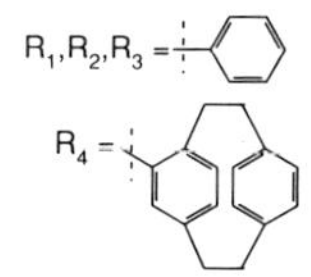	**47**	[118]
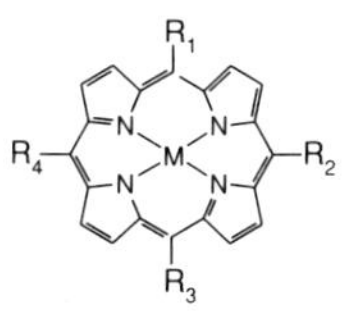	2H, Mn, Fe, Co, Ni	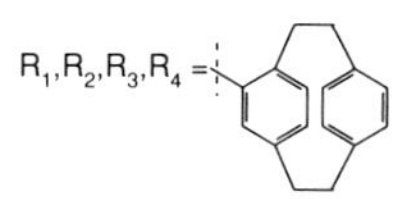	**48**	[118]
	Cu, Fe, Pd Co, Ni, Mn, (t-BuPy)–Ru–CO,	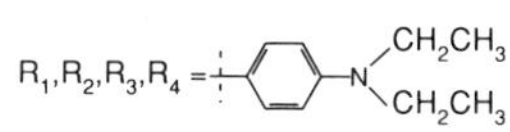	**49**	[118]
	Zn(II), Co(II)	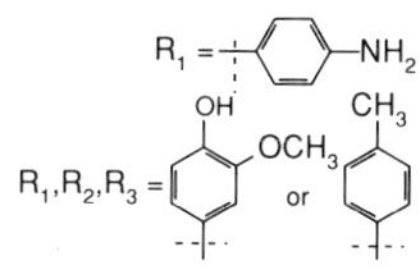	**50**	[119]
	Co(II)	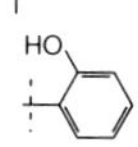	**51**	[58, 105]
	Co(II)	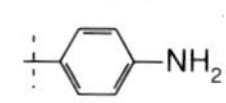	**52**	[120]
	Co(II)	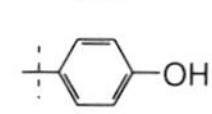	**53**	[121]
	Zn(II)		**53**	[122–123]
	Zn(II), Pd(II)		**53**	[124]
	2H	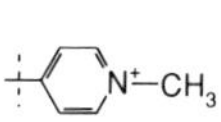	**54**	[124–126]
	Zn(II), 2H	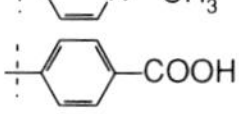	**55**	[125]

Figure 8.13. Examples of porphyrin and phthalocyanine derivatives bearing electropolymerizable moieties different pyrrole and thiophene units.

held together by cross-linking with glutaraldehyde. It was shown later that simple polymeric films[116] and ion-selective membranes[117] could be formed on smooth platinum and glassy carbon electrodes, respectively, by oxidative electropolymerization of Co(II)-**44**). More elaborate structures such as (Co(II)-**47**) and M-**48** (with M = Fe, Co, Mn and Ni as central metals) were also successfully electrodeposited via electropolymerization, as well as free-base and M-**49** (with M = Cu, Fe, Co, Mn, Ni and Pd as central metals) and the Ru carbonyl pyridynate analogue of this latter porphyrin group with carbonyl and *tert*-butylpyridine (*t*-Bupy)

Porphyrins	Metal	R_1, R_2, R_3, R_4	Identification number	References
	Zn(II)		**53**	[125–127]
	Zn(II), 2H	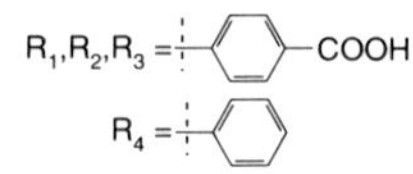	**56**	[125]
	Zn(II), 2H		**1**	[125]
	Zn(II)		**51**	[126]
	Zn(II)		**51**	[127]
	Zn(II)	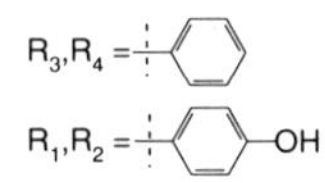	**57**	[127]
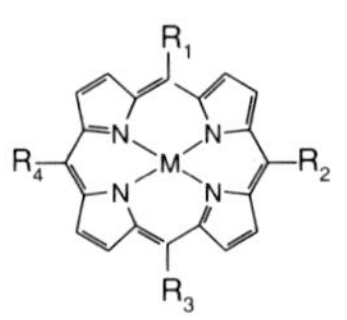	Zn(II)	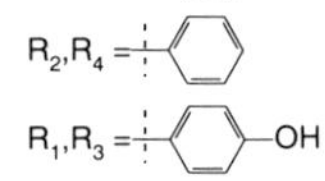	**58**	[127]
	Zn(II)	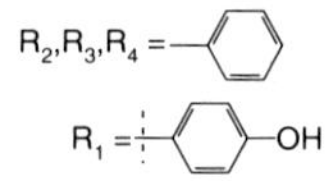	**59**	[128]
	2H, Fe(III), Zn(II)	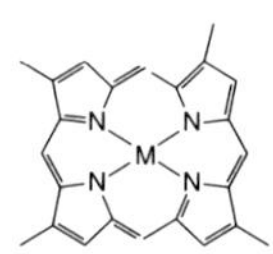	**60**	[129]
	Co(II)		**53**	[130–131]
		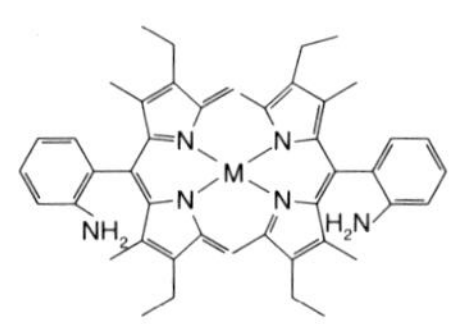	**61**	[134]
	Co(II), Ni(II), Zn(II), Cu(II)		**44**	[135]
	Pt		**1**	[136]

Figure 8.13. *Continued*

as axial ligands[118]. This work focused mainly on the surface morphology of the obtained materials and their catalytic activity towards the electrooxidation of hydrazine. A typical continuous scan cyclic voltammogram of [Ru(*t*-Bupy)(CO)-**49**] obtained during film formation is presented in Figure 8.14[118]. It demonstrates that the current of the peaks related to the porphyrin redox processes grows with each successive scan of the potential.

A scanning electron micrograph of the formed polymer, Figure 8.15, reveals a relatively smooth, caramel-like looking surface, containing many craters[118]. A closer look inside the crater reveals a comparatively even compact microspheroid surface morphology.

Malinski's group showed in early 1990s pioneering works on the electrochemical synthesis and characterization of conductive porphyrinic polymeric and

Porphyrins	Metal	R_1, R_2, R_3, R_4	Identification number	References
	Pt	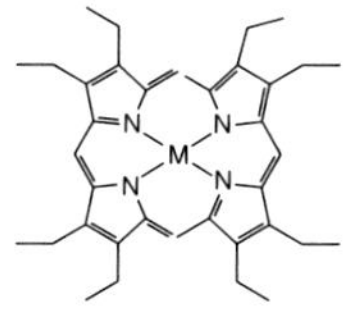	**62**	[136]
	2H, Fe, Cu(II), Ni(II)		**44, 52**	[137, 138]
	2H, Fe, Cu(II), Ni(II)	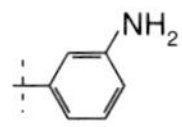	**63**	[137, 138]
	Ru(II)-CO	R_1, R_2, R_3, R_4 = 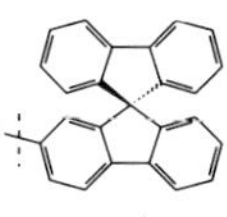	**64**	[139]
	Fe(III), Cu(II)	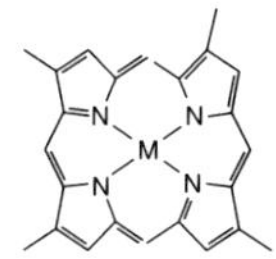	**65**	[140]
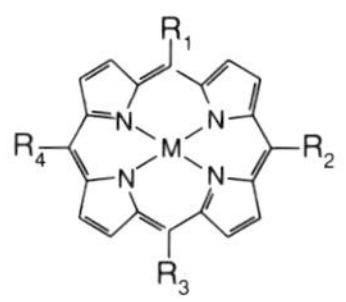	Fe(III)-Cl	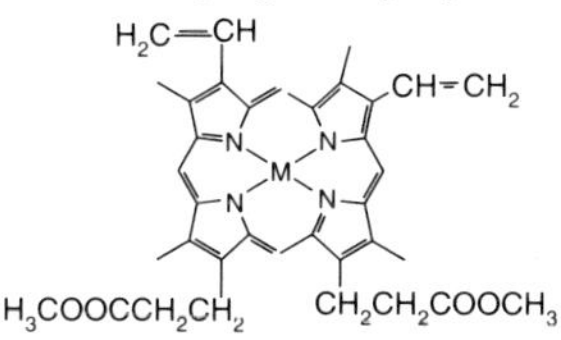	**66**	[141]
	Fe(III)–OH, Fe(III)–CKI	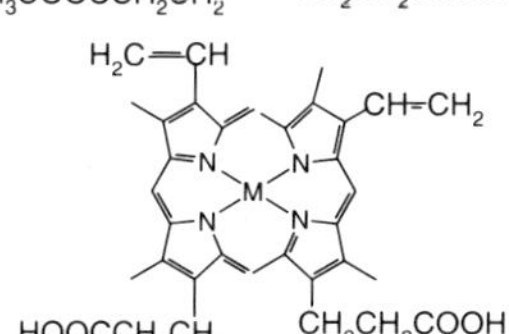	**67**	[46]
	Zn(II)	R_2, R_3, R_4 = 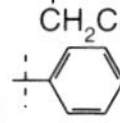 R_1 = 	**68**	[142]
	Zn(II), Co(II), Ni(II), Cu(II)		**66**	[143]

Figure 8.13. *Continued*

copolymeric materials obtained from several free-base and metalated porphyrins bearing 3-methoxy-4-hydroxyphenyl moieties as electropolymerizable groups. Metalloporphyrins **50** with Zn(II) and Co(II) were studied[119].

But the most studied electropolymerizable porphyrins are the tetra-amino- and tetrahydroxyphenyl-substituted ones. Indeed, both of them were developed to elaborate electrodes having potentiometric responses to several kinds of anions, such as iodide for example[120]. Co(II) tetrakis(*p*-hydroxyphenyl)porphyrin **53**-based films were used for the elaboration of potentiometric and fibre optic pH sensors with minimal interferences from anions[121]. Zn(II)-**53**-based films were also used and characterized for kinetics studies of hydrogen evolution at their surface[122, 123]. Zn(II)-**53** as well as Pd(II)-**53** and free base tetra(4-methyl pyridinium) porphyrin (H_2-**54**) were electropolymerized on indium tin oxide substrates, leading thus to starting materials for the making of donor/acceptor

Phthalocyanines	Metal	R_1, R_2, R_3, R_4	Identification number	References
	Co(II), Ni(II)	$-NH_2$	**68**	[144]
	Co(II), Ni(II), Pd(II)		**68**	[145]
	Lu(III)		1/2 **68**	[147,148]
	Co(II)		**68**	[149]
	Co(II), Cu(II)		**68**	[150]
	Ni(II)		**68**	[151]
	2H, Co(II)		**68**	[152,153]
	Co(II), Ni(II), Zn(II)		**68**	[154]
	Co(II)		**68**	[146,155]
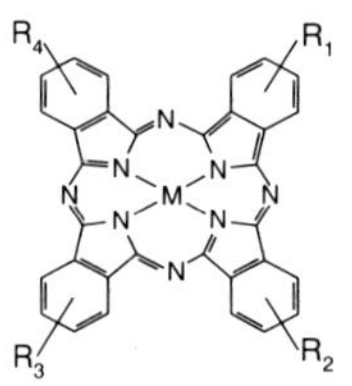	Ni(II)		**68**	[156,157]
	Co(II)		**68**	[158,159]
	2H		**68**	[160]
	2H, Fe(III)		**68**	[161]
	Co(II)		**68**	[162]
	Co(II), Ni(II), Cu(II)		**68**	[163,164]
	Ni(II)		**68**	[165]
	Co(II)		**68**	[166]
	Cu(II)		**68**	[167]
	Co(II) (metallisation)		**68**	[168]
	Mn(II)		**68**	[169]
	Co(II), Zn(II)		**68**	[170]
	2H		**68**	[81]
	Co(II)		**68**	[171]
	Mn(II)		**68**	[172]

Figure 8.13. *Continued*

heterojunctions in which photocurrents were produced and observed successfully[124]. Charge carrier photogeneration was also investigated in various electrochemically generated solid porphyrin films using optical and photocurrent action spectroscopy[125–127]. The porphyrin derivatives used therein were again Zn(II)-**53**, Zn(II)-**54** as well as tetra(*p*-carboxyphenyl) porphyrin **2**. Zn(II)-**53**-based films were examined by steady state absorption and emission spectroscopy in order to determine the type of interactions between the different porphyrin molecules present in the polymeric thin films[128]. Porphyrin-based films were also characterized using scanning probe techniques such as atomic force microscopy (AFM) and scanning tunnelling microscopy (STM) for their subsequent use in the electrochemical detection of the neurotransmitter, dopamine[129]. Only recently poly[Co(II)-**53**] films were shown to have catalytic properties towards the oxidation of thiols[130,131].

Electropolymerized films of [Co(II)-**44**] were investigated for their electrocatalytic behaviour towards dioxygen reduction[132,133] and thiol oxidation. Electropolymerization of Co(II)-**44** can be achieved by repeated potential scans

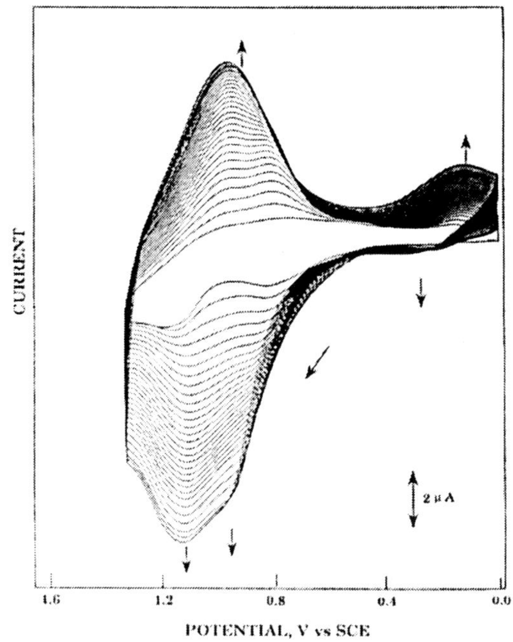

Figure 8.14. Continuous-scan cyclic voltammogram of [Ru(III)(*t*-BuPy)(CO)-**49**] from 0.0 to 1.5 V in CH_2Cl_2(0.1 M TBAP). Reprinted from ref.[118]. Copyright (1991) American Chemical Society.

between 0 and +1.1 V, respectively (at 0.2 Vs^{-1}). The repeated cycling around the anodic limit of the potential range results in the oxidation of the aminophenyl groups, leading to the deposition of the corresponding electroactive polymers on the electrode surface (see Figure 8.16)[131]. Thus, different polymer thickness can be obtained by controlling the number of the electropolymerizing scans.

Simultaneous piezoelectric microgravimetry and cyclic voltammetry measurements at an EQCM have been used to monitor the growth of poly[Co(II)-**44**] films, during potential cycling. Thus, formation of the electropolymerized films can be followed by microgravimetry as illustrated in Figure 8.17[131]. The mass uptake data, from the resonant frequency shift according to the Sauerbrey equation, give surface coverages of 620 ng cm^{-2} per scan of Co(II)-**44**. By taking into account the size and the shape of the square planar complexes (20 × 20$Å^2$ for the porphyrin) and the average distance between two stacked moieties (4 Å), one can estimate the thickness of the deposit to be *ca.* 8.4 nm per scan, or the equivalent of 21 monolayers of complex per scan.

The formation of the complex-based films on ITO transparent electrodes was confirmed by UV–visible spectrophotometry. This is exemplified in the case of poly[Co(II)-**44**], as shown in Figure 8.18[131]. The absorption spectrum of Co(II)-**44** in ACN solution is compared with that of the films, prepared by 24, 48, 80 and 160 electropolymerizing scans on a transparent ITO electrode. The intensity of the absorption bands does increase linearly over the first 100

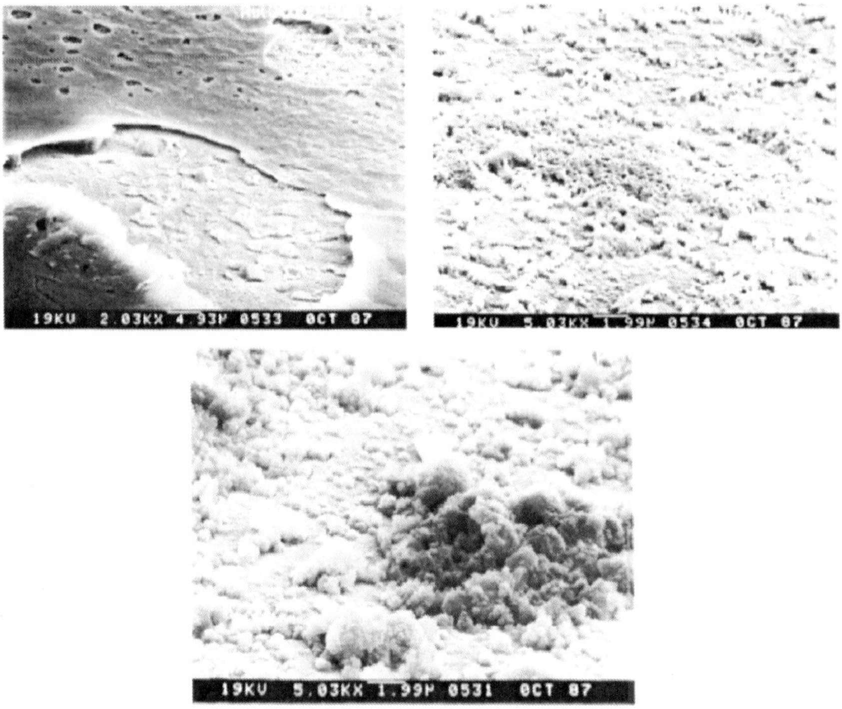

Figure 8.15. Scanning electron micrographs of [Ru(III)(*t*-BuPy)(CO)-**49**] film grown electrochemically on a Pt electrode at a constant potential of 1.45 V for 10 min. Magnification: (A) 2000:1, (B) 5000:1, from inside the upper right crater that appears in (A). Reprinted from ref. [118]. Copyright (1991) American Chemical Society.

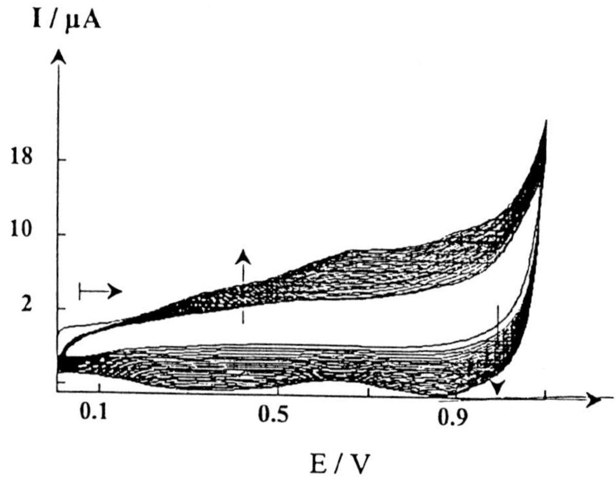

Figure 8.16. Cyclic voltammetry at vitreous carbon disk electrode of Co(II)-**44** (1 mM) in ACN + 0.1 M $TBABF_4$. Scan rate 0.2 V s^{-1}. Reprinted from ref.[131]. Copyright (2002), reproduced by permission of The Royal Society of Chemistry.

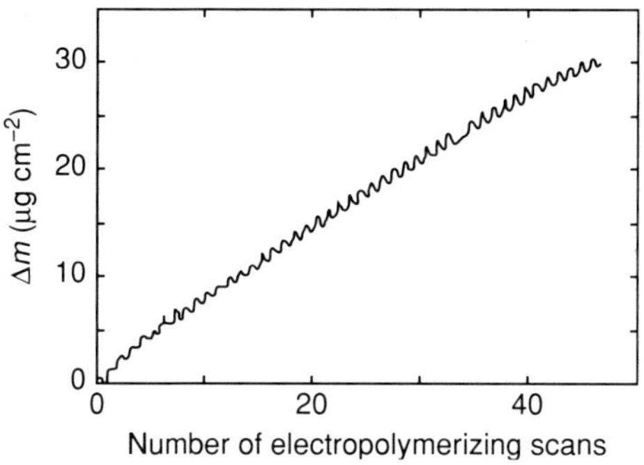

Figure 8.17. Evolution of deposited mass of Co(II)-**44** vs. the number of electropolymerizing scans (same electropolymerization conditions as Figure 8.16. Reprinted from ref.[131]. Copyright (2002), reproduced by permission of The Royal Society of Chemistry.

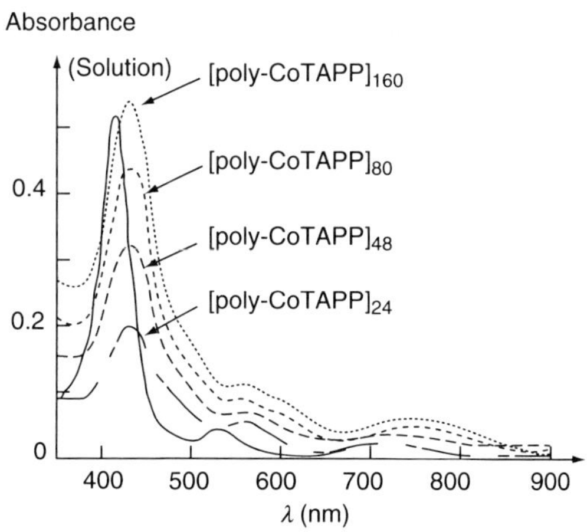

Figure 8.18. Absorption spectra of poly[Co(II)-**44**] films on ITO electrodes immersed in ACN. The straight line curve corresponds to Co(II)-**44** monomer in ACN solution. Reprinted from ref.[131]. Copyright (2002), reproduced by permission of The Royal Society of Chemistry.

electropolymerizing scans (and then increases but not linearly), indicating that the electropolymerization process progresses during the cycling of the potential. The similarity between the spectra of the monomer and that of the polymer indicates that the structure of the complex, upon electropolymerization, remains almost the same. In the case of the Soret band, a peak broadening is observed with a hyperchromic shift. Considerable broadness of the Soret band was previously observed in the case of several supported porphyrins and was attributed to some degree of aggregation and stacking of the porphyrin molecules on the electrode surface. The similarity of the polymer and monomer spectra indicates that the structure of the complex unit has remained intact upon electropolymerization.

The remarkable results described above show that the EQCM, UV–visible and cyclic voltammetry data demonstrate that a limited amount of cobalt complexes is electroactive, in the various electropolymerized films. This confirms previously reported results indicating that as the films grow, a considerable fraction remains electrochemically inactive. All this suggests that the cobalt complex-based films are very dense and then only very few monolayers, probably those located at the external film surface/solution interface, are electrochemically active (i.e. accessible).

Among the various reported studies related to electropolymerized porphyrin films of this type, the following ones should be emphasized since they constitute timely achieved contributions in this field, dealing either with the elaboration of potentiometric sensors aimed at the detection of numerous and common anions built from electropolymerized films of **61**, a 5,15-bis(2-aminophenyl) porphyrin[134] or the coupling of cyclic voltammetry, electrochemical impedance spectroscopy (EIS), infrared spectroscopy and AFM to demonstrate that the electropolymerization mechanism of *ortho*-substituted phenyl porphyrins was highly related to the nature of the solvent (organic or aqueous)[135]. Poly[Pt(II)-**1**] and poly[Pt(II)-**62**] films having luminescence properties were also used for elaborating fast optical sensors for oxygen[136] and various metalated (Fe, Cu, Ni) and non-metalated tetra-aminophenylporphyrins (**44**, **52**, **63**) with several amino groups (2–4) located in different positions of the phenyl substituents were prepared and polymerized electrochemically in order to determine the effect of the amino groups' position on the properties of the resulting modified electrodes[137, 138].

Recently, a novel porphyrin-based polymer, namely a poly(CO–Ru(II)-**64**) (with CO–Ru(II)–**64** = ruthenium carbonyl spirobifluorenylporphyrin), was prepared electrochemically and used for the transfer of carbene to olefins and sulphides in a solid-state reaction[139]. In another original study, a bimetallic porphyrin film using **65** was studied as electrode modifier with catalytic activity for molecular oxygen reduction and hydrogen peroxide reduction[140].

Finally, Meyer's group reported for the first time in 1992 on the electrochemical formation of thin polymeric films from [Fe(III)-**66**] (with **66** = protoporphyrin IX dimethyl ester complex) on glassy carbon or optically transparent SnO_2 electrodes[141]. The obtained molecular material was shown to be effective catalyst for the electroreduction of nitrite or NO to N_2O, N_2, NH_2OH and NH_3. Later, [HO–Fe(III)-**67**] and [Cl–Fe(III)-**67**] two iron porphyrins bearing vinyl groups (hematin and hemin, respectively) both bearing vinyl groups were electrochemically immobilized on vitreous carbon disk electrodes[46]. The electrochemical behaviour of a vinyl-substituted metalloporphyrin, Zn(II) 5-vinyl-10,15, 20-triphenylporphyrin (Zn(II)-**67**) was investigated in detail in order to confirm its deposition on electrode surfaces via electrochemically initiated polymerization[142]. Gold-coated quartz crystal of EQCM electrochemically coated with [M-**66**] (M = Zn(II), Co(II), Ni(II) and Cu(II)) was used for the elaboration of gas sensors[143].

3.4.2. Significant Examples of Electropolymerized Tetra-Amino Phthalocyanines

Tetra-aminophthalocyanine **68** was exclusively developed for the formation of electropolymerized phthalocyanine-based films. Indeed, **68** was successfully electrochemically oxidized and polymerized in dimethyl sulphoxide or dimethylformamide at a glassy carbon electrode[144](M = Co or Ni) or at optically transparent indium/tin oxide-coated glass electrodes (M = Pd, Co or Ni)[145]. In the case of the cobalt complex, it has been shown that the consecutive cyclic voltammograms of a vitreous carbon electrode in DMF + 0.1 mol L^{-1} $TBABF_4$ solution containing 1 mmol L^{-1} [Co(II)-**68**] (with a scan rate of 0.2 V s^{-1}) allows the build up of the electropolymerized poly[Co(II)-**68**] film]. Thus, repeated cycling around the Co(II)-**68** ligand oxidation results in the polymerization of the complex and the deposition of an electroactive polymer on the electrode surface and different polymer thickness can be obtained by controlling the number of the electropolymerizing scans. Similar films were also obtained on tin-doped In_2O_3 optically transparent electrodes (ITO). This allows the monitoring of the growth of the film by the changes observed in the UV–visible spectra. Typical spectra for three films of poly[Co(II)-**68**] of different thickness on ITO (prepared by 24, 48 and 80 electropolymerizing scans and noted as poly[Co(II)-**68**]$_{24}$, poly[Co(II)-**68**]$_{48}$, poly[Co(II)-**68**]$_{80}$, respectively, the subindex indicates the number of electropolymerizing scans), immersed in $TBABF_4$ 0.1 mol L^{-1} DMF solution, are illustrated in Figure 8.19[146]. The dashed line corresponds to the absorption spectra of [Co(II)-**68**] monomer in a 10^{-5} mol L^{-1} DMF solution and shows bands at 350 and 720 nm. The UV–visible spectra obtained for the polymer films are similar to monomer species, but exhibit a notorious broadening of the peaks which appear at 338 and 717 nm which increase with thickness. Typical thicknesses of films grown on ITO, measured by scanning electron microscopy, are 60, 113 and 214 nm for poly[Co(II)-**68**]$_{24}$, poly[Co(II)-**68**]$_{48}$ and poly[Co(II)-**68**]$_{80}$, respectively. Even though these measurements are conducted with dry films, the results indicate clearly that repeated electropolymerizing scans lead to continuous growth of the film, by nearly 2.5 nm per scan, and that the obtained films are conductive enough to ensure further electro-oxidation of Co(II)-**68** monomers from the solution.

Among the several recently reported examples[147–172], it is important to mention the following works:

- The rather rarely used lutetium diphthalocyanine [Lu(III)$_2$-**68**] was shown to be able to undergo a successful but very slow oxidative electropolymerization process under a continuous potential cycling carried out during several hours[147, 148].
- Electropolymerized cobalt phthalocyanine films showed an electro catalytic activity that allowed the electrochemical detection of 2-thiothiazolidine-4-carboxylic acid in urine samples with a detection limit of 7 pmol without any sample preparation[149].

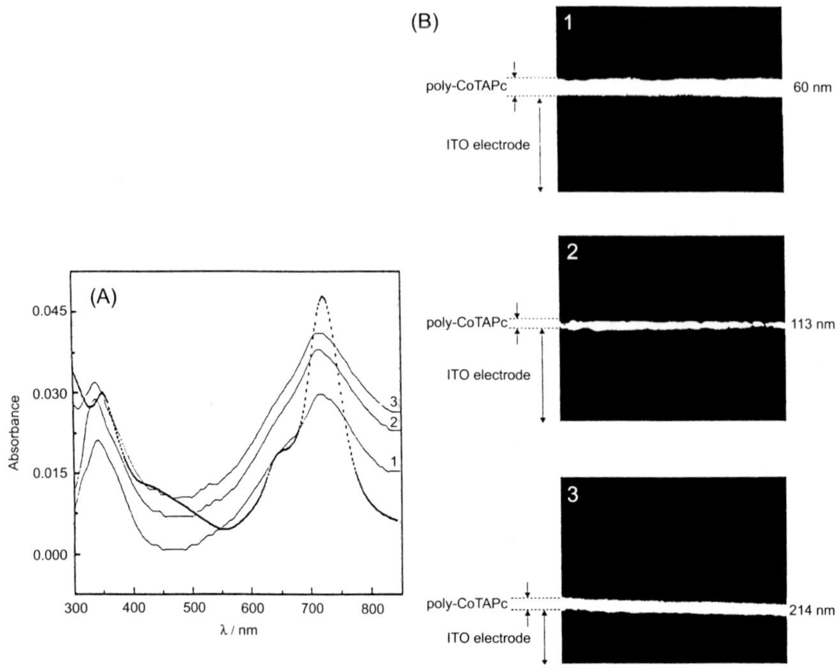

Figure 8.19. (A) UV–visible spectra of poly[Co(II)-**68**]$_{24}$ (curve 1), poly[Co(II)-**68**]$_{48}$ (curve 2) and poly[Co(II)-**68**]$_{80}$ (curve 3) on ITO electrodes. Dashed curve corresponds to monomer in DMF solution. (B) SEM micrographs of poly[Co(II)-**68**]$_{24}$ (micrograph 1), poly[Co(II)-**68**]$_{48}$ (micrograph 2), poly[Co(II)-**68**]$_{80}$ (micrograph 3) on ITO electrodes. Reprinted from ref.[146]. Copyright (2001), with permission from Elsevier.

- Polymetallophthalocyanine film and a polypyrrole film incorporating glucose oxidase (PPy-Gox) were prepared on glassy carbon electrode by the successive electrochemical deposition of the two polymers, allowing thus the elaboration of a bilayer conducting polymer electrode showing catalytic behaviour[150].
- Nickel tetra-aminophthalocyanine polymer was ingeniously tested for its electronic conductivity via kinetics studies of heterogeneous electron transfer taking place at the interface between them and the liquid phase[151].
- Thin layers of polymeric tetra-aminophthalocyanines were suggested as interesting electrode modifiers for the electrocatalytic oxidation of hydrazine in basic solution (M = Co)[152,153], the electrocatalytic oxidation of NADH (M = Co, Ni, Zn)[154], the electrochemical detection of sulphide anions and 2-mercaptoethanol (M = Co)[146,155], the electrochemical oxidation of dopamine (M = Zn[156], Ni[157]), the electrocatalytic reduction of dioxygen (M = Co[133,157,158], H_2[160] and Fe[160]), the electrooxidation of nitric oxide and nitrite (M = Co[161], Co, Ni, Cu[163,164] and Ni[165]), or the electroanalysis of 2-mercaptoethanesulphonic acid, reduced glutathione and L-cysteine (M = Co)[166].

3.5. Electrodeposited Films of Nickel N_4-Complexes in Alkaline Solution

Although the electrochemistry and the electrocatalytic properties of nickel macrocyclic complexes, phthalocyanines and porphyrins have been well studied in various solvents, few data exist on their special and typical behaviour when "electropolymerized" films in aqueous alkaline solution. During the last decade, it has been shown that nickel tetraazamacrocyclic complexes (examples shown in Figure 8.20) can be easily deposited onto an electrode surface in alkaline solutions

Porphyrins	Metal	R_1, R_2, R_3, R_4	Identification number	References
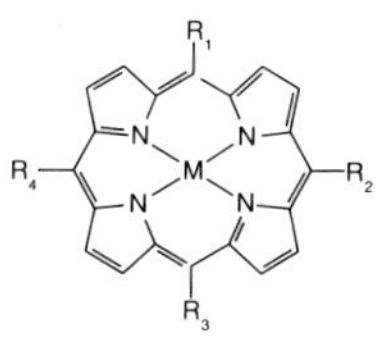	Ni(II)	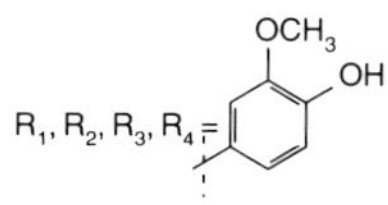	69	[173, 174]
	Ni(II)	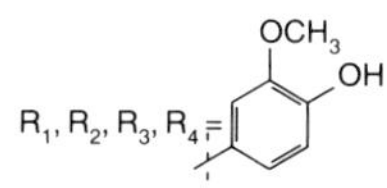	69	[175, 179]
	Ni(II)	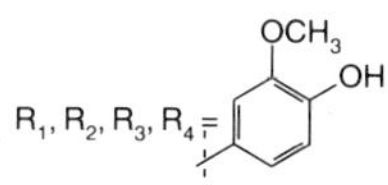	69	[176]
	Ni(II)	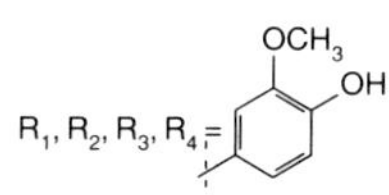	69	[177]
	Ni(II)	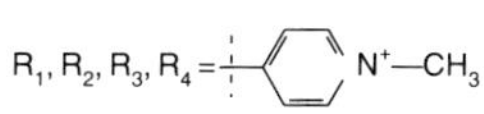	70	[178]
	Ni(II)	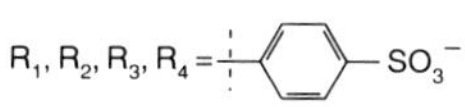	71	[178]
	Ni(II)	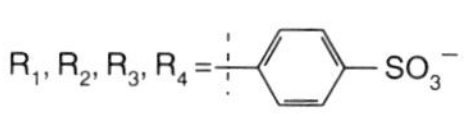	71	[180]
	Ni(II)	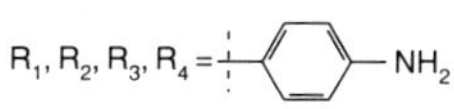	72	[180]

Phthalocyanines	Metal	R_1, R_2, R_3, R_4	Identification number	References
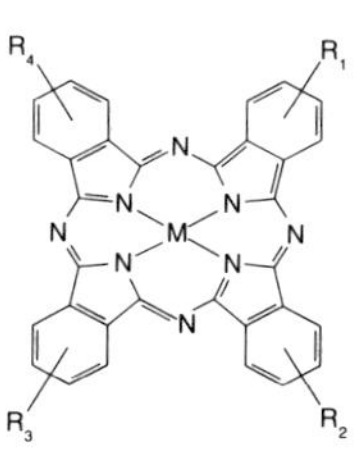	Ni(II)	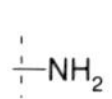	73	[182]

Figure 8.20. Examples of Ni(II)–porphyrin and phthalocyanine derivatives.

Salen derivatives	Metal	R_1, R_2, R_3, R_4	Identification number	References
	Ni(II)	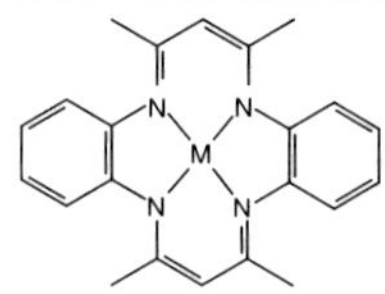	**74**	[183]
	Ni(II)	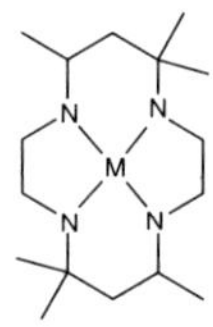	**75**	[184]
	Ni(II)	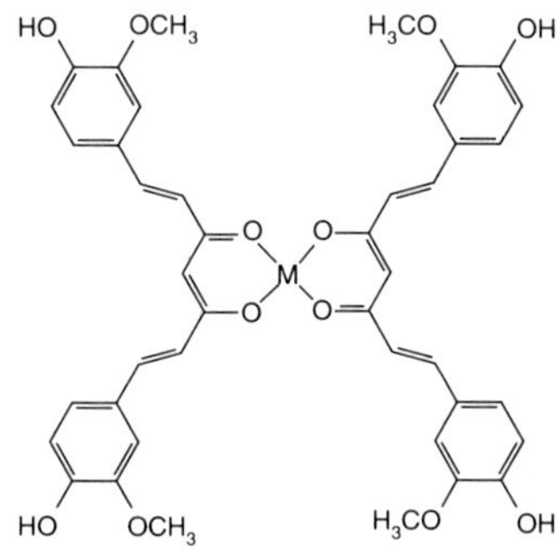	**76**	[185]

Figure 8.20. *Continued*

to form stable modified electrodes that catalyse the oxidation of several substrates. But the explanation of why the voltammetric behaviour of the films formed from various kinds of nickel planar macrocyclic complexes is nearly the same in aqueous basic solution has not been fully understood and has been a point of discussion. Two postulates have been reported: one of them considers that upon the electrochemical polymerization of the nickel complex in alkaline media, the nickel–nitrogen tetracoordination of the complex incorporated in the polymeric film is lost and the films behave in a similar manner than nickel hydroxide electrode. Even though the coordination of the nickel centres was probably lost, they remain entrapped inside the polymeric skeleton and exhibit a stable redox behaviour. The second postulate considers that the complexes do not decompose upon the electropolymerization procedure and they are attached to the electrode surface and interconnected via oxo-bridges.

The typical electrodeposition process observed with nickel derivatives was first investigated by Milczarek using a Ni(II) tetrakis(3-methoxy-4-hydroxyphenyl) porphyrin as a monomer. The resulting conductive film was used for the electrocatalytic oxidation of methanol and other simple alcohols[173,174], nitric oxide[175], formaldehyde[176], or ascorbic acid and NADH[177]. Later, [Ni(II)-**69**], [Ni(II)-**70**] and [Ni(II)-**71**] were shown to provide a large benefit in the electrochemical determination of nitric oxide in aqueous solution[178,179] or for the electrocatalytic

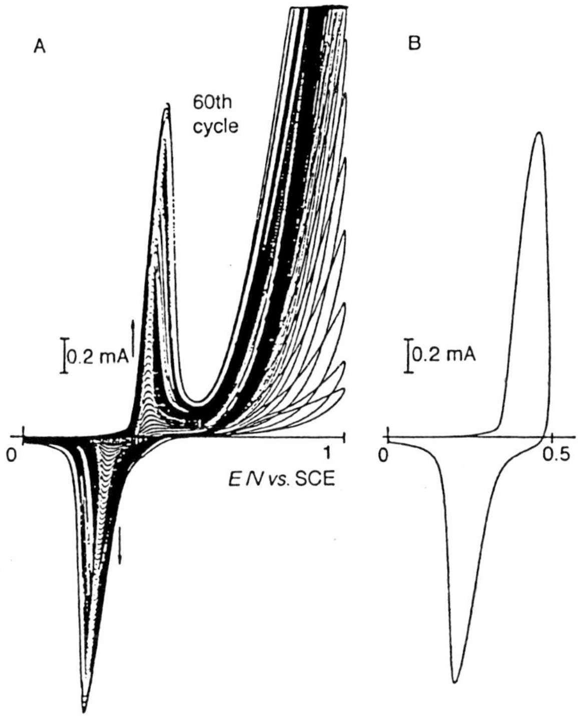

Figure 8.21. (A) Evolution of repeated cyclic voltammograms of ITO electrode in 0.1 M NaOH aqueous solution + Ni(II)-**71** monomer (2 mM) (60 scans are shown); (B) Cyclic voltammogram of poly[Ni(II)-**71**] film-coated electrode prepared by 60 electropolymerizing scans, as indicated in (A) potential scan rate 100 mV s^{-1} Reprinted from ref.[180]. Copyright (1997), reproduced by permission of The Royal Society of Chemistry.

oxidation of methanol, ethanol and hydrazine[180]. Most of this work on nickel derivatives of porphyrin had been reviewed recently[181]. In brief, a typical study of the electrodeposition of a nickel(II) N_4-complex can be analysed as follows, with the example of [Ni(II)-**71**][180]. Figure 8.21A shows the evolution of the cyclic voltammogram obtained for [Ni(II)-**71**] porphyrin film growth, denoted as poly[Ni(II)-**71**], on an ITO electrode. The potential was continuously cycled between 0 and 1.0 V at 100 mV s^{-1} while the ITO electrode was immersed in monomer solution of 2 mM Ni(II)-**71** in aqueous 0.1 M NaOH. It can be seen that an ill-defined redox couple appears during the first scan which increases and becomes well defined when the potential scans are repeated. The continuous increase in the amplitude of the voltammometric peaks, around 0.34 V, indicates that the film has been formed as a result of the anodic deposition of the porphyrin complex. The mechanism of electrooxidative deposition of the nickel complex is not fully clear at present. Several authors have proposed a tentative explanation for the electrochemical deposition process by suggesting that the attachment of the nickel complexes to the electrode surface is connected with oxidation of

OH^- anions. Indeed, the pH value of 13 is critical for film formation and one can imagine that the oxidation of OH^- creates a variety of functional groups on the electrode surface. It is important to note that no electrodeposition process occurs when the potential scan was limited to a range before the beginning of the oxidation of OH^- (i.e. $E < 0.8$ V). Thus, coupling the complexes to the electrode surface via an –O–Ni(II) bond seems to be probable. During further scans, the attached Ni(II) complexes undergo oxidation to Ni(III) and bind to a new coming Ni complex, supplied to the electrode surface by diffusion, via oxo-bridges. Anyhow, it appears clearly that the redox process exhibited by the nickel complex-based films in alkaline solution is similar to that shown by electrochemically formed $Ni(OH)_2$ electrodes in which the observed redox process is usually related to α-Ni(OH)2/γ-NiOOH transformation (for hydrated Ni(II) structures).

After a period of scanning, the electrode was transferred, after careful rinsing with water, to a 0.1 M NaOH aqueous solution (without the monomer in it). The cyclic voltammogram of the poly[Ni(II)-**71**] film is shown in Figure 8.21B[180]. It exhibits the previously observed electrochemical behaviour assigned to the Ni(III)/Ni(II) process at 0.34 V. Similar results are obtained on vitreous carbon and platinum electrodes. The apparent surface coverages of the poly[Ni(II)-**71**] film electrodes by nickel porphyrin can be calculated from the cyclic voltammograms of these films, and the calculation is based on the charge under the Ni(III)/Ni(II) oxidative or reductive peak. The formation of poly[Ni(II)-**71**] films on ITO transparent electrodes was confirmed by UV–visible spectrophotometry. The absorption spectrum of [Ni(II)-**71**] in solution (Figure 8.22A; curve 1) is compared with that of the poly[Ni(II)-**71**] film on transparent electrode (Figure 8.22A; curve 2). This clearly indicates that a nickel porphyrin film is present on the electrode surface. In the case of the Soret band, a peak broadening is observed with an hypsochromic shift. In Figure 8.22B are reported the absorbances of films (at $\lambda = 403$ nm, the wavelength of the Soret band) as a function of the number of electropolymerizing potential scans n. The electrodeposition of the porphyrin seems to occur regularly for $n < 100$ since absorbance of the Soret band is fairly proportional to the number of scans. But a decrease of the absorbance is observed when the thickness of the film increases. The origin of such a behaviour still is not understood.

In order to further elucidate the origin of the redox process displayed by all the studied nickel complex-based films in alkaline aqueous solutions, the voltammetric behaviour of nickel porphyrin-based film is prepared in acetonitrile solution. Figure 8.23A shows the polymerization of [Ni(II)-**72**] porphyrin performed on ITO electrode by cyclic voltammetry between -1.6 and 1.2 V in acetonitrile + 0.1 M Bu_4NBF_4, at an scan rate of 100 mV s^{-1}[180]. The cyclic voltammograms exhibit the expected well-known reversible one-electron porphyrin ring oxidation $NiP/NiP^{+\circ}$ and reversible one-electron ring reduction $NiP/NiP^{-\circ}$ at +0.77 V and -1.36 V, respectively. The evolution of the cyclic voltammograms during repeated potential scans shows a continuous increase in the amplitude of the cyclic voltammogram peaks for the previously described systems and this indicates that a film was formed on the electrode as a consequence of electrochemical oxidative polymerization of the attached aminophenyl groups. The characterization of the

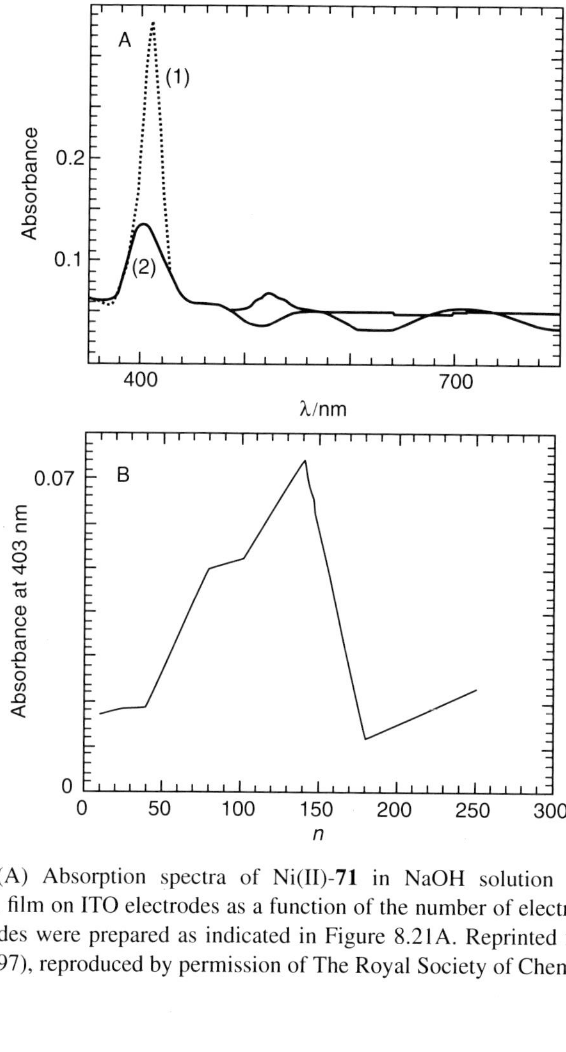

Figure 8.22. (A) Absorption spectra of Ni(II)-**71** in NaOH solution (curve 1) and poly[Ni(II)-**71** film on ITO electrodes as a function of the number of electropolymerizing scans. Electrodes were prepared as indicated in Figure 8.21A. Reprinted from ref.[180]. Copyright (1997), reproduced by permission of The Royal Society of Chemistry.

poly[Ni(II)-**72**] film was first carried out in acetonitrile + 0.1M Bu_4NBF_4. The cyclic voltammogram of the film exhibited similar electrochemical responses as the starting monomer which indicates that the polymeric film contains the nickel porphyrin. When the poly[Ni(II)-**72**] film was characterized in 0.1 M NaOH solution, different features were observed (Figure 8.23B)[180]. During the first scan, a huge irreversible anodic peak was observed at $E > +0.8$ V showing that a transformation of the film was occurring. As soon as the second scan was performed, the intensity of the previously reported anodic peak became significantly smaller showing the end of the irreversible transformation of the film. In addition, it appeared a new well-defined redox system at $E_{1/2} = 0.4$ V that grows continuously until it reaches a steady state. The shape of the final cyclic voltammogram (Figure 8.23C) is very similar to those obtained with the electropolymerized nickel macrocyclic-based films obtained in alkaline solution, which are also similar to

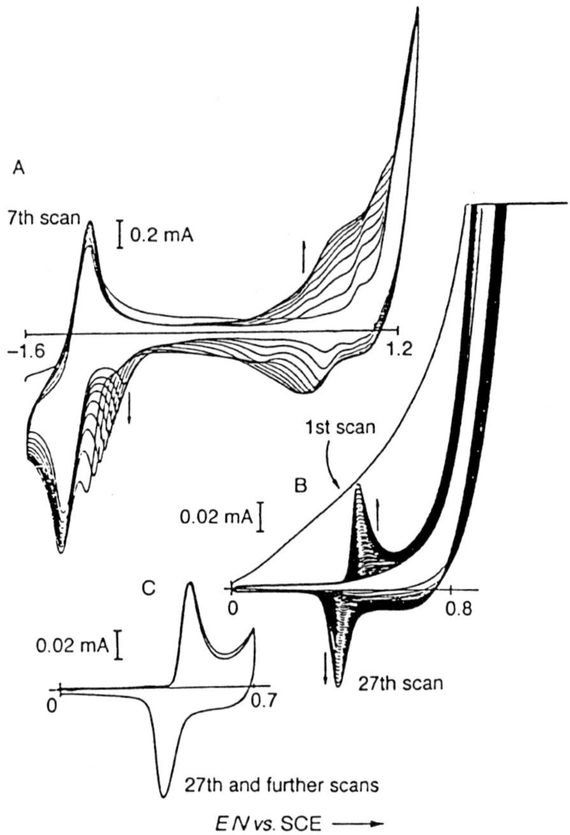

Figure 8.23. (A) Repeated cyclic voltammograms of Ni(II)-**72** (2 mM) in acetonitrile + 0.1M Bu_4NBF_4 at an ITO electrode (seven cycles are shown). (B) Evolution of the cyclic voltammograms of the ITO electrode-coated by poly[Ni(II)-**72**] (prepared as indicated in (A)) and transferred to fresh 0.1 M NaOH aqueous solution. (C) 27th and further scans. Potential scan rate 100 mV s^{-1}. Reprinted from ref.[180]. Copyright (1997), reproduced by permission of The Royal Society of Chemistry.

those of $Ni(OH)_2$ electrode. The redox process observed at 0.40 V remains stable and tends to show that during the electrochemical treatment of poly[Ni(II)-**72**] in alkaline media, the nickel porphyrin (or macro-complex, in general) incorporated into the polymeric film behaves in a similar manner than nickel hydroxide electrode. It should also be noted that during this process the colour of the film (observed on ITO electrodes) goes from deep blue to dark gold. This may be explained by the changes in the axial occupation of the Ni sites as a consequence of the formation of the O–Ni oxo bridges in alkaline solution.

This fundamental observation confirms the one reported by Cataldi and coworkers[183] in the case of polyNi(II)-tetramethyldibenzo-teraaza[14]annulene formed in dichloromethane and transformed in alkaline aqueous solution. It tends to show that upon the electrochemical polymerization of the nickel complex in organic solvent and the treatment of the film in alkaline solution, the complex incorporated in the polymeric film behaves in a similar manner than nickel hydroxide electrode. As we mentioned earlier, two postulates are reported in the literature: one considers that the coordination of the nickel centres may be lost and the second postulate considers that the complexes do not decompose and remain interconnected via oxo-bridges. But it is also important to note that (i) the redox process exhibited by all the nickel-based films (either directly prepared by electropolymerization in alkaline solution or in organic solvent followed by treatment in alkaline solution) and assigned to Ni(II)/Ni(III), is similar to that observed with nickel hydroxide films and shows the same pH dependence, and (ii) the UV–visible spectra of the films (particularly in the case of nickel porphyrins) indicate very clearly that the macrocyclic configuration of the nickel complexes remains intact upon the electrochemical polymerization.

Besides the largely reported case of the nickel tetrasulphonated phthalocyanine, only one example dealing with the electrodeposition of nickel tetraaminophthalocyanine via the formation of O–Ni–O oxo bridges can be found in the literature[182]. The resulting polymeric material was investigated as a likely electrocatalyst towards dopamine redox process. In this study, it was shown that the formation of these oxo bridges is beneficial for the electrocatalytic activity of the polymeric material besides the electropolymerization process based on the electro-oxidative N–N coupling of the amino groups.

As specified above, it has also been shown that tetraaza[14]annulene **74** (5,7,12,14-tetramethyl-dibenzo-1,4,8,11-tetraaza[14]annulene) can be electrodeposited on common conducting substrates and electrochemically treated so as to display a nickel-based catalytic deposit towards the oxidation of carbohydrates in alkaline solutions which shows strong similarities to the nickel hydroxide electrode[183]. Similar observations were reported for Ni complex **75** although they were rather oriented towards the oxidation of hydrogen peroxide[184] and for Ni curcumin films towards the electrocatalytic oxidation of aliphatic alcohols[185].

3.6. Electropolymerized Films of Salen Complexes

Several Schiff base complexes (especially salen = ethylenebis(salicylideneaminato)) display an ability for electrodeposition process that does not allow their classification under "electropolymerization" nomination, in the sense that the salen ligand is not bearing an electropolymerizable group. Nevertheless, such a process relies essentially on the parent structure itself of the salen derivatives and it deserves attention, since it is now largely involved in designing salen-based polymers via electrochemical processing (called here as "electropolymerization"). Some significant examples reported in the literature are described below (see examples in Figure 8.24).

Thin films of σ-bonded tetraazaannulenic complexes of cobalt were electrochemically synthesized. From a potential cycling applied to a platinum electrode in an electrolytic solution containing the monomer, the electrodeposition process occurred via a 6-6′σ polymerization of the monomers[186]. Also, the electrochemical homopolymerization of M-**78** (M = Ni, Cu) derivatives was thoroughly investigated[187], some of which were functionalized at the γ-position[188].

Salen derivatives	Metal	Identification number	References
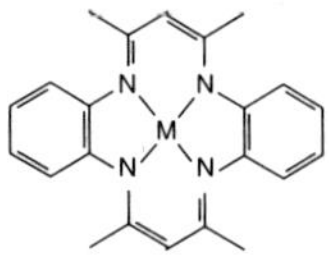	Co(III)-R (R = CH_3, C_2H_5 or C_6H_5)	**77**	[186]
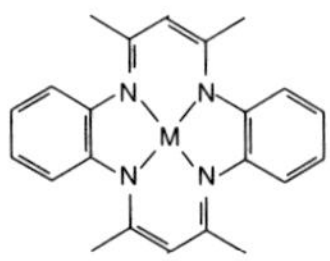	Cu(II), Ni(II)	**78**	[187]
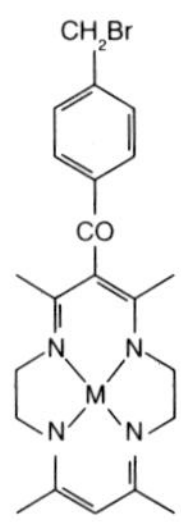	Ni(II)	**79**	[188]
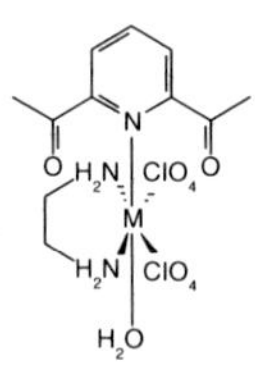	Cu(II)	**80**	[189]
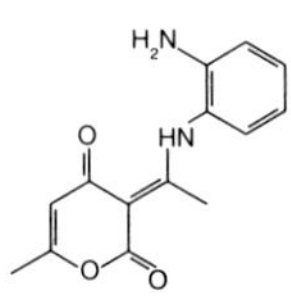		**81**	[190, 191]
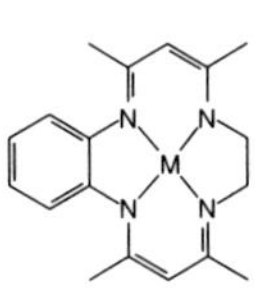	Ni(II)	**82**	[192]
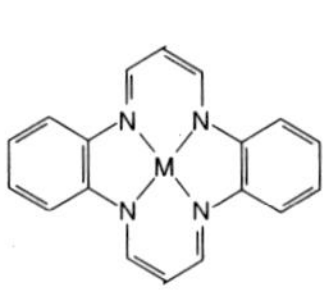	Co(II)	**83**	[193]

Figure 8.24. Examples of electropolymerizable salen derivatives bearing no pyrrole and thiophene units.

Salen derivatives	Metal	Identification number	References
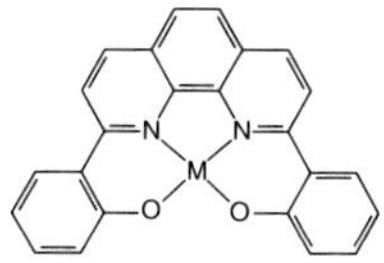	Ni(II), Cu(II)	**84**	[194]
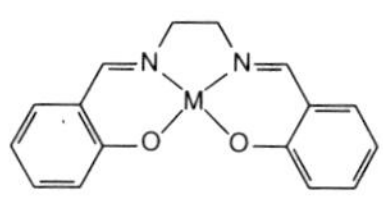	Ni(II)	**85**	[195]
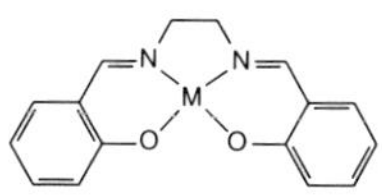	Co(II), Fe(II), Cu(II), Mn(II)	**85**	[196]
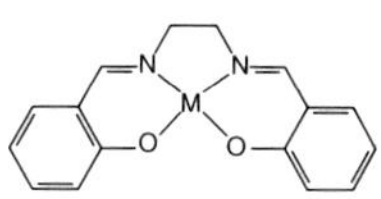	Fe(III)	**85**	[197]
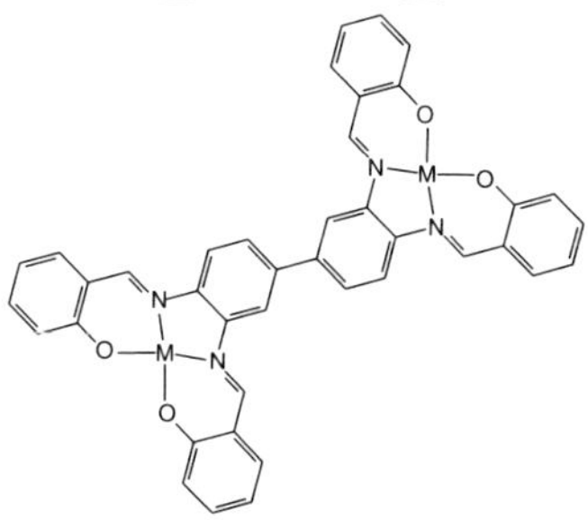	Ni(II)	**86**	[198]
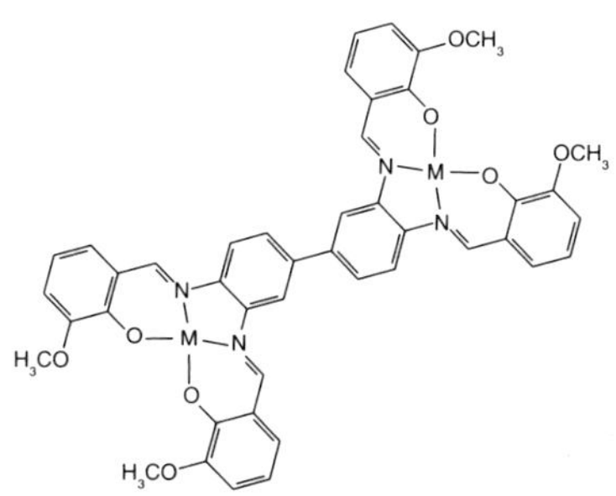	Ni(II)	**87**	[198]

Figure 8.24. *Continued*

By the same manner, the electropolymerization of the copper complex of **80** (half-unit = 2,6-diacetylpyridine-mono(ethylenediamine)) was performed, which led to an electroactive polymeric film whose electrocatalytic properties towards biodegradable agrochemicals were evaluated[189]. One can cite the example of the electrooxidative polymerization of **81**, a 3-[1-(2-amino-phenylimino)-ethyl]-6-methylpyran-2,4-dione Schiff base. The corresponding films are insoluble in most solvents, electroactive in water but electro-inactive in acetonitrile and they display electrocatalytic properties towards the oxidation of hydroquinone[190,191].

Salen derivatives	Metal	Identification number	References
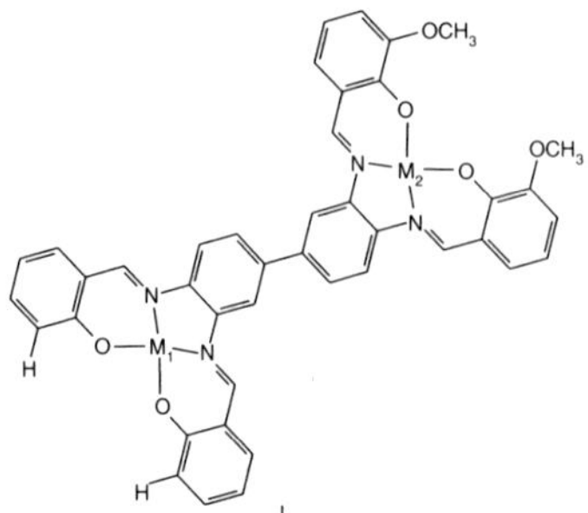	Ni(II)	**88**	[198]
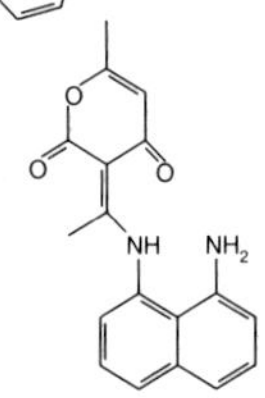		**89**	[199]

Figure 8.24. *Continued*

Finally, **82**, an asymmetric Ni-macrocyclic complex containing dianionic benzotetraaza[14]annulene ligand was electropolymerized on glassy carbon electrodes so as to develop the electrocatalytic properties of the resulting films towards the reduction of dioxygen and the oxidation of oxalate and L-ascorbic acid[192]. Poly(cobalt tetraazaanulene) was electrodeposited at glassy carbon electrode surface, providing thus a modified electrode aimed at the electrocatalytic reduction of oxygen[193].

In the case of salen complexes, phenanthroline-bridged salen compounds have been electrochemically studied and electropolymerized, allowing then the exchange of the inserted metal ion[194]. Similar studies dealing with the nickel salen complex and coupled with *in situ* FTIR and UV–visible spectroscopies revealed ultimately that electropolymerization process is a ligand-based process that takes place through a mixture of *o*- and *p*-linking of the phenol rings. It also indicated that the resulting films exhibit physical/chemical properties that cannot be attributed to an aggregation of individual complexes, behaving rather like a polyphenylene compound, with the metal ion acting as a bridge between biphenylene moieties[195]. Electropolymerized films obtained from the complexes of this same salen with cobalt, iron, copper and manganese were utilized as materials for development of an electrochemical sensor for the determination of NO in solution[196]. The electrochemical behaviour of Fe(III) salen complex was reported, as well as its electropolymerization. Although this latter appears more difficult than for several other poly(metal-salen) films (e.g. with nickel or copper), poly(Fe-salen) constitutes an efficient electrocatalyst towards the reduction of hydrogen peroxide and oxygen even if a different kinetic behaviour has been observed for both substrates, with a much greater efficiency in the case of oxygen[197]. Indeed, poly(Fe-salen) film displays a stable electroactivity in aqueous medium, as shown

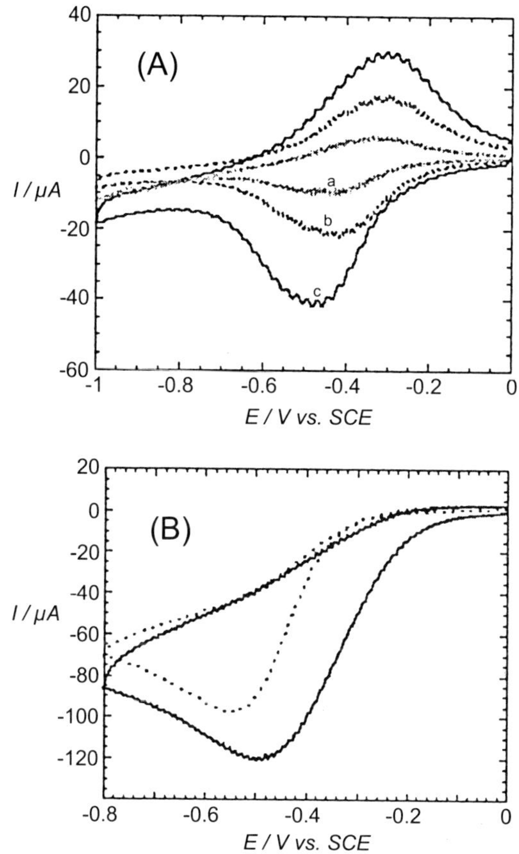

Figure 8.25. (A) CVs of poly[Fe(III)-**85** on an ITO electrode in phosphate buffer (pH 7.2) at various scan rates (mV s^{-1}): (a) 20, (b) 50, (c) 100. (B) Reduction of hydrogen peroxide (*ca.* 0.4 mM in phosphate buffer (pH 7.2) on poly[Fe(III)-**85**]-ITO electrode (—) and bare ITO electrode (- - -). Potentials are referred to SCE. Reprinted from ref.[197]. Copyright (2001), with permission from Elsevier.

in Figure 8.25a[197]. The surface coverage of the electrode, estimated from the charge under the voltammetric peaks was found to be *ca.* 1.1×10^{-9} mol/cm^{-2} and the thickness of the film was estimated to *ca.* 2.2 nm. The reduction of hydrogen peroxide on poly(Fe-salen) has been investigated mainly in phosphate buffer media (pH 7.2). The electrocatalysis phenomenon was evidenced by the fact that the reduction of H_2O_2 occurs at a slightly more positive potential in presence of poly(Fe-salen), with respect to the same reduction on bare electrode (see Figure 8.25b).

Finally, innovative bimetallic complexes as dissymmetric ligands can easily be built, giving birth to an unprecedented class of metal Schiff base salen complexes that can be successfully electropolymerized concomitantly with an overoxidation process[198]. The electrooxidation of 8-(3-acetylimino-6-methyl-2,

4-dioxopyran)-1-amino-naphthalene Schiff base in acidic aqueous solution yielded a polymer film on glassy carbon electrodes that showed a reversible well-defined redox system in aqueous acidic solution, as well as electrocatalytic activity towards oxidation of catechol and hydroquinone[199].

3.7. Miscellaneous Electrodeposition Processes

Besides the list of methods allowing the electrodeposition of porphyrinic compounds that was revealed so far in this chapter, miscellaneous processes have been reported, most of which are either ingenious or especially dedicated to precise applications. For example, Mizutani and coworkers achieved the immobilization of Co(II) *meso*-tetraphenylporphyrin as catalytic sites in a conducting polymer film electrode consisting of polypyrrole-poly(vinyl chloride) alloy[200]. To this end, the catalyst was first immobilized in poly(vinyl chloride) (PVC) film and, second, the insulating PVC matrix is converted into a conducting PPy–PVC alloy via the electrochemical polymerization of pyrrole at the surface of a glassy carbon electrode coated with the catalyst–PVC composite. Another ingenious work was published in 2000, which relies on processes frequently used for the deposition of non-electropolymerizable compounds on electrode surfaces, such as dip coating or electrostatic assembly. In that work, the relevant series of symmetric supramolecular porphyrins that was used had been obtained by attaching four [Ru(II)(bipy)$_2$Cl] groups to the bipyridyl substituents of *meso*-tetra (4-pyridyl)porphyrin and its metalated derivatives[201]. In a later published work, porphyrinic compounds were cast as porphyrin–fullerene dyad cluster films on nanostructured SnO_2 films, providing thus photoelectrochemically active materials that can generate anodic photocurrent under visible light excitation[202]. More recently, in a quite innovative work, magnetic nickel nanowires had been functionalized with fluorescent hematoporphyrin IX, [8,13-bis(1-hydroxyethyl)-3,7,12,17-tetra-methyl-21*H*,23*H*-porphyrin-2,18-dipropionic acid], which has two intense, red fluorescence bands and two carboxylic acid groups that bind strongly to metal oxides and to the native oxide films on metals, such as nickel[203]. Finally, another method consisting in electropolymerizing a molecular film, previously positioned on the electrode surface by the well-known method of dip-coating, was successfully reported. This strategy was employed with [H_2(4-TPyP){Ru(Clphen)$_2$Cl}$_4$], a tetraruthenated porphyrin and applied to the electrocatalytic oxidation of nitrite ion[204].

Various original methods were also proposed for the electrochemical deposition of phthalocyanines. For example, "cobalt phthalocyanine-on-carbon" was prepared from the precipitation of cobalt phthalocyanine CoPc on carbon black by pouring the slurry into cold water[205]. Once filtered and dried, the material was pyrolysed for 2 h in a tubular furnace at temperatures ranging from 400 to 1,100°C under flowing argon atmosphere. Both half- and full-cell measurements were carried out 2–3 weeks after pyrolysis of the CoPc/C electrocatalyst at various temperatures. In a completely different procedure, glassy carbon electrodes were modified with thin films of CoPc, prepared by spreading $CCl_3COOH + CHCl_3$ solution containing the complex and allowing thus its deposition of the order of a monolayer[206].

Another interesting electrodeposition procedure was reported: it consists in carrying out the cathodic electrolytic deposition of Mg(II)TCPc (TCPc = 2,9,16,23-tetracarboxyphthalocyanine) on indium/tin oxide-coated glass electrodes in the presence of bridging metal ions provided by $MgCl_2$, $CaCl_2$, $CoCl_2$ or $ZnCl_2$ used as a supporting electrolyte[207]. Also, the electrochemical oxidation of Li_2Pc (lithium phthalocyanine) was achieved to allow the electrodeposition (electrocrystallization) of Li_2Pc onto indium tin oxide substrate and detailed studies of the influence of the electrodeposition parameters such as the electrolysis time, the nature of the solvent or the oxidation potential on the thin film morphology[208,209] and magnetic properties[210,211] were carried out. Cathodic electrodeposition from an aqueous mixed solution of zinc nitrate and water soluble tetrasulfonated metallophthalocyanine (M = Zn, Al, Si) was also reported. It resulted in a self-assembled growth of zinc oxide (ZnO) thin films whose surface is modified by the phthalocyanine moieties[212–214]. In a similar approach, electro-codeposited and homogeneous film was immobilized on indium tin oxide-coated glass electrode by cathodic polarization of a solution of 2,9,16,23-tetracarboxyphthalocyanatomagnesium, *N*,*N*′-4-hydroxyphenyl-3,4,9,10-perylenetetracarboxylic-diimide and $ZnCl_2$ as supporting electrolyte in DMF[215]. A device based on this film displayed a number of p–n junction interfaces, as suggested by its rectification and photoconduction properties.

One can also mention the case of composites-based conducting polymers electrodeposited and characterized on anodes of platinum- or carbon black- filled polypropylene from a stirred electrolyte with dispersed copper phthalocyanine[216]. The electrolytic solution contained, besides the solvent (water or acetonitrile), the monomer (pyrrole or thiophene) and a supporting electrolyte. Patterned thin films were obtained from phthalocyanine derivatives, as reported in the case of (2,3,9,10,16,17,23,24-oktakis((2-benzyloxy)ethoxy)phthalocyaninato) copper[217]. Such films were prepared by means of capillary flow of chloroform solutions into micrometer-dimension hydrophobic/hydrophilic channels initially created by a combination of microcontact printing of octadecylmercaptan (C_{18}–SH) layers on gold electrodes. These latter gave birth to a hydrophobic channel bottom while oxidative electropolymerization of *m*-aminophenol (at pH 4) led to hydrophilic channel walls.

Finally, vacuum-deposited organic photovoltaic cells were elaborated from a phthalocyanine/C_{60} multilayer configuration[218]. Such multilayer cells were constructed in a vacuum chamber coupled to an Ar-glove-box/characterization chamber. Sublimation purified copper phthalocyanine, C_{60}, and bathocuprine were vacuum deposited on various cleaned and small molecule-modified ITO substrates, over which either PEDOT layers were spin casted, or were electrochemically grown.

Interestingly, the reaction between π-cation radicals generated by electrochemical or chemical oxidation of pyrrole and metallo-organic complexes (such as tetramethyldibenzo tetraazaannulene nickel (Nitmdbtaa) or tetramethyl dinaphtho tetrazaannulene nickel (Nitmdntaa)) was investigated. In the case of the first compound, it appeared from voltammetric evidence that pyrrolyl units insert as organic spacers between nickel-macrocycle molecules, leading therefore to thin conducting films growing on common types of electrode surfaces[219]. The electrodeposition of pyrrole-free salen derivatives via an electrochemical copolymerization process with pyrrole was also reported in the case of 6, 6′bis(2-hydroxyphenyl)bipyridine in the complex (Ni-bpsalen)[220]. In an original procedure, a new coordination polymer was prepared through the polycondensation reaction of α, ω-bis(chloromethyl)polymethylphenylsilane with Ni-salen complex, necessitating a new preparation method of the chloromethylated polysilane[221].

4. Electrocatalytic and Electroanalytic Applications of Electropolymerized N_4-Macrocyclic-Based Films

The most relevant and promising applications of electropolymerized N_4-macrocyclic complexes are related to the biomimetic activation of small molecules of biological importance such as molecular oxygen, nitrite, nitric oxide, thiols, neurotransmitters etc. By now, the authors are making profit from the extensively well-developed electroassisted biomimetic activation of small molecules, to elaborate sensing materials for electroanalysis. Such an approach combines several benefits coming from the particular versatile reactivity of the N_4-macrocyclic complexes with their ability to form adherent electroactive and well-characterized thin films on electrodes of various shapes, sizes and geometries. We focus in this section on two major applications reported during the last decade: the electroassisted biomimetic reduction of molecular oxygen by electropolymerized manganese porphyrin films and the electroanalytical devices developed for nitric oxide produced by biological systems.

4.1. Electroassisted Biomimetic Reduction of Molecular Oxygen: Mechanistic and Electrochemical Approach

Achieving biomimetic hydrocarbon oxidations with molecular oxygen and synthetic porphyrin models of cytochrome P-450 is a field where electropolymerized metalloporphyrin films played a significant role. Efficient high-valent metal-oxo porphyrin complexes, the reactive oxygenating intermediates, have been postulated in numerous systems[4, 222–239]. Electroassisted formation of the oxo species requires the enzyme model plus molecular oxygen, an activator (such as anhydride or acid) and one or several axial ligands (such as imidazole) as shown in Figure 8.26 and as it has been described by several groups[229–234, 236, 238, 240].

We and other groups have shown the first examples in which a manganese porphyrin supported on a polypyrrole film[229, 234, 236] acts as an efficient model. Figure 8.27 shows the examined test reactions. In brief, molecular oxygen was electro-activated by the electropolymerized film of a manganese porphyrin deposited on graphite tissue and carbon felt electrodes. The supported catalyst was regenerated by controlled-potential electrolysis at $E = -0.5$V/SCE.

The essential observations that can be summarized from the reported studies allows to state that the catalytic efficiency and stability of the complexes are

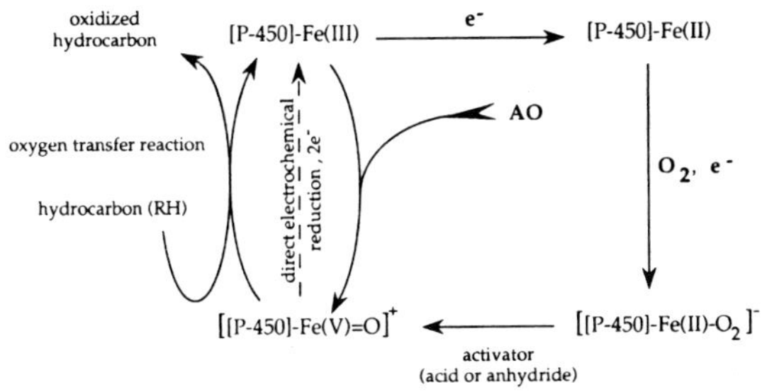

Figure 8.26. Simplified catalytic cycle for the oxidation of hydrocarbons (RH) by molecular oxygen (long cycle) or by oxygen atom donors AO (short cycle) and cytochrome P-450 (or its synthetic models). Reprinted with permission from *Acc. Chem. Res.* 1995, **28**, 30–36. Copyright 1995 American Chemical Society.

obviously enhanced when it is deposited within the polymer network, compared to those of the same complexes in homogeneous phase. Indeed:

- The oxidation products are formed with a very good efficiency by the supported catalyst, up to 500 turnovers of the catalyst per hour;
- The faradaic efficiency expressed as the ratio of moles of oxidation products analyzed to the electrochemical charge passed is relatively poor, but reached 99% in some cases;
- The nature of the electrode material (graphite tissue or carbon felt) has a small but significant influence on the chemical and faradic efficiencies of the supported catalyst (for the *cis*-cyclooctene epoxidation reaction, for example). This is probably due to either a preferred orientation of the supported catalyst for approach of the hydrocarbon to the presumed active manganese site, or to the high local concentration of the catalyst. However, the low faradaic yields obtained in some cases are certainly due to the direct electrochemical reduction of the active oxidant, a high-valent manganese "oxo" complex, at the potential of electrolysis according to the electroactivity study of this "oxo" form of the catalyst previously reported[232,236,240]. Thus, the fixation of the catalyst onto the electrode promotes the electroreductive consumption of the active oxidant. Anyhow, it is important to note that with the electropolymerized films, the products are formed using a substrate/catalyst ratio higher than 6,000, with similar yields compared with those obtained in homogeneous solution (with a substrate/catalyst ratio up to 400). It was also found that the recovered porphyrin polymer electrodes (stored in air, without precaution) retained 95% of their initial catalytic activity during the second run, and 70% of their initial activity during the third run[236]. It also appears from these results that it is not clear to bring out any valid information on the stereospecificity of the supported catalyst. More experiments with well-adapted test reactions are needed to reflect the importance of this aspect.

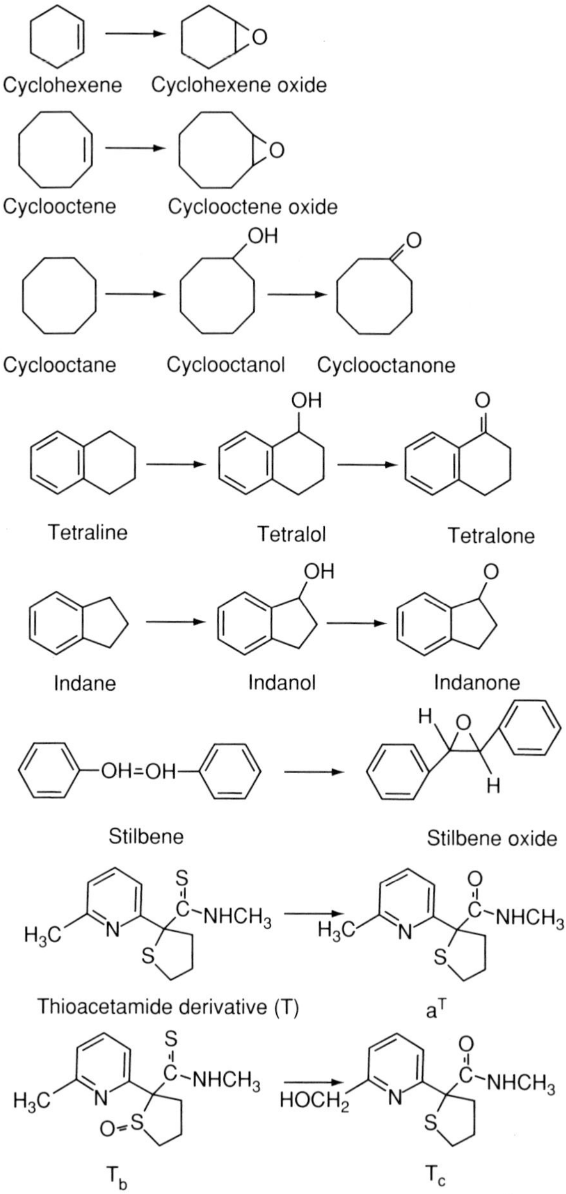

Figure 8.27. Studied examples of oxidizable hydrocarbons.

We recently reported for the first time on the electroassisted-biomimetic activation of molecular oxygen by a newly prepared electropolymerized polypyrrole-manganese phthalocyanine film[241]. The prepared films and their intervention in the electroassisted-catalytic reduction of molecular oxygen were analyzed by cyclic voltammetry and UV–visible spectrophotometry on optically transparent electrodes. The obtained results demonstrate the probable existence of the key- steps responsible for the suggested formation of the highly reactive manganese oxo intermediate. Figure 8.28a (curve 1) shows the cyclic voltammogram of poly[Mn(III)-**32**] film in deaerated DCM + 0.1 mol L^{-1} Bu_4NBF_4 solution (curve 1) and in presence of molecular oxygen, without (curve 2) and with benzoic anhydride (curve 3)[241]. It appears that in presence of molecular oxygen, the Mn(III)/Mn(II) redox process is subjected to a major change. Indeed, its associated re-oxidation peak disappears (curve 2). Such modification of the cyclic voltammogram suggests that, in a first step, oxygen binds to Mn(II) leading to a formulated Mn(III)-superoxide adduct that accepts a second electron to form a doubly reduced superoxo intermediate (see below). Curve 3 (Figure 8.28a) shows the cyclic voltammogram of poly[Mn(III)-**32**] film in DCM + 0.1 mol L^{-1} Bu_4NBF_4 solution containing benzoic anhydride (activator), upon addition of molecular oxygen. The anhydride function is to cleave the O–O bond once the oxygen is coordinated to and activated by the Mn centre. A large enhancement of the Mn(III)/Mn(II) reduction current peak is clearly observed.

With reference to previously reported studies of porphyrins and Schiff base complexes, the modification of the cyclic voltammogram can be explained by the occurrence of the expected formation of the high-valent manganese-oxo $[Mn(V){=}O]^+$ intermediate, followed by its reduction and the steady-state electrocatalytic regeneration of Mn(III) form[242–249] according to the following set of reactions (where $(PhCO)_2O$ is benzoic anhydride):

$$[Mn(III)]^{+} + e^{-} \rightarrow [Mn(II)] \quad (8.1)$$

$$[Mn(II)] + O_2 \rightarrow [Mn(III){-}O_2^{-}] \quad (8.2)$$

$$[Mn(III){-}O_2^{\,-}] + e^{-} \rightarrow [Mn(III){-}O_2^{\,-}]^{-} \text{ (or } [Mn(II){-}O_2^{\,-}]^{\cdot}) \quad (8.3)$$

$$[Mn(III){-}O_2^{\,-}]^{-} + (PhCO)_2O \rightarrow [Mn(V){=}O]^{+} + 2PhCOO^{-} \quad (8.4)$$

$$[Mn(V){=}O]^{+} + 2e^{-} + 2H^{+} \rightarrow [Mn(III)]^{+} + H_2O \quad (8.5)$$

Note that the direct reduction of molecular oxygen does not take place in the potential range investigated in this study, and that the presence of benzoic anhydride does not induce any electrochemical interference. Rotating disk voltammetry confirms all the above-described observations as it is shown in Figure 8.28b. Also, experiments performed with dissolved monomer in solution provided information similar to that obtained with the supported complex, thus confirming that the Mn complex behaves according the same way in both phases.

Spectroelectrochemistry experiments using poly[Mn(III)-**32**] film deposited on optically transparent tin oxide electrode, flattened against one of the walls of the spectrophotometer cuvette orthogonal to the light beam pathway and mounted as an electrochemical cell, were performed to provide further insights into the nature of the redox activity observed by cyclic voltammetry[241]. Figure 8.29a shows spectral changes observed during the controlled-potential reduction of the film immersed in DCM + 0.1 mol L^{-1} Bu_4NBF_4 solution at −0.7 V/Ag–AgCl, a value more negative than the reduction potential of Mn(III)/Mn(II) redox process. The Q band shifts from 734 to 685 nm. This observed spectral change is consistent with the reduction of the Mn(III) sites in the film. Also, the spectral changes upon reduction at −0.7 V were similar with or without benzoic anhydride, and as long as molecular oxygen is not present in the solution.

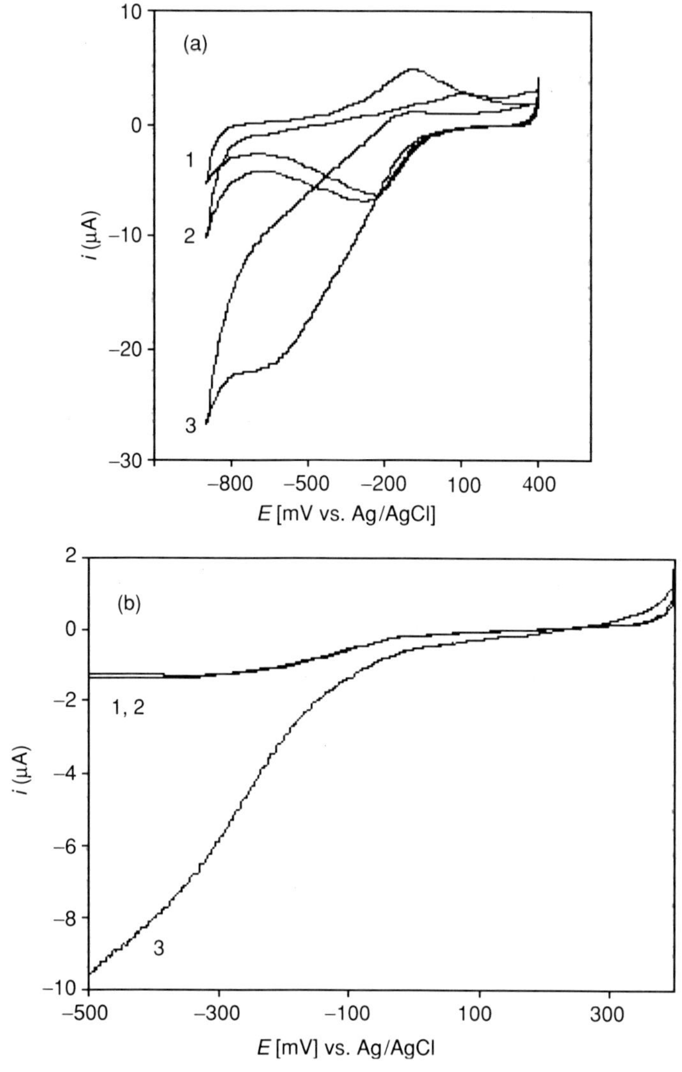

Figure 8.28. (a) Cyclic voltammograms (scan rate = 100 mV s^{-1}) and (b) hydrodynamic voltammograms (scan rate = 10 mV s^{-1}, electrode rotation rate = 400 rpm) of a vitreous carbon disk electrode modified with a film of poly[Mn(III)-**32**] in DCM + 0.1 mol L^{-1} Bu_4NBF_4 under argon (curve 1), in the presence of dissolved oxygen (curve 2) and in the presence of dissolved oxygen and 0.1 M benzoic anhydride (curve 3). Adapted from ref. [241]

Figure 8.29b shows the changes in the Q band of the UV–visible spectra of the reduced poly[Mn(II)-**32**] film (curve 1, same as the final spectrum in Figure 8.29a) upon addition of molecular oxygen to the DCM solution, without benzoic anhydride, and maintaining the reductive electrolysis (curve 2). It appears that O_2 provokes two modifications of the spectral features: the spontaneous decrease in the Mn(II) Q band that is also red shifted by 3 nm and the apparition of a

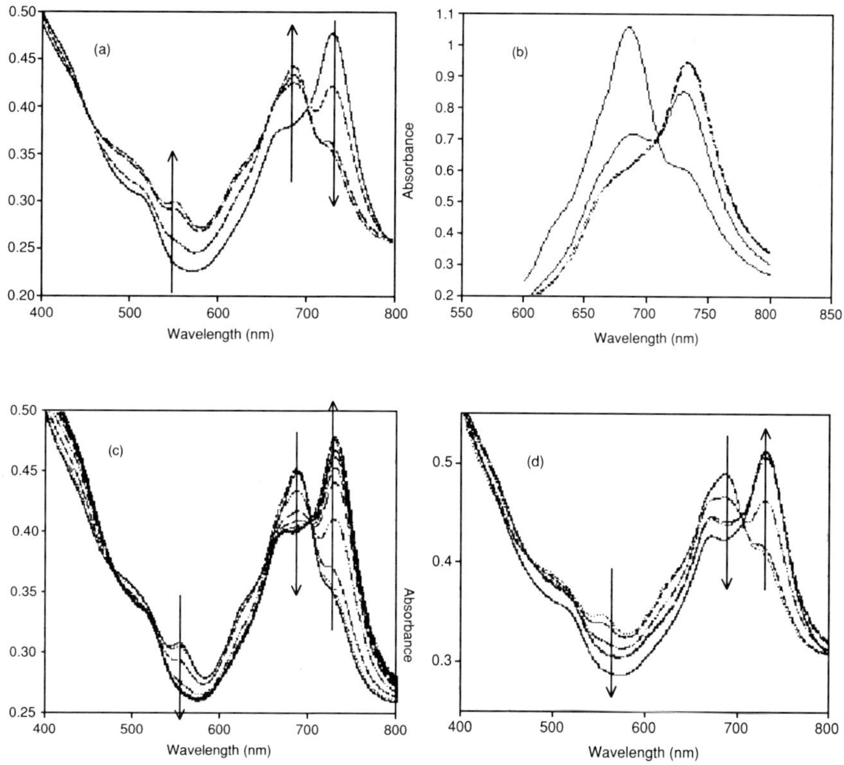

Figure 8.29. (a) UV–visible spectroscopic changes of a poly[Mn(III)-**32**] film on ITO in oxygen-free DCM + 0.1 mol L^{-1} Bu_4NBF_4 during constant potential reductive electrolysis at −0.7 V vs. Ag/AgCl. (b) UV–visible spectra of (curve 1) reduced poly[Mn(III)-**32**] film on ITO in oxygen-free DCM + 0.1 mol L^{-1} Bu_4NBF_4 obtained as in Figure 8.4a; (curve 2) = upon addition of oxygen without stopping the reductive electrolysis and (curve 3) = after stopping the electrolysis. Dashed curve shows the spetra of the native poly[Mn(III)-**32**]. (c) Evolution of the UV–visible spectra of the reduced poly[Mn(III)-**32**] film (obtained as in Figure 8.4a) on ITO in DCM + 0.1 mol L^{-1} Bu_4NBF_4 upon addition of anhydride benzoic and oxygen (electrolysis at −0.7 V was maintained during addition of oxygen) and (d) electrolysis was stopped during addition of oxygen. Adapted from ref.[241].

Mn(III)-like Q band blue shifted by 4 nm compared to the native Mn(III) one. When the reductive electrolysis is stopped (curve 3), there is a complete vanishing of the band at 688 nm to the detriment of the Mn(III)-like one at 729 nm. The Q band for the native Mn(III) film is reported for comparison (dashed curve). These observations may be explained by the formation of the doubly reduced superoxo intermediate (steps 1–3) that exhibits a Q band at 688 nm, while the formulated Mn(III)-superoxide adduct exhibits a Q band at 729 nm. When molecular oxygen was added to the reduced poly[Mn(II)-**32**] film in a DCM solution containing benzoic anhydride, with or without stopping the reductive electrolysis, it appears that there is a total restitution of the native film within 10 min, as it can be seen in

Figures 8.29c,d, respectively. No more changes in the spectra were observed when maintaining the electrolysis for more than 40 min, as far as molecular oxygen and benzoic anhydride were present in the solution. Indeed, by following the evolution of the measured current during the performed chronoamperometry, a decrease of the current during the reduction of Mn(III) to Mn(II) sites in the film was first reported. Then, the introduction of benzoic anhydride and molecular oxygen (air saturation) in the solution produced an increase in the measured cathodic current that reached a steady state indicating the establishment of the catalytic process. This can be explained by the occurrence of reactions 2–5, with steps 4 and 5 which are extremely fast compared to the UV–visible timescale.

Table 8.1 recapitulates the most significant examples of use of electropolymerized metallophthalopcyanines and metalloporphyrins for the activation of molecular oxygen. The reported fields of application are mainly devoted to electrosynthesis involving O_2, mechanistic electroanalysis of O_2 reduction and assessment of the feasibility of the O_2 activation.

4.2. Electropolymerized N_4-Macrocyclic Films as Electrochemical Sensors for Nitric Oxide in Solution

Detection of nitric oxide remains a challenge, especially because of the difficult dual requirements for specificity and sensitivity. Among the thousands of publications investigating the biosynthesis of NO and its physiological effects, only few percentage includes estimations of its levels. Most methods currently available for detecting NO are indirect and some of them prone to severe interferences and artefacts. The most commonly developed techniques are chemiluminescence, UV–visible spectroscopy, fluorescence, electron paramagnetic resonance spectroscopy (EPR) and electrochemistry. While each technique has certain individual advantages, all of them have disadvantages that limit their application. These disadvantages range from lack of sensitivity or specificity to interference from factors commonly present in biological systems. The choice of a technique is therefore dependent upon the desired applications. Exception made for the electrochemical techniques, most of the approaches use indirect methods for estimating endogenous NO, rely on measurements of secondary species such as nitrite, nitrate or NO-adducts.

The reality is that surface electrode modification is needed to make the ultramicroelectrode material selective for NO. Therefore, the design of modified electrode surfaces using organized layers is very attractive and provides the ideal strategy. In the general case, the chemical modification of electrode surfaces with polyelectrolytes and metal complex-based polymer films has expanded the scope of application of such designed electrodes and provided a lot of options for their use in various experimental conditions. In addition to their electrocatalytic applications, such electrodes showed a great promise for electroanalysis. As far as this aspect is concerned, substantial improvements in selectivity, sensitivity, versatility and reproducibility can be achieved.

Malinski and Taha reported in 1992 on a very interesting application of an electropolymerized nickel porphyrin film electrode for *in situ* amperometric detection of nitric oxide in biological systems[250]. The authors coated carbon fibres

Table 8.1. Applications of Electropolymerized N_4-Macrocycles in Electroassisted-Molecular Oxygen Activation

Complex	Metal	References
Porphyrin derivatives		
2	Co(III), Fe(III), Mn(III)	[31]
3	Co(III), Fe(III), Mn(III)	[31]
3	Fe(III), Co(II), 2H	[47]
4	Fe(III)	[48]
3	Co(III), Fe(III)	[49]
7	Mn(III)	[64]
18	Cu(II), Mn(III), Fe(III)	[67]
20, 21	Fe(III)–Cl, Mn(III)–Cl, Co(II)	[70]
28	2H, Fe(III)–Cl, Mn(III)–Cl	[77]
44	2H, Co(II), Fe(II), Mn(II)	[104]
44	Co(II)	[115]
44	Co(II)	[132, 133]
50	Zn(II), Co(II)	[119]
1, 62	Pt	[136]
44, 52, 63	2H, Fe, Cu(II), Ni(II)	[137]
65	Fe(III), Cu(II)	[140]
Phthalocyanine derivatives		
5	Fe(II), Co(II)	[51]
5	Fe(III), Co(II)	[52]
68	Co(II)	[133, 158, 159]
68	2H, Fe(III)	[160, 161]
68	2H, Fe(III)	[241]
Schiff base derivatives		
35	Co(II)	[85]
37	Cu(II)	[87]
82	Ni(II)	[192]
83	Co(II)	[193]
85	Fe(III)	[197]

(2–6 μm length; 0.8 μm diameter) with thin polymeric nickel porphyrin layers and Nafion® to minimize detection of nitrite. They successfully reported that the sensor, which can be operated in either the amperometric or voltammetric mode, can operate within a linear response up to 300 μM of NO and offers a detection limit of 10 nM (by using differential normal pulse amperometry). This important result allowed, at that time, the possible use of this nickel porphyrin and Nafion®-based sensor for most real-time assays in biomedical applications. By doing so, Malinski and coworkers reported the very pioneering and promising researches for successful *in vivo* detection of NO production from vascular endothelial cells of normotensive or hypertensive animals[251–254], rat cerebral tissue[255], platelets in whole human blood or washed in suspension[256], rat cerebral artery after ischaemia[257], human blood vessels of healthy volunteers during infusion of acetylcholine or bradykinin[258], vascular endothelial cells under sheer stress[259] etc. The intense research activity of this group during the last decade has been over-reviewed in a recent report[260].

Devynck and Bedioui's group provided the first example of a calibration of the electrochemical nickel porphyrin and Nafion® sensor[261] for the sensitive and selective determination of NO with concentrations ranging from 1.5 nM to 40 μM. This was achieved by using standards from NO gas-saturated stock solution and differential normal pulse amperometry technique at 0.75 V vs. saturated calomel reference electrode, in aerobic and anaerobic biological phosphate buffer solution model (pH = 7.4). The calculated detection limit at a signal-to-noise ratio of 3 was determined equal to 1.5 nM. Figure 8.30 illustrates an example of the obtained amperometric output. Since slight variations from one sensor to another can be observed, it is necessary to calibrate each electrode before use. Once calibrated, each sensor maintains its performance over a 24-hr period in the same medium. Finally, it is also important to mention that the sensitivity of the nickel porphyrin and Nafion®-based sensor depends significantly on the quality of the polymeric film, the purity of the starting monomer porphyrin and the potential range of polymerization.

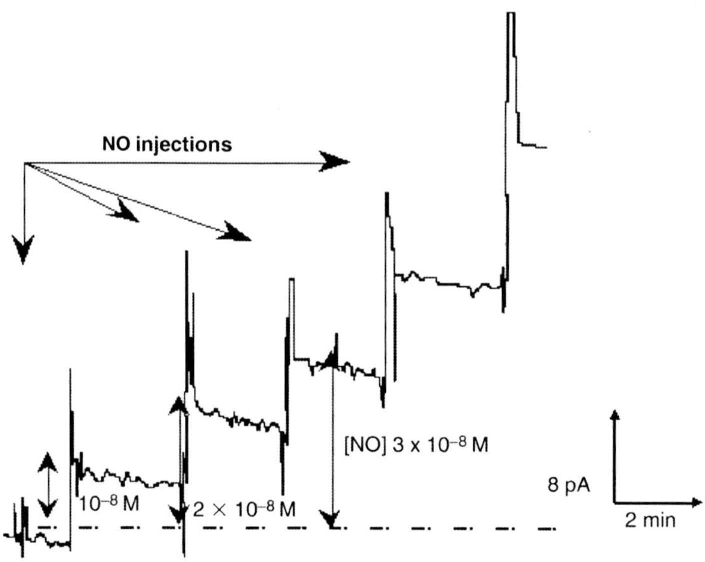

Figure 8.30. Differential pulse amperogram of a Nafion® and nickel porphyrin-coated carbon microfiber electrode in anaerobic PBS with different NO concentrations. Adapted from ref. [262].

Various interesting examples of new molecular materials are now being studied extensively with the aim of investigating the electrocatalytic oxidation of nitric oxide in aqueous solutions. These materials constitute a huge pool of possible molecular materials suitable for electrode surface modification to design new electrochemical microsensors. These approaches provide elegant ways to build up multilayered structures and architectures that result in NO-sensors with high performances. It is clear, from the various described examples reported in a recent review by Villeneuve and Bedioui[262], that designing microelectrode device with specific behaviour towards NO provides a unique approach for the direct and *in vivo* determination of this molecule in biological systems. The use of chemically modified microelectrodes provides an elegant way to build up multilayered structures and architectures that result in NO-sensors with high-performance characteristics (sensitivity and selectivity). Although the commercially available sensors are very attractive due to their "apparent facility of use", they do not offer any flexibility to ensure their adaptability to complex systems. Also, their claimed performances have to be systematically checked and validated before use, and their calibration carefully reconsidered with *true* standard NO solutions. It is important to give special attention to the electroformed nickel porphyrin (or phthalocyanine)-based sensor which appears to be a powerful tool. This is principally due to the fact that users can ingeniously adapt the configuration of this homemade sensor to every special case they have to investigate.

4.3. Miscellaneous Significant Examples

Table 8.2 recapitulates the most significant examples of use of electropolymerized N_4-macrocyclic-based films for the electrocatalysis and electroanalysis of several biologically relevant molecules. Among them, one should mention that Younathan *et al.*[113] reported the pioneering study which involved the application of electropolymerized iron protoporphyrin films to model the activity of the nitrite reductase enzymes. The authors have transferred the reactivity of an iron porphyrin to the electrode–solution interface by electropolymerizing the catalyst to form a thin polymeric film. Their approach was based on the fact that the polymeric structure may possess some of the properties of the biological catalyst. Thus, it was shown that the electropolymerized films of iron protoporphyrin-IX dimethylester are effective catalysts for the electroreduction of $HONO/NO_2^-$ or NO to N_2O, N_2, N_2OH and NH_3[113]. The yield of N_2 was enhanced in the compact environment of the redox polymer relative to that for comparable monomeric catalyst in homogeneous solution. The authors also suggested that the sensitivity of the polymeric catalyst to millimolar quantities of $HONO/NO_2^-$ could be used as an analytical tool for the detection of these species. In the same way, we have recently reported[91] the reactivity of a poly(pyrrole–alkylammonium) film electrode containing iron(III) porphyrin towards nitric oxide in acidic aqueous solution. The formation of the iron–nitrosyl intermediate within the polypyrrole film was proven by cyclic voltammetry and spectrophotometric measurements. Malinski *et al.*[106] also reported a similar approach by using an electropolymerized nickel porphyrin film as a new electrode material in the electrooxidation process of water, methanol, hydrazine to O_2, CO_2 and N_2, respectively. The authors

Table 8.2. Applications of N_4-Macrocycles in Electrocatalysis or Electroanalysis

Complex	Metal	Catalysis	Analysis	References
Porphyrin derivatives				
2	Co(II)	*N*-acetyl-L-cysteine		[90]
2	Mn(III), Zn(II), Fe(III)		NO	[91]
3	Co(II)	*N*-acetyl-L-cysteine		[90]
3	Mn(III), Zn(II), Fe(III)		NO	[91]
3	Fe(III)–OH, Fe(III)–Cl	NO_2^-, NO		[46]
4	Fe(III)	Cyclopent-2-ene-1-acetic acid		[48]
8	Co(II)	Hydrocarbons, olefins, thioacetamide derivative		[65]
9	Co(II)	Hydrocarbons, olefins, thioacetamide derivative		[65]
10	Mn(III)	Hydrocarbons, olefins, thioacetamide derivative		[65]
11	Co(II)	*N*-acetyl-L-cysteine		[90]
12, **13**	Mn(III)	Hydrocarbons, olefins, thioacetamide derivative		[65]
18	Cu(II), Mn(III), Fe(III)	Benzoic anhydride	CN^-	[67]
18	Co(II)		NO_2^-	[68]
19	Mn(II)		NO	[69]
20, **21**	Fe(III)–Cl, Mn(III)–Cl, Co(II)	CN^-		[70]
40, **41**	Co(II)	*N*-acetyl-L-cysteine		[90]
40, **41**	Mn(III), Zn(II), Fe(III)		NO	[91]
43	Fe(III)–OH, Fe(III)–Cl	NO_2^-, NO		[46]
45	Fe(III)–Cl	NO_2^-, NO		[109, 113]

Table 8.2. *Continued.*

Complex	Metal	Catalysis	Analysis	References
44	Co(II)		SCN^-, ClO_4^-, I^-, NO_2^-, salicylate, Br^-, Cl^-, HCO_3^-, PO_4^{3-}	[117]
44	Co(II)	2-Mercaptoethanol		[131]
47	2H, Co(II)	Hydrazine		[118]
48	2H, Mn, Fe, Co, Ni	Hydrazine		[118]
49	Cu, Fe, Pd Co, Ni, Mn, *t*-BuPy-Ru-CO	Hydrazine		[118]
50	Zn(II), Co(II)	Methanol, hydrazine		[119]
51	Co(II)		H^+	[58, 105]
52	Co(II)		I^-, Cl^-, Br^-, F^-, SCN^-, NO_3^-, SO_4^{2-}, ClO_4^-, $HCOO^-$, salicylate	[120]
53	Co(II)		H^+	[121]
60	2H, Fe(III), Zn(II)		Dopamine, ascorbic acid	[129]
53, 44	Co(II)	2-Mercaptoethanol		[130, 131]
61			I^-, Cl^-, F^-, SCN^-, NO_3^-, SO_4^{2-}, ClO_4^-, CH_3COO^-, salicylate	[134]
44	Co(II), Ni(II), Zn(II), Cu(II)		I^-, Br^-, Cl^-, SCN^-, NO_2^-, NO_3^-, ClO_4^-, OH^-, CH_3COO^-, SO_3^{2-}, $S_2O_3^{2-}$, $C_2O_4^{2-}$	[135]
44, 52, 63	2H, Fe, Cu(II), Ni(II)	NO_2^-		[138]
64	Ru(II)–CO	Ethyl diazoacetate		[139]
65	Fe(III), Cu(II)	H_2O_2	Oxidase	[140]
66	Fe(III)–Cl	NO_2^-, NO		[141]
67	Fe(III)–OH, Fe(III)–Cl	NO_2^-, NO		[46]
66	Zn(II), Co(II), Ni(II), Cu(II)		Triethylamine, ethanol, acetic acid, toluene	[143]
69	Ni(II)	Methanol, ethanol, propanol, butanol		[173, 174]

Table 8.2. *Continued.*

Complex	Metal	Catalysis	Analysis	References
69	Ni(II)	NO	NO	[175, 179]
69	Ni(II)	Formaldehyde		[176]
69	Ni(II)	Ascorbic acid	NADH	[177]
70, 71	Ni(II)		NO	[178]
71, 72	Ni(II)	Methanol, ethanol, hydrazine		[180]
Phthalocyanine derivative				
32	Co(II), Fe(II), Mn(III), Ni(II), Zn(II)	L-cysteine		[82]
68	Co(II)	2-Thiothiazolidine-4-carboxylic acid		[149]
68	Co(II), Cu(II)		Glucose	[150]
68	2H, Co(II)	Hydrazine		[152, 153]
68	Co(II), Ni(II), Zn(II)	NADH		[154]
68	Co(II)		2-Mercapto-ethanol, S^{2-}	[155]
68	Co(II)	2-Mercapto-ethanol		[146]
68	Ni(II)	Dopamine		[156, 517]
68	Co(II)		NO, NO_2^-	[162]
68	Co(II), Ni(II), Cu(II)		NO	[163, 164]
68	Ni(II)		NO_2^-	[165]
68	Co(II)	2-Mercapto-ethanol, 2-mercapto-ethane-sulphonic acid, reduced glutathione, L-cysteine		[166]
68	Co(II)	L-cysteine		[168]
68	Mn(II)		O=N–O–O$^-$	[169]
68	Co(II)	2-Mercapto-ethanol		[171]
68	Mn(III)	Glycine		[172]
73	Ni(II)	Dopamine		[182]
35	Co(II)	CO_2		[85]

Table 8.2. *Continued.*

Complex	Metal	Catalysis	Analysis	References
Salen-type N_4 derivatives				
36	Ni(II)	Iodoethane, methanol, hydrazine		[86]
38		Hydroquinone		[88]
39		Hydroquinone, catechol		[89]
74	Ni(II)	Arabinose, glucose, fructose, lactose, sucrose, maltose, maltotriose		[183]
75	Ni(II)	H_2O_2		[184]
76	Ni(II)	Methanol, ethanol, propanol, butanol		[185]
80	Cu(II)	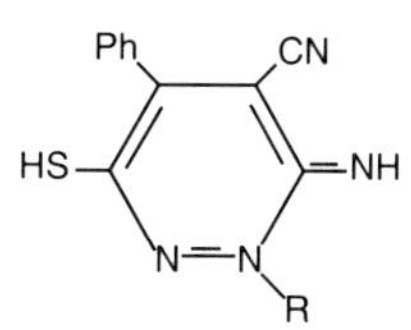 R = C_6H_5, $C_6H_4CH_2$, $C_6H_4NO_2$, or C_6H_4Cl		[189]
81		Hydroquinone oxalate anion,		[190, 191]
82	Ni(II)	L-ascorbic acid		[192]
85	Co(II), Fe(II), Cu(II), Mn(II)		NO	[196]
85	Fe(III)	H_2O_2		[197]
89		Catechol, hydroquinone		[199]

demonstrated high catalytic activity and good effects in the electrooxidation of methanol and hydrazine with overpotentials of 110 and 360 mV, respectively.

Several authors have also reported the potential use of electropolymerized metalloporphyrin films as new electrode materials for anion detection and pH measurements. For example, Daunert *et al.*[102] reported that anion-selective membrane electrodes can be prepared by electropolymerizing aniline-substituted cobalt tetraphenylporphyrin complexes onto a glassy carbon surface. It was shown that the resulting electrodes are highly selective towards the detection of thiocyanate and nitrite anions. The mechanism by which these metalloporphyrin electrode sensors operate is probably related to the well-known axial complexation capability of the central metal cation of the porphyrin towards different kind of anions and bases. The same authors also reported a detection limit of 5×10^{-7} M for these ion-selective electrodes which have lifetimes of more than 2 months[102].

Meyerhoff *et al.*[103] have reported the use of a non-metallated porphyrin film as an ion-selective electrode. They have shown that the electropolymerized films offer significant selectivity for the detection of iodide over a wide range of anions. The lifetime of these electrodes was estimated by analyzing some of their characteristics such as sensitivity and detection limits. The effect of pH on the detection of iodide by the film-supported electrodes was also established. For pH values from 3 to 8, there is relatively little effect of the pH on the electrochemical response of the electrodes. At pH values higher than 8, the electrode response is affected by the interference by hydroxide ions.

More recently, hydroxy-substituted cobalt tetraphenylporphyrins, electropolymerized onto an optically transparent electrode and adapted onto the tip of an optic fibre were used as sensing materials for the development of optic fibres as pH-sensors[105]. The pH values by this chemically modified optic fibre were monitored by spectrophotometric measurement of the Soret band wavelength for the supported porphyrin.

5. Conclusions

It appears from the studies on the electrochemical design of N_4-macrocycle complexes films described above that the idea of building new molecular devices having catalytic and analytical properties is realistic. The use of electropolymerization techniques provides an elegant way to build up multilayered structures and architectures. In the case of biomimetic catalysis, the polymer matrix, which can be compared to a protein, prevents supported metalloporphyrins degradation and controls the access of substrate molecules to the reactive centres. In most cases, the reactivity of the supported catalysts is largely enhanced. The electropolymerized metalloporphyrin films also appear to be powerful new analytical tools, especially for biosensoring. The significant developments in biosensor science and technology, such as the design of ultramicroelectrodes, should stimulate intense exploration of this field of research.

Acknowledgements

We would like to thank our co-workers whose names appear in the references of this paper. This research was partially supported by CNRS, EU (Cellsens program) and ECOS-Sud (C03E02).

References

1. Bein, T. (ed) (1992). *Supramolecular Architecture.* ACS Publishers, Washington, DC.
2. Morgan, B., D. Dolphin (1985). Synthesis and structure of biomimetic porphyrins. *Struct. Bond.* **64**, 115.
3. Momenteau, M. (1989). Aspect biologique du transport de l'oxygène: modélisation. *L'Actualité Chimique* **4**, 95–109.
4. Mansuy, D. and P. Battioni (1993). In J. Reedijk (ed), *Bioorganic Catalysis.* Marcel Dekker, New York, p. 395.
5. Shilov, A.E. Chemical-models of metallo-enzymes (1988). *J. Molecular Catal.* **47**, 351–362.
6. Cook, B.R., T.J. Reinert, and K.S. Suslick (1986). Shape selective alkane hydroxylation by metalloporphyrin catalysts. *J. Am. Chem. Soc.* **108**, 7281–7286.
7. Suslick, K., M. Cook, and J. Fox (1985). Shape-selective alkane hydroxylation. *J. Chem. Soc. Chem. Commun.* 580–582.
8. Collman, J.P., X. Zhang, V.J. Lee, U.S. Uffelman, and J.I. Brauman (1993). Regioselective and enantioselective epoxidation catalyzed by metalloporphyrins. *Science* **261**, 1404–1411 and references cited therein.
9. Herron, N. (1988). The selective partial oxidation of alkanes using zeolite based catalysts—phthalocyanine (PC) ship-in-bottle species. *J. Coord. Chem.* **19**, 25–38.
10. Herron, N. (1989). Towards Si-based life—zeolites as enzyme mimics. *Chemtech.* **19**(9), 542–548.
11. Mitchell, P.C.H. (1991). Zeolite-encapsulated metal-complexes—biomimetic catalysts. *Chem. Ind.* 308–311.
12. Mansuy, D. (1993). Activation of alkanes—the biomimetic approach. *Coord. Chem. Rev.* **125**, 129–141.
13. Karlin, K.D. (1993). Metalloenzymes, structural motifs, and inorganic models. *Science* **261**, 701–708.
14. Parton, R., D. De Vos, and P.A. Jacobs (1992). In E.G. Derouane (ed), *Zeolite Microporous Solids: Synthesis, Structure, and Reactivity.* Kluwer Academic Publishers, Amsterdam, p. 555.
15. Bedioui, F. (1995). Zeolite-encapsulated and clay-intercalated metal porphyrin, phthalocyanine and Schiff-base complexes as models for biomimetic oxidation catalysts—an overview. *Coord. Chem. Rev.* **144**, 39–68.
16. Kojima, T., Y. Urata, and H. Yonekura (1990). Synthesis of polymer with Fe(III) porphyrin rings. *J. Polym. Sci., Part C, Polym. Lett.* **28**, 129–132.
17. Traylor, T.G., Y.S. Byun, P.S. Traylor, P. Battioni, and D. Mansuy (1991). Polymeric polyhalogenated metalloporphyrin catalysts for hydroxylation of alkanes and epoxidation of alkenes. *J. Am. Chem. Soc.* **113**, 7821–7823.
18. Soloviera, A.B., E.A. Lukashova, A.V. Vorobiev, and S.F. Timashev (1991/1992). Polymer sulfofluoride films as carriers for metalloporphyrin catalysts. *React. Polym.* **16**, 9–17.

19. Linford, R.G. (ed) (1987). *Electrochemical Science and Technology of polymers.* Elsevier, London and New York.
20. Heinze, J. (1990). Electronically conducting polymers. *Top. Curr . Chem.* **152**, 1.
21. Merz, A. (1990). Chemically modified electrodes. *Topics Curr. Chem.* **152**, 49.
22. Murray, R.W. (ed) (1992). *Molecular Design of Electrode Surfaces.* Wiley Interscience, New York.
23. Gratzel, M. and K. Kalyanasundaram (ed) (1991). *Kinetics and Catalysis in Microheterogeneous Systems.* Marcel Dekker, New York.
24. Skotheim, T.A. (ed) (1986). *Handbook of Conducting Polymers*, Vols. 1 and 2. Marcel Dekker, New York.
25. Deronzier, A. and J.C. Moutet (1989). Functionalized polypyrroles—new molecular materials for electrocatalysis and related applications. *Acc. Chem. Res.* **22**, 249–255.
26. Curran, D., J. Grimshaw, and S.D. Perera (1991). Poly(pyrrole) as a support for electrocatalytic materials. *Chem. Soc. Rev.* **20**, 391–404.
27. Roncali, J. (1992). Conjugated poly(thiophenes)—synthesis, functionalization, and applications. *Chem. Rev.* **92**, 711–738.
28. Okabayashi, K., O. Ikeda, and H. Tamura (1983). Electrochemical doping with *meso*-tetrakis(4-sulphonatophenyl)-porphyrincobalt of a polypyrrole film electrode. *J. Chem. Soc., Chem. Commun.* 684–685.
29. Bedioui, F., C. Bongars, C. Hinnen, C. Bied-Charreton, and J. Devynck (1985). Insitu characterization of redox properties of *meso*tetrakis (4-carboxyphenyl) porphine cobalt in a polypyrrole film electrode using the electroreflectance technique. *Bull. Soc. Chim. Fr.* 679–682.
30. Ikeda, O., O. Okabayashi, N. Yoshida, and H. Tamura (1985). Spectroelectrochemical study of oxygen reduction at metalloporphyrin-doped polypyrrole film electrodes. *J. Electroanal. Chem.* **191**, 157–174.
31. Bedioui, F., C. Bongars, J. Devynck, C. Bied-Charreton, and C. Hinnen (1986). Metalloporphyrin polypyrrole film electrode—characterization and catalytic application. *J. Electroanal. Chem.* **207**, 87–99.
32. Bull, R.A., F.R. Fan, and A.J. Bard (1984). Polymer-films on electrodes .13. Incorporation of catalysts into electronically conductive polymers—Iron phthalocyanine in polypyrrole. *J. Electrochem. Soc.* **131**, 687.
33. Skotheim, T., M. Velasquez Rosenthal, and C.A. Linkous (1985). Polypyrrole complexed with cobalt phthalocyanine—A conducting polymer which is stable to air and moisture. *J. Chem. Soc., Chem. Commun.* 612–613.
34. Mizutani, F., S.I. Lijima, Y. Tanabe, and K. Tsuda (1985). Conducting polymer film electrodes with immobilized catalytic sites. *J. Chem. Soc., Chem. Commun.* 1728–1729.
35. Velasquez Rosenthal, M., T.A. Skotheim, and C.A. Linkous (1986). Polypyrrole phthalocyanine. *Synth. Met.* **15**, 219–227.
36. Elzing, A., A. Van Der Putten, W. Visscher, and E. Barendrecht (1987). The mechanism of oxygen reduction at iron tetrasulfonato-phthalocyanine incorporated in polypyrrole. *J. Electroanal. Chem.* **233**, 113–123.
37. Jiang, R. and S. Dong (1988). Research on chemically modified electrodes—catalytic reduction of dioxygen at a cobalt phthalocyanine-doped polyaniline film electrode. *J. Electroanal. Chem.* **246**, 101–117.
38. Choi, C.S. and H. Tachikawa (1990). Electrochemical behavior and characterization of polypyrrole copper phthalocyanine tetrasulfonate thin-film—cyclic voltammetry and in situ Raman spectroscopic investigation. *J. Am. Chem. Soc.* **112**, 1757–1768.
39. Walton, D.J., D.M. Hadingham, C.E. Hall, I.V.F. Viney, and A. Chyla (1991). Competitive doping in polypyrrole. *Synth. Met.* **41–43**, 295–299.

40. Saunders, B.R., K.S. Murray, and R.J. Fleming (1992). Physical properties of polypyrrole films containing sulfonated metallophthalocyanine anions. *Synth. Met.* **47**, 167–178.
41. Kanazawa, K.K., AF. Diaz, W.D. Gill, P. Grant, G.B. Street, G.P. Gardini, and J.F. Kwak (1980). Polypyrrole—an electrochemically synthesized conducting organic polymer. *Synth. Met.* **1**, 329–336.
42. Diaz, A.F., J.M. Vasquez Vallejo, and A. Martinez Duran (1981). [Pt] polypyrrole—a new organic electrode material. *IBM J. Res. Develop.* **25**, 42–50.
43. Noufi, R., D. Trench, and L.F. Warren (1981). Protection of semiconductor photoanodes with photoelectrochemically generated polypyrrole films. *J. Electrochem. Soc.* **128**, 2596–2599.
44. Zinger, B. and L.L. Miller (1984). Timed release of chemicals from polypyrrole films. *J. Am. Chem. Soc.* **106**, 6861–6863.
45. Miller, L.L., B. Zinger, and Q.N. Zhou (1987). Electrically controlled release of $Fe(CN)_6^{(4-)}$ from polypyrrole. *J. Am. Chem. Soc.* **109**, 2267–2272.
46. Bedioui, F., S. Trevin, V. Albin, M. Guadalupe Gomez Villegas, and J. Devynck (1997). Design and characterisation of chemically modified electrodes with iron(III) porphyrinic-based polymers: Study of their reactivity toward nitrites and nitric oxide in aqueous solution. *Anal. Chim. Acta* **341**, 177–185.
47. Song, E.-H. and W.-K. Paik (1998). Polypyrrole doped with sulfonate derivatives of metalloporphyrins: Use in cathodic reduction of oxygen in acidic and basic solutions. *Bull. Korean Chem. Soc.* **19**(2), 183–188.
48. Liu, M.-H. and Y.O. Su (1999). Electrocatalytic reactions by an iron porphyrin/polypyrrole modified electrode monitored by electrochemical quartz crystal microbalance. *J. Chin. Chem. Soc.* **46**, 115–119.
49. Song, E.H., I.-H. Yeo, and W.-K. Paik (1999). Poly(3,4-ethylenedioxythiophene) electrodes doped with anionic metalloporphyrins. *Bull. Korean Chem. Soc.* **20**(11), 1303–1308.
50. Paik, W., I.-H. Yeo, H. Suh, Y. Kim, and E. Song (2000). Ion transport in conducting polymers doped with electroactive anions examined by EQCM. *Electrochim. Acta* **45**, 3833–3840.
51. Elzing, A., A. Van Der Putten, W. Visscher, and E. Barendrecht (1987). The mechanism of oxygen reduction at iron tetrasulfonato-phthalocyanine incorporated in polypyrrole. *J. Electroanal. Chem.* **233**, 113–123.
52. Coutanceau, C., A. El Hourch, P. Crouigneau, J.M. Leger, and C. Lamy (1995). Conducting polymer electrodes modified by metal tetrasulfonated phthalocyanines: Preparation and electrocatalytic behaviour towards dioxygen reduction in acid medium. *Electrochim. Acta* **40**(17), 2739–2748.
53. Čabala, R., J. Škarda, and K. Potje-Kamloth (2000). Spectroscopic investigations of thermal treatment of doped polypyrrole. *Phys. Chem. Chem. Phys.* **2**, 3283–3291.
54. Mikalo, R.P., G. Appel, P. Hoffmann, and D. Schmeißer (2001). Band bending in doped conducting polypyrrole: Interaction with silver. *Synth. Met.* **122**, 249–261.
55. Aguilar-Hernández, J. and K. Potje-Kamloth (2001). Evaluation of the electrical conductivity of polypyrrole polymer composites. *J. Phys. D: Appl. Phys.* **34**, 1700–1711.
56. Nguyen Van, C. and K. Potje-Kamloth (2001). Electrical and NO_x gas sensing properties of metallophthalocyanine-doped polypyrrole/silicon heterojunctions. *Thin Solid Films* **392**, 113–121.
57. Deronzier, A. and J.M. Latour (1987). A poly(pyrrole nickel(II) pyridiniumporphyrin)-modified electrode. *J. Electroanal. Chem.* **224**, 295–301.
58. Bettelheim, A., B.A. White, S.A. Raybuck, and R.W. Murray (1987). Electrochemical polymerization of amino-substituted, pyrrole-substituted, and hydroxy-substituted tetraphenylporphyrins. *Inorg. Chem.* **26**, 1009–1017.

59. Bedioui, F., A. Merino, J. Devynck, C.E. Mestres, and C. Bied-Charreton (1988). Poly(pyrrole-manganese tetraphenylporphyrin) film electrodes in acetonitrile solution. *J. Electroanal. Chem.* **239**, 433–439.
60. Armengaud, C., P. Moisy, F. Bedioui, J. Devynck, and C. Bied-Charreton (1990). Electrochemistry of conducting polypyrrole films containing cobalt porphyrin. *J. Electroanal. Chem.* **277**, 197–211.
61. Bedioui, F., M. Voisin, J. Devynck, and C. Bied-Charreton (1991). Electrochemistry of conducting polypyrrole films containing cobalt porphyrin. 2. New developments and inclusion of metallic aggregates in the coordination polymer. *J. Electroanal. Chem.* **297**, 257–269.
62. Ramachandraiah, G., F. Bedioui, J. Devynck, M. Serrar, and C. Bied-Charreton (1991). Electrochemical preparation and characterization of zinc porphyrin-coated electrodes. *J. Electroanal. Chem.* **319**, 395–402.
63. Deronzier, A., R. Devaux, D. Limozin, and J.M. Latour (1992). Poly(pyrrole-metallotetraphenylporphyrin)-modified electrodes. 2. *J. Electroanal. Chem.* **324**, 325–337.
64. Deronzier, A. (1996). Des électrodes modifiées par des complexes de métaux de transitions. Comment et pour quoi faire ? *J. Chim. Phys.* **93**, 611–619.
65. Bedioui, F., J. Devynck, and C. Bied-Charreton (1996). Electropolymerized manganese porphyrin films as catalytic electrode materials for biomimetic oxidations with molecular oxygen. *J. Mol. Catalysis* **113**, 3–11.
66. Carvalho de Medeiros, M.A., S. Cosnier, A. Deronzier, and J.-C. Moutet (1996). Synthesis and characterisation of a new series of nickel(II) *meso*-tetrakis-(polyfluorophenyl)porphyrins functionalized by pyrrole groups and their electropolymerized films. *Inorg. Chem.* **35**, 2659–2664.
67. Cosnier, S., C. Gondran, R. Wessel, F.-P. Montforts, and M. Wedel (2000). Poly (pyrrole-metallodeuteroporphyrin)electrodes: Towards electrochemical biomimetic devices. *J. Electroanal. Chem.* **488**, 83–91.
68. Cosnier, S., C. Gondran, R. Wessel, F.-P. Montforts, and M. Wedel (2003). A poly (pyrrole-cobalt(II)deuteroporphyrin) electrode for the potentiometric determination of nitrite. *Sensors* **3**, 213–222.
69. Diab, N. and W. Schuhmann (2001). Electropolymerized manganese porphyrin/ polypyrrole films as catalytic surfaces for the oxidation of nitric oxide. *Electrochim. Acta* **47**, 265–273.
70. Wedel, M., A. Walter, and F.-P. Montforts (2001). Synthesis of metalloporphyrins and metallochlorins for immobilization on electrode surfaces. *Eur. J. Org. Chem.* 1681–1687.
71. Cosnier, S., C. Gondran, K. Gorgy, R. Wessel, F.-P. Montforts, and M. Wedel (2002). Electrogeneration and characterisation of a poly(pyrrole-nickel (II) chlorin) electrode. *Electrochem. Commun.* **4**, 426–430.
72. Shimidzu, T. (1995). Functionalized conjugating polymers: From molecule and ions transporting membranes to advanced electronic and photonic materials. *Pure & Appl. Chem.* **67**(12), 2039–2046.
73. Shimidzu, T., T. Iyoda, and H. Segawa (1996). Advanced materials by functionalization of conjugated polymers. *Macromol. Symp.* **101**, 207–218.
74. Shimidzu, T. (1996). Approaches to polymer superlattice and molecular device. *Macromol. Symp.* **104**, 127–135.
75. Shimidzu, T. (1996). Approaches to ultimate functional polymer materials. *Polym. Adv. Technol.* **8**, 275–280.
76. Schäferling, M. and P. Bäuerle (1999). Synthesis and properties of porphyrin-functionalized poly(bithiophenes). *Syn. Met.* **101**, 38–39.

77. Schäferling, M. and P. Bäuerle (2001). Electrochemical properties of porphyrin-functionalized polythiophenes. *Syn. Met.* **119**, 289–290.
78. Too, C.O., and G.G. Wallace, A.K. Burell, G.E. Collis, D.L. Officer, E.W. Boge, S.G. Brodie, and E.J. Evans (2001). Photovoltaic devices based on polythiophenes and substituted polythiophenes. *Syn. Met.* **123**, 53–60.
79. Yamashita, K., M. Ikeda, M. Takeuchi, and S. Shinkai (2003). Electropolymerization of bithienyl-appended cerium(III) triple decker porphyrin complex. *Chem. Lett.* **32**(3), 264–265.
80. Trombach, N., O. Hild, D. Schlettwein, and D. Wöhrle (2002). Synthesis and electropolymerisation of pyrrol-1-yl substituted phthalocyanines. *J. Mater. Chem.* **12**, 879–885.
81. Wöhrle, D., O. Hild, N. Trombach, R. Benters, G. Schnurpfeil, and O. Suvorova (2002). Conjugated polymeric phthalocyanines by poly-cyclotetramerization and electropolymerization. *Macromol. Symp.* **186**, 99–104.
82. Obirai, J., N.P. Rodrigues, F. Bedioui, and T. Nyokong (2003). Synthesis, spectral and electrochemical properties of a new family of pyrrole substituted cobalt, iron, manganese, nickel and zinc phthalocyanine complexes. *J. Porphyrins Phthalocyanines* **7**, 508–520.
83. King, G., S.J. Higgins, and N. Price (1992). Novel conducting polymers incorporating covalently bound metal-tetraazamacrocycle complexes. *Analyst* **117**, 1243–1246.
84. Higgins, S.J., T.J. Pounds, and P.A. Christensen (2001). Synthesis and electro (co)polymerization of novel thiophene- and 2,2: $5'$, $2''$-terthiophene-functionalized metal-tetraazamacrocycle complexes, and electrochemical and spectroelectrochemical characterization of the resulting polythiophenes. *J. Mater. Chem.* **11**, 2253–2261.
85. Losada, J., I. Del Peso, L. Beyer, J. Hartung, V. Fernandez, and M. Möius (1995). Electrocatalytic reduction of O_2 and CO_2 with electropolymerised films of polypyrrole cobalt(II)Schiff-base complexes. *J. Electroanal. Chem.* **398**, 89–93.
86. Losada, J., I. Del Peso, and L. Beyer (1998). Redox and electrocatalytic properties of electrodes modified by films of polypyrrole nickel(II) Schiff-base complexes. *J. Electroanal. Chem.* **447**, 147–154.
87. Losada, J., I. Del Peso, and L. Beyer (2001). Electrochemical and spectroelectrochemical properties of copper(II) Schiff-base complexes. *Inorg. Chim. Acta* **321**, 107–115.
88. Diab, A.S., A.A. Hathoot, M. Abdel-Azzem, and A. Merz (2000). Preparation of a novel conducting polymer by electropolymerization of thiophenylidine 8-naphthylamine Schiff-base. *Eur. Polym. J.* **36**, 1959–1965.
89. Hathoot, A.A. (2000). Electro-oxidative, polymerization of Schiff-base of 1,8-diaminonaphthaline and 3-acetylthiophene. I. Preparation and study the redox behaviour of the resulting polymer. *Eur. Polym. J.* **36**, 1063–1071.
90. De Gregori, I., M. Carrier, A. Deronzier, J.C. Moutet, F. Bedioui, and J. Devynck (1992). Incorporation of anionic cobalt porphyrin by anion-exchange into polypyrrole films containing alkylammonium groups. *J. Chem. Soc. Faraday Trans.* **88**, 1567–1572.
91. Bedioui, F., Y. Bouhier, C. Sorel, J. Devynck, L. Coche-Guerrente, A. Deronzier, and J.C. Moutet (1993). Incorporation of anionic metalloporphyrins into poly(pyrrole-alkylammonium) films. 2. Characterization of the reactivity of the iron(III) porphyrinic-based polymer. *Electrochim. Acta* **38**, 2485–2491.
92. Cosnier, S., A. Deronzier, J.C. Moutet, and J.F. Roland (1989). Alkylammonium and pyridinium group-containing polypyrroles, a new class of electronically conducting anion-exchange polymers. *J. Electroanal. Chem.* **271**, 69–81.
93. Moutet, J.C. and C.J. Pickett (1989). Iron-sulfur clusters in ionic polymers on electrodes. *J. Chem. Soc. Chem. Commun.* **3**, 188–190.

94. Keita, B., D. Bouaziz, L. Nadjo, and A. Deronzier (1990). Surface functionalization with oxometallates entrapped in polymeric matrices. 2. Substituted pyrrole-based ion-exchange polymers. *J. Electroanal. Chem.* **279**, 187–203.
95. Coche-Guerrente, L., A. Deronzier, B. Galland, P. Labbé, J.C. Moutet, and G. Reverdy (1991). Immobilization of redox anions in poly(amphiphilic pyrrolylalkylammonium) using a simple and monomer-saving one-step procedure in pure water-electrolyte. *J. Chem. Soc. Chem. Commun.* 386–388.
96. De Olivera, I.M., J.C. Moutet, and S. Hamar-Thibault (1992). Electrocatalytic hydrogenation activity of palladium and rhodium microparticles dispersed in alkylammonium-substituted and pyridinium-substituted polypyrrole films. *J. Mater. Chem.* **2**, 167–173.
97. Stéphan, O., M. Carrier, M. Le Bail, A. Deronzier, and J.-C. Moutet (1995). Ion binding by poly[4-(pyrrol-1-ylmethyl)benzoic acid] thin films. *J. Chem. Soc. Faraday Trans.* **91**(8), 1241–1246.
98. D'Souza, F., Y.-Ying Hsieh, H. Wickman, and W. Kutner (1997). β-cyclodextrin and carboxymethylated β-cyclodextrin polymer film modified electrodes, hosting cobalt porphyrins, as sensors for electrocatalytic determination of oxygen dissolved in solution. *Electroanalysis* **9**(14), 1093–1101.
99. White, B.A. and R.W. Murray (1985). Electroactive porphyrin films from electropolymerized metallotetra(ortho-aminophenyl)porphyrins. *J. Electroanal. Chem.* **189**, 345–352.
100. Macor, K.A., Y.O. Su, L.A. Miller, and T.G. Spiro (1987). Electrochemical and resonance Raman-spectroscopic characterization of polyaniline and polyaniline metalloporphyrin electrode films. *Inorg. Chem.* **26**, 2594–2598.
101. White, B.A. and R.W. Murray (1987). Kinetics of electron self-exchange reactions between metalloporphyrin sites in submicrometer polymeric films on electrodes. *J. Am. Chem. Soc.* **109**, 2576–2581.
102. Daunert, S., S. Wallace, A. Florido, and L.G. Bachas (1991). Anion-selective electrodes based on electropolymerized porphyrin films. *Anal. Chem.* **63**, 1676–1679.
103. Kliza, D.M. and M.E. Meyerhoff (1992). Potentiometric anion response of poly(tetrakis(p-aminophenyl)porphyrin) film-modified electrodes. *Electroanalysis* **4**, 841–849.
104. Hayon, J., A. Raveh, and A. Bettelheim (1993). Electrocatalytic properties of chemically polymerized films of cobalt, iron and manganese tetrakis(o-aminophenyl)porphyrins. *J. Electroanal. Chem.* **359**, 209–221.
105. Blair, T.L., J.R. Allen, S. Daunert, and L.G. Bachas (1993). Potentiometric and fiber optic sensors for pH based on an electropolymerized cobalt porphyrin. *Anal. Chem.* **65**, 2155–2158.
106. Malinski, T., A. Ciszewski, J. Bennett, J.R. Fish, and L. Czuchajowski (1991). Characterization of conductive polymeric nickel(II) tetrakis(3-methoxy-4-hydroxyphenyl)porphyrin as an anodic material for electrocatalysis. *J. Electrochem. Soc.* **138**, 2008–2015.
107. Malinski, T., A. Ciszewski, J. Fish, E. Kubaszewski, and L. Czuchajowski (1992). Conductive polymeric Cu(II) tetrakis(3-methoxy-4-hydroxyphenyl)porphyrin as a photosensitizer in a photoelectrochemical cell. *Adv. Mater.* **4**, 354–357.
108. Fish, J.R., E. Kubaszewski, A. Peat, T. Malinski, J. Kaczor, P. Kus, and L. Czuchajowski (1992). Synthesis and electrochemistry of conductive copolymeric porphyrins. *Chem. Mater.* **4**, 795–803.
109. Macor, K.A. and T.G. Spiro (1983). Porphyrin electrode films prepared by electrooxidation of metalloprotoporphyrins. *J. Am. Chem. Soc.* **105**, 5601–5607.
110. Macor, K.A. and T.G. Spiro (1984). Oxidative electrochemistry of electropolymerized metalloprotoporphyrin films. *J. Electroanal. Chem.* **163**, 223–236.

111. Basu, J. and K.K. Rohatgi-Mukherjee (1988). Photoelectrochemical studies of metalloporphyrins. *Photochem. Photobiol.* **48**, 417–422.
112. Basu, J. and K.K. Rohatgi-Mukherjee (1991). Photoelectrochemical characterization of porphyrin-coated electrodes. *Solar Energ. Mater.* **21**, 317–325.
113. Younathan, J., K.S. Wood, and T.J. Meyer (1992). Electrocatalytic reduction of nitrite and nitrosyl by iron(III) protoporphyrin-IX dimethyl ester immobilized in an electropolymerized film. *Inorg. Chem.* **31**, 3280–3285.
114. Czuchajowski, L., J.E. Bennett, S. Goszczynski, D.E. Wheeler, A.K. Wisor, and T. Malinski (1989). *meso*-[2.2]paracyclophanyltriphenylporphyrin—electronic consequences of linking paracyclophane to porphyrin. *J. Am. Chem. Soc.* **111**, 607–616.
115. Oyama, N., T. Ohsaka, M. Mizunuma, and M. Kobayashi (1988). Electropolymerized cobalt tetrakis(*ortho*-aminophenyl) porphyrin film mediated enzyme electrode for amperometric determination of glucose. *Anal. Chem.* **60**(22), 2536–2537.
116. Pressprich, K.A., S.G. Maybury, R.E. Thomas, R.W. Linton, E.A. Irene, and R.W. Murray (1989). Molecular sieving by electropolymerized porphyrin films only a few monolayers thick. *J. Phys. Chem.* **93**, 5568–5574.
117. Daunert, S., S. Wallace, A. Florido, and L.G. Bachas (1991). Anion-selective electrodes based on electropolymerized porphyrin films. *Anal. Chem.* **63**, 1676–1679.
118. Bennett, J.E. and T. Malinski (1991). Conductive polymeric porphyrin films: Application in the electrocatalytic oxidation of hydrazine. *Chem. Mater.* **3**, 490–495.
119. Fish, J.R., E. Kubaszewski, A. Peat, T. Malinski, J. Kaczor, P. Kus, and L. Czuchajowski (1992). Synthesis and electrochemistry of conductive copolymeric porphyrins. *Chem. Mater.* **4**, 795–803.
120. Kliza, D.M. and M.E. Meyerhoff (1992). Potentiometric anion response of poly(tetrakis(p-aminophenyl)porphyrin) film-modified electrodes. *Electroanalysis* **4**, 841–849.
121. Blair, T.L., J.R. Allen, S. Daunert, and L.G. Bachas (1993). Potentiometric and fiber optic sensors for pH based on an electropolymerised cobalt porphyrin. *Anal. Chem.* **65**, 2155–2158.
122. Marée, C.H.M., A.M. Vredenberg, and F.H.P.M. Habraken (1996). Mechanism of MeV induced hydrogen depletion from organic layers. *Mat. Chem. Phys.* **46**, 198–205.
123. Marée, C.H.M., A. Kleinpenning, A.M. Vredenberg, and F.H.P.M. Habraken (1996). Ion beam analysis of electropolymerized porphyrin layers. *Nucl. Instr. Meth. Phys. Res. B* **118**, 301–306.
124. Marée, C.H.M., S.J. Roosendaal, T.J. Savenije, R.E.I. Schropp, T.J. Schaafsma, and F.H.P.M. Habraken (1996). Photovoltaic effects in porphyrin polymer films and heterojunctions. *J. Appl. Phys.* **80**(6), 3381–3389.
125. Schaafsma, T.J., T.J. Savenije, R.B.M. Koehorst, F.J. Vergeldt, and J. Wienke (1997). Luminescent properties and photogeneration of charge carriers in porphyrin dimers and molecular photo-diodes. *J. Lumin.* **72–74**, 81–82.
126. Savenije, T.J., E. Moons, G.K. Boschloo, A. Goossens, and T.J. Schaafsma (1997). Photogeneration and transport of charge carriers in a porphyrin p/n heterojunction. *Phys. Rev. B* **55**(15), 9685–9692.
127. Savenije, T.J., R.B. Koehorst, and T.J. Schaafsma (1997). Spectroelectrochemical measurement of charge transport properties of electropolymerized tetrakis (hydroxyphenyl)porphyrins. *J. Phys. Chem. B* **101**, 720–725.
128. Balasubramaniam, E. and P. Natarajan (1997). Photoelectrochemical studies on electrodes coated with polymer-bound porphyrin systems. *J. Photochem. Photobiol. A: Chem.* **109**, 39–46.

129. Duong, B., R. Arechabaleta, and N.J. Tao (1998). In situ AFM/STM characterization of porphyrin electrode films for electrochemical detection of neurotransmitters. *J. Electroanal. Chem.* **447**, 63–69.
130. Griveau, S. and F. Bedioui (2001). Electrocatalytic oxidation of 2-mercaptoethanol by electropolymerized cobalt porphyrin film on vitreous carbon electrodes. *Electroanalysis* **13**(3), 253–256.
131. Griveau, S., V. Albin, T. Pauporté, J.H. Zagal, and F. Bedioui (2002). Comparative study of electropolymerized cobalt porphyrin and phthalocyanine based films for the electrochemical activation of thiols. *J. Mater. Chem.* **12**, 225–232.
132. El Mouahid, O., C. Coutanceau, E.M. Belgsir, P. Crouigneau, J.M. Léger, and C. Lamy (1997). Electrocatalytic reduction of dioxygen at macrocycle conducting polymer electrodes in acid media. *J. Electroanal. Chem.* **426**, 117–123.
133. El Mouahid, O., A. Rakotondrainibe, P. Crouigneau, J.M. Léger, and C. Lamy (1998). A UV-visible study of the electropolymerization of CoTAPP at vitreous carbon and investigation of its catalytic activity towards the electroreduction of dioxygen. *J. Electroanal. Chem.* **455**, 209–222.
134. Volf, R., T.V. Shishkanova, P. Matejka, M. Hamplavá, and V. Král (1999). Potentiometric anion response of poly(5,15-bis(2-aminophenyl)porphyrin) electropolymerized electrodes. *Anal. Chim. Acta* **381**, 197–205.
135. Bruti, E.M., M. Giannetto, G. Mori, and R. Seeber (1999). Electropolymerization of tetrakis(*o*-aminophenyl)porphyrin and relevant transition metal complexes from aqueous solution. The resulting modified electrodes as potentiometric sensors. *Electroanalysis* **11**(8), 565–572.
136. Holmes-Smith, A.S., A. Hamill, M. Campbell, and M. Uttamlal (1999). Electropolymerised platinum porphyrin polymers for dissolved oxygen sensing. *Analyst* **124**, 1463–1466.
137. Cornejo, G., G. Ramírez, M. Villagrán, J. Costamagna, E. Trollund, and M.J. Aguirre (2003). Electropolymerization and electrocatalytic behavior of electrodes modified with Fe and non-metalled tetraaminophenylporphyrins: Effect of the position of the amino group on the ligand. *J. Chil. Chem. Soc.* **48**(1), 49–55.
138. Armijo, F., E. Trollund, M. Reina, M.C. Arévalo, and M.J. Aguirre (2003). Preparation and characterization of electrodes modified with metalloporphyrins. Application to reduction of nitrite. *Collect. Czech. Chem. Commun.* **68**, 1723–1735.
139. Poriel, C., Y. Ferrand, P. Le Maux, C. Paul, J. Rault-Berthelot, and G. Simonneaux (2003). Poly(ruthenium carbonyl spirobifluorenylporphyrin): A new polymer used as a catalytic device for carbene transfer. *Chem. Commun.* 2308–2309.
140. Vago, J.M., V. Campo Dall'Orto, E. Forzani, J. Hurst, and I.N. Rezzano (2003). New bimetallic porphyrin film: An electrocatalytic transducer for hydrogen peroxide reduction, application to first-generation oxidase-based biosensors. *Sens. Actuators B* **96**, 407–412.
141. Younathan, J.N., K.S. Wood, and T.J. Meyer (1992). Electrocatalytic reduction of nitrite and nitrosyl by iron(III) protoporphyrin IX dimethyl ester immobilized in an electropolymerized film. *Inorg. Chem.* **31**, 3280–3285.
142. Kuester, S.N., M.M. McGuire, and S.M. Drew (1998). Electrochemically initiated polymerization of zinc(II)5-vinyl-10,15,20-triphenyl porphyrin. *J. Electroanal. Chem.* **452**, 13–18.
143. Paolesse, R., C. Di Natale, V. Campo Dall'Orto, A. Macagnano, A. Angelaccio, N. Motta, A. Sgarlata, J. Hurst, I. Rezzano, M. Mascini, and A. D'Amico (1999). Porphyrin thin films coated quartz crystal microbalances prepared by electropolymerisation technique. *Thin Solid Films* **354**, 245–250.

144. Li, H. and T.F. Guarr (1989). Formation of electronically conductive thin films of metal phthalocyanines via electropolymerisation. *J. Chem. Soc., Chem. Commun.* 832–834.
145. Li, H. and T.F. Guarr (1999). Reversible electrochromism in polymeric metal phthalocyanine thin films. *J. Electroanal. Chem.* **297**, 169–183.
146. Griveau, S., J. Pavez, J.H. Zagal, and F. Bedioui (2001). Electro-oxidation of 2-mercaptoethanol on adsorbed monomeric and electropolymerized cobalt tetra-aminophthalocyanine films. Effect of film thickness. *J. Electroanal. Chem.* **497**, 75–83.
147. Moore, D.J. and T.F. Guarr (1991). Electrochromic properties of electrodeposited lutetium diphthalocyanine thin films. *J. Electroanal. Chem.* **314**, 313–321.
148. Mortimer, R.J. (1999). Organic electrochromic materials. *Electrochim. Acta* **44**, 2971–2981.
149. Ciucu, A. and R.P. Baldwin (1992). Determination of 2-thiothiazolidine-4-carboxylic acid in urine by liquid chromatography with electrochemical detection. *Electroanalysis* **4**, 515–519.
150. Sun, Z. and H. Tachikawa (1992). Enzyme-based bilayer conducting polymer electrodes consisting of polymetallophthalocyanines and polypyrrole-glucose oxidase thin films. *Anal. Chem.* **64**, 1112–1117.
151. Mu, X.H. and F.A. Schultz (1993). Heterogeneous electron transfer at electrodes coated with electronically conducting nickel-tetraaminophthalocyanine polymer films. *J. Electroanal. Chem.* **361**, 49–56.
152. Peng, Q.Y. and T.F. Guarr (1994). Electro-oxidation of hydrazine at electrodes modified with polymeric cobalt phthalocyanine. *Electrochim. Acta* **39**(17), 2629–2632.
153. Trollund, E., P. Ardiles, M.J. Aguirre, S.R. Biaggio, and R.C. Rocha-Filho (2000). Spectroelectrochemical and electrical characterization of poly(cobalt-tetraaminophthalocyanine)-modified electrodes: Electrocatalytic oxidation of hydrazine. *Polyhedron* **19**, 2303–2312.
154. Xu, F., H. Li, S.J. Cross, and T.F. Guarr (1994). Electrocatalytic oxidation of NADH at poly(metallophthalocyanine)-modified electrodes. *J. Electroanal. Chem.* **368**, 221–225.
155. Tse, Y.H., P. Janda, H. Lam, and A.B.P. Lever (1995). Electrode with electropolymerized tetraaminophthalocyanatocobalt(II) for detection of sulfide ion. *Anal. Chem.* **67**, 981–985.
156. Kang, T.F., G.-L. Shen, and R.-Q. Yu (1997). Voltammetric behaviour of dopamine at nickel phthalocyanine polymer modified electrodes and analytical applications. *Anal. Chim. Acta* **356**, 245–251.
157. Kang, T.F., G.-L. Shen, and R.-Q. Yu (1997). Voltammetric behaviour of dopamine at nickel phthalocyanine polymer modified electrodes and analytical applications. *Anal. Chim. Acta* **354**, 343–349.
158. Tse, Y.-H., P. Janda, H. Lam, J. Zhang, W.J. Pietro, and A.B.P. Lever (1997). Monomeric and polymeric tetra-aminophthalocyanatocobalt(II) modified electrodes: Electrocatalytic reduction of oxygen. *J. Porphyrins Phthalocyanines* **1**, 3–16.
159. El Mouahid, O., C. Coutanceau, E.M. Belgsir, P. Crouigneau, J.M. Léger, and C. Lamy (1997). Electrocatalytic reduction of dioxygen at macrocycle conducting polymer electrodes in acid media. *J. Electroanal. Chem.* **426**, 117–123.
160. Ramírez, G., E. Trollund, J.C. Canales, M.J. Canales, F. Armijo, and M.J. Aguirre (2001). Effect of the conditions of electropolymerization on the electrocatalytic response of non-metalled-poly-tetraaminophthalocyanine-modified electrodes toward the reduction of oxygen. *Boletin de la sociedad chilena de quimica* **46**(3), 247–255.

161. Ramírez, G., E. Trollund, M. Isaacs, F. Armijo, J. Zagal, J. Costamagna, and M.J. Aguirre (2002). Electroreduction of molecular oxygen on poly-iron-tetraaminophthalocyanine modified electrodes. *Electroanalysis* **14**(7–8), 540–545.
162. Mao, L., G. Shi, Y. Tian, H. Liu, L. Jin, K. Yamamoto, S. Tao, and J. Jin (1998). A novel thin-layer amperometric detector based on chemically modified ring–disc electrode and its application for simultaneous measurements of nitric oxide and nitrite in rat brain combined with in vivo microdialysis. *Talanta* **46**, 1547–1556.
163. Jin, J., T. Miwa, L. Mao, H. Tu, and L. Jin (1999). Determination of nitric oxide with ultramicrosensors based on electropolymerized films of metal tetraaminophthalocyanines. *Talanta* **48**, 1005–1011.
164. Tu, H., J. Xue, X. Cao, W. Zhang, and L. Jin (2000). A novel electrochemical microsensor for the determination of NO and its application to the study of the NO donor *S*-nitrosoglutathione. *Analyst* **125**, 163–167.
165. Wen, Z.-H. and T.-F. Kang (2004). Determination of nitrite using sensors based on nickel phthalocyanine polymer modified electrodes. *Talanta* **62**, 351–355.
166. Griveau, S., M. Gulppi, J. Pavez, J.H. Zagal, and F. Bedioui (2003). Cobalt phthalocyanine-based molecular materials for the electrocatalysis and electroanalysis of 2-mercaptoethanol, 2-mercaptoethanesulfonic acid, reduced glutathione and L-cysteine. *Electroanalysis* **15**(9), 779–785.
167. Brown, K.L. and H.A. Mottola (1998). Voltammetric, chronocoulometric, and spectroelectrochemical studies of electropolymerized films based on Cu(II/I)-4,9,16,23-tetraaminophthalocyanine. *Langmuir* **14**, 3411–3417.
168. Todd, W.J., F. Bailly, J. Pavez, P.W. Faguy, R.P. Baldwin, and R.M. Buchanan (1998). Electrochemically induced metalation of polymeric phthalocyanines. *J. Am. Chem. Soc.* **120**, 4887–4888.
169. Xue, J., X. Ying, J. Chen, Y. Xian, L. Jin, and J. Jin (2000). Amperometric ultramicrosensors for peroxynitrite detection and its application toward single myocardial cells. *Anal. Chem.* **72**, 5313–5321.
170. Brown, K.l., J. Shaw, M. Ambrose, and H.A. Mottola (2002). Voltammetric, chronocoulometric and spectroelectrochemical studies of electropolymerized films based on Co(III/II)- and Zn(II)-4,9,16,23-tetraaminophthalocyanine: Effect of high pH. *Microchem. J.* **72**, 285–298.
171. Griveau, S., V. Albin, T. Pauporté, J.H. Zagal, and F. Bedioui (2002). Comparative study of electropolymerized cobalt porphyrin and phthalocyanine based films for the electrochemical activation of thiols. *J. Mater. Chem.* **12**, 225–232.
172. Obirai, J. and T. Nyokong (2004). Electrochemical studies of manganese tetraamminophthalocyanine monomer and polymer. *Electrochim. Acta* **49**, 1417–1428.
173. Ciszewski, A. and G. Milczarek (1996). Electrocatalytic oxidation of alcohols on glassy carbon electrodes electrochemically modified by conductive polymeric nickel(II)tetrakis(3-methoxy-4-hydroxyphenyl)porphyrin film. *J. Electroanal. Chem.* **413**, 137–142.
174. Ciszewski, A. and G. Milczarek (1997). Glassy carbon electrode modified by conductive, polymeric nickel(II) porphyrin complex as a 3D homogeneous catalytic system for methanol oxidation in basic media. *J. Electroanal. Chem.* **426**, 125–130.
175. Ciszewski, A., G. Milczarek, E. Kubaszewski, and M. Lozynski (1998). Oxidation of nitric oxide at a porphyrinic-based sensor. New results from rotating disk experiments. *Electroanalysis* **10**(9), 628–632.
176. Ciszewski, A. and G. Milczarek (1999). Kinetics of electrocatalytic oxidation of formaldehyde on a nickel porphyrin-based glassy carbon electrode. *J. Electroanal. Chem.* **469**, 18–26.

177. Milczarek, G. and A. Ciszewski (2001). Voltammetric detection of NADH free from interference of ascorbic acid using a glassy carbon electrode modified with an electropolymerised porphyrin film. *Electroanalysis* **13**(2), 164–166.
178. Trévin, S., F. Bedioui, and J. Devynck (1996). New electropolymerised nickel porphyrin films. Application to the detection of nitric oxide in aqueous solution. *J. Electroanal. Chem.* **408**, 261–265.
179. Trévin, S., F. Bedioui, and J. Devynck (1996). Electrochemical and spectrophotometric study of the behavior of electropolymerised nickel porphyrin films in the determination of nitric oxide in solution. *Talanta* **43**, 303–311.
180. Trévin, S., F. Bedioui, M. Guadalupe Gomez Villegas, and C. Bied-Charreton (1997). Electropolymerised nickel macrocyclic complex-based films: Design and electrocatalytic application. *J. Mater. Chem.* **7**(6), 923–928.
181. Ciszewski, A. and G. Milczarek (2003). Electrochemical detection of nitric oxide using polymer modified electrodes. *Talanta* **61**, 11–26.
182. Goux, A., F. Bedioui, L. Robbiola, and M. Pontié (2003). Nickel tetraaminophthalocyanine based films for the electrocatalytic activation of dopamine. *Electroanalysis* **15**(11), 969–974.
183. Cataldi, T.R.I., E. Desimoni, G. Ricciardi, and F. Lelj (1995). Study of the nickel-based chemically modified electrode obtained by electrochemical deposition of an Ni^{II}–tetramethyl-dibenzo-tetraaza[14] annulene complex. Redox catalysis of carbohydrates in alkaline solutions. II. *Electroanalysis* **7**(5), 435–441.
184. Ganesan, V. and R. Ramaraj (2001). Electrocatalytic oxidation of hydrogen peroxide by poly(Ni^{II}teta) modified electrodes. *J. Appl. Electrochem.* **31**, 585–590.
185. Ciszewski, A., G. Milczarek, B. Lewandowska, and K. Krutowski (2003). Electrocatalytic properties of electropolymerized Ni(II)curcumin complex. *Electroanalysis* **15**(5–6), 518–523.
186. Lelj, F., G. Morelli, G. Ricciardi, and A. Rosa (1990). The electrochemical behaviour and electrooxidative polymerization of tetraazannulenic alkyl- and aryl-cobalt complexes. *Inorg. Chim. Acta* **176**, 189–194.
187. Deronzier, A. and M.-J. Marques (1994). Thin films of metal-dibenzotetraaza [14]annulene complexes–Part 4. Mechanistic reinvestigations of the electropolymerization of Ni((II) and Cu(II) complexes at an electrode surface. Sequential electrochemical formation of dimers, oligomers and polymers. *Electrochim. Acta* **39**(10), 1377–1383.
188. Deronzier, A. and M.-J. Marques (1994). Thin films of metal dibenzotetraaza[14]annulene complexes. Part 3. Dimerization and alternated dimerizations of γ-substituted Ni^{2+} complexes. *J. Electroanal. Chem.* **370**, 151–158.
189. Abdel Azzem, M., Z.F. Mohamed, and H.M. Fahmy (1995). Electrocatalytic reduction of some imino compounds on a glassy carbon electrode electrochemically modified with a new copper-salen complex. *J. Electroanal. Chem.* **399**, 121–125.
190. Yousef, U.S. Electrooxidative polymerization of 3-[1-(2-amino-[henylimino)-ethyl]-6-methylpyran-2,4-dione Schiff base in acetonitrile and redox properties of the resulting platinum-film modified electrodes. *Eur. Polym. J.* **35**, 133–143.
191. Yousef, U.S. (2000). A novel conducting polymer film by electrochemical oxidation of 3-[1-(2-aminophenylimino)-ethyl]-6-methylpyran-2,4-dione Schiff base in aqueous medium. *Eur. Polym. J.* **36**, 1629–1644.
192. Bae, Z.-U., Y.-C. Park, J.-H. Lee, H.-Y. Chang, and S.-H. Lee (2000). Electrocatalytic properties of a modified electrode with an asymmetric nickel(II)-tetraaza[14]annulene complex. *Bull. Korean Chem. Soc.* **21**(7), 749–751.
193. Convert, P., C. Coutanceau, P. Crouïgneau, F. Gloaguen, and C. Lamy (2001). Electrodes modified by electrodeposition of CoTAA complexes as selective oxygen cathodes in a direct methanol fuel cell. *J. Appl. Electrochem.* **31**, 945–952.

194. Maumy, M., P. Capdevielle, P.H. Aubert, P. Audebert, and M. Roche (1997). Synthesis and electrochemical properties of salens and poly(salens) where the nitrogens are bridged by a phenanthroline structure. *New J. Chem.* **21**, 621–626.
195. Vilas-Boas, M., C. Freire, B. De Castro, P.A. Christensen, and A.R. Hillman (1997). New insights into the structure and properties of electroactive polymer films derived from [Ni(salen)]. *Inorg. Chem.* **36**, 4919–4929.
196. Mao, L., K. Yamamoto, W. Zhou, and L. Jin (2000). Electrochemical nitric oxide sensors based on electropolymerized film of M(salen) with central ions of Fe, Co, Cu, and Mn. *Electroanalysis* **12**(1), 72–77.
197. Miomandre, F., P. Audebert, M. Maumy, and L. Uhl (2001). Electrochemical behaviour of iron (III) salen and poly(iron-salen). Application to the electrocatalytic reduction of hydrogen peroxide and oxygen. *J. Electroanal. Chem.* **516**, 66–72.
198. Aubert, P.-H., P. Audebert, M. Roche, P. Capdevielle, M. Maumy, and G. Ricart (2001). Synthesis and electrochemical investigations of bis(salen) complex precursors allowing the formation of a ladder-type polymer. *Chem. Mater.* **13**, 2223–2230.
199. Yousef, U.S., A.A. Hathoot, and M. Abdel-Azzem (2001). Electrochemical synthesis and characterization of poly[8-(3-acetylimino-6-methyl-2,4-dioxopyran)-1-aminonaphthalene] film electrodes in acidic aqueous medium. *Eur. Polym. J.* **37**, 1267–1276.
200. Mizutani, F., S.-I. Lijima, Y. Tanabe, and K. Tsuda (1985). Conducting polymer film electrodes with immobilized catalytic sites. *J. Chem. Soc., Chem. Commun.* 1728–1729.
201. Araki, K., S. De Souza Lima, and H. Winnischofer (2000). Thin molecular films of supramolecular porphyrins. *An. Acad. Bras. Ci.* **72**(1), 27–32.
202. Imahori, H., T. Hasobe, H. Yamada, P.V. Kamat, S. Barazzouk, M. Fujitsuka, O. Ito, and S. Fukuzumi (2001). Spectroscopy and photocurrent generation in nanostructured thin films of porphyrin-fullerene dyad clusters. *Chem. Lett.* 784–785.
203. Chien, C.L., L. Sun, M. Tanase, L.A. Bauer, A. Hultgren, D.M. Silevitch, G.J. Meyer, P.C. Searson, and D.H. Reich (2002). Electrodeposited magnetic nanowires: Arrays, field-induced assembly, and surface functionalization. *J. Magn. Magn. Mater.* **249**, 146–155.
204. Winnischofer, H., S. De Souza Lima, K. Araki, and H. Eisi Toma (2003). Electrocatalytic activity of a new nanostructured polymeric tetraruthenated porphyrin film for nitrite detection. *Anal. Chim. Acta* **480**, 97–107.
205. Tamizhmani, G., J.P. Dodelet, D. Guay, G. Lalande, and G.A. Capuano (1994). Electrocatalytic activity of Nafion-impregnated pyrolyzed cobalt phthalocyanine. A correlative study between rotating disk and solid polymer electrolyte fuel-cell electrodes. *J. Electrochem. Soc.* **141**(1), 41–45.
206. Böttger, B., U. Schindewolf, J.L. Avila, and R. Rodríguez-Amaro (1997). Catalytic electrodeposition of silver on glassy carbon electrodes modified with films of cobalt phthalocyanine. *J. Electroanal. Chem.* **432**, 139–144.
207. Kudo, T., M. Kimura, K. Hanabusa, and H. Shirai (1998). Electrical properties of junctions between electrodeposited magnesium phthalocyanine and aluminium. *Polym. J.* **30**(3), 177–180.
208. Brinkmann, M., S. Graff, C. Chaumont, and J.-J. André (1999). Electrodeposition of lithium phthalocyanine thin films: Part I. Structure and morphology. *J. Mater Res.* **14**(5), 2163–2172.
209. Hangovan, G., J.L. Zweier, and P. Kuppusamy (2000). Electrochemical preparation and EPR studies of lithium phthalocyanine: Evaluation of the nucleation and growth mechanism and evidence for potential-dependent phase formation. *J. Phys. Chem. B* **104**, 4047–4059.

210. Brinkmann, M. and J.-J. André (1999). Electrodeposited lithium phthalocyanine thin films. Part II: Magnetic properties and mesoscopic effects. *J. Mater. Chem.* **9**, 1511–1520.
211. André, J.-J. and M. Brinkmann (2001). Electrodeposited thin films of lithium phthalocyanine: Morphology, structure and magnetic properties. *Synth. Met.* **121**, 1359–1360.
212. Yoshida, T., M. Tochimoto, D. Schlettwein, D. Wöhrle, T. Sugiura, and H. Minoura (1999). Self-assembly of zinc oxide thin film modified with tetrasulfonated metallophthalocyanines by one-step electrodeposition. *Chem. Mater.* **11**, 2657–2667.
213. Schlettwein, D., T. Oekermann, T. Yoshida, M. Tochimoto, and H. Minoura (2000). Photoelectrochemical sensitisation of ZnO–tetrasulfophthalocyaninatozinc composites prepared by electrochemical self-assembly. *J. Electroanal. Chem.* **481**, 42–51.
214. Oekermann, T., T. Yoshida, D. Schlettwein, T. Sugiura, and H. Minoura (2001). Photoelectrochemical properties of ZnO/tetrasulfophthalocyanine hybrid thin films prepared by electrochemical self-assembly. *Phys. Chem. Chem. Phys.* **3**, 3387–3392.
215. Kudo, T., M. Kimura, K. Hanabusa, and H. Shirai (1999). Preparation of electrocodeposited film consisting of phthalocyanine and perylene derivatives. *J. Porphyrins Phthalocyanines* **3**, 310–315.
216. Beck, F., E. Abdelmula, and M. Dahlhaus (2000). Anodic deposition of composites with a matrix of intrinsically conducting polymers. *Electrochim. Acta* **45**, 3423–3429.
217. Zangmeister, R.A.P., D.F. O'Brien, and N.R. Armstrong (2002). Selective deposition of rod-like phthalocyanine aggregates on Au surfaces patterned with a combination of microcontact printing and electropolymerization. *Adv. Func. Mater.* **12**(3), 179–186.
218. Armstrong, N.R., C. Carter, C. Donley, A. Simmonds, P. Lee, M. Brumbach, B. Kippelen, B. Domercq, and S. Yoo (2003). Interface modification of ITO thin films: Organic photovoltaic cells. *Thin Solid Films* **445**, 342–352.
219. Ricciardi, G. and F. Lelj (1992). Reaction of organic and metallorganic π-cation radicals: Electrical and electrochemical properties of polymeric materials obtained by oxidation of pyrrole and tetraaza[14]annulenic nickel(II) complexes. *Polyhedron* **11**(16), 2089–2097.
220. Aubert, P.-H., A. Neudeck, L. Dunsch, P. Audebert, P. Capdevielle, and M. Maumy (1999). Electrochemical synthesis and structural studies of copolymers based on the electrooxidation of pyrrole and some salen compounds. *J. Electroanal. Chem.* **470**, 77–78.
221. Sacarescu, G., R. Ardeleanu, L. Sacarescu, and M. Simionescu (2003). Synthesis of polysilane-bis(salicyliden)ethylenediamine Ni(II) complex. *J. Organomet. Chem.* **685**, 202–206.
222. Tabushi, I. and N. Koga (1979). P-450 type oxygen activation by porphyrin–manganese complex. *J. Am. Chem. Soc.* **101**, 6456–6458.
223. Perrée-Fauvet, M. and A. Gaudemer (1981). Manganese porphyrin-catalyzed oxidation of olefins to ketones by molecular-oxygen. *J. Chem. Soc., Chem. Commun.* **17**, 874–875.
224. Tabushi, I. and A. Yazaki (1981). P-450-type dioxygen activation using H-2 colloidal Pt as an effective electron-donor. *J. Am. Chem. Soc.* **103**, 7371–7373.
225. Fontecave, M. and D. Mansuy (1984). Monooxygenase-like oxidations of olefins and alkanes catalyzed by manganese porphyrins—comparison of systems involving either O_2^- and ascorbate or iodosylbenzene. *Tetrahedron* **40**, 4297–4311.
226. Tabushi, I., M. Kodera, and M. Yokoyama (1985). Kinetics and mechanism of reductive dioxygen activation catalyzed by the P-450 model system—iron picket fence as a catalytic center. *J. Am. Chem. Soc.* **107**, 4466–4473.

227. Creager, S.E., A.A. Raybuck, and R.W. Murray (1986). An efficient electrocatalytic model cytochrome-P-450 epoxidation cycle. *J. Am. Chem. Soc.* **108**, 4225–4227.
228. Battioni, P., J.F. Bartoli, P. Leduc, M. Fontecave, and D. Mansuy (1987). A new and efficient biomimetic system for hydrocarbon oxidation by dioxygen using manganese porphyrins, imidazole, and zinc. *J. Chem. Soc., Chem. Commun.* 791–792.
229. Moisy, P., F. Bedioui, Y. Robin, and J. Devynck (1988). Epoxidation of *cis*-cyclooctene by molecular-oxygen electrocatalyzed by polypyrrole manganese porphyrin film modified electrodes. *J. Electroanal. Chem.* **250**, 191–199.
230. Leduc, P., P. Battioni, J.F. Bartoli, and D. Mansuy (1988). A biomimetic electrochemical system for the oxidation of hydrocarbons by dioxygen catalyzed by manganese-porphyrins and imidazole. *Tet. Lett.* **29**, 205–208.
231. Nishihara, H., K. Pressprich, R.W. Murray, and J.P. Collman (1990). Electrochemical olefin epoxidation with manganese *meso*-tetraphenylporphyrin catalyst and hydrogen-peroxide generation at polymer-coated electrodes generation at polymer-coated electrodes. *Inorg. Chem.* **29**, 1000–1006.
232. Horwitz, C.P., S.E. Creager, and P.W. Murray (1990). Electrocatalytic olefin epoxidation using manganese Schiff-base complexes and dioxygen. *Inorg. Chem.* **29**, 1006–1011.
233. Tsang, P.K.S. and D.T. Sawyer (1990). Electron-transfer thermodynamics and bonding for the superoxide ($O_2^{\cdot -}$), dioxygen (O_2), and hydroxyl (OH) adducts of (tetrakis(2,6-dichlorophenyl)porphinato)iron, manganese, and cobalt in dimethylformamide. *Inorg. Chem.* **29**, 2848–2855.
234. Bedioui, F., P. Moisy, C. Bied-Charreton, and J. Devynck (1989). Preparation and catalytic applications of poly(pyrrole manganese porphyrin) films modified electrodes). *J. Chim. Phys.* **86**, 235–240.
235. Arasasinghan, R.D. and T.C. Bruice (1990). Reaction of hydroxide ion with manganese(III) tetramesitylporphyrin and the oxidation-states of manganese tetramesitylporphyrins. *Inorg. Chem.* **29**, 1422–1427.
236. Bedioui, F., S. Guterriez Granados, C. Bied Charreton, and J. Devynck (1991). Biomimetic oxidation of hydrocarbons by dioxygen—electrocatalysis using polypyrrole–manganese porphyrin film modified electrodes. *New J. Chem.* **15**, 939–941.
237. Lu, W.Y., J.F. Bartoli, P. Battioni, and D. Mansuy (1992). Selective oxygenation of hydrocarbons and sulfoxidation of thioethers by dioxygen with a Mn-porphyrin-based cytochrome P450 model system using Zn as electron-donor. *New J. Chem.* **16**, 621–628.
238. Cauquis, G., S. Cosnier, A. Deronzier, B. Galland, D. Limosin, J.C. Moutet, J. Bizot, D. Deprez, and J.P. Pulicani (1993). Poly(pyrrole-manganese porphyrin)—a catalytic electrode material as a model system for olefin epoxidation and drug-metabolism with molecular-oxygen. *J. Electroanal. Chem.* **352**, 181–195.
239. Meunier, B. (1992). Metalloporphyrins as versatile catalysts for oxidation reactions and oxidative DNA cleavage. *Chem. Rev.* **92**, 1411–1456.
240. Gutierrez Granados, S., F. Bedioui, and J. Devynck (1993). Electroanalytical study of the activation of dioxygen in acetonitrile solution by manganese porphyrin films deposited onto carbon electrodes. *Electrochim. Acta* **38**, 1747–1751.
241. Pereira Rodrigues, N., J. Obirai, T. Nyokong, and F. Bedioui (2005). Electropolymerized pyrrole-substituted manganese phthalocyanine films for the electroassisted biomimetic catalytic reduction of molecular oxygen. *Electroanalysis* **17**, 186–190.
242. Gaillon, L., F. Bedioui, P. Battioni, and J. Devynck (1993). Electroassisted biomimetic oxidation of hydrocarbons by molecular-oxygen catalyzed by manganese porphyrin complexes intercalated into montmorillonite. *J. Mol. Catal.* **78**(2), L23–L26.

243. Bedioui, F., S. Gutierrez Granados, L. Gaillon, C. Bied-Charreton, and J. Devynck (1991). Electroassisted oxidation of *cis*-cyclooctene and adamantane by molecular oxygen catalyzed by polypyrrole manganese porphyrin films. *Stud. Surf. Sci. Catal.* **66**, 221–228.
244. Bedioui, F., P. Moisy, J. Devynck, L. Salmon, and C. Bied-Charreton (1989). Poly(pyrrole manganese porphyrin) film electrode as a catalyst in electro-assisted oxidation reactions using molecular-oxygen. Comparison with described homogeneous systems. *J. Mol. Catal.* **56**(1–3), 267–275.
245. Creager, S.E., S.A. Raybuck, and R.W. Murray (1986). An efficient electrocatalytic model cytochrome-P-450 epoxidation cycle. *J. Amer. Chem. Soc.* **108**(14), 4225–4227.
246. Creager, S.E. and R.W. Murray (1987). Electrochemical reactivity of manganese(II) porphyrins-effects of dioxygen, benzoic anhydride, and axial ligands. *Inorg. Chem.* **26**(16), 2612–2618
247. Gaillon, L., F. Bedioui, J. Devynck, P. Battioni, L. Barloy, and D. Mansuy (1991). Electrochemistry of manganese porphyrin intercalated into montmorillonite. *J. Electroanal. Chem.* **303**(1–2), 283–287.
248. Gaillon, L. and F. Bedioui (2001). First example of electroassisted biomimetic activation of molecular oxygen by a (salen)Mn epoxidation catalyst in a room-temperature ionic liquid. *Chem. Commun.* (16), 1458–1459.
249. Albin, V. and F. Bedioui (2003). First electrochemical evidence of existence of an oxomanganese(V) porphyrin intermediate in the reaction of manganese(III) porphyrin and hydrogen peroxide as an intermediate in the reaction of manganese(III) porphyrin and hydrogen peroxide as a model of enzyme mimetics. *Electrochem. Commun.* 5, 129–132.
250. Taha, Z. and T. Malinski (1992). Nitric-oxide release from a single cell measured in-situ by a porphyrinic-based microsensor. *Nature* **358**, 676–678.
251. Malinski, T., Z. Taha, S. Grunfeld, A. Burewicz, P. Tomboulian, and F. Kiechle (1993). Measurements of nitric-oxide in biological materials using a porphyrinic microsensor. *Anal. Chim. Acta* **279**(1), 135–140.
252. Taha, Z., F. Kiechle, and T. Malinski (1992). Oxidation of nitric-oxide by oxygen in biological systems monitored by porphyrinic sensor. *Biochem. Biophys. Res. Commun.* **188**(2), 734–739.
253. Malinski, T., Z. Taha, S. Grunfeld, S. Patton, M. Kapturczak, and P. Tomboulian (1993). Diffusion of nitric-oxide in the aorta wall monitored in-situ by porphyrinic microsensors. *Biochem. Biophys. Res. Commun.* **193**(3), 1076–1082.
254. Malinski, T., M. Kapturczak, J. Dayharsh, and D. Bohr (1993). Nitric-oxide synthase activity in genetic hypertension. *Biochem. Biophys. Res. Commun.* **194**(2), 654–658.
255. Malinski, T., F. Bailey, Z. G. Zhang, and M. Chopp (1993). Nitric-oxide measured by a porphyrinic microsensor in rat-brain after transient middle cerebral–artery occlusion. *J. Cereb. Blood Flow Metab.* **13**(3), 355–358.
256. Malinski, T., M. W. Radomski, Z. Taha, and S. Moncada (1993). Direct electrochemical measurement of nitric-oxide released from human platelets. *Biochem. Biophys. Res. Commun.* **194**(2), 960–965.
257. Zhang, Z.G., M. Chopp, F. Bailey, and T. Malinski (1995). Nitric-oxide changes in the rat-brain after transient middle cerebral–artery occlusion. *J. Neur. Sci.* **28**(1), 22–27.
258. Vallance, P., S. Patton, K. Bhagat, R. MacAllister, M. Radomski, S. Moncada, and T. Malinski (1995). Direct measurement of nitric-oxide in human-beings. *The Lancet* **346**, 153–154.

259. Kanai, A.J., H.C. Strauss, G.A. Truskey, A. L. Crews, S. Grunfeld, and T. Malinski (1995). Shear-stress induces ATP-independent transient nitric-oxide release from vascular endothelial-cells, measured directly with a porphyrinic microsensor. *Cir. Res.* **77**(2), 284–293.
260. Malinski, T. (1999). In K.M. Kadish, K.M. Smith, R. Guilard (eds), *The Porphyrin Handbook*, Vol. 6. Academic Press, New York.
261. Lantoine, F., S Trévin, F. Bedioui, and J. Devynck (1995). Selective and sensitive electrochemical measurement of nitric-oxide in aqueous solution–Discussion and new results. *J. Electroanal. Chem.* **392**(1–2), 85–89.
262. Villeneuve, N. and F. Bedioui (2003). Electrochemical nitric oxide sensors for biological samples. Principle, selected examples and applications. *Electroanalysis* **15**, 5–18.

9

Electron Transfer Processes of β-Pyrrole Brominated Porphyrins: Structural vs. Electronic Effects

Francis D'Souza and Karl M. Kadish

1. Introduction

The electron transfer properties of metalloporphyrin complexes (MN_4) have been a subject of intense research over the last three decades because of their importance in biomimetic, catalytic, or sensor applications[1]. To date, MN_4 complexes bearing *meso* and β-pyrrole substituents as well as extended conjugated structures containing over three dozen metal ions (M) have been probed electrochemically to determine their redox behavior[2]. The oxidation state of the central metal ion, M, can sometimes be varied between +1 and +4 and this may or may not depend upon the structure of the N_4 porphyrin macrocycle. The value of the oxidation and reduction potentials and the resulting HOMO–LUMO gap (HOMO stands for the highest occupied molecular orbital, and LUMO for the lowest unoccupied molecular orbital) can vary substantially for a given complex depending upon the substituents on the macrocycle and the type and number of axial ligands coordinated to the metal center. A large variation of the electrochemical HOMO–LUMO gap, ranging between 1.7 and 2.4 V, has been achieved by appropriate substitution at the periphery of the porphyrin macrocycle[2,3].

Studies on β-pyrrole halogenated *meso*-tetraarylmetalloporphyrins have witnessed a rapid growth in recent years since some of these complexes have been shown to have a high catalytic efficiency for several specific reactions (such as oxygen transfer reactions) as compared to non-halogenated porphyrins having the same macrocyclic structure. This property has been attributed to the presence of electron-withdrawing substituents on the macrocycle and the nonplanarity of the porphyrin ring which results from β-pyrrole halogenation (generally chlorination

---

Francis D'Souza • Department of Chemistry, Wichita State University, 1845 Fairmount, Wichita, KS 67260-0051.
Karl M. Kadish • Department of Chemistry, University of Houston, Houston, TX 77205-2004.

N_4-Macrocyclic Metal Complexes, edited by José H. Zagal, Fethi Bedioui and Jean-Pol Dodelet. Springer Science+Business Media, Inc., New York, 2006.

and bromination but sometimes fluorination)[4] which prevents formation of μ-oxo species in solution. The electron-withdrawing substituents make the macrocycle more electron deficient which in turn makes the porphyrin π-ring system harder to oxidize, thus stabilizing it against oxidative degradation. Among the known β-pyrrole halogenated porphyrins, β-pyrrole brominated porphyrins have been the most widely studied, primarily because of their synthetic ease and the fact that the number of bromo substituents can be controlled with a high degree of certainty.

Three series of β-pyrrole brominated porphyrins have been reported in the literature[5–29]. The first type-1 is represented as $(TPPBr_x)M$ where $TPPBr_x$ is the dianion of β-pyrrole brominated tetraphenylporphyrin, $x = 0$–8 and M = 2H, Mn, Fe, Co, Ni, Cu, or Zn[7,12–16] (Figure 9.1). The $TPPBr_x$ derivatives have been studied in order to ascertain how distortion of the porphyrin ring will quantitatively affect the redox and spectral properties of the porphyrin as a function of the number of halogen substituents at the β-pyrrole positions which can be varied from 0 to 8.

The second series of compounds (type-2) are water-soluble porphyrins, and most are represented by $(TMPyPBr_8)M$ where $TMPyPBr_8$ is the dianion of the tetrachloro salt of octa-β-pyrrole brominated N-methylpyridylporphyrin, M = 2H, Zn, Co, Mn, or Cu or $(TPPSBr_8)M$, where $TPPSBr_8$ is the dianion of the tetrasodium salt of octa β-pyrrole brominated (4-sulfonatophenyl)porphyrin, M = 2H, Zn or Co[27–30] (Figure 9.2). Studies on these water-soluble porphyrins have provided information as to how the porphyrin ring basicity, aggregation, redox properties, spectroscopic properties, and axial ligand-binding reactions of these compounds vary in comparison with the unbrominated analogs[29,30]. The cobalt(II) derivatives of these water-soluble porphyrins can catalyze the two-electron reduction of dioxygen to hydrogen peroxide in aqueous solutions[27–29], similar to the unbrominated, monomeric, (TMPyP)Co which is known to reduce

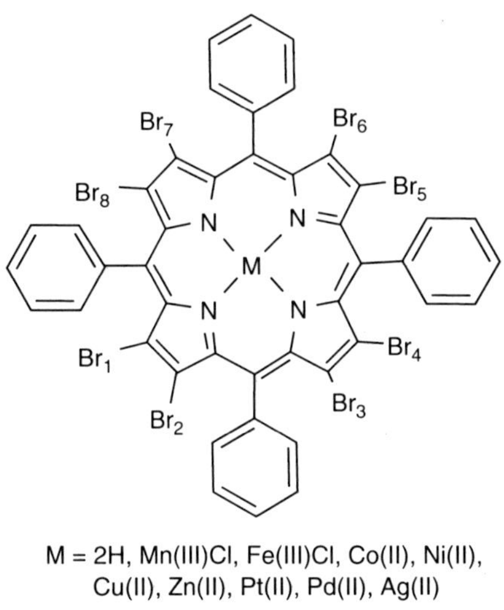

Figure 9.1. Structural formula of "type-1" β-pyrrole brominated metalloporphyrins.

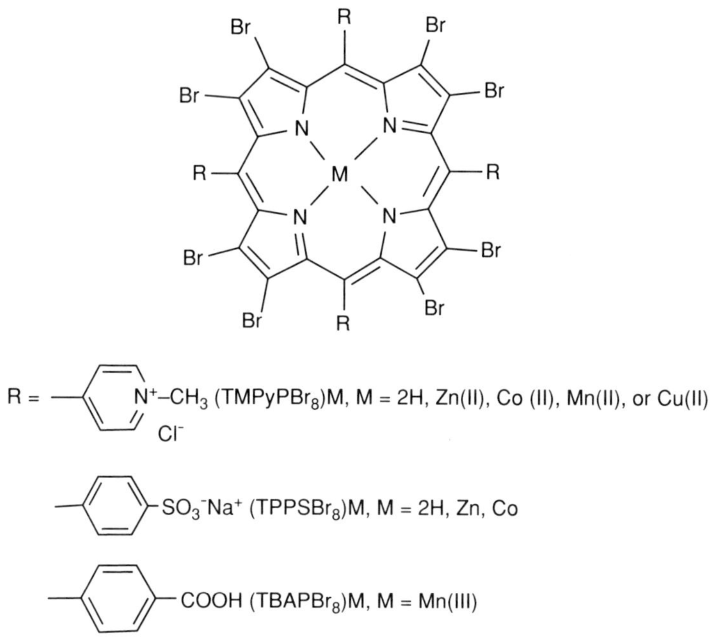

Figure 9.2. Structures of "type-2" β-pyrrole brominate water-soluble porphyrins.

oxygen by two electrons in non-aqueous media[91]. Studies have also shown that the manganese derivative of the same water-soluble β-pyrrole brominated porphyrin, $(TPPSBr_8)Mn^{II}$, behaves as a potent superoxide dismutase mimic[31,32].

The third series of metalloporphyrin catalysts for oxygenation of organic substrates (type-3), the β-pyrrole halogenated complexes possessing additional bulky substituents on the *meso*-aryl groups (Figure 9.3), has received much interest due to the pioneering work of several groups around the world[17,33–36]. The first report on the catalytic activity of β-pyrrole brominated porphyrins was by Traylor and Tsuchiya[17] who showed an improved yield for the hydroxylation reactions and a good chemical stability of the catalyst, which was $(TDCPPBr_8)Fe^{III}Cl$, where $TDCPPBr_8$ is the dianion of 2,3,7,8,12,13,17,18-octabromo-*meso*-tetra(2,6- dichlorophenyl)porphyrin. The $(TMPBr_8)Mn^{III}Cl$ derivative, where $TMPBr_8$ is the dianion of 2,3,7,8,12,13,17,18-octabromo-*meso*-tetra-(2,4,6-trimethylphenyl) porphyrin, was found to be a moderate catalyst for the oxidation of organic substrates when H_2O_2 was used but gave good results when $KHSO_5$ or magnesium monoperoxyphthalate was used as the oxygen source[36]. In a similar study, $(TDCPBr_8)Mn^{III}Cl$, when used as a catalyst for the oxidation of an aromatic substrate, anisole by H_2O_2 yielded better results, with the final products of the reaction being 2-and/or 4-methoxy phenol[23]. X-ray structural characterization of these fully halogenated porphyrins[5,9] have revealed a severe distortion of the macrocyclic ring due to steric interactions between the halogens and the phenyl groups. The saddle-shaped conformations have been related to a nonlinear dependence of the oxidation potentials on the number of halogen atoms[5,9,14]. Boschi and coworkers[37] utilized β-tetrabrominated porphyrins in the reaction of H_2O_2-dependent oxidation of cyclooctene and adamantane and demonstrated better catalytic activity as compared with the octabrominated derivatives. Several iron porphyrins of the third series are also known to catalyze the O_2-dependent oxidation of alkanes, even at room temperature[38,39]. For example, $(TF_5PPBr_8)Fe^{III}Cl$, where TF_5PPBr_8 is the dianion of *meso*-tetra(pentafluoro phenyl)-2,3,7,8,12,13,17,18-octabromoporphyrin is able to oxidize isobutane to *t*-BuOH with 91% selectivity and 890 turnovers[38,39].

The above discussions clearly demonstrates the importance of β-pyrrole brominated porphyrins in both fundamental and applied fields of porphyrin chemistry. While one aim of these studies has been to gain insights into the role of macrocycle distortion[5,29] in relation to biological systems containing porphyrins or related macrocycles[40,41], another has been to determine which factors influence the robustness of such catalysts in the oxidation of organic substrates[24,28,42–59]. Both measurements of electrochemical redox potentials and spectral properties of the porphyrins are needed to understand these properties and both are reviewed in this chapter on β-pyrrole brominated porphyrins.

In general, the properties of β-halogenated porphyrins have been explained in terms of a combination of two effects: one is the electron-withdrawing properties of the halogen substituents at the β-pyrrole positions of the macrocycle and the other is the nonplanarity of the porphyrin macrocycle due to steric hindrance of the peripheral groups[60–65]. In order to separate these two factors and to obtain an estimate of their magnitude, different series of β-pyrrole brominated porphyrins have been examined, some of which are water soluble. Most results on these compounds have been reviewed in the literature[12,16,27–30,67], and our discussion in this chapter mainly covers: (i) synthetic routes needed to obtain porphyrins bearing different numbers of β-pyrrole bromo substituents, (ii) the effect of Br substituents and solvent interactions on spectral properties of the neutral, oxidized, or reduced complexes, (iii) electrochemical reduction and oxidation behavior and HOMO–LUMO energy gap changes with changes in the number of bromo substituents, and (iv) the effect of Br substituents on axial ligation properties.

2. Synthesis

2.1. Synthesis and Characterization of β-Pyrrole Brominated *meso*-Tetraphenylporphyrins

Although the present chapter deals mainly with the electrochemical behavior of β-pyrrole brominated porphyrins, it is worthwhile for the sake of completeness to discuss the synthetic methodology adopted by different research groups. In general, the bromination of porphyrins has been carried out either by direct bromination using liquid bromine in an organic solvent or by reacting the porphyrins with a brominating agent such as *N*-bromosuccinamide (NBS). Depending upon the reaction conditions and substituents on the *meso*-tetraphenyl moieties (Figures 9.1 and 9.3), partial-to-complete β-pyrrole bromination was achieved. Callot[68] reported the first synthesis of partially β-pyrrole brominated *meso*-tetraphenylporphyrins. The substrate, free-base porphyrin, was reacted with increasing amounts of NBS as the halogen donor in boiling chloroform and atmospheric oxygen was used as a radical initiator. After chromatographic separation and purification, the mono-, bis- (two isomers), tri- and tetra-β-pyrrole brominated derivatives were isolated, characterized, and also used as starting materials for preparation of other β-substituted compounds and their metal derivatives[24, 57]. The author observed an approximate 6 nm red shift of the Soret band for every added bromine atom[68]. The stereochemistry of the tetra-β-pyrrole brominated porphyrins was ascertained by Crossley *et al.*[35] using ^{1}H NMR spectroscopy. These studies showed that the two double bonds which were not directly part of the aromatic conjugation could be completely tetrabrominated because they possess a higher electron density than at the other positions in the aromatic conjugation. Molecular structures of several tetrahalogenated porphyrins have also been obtained and compared to structures of the octahalogenated compounds[5]. The first report for a successful β-pyrrole perbromination reaction using tetraphenylporphyrin was provided by Traylor and Tsuchiya[17] who showed the possibility to access a new class of compounds having a lower electron density on the macrocyclic ring.

The bromination of (TDCPP)Zn, where TDCPP is the dianion of *meso*-tetra(2,6-chloro)phenylporphyrin in CCl_4 by NBS, resulted in bromination at the eight β-pyrrole positions of the macrocycle. The reproducibility of such a reaction was very difficult to obtain because of a low solubility of the starting compound in CCl_4 and better yields (70–75%) were obtained when chloroform or tetrachloroethene was used as the reaction solvent[36]. Several articles have reported the synthesis of zinc porphyrins completely halogenated at the β-pyrrole positions, using NBS in chlorinated solvents or methanol[9, 33, 34, 70]. In some cases, different synthetic procedures using Lewis acids were proposed in order to enhance

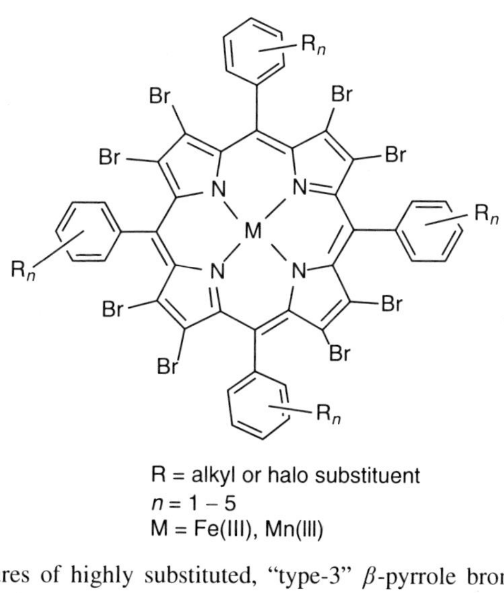

Figure 9.3. Structures of highly substituted, “type-3” β-pyrrole brominated porphyrins used in catalytic applications.

the halogenation in chlorinated solvents. The choice of procedure was based on the different mechanisms, radical or ionic, operating in polar or nonpolar solvents[33,34].

A complete bromination of (TPP)Cu was achieved by Bhyrappa and Krishnan[7] using liquid bromine in $CHCl_3/CCl_4$. The resulting porphyrin was demetalated in perchloric acid, giving the desired free-base porphyrin in 75% yield. Similar experimental procedures, using liquid bromine as a reagent[38], led to completely β-pyrrole brominated porphyrins containing iron as a central metal ion. Interestingly, partial bromination of (TPP)Zn has been achieved using NBS in CCl_4[12,14]. Chromatographic separation of the reaction products afforded the penta-, hexa-, and heptabromo derivatives. Characterization of the above free-base complexes by positive ion fast atom bombardment (FAB)-mass spectrometry showed dehalogenation process to occur during formation of the π-radical cation and this reaction was attributed to the high acidity of the matrix[36]. Similar reactions were also observed during electroreduction of $(TPPBr_x)Co$ and $(TPPBr_x)FeCl$ in benzonitrile solvent[12,14].

Recently, different protocols have been reported for the regioselective synthesis of partially brominated porphyrins. Tse *et al.*[71] utilized 5,10,15,20-tetrakis-(trifluoromethyl)-porphyrinatozinc, **I**, and showed the occurrence of a monobromination reaction to yield compound **II**, and regioselective dibromination reaction to yield compounds **III** and **IV** (Figure 9.4). Subsequently, the free-base β-pyrrole bromoporphyrins were converted to aryl porphyrins through Suzuki cross-coupling reaction. Bhyrappa and coworkers[72] reported an improved protocol for regioselective synthesis of 2,3,12,13-tetrabromo-5,10,15, 20-tetraphenylporphyrin, $(TPPBr_4)H_2$.

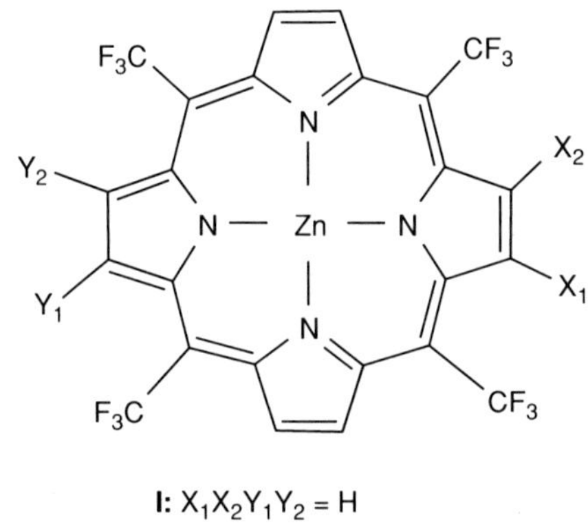

I: $X_1X_2Y_1Y_2 = H$

II: $X_1 = Br$, $X_2Y_1Y_2 = H$

III: $X_1X_2 = Br$, $Y_1Y_2 = H$

IV: $X_1X_2Y_1Y_2 = Br$

Figure 9.4. Partially β-pyrrole brominated porphyrins obtained by regioselective synthetic procedure.

2.2. Synthesis of Water-Soluble β-Pyrrole Brominated Porphyrins

Water-soluble β-pyrrole porphyrins can be synthesized by either direct bromination of functionalized water-soluble porphyrins or by introducing ionic groups at the tetraphenyl positions of the brominated porphyrin. $(TMPyPBr_8)Zn$ was prepared by direct bromination of (TMPyP)Zn using liquid bromine[25]. The $(TMPyPBr_8)Co$ derivative was synthesized by metathesis of $(TMPyPBr_8)Zn$ with cobalt(II) chloride[28]. The free-base porphyrin was isolated in its protonated form, $(TMPyPBr_8)H_3^+$, by demetallating the zinc(II) derivative, $(TMPyPBr_8)Zn$, and then isolating the product as a hexafluorophosphate salt[25,27].

Using an approach similar to the above, $(TPPSBr_8)Zn$ was synthesized by direct bromination of (TPPS)Zn[29]. The free-base derivative, $(TPPSBr_8)H_2$, was obtained by demetalating $(TPPSBr_8)Zn$ in excess sulfuric acid at 0° C, followed by slow addition and neutralization of the reaction product with 1 M NaOH[29,30]. Alternatively, Tabata *et al.*[26] reported the synthesis of $(TPPSBr_8)H_2$ by bromination of $(TPP)H_2$ using NBS, followed by sulfonation using concentrated sulfuric acid at 70° C for 2 days. Here, the product was isolated in the tetrasulfonic acid form.

Two cationic Mn(II) and Cu(II) metal derivatives of octabrominated 5, 10,15,20-tetrakis-(*N*-methylpyridinium-4-yl) porphyrins, $(TMPyPBr_8)Mn^{II}$ and $(TMPyPBr_8)Cu^{II}$, were prepared[31] using the literature procedure for bromination of water soluble porphyrins[25]. Unlike the TMPyP derivative, where Mn^{III} is stable, the $(TMPyPBr_8)Mn$ derivative was isolated in its Mn^{II} form. Another anionic porphyrin, Mn(III) β-octabromo-*meso*-tetrakis(4-carboxyphenyl) porphyrin, $(TBAPBr_8)Mn^{III}$, was recently prepared with an overall yield of 50% starting from the methyl ester derivative of the free-base β-octabromo-*meso*-tetrakis(4-carboxyphenyl)porphyrin, $(TBAPBr_8)H_2$[32].

3. Effect of the Br Substituents and Solvent Interactions on the UV–Visible Spectra

Starting from the first observation of a red shift of the Soret band upon β-pyrrole bromination[68], several studies dealing with solvation behavior of brominated porphyrins have appeared in literature[7,12–15,73–76]. The solvation of $(TPPBr_8)Zn$ in 12 different nonaqueous solvents resulted in large red shifts of the B and Q bands of the porphyrin accompanied by enhanced absorbance ratios of the Q bands[7,8]. These observations were ascribed to a destabilization of the HOMO a_{2u} of the porphyrin arising from a flow of charge from the axial ligand to the porphyrin ring through the Zn(II) ion. More recently, the effect of solvent polarity on the optical absorption spectra of perhalogenated porphyrins was reported by Bhyrappa coworkers[75,76]. In contrast to the unbrominated tetraphenylporphyrins, the perbrominated porphyrins revealed strong solvent-dependent absorption spectral features. Red shifts of more than 25 nm for the Soret band and 50 nm for the visible bands were observed. The authors[75,76] ascribed these effects to an enhanced distortion of the macrocyclic ring induced by a hydrogen-bonding interaction between the porphyrin core pyrrolic N–H or pyrrolenine nitrogens and the solvent molecules.

Electronic absorption spectra of Ni(II) and Cu(II) complexes of 2,3, 7,8,12,13,17,18-octa(bromo and chloro)-5,10,15,20-tetraphenylporphyrins (MOXTPP; X = Br, Cl) in various solvents were recently examined[93]. Both complexes exhibited dramatic shifts in the peak maxima and peak intensity (ε) as compared to their unhalogenated analogs. The solvent-dependent absorption spectral features of the electron deficient metal(II) perhaloporphyrins were attributed to a coordinative interaction of the solvent with the metal ion of the porphyrin cavity[93].

Quantitative analysis of the effect of bromo substitution on the geometric and electronic structures was possible by systematically examining different metal complexes with 0–8 number bromo substituents. A detailed study on the HOMO–LUMO gap of $(TPPBr_x)Zn$ as a function of the number of Br substituents ($x = 0$–8) was performed by optical absorption, semiempirical AM1-CI calculations and electrochemical methods[73]. Geometry optimizations by AM1-CI calculations revealed little or no porphyrin ring ruffling for compounds with $x = 0$–4 while for compounds with $x = 5$–8, increased ruffling with respect to Br substituents was predicted. A plot of the AM1-CI estimated HOMO–LUMO energy gap (energy of Q_2 band) vs. the number of Br groups showed two linear segments with different slopes (Figure 9.5, curve marked as AM1-CI). This study also suggested that the Soret and visible absorption bands of $(TPPBr_x)Zn$ are sensitive to distortion of the porphyrin macrocycle which changes as a function of the number of Br groups on the complex[73]. Plots constructed by using optical data (energy of the low energy visible band) and electrochemical data (difference between the first oxidation potential and the first reduction potential) for different $(TPPBr_x)Zn$ derivatives, were indeed found to follow the predicted trend (Figure 9.4). Both plots showed two segments, one corresponding to the planar, less brominated porphyrins while the other was for nonplanar, higher brominated porphyrins, thus providing a quantitative estimation of the effects caused by structural distortions as opposed to the electron-withdrawing effects of bromo substituents.

The spectral features of the Fe(III), Mn(III), Co(II), and Zn(II) β-pyrrole brominated porphyrins in various solvents have been investigated as a function of the number of β-bromo substituents. In general, the Soret and visible bands in each series of compounds are all increasingly red shifted with increase in the number of Br groups on the macrocycle[7,12–15,73,74]. Moreover, as described in subsequent sections of this review, the utilization of spectroelectrochemistry has provided an opportunity to study the spectral behavior of different brominated porphyrins as a function of their different oxidation states.

4. Electrochemical Behavior of β-Pyrrole Brominated Metalloporphyrins

Initial electrochemical studies on partially β-pyrrole brominated porphyrins ($x = 0$–4) were carried out by the groups of Gross and Callot[95,97] and others[96]. In

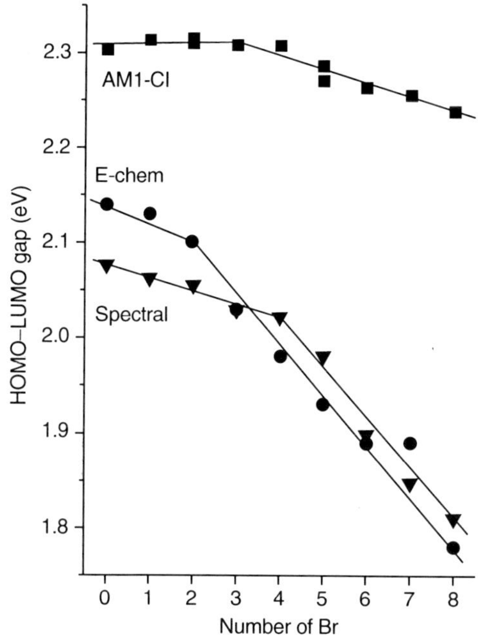

Figure 9.5. Correlations between the calculated or experimental HOMO–LUMO gap of each compound and the number of Br substituents on the macrocycle of (TPP)Br_x)Zn. (Reproduced with permission from ref. [73].)

their analysis of half-wave potentials against "sigma" plots, the authors observed differences in the electrochemical behavior of β-pyrrole substituted porphyrins. This was initially rationalized in terms of inductive effects of the lone pair of electrons of the pyrrolic nitrogens. The lone pair of electrons located on the pyrrole nitrogen were suggested as the site of electron transfer for β-pyrrole substituted porphyrins[94–97]. Studies performed later on a series of β-pyrrole brominated metalloporphyrins bearing 0–8 bromo groups demonstrated the significance of porphyrin ring distortions in the observed differences of electrochemical behavior.

The electrochemistry of ($TPPBr_8$)M, where M = Co, Ni, Cu, Zn, Pt, Pd, and Ag generally revealed the expected easier reduction for both metal-centered and ring-centered reactions as a result of the electron withdrawing bromo substituents. Quantitative work carried out on the ($TPPBr_x$)M and ($TPPBr_x$)MCl series[12,14,15,73,74] yielded a wealth of information and allowed for a separation of geometric (steric) from electronic effects caused by the complete or partial β-pyrrole bromination. As pointed out earlier, investigations performed on the ($TPPBr_x$)Zn series of compounds revealed two one-electron oxidations and two one-electron reductions, as expected for porphyrins bearing electroinactive metal ions[2]. Since the first oxidation and first reduction both involved the porphyrin ring, it was possible to estimate the HOMO–LUMO gap (potential difference between the first reversible oxidation and first reversible reduction of the macrocycle) and compare the results with data from computational and spectral studies. As discussed earlier, the computational and experimental plots of

HOMO–LUMO gap vs. number of bromo substituents plots yielded two segments corresponding, respectively, to the less brominated, planar and more brominated, nonplanar porphyrins (Figure 9.5) (*vide supra*).

The investigated ($TPPBr_x$)MnCl, ($TPPBr_x$)FeCl, and ($TPPBr_x$)Co derivatives revealed both metal-centered and ring-centered reduction and oxidation processes[7,12,14,15,73,74]. Detailed discussions of the electrochemical reduction and oxidation behavior of these complexes along with the resulting electrochemical trends with change in the number of β-pyrrole bromo substituents on the macrocycle are discussed separately in the following sections.

4.1. Electroreduction Behavior of β-Pyrrole Brominated Metalloporphyrins

The reduction of ($TPPBr_x$)MCl and ($TPPBr_x$)M, where $x = 0$–8, $TPPBr_x =$ the dianion of β-pyrrole brominated tetraphenylporphyrin and M = Fe, Mn, Co or Zn, was investigated in benzonitrile containing 0.1 M tetra-*n*-butylammonium perchlorate as supporting electrolyte[12,14,15,73,74]. Table 9.1 lists the reduction and oxidation potentials of these compounds and Figure 9.5 shows the trends in first oxidation and first reduction potentials as a function of the number of bromo substituents on the investigated compounds. As mentioned earlier, two well-defined reductions are observed for the ($TPPBr_x$)Zn complexes[73] as well as for the free-base, Cu, Ni, Pd, Pt, and VO complexes of $TPPBr_8$[7]. In the case of ($TPPBr_x$)Zn, the potentials for both ring-centered reductions are progressively shifted in a more positive direction with increase in the number of Br substituents. A linear relationship is seen between $E_{1/2}$ for the first reduction and the number of Br groups (slope = 61 mV/Br, Figure 9.6), but two linear segments are seen in the case of the second reduction[73]. The absolute potential difference between the two reductions of ($TPPBr_x$)Zn in PhCN thus decreases from 0.42 to 0.32 V with increase of the number of Br groups from 0 to 8.

The first reductions of ($TPPBr_x$)FeCl, ($TPPBr_x$)MnCl, and ($TPPBr_x$)Co are all metal centered. In each case, these reductions shift in a positive direction with increase in the number of Br substituents and a single linear free-energy relationship can be observed between $E_{1/2}$ and the number of Br groups. The slope of the plot of $E_{1/2}$ vs. the number of Br groups is 51 mV for ($TPPBr_x$)FeCl, 47 mV for ($TPPBr_x$)MnCl, and 61 mV for both ($TPPBr_x$)Co (Figure 9.6). Reduction of the cobalt complexes leads to a Co(I) porphyrin and this $Co^{II/I}$ process has the same slope (61 mV) as the first ring-centered reduction of (TPP)Zn, i.e., 61 mV. Computational studies predicts that porphyrin ring deformation should have a minimum effect on the reduction potentials[14,15] and hence, the observed linear trend for all four series of complexes in Figure 9.6 can be simply explained as due to an inductive effect of the electron-withdrawing halogen substituents.

The first reduction of ($TPPBr_x$)FeCl in PhCN results in an equilibrium mixture of $[(TPPBr_x)Fe^{II}Cl]^-$ and $(TPPBr_x)Fe^{II}$ [15,78]. The UV–visible spectral data suggest that a dissociation of the Cl^- ligand from $[(TPPBr_x)Fe^{II}Cl]^-$ is favored for compounds with 0–4 Br groups and rapidly occurs for the complexes with 0 or 1 Br group. However, the dominant species in solution is $[(TPPBr_x)Fe^{II}Cl]^-$ for compounds with 6–8 Br groups[15]. Spectroelectrochemical and ESR data indicate

Table 9.1. Electrochemical Potentials ($E_{1/2}$ vs. SCE) for ($TPPBr_x$)M (where M = Zn, Co, FeCl, or MnCl) in Benzonitrile Containing 0.1 M (TBAP)PF_6 (for Zn and FeCl), or CH_2Cl_2 containing 0.1 M (TBAP)PF_6 (for MnCl), or 0.2 M (TBAP)PF_6 (for Co). Scan rate = 100 mV s^{-1}. (Data taken from refs. [12, 14, 15, 73, 74].)

Compound	3rd oxidation	2nd oxidation	1st oxidation	1st reduction	2nd reduction	3rd reduction
($TPPBr_0$)Zn		1.14	0.82	−1.32	−1.74	
($TPPBr_1$)Zn		1.18	0.88	−1.25	−1.68	
($TPPBr_2$)Zn		1.17	0.91	−1.19	−1.58	
($TPPBr_3$)Zn		1.16	0.93	−1.10	−1.42	
($TPPBr_4$)Zn		1.19	0.95	−1.03	−1.32	
($TPPBr_5$)Zn		1.15	0.96	−0.97	−1.28	
($TPPBr_6$)Zn		1.11	0.96	−0.93	−1.20	
($TPPBr_7$)Zn		1.14	0.97	−0.92	−1.15	
($TPPBr_8$)Zn		1.15	0.96	−0.82	−1.15	
($TPPBr_0$)Co	1.30	1.10	0.90	−0.84	−1.97	
($TPPBr_1$)Co	1.31	1.12	0.91	−0.79		
($TPPBr_2$)Co	1.33	1.14	0.92	−0.73		
($TPPBr_3$)Co	1.33	1.15	0.94	−0.67		
($TPPBr_4$)Co	1.36	1.20	0.95	−0.61		
($TPPBr_5$)Co	1.37	1.22	0.99	−0.52		
($TPPBr_6$)Co	1.40	1.27	0.99	−0.46		
($TPPBr_7$)Co	1.44	1.32	0.97	−0.42		
($TPPBr_8$)Co	1.47	1.33	0.98	−0.37		
($TPPBr_0$)FeCl		1.52	1.20	−0.29	−1.06	−1.73
($TPPBr_1$)FeCl		1.52	1.24	−0.26	−1.07	−1.73
($TPPBr_2$)FeCl		1.51	1.28	−1.18	−1.01	−1.61
($TPPBr_3$)FeCl		1.53	1.27	−0.13	−0.97, −1.26[a]	−1.60
($TPPBr_4$)FeCl		1.56	1.26	−0.07	−0.86, −1.18[a]	−1.37
($TPPBr_5$)FeCl		1.58	1.26	−0.02	−0.83, −1.14[a]	−1.36
($TPPBr_6$)FeCl		1.60	1.24	0.04	−0.73, −1.17[a,b]	−1.28
($TPPBr_7$)FeCl		1.62	1.21	0.06	−0.67, −1.13[a,b]	−1.24
($TPPBr_8$)FeCl		1.64	1.19	0.10	−0.62, −0.97[a,b]	−1.20
($TPPBr_0$)MnCl			1.17	−0.29		
($TPPBr_2$)MnCl			1.26	−0.26		
($TPPBr_3$)MnCl			1.29	−0.09		
($TPPBr_4$)MnCl			1.30	−0.06		
($TPPBr_6$)MnCl			1.37	0.02		
($TPPBr_7$)MnCl			1.36	0.05		
($TPPBr_8$)MnCl			1.38	0.09		

[a]Corresponding to the reduction of ($TPPBr_x$)Fe^{II}/[($TPPBr_x$)Fe]$^-$ and [($TPPBr_x$)Fe^{II}Cl]$^-$/[($TPPBr_x$)Fe]$^-$ forms of the complex.
[b]E_{pc} at 100 mV s^{-1}.

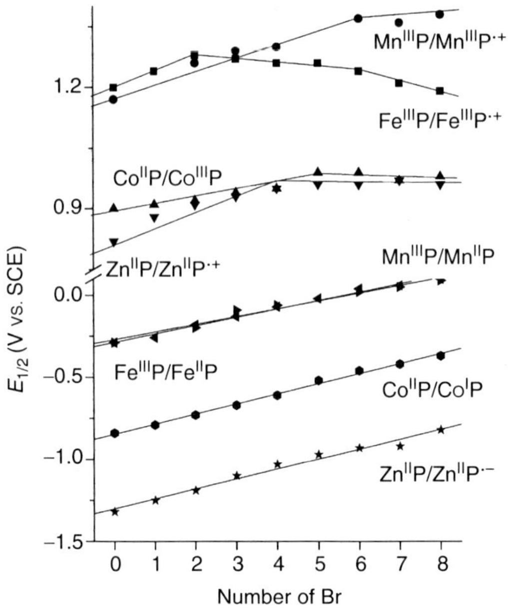

Figure 9.6. Plots of the first oxidation potentials and first reduction vs. the number of Br groups on the macrocycle of $(TPPBr_x)FeCl$, and $(TPPBr_x)Zn$ in PhCN, 0.1 M TBAP, $(TPPBr_x)MnCl$ in CH_2Cl_2, 0.1 M TBAP, and $(TPPBr_x)Co$ in CH_2Cl_2, 0.2 M $(TBA)PF_6$. (Data taken from ref. [12, 14, 15, 73, 74] for construction of the plot.)

that the second reduction of $(TPPBr_x)FeCl$ leads to a loss of the chloride axial ligand and $[(TPPBr_x)Fe^I]^-$ is then the only porphyrin present in solution[15]. The third electroreduction of $(TPPBr_x)FeCl$ has been assigned as a reduction at the macrocycle to give an Fe(I) porphyrin π-anion radical[15]. The first reduction of each $(TPPBr_x)Mn^{III}Cl$ complex is also reversible. Spectroelectrochemical data indicate that Mn(II) porphyrins are formed after the first reduction in PhCN.

The first reduction of $(TPPBr_x)Co^{II}$ is reversible and $[(TPPBr_x)Co^I]^-$ is generated after the first reduction. A further reduction at the macrocycle leads to a stepwise elimination of Br^-, thus giving a series of $[(TPPBr_x)Co^I]^-$ complexes with smaller and smaller values of x and $(TPP)Co^I$ is ultimately formed as the final porphyrin electroreduction product in PhCN. These results were confirmed by experiments carried out at a rotating ring disk electrode and also by thin-layer spectroelectrochemistry (*vide infra*)[12].

4.2. Electrochemical Oxidation Behavior of β-Pyrrole Brominated Metalloporphyrins

Studies on the oxidation of brominated porphyrins revealed the most interesting behavior. A nonlinear trend in the $E_{1/2}$ for oxidation was first observed by Callot for porphyrins containing 1-4 β-pyrrole substituents[68], and later systematic work on the full series of brominated porphyrins confirmed the initial reports[12, 14, 15, 73, 74]. Nonplanar distortions are believed to be responsible for these nonlinear trends. As shown in Figure 9.6, the shift in $E_{1/2}$ as a function of Br groups consists of two segments for first oxidation of $(TPPBr_x)Zn$ in PhCN, 0.1 M TBAP. The first slope comprises compounds with 0–4 Br groups and second those with 4–8 Br groups. This result parallels the spectroscopic data (*vide supra*) and is similar to what is seen for Fe, Mn, and Co porphyrins which have the same set of β-pyrrole Br-substituted macrocycles (Table 9.1). The two one-electron oxidations of $(TPPBr_x)Zn$ are both porphyrins ring-centered and lead to formation of π-cation radicals and dications[73]. The second oxidation of $(TPPBr_x)Zn$ shifts in a negative direction with increase in the number of Br substituents but this reaction exhibits a rather small substituent effect from the Br groups for the derivatives with $x = 5, 6$, or 8. Because of the non-linear shift in the first oxidation, the absolute potential difference between the two oxidations of $(TPPBr_x)Zn$ in PhCN decreases from 0.32 to 0.19 V with increase in the number of Br groups from 0 to 8[73]. For the same reason, the HOMO–LUMO gap of $(TPPBr_x)Zn$ also decreases with increase of the number of Br groups. The HOMO–LUMO gap in PhCN is 2.14 V when $x = 0$ but decreases to 1.78 V when $x = 8$[73]. The electrochemical results parallel data from theoretical calculations using the AM1-CI method and are consistent with earlier calculations performed by other groups on nonplanar porphyrins (*vide supra*)[79–81].

The Co(II) and Ag(II) derivatives show three one-electron oxidations because of the additional metal-centered, $M^{II/III}$, oxidation process[7,12,14,73,74] in PhCN or CH_2Cl_2 containing 0.1 M TBAP. As predicted from computational studies, the porphyrin ring distortion of the brominated compounds results in a decreased stability of the HOMO level and has a large effect on the oxidations. For this reason a nonlinear relationship was observed between $E_{1/2}$ for the first oxidation and the number of Br groups on all of the studied compounds in Figure 9.6.

In the case of $(TPPBr_x)FeCl$ in PhCN containing 0.1 M TBAP, each porphyrin undergoes two reversible one-electron oxidations[15], leading stepwise to an iron(III) π-cation radical and dication. The first reversible oxidation potential initially shifts in the expected anodic direction and reaches a maximum positive value for $(TPPBr_2)FeCl$ after which $E_{1/2}$ changes direction and then shifts negatively upon going from $x = 3$ to $x = 8$. In contrast, the second oxidation of the same compounds is shifted negatively upon going from $x = 0$ to $x = 2$ and then positively from $x = 3$ to $x = 8$; thus, the absolute potential difference ($\Delta E_{1/2}$) between the two ring-centered oxidations of $(TPPBr_x)FeCl$ varies with the number of Br groups on the macrocycle and increases (linearly) from 230 mV in the case of $(TPPBr_2)FeCl$ to 450 mV in the case of $(TPPBr_8)FeCl$[15].

(TPP)MnCl also undergoes two one-electron oxidations in CH_2Cl_2 or PhCN, 0.1 M TBAP. The first oxidation leads to formation of a Mn(III) π-cation radical at room temperature[2] but Mn(IV) is formed at low temperature[82]. The same assignment was proposed[77] for each $(TPPBr_x)MnCl$ complex, based on UV-visible and ESR spectra of the singly oxidized species.

Three one-electron oxidations are observed for $(TPPBr_x)Co$ in PhCN containing 0.1 M TBAP[12]. The first oxidation of cobalt(II) porphyrins can occur at either the metal center or at the conjugated macrocycle depending upon the strength of the counteranion and/or the presence of any coordinating ligands in solution[83] and a differentiation of these two processes can often be made on the basis of IR or UV–visible spectroscopic data[84–90]. A Co(III) complex is formed after the first oxidation of (TPP)Co in PhCN[83] and this is also the case after the abstraction of one electron from ($TPPBr_x$)Co in the same solvent. The second and third oxidations are macrocycle centered, leading to a Co(III) porphyrin π-cation radical and dication[12].

Three oxidations are also seen for ($TPPBr_x$)Co in CH_2Cl_2 containing 0.2 M (TBA)PF_6[74]. The first oxidation linearly shifts in a positive direction for compounds with $x = 0$–5 but shifts in a slightly negative direction for compounds with 6–8 Br groups. Both the second and third oxidations shift in a positive direction as the number of Br groups on the macrocycle is increased and there is an increased substituent effect for compounds with 6–8 Br groups. Consequently, the absolute potential difference between the first two oxidations of ($TPPBr_x$)Co remains virtually constant upon going from (TPP)Co to ($TPPBr_5$)Co but increases by 120 mV upon going from ($TPPBr_5$)Co to ($TPPBr_8$)Co. Likewise, the potential separation between the second and third oxidations of ($TPPBr_x$)Co decreases from 200 mV in the case of (TPP)Co to 140 mV in the case of ($TPPBr_8$)Co.

Two linear segments are seen in plots of $E_{1/2}$ for the first oxidation of ($TPPBr_x$)Co in CH_2Cl_2, 0.2 M $TBAPF_6$ vs. the number of Br groups on the porphyrin (Figure 9.6). The slope of the first segment is 17 mV (for ($TPPBr_x$)Co where $x = 0$–5) while that of the second is −5 mV ($x = 5$–8). The electrochemical data suggest that cobalt porphyrins with > 5 Br groups are oxidized at the porphyrin π-ring system; hence, the same substituent effect is observed for oxidation of ($TPPBr_x$)FeCl with $x = 2$–5 and ($TPPBr_x$)Co with $x = 6$–8. The spectroelectrochemical data[74] are consistent with the electrochemical data and suggest that the initial site of electron transfer for the first oxidation of ($TPPBr_x$)Co in CH_2Cl_2 containing 0.2 M (TBA)PF_6 shifts from the central metal to the macrocycle as x is increased above 5 (*vide infra*).

In another interesting study[69], the electrochemical oxidation and reduction of Ni(II) and Cu(II) complexes of a series of β-octahalogenao-*meso*-tetrakis(*p*-X-phenylporphyrins, (where halo = Cl or Br and X = CH_3, H, F, Br, COOMe, CF_3, NO_2) were investigated and the data were analyzed by means of the Hammett equation. Significant differences in electrochemical behavior were observed for the brominated porphyrins. Density functional calculations (DFT) suggested that these differences might be due to the differences in saddling-induced metal ($d_{x_2-y_2}$) to porphyrin (a_{2u}) orbital interaction between the Ni and Cu porphyrins.

4.3. Electrochemical Behavior of Water-Soluble β-Pyrrole Brominated Porphyrins

The electrochemical behavior of water-soluble β-pyrrole brominated porphyrins is more complex than that of their water insoluble analogs. The metal-centered redox reactions of (TMPyP)Mn^{III} and ($TMPyPBr_8$)Mn^{II} are reversible while the majority of porphyrin ring-centered redox reactions of the free-base, Cu, and Mn derivatives of $TMPyPBr_8$ are irreversible[31]. The metal-centered oxidation of ($TMPyPBr_8$)Mn^{II} is anodically shifted by $\sim$ 420 mV compared to $E_{1/2}$ for the corresponding reaction of (TMPyP)Mn^{II} (Table 9.2). The metal-centered

Table 9.2. Redox Potentials (V vs. NHE) for the Water-Soluble, β-Pyrrole Brominated Porphyrins in Aqueous Media

Compound	1st ox	1st red	2nd red	Ref
(TMPyP)Co[a]	0.45	−0.68		28–30
$(TMPyPBr_8)Co$[a]	0.53	−0.31		28–30
(TPPS)Co[a]	0.35	−0.90		28–30
$(TPPSBr_8)Co$[a]	0.50	−0.49		28–30
$(TMPyP)H_2$[b]		−0.33, −1.14[c]		31
$(TMPyPBr_8)H_2$[b]		−0.01, −1.09[c]		31
$(TMPyP)Mn^{III}$[b]		−0.51, −1.19[c]	0.06	31
$(TMPyPBr_8)Mn^{II}$[b]		−0.36, −1.11[c]	0.48	31
(TMPyP)Cu[b]		−0.48, −0.63[c]		31
$(TMPyPBr_8)Cu$[b]		−0.13, −0.35[c]		31
$(TBAP)Mn^{III}$			−0.19	32, 67
$(TBAPBr_8)Mn^{III}$			0.13	32, 67

[a]In phosphate buffer (pH = 7) containing 0.1 M NaCl, reference = Ag/AgCl.
[b]Conditions: 0.5 mM porphyrins, 0.05 M phosphate buffer, 0.1 M NaCl, pH = 7.80, 2 mM *N*-methyl imidazole, 0.5 mM $Na_2S_2O_4$.
[c]Cathodic peak potentials corresponding to the first two ring-centered reductions.

reduction of the negatively charged water-soluble $(TBAPBr_8)Mn^{III}$ is located at 0.13 V vs. NHE and this can be compared with an $E_{1/2} = -0.19$ V vs. SHE for $(TBAP)Mn^{III}$ [32]. Thus, a shift of ~ 320 mV in the $Mn^{III/II}$ reaction is observed as a result of the eight bromine substituents at the β-pyrrole positions of the porphyrin macrocycle. The reduction potential of $(TBAPBr_8)Mn^{III}$ is more negative than that reported for the $(TMPyPBr_8)Mn^{II}$(Table 9.2).

A detailed electrochemical characterization of two water-soluble cobalt porphyrins, $(TMPyPBr_8)Co^{II}$ and $(TPPSBr_8)Co^{II}$, was carried out under different pH (pH = 2–13) conditions[30]. As predicted[73], the investigated nonplanar porphyrins were found to be electron deficient owing to the presence of the eight Br groups at the β-pyrrole positions of the macrocycle. The half-wave potentials corresponding to both the first oxidation and first reduction were shifted positively as compared with redox potentials of the respective unbrominated porphyrin derivatives (Table 9.2) and such changes were found to be larger for the reductions, a trend that generally agrees with other $TPPBr_8$ derivatives that have been examined in nonaqueous solutions. The peak potentials for the first irreversible oxidation of both $(TMPyPBr_8)Co^{II}$ and $(TPPSBr_8)Co^{II}$ were pH dependent[30].

5. Spectroelectrochemical Studies of β-Pyrrole Brominated Metalloporphyrins

Utilization of spectroelectrochemistry involving UV–visible, FT-IR, and ESR techniques along with the electrochemical methods, has yielded information about the site of electron transfer and provided an opportunity to obtain the spectrum of oxidized and reduced forms of β-pyrrole brominated porphyrins. In the absence of X-ray structural data on oxidized and reduced brominated porphyrins, the spectral data should be helpful in predicting the structural and electronic effects of the substituents on the properties of the compounds. The following paragraphs summarize the spectral features of brominated metalloporphyrins, determined by spectroelectrochemical methods.

Electrochemical oxidation of cobalt(II) porphyrins can occur either at the metal center to yield a cobalt(III) porphyrin complex or at the porphyrin macrocycle to yield a cobalt(II) porphyrin π-cation radical depending upon the solution conditions[2]. Spectroelectrochemical studies on water-soluble $(TMPyPBr_8)Co^{II}$ and $(TPPSBr_8)Co^{II}$ revealed the formation of a cobalt(III) complex during the first oxidation. Both the Soret and visible bands were red shifted without significant loss of peak intensity. Moreover, no new absorption bands corresponding to the formation of π-cation radicals were observed. Similarly, the first reduction of $(TPPSBr_8)Co^{II}$ results in formation of a cobalt(I) complex as revealed by a split Soret band[30], but the site of electroreduction in the case of the $(TMPyPBr_8)Co^{II}$ could not be confirmed. Involvement of either a $Co^{II/I}$ or the peripheral *N*-methylpyridiniumyl groups was suggested[30] and this is in contrast to $(TMPyP)Co^{II}$ where a $Co^{II/I}$ reaction during the first electroreduction in dimethylformamide was confirmed spectroelectrochemically[91]. Debromination of the β-pyrrole Br substituents of $(TMPyPBr_8)Co$ was observed to occur at more negative potentials (> -0.65 V)[30]. The UV–visible spectrum obtained after bulk electrolysis of $(TMPyPBr_8)Co$ at -1.0 V vs. Ag/AgCl followed by reoxidation at 0.2 V to neutralize the reduced product indicated complete elimination of the Br substituents from the porphyrin periphery, thus giving (TMPyP)Co as a final product. This observation, along with observations discussed earlier for $(TPPBr_x)Co$ in benzonitrile[12], collectively suggest that the brominated cobalt porphyrins undergo debromination reactions upon adding a second electron to the porphyrin macrocycle.

In the investigated $(TPPBr_x)Co$ series, the Soret band of the singly reduced or singly oxidized cobalt porphyrin is also red shifted upon going from (TPP)Co to $(TPPBr_x)Co$ in PhCN containing 0.1 M TBAP. For Co(II) porphyrins, the Soret band shifts by 41 nm upon going from TPP to $TPPBr_8$ of the macrocycle but the Soret bands of the Co(I) and Co(III) porphyrins shift by 36 and 39 nm, respectively, when the number of Br groups is increased from 0 to 8[12]. However, the magnitude of the Br substituent effect depends upon the oxidation state of the cobalt ion. This is shown in Figure 9.7 as an example for the cobalt(II) and cobalt(III) complexes as a function of number of bromo substituents on the porphyrin ring. While a linear plot is obtained for the cobalt(II) complexes, a similar plot for the cobalt(III) complexes revealed two segments, suggesting different degrees of nonplanarity of the β-pyrrole brominated porphyrin macrocycle depending upon the oxidation state of the central metal ion. Interestingly, the UV-visible spectrum of singly oxidized $(TPPBr_x)Co$ ($x = 6$–8) in CH_2Cl_2 containing 0.2 M $TBAPF_6$ differs from the spectra of the same singly oxidized compounds in PhCN containing 0.1 M TBAP[74]. That is, the spectra of the singly oxidized porphyrins

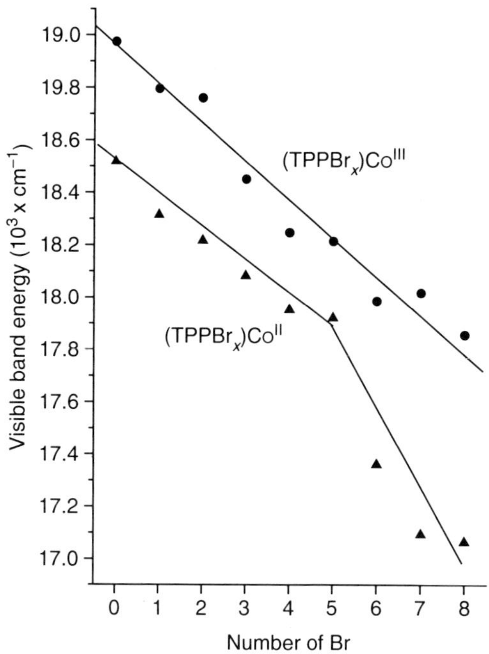

Figure 9.7. Plots of visible band energy vs. number of bromo substituents for the $(TPPBr_x)Co^{II}$ and $(TPPBr_x)Co^{III}$ complexes. (Data taken from Ref. 12 and 74 for construction of the plot.)

in CH_2Cl_2 containing 0.2 M $TBAPF_6$ exhibit broad, low-intensity Soret bands along with broad visible bands and were assigned as representing Co(II) porphyrin π-cation radicals. In contrast, the spectra in PhCN containing 0.1 M TBAP are characterized by a sharp Soret band with well-defined visible bands which were assigned to a cobalt(III) complex with an uncharged porphyrin ring[12].

The Soret band of singly oxidized $(TPPBr_x)FeCl$ in PhCN decreases in intensity upon the first oxidation as compared to neutral $(TPPBr_x)FeCl$ and there is also a 1–14 nm blue shift of the bands, with the exact magnitude of the shift depending upon the number of Br groups on the macrocycle[14,15]. This result indicates that the first oxidation of each $(TPPBr_x)FeCl$ complex is porphyrin ring centered. The Soret band of reduced $(TPPBr_x)FeCl$ is red shifted as compared to the neutral complexes. The magnitude of the shift ranges from 13 nm for (TPP)FeCl to 22 nm for $(TPPBr_7)FeCl$. There is also an increase in intensity of the Soret band upon the first reduction of $(TPPBr_x)FeCl$ to give an Fe(II) complex. The Soret band of neutral $(TPPBr_x)FeCl$ in PhCN shifts by 44 nm as the number of Br groups is increased from 0 to 7. However, the Soret bands of reduced and oxidized $(TPPBr_x)FeCl$ shift to the red by 53 and 29 nm, respectively[15]. Figure 9.8 illustrates the spectral trends observed for the ferrous and ferric forms of $(TPPBr_x)Fe$ as a function of number of Br substituents. Unlike what is observed in the cobalt series, linear trends were observed for both ferrous

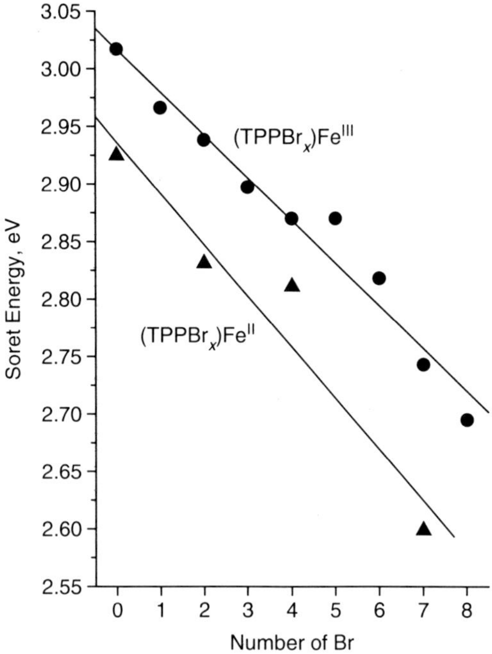

Figure 9.8. Plots of visible band energy vs. number of bromo substituents for the (TPPBr$_x$)Fe^{II} and (TPPBr$_x$)Fe^{III}Cl complexes. (Data taken from ref. 14 and 15 for construction of the plot).

and ferric porphyrins. However, the slope of the straight line plot was found to be higher for the ferrous form compared to the ferric form of the porphyrin complexes. This result indicates that the effect of Br substituents on the spectral data depends on the oxidation state of the metal porphyrin and follows the order: (TPPBr$_x$)Fe^{II} > (TPPBr$_x$)Fe^{III}Cl > [(TPPBr$_x$)Fe^{III}Cl]$^{+\cdot}$.

6. Effect of the Br Substituents on Axial Ligand Binding

The binding constants for complexation of nitrogenous bases to (TPPBr$_8$)Zn are an order of magnitude higher as compared to the binding constants for the corresponding adducts of (TPP)Zn and vary in the following order: piperidine > imidazole > pyridine > 3-methylpyridine > pyridine-3-carbaldehyde[8]. The enhanced binding constants and large spectral shifts were interpreted in terms of the increased electrophilicity of (TPPBr$_x$)Zn induced by the electron-withdrawing Br substituents on the porphyrin core.

Symmetrically substituted perhaloporphyrins, 2,3,7,8,12,13,17,18-octahalo (bromo and chloro)-5,10,15,20-tetraphenylporphyrins (TPPBr$_x$)H$_2$ and (TPPCl$_8$) H$_2$ have been examined as receptors for binding to Lewis bases[92]. Hydrogen bonding between the pyrrolic NH protons with solvent/bases as a binding mechanism was suggested[92]. The base binding to these halogenated porphyrins revealed enhanced binding constants for (TPPBr$_x$)H$_2$ relative to (TPPCl$_8$)H$_2$ and followed a linear trend with increase in pKa values of the bases. The binding of Lewis bases to the halogenated porphyrin core was more influenced by the extent of nonplanarity of the porphyrin core than the electron-withdrawing effect of the substituents. The higher binding constants of $(TPPBr_8)H_2$ relative to $(TPPCl_8)H_2$ were interpreted in terms of a greater nonplanarity of the former complexes as compared to the latter[92].

The pyridine-binding reaction of $(TPPBr_x)$Co and $(TPPBr_x)$Zn in CH_2Cl_2 were monitored by electrochemical methods (for cobalt) and by UV–visible spectroscopy (for zinc)[77]. Only a mono-pyridine adduct was formed for both series of complexes under the utilized experimental conditions. In general, higher values of binding constants are observed for brominated porphyrins than nonbrominated porphyrins and, as mentioned earlier, this is in large part due to the increased electrophilicity of the porphyrin ring as a result of electron-withdrawing bromo substituents. It was seen that the pyridine-binding constants for both $(TPPBr_x)$Zn and $(TPPBr_x)$Co show a nonlinear relationship with the number of Br groups on the macrocycle but a linear relationship is seen between log K for pyridine binding and the number of Br groups on the macrocycle of $(TPPBr_x)$Zn[77]. Figure 9.9

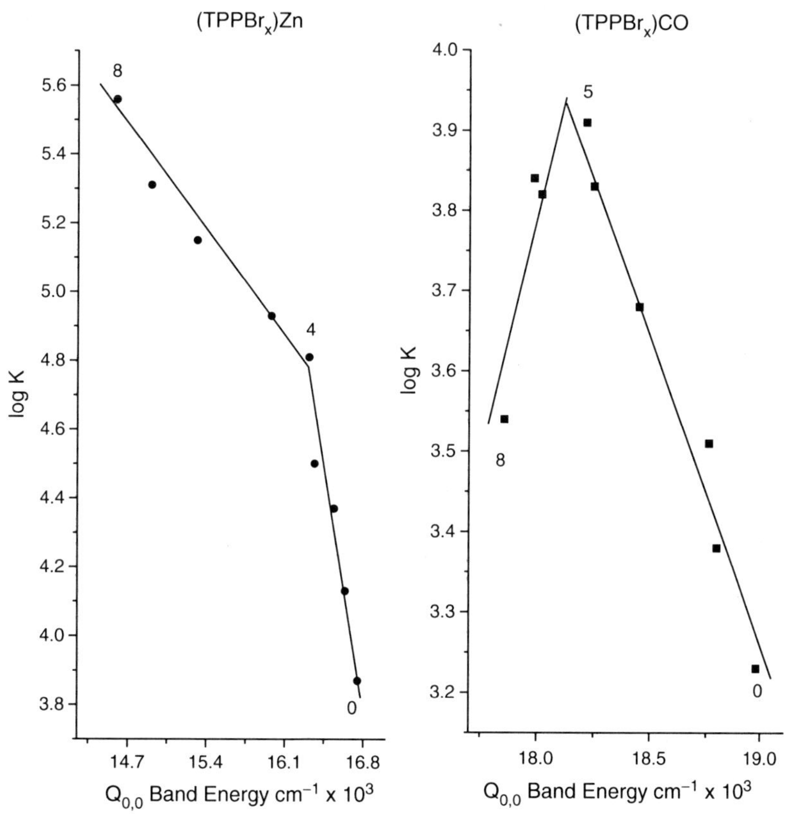

Figure 9.9. Plots of log K vs. $Q_{0,0}$ for pyridine binding of different β-pyrrole brominated zinc and cobalt porphyrins. (Data taken from ref. [77] for construction of the plot).

shows the dependence of log *K* on the low-energy visible band of these two series of porphyrins. Two segments are clearly observed. One is attributed to the planar porphyrins of the partially brominated complexes and the other to the nonplanar porphyrins with a larger number of bromo groups. These results suggest that the ring nonplanarity reduces the axial-binding ability to some extent although with increased bromination the values of *K* for base binding increase due to the increased electrophilicity of the porphyrin ring.

The effect of increased electrophilicity of the porphyrin ring as a result of bromination and the resulting nonplanarity effects are better shown in the case of Lewis acid type ligand (*viz.* carbon monoxide) binding to electrogenerated Fe(II) and Co(III) brominated porphyrins[15,74]. In the case of the iron complexes, the FTIR spectra were taken after the first reduction of $(TPPBr_x)FeCl$ (x = 0–8) in CH_2Cl_2 containing 0.2 M TBAP under a CO atmosphere. The spectra of the compounds with 0–6 Br groups show two IR bands at 1962–2043 cm^{-1} whose relative intensity depends on the number of Br groups. Under the same solution conditions, the spectrum of the electroreduced $(TPPBr_7)FeCl$ exhibits only a single band at 1980 cm^{-1} and no IR band is seen for the spectrum of electroreduced $(TPPBr_8)FeCl$, indicating a lack of CO binding.

The bands located between 1962 and 1980 cm^{-1} are assigned to a mono-CO adduct while the higher energy bands, located between 2028 and 2043 cm^{-1}, are assigned the asymmetric stretches of a bis-CO adduct. The intensity of the absorption peak corresponding to the bis-CO adduct decreases in intensity with increasing the bromo substituents and then vanishes completely for Br_7 and Br_8 porphyrin derivatives. However, for the Br_8 derivative, no evidence for either a mono- or bis-CO adduct was observed. Figure 9.10 illustrates ν_{CO} of mono- and bis-adducts of CO vs. the number of Br groups on the porphyrin ring. Two segments for both the mono- and bis-adducts were observed although $(TPPBr_8)FeCl$ did not show any evidence of CO binding as indicated above.

The electrogenerated $[(TPPBr_x)Co^{III}]^+$ derivatives with 0–5 Br groups on the macrocycle show two IR bands between 2109 and 2148 cm^{-1}. The higher energy bands at 2144–2148 cm^{-1} are assigned to the bis-CO adduct, $[(TPPBr_x)Co(CO)_2]^+$, while those at 2109–2112 cm^{-1} are assigned to the mono-CO adduct, $[(TPPBr_x)Co(CO)]^+$. The derivatives having 6–8 Br groups show no IR bands at all between 2100 and 2200 cm^{-1}. These results suggest that the electron deficiency of the porphyrin macrocycle caused by the Br substituents lowers the binding ability for CO binding[15].

7. Summary

The β-pyrrole brominated porphyrins have proven to be highly versatile N_4 ligands to study structural and electronic effects of the compounds as a function of both the number of bromo substituents on the macrocycle and the type or oxidation state of the metal ion in the porphyrin central cavity. As outlined in this chapter, the studies performed so far on these compounds have revealed pronounced spectral shifts as a function of changes in degree of bromo substitution or nature of the solvent. For all the studied brominated MN_4 complexes, the electrochemical

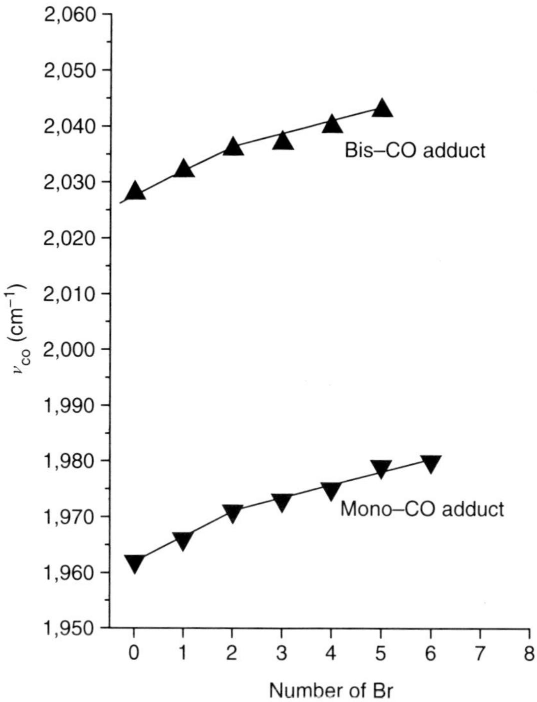

Figure 9.10. Plots illustrating ν_{CO} of mono- and bis-adducts of CO binding to $(TPPBr_x)FeCl$ vs. the number of Br groups on the macrocyle. (Data taken from ref. [15].)

reduction is linearly related to the number of β-pyrrole bromo substituents and there is no obvious effect on the reduction potentials as a result of the porphyrin ring distortion. Interestingly, as predicted from computational studies, the electrochemical oxidation shows a nonlinear dependence of $E_{1/2}$ on the number of bromo substituents at the β-pyrrole position of the macrocycle. In general, for a given MN_4 complex, the site of electron transfer was found to be similar to that observed for unbrominated analogs, a key piece of information that is useful to arrive at mechanistic details of catalytic reactions involving these MN_4 complexes. Spectroelectrochemical studies have also shown different spectral trends with respect to the oxidation state of the brominated porphyrins. However, some of the investigated brominated porphyrins have been found to be unstable at extremely cathodic potentials and undergo debromination. The brominated MN_4 complexes also display higher axial ligand-binding ability toward Lewis base addition as a result of electrophilicity of the porphyrin ring although the nonplanarity of the porphyrin ring seems to reduce the K values to some extent. Interestingly, an opposite trend was observed for the binding of a Lewis acid, carbon monoxide as axial ligand. A metalloporphyrin-CO adduct was never observed for either Co(III) or Fe(III) octabrominated porphyrins. These properties might be useful in gauging the substrate-binding properties in both catalytic and electrocatalytic applications of β-pyrrole brominated porphyrins.

Acknowledgments

The support of the Robert A. Welch Foundation (Grand E-680, KMK), the Petroleum Research Funds administered by the American Chemical Society, and National Institutes of Health (to FD) is gratefully acknowledged.

References

1. Kadish, K.M., K.M. Smith, R. Guilard (eds) (2001, 2003). *The Porphyrin Handbook*, Vol. 1–20. Academic Press, New York.
2. Kadish, K.M., E. Van Caemelbecke, and G. Royal (2000). In K.M. Kadish, K.M. Smith, and R. Guilard (eds), *The Porphyrin Handbook*, Vol. 8, Chapter 55. Academic Press, New York, pp. 1–114.
3. Kadish, K.M., G. Royal, E. Van Caemelbecke, and L. Gueletti (2000). In K.M. Kadish, K.M. Smith, and R. Guilard (eds), *The Porphyrin Handbook*, Vol. 9, Chapter 59. Academic Press, New York, pp. 1–219.
4. Senge, M.O. (2001). In K.M. Kadish, K.M. Smith, and R. Guilard (eds), *The Porphyrin Handbook*, Vol. 1, Chapter 6. Academic Press, New York, pp. 240–347. (b) Shelnutt, J.A., X.-Z. Song, J.-G. Ma, S.-L. Jia, W. Walter, and C.J. Medforth (1998). Nonplanar porphyrins and their significance in proteins. *Chem. Soc. Rev.* **27**, 31–42.
5. Ochsenbein, P., K. Ayougou, D. Mandon, J. Fischer, R. Weiss, R.N. Austin, K. Jayaraj, A. Gold, J. Terner, and J. Fajer (1994). Effect of conformation on the redox potentials of porphyrins halogenated in the β-pyrrole positions. *Angew Chem., Int. Ed. Engl.* **106**, 348–350.
6. Bhyrappa, P., M. Nethaji, and V. Krishnan (1993). Structure of nonplanar octabromo tetraphenylporphyrin and kinetics of rapid metalation reactions. *Chem. Lett.* **5**, 869–872.
7. Bhyrappa, P. and V. Krishnan (1991). Octabromotetraphenylporphyrin and its metal derivatives: Electronic structure and electrochemical properties. *Inorg. Chem.* **30**, 239–245.
8. Bhyrappa, P., V. Krishnan, and M. Nethaji (1993). Solvation and axial ligation properties of (2,3,7,8,12,13,17,18-octabromo-5,10,15,20-tetraphenylporphyrinato)zinc(II). *J. Chem. Soc., Dalton Trans.* **12**, 1901–1906.
9. Mandon, D., P. Oschsenbein, J. Fischer, R. Weiss, K. Jayaraj, R.N. Austin, A. Gold, P.S. White, O. Brigaud, P. Battioni, and D. Mansuy (1992). Beta-halogenated-pyrrole porphyrins. Molecular structures of 2,3,7,8,12,13,17,18-octabromo-5,10,15,20-tetramesitylporphyrin, nickel(II) 2,3,7,8,12,13,17,18-octabromo-5,10,15,20-tetramesitylporphyrin, and nickel(II) 2,3,7,8,12,13,17,18-octabromo-5,10,15,20-tetrakis (pentafluorophenyl)porphyrin. *Inorg. Chem.* **31**, 2044–2049.
10. Scheidt, W.R. and Y.R. Lee (1987). Recent advances in the stereochemistry of metallotetrapyrroles. *Struct. Bond. (Berlin)* **64**, 1–70.
11. Ravikanth, M. and T.K. Chandrashekar (1995). Nonplanar porphyrins and their biological relevance: Ground and excited state dynamics. *Struct. Bond. (Berlin)* **82**, 105–188.
12. D'Souza, F., A. Villard, E. Van Caemelbecke, M. Franzen, T. Boschi, P. Tagliatesta, and K.M. Kadish (1993). Electrochemical and spectroelectrochemical behavior of cobalt(III), cobalt(II), and cobalt(I) complexes of meso-tetraphenylporphyrinate bearing bromides on the beta-pyrrole positions. *Inorg. Chem.* **32**, 4042–4048.

13. Kadish, K.M., E. Van Caemelbecke, F. D'Souza, C. Medforth, K.M. Smith, A. Tabard, and R. Guilard (1995). Electrochemistry and spectroelectrochemistry of sigma-bonded iron(III) porphyrins with nonplanar porphyrin rings. Reactions of (OETPP)Fe(R) and (OETPP)FeCl, where R = C_6H_5, C_6F_4H, or C_6F_5 and OETPP is the dianion of 2,3,7,8,12,13,17,18-octaethyl-5,10,15,20- tetraphenylporphyrin. *Inorg. Chem.* **34**, 2984–2989.
14. Kadish, K.M., F. D'Souza, A. Villard, M. Autret, E. Van Caemelbecke, P. Bianco, A. Antonini, and P. Tagliatesta (1994). Effect of porphyrin ring distortion on redox dotentials of beta-brominated-pyrrole Iron(III) tetraphenylporphyrins. *Inorg. Chem.* **33**, 5169–5170.
15. Tagliatesta, P., J. Li, M. Autret, E. Van Caemelbecke, A. Villard, F. D'Souza, and K.M. Kadish (1996). Electrochemistry and spectral characterization of oxidized and reduced ($TPPBr_x$)FeCl where $TPPBr_x$ is the dianion of β-brominated-pyrrole tetraphenylporphyrin and x varies from 0 to 8. *Inorg. Chem.* **35**, 5570–5576.
16. Baciocchi, E., T. Boschi, L. Cassioli, C. Galli, L. Jaquinod, A. Lapi, R. Paolesse, K.M. Smith, P. Tagliatesta (1999). Electronic effects on the stereoselectivity of epoxidation reactions catalyzed by manganese porphyrins. *Eur. J. Org. Chem.* **12**, 3281–3286.
17. Traylor, T.G. and S. Tsuchiya (1987). Perhalogenated tetraphenylhemins: Stable catalysts of high turnover catalytic hydroxylations. *Inorg. Chem.* **27**, 1338–1339.
18. Wijesekara, T., A. Matsumoto, D. Dolphin, and D. Lexa (1990). Highly chlorinated and perchlorinated meso-tetraphenylporphyrins, *Angew Chem., Int. Ed. Engl.* **102**, 1073–1074.
19. Watanabe, E., S. Nishimura, H. Ogoshi, and Z. Yoshida (1975). Orientation of electrophilic meso-substitution in metallooctaethylporphyrins. *Tetrahedron* **31**, 1385–1390.
20. Gong, L.G. and D. Dolphin (1985). Nitrooctaethylporphyrins: Synthesis, optical and redox properties. *Can. J. Chem.* **63**, 401–405.
21. Gong, L.G. and D. Dolphin (1985). Nucleophilic substitution of meso-nitrooctaethylporphyrins. *Can. J. Chem.* **63**, 406–411.
22. Wu, G.Z., W.X. Gan, and H.K. Leung (1991). Porphyrin chemistry. 6. Photophysical properties of meso-substituted octaethylporphines and their zinc complexes. *J. Chem. Soc., Faraday Trans.* **87**, 2933–2937.
23. Carrier, M.N., C. Scheer, P. Gouvine, J.F. Bartoli, P. Battioni, and D. Mansuy (1990). Biomimetic hydroxylation of aromatic compounds: Hydrogen peroxide and manganese-polyhalogenated porphyrins as a particularly good. *Tetrahedron Lett.* **31**, 6645–6648.
24. Giraudeau, H.J., J.J. Callot, I. Ezaher, and M. Gross (1979). Substituent effects in the electroreduction of porphyrins and metalloporphyrins. *J. Am. Chem. Soc.* **101**, 3857–3862.
25. Richards, R.A., K. Hammons, M. Joe, and G.M. Miskelly (1996). Observation of a stable water-soluble lithium porphyrin. *Inorg. Chem.* **35**, 1940–1944.
26. Tabata, M., J. Nishimoto, A. Ogata, T. Kusano, and N. Nahar (1996). Metalation of water-soluble octabromoporphyrin with lithium(I), cadmium(II), and mercury(II). *Bull. Chem. Soc. Jpn* **69**, 673–677.
27. D'Souza, F., G.R. Deviprasad, and M.E. Zandler (1997). Aggregation and axial ligand exchange behavior of water-soluble pyrrole-β brominated porphyrins. *J. Chem. Soc., Dalton Trans.* 3699–3703.
28. D'Souza, F., G.R. Deviprasad, and Y.Y. Hsieh (1996). Synthesis and studies on the electrocatalytic reduction of molecular oxygen by non-planar cobalt(II) tetrakis(N-methylpyridyl)-β-octabromoporphyrin. *J. Electroanal. Chem.* **411**, 167–171.

29. D'Souza, F., Y.Y. Hsieh, and G.R. Deviprasad (1997). Electrocatalytic reduction of molecular oxygen using non-planar cobalt tetrakis(4-sulfonatophenyl)-β-octabromoporphyrin. *J. Electroanal. Chem.* **426**, 17–21.
30. D'Souza, F., Y.Y. Hsieh, and G.R. Deviprasad (1998). Electrochemical and spectroelectrochemical characterization of water-soluble, β-pyrrole-brominated cobalt porphyrins. *J. Porphyrins Phthalocyanines* **2**, 429–437.
31. Batinic-Haberle, I., Liochev, I. Spasojevic, and I. Fridovich (1997). A potent superoxide dismutase mimic: Manganese β–octabromo-meso-tetrakis-(*N*-methylpyridinium-4-yl) porphyrin. *Arch. Biochem. Biophys.* **343**, 225–233.
32. Kachadourian, R., M.M. Flaherty, A.L. Crumbliss, M. Patel, and B.J. Day (2003). Synthesis and in vitro antioxidant properties of manganese(III) β-octabromo-meso-tetrakis(4-carboxyphenyl)porphyrin. *J. Inorg. Biochem.* **95**, 240–248.
33. Mansuy, D. (1993). Activation of alkanes: The biomimetic approach. *Coord. Chem. Rev.* **125**, 129–141.
34. Meunier, B. (1992). Metalloporphyrins as versatile catalysts for oxidation reactions and oxidative DNA cleavage. *Chem. Rev.* **92**, 1411–1456.
35. Crossley, M.J., P.L. Burn, S.S Chew, F.B. Cuttance, and I.A. Newsom (1991). Regiospecific introduction of four substituents to porphyrin systems at antipodal pyrrolenic positions. *J. Chem. Soc., Chem. Commun.* **21**, 1564–1566.
36. Hoffmann, P., A. Robert, and B. Meunier (1992). Preparation and catalytic activities of the manganese and iron derivatives of Br_8TMP and $Cl_{12}TMP$, two robust porphyrin ligands obtained by halogenation of tetramesitylporphyrin. *Bull. Soc. Chim. Fr.* **129**, 85–97.
37. Baciocchi, E., T. Boschi, L. Cassioli, C. Galli, A. Lapi, and P. Tagliatesta (1997). Epoxidation and hydroxylation reactions catalyzed by β-tetrahalo and β-octahalo manganese porphyrins. *Tetrahedron Lett.* **38**, 7283–7286.
38. Lyons, J.E. and P.E. Ellis, Jr. (1991). Selective low temperature hydroxylation of isobutane by molecular oxygen catalyzed by an iron perhaloporphyrin complex. *Catal. Lett.* **8**, 45–51.
39. Ellis, Jr., E. and J.E. Lyons (1989). Halogen substituent effects on the catalytic activity of iron porphyrin complexes for selective air-oxidation of alkanes in the liquid phase. *Catal. Lett.* **3**, 389–397.
40. Barkigia, K.M., L. Chantranupong, K.M. Smith, and J. Fajer (1988). Structural and theoretical models of photosynthetic chromophores. Implications for redox, light-absorption properties and vectorial electron flow. *J. Am. Chem. Soc.* **110**, 7566–7567.
41. Fajer, J., K.M. Barkigia, K.M. Smith, E. Zhong, E. Gudowska-Nowak, and M. Newton (1990). In Micheal Beyerle M.E. (ed), *Reaction Centers of Photosynthetic Bacteria*. Springer-Verlag, Berlin, p. 367.
42. Lyons, J.E. and P.E. Ellis (1994). *Metalloporphyrins in catalytic oxidations*. M and Dekker, New York, p. 297.
43. Mansuy, D. (1993). The activation of dioxygen and homogeneous catalytic oxidation. Plenum Press, New York and London, p. 347.
44. Giraudeau, H.J. Callot, and M. Gross (1979). Effects of electron-withdrawing substituents on the electrochemical oxidation of porphyrins. *Inorg. Chem.* **18**, 201–206.
45. Grinstaff, M.W., M.G. Hill, J.A. Labinger, and H.B. Gray (1994). Mechanism of catalytic oxygenation of alkanes by halogenated iron porphyrins. *Science* **264**, 1311–1313.
46. Ellis, Jr., P.E. and J.E. Lyons (1990). Selective air oxidation of light alkanes catalyzed by activated metalloporphyrins — the search for a suprabiotic system. *Coord. Chem. Rev.* **105**, 181–193.

47. Lyons, J.E., P.E. Ellis, Jr., R.W. Wagner, P.B. Thompson, H.B. Gray, M.E. Hughes, and J.A. Hodge (1992). Proc. American Chemical Socicty Division of Petroleum Chemistry, Symposium on Natural Gas Upgrading II. American Chemical Society, Washington, DC, p. 307.
48. Bartoli, J.F., O. Brigaud, P. Battioni, and D. Mansuy (1992). Hydroxylation of linear alkanes catalyzed by iron porphyrins: Particular efficacy and regioselectivity of perhalogenated porphyrins. *J. Chem. Soc., Chem. Commum.* **6**, 440–442.
49. Traylor, T.G., K.W. Hill, W. Fann, S. Tasuchiya, and B.E. Dunlap (1992). Aliphatic hydroxylation catalyzed by iron(III) porphyrins. *J. Am. Chem. Soc.* **114**, 1308–1312.
50. Bruice, T.C. and G. He (1991). Nature of the epoxidizing species generated by reaction of alkyl hydroperoxides with iron(III) porphyrins. Oxidations of cis-stilbene and (Z)-1,2-bis(trans-2, trans-3-diphenylcyclopropyl)ethene by tert-BuOOH in the presence of [meso-tetrakis(2,4,6-trimethylphenyl)porphinato]-, [meso-tetrakis(2,6-dichlorophenyl)-porphinato]-, and [meso-tetrakis(2,6-dibromophenyl)porphinato]iron(III) chloride. *J. Am. Chem. Soc.* **113**, 2747–2753.
51. Bruice, T.C. (1988). Mechanistic *Principles of Enzyme Activity*, Chapter 8. VCH Publishers, New York.
52. Ellis, Jr., P.E. and J.E. Lyons (1989). Effect of fluorination of the meso-phenyl groups on selective tetraphenylporphyrinatoiron(III)-catalyzed reactions of propane with molecular oxygen. *J. Chem. Soc Chem. Commun.* **41**, 1315–1316.
53. Renoud, J.P., P. Battioni, J.F. Bartoli, and D. Mansuy (1985). A very efficient system for alkene epoxidation by hydrogen peroxide: catalysis by manganese-porphyrins in the presence of imidazole. *J. Chem. Soc., Chem. Commun.* **13**, 888–889.
54. Bruice, T.C. and D. Ostovic (1988). Transition state geometry in epoxidation by iron-oxo porphyrin at the compound I oxidation level. Epoxidation of alkenes catalyzed by a sterically hindered meso-tetrakis(2,6-dibromophenyl)porphinato iron(III) chloride. *J. Am. Chem. Soc.* **110**, 6906–6908.
55. Meunier, B. (1983). Homogeneous-phase oxidations catalyzed by transition metals: Recent advances. *Bull. Soc. Chim. Fr.* **11-12**, 345–366.
56. Hoffmann, P., G. Labat, A. Robert, and B. Meunier (1990). Highly selective bromination of tetramesitylporphyrin: An easy access to robust metalloporphyrins, M-Br_8TMP and M-Br_8TMPS. Examples of application in catalytic oxygenation and oxidation reactions. *Tetrahedron Lett.* **31**, 1991–1994.
57. Gonsalves, A.M., R.A.W. Johnstone, M.M. Pereira, J. Shaw, and A.J.F. do N. Sobral (1991). Metal-assisted reactions. Part 22. Synthesis of perhalogenated porphyrins and their use as oxidation catalysts. *Tetrahedron Lett.* **32**, 1355–1358.
58. Labinger, J.A. (1994). A simplified model for catalyzed isobutane autoxidation: Implications for the mechanism of catalysis by halogenated porphyrin complexes. *Catal. Lett.* **26**, 95–99.
59. Dolphin, D., T.G. Traylor, and L.Y. Xie (1997). Polyhaloporphyrins: Unusual ligands for metals and metal-catalyzed oxidations. *Acc. Chem. Res.* **30**, 251–259.
60. George, R.G. and M.P.D.H. Padmanabhan (2003). Studies on iron(III) and manganese(III) derivatives of new meso-aryl substituted and brominated porphyrins. *Transit Met. Chem.* **28**, 858–863.
61. Grinstaff, M.W., M.G. Hill, E.R. Birnbaum, W.P. Schaefer, J.A. Labinger, and H.B. Gray (1995). Structures, electronic properties, and oxidation–reduction reactivity of halogenated iron porphyrins. *Inorg. Chem.* **34**, 4896–4902.
62. Birnbaum, E.R., W.P. Schaefer, J.A. Labinger, J.E. Bercaw, and H.B. Gray (1995). Electronic structures of halogenated ruthenium porphyrins. Crystal structure of $RuTFPPCl_8(CO)H_2O$ ($TFPPCl_8$ = Octa-beta-chlorotetrakis(pentafluorophenyl) porphyrin). *Inorg. Chem.* **34**, 1751–1755.

63. Sen, A. and V. Krishnan (2001). Novel covalently linked porphyrin trimers with redox-distinct properties. *Tetrahedron Lett.* **39**, 6539–6542. (b) Bhyrappa, P. B. Purushothaman, and P. Bhavana (2001). Perbrominated 2-nitrotetraphenylporphyrins: Electrochemical and axial ligation properties. *J. Chem. Soc., Chem. Perkin Trans.* **2**, 238–242.
64. DiMagno, S.G., A.K. Wertsching, and C.R. Ross, II (1995). Electronic consequences of nonplanar core conformations in electron-deficient porphyrins: The structure and spectroscopic properties of [5,10,15,20-tetrakis(heptafluoropropyl)porphinato] cobalt(II). *J. Am. Chem. Soc.* **117**, 8279–8280.
65. Ryeng, H. and A. Ghosh (2002). Do nonplanar distortions of porphyrins bring about strongly red-shifted electronic spectra? Controversy, consensus, new developments, and relevance to chelatases. *J. Am. Chem. Soc.* **124**, 8099–8103.
66. Bailey, S.L. and P. Hambright (2003). Kinetics of the reactions of divalent copper, zinc, cobalt, and nickel with a deformed water soluble centrally monoprotic porphyrin. *Inorg. Chem. Acta* **344**, 43–48.
67. Spasojevic, I. and I. Batinic-Haberle (2001). Manganese(III) complexes with porphyrins and related compounds as catalytic scavengers of superoxide. *Inorg. Chim. Acta.* **317**, 230–242.
68. Callot, H.J. (1974). Bromination of m-tetraphenylporphine. Preparation of alkyl derivatives and polycyanoporphines. *Bull. Soc. Chim. Fr.* **7–8**, 1492–1496.
69. Ghosh, I. Halvorsen, H.J. Nilsen, E. Steene, T. Wondimagegen, R. Lie, E.V. Caemelbecke, N. Guo, Z. Ou, and K.M. Kadish (2001). Electrochemistry of nickel and copper β-octahalogeno-meso-tetraarylporphyrins. Evidence for important role played by saddling-induced metal (d_{x2-y2})-porphyrine("a_{2u}") orbital interactions. *J. Phys. Chem. B* **105**, 8120–8124.
70. Boschi, T., G. D'Arcangelo, and P. Tagliatesta (1995). Fast atom bombardment mass spectral observations on β-pyrrole-substituted tetraphenylporphyrins. *J. Chem. Res. (s)* **8**, 326–327.
71. Tse, M.K., Z.Y. Zhou, T.C.W. Mak, and K.S. Chan (2000). Regioselective bromination and subsequent Suzuki cross-coupling of highly electron deficient 5,10,15,20-tetrakis(trifluoromethyl)porphyrin. *Tetrahedron* **56**, 7779–7783.
72. Kumar, P.K., P. Bhyrappa, and B. Varghese (2003). An improved protocol for the synthesis of antipodal β-tetrabromo-tetraphenylporphyrin and the crystal structure of its Zn(II) complex. *Tetrahedron Lett.* **44**, 4849–4851.
73. D'Souza, F., M.E. Zandler, P. Tagliatesta, Z. Ou, J. Shao, E. Van Caemelbecke, and K.M. Kadish (1998). Electronic, spectral, and electrochemical properties of ($TPPBr_x$)Zn where $TPPBr_x$ is the dianion of β-brominated-pyrrole tetraphenylporphyrin and x varies from 0 to 8. *Inorg. Chem.* **37**, 4567–4572.
74. Kadish, K.M., J. Li, E. Van Caemelbecke, Z. Ou, N. Guo, M. Autret, F. D'Souza, and P. Tagliatesta (1997). Electrooxidation of cobalt(II) β-brominated-pyrrole tetraphenylporphyrins in CH_2Cl_2 under an N_2 or a CO atmosphere. *Inorg. Chem.* **36**, 6292–6298.
75. Bhyrappa, P., and P. Bhavana (2001). Unusual solvent dependent optical absorption spectral properties of free-base perhaloporphyrins. *Chem. Phys. Lett.* **342**, 39–44.
76. Bhyrappa, P., B. Purushothaman, P. Bhavana, and J.J. Vittal (2003). Highly brominated porphyrins: Synthesis, structure and their properties. *J. Porphyrins Phthalocyanines* **7**, 682–692.
77. Ou, Z., J. Shao, F. D'Souza, P. Tagliatesta, and K.M. Kadish (2004). β-pyrrole brominated meso-tetraphenylporphyrins: Synthesis, spectral and electrochemical properties. *J. Porphyrins Phthalocyanines* **8**, 201–214.
78. Kadish, K.M. and R.K. Rhodes (1983). Thin-layer spectroelectrochemical evidence of anion binding to (tetraphenylporphinato)iron(II) in nonaqueous media. *Inorg. Chem.* **22**, 1090–1094.

79. Takeuchi, T., H.B. Gray, and W.A. Goddard, III (1994). Electronic structures of halogenated porphyrins: Spectroscopic properties of ZnTFPPX8 (TFPPX8 = Octa-beta-halotetrakis (pentafluorophenyl)porphyrin; X = Cl, Br). *J. Am. Chem. Soc.* **116**, 9730–9732.
80. Brigaud, O., P. Battioni, and D. Mansuy (1992). Structure of meso-tetraaryl-β-octahalogeno porphyrins: A semi-empirical quantum-mechanical investigation. *New. J. Chem.* **16**, 1031–1038.
81. Hariprasad, G., S. Dahal, and B.G. Maiya (1996). meso-Substituted octabromoporphyrins: Synthesis, spectroscopy, electrochemistry and electronic structure. *J. Chem. Soc., Dalton Trans.* 3429–3436. (b) Hodge, J.A., M.G. Hill, and H.B. Gray (1995). Electrochemistry of nonplanar zinc(II) tetrakis(pentafluorophenyl)porphyrins. *Inorg. Chem.* **34**, 809–817.
82. Rodgers, K.R. and H.M. Goff (1988). Generation and characterization of highly reactive oxo-transfer intermediates and related species derived from (tetraarylporphinato) manganese(III) complexes. *J. Am. Chem. Soc.* **110**, 7049–7060.
83. Kadish, K.M. (1986). The electrochemistry of metalloporphyrins in nonaqueous media. *Prog. Inorg. Chem.* **34**, 435–605.
84. Hu, Y., B.C. Han, L.Y. Bao, X.H. Mu, and K.M. Kadish (1991). Electrochemistry of (octaethylporphinato)cobalt(II), (OEP)Co, under a carbon monoxide atmosphere. Electrogeneration and characterization of $[(OEP)Co^{III}(CO)]^+$. *Inorg. Chem.* **30**, 2444–2446.
85. Hinman, A.S., B.J. Pavelich, and K. McGarty (1988). In situ Fourier-transform infrared spectroelectrochemical studies of the oxidation of some tetraphenylporphyrin complexes. *Can. J. Chem.* **66**, 1589–1595.
86. Jones, D.H. and A.S. Hinman (1992). In situ infrared spectroelectrochemical studies of tetraphenylporphyrin complexes containing manganese, iron and cobalt. *J. Chem. Soc., Dalton Trans.* 1503–1508.
87. Mu, X.H., X.Q. Lin, and K.M. Kadish (1989). Microvoltammetric and spectroelectrochemical studies of (tetraphenylporphyrinato)cobalt (TPP)Co) oxidation/reduction in toluene and benzene solutions. *Electroanalysis* **1**, 113–116.
88. Araullo-McAdams, C. and K.M. Kadish (1990). Electrochemistry, spectroscopy, and reactivity of (meso-tetrakis(1-methylpyridinium-4-yl)porphinato)cobalt(III,II,I) in nonaqueous media. *Inorg. Chem.* **29**, 2749–2757.
89. Mu, X.H. and K.M. Kadish (1989). Oxidative electrochemistry of cobalt tetraphenylporphyrin under a CO atmosphere. Interaction between carbon monoxide and electrogenerated $[(TPP)Co]^+$ in nonbonding media. *Inorg. Chem.* **28**, 3743–3747.
90. Scholz, W.F., C.A. Reed, Y.J. Lee, W.R. Scheidt, and G.J. Lang (1982). Magnetic interactions in metalloporphyrin pi-radical cations of copper and iron. *J. Am. Chem. Soc.* **104**, 6791–6793.
91. Araullo-McAdams, C. and K.M. Kadish (1990). Electrochemistry, spectroscopy, and reactivity of (meso-tetrakis(1-methylpyridinium-4-yl)porphinato)cobalt(III,II,I) in nonaqueous media. *Inorg. Chem.* **29**, 2749–2757.
92. Bhyrappa, P. and P. Bhavana (2002). Lewis-base binding properties of free-base β-octahalotetraphenylporphyrins. *Chem. Phys. Lett.* **357**, 108–112.
93. Sankar, M., C. Arunkumar, and P. Bhyrappa (2004). Unusual solvent dependent electronic absorption spectral properties of nickel(II) and copper(II) perhaloporphyrins. *J. Porphyrins Phthalocyanines* **8**, 1343–1355.
94. Giraudeau, A., I. Ezhar, M. Gross, H.J. Callot, and J. Jordan (1976). Effect of electrophilic substituents on the polarographic behavior of porphyrins. *J. Bioelectrochem. Bioener.* **3**, 519–527.
95. Callot, H.J., A. Giraudeau, and M. Gross (1975). Cyanoporphyrins. Coordinating and electrochemical properties. *J. Chem. Soc., Perkin Trans. 2*, 1321–1324.

96. Arnold, D.P., G.A. Heath, and D.A. James (1998). Conjugated dimers of nickel(II) octaethylporphyrin linked by extended meso,meso-alkynyl bridges. II. Redox properties and electronic spectra of electrogenerated anions and dianions. *New J. Chem.* **22**, 1377–1387.
97. Giraudeau, A., H.J. Callot, J. Jordan, I. Ezhar, and M. Gross (1976). Nitrogen-15 nuclear magnetic resonance of organophosphorus compunds. Experimental determination and SCF-MO finite perturbation calculation of nitrogen-15-phosphorus-31 nuclear spin coupling constants in tris(dimethylamino)phosphine, -phosphine sulfide, and -phosphine oxide. *J. Am. Chem. Soc.* **101**, 3857–3862.

10

Photoelectrochemical Reactions at Phthalocyanine Electrodes

Derck Schlettwein

1. Introduction

One major driving force for the study of photoelectrochemically active electrode materials is the search for commercially viable solar cells. Their principles of operation have been explored mainly between 1960 and 1980 and a number of strategies were developed that worked quite efficiently for single-crystalline inorganic semiconductors. Subsequent research also tried to explore new systems of acceptable efficiency and stability at a more affordable price. The use of amorphous inorganic semiconductors or materials grown as thin films was one strategy, the use of organic semiconductors another. In all cases, chemical stability in particular under photoelectrochemical working conditions turned out to be a major problem. Photogenerated high-energy electron vacancies (holes) that are supposed to be harvested externally often also led to irreversible oxidation reactions at the semiconductor surface. Electron vacancies in the valence band of the semiconductor electrodes therefore had to be avoided. This could be achieved by the use of n-doped semiconductors with a bandgap large enough to not absorb the incident light. This latter main task, however, is of course crucial for a working solar cell. Organic dye molecules can serve to efficiently absorb light in the visible range of the spectrum. A concept of sensitization was therefore developed that combined chemically stable inorganic n-type wide bandgap semiconductors with adsorbed dye molecules. To provide a stable and efficiently working electrode, the excited state of the dye molecule has to inject an electron into the conduction band of the semiconductor and the remaining electron vacancy in the (then oxidized) dye has to be transferred to a redox electrolyte at a sufficiently high rate to avoid subsequent irreversible reactions at the dye molecule. Oxide electrodeslike

---

Derck Schlettwein • Institute of Applied Physics, Justus-Liebig-University Gießen, Heinrich-Buff-Ring 16, D-35392 Gießen, Germany.

N_4-Macrocyclic Metal Complexes, edited by José H. Zagal, Fethi Bedioui and Jean-Pol Dodelet. Springer Science+Business Media, Inc., New York, 2006.

TiO_2, ZnO, etc. sensitized by covalently attached dye molecules turned out to be suitable candidates for such reactions. To avoid nonradiative decay, dye aggregates have to be avoided in most cases, asking for at the most monolayer coverage of the semiconductor electrodes. Even at the high absorption coefficients reached by organic dyes, porous semiconductor electrodes with high internal surface area still accessible by dye molecules and redox electrolytes are needed to provide a high overall dye concentration in the film, maximize the active semiconductor/dye interface, and still avoid dye aggregation. Based on this concept, a good chance is seen to develop photoelectrochemical solar cells since semiconducting electrodes can be prepared that absorb visible light and still provide sufficient chemical stability to convert solar energy at attractive efficiencies[1–14].

In photoelectrochemical studies on phthalocyanines the chromophore is either adsorbed as a dye to an inorganic oxide surface as outlined above or used in a pure pigment thin film, thereby also utilizing the semiconducting properties of solid phthalocyanines. The experiments with organic pigment thin films are performed to evaluate the feasibility of such an approach for solar energy conversion and also to investigate the role of dye aggregates at phthalocyanine-sensitized oxide electrodes. Such studies also provided quite fundamental insight. One major goal was to probe molecular materials in contact with electrolytes in order to establish aspects which are common and different, respectively, compared with classical semiconductors or insulators[15] on the one hand and in contact with a second solid on the other hand[16–20]. An important aspect in these studies was the use of photoelectrochemical reactions to characterize the electrical properties in thin films of molecular semiconductors and to analyze their surface properties. Mainly unsubstituted phthalocyanines Pc were studied as organic semiconductor photoelectrodes[21–55], or, with a chemical anchor group added, in the sensitization of oxide semiconductors[21,56–60].

2. Essentials of Photoelectrochemical Reactions

Changes in the electrochemical currents or the established potentials as a consequence of illumination of an electrode/electrolyte interface can be observed if illumination leads to changes in the occupation of electronic levels that have a lifetime longer than the timescale of electrochemical reactions[1]. Such changes in principal can occur either in the electrode or in the electrolyte. Electrodes in contact with dye solutions, e.g., were studied to harvest electrochemically the energy of the absorbed light stored in the excitation energy of the molecules[61]. Much higher efficiencies, however, can be reached by electronic changes in an absorbing electrode. If a semiconductor electrode, e.g., absorbs light of energy higher than its band gap, excess electrons are created in the conduction band and electron vacancies are created in the valence band of the semiconductor[1,2,62]. Both these changes can be detected in intrinsic semiconductors. In a doped semiconductor,

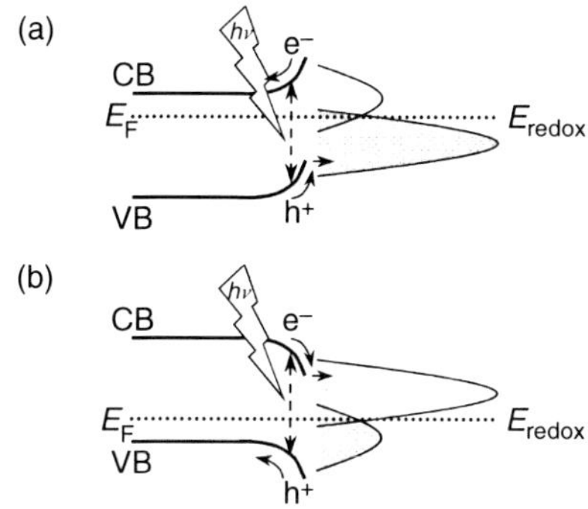

Figure 10.1. Schematic representation of photoelectrochemically active (a) n-type semiconductors and (b) p-type semiconductors, each in contact with a redox electrolyte that allows facile transfer of the photogenerated minority carriers from the semiconductor electrode.

only the concentration of minority carrier is changed significantly because of the high concentration of majority carriers anyway present in a doped semiconductor. N-doped electrodes are therefore characterized by anodic photocurrents, and p-doped electrodes by cathodic photocurrents. To observe these currents, reduced redox levels at energies in the range of the valence band of n-doped electrodes or oxidized redox levels at energies in the range of the conduction band of p-doped electrodes are needed[1,2]. In order to further provide transportation of the photogenerated minority charge carriers from the site of light absorption also inside the electrode to the interface with the electrolyte, the redox potential of the electrolyte has to be chosen appropriately. When brought into contact with an electrolyte, a semiconductor electrode will establish a contact similar to a semiconductor/metal contact because of the higher ionic concentration of charges in the electrolyte when compared with the charge density in a semiconductor and the space charge to compensate initial energy differences will be established mainly in the semiconductor surface and hence will lead to band bending there, as also known in semiconductor/metal contacts[1,2,62].

Roughly speaking, to observe the anodic photocurrents at n-doped electrodes a redox potential of the electrolyte below the semiconductor Fermi energy has to be chosen, for the cathodic photocurrents at p-doped electrodes, a redox potential above the semiconductor Fermi energy is needed[1,2]. Optimum conditions are summarized in Figure 10.1. A more detailed discussion, however, would have to include the quasi-Fermi-levels of the light-induced minority carriers[2]. Changes in an applied external potential will lead to changes in the space–charge region since this is the zone of the lowest charge density. In the context of this chapter it has to be pointed out that applicability of a semiconductor band model for the transport of charge carriers is not a prerequisite to establish a space-charge region nor is it a necessary condition for the observation of photoelectrochemical currents (see below).

3. Photoelectrochemical Experiments at Phthalocyanine Thin Films

3.1. Preparation of Thin Films

Film preparation plays a crucial role in determining the photoelectrochemical properties of phthalocyanine electrodes. Since the coupling of individual chromophores strongly depends on their relative orientation, the position of the absorption maximum and its width shows a clear dependence on the structure of thin films. Also the charge transport within phthalocyanine films, a fundamental necessity for the films to work as electrodes, depends upon the overlap of the frontier orbital wave functions. Beyond the microscopic structure of films also the morphology of films plays an important role. In the case of crystalline films, the orientation of crystallites relative to the electrode surface will be relevant because of anisotropies in optical absorption and charge transport. The size of the observed photocurrent directly depends on the real electrode surface area accessible by the electrolyte and this leads to a strong dependence on the porosity of the films.

In view of the high thermal stability of a number of phthalocyanines, *physical vapor deposition (PVD)*[20,63–69] has become one of the major preparation routes for thin films. The method is based on sublimation under at least high vacuum conditions (HV, base pressures below 10^{-5} mbar) or, better, ultrahigh vacuum conditions (UHV, base pressures below 10^{-8} mbar). Thin film preparation by PVD is, compared to the methods from solutions, slightly more demanding with respect to the equipment involved in the laboratory because vacuum technology is needed. On the material side, however, PVD is a less demanding technique because a number of molecules are more easily accessible as a vacuum-compatible pigment than they are as a soluble dye. For film preparation by PVD, the respective phthalocyanine is placed in a crucible, which can be heated resistively to temperatures typically between 200 and 400°C. Substrates on which the films are deposited are mounted on a substrate holder which is placed approximately 10–50 cm above the source. This distance and the temperature of a given source serve to control the rate of deposition. Aside from the rate of deposition the substrate temperature is another important parameter for film growth. Since the parameters of film growth and the conditions of the substrate and the source can be rather well controlled in a vacuum environment, the reproducibility of film deposition and hence film properties is also rather good relative to many other methods of film preparation. The films can be prepared in a wide thickness range from sub-monolayer coverage up to a few μm. Due to the vacuum environment inherent to the film preparation by PVD, the mechanism of film growth and its crystallization are perfectly suited to be studied *in situ*. This is in particular the case for a number of characterization tools of organic thin films which are surface science methods like electron diffraction, and electron spectroscopies, and also for more general methods like conductivity measurements, optical spectroscopy, and surface probe microscopies which can be performed in vacuum after proper installation but which in general could also be performed in air. A general advantage of *in situ* studies as opposed to studies at films, which have been prepared by methods at ambient atmosphere or have been transported through it, is given by the good shielding of the *in situ* experiments from uncontrolled reactions of the films with ambient atmosphere prior to the measurements. For photoelectrochemical experiments at phthalocyanine surfaces this advantage is more difficult to use since a direct transfer of the thin film from a vacuum environment to an electrochemical cell through a controlled atmosphere of a load lock or a glove box is needed. Electrochemical experiments have been interfaced with such vacuum environments before[70–75] and work is on the way to realize this also for the photoelectrochemical characterization of vapor-deposited phthalocyanine films. A special case of PVD can be seen in the technique of organic molecular beam epitaxy (OMBE). It makes use of the high purity of substrates, which can be achieved in UHV and can lead to highly ordered, crystalline organic films of well-defined chromophore interactions[63–69].

The methods of *film preparation based on solutions of phthalocyanines* differ strongly in the kind of films that can be obtained, especially with respect to molecular order, choice of substrates, and achievable film thickness. Film preparation from solution further carries the attractive perspective of using continuous ("roll-to-roll") processing technologies, a key toward affordable production of solar cells. Preparation of films consisting of phthalocyanine liquid crystals[76–79], the self-assembly of monolayers (SAM)[78,80], and film preparation by the Langmuir–Blodgett (LB) technique[76,77,80–85] allow to produce highly ordered films whereas spin coating and drop coating are examples of methods which aim at the deposition of bulk materials which typically are amorphous or polycrystalline. SAM and LB techniques on the other hand are normally restricted to the deposition of only one or a few monolayers or at least yield optimum results for such thin films and the deposition is restricted to specifically chosen molecules and substrates, whereas films up to a few μm average thickness of a soluble molecule can be prepared on almost any substrate by spin coating and drop coating. Films of pure molecular compounds, materials embedded in a polymer matrix and/or in a mixture with other components can be prepared. The matrix materials in general influence the formation of semiconductor aggregates, their structure, and hence chromophore interaction. Beyond this the matrix can also play an active role in the exciton dissociation and charge-carrier transport and thereby can be used to optimize the performance of electrodes[51]. Wetting of the substrate, the concentration of the components in the solution, and adjustment of the evaporation conditions are the main tools to control the film morphology and structure.

3.2. Semiconductor Characteristics of Solid Phthalocyanine Films

Since semiconducting properties can be regarded somewhat extraordinary characteristics when organic materials are discussed, and since the photoelectrochemical reactions discussed here heavily depend on these properties, some remarks about semiconductor characteristics of phthalocyanines appear appropriate. One of the most characteristic properties of phthalocyanines is their intense blue-green color caused by strong absorption at wavelengths from 500 to 750 nm (Figure 10.2) corresponding to about 1.6–2.5 eV. This transition in the Q-band (HOMO $\rightarrow$ LUMO) leads to additional charge carriers in the HOMO (electron

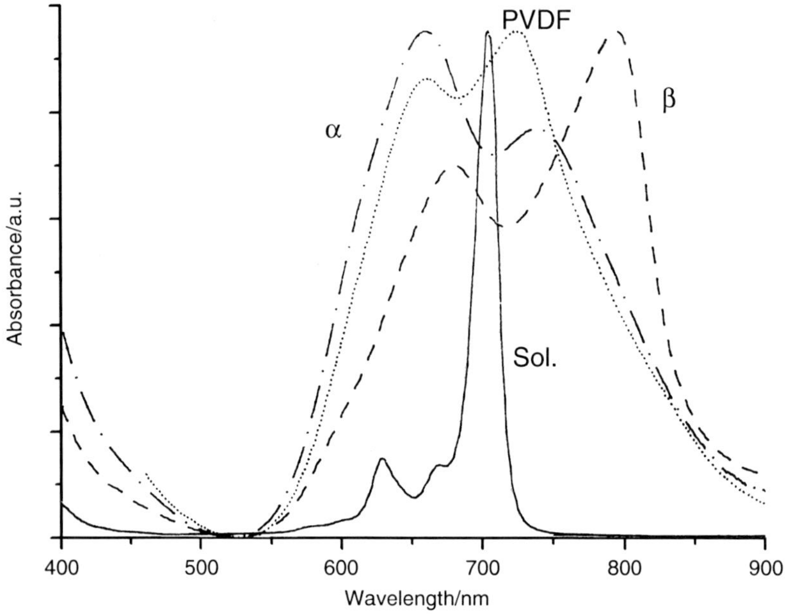

Figure 10.2. Optical absorption spectra of a solution of PcZn in *N*,*N*-dimethylacetamide (—) compared with thin films of PcZn drop coated from a mixed solution of PcZn and polyvinylidenefluoride (PVDF; · · ·), and PcZn vapor deposited at room temperature (α phase; -·-·-) or after annealing at 573 K for 3 hr (β phase; - - -).

vacancies) and LUMO (excess electrons) in a very simplified view. At second sight delocalization of the electrons in the excited state cannot be assumed in the films but the molecular excited state has to be looked at as a localized state. This corresponds to an exciton in the language of semiconductor physics. The dissociation of these excitons into electrons and electron vacancies (holes) that can separately contribute to charge transport in the films, needs an additional exciton dissociation (or binding) energy. It was determined to be in the order of 0.5–1.5 eV[86–88], a significant energy when compared with the estimated difference of the transport level from the Fermi energy of 0.2–1.0 eV. Only from separated charge carriers we can expect photoelectrochemical activity for a phthalocyanine film.

From a number of studies at thin films of unsubstituted phthalocyanines which have been reviewed earlier[18,89–94], it has become clear that the films behave as semiconductors. The photoconductivity generally observed for films of phthalocyanines clearly showed the possibility to dissociate excitons into mobile carriers. Under most conditions films of unsubstituted phthalocyanines were found to be partly oxidized and hence showed p-type characteristics. This was seen from measurements of the electrical conduction, the thermopower established by a temperature gradient across a sample, and by the rectifying characteristics in junctions. A generally quite low mobility of about 10^{-4}–10^{-1} $cm^2\,V^{-1}\,s^{-1}$ was found for the electron vacancies in phthalocyanines, barely allowing the applicability of a band model which dwells upon the assumption of delocalized electrons in a periodic potential of a solid. A rather localized character of the electronic system in phthalocyanines, however, is not at all surprising given the fact of rather loose bonds (van der Waals) among the molecules and very strong (covalent) bonds within the molecules as opposed to a three- dimensional covalent network of bonds in classical semiconductors. The materials are therefore generally referred to as molecular semiconductors, taking into account the similarities and also the differences to classical semiconductors. Models of charge-carrier transport based on hopping or tunneling of charge carriers from site to site in most cases are more appropriate than the band model of delocalized transport[68].

The rather small coupling of the π-electrons of phthalocyanine molecules in films is also directly reflected in a comparison of the absorption spectra of solutions and thin films (Figure 10.2). As opposed to a band edge and almost constant absorption beyond it as characteristic for semiconductors, the well-defined molecular absorption band detected in solutions of phthalocyanines is preserved also in the solid state, just split and broadened by the rather weak chromophore interactions in the solid. Nevertheless characteristically different spectra are obtained for different intermolecular arrangements in the solid state. This is seen in Figure 10.2 for the two different crystalline modifications (α- and β- form) of PcZn and for PcZn crystallized in a polymer matrix[51]. Such changes were reported for a number of different phthalocyanines and their different crystalline modifications, as reviewed earlier[65,67,68]. The differences in band splitting can be explained based on dipole–dipole interactions of the transition dipole that can be assigned to the optical transition.

Corresponding optical and electrical measurements at phthalocyanines with electron- withdrawing substituents in the ligand-like octacyanophthalocyanines $(CN)_8Pc$, tetrapyridotetraazaporphyrines TPyTAP, tetrapyrazinotetraazaporphyrins TPzTAP, or hexadecafluorophthalocyanines $F_{16}Pc$ showed quite comparable optical characteristics but the important difference that n-type conduction was observed[68,95–99], caused by the easy reduction of these molecules. Whereas the mobility in unsubstituted phthalocyanines was often found to be sufficiently high for a limited use of the band model, hopping transport had to be assumed in most cases for the substituted materials. Taking into account the observed conductivity in the dark and under illumination[68,91], and comparing them to the observed photoelectrochemical currents (see below), it is seen that the conductivity is sufficiently high to provide sufficiently fast transport of light-induced charge carriers to the electrode surface. The observed currents in photoelectrochemical experiments are hence limited by the kinetics of charge transfer at the electrode surface. This is also confirmed from the higher photocurrents in organic/organic′heterojunctions when compared with the photocurrents for the same films in photoelectrochemical cells[45].

3.3. Position of Frontier Energy Levels in Phthalocyanines

Since the rate and direction of light-induced charge transfer is governed by the relative position of occupied and unoccupied electronic states in the electrode and the electrolyte, this aspect of semiconductor characteristics of phthalocyanines is reviewed briefly. Knowledge of the position of the highest occupied molecular orbital (HOMO) and the lowest unoccupied molecular orbital (LUMO), the so-called frontier orbitals, of a given molecular electrode is crucial to discuss the photoelectrochemical characteristics. For the observed photoelectrochemical response of a semiconducting electrode the sign of the minority carrier in an electrode material is of uttermost importance since it defines the observed direction of the light-induced changes in the interfacial charge transfer. The sign of the charge carriers is defined by redox interactions of the semiconductor with dopant molecules defined by the ambient conditions.

The experimental methods to elucidate the position of the electronic energy levels in molecular materials aim at ionization of the sample and analysis of the energetic change involved. Two different sets of methods are frequently used, either based on redox electrochemistry as described in a number of chapters in this book or based on ionization in vacuum. Here, ultraviolet photoelectron spectroscopy (UPS) is a very direct method to determine occupied electronic states in a given material. The kinetic energy of photoelectrons following ionization by irradiation in the vacuum UV (e.g., He_I 21.2 eV) or soft X-ray (e.g., synchrotron) is measured. Molecules can either be in the gas phase or in a thin film on a conductive substrate. Inverse photoelectron spectroscopy, IPES (irradiation of the sample by an electron beam, spectral analysis of emitted photons) is an extension to the direct analysis of unoccupied electronic states. Electrochemical redox potentials reflect free enthalpies of reactions which involve ionization of the molecule of interest either in a solvent environment or as a thin film, then accompanied by intercalation of charge-balancing counter ions. In electrochemical experiments the accessible potential range of the electrolyte determines the experimental range. For a number of phthalocyanines, only the redox potential of reduction (characteristic for unoccupied levels) or oxidation (characteristic for occupied levels) can be obtained. To open a discussion about both occupied and unoccupied frontier levels of a material despite these experimental limits, optical spectroscopy (absorption or emission) or electron energy loss spectroscopy (EELS) was frequently used to estimate the frontier orbital gap. A rather constant gap of about 1.6 eV for most Pc (with the exception perhaps of cofacially stacked polymorphs[100] was determined[52,95,101,102]. In addition to this, however, an additional exciton binding energy (see above) has to be considered to estimate the energy needed to create mobile carriers.

UPS is also well suited to analyze details of the energetic alignment in heterojunctions beyond phenomenological observations[103]. The analysis of such heterojunctions[19,20,104–116] is of considerable interest in organic photovoltaic cells, organic light emitting diodes, and also for the characteristics of thin films in photoelectrochemically active junctions. Aside from the expected and in many cases desired establishment of a space charge in the films an interface reaction occurs leading to surface defects, a surface dipole and hence a shift of the vacuum level across the junction.

3.4. Photocurrent Direction at Phthalocyanine Electrodes

Following absorption of a photon in a molecule, this molecule in general is a stronger reducing agent due to an electron in a state of higher energy (LUMO in the case of the Q band of phthalocyanines), and also a stronger oxidizing agent due to the electron vacancy in a state of lower electron energy (HOMO in the case of the Q band of phthalocyanines) when compared with the molecule in the ground state[117,118]. From this point of view, an increased driving force for reduction reactions and oxidation reactions is expected for molecules in homogeneous solution. The same should then also be valid for reactions occurring at the surface of a solid electrode. At a closer look, however, it becomes clear that this principle can only be applied for intrinsic (or fully compensated[62]) electrode materials, since for doped (partially reduced or oxidized) materials the *relative* changes in the concentrations of excess electrons (LUMO) and electron vacancies (HOMO) will occur unsymmetrical, leading to preferentially anodic photocurrents at n-type electrodes and cathodic photocurrents at p-type electrodes (see above).

In a comparison of the catalytic activity of thin films of PcH_2, PcCu, or PcFe in the dark and under illumination[42], it was found that the photocatalytic activity was barely correlated to the catalytic activity in the dark and this difference was discussed as arising from a photoconductivity effect in the bulk of the phthalocyanine films. A photocurrent and also photovoltage observed at H_2Pc[41] was discussed to be due to a photovoltaic effect at the phthalocyanine film. Phthalocyanine thin films used in early photogalvanic cells[40] were described as p-type semiconducting electrodes that established a region of space charge at the Pc/electrolyte interface. Nevertheless also small anodic photocurrents were observed and explained by light-induced changes in the concentration of both types of charge carriers in a weakly doped semiconductor[38,39] or by a weak sensitization effect (see also below, Section 4) at the back electrode in the case of SnO_2 substrates[21,37]. This reaction at the SnO_2 back contact was characterized in detail by the analysis of experiments at varied illumination conditions and a thickness of the photoactive layer on the electrolyte side of 35 nm was established[33], in agreement with potential-dependent impedance measurements at electrochemical contacts with films of PcH_2, PcNi, and PcZn[119]. It was noticed that the role of acceptor molecules in the volume of the films is crucial for the observed characteristics and that oxygen from air in most cases caused the observed p-doping and hence cathodic photocurrents. Also attempted was a more controlled doping by reactions with halogen molecules like iodine, or organic molecules like *ortho*-chloranil, tetracyanoquinodimethane, tetrafluorenone, etc. This work quickly led to the need of a detailed investigation of the electrical properties in the solid state which are sketched above but which have been reviewed in great detail earlier[91] and also recently[68].

Utilization of the photoelectrochemical reactions purely based on the p-type character of phthalocyanine electrodes and optimization of the electrolyte led to photoelectrochemical cells with a power-conversion efficiency of 0.07% under illumination with $75\,\mathrm{mW\,cm^{-2}}$ white light[34]. Electrodes consisting of phthalocyanines in polymer binders were used. A space- charge region of 30 nm thickness and the presence of interfacial states at the electrolyte were detected in capacitance measurements also for these electrodes. The interfacial states served as mediators of charge to the electrolyte, but they were also thought to limit the open-circuit photovoltage and lead to surface recombination of carriers (see below). Doping of phthalocyanine films with organic electron acceptors was found to significantly increase the photoelectrochemical activity of evaporated phthalocyanine thin films and a quantum efficiency of 4.6% was reached for doped PcH_2 corresponding, however, to only 0.04% power conversion efficiency[33]. In a comparison of different methods of film preparation of divalent phthalocyanines, it was found that vapor-deposited films showed higher photocurrents when compared with drop-coated films[48]. This was explained by an increased trap density at grain surfaces of the drop-coated films leading to a decreased mobility across grain boundaries and an increased probability for the recombination of charge carriers. An electrophoretic deposition of $PcH_2$from suspensions of the pigment by use of surfactants was also used to obtain electrodes with an increased porosity, which led to an efficiency of 0.06% under illumination with 6 mW cm^{-2} white light[24].

The photoelectrochemical characteristics of thin films of phthalocyanines with higher- valent central metals carrying additional electron-withdrawing axial ligands like PcAl(OH), PcAl(Cl), PcGa(F), PcGa(Cl), PcIn(Cl), PcTi(O), PcV(O) were also investigated in great detail[22,27–29,31,35,120–122]. For this group of materials, although all films were vapor deposited, large differences in film characteristics were observed with either anodic or cathodic photocurrents dominating the photoelectrochemical behavior. The role of oxygen as a dopant leading to the typically observed p-type characteristics of phthalocyanines could be clearly elucidated in these studies[27,120,121]. The growth mode and morphology of films turned out to be crucial for interactions with the gas phase in vacuum as well as under ambient conditions and led to different levels of doping and trap sites[22,28,121]. An increased crystallinity and optimized orientation of crystals further led to larger exciton diffusion lengths and charge-carrier mobilities as shown in an increased photoelectrochemical efficiency[122]. Films consisting of rather large crystallites with a small surface area provided conditions of almost defect-free surfaces as reflected, e.g., in a proportionality between the observed photopotential and the electrolyte redox potential (no Fermi-level-pinning, see below)[31] and photovoltaic efficiencies in the range of 0.1% could be reached under illumination with 75 mW cm^{-2} in the visible range (470–900 nm).

In the case of the aluminum complexes, the influence of different methods of film preparation and different chemical film treatment on the photoelectrochemical characteristics can be discussed. By attachment of a surfactant chain to Al, electrophoretic deposition of very porous active thin films could be achieved which showed, however, a rather low efficiency of 0.01% under illumination with white light of 0.5–100 mW cm^{-2} due to an increased recombination of photogenerated carriers[32]. Spin coating of PcAl(OH) from DMSO was also used and gave active electrodes, however, with a low efficiency of only 0.006% at 10 mW cm^{-2} white light if a stable counter electrode was used[123]. Treatment of vapor- deposited thin films of PcAl(OH) in aqueous solutions of phthalic acid induced a structural transformation of the films and led to an increased efficiency of the films of 0.02% under 10 mW cm^{-2} white light[26]. Embedding of PcAl(OH) particles in a poly(vinylpyridine) binder led to an efficiency of 0.04% (10 mW cm^{-2} white light). A detailed analytical investigation of different ways of chemical treatment of PcAl(Cl) as compared to PcGa(Cl) and PcIn(Cl) revealed the hydrolysis of

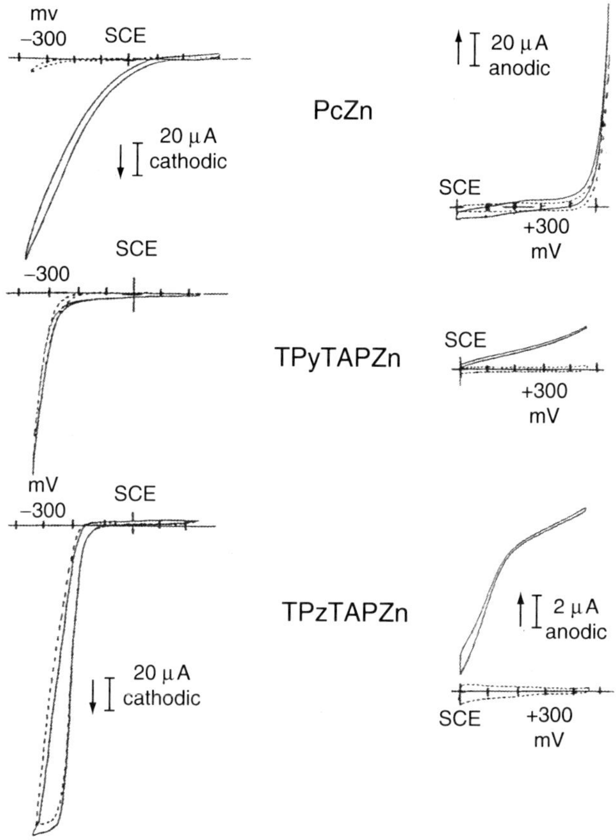

Figure 10.3. Cyclic voltammograms measured at thin film electrodes of PcZn, TPyTAPZn, and TPzTAPZn in the reduction of 1.3×10^{-3}M O_2 (left-hand side) or in the oxidation of 0.1 M EDTA (right-hand side). Experiments were performed in 0.5 M KNO_3 in the dark (- - -) or under illumination with 400 mW cm^{-2} white light of a Xenon arc lamp (——) (Reproduced from ref. [96] with permission from the American Chemical Society.)

the Al–Cl bond as the key step towards the formation of more efficient electrode materials[22].

Chemical substitution at the phthalocyanine ligand with electron-withdrawing substituents led to a dominance of anodic photocurrents as opposed to that of cathodic photocurrents observed at the unsubstituted materials. This was shown for $(CN)_8$PcZn[52,124], TPyTAPZn[48,52,125] and TPzTAPZn[48,52]. To restrict the observed differences in the photoelectrochemical characteristics to the influence of the different ligands, the central metal was kept constant in these studies and films were prepared by vapor deposition if applicable and also by drop coating for comparison purposes. Figure 10.3 shows the dominance of cathodic photocurrents (high anodic dark currents) at PcZn and the dominance of anodic photocurrents (high cathodic dark currents) at electrodes of

TPyTAPZn and TPzTAPZn, also seen for $(CN)_8PcZn$[52,96]. Based on these results, n-type conductivity was assigned to these materials of molecules with electron-withdrawing substituents based on the argument outlined above[48,52,96,124–128]. Electrodes of $F_{16}PcZn$ showed photocurrents of almost identical significance in both photocurrent directions[128–130]. Since films of $F_{16}PcZn$ investigated directly following the preparation still showed n-type characteristics, the photoelectrochemical behavior was explained by compensation of the dopants by air. The assignments were supported by studies of the electrical properties of the materials[47,95–98,125] and in solid heterojunctions[19,20,107]. In principle the same influence of an electron-withdrawing ligand was observed also when electrodes of naphthalocyaninatozinc (cathodic photocurrents, p-conduction) were compared with electrodes of tetraquinoxalinotetraazaporphyrinatozinc (anodic photocurrents, n-conduction)[126].

Thin films of TPyTAPZn have also been prepared by electrochemical deposition[131,132]. The n-type character was preserved also for films prepared in this way as shown by capacitance measurements. No clear preference of a photocurrent direction could be established, however, and the photocurrent action spectrum had no resemblance with the absorption spectrum of the material. It had to be assumed that the effect is caused by impurities in the film, which are incorporated during electrochemical film deposition.

Electrodes of two-dimensional sheet polymers of phthalocyanines prepared by *in situ* reaction of tetracyanobenzene with thin metal films were prepared and characterized in their photoelectrochemical characteristics[133]. Anodic photocurrents were detected, characterizing these films as n-type semiconducting materials. It was also found that such films could be reduced at more positive potentials than unsubstituted phthalocyanines. This change when compared to films of divalent unsubstituted phthalocyanines is caused by unreacted CN^- groups that were found in the films[134,135]. Such films therefore can be looked at as substituted phthalocyanines with electron-withdrawing CN^- groups which explains their photoelectrochemical characteristics.

3.5. Role of Higher Excited States

Caused by the possibility to separately absorb light in different absorption bands of a molecular semiconductor and by the separate character of the resulting excited electronic states, phthalocyanine electrodes show the specific characteristics of a switchable photocurrent direction which is demonstrated in this section[128]. Phthalocyanines show two distinct absorption bands in the visible range, namely the Q band around 650–700 nm and the B or Soret bands around 300–350 nm. In the solid state these optical transitions of Pc molecules are preserved, just slightly shifted and broadened considerably[136]. Most of the light of a white light source that leads to the observed photocurrents described above will be absorbed in the Q bands. Excitation in the B band of PcZn (≅340 nm), however, led to a strongly contrasting behavior to the situation described above for Q band illumination and the origin of bands should therefore be introduced briefly.

MO calculations for Pc showed that the Q band basically corresponds to the excitation of an electron from the highest occupied molecular orbital (HOMO,

$2a_{1u}$) to the lowest unoccupied molecular orbital (LUMO, $6e_g$) and the B band to the transition from the second-highest occupied molecular orbital (SHOMO, $4a_{2u}$) to the LUMO as in the case of porphyrins[136–138]. The higher excited singlet state S_2 following B-band excitation as opposed to the first excited singlet state S_1 which was populated by Q band absorption is of sufficient lifetime to be transferred to reactants of the electrolyte despite the competing relaxation into S_1 and to the ground state S_0. The assignment of the well-defined B and Q bands in the solid state to transitions between distinct molecular orbitals was proven to be of relevance also for the observed charge transfer in photoelectrochemical reactions at thin films. If a material of a suitable position of energy levels was chosen the direction of charge transfer could be switched by illumination with light of the two different wavelengths (Figurc 10.4). A rather large ratio of dark currents to photocurrents was seen for PcZn in contact with ethylenediaminetetraacetate EDTA (Figure 10.4a) but a relative small ratio was observed for PcZn in contact with O_2 as an oxidant in the cathodic direction (Figure 10.4b). A considerable concentration of electron vacancies (holes) in the HOMO of PcZn is thereby indicated already in the dark stemming from partial ionization (doping) of the films. The sign of the respective carrier confirms the characterization of PcZn as p-type material. Illumination in the Q band (620 nm) led to significant photocurrents in the reduction of O_2 (Figure 10.4b) by photogenerated electrons in the PcZn LUMO but no photocurrents were added to the dark currents in the oxidation of EDTA (Figure 10.4a) showing the absence of a significantly changed hole concentration in the PcZn HOMO, again consistent with the conduction type of the material.

Under B band illumination of PcZn as opposed to the situation under Q band illumination, also anodic photocurrents could be observed (Figure 10.4a) aside from the still dominating cathodic photocurrents. The hole generated in the SHOMO obviously was stable enough to lead to considerable changes in the electrode characteristics which at first sight appeared rather surprising in view of a presumably high concentration of surface defects and a number of possible relaxation reactions to S_1 or even recombination to S_0 which should lead to generally rather short lifetimes of the S_2 excited state. The electron generated in the LUMO leads to the cathodic photocurrent in contact with O_2 (Figure 10.4b) as also observed under Q band illumination. The validity of the band assignment (Figure 10.4c) as well as the partial oxidation of the PcZn HOMO (p-conductor) was clearly confirmed by this series of experiments and the possibility to switch the direction of interfacial charge transfer was demonstrated[128].

Anodic photocurrents dominated at electrodes of TPyTAPZn and F_{16}PcZn as described above, characterizing those materials as n-doped molecular semiconductors[128]. For these materials no such switching of the photocurrent direction as observed at PcZn could be observed. This is in good accordance with the assignment of the optical bands considering their conduction type because both holes in the HOMO (following Q band excitation) or holes in the SHOMO (following B band excitation) would represent minority carriers leading to significant photocurrents in the same direction. The independence of the two paths of charge-carrier generation was also proven in these cases, however, by different quantum efficiencies of photocurrent generation[128]. For molecular electrodes it is therefore appropriate to think of the transitions to higher excited states not only as

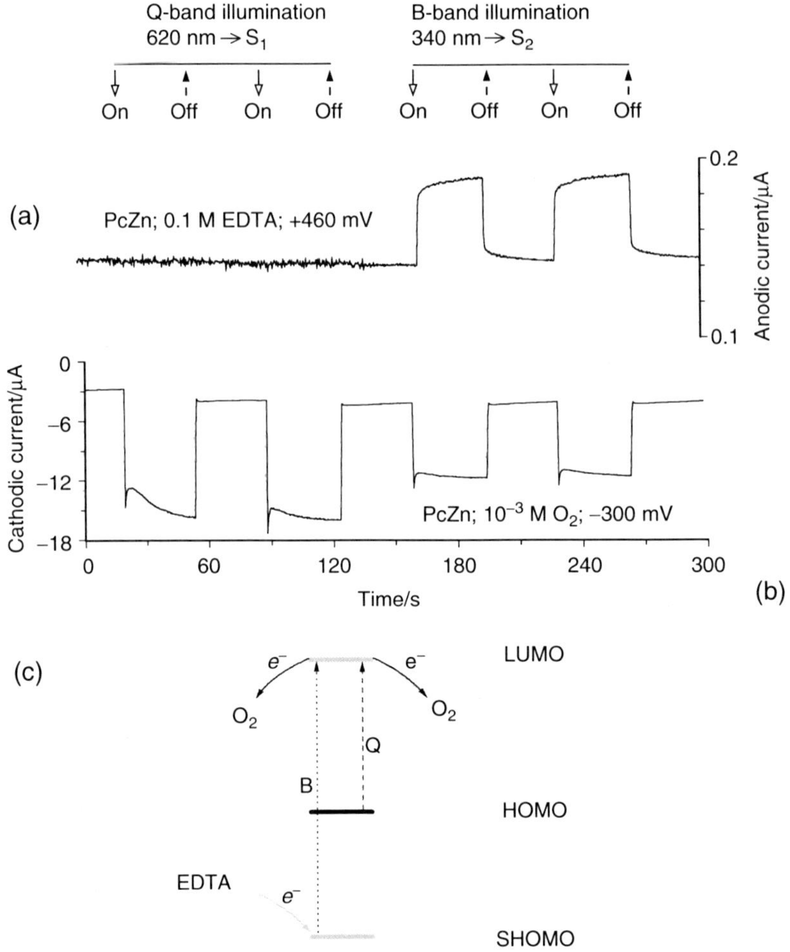

Figure 10.4. Current at films of PcZn (100 nm) vapor deposited on ITO (1 cm^2) observed in contact with aqueous electrolytes during potentiostatic polarization. Illumination occurred either in the B band (3×10^{15} photons cm^{-2} s^{-1}) or in the Q band (7×10^{15} photons cm^{-2} s^{-1}). (a) In the presence of 0.1 M EDTA (+460 mV), (b) in the presence of 10^{-3} M O_2 (−300 mV vs. SCE), (c) schematic representation of frontier energy levels and observed photocurrents (adapted and in part reprinted from ref.[128] with permission from Elsevier Science.)

contributors to an increased light absorption but also to consider charge transfer out of the higher excited states. This led to the observed switching in the direction of interfacial charge transfer by different illumination conditions[128], but also opens new reaction pathways utilizing the higher energy of the higher excited state as desired in the context of third generation solar cells[139].

3.6. Reactant Adsorption

In studies of electrochemical photocurrents obtained at molecular semiconductor thin films (PcZn, $(CN)_8PcZn$, TPyTAPZn, MePTCDI) and their dependence on the concentration of the reactant in the electrolyte (O_2, ethylthiolate (RS^-), hydroquinone/benzoquinone (HQ/BQ), $Fe(CN)_6^{3-/4-}$, $Ce^{3+/4+}$), a saturation behavior of the photocurrents was found arising from reactant adsorption prior to charge transfer[45,50,124,127]. Often functions according to Langmuir's adsorption isotherm based on the presence of only one kind of independent adsorption sites and a maximum coverage of a monolayer were found to give a reasonable fit to the observed photocurrent data. Assuming that only adsorbed species react, the reaction products desorb quickly, and the photocurrent density i shows a first-order dependence on the concentration c^* of minority carriers at the surface and on the surface coverage Γ by the reactant R, the photocurrent is obtained as

$$i = n\ F\ k_f c^* \Gamma \tag{10.1}$$

where n is the number of charge carriers transferred to each molecule of reactant, F is the Faraday constant, and k_f is the rate constant of the electrochemical reaction. The value of c^* is assumed to be constant at a fixed potential and light intensity. At increasing light intensity, however, the photocurrent increased, excluding diffusion limitation to be a reason of the observed saturation with the reactant concentration[50,53,130]. Assuming Langmuir adsorption equilibrium[140] for R leads to

$$d\Gamma/dt = k(\Gamma_{max} - \Gamma)c_R - k'\Gamma - k_f c^* \Gamma \tag{10.2}$$

where k is the rate constant of adsorption, k' is the rate constant of desorption, c_R is the reactant concentration in the electrolyte, and Γ_{max} is the maximum coverage arising from the occupation of all available sites. For the initial photocurrent ($i = i_{in}$) for which an unperturbed adsorption equilibrium can be assumed and for the steady state ($i = i_{stead}$) follows from Equations (10.1) and (10.2)

$$c_R/i_{in} = c_R/i_{max} + (k'/k)/i_{max} \tag{10.3a}$$

$$c_R/i_{stead} = c_R/i_{max} + [(k'/k) + (k_f c^*/k)]/i_{max} \tag{10.3b}$$

with i_{max} as the photocurrent arising from Γ_{max}[45,50,124,127]. Assumption of a more complex adsorption equilibrium considering rather a distribution (Freundlich isotherm) of adsorption constants $k_a = k/k'$ than a fixed value (Langmuir isotherm) provided a better model[124]. In many cases, however, the Langmuir isotherm allowed a successful discussion of the observed data. The accommodation coefficients k_a can be discussed as a parameter characteristic for each pair of electrode material and reactant. Different tendencies of reactant adsorption were observed: a rather weak driving force, e.g., was observed for adsorption of the anionic electron acceptor $Fe(CN)_6^{3-}$ at PcZn, and an equilibrium clearly on the side of adsorbed reactants was seen for the anionic hole acceptor RS^- at $(CN)_8PcZn$ and also for the cationic Ce^{3+} at $F_{16}PcZn$. No general trend was seen within the series of experiments but specific interactions seemed to dominate the strength of adsorption.

The time dependence of photocurrents on the scale of seconds to minutes showed a decay which was caused by changes in the surface coverage of the reactant as a result of the electrochemical reaction (term $k_f c^*/k$ in Equation 10.3b)[50,124,127] and a monoexponential decrease was expected within the Langmuir model:

$$i(t) = i_{\text{stead}} + (i_{\text{in}} - i_{\text{stead}})^* \exp[-((k\, i_{\max} c_{\text{R}})/i_{\text{stead}})t] \qquad (10.4)$$

Monoexponential decay functions were found to fit the transients in the light- induced reduction of O_2 at electrodes of PcZn showing the validity of the model[50]. Qualitatively similar transients were also observed, e.g., in the light-induced oxidation of RS^-[124,127], but its detailed analysis showed that a monoexponential decrease did not fit these data showing the limits of this approach. If again a Freundlich isotherm was assumed, however, a multi-exponential behavior would be expected and already the assumption of just two adsorption sites (bi-exponential fit) gave a good fit to the experimental data[124].

3.7. Surface Defects

3.7.1. Fermi-Level Pinning

For many electrode materials, surface defects were found to play an important role for the surface recombination of charge carriers[141–143], leading to a loss in the conversion efficiency, but also for the charge transfer to the electrolyte[4,144,145]. These surface defects may be already present without electrolyte contact, caused, e.g., by crystal defects, or are introduced by adsorbed species from the electrolyte[146]. Also charge transfer via surface defects was suggested for PcH_2[34].

Surface defects can cause a potential drop across the electrode surface, which does not contribute to the potential drop in the space–charge layer of the semiconductor and hence the photovoltage. This potential drop at the surface therefore leads to a complete or partial Fermi-level pinning (FLP)[147,148]. The extent of FLP can be measured by the difference between the rest potential under illumination and in the dark (V_{ph}) of the electrode using electrolytes of different redox potentials. Since transfer occurred to adsorbed reactants, the redox potential of the adsorbed reactant would be the relevant entity. Since these, however, were not accessible in the experimental work, differences in the redox potential of the solution were used as an approximation. For an ideal behavior (no pinning), V_{ph} should change directly with the redox potential, leading to a slope of –1 for a p-type semiconductor (1 for an n-type semiconductor) in a plot of V_{ph} against the dark rest potential of the electrode[149]. Total FLP (no potential drop across the space-charge region), is characterized by a slope of 0, while slopes between –1 and 0 (p-type) and between 0 and 1 (n-type) indicate partial FLP, a potential drop across both, the space-charge region and the layer of surface defects.

Illumination of electrodes of PcZn in contact with a variety of redox electrolytes under open circuit conditions showed positive photovoltages as expected for a p-type electrode, when the rest potential in the dark is negative of its flat band potential. Measurements in KCl solutions containing various electroactive species led to rest potentials in the dark between 0 and 300 mV vs. SCE[44]. The observed photovoltage V_{ph} was proportional to changes in the dark rest potential. The more positive the equilibrium potential of the electrolyte, the smaller the potential drop in the space–charge layer and hence the observed photovoltage. The quantitative comparison of the photovoltages observed at different equilibrium potentials indicated partial FLP at PcZn electrodes with a proportionality coefficient clearly smaller than unity. The obtained behavior of V_{ph} was quite independent of the redox pair, i.e., the surface defects were obviously not introduced by contact with the electrolytes but are a property of the electrode as prepared and transferred to the cell. The density of surface defects can be inferred from a plot of V_{ph} against the rest potential of the electrode in the dark from a series of experiments where the redox potential of the electrolyte is varied by changing the concentration ratio of the redox pair (ferri/ferrocyanide) rather than the kind of redox pair and where the rate of hole transfer from the electrode HOMO (majority carriers of this electrode material) was held at a constant level by a constant concentration of ferrocyanide[44]. A linear behavior was obtained for the plot of V_{ph} against the rest potential in the dark, with a slope of −0.35 for a series of experiments involving a PcZn electrode which was exposed to air for 5 days after preparation[44]. For an otherwise identical electrode, which had been exposed, to air for 5 months a smaller slope of −0.21 was obtained[44]. The smaller slope after extensive exposure to air clearly showed a higher surface defect density. It was therefore concluded that the surface defects at PcZn are caused by interaction with oxygen from air.

Under the assumption of a uniform energy distribution of the surface defects in the examined potential region, a semiquantitative model was introduced to interpret partial FLP[149] according to the equation

$$V_{ph} = a\, E_{redox} + b \tag{10.5}$$

where b is the photovoltage observed at reference zero. A plot of V_{ph} against E_{redox} should therefore yield a linear dependence with a slope $-1 < a < 0$, as it was the case for PcZn. From a the surface defect density D_s can be calculated according to

$$D_s = \frac{-\varepsilon_0\, \varepsilon_i\, (1 + a)}{e_0^2\, \delta\, a} \tag{10.6}$$

where ε_0 is the vacuum permittivity, ε_i the dielectric constant, e_0 the elementary charge, and δ the thickness of the interfacial layer, where the surface defects lead to the potential drop that causes the partial FLP. The value of ε_i is assumed to be close to 1 since it relates to the spatial region where surface defects are exposed at the semiconductor phase boundary with the electrolyte. Assuming that only surface molecules can participate in the formation of surface defects, a value of 0.8 nm was estimated for δ in the case of PcZn, as the thickness of a monolayer of PcZn would vary from 0.36 to 1.2 nm, depending on the orientation of the molecules. According to these geometrical assumptions a surface defect density $D_s = 4.2 \times 10^{13}$ cm^{-2} eV^{-1} was calculated from the slope of $a = -0.21$ obtained for an electrode stored in air for 5 months. For an electrode exposed to air for only 5 days prior to the experiments, a smaller $D_s = 2.1 \times 10^{13}$ cm^{-2} eV^{-1} was obtained from the slope of $a = -0.35$[44].

Assuming further that D_s is constant over the whole range of the energy gap (1.65 eV for PcZn), the overall number of surface defects was estimated to about 0.7×10^{14} cm^{-2} based on $a = -0.21$. This value is of a similar size as the calculated number of 0.6 to 2×10^{14} cm^{-2} molecules of PcZn present at the surface, if all possible orientations of the molecules are considered equally. According to this estimation a considerable part of the surface molecules represents surface defects, which can be filled with electrons.

3.7.2. Photoelectrochemical Electrode Kinetics

Photocurrent transient methods[44, 150–155] and intensity modulated photocurrent spectroscopy (IMPS)[145, 151, 156–161] have been established as good techniques for the investigation of semiconductor surfaces[162]. By these methods the working electrode is illuminated with either a light pulse or with sinusoidal modulated light, and the resulting photocurrent is monitored. Both techniques have been used at phthalocyanine electrodes to analyze in detail the kinetics of the photoelectrochemical reactions.

Photocurrent Transients

In addition to the perturbations of the reactant adsorption equilibrium, which lead to a transient behavior of the photocurrent in the second-to minute-regime, charging and discharging photocurrents were observed in the millisecond-regime, caused by relaxation of light-induced charge carriers in surface defects and their subsequent recombination[151]. Currents dropping below the steady-state current in the dark shortly after the end of illumination indicated the discharge of surface trap states. Such charging and discharging of surface defects were analyzed in detail for PcZn in contact with O_2, BQ/HQ and $Fe(CN)_6^{3-/4-}$ and for F_{16} PcZn in contact with $Ce^{3+/4+}$ and $Fe(CN)_6^{3-/4-}$ on a time scale of milliseconds[44, 129, 163].

A typical result for the example of PcZn electrodes after different times of exposure to air is displayed in Figure 10.5[44]. As expected for a p-type semiconductor, cathodic photocurrents were observed at PcZn electrodes as also seen earlier in steady-state photocurrent measurements[37, 44, 119, 164]. During the opening time of the photographic shutter used in these experiments (about 10 ms), the photocurrent increased beyond the value i_{in} (Equation (10.3a)) to a peak of the charging current (i_{peak}) and then decayed to i_{in} within about 30–40 ms. A reverse discharging current was observed upon closure of the shutter before the dark current was re-established. The steady-state photocurrents as well as the charging and discharging currents increased with the exposure time of the electrode to air. The increased steady-state photocurrent can be explained by the doping of the electrode bulk with oxygen from air. The increasing charging and discharging currents are consistent with the increasing surface defect density D_s found under the influence of oxygen.

Anodic photocurrents should be observed in the case of an n-type semiconductor like F_{16}PcZn. Figure 10.6, however, reveals both anodic and cathodic

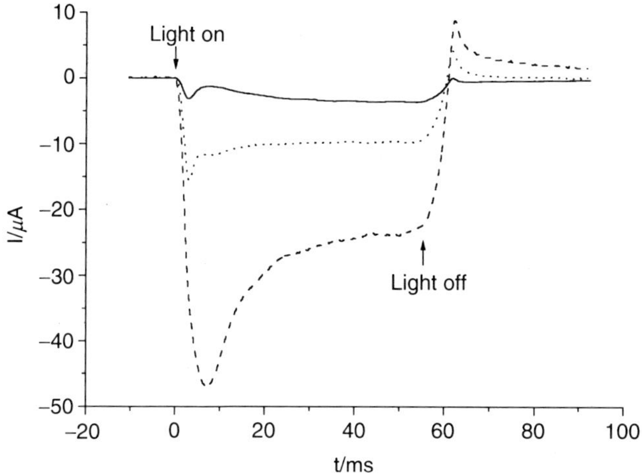

Figure 10.5. Time-resolved photocurrent response of PcZn electrodes to a 50 ms flash of white light in an oxygen-saturated aqueous 1 mol l^{-1} KCl solution at 0 mV vs. SCE. All electrodes were prepared in the same process and were exposed to air for 30 min (—), 1 day (···) and 6 days (- - -) after preparation. (Reprinted from ref.[44] with permission from Elsevier Science.)

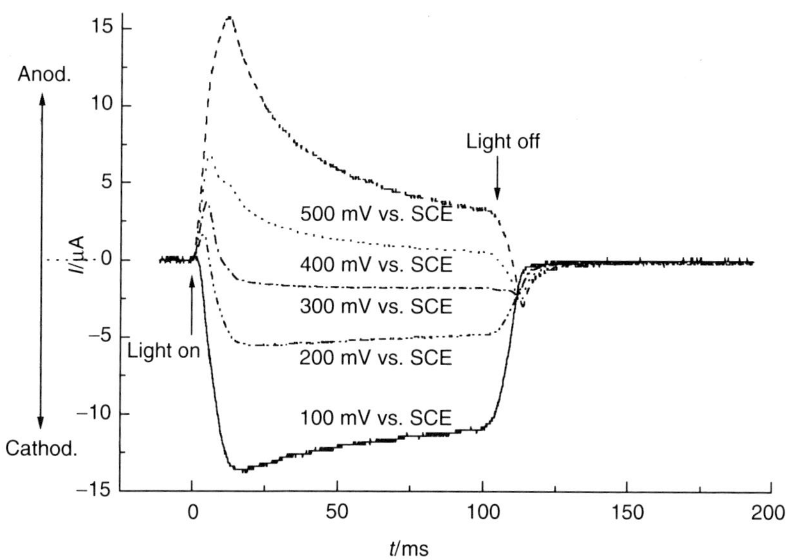

Figure 10.6. Time-resolved photocurrent responses of F_{16}PcZn electrodes to a 100 ms flash of white light at different electrode potentials between 100 and 500 mV vs. SCE as indicated. The electrolyte was an aqueous KCl (1 mol l^{-1}) solution with 10^{-3} mol l^{-1} $Ce^{3+/4+}$. (Reproduced from ref.[129] with permission from the Society of Porphyrins and Phthalocyanines.)

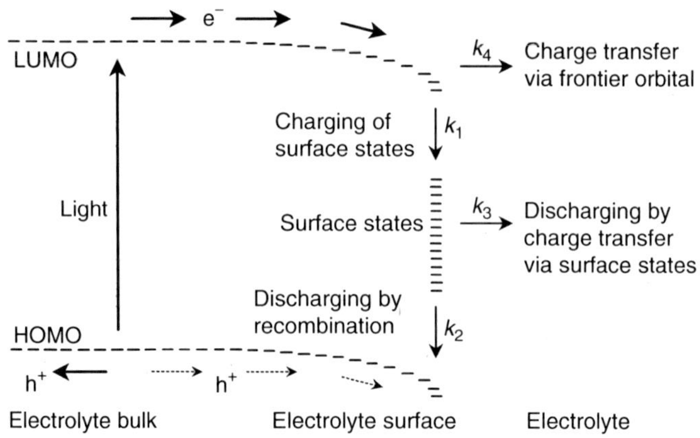

Figure 10.7. Qualitative model showing the involvement of surface defects in the charging and discharging of electrode surfaces and in charge transfer. The k_i indicate rate constants, and the respective rates are referred to as r_i in the text. (Reproduced from ref.[163] with permission from the American Chemical Society.)

photocurrents in dependence on the electrode potential[125, 129]. The dark currents were generally quite small. The concentration of both electrons and holes was small in the dark and could be increased significantly under illumination. Such a behavior is characteristic of an intrinsic or compensated semiconductor electrode. A compensation of donor sites in films of $F_{16}PcZn$ by oxygen had been seen earlier in conductivity measurements in vacuum and under air[97].

The concentration dependence for i_{in} as discussed above was confirmed for these photocurrent transient measurements, and a Langmuir adsorption isotherm according to Equation (10.3a) was successfully used to fit the data, e.g., for the photoreduction of benzoquinone at the surface of PcZn[44] and for the photo oxidation of Ce^{3+} at the surface of $F_{16}PcZn$[129]. In these studies quite remarkable differences were seen in the dependence or independence of charging and discharging currents on the reactant concentration. In the case of the photo reduction of benzoquinone at the surface of PcZn a clear inverse dependence on the coverage of adsorbed reactants was observed for both the charging and the discharging currents[44]. For the photo oxidation of Ce^{3+} at $F_{16}PcZn$, however, charging and discharging currents were found to be independent on the Ce^{3+} concentration[129] in spite of the high similarity in the concentration dependence of i_{in} as discussed above. A model involving the rate constants k_1 for surface defect charging by light-induced minority carriers, k_2 for their recombination with majority carriers, k_3 for the transfer of trapped minority carriers to the electrolyte, and k_4 for the direct transfer of minority carriers to the electrolyte as shown in Figure 10.7 for a p-type semiconductor was used to discuss these differences[44, 129, 163].

The back reactions corresponding to the detrapping from surface defects (k_{-1}) and dissociation of electron–hole pairs to a trapped minority carrier and a majority carrier (k_{-2}) were energetically not favored at room temperature ($kT = 0.048$ eV as compared to a frontier orbital gap of about 1.65 eV) and were therefore assumed to be negligible in first approximation. Electron injection from the electrolyte into the electrode (k_{-3} and k_{-4}) had also not been observed and hence all back reactions were assumed to be negligible. The charging and discharging of surface defects at molecular electrodes following the charging events but in a still unperturbed adsorption equilibrium could then be described by $r_1 = r_2 + r_3$ or, in more detail, by

$$k_1\, c_{\min}\, N_{ss}\, (1 - \Theta) = k_2\, c_{\mathrm{maj}}\, N_{ss}\, \Theta + k_3\, c_{\mathrm{redox,ad}}\, N_{ss}\Theta \tag{10.7}$$

where $c_{\min}$ and c_{maj} are the concentrations of light-induced minority carriers and majority carriers at the surface, $c_{\mathrm{redox,ad}}$ is the concentration of the electroactive species adsorbed at the electrode surface, N_{ss} is the overall number of chargeable surface defects, and Θ the occupation of these states ($0 \leq \Theta \leq 1$). Therefore Θ in this situation could be described by

$$\Theta = \frac{K}{1 + K} \tag{10.8}$$

with

$$K = \frac{k_1\, c_{\min}}{k_2\, c_{\mathrm{maj}} + k_3\, c_{\mathrm{redox,ad}}} \tag{10.9}$$

Equations (10.8) and (10.9) predicted a decreasing Θ when $c_{\mathrm{redox,ad}}$ is increased. Both, the amount of surface defect charging upon opening of the shutter and their discharging upon closure of the shutter were expected to be proportional to Θ. As shown before, $c_{\mathrm{redox,ad}}$ could be described by a Langmuir isotherm. Therefore K, Θ, and also the charging and discharging currents should behave antipodal to i_{in}. This could clearly be seen for the photo reduction of BQ at PcZn in a decrease of the peak currents i_{peak} at increasing electrolyte concentration[44]. Consequently, i_{peak} of the charging and discharging reactions could be approximated by a modified Langmuir isotherm:[44]

$$i_{\mathrm{peak}} = i_{\mathrm{max,peak}} - \frac{i_{\mathrm{max,peak}} - i_{\mathrm{min,peak}}}{1 + \frac{1}{k'_{\mathrm{a}} c_{\mathrm{redox}}}} \tag{10.10}$$

where $i_{\mathrm{max,peak}}$ and $i_{\mathrm{min,peak}}$ at a given electrode potential were the maximum peak current for slow faradaic charge transfer (low values of c_{redox}) and the minimum peak current for fast faradaic charge transfer (high values of c_{redox}), respectively. The constant $k'_{\mathrm{a}} = f\, k_{\mathrm{a}}$ can be considered to be a corrected accommodation coefficient with f taking into account the strength of coupling between the coverage by adsorbed reactants and the size of charging and discharging currents. The accommodation coefficients and other constants according to Equations (10.3a) and (10.10) are listed in Table 10.1 for PcZn and F_{16}PcZn electrodes for a variety of redox electrolytes. In spite of a stronger interaction (higher k_{a}) of F_{16}PcZn with, e.g., $Fe(CN)_6^{3-/4-}$ compared to that of PcZn a smaller $i_{\max}$ is reached at F_{16}PcZn under identical illumination conditions, indicating a smaller concentration of light-induced charges at the surface or a smaller charge-transfer rate.

As mentioned above, the anodic charging currents and cathodic discharging currents at F_{16}PcZn in aqueous $Ce^{3+/4+}$ solution were found to be independent of the electrolyte concentration and could therefore not be analyzed by Equation (10.10), although i_{in} did show a saturation behavior comparable to PcZn.

Table 10.1. Values of Accommodation Coefficients and Other Constants[130, 163] According to Equations (10.3) and (10.10) for PcZn and $F_{16}PcZn$ Electrodes in Different Electrolytes as Indicated in Parentheses. The Indices "cath" and "an" Stand for the Observed Cathodic or Anodic Photocurrents, Respectively

Measured photocurrent	Constant	PcZn ($Fe(CN)_6^{3-/4-}$)	PcZn (O_2)	$F_{16}PcZn$ ($Ce^{3+/4+}$)	$F_{16}PcZn$ ($Fe(CN)_6^{3-/4-}$)
i_{in}	$i_{max,in}/\mu A$	55.7	4.1	2.2	3.6
	$k_a/l\,mmol^{-1}$	0.7	11.3	32.1	38.7
Charging current	$i^{cath}_{max,peak}/\mu A$	17.6	13.6	4.5	8.3
	$i^{cath}_{min,peak}/\mu A$	3.1	5.8	4.5	0.4
	$k'_a{}^{cath}/l\,mmol^{-1}$	2.7	4.3	—	19.9
Discharging current	$i^{an}_{max,peak}/\mu A$	8.3	5.2	2.6	1.7
	$i^{an}_{min,peak}/\mu A$	3.3	0.9	2.6	0.2
	$k'_a{}^{an}/l\,mmol^{-1}$	8.7	27.1	—	7.3

This difference was discussed to arise from a missing pathway of minority carrier transfer from the surface traps to the electrolyte and hence a negligible k_3 pointing at the presence of traps in the surface region which are not, however, accessible, even by adsorbed electrolyte species and were therefore referred to as "near-surface defects"[129].

Since cathodic as well as anodic photocurrents were seen at $F_{16}PcZn$ electrodes the question arose whether electron surface traps might also be detectable. These electrodes were therefore also investigated at considerably more negative electrode potentials, using $Fe(CN)_6^{3-/4-}$ in the electrolyte. Indeed, not only cathodic steady-state photocurrents, but also cathodic charging currents and anodic discharging currents could be found under these conditions[163]. Their dependence on the electrolyte concentration could be analyzed according to Equation (10.10), with the accommodation coefficients and other constants shown in Table 10.1. Such behavior is opposite of that observed for the anodic charging and cathodic discharging currents[129] at $F_{16}PcZn$ with $Ce^{3+/4+}$ and resembled the characteristics found for cathodic charging and anodic discharging currents at PcZn[44, 50]. It was concluded that, although holes were trapped in "near-surface defects" for which no charge transfer to the reactant in the electrolyte could be observed, photogenerated electrons in $F_{16}PcZn$ were trapped in states at the very surface and could be transferred to acceptors in the electrolyte[163].

Intensity-Modulated Photocurrent Spectroscopy (IMPS)

To quantitatively determine the rate constants, IMPS measurements were performed at phthalocyanine electrodes when in contact with different redox electrolytes. The Pc working electrode was illuminated by a modulated light-emitting diode as light source and the phase shift φ of the resulting photocurrent and its amplitude A were measured for different modulation frequencies ω. These served to calculate kinetic data of the photo–induced processes in a model of doped semiconductor electrodes, which considers surface recombination and charge transfer from the frontier orbital levels as well as from surface defects[151,162,165]. According to the following reactions for a p-type semiconductor, which are also shown in Figure 10.7, pseudo-first order rate constants k'_i are introduced:

$$SS_{\text{vacant}} + e^- \xrightarrow{k_1} SS^-_{\text{occupied}} \quad k'_1 = k_1 N_{\text{ss}}(1 - \Theta) \tag{10.11a}$$

$$SS^-_{\text{occupied}} + h^+ \xrightarrow{k_2} SS_{\text{vacant}} \quad k'_2 = k_2[h] \tag{10.11b}$$

$$O + SS^-_{\text{occupied}} \xrightarrow{k_3} R + SS_{\text{vacant}} \quad k'_3 = k_3[O] \tag{10.11c}$$

$$O + e^- \xrightarrow{k_4} R \quad k'_4 = k_4[O] \tag{10.11d}$$

where SS_{vacant} and SS^-_{occupied} are vacant and occupied surface defects, respectively, O is the oxidized and R is the reduced form of the electroactive species in the electrolyte, k_i are the rate constants, N_{SS} is the density of surface defects, Θ is their occupation, $[h]$ is the concentration of electron vacancies (holes) in the HOMO at the surface, and $[O]$ is the concentration of the oxidized form of the electroactive species in the electrolyte. The amplitudes A and phase shifts φ for different modulation frequencies $\omega = 2\pi f$ are presented in the complex plane as shown schematically in the first part of Figure 10.8. Kinetic data of charge transfer and surface recombination can be extracted from the first quadrant in the complex plane plot by use of the intersections of the plot with the real axis I_1 and I_2[165]. The low-frequency limit I_1 represents the differential steady-state photocurrent increase due to a differential increase in the light intensity[162]. The high-frequency intercept I_2 represents the amplitude of the Gärtner flux g, which is the flux of minority carriers to the electrode surface.

The values k'_2 and k'_3 can be calculated from the frequency of the highest imaginary part of the photocurrent (f_{max}):

$$f_{\text{max}} = \frac{k'_2 + k'_3}{2\pi} \tag{10.12}$$

If no electroactive species is present in the electrolyte ($k'_3 = 0$), k'_2 can be directly calculated with this equation, given that photo corrosion can be neglected. This is normally the case for phthalocyanine electrodes even at very low electrolyte concentrations[45,49]. After adding electroactive species, k'_3 can be determined if k'_2 is known. If both k'_2 and k'_3 are known, the ratio k'_4/k'_1 can be calculated from I_1 and I_2[165]:

$$\frac{k'_4}{k'_1} = -\frac{I_2}{I_1 - I_2}\frac{k'_2}{k'_2 + k'_3} - 1 \tag{10.13}$$

At higher frequencies the effects of the $R_{\text{el}}C_{\text{sc}}$ time constant of the cell, where R_{el} is the resistance of the electrolyte and C_{sc} is the capacitance of the space–charge layer, become evident and the photocurrent is attenuated. The frequency response then describes a semicircle in the fourth quadrant of the complex plain. For an n-type semiconductor the semicircle caused by surface recombination will

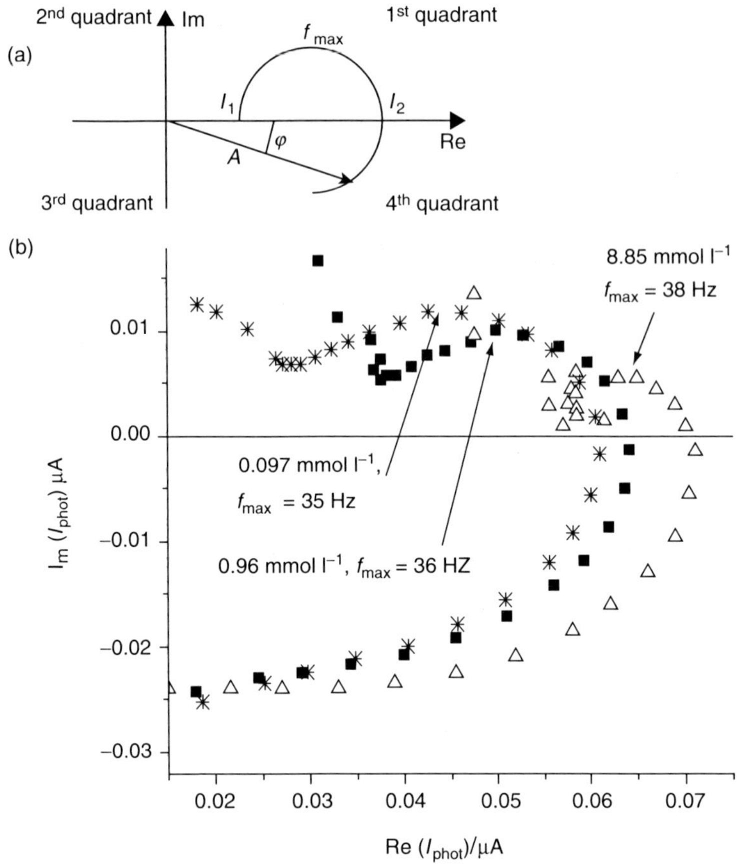

Figure 10.8. (a) Principal shape of an IMPS plot in the complex plane, as found in the presence of surface recombination. I_1 and I_2 are the intersections of the plot with the real axis. The modulation frequency with the highest imaginary value is f_{max}. (b) IMPS plots of a PcZn electrode using different $Fe(CN)_6^{3-}$ concentrations in the electrolyte as indicated, measured at an electrode potential of 0 mV vs. SCE. (Reproduced from ref.[163] with permission from the American Chemical Society.)

appear in the third and the semicircle due to the $R_{el}C_{sc}$ time constant in the second quadrant.

IMPS at PcZn electrodes: Although the principal characteristics of IMPS were observed for electrodes of PcZn in contact with various electrolytes (Figure 10.8)[163], flattened semicircles were observed in the first quadrant instead of a true semicircle expected by the model[165]. This pointed toward an energy distribution of the surface defects, since model calculations had shown that a Gaussian distribution of energies and the resulting Gaussian distribution of the rate constants k_i would lead to such characteristics[156,166]. Further, however, a kinetic limitation by diffusion in the electrolyte cannot be excluded, especially in those cases of rather large photocurrents and low concentration of reactive species in the electrolyte.

Another deviation from the ideal behavior was found for PcZn and can also be seen in Figure 10.8: for very low modulation frequencies, $\mathrm{Im}(i_{\mathrm{phot}})$ increased again at 0 and –200 mV vs. SCE instead of approaching 0 as it would be expected from the above model, indicating the presence of a further, very slow reaction at the electrode surface[163]. In a series of experiments in which oxidized species were present in the electrolyte this increase in $\mathrm{Im}(i_{\mathrm{phot}})$ for very low frequencies was observed to a much higher extent compared to experiments without electroactive species added. Since the plots tend to new maxima at frequencies below 0.2 Hz (the lowest frequency applied in these studies), a slow process occurring in the second or minute regime was indicated. In this time regime, the adsorption equilibrium of reactants at molecular semiconductor electrodes (see above) was found to change and establish a new equilibrium under illumination. This change in the adsorption equilibrium could therefore also be detected in the IMPS plots at very low modulation frequencies.

Figure 10.8 depicts a typical series of IMPS experiments in which the concentration of an electroactive species has been changed, in this case the concentration of $Fe(CN)_6^{3-}$. With increasing electrolyte concentration an increasing I_2 and a decreasing difference between I_2 and I_1 was observed, indicating a higher charge-transfer rate from the LUMO (k_4'). On the other hand, an increasing $f_{\max}$ also showed a faster charge transfer from surface defects (k_3'). Similar measurements were also performed using different *p*-benzoquinone (BQ) concentrations and an oxygen-saturated solution. The results for $Fe(CN)_6^{3-}$ and O_2 given in Table 10.2 show that charge transfer to $Fe(CN)_6^{3-}$ mainly occurred from the PcZn LUMO, while charge transfer from surface defects also took place but only to a minor extent. Charge transfer to oxygen, however, mostly occurred via surface defects. This was explained by an additional introduction of surface defects by oxygen, which also could be observed in higher charging- and discharging currents, especially at freshly prepared PcZn electrodes, in oxygen-saturated as compared to nitrogen-saturated KCl solution[163]. The overall charge transfer to oxygen was less efficient compared with $Fe(CN)_6^{3-}$ and *p*-benzoquinone, which was attributed to the low charge-transfer rate from the LUMO to oxygen.

Table 10.2 also showed that k_3' and k_4' obviously did not depend linearly on the electrolyte concentration $[O]$, as it would have been expected from Equations (10.11c) and (10.11d). This result confirmed that charge transfer from PcZn electrodes could only occur to adsorbed electroactive species (see above). If $[O]$ in Equations (10.11c) and (10.11d) is replaced by the concentration of adsorbed oxidized species $[O]_{\mathrm{ad}}$, which should show a saturation behavior toward high values of $[O]$ when a Langmuir isotherm is assumed, the observed dependence of k_3' and k_4' on $[O]$ is explained.

IMPS at F_{16} *PcZn electrodes:* IMPS plots for F_{16}PcZn in different electrolytes are shown in Figure 10.9[163]. In aqueous KCl solution without electroactive species a semicircle appeared in the third quadrant, indicating the recombination of photogenerated holes with electrons from the LUMO. It was mentioned before that photogenerated holes in F_{16}PcZn are trapped in near-surface defects, which

Table 10.2. Kinetic Data for a PcZn Electrode in Different Electrolytes of Different Concentrations at an Electrode Potential of 0 mV vs. SCE. The Three Columns on the Right Show the Percentages of Electrons that Reach the Surface (100%) that will Eventually Recombine or Undergo Charge Transfer to the Electrolyte from Either the LUMO or from Surface Defects[163]

Electrolyte	$c/(\text{mmol l}^{-1})$	$k'_3/(\text{s}^{-1})$	k'_4/k'_1	Charge transfer via LUMO (%)	Charge transfer via surface defects (%)	Recombination (%)
$Fe(CN)_6^{3-}$	0.097	6.3	0.975	49.4	1.4	49.2
	0.96	12.6	1.518	60.3	2.2	37.5
	8.85	25.1	3.538	78.0	2.3	19.7
O_2	1.2	75.4	0.086	7.9	24.0	68.1

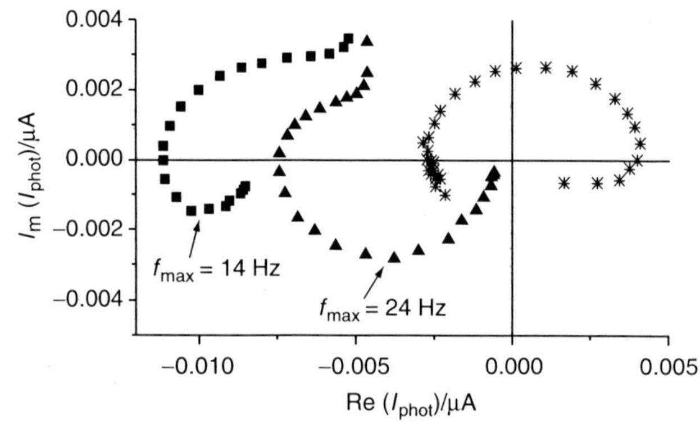

Figure 10.9. IMPS plots of a F_{16}PcZn electrode at 400 mV vs. SCE. The electrolytes were 1 moll^{-1} aqueous KCl solution (▲), aqueous KCl with 1 mmol l^{-1} $Fe(CN)_6^{4-}$ (∗) and aqueous KCl with 1 mmol l^{-1} $Fe(CN)_6^{3-}$ and 1 mmol l^{-1} $Fe(CN)_6^{4-}$ (■). (Reproduced from ref.[163] with permission from the American Chemical Society.)

represent recombination centers[129]. With $Fe(CN)_6^{4-}$ added to the electrolyte, the IMPS plot appeared mainly in the first quadrant, indicating surface charging with photogenerated electrons and their recombination with holes from the HOMO (Figure 10.9). The charging of the surface with photogenerated holes was also evident from a small cathodic charging current observed at the beginning of the illumination under these conditions[163].

Taking this into account, the IMPS plots at F_{16}PcZn electrodes were discussed in a model shown in Figure 10.10[129, 163]. As already seen in the photocurrent transient measurements, the n-type behavior of F_{16}PcZn is compensated by oxygen after storage in air. Both electrons and holes are photogenerated in significant amounts compared to their concentration in the dark. At +400 mV vs.

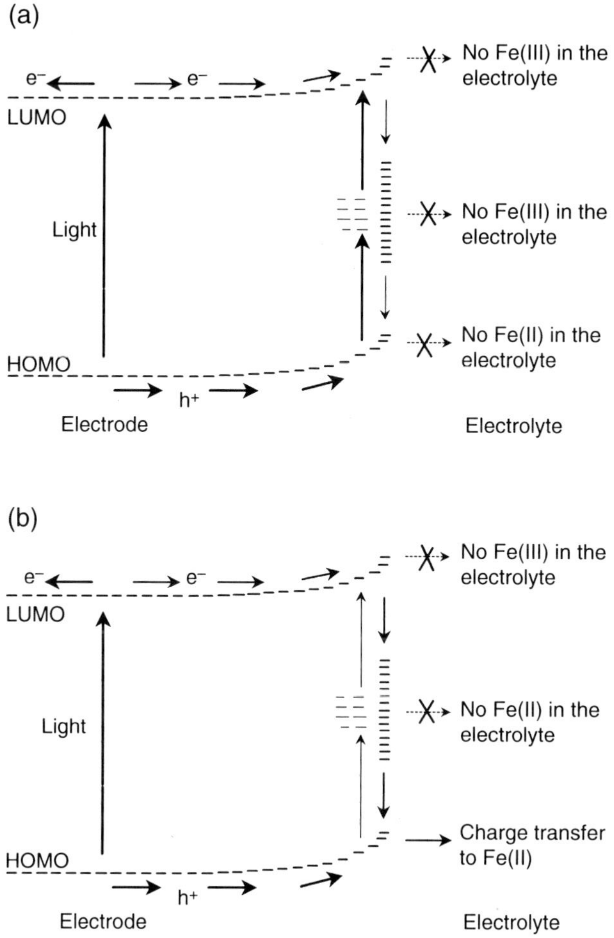

Figure 10.10. Situation at the interface between the F_{16}PcZn electrode and the electrolyte (1 mol l^{-1} KCl) at about +400 mV vs. SCE (a) before and (b) after the addition of 1 mmol l^{-1} $Fe(CN)_6^{4-}$. (Reproduced from ref.[163] with permission from the American Chemical Society.)

SCE, the transport of holes to the surface is favored as compared to the transport of electrons. Charging of surface defects with electrons may also occur, but to a smaller extent compared to the charging of near-surface defects with holes (Figure 10.10a). Following the addition of $Fe(CN)_6^{4-}$, photogenerated holes from the HOMO can undergo charge transfer to the electrolyte, and the charging of near-surface defects is significantly reduced. The charging of surface defects with electrons and their recombination can therefore be observed in the photocurrent transients and in IMPS (Figure 10.10b).

With both $Fe(CN)_6^{4-}$ and $Fe(CN)_6^{3-}$ added to the KCl solution in the same concentrations, the semicircle of the IMPS plot again appeared in the third quadrant (Figure 10.9), since electrons now could also undergo charge transfer from the LUMO, so that anodic charging currents again became higher than the cathodic charging currents. The value of f_{max} (14 Hz) was now lower than it had been in KCl solution (24 Hz). The quantitative IMPS model presented above, because it applies to doped semiconductors only and the concentration of majority carriers in a doped semiconductor is not significantly changed under illumination, does not predict this change in f_{max}. The experimental results obtained at $F_{16}PcZn$ electrodes, on the contrary, clearly showed that $F_{16}PcZn$ is a compensated semiconductor after storage in air and that photocurrents involving both types of charge carriers have to be considered.

4. Sensitization of Oxide Semiconductors by Phthalocyanines

4.1. Sensitization of Nanoparticulate Semiconductors

As already mentioned above, the investigation of the photoelectrochemical properties of phthalocyanine thin films originated from studies in which phthalocyanine films had been deposited on inorganic oxide electrodes to investigate their role as sensitizers and protecting layers against photo corrosion. Anodic sensitization could be achieved by vapor-deposited PcCu and by surface-bound tetrasulfonated copper phthalocyanine on n-SnO_2 conducting glass[56] and also by vapor-deposited thin films of PcH_2 on single-crystal electrodes of n-TiO_2, n-$SrTiO_3$, n-WO_3, n-ZnO, n-CdS, n-CdSe, n-Si, n-GaP, or slides of SnO_2 conducting glass[58]. The sensitization capability of the phthalocyanine layers could clearly be shown by anodic photocurrents that followed the absorption spectrum in their spectral dependence but the quantum efficiency of the sensitization was generally low and photo corrosion of the semiconductor substrate electrodes (evident for those materials of smaller band gap) could not be suppressed sufficiently. The same conclusion was drawn from an extended study in which also films of PcMg, PcZn, PcAl(Cl), PcTi(O), PcCo, and PcFe on single crystalline n-TiO_2 or n-WO_3 were investigated[57]. In more recent studies it had been found that although surface defects on TiO_2 quenched the photocurrent from light absorption in the oxide electrode, the (sensitized) photocurrents following light absorption in the phthalocyanine were left widely unchanged[59] and details of the energy level alignment were discussed. It was also shown that the film morphology of PcTi(O) or PcV(O) played a crucial role. Closed layers of the materials led to the occurrence of cathodic photocurrents originating from the phthalocyanine as active semiconductor material, whereas islands of the materials led to sensitized anodic photocurrents[167].

The photoelectrochemical properties of PcIn(Cl) thin films were also investigated on single-crystalline layered semiconductor electrodes of SnS_2 or MoS_2[168,169]. These surfaces turned out to provide a suitable substrate to deposit highly ordered epitaxial phthalocyanine films by OMBE and to also allow the study of such films as sensitizers for the n-type semiconductors. Very narrow absorption spectra and also photocurrent action spectra could be obtained speaking for a well-defined crystallization of PcIn(Cl) on n-SnS_2 and injection of electrons from the excited state of the dyes in the crystallites into SnS_2. A quite constant quantum efficiency of about 10% was found when such highly ordered films were compared with polycrystalline films indicating a direct injection of charge from molecules adjacent to SnS_2 since otherwise more pronounced differences should have been observed among the different crystalline modifications[169]. An increased quantum efficiency of up to 44% was obtained for films of phthalocyanines carrying eight dodecylamide or dodecylester functions drop coated from solutions on SnS_2[170]. These films showed liquid–crystalline properties and small chromophore interaction. The high quantum efficiency was reached for films of submonolayer coverage and decreased for thicker films pointing toward efficient injection again only from molecules close to the interface with SnS_2. Consequently the incident photon-to-current conversion efficiency for submonolayer or thicker films (up to 30 equivalent monolayers) was found quite constant at only 0.8%.

Higher efficiencies of phthalocyanine-modified oxide semiconductor electrodes could be achieved if nanoparticulate oxide films were used since the electrode surface and hence the interface area of the phthalocyanine with the oxide could be increased significantly. In these cases a monolayer of phthalocyanine on the nanoparticles provided sufficient light absorption. A high quantum efficiency can be maintained if aggregation of the phthalocyanines can be suppressed. Vapor deposition of the phthalocyanines therefore turned out to be not appropriate. The chromophores were typically adsorbed from solutions and preferably of phthalocyanines that carried a chemical substituent that provided good anchoring to the oxide surface. Tetrasulfonated metal-free phthalocyanine and the complexes of Zn, Ga(OH), Co, In(OH), and Ti(O) were adsorbed to nanocrystalline films of TiO_2 prepared by the sol–gel technique and their photoelectrochemical activity was measured under monochromatic illumination[171]. Among these, the zinc complex showed the highest IPCE at a sensitization quantum yield of still only 5.7%. Also investigated was the aluminum complex of tricarboxymonoamidephthalocyanine adsorbed on commercial TiO_2 particles and photocatalytic oxidations of organic molecules like phenols or hydroquinone could be achieved[172]. A composite electrode of TiO_2/phthalocyanine could also be prepared by spray pyrolysis of a mixed solution of titanium-oxy-acetylacetonate and the Cu complex of a tetrasulfonated phthalocyanine with a sensitized photocurrent obtained in the phthalocyanine Q band absorption range. However, the photocurrents reached only a few $\mu A\,cm^{-2}$ under illumination with 100 $mW\,cm^{-2}$ of white light[173]. Also attempts have been made to prepare very complex materials of TiO_2 particles modified by quantum-sized CdS on its surface and then further modified by Ga(OH), Zn, In(OH), or V(O) complexes of tetrasulfonated phthalocyanines and efficient sensitization by monomers of the Ga(OH) complex was claimed[174] at IPCE of up to 10%, but photo corrosion will be a severe problem for such electrodes as reported earlier. By far the most efficient electrodes sensitized by phthalocyanines up to now have been prepared by use of a Ru complex of octamethylphthalocyanine with two additional axial ligands of pyridine–biscarboxylic acid to bind to the oxide surface grafted to nanoparticulate TiO_2[175]. An IPCE of 60% and photocurrents of 10 mA cm^{-2} were reached under illumination with white light under AM 1.5 conditions. In a study using a number of differently carboxylated or sulfonated phthalocyanine complexes of Zn or Al(OH) similar values were reached with the zinc complexes of tetracarboxyphthalocyanine or tetrasulfophthalocyanine showing the best performance within this group of materials reaching an IPCE of 30–45% and a conversion efficiency of about 1% under AM 1.5 (100 mW cm^{-2}) conditions[60]. For phthalocyanines these are quite promising values, but it should be kept in mind that around 10% efficiency is typically reached by dyes of the Ru-trisbipyridyl-class when adsorbed at TiO_2.

4.2. Sensitization of Electrodeposited Semiconductor Thin Films

Aside from TiO_2, ZnO is one of the most promising semiconductor materials[176–179]. On the way towards both improved electrode properties and more economically feasible cells, new ways of preparing semiconductor–dye composites are sought and ZnO offers new perspectives in this respect. Crystalline films can be obtained electrochemically on the cathode during electrolysis of aqueous $Zn(NO_3)_2$ or oxygen-containing $ZnCl_2$ solutions.

4.2.1. Electrodes Deposited in the Presence of Phthalocyanines

The successful preparation of dye-loaded porous but crystalline ZnO semiconductor thin films in the presence of water soluble dyes[180] such as metal complexes of tetrasulfonated phthalocyanines (TSPcMt)[181,182] was reported. This one-step synthesis by dissolving the dye in the bath for the ZnO electrodeposition was found to allow *simultaneous self-assembly of ZnO and the adsorbed dye*, thus yielding homogeneously colored crystalline ZnO films with porous morphology. The dyes showed good compatibility with the aqueous deposition conditions and a suitable range of redox potentials to allow proper alignment of the energy levels in the contact with ZnO. Adsorption of dye molecules onto the growing surface of ZnO strongly affected the crystal growth of the ZnO and led to a significantly higher surface area when compared to pure ZnO films formed without addition of the dyes, opening up a new synthetic route to photoactive materials for dye-sensitized semiconductor electrodes. Strongly differing crystal sizes, morphologies, and porosities of the crystalline ZnO as well as different degrees of aggregation of the dye molecules could be obtained dependent upon the adsorbed dye molecules and proper choice of the deposition conditions[182,183].

Examples of different film morphology are shown in Figure 10.11. It was seen how for a given phthalocyanine ligand, the choice of the central metal group allows to grow films of quite remarkably different morphology and relative orientation of the ZnO nanocrystals[182]. The formation of ester-like bonds through the sulfonic acid groups of a TSPcMt molecule to the surface of ZnO probably plays a decisive role in this assembly, since a formation of such bonds was also seen in TSPcCo/TiO_2 composites[184]. The adsorption of TSPcMt takes place preferentially onto the a-/b-plane of ZnO crystals, since crystal growth predominantly along the *a* and *b* axes was observed. The overall growth direction is defined vertical to the electrode. Therefore those ZnO crystallites with their *a* and *b* axes

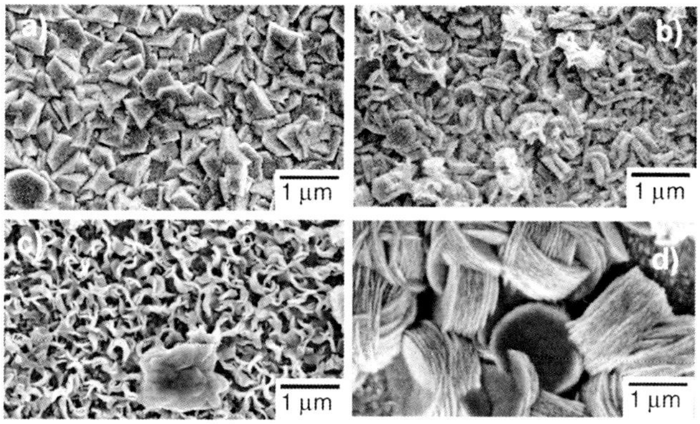

Figure 10.11. Influence of the adsorbed dye molecules on the morphology of the films deposited at −0.9 V vs. SCE from 0.1 M $Zn(NO_3)_2$ of (a) pure ZnO without dye and in the presence of: (b) 50 μM TSPcZn, (c) 50 μM TSPcAl(OH), and (d) 50 μM $TSPcSi(OH)_2$ as revealed by scanning electron microscopy. (Reproduced from ref.[190] with permission from The International Society for Optical Engineering.)

perpendicular to the substrate grew faster and represented the preferential crystal orientation of the final film[182]. Differences in the stability of TSPcMt adsorption to different crystallographic faces of ZnO led to the anisotropy of the crystal growth (Figure 10.11).

Dye molecules at the interface not only react with ZnO but also form well-defined intermolecular structures. The intermolecular electronic interaction can be nicely studied by measuring UV–Vis absorption spectra, because the interaction of the chromophores in condensed dye assemblies leads to specific changes of the electronic structure[136, 185, 186]. This was discussed in detail for the electrodeposited ZnO/TSPcZn hybrid thin films[182, 187]. Films electrodeposited at −0.7 V (slow growth) or −0.9 V vs. SCE (faster growth) showed characteristic differences. Whereas both films were blue and less scattering than the pure ZnO, the film deposited at −0.7 V obviously contained a higher amount of dye and showed absorption maxima at 337, 639 nm, and a shoulder at around 680 nm. The film deposited at –0.9 V was pale blue and had a Q-band absorption peak at 690 nm and only a shoulder at around 630 nm. The absorption spectra of the ZnO/TSPcZn films deposited at −0.7 V indicated formation of π-stacking aggregates for the TSPcZn molecules in the ZnO/TSPcZn film. Formation of such ordered dye aggregates on inorganic surfaces was also reported for merocyanine dyes adsorbed on TiO_2, Al_2O_3, and ZrO_2, as a consequence of chemical interactions between the dye molecules and the ordered surface of inorganic materials as well as that among the neighboring dye molecules[188]. Films deposited at −0.9 V showed dye loading as monomers noticed from the dominance of the 690 nm Q-band absorption[186]. Washing the ZnO/TSPcZn hybrid thin film deposited at −0.7 V with a surfactant solution that leads to optimum solubility of TSPcZn led to partial desorption of TSPcZn molecules from the composite films and converted their absorption spectra to those very similar to those of the hybrid thin film deposited at -0.9 V[182]. Such a change clearly indicated formation of dye multilayers due to dye/dye interaction in the films deposited at -0.7 V. At the bottom of such dye multilayers, however, TSPcZn molecules are present that are chemically attached to ZnO and that cannot be washed away by the CTAC treatment.

The *sensitization* of such films were studied in detail by photocurrent spectra, time-resolved photocurrent measurements, and intensity-modulated photocurrent spectroscopy (IMPS)[187,189]. Visible light absorbed in the dyes led to the sensitization of ZnO in contact with an organic I_2/I_3^- electrolyte as shown in Figure 10.12. Many experiments were performed in a conventional three-electrode arrangement with the deposited hybrid thin films as working electrode, a Pt counter electrode, and a $Ag/AgNO_3$ reference electrode. 0.5 M KI in acetonitrile/ethylenecarbonate (1:4 by volume) was used as the electrolyte. Among TSPcMt of different central groups, significantly higher conversion efficiencies were found for monomeric as compared to aggregated TSPcMt adsorbed on the ZnO surface[187,189,190]. This was also the case for electrodes having mostly aggregated TSPcZn, although these aggregates dominated the absorption spectra. A considerably higher quantum efficiency of monomeric dye when compared with aggregated dye was thereby shown. When the values of the IPCE are considered it has to be noted that the values reached by ZnO–TSPcMt hybrid thin films were generally low. Among the materials, however, clear differences could be detected showing the strong influence of the central group. $TSPcSi(OH)_2$ proved to be the most efficient sensitizer since a peak IPCE of 0.4% was reached at an absorbance of only 0.04. This indicated a quantum efficiency about a factor of 3 higher when compared to monomers of TSPcZn (0.12% at similar absorbance) and about a factor of 17 higher when compared to TSPcAl(OH). The higher photocurrents in this comparison for $TSPcSi(OH)_2$ could be explained by quite efficient electron injection from the excited state of the dye to the conduction band of ZnO. A suitable relative position of the electron energy levels combined with a strong chemical interaction of $TSPcSi(OH)_2$ with ZnO was found to provide the basis for this increased efficiency when compared with other TSPcMt[189,190]. In a direct comparison of monomeric vs. aggregated zinc complex TSPcZn on ZnO (Figure 10.12) the higher injection efficiency from monomers was confirmed by an increased quantum efficiency. In spite of a smaller coverage by the monomers and a hence decreased absorption even the incident IPCE was higher for the films with monomeric TSPcZn when compared to those with aggregated TSPcZn, in which photogenerated charge carriers mostly recombine within the aggregates[187,189].

Time-resolved photocurrent measurements in the ms regime were performed to discuss the efficiency of monomeric vs. aggregated forms of TSPcMt and the role of the central group in TSPcMt (Figure 10.13)[190]. Electrodes with monomers of TSPcZn on ZnO led to a more pronounced overshoot of the photocurrent above the stationary value when the light was switched on and significant cathodic currents when the beam was shut, whereas electrodes with aggregates of TSPcZn showed a more rectangular response consisting of anodic currents only. A more facile transfer of electrons explained this behavior from I_3^- to the aggregates compared with the monomers[187]. The slower regeneration of

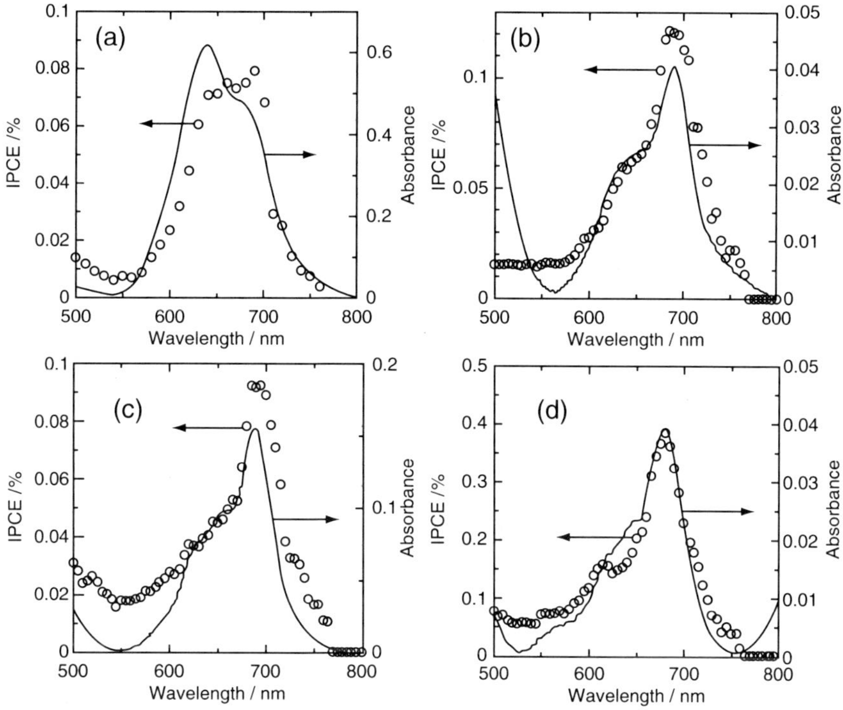

Figure 10.12. Optical absorption spectra (lines) and photocurrent action spectra in the oxidation of iodide at −0.2 V vs. $Ag/AgNO_3$ (circles) measured at films of TSPcZn/ZnO grown at −0.7 V vs. SCE (a) and at −0.9 V vs. SCE (b), TSPcAl(OH)/ZnO (c), and $TSPcSi(OH)_2/ZnO$ (d), both grown at −0.9 V vs. SCE. (Reproduced from ref.[190] with permission from The International Society for Optical Engineering.)

the neutral dye following electron transfer from the excited dye to the conduction band of ZnO led to a significantly higher concentration of oxidized dye in the stationary state with the consequence of the observed cathodic currents by electrons from the ZnO valence band when the light beam was shut. The low efficiency of TSPcZn aggregates was not caused by slow electrode kinetics but most likely by fast radiationless decay within TSPcZn aggregates or perhaps by a slower electron injection rate into the conduction band of ZnO. Although electron transfer from I_3^- to the aggregates occurred significantly faster, the electrodes were significantly less efficient.

The different TSPcMt monomers showed transients of similar shape indicating similar relative rates of electron injection and regeneration. As compared to $TSPcSi(OH)_2$, however, smaller ratios of charging/steady-state and discharging/steady-state currents were observed at TSPcZn and TSPcAl(OH) on ZnO (Figure 10.13)[187, 189, 190]. A faster charge transfer from TSPcZn or TSPcAl(OH) to the electrolyte was thereby indicated relative to the injection rate to ZnO. Since lower steady-state values of the photocurrents were observed

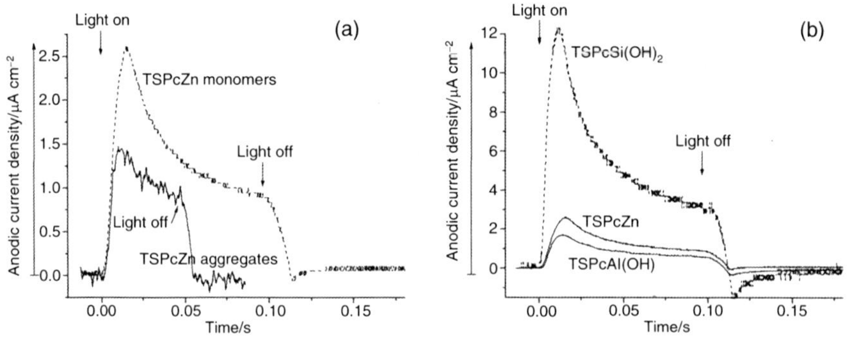

Figure 10.13. Photocurrent transients observed at −200 mV vs. $Ag/AgNO_3$ in contact with 0.5 M KI in acetonitrile/ethylenecarbonate (1:4 by volume). Comparison between aggregated and monomeric forms of TSPcZn (a), and of the influence of the central group in monomeric TSPc (b). (Reproduced from ref.[190] with permission from The International Society for Optical Engineering.)

when compared with $TSPcSi(OH)_2$ it was concluded that $TSPcSi(OH)_2$ provided faster and more efficient electron injection into ZnO compared with TSPcZn or TSPcAl(OH).

The interplay of different dye molecules present during the electrodeposition reactions of dye-modified ZnO and also under photoelectrochemical working conditions were studied for zinc complexes of tetrasulfonated phthalocyanine (TSPcZn) and tetraphenylporphyrine (TSTPPZn). Both dyes were simultaneously adsorbed to ZnO during the deposition[191]. The typical absorption bands for both dyes were detected. Films of TSPcZn/ZnO consisted of larger particulate domains when compared with TSTPPZn/ZnO or (TSTPPZn+TSPcZn)/ZnO as also revealed by atomic force microscopy[191]. The composite film with both dyes present (TSTPPZn+TSPcZn)/ZnO absorbed 48% of the light at 420 nm and 31% at 680 nm. The presence of the porphyrin obviously stabilized the phthalocyanine on the ZnO since a higher amount of TSPcZn was adsorbed. It is relevant to note, however, that the aggregation tendency of TSPcZn was even increased in the films with both dyes present as seen in the increase of the aggregate absorption band (640 nm) relative to the monomer band (680 nm) of TSPcZn[191]. Both, TSPcZn and TSTPPZn worked in parallel and the contribution of each dye to the overall photocurrent was collected[192]. Beyond this, the presence of TSPcZn in the film together with TSTPPZn improved the photoelectrochemical properties of TSTPPZn since a threefold increase of the photoelectrochemical quantum efficiency was found at 425 nm[192]. In this case the increased rate constant of the electron transfer to oxidized TSPcZn could be efficiently utilized to suppress recombination of TSTPPZn and TSPcZn took over a role as charge mediator in the regeneration reaction of oxidized TSTPPZn. Nevertheless only poor IPCE in the 0.1% range could be reached for these electrodes in which TSPcZn was present during electrodeposition of ZnO leading to aggregates of TSPcZn and a small surface area of the hybrid material.

4.2.2. Sensitization by Subsequently Adsorbed Phthalocyanines

Clearly improved electrode characteristics were obtained when the ZnO was prepared in a separate reaction step in the presence of a more favorable structure-directing agent before the phthalocyanine chromophores were adsorbed in a subsequent step[192,193]. Films of ZnO with a high porosity at preserved crystallinity were obtained in the presence of Eosin Y[194,195]. Since Eosin Y could be desorbed from such films an efficient ZnO semiconductor matrix could be provided by this route that can subsequently be loaded with a variety of sensitizer molecules by adsorption from solution ("readsorption method")[193]. Using this method, TSPcZn was re-adsorbed on ZnO. By analysis of these films it was clearly observed that TSPcZn could perform at much better efficiency when compared with the films deposited in its presence. In the Q-band an absorbance of about 0.7 above the scattering background was observed. This is about a factor of 5–10 higher than that in the films deposited in the presence of TSPcZn speaking for an efficient adsorption of the sensitizer at ZnO also when the latter was grown in the presence of Eosin. The shape of the band speaks for a widely monomeric adsorption of TSPcZn. The spectral dependence of the photocurrent (IPCE) basically follows the optical absorbance spectrum and shows the high efficiency of adsorbed monomers of TSPcZn. The efficiency reaches 13% at the absorption maximum of 680 nm leading to a photocurrent density of 1.8 mA cm^{-2} calculated for the geometric surface area. When these values of the IPCE or the current density are compared to the values reached for the films deposited in the presence of TSPcZn, an increase by a factor of about 100–200 was observed. By normalizing for the different dye content, a more than 1000 times higher sensitization efficiency can be estimated. These improvements of the photoelectrochemical properties for the TSPcZn/ZnO films can be understood by a mainly monomeric adsorption of the dye on the surface of ZnO, by the preserved crystallinity of ZnO and by the large overall surface area[192].

5. Technology Outlook

Photoelectrochemical properties of molecular semiconductors have recently regained new interest in view of their possible application in photovoltaic cells. Phthalocyanines are of interest for this application due to their high absorption coefficients for visible light, their chemical and thermal stability, and their commercial availability. They already had been tested successfully for applications in all-solid state photovoltaic cells[16,18,196,197], where the best results with molecular semiconductors were obtained for p–n-cells containing a phthalocyanine as p-type semiconductor and perylenetetracarboxylic acid diimides (PTCDI) as n-type semiconductor in a configuration ITO/PTCDI/Pc/Au[18,196]. Efficiencies of up to about $\eta = 2\%$ were found for such cells of PcZn and N,N'-dimethyl-3,4,9,10- perylenetetracarboxylic acid diimide (MePTCDI)[196]. The efficiency of such cells can be increased significantly in interpenetrating networks of the two electrode materials leading to an increased interfacial area[197]. This has also been successfully demonstrated in mixtures of PcZn with buckminsterfullerence C_{60} as an acceptor where efficiencies of up to 3.8% could be reached[201]. Crystallographic strain can be expected in the interface of two crystalline molecular semiconductors, leading to a small interfacial contact area and efficiency losses due to chemical and physical defect centers. Such interfacial states have been directly observed as surface dipoles in photoelectron spectroscopy studies of organic/organic′ heterojunctions[19,20,103,107,113,116]. PcZn and MePTCDI were therefore also investigated in photoelectrochemical p–n-cells to increase the effective surface area by an electrolyte between the two semiconductor thin films (e.g., ITO/PcZn/$Fe(CN)_6^{3-/4-}$(aq)/MePTCDI/ITO)[45]. However, although the semiconductor/electrolyte interface avoids the organic/organic′ solid contact and should provide larger contact areas at the surfaces, much smaller efficiencies ($\eta = 0.00032\%$) were found for these cells. Especially the very low photocurrent ($i_{SC} = 10.5\ \mu A\ cm^{-2}$) led to such a small efficiency[45]. This small photocurrent, however, could be explained by charge-transfer limitation due to a necessity of reactant adsorption prior to the charge-transfer step, which was generally found for electrodes of phthalocyanines,[45,50,124] and also for thin films of perylene derivatives[45]. The surface defects observed at thin films of phthalocyanines and perylene derivatives[163] play a negative role, since they mainly act as recombination centers, since charge transfer from surface defects is slow. The rather small surface area of the Pc and PTCDI thin films (roughness factor only about 2)[45] could by far not compensate these current limiting effects of reactant adsorption and surface recombination. Rather low efficiencies of only up to 0.08% were therefore also reported for photoelectrochemical solar cells with Pc electrodes and metal counter electrodes, corresponding to all-solid state Schottky cells[121,198].

The problems of a low surface area and surface recombination could alternatively be solved by the development of nanocrystalline dye-sensitized solar cells, using organic dyes as sensitizers for inorganic wide bandgap semiconductors like TiO_2 or ZnO. Especially for the Grätzel cell[3], which uses nanocrystalline TiO_2 films of very high surface area and various trisruthenium complexes as sensitizers, high efficiencies of up to $\eta = 10.4\%$ were already reported. When photoelectrochemical cells are considered, the concept of dye-sensitized oxide electrodes therefore provides the best perspective. Although good oxide–dye combinations have already been presented, further work will reveal new composite materials, prepared by different, preferably simpler, preparation techniques and phthalocyanines might very well play a role in this context. One should keep in mind, however, the mostly low efficiency of phthalocyanine aggregates in the sensitization of oxidic semiconductors as opposed to considerably more efficient monomers of phthalocyanines on the one hand and the high aggregation tendency of phthalocyanines on the other hand. A goal therefore will be to prepare tailored phthalocyanines with specifically designed anchoring groups to bring the chromophore π-system close toward the oxide surface, optimize the relative energetic and local orbital overlap with the oxide, and at the same time hinder the chromophores from aggregate formation in spite of a high coverage of monomers on the surface. The photoelectrochemical properties of phthalocyanine surfaces are also of interest in interactions with gas environments. Chemical reactions with redox-active gas molecules lead to similar changes of the surface trap distribution as discussed above for the reactions in contact with a liquid redox-active electrolyte. Changes of the electrical conductivity as a consequence of redox interactions with gas molecules were therefore often used to test phthalocyanine thin films as sensitive layers in chemical sensors. Vapor-deposited phthalocyanine thin films and thin films prepared by the LB technique were extensively studied and discussed in review articles[68, 80, 92–94, 199]. Photoelectrochemical experiments in contact with solutions of the analytes or their model compounds can be used to study the surface reactions at the electrodes in great detail and hence assist to broaden the understanding of chemical sensing at phthalocyanine surfaces. Of particular interest in this context is the finding that the photoconductivity of phthalocyanine films showed a considerably higher sensitivity to the presence of redox-active gas molecules than the dark conductivity[200]. A clear relevance of surface defects as centers of charge carrier dissociation and traps for charge-carriers is thereby indicated. A close link of the photoelectrochemical research at phthalocyanine thin films and research in view of their application as chemical sensors should therefore be maintained and used to further develop both fields, although they appear quite independent judged from their technological aspects.

6. Acknowledgment

The author is grateful to the VolkswagenFoundation (I/80314) for financial support. A fruitful joint research and numerous discussions on a variety of subjects in the context of this chapter with N. Jaeger and D. Wöhrle (University of Bremen, Germany), T. Oekermann (University of Hannover, Germany), and T. Yoshida (Gifu University, Japan) are gratefully acknowledged as are the contributions of E. Karmann, T. Oekermann, and K. Nonomura to the experimental work in the author's group.

References

1. Gerischer, H. and F. Willig (1976). Reactions of excited dye molecules at electrodes. In F.L. Boschke (ed), *Physical and Chemical Applications of Dyestuffs. Topics in Current Chemistry*, Vol. 61. Springer, Berlin, Heidelberg, pp. 31–84.
2. Memming, R. (1996). *Semiconductor Electrochemistry*. Wiley-VCH, Weinheim.
3. Grätzel, M. (2001). Photoelectrochemical cells. *Nature* **414**, 338–344.
4. Nozik, A.J. and R. Memming (1996). Physical chemistry of semiconductor–liquid interfaces. *J. Phys. Chem.* **100**, 13061–13078.
5. Hannappel, T., B. Burfeindt, W. Storck, and F. Willig (1997). Measurement of ultrafast photoinduced electron transfer from chemically anchored Ru-dye molecules into empty electronic states in a colloidal anatase TiO_2 film. *J. Phys. Chem. B* **101**, 6799–6802.
6. Grünwald, R. and H. Tributsch (1997). Mechanisms of instability in Ru-based dye sensitization solar cells. *J. Phys. Chem. B* **101**, 2564–2575.
7. Dloczik, L., O. Ileperuma, I. Lauermann, L.M. Peter, E.A. Ponomarev, G. Redmond, N.J. Shaw, and I. Uhlendorf (1997). Dynamic response of dye-sensitized nanocrystalline solar cells: Characterization by intensity-modulated photocurrent spectroscopy. *J. Phys. Chem. B* **101**, 10281–10289.

8. Huang, S. Y., G. Schlichthörl, A.J. Nozik, M. Grätzel, and A.J. Frank (1997). Charge recombination in dye-sensitized nanocrystalline TiO_2 solar cells. *J. Phys. Chem. B* **101**, 2576–2582.
9. de Jongh, P.E. and D. Vanmaekelbergh (1997). Investigation of the electronic transport properties of nanocrystalline particulate TiO_2 electrodes by intensity-modulated photocurrent spectroscopy. *J. Phys. Chem. B* **101**, 2716–2722.
10. Solbrand, A., H. Lindström, H. Rensmo, A. Hagfeldt, S.-E. Lindquist, and S. Södergren (1997). Electron transport in the nanostructured TiO2-electrolyte system studied with time-resolved photocurrents. *J. Phys. Chem. B* **101**, 2514–2518.
11. Hilgendorff, M. and V. Sundström (1998). Dynamics of electron injection and recombination of dye-sensitized TiO_2 particles. *J. Phys. Chem. B* **102**, 10505–10514.
12. Ellingson, R.J., J.B. Asbury, S. Ferrere, H.N. Ghosh, J.R. Sprague, T. Lian, and A.J. Nozik (1998). Dynamics of electron injection in nanocrystalline titanium dioxide films sensitized with [Ru(4,4′-dicarboxy-2,2′-bipyridine)(2)(NCS)(2)] by infrared transient absorption. *J. Phys. Chem. B* **102**, 6455–6458.
13. Ghosh, H.N., J.B. Asbury, and T. Lian (1998). Direct observation of ultrafast electron injection from coumarin 343 to TiO_2 nanoparticles by femtosecond infrared spectroscopy. *J. Phys. Chem. B* **102**, 6482-6486.
14. Haque, S.A., Y. Tachibana, D.R. Klug, and J.R. Durrant (1998). Charge recombination kinetics in dye-sensitized nanocrystalline titanium dioxide films under externally applied bias. *J. Phys. Chem. B* **102**, 1745–1749.
15. Willig, F. (1981). Electrochemistry at the organic molecular crystal/aqueous electrolyte interface. In H. Gerischer and C.W. Tobias (eds), *Advances in Electrochemistry and Electrochemical Engineering*. John Wiley & Sons, New York, pp. 1–111.
16. Wöhrle, D. and D. Meissner (1991). Organic solar cells. *Adv. Mater.* **3**, 129–138.
17. Wöhrle, D., J. Elbe, L. Kreienhoop, G. Schnurpfeil, B. Tennigkeit, S. Hiller, and D. Schlettwein (1995). Investigation of n/p junction photovoltaic cells of perylenetetracarboxylic acid diimides and phthalocyanines. *J. Mater. Chem.* **5**, 1819–1829.
18. Wöhrle, D., L. Kreienhoop, and D. Schlettwein (1996). Phthalocyanines and related macrocycles in organic photovoltaic junctions. In C.C. Leznoff and A.B.P. Lever (eds), *Phthalocyanines—Properties and Applications*. Vol. **4**. VCH, New York, Weinheim, Cambridge, pp. 219–284.
19. Schlettwein, D., T. Oekermann, N.I. Jaeger, N.R. Armstrong, and D. Wöhrle (2003). Interfacial trap states in junctions of molecular semiconductors. *Chem. Phys.* **285**, 103–112.
20. Schlettwein, D., K. Hesse, N. Gruhn, P.A. Lee, K.W. Nebesny, and N.R. Armstrong (2001). Electronic energy levels in individual molecules, thin films, and organic heterojunctions of substituted phthalocyanines. *J. Phys. Chem. B* **105**, 4791–4800.
21. Minami, N., T. Watanabe, A. Fujishima, and K.-I. Honda (1979). Photoelectrochemical study on copper phthalocyanine films. *Ber. Bunsenges. Phys. Chem.* **83**, 476–481.
22. Santerre, F., R. Cote, G. Veilleux, R.G. Saint-Jacques, and J.P. Dodelet (1996). Highly photoactive molecular semiconductors: Determination of the essential parameters that lead to an improved photoactivity for modified chloroaluminum phthalocyanine thin films. *J. Phys. Chem.* **100**, 7632–7645.
23. Harima, Y. and K. Yamashita (1989). Electrochemical characterization of phthalocyanine thin films prepared by the electrolytic micelle disruption method. *J. Phys. Chem.* **93**, 4184–4188.
24. Harima, Y. and K. Yamashita (1988). Phthalocyanine photoelectrochemical cell prepared by a micelle disruption method. *Appl. Phys. Lett.* **52**, 1542–1543.
25. Schuhmann, W., H.-P. Josel, and H. Parlar (1987). A new photosynthesis-like system for the light-induced reduction of water to molecular hydrogen. *Angew. Chem. Int. Ed.* **26**, 241–243.

26. Perrier, G. and L.H. Dao (1987). Improvement of the performance of a hydroxyaluminum phthalocyanine photoelectrochemical cell by a phthalic acid treatment. *J. Electrochem. Soc.* **134**, 1148–1152.
27. Klofta, T.J., T.D. Sims, J.W. Pankow, J. Danziger, K.W. Nebesny, and N.R. Armstrong (1987). Spectroscopic and photoelectrochemical studies of trivalent metal phthalocyanine thin films: The role of gaseous dopants (oxygen and hydrogen) in determining photoelectrochemical response. *J. Phys. Chem.* **91**, 5651–5659.
28. Klofta, T., P. Rieke, C. Linkous, W.J. Buttner, A. Nanthakumar, T.D. Mewborn, and N.R. Armstrong (1985). Tri- and tetravalent phthalocyanine thin film electrodes: Comparison with other metal and demetallated phthalocyanine systems. *J. Electrochem. Soc.* **132**, 2134-2143.
29. Buttner, W.J., P.C. Rieke, and N.R. Armstrong (1985). The gold/GaPc-Cl/ferri, ferrocyanide/GaPc-Cl/platinum photoelectrochemical cell. *J. Am. Chem. Soc.* **107**, 3738–3739.
30. Harima, Y. and K. Yamashita (1985). Organic photoelectrodes based on p–p iso-type junctions. *J. Phys. Chem.* **89**, 5325–5327.
31. Rieke, P.C. and N.R. Armstrong (1984). Light-assisted, aqueous redox reactions at chlorogallium phthalocyanine thin-film photoconductors: Dependence of the photopotential on the formal potential of the redox couple and evidence for photoassisted hydrogen evolution. *J. Am. Chem. Soc.* **106**, 47–50.
32. Belanger, D., J.P. Dodelet, L.H. Dao, and B.A. Lombos (1984). Photoelectrochemical characteristics and behavior of a surfactant aluminum phthalocyanine cell. *J. Phys. Chem.* **88**, 4288–4295.
33. Leempoel, P., F.-R.F. Fan, and A.J. Bard (1983). Semiconductor electrodes. 50. Effect of mode of illumination and doping on photochemical behavior of phthalocyanine films. *J. Phys. Chem.* **87**, 2948–2955.
34. Loutfy, R.O. and L.F. McIntyre (1983). Phthalocyanines: Electrolyte Schottky-junction devices. *Can. J. Chem.* **61**, 72–77.
35. Mezza, T.M., C.L. Linkous, V. R. Shepard, and N.R. Armstrong (1981). Improved photoelectrochemical efficiencies at methalocyanine-modified SnO_2 electrodes. *J. Electroanal. Chem.* **124**, 311–320.
36. Meier, H., W. Albrecht, U. Tschirwitz, N. Geheeb, and E. Zimmerhackl (1979). Organische Halbleiter in Photoelektroden. *Chem. Ing. Tech.* **51**, 653–656.
37. Tachikawa, H. and L.R. Faulkner (1978). Electrochemical and solid state studies of phthalocyanine thin film electrodes. *J. Am. Chem. Soc.* **100**, 4379–4385.
38. Meshitsuka, S. and K. Tamaru (1977). Photoelectrocatalysis by metal phthalocyanine evaporated films in the oxidation of oxalate ion. *J. Chem. Soc. Faraday Trans. 1*, **73**, 236–242.
39. Meshitsuka, S. and K. Tamaru (1977). Spectral distributions of photoelectrochemical reactions over metal phthalocyanine electrodes. *J. Chem. Soc. Faraday. Trans. 1*, **73**, 760–767.
40. Meier, H., W. Albrecht, U. Tschirwitz, E. Zimmerhackl, and N. Geheeb (1977). Zum photovoltaischen Effekt am System Organischer Halbleiter/Elektrolyt. *Ber. Bunsenges. Phys. Chem.* **81**, 592–597.
41. Sevastyanov, V.I., G.A. Alferov, A.N. Asanov, and G.G. Komissarov (1975). Photovoltaiv effect in the films of pigments contacting with electrolyte. *Biofizika* **20**, 1004–1009.
42. Shumov, Y.S. and M. Heyrovsky (1975). The relation between catalytic and photoelectro-chemical properties of phthalocyanine films. *J. Electroanal. Chem.* **65**, 469–471.

43. Schlettwein, D., H. Graaf, J.-P. Meyer, T. Oekermann, and N.I. Jaeger (1999). Molecular interactions in thin films of hexadecafluorophthalocyaninatozinc (F16PcZn) as compared to islands of *N,N′*-dimethylperylene-3,4,9,10-bicarboximide (MePTCDI). *J. Phys. Chem. B* **103**, 3078–3086.
44. Oekermann, T., D. Schlettwein, and N.I. Jaeger (1999). Role of surface states and adsorbates in time-resolved photocurrent measurements and photovoltage generation at phthalocyaninatozinc(II)-photocathodes. *J. Electroanal. Chem.* **462**, 222–234.
45. Oekermann, T., D. Schlettwein, and D. Wöhrle (1997). Characterization of *N,N′*-dimethyl-3,4,9,10-perylenetetracarboxylic acid diimide and phthalocyaninatozinc(II) in electrochemical photovoltaic cells. *J. Appl. Electrochem.* **27**, 1172–1178.
46. Wöhrle, D., D. Schlettwein, G. Schnurpfeil, G. Schneider, E. Karmann, T. Yoshida, and M. Kaneko (1995). Phthalocyanines and related macrocycles for multi-electron transfer in catalysis, photochemistry and photoelectrochemistry. *Polym. Adv. Technol.* **6**, 118–130.
47. Schlettwein, D., D. Wöhrle, E. Karmann, and U. Melville (1994). Conduction type of substituted tetraazaporphyrins and perylene tetracarboxilic acid diimides as detected by thermoelectric power measurements. *Chem. Mater.* **6**, 3–6.
48. Yanagi, H., K. Tsukatani, H. Yamaguchi, M. Ashida, D. Schlettwein, and D. Wöhrle (1993). Semiconducting behavior of substituted tetraazaporphyrin thin films in photoelectrochemical cells. *J. Electrochem. Soc.* **140**, 1942–1948.
49. Sobbi, A.K., D. Wöhrle, and D. Schlettwein (1993). Stability of various porphyrins in solution and as thin film electrodes. *J. Chem. Soc., Perkin Trans. 2* 481–488.
50. Schlettwein, D. and N.I. Jaeger (1993). Identification of the mechanism of the photoelectrochemical reduction of oxygen on the surface of a molecular semiconductor. *J. Phys. Chem.* **97**, 3333–3337.
51. Schlettwein, D., N.I. Jaeger, and D. Wöhrle (1992). Influence of polymer matrices on the photoelectrochemical properties of a molecular semiconductor by structural modification. *Makromol. Chem., Macromol. Symp.* **59**, 267–279.
52. Schlettwein, D., N.I. Jaeger, and D. Wöhrle (1991). Photoelectrochemical investigations of molecular semiconductors: Characterization of the conduction type of various substituted porphyrins. *Ber. Bunsenges. Phys. Chem.* **95**, 1526–1530.
53. Schlettwein, D., M. Kaneko, A. Yamada, D. Wöhrle, and N.I. Jaeger (1991). Light-induced dioxygen reduction at thin film electrodes of various porphyrins. *J. Phys. Chem.* **95**, 1748–1755.
54. Wöhrle, D., D. Schlettwein, M. Kirschenmann, M. Kaneko, and A. Yamada (1990). The combination of phtha-locyanines and polymers for electrochemically induced processes. *J. Macromol. Sci. Chem. A* **27**, 1239–1260.
55. Kaneko, M., D. Wöhrle, D. Schlettwein, and V. Schmidt (1988). Dioxygen sensitivity of a photoexcited thin film of phthalocyanine dispersed in poly(vinylcarbazole). *Makromol. Chem.* **189**, 2419–2425.
56. Shepard, V.R. and N.R. Armstrong (1979). Electrochemical and photoelectrochemical studies of copper and cobalt phthalocyanine–tin oxide electrodes. *J. Phys. Chem.* **83**, 1268–1276.
57. Giraudeau, A., F.-R.F. Fan, and A.J. Bard (1980). Semiconductor electrodes. 30. Spectral sensitization of the semiconductors titanium oxide (n-TiO_2) and tungsten oxide (n-WO_3) with metal phthalocyanines. *J. Am. Chem. Soc.* **102**, 5137–5142.
58. Jaeger, C.D., F.-R.F. Fan, A.J. Bard (1980). Semiconductor electrodes. 26. Spectral sensitization of semiconductors with phthalocyanine. *J. Am. Chem. Soc.* **102**, 2592–2598.

59. Yanagi, H., S. Chen, P.A. Lee, K.W. Nebesny, N.R. Armstrong, and A. Fujishima (1996). Dye-sensitizing effect of TiOPc thin film on n-TiO_2 (001) surface. *J. Phys. Chem.* **100**, 5447–5451.
60. Nazeeruddin, M.K., R. Humphry-Baker, M. Grätzel, D. Wöhrle, G. Schnurpfeil, G. Schneider, A. Hirth, and N. Trombach (1999). Efficient near-IR sensitization of nanocrystalline TiO_2 films by zinc and aluminum phthalocyanines. *J. Porphyrins Phthalocyanines* **3**, 230–237.
61. Gerischer, H. (1978). Electrochemistry of the excited electronic state. *J. Electrochem. Soc.* **125**, 218C–226C.
62. Sze, S.M. (1981). *Physics of Semiconductor Devices*. Wiley, New York.
63. Rompf, C., D. Ammermann, and W. Kowalsky (1995). Deposition and characterization of crystalline organic semiconductors for photonic devices. *Mater. Sci. Technol.* **11**, 845–848.
64. Koma, A. (1995). Molecular-beam epitaxial-growth of organic thin films. *Prog. Crystal Growth Charact.* **30**, 129–152.
65. Schmidt, A., L.K. Chau, A. Back, and N.R. Armstrong (1996). Epitaxial phthalocyanine ultrathin films grown by organic molecular beam epitaxy (OMBE). In C.C. Leznoff and A.B.P. Lever (eds), *Phthalocyanines: Properties and Applications*, Vol. 4. VCH, New York, Weinheim, Cambridge, pp. 307–341.
66. Yamashita, A. and T. Hayashi (1996). Organic molecular beam deposition of metallophthalocyanines for opto-electonics applications. *Adv. Mater.* **8**, 791–799.
67. Schlettwein, D., D. Alloway, A. Back, K.W. Nebesny, P.A. Lee, and N.R. Armstrong (2002). Organic molecular beam epitaxy (OMBE): Creation of ordered organic thin films and organic/organic$'$ heterojunctions. In A. Hubbard (eds), *Encyclopedia of Surface and Colloid Science*, Marcel Dekker, New York, pp. 3842–3857.
68. Schlettwein, D. (2001). Electronic properties of molecular organic semiconductor thin films. In H. Nalwa (ed), *Supramolecular Photosensitive and Electroactive Materials*, Academic Press, San Diego 2001, pp. 211–338.
69. Hooks, D.E., T. Fritz, and M.D. Ward (2001). Epitaxy and molecular organization on solid substrates. *Adv. Mater.* **13**, 227–241.
70. Jaegermann, W. and T. Mayer (1995). What do we learn from model experiments of semiconductor/electrolyte interfaces in UHV-coadsorption of Br_2 with Na and H_2O on WSe(0001)? *Surf. Sci.* **335**, 343–352.
71. Svensson, S., J.-O. Forsell, H. Siegbahn, A. Ausmees, G. Bray, S. Södergren, S. Sundin, S.J. Osborne, S. Aksela, E. Nommiste, J. Jauhiainen, M. Jurvansuu, J. Karvonen, P. Barta, W.R. Salaneck, A. Evaldsson, M. Lögdlund, and A. Fahlmann (1996). New end station for the study of gases, liquids, and solid films at the MAX laboratory. *Rev. Sci. Instrum.* **67**, 2149–2156.
72. Lindström, H., H. Rensmo, S.-E. Lindquist, A. Hagfeldt, A. Henningsson, S. Södergren, and H. Siegbahn (1998). Redox properties of nanoporous TiO2 (anatase) surface modified with phosphotungstic acid. *Thin Solid Films* **323**, 141–145.
73. Bansal, A. and N.S. Lewis (1998). Stabilization of Si photoanodes in aqueous electrolytes through surface alkylation. *J. Phys. Chem. B* **102**, 4058–4060.
74. Sturzenegger, M., N. Prokopuk, C.N. Kenyon,W.J. Royea, and N.S. Lewis (1999). Reactions of etched, single crystal (111)B-oriented InP to produce functionalized surfaces with low electrical defect densities. *J. Phys. Chem. B* **103**, 10838–10849.
75. Beerbom, M., O. Henrion, A. Klein, T. Mayer, and W. Jaegermann (2000). XPS analysis of wet chemical etching of GaAs(110) by Br-2-H2O: Comparison of emersion and model experiments. *Electrochim. Acta* **45**, 4663–4672.

76. Smolenyak, P.E., R.A. Peterson, D.R. Dunphy, S. Mendes, K.W. Nebesny, D.F. O'Brien, S.S. Saavedra, and N.R. Armstrong (1999). Formation and spectroelectrochemical characterization of multilayer and submonolayer thin films of 2,3,9,10,16,17,23,24-octa(2-benzyloxyethoxy) phthalocyninato copper (CuPc(OC-(2)OBz)(8)). *J. Porphyrins Phthalocyanin* **3**, 620–633.
77. Smolenyak, P.E., R.A. Peterson, K.W. Nebesny, M. Törker, D.F. O'Brien, and N.R. Armstrong (1999). Highly ordered thin films of octasubstituted phthalocyanines. *J. Am. Chem. Soc.* **121**, 8628–8636.
78. Fox, M.-A. (1999). Fundamentals in the design of molecular electronic devices: Long-range charge carrier transport and electronic coupling. *Acc. Chem. Res.* **32**, 201–207.
79. Eichhorn, H., D.W. Bruce, and D. Wöhrle (1998). Amphitropic mesomorphic phthalocyanines—A new approach to highly ordered layers. *Adv. Mater.* **10**, 419–422.
80. Cook, M.J. (1991). Phthalocyanine thin films. *Pure Appl. Chem.* **71**, 2145–2151.
81. Wegner, G. (1991). Ultrathin films of polymers. *Ber. Bunsenges. Phys. Chem.* **95**, 1326–1333.
82. Wegner, G. (1992). Ultrathin films of polymers: Architecture, characterization, properties. *Thin Solid Films* **216**, 105–116.
83. Wu, J., G. Lieser, and G. Wegner (1996). Direct imaging of individual shape-persistent macromolecules and their interaction by TEM. *Adv. Mater.* **2**, 151–154.
84. Silerova (Back), R., L. Kalvoda, D. Neher, A. Ferencz, J. Wu, and G. Wegner (1998). Electrical conductivity of highly organized Langmuir–Blodgett films of phthalocyaninatopolysiloxane. *Chem. Mater.* **10**, 2284–2292.
85. Gattinger, P., H. Rengel, D. Neher, M. Gurka, M. Buck, A.M. van de Craats, and J.M. Warman (1999). Mechanism of charge transport in anisotropic layers of a phthalocyanine polymer. *J. Phys. Chem. B* **103**, 3179–3186.
86. Sato, N., H. Yoshida, and K. Tsutsumi (2003). Unoccupied electronic states in phthalocyanine thin films studied by inverse photoemission spectroscopy. *Synth. Metals* **133**, 673–674.
87. Hill, I.G., A. Kahn, Z.G. Soos, and R.A. Pascal, Jr. (2000). Charge-separation energy in films of pi-conjugated organic molecules. *Chem. Phys. Lett.* **327**, 181–188.
88. Wu, C.I., Y. Hirose, H. Sirringhaus, and A. Kahn (1997). Electron–hole interaction energy in the organic molecular semiconductor PTCDA. *Chem. Phys. Lett.* **272**, 43–47.
89. Meier, H. and W. Albrecht (1965). Zum Problem der pn-Übergänge zwischen organischen und anorganischen Photoleitern. *Ber. Bunsenges. Phys. Chem.* **69**, 160–167.
90. Meier, H. (1974). *Organic Semiconductors*. VCH, Weinheim.
91. Simon, J. and J.-J. Andre (1985). *Molecular Semiconductors: Photoelectrical Properties and Solar Cells*. Springer, Berlin.
92. Wright, J.D. (1989). Gas adsorption on phthalocyanines and its effects on electrical properties. *Prog. Surf. Sci.* **31**, 1–60.
93. Snow, A.W. and W.R. Barger (1989). Phthalocyanine films in chemical sensors. In C.C. Leznoff and A.B.P. Lever (eds), *Phthalocyanines: Properties and Applications*, Vol. 1. VCH, New York, Weinheim, Cambridge, pp. 341–392.
94. Guillaud, G., J. Simon, and J.P. Germain (1998). Metallophthalocyanines—Gas sensors, resistors and field effect transistors. *Coord. Chem. Rev.* **178**, 1433–1484.
95. Schlettwein, D., N.R. Armstrong, P.A. Lee, and K.W. Nebesny (1994). Factors which control the n-type or p-type behavior of molecular semiconductor thin films. *Mol. Cryst. Liq. Cryst.* **253**, 161–171.

96. Schlettwein, D., and N.R. Armstrong (1994). Correlation of frontier orbital positions and conduction type of molecular semiconductors as derived from UPS in combination with electrical and photoelectrochemical experiments. *J. Phys. Chem.* **98**, 11771–11779.
97. Meyer, J.-P., D. Schlettwein, D. Wöhrle, and N.I. Jaeger (1995). Charge transport in thin films of molecular semiconductors as investigated by measurements of thermoelectric power and electrical conductivity. *Thin Solid Films* **258**, 317–324.
98. Meyer, J.-P. and D. Schlettwein (1996). Influence of central metal and ligand system on conduction type and charge carrier transport in phthalocyanine thin films. *Adv. Mat. Opt. Electron.* **6**, 239–244.
99. Schlettwein, D., J.-P. Meyer, and N.I. Jaeger (1999). Intermolecular interactions and electrical properties in thin films of tetrapyridotetraazaporphyrinatozinc(II). *J. Porphyrins Phthalocyanin* **3**, 611–619.
100. Schmidt, A., R. Schlaf, D. Louder, L.K. Chau, S.-Y. Chen, T. Fritz, M.F. Lawrence, B.A. Parkinson, and N.R. Armstrong (1995). Epitaxial growth of the ionic polymer fluoroaluminum phthalocyanine on the basal plane of singlecrystal tin disulfide. *Chem. Mater.* **7**, 2127–2135.
101. Chau, L.K., C.D. England, S.-Y. Chen, and N.R. Armstrong (1993). Visible absorption and photocurrent spectra of epitaxially deposited phthalocyanine thin films: Interpretation of exciton coupling effects. *J. Phys. Chem.* **97**, 2699–2706.
102. Bufler, J. (1993). *Dissertation: Vergleichende elektrische, elektrochemische und oberflächenspektroskopische Untersuchungen an den radikalischen Phthalocyaninen* $LuPc_2$ *und LiPc*. Universität Tübingen, Tübingen.
103. Ishii, H., K. Sugiyama, E. Ito, and K. Seki (1999). Energy level alignment and interfacial electronic structures at organic metal and organic organic interfaces. *Adv. Mater.* **11**, 605–625.
104. Ishii, H., K. Sugiyama, D. Yoshimura, E. Ito, Y. Ouchi, and K. Seki (1998). Energy-level alignment at model interfaces of organic electroluminescent devices studied by UV photoemission: Trend in the deviation from the traditional way of estimating the interfacial electronic structures. *IEEE J. Select. Topics Quant. Chem.* **4**, 24–33.
105. Tamoto, N., C. Adachi, and K. Nagai (1997). Electroluminescence of 1,3,4-oxadiazole and triphenylamine-containing molecules as an emitter in organic multilayer light emitting diodes. *Chem. Mater.* **9**, 1077–1085.
106. Hung, L.S. and C.W. Tang (1999). Interface engineering in preparation of organic surfaceemitting diodes. *Appl. Phys. Lett.* **74**, 3209–3211.
107. Hiller, S., D. Schlettwein, N.R. Armstrong, and D. Wöhrle (1998). Influence of surface reactions and ionization gradients on junction properties of $F_{16}PcZn$. *J. Mater. Chem.* **8**, 945–954.
108. Karl, N. and N. Sato (1992). UV-photoelectron spectroscopy of one- and two-component organic crystals. *Mol. Cryst. Liq. Cryst.* **218**, 79–84.
109. Sato, N. and M. Yoshikawa (1996). Valence electronic structure at the interface of organic thin films. *J. Electron Spectrosc. Relat. Phenom.* **78**, 387–390.
110. Schlaf, R., B.A. Parkinson, P.A. Lee, K.W. Nebesny, and N.R. Armstrong (1998). Determination of frontier orbital alignment and band bending at an organic semiconductor heterointerface by combined X-ray and ultraviolet photoemission measurements. *Appl. Phys. Lett.* **73**, 1026–1028.
111. Rajagopal, A. and A. Kahn (1998). Molecular-level offset at the PTCDA/Alq(3) heterojunction. *Adv. Mater.* **10**, 140–144.
112. Hill, I.G. and A. Kahn (1998). Energy level alignment at interfaces of organic semiconductor heterostructures. *J. Appl. Phys.* **84**, 5583–5586.
113. Rajagopal, A., C.I. Wu, and A. Kahn (1998). Energy level offset at organic semiconductor heterojunctions. *J. Appl. Phys.* **83**, 2649–2655.

114. Hill, I.G. and A. Kahn (1999). Combined photoemission/in vacuo transport study of the indium tin oxide/copper phthalocyanine/*N*, *N*′-diphenyl-*N*, *N*′-bis(l-naphthyl)-1,1′ biphenyl-4,4″ diamine molecular organic semiconductor system, *J. Appl. Phys.* **86**, 2116–2122.
115. Lee, S.T., Y.M. Wang, X.Y. Hou, and C.W. Tang (1999). Interfacial electronic structures in an organic light-emitting diode. *Appl. Phys. Lett.* **74**, 670–672.
116. Schlaf, R., B.A. Parkinson, P.A. Lee, K.W. Nebesny, and N.R. Armstrong (1999). HOMO/LUMO alignment at PTCDA/ZnPc and PTCDA/ClInPc heterointerfaces determined by combined UPS and XPS measurements. *J. Phys. Chem. B* **103**, 2984–2992.
117. Darwent, J.R., P. Douglas, A. Harriman, G. Porter, and M.-C. Richoux (1982). Metal phthalocyanines and porphyrins as photosensitizers for reduction of water to hydrogen. *Coord. Chem. Rev.* **44**, 83–126.
118. Ferraudi, G. (1989). Photochemical properties of metallophthalocyanines in homogeneous solution. In C.C. Leznoff and A.B.P. Lever (eds), *Phthalocyanines, Properties and Applications*, Vol. 1. VCH, New York, Weinheim, Cambridge, pp. 291–340.
119. Fan, F.-R. and L.R. Faulkner (1979). Phthalocyanine thin films as semiconductor electrodes, *J. Am. Chem. Soc.* **101**, 4779–4787.
120. Klofta, T., W.J. Buttner, and N.R. Armstrong (1986). Effect of crystallite size and hydrogen and oxygen uptake in the photoelectrochemistries of thin films of chlorogallium phthalocyanine. *J. Electrochem. Soc.* **133**, 1531–1532.
121. Klofta, T.J., J. Danziger, P.A. Lee, J. Pankow, K.W. Nebesny, and N.R. Armstrong (1987). Photoelectrochemical and spectroscopic characterization of thin films of titanyl phthalocyanine: Comparisons with vanadyl phthalocyanine. *J. Phys. Chem.* **91**, 5646–5651.
122. Yanagi, H., S. Douko, Y. Ueda, M. Ashida, and D. Wöhrle (1992). Improvement of photoelectrochemical properties of chloroaluminum phthalocyanine thin films by controlled crystallization and molecular orientation. *J. Phys. Chem.* **96**, 1366–1372.
123. Perrier, G. and L.H. Dao (1986). Cellules photoélectrochimiques de phtalocyanine d'hydroxyaluminium déposées par rotation. *Can. J. Chem.* **64**, 2431–2439.
124. Karmann, E., D. Schlettwein, and N.I. Jaeger (1996). Photoelectrochemical oxidation of 2-mercaptoethanol at the surface of octacyanophthalocyanine thin film electrodes. *J. Electroanal. Chem.* **405**, 149–158.
125. Karmann, E., J.-P. Meyer, D. Schlettwein, N.I. Jaeger, M. Anderson, A. Schmidt, and N.R. Armstrong (1996). Photoelectrochemical effects and (photo)conductivity of "n-type" phthalocyanines. *Mol. Cryst. Liq. Cryst.* **283**, 283–291.
126. Yanagi, H., Y. Kanbayashi, D. Schlettwein, D. Wöhrle, and N.R. Armstrong (1994). Photochemical investigations on naphthalocyanine derivatives in thin films. *J. Phys. Chem.* **98**, 4760–4766.
127. Karmann, E. (1996). *Dissertation: Photoelektrochemische Untersuchungen an dünnen Filmen molekularer n-Halbleiter mit Porphyrinstruktur*. Universität Bremen, Bremen.
128. Schlettwein, D., E. Karmann, T. Oekermann, and H. Yanagi (2000). Wavelength-dependent switching of the photocurrent direction at the surface of molecular semiconductor electrodes based on orbital-confined excitation and transfer of charge carriers from higher excited states. *Electrochim. Acta* **45**, 4679–4704.
129. Oekermann, T., D. Schlettwein, N.I. Jaeger, and D. Wöhrle (1999). Influence of electronwithdrawing substituents on photoelectrochemical surface phenomena at phthalocyanine thin film electrodes. *J. Porphyrins Phthalocyanin* **3**, 444–452.
130. Oekermann, T. (2000). *Die Rolle von Oberflächenzuständen in der Kinetik photoelektrochemischer Reaktionen an Elektroden molekularer Halbleiter*. Shaker Verlag, Aachen.

131. van Vlierberge, B., M.Z. Yang, F.X. Sauvage, M.-G. de Backer, and A. Chapput (1986). The photoreduction of zinc tetra-2,3-pyridino porphyrazine: A photochemical, electrochemical and spectroscopic study. *Spectrochim. Acta* **42**, 1133–1139.
132. Yang, M.Z., M.G. de Backer, and F.X. Sauvage (1990). Electrochemical and photoelectrochemical characterizations of electrodes covered by zinc tetra 2,3 pyridinoporphyrazine layers. *New J. Chem.* **14**, 273–277.
133. Wöhrle, D., R. Bannehr, B. Schumann, G. Meyer, and N.I. Jaeger (1983). Synthesis, electrochemical and photoelectrochemical properties of polyphthalocyanine coated electrodes. *J. Mol. Cat.* **21**, 255–262.
134. Knothe, G. and D. Wöhrle (1989). Polymeric phthalocyanines and their precursors, 16: A structure model for polymeric phthalocyanines. *Makromol. Chem.* **190**, 1573–1586.
135. Knothe, G. (1993). Isomerism and symmetry of bridged polymeric phthalocyanines. *Macromol. Theor. Simul.* **2**, 503–516.
136. Stillman, M.J. and T. Nyokong (1989). Absorption and magnetic circular dichroism spectral properties of phthalocyanines. In C.C. Leznoff and A.B.P. Lever (eds), *Phthalocyanines: Properties and Applications*, Vol. 1. VCH, New York, Weinheim, Cambridge, pp. 133–290.
137. Gouterman, M. (1978). Optical spectra and atomic structure. In D. Dolphin (ed), *The Porphyrins*, Part A Vol. III, Academic Press, New York pp. 1–159.
138. Kobayashi, N. and H. Konami (1996). Molecular orbitals and electronic spectra of phthalocyanine analogues. In C.C. Leznoff and A.B.P. Lever (eds), *Phthalocyanines: Properties and Applications*, Vol. 4, VCH, New York, Weinheim, Cambridge, pp. 343–404.
139. Green, M.A. (2003). *Third Generation Photovoltaics: Advanced Solar Energy Conversion*. Springer, Berlin.
140. Bard, A.J. and L.R. Faulkner (1980). *Electrochemical Methods*. Wiley, New York.
141. Abrantes, L.M. and L.M. Peter (1983). Transient photocurrents at passive iron electrodes. *J. Electroanal. Chem.* **150**, 593–601.
142. Gerischer, H. (1991). Electron-transfer kinetics of redox reactions at the semiconductor/electrolyte contact. A new approach. *J. Phys. Chem.* **95**, 1356–1359.
143. Hagfeldt, A. and M. Grätzel (1995). Light-induced redox reactions in nanocrystalline systems. *Chem. Rev.* **95**, 49–68.
144. Wilson, R.H. (1977). A model for the current–voltage curve of photoexcited semiconductor electrodes. *J. Appl. Phys.* **48**, 4292–4297.
145. Chazalviel, J.-N. (1982). Electrochemical transfer via surface states: A new formulation for the semiconductor/electrolyte interface. *J. Electrochem. Soc.* **129**, 963–969.
146. Batchelor, R.A. and A. Hamnett (1992). Surface states on semiconductors. In J.O. Bockris, B.E. Conway, and R.E. White (eds), *Modern Aspects of Electrochemistry*, Vol. 22. Kluwer Academic Publishers/Plenum Publishers, Boston, pp. 265–415.
147. Bard, A.J., A.B. Bocarsly, F.-R.F. Fan, E.G. Walton, and M.S. Wrighton (1980). The concept of Fermi level pinning at semiconductor/liquid junctions. Consequences for energy conversion efficiency and selection of useful solution redox couples in solar devices. *J. Am. Chem. Soc.* **102**, 3671–3677.
148. Bocarsly, A.B., D.C. Bookbinder, R.N. Dominey, N.S. Lewis, and M.S. Wrighton (1980). Photoreduction at illuminated p-type semiconducting silicon photoelectrodes. Evidence for Fermi level pinning. *J. Am. Chem. Soc.* **102**, 3683–3688.
149. Lewerenz, H.J. (1993). Surface states and Fermi level pinning at semiconductor/electrolyte junctions. *J. Electroanal. Chem.* **356**, 121–143.
150. Albery, W.J., N.L. Dias, and C.P. Wilde, (1987). The photoelectrochemical kinetics of n-type cadmium sulfide I. The hydroquinone system. *J. Electrochem. Soc.* **134**, 601–609.

151. Peter, L.M. (1990). Dynamic aspects of semiconductor photoelectrochemistry. *Chem. Rev.* **90**, 753–769.
152. Rajeshwar, K. (1992). Charge transfer in photoelectrochemical devices via interface states: Unified model and comparison with experimental data. *J. Electrochem. Soc.* **129**, 1003–1008.
153. Kelly, J.J. and R. Memming (1992). The influence of surface recombination and trapping on the cathodic photocurrent at p-type III–V-electrodes. *J. Electrochem. Soc.* **129**, 730–738.
154. Schwarzburg, K. and F. Willig (1997). Modeling of electrical transients in the semiconductor/electrolyte cell for photogeneration of charge carriers in the bulk. *J. Phys. Chem. B* **101**, 2451–2458.
155. Smith, B.B. and A.J. Nozik, (1997). Theoretical studies of electron transfer and electron localization at the semiconductor–liquid interface. *J. Phys. Chem. B* **101**, 2459–2475.
156. Albery, W.J., P.N. Bartlett and C.P. Wilde (1987). Modulated light studies of the electrochemistry of semiconductors. Theory and experiment. *J. Electrochem. Soc.* **134**, 2486–2490.
157. Schefold, J. (1992). Impedance and intensity modulated photocurrent spectroscopy as complementary differential methods in photoelectrochemistry. *J. Electroanal. Chem.* **341**, 111–136.
158. Modestov, A.D., G.-D. Zhou, H.-H. Ge, and B.H. Loo (1994). A study of copper electrode behavior in alkaline solutions containing benzotriazole-type inhibitors by the photocurrent response method and intensity-modulated photocurrent spectroscopy. *J. Electroanal. Chem.* **375**, 293–299.
159. Goossens, A. (1996). Intensity-modulated photocurrent spectroscopy of thin anodic films on titanium. *Surf. Sci.* **365**, 662–671.
160. Oskam, G., J.C. Schmidt, and P.C. Searson (1996). Electrical properties of n-type (111)Si in aqueous $K_4Fe(CN)_6$ solution. 2. Intensity modulated photocurrent spectroscopy. *J. Electrochem. Soc.* **143**, 2538–2543.
161. Schlichthörl, G., N.G. Park, and A.J. Frank (1999). Estimation of the charge-collection efficiency of dye-sensitized nanocrystalline TiO_2 solar cells. *Z. Phys. Chem.* **212**, 45–50.
162. Peter, L.M. and D. Vanmaekelbergh (1999). Time and frequency resolved studies of photoelectrochemical kinetics. In R.C. Alkire and D.M. Kolb (eds), *Advances in Electrochemical Science and Engineering*, Vol. 6. Wiley-VCH, Weinheim, pp. 77–163.
163. Oekermann, T., D. Schlettwein, and N.I. Jaeger (2001). Charge transfer and recombination kinetics at electrodes of molecular semiconductors investigated by intensity modulated photocurrent spectroscopy. *J. Phys. Chem. B* **105**, 9524–9532.
164. Meier, H. (1965). Organic dyes as photoelectric semiconductors. *Angew. Chem. Int. Ed.* **4**, 619–635.
165. Peat, R. and L.M. Peter (1987). A study of the passive film on iron by intensity modulated photocurrent spectroscopy. *J. Electroanal. Chem.* **228**, 351–364.
166. Li, J. and L.M. Peter (1986). Surface recombination at semiconductor electrodes: Part iv. Steady-state and intensity modulated photocurrents at n-GaAs electrodes. *J. Electroanal. Chem.* **199**, 1–26.
167. Taira, S., T. Miki, and H. Yanagi (1999). Dye-sensitization of n-TiO_2 single-crystal electrodes with vapor-deposited oxometal phthalocyanines. *Appl. Surf. Sci.* **143**, 23–29.
168. Armstrong, N.R., K.W. Nebesny, G.E. Collins, P.A. Lee, L.K. Chau, C. Arbour, and B.A. Parkinson (1991). O/I-MBE: Formation of highly ordered phthalocyanine/semiconductor junctions by molecular-beam epitaxy: Photoelectrochemical

characterization. In R.A. Lessard (ed), *Photopolymer Device Physics, Chemistry, and Applications II*. Proceedings of SPIE Volume 1559. The International Society for Optical Engineering, Bellingham, WA, pp. 18–26.

169. Chau, L.K., C. Arbour, G.E. Collins, K.W. Nebesny, P.A. Lee, C.D. England, N.R. Armstrong, and B.A. Parkinson (1993). Phthalocyanine aggregates on metal dichalcogenide surfaces: Dye sensitization on tin disulfide semiconductor electrodes by ordered and disordered chloroindium phthalocyanine thin films. *J. Phys. Chem.* **97**, 2690–2698.
170. Chau, L.K., E.J. Osburn, N.R. Armstrong, D.F. O'Brien, and B.A. Parkinson (1994). Dye sensitization with octasubstituted liquid crystalline phthalocyanines. *Langmuir* **10**, 351–353.
171. Deng, H., H. Mao, B. Liang, Y. Shen, Z. Lu, and H. Xu (1996). Aggregation and the photoelectric behavior of tetrasulfonated phthalocyanine adsorbed on a TiO_2 microporous electrode. *J. Photochem. Photobiol. A* **99**, 71–74.
172. Hodak, J., C. Quinteros, M.I. Litter, and E. San Roman (1996). Sensitization of TiO_2 with phthalocyanines. 1. Photo-oxidations using hydroxoaluminium tricarboxymonoamidephthalocyanine adsorbed on TiO_2. *J. Chem. Soc. Faraday Trans.* **92**, 5081–5088.
173. Yanagi, H., Y. Ohoka, T. Hishiki, K. Ajito, and A. Fujishima (1997). Characterization of dyedoped TiO_2 films prepared by spray-pyrolysis. *Appl. Surf. Sci.* **113**, 426–431.
174. Fang, J., J. Wu, X. Zhang, H. Mao, Y. Shen, and Z. Lu (1997). Fabrication, characterization and photovoltaic study of a GaTSPc-CdS/TiO_2 particulate film. *J. Mater. Chem.* **7**, 737–740.
175. Nazeeruddin, M.K., R. Humphry-Baker, M. Grätzel, and B.A. Murrer (1998). Efficient near IR sensitization of nanocrystalline TiO2 films by ruthenium phthalocyanines. *Chem. Commun.* 719–720.
176. Rensmo, H., K. Keis, H. Lindström, S. Södergren, A. Solbrand, A. Hagfeldt, S.-E. Lindquist, L.N. Wang, and M. Muhammed (1997). High light-to-energy conversion efficiencies for solar cells based on nanostructured ZnO electrodes. *J. Phys. Chem. B* **101**, 2598–2601.
177. Bedja, I., P.V. Kamat, X. Hua, A.G. Lappin, and S. Hotchandani (1997). Photosensitization of nanocrystalline ZnO films by bis(2, 2′-bipyridine)(2, 2′-bipyridine-4, 4′-dicarboxylic acid)ruthenium(II). *Langmuir* **13**, 2398–2403.
178. Rao, T.N. and L. Bahadur (1997). Photoelectrochemical studies on dye-sensitized particulate ZnO thin-film photoelectrodes in nonaqueous media. *J. Electrochem. Soc.* **144**, 179–185.
179. Fessenden, R.W. and P.V. Kamat (1995). Rate constants for charge injection from excited sensitizer into SnO_2, ZnO, and TiO_2 semiconductor nanocrystallites. *J. Phys. Chem.* **99**, 12902–12906.
180. Yoshida, T. and H. Minoura (2000). Electrochemical self-assembly of dye-modified zinc oxide thin films. *Adv. Mater.* **12**, 1219–1222.
181. Yoshida, T., K. Miyamoto, N. Hibi, T. Sugiura, H. Minoura, D. Schlettwein, T. Oekermann, G. Schneider, and D. Wöhrle (1998). Self assembled growth of nano particulate porous ZnO thin film modified by 2,9,16,23-tetrasulfophthalocyanatozinc(II) by one-step electrodeposition. *Chem. Lett.* **7**, 599–600.
182. Yoshida, T., M. Tochimoto, D. Schlettwein, G. Schneider, D. Wöhrle, T. Sugiura, and H. Minoura (1999). Self-assembly of zinc oxide thin films modified with tetrasulfonated metallophthalocyanines by one-step electrodeposition. *Chem. Mater.* **11**, 2657–2667.
183. Yoshida, T., K. Terada, D. Schlettwein, T. Oekermann, T. Sugiura, and H. Minoura (2000). Electrochemical self-assembly of nanoporous ZnO/eosin Y thin films and their sensitized photoelectrochemical performance. *Adv. Mater.* **12**, 1214–1217.

184. Schubert, U., A. Lorenz, N. Kundo, T. Stuchinskaya, L. Gogina, A. Salanov, V. Zaikonovskii, V. Maizlish, and G.P. Shaposhnikov (1997). Cobalt phthalocyanine derivatives supported on TiO_2 by sol-gel processing. 1. Preparation and microstructure. *Chem. Ber./Recl.* **130**, 1585–1589.
185. Hunter, C.A. and K.M. Sanders (1990). The nature of π–π interactions. *J. Am. Chem. Soc.* **112**, 5525–5534.
186. Schneider, G., D. Wöhrle, W. Spiller, J. Stark, and G. Schulz-Ekloff (1998). Photooxidation of 2-mercaptoethanol by various water soluble phthalocyanines in aqueous alkaline solution under irradiation with visible light. *Photochem. Photobiol.* **60**, 333–342.
187. Schlettwein, D., T. Oekermann, T. Yoshida, M. Tochimoto, and H. Minoura (2000). Photoelectrochemical sensitisation of ZnO–tetrasulfophthalocyaninatozinc composites prepared by electrochemical self-assembly. *J. Electroanal. Chem.* **481**, 42–51.
188. Nüesch, F., J.E. Moser, V. Shklover, and M. Grätzel (1996). Merocyanine aggregation in mesoporous networks. *J. Am. Chem. Soc.* **118**, 5420–5431.
189. Oekermann, T., T. Yoshida, D. Schlettwein, T. Sugiura, and H. Minoura (2001). Photoelectrochemical properties of ZnO/tetrasulfophthalocyanine hybrid thin films prepared by electrochemical self-assembly. *Phys. Chem. Chem. Phys.* **3**, 3387–3392.
190. Schlettwein, D., T. Oekermann, T. Yoshida, T. Sugiura, H. Minoura, and D. Wöhrle (2002). Electrochemically self-assembled ZnO/dye electrodes: Preparation and time-resolved photoelectrochemical measurements. In Z.H. Kafafi (ed), *Organic Photovoltaics II*. Proceedings of SPIE Volume 4465. The International Society for Optical Engineering, Bellingham, WA, pp. 113–122.
191. Michaelis, E., K. Nonomura, D. Schlettwein, T. Yoshida, H. Minoura, and D. Wöhrle (2004). Hybrid thin films of ZnO with porphyrins and phthalocyanines prepared by one-step electrodeposition. *J. Porphyrins Phthalocyanin* **8**, 1366–1375.
192. Nonomura, K., T. Loewenstein, E. Michaelis, D. Wöhrle, T. Yoshida, H. Minoura, and D. Schlettwein (2006). Photoelectrochemical characterisation and optimisation of electrodeposited ZnO thin films sensitized by porphyrins and phthalocyanines. *Phys.Chem.Chem.Phys.* submitted.
193. Yoshida, T., M. Iwaya, H. Ando, T. Oekermann, K. Nonomura, D. Schlettwein, D. Wöhrle, and H. Minoura (2004). Improved photoelectrochemical performance of electrodeposited ZnO/EosinY hybrid thin films by dye re-adsorption. *Chem. Commun.* 400–401.
194. Yoshida, T., K. Terada, D. Schlettwein, T. Oekermann, T. Sugiura, and H. Minoura (2000). Electrochemical self-assembly of nanoporous ZnO/eosin Y thin films and their sensitized photoelectrochemical performance. *Adv. Mater.* **12**, 1214–1217.
195. Yoshida, T., T. Oekermann, K. Okabe, D. Schlettwein, K. Funabiki, and H. Minoura (2002). Cathodic electrodeposition of ZnO/eosinY hybrid thin films from dye added zinc nitrate bath and their photoelectrochemical characterizations. *Electrochemistry* **70**, 470–487.
196. Rostalski, J. and D. Meissner (2000). Photocurrent spectroscopy for the investigation of charge carrier generation and transport mechanisms in organic p/n-junction solar cells. *Sol. Energy Mater. Sol. Cells* **63**, 37–47.
197. Brabec, C.J. N.S. Sariciftci, and J.C. Hummelen (2001). Plastic solar cells. *Adv. Mater.* **13**, 15–26.
198. Yanagi, H., M. Ashida, Y. Harima, and K. Yamashita (1990). Photoelectrochemical properties of orientation-controlled thin film for 5,10,15,20-tetraphenylporphyrin. *Chem. Lett.* 385–388.

199. Zhou, R., F. Josse, W. Göpel, Z.Z. Öztürk, and Ö. Bekaroglu (1996). Phthalocyanines as sensitive materials for chemical sensors. *Appl. Organomet. Chem.* **10**, 557–577.
200. Waite, S., J. Pankow, G.E. Collins, P.A. Lee, and N.R. Armstrong (1989). Interactions of ammonia and oxygen with the surfaces of chlorogallium phthalocyanine thin films: Microcircuit photoconductivity and quartz-crystal microgravimetry studies. *Langmuir* **5**, 797–805.
201. Drechsel, J, B. Männig, F. Kozlowski, M. Pfeiffer, K. Leo, and H. Hoppe (2005). Efficient organic solar cells based on a double p-i-n architecture using doped wide-gap transport layers. *Appl. Phys. Lett.* **86**, 244102.

11

Organisation And Photoelectrochemical Reactivity of Water-Soluble Metalloporphyrins at the Liquid/Liquid Interface

David J. Fermín and Nicolas Eugster

1. Introduction

The present chapter discusses recent accounts on the organisation and photoreactivity of water-soluble metalloporphyrins at interfaces between two immiscible electrolyte solutions (ITIES). This class of molecular interfaces provides an interesting framework for studying heterogeneous photoredox processes between water-soluble porphyrins and hydrophobic redox species employing photoelectrochemical techniques. These experimental approaches are based on ideally polarisable liquid/liquid junctions, in which the Galvani potential difference can be controlled potentiostatically[1–3]. By contrast to photoelectrochemistry at porphyrin sensitised electrodes, processes at these interfaces are free from surface defects which can act as charge traps and recombination centres[4]. Consequently, the redox energy levels involved in the photoelectrochemical reactions can be well defined as in the case of classical condensed phase photochemistry.

The interfacial reactivity of metalloporphyrins and chlorins is critically determined by their specific adsorption at the liquid/liquid junctions. The affinity of the water-soluble porphyrins for these interfaces is associated with their complex solvation structure arising from the hydrophobic central ring and peripheral ionisable groups. This chapter will also feature a brief overview on recent studies of the molecular organisation of water-soluble porphyrins as probed by a variety

David J. Fermín • Departement für Chemie und Biochemie, Universität Bern, Freiestrasse 3, Bern CH-3012, Switzerland

Nicolas Eugster • Laboratoire d'Electrochimie Physique et Analytique, Institut des Sciences et Ingénierie Chimiques, Ecole Polytechnique Fédérale de Lausanne, Lausanne CH-1015, Switzerland

N_4-Macrocyclic Metal Complexes, edited by José H. Zagal, Fethi Bedioui and Jean-Pol Dodelet. Springer Science+Business Media, Inc., New York, 2006.

of techniques including quasi-elastic laser scattering, second harmonic generation and impedance spectroscopy.

Under potentiostatic conditions, photoinduced heterogeneous electron transfer between specifically adsorbed porphyrins and redox couples confined to the organic phase manifests itself by photocurrent responses[5]. As in the case of dynamic photoelectrochemistry, these photoresponses provide information on the dynamics of heterogeneous electron transfer and recombination processes. In addition, we shall demonstrate that photocurrent measurements can be used to characterise the interfacial coverage of the specifically adsorbed porphyrins as well as their molecular orientation.

2. The Polarisable Liquid/Liquid Interface

2.1. The Structure of the Neat Liquid/Liquid Boundary

The structure of the interfacial boundary between two immiscible electrolyte solutions has been subject to numerous studies and a comprehensive discussion on this topic is beyond the scope of this chapter. Several review articles provide an excellent background on the structure and dynamics at the molecular level[6–9]. Important progress has been made in recent years, yet key information on the nature of the liquid/liquid junction remain controversial. A major obstacle is that high-resolution scanning probe techniques cannot be directly employed for accessing information on the structure of the interface. One of the few attempts reported so far was done by De Serio *et al.* who developed a method combining a near-field optical probe and Raman spectroscopy for studying the structure of the junction between *p*-xylene and ethylene glycol[10]. Initial reports revealed changes in the intensity of the *p*-xylene signal over a region as far as 10 μm from the interfacial boundary. The authors recognised that this distance is significantly larger than the expected density profiles at the boundary between two immiscible liquids and several aspects were considered to influence the results, including far-field contributions. On the other hand, a wealth of information about the structure of the interface has been obtained by other techniques such as non-linear optics[11–16], X-ray[17–21] and neutron reflectivity[22,23].

Molecular dynamics simulations have consistently provided pictures of microscopically corrugated liquid/liquid boundaries at the picosecond time scale, as exemplified in Figure 11.1a[6,24–27]. The characteristic width of the neat interface involves two contributions: (i) the intrinsic width determined by the molecular structure of solvents and (ii) the broadening induced by capillary wave fluctuations. The snapshot picture in Figure 11.1a shows that the interface is molecularly sharp, indicating that the intrinsic width is approximately 1 Å thick or less[27]. A very important aspect illustrated in Figure 11.1a is the dynamic surface roughness induced by thermally excited capillary waves. The relevance of these features towards heterogeneous electron transfer becomes more evident as the transverse density fluctuations are time averaged. Figure 11.1b shows the average of trajectories over hundreds of picoseconds, resulting in monotonic changes in the density profiles for both solvents in the range of 10 Å[6]. According to capillary waves

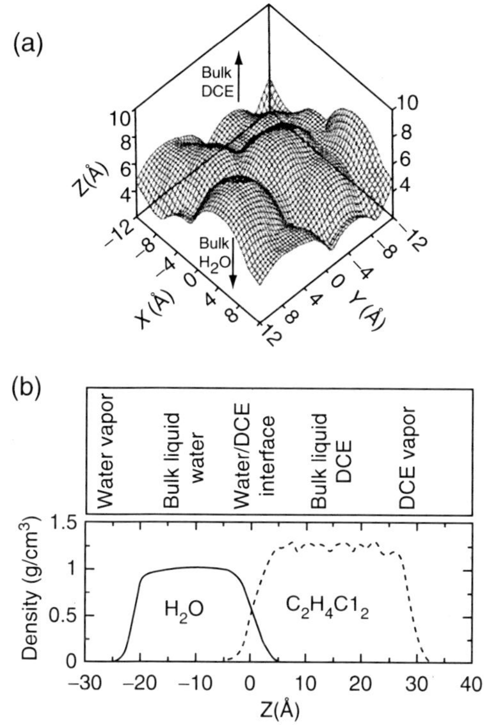

Figure 11.1. Conolly surface plot of the oxygen atoms obtained from molecular dynamic simulation of the water/1,2-dichloroethane (DCE) junction (a). A flexible SPC model was employed for the 343 water molecules, while the 108 DCE molecules were described by a four-centre, simple charge, flexible model. A spherical probe was used with a radius of 5Å and the Gibbs dividing surface is located at z = 4Å. Density variations of the water/vapour, water/DCE and DCE/vapour junction as obtained from the average of trajectories over 200 ps at 300 K (b). Reprinted with permission from ref.[6]. Copyright (1996) American Chemical Society.

theory[28], the width of the density profile decay is dependent on the surface tension. In the case of the water/1,2-dichloroethane (DCE) interface, the width of the interfacial region is larger than that of the corresponding liquid/vapour junctions, which feature larger surface tensions. Estimations of the interfacial width associated with capillary wave fluctuations by second-harmonic generation (SHG)[15] and neutron reflectivity[22,23,29] also provide consistent values between 7 and 15 Å. On the other hand, analysis of X-ray reflectograms at the water/alkane interface appears to show certain deviations from the capillary-wave predicted widths[21]. Despite the slight variations in the reported values for the interfacial width, the different approaches appear to converge to values just under 10 Å. Furthermore, theoretical studies based on capillary wave theory including non-linear polarisation of the interface predict strong dependence of the interfacial corrugation on the potential drop across the liquid/liquid boundary[30].

As mentioned previously, the photoreactivity of porphyrin molecules at this class of interfaces strongly depends on their specific adsorption. The organisation of adsorbed species is determined by their solvation properties as well as local interfacial properties such as pH and H-bonding. SHG studies have demonstrated that the concentration of protons in the vicinity of the interface can differ by orders of magnitude from that in the bulk water phase[7,31]. Sum frequency generation (SFG) studies have also shown an increase in the concentration of H-bonded water molecules at the interfacial region[13,14,32–34]. In the case of water soluble metalloporphyrins, not only interfacial protonation and H-bonding but also aggregation phenomena do play an important role on the molecular organisation[35–38]. Specific aspects on the correlation between porphyrin organisation and reactivity at the liquid/liquid interfaces will be further discussed in Section 4 of this chapter.

2.2. The Potential Distribution Across the Polarisable Liquid/Liquid Interface

Let us consider the case where a porphyrin molecule located in the water phase undergoes a photoinduced heterogeneous electron transfer with a redox couple located in the organic phase as schematically depicted in Figure 11.2. The driving force for the electron-transfer process is dependent on the Fermi level of the electrons in the redox couples as well as on the potential drop across the interface. The potential distribution across the liquid/liquid boundary has been subject to intense investigations for the last 20 years and several review articles have been published on this issue[2,3,39,40]. One of the main points of discussion is whether a significant portion of the potential drop is developed between the reaction planes where the redox species are located. In principle, the reaction planes can be related to the onset of the average solvent density profiles generated by the thermal fluctuations. Two limiting cases can be distinguished in this respect, (A) the distance between the two planes is negligible and the potential is distributed in two back-to-back diffuse layers and (B) the potential drop occurs essentially between the two reaction planes. It can be envisaged that in case (A) changes in the Galvani potential difference will only affect the concentration of the ionic species but not the rate constant of electron transfer (concentration polarisation). On the other hand, case (B) will determine a "Butler–Volmer" type behaviour of the current density associated with the electron-transfer process.

As a first approximation, the distribution of the electrostatic potential can be analytically solved by describing the interface as back-to-back diffuse layers. Employing the Gouy–Chapman model for 1:1 electrolytes in both phases, the electric field is given by[3,41],

$$\left(\frac{\partial \phi}{\partial x}\right)_{\mathrm{w}<x\leq x_{\mathrm{s}}} = \left(\frac{8RTc_{\mathrm{w}}}{\varepsilon_{\mathrm{w}}}\right)^{1/2} \sinh\left(\frac{F}{2RT}\left(\phi(x)-\phi^{\mathrm{w}}\right)\right) \tag{11.1}$$

and

$$\left(\frac{\partial \phi}{\partial x}\right)_{x_{\mathrm{s}}<x\leq \mathrm{o}} = -\left(\frac{8RTc_{\mathrm{o}}}{\varepsilon_{\mathrm{o}}}\right)^{1/2} \sinh\left(\frac{F}{2RT}\left(\phi(x)-\phi^{\mathrm{o}}\right)\right) \tag{11.2}$$

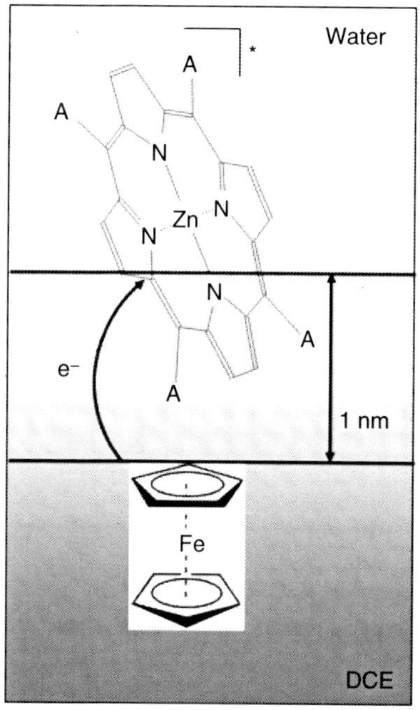

Figure 11.2. Schematic representation of the heterogeneous photo-oxidation of ferrocene by specifically adsorbed water-soluble metalloporphyrin at the water/DCE interface. It is postulated that the distance of closest approach is determined by the onset of the solvent density profiles generated by capillary wave fluctuations.

where $\phi(x)$ is the Galvani potential at a given distance from the interface, ε is the dielectric permittivity, c is the concentration of the supporting electrolyte, x_S is the position of the liquid/liquid boundary and the indexes "w" and "o" relate to the bulk of the water and organic phases, respectively. Neglecting the contribution from surface dipoles and in the absence of specifically adsorbed species, the continuity of the electric field establishes;

$$\varepsilon_W \left(\frac{\partial \phi}{\partial x}\right)_{x_s} = \varepsilon_O \left(\frac{\partial \phi}{\partial x}\right)_{x_s} \tag{11.3}$$

and the fraction of the applied potential developed in the organic side of the interface $\left(\Delta_o^s\phi = \phi^s - \phi^o\right)$ can be estimated as:

$$\Delta_o^s\phi = \phi^s - \phi^o = \frac{2RT}{F}\tanh^{-1}\left(\frac{\sqrt{\varepsilon_W c_W}\sinh\left(\frac{F}{2RT}\Delta_o^w\phi\right)}{\sqrt{\varepsilon_W c_W}\cosh\left(\frac{F}{2RT}\Delta_o^w\phi\right) + \sqrt{\varepsilon_O c_O}}\right) \tag{11.4}$$

where $\Delta_o^w\phi = \phi^w - \phi^o$ is the Galvani potential difference. Taking the Galvani potential of the organic phase as reference, the potential drop in the aqueous phase $\left(\Delta_s^w\phi\right)$ can be simply obtained from the difference between Equation (11.4) and $\Delta_o^w\phi$. The dependence of $\Delta_s^w\phi$ on $\Delta_o^w\phi$ is exemplified in Figure 11.3a for the water/DCE interface($\varepsilon_o = 10$). As expected, the fraction of the potential drop in

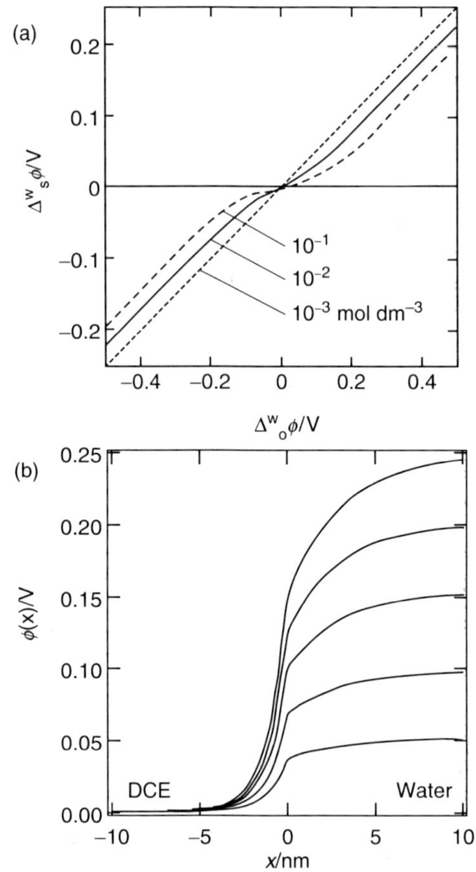

Figure 11.3. Fraction of the potential drop in the aqueous side as a function of the Galvani potential difference at the water/DCE interface as estimated by the back-to-back diffuse layer model (a). Different concentration of 1:1 electrolytes are introduced in the aqueous phase, while the concentration in DCE is 10^{-2} mol dm^{-3}. The relative permittivities of water and DCE were taken as 78 and 10, respectively. Potential distribution across the liquid/liquid interface for 10^{-2} mol dm^{-3} concentration of 1:1 electrolytes in both phases (b).

the aqueous phase decreases as the concentration of the aqueous supporting electrolyte increases. For electrolyte concentrations of 10^{-2} mol dm^{-3} in both phases, the fraction of the potential in the aqueous phase varies between 25 and 45%.

Based on the potential distribution estimated from Equation (11.4), the electrostatic potential profile across the water/DCE interface can be obtained as shown in Figure 11.3b. This simplified approach suggests that the diffuse layers extend over a region of 10 nm for electrolyte concentrations of 10^{-2} mol dm^{-3}. As mentioned in Section 2.1, this distance is larger than the average solvent density profiles. If the distance separating the redox species is determined by the solvent profile at the interface, it would be expected that the driving force for heterogeneous electron transfer will be little dependent on changes in the Galvani potential difference. This point has been raised by a number of authors[42,43], and in particular by Schmickler on the basis of a model employing a three-dimensional lattice gas[44]. However, studies based on techniques such as ac impedance[45–47] and scanning electrochemical microscopy[48–51] have demonstrated a dependence of the electron-transfer rate constant on the Galvani potential difference. In Section 5, we shall review a series of photoelectrochemical experiments involving self-assembled metalloporphyrin that provide clear evidence that the Galvani potential difference does affect the phenomenological rate constant of electron transfer. Depending on the nature of the specific adsorption of the porphyrin species, the applied potential can also induce changes in the interfacial concentration and orientation of the dye. This behaviour will be discussed in Section 4.

3. Basic Photophysics of Metalloporphyrins and Chlorins

3.1. Electronic Transitions and Lifetime of Excited States in Porphyrin-Based Compounds

The photophysics of porphyrin molecules have been the subject of intensive research due to the central role played by these molecules in life-sustaining processes[52]. In this section, we shall briefly introduce basic aspects about electronic transitions which are directly connected to their photoredox behaviour at interfaces. The absorption properties of regular porphyrins in the UV–visible region are determined by electronic transitions from the ground state S_0 to the two lowest singlet excited states S_1 (Q state) and S_2 (B state). The $S_0 \rightarrow S_1$ transition gives rise to the weak Q bands in the visible region (550–650 nm), while the $S_0 \rightarrow S_2$ transition produces the strong B band (or Soret band) in the near UV (380–450 nm). The absorption spectra of regular metalloporphyrins usually feature two Q bands between 500 and 600 nm. The lower energy band is the electronic origin Q(0,0) of the lowest singlet excited state S_1. The second band is its vibrational overtone and is labelled Q(1,0). Similar transitions are also observed in chlorin-type of molecules, in which one of the four pyrrole ring is partially saturated.

The types of porphyrin molecules to be studied throughout this chapter are displayed in Figure 11.4. These species are water soluble due to ionised groups located in the periphery of the tetrapyrrole ring. As discussed in Section 4, these porphyrin-based molecules can be separated into those featuring symmetrical or asymmetrical distribution of ionised groups. The absorption spectra of chlorin e-6, protoporphyrin IX (protoIX) and Sn(IV) *meso*-tetra(4-carboxyphenyl) porphyrin dichloride (SnTPPC) are contrasted in Figure 11.5. The Soret band is the most prominent features in this spectral range. Most of the photoelectrochemical studies reviewed in the next sections involve excitation to the S_2 state, which features a higher capture cross-section.

The dynamics of relaxation for single porphyrin molecules are illustrated by the Jablonski diagram in Figure 11.6. According to the Frank–Condon approximation, light absorption leads to electronic transitions in which the momentum

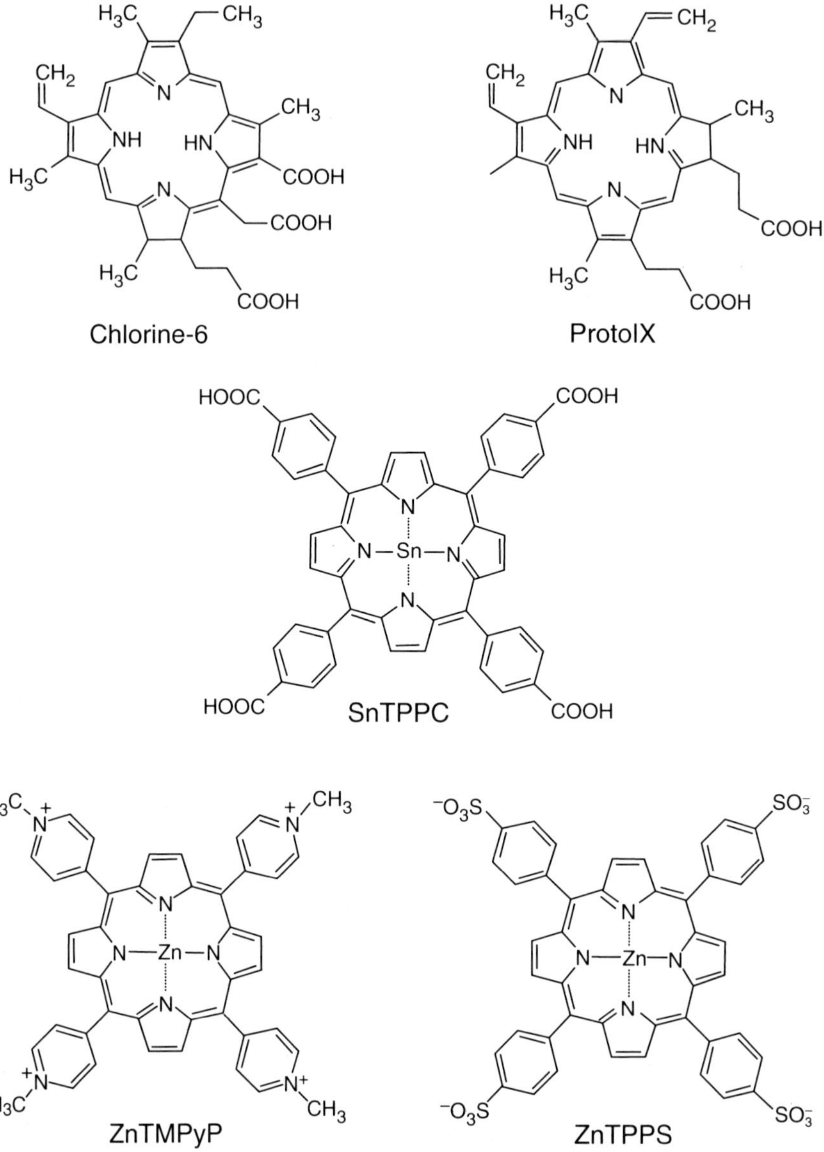

Figure 11.4. Molecular structure of the various water-soluble porphyrins and chlorins studied by photoelectrochemistry at the liquid/liquid interfaces: chlorin e-6, protoporphyrin IX (protoIX), Sn(IV) *meso*-tetra(4-carboxyphenyl) porphyrin dichloride (SnTPPC), Zn(II) *meso*-tetra(*N*-methyl-4-pyridinium) porphyrin (ZnTMPyP) and Zn(II) *meso*-tetra (4-sulphonatophenyl) porphyrin (ZnTPPS). The Zn(II) derivatives of TPPC have also been extenively studied.

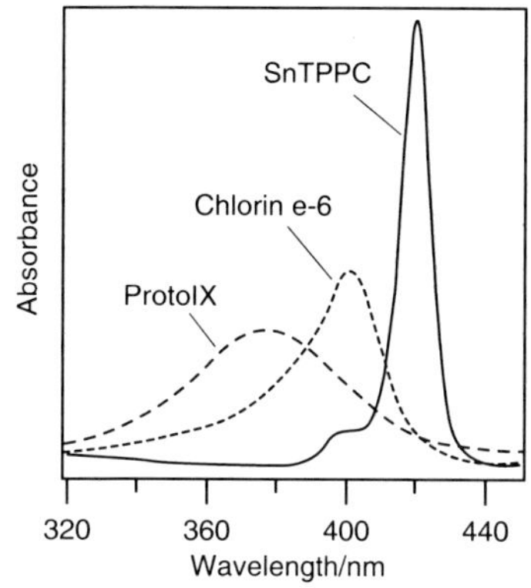

Figure 11.5. Typical absorption spectra of the Soret band of SnTPPC, chlorin e-6 and protoIX in aqueous phase.

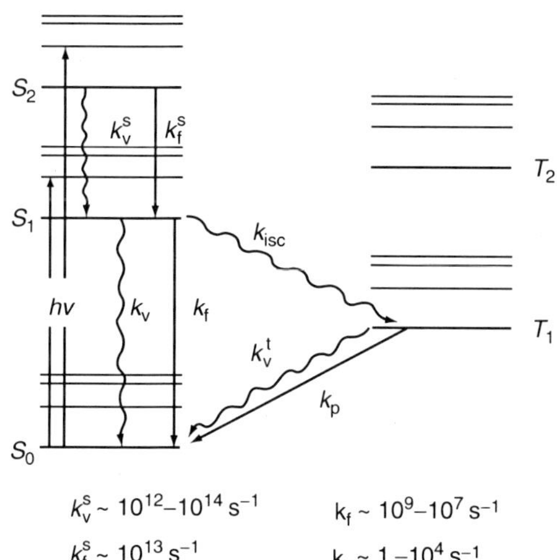

Figure 11.6. Jablonsky energy level diagram of porphyrins and chlorins. Characteristic rate constants for various radiative and non-radiative relaxations are also indicated.

is preserved. The "hot" excited state readily relaxes to thermally equilibrated excited states within picoseconds. Further relaxation phenomena involve non-radiative as well as radiative transitions between states of the same spin multiplicity (fluorescence) or different multiplicity (phosphorescence). Excited states can also undergo intersystem crossing, where the system changes from a state of low- to high-spin multiplicity. In the case of the porphyrins shown in Figure 11.4, the efficiency of intersystem crossing is rather high and the generated triplet state can live for hundreds of microseconds. The decays of the triplet states of chlorin e-6, protoIX and SnTPPC as monitored by flash photolysis are contrasted in Figure 11.7[53].

The efficiency of heterogeneous photoinduced electron transfer will be determined by the ratio between the pseudo-first-order rate constant of electron transfer and the decay of the excited state. Considering the distance separating

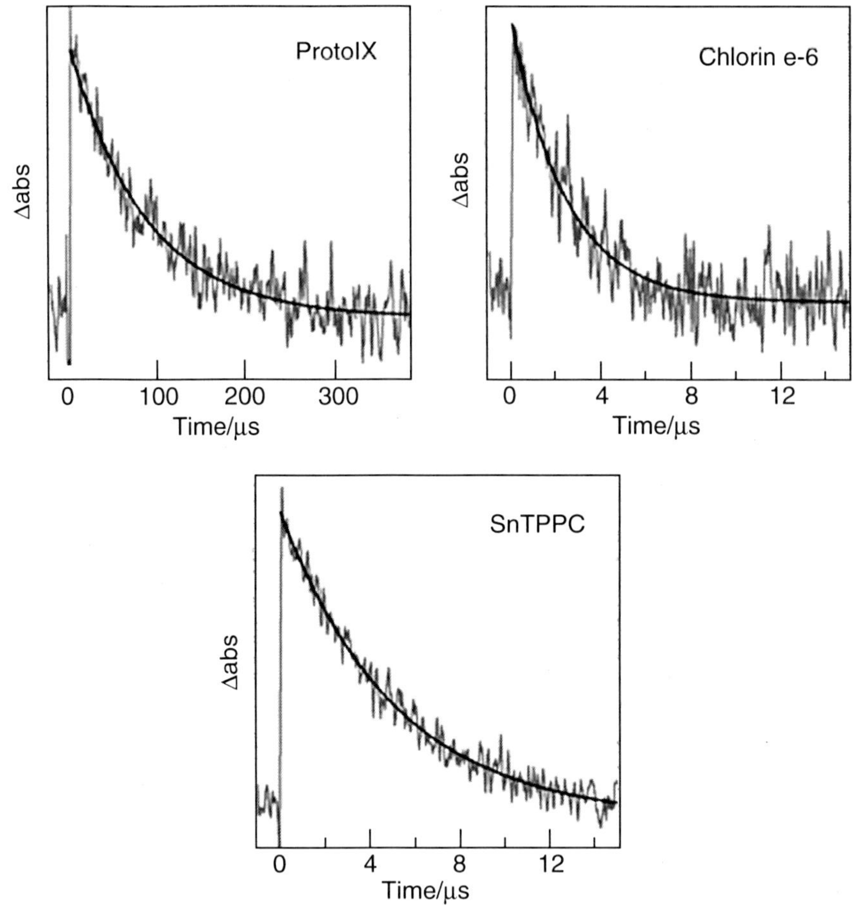

Figure 11.7. Characteristic decay of the triplet state of protoIX, chlorin e-6 and SnTPPC in aqueous phase monitored by flash photolysis. The composition of the electrolyte is identical to the one in photocurrent measurement at the water/DCE interface. Photoinduced heterogeneous electron transfer is effectively in competition with their relaxation process.

the redox couples from the porphyrin species at the liquid/liquid interface, it is expected that the average rate for heterogeneous electron transfer can only compete with the slowest decay path. Consequently, the characteristic decay times illustrated in Figure 11.7 provide an indication of the time scale involved in the electron-transfer process.

3.2. Ultrafast Relaxation in Porphyrin Ion Pairs

Another class of metalloporphyrin species of remarkable photoactivity at the liquid/liquid interface is formed by the electrostatic association of oppositely charged species[54–58]. The electrostatic attraction between the charged substituents and the hydrophobic interaction of the aromatic macrocycles cooperate in holding the individual species in close proximity such that extensive orbital overlap can occur. For instance, the negatively charged Zn(II) *meso*-tetrakis(*p*-sulphonatophenyl) porphyrin (ZnTPPS) and the positively charged Zn(II) *meso*-tetrakis(*N*-methylpyridyl) porphyrin (ZnTMPyP) undergo spontaneous association[59]. The resulting ion pair, or heterodimer, features a strong affinity to the water/DCE interface. Analysis of the association constant on the basis of the Bjerrum model suggests that the distance separating the centres of the metalloporphyrins in the complex is approximately 6 Å[59]. As discussed in subsequent sections, the ZnTPPS–ZnTMPyP complex is rather reactive towards heterogeneous photoreductions and photo-oxidations.

Evidence of the strong association of the two porphyrins in the aqueous phase can be obtained from the evolution of the absorption spectrum. Figure 11.8 displays the absorption spectra of the ZnTPPS and ZnTMPyP monomers (10^{-5} mol dm^{-3}), as well as that of an equimolar mixture of both porphyrins[60]. It can be observed that the Soret and Q bands of the ZnTPPS–ZnTMPyP ion pair are not linear combinations of the bands of the isolated porphyrin. Indeed, the absorption bands of the ion pair are slightly broadened and red shifted in comparison to the bands of the single porphyrins. Figure 11.9a also illustrates a strong quenching of the $S_1 \rightarrow S_0$ fluorescence signal upon the formation of the complex. Looking closer to the spectral features of the complex emission (Figure 11.9b), it is also observed that the fluorescence spectrum is not a linear combination of the spectra of the free monomers. These spectral properties can be interpreted in terms of the formation of Frenkel-type excitons as a result of the face-to-face conformation of the ion pair[54,59,61]. These considerations are consistent with the recent Raman

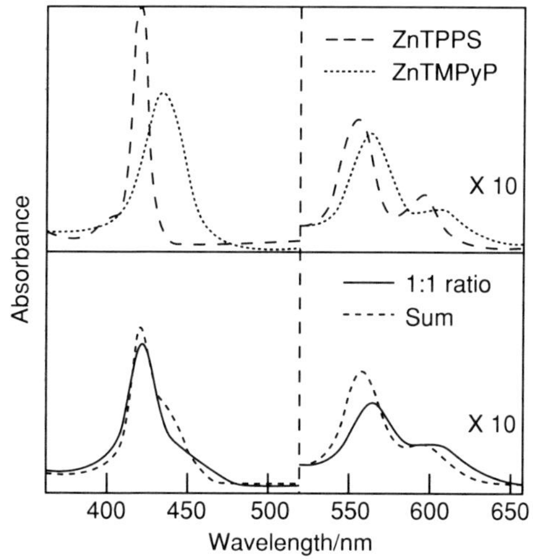

Figure 11.8. Absorption spectra of ZnTPPS, ZnTMPyP and the equimolar concentration of both porphyrins in aqueous phase. The spectral features of the 1:1 ratio do not correspond to a linear combination of the spectra of the individual porphyrins. These results are taken as evidence for the formation of exciton type of structures in the ZnTPPS-ZnTMPyP ion pair. Reprinted with permission from ref.[60]. Copyright(2003) American Chemical Society.

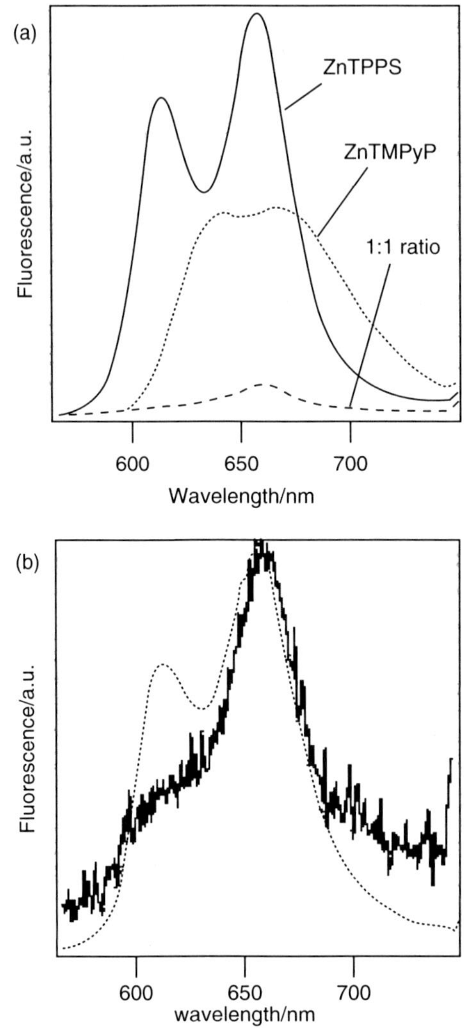

Figure 11.9. Fluorescence spectra of ZnTPPS-ZnTMPyP and the ZnTPPs–ZnTMPyP ion pair in aqueous solution (a). As for the absorption spectra in Figure 11.8, the fluorescence spectrum of the ion pair cannot be interpreted in term of a linear combination of the individual porphyrin spectra (b).

studies by Chen *et al.*, which revealed that CuTPPS and CuTMPyP undergo very slight changes of the porphyrin skeleton upon ion pairing[55].

The dynamics of the photochemical process involving the ZnTPPS–ZnTMPyP ion pair has been recently investigated by pump-probe transient absorption[60]. The transient absorption responses shown in Figure 11.10 were obtained by pumping the porphyrins at 560 nm and probing at 510 nm. It should be indicated that the depopulation of the S_1 state for ZnTPPS and ZnTMPyP in their monomeric form exhibits time constants of 1.6 and 0.6 ns, respectively. As mentioned before, this relaxation mainly contains contribution from radiative

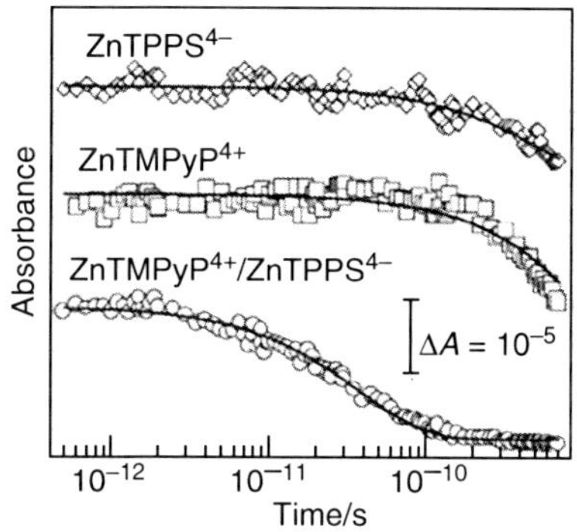

Figure 11.10. Femtosecond resolved transient absorption of ZnTPPS, ZnTMPyP and the ion pair in aqueous solution. The 40 ps decay observed in the presence of the ion pair suggests the formation of an intermediate charge separated state. Reprinted from ref.[60]. Copyright (2003) American Chemical Society.

(fluorescence) and non-radiative (intersystem crossing) processes. These values are within the range typically observed for the relaxation of the S_1 state of zinc tetraphenyl porphyrins[52,62,63]. By contrast, the decay of the excited state for the ZnTPPS–ZnTMPyP ion-pair occurs within 40 ps with no response associated with the absorption of the triplet state. Considering that the driving force for electron transfer from ZnTMPyP to ZnTPPS is 0.4 eV[60,64], excluding the electrostatic energies associated with the charge of the molecules, it is expected that the excitation of the ion pair leads to the formation of an intermolecular charge transfer state of the form,

$$\left[\left(\mathrm{ZnTPPS}^{4-}\right)\left(\mathrm{ZnTMPyP}^{4+}\right)\right]^{*} \xrightarrow{k_q} \left[\left(\mathrm{ZnTPPS}^{3-}\right)\left(\mathrm{ZnTMPyP}^{3+}\right)\right]^{\mathrm{CS}} \tag{11.5}$$

Although Equation (11.5) appears reasonable from the thermodynamic point of view, the nature of the charge separated state should be considered cautiously. As recognised by Holten *et al.*[62], the spectroscopic signatures of ground and different excited states of porphyrin can be convoluted in the visible region and the identification of the intermediate species by transient absorption becomes non-trivial. However, the strong photoreactivity of this ion-pair at the liquid/liquid interface provides clear evidence that an intermediate species with lifetime extending into the microsecond range is generated after 40 ps[60].

3.3. Interfacial Photoelectrochemistry vs. Time-Resolved Spectroscopy

The photoredox properties of porphyrin molecules at interfaces can be studied by time-resolved spectroscopic and photoelectrochemical techniques. A key difference between both approaches is that the latter only probes molecules located at distances from the interface not larger than the characteristic diffusion length of the excited state. According to the Einstein–Smoluchowski equation, the diffusion length (δ) is determined by the lifetime of the excited state (τ) and the diffusion coefficient ($D \sim 10^{-5}$ cm^{-2}s^{-1}),

$$\delta = \sqrt{2D\tau} \tag{11.6}$$

From this equation, it is clear that photoelectrochemical responses featuring long-living triplet states as exemplified in Figure 11.7 ($\tau \sim 10^{-5}$ s) involves species located within 100 nm from the interfacial boundary. Obviously, the shortest is the lifetime of the excited state, the closest the excited state should be to the reaction plane. For instance, photoreactions featuring vibrationally excited singlet (hot) states only occur in systems with large orbital coupling, e.g. photoactive species covalently bonded to metal oxides[65,66].

In the case of liquid/liquid interfaces, the fact that only porphyrins located within few nanometers from the molecular junction can react heterogeneously introduces large complexities for fully spectroscopic approaches. Indeed, linear time-resolved optical techniques are overwhelmed by the dynamics of the photophysical processes of the species in the bulk phase. Transient absorption in total internal reflection (TIR) has been employed in order to enhance the spectroscopic signal arising from the porphyrin species involved in the heterogeneous electron transfer. Figure 11.11 shows that the triplet state lifetime of ZnTPPC located within 150 nm from the water/toluene boundary (penetration depth of the evanescent wave) decreases with increasing concentrations of tetracyanoquinodimethane in the organic phase[67]. A plausible analysis of these data can be performed in terms of a variation of the Stern–Volmer equation for heterogeneous quenching reaction. However, there is no "direct" proof that the decrease of the triplet state lifetime is associated with heterogeneous rather than homogeneous

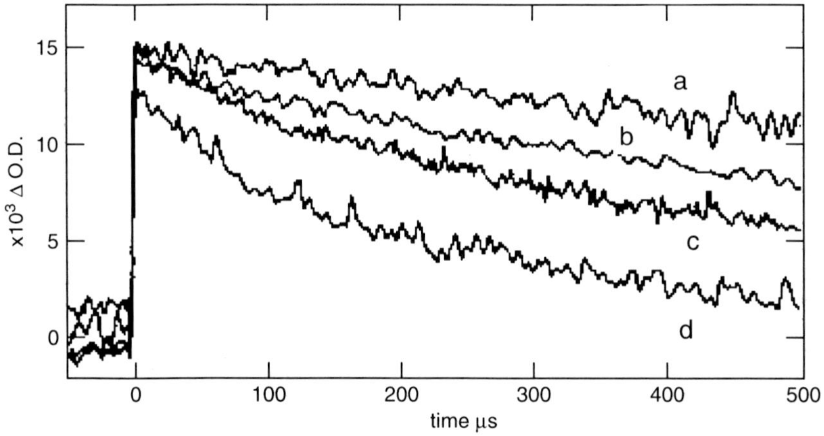

Figure 11.11. Flash photolysis measurement under total internal reflection (TIR) from the water/toluene interface in the presence of ZnTPPC ($5.0 \cdot 10^{-5}$ mol dm^{-3}) and different concentrations of the redox acceptor TCNQ: 0 (a), $0.5 \cdot 10^{-3}$ (b), $1.0 \cdot 10^{-3}$ (c) and $1.5 \cdot 10^{-3}$ mol dm^{-3} (d). The decrease in the lifetime of the triplet state was interpreted in terms of heterogeneous quenching by the redox acceptor in the organic phase. Reprinted from ref.[67] with permission from Elsevier Science.

photoprocesses, as partition of TCNQ cannot be entirely excluded from these experiments.

By contrast to linear optics, electrochemistry at interfaces between two immiscible electrolyte solutions is very sensitive to heterogeneous redox processes. We shall demonstrate that photocurrent responses occur when photoexcited porphyrin species in one solution undergo heterogeneous redox quenching by species located in the other electrolyte (see Figure 11.2) There are two key requirements for measuring photocurrents under potentiostatic conditions: (i) the rate of heterogeneous electron transfer should be comparable to the rate of excited state decay, and (ii) the time constant for double layer charging should be shorter than other short-circuiting events such as ion transfer or back electron transfer. Indeed, stable photocurrent signals can be observed in the potential range where neither charged photoproducts nor the supporting electrolyte transfer from one electrolyte solution

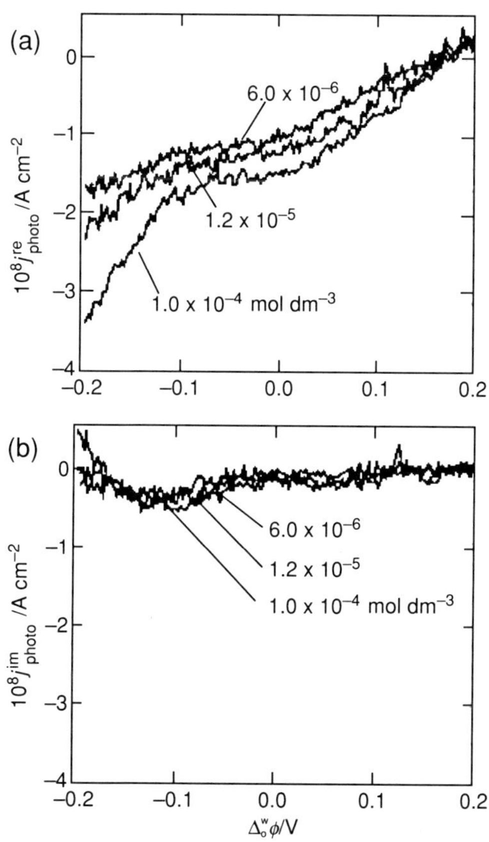

Figure 11.12. In-phase (a) and quadrature (b) components of the photocurrent associated with the photoreduction of TCNQ by ZnTPPC at the water/DCE interface. The photoresponses were recorded for various concentrations of ZnTPPC under chopped illumination and lock-in detection as a function of the Galvani potential difference. From ref.[68]. Reproduced by permission of the Royal Society of Chemistry.

to the other. For instance, photocurrent signals at the water/DCE interface in the presence of ZnTPPC and TCNQ can be observed in Figure 11.12[68]. As discussed later, these results unequivocally demonstrate that photoinduced heterogeneous electron transfer can take place within a narrow range of Galvani potential differences. Under these conditions, photoelectrochemical signals can be regarded as the ultimate surface specific approach.

As in all potentiostatic techniques, the double layer charging is a "parallel" process to the faradaic reaction that can substantially attenuate the photocurrent signal at "short-time scale" (see Section 5.3)[4,69]. This element introduces another important difference between fully spectroscopic and electrochemical techniques. Commercially available optical instrumentation can currently deliver time resolution of 50 fs or less for conventional techniques such as transient absorption. On the other hand, the resistance between the two reference electrodes commonly employed in electrochemical measurements at the liquid/liquid interfaces and the interfacial double layer capacitance provide time constants of the order of hundreds of microseconds. Consequently, direct information on the rate of heterogeneous electron injection from/to the excited state is not accessible from photocurrent measurements. These techniques do allow sensitive measurements of the ratio between electron injection and decay of the excited state under photostationary conditions. Other approaches such as photopotential measurements, i.e. relative changes in the Fermi levels in both phases, can provide kinetic information in the nanosecond regime.

These considerations emphasize that the information extracted from time resolved spectroscopy and photoelectrochemical measurements are not alternative but complementary. As discussed in Section 5, the spectroscopic information reviewed so far will allow rationalising the photocurrent signals obtained in a considerably longer time scale. Before analysing the dynamic photocurrent responses, we shall demonstrate that the photoreactions are strongly connected to the specific adsorption of the porphyrin at the liquid/liquid interface.

4. Organisation of Water-Soluble Porphyrins at the Liquid/Liquid Interface

The specific adsorption of molecules at liquid/liquid interfaces has generated a lot of interest since the pioneering work of Koryta *et al.* on phospholipid monolayers at the ITIES[70–72]. The formation of condensed monolayers has been considered as a convenient strategy for studying physical aspects of charge transfer processes in biological membranes. The surface concentration of adsorbates at these interfaces can be accurately approached by surface tension measurements. Indeed, the formation of phospholipid monolayers brings about a substantial drop of the surface tension depending on the applied Galvani potential difference. Several authors have employed drop-size imaging techniques for monitoring this behaviour statically[73–75] and dynamically[76–78]. Surface tension measurements have been complemented by spectroscopic techniques specially sensitive to the electronic structure and organisation of adsorbates such as SHG and SFG[79], X-ray and Neutron Reflectivity[8], quasi-elastic laser scattering (QELS)[80–82], and potential modulated fluorescence (PMF)[83,84]. Theoretical analysis of these phenomena using the lattice-gas model has also been reported[85–87].

As discussed in Section 2, capillary wave fluctuations establish solvent density profiles across the liquid/liquid boundary extending through a distance of less than 1 nm. As adsorption at these interfaces is linked to solvation forces, the onset of the solvent density profiles can be considered as "adsorption planes". Based on this picture, the interface is characterised by two adsorption planes and the surface concentration of the adsorbates is in equilibrium with the bulk concentration in each phase. Furthermore, the bulk concentrations are also correlated by the partitioning equilibrium[83,88]. The porphyrin derivatives shown in Figure 11.4 exhibit very low solubility in organic solvents such as DCE, manifesting itself by positive values for the Gibbs free energy of transfer ($\Delta G_{\mathrm{tr}}^{\circ,\mathrm{w}\to\mathrm{o}}$). In the case of ionic species with a charge z, the Gibbs free energy of ion transfer determines the standard ion transfer potential ($\Delta_{\mathrm{o}}^{\mathrm{w}}\phi_{\mathrm{tr}}^{\circ}$)[89],

$$\Delta_{\mathrm{o}}^{\mathrm{w}}\phi_{\mathrm{tr}}^{\circ} = \frac{\Delta G_{\mathrm{tr}}^{\circ,\mathrm{w}\to\mathrm{o}}}{zF} \tag{11.7}$$

Depending on the charge of the porphyrin species, polarisation of the liquid/liquid boundary to either positive or negative potentials will affect the concentration ratio of the ionic species in both phases,

$$\Delta_{\mathrm{o}}^{\mathrm{w}}\phi = \Delta_{\mathrm{o}}^{\mathrm{w}}\phi_{\mathrm{tr}}^{\circ} + \frac{RT}{zF}\ln\left(\frac{a^{\mathrm{o}}}{a^{\mathrm{w}}}\right) \tag{11.8}$$

PMF and SHG measurements have provided clear evidence that the surface concentration of ZnTMPyP or ZnTPPC increases in both adsorption planes as the Galvani potential difference approaches $\Delta_{\mathrm{o}}^{\mathrm{w}}\phi_{\mathrm{tr}}^{\circ}$[36,37,83,84]. As discussed further below, photocurrent measurements confirmed that the coverage of ionic porphyrins at the liquid/liquid boundary is affected by $\Delta_{\mathrm{o}}^{\mathrm{w}}\phi$.

As illustrated by Figure 11.3, the potential profile is distributed between both electrolyte phases. Depending on the charge of the porphyrin molecule, the Gibbs free energy of adsorption (ΔG_{ads}) of water-soluble metalloporphyrins can be affected by the fraction of the potential drop in the aqueous phase $\Delta_{\mathrm{s}}^{\mathrm{w}}\phi$[83,90],

$$\Delta G_{\mathrm{ads}} = \Delta G_{\mathrm{ads}}^{\circ} - \frac{zF}{RT}\Delta_{\mathrm{s}}^{\mathrm{w}}\phi \tag{11.9}$$

The fugacity of porphyrins at the interface $f(\theta_{\mathrm{p}})$ and the activity fraction in bulk water are linked by ΔG_{ads},

$$\frac{a_{\mathrm{p}}}{a_{\mathrm{H_2O}}} = f(\theta_{\mathrm{p}})\exp\left(\frac{\Delta G_{\mathrm{ads}}}{RT}\right) = f(\theta_{\mathrm{p}})\left\{\exp\left(\frac{\Delta G_{\mathrm{ads}}^{\circ}}{RT}\right)\exp\left(-\frac{zF}{RT}\Delta_{\mathrm{s}}^{\mathrm{w}}\phi\right)\right\} \tag{11.10}$$

the parameter $f(\theta_{\mathrm{p}})$ is a function of the relative surface coverage θ_{p}[91].

Different models describing the adsorbate–adsorbate and adsorbate–surface interactions lead to various forms of $f(\theta_{\mathrm{p}})$. The isotherm most frequently used to describe experimental results is the Langmuir isotherm, in which no interaction between adsorbates is considered,

$$f\left(\theta_{\mathrm{p}}\right) = \frac{\theta_{\mathrm{p}}}{1-\theta_{\mathrm{p}}} \tag{11.11}$$

Other isotherms may empirically introduce parameters associated with adsorbate–adsorbate interactions such as the Frumkin isotherm,

$$f\left(\theta_{\mathrm{p}}\right) = \frac{\theta_{\mathrm{p}}}{1-\theta_{\mathrm{p}}} \exp\left(-2\alpha\theta_{\mathrm{p}}\right) \tag{11.12}$$

where α corresponds to the interaction parameter. Equations (11.10)–(11.12) are special cases arising from the generalised Frumkin isotherm developed by Markin and Volkov for adsorption at the liquid/liquid interfaces[92].

As discussed in Section 4.1, the specific adsorption of the ionic porphyrins at the liquid/liquid interface manifests itself by changes in the distribution of ions at the interface as well as in the interfacial tension. However, these two parameters do not provide direct information on the organisation of the adsorbed species in terms of lateral interactions and average molecular orientation. These aspects will be reviewed in Section 4.2 based on SHG and photocurrent light polarisation anisotropy studies of metalloporphyrins at the polarised water/DCE interface.

4.1. Excess Charge Associated with the Specific Adsorption of Ionic Porphyrins

At constant temperature and pressure, the changes in surface tension γ are related to the surface excess (Γ) of the various species in the system by the Gibbs adsorption equation[3,93],

$$-\mathrm{d}\gamma = \sum_i \Gamma_i \,\mathrm{d}\tilde{\mu}_i \tag{11.13}$$

where $\tilde{\mu}_i$ is the electrochemical potential of the species "i". Separating the chemical and electrostatic components of the electrochemical potential, Equation (11.13) can be written similarly to the electrocapillary equation for mercury/electrolyte solution interfaces,

$$-\mathrm{d}\gamma = q^{\mathrm{w}}\mathrm{d}\left(\Delta_{\mathrm{o}}^{\mathrm{w}}\phi\right) + \sum_i \Gamma_i \,\mathrm{d}\mu_i \tag{11.14}$$

where q^{w} is the surface excess charge density on the aqueous side of the interface ($q^{\mathrm{w}} = -q^{\mathrm{o}}$). The surface excess charge density in one phase corresponds to the sum of the excess charges carried by the ions,

$$q^{\mathrm{w}} = \sum_i z_i F \Gamma_i^{\mathrm{w}} = -\sum_i z_i F \Gamma_i^{\mathrm{o}} \tag{11.15}$$

and is defined by the following equation:

$$q^{\mathrm{w}} = -\left(\frac{\partial\gamma}{\partial\left(\Delta_{\mathrm{o}}^{\mathrm{w}}\phi\right)}\right)_{T,P,\mu_i} \tag{11.16}$$

In order to semi-quantitatively illustrate the effect of porphyrin adsorption on the differential capacitance at the polarisable liquid/liquid interface, let us redefine the potential distribution as discussed in Section 2.2. We shall simply assume one adsorption plane located at the interfacial boundary and a given concentration of 1:1 electrolytes in both phases. It follows[3,41],

$$q^{\mathrm{w}} = -\sqrt{(8RT\varepsilon_{\mathrm{w}}c_{\mathrm{w}})}\,\sinh\left(\frac{F\Delta_{\mathrm{s}}^{\mathrm{w}}\phi}{2RT}\right) - q^{\mathrm{ads}} = -q^{\mathrm{o}} \tag{11.17}$$

where the charge introduced by the specifically adsorbed porphyrin is described in terms of the relative surface coverage θ_{p} and the maximum surface concentration $\Gamma_{\mathrm{p}}^{\mathrm{max}}$:

$$q^{\mathrm{ads}} = zF\Gamma_{\mathrm{p}} = zF\theta_{\mathrm{p}}\Gamma_{\mathrm{p}}^{\mathrm{max}} \tag{11.18}$$

Furthermore, the coverage θ_{p} can be computed by combining Equations (11.10) and (11.11) in the case of a Langmuir isotherm,

$$\theta_{\mathrm{p}} = \frac{1}{1 + \frac{a_{\mathrm{p}}}{a_{\mathrm{H_2O}}}\exp\left(-\frac{\Delta G_{\mathrm{ads}}}{RT}\right)} \tag{11.19}$$

The differential capacitance is simply defined as the first derivative of the excess charge with respect to the potential[3]:

$$C_{\mathrm{w}} = \frac{F}{RT}(2RT\varepsilon_{\mathrm{w}}c_{\mathrm{w}})^{\frac{1}{2}}\cosh\left(\frac{F\Delta_{\mathrm{s}}^{\mathrm{w}}\phi}{2RT}\right) \tag{11.20}$$

$$C_{\mathrm{o}} = \frac{F}{RT}(2RT\varepsilon_{\mathrm{o}}c_{\mathrm{o}})^{\frac{1}{2}}\cosh\left(\frac{F\Delta_{\mathrm{o}}^{\mathrm{s}}\phi}{2RT}\right) \tag{11.21}$$

Assuming that the porphyrin monolayer does not affect the dielectric properties of the liquid/liquid junction, the differential double layer capacitance can be simply obtained from

$$\frac{1}{C} = \frac{1}{C_{\mathrm{o}}} + \frac{1}{C_{\mathrm{w}}} \tag{11.22}$$

The potential dependent capacitance as estimated from Equations (11.17)–(11.22) for a molecule charged −4 (as in the case of ZnTPPC in neutral aqueous phase) is exemplified in Figure 11.13. The Gibbs energy of adsorption was taken as a constant value of $\Delta G_{\mathrm{ads}} = -38\ \mathrm{kJ\ mol^{-1}}$, and the maximum surface concentration was $\Gamma_{\mathrm{p}}^{\mathrm{max}} = 10^{-6}\ \mathrm{mol\ m^{-2}}$[94]. This simplified model predicts an increase of the capacitance at negative potential and a shift of the potential of zero charge towards more positive potentials as the concentration of porphyrin molecule increases. Qualitatively similar results were obtained by Frank and Schmickler employing a Lattice Gas model and Monte Carlo simulations[87].

The differential capacitance curves illustrated in Figure 11.14a were recorded at the water/DCE interface with various concentrations of ZnTPPC[94]. In agreement with the simulations (Figure 11.13), the experimental data do exhibit an increase of the capacitance at negative $\Delta_{\mathrm{o}}^{\mathrm{w}}\phi$ and a shift of the minimum of the capacitance towards positive potentials as the concentration of ZnTPPC increases. However, it is clear that the model described in this section is not sufficiently

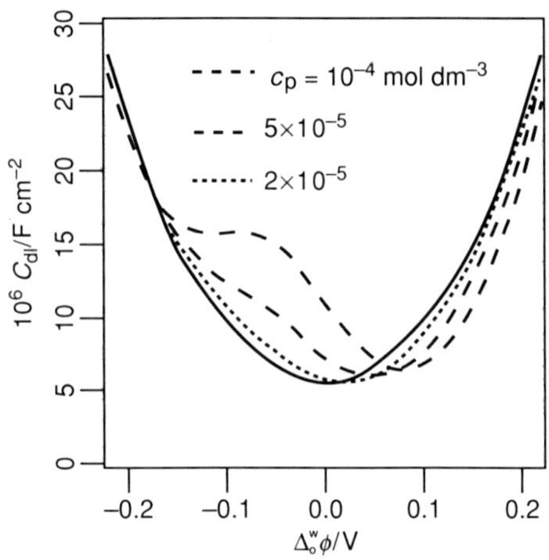

Figure 11.13. Simulations of the capacitance–potential curves employing Equations (11.17)–(11.19) taking $\Delta G_{ads} = -38$ kJ mol^{-1}, $\Gamma_p^{max} = 10^{-6}$ mol m^{-2} and various bulk concentrations of an adsorbate with a nominal charge −4.

accurate to quantitatively describe the various features observed experimentally. These deviations are even more evident in the case of Cu-chlorophyllin where the overall capacitance decreases and the potential dependence becomes weaker with increasing concentration of the dye (Figure 11.14b)[95]. Similar behaviour of the interfacial capacitance have been observed for the metalloporphyrin and chlorin molecules depicted in Figure 11.4[38]. The capacitance curves clearly show that water-soluble porphyrins generally exhibit strong specific adsorption, introducing substantial changes in the potential distribution across the interface. As discussed in Section 4.3.1, aggregation phenomena take place at the interface as the coverage of the porphyrin increases. Interporphyrin interactions as well as interfacial protonation also play an important role in the potential distribution and should be considered in order to extract quantitative information from the capacitance data. However, these aspects may be rather complex to model as aggregation may introduce changes in the solubility and the formation of condensed layers at the interface[96,97]. In the case of ZnTPPC, it has been reported that the self-assembly at the water/DCE junction is extremely sensitive to the pH in the aqueous phase[35]. Photocurrent and impedance data showed that the adsorption of ZnTPPC only takes place at pH lower than 9, indicating that the species adsorbed at the interface are not fully deprotonated. These results suggest that a key phenomenon involved in the stabilisation of the carboxyphenyl-derivatised porphyrin is the formation of cooperative hydrogen bonds. In Section 4.3.2, we shall provide links between the cooperative hydrogen bonding and the average molecular orientation of ZnTPPC and SnTPPC at the liquid/liquid boundary.

4.2. Electrocapillary Curves of the Liquid/Liquid Interface

The specific adsorption of porphyrins at the liquid/liquid boundary has also been addressed by surface tension measurements. Quasi-elastic laser scattering

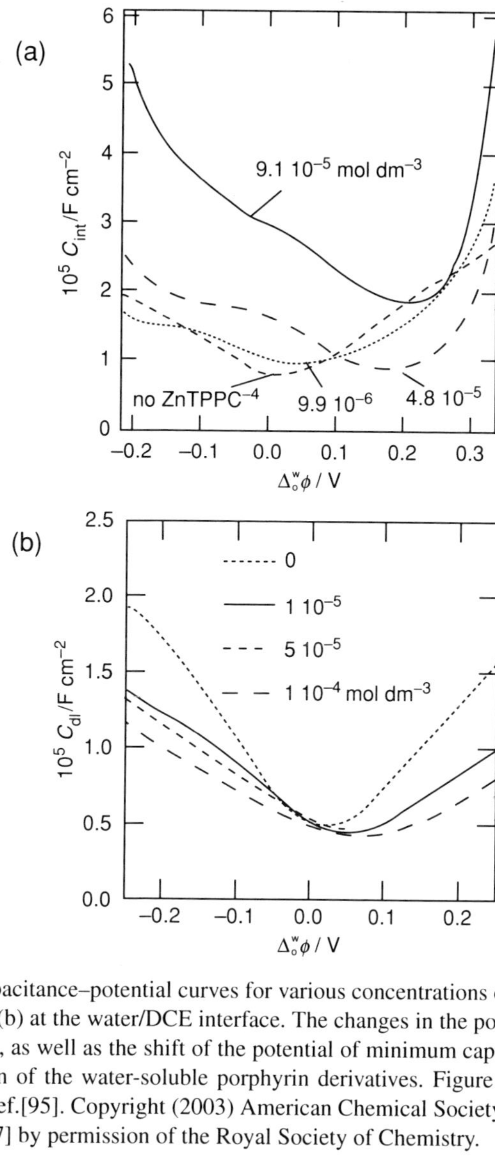

Figure 11.14. Capacitance–potential curves for various concentrations of ZnTPPC (a) and Cu-chlorophyllin (b) at the water/DCE interface. The changes in the potential dependence of the capacitance, as well as the shift of the potential of minimum capacitance reflect the specific adsorption of the water-soluble porphyrin derivatives. Figure (a) reprinted with permission from ref.[95]. Copyright (2003) American Chemical Society. Figure (b) reproduced from ref.[87] by permission of the Royal Society of Chemistry.

(QELS) has proved to be a valuable technique for monitoring changes in the frequency of capillary waves at liquid interfaces[80,81,98]. The principle of this measurement is based on the determination of the Doppler shift associated with the quasi-elastic scattering of the laser light by the capillary waves[80]. The scattered beam is optically mixed with a diffracted portion of the light from a grating introduced in the optical path. The mixing of the scattered and diffracted components generates an optical beat which frequency corresponds to the frequency of the capillary waves (f_0). The relationship between f_0 and the surface tension is given by the Lamb equation[28]:

$$f_0 = \frac{1}{2\pi}\left(\frac{\gamma}{\rho_w + \rho_o}\right)^{1/2} k^{3/2} \tag{11.23}$$

where f_0 is a characteristic frequency of the capillary waves, and ρ_w and ρ_o are the densities of the aqueous and organic solvents, respectively. For small scattering angles θ, the circular wavenumber k is related to the grating constant d by[80],

$$k = \frac{2\pi n}{d} \tag{11.24}$$

where n is the order of the diffraction spot. The angle θ is defined as the angle between the diffracted and fundamental beams at the detector plane. The relationship between θ and d is given by[80]:

$$d \sin(\theta) = n\lambda \tag{11.25}$$

where λ is the wavelength of the laser beam. Experimentally, the mixed optical signal is fast-Fourier transformed in order to obtain the capillary wave frequency. This non-contact method provides a fast and simple mean to estimate the surface tension at liquid/liquid interfaces under potentiostatic control[37,82].

For constant electrolyte composition and Galvani potential difference, the surface concentration of the adsorbed porphyrins can be evaluated from the Gibbs surface tension expression described by Equation (11.14)[93],

$$\Gamma_p = -\frac{1}{RT}\frac{d\gamma}{d\ln(a_p)} \tag{11.26}$$

where a_p is the bulk activity of the porphyrins. Nagatani *et al.* employed QELS to study the adsorption of ZnTPPC at the water/DCE interface as a function of the Galvani potential difference[37]. Figure 11.15a illustrates the effect of the ZnTPPC concentration on the power spectrum recorded at −0.20V. According to Equation (11.23), the decrease in the frequency associated with the capillary wave fluctuations indicates a decrease in the surface tension at the liquid/liquid boundary. The potential dependence of the surface tension for this system is shown in Figure 11.15b, where a substantial decrease of the surface tension in the presence of ZnTPPC is observed at negative potentials. On the other hand, the surface tension is little dependent on the ZnTPPC concentration at positive Galvani potential differences. Estimations of the surface concentration employing Equation (11.26) indicated that for concentrations of 5.0×10^{-6} and 5.0×10^{-5} mol dm^{-3} of ZnTPPC, Γ_p is of the order of 0.3×10^{-11} and 2.8×10^{-11} mol cm^{-2} at 0 V, respectively[37]. The electrocapillary curves in Figure 11.15b are consistent with the differential capacitance data described in Figure 11.14a. Both set of data clearly show that the coverage of ZnTPPC increases as the Galvani potential difference is shifted negatively with respect to the Galvani potential in the organic phase.

The potential dependence of the ZnTPPC coverage illustrates the typical trend observed for other anionic metalloporphyrin and chlorins as those shown in Figure 11.4. For instance, studies based on potential modulated fluorescence and capacitance curves have shown that ZnTPPS specifically adsorbs at the

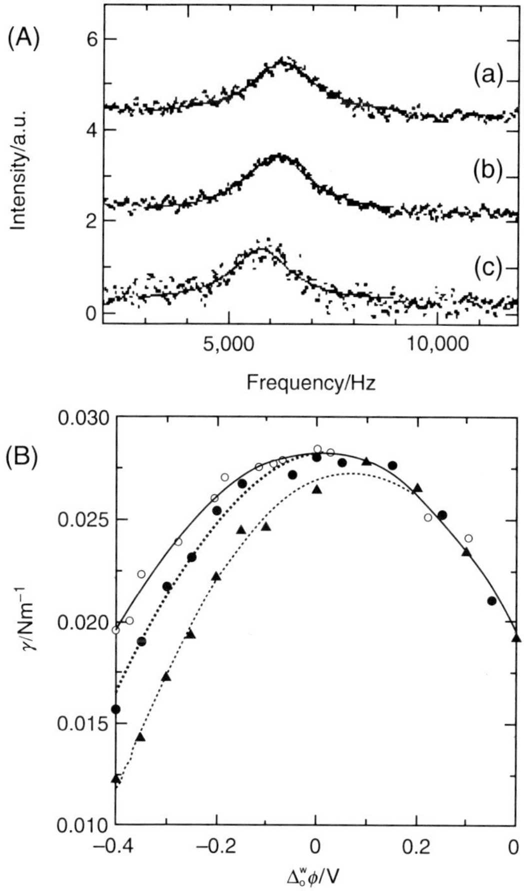

Figure 11.15. QELS power spectra of the third-order diffracted spot from the water/DCE interface for various concentrations of ZnTPPC: 0 (a), 5.0×10^{-6} and (b) 5.0×10^{-5} mol dm^{-3} (c). The applied Galvani potential difference was −0.20V (A). Electrocapillary curves at the water/DCE interface for various concentrations of ZnTPPC: 0 (hollow circle), 5.0×10^{-6} (full circle) and (b) 5.0×10^{-5} mol dm^{-3} (full triangle) (B). The decrease in the surface tension at negative potentials indicates an increase of the ZnTPPC surface coverage. Reprinted with permission from ref.[37]. Copyright (2003) American Chemical Society.

water/DCE boundary within a narrow potential range close to the formal potential of transfer to the organic phase[84]. In addition, concentration dependence of the photocurrent (further discussed in Section 5.2.) shows that the coverage of Cu-chlorophillin[95], chlorin e-6, protoIX and SnTPPC decreases towards negative potentials[38,53]. By contrast, the metalloporphyrin ion pair ZnTPPS–ZnTMPyP exhibits a potential independent surface coverage, suggesting that this supramolecular structure is strongly stabilised at the liquid/liquid boundary[59]. As discussed in Section 5.2, the analysis of the photocurrent as a function of the Galvani potential difference is strongly connected to the behaviour of the surface coverage.

4.3. Interfacial Molecular Orientation and Lateral Porphyrin Interactions

The molecular orientation of photoactive porphyrin species assembled at the liquid/liquid interface has been addressed by three methods based on illumination in TIR employing linearly polarised light, namely potential modulated fluorescence (PMF)[84], SHG[36] and photocurrent measurements[95]. Although different responses are measured from these techniques, in all cases the appropriate selection of experimental conditions allows specifically addressing the properties of the assembled porphyrins. The surface specificity of PMF is based on the fact that ac-fluorescence responses arise from the changes in the coverage of the porphyrin molecules with the applied potential. These studies have led to a series of interesting conclusions such as the presence of different adsorption planes at the interface depending on the applied potential[83, 84]. However, in this section we shall make more emphasis on studies based on SHG and photocurrent measurements.

4.3.1. SHG Studies of Metalloporphyrins Adsorption at the Water/DCE Interface

The principle of SHG resides in a resonance process whereby two photons of a given frequency ω are converted into one photon at the harmonic frequency 2ω. In the electric dipole approximation, the SHG response is restricted to anisotropic environments, e.g. to the liquid/liquid boundary[7, 79, 99]. The intensity of the SHG response ($I^{2\omega}$) measured in TIR is given by the expression[7],

$$I^{2\omega} = \frac{\omega^2}{8\varepsilon_0 c} \frac{\left(\varepsilon_1^{2\omega}\right)^{\frac{1}{2}}}{\varepsilon_1^{\omega}\left(\varepsilon_{\mathrm{m}}^{2\omega} - \varepsilon_1^{\omega}\sin^2\theta_1^{\omega}\right)} \left|\chi^{(2)}\right|^2 \left(I^{\omega}\right)^2 \tag{11.27}$$

where $I^{\omega}$and θ_1^{ω} are the intensity and angle of incidence of the fundamental beam, and the parameters ε_1 and ε_m corresponds to the relative permittivity of the incident phase and the interfacial region. The macroscopic second-order surface susceptibility tensor $\chi^{(2)}$ is antisymmetric with respect to inversion, consequently $I^{2\omega}$ is effectively zero in isotropic media, e.g. bulk liquids. The resonant component of $\chi^{(2)}$ is proportional to the microscopic hyperpolarisability β[15],

$$\chi^{(2)} \propto N_{\mathrm{p}} \sum_{\mathrm{k,e}} \frac{\left\langle \mu_{\mathrm{gk}}\mu_{\mathrm{ke}}\mu_{\mathrm{eg}}\right\rangle}{\left(\omega_{\mathrm{gk}} - \omega - i\Gamma\right)\left(\omega_{\mathrm{eg}} - 2\omega + i\Gamma\right)} \propto N_{\mathrm{p}}\left\langle\beta\right\rangle \tag{11.28}$$

where N_{p} is the number of molecules contributing to the spectroscopic responses, μ_{ij} is the transition matrix element between states i and j, while Γ is the line width term. The subscripts g, k and e stands for ground state, intermediate (virtual) state and first excited state, respectively. According to Equations (11.27) and (11.28), the SHG intensity strongly increases when 2ω is resonant to the ω_{eg}. This principle allows measuring the effective excitation spectrum of the organised porphyrin

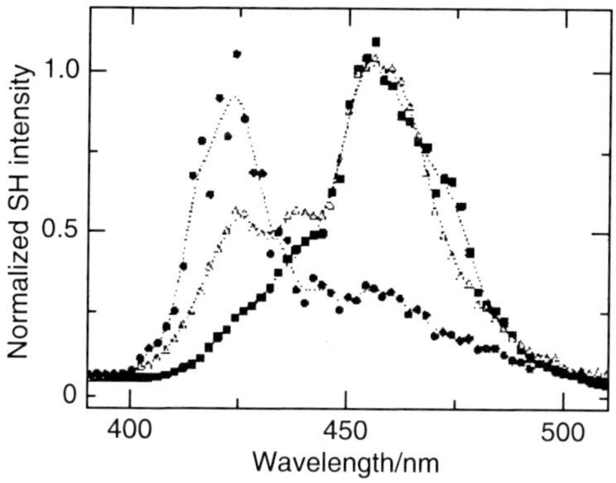

Figure 11.16. SHG spectra associated with specifically absorbed ZnTPPC at the water/DCE interface at −0.15 (full circles), −0.20 (hollow triangles) and −0.25V (full squares). The shift of the resonant transition to longer wavelength was interpreted in terms of the formation of *J*-aggregates at the liquid/liquid boundary as the coverage of ZnTPPC increases. Reprinted with permission from ref.[37]. Copyright (2003) American Chemical Society.

species. In addition, the matrix element in brackets corresponds to the average orientational distribution from which the molecular orientation at the interface can be extracted.

Nagatani *et al.* have recorded the SHG spectra of ZnTPPC at the water/DCE interface as a function of the applied Galvani potential difference[37]. In Figure 11.16, it is shown that the SHG signal in resonance with the Soret band strongly shifts to the red as the potential is tuned towards negative values. The SHG maximum at −0.15V is located at wavelengths very close to the Soret band observed in bulk water phase (for comparison, see the absorption spectra of SnTPPC in Figure 11.5). At more negative potentials, this band decreases and other SHG bands are observed with the most intense contribution located at 460 nm, i.e. approximately 40 nm red-shifted with respect to the signal measured at −0.15 V. This behaviour of the SHG signal has been associated with the formation of *J*-aggregates at the water/DCE interface at negative potentials. As discussed in Section 4.2, the coverage of ZnTPPC increases in this potential range, increasing the interporphyrin interactions. This aggregation process is rather unusual in the sense that it is fully reversible, i.e. the SHG intensity decreases and shifts towards shorter wavelength as the potential is swept back towards positive values[37].

SHG studies have also provided interesting insights into the adsorption of other porphyrins including H_2TMPyP, ZnTMPyP[36], protoIX[36], as well as the proteins myoglobin and haemoglobin[100]. Comparison between the SHG and visible spectra not only offers information on aggregation and interporphyrin interactions, but also on the environment of the porphyrin probe in the adsorption plane. Determination of the average molecular orientation is rather complex in these systems due to the enhancement of the SHG signal by the formation of aggregates. This aspect will be further discussed from the photoelectrochemical point of view.

4.3.2. Molecular Orientation Studied by Polarisation Angle Photocurrent Anisotropy

In the case that specifically adsorbed metalloporphyrins exhibit an average orientation at the adsorption plane, the excited state population will be determined by the relative orientation of the electromagnetic wave of the excitation beam with respect to the molecular transition dipole. As the photocurrent responses are associated with adsorbed porphyrins, changes in the angle of polarisation will affect the overall intensity of the photocurrent. In order to quantitatively analyse this correlation, the transition dipole moment M must be expressed in terms of the x, y, z coordinates, corresponding to the referential of the interface,

$$\begin{cases} M_x = M \sin\xi \cos\phi \\ M_y = M \sin\xi \sin\phi \\ M_z = M \cos\xi \end{cases} \tag{11.29}$$

where ξ corresponds to the tilting angle of the transition dipoles with respect to the surface normal as schematically represented in Figure 11.17. The angle ϕ denotes the orientation in the xy plane. The absorption intensity in each direction is proportional to the square of the dipole moment integrated over all of ϕ, i.e. the molecule is considered to have a fixed angle from the z axis (ξ constant), but can rotate around z.

$$A_i = C \int_0^{2\pi} |M_i|^2 \mathrm{d}\phi \tag{11.30}$$

where C is a proportionality constant involving the density of transition dipoles. The absorption constant in the spatial coordinates is obtained by combining Equations (11.29) and (11.30),

$$\begin{cases} A_x = C\pi \sin^2\xi \\ A_y = C\pi \sin^2\xi \\ A_z = C2\pi \cos^2\xi \end{cases} \tag{11.31}$$

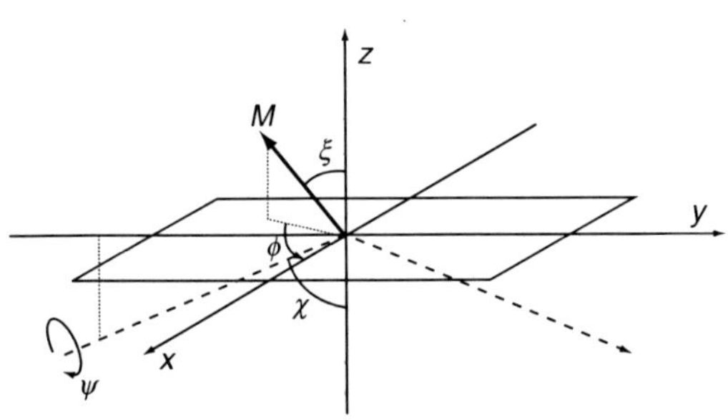

Figure 11.17. Schematic representation of the angles ξ, χ and ϕ defining the orientation of the prophrin transition dipole M with respect to the liquid liquid boundary. To simplify the illustration a single dipole is shown (M), however, porphyrin molecules exhibit two orthogonal transition dipoles located in the plane of the ring. The z axis is perpendicular to the plane of liquid/liquid boundary, and it is assumed that the excitation beam (dashed lines) follows the y direction. The angle χ denotes the angle between the laser beam and the interface, while Ψ represents the angle of polarisation of the incoming radiation.

and the total photocurrent can be expressed as,

$$\frac{j_{\rm ph}}{N} = |E_x|^2 A_x + |E_y|^2 A_y + |E_z|^2 A_z \tag{11.32}$$

where E_x, E_y and E_z are the components of the electric field, and N is a constant dependent on C as well as on the efficiency of the electron transfer process. The photocurrent anisotropy has been measured under TIR conditions, where the field at the reflection plane corresponds to[53],

$$\begin{cases} |E_x| = \dfrac{2n^{\rm o}\cos\chi}{\left((n^{\rm o})^2 - (n^{\rm w})^2\right)^{\frac{1}{2}}} E_{\rm i}\cos\Psi \\ |E_y| = \dfrac{2\,(n^{\rm w})^2\cos^2\chi}{\left((n^{\rm o})^2 - (n^{\rm w})^2\right)^{\frac{1}{2}}\left(\sin^2\chi\left((n^{\rm o})^2 + (n^{\rm w})^2\right) - (n^{\rm w})^2\right)^{\frac{1}{2}}} E_{\rm i}\sin\Psi \\ |E_z| = \dfrac{2n^{\rm o}\sin\chi\left((n^{\rm o})^2\sin^2\chi - (n^{\rm w})^2\right)^{\frac{1}{2}}}{\left((n^{\rm o})^2 - (n^{\rm w})^2\right)^{\frac{1}{2}}\left(\sin^2\chi\left((n^{\rm o})^2 + (n^{\rm w})^2\right) - (n^{\rm w})^2\right)^{\frac{1}{2}}} E_{\rm i}\sin\Psi \end{cases} \tag{11.33}$$

where Ψ is the angle of polarisation of the incident beam and χ is the angle of illumination with respect to the surface normal. E_i is the magnitude of the electric field associated with the incident beam. Considering the interaction of the incident beam with the adsorbed molecules, the photocurrent in Equation (11.32) is simplified to

$$\frac{j_{\rm ph}}{N} = \cos^2\Psi\sin^2\xi + \sin^2\Psi\left(\cos^2\chi\sin^2\xi + 2\sin^2\chi\cos^2\xi\right) \tag{11.34}$$

Jensen *et al.* have demonstrated that the photocurrent dependence on the angle of polarisation of the incoming light in TIR in the presence of ZnTPPC is affected by the applied Galvani potential difference, as well as by the bulk concentration of the porphyrin as illustrated in Figure 11.18. The photocurrent anisotropy confirms that the transition dipoles exhibit a well-defined orientation with respect to the interfacial plane. In the case of symmetrically substituted porphyrins, the transition dipoles are orthogonal and located in the plane of the porphyrin ring. The minimum in the photocurrent for *p*-polarised light suggests that the porphyrin species adopt a rather flat orientation with respect to the interfacial plane. As the Galvani potential difference is shifted to potentials closer to 0 V, the photocurrents exhibit a weaker dependence on the polarisation angle, suggesting a change in the orientation from a coplanar to a more upright position. Similar trends have been measured for SnTPPC under identical experimental conditions[38].

The average tilting angles obtained for ZnTPPC, SnTPPC, protoIX and chlorin e-6 are contrasted in Figure 11.19. The symmetrically substituted porphyrins show an increase in the angle ξ as the surface coverage decreases (Figures 11.19a and b). Indeed, the results obtained from capacitance (Figure 11.14), QELS (Figure 11.15) and SHG (Figure 11.16) clearly demonstrate that the surface coverage of ZnTPPC decreases as the applied potential increases. Under these conditions, the ZnTPPC molecules adopt a more coplanar orientation with

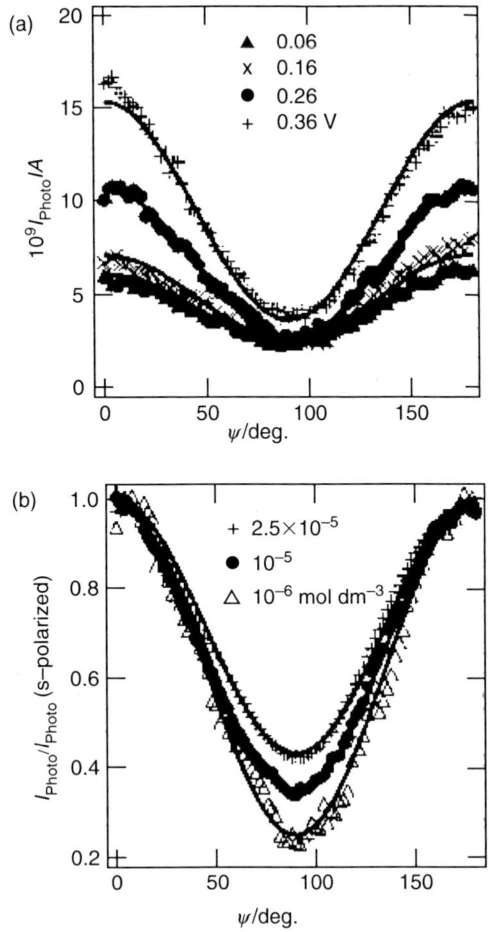

Figure 11.18. Photocurrent dependence on the angle of polarisation of the incoming light in TIR in the presence of ZnTPPC and Fc at the water/DCE interface. The effect of the Galvani potential difference for a bulk concentration of ZnTPPC of 1.0×10^{-6}mol dm^{-3} is illustrated in (a). The photocurrent anisotropy at 0.26V as a function of the ZnTPPC concentration is shown in (b). Reprinted with permission form ref.[35]. Copyright (2003) American Chemical Society.

respect to the liquid/liquid boundary enhancing the formation of cooperative H-bonds via the peripheral carboxyphenyl groups[35]. Increasing the bulk concentration or applying more negative potentials, the surface coverage is increased and the porphyrins adopt a more upright position (lower values of ξ). This behaviour suggests that the adsorbed ZnTPPC or SnTPPC is organised in a fashion minimising the lateral electrostatic repulsion and enhancing the interaction between the porphyrin rings. The formation of porphyrin aggregates revealed by SHG studies (see Section 4.3.1) is consistent with the molecular orientation at high coverage revealed by the photocurrent analysis.

The behaviour observed for asymmetrically derivatised porphyrins such as chlorin e-6, protoIX[38] and Cu-chlorophyllin[95] is remarkably different to the ZnTPPC and SnTPPC cases. The results in Figure 11.19c indicate that the orientation of chlorin e-6 and protoIX are effectively independent of the bulk concentration of the porphyrin. Furthermore, no dependence on the Galvani potential difference was observed either. These results appear to indicate that lateral interactions between the adsorbed species do not play an important role in the molecular orientation of these species. The asymmetric distribution of the carboxyl groups around the chromophore ring appears to introduce amphiphillic properties to these dyes. This interpretation is consistent with the fact that chlorin e-6 and protoIX exhibit a more upright orientation than SnTPPC[101].

5. Photoelectrochemical Reactivity at Porphyrin-Sensitised Liquid/Liquid Interfaces

The previous analysis has so far shown that photocurrent responses are directly linked to specifically adsorbed metalloporphyrin at the liquid/liquid boundary. The interfacial organisation of these species is affected by a series of parameters such as the bulk concentration and the Galvani potential difference. In this section, we shall concentrate on the dynamic and mechanistic aspects associated with the photocurrent signal. The information discussed in the previous sections concerning lifetimes of excited states and surface coverage will be relevant to the estimation of the heterogeneous electron-transfer rate constant from the photocurrent data.

5.1. The Origin of the Photocurrent Responses

Under potentiostatic conditions, charge transfer (either electrons or ions) from one electrolyte phase to the other manifests itself by current responses. The charge redistribution as a result of the faradaic process is counterbalanced by electrochemical processes at the two secondary electrodes. As in conventional electrochemical measurements, the flux of charges across the interfaces is matched by the flux of species to and from each electrolyte phases. In the particular context of porphyrin-sensitised liquid/liquid interfaces, photocurrents may arise from: (I) heterogeneous electron injection involving the porphyrin excited state and the hydrophobic redox species, eventually followed by the regeneration of the ground state by redox species in solution (co-sensitiser) or (II) "homogeneous" photoredox process followed by heterogeneous electron transfer to the oxidised porphyrin. For the photo-oxidation of an organic phase donor Q_o, these processes can be schematically represented as:

$$\mathrm{P_w} \xrightarrow{h\nu} \mathrm{P_w^*} \qquad \text{Photoexcitation} \qquad (11.35)$$

$$\mathrm{P_w^*} \xrightarrow{k_d} \mathrm{P_w} \qquad \text{Excited state relaxation} \qquad (11.36)$$

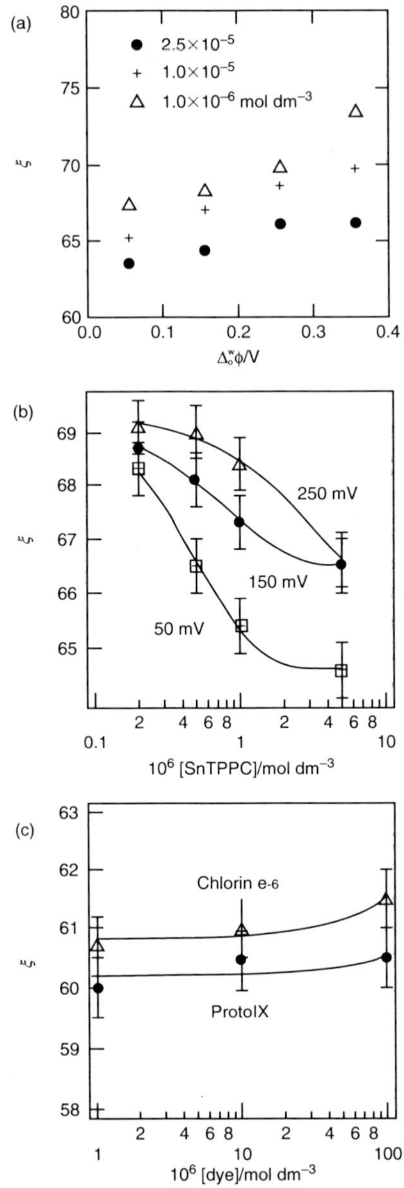

Figure 11.19. Potential dependence of the ZnTPPC orientation with respect to the normal of the water/DCE interface at various concentrations as estimated from Equation (11.34) and the results on Figure 11.18(a). Reprinted with permission form ref.[35]. Copyright (2003) American Chemical Society. Orientation dependence of SnTPPC (b), chlorin e-6 and protoIX (c) as a function of the bulk concentration of the photoactive species in the water phase. Similarly to the case of ZnTPPC, a substantial dependence of the SnTPPC orientation on the applied potential is observed. On the other hand, the orientation of chlorin e-6 and protoIX is effectively independent of the concentration and Galvani potential difference. Reprinted from ref.[38] with permission from Elsevier Science.

Mechanism I

$$P_w^* + Q_o \xrightarrow{k_{et}} [P_w^- \cdots Q_o{}^+] \qquad \text{Heterogeneous electron injection} \qquad (11.37)$$

$$[P_w^- \cdots Q_o^+] \xrightarrow{k_{ps}} P_w^- + Q_o^+ \qquad \text{Product Separation} \qquad (11.38)$$

$$[P_w^- \cdots Q_o^+] \xrightarrow{k_b} P_w + Q_o \qquad \text{Back electron transfer} \qquad (11.39)$$

In the presence of a co-sensitiser in the aqueous phase, the following additional reactions can occur,

$$[P_w^- \cdots Q_o^+] + CS_w \xrightarrow{k_{CS}} P_w + CS_w^- + Q_o^+ \qquad \text{Co-sensitisation} \qquad (11.40)$$

$$P_w^- + CS_w \xrightarrow{k_{CS}} P_w + CS_w^- \qquad \text{Ground state regeneration} \qquad (11.41)$$

Mechanism II

$$P_w^* + SS_w \xrightarrow{k_{ss}} P_w^+ + SS_w^- \qquad \text{Supersensitisation} \qquad (11.42)$$

$$P_w^+ + Q_o \xrightarrow{k_{set}} P_w + Q_o^+ \qquad \text{Heterogeneous electron transfer} \qquad (11.43)$$

$$P_w^+ + SS_w^- \xrightarrow{k_{sb}} P_w + SS_w \qquad \text{Back electron transfer} \qquad (11.44)$$

where P_w, CS_w and SS_w represent the porphyrin, co-sensitiser and supersensitiser confined to the water phase. The photocurrent responses arise from the heterogeneous electron transfer steps (11.37) or (11.43). These steps are in competition with the relaxation of the excited state (11.36). In mechanism I, the heterogeneous electron transfer generates an intermediate species $[P_w{}^- \cdots Q_o{}^+]$ that can separate to yield the final products (Equation (11.38)) or undergo back electron transfer (Equation (11.39)). The latter step involves a flux of charges with a sign opposite to that of the forward heterogeneous electron injection (11.37). As discussed later, the heterogeneous back electron transfer manifests itself by the relaxation of the photocurrent responses. This feature is phenomenologically similar to the interfacial electron–hole recombination processes in semiconductor/electrolyte junctions. However, while recombination in semiconductors is a second-order process, the photocurrent decay associated with step (11.39) follows first-order kinetics[53, 68, 102].

Typical examples of photocurrent responses arising from mechanism I are illustrated in Figure 11.20. The photo-oxidation of decamethyl ferrocene ($Me_{10}Fc$) by the ion pair ZnTPPS–ZnTMPyP (Figure 11.20a) exhibits quasi-quadratic responses to on–off light perturbations[101]. As discussed in the following section, the magnitude of the photocurrent signal is determined by a series of parameters including the photon flux, the coverage of the heterodimer and the rate of heterogeneous electron transfer k_{et}. The small decay of the photocurrent after the initial value is related to the diffusion of the ferrocene derivatives from the bulk of the organic phase to the interface. On the other hand, experiments performed over a wide range of light intensities and heterodimer concentrations indicate that the ground state regeneration (11.41) is rather efficient and no depletion of the

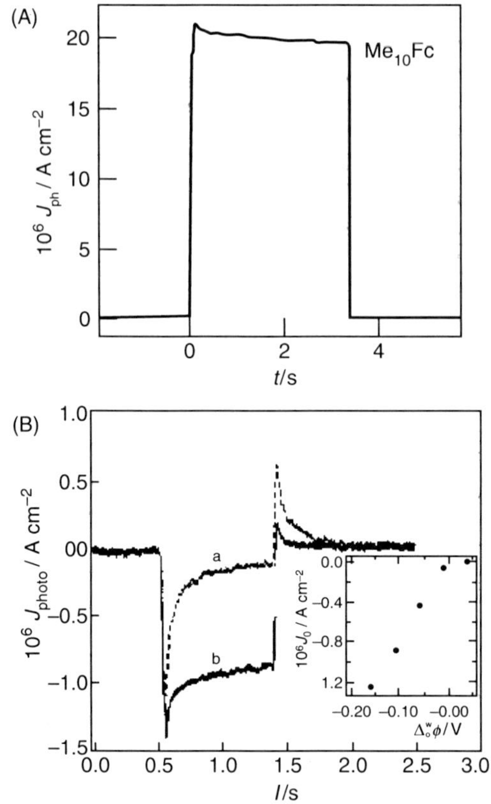

Figure 11.20. Typical photocurrent transient response associated with the photo-oxidation of decamethylferrocene by the ZnTPPS–ZnTMPyP ion pair at the water/DCE interface (A). Reprinted with permission from ref.[101]. Copyright (2003) American Chemical Society. Photocurrent transients originating from the photoreduction of TCNQ in the absence (a) and in the presence (b) of the co-sensitiser hexacyanoferrate (B). The inset shows the increase of the initial photocurrent as the Galvani potential difference is shifted towards negative values. Reprinted from ref.[103] with permission from Elsevier Science.

heterodimer is observed[53]. The important point remaining to be addressed is the process behind the porphyrin regeneration as no other electro-active species are present in the aqueous phase. Preliminary results appear to indicate that dissolved oxygen does not play a significant role in reaction (11.41).

The photoreduction of TCNQ by the heterodimer in Figure 11.20b exhibits negative photocurrents, which by convention indicate that electrons are transferred from the aqueous to the organic phase[103]. In this case, strong relaxation of the photocurrent is observed followed by a positive overshoot in the off transient. Taking into account the appropriate sign reversal, the photocurrent decay and overshoot are associated with the back electron transfer step (11.39). Upon addition of the ferri/ferrocyanide couple in the organic phase, the oxidised ion pair is regenerated prior to the back electron transfer, and the photocurrent transient approaches the same behaviour as in the case of $Me_{10}Fc$. This process, commonly

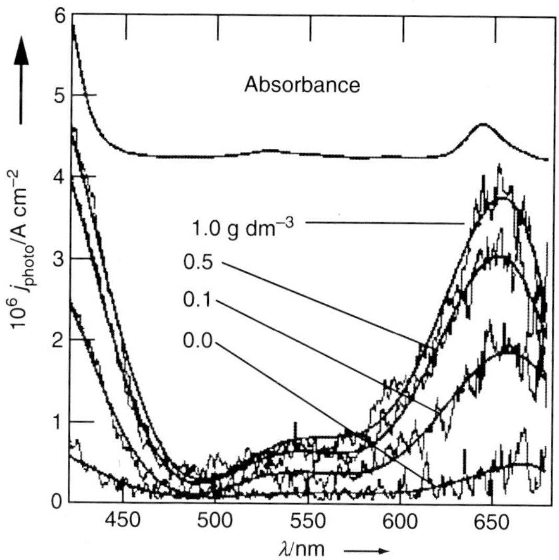

Figure 11.21. Photocurrent action spectra associated with the heterogeneous photo-oxidation of ferrocene by chlorin e-6 sensitised TiO_2 nanoparticles assembled at the water/DCE interface. The photocurrent significantly increases as the concentration of TiO_2 particles increases in the aqueous phase. The photocurrent spectra exhibit the same features as the chlorin e-6 spectrum superimposed in the figure. Reprinted from ref.[104] with permission from Wiley-VCH.

referred to as co-sensitisation, is rather difficult to distinguish from the regeneration of the ground state (Equation 11.41) solely on the basis of photocurrent analysis. Both processes occur in homogeneous phase and no photoresponses are directly associated with them. The fact that the co-sensitiser decreases the rate of first-order back electron transfer as shown in Figure 11.20b indicates that the reaction involves the intermediate complex $\left[P_w^- \cdots Q_o^+\right]$. On the other hand, there is no direct experimental evidence that the regeneration of the ground state occurs before or after a product separation takes place. In any case, these mechanistic details will not affect the most important discussions in this chapter which are related to the photoelectrochemistry of metalloporphyrin sensitised liquid/liquid junctions.

An example of mechanism II can be recalled from dye sensitised TiO_2 particles assembled at the water/DCE interface[104]. Figure 11.21 illustrates photocurrent spectra for the photo-oxidation of ferrocene (Fc) by chlorin e-6 for various concentrations of TiO_2 colloids. Dye species featuring peripheral carboxylic groups strongly bind to oxide surfaces and ultrafast electron injection into the conduction band takes place upon illumination[65, 66, 105–108]. The fact that photocurrent spectra show the same features as the absorption spectrum of chlorin e-6 indicates that the process is initiated by electron injection from the dye to the conduction band of the TiO_2 colloid. As in mechanism I, the efficiency of the photocurrent is determined by the competition between the heterogeneous electron transfer from ferrocene to the oxidised dye and electron capture from TiO_2

conduction band (these steps can be considered equivalent to reactions (11.43) and (11.44)). It should also be mentioned that photocurrent relaxations as exemplified in Figure 11.20b are not observed in this kind of mechanism. This is due to the fact that the back electron transfer is not heterogeneous, consequently no photoresponses are recorded in the potentiostatic circuit[104].

The remaining sections of this chapter will be mostly concerned with reactions following mechanism I. In this case, the photocurrent magnitude is directly connected to the ratio between the rates of heterogeneous electron transfer and excited state decay. In addition, the effect of the Galvani potential difference on the photocurrent magnitude provides valuable information on the reactivity of the metalloporphyrin at the liquid/liquid boundary.

5.2. Photocurrent Responses as a Function of the Galvani Potential Difference

The photocurrent anisotropy on the angle of light polarisation (see Figure 11.18) clearly demonstrated that only the metalloporphyrins self-assembled at the liquid/liquid boundary are responsible for the photoeffects observed. As discussed in Section 4, the interfacial coverage of the ionic porphyrins is determined not only by the bulk concentration but also by the Galvani potential difference. Consequently, changes in the photocurrent responses with $\Delta_o^w\phi$ can be brought about by changes in the surface coverage of porphyrins.

Neglecting the back electron transfer step and the diffusion of redox active species to the liquid/liquid junction, the photocurrent signal can be expressed in terms of the surface concentration of porphyrins (see Equations 11.9 and 11.19) and the competition between electron transfer and excited state decay[94],

$$j_{\text{photo}} = eI_0\sigma\left(\frac{k_{\text{et}}}{k_{\text{et}}+k_{\text{d}}}\right)\theta_{\text{p}}\Gamma_{\text{p}}^{\text{max}} \tag{11.45}$$

where I_0 is the photon flux arriving to the liquid/liquid boundary, σ is the photon capture cross section. In this approximation, the surface coverage of porphyrins is described by a potential dependent Langmuir isotherm (Equations 11.9 and 11.19)

$$\theta_{\text{p}} = \frac{1}{1+\frac{a_{\text{H}_2\text{O}}}{a_{\text{p}}}\exp\left(\frac{\Delta G^{\circ}_{\text{ads}}}{RT}\right)\exp\left(-\frac{zF}{RT}\Delta_{\text{s}}^{\text{w}}\phi\right)} \tag{11.46}$$

The fraction of the potential drop in the aqueous phase $\Delta_s^w\phi$ can be estimated from similar analysis as those presented in Section 2. At high porphyrin coverage, the surface charge density is expected to affect substantially the potential distribution across the interfacial region.

The photocurrent dependence on the porphyrin concentration is illustrated in Figure 11.22 for ZnTPPC[94] and the ion pair ZnTPPS–ZnTMPyP[59]. In these experiments, the photo-oxidation of ferrocene derivatives were examined under the conditions where responses associated with back electron transfer are negligible. For both set of results, it is observed that the photocurrent increases with increasing Galvani potential difference. The lines also indicate that the concentration dependence can be rationalised in terms of expression (11.46). In Figure 11.22a, the photocurrent responses exhibit a stronger dependence on the ZnTPPC

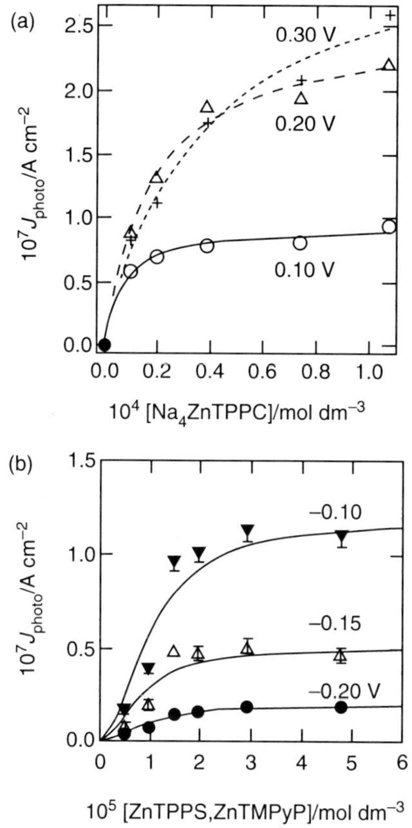

Figure 11.22. Photocurrent dependence on the concentration of ZnTPPC (a) and ZnTPPS–ZnTMPyP ion pair (b) in the aqueous phase at various Galvani potential differences. The photoresponses originate from the photo-oxidation of Fc at the water/DCE interface. Reprinted with permission from refs.[94] and [59]. Copyright (1998/1999) American Chemical Society.

concentration as the Galvani potential difference is increased. On the other hand, the photocurrent dependence on the concentration of ZnTPPS–ZnTMPyP appears unaffected by the applied potential. These results have been interpreted in terms of the potential dependence of the surface coverage. In Section 4, it was concluded that the surface concentration of ZnTPPC decreases with increasing $\Delta_o^w\phi$. On the other hand, the same experimental approaches have shown that the surface coverage of the heterodimer ZnTPPS–ZnTMPyP is effectively independent of the Galvani potential difference[53].

The analysis of the photocurrent responses on the basis of Equations (11.45) and (11.46) appears consistent with the results obtained by other spectroscopic techniques in relation to the potential dependence of the metalloporphyrin coverage. However, the physical meaning of the parameters associated with the Langmuir isotherm must be considered cautiously. The contribution of aggregation, interfacial protonation and cooperative H-bonding associated with the adsorption of ZnTPPC (Section 4) are implicitly included in the phenomenological term ΔG°_{ads} in Equation (11.46).

The two examples illustrated in Figure 11.22 clearly show that the photocurrent responses originating from heterogeneous reductive quenching increases as $\Delta_o^w\phi$ is shifted towards positive values. These results demonstrate that the apparent rate constant of electron transfer k_{et} is affected by the applied potential. This evidence does not support one of the arguments presented in Section 2, i.e. changes in $\Delta_o^w\phi$ mainly perturb the distribution of ionic species rather than the electron transfer rate constant. In the next section, we shall examine more quantitative relationships between the thermodynamic electron transfer driving force and k_{et} with particular emphasis on the case of the heterodimer ZnTPPS–ZnTMPyP. As mentioned previously, the surface coverage of this complex is independent of $\Delta_o^w\phi$, allowing a direct correlation between photocurrent magnitude and k_{et}.

5.2.1. Correlation between Photocurrent and the Gibbs Free Energy of Electron Transfer

The Gibbs free energy of electron transfer is determined by the formal electron transfer potentials associated with the excited state of the metalloporphyrin ($E^{\circ',w}_{P*}$) and the redox couple in the organic phase ($E^{\circ',o}_{Q}$)[101],

$$\Delta G^{\circ'}_{et} = -F\left(E^{\circ',w}_{P*} - E^{\circ',o}_{Q} + \Delta_o^w\phi\right) \tag{11.47}$$

This expression is defined for an electron transfer from donors in the organic phase to redox species in the aqueous phase. In the case that the species in the organic phase are reduced, the sign of Equation (11.47) must be reversed. As we shall describe here, a key distinctive point of carrying out heterogeneous photoelectrochemical studies at the polarisable ITIES is that $\Delta G^{\circ'}_{et}$ can also be changed by tuning the Galvani potential difference. In order to evaluate the Gibbs free energy as formulated in Equation (11.47), the formal redox potentials $E^{\circ',w}_{P*}$ and $E^{\circ',o}_{Q}$ must be formally referred to a common reference state. Taking the redox couple ferrocene/ferricinium as an example, the formal redox potential in DCE vs. SHE in water is defined in terms of the reaction[47,101],

$$\frac{1}{2}H_2\,(w) + Fc^+\,(DCE) \rightarrow H^+\,(w) + Fc\,(DCE) \tag{11.48}$$

The formal transfer potential for reaction (11.48) $\left(\left[E^{\circ',o}_{Fc^+/Fc}\right]_{SHE}\right)$ can be evaluated by the following thermodynamic cycle,

$$Fc^+\,(DCE) \rightarrow Fc^+\,(w) \qquad -F\Delta_o^w\phi^{\circ'}_{Fc^+} \tag{11.49}$$

$$\frac{1}{2}H_2\,(w) + Fc^+\,(w) \rightarrow H^+\,(w) + Fc\,(w) \qquad -F\left[E^{\circ',w}_{Fc^+/Fc}\right]_{SHE} \tag{11.50}$$

$$Fc\,(W) \rightarrow Fc\,(DCE) \qquad \Delta G^{w\rightarrow o}_{Fc} \tag{11.51}$$

From Equations (11.49)–(11.51), it follows that

$$\left[E^{o',o}_{Fc^+/Fc}\right]_{SHE} = \Delta^w_o \phi^{o'}_{Fc^+} + \left[E^{o',w}_{Fc^+/Fc}\right]_{SHE} - \frac{\Delta G^{w\to o}_{Fc}}{F} \quad (11.52)$$

Independent measurements of each of these parameters have allowed the evaluation of Equation (11.52) giving as a result $\left[E^{o',o}_{Fc^+/Fc}\right]_{SHE} = 0.64 \pm 0.02\,V$[47]. This value can be used as a reference for estimating the formal redox potential of other species in DCE as summarised in Table 11.1[60,101]. The redox potentials estimated for the intermediate charge separated state associated with the ZnTPPS–ZnTMPyP heterodimer (see reaction 11.5) are also included.

Table 11.1. Formal Redox Potentials and Gibbs Free Energy for Heterogeneous Electron Transfer Between Photoexcited ZnTPPS–ZnTMPyP Ion Pair (P*) in Water and Redox Species in 1,2-Dichloroehtane

Redox couple	$E^{o'}/V$ vs. SHE	$\Delta G^{o'}_{et}/eV^{\ddagger}$
P^*/P^-	1.00	/
P^+/P^*	−0.35	/
Ferrocene Fc^+/Fc	0.64	−0.36
Dimethylferrocene Me_2Fc^+/Me_2Fc	0.55	−0.45
Diferrocenylethane Fc_2Et^+/Fc_2Et	0.55	−0.45
Butylferrocene $BuFc^+/BuFc$	0.56	−0.44
Decamethylferrocene $Me_{10}Fc^+/Me_{10}Fc$	0.07	−0.93
Tetracyanoquinodimethane TCNQ / $TCNQ^-$	0.29	−0.64
Tetrachlorobenzoquinone TCBQ / $TCBQ^-$	0.17	−0.52
Dichlorobenzoquinone DCBQ / $DCBQ^-$	−0.02	−0.33
Benzoquinone BQ/BQ^-	−0.43	0.08

† Data reproduced from refs. [54] and [93].
‡ Estimations based on Equation (11.47) for $\Delta^w_o\phi = 0\,V$. The appropriate sign reversal should be considered for photoreduction processes.

From the data in Table 11.1 and Equation (11.47), it is expected that the photocurrent responses at a given Galvani potential difference will be affected by the redox potential of the quencher in the organic phase. This behaviour is illustrated in Figure 11.23 for a series of electron donors and acceptors[60,101]. The transient responses are rather different for both set of data, showing quasi-quadratic responses in the photo-oxidation (Figure 11.23a) and strong photocurrent relaxation in the case of the electron acceptors (Figure 11.23b). As discussed in Section 5.1, the relaxation phenomena are associated with back electron transfer processes and diffusion of the redox active species to the interfacial region. The contribution of diffusion responses is rather small for the photocurrent magnitudes shown in this figure. Indeed the small relaxation obtained in the presence of $Me_{10}Fc$ (see Figure 11.20a) illustrates the limited effect of diffusion under these experimental conditions[53]. In the following discussion, the initial value of

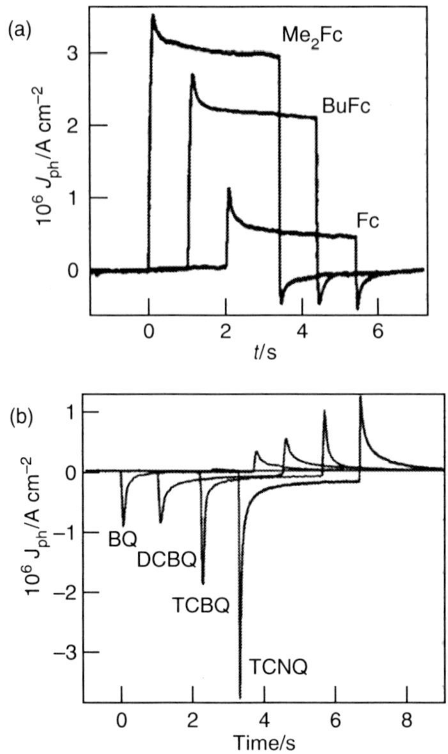

Figure 11.23. Photocurrent transient responses associated with the photo-oxidation of various ferrocene derivatives (a) and photoreduction of electron acceptors (b) by the ion pair ZnTPPS–ZnTMPyP. Experiments were performed at the water/DCE interface at 0 V (a) and −0.12 V(b). The difference in photocurrent magnitude is connected to the Gibbs free energy of electron transfer (see Table 11.1). Reprinted with permission from refs.[101] and [60]. Copyright (2002/2003) American Chemical Society.

the photocurrent (j^0_{ph}) will be employed to extract information on the competition between the heterogeneous electron transfer and the relaxation of the charge separated state. Assuming that the contribution from back electron transfer and other competing processes to the measuring current is negligible at short times[101], the parameter j^0_{ph} can be simply described by Equation (11.45). However, the estimation of j^0_{ph} from the photocurrent transient could be non-trivial in cases where the photocurrent relaxations are fast. This aspect will be discussed in more detail in Section 5.3. The results in Figure 11.23 show that the initial photocurrent is clearly dependent on $\Delta G^{o'}_{et}$, confirming the correlation between the thermodynamic driving force and the phenomenological electron-transfer rate constant.

The effect of the Galvani potential difference on the photocurrent responses is illustrated in Figure 11.24 for Fc and tetrachloro benzoquinone (TCBQ). In the case of the electron donor, the photocurrent increases as $\Delta^w_o\phi$ increases (Figure 11.24a). The opposite dependence is observed for the acceptor TCBQ. The behaviour of j^0_{ph} in Figures 11.23 and 11.24 is self-consistent within the correlation

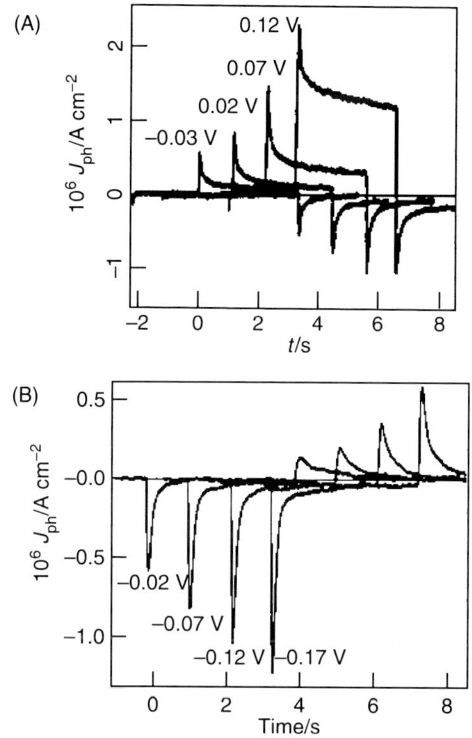

Figure 11.24. Potential dependence of the transient photocurrent for the photo–oxidation of Fc (a) and photoreduction of TCBQ (b) at the water/DCE interface sensitised by ZnTPPS–ZnTMPyP. The magnitude of the photocurrent increases with increasing Galvani potential difference in the heterogeneous photo-oxidation process, while the opposite trend is observed for the photoreduction. These results unequivocally demonstrate that the heterogeneous photoinduced electron transfer is affected by the Galvani potential difference across the liquid/liquid boundary. Reprinted with permission from refs.[101] and [60]. Copyright (2002/2003) American Chemical Society.

established between $\Delta G_{\text{et}}^{\circ\prime}$ and $\Delta_{\text{o}}^{\text{w}}\phi$ in Equation (11.47). This correlation allows gathering the trends associated with the applied voltage and the formal potential of the redox species as illustrated in Figure 11.25[60, 101]. The photocurrent values show a marked increase as $\Delta G_{\text{et}}^{\circ\prime}$ becomes more negative, showing a weaker dependence for values below -1 eV. This behaviour not only confirms that contributions from diffusion are negligible under the photon flux employed, but also that the dynamics of electron transfer for these redox species can be described in terms of a common set of physical parameters. These parameters include the reorganisation energy and electronic coupling between the excited metalloporphyrin dimer and the electron donor.

A similar behaviour is also observed for the electron acceptors (Figure 11.25b), except for the case of 1,4-benzoquinone (BQ). For this quencher, $\Delta G_{\text{et}}^{\circ\prime}$ is positive within a certain potential range and electron transfer from the populated high energy level of the heterodimer should not be observed under thermodynamic

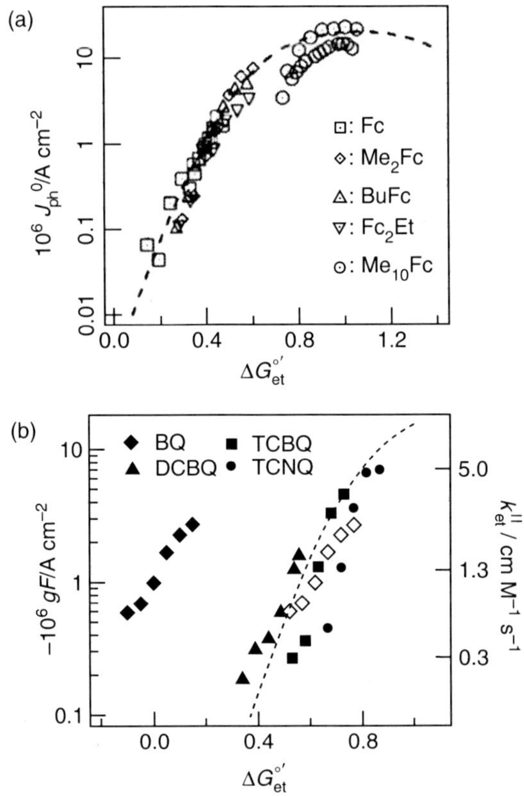

Figure 11.25. Photocurrent dependence on the Gibbs free energy of electron transfer for the photo-oxidation of ferrocene derivatives (a) and photoreduction of quinone-type molecules (b) at the water/DCE interface. $\Delta G^{o'}_{et}$ is evaluated from Equation (11.47), employing the formal redox potentials summarised in Table 11.1 and the applied Galvani potential difference. A deconvolution of the photocurrent relaxation in the presence of the electron acceptors was performed in order to estimate the flux of electron injection g. The second-order rate constant for the photoninduced heterogeneous electron transfer is also calculated assuming values of 1 nm for d_{cc} and $5 \times 10^{-5}\ s^{-1}$ for k_d. The trends observed in both set of data were rationalised in terms of a single solvent reorganisation energy and activation-less limit for the rate constant. Reprinted with permission from refs.[101] and [60]. Copyright (2002/2003) American Chemical Society.

grounds. This behaviour has been rationalised in terms of shifts of the effective redox potential of BQ towards more positive potentials via hydrogen bonding and coupled protonation of the radical semiquinone[60, 109–112]. The formal redox potentials in Table 11.1 are calculated from the redox potential measured with a Pt micro-electrode in dry DCE with respect to the Fc couple[60]. Therefore, it can be concluded that the apparent redox potential of BQ is shifted towards positive values in the proximity of the water/DCE interface. Experimental analysis reviewed in Section 5.3 provides evidence that the shift of the interfacial redox potential is connected to interfacial hydrogen bonding.

5.2.2. The Electron Transfer Activation Energy and Solvent Reorganisation Term

As we mentioned earlier, the surface coverage of the ZnTPPS–ZnTMPyP complex is effectively independent of the Galvani potential difference. Estimations of the interfacial tension based on QELS studies have shown that the density number of ZnTPPS–ZnTMPyP is of the order of 10^{14} cm^{-2} for a bulk concentration of 5×10^{-5} mol dm^{-3} [53]. From the results illustrated in Figure 11.22b, this value corresponds to the maximum surface concentration at the water/DCE interface. According to Equation (11.45), the dependence of j^0_{ph} on $\Delta G^{o\prime}_{\text{et}}$ is determined by the effective quantum yield (Ω_{inj}) for the photoinduced heterogeneous electron transfer:

$$\Omega_{\text{inj}} = \frac{j^0_{\text{ph}}}{eI_0\sigma N_s} = \frac{j^0_{\text{ph}}}{j^{\max}_{\text{ph}}} = \left(\frac{k_{\text{et}}}{k_{\text{et}}+k_{\text{d}}}\right) \tag{11.53}$$

As we mentioned previously, k_{et} increases as the redox potential of the electron donor is more negative and $\Delta^{\text{w}}_{\text{o}}\phi$ is tuned towards positive values. From these relationships, we can extract information on the reorganisation energy and electronic coupling between the redox species.

Within the framework of Marcus theory of electron transfer, the activation energy for a non-adiabatic electron transfer process is given by[113–116],

$$\Delta G_{\text{act}} = w_{\text{r}} + \frac{\left(\lambda + \Delta G^{o\prime}_{\text{et}} + w_{\text{p}} - w_{\text{r}}\right)^2}{4\lambda} \approx \frac{\left(\lambda + \Delta G^{o\prime}_{\text{et}}\right)^2}{4\lambda} \tag{11.54}$$

where w_{r} and w_{p} are the work terms related to reactants encounter and product separation. Analysis performed in other publications have shown that the terms w_{r} and w_{p} are approximately an order of magnitude smaller than $\Delta G^{o\prime}_{\text{et}}$ and the reorganisation energy λ,[43] consequently these parameters will be neglected as a first approximation. The parameter λ involves contributions from internal conformation changes of the reactants as well as the solvation structure during the electron-transfer step. Assuming a sharp liquid/liquid boundary which cannot be penetrated by the reactants, the relationship between the activation energy and the bimolecular electron transfer rate constant $k^{\text{II}}_{\text{et}}$ (units cm^4 s^{-1}) corresponds to[116]

$$k^{\text{II}}_{\text{et}} = 2\pi\kappa\nu\left(r_{\text{p}} + r_{\text{q}}\right)\Delta R^3 \exp\left(\frac{-\Delta G_{\text{act}}}{k_{\text{B}}T}\right) \tag{11.55}$$

where κ is the Landau–Zener transmission coefficient, ν is the frequency of nuclear motion and ΔR is the characteristic distance where the volume integrand is maximised. The parameters r_{p} and r_{q} correspond to the molecular radii of the porphyrin and the redox quencher. The dimensionality of the rate constants k_{et} and $k^{\text{II}}_{\text{et}}$ should be clearly differentiated. The former corresponds to an average of the excited state quenching rate constant, in which the characteristic tunnelling distance and the concentration of the redox quencher are included. The unit of k_{et} is s^{-1}, which allows to be compared with the rate constant of relaxation of the metalloporphyrin excited state (k_d). On the other hand, k_{et}^{II} is independent of the quencher concentration and is defined for a given average distance between the redox species. Obviously, both parameters are closely related as discussed further below. From Equations (11.53), (11.54) and (11.55), the relationship between j_{ph}^0 and $\Delta G_{et}^{o'}$ can be expressed as[101],

$$j_{ph}0 = j_{ph}^{max} \frac{\exp\left(\frac{-\left(\lambda+\Delta G_{et}^{o'}\right)^2}{4\lambda k_B T}\right)}{\exp\left(\frac{-\left(\lambda+\Delta G_{et}^{o'}\right)^2}{4\lambda k_B T}\right) + \frac{k_d}{Z}} \tag{11.56}$$

where Z corresponds to the maximum value of the electron-transfer rate constant expressed in units of a homogeneous pseudo-first-order reaction with respect to the concentration of the excited state (s^{-1}). This parameter is determined by the pre-exponential term in Equation (11.55), the average centre-to-centre distance between the redox species at the interface (d_{cc}) and the concentration of the quencher (c_Q)[101],

$$Z = \frac{2\pi \kappa \nu \left(r_p + r_q\right) \Delta R^3 N_a c_Q^o}{d_{cc}}, \tag{11.57}$$

where N_a is the Avogadro constant. The ratio k_d/Z as well as λ were evaluated from non-linear least square fitting of Equation (11.56) to the curve in Figure 11.25a, taking $j_{ph}^{max} = 3 \times 10^{-5}$ A cm^{-2}. The dashed line in Figure 11.25a was obtained for $k_d/Z = 0.5$, and a reorganisation energy $\lambda = 1.05 \pm 0.05$ eV[101]. The former value confirms that the characteristic transient time of the electron transfer is of the same order of magnitude than the decay of the excited state.

The same analysis can be employed for rationalising the behaviour of j_{ph}^0 as a function $\Delta G_{et}^{o'}$ for the set of electron acceptors (Figure 11.25b). As we mentioned before, the strong photocurrent relaxation affect the apparent value of j_{ph}^0 and deconvolution of the various components involved in the photocurrent transients should be performed in order to have accurate estimations of k_{et}. The result obtained from the deconvolution (further described in Section 5.3.) is denoted as g. The dashed lines in Figure 11.25b illustrate the fit of j_{ph}^0 to Equation (11.56), resulting in identical values for k_d/Z and λ as in the case of the ferrocene derivatives[60]. The fact that the photocurrent dependence on the driving force for TCNQ, TCBQ, DCBQ and the various ferrocene derivatives can be described by a single value of Z and λ reveals that both parameters are mainly determined by the structure of the interface and the solvent reorganisation energy (λ_{solv}).

Considering an impenetrable sharp liquid/liquid boundary, the solvent reorganisation energy can be expressed as[113, 114, 117, 118],

$$\lambda_{\text{solv}} = \frac{(\Delta e)^2}{4\pi\varepsilon_0}\left(\frac{1}{\varepsilon^{\text{o}}_{\text{opt}}} - \frac{1}{\varepsilon^{\text{o}}_{\text{stat}}}\right)\left(\frac{1}{2r_{\text{q}}}\right) + \frac{(\Delta e)^2}{4\pi\varepsilon_0}\left(\frac{1}{\varepsilon^{\text{w}}_{\text{opt}}} - \frac{1}{\varepsilon^{\text{w}}_{\text{stat}}}\right)\left(\frac{1}{2r_{\text{p}}}\right)$$
$$+\frac{(\Delta e)^2}{4\pi\varepsilon_0}\left(\frac{1}{\varepsilon^{\text{o}}_{\text{opt}}}\left(\frac{\varepsilon^{\text{o}}_{\text{opt}} - \varepsilon^{\text{w}}_{\text{opt}}}{\varepsilon^{\text{o}}_{\text{opt}} + \varepsilon^{\text{w}}_{\text{opt}}}\right) - \frac{1}{\varepsilon^{\text{o}}_{\text{stat}}}\left(\frac{\varepsilon^{\text{o}}_{\text{stat}} - \varepsilon^{\text{w}}_{\text{stat}}}{\varepsilon^{\text{o}}_{\text{stat}} + \varepsilon^{\text{w}}_{\text{stat}}}\right)\right)\left(\frac{1}{4d_{\text{o}}}\right)$$
$$+\frac{(\Delta e)^2}{4\pi\varepsilon_0}\left(\frac{1}{\varepsilon^{\text{w}}_{\text{opt}}}\left(\frac{\varepsilon^{\text{w}}_{\text{opt}} - \varepsilon^{\text{o}}_{\text{opt}}}{\varepsilon^{\text{o}}_{\text{opt}} + \varepsilon^{\text{w}}_{\text{opt}}}\right) - \frac{1}{\varepsilon^{\text{w}}_{\text{stat}}}\left(\frac{\varepsilon^{\text{w}}_{\text{stat}} - \varepsilon^{\text{o}}_{\text{stat}}}{\varepsilon^{\text{o}}_{\text{stat}} + \varepsilon^{\text{w}}_{\text{stat}}}\right)\right)\left(\frac{1}{4d_{\text{w}}}\right)$$
$$-\frac{(\Delta e)^2}{4\pi\varepsilon_0}\left(\frac{1}{\varepsilon^{\text{o}}_{\text{opt}} + \varepsilon^{\text{w}}_{\text{opt}}} - \frac{1}{\varepsilon^{\text{o}}_{\text{stat}} + \varepsilon^{\text{w}}_{\text{stat}}}\right)\left(\frac{2}{d_{\text{cc}}}\right) \quad (11.58)$$

where d_{o} and d_{w} are the characteristic distances between the interface boundary and the centre of the reactive species. Taking the various parameters in Equation (11.58) as, $r_{\text{q}} \sim 0.2$ nm, $r_{\text{p}} \sim 0.7$ nm, $d_{\text{o}} = d_{\text{w}} = d_{\text{cc}}/2$, $\varepsilon^{\text{w}}_{\text{opt}} = 1.78$, $\varepsilon^{\text{o}}_{\text{opt}} = 2.09$, $\varepsilon^{\text{w}}_{\text{stat}} = 78.4$ and $\varepsilon^{\text{o}}_{\text{stat}} = 10$, the estimated solvent reorganisation energy of 1.05 eV corresponds to an average distance d_{cc} of 0.8 nm[101].

As discussed in Section 2.1, the value of d_{cc} estimated from the solvent reorganisation energy is comparable to the characteristic density profiles generated by thermally activated capillary wave fluctuations. We believe that these experiments provide sound proof that the dynamics of heterogeneous electron transfer are determined by the average solvent distribution at the liquid/liquid boundary. A point that should be reiterated is that the photoreactivity of adsorbed metalloporphyrin is clearly affected by $\Delta^{\text{w}}_{\text{o}}\phi$ despite the fact that the distance separating the redox species is less than 1 nm. This conclusion confirms that the electrostatic potential exhibits a distribution considerably sharper than expected from the simplified Gouy–Chapman model (see Figure 11.3). However, it remains a matter of controversy whether the use of a step function for the change in dielectric permittivity is self-consistent with the microscopic description of the liquid/liquid boundary.

5.2.3. Comparison Between Porphyrin Photoreactivity in Bulk Solutions and at the ITIES

In order to rationalise the dynamics of heterogeneous vs. homogeneous photoinduced electron transfer, values of $k^{\text{II}}_{\text{et}}$ must be estimated. However, the extraction of absolute values for k_{et} from the photocurrent data illustrated in Figure 11.25 requires knowledge of the lifetime of the ZnTPPS–ZnTMPyP charge separated state (k_{d}). As mentioned in Section 3.3, a limitation of the photoelectrochemical approach is that the dynamics of electron transfer cannot be time resolved as in the case of fully spectroscopic techniques. Previous studies in bulk solution suggested that long living transient states in porphyrin ion pairs exhibit decay rate constants of the order of $k_{\text{d}} = 5 \times 10^5\ \text{s}^{-1}$[53,54]. Estimations of $k^{\text{II}}_{\text{et}}$ based on this value of k_{d} are also illustrated in Figure 11.25b for the case of electron accepting species. Extrapolation to the activation-less limit of electron-transfer rate constant provide values of the order of $120\ \text{M}^{-1}\ \text{cm s}^{-1}$, which is consistent with the non-adiabatic limit calculated by Smith *et al.*[119]. This non-adiabatic character of the heterogeneous quenching of the porphyrins comes as a result of the distance separating the redox species at the interface.

The data shown in Figure 11.25b reveals that the maximum rate constant obtained for the acceptor TCNQ was close to $5\,M^{-1}\,cm\,s^{-1}$. This value was obtained at the negative edge of the polarisable window i.e. $\Delta_o^w\phi \sim -0.20\,V$. Taking an average separation distance of 1 nm, this rate constant remains two to three orders of magnitude smaller than homogeneous quenching rate constants of ZnTMPyP by TCNQ ($k_q \sim 10^{10}\,M^{-1}\,s^{-1}$)[67]. The estimates from photocurrent measurements are also considerably lower than those reported from nanosecond flash photolysis in total internal reflection at the water/toluene interface (see Figure 11.9)[67]. In this case the quenching rate was of the order of $100\,M^{-1}\,cm\,s^{-1}$. The substantial difference between both approaches may be connected to the specificity of the detection method rather than to physical aspects associated with the heterogeneous electron transfer process. While flash photolysis provides the time resolution, there is no possibility to distinguish between homogeneous and heterogeneous electron transfer. In addition, photocurrent responses have clearly demonstrated that heterogeneous electron transfer only involves specifically adsorbed metalloporphyrins. The values of the optical density in the interfacial flash photolysis experiment indicate that more than the specifically adsorbed porphyrins are probed by this method.

Another interesting comparison can be made with the charge transfer dynamics observed in porphyrin–ferrocene dyads. These supramolecular structures feature porphyrin and ferrocene units covalently linked. Imahori *et al.* have shown that the quenching of zinc tetraphenylporphyrin–ferrocene dyads takes place via the singlet state with a rate constant in the range of $10^8\,s^{-1}$ [120,121]. This rate constant is two to three orders of magnitude larger than the value of k_{et} estimated for ferrocene from the photocurrent responses. This difference confirms that the large value of λ is connected to the solvent reorganisation energy rather than changes in the internal conformation of the redox species.

5.3. Dynamics of Photocurrent Relaxation

In Section 5.1, we have shown that the relaxation of photocurrent responses can be connected to heterogeneous back electron transfer processes (mechanism I). In the case of ferrocene, detailed analysis based on intensity modulated photocurrent spectroscopy[68] and photocurrent transients (Figure 11.24a)[53] have shown that the back electron transfer process follows first-order kinetics. These experimental evidences support the mechanism describing back electron transfer "within" the interfacial intermediate species, and not by the separated ionic photoproducts. According to mechanism I, a quasi-steady-state photocurrent (j_{ph}^{ss}) is developed in the case that the regeneration of the metalloporphyrin ground state is fast and the interfacial concentration of the organic phase redox couple is not depleted. j_{ph}^{ss} is determined by the competition between product separation (Equation 11.38) and back electron transfer (Equation 11.39)[68],

$$j_{ph}^{ss} = \Omega_{inj}\, j_{ph}^{max}\left(\frac{k_{ps}}{k_{ps}+k_b}\right) = j_{ph}^{0}\left(\frac{k_{ps}}{k_{ps}+k_b}\right) \tag{11.59}$$

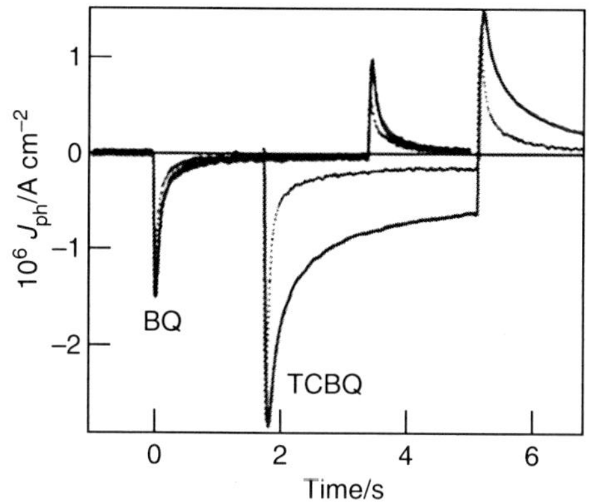

Figure 11.26. Effect of the introduction of a co-sensitiser on the photocurrent transients for the photoreduction of TCBQ and BQ by the ion pair ZnTPPS–ZnTMPyP. The full lines correspond to the photoresponses in the presence of the hexacyanoferrate redox couple in the aqueous phase. The relaxation in the photoreduction of BQ appears little affected by the co-sensitiser in comparison to TCBQ and TCNQ (see Figure 11.20). Reprinted with permission from ref.[60]. Copyright (2003) American Chemical Society.

Despite the fact that Equation (11.59) can describe phenomenologically the experimental responses observed in a variety of systems, it should be clarified that formally no "steady-state" photocurrent can be defined for bulk liquid/liquid junctions in the absence of convection forces. Even assuming a fast regeneration of the metalloporphyrins at the interface, the photocurrent flux will be controlled by the diffusion of the redox species in the organic phase. However, the depletion of the redox species is not significant for moderate light intensities as illustrated in the photocurrent transients in Figures 11.20, 11.23 and 11.24. In this section, we shall describe the photocurrent transients observed in the presence of BQ, where other processes such as coupled proton transfer can contribute to the photorelaxation.

The formation of the BQ radical anion by heterogeneous electron injection from the ZnTPPS–ZnTMPyP complex is characterised by negative photocurrent responses followed by a strong decay in the millisecond time domain (see Figure 11.23b). According to mechanism I and Equation (11.59), this decay can be attributed to an efficient back electron process, i.e. $k_b >> k_{ps}$. A similar behaviour is also observed for the acceptors TCNQ and TCBQ. As illustrated in Figure 11.20b, the photocurrent decay in the photoreduction of TCNQ is substantially affected upon addition of the hexacyanoferrate couple in the aqueous phase. This couple acts as a co-sensitiser competing with the back electron transfer as well as regenerating the initial ground state of the complex as in Equation 11.40. Experiments under similar conditions have also been reported in the presence of BQ and TCBQ as shown in Figure 11.26[60]. It is observed that the photocurrent decay in the case of TCBQ is significantly affected by the co-sensitiser, while very small changes are observed for the BQ photoreduction. Previous studies based on intensity modulated photocurrent spectroscopy show that increasing concentrations of the hexacyanoferrate couple do affect the photocurrent relaxation; however, the "steady-state" photocurrent consistently remains close to zero[102]. These experimental evidences suggest that other processes may be involved in the dynamics of photocurrent decay in the case of BQ.

The fast relaxation in the case of BQ has been associated with an interfacial protonation reaction[60],

$$Q_o^{\cdot -} + H_w^+ \xrightarrow{k_{tr}} QH_o \qquad \text{Heterogeneous proton transfer} \tag{11.60}$$

This reaction involves the transfer of a proton from the aqueous to the organic phase, which generates a positive current response. Taking into account the processes contributing to the photocurrent responses, Equations (11.35)–(11.39) and (11.60), the differential equations associated with the concentrations of the intermediate species ([PQ]) and the semiquinone radical $\left([Q^{\cdot -}]\right)$ are given by

$$\frac{d[PQ]}{dt} = g - \left(k_b + k_{ps}\right)[PQ] \tag{11.61}$$

$$\frac{d[Q^{\cdot -}]}{dt} = k_{ps}[PQ] - \left(k_{tr} + k_{di}\right)[Q^{\cdot -}] \tag{11.62}$$

where the parameter g is given by Equation (11.63),

$$g = k_{et}\left[P^*\right] = j_{ph}^{max}\left(\frac{k_{et}}{k_{et} + k_d}\right) \tag{11.63}$$

In this treatment, we have expressed the disappearance of the semiquinone radical from the interface as a first-order process (k_{di}) in order to simplify the analysis. A more rigorous treatment involves the computing of the concentration profiles of the products. However, considering that the steady-state photocurrents are rather small, this approximation has very little effect on the present analysis of the photocurrent transients.

The resolution of the differential equations in order to obtain the time dependent photocurrent has been performed via the Laplace method. In this approach, the concentration of the intermediates and products can be expressed as[60]

$$\overline{[PQ]} = \frac{g}{s}\left(\frac{1}{s + k_b + k_{ps}}\right) \tag{11.64}$$

$$\overline{[Q^{\cdot -}]} = \frac{g}{s}\left(\frac{k_{ps}}{s + k_b + k_{ps}}\right)\left(\frac{1}{s + k_{tr} + k_{di}}\right) \tag{11.65}$$

where s is the Laplace variable. Taking into account that steps (11.37), (11.39) and (11.60) involve heterogeneous charge transfer processes, the overall photocurrent is given by,

$$j_{ph} = -F\left(k_{et}\left[P^*\right] - k_b[PQ] - k_{tr}\left[Q^{\cdot -}\right]\right) \tag{11.66}$$

which can be expressed in the Laplace plane as

$$\overline{j_{ph}} = -F\left(k_{et}\overline{[P^*]} - k_b\overline{[PQ]} - k_{tr}\overline{[Q^{\cdot -}]}\right) \tag{11.67}$$

As mention in section 3.3, the photocurrent responses are attenuated by the RC component of the cell at short times. This parameter can be introduced into expression (11.67) yielding[68,69],

$$\overline{j_{\mathrm{ph}}} = -F\left(k_{\mathrm{et}}\overline{[\mathrm{P}^*]} - k_{\mathrm{b}}\overline{[\mathrm{PQ}]} - k_{\mathrm{tr}}\overline{[\mathrm{Q}^{\cdot -}]}\right)\left(\frac{1}{1+RCs}\right) \tag{11.68}$$

The inverse Laplace transform of Equation (11.68) provides the following relation for the photocurrent as a function of time under constant illumination[60]:

$$j_{\mathrm{ph}} = -Fg\,(\alpha - \beta - \gamma) \tag{11.69}$$

where

$$\alpha = 1 - \exp\left(-\frac{t}{RC}\right) \tag{11.70}$$

$$\beta = k_{\mathrm{b}}\left(\frac{1-\exp\left(-\left(k_{\mathrm{b}}+k_{\mathrm{ps}}\right)t\right)}{k_{\mathrm{b}}+k_{\mathrm{ps}}} - RC\frac{\exp\left(-\left(k_{\mathrm{b}}+k_{\mathrm{ps}}\right)t\right)-\exp\left(-t/RC\right)}{1-RC\left(k_{\mathrm{b}}+k_{\mathrm{ps}}\right)}\right) \tag{11.71}$$

$$\begin{aligned}\gamma = \frac{k_{\mathrm{ps}}k_{\mathrm{tr}}}{k_{\mathrm{b}}+k_{\mathrm{ps}}-k_{\mathrm{tr}}-k_{\mathrm{di}}}\Bigg(&\frac{1-\exp\left(-\left(k_{\mathrm{b}}+k_{\mathrm{ps}}\right)t\right)}{k_{\mathrm{b}}+k_{\mathrm{ps}}} - \frac{1-\exp\left(-\left(k_{\mathrm{tr}}+k_{\mathrm{di}}\right)t\right)}{k_{\mathrm{tr}}+k_{\mathrm{di}}}\\ &-RC\frac{\exp\left(-\left(k_{\mathrm{b}}+k_{\mathrm{ps}}\right)t\right)-\exp\left(-t/RC\right)}{1-RC\left(k_{\mathrm{b}}+k_{\mathrm{ps}}\right)}\\ &+RC\frac{\exp\left(-\left(k_{\mathrm{tr}}+k_{\mathrm{di}}\right)t\right)-\exp\left(-t/RC\right)}{1-RC\left(k_{\mathrm{tr}}+k_{\mathrm{di}}\right)}\Bigg)\end{aligned} \tag{11.72}$$

The term α is associated with the initial rise of the photocurrent response which is determined by the RC constant of the cell. As the uncompensated resistance tends to zero, this parameter approaches unity and the RC attenuation vanishes. This effect is illustrated in the simulations in Figure 11.27a. The terms β and γ describe the exponential decays associated with the back electron transfer and the heterogeneous protonation, respectively. In Figure 11.27b, the transient evolution of the parameters α, β and γ are contrasted to the photocurrent as estimated from Equation (11.69), taking the RC constant as 30 ms. This figure clearly indicates that to estimate the flux of electron injection g from the photocurrent response, the contributions from the fast relaxation processes and the RC constant should be deconvoluted. On the other hand, if the photocurrent decay is not significant (e.g. Figure 11.20b), then j^0_{ph} will be approximately equal to g. Fitting of Equation (11.69) to photocurrent transients for the various electron acceptors and Galvani potential differences provide consistent values for g, k_{ps}, k_{b}, k_{tr} and k_{di}. A typical fitting of the photocurrent transients is exemplified in Figure 11.27c. The values for RC were estimated from impedance measurements. The extraction of the parameter g by this approach allowed accurate analysis of the rate of electron injection as described in the previous section.

The analysis of photocurrent transients employing Equation (11.72) also provided values for k_{b} in the range of 10–50 s^{-1}, while k_{tr} was of the order of

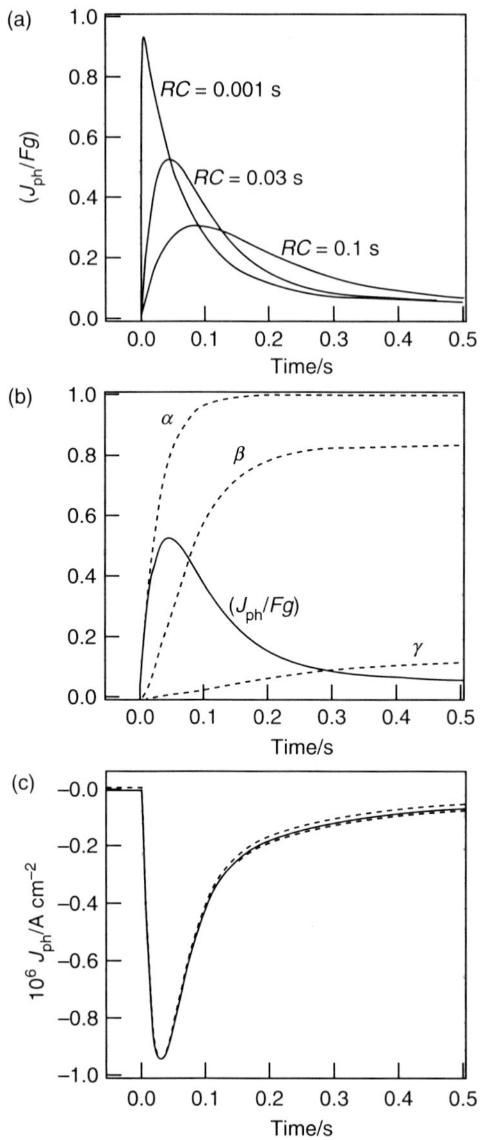

Figure 11.27. Analysis of the photocurrent transient responses including the characteristic *RC* attenuation, back electron transfer and coupled ion transfer. The effect of the *RC* constant as evaluated from Equation (11.69)–(11.72) is illustrated in (a) taking $k_b = 15\,s^{-1}, k_{ps} = 3\,s^{-1}, k_{tr} = 5\,s^{-1}$ and $k_{di} = 2\,s^{-1}$. The time evolution of the terms α, β and γ are contrasted with the normalised photocurrent response taking RC=0.03 s (b). Photocurrent response in the presence of BQ at −0.12 V and the corresponding fitting to Equation (11.69) are shown in (c). The parameters employed in the numerical analysis were: $k_b = 32\,s^{-1}, k_{ps} = 4.7\,s^{-1}, k_{tr} = 4.1\,s^{-1}, k_{di} = 0.4\,s^{-1}, RC = 0.025\,s$ and $g = 2.2 \times 10^{-11}\,mol\,s^{-1}\,cm^{-2}$. Reprinted with permission from ref.[60]. Copyright (2003) American Chemical Society.

1–5 s^{-1}. The interpretation of these results on the basis of the physics of heterogeneous electron and proton transfer remains challenging. Preliminary comparisons between the potential dependence of k_{et} and k_b for ferrocene derivatives suggest that the activation energy is comparable for both processes[68]. In the case of proton-transfer processes, the situation is even more complex as the estimated rate constant is several orders of magnitude smaller than those estimated for homogeneous processes. How is the activation energy defined for the H^+ transfer process at these interfaces? Can this reaction be treated as a conventional ion transfer reaction? What is the effective proton concentration at the interface and what is its dependence on the water phase pH and $\Delta_o^w \phi$? Our interpretation of the phenomenological data obtained from the photocurrent measurements will be rather limited until these questions are properly addressed.

The apparent slow proton transfer to the semiquinone radical brings about some interesting considerations with respect to the interfacial redox potential of BQ. As illustrated in Figure 11.25b, the photocurrent dependence on $\Delta G_{et}^{o'}$ shows a consistent trend for all the redox acceptors except BQ. The hollow diamonds in Figure 11.25b are obtained by shifting this redox potential by 0.6 V towards more positive values. In principle, substantial displacement of the reduction potential of BQ can take place via the formation of H-bonds as well as interfacial protonation processes. However, the interfacial proton transfer does manifest itself as a positive current response. If the proton transfer step occurs at time scales shorter than the relaxation associated with the *RC* component of the cell, the charge transfer originating from the quinone reduction is effectively counterbalanced and no net photocurrent can be measured under potentiostatic conditions. The analysis of the photocurrent relaxation in Figure 11.27c clearly shows that the proton transfer occurs in the sub-second time domain, i.e. over four orders of magnitude slower than the electron transfer step to BQ. Consequently, the protonation equilibrium cannot be invoked as the origin of the positive shift of the effective reduction potential of BQ as estimated from the flux of electron injection from the charge-separated intermediate state.

Although potential shifts of the redox potential by 600 mV due to H-bonding appears somewhat large, it should be mentioned that theoretical estimations of the electron affinity for BQ hydrogen bonded to four water molecules predicted increments of the order of 1.1 eV[109]. An interesting example of the role of interfacial H-bonding on the organisation and properties of molecules at the liquid/liquid interface is also found in the work of Steel and Walker[15]. SHG measurements employing ingeniously designed chromophore surfactants self-assembled at the water/1-octanol interface have shown that a region with a dielectric permittivity lower than bulk 1-octanol can be probed at the interface. This region is formed as a result of the orientation adopted by the first layer of 1-octanol to maximise H-bonding with water molecules. MD simulations by Benjamin have recently confirmed the non-monotonic changes of polarity and solvent density at the 1-octanol/water interface[122]. In addition, other MD simulations by Jedlovszky *et al.* have concluded that H-bonding and water orientation undergo substantial changes as function of the position at the interface[122]. For instance, it is proposed that in the region where the density of water is less than 10%, water molecules adopt an orientation with the molecular plane perpendicular to the interface with an O–H bond pointing towards the organic phase.

6. Concluding Remarks

This chapter has illustrated the wealth of information about structure and reactivity at polarisable liquid/liquid interfaces as probed by water-soluble metalloporphyrins. The multidisciplinary approach based on spectroscopic and photoelectrochemical techniques has allowed unravelling the specific interaction between the porphyrins and the liquid/liquid boundary, their molecular organisation and the factors governing their reactivity towards photoinduced heterogeneous electron transfer. Electrochemical impedance, QELS and SHG have demonstrated that metalloporphyrins specifically adsorb at the interface as a results of their complex solvation structure. The characteristic hydrophobic ring and ionisable peripheral groups are responsible for a variety of interactions such as ion pairing, H-bonding, aggregation and surfactant-like ordering at the liquid/liquid boundary. The coverage and molecular orientation are dependent on the Galvani potential difference between the two liquid phases.

In the presence of redox couples confined to the hydrophobic liquid phase, photoinduced heterogeneous electron transfer can be effectively monitored by photoelectrochemical techniques under potentiostatic conditions. The photocurrent responses are uniquely related to specifically adsorbed porphyrins, as demonstrated by the photocurrent anisotropy to the angle of polarisation of the incident illumination (Section 4.3). Systematic studies of the photocurrent intensity as a function of the formal potential of the redox couple and the Galvani potential difference revealed that the dynamics of electron transfer are determined by the distance separating the redox species at the interface. Other processes including decay of the electronically excited state, back electron transfer, porphyrin regeneration and coupled ion transfer play important role on the dynamics of the photocurrent responses.

A key remaining challenge in this direction is the possibility of manipulating the distance separating the redox couples and porphyrins at the liquid/liquid boundary. This can be achieved by designing molecular dyads in which both units are connected by a linker acting as a molecular wire. Examples of this kind of structures are widespread in the literature[120, 123–126]. The rationale behind this approach is to increase the electronic coupling between the excited state of the porphyrin and the redox species (pre-exponential term in Equation 11.55), as well as decreasing the reorganisation energy term (Equation 11.58). However, optimisation of these two parameters will undoubtedly increase the rate of back electron transfer and decrease the overall photocurrent quantum efficiency. A systematic study of photocurrent responses as a function of the length and nature of the linker could provide not only an optimum porphyrin-redox distance, but also a mean to probe the fraction of the applied potential acting on the electron transfer process.

Acknowledgements

We gratefully acknowledge the financial support by the Swiss National Science Foundation (Projects 2000-067050.01 and PP002—68708). We would like to express our gratitude to Dr. Virginie Mengeaud for her valuable support in editing and final touches of this chapter.

References

1. Samec, Z., V. Marecek, and J. Weber (1978). Detection of an electron transfer across the interface between two immiscible electrolyte solutions by cyclic voltammetry with four-electrode system. *J. Electroanal. Chem.* **96**, 245–247.
2. Samec, Z. and T. Kakiuchi (1995). Charge transfer kinetics at water–organic solvent phase boundaries. In H. Gerischer and C.W. Tobias (eds), *Advances in Electrochemical Science and Engineering*. VCH, Weinheim, pp. 297–361.
3. Samec, Z. (1988). Electrical double layer at the interface between two immiscible electrolyte solutions. *Chem. Rev.* **88**, 617–632.
4. Peter, L.M. and D. Vanmaekelbergh (1998). Time and frequency resolved studies of photoelectrochemical kinetics. In R. Alkire and D.M. Kolb, (eds), *Advances in Electrochemical Science and Engineering*. VCH, Weinheim, pp. 77–163.
5. Fermín, D.J., Z.F. Ding, H.D. Duong, P.F. Brevet, and H.H. Girault (1998). Photocurrent responses associated with heterogeneous electron-transfer at liquid/liquid interfaces. *Chem. Commun.* 1125–1126.
6. Benjamin, I. (1996). Chemical-reactions and solvation at liquid interfaces—a microscopic perspective. *Chem. Rev.* **96**, 1449–1475.
7. Eisenthal, K.A. (1996). Liquid interfaces probed by secon-harmonic and sum-frequency spectroscopy. *Chem. Rev.* **96**, 1343–1360.
8. Schlossman, M.L. (2002). Liquid–liquid interfaces: Studied by X-ray and neutron scattering. *Curr. Opin. Colloid Interface Sci.* **7**, 235–243.
9. Richmond, G.L. (2001). Structure and bonding of molecules at aqueous surfaces. *Annu. Rev. Phys. Chem.* **52**, 357–389.
10. De Serio, M., A.N. Bader, M. Heule, R. Zenobi, and V. Deckert (2003). A near-field optical method for probing liquid–liquid interfaces. *Chem. Phys. Lett.* **380**, 47–53.
11. Wang, H.F., E. Borguet, and K.B. Eisenthal (1998). Generalized interface polarity scale based on second harmonic spectroscopy. *J. Phys. Chem. B* **102**, 4927–4932.
12. Wang, H.F., E. Borguet, and K.B. Eisenthal (1997). Polarity of liquid interfaces by second harmonic generation spectroscopy. *J. Phys. Chem. A* **101**, 713–718.
13. Scatena, L.F., and G.L. Richmond (2001). Orientation, hydrogen bonding, and penetration of water at the organic/water interface. *J. Phys. Chem. B* **105**, 11240–11250.
14. Scatena, L.F., M.G. Brown, and G.L. Richmond (2001). Water at hydrophobic surfaces: Weak hydrogen bonding and strong orientation effects. *Science* **292**, 908–912.
15. Steel, W.H. and R.A. Walker (2003). Measuring dipolar width across liquid–liquid interfaces with 'molecular rulers'. *Nature* **424**, 296–299.
16. Steel, W.H. and R.A. Walker (2003). Solvent polarity at an aqueous/alkane interface: The effect of solute identity. *J. Am. Chem. Soc.* **125**, 1132–1133.
17. Roser, S.J., R. Felici, and A. Eaglesham (1994). X-ray reflection from a liquid–liquid Interface. *Langmuir* **10**, 3853.

18. Zhang, Z.J., D.M. Mitrinovic, S.M. Williams, Z.Q. Huang, and M.L. Schlossman (1999). X-ray scattering from monolayers of F(CF2)(10)(CH2)(2)OH at the water–(hexane solution) and water–vapor interfaces. *J. Chem. Phys.* **110**, 7421–7432.
19. Mitrinovic, D.M., Z.J. Zhang, S.M. Williams, Z.Q. Huang, and M.L. Schlossman (1999). X-ray reflectivity study of the water–hexane interface. *J. Phys. Chem. B* **103**, 1779–1782.
20. Tikhonov, A.M., D.M. Mitrinovic, M. Li, Z.Q. Huang, and M.L. Schlossman (2003). An X-ray reflectivity study of the water–docosane interface. *J. Phys. Chem. B* **104**, 6336–6339.
21. Mitrinovic, D.M., A.M. Tikhonov, M. Li, Z.Q. Huang, and M.L. Schlossman (2000). Noncapillary-wave structure at the water–alkane interface. *Phys. Rev. Lett.* **85**, 582–585.
22. Zarbakhsh, A., J. Bowers, and J.R.P. Webster (1999). A new approach for measuring neutron reflection from a liquid/liquid interface. *Meas. Sci. Technol.* **10**, 738–743.
23. Strutwolf, J., A.L. Barker, M. Gonsalves, D.J. Caruana, P.R. Unwin, D.E. Williams, and J.R.P. Webster (2000). Probing liquid/liquid interfaces using neutron reflection measurements and scanning electrochemical microscopy. *J. Electroanal. Chem.* **483**, 163–173.
24. Michael, D. and I. Benjamin (1998). Molecular dynamics simulation of the water/nitrobenzene interface. *J. Electroanal. Chem.* **450**, 335–345.
25. Michael, D. and I. Benjamin (2001). Molecular dynamics computer simulations of solvation dynamics at liquid/liquid interfaces. *J. Chem. Phys.* **114**, 2817–2824.
26. Fernandes, P.A., M.N.D.S. Cordeiro, and J.A.N.F. Gomes (1999). Molecular dynamics simulations of the water/2–heptanone liquid/liquid interface. *J. Phys. Chem. B* **103**, 6290–6299.
27. Senapati, S. and M.L. Berkowitz (2001). Computer simulation study of the interface width of the liquid/liquid interface. *Phys. Rev. Lett.* **87**, 176101–176101.
28. Lamb. H. (1945). *Hydrodynamics*. Dover, New York.
29. Bowers, J., A. Zarbakhsh, J.R.P. Webster, L.R. Hutchings, and R.W. Richards (2001). Neutron reflectivity studies at liquid/liquid interfaces: Methodology and analysis. *Langmuir* **17**, 2548.
30. Daikhin, L.I., A.A. Kornyshev, and M. Urbakh (2001). Ion penetration into an 'unfriendly medium' and the double layer capacitance of the interface between two immiscible electrolytes. *J. Electroanal. Chem.* **500**, 461–470.
31. Naujok, R.R., D.A. Higgins, D.G. Hanken, and R.M. Corn (1995). Optical 2nd-harmonic generation measurements of molecular adsorption and orientation at the liquid–liquid electrochemical interface. *J. Chem. Soc. Faraday Trans.* **91**, 1411–1420.
32. Gragson, D.E. and G.L. Richmond (1998). Potential dependent alignment and hydrogen bonding of water molecules at charged air/water and CCl4/water interfaces. *J. Am. Chem. Soc.* **120**, 366–375.
33. Gragson, D.E. and G.L. Richmond (1998). Investigations of the structure and hydrogen bonding of water molecules at liquid surfaces by vibrational sum frequency spectroscopy. *J. Phys. Chem. B* **102**, 3847–3861.
34. Gragson, D.E. and G.L. Richmond (1997). Probing the intermolecular hydrogen bonding of water molecules at the CCl4/water interface in the presence of charged soluble surfactant. *J. Chem. Phys.* **107**, 9687–9690.
35. Jensen, H., J.J. Kakkassery, H. Nagatani, D.J. Fermín, and H.H. Girault (2000). Photoinduced electron transfer at liquid/liquid interfaces. Part IV. Orientation and reactivity of zinc tetra(4–carboxyphenyl) porphyrin self–assembled at the water/1,2–dichloroethane junction. *J. Am. Chem. Soc.* **122**, 10943–10948.

36. Nagatani, H., A. Piron, P.F. Brevet, D.J. Fermín, and H.H. Girault (2002). Surface second harmonic generation of cationic water-soluble porphyrins at the polarized water/1,2-dichloroethane interface. *Langmuir* **18**, 6647–6652.
37. Nagatani, H., Z. Samec, P.F. Brevet, D.J. Fermín, and H.H. Girault (2003). Adsorption and aggregation of meso–tetrakis(4- carboxyphenyl)porphyrinato zinc(II) at the polarized water/1,2-dichloroethane interface. *J. Phys. Chem. B* **107**, 786–790.
38. Eugster, N., H. Jensen, D.J. Fermín, and H.H. Girault (2003). Photoinduced electron transfer at liquid/liquid interfaces. Part VII. Correlation between self–organisation and structure of water–soluble photoactive species. *J. Electroanal. Chem.* **560**, 143–149.
39. Girault, H.H. (1993). Charge transfer across liquid–liquid interfaces. In J.O.M. Bockris, B.E. Conway, and R.E. White (eds), *Modern Aspects of Electrochemistry*. Plenum Press, N Y pp. 1–62.
40. Koczorowski, Z. (2001). Interfacial potentials and cells. In A. Volkov (ed), *Liquid Interfaces in Chemical, Biological, and Pharmaceutical Applications*. Marcel Dekker, Boca Raton, pp. 1–22.
41. Gavach, C., P. Seta, and B. D'Epenoux (1977). The double layer and ion adsorption at the interface between two non miscible solutions. Part I. Interfacial tension measurements for the water–nitrobenzene tetraalkylammonium bromide systems. *J. Electroanal. Chem.* **83**, 225–235.
42. Samec, Z., V. Marecek, J. Weber, and D. Homolka (1981). Charge transfer between two immiscible electrolyte solutions. Part VII. Convolution potential sweep voltammetry of cesium(1+) ion transfer and of electron transfer between ferrocene and hexacyanoferrate(III) ion across the water/nitrobenzene interface. *J. Electroanal. Chem.* **126**, 105–119.
43. Fermín, D.J., H. Jensen, and H.H. Girault (2003). Theoretical aspects associated with charge-transfer kinetics across interfaces between two immiscible electrolyte solutions. In E.J. Calvo (ed), *Encyclopedia of ELectrochemistry. Vol. 2. Interfacial Kinetics and Mass Transport*. Wiley-VCH, New York, pp. 360–390.
44. Schmickler, W. (1997). Electron-transfer reactions across liquid/liquid interfaces. *J. Electroanal. Chem.* **428**, 123–127.
45. Cheng, Y., and D.J. Schiffrin (1993). AC impedance study of rate constants for two-phase electron-transfer reactions. *J. Chem. Soc. Faraday Trans.* **89**, 199–205.
46. Ding, Z.F., D.J. Fermín, P.F. Brevet, and H.H. Girault (1998). Spectroelectrochemical approaches to heterogeneous electron-transfer reactions at the polarized water/1,2-dichloroethane interfaces. *J. Electroanal. Chem.* **458**, 139–148.
47. Fermín, D.J. and R. Lahtinen (2001). Dynamic aspects of heterogeneous electron transfer reactions at liquid/liquid interfaces. In A. Volkov, (ed), *Liquid Interfaces in Chemical, Biological, and Pharmaceutical Applications*. Marcel Dekker, Boca Raton, pp. 179–228.
48. Wei, C., Bard, A.J. and M.V. Mirkin (1995). Scanning electrochemical microscopy. 31. Application of secm to the study of charge-transfer processes at the liquid–liquid interface. *J. Phys. Chem.* **99**, 16033–16042.
49. Tsionsky, M., A.J. Bard, and M.V. Mirkin (1996). Scanning electrochemical microscopy. 34. Potential dependence of the electron-transfer rate and film formation at the liquid/liquid interface. *J. Phys. Chem.* **100**, 17881–17888.
50. Barker, A.L., P.R. Unwin, S. Amemiya, J.F. Zhou, and A.J. Bard (1999). Scanning electrochemistry microscopy (SECM) in the study of electron-transfer kinetics at liquid/liquid interfaces—beyond the constant composition approximation. *J. Phys. Chem. B* **103**, 7260–7269.

51. A.L. Barker, J.V. Macpherson, C.J. Slevin, and P.R. Unwin (1998). Scanning electrochemical microscopy (Secm) as a probe of transfer processes in 2-phase systems—theory and experimental applications of secm-induced transfer with arbitrary partition-coefficients, diffusion-coefficients, and interfacial kinetics. *J. Phys. Chem. B* **102**, 1586–1598.
52. Kalyanasundaram, K. (1992). *Photochemistry of Polypyridine and Porphyrin Complexes*. Academic Press, London.
53. Eugster, N. (2004). *PhD Thesis*. Ecole Polytechnique Fédérale de Lausanne.
54. Hugerat, M., H. Levanon, E. Ojadi, L. Biczok, and H. Linschitz (1991). Multiple decay pathways and electron transfer in excited ion-paired zinc–copper porphyrins: Laser photolysis and time-resolved EPR spectroscopy. *Chem. Phys. Lett.* **181**, 400–406.
55. Chen, D.-M., Y.-H. Zhang, T.-J. He, and F.-C. Liu (2002). Raman and UV–visible absorption spectra of ion-paired aggregates of copper porphyrins. *Spectrochim. Acta Part A* **58**, 2291–2297.
56. Van Willigen, H., U. Das, E. Ojadi, and H. Linschitz (1985). Triplet ESR study of dimerization of cationic and anionic water-soluble porphyrins. *J. Am. Chem. Soc.* **107**, 7784–7785.
57. Hofstra, U., R.B.M. Koehorst, and T.J. Schaafsma (1986). Excited-state properties of water-soluble porphyrin dimers. *Chem. Phys. Lett.* **130**, 555–559.
58. Vergeldt, F.J., R.B.M. Koehorst, T.J. Schaafsma, J.C. Lambry, J.L. Martin, D.G. Johnson, and M.R. Wasiliewski (1991). Sub-picosecond photoinduced electron transfer in water-soluble porphyrin dimers. *Chem. Phys. Lett.* **182**, 107–113.
59. Fermín, D.J., H.D. Duong, Z.F. Ding, P.F. Brevet, and H.H. Girault (1999). Photoinduced electron-transfer at liquid/liquid interfaces—part III—Photoelectrochemical responses involving porphyrin ion-pairs. *J. Am. Chem. Soc.* **121**, 10203–10210.
60. Eugster, N., D.J. Fermín, and H.H. Girault (2003). Photoinduced electron transfer at liquid/liquid interfaces. Dynamics of the heterogeneous photoreduction of quinones by self-assembled porphyrin ion-pairs. *J. Am. Chem. Soc.* **125**, 4862–4869.
61. Ojadi, E., R. Selzer, and H. Linschitz (1985). Properties of porphyrin dimers, formed by pairing cationic and anionic porphyrins. *J. Am. Chem. Soc.* **107**, 7783–7784.
62. Rodriguez, J., C. Kirmaier, and D. Holten (1989). Optical properties of metalloporphyrin excited states. *J. Am. Chem. Soc.* **111**, 6500–6506.
63. Yu, H.-Z., S. Baskin, and A.H. Zewail (2002). Ultrafast dynamics of poprhyrins in the condensed phase: II. Zinc tetraphenylporphyrins. *J. Phys. Chem. A* **106**, 9845–9854.
64. Kalyanasundaram, K. and M. Neumann-Spallart (1982). Photophysical and redox properties of water-soluble porphyrins in aqueous media. *J. Phys. Chem.* **86**, 5163–5169.
65. Huber, R., J.E. Moser, M. Gratzel, and J. Wachtveitl (2002). Real-time observation of photoinduced adiabatic electron transfer in strongly coupled dye/semiconductor colloidal systems with a 6 fs time constant. *J. Phys. Chem. B* **106**, 6494–6499.
66. Huber, R., S. Sporlein, J.E. Moser, M. Gratzel, and J. Wachtveitl (2000). The role of surface states in the ultrafast photoinduced electron transfer from sensitizing dye molecules to semiconductor colloids. *J. Phys. Chem. B* **104**, 8995–9003.
67. Duong, H.D., P.F. Brevet, and H.H. Girault (1998). Heterogeneous electron transfer reactions at liquid/liquid interfaces studied by time resolved absorption spectroscopy. *J. Photochem. Photobiol. A* **117**, 27–33.
68. Fermín, D.J., H.D. Duong, Z.F. Ding, P.F. Brevet, and H.H. Girault (1999). Photoinduced electron-transfer at liquid/liquid interfaces—Part II—A study of the electron-transfer and recombination dynamics by intensity-modulated photocurrent spectroscopy (IMPS). *Phys. Chem. Chem. Phys.* **1**, 1461–1467.

69. Peter, L.M. (1990). Dynamic aspects of semiconductor photoelectrochemistry. *Chem. Rev.* **90**, 753–769.
70. Koryta, J., M. Brezina, A. Hofmanova, D. Homolka, H. Le Quoc, W. Khalil, V. Marecek, Z. Samec, S.K. Sen, et al. (1980). A new model of membrane transport: Electrolysis at the interface of two immiscible electrolyte solutions. *Bioelectrochem. Bioenerg.* **7**, 61–68.
71. Koryta, J., L.Q. Hung, and A. Hofmanova (1982). Biomembrane transport processes at the ITIES with an adsorbed phospholipid monolayer. *Stud. Biophys.* **90**, 25–29.
72. Koryta, J. (1987). Electrochemistry of liquid membranes: interfacial aspects. *Electrochim. Acta* **32**, 419–424.
73. Kakiuchi, T., M. Nakanishi, and M. Senda (1989). The electrocapillary curves of the phosphatidylcholine monolayer at the polarized oil–water interface. II. Double layer structure of dilauroylphosphatidylcholine monolayer at the nitrobenzene–water interface. *Bull. Chem. Soc. Jpn* **62**, 403–409.
74. Kakiuchi, T., M. Nakanishi, and M. Senda (1988). The electrocapillary curves of a phosphatidylcholine monolayer at a polarized oil–water interface. I. Measurement of interfacial tension using a computer-aided pendant-drop method. *Bull. Chem. Soc. Jpn* **61**, 1845–1851.
75. Girault, H.H.J. and D.J. Schiffrin (1984). Adsorption of phosphatidylcholine and phosphatidylethanolamine at the polarized water/1,2-dichloroethane interface. *J. Electroanal. Chem.* **179**, 277–284.
76. Roozeman, R.J., P. Liljeroth, C. Johans, D.E. Williams, and K. Kontturi (2002). Dynamic interfacial tension at electrified liquid/liquid interfaces. *Langmuir* **18**, 8318–8323.
77. Allen, R.M., K. Kontturi, and D.E. Williams (2000). Interfacial tension measurements at the interface between two highly immiscible electrolyte solutions: The trapped air bubble method. *Electrochem. Commun.* **2**, 703–706.
78. Allen, R.M., K. Kontturi, L. Murtomaki, and D.E. Williams (2000). Probing adsorption reactions at the liquid/liquid interface by area-step experiments. *J. Electroanal. Chem.* **483**, 57–67.
79. Brevet, P.F. (2001). Nonlinear optics at liquid–liquid interfaces. In A. Volkov (ed), *Liquid Interfaces in Chemical, Biological, and Pharmaceutical Applications*. Marcel Dekker, Boca Raton, pp. 123–151.
80. Tsuyumoto, I. and T. Sawada (2001). Dynamic behaviors of molecules at liquid–liquid interfaces using the time-resolved quasi-elastic laser scattering method. In A. Volkov, (ed), *Liquid Interfaces in Chemical, Biological, and Pharmaceutical Applications*. Marcel Dekker, Boca Raton, pp. 229–239.
81. Trojanek, A., P. Krtil, and Z. Samec (2001). Quasi-elastic laser light scattering from thermally excited capillary waves on polarised liquid/liquid interfaces Part 1: Effects of adsorption of hexadecyltrimethylammonium chloride at the water/1,2-dichloroethane interface. *J. Electroanal. Chem.* **517**, 77–84.
82. Su, B., J.P. Abid, D.J. Fermín, H.H. Girault, H. Hoffmannova, P. Krtil, and Z. Samec (2004). Reversible voltage-induced assembly of Au nanoparticles at liquid/liquid interfaces. *J. Am. Chem. Soc.* **126**, 915–919.
83. Nagatani, H., D.J. Fermín, and H.H. Girault (2001). A kinetic model for adsorption and transfer of ionic species at polarized liquid/liquid interfaces as studied by potential modulated fluorescence spectroscopy. *J. Phys. Chem. B* **105**, 9463–9473.
84. Nagatani, H., R.A. Iglesias, D.J. Fermín, P.F. Brevet, and H.H. Girault (2000). Adsorption behavior of charged zinc porphyrins at the water/1,2-dichloroethane interface studied by potential modulated fluorescence spectroscopy. *J. Phys. Chem. B* **104**, 6869–6876.

85. Frank, S. and W. Schmickler (2000). A lattice-gas model for ion pairing at liquid/liquid interfaces. *J. Electroanal. Chem.* **483**, 18–21.
86. Huber, T., O. Pecina, and W. Schmickler (1999). The influence of the ions on the capacity of liquid/liquid interfaces. *J. Electroanal. Chem.* **467**, 203–206.
87. Frank, S. and W. Schmickler (2001). A lattice-gas model for specific ion adsorption at liquid/liquid interfaces. *J. Electroanal. Chem.* **500**, 491–497.
88. Kakiuchi, T. (2001). Adsorption at polarized liquid–liquid interfaces. In A.G. Volkov, (ed), *Liquid Interfaces in Chemical, Biological and Pharmaceutical Applications*. Marcel Dekker, New York, pp. 105–121.
89. Gavach, C. and N. Davion (1973). Interfacial potential and ion exchange equilibrium between a potassium halide aqueous solution and an alkyltrimethyl ammonium picrate solution in nitrobenzene. *Electrochim. Acta* **18**, 649–656.
90. Higgins, D.A. and R.M. Corn (1993). Second harmonic generation studies of adsorption at a liquid/liquid interface. *J. Phys. Chem.* **97**, 489–493.
91. Kakiuchi, T. (2001). *J. Electroanal. Chem.* **496**, 137–142.
92. Markin, V.S. and A.G. Volkov (1996). Adsorption isotherms and the structure of oil/water interfaces. In A.G. Volkov and D.W. Deamer (eds), *Liquid–Liquid Interface Theory and Methods*. Marcel Dekker, New York, pp. 63–75.
93. Parsons, R. (1980). Thermodynamic methods for the study of interfacial regions in electrochemical systems. In J.O.M. Bockris, B.E. Conway, and E. Yeager (eds), *Comprehensive Treatise of Electrochemistry*, Vol. 1. Plenum Press, New York, pp. 1–44.
94. Fermín, D.J., Z.F. Ding, H.D. Duong, P.F. Brevet, and H.H. Girault (1998). Photoinduced electron-transfer at liquid/liquid interfaces—I—Photocurrent measurements associated with heterogeneous quenching of zinc porphyrins. *J. Phys. Chem. B* **102**, 10334–10341.
95. Jensen, H., D.J. Fermín, and H.H. Girault (2001). Photoinduced electron transfer at liquid/liquid interfaces. Part V. Organisation of water-soluble chlorophyll at the water/1,2-dichloroethane interface. *Phys. Chem. Chem. Phys.* **3**, 2503–2508.
96. Cunderlikova, B., L. Gangeskar, and J. Moan (1999). Acid–base properties of chlorin e(6): Relation to cellular uptake. *J. Photochem. Photobiol. B Biol.* **53**, 81–90.
97. Cunderlikova, B., M. Kongshaug, L. Gangeskar, and J. Moan (2000). Increased binding of chlorin e(6) to lipoproteins at low pH values. *Int J. Biochem. Cell Biol.* **32**, 759–768.
98. Tsuyumoto, I., N. Noguchi, T. Kitamori, and T. Sawada (1998). Density-estimation of liquid/liquid interfacial regions using a quasi-elastic laser scattering method. *J. Phys. Chem. B* **102**, 2684–2687.
99. Corn, R.M. and D.A. Higgins (1994). Optical 2nd-harmonic generation as probe of surface chemistry. *Chem. Rev.* **94**, 107–125.
100. Perrenoud-Rinuy, J., P.F. Brevet, and H.H. Girault (2002). Second harmonic generation study of myoglobin and hemoglobin and their protoporphyrin IX chromophore at the water/1,2- dichloroethane interface. *Phys. Chem. Chem. Phys.* **4**, 4774–4781.
101. Eugster, N., D.J. Fermín, and H.H. Girault (2002). Photoinduced electron transfer at liquid/liquid interfaces. Part VI. On the thermodynamic driving force dependence of the phenomenological electron transfer rate constant. *J. Phys. Chem. B* **106**, 3428–3433.
102. Lahtinen, R., D.J. Fermín, K. Kontturi, and H.H. Girault (2000). Artificial photosynthesis at liquid/liquid interfaces: photoreduction of benzoquinone by water soluble porphyrin species. *J. Electroanal. Chem.* **483**, 81–87.
103. Fermín, D.J., H.D. Duong, Z.F. Ding, P.F. Brevet, and H.H. Girault (1999). Solar energy conversion using dye-sensitised liquid/liquid interfaces. *Electrochem. Commun.* **1**, 29–32.

104. Fermín, D.J., H. Jensen, J.E. Moser, and H.H. Girault (2003). Organisation and reactivity of nanoparticles at molecular interfaces. Part II. Dye sensitisation of TiO2 nanoparticles assembled at the water/1,2-dichloroethane interface. *Chem. Phys. Chem.* **4**, 85–89.
105. Moser, J.E. and M. Gratzel (1998). Excitation-wavelength dependence of photo-induced charge injection at the semiconductor-dye interface: Evidence for electron transfer from vibrationally hot excited states. *Chimia* **52**, 160–162.
106. Moser, J.E., M. Wolf, F. Lenzmann, and M. Gratzel (1999). Photoinduced charge injection from vibronically hot excited molecules of a dye sensitizer into acceptor states of wide-bandgap oxide semiconductors. *Z. Phys. Chemie—Int. J. Res. Phys. Chem. Chem. Phys.* **212**, 85–92.
107. Durrant, J.R., Y. Tachibana, I. Mercer, J.E. Moser, M. Gratzel, and D.R. Klug (1999). The excitation wavelength and solvent dependance of the kinetics of electron injection in Ru(dcbpy)(2)(NCS)(2) sensitized nanocrystalline TiO2 films. *Z. Phys. Chemie—Int. J. Res. Phys. Chem. Chem. Phys.* **212**, 93–98.
108. Durrant, J.R., Y. Tachibana, J.E. Moser, M. Gratzel, and D.R. Klug (1997). Interfacial electron transfer in dye sensitised nanocrystalline TiO2 films. *Proc. Indian Acad. Sci. Chem. Sci.* **109**, 411–414.
109. O'Malley, P.J. (1997). A density functional study of the effect of resuction on the geometry and electron affinity of hydrogen bonded 1,4-benzoquinone. Implications for quinone reduction and protonation in photosynthetic reaction centres. *Chem. Phys. Lett.* **274**, 251–254.
110. Gupta, N. and H. Linschitz (1997). Hydrogen-bonding and protonation effects in electrochemistry of quinones in aprotic solvents. *J. Am. Chem. Soc.* **119**, 6384–6391.
111. Kim, H.S., T.D. Chung, and H. Kim (2001). Voltammetric determination of the pKa of various acids in polar aprotic solvents using 1,4-benzoquinone. *J. Electroanal. Chem.* **498**, 209–215.
112. Kim, J., T.D. Chung, and H. Kim (2001). Determination of biologically active acids based on the electrochemical reduction of quinone in acetonitrile + water mixed solvent. *J. Electroanal. Chem.* **499**, 78–84.
113. Marcus, R.A. (1990). Theory of electron-transfer rates across liquid–liquid interfaces. *J. Phys. Chem.* **94**, 4152–4155.
114. Marcus, R.A. (1990). Theory of electron-transfer rates across liquid–liquid interfaces [Erratum to document cited in CA112(24):225352x]. *J. Phys. Chem.* **94**, 7742.
115. Marcus, R.A. (1990). Reorganization free energy for electron transfers at liquid–liquid and dielectric semiconductor–liquid interfaces. *J. Phys. Chem.* **94**, 1050–1055.
116. Marcus, R.A. (1991). Theory of electron-transfer rates across liquid–liquid interfaces. 2. Relationships and application. *J. Phys. Chem.* **95**, 2010–2013.
117. Girault, H.H. (1995). Solvent reorganization energy for heterogeneous electron-transfer reactions at liquid–liquid interfaces. *J. Electroanal. Chem.* **388**, 93–100.
118. Benjamin, I. and Y.I. Kharkats (1998). Reorganization free energy for electron transfer reactions at liquid/liquid interfaces. *Electrochim. Acta* **44**, 133–138.
119. Smith, B.B., J.W. Halley, and A.J. Nozik (1996). On the Marcus model of electron transfer at immiscible liquid interfaces and its application to the semiconductor/liquid interface. *Chem. Phys.* **205**, 245–267.
120. Imahori, H., K. Tamaki, D.M. Guldi, C.P. Luo, M. Fujitsuka, O. Ito, Y. Sakata, and S. Fukuzumi (2001). Modulating charge separation and charge recombination dynamics in porphyrin fullerene linked dyads and triads: Marcus-normal versus inverted region. *J. Am. Chem. Soc.* **123**, 2607–2617.
121. Imahori, H., H. Yamada, Y. Nishimura, I. Yamazaki, and Y. Sakata (2000). Vectorial multistep electron transfer at the gold electrodes modified with self-assembled monolayers of ferrocene–porphyrin–fullerene triads. *J. Phys. Chem. B* **104**, 2099–2108.

122. Benjamin, I. (2004). Polarity of the water/octanol interface. *Chem. Phys. Lett.* **393**, 453–456.
123. Schuster, D.I., P. Cheng, P.D. Jarowski, D.M. Guldi, C.P. Luo, L. Echegoyen, S. Pyo, A.R. Holzwarth, S.E. Braslavsky, R.M. Williams, and G. Klihm (2004). Design, synthesis, and photophysical studies of a porphyrin–fullerene dyad with parachute topology: Charge recombination in the Marcus inverted region. *J. Am. Chem. Soc.* **126**, 7257–7270.
124. Guldi, D.M., B. Nuber, P.J. Bracher, C.A. Alabi, S. MacMahon, J.W. Kukol, S.R. Wilson, and D.I. Schuster (2003). Synthesis and photophysics of a copper–porphyrin–styrene-C-60 hybrid. *J. Phys. Chem. A* **107**, 3215–3221.
125. Guldi, D.M., C.P. Luo, M. Prato, A. Troisi, F. Zerbetto, M. Scheloske, E. Dietel, W. Bauer, and A. Hirsch (2001). Parallel (face-to-face) versus perpendicular (edge-to-face) alignment of electron donors and acceptors in fullerene porphyrin dyads: The importance of orientation in electron transfer. *J. Am. Chem. Soc.* **123**, 9166–9167.
126. Guldi, D.M., I. Zilbermann, G. Anderson, A. Li, D. Balbinot, N. Jux, M. Hatzimarinaki, A. Hirsch, and M. Prato (2004). Multicomponent redox gradients on photoactive electrode surfaces. *Chem. Commun.* 726–727.

12

Theoretical Insights on the Chemical Reactivity of MetalloPorphyrins Using Density Functional Theory

Ilaria Ciofini, Laurent Joubert, Michele Pavone, Vincenzo Barone and Carlo Adamo

1. Introduction

Metal complexes of N_4-ligands, such as porphyrins, are widely studied as biomimetic models for several biological redox processes, in particular for molecular oxygen transport and activation[1]. They are also well known as efficient catalysts for oxidative degradation of various types of pollutants and residual wastes[2]. In the last decade, related complexes such as metallophthalocyanines were also reported to be efficient catalysts[3], especially for the electrochemical oxidation of thiols[4–6]. For this reason, modified electrodes coated by adsorbed or electropolymerized films of these complexes have been extensively used since they act as electrocatalysts lowering the overpotential of oxidation (or reduction) of the target molecules[3–7]. Studies, reported in literature, concerning the electro-oxidation of thiols have shown that the catalytic activity of phthalocyanines-coated electrodes strongly depends on the nature of the central metal, with cobalt derivatives giving the best activity[4–6]. Nevertheless, the mechanism of the electrocatalytic process is still not well understood. In a recent study, Bedioui and collaborators have extended the use of electropolymerized cobalt porphyrin film-coated electrode for the electrocatalysis of the oxidation of thiols[8]. For the first time, it was shown that the cobalt porphyrin-modified electrode possesses a potential electrocatalytic activity for 2-mercaptoethanol electro-oxidation. Anyway, its activity is significantly lower than that of the phthalocyanine-based one. Further extension of the

Ilaria Ciofini, Laurent Joubert, and Carlo Adamo • Ecole Nationale Supérieure de Chimie de Paris, Laboratoire d'Electrochimie et Chimie Analytique, UMR 7575, 11 rue P. et M. Curie, F-75231 Paris cedex 05.

Michele Pavone and Vincenzo Barone • Laboratorio di Struttura e Dinamica Molecolare, Dipartimento di Chimica, Complesso Universitario Monte Sant'Angelo, Via Cintia, I-80126 Napoli, Italy.

N_4-Macrocyclic Metal Complexes, edited by José H. Zagal, Fethi Bedioui and Jean-Pol Dodelet. Springer Science+Business Media, Inc., New York, 2006.

use of various types of macrocyclic complexes, such as tetrabenzoporphyrin and porphyrazine, for the electrocatalytic activation of thiols may offer new alternatives in this field. As a matter of fact, under experimental conditions, several effects, such as the role of the central metal, functionalisation, solvent, the substrate (here 2-mercaptoethanol), and the absorption, sum up to give the final electrochemical activity of the device. Indeed, these effects are strongly interconnected and very difficult to analyse on experimental basis only. Therefore, especially to tune and guide the experimental work, a theoretical approach could be of great help, if not mandatory.

Theoretical (and computational) approaches offer the great advantage of being able not only to simulate the overall process, i.e. to perform an "ideal" reaction, but to provide tools of analysis able to separate and quantify each of the effects playing a role. In this respect, two main approaches can be followed: (1) perform a "chemical experiment" using computational approaches, i.e. simulate the overall reaction taking place (direct approach); (2) use reactivity indexes computed for isolated reactants/products in order to qualitatively and quantitatively evaluate the reaction yields (indirect approach). The first approach offers the advantage of giving information not only on the thermodynamics of the reaction but also on its kinetics and more generally on its microscopic steps. In other words the direct approach gives insights on the mechanism of the reaction, thus providing a global description of the overall process via its elementary steps. Nevertheless, if we limit to quantum *ab initio* methods (i.e. quantum approaches that do not use any parametrization) the direct approach becomes very expensive as the size of the systems increases and difficult to be efficiently used to simulate reactions, including a great variety of phases and interactions, such as in the case of the electrocatalytic process we are interested in. Furthermore, some theoretical tools to analyse the computed reaction profiles and rationalize them in terms of simple chemical concepts will, indeed, be needed. Therefore, even if we are currently studying the possibility of following the oxidation of 2-mercaptoethanol using a direct approach, that is performing an *ab initio* molecular dynamic simulation for the approach of the substrate to a Co(II)–N_4 complex, it is not surprising that simpler, yet efficient, methods are currently used to rationalise the oxidative power of N_4 complexes. In this context, the global descriptors of reactivity, such as the electronic chemical potential (μ) and chemical hardness (η)[9–11], represent a simple way to rationalise the different chemical behaviour of similar species. In particular, μ characterises the escaping tendency of electrons from the equilibrium system while η can be seen as a resistance to charge transfer. These two entities are global properties of the investigated systems and the characterisation of their profiles along a reaction coordinate has been shown to be useful to study their chemical reactivity[12–17]. More recently, Parr and co-workers[18] have introduced a new index, the electrophilicity (ω), as a convenient parameter to assess the electrophilicity power of an atom or molecule. This index can be defined in terms of hardness and chemical potential[18].

From a more theoretical point of view, the chemical potential, and hardness (and softness) concepts can be rigorously defined in the framework of density functional theory (DFT), allowing their non-empirical evaluation and accurate calculation (see for instance ref.[12] and [13]). DFT is an *ab initio* approach including both exchange and electron correlation contributions to the total energy with such a favourable scaling that large size systems can be routinely treated[19].

As concerns the calculation of chemical reactivity index, it can easily be shown that within the DFT formalism, being the total energy of a system a function of the density and external potential ($E[\rho(r), \upsilon(r)]$), the chemical potential can be defined as:

$$\mu = \left(\frac{\partial E}{\partial N}\right)_{\upsilon(r)}, \qquad (12.1)$$

where N is the total number of electrons.

Analogously the hardness is the second derivative of the total energy with respect to the number of electrons at constant external potential:

$$\eta = \left(\frac{\partial^2 E}{\partial N^2}\right)_{\upsilon(r)}. \qquad (12.2)$$

This rigorous definition given by Parr and Pearson in 1983[9,10], formalizes the concept expressed by Pearson in 1960 to rationalize acid–base reactions and has been widely used since then[20]. Furthermore, Parr and Pearson[9] using a three-point finite difference approximation, derived the following working definition (here we omit the 1/2 factor arbitrarily introduced in the original definition):

$$\eta = (I - A), \qquad (12.3)$$

where I and A are the first vertical ionisation potential and electron affinity of the neutral molecule, respectively. Analogously, the electronic chemical potential (μ) can be defined as:

$$\mu = \frac{1}{2}(I + A). \qquad (12.4)$$

Starting from these two quantities, the electrophilicity index (ω) was successively defined by Parr and co-workers[18] as:

$$\omega = \frac{\mu^2}{2\eta}. \qquad (12.5)$$

This expression can be regarded as a quantitative formulation of the model of Maynard et al. who introduced the concept of "electrophilicity power" of a ligand as a measure of its electron saturation[21]. This index quantifies the tendency of a molecule to accept an electron from a generic donor. Therefore, the higher is the electrophilicity index, the greater is the propensity of the complex to attract electrons from a generic donor molecule.

Indeed, all these three indexes are related to the properties of an isolated reactant, that is they do not take into account the "selectivity" for other species. To this end, a new intermolecular index can be introduced, the donor–acceptor intermolecular hardness (η_{DA}) defined as:

$$\eta_{DA} = (I_D - A_A), \qquad (12.6)$$

where A_A is the electron affinity of the acceptor A (for instance the metal–N_4 complexes) and I_D the vertical ionisation energy of the donor molecule D (for instance the thiol anion, namely 2-mercaptoethanol)[22,23]. This index compactly conveys the classical principle of electron transfer reaction: "transfer of electrons from D to A is faster the closer in energy the highest occupied molecular orbital (HOMO) of D is to the lowest unoccupied molecular orbital (LUMO) of A"[24]. It is important to mention that η_{DA} represents the hardness of the initial system (A+D) by considering the isolated acceptor and donor species, where the charge transfer has still not occurred and therefore gives a general idea on their reactivity. Nevertheless, the intramolecular hardness as reactivity index has been successfully used for the study of intermolecular charge transfer reactions of electrochemical interest (see for instance ref.[25]).

The accurate calculation of chemical potential, hardness or electrophilicity requires (only) an accurate evaluation of the ionisation energy and electron affinity of the systems. As for other molecular properties, DFT is an invaluable tool, if an adequate exchange-correlation functional is used. In fact, within the Kohn–Sham (KS) approach to DFT the total (exact) energy of a system reads:[26]

$$E_{\mathrm{KS}}[\rho, \nu] = T_{\mathrm{s}}[\{\varphi_{\mathrm{i}}\}] + J[\rho] + \int \rho(r)\upsilon(r)\mathrm{d}r + E_{\mathrm{xc}}[\rho], \qquad (12.7)$$

where $\{\varphi_i\}$ are the spin orbitals. The first term in Equation (12.7) is the kinetic energy of a system of non-interacting particle, the second is the Coulomb interaction and the third is the interaction energy between the electron density $\rho(\mathbf{r})$ and the external potential $\nu(\mathbf{r})$. All terms are known exactly, but the last one E_{xc}, containing all the remaining contribution to both the kinetic energy and the electron–electron interaction, is approximated by some functional form, the so-called exchange-correlation functionals.

A number of functionals based on the density and its gradient (the generalized gradient approximation, GGA) have been developed in the last few years[19,27,28] either by fitting parameters to some set of experimental data or by imposing the fulfilment of physical constraints[28]. In this context, the Perdew–Burke–Erzerhof (PBE) exchange-correlation functional[29] is a peculiar approach since it does not contain any parameter fitted to experimental data. The PBE generally provides results that are at least comparable to those obtained with more empirical functionals[30,31]. In order to improve GGA results and to partially correct some of their intrinsic failures (such as self-interaction error) a new class of functionals (called hybrid functionals) was developed in the late 1980s self-consistently adding a predefined amount of Hartree–Fock (HF) exchange to the DFT contribution[25–27]. Hybrid functionals lead to a great improvement of structural and electronic properties and cure many of the most striking failures of GGA[27,32,33]. A number of them were developed, the use of some being particularly widespread, such as for B3LYP[34]. In particular, casting the PBE functional in hybrid model, leads to an extremely accurate functional[35–38]. A number of tests have shown that by using this hybrid model (hereafter referred to as PBE0), the calculated ionisation potentials, electronic affinities and a number of other electronic properties are remarkably close to experimental values[37,39,40]. For these reasons, the PBE0 functional will be mainly used for the calculations reported in this review.

If the use of the DFT machinery to compute reactivity indexes is not new, very few studies have been performed to understand the role of the environment (i.e. solvent or surface adsorption) on the reactivity descriptors[16,17,41]. Indeed, the environment plays an important role in many of the reactions. Here, the solvent will be introduced via simplified, yet reliable, method based on the continuum model of solvent (Polarizable Continuum Model, PCM[42]) neglecting short-range (specific) solute–solvent interactions. The first continuum models were elaborated at the beginning of last century[43–47] and their success relies in their simplicity and in their straightforward link to physical concepts. Nevertheless current continuum models have been developed much from the first simplified schemes. An extensive review of continuum models up to the most recent developments and applications can be found in several studies[48–53]. The basic idea of all continuum models is the partitioning of the solution into two subunits: the solute (M) and the solvent (S). The solute (M) is described fully at a microscopic (here DFT) level. The solvent, on the other hand, is incorporated as a continuum medium (a continuous electric field, the so-called reaction field), representing a statistical average over all solvent degrees of freedom at thermal equilibrium. The main solute–solvent interaction is electrostatic although recent developments of continuum models included have also non-electrostatic terms[54].

Finally, in order to fully analyse the effect of the environment the adsorption of the acceptors on the electrode will be also modelled. Nowadays, there are several possibilities for a theoretical treatment of adsorption in large systems, for instance the super-molecule technique, simulating total complexes (substrate + adsorbed species) by appropriate finite clusters, and the embedded-molecule technique, through the use of infinite periodic lattices. A different approach may be found using composite methodologies with different theoretical levels of description for the different parts of the total system[55,56] but the question of taking into account electron correlation for an (in principle) infinite number of electrons raises technical problems (nowadays in course of solution[57–59].) Here, the super-molecule approach has been preferred and the surface of the electrode has been modelled by a monolayer of graphite where the border carbon atoms have been saturated by hydrogens. A similar model has been already used, for instance, in literature to investigate the adsorption of hydrogen molecule on pregraphitic surfaces[60].

Here we review our work aimed at correlating the reactivity of a series of M(II) N_4-ligands, see Figure 12.1, (M = Co, Fe, Mn and N_4 = porphyrin (P), phthalocyanine (Pc), teraphenylporphyrin (TPP), tetrabenzoporphyrin (TBP) and tetraazaporphyrin (TAP)) towards the electrocatalytic oxidation of 2-mercaptoethanol. Different effects will be analysed, namely the role of the metal atoms, the role of the N_4 functionalisation, solvent and the impact of the adsorption on the electrode on the electrochemical activity. The whole machinery of DFT and the notions of hardness, chemical potential, intramolecular hardness and electrophilicity are used to better quantify these effects and discriminate between the examined molecular complexes.

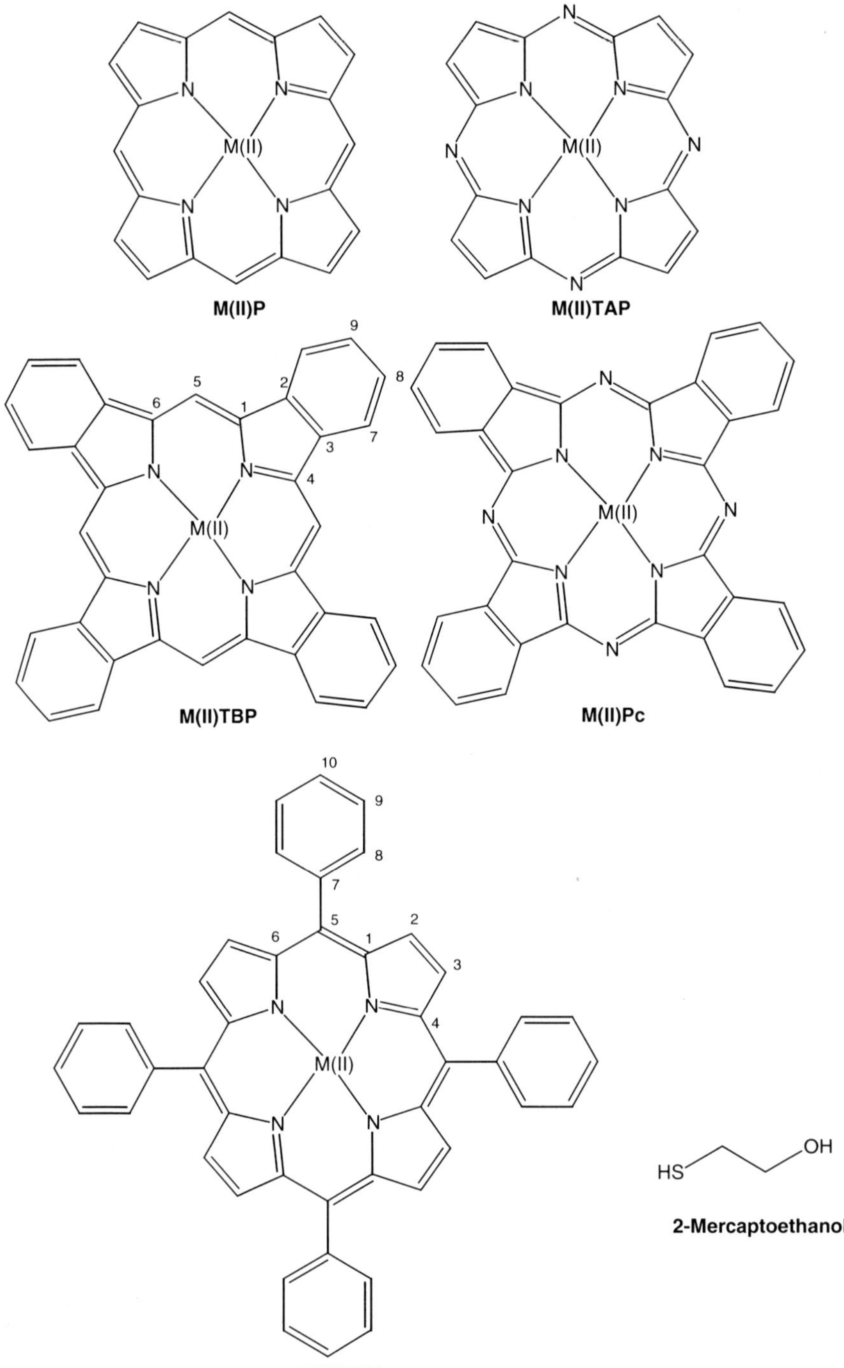

Figure 12.1. Sketches and labelling of the systems considered: M(II) porphyrin (M(II)P), M(II) tetraazaporphyrin (M(II)TAP), M(II) tetrabenzoporphyrin (M(II)TBP), M(II) phthalocyanine (M(II)Pc), M(II) tetraphenylporphyrin (M(II)TPP) and 2-mercaptoethanol.

2. Computational Details

All calculations were carried out with the Gaussian code[61], using a recent hybrid Kohn–Sham/Hartree–Fock (KS/HF) model hereafter referred to as PBE0[38]. This approach is obtained casting the PBE exchange and correlation functional[29] in a hybrid scheme HF/DFT, where the HF exchange ratio (1:4) is fixed *a priori*[62].

Two different basis sets have been used for geometry optimization. The small set is composed of the CEP-121 pseudo-potentials and basis set (contraction [8s8p6d/4s4p3d]) for the metal atom, and by the 6-31G(d) basis for C, N, O, S and H[63]. It has been shown that such a basis set provides accurate geometrical parameters for metal porphyrins[64]. To test basis set effect, a larger basis set has been also used, where the lighter atoms are described via a 6-311G(d,p) basis sets and the metal basis set is enhanced adding one p polarization function (exp $= 0.08$) to the CEP-121 basis[65]. Only for the study of the effect on reactivity of the metal atom, that is for the calculation of Co(II), Fe(II) and Mn(II) P and Pc complexes, a smaller basis set was used for all atoms, namely the Los Alamos double zeta valence (LANL2DZ)[66] and corresponding pseudo-potentials[67]. Furthermore, only in this case the hybrid Becke-three parameter exchange correlation functional (B3LYP) was applied[34].

Each stationary point found was characterized as minimum or first-order saddle point by computing the harmonic vibrational frequencies.

Solvent effects were evaluated using the Polarizable Continuum Model (PCM)[48]. In particular, optimised structures and solvation energies were computed by a cavity model, namely the United Atoms Topological Model (UATM)[68], coupled to the conductor-like Polarizable Continuum Model (CPCM)[69]. This approach provides results very close to those obtained by the original dielectric model for high dielectric constant solvents, but it is significantly more effective in geometry optimizations, and less prone to numerical errors arising from the small part of the solute electron cloud lying outside the cavity (escaped charge effects)[69].

Two global reactivity descriptors are discussed: the inter-molecular hardness and the electrophilicity index. Although originally developed for closed-shell systems, the extension of these indexes to the study of reactions involving open-shell systems (like those considered here) is straightforward and already reported in literature (see for instance ref.[10b,12, 13]). Indeed, the Koopmans theorem cannot be used unambiguously to determine η and ω in terms of the energies of the HOMO and LUMO for unrestricted wave functions. In order to avoid artefacts for all calculations (except the ones concerning the effect of metal) we have calculated the ionisation potentials and the electron affinities as the differences between the energies of the reference molecule (N electron system) and the corresponding $N + 1$ or $N - 1$ electron systems, at the geometry of the reference species. All these latter calculations, i.e. the evaluations of the energies of the N, $N + 1$ and $N - 1$ electron systems, have been carried out using an even larger basis, i.e. the 6-311 + G(d,p) basis set for light atoms and the polarized CEP-121G basis for Co.

Finally, in order to give a chemical sounding description of the electronic structure of these molecules the natural bond orbital (NBO) approach and the related natural population analysis (NPA)[70,71] have been computed and discussed. The NPA approach is particularly effective in the case of inorganic complexes, since it gives a description of the electronic distribution less sensitive to the computational parameters (e.g. basis set) with respect to other more commonly used population analysis such as the Mulliken one [70].

In the case of open-shell systems, unrestricted calculations were performed and spin contamination was monitored by the expectation values of S^2.

3. The Chemical Effects Tuning the Reactivity of M(II)–N_4 Complexes

One of the major advantages of theoretical approaches is, as mentioned before, the possibility of decomposing a complex chemical event in its basic constituents. The net result is the weighting of each elementary brick with respect to the global, experimentally observed reaction. Therefore, following this "ideal" approach, we have identified the main components of our phenomenon: the central metal atom, the macrocycle, the solvent and the graphite surface. The effect of each of these units on the observed reactivity is discussed in the following.

In all the cases considered, 2-mercaptoethanol ($HSCH_2CH_2OH$) is used as electron donor. Moreover, since the experiments are usually carried out in a basic solution[6,7], the de-protonated anion, namely $[HOCH_2CH_2S]^-$, is our reference species. We do not discuss in detail the results concerning this molecule, since we are interested in differences in reactivity due to changes in the electron acceptor, i.e. the N_4-macrocycles. Here, we want only to recall that the HOMO of the anion is a p orbital localised on the sulphur atom, with the correct symmetry for an interaction with a d_π orbital of the central metal atom of the acceptor. This orbital is strongly stabilized by interaction with a surrounding medium (solvent), as expected in the case of a small, charged species.

3.1. Influence of the Metal Atom

Preliminary studies were carried out in order to justify, using the reactivity index machinery, the higher reactivity of Co(II) derivatives with respect to other M(II) transition metal complexes, in particular when M = Mn(II) or Fe(II). Several *ab intio* studies of the ground state properties of M–N_4 complexes can be found in literature, especially concerning the relative stability of the different spin states (for instance in the case of Fe(II) derivatives). Here we consider only the most stable spin state for each metal complex and analyse the effect of the metal on the reactivity indexes (i.e. hardness, softness and electrophilicity). As already mentioned, and contrary to all other calculations reported in this review, these computations were performed using the parametrized hybrid Becke three-parameter exchange correlation functional (B3LYP[34]) and a smaller basis set. The same level of theory was used to compute the donor molecule, i.e. the anionic form of 2-mercaptoethanol.

From a qualitative point of view the ground state molecular orbital splitting of M(II)–N_4 can be deduced from very simple, symmetry-based, orbital interaction diagrams[72]. Since these complexes are the typical prototypes for a square planar coordination of a transition metal, the d manifold of the metal is split into a set of three practically non-bonding orbitals (d_{x2-y2}, d_{xz}, d_{yz}), a slightly anti-bonding orbital (d_{z2}) and a strongly anti-bonding orbital (d_{xy}, directly pointing vs. the nitrogen atoms of the N_4) thus much higher in energy. As a consequence of such an orbital splitting, the ground electronic state in the case of d^7 systems (such as Co(II)) will be a doublet. More difficult is the definition of the spin state for Mn(II) (d^5) and Fe(II) (d^6) systems. Indeed, in the case of Pc and P derivatives we found the intermediate spin state (S = 1) for Fe and the high spin state (S = 5/2) for Mn to be the most stable, in agreement with the available experimental magnetic susceptibility data and previous DFT calculations[73–78].

Before discussing in detail the results, some remarks on the spin-contamination problems when dealing with open-shell systems, should be recalled. Indeed, it is well known that the interpretation of the expectation value for S^2, $<S^2>$, is not straightforward in the framework of DFT methods. Nevertheless this value still indicates, at least roughly, the extent of contamination by higher spin states. In our case, the computed B3LYP values are, in the case of Co(II) complexes, always $<S^2> < 0.77$, thus suggesting that we are dealing with essentially pure doublet states ($<S^2> = 0.75$) while for Mn(II) systems the $<S^2>$ value never exceeds 8.78 (pure sextuplet state: $<S^2> = 8.75$) and for Fe(II) systems the $<S^2>$ value is always around 2.04 (pure triplet: $<S^2> = 2.00$).

In Table 12.1 the computed structural parameters for two series of M(II)–N_4 complexes, namely Pc or P, are reported following the labelling scheme depicted in Figure 12.1. Although slight variations in the metal-to-nitrogen distances are computed, mainly due to the different size of the metal ions, no significant difference in the overall structure can be noticed. Therefore the difference in reactivity between the different metals cannot be derived on the basis of simple geometrical considerations (such as greater exposure of the metal to the donor).

Table 12.1. Structural Parameters (in Å and degrees) for Selected Metal-Phthalocyanines (M(II)Pc) and 1 Metal-Porphyrines (M(II)P) as a Function of the Central Metal Ion Computed at B3LYP level Using a LANL2DZ Basis

	M(II)Pc			M(II)P		
	Co(II)	Fe(II)	Mn(II)	Co(II)	Fe(II)	Mn(II)
M–N	1.952	1.961	1.962	1.998	2.011	2.042
N–C1	1.397	1.399	1.417	1.396	1.396	1.391
C1–C2	1.461	1.462	1.446	1.452	1.455	1.459
C2–C3	1.414	1.415	1.420	1.373	1.372	1.381
C1–X5	1.333	1.333	1.342	1.392	1.392	1.410
M–N–C1	126.4	126.2	126.3	127.4	127.4	126.1
N–C1–X5	127.2	127.1	127.4	125.5	125.4	125.2

In order to define the relative reactivity of the M(II)Pc and M(II)P towards the 2-mercaptoethanol we considered two different reactivity indexes: the electrophilicity (ω) and the intermolecular hardness (η_{AD}). The first one is an "intramolecular" parameter, depending only on the electronic characteristic of the acceptor species, while the second one, defined in terms of both acceptor and donor properties, is an "intermolecular" parameter. As a consequence, the combined use of these two indexes should give a complete picture of the overall oxidation process. In Figure 12.2, the relative electrophilicity ($\Delta\omega$) and intermolecular hardness ($\Delta\eta_{DA}$) with respect to those of the analogous Co–N_4 systems are reported.

From the analysis of the electrophilicity index only a different behaviour is predicted for Pc and P derivatives. In fact, while in the case of porphyrine complexes, the Mn(II) one is computed to have the largest electrophilicity and, thus, the highest capability to reduce 2-mercaptoethanol, the contrary holds for phthalocyanine derivatives. Furthermore, if for porphyrine systems a quite large difference in reactivity is predicted between Mn ($\Delta\omega = 0.5\,\text{eV}$), for phthalocyanine complexes smaller differences are computed, the Co(II) complex being always predicted to be the least reactive. These results are clearly contrary to the experimental findings and, moreover, do not allow to derive common trends between Pc and P derivatives. The only general feature appearing from electrophilicity index is that Co and Fe derivatives show a very similar reactivity. Nevertheless, it should be kept in mind that the electrophilicity is an intramolecular index and that it tells nothing about the possible selectivity for a given donor. A (intrinsic) measure of the selectivity of an acceptor–donor couple is best represented, in terms of reactivity indexes, by the intermolecular donor–acceptor hardness since its values practically quantify the very intuitive concept that the interaction between a donor and an acceptor will be maximised by the couple that minimises the HOMO(donor)–LUMO(acceptor) energy difference.

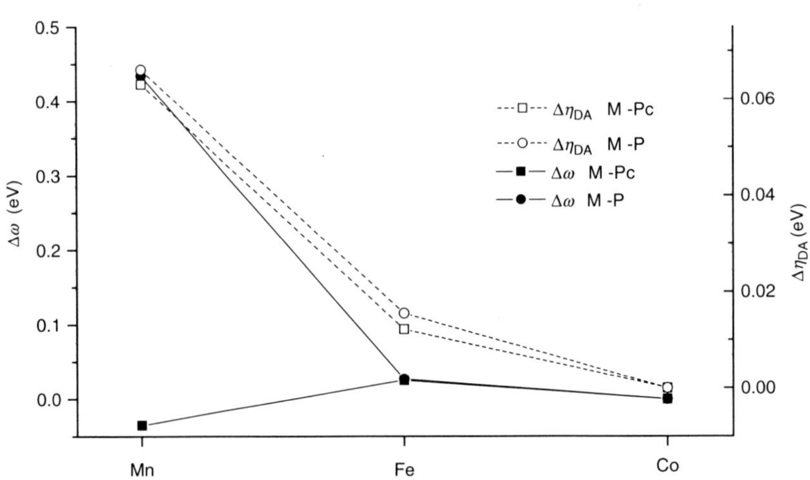

Figure 12.2. Computed variations of electrophilitcity index ($\Delta\omega$, eV) and intramolecular hardness ($\Delta\eta_{DA}$, eV) for selected M(II)P and M(II)Pc (M = Fe, Mn and Co) with respect to the corresponding Co(II) complex.

When analysing the variation of the computed η_{AD} a completely different and clear trend appears: Mn(II) complexes possess the highest hardness, the corresponding Fe(II) and Co(II) complexes showing very similar reactivity. Indeed, in agreement with experiments, Co(II) systems are predicted to be the most reactive[6]. It is worthy to underline that these results have to be considered as qualitative since the possibility of axial coordination of a solvent molecule cannot be neglected. In particular, especially for Fe(II) and Mn(II) systems, axial coordination alters the ground spin state, and, in principle should not be neglected in the study of their reactivity. Nevertheless we do not expect that the relative reactivity order will be significantly altered. Furthermore, in the case of Co systems (the one that we analyse in the following) the eventual axial coordination cannot change the ground spin state of the complex and the effect on reactivity can be considered as negligible (see *infra*).

3.2. Influence of the N_4 Functionalisation

A series of different Co(II)–N_4 macrocycles, acting as electron acceptor and characterised by different substitution on position 5 (methyne, nitrogen or phenyl), were considered. The optimised geometrical parameters of all acceptor molecules (refer to Figure 12.1) are reported in Table 12.2. All the complexes, but TPP, have a planar arrangement of the porphyrin ring and an overall D_{4h} symmetry, in agreement with experimental data[79–81]. In the case of TPP, the cycle is significantly distorted from planarity, with the typical ruffling of the skeleton ring[82] (see Figure 12.3), due to the strong steric interactions between the phenyl rings[83]. The optimised gas-phase structural parameters are in excellent agreement with the experimental data (error on bonds $\leq$ 0.01 Å, error on valence angle $\leq 0.9°$). Furthermore, our calculations well reproduce the experimental trends on the geometries observed along the series. For instance we can notice that the cobalt–pyrrole distance increases when replacing the nitrogen atoms bridged to pyrrole by a methine group. The enlargement of the central cavity upon substitution is directly related to the greater steric effect of the lone pairs of the nitrogen atom (in CoTAP and CoPc) with respect to the methyne hydrogen atom (in CoP and CoTBP)[81,84]. The only peculiar case is represented by CoTPP, due its twisted structure.

From a more technical point of view, no basis set dependence of the geometrical parameter is noticed, the maximum variation in bond lengths being 0.004 Å and $+0.4°$ on valence angles when going from the small to the large basis set. In Table 12.3 are collected the Natural Population Analysis (NPA) charges for the whole series of Co(II)–N_4 complexes. The covalent character of the metal–ligand bonding is highlighted by the relatively small charges computed for Co(II) (around 1.0 $|e^-|$) quite far from the formal charge of $+2$ associated to neutral N_4-complex. At the same time there is a striking difference between the macrocycles possessing a porphyrin-like ring (such as CoP, CoTBP and CoTPP) and the "tetra-azo porphyrins" ones (CoTAP and CoPc). In fact, the substitution of the methylene carbon atoms in the pyrrole-bridging position with the more electronegative nitrogen atoms significantly alters the electronic structure. The largest effects are computed for the atoms in position 5 (C_5 or N_5) which have

Table 12.2. Main Geometrical Parameters (Å and degrees) of the Co(II) Tetraphenylporphyrin (CoTPP), Co(II) Tetrabenzoporphyrin (CoTBP) and of Co(II) Phthalocyanin (CoPc)

	CoP			CoTAP		CoTPP			CoTBP			CoPc		
	Small basis	Large basis	Exp.[a]	Small basis	Large basis	Small basis	Large basis	Exp.[b]	Small basis	Large basis	Exp.[c]	Small basis	Large basis	Exp.[d]
CoN	1.979	1.978	1.967(3)	1.901	1.902	1.965	1.964	1.949(1)	2.004	2.003	1.984(5)	1.923	1.923	1.919(1)
NC1	1.371	1.369	1.371(3)	1.368	1.366	1.374	1.372	1.380(2)	1.370	1.370	1.376(7)	1.371	1.369	1.380(3)
C1C2	1.436	1.435	1.435(4)	1.449	1.447	1.442	1.440	1.433(2)	1.448	1.445	1.447(8)	1.450	1.449	1.457(7)
C2C3	1.356	1.354	1.357(4)	1.353	1.350	1.352	1.350	1.354(2)	1.401	1.398	1.401(8)	1.398	1.396	1.398(1)
C1X5	1.381	1.379	1.383(3)	1.321	1.319	1.391	1.388	1.386(2)	1.373	1.375	1.383(8)	1.317	1.314	1.325(1)
C3C7	—	—	—	—	—	—	—	—	1.399	1.396	1.393(8)	1.392	1.390	1.397(1)
C5C7	—	—	—	—	—	1.490	1.490	1.488(2)	—	—	—	—	—	—
C7C8	—	—	—	—	—	1.400	1.396	1.395(2)	1.390	1.384	1.372(9)	1.390	1.387	1.396(1)
C8C9	—	—	—	—	—	1.392	1.390	1.387(2)	1.401	1.404	1.389(9)	1.406	1.403	1.410(1)
C9C10	—	—	—	—	—	1.395	1.392	1.388(2)	—	—	—	—	—	—
CoNC1	127.5	127.5	127.47(17)	126.9	126.8	127.4	127.3	127.7(1)	126.6	126.8	127.5(4)	126.3	126.2	126.50(78)
NC1X5	125.3	125.4	125.18(20)	127.5	127.5	126.2	126.2	126.5(1)	126.0	125.8	124.7(6)	127.8	127.7	127.93(24)
NC1C2	111.0	110.9	110.64(22)	110.1	110.0	110.6	110.6	110.4(1)	110.5	110.7	111.7(5)	110.0	109.9	110.14(21)
C1C2C3	106.5	106.6	106.75(25)	106.8	106.8	106.8	106.8	107.0(1)	106.0	106.0	105.8(5)	106.3	106.3	106.38(8)
C1X5C6	124.3	124.3	124.73(22)	121.3	121.3	122.8	123.0	123.4(1)	124.8	125.0	125.6(6)	122.1	122.1	107.00(9)
C4C3C7	—	—	—	—	—	—	—	—	132.8	133.1	133.9(6)	132.3	132.3	132.06(57)
C3C7C8	—	—	—	—	—	—	—	—	117.7	118.1	118.1(6)	117.4	117.4	117.06(17)
C7C8C9	—	—	—	—	—	120.6	120.5	120.6(1)	121.4	121.1	121.7(6)	121.3	121.3	121.36(27)
C8C9C10	—	—	—	—	—	120.1	120.1	120.0(1)	—	—	—	—	—	—
C9C10C9	—	—	—	—	—	119.8	119.8	119.9(1)	—	—	—	—	—	—

All Values Have Been Computed at the PBE0 Level Using Two Different Basis Sets: (a) CEP-121G on Co and 6-31G(d) on C,N and H (small basis), (b) CEP-121G on Co and 6-311G(d,p) on C,N and H (large basis)

[a]ref.[80a].

[b]ref.[82].

[c]ref.[81].

[d]ref.[80b].

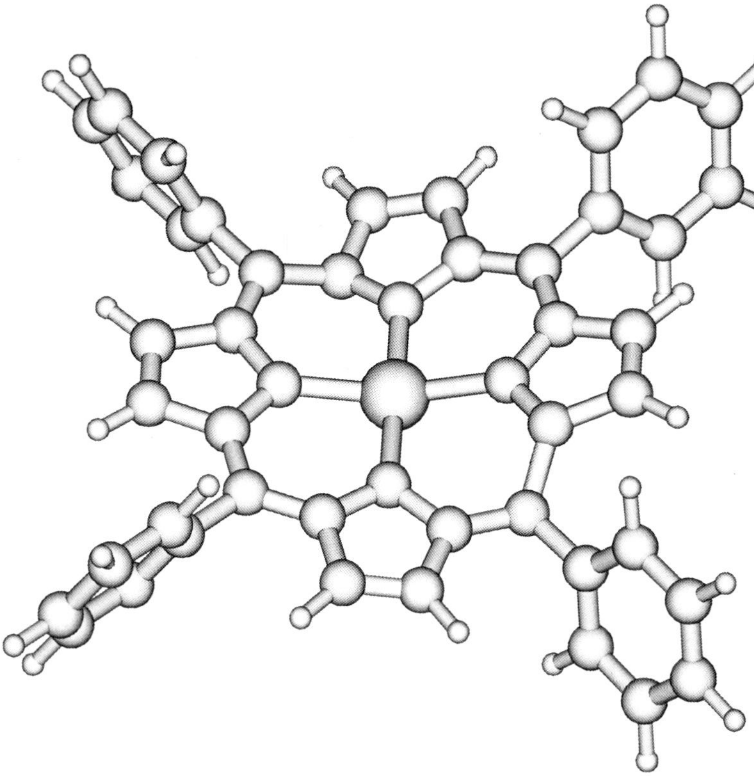

Figure 12.3. Optimized structure of the Co(II) tetraphenylporphyrin (CoTPP).

a charge of about −0.25 in the first case and −0.46 in the second. Indeed, also the C1 atoms are significantly affected (+0.40 vs. +0.16/0.19$|e^-|$) while only small variations are found on the N atoms bonded to the metal or even on the Co atom (about +1.20 vs. +1.08$|e^-|$). As concerns the metal atom, the small differences found within the series area a consequence of a slightly different occupation of the d orbitals. In particular, while the d_{z2} has a small excess of electron density ($d_{z2}^{1.02}$) in all the complexes, differences are found for the occupation of the in-plane $d_{x_2-y_2}$ orbital. This latter has a smaller occupation in the methyne-substituted molecule (CoP, CoTBP) than in the nitrogen-substituted complexes (CoTAP, CoPc). Interestinly CoTPP is closer to the nitrogen analogous, its configuration being $d_{z_2}^{1.03}d_{x_2-y_2}^{0.74}$.

Neutron diffractions studies allow for an experimental estimation of the charge density on the central Co and, in particular, of the d-orbital electron population[79,82,85]. Anyway, a quantitative comparison between our NBO analysis and the experimental results (obtained through a least-square fitting of the density) is not strictly rigorous, since the procedure of partitioning the total electron density is different[79]. Nevertheless, the agreement of the PBE0 calculations with the experiments is qualitatively satisfactory. For instance, the X-ray analysis of the CoTPP gives a density corresponding to $d_{z_2}^{0.92}d_{x_2-y_2}^{0.83}$, close to our values[83].

Table 12.3. Natural Population Analysis for Selected Co(II)–N_4 Complexes (N_4 = P, TAP, TPP, TBP, Pc) Computed in Gas Phase, Aqueous Solution and Adsorbed on Surface at the PBE0 Level, Using the CEP-121/6-31G(d) Basis Set and the Corresponding Optimised Geometries

	Gas-phase					Solution					Adsorbed	
	CoP	CoTAP	CoTBP	CoPc	CoTPP	CoP	CoTAP	CoTBP	CoPc	CoTPP	CoPc[a]	CoPc[b]
Co	1.116	1.076	1.119	1.097	1.093	1.112	1.096	1.132	1.114	1.109	1.080	1.012
N	−0.596	−0.641	−0.576	−0.629	−0.603	−0.609	−0.634	−0.582	−0.634	−0.613	−0.620	−0.638
C1	0.160	0.402	0.189	0.441	0.185	0.149	0.400	0.181	0.438	0.182	0.437	0.442
C2	−0.264	−0.252	−0.068	−0.089	−0.253	−0.260	−0.260	−0.077	−0.096	−0.259	−0.084	−0.084
X5	−0.254	−0.458	−0.264	−0.487	−0.059	−0.250	−0.481	−0.270	−0.507	−0.063	−0.484	−0.486
C7	—	—	−0.215	−0.164	−0.061	—	—	−0.226	−0.171	−0.062		
C8	—	—	−0.242	−0.197	−0.229	—	—	−0.249	−0.202	−0.233		
C9	—	—	−0.242	−0.197	−0.241	—	—	−0.249	−0.202	−0.244		
C10	—	—	—	—	−0.244	—	—	—	—	—0.244		

[a]Small cluster model.
[b]Large cluster model.

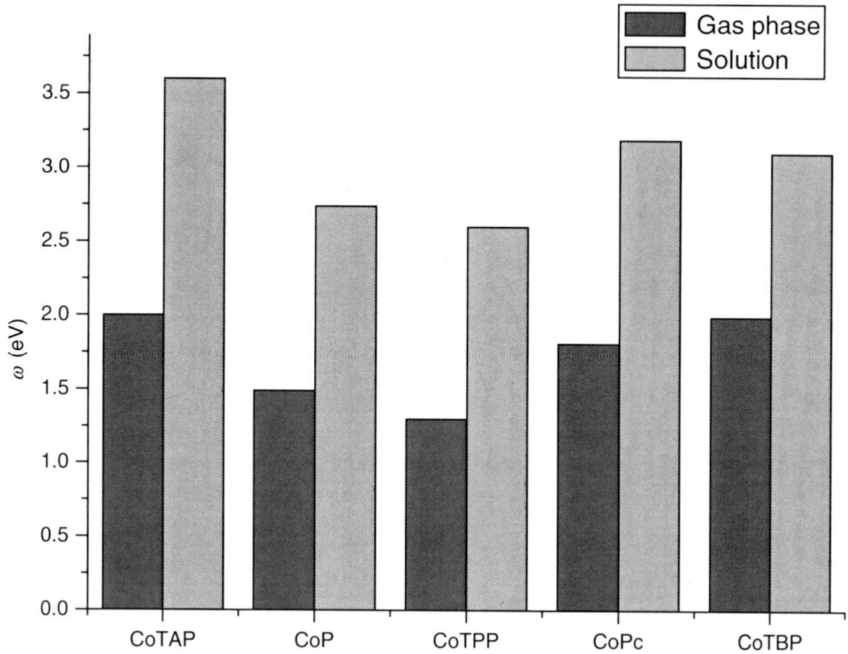

Figure 12.4. Plots of the electrophilicity power (ω, eV) in the gas phase and in aqueous solution for selected Co(II)–N_4 complexes (N_4 = P, TAP, TBP, Pc and TPP).

Slightly different values are found from neutron experiments, $d_{z^2}^{1.21}d_{x^2-y^2}^{0.21}$. The apparent discrepancy observed for the $d_{x^2-y^2}$ orbital derives from the low sensitivity of neutron experiments to the population of this orbital where electron spins are mostly paired[84].

As previously done to analyse the effect of the metal, two global reactivity indexes will be discussed: ω and η_{DA}. The values computed for the electrophilicity index are reported in Figure 12.4. All the Co(II)–N_4 complexes show comparable ω values, ranging from 2.0 (CoTAP) to 1.3 eV (CoTPP). In particular, CoTAP has a ω value significantly higher than the corresponding porphyrins (CoP and CoTPP), due to the electron-withdrawing effect of the nitrogen atoms in position 5. In contrast, CoPc and CoTBP have close ω values, thus underlying the major role played by the benzene rings in storing electrons. The calculated electrophilicity indexes suggest that aza-porphyrins have a greater capacity to accept electrons from a generic electron donor and, consequently, they would be more active in the oxidation reaction of 2-mercaptoethanol. From a very qualitative picture we can figure that the transfer of electrons from a donor D to an acceptor A is facilitated when the HOMO of D is close in energy to the LUMO of A. In our case, D is the anion of the 2-mercaptoethanol and the acceptors are the complexes of Co(II). The orbitals involved in this interaction are a p orbital localized on sulphur atom for D and a π^* for all the macrocycles (see Figure 12.5). This latter has a significant contribution from the d_π orbital of the Co and possesses the appropriate symmetry to interact with the HOMO of the donor.

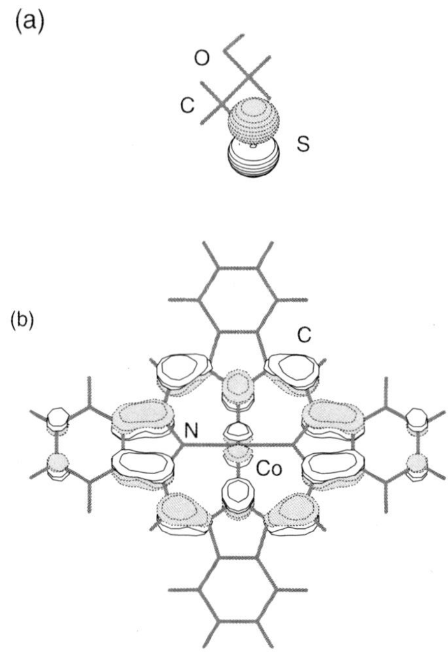

Figure 12.5. Plot of the frontier molecular orbitals involved in the oxidation mechanism: (a) HOMO of the mercaptoethanol anion; (b) LUMO of the Co(II) phthalocyanine.

Table 12.4. Computed Donor–Acceptor Intermolecular Hardness (η_{AD}, eV) for selected Co(II)–N_4 Complexes

	Gas phase	Solution[a]	Solution[b]
CoTAP	3.26	2.36	2.34
CoP	2.36	3.20	3.18
CoTPP	2.34	3.22	3.28
CoPc	2.98	2.52	2.50
CoTBP	2.34	3.22	3.26

[a]Computed using gas-phase geometry
[b]Computed using the structure optimised in solution.

The intermolecular donor–acceptor hardness′ (η_{DA}, as defined by Equation (12.6)), provides a more rigorous description of these nucleophile/electrophile interactions. The η_{DA} values for the whole series of N_4 macrocycles are collected in Table 12.4 and plotted in Figure 12.6.

We recall that a low value for the hardness implies greater interactions between donor and acceptor species, corresponding to a small energy gap between the donor HOMO orbital and the LUMO of acceptor. Some trends appear from the

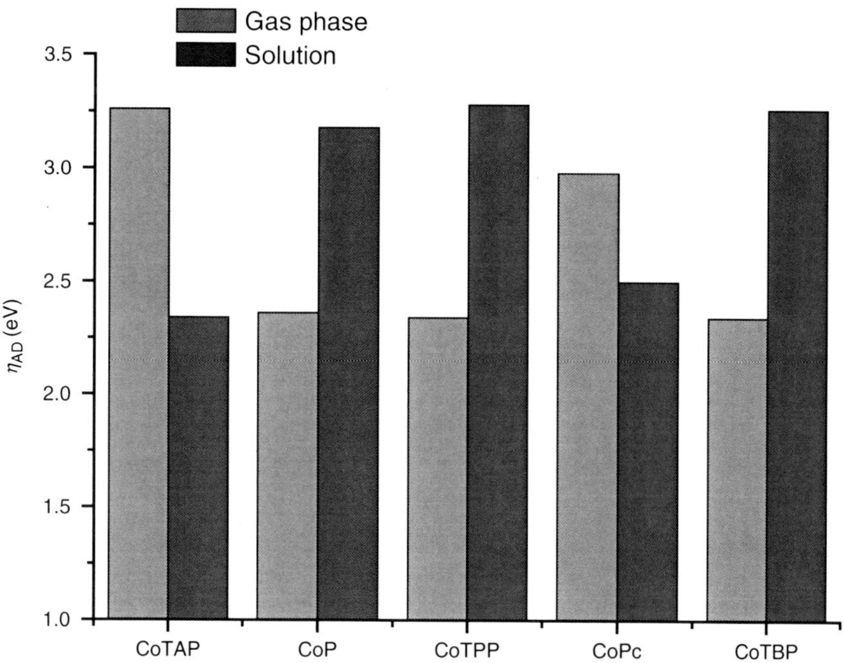

Figure 12.6. Plots of the intermolecular donor–acceptor hardness (η_{DA}, eV) in the gas phase and in aqueous solution for selected Co(II)–N_4 complexes (N_4 = P, TAP, TBP, Pc and TPP).

analysis of our gas-phase values (first column of Table 12.5): the porphyrin-like macrocycles (CoP, CoTBP, and CoTPP) are predicted as the most reactive species, all showing similar η_{DA} (about 2.4 eV). Higher hardness′ are instead found for the aza-compounds, CoPc and CoTAP, (3.0 and 3.2 eV, respectively). This result can be rationalized in terms of the relative energies of the donor and acceptor orbitals. In fact, the electron-withdrawing substituent (nitrogen atom in aza-porphyrins) in position 5 of the macrocycles significantly stabilizes the SOMO, whereas the contrary is true for electron-donating substituent (methylene group). Since the HOMO of the thiol is quite high in energy, CoTAP and CoPc show higher hardness. It is also interesting to note that the main contributions to the LUMO come from the atoms belonging to the inner porphyrin ring, so that substituents on pyrrole have practically no effect. As a consequence, CoP and COTBP have very similar hardness and the same holds for CoTPP, since there is no contribution from the orbitals of the phenyl groups, almost orthogonal to the N_4 plane.

3.3. Influence of the Environment: Solvent Effects

As mentioned above, we have considered the effect of bulk solvent (here water) on the electronic properties of Co(II) complex by the CPCM model[68]. While this approach well reproduces non-specific solute–solvent interactions,

Table 12.5. Computed Structural Parameters (in Å and degrees) for Selected Co(II)–N_4 Complexes

	CoP	CoTAP	CoTBP	CoPc	CoTPP
CoN	1.983	1.899	2.005	1.921	1.977
NC1	1.370	1.362	1.368	1.365	1.373
C1C2	1.438	1.454	1.446	1.453	1.440
C2C3	1.361	1.355	1.400	1.402	1.352
C1X5	1.382	1.316	1.378	1.311	1.389
C3C7	—	—	1.399	1.391	—
X5C7	—	—	—	—	1.492
C7C8	—	—	1.389	1.387	1.401
C8C9	—	—	1.407	1.402	1.392
C9C10	—	—	—	—	1.395
CoNC1	127.3	127.1	126.4	126.7	127.5
NC1X5	126.0	126.4	125.9	126.6	126.2
NC1C2	110.7	110.8	110.3	109.6	110.8
C1C2C3	106.6	106.3	106.1	105.9	106.7
C1X5C6	123.5	123.0	125.3	123.0	122.6
C4C3C7	—	—	132.9	132.0	—
C3C7C8	—	—	117.9	117.5	—
C7C8C9	—	—	121.1	121.0	120.6
C8C9C10	—	—	—	—	120.1
C9C10C9	—	—	—	—	119.8

N_4 = P, TAP, TPP, TBP and Pc in Aqueous Solution Using the CEP-121G/6-31G(d) Basis Set Solution.

specific (e.g. H-bond) effects are only partially mimicked[86,87]. In our case, solvent (water) molecules strongly coordinated to the cobalt atom might significantly affect the electronic properties of the solute. Nevertheless, we thought that no significant changes in the general trends can be expected from these interactions, due to structural similarities along the molecular series. Furthermore, it should be noticed that even if the electrophile/nucleophile interactions, like that reported in the present study, might be assumed to be preceded by a desolvation step, it is expected that this desolvation process may not be complete and that partial solvation may affect these interactions to some extent[88]. In conclusion, the PCM model should give an idea about the effects of bulk solvent on the electronic properties of such macrocycles and similar solvent models have been recently applied to rationalise the oxidation potential of substituted anilines[89].

In Table 12.5 are reported the structural parameters computed at PBE0 level in aqueous solution, using the small basis set. From these data it is clear that solvent has a marginal effect on the geometries, the largest differences with respect to the gas-phase parameters being ±0.05 Å. A larger effect can be expected on the electronic distributions. In particular, there is an increase of the charge separation within the molecule, with an increase of the positive charge brought by the metal atom and of the negative charge over the ring. This effect, which corresponds to a stabilization of the LUMO, is larger in the more polarizable tetra-azo complexes than in the methyne series.

If we analyse the effect of solvent on reactivity indexes, we can notice that the solvent strongly rules the magnitude of the calculated electrophilicities, but does not significantly alter the overall trend. In fact, in going from the gas phase to aqueous solution, CoTAP still has the highest ω value (3.6 eV) and CoTPP the lowest one (2.6 eV). The only remarkable difference with respect to the gas phase findings is, in aqueous solution, CoPc is computed to be slightly more electrophilic than CoTBP. The variations observed in solution are related to a strong change of the chemical hardness, significantly decreasing from gas phase to solution, while the chemical potential is practically constant[41]. In conclusion the calculated electrophilicity indexes in gas phase suggest that aza-porphyrins have a greater capacity to accept electrons from a generic electron donor and this trend is significantly enhanced, but not changed, in the presence of a polar solvent.

On the other hand, the solvent has a drastic effect on the intermolecular hardness (Table 12.4). In particular, the η_{DA} for CoP and CoTBP changes from 2.4 to 3.2 eV, in going from gas phase to aqueous solution. Similar variations, but in the opposite directions, are also found for the aza-compounds. It is also noteworthy that CoTAP is predicted to be slightly more reactive than CoPc (2.4 vs. 2.6 eV).

These are two main results. First of all the anion of 2-mercaptiol, being a charged species, is strongly stabilized by the electrostatic interactions with the solvent and the energy of its HOMO significantly lowers. At the same time, aza-porphyrins are more polarisable than porphyrins (having higher dipole moments), so that they strongly experience the reaction field generated by the continuum. The net effect is a complete inversion of the prediction in going from the gas phase to the aqueous solution. It is interesting to note that electron-withdrawing groups, even though they decrease the electron density on the cobalt centre (see Table 12.3), also contribute to reduce the gap between the energy of the SOMO of the macrocycle and the HOMO of the thiol, thus decreasing the reactivity index. Finally, similar results are obtained using gas phase or solution geometries, thus confirming the electronic nature of the effect induced by solvent.

It is clear from our results that bulk solvent effects control to some extent the reactivity of Co(II)–N_4 species, even in absence of any specific solute–solvent interactions. Work is in progress in order to obtain even more reliable results, aiming to investigate the role of the first solvation shell in the electro-oxidation reactions[90].

3.4. Influence of the Environment: Effect of the Absorption on the Electrode

Up to now, we have shown how the reactivity towards the oxidation of 2-mercaptoethanol changes within the different members of a Co(II)–N_4 complex family (porphyrin, phthalocyanine, tetraphenylporphyrin, tetrabenzoporphyrin and tetraazaporphyrin) and due to solvation (gas phase vs. solution). In other words we have investigated how the nearest chemical environment (nature of the N_4 cycle) and the far chemical environment (bulk solvent) can influence their

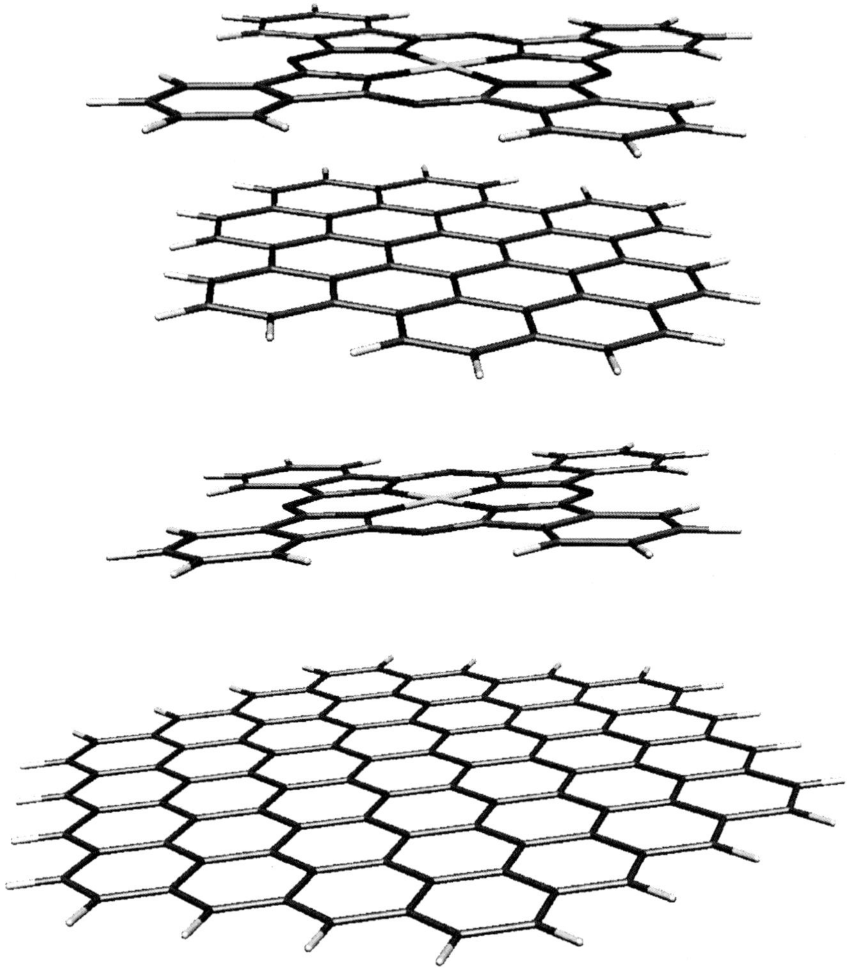

Figure 12.7. Sketch of Co(II)-phthalocyanine complex adsorbed over the small (46 carbon atom, up) and large (106 carbon atom, down) clusters.

reactivity[91]. Here, we focus the variation in reactivity induced by the absorption of the complexes on the electrode surface[92].

In particular, we investigated the effect of the electrode surface on the redox property of CoPc. To this end we have first modelled the electrode graphite surface as a single carbon layer composed of 15 (small cluster) and 38 (large cluster) benzene rings. To saturate the valence of border carbon atoms, hydrogen atoms were added. The resulting clusters are sketched in Figure 12.7. Similar cluster approaches have already been used in literature to model surface effects on adsorbate properties (see for instance ref.[93]). The internal structural parameters of CoPc as well as the complex to surface distance have been fully optimised. On the other hand, no surface reconstruction has been taken into account (i.e.

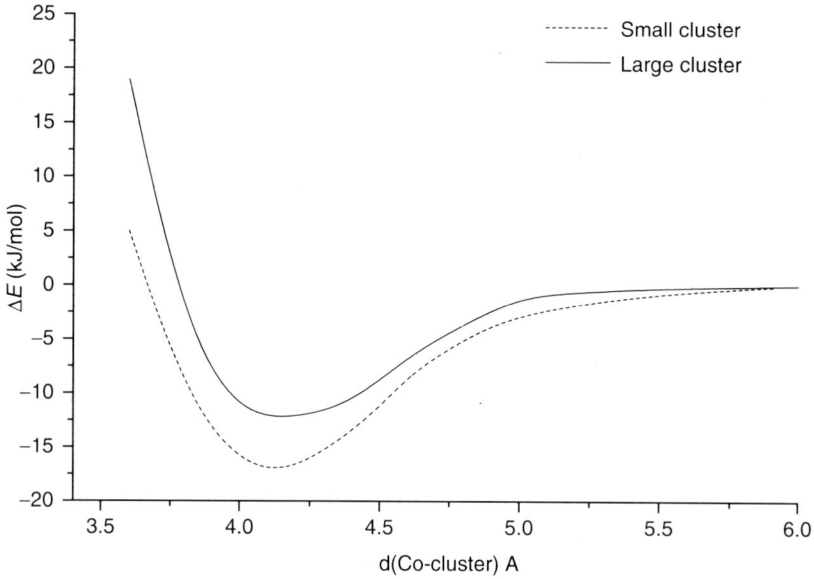

Figure 12.8. Computed potential energy profiles for the small (line) and large (dotted line) clusters as a function of the Co-graphite distances.

the graphite cluster representing the surface has not been relaxed). The internal structural parameters of CoPc are not significantly altered due to the adsorption. The reason of the negligible perturbation in geometry of the CoPc is the large computed metal–graphite distance (4.3 Å). This result is in agreement with recent X-ray investigation of iron-phthalocyanine adsorbed on carbon electrode, where no significant structural changes were found upon adsorption[94]. In order to have an estimation of the binding energy of CoPc on the graphite surface, a scan of the potential energy surface along the cluster-CoPc distance has been carried out freezing the CoPc geometry at the resulting minimum energy structure. The results, corrected for Basis Set Superimposition Error (BSSE)[95], are reported in Figure 12.8. The obtained profile, calculated for the small cluster, shows a minimum at around 4.3 Å, corresponding to an interaction energy of about 17.0 kJ/mol. For the larger cluster a smaller interaction energy is computed (12.3 kJ/mol). Unfortunately to the best of our knowledge line information is available about the interaction of metallophthalocyanines with the graphite surface. In particular, some electrochemical evidence suggests a very weak interaction[4], while X-ray photoelectron spectroscopy studies of spectrum variations in function of the temperature, for nickel-phthalocyanine complex over graphite, indicate a desorption energy of about 6 kJ/mol[96]. This latter value was roughly estimated, since the authors were not able to unambiguously assign this energy to the phthalocyanine–phthalocyanine or to the phthalocyanine–surface interaction[96].

As for the isolated molecule, the first step in understanding the role of the environment (here the electrode surface) on the redox properties of CoPc is the

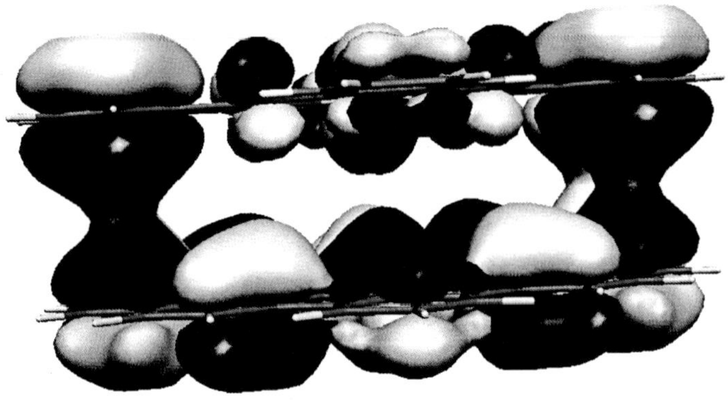

Figure 12.9. Plot of one of the orbitals associated with the $\pi - \pi$ interaction between CoPc and the graphite surface.

analysis of the electronic distribution of the complex. In the last two columns of Table 12.3 are reported the NPA charges computed for CoPc adsorbed over the small and the large clusters. A decrease of the positive charge on the cobalt atom is observed (1.08 $|e^-|$) while the nitrogens close to Co become more positive and the peripheral carbon atoms significantly more negative ($-0.04|e^-|$). The whole picture corresponds to $\pi - \pi$ stacking interaction between the CoPc and the graphite sheet, as nicely depicted by one of the involved orbitals reported in Figure 12.9. Both the interaction energy and the CoPc-graphite distances are typical of a $\pi - \pi$ stacking interaction, as in the case of the CoPc dimer or in the CoPc crystals[97,98]. Despite the large electronic rearrangement, the interaction with the thiol is still likely, since the LUMO of the adsorbed CoPc has the same shape as that in the bare molecule: a π^* orbital of the macrocycle with a significant contribution from the d_π orbital of the Co (see Figure 12.5). In a plain scheme, these results suggest that the graphite surface electronically buffers the phthalocyanine and makes it less reactive towards an electron donor. As a consequence, there is a decrease not only of the positive charge localized on the Co atom but also of the electrophilicity ($\Delta\omega = -0.3$ eV, see Table 12.6). The decrease of ω is related to the change in IP in going from isolated to absorbed CoPc (from 8.3 to 7.1 eV), due to the Pauli repulsion between the filled orbitals of the graphite surface and of the macrocycle. The EA is, instead, almost constant (around 1.5 eV). So both indexes, charges and electrophilicity, suggest a decrease of the reactivity. In contrast a large decrease of the η_{DA}, with respect to the gas-phase value, is found ($\Delta\eta_{DA} = -1.2$ eV), suggesting higher reactivity. As before, the repulsive Pauli interaction between the surface and CoPc pushes the LUMO energy of this latter quite high in energy, thus reducing the energy gap with the HOMO of the electron donor. The net result is an increase of the reactivity, thus well underlining how the surface effects rule the observed reactivity[99].

In going from the small to the large clusters, a small variation of both ω and η_{DA} is observed: the first becoming more positive and the second slightly more negative (see Table 12.2), but the overall trends are preserved. This suggests that all main features of the interaction CoPc/graphite surface are already

Table 12.6. Gas-Phase Electrophilities (ω, eV) and Donor–Acceptor Intermolecular Hardness (η_{DA}, eV) of the CoPc–N_4 System and its Variations (Δ) with Respect to the Gas Phase Values

Molecule	Gas phase	Δ (solution)	Δ (small cluster)	Δ (large cluster)	Best estimation[a]
ω	1.81	+1.38	−0.29	−0.15	3.04
η_{DA}	2.98	−0.46	−1.15	−1.25	1.27

[a]Column 2 + 3 + 5.

present in the small cluster model. At the the same time, however, the interaction energy slightly decreases (as one can see from the plot of Figure 12.3), due to a stronger Pauli repulsion between the two moieties. The calculated interaction energy, 10.5 kJ/mol, is significantly lower than that found for the smaller cluster, even if it is quite larger than the experimental estimation.

In summary, if the different contributions are considered as additive, one can see that the solvent strongly rules the capacity of CoPc to accept electrons from a generic donor, since ω values almost double in going from gas phase to solution, while the surface plays a minor role. On the other hand, both solvent and surface effects increase the reactivity of the adsorbed species, the total intermolecular hardness being less than one-half of the gas-phase values (3.0 vs. 1.3 eV, respectively).

4. Conclusions

In this review we have reported a detailed analysis of the different effects (metal, ligand, solvent and surface) determining the reactivity of a family of metal–N_4 macrocycles. We have shown how they can be identified and rationalised using the reactivity index, rigorously provided by the DFT model.

In particular, the two selected reactivity indexes, hardness and electrophilicity, provide coherent and complementary informations about the reactivity of the different species. In fact, both indexes point out that aza-porphyrins are more reactive than the corresponding porphyrins, due to the presence of the electronegative (and more polarizable) nitrogen atoms in the inner ring. Nevertheless, the electrophilicity index alone, does not allow to unambiguously distinguishing between the different species. Moreover, bulk solvent effects control to some extent the reactivity of such species, even in absence of any specific solute–solvent interactions. Work is in progress to investigate in more detail the role of the first solvation shell in the electro-oxidation reactions[61].

From a more general point of view, our results show how only an integrated approach including all the elementary bricks, even in an approximate way, should be used to obtain meaningful theoretical simulations, since their weights are of comparable magnitude.

Acknowledgements

Part of this work has been carried out within the framework of the ECOS project N. C03E02. Dr. Fethi Bedioui, Prof. José Zagal, Prof. Gloria Cardenas-Jiron, Dr. Claudia Caro, Dr. Sophie Griveau and Prof. Toro-Labbé are acknowledged for fruitful collaborations and discussions.

References

1. Tabushi, I., (1988). *Coord. Chem. Rev.* **86**, 1.
2. Meunier, B., A. Robert, G. Pratviel, and J. Bernadou, in *The Porphyrin*, vol. 4. Kadish, K.M., Smith K.M., Guilard, R. (eds); Academic Press: San Diego, 2000.
3. Leznoff, C.C., A.B. Lever, *Phthalocyanines: Properties and Applications*, Verlag VCH: New York, 1989.
4. Zagal, J.H., (1992). *Coord. Chem. Rev.* **119**, 89.
5. Zagal, J.H., M. Gulppi, M. Isaacs, G. Cárdenas-Jíron, and M.J. Aguirre, (1998). *Electrochim. Acta* **44**, 1349.
6. Griveau, S., J. Pavez, J.H. Zagal, and F. Bedioui, (2001). *J. Electroanal. Chem.* **497**, 75.
7. Bedioui, F., J. Devynck, and C. Bied-Charreton, (1995). *Acc. Chem. Res.* **28**, 30.
8. Griveau, S., and F. Bedioui, (2001). *Electroanalysice* **13**, 253.
9. Parr, R., and R.G. Pearson, (1983). *J. Am. Chem. Soc.* **105**, 7512.
10. (a) Pearson, R.G. *Chemical Hardness*, VCH, Weinheim (1997). (b) Pearson, R.G., (1989). *J. Org. Chem.* **54**, 1423.
11. Chermette, H., (1999). *J. Comput. Chem.* **20**, 129.
12. De Proft, F., and Geerlings, P. (1997). *J. Chem. Phys.* **106**, 3272.
13. Kar, T., S. Scheiner, and A.B. Sannigrahi, (1998). *J. Phys. Chem.* A **102**, 5967.
14. Ghanty, T., and S.K. Gosh, (1996). *J. Phys. Chem.* **100**, 2295.
15. Cardenas-Jiron, G.I., S. Gutierrez-Oliva, J., Melin, and A. Toro-Labbé, (1999). *J. Phys. Chem. A* **101**, 61.
16. Chattaraj, P.K., P. Pérez, J. Zevallos, and A. Toro-Labbé, (2001). *J. Phys. Chem.* **105**, 4272.
17. Safi, B., K. Choho, and P. Geerlings, (2001). *J. Phys. Chem.* **105**, 591.
18. Parr, R.G., L. Szentpaly, and S. Liu, (1991). *J. Am. Chem. Soc.* **121**, 1922.
19. Kock, W., M.C. Holthausen, (2000). *A Chemist's Guide to Density Functional Theory*, Wiley-VCH Weinheim.
20. Pearson, R.G., (1963). *J. Am. Chem. Soc.* **85**, 3533.
21. Maynard, A.T., M. Huang, W.G. Rice, and D.G. Covel, (1998). *Proc. Natl Acad. Sci. USA*, **95**, 11578.
22. Zhou, Z., and R.G. Parr, (1990). *J. Am. Chem. Soc.* **112**, 5720.
23. Parr, R.G., and Z. Zhou, (1993). *Acc. Chem. Res.* **26**, 256.
24. Klopman, G., (1968). *J. Am. Chem. Soc.* **90**, 223.
25. Zagal, J.H., and G.I. Cárdenas-Jíron, (2001). *J. Electroanal. Chem.* **497**, 55.
26. Parr, R.G., W. Yang, (1989). *Density-Functional Theory of Atoms and Molecules*. International Series of Monographs on Chemistry, Vol. 16. Oxford University Press/Clarendon Press, Oxford.
27. Cramer, C.J., (2002). *Essentials of Computational Chemistry: Theories and Models*, Wiley and sons, New york.
28. Adamo, C., A. di Matteo, and V. Barone, (1999). *Adv. Quantum Chem.* **36**, 4.

29. Perdew, J.P., K. Burke, and M. Ernzerhof, (1996). *Phys. Rev. Lett.* **77**, 3865.
30. Novikov, D.L., A.J. Freeman, N.E. Christensen, A. Svane, and C.O. Rodriguez, (1997). *Phys. Rev. B* **56**, 7206.
31. Jaffe, J.E., Z. Lin, and A.C. Hess, (1998). *Phys. Rev. B* **57**, 11834.
32. Bally, T., and G.N. Sastry, (1997). *J. Phys. Chem. A.* **101**, 7923.
33. Jacquemin, D., E.A. Perpète, I. Ciofini, and C. Adamo, *Chem. Phys. Lett.* in press
34. Becke, A.D. (1993). *J. Chem. Phys.* **98**, 5648.
35. Adamo, C., and V. Barone, *J. Chem. Phys.* **108**, 664.
36. Adamo, C., and V. Barone, *Chem. Phys. Lett.* **274**, 242.
37. Ernzerhof, M., and G.E. Scuseria, (1999). *J. Chem. Phys.* **110**, 5029.
38. Adamo, C., and V. Barone, (1999). *J. Chem. Phys.* **110**, 6158.
39. Adamo, C., V. Barone, and G.E. Scuseria, (1999). *J. Chem. Phys.* **111**, 2889.
40. Adamo, C., M. Cossi, G. Scalmani, and V. Barone, (1999). *Chem. Phys. Lett.* **307**, 265.
41. Pérez, P., A. Toro-Labbé, and R. Contreras, (2001). *J. Am. Chem. Soc.* **123**, 5527.
42. Miertuš, S., E. Scrocco, and J. Tomasi, (1981). *Chem. Phys.* **55**, 117; Cammi, R., J. Tomasi, (1995). *J. Comput. Chem.* **16**, 1449.
43. Born, M.Z., (1920). *Phys.* **1**, 45.
44. Bell, R.P., (1931). *Trans. Faraday Soc.* **27**, 797.
45. Kirkwood, J.G., (1934). *Chem. Phys.* **2**, 351.
46. Onsager, L., (1936). *J. Am. Chem. Soc.* **58**, 1486.
47. Debye, P., and E. Hückel, (1923). *Phys. Z.* **24**, 185.
48. Tomasi, J., and M. Persico, (1994). *Chem. Rev.* **94**, 2027.
49. Rivail, J.-L., D. Rinaldi, (1995). In J. Leszczynki (ed), *Computational Chemistry, Review of Current Trends*, World Scientific, New York, p. 139; Cramer, C.J., D.G. Truhlar In K.B. Lipkowitz, D.B. Boyd, (eds), *Reviews in Computational Chemistry*, VCH, New York, Vol. 6, p. 1; Cramer, C.J., D.G. Truhlar, (1996). In *Solvent Effects and Chemical Reactivity*, O. Tapia, J. Bertran, (eds), Kluwer, Dordrecht, p. 1.
50. Cramer, C.J., D.G. Truhlar, (1999). *Chem. Rev.* **99**, 2160.
51. Tomasi, J., B. Mennucci, R. Cammi, and M. Cossi, (1997). In G. Náray-Szabo, A. Warshel, (eds), *Theoretical Aspects of Biochemical Reactivity*, Kluwer, Dordrecht p. 1.
52. Amovilli, C., V. Barone, R. Cammi, E. Cancès, M. Cossi, B. Mennucci, C.S. Pomelli, and J. Tomasi, (1998). *J. Adv Quantum Chem* **32**, 227.
53. Tomasi, J., R. Cammi, B. Mennucci, C. Cappelli, and S. Corni, (2002). *Phys. Chem. Chem. Phys.* **4**, 5697.
54. (a) Amovilli, C., and B. Mennucci, (1997). *J. Chem, Phys. B* **101**, 1051. (b) Pierotti, R.A., (1976). *Chem. Rev.* **76**, 717. (c) Claverie, P. (1978). *Perspect. Quantum Chem. Biochem.* **2**, 69. (d) Colominas, C, F.J. Luque, and J. Texidó, (1999). M. Orozco, *Chem. Phys.* **240**, 253. (e) Floris, F.M., J. Tomasi, and J.L. Pascual-Ahuir, (1991). *J. Comput. Chem.* **12**, 784. (f) Tomasi, J., B. Mennucci, and E. Cancès, (1999). *J. Mol. Struct. (Theochem)* **211**, 464.
55. Whitten, J.L., and T.A. Pakkonen, (1981). *Phys. Rev. B* **21**, 4357.
56. Maseras, F., and K. Morokuma, *J. Comput. Chem.* **16**, 1170.
57. Kudin, K.N., and G.E. Scuseria, (2000). *Phys. Rev. B* **61**, 16440.
58. Albrecht, M., and P. Fulde, (2002). *Phys. Stat. Sol.* **234**, 313.
59. Pisani, C., M. Busso, G. Capecchi, S. Casassa, R. Dovesi, L. Maschio, V. Saunders, C. Zicovich, M. Schutz, *in press*
60. Tran, F., J. Weber, T. Wesolowski, F. Cheikh, Y. Ellinger, and F. Pauzat, (2002). *J. Phys. Chem. A* **106**, 8696.

61. Frisch, M.J., G.W. Trucks, H.B. Schlegel, G.E. Scuseria, R.E. Stratmann, J.C. Burant, S. Dapprich, J.M. Millam, A.D. Daniels, K.N. Kudin, M.C. Strain, O. Farkas, J. Tomasi, V. Barone, M. Cossi, R. Cammi, B. Mennucci, C. Pomelli, C. Adamo, S. Clifford, J. Ochterschi, Q. Cui, P.M.W. Gill, B.G. Johnson, M.A. Robb, J. R. Cheeseman, T. Keith, M.K. Petersson, D.K. Malick, A.D. Rabuck, G.A., J.A. Montgomery, K. Raghavachari, M.A. Al-Laham, V.G. Zakrewski, J.V. Ortiz, J.B. Foresman, J. Cioslowski, B.B. Stefanov, A. Nanayakkara, J. Liu, A. Liashenklo, P. Piskorz, I. Komaromi, M. Challacombe, C.Y. Peng, P.Y. Ayala, W. Chen, M.W. Wong, J.L. Andres, E.S., Replogle, R. Gomperts, R.L. Martin, D.J. Fox, J.S. Binkley, D.J. DeFrees, J. Baker, J.P. Stewart, M. Head-Gordon, C. Gonzalez, and J.A. Pople, (1998). *Gaussian 98 (Revision A.7)*, Gaussian Inc., Pittsburgh, PA.
62. Perdew J.P., and M. Ernzerhof, (1996). *J. Chem. Phys.* **105**, 9982.
63. Frisch, Æ., M.J. Frisch, (1998). *Gaussian 98 User's Reference*, Gaussian, Inc., Pittsburgh, and references therein.
64. Liu, Y.P., (2001). *J. Chem. Inf. Comput. Scien.* **40**, 1337.
65. Couty, M., and M.B. Hall, (1996). *J.Comput.Chem.* **17**, 1359.
66. Dunning, T.H. Jr, P.J. Hay (1976). In H.F. Schaefer III (ed), *Modern Theoretical Chemistry*, Plenum, New York, pp. 1–28.
67. Hay, J., and W.R. Wadt, (1985). *J. Chem. Phys.* **82**, 299.
68. Barone, V., M. Cossi, and J. Tomasi, (1998). *J.Comput.Chem.* **19**, 407.
69. Barone, V., and M. Cossi, (1998). *J. Phys. Chem. A* **102**, 1995.
70. Reed, A.E., L.A. Curtiss, and F. Weinhold, (1988). *Chem. Rev.* **88**, 899.
71. Glendening, E.D., J.K. Badenhoop, A.E. Reed, J.E. Carpenter, F. Weinhold (1996). NBO 4.0 program. Theoretical Chemistry Institute, University of Wisconsin, Madison.
72. Albright, T.A., J.K. Burdett, and M.-H. Whangbo (1985). *Orbital Interaction in Chemistry*, Wiley and sons.
73. (a) Da Silva, C., L. Bonomo, E. Solari, R. Scopelliti, C. Floriani, and N. Re, (2000). *Chem. Eur. J.* **6**, 4518. (b) Baerends, E.J., G. Ricciardi, A. Rosa, and S.J.A. van Gisbergen, (2002). *Coord. Chem. Rev.* **230**, 5.
74. Scheidt, W.R., C.A. Reed, (1981). *Chem. Rev.* **81**, 543.
75. Boyd, P.D., D.A. Buckingham, R.F. McMeeking, and S. Mitra, (1979). *Inorg. Chem.* **18**, 3585.
76. Lin, W.C., (1976). *Inorg. Chem.* **15**, 1114.
77. Reynolds, P.A., and B.N. Figgis, (1991). *Inorg. Chem.* **30**, 2294.
78. Ricciardi, G., A. Rosa, I. Ciofini, and A. Bencini, (1999). *Inorg. Chem.* **38**, 1422.
79. Figgis, B.N., E.S. Kucharski, and P.A. Reynolds, (1989). *J. Am. Chem. Soc.* **111**, 1683.
80. (a) Scheidt, W.R., and Turowska-Tyrk, I. (1994). *Inorg. Chem.* **33**, 1314. (b) Wang, J., A.D. Becke, and V.H. Smith, (1995). *J. Chem. Soc.* **102**, 3477.
81. Liou, K., T.P. Newcomb, M.D. Heavy, J.A. Thompson, W.B. Heuer, R.L. Musselman, C.S. Jacobsen, B.M. Hoffman, and J.A. Ibers, (1992). *Inorg. Chem.* **31**, 4517.
82. Stevens, E.D., (1981). *J. Am. Chem. Soc.* **103**, 5087.
83. Madura, P., and W.R. Scheidt, (1976). *Inorg. Chem.* **15**, 3182.
84. Hoard J.L. (1975). In K.M. Smith, (ed), *Porphyrins and Metalloporphyrins*, Elsevier, New York.
85. Coppens, P., A. Holladay, and E.D., Stevens, (1982). *J. Am. Chem. Soc.* **104**, 3546.
86. Barone, V., and C. Adamo, (1995). *J. Phys. Chem.* **99**, 15062.
87. Hassan, S.A., F. Guarnieri, and E.L., Mehler, (2000). *J. Phys. Chem. B* **104**, 6490.
88. Reichardt, C., (1988). *Solvents and Solvents Effects in Organic Chemistry*, 2nd edn. VCH, Weinheim.
89. Winget, P., E.J. Weber, C.J. Cramer, and D.G. Truhlar, (2000). *Phys. Chem. Chem. Phys.* **2**, 1231.

90. Adamo, C., L. Joubert, J. Bloino, work in progress.
91. Griveau, S., F. Bedioui, and C. Adamo, (2001). *J. Phys. Chem. A* **105**, 11304.
92. Ciofini, I., F. Bedioui, J. Zagal, and C. Adamo, (2003). *Chem. Phys. Lett.* **376**, 690.
93. Lou, L., L. Österlund, and B. Hellsing, (2000). *J. Chem. Phys.* **112**, 4788.
94. Kim, S., T. Ohta, and G. Kwag, (2000). *Bull. Korean Chem. Soc.* **21**, 588.
95. Boys, S.F., and F. Bernardi, (1970). *Mol. Phys.* **19**, 553.
96. Ottaviano, L., S. Di Nardo, L. Lozzi, M. Passacantando, P. Picozzi, and S. Santucci, (1997). *Surf. Science* **373**, 318.
97. Orti, E., J.L. Bredas, and C. Clarisse, (1990). *J. Chem. Phys.* **92**, 1228.
98. Mason, R., G.A. Williams, and P.E. Fielding, (1979). *J. Chem. Soc. Dalton Trans.* 676.
99. R.O., Lezna, S. Juanto, and J.H. Zagal, (1998). *J. Electroanal. Chem.* **452**, 221.

13

Organized Multiporphyrinic Assemblies for Photoconduction and Electroconduction

Jean Weiss and Jennifer Wytko

After a brief recall of the physics principles involved in the generation of light-induced excited states and their evolution in time, this chapter surveys the properties of organized multiporphyrin species that have been reported from the end of 2001 until the beginning of 2004. Once the properties of the chromophores have been established as electron or energy donors or acceptors, the structures described mainly differ by the synthetic approaches from which they originate. While extremely well-defined geometries can be obtained by covalent synthesis, providing useful structure–activity information, self-assembly of photo and electroactive species is attractive because of its simplicity. This chapter is meant to demonstrate the complementarity of both approaches.

1. Introduction

Energy and electron transfer processes are essential to the development of bottom-up approaches of challenging modern technologies. So far, in terms of efficiency, Nature provides the most carefully programmed systems for performing both processes. Energy-transfer reactions are ultimately mastered in the collection of visible light in antennae systems of the photosynthetic apparatus until the injection of photons in the charge separation process at the photosynthetic reaction center. After excitation of the special pair that is responsible for the initial charge separation, electron-transfer reactions are optimized to allow an energetic input into chemical transformations associated with the terminal stage of photosynthesis. The efficiency of these processes and the perfect spatial and functional adjustments of the components involved are the result of billions of years of evolution. In comparison, the development of synthetic photo- and electroactive

---

Jean Weiss and Jennifer Wytko • Laboratoire de Chimie des Ligands à Architecture Contrôlée, Institut de Chimie, Université Louis Pasteur, 4 rue Blaise Pascal, 67070 Strasbourg, France. Telephone: (33)–390–24–14–23, Fax: (33)–390–24–14–25, e-mail: jweiss@chimie.u-strasbg.fr

N_4-Macrocyclic Metal Complexes, edited by José H. Zagal, Fethi Bedioui and Jean-Pol Dodelet. Springer Science+Business Media, Inc., New York, 2006.

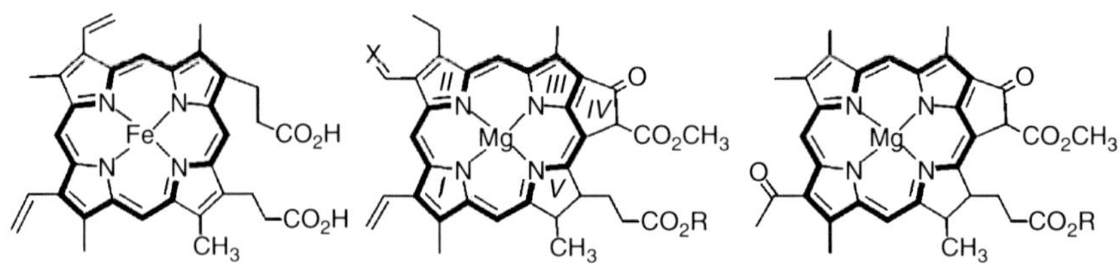

Protopophyrin IX

Chlorophyll a: X = H_2
Chlorophyll b: X = O

Bacterichlorophyll a:
R = Phytyl of Geranyl

Figure 13.1. Most abundant porphyrin derivatives in Nature.

molecular systems is still in its genesis stage. Yet, intuitively or rationally, chemists take advantage of the information provided by the study of biological systems.

Nature selected tetrapyrrolic macrocycles (Figure 13.1) as the most widespread photo- and electroactive components. As a consequence, porphyrins have been a first class choice of molecular building blocks in synthetic energy and electron transfer systems. Although the involvement of porphyrins in natural photochemical and redox events has long been known, models and applications based on multiporphyrin assemblies is a field that really exploded in the last 15 years. Comprehension of natural systems arose from the combination of fundamental rules of energy and electron transfers established almost 5 decades ago. The increased availability of crystallographic data on genuine or mutant reaction centers and light harvesting systems has also improved our understanding of the topographic arrangement of porphyrin derivatives.

Thus, many synthetic photoactive devices performing electronic or energetic transfers appear frequently in the literature. Due to the importance of this relatively young research field, an exhaustive coverage of the literature up until the end of 2001 (over 400 references, by Pierre Harvey, Sherbrooke, Québec) has appeared in the second set of the Porphyrin Handbook[1]. Our coverage will start from the end of 2001 and end in February 2004. A fewer number of references complementary to those analyzed by Harvey will be added. It should be noted, as reflected by the title of this chapter, that only systems involving more than one porphyrin will be covered in this chapter. Donor–acceptor (D–A) dyads involving only one porphyrin as either D or A will be omitted.

This chapter will thus begin with a rapid overview of the basic principles of energy and electron transfers, and the description of elementary events in photosynthesis. Once the source of inspiration has been set, recent advances in the use of synthetic multiporphyrinic assemblies for these purposes will be arranged in sections defined according to the synthetic procedures for assembling. For energy transfer systems, well-defined species obtained through stepwise covalent construction will be treated first. The systems presented may comprise either exclusively multiporphyrinic species, or multiporphyrins associated with other components such as fullerenes or quinones. Polymer structures involving functionalized porphyrin monomers will be categorized into structures incorporating the tetrapyrrolic macrocycles in the polymer chain or structures bearing pendant chromophores. A large section will cover the field of self-assembled

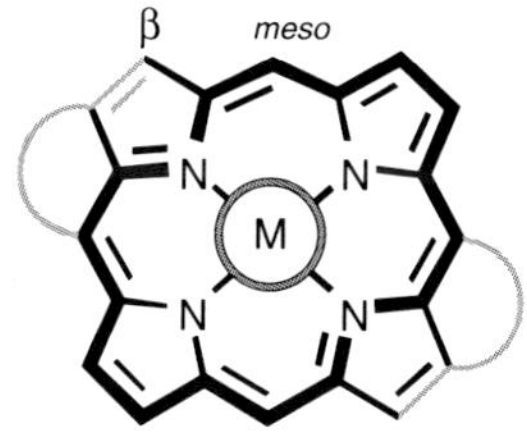

Figure 13.2. The porphyrin chromophore.

multiporphyrinic species, and will be arranged according to the type of non-covalent interactions controlling the assembly (axial coordination, peripheral co-ordination, H-bonding, etc.). A similar arrangement of sections will follow for synthetic systems performing electron transfer and extensively delocalized porphyrinic assemblies.

2. The Porphyrinic Chromophore

As a chromophore (from the Greek: *chromo* = color, *phero* = to carry along, to bear), porphyrin (Figure 13.2) can be described by its characteristic signature observed by UV–visible spectrophotometry. The intense color of porphyrin derivatives arises from a highly delocalized π system involving 18 electrons (bold lines) which, because of its subsequent aromatic character, partially explains the survival of this structure through evolution.

In natural chromophores, additional cyclic structures may be found at the periphery of the macrocycle. Partial reduction of π bonds not involved in aromaticity is also observed (see Figure 13.1). The energetic positioning of the corresponding molecular orbitals, HOMO and LUMO, may be adjusted either by adequate substitution of the available β or *meso* positions, or by the coordination of a metal within the N4 core of the macrocycle. The global behavior of the chromophore remains roughly unaffected in terms of absorption, however. As it will be explained later, the nature of the metal mostly affects the characteristics of the excited state. This persistent photophysical behavior of the porphyrin core allows the use of various peripheral functions in order to connect chromophores in more or less defined arrangements without perturbing the generation of excited states. Thus, the following general description of excited states and their ability to transfer energy or electrons will apply throughout this chapter unless specified otherwise.

2.1. Absorption of Light

Light absorption promotes an electron from the ground state to an excited state. The presence of 18, π electrons in the aromatic core of the porphyrin consequently generates a UV–visible spectrum that displays several absorption bands corresponding to rather energetically inexpensive π–π* transitions. Our attention should focus mainly on two types of absorption bands, the Soret or B band, and

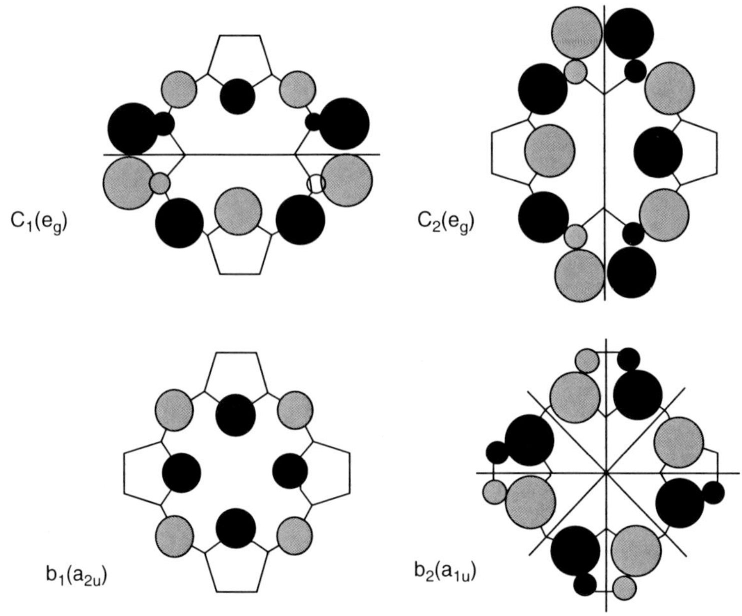

Figure 13.3. HOMO–LUMO frontier orbitals for a D_{4h} porphyrin.

the Q bands. These absorptions are due to transitions between frontier orbitals, the two HOMOs and LUMOs defined by the four orbital model of Gouterman[2] (Figure 13.3).

Typical molar extinctions are in the range of 5,000–35,000 $M^{-1}cm^{-1}$ for two (D_{4h}) to four (D_{2h}) Q bands ($S_0 \rightarrow S_1$) appearing between 450 and 750 nm, and around 250,000 $M^{-1}cm^{-1}$ for the more energetic B band ($S_0 \rightarrow S_2$) which is situated between 360 end 450 nm. In metal porphyrins which are closer to D_{4h} symmetry, the B band usually originates from two degenerate transitions of perpendicular polarization, and thus appears as a single absorption barely affected by vibrational modes. Only a slight bandwidth increase is observed for the Soret band in the corresponding D_{2h} free base porphyrins, while the appearance of less energetic Q bands can be strongly affected by vibrational modes.

The promotion of one electron to an excited state upon light absorption generates an electron–hole pair in the frontier orbital system of the porphyrinic chromophore. This electron–hole pair is an exciton, and its generation introduces a nonpermanent (transitory) dipole in the chromophore. The energy transfer process is strongly related to the interactions between excitons in multiple chromophore assemblies.

2.1.1. Transition Moments

Transition dipole moments, generated by electronic transitions, are involved in both electronic interactions between chromophores (excitonic coupling) and, as we will see later in this chapter, energy transfer between donors and acceptors. In porphyrinic structures, the orientation of the dipoles existing in excited states are associated with symmetry considerations of the HOMOs and LUMOs involved in the electronic transition. In general, it is admitted that transition moments due to π–π* electronic transitions are in plane, and n–π* transitions are out of plane of the macrocyclic structure. For both B and Q transitions, in an isolated D_{4h} porphyrin, the four orbital model of Gouterman leads to the orientation of two perpendicular transition dipole moments (B_x, B_y, Q_x and Q_y) along the axis of opposite pyrrole nitrogen atoms. Substitution, oligomerization, or aggregation will influence the orientation of the transition dipoles of the excited states. In a simplified approach, dipole transition moments are often considered as being oriented along the highest possible symmetry axis in the chromophore. In this way, most of the spectral changes observed in combinations of porphyrin chromophores may be explained. Although not very common, experimental determination of the orientation of dipole transition moments can be performed via several techniques such as fluorescence anisotropy spectroscopy or linear dichroism[3]. Time resolved EPR measurements of the reaction center in *Rhodbacter Sphaeroides* R-26 primary donor triplet state generated by polarized light, and analyzed in the light of X-ray data have also been used to determine the orientation of the Q_y optical transition moment[4].

The way that transition dipole moments interact together in multichromophore assemblies was initially examined by Kasha in his pioneering work on excitonic coupling[5]. The original point dipole treatment was later refined by Hunter and Sanders who have developed a transition monopole treatment that allows quantitative prediction of the effects of excitonic interaction[6]. Figure 13.4 qualitatively summarizes the nature of the absorption band shifts depending on the

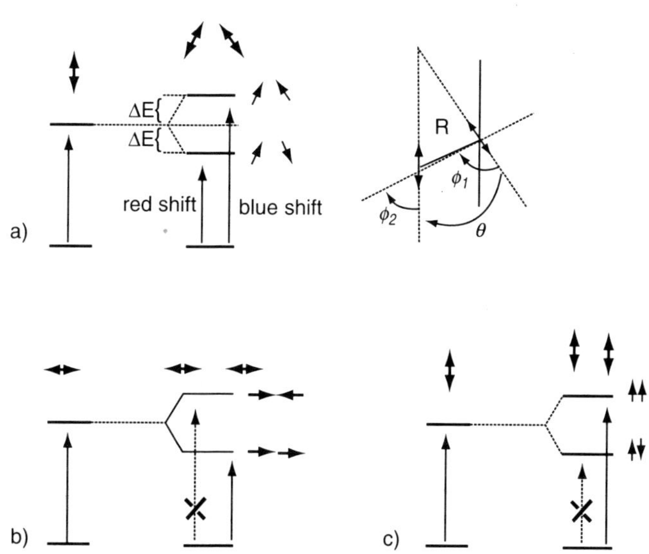

Figure 13.4. Point dipole treatment of chromophore dimers: (a) general case of oblique dipoles; (b) extreme case of aligned dipoles; (c) extreme case of stacked dipoles (double arrow: dipole moment, single arrow: electronic transition).

respective orientations of the dipoles, and extreme cases in which dipole moments are parallel or aligned. Oblique interaction between transition dipoles will result in an angular and distance dependent band splitting, for which ΔE is proportional to $(\cos\theta - 3\cos\phi_1\cos\phi_2)R^3$. In tetrapyrrolic chromophores, the situation is complicated by the presence of two perpendicular dipole moments. In a first approach they can be treated separately according to Figure 13.4 for each absorption band, but the choice of orientation sometimes remains difficult.

In some cases, the orientation of transition dipole moments according to the Gouterman model along the opposite N–N axis will give a qualitatively correct explanation of experimental spectral changes. In other cases, the orientation of these dipoles according to symmetry considerations will prevail. Ribó and coworkers have proposed a helpful detailed examination of two experimental examples[7]. Figure 13.5 summarizes the possible spectral changes that can be predicted from different dipole orientations in side to side, perpendicularly arranged tetrapyrrolic chromophores. In the case of the Osuka type C_{meso}–C_{meso} coupled porphyrin examined[8], the situation (Figure 13.5c) where dipoles are oriented according to the highest possible symmetry axis explains experimental data.

A similar treatment of slipped dimers, from the extreme situations of side-to-side arrangement to stacked species, shows that for slipped aggregates of tetrasulfonated tetraphenyl porphyrin dications[9] only the Gouterman orientation allows a consistent interpretation of experimental spectral changes. The treatment of these two cases illustrates the context of validity of Kasha's point dipole approach, which considers two independent electronic transitions occurring in

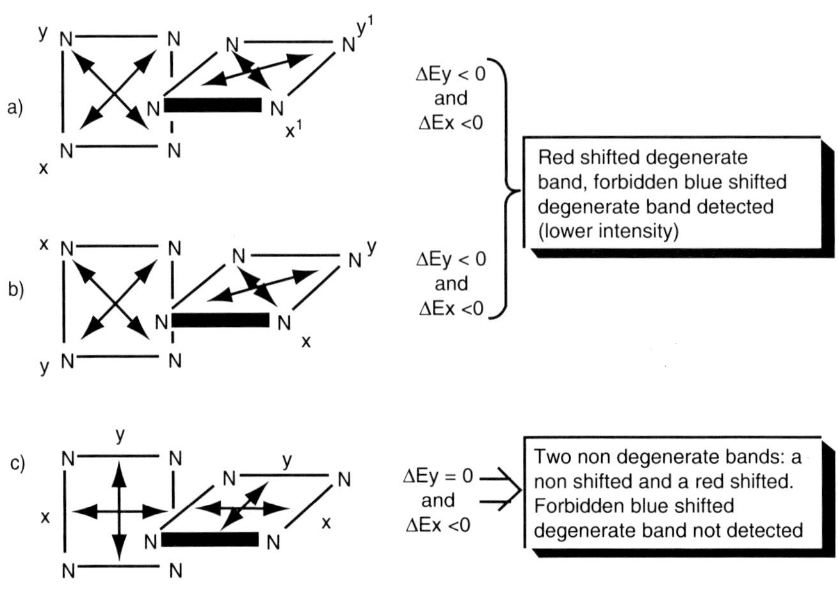

Figure 13.5. Expected spectral changes predicted as a function of the dipole moment orientation. In the case examined in ref.[7], only (c) can explain the experimental data.

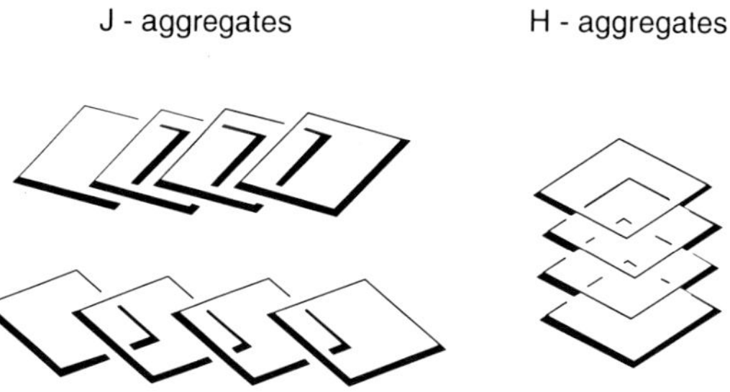

Figure 13.6. Schematic representation of J and H aggregates.

electronically distinct chromophores. When electronic communication between the chromophores is present, the frontier orbitals involved in the transitions are mixed. A refined experimental and theoretical comparative analysis by Crossley and Reimers[10a] for bis-porphyrins fused with tetraaza anthraquinones or tetraaza anthracene illustrates the complexity of the excitonic coupling analysis in fully conjugated multiporphyrin species. In dimers for which the orientation of the two chromophores can be tuned, Osuka and Kim[10b] have investigated the dihedral angle charge resonance dependence by Raman resonance spectroscopy and MO approaches. Strong orbital mixing is found as the dihedral angle decreases in such systems.

In noncovalent multiporphyrinic assemblies, aggregated or self-assembled in a controlled manner, the two main types of arrangements depicted in Figure 13.6 are observed. J-aggregates correspond to tape or ribbon shaped arrangements of chromophores, which in terms of transition dipole moments, leads to repetition of the slipped dimer situation. Red shifts of B and Q bands are observed in the absorption spectrum of J-aggregates. H-aggregates correspond to a repeated stacked dimer for which, together with a blue shift, broadening and/or hypochromism may be observed. Thus, the excitonic coupling between chromophores is responsible for significant changes in absorption spectra that are a useful source of information on the respective orientation of the chromophores involved. Because of their intensity, spectral variations induced by excitonic coupling in the B band are more likely to be used as a diagnosis concerning the respective orientation of chromophores. As far as energy transfer between chromophores is concerned, respective orientation and coupling of transition dipole moments of the Q bands is a determining parameter. In the case of a strong red shift of the B bands, or a large splitting due to excitonic interaction, it should be noted that an intensity transfer may be observed from the coupled B bands to the Q bands[11].

Although not within the scope of this chapter, it should be noted that excitonic coupling between B or Soret bands in achiral metalloporphyrin hosts or bis-metalloporphyrin tweezers are extensively used as an analytical tool for assignment of absolute configuration of chiral guests, by taking advantage of the induction of CD features in the achiral hosts[12].

2.1.2. Excited States

All excited states occur by absorption of light energy by one electron in the fundamental electronic state S_0 or the vibrational levels associated with it, and its relocation to a higher energy level of the same spin parity S_1, S_2, etc. Figure 13.7 summarizes the multiple pathways available for the return of the promoted electron to the ground state.

Vibrational relaxation and internal conversion are nonradiative deactivation processes which correspond to a waste of energy and will not be examined in detail; however, these processes are always in competition with either energy or electron transfer.

Intersystem crossing (isc) is a nonradiative process which may be part of deactivation. It characterizes the conversion of a singlet excited state into a triplet excited state. Because of Hund's rule, T_n states are of lower energy than the corresponding S_n state, and isc efficiency is strongly related to the spin–orbital coupling. Therefore, heavy atoms usually favor intersystem crossing. Reverse isc is possible if T_n and S_n states are close in energy, and may be enhanced thermally or by collision between two T_n states.

Internal conversion converts S_n states into S_1, fluorescence usually occurs from S_1 to S_0. Thus, it is a spin allowed transition that occurs very fast, causing rather short lifetimes for excited singlet states (10^{-10}–10^{-7} s). Since some of the absorbed energy is lost in the faster vibrational relaxation (10^{-12}–10^{-10} s), the photons are emitted at lower energy. The difference between the absorption and

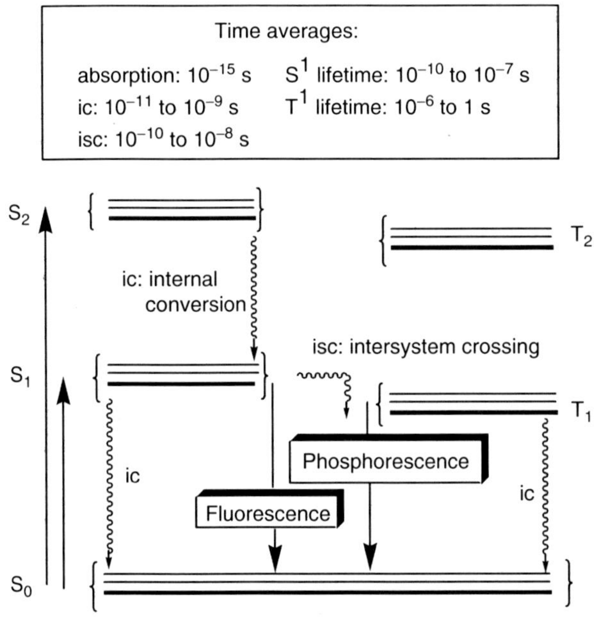

Figure 13.7. Deactivation pathways for excited states (adapted from ref.[13]).

the emission wavelength is called the Stoke's shift. If reverse isc is possible, and if the lifetime of T_1 is long enough, a delayed fluorescence will be observed at identical wavelengths but with a significantly longer decay time constant. Phosphorescence is a spin forbidden transition, and is therefore a much slower process (10^{-6}–1 s). A more detailed description of these phenomena can be found in a book devoted to molecular fluorescence and its applications by Prof. Bernard Valeur (CNAM, Paris, France)[13].

Lifetime of the Excited State

For a given S_1 singlet excited state $^{S}P*$ of a porphyrin P, a rate constant is associated with each deactivation process, k_F, k_{ic}, and k_{isc}, which are, respectively, the rate constants for fluorescence, internal conversion, and intersystem crossing. As these pathways contribute to the disappearance of the singlet excited state, the rate of deactivation can be expressed as

$$-d[^{S}P*]/dt = (k_F + k_{ic} + k_{isc})[^{S}P*]$$

which after integration gives

$$[^{S}P*] = [^{S}P*]_0 \exp(-1/\tau_S)$$

where $[^{S}P*]_0$ is the initial concentration of excited molecules of P.

The lifetime of the excited state S_1 is given by

$$\tau_S = 1/(k_F + k_{ic} + k_{isc})$$

If we consider the complete process of absorption (excitation) and return (emission) to the ground state in terms of number of photons, a fluorescence quantum yield Φ_F corresponding to the ratio of the number of emitted photons (fluorescence) vs. the number of absorbed photons can be defined. It is expressed as

$$\Phi_F = k_F/(k_F + k_{ic} + k_{isc}) = k_F\ \tau_S$$

When P is considered as an isolated species, the absorbed energy is rapidly wasted in deactivation pathways. If other species are present, interactions with the excited states may lead to other deactivation pathways, such as energy or electron transfer. In that case, the energy stored in the excited state is quenched more or less efficiently.

Direct connection between a fluorescence quenching moiety and the excited species may lead to a loss of fluorescence due to two distinct, and eventually competing mechanisms, energy transfer (ET) or electron transfer (eT). In this introduction and the following chapter, we will not consider dynamic systems in which the association of the quencher with the excited species is not controlled. In this regard, the Stern–Volmer method for the study of fluorescence quenching will not be introduced. Information relevant to this method can be found in the literature[1].

Electron Transfer (eT)

One of the most important features of the excited states of a given molecule is that it is both a better reducing agent and a better oxidizing agent than the corresponding ground state. In a very simplistic view, after a photoinduced HOMO–LUMO electronic transition, the former HOMO, now electron deficient, is ready to oxidize an electron donor, while the former LUMO, now half filled with a high energy electron, is prompted to reduce an electron acceptor. Thus, in the presence of an electron acceptor or an electron donor, photoinduced electron transfer can be viewed as a process competing with fluorescence and phosphorescence decays. The simplest approach considers the electron transfer between a donor and an acceptor, one of them being in its excited state. Figure 13.8 describes the possible evolution of a donor–acceptor ensemble after photoexcitation. The fluorescence, phosphorescence, and nonradiative internal conversion processes are combined into one global deactivation rate constant k_{d}, and the terms CS (k_{CS}, and τ_{CS}), and CR (k_{CR}, $k_{-\mathrm{CR}}$, and τ_{CR}) are, respectively, assigned to charge separation and charge recombination processes.

As stated in an excellent review article by Michael Paddon-Row[14], optimization of charge separation, as the initial electron transfer in a photoconduction or photoconversion process, will target a high value of k_{CS} compared to the competing k_{d}, a long lifetime of the charge separated states (small value of k_{CR}), and the highest energy content for the CS state. However, if the CS state (D^+–A^-) is too close in energy (*ca.* $< 0.2\,\mathrm{eV}$) to the *(D–A) state, back electron transfer ($k_{-\mathrm{CR}}$) may also be observed.

The term "long-range electron transfer" is generally applied to processes in which the redox centers involved in the electron transfer are separated by a distance superior to the van der Waals separation. This type of distance separation limits the occurrence of the charge recombination process. Given a few approximations concerning the decoupling of nuclear and electronic motions, the absence of bond rupture or formation, and the existence of a solvent dielectric continuum, long-range electron transfer can be expressed according to the Marcus theory[15] as

$$\Delta G^{\neq} = (\lambda/4)(1 + \Delta G_{\mathrm{cs}}/\lambda)^2 \qquad \text{with} \quad \lambda = \lambda_{\mathrm{int}} + \lambda_{\mathrm{s}}$$

in which λ is the reorganization energy, and represents the energy necessary to distort the solvated initial redox state (D–A) into the solvated charge separated state

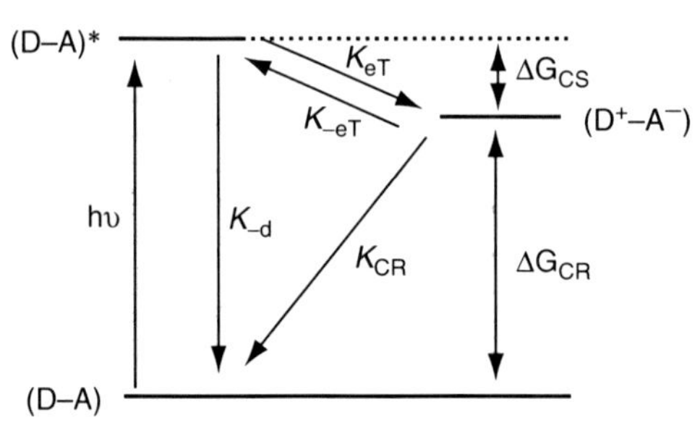

Figure 13.8. Deactivation pathways for a photoexcited dyad D–A.

(D^+–A^-). λ is the sum of the inner sphere reorganization energy (λ_{int}) during eT, and the outer sphere reorganization energy of the solvent during the process (λ_s). The value of λ_{int} takes into account the structural changes during a redox process. If the reduced and oxidized species have similar structures, then λ_{int}, and thus $\Delta G^{\neq}$, will be small. The value of λ_s mostly takes into account the dielectric constant of the solvent separating the redox active species.

Quite logically, k_{eT} will increase ($\Delta G^{\neq}$ decrease) if ΔG_{cs} decreases. When considering a constant value of λ, this trend is respected only if $|\Delta G_{cs}| < \lambda$, which corresponds to the *normal* region of electron transfer. If $|\Delta G_{cs}| > \lambda$, the decrease of ΔG_{cs} will have a reverse effect on k_{eT}, and the rate of eT will decrease with ΔG_{cs} in the region thus named the *inverse* region.

Again, given a few approximations on temperature and vibrational levels, k_{eT} can be expressed as

$$k_{eT} = (2\pi/h)(H_{AB})^2(4\pi\lambda RT)^{-1/2}\ \exp -((\Delta G_{cs} + \lambda)^2/4\lambda RT)$$

where h is Planck's constant, and H_{AB} defines the electronic interaction (Figure 13.8) between the redox sites $D+/D$ and A/A^- at the time of the transfer. H_{AB} and thus k_{eT} decrease exponentially with increasing D–A separation moderated by damping factors that are significant of the bridge "conductivity".

The nature of the bridge, if any, between D and A will define whether electron transfer or electron transport is responsible for charge separation[14]. Electron transfer occurs in species where either no bridge or electronically insulating bridges are separating D and A. Thus, the electron will never be localized on the bridge. Electron transport takes place in D–A species where the bridging part is conducting, whether it be a metallic particle, or a conjugated, or aromatic linker. A combination of both situations will be found in species containing multiple intermediate acceptors for example. In these systems, charge hopping from the donor to the ultimate acceptor will temporarily localize the charges on the intermediate acceptors.

Hereafter, charge-separating devices will be classified according to the nature of the bridge: noncovalent, for which charge separation proceeds mostly via electron transfer, and covalent, for which subsections will consider charge transfer and charge transport, respectively. If possible, distinctions will be made between eT proceeding via superexchange mechanism, and transfers occurring through tunneling, which happens mostly for eT reactions in the inverse Marcus region.

Energy Transfer (ET)

Nonradiative excitation energy transfer from a donor D to an acceptor A summarized in Figure 13.9 requires interactions between them. These interactions can be of the Coulombic type, or due to an orbital overlap between the two species, or be a combination of both types. Two mechanisms can explain nonradiative energy transfer in donor–acceptor systems. As is the case of electron transfer, energy transfer can also happen in cascade processes. For clarity reasons, however, we will only describe the basic theoretical background for energy dyads comprising one donor D and one acceptor A.

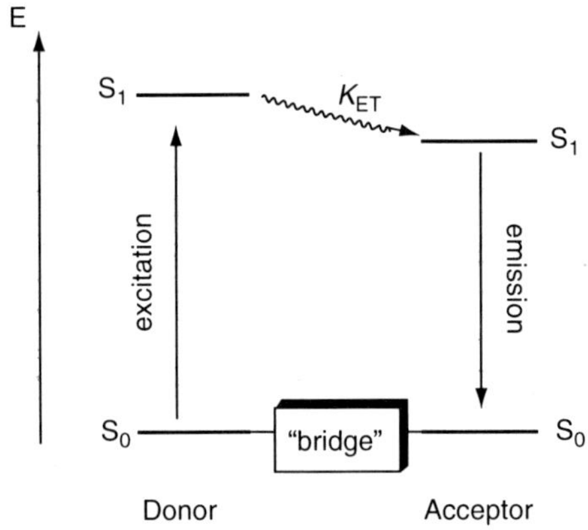

Figure 13.9. Singlet–singlet energy transfer from D to A.

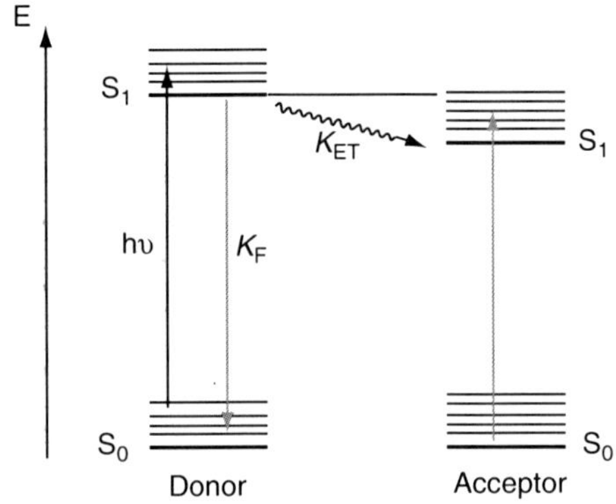

Figure 13.10. Resonant transitions in energy transfer.

The nature of the bridge, if any, between D and A will usually define the mechanism by which the energy transfer is more likely to proceed. It should be pointed out that in most cases detailed in this chapter, weakly coupled systems will be considered. A weakly coupled dyad corresponds to a D–A pair in which Coulombic interactions in the ground state are small compared to the energy of the electronic transitions, and thus, the absorption spectra of separated species are very similar to the absorption spectrum of the dyad.

The Förster Mechanism The Förster mechanism is related to Coulombic interactions between dipolar transition moments and implies that due to resonant transitions (see Figure 13.10), the promoted electron (excited) of the donor returns to its original ground state during simultaneous promotion of an electron on the acceptor[16].

The energy is transferred through space via dipolar interactions to the acceptor and can occur over rather long distances of up to several nanometers[13]. Consequently, in the Förster mechanism, for a D–A dyad in a given medium, the rate of energy transfer will be dependent on the distance r separating D and A, and is expressed as

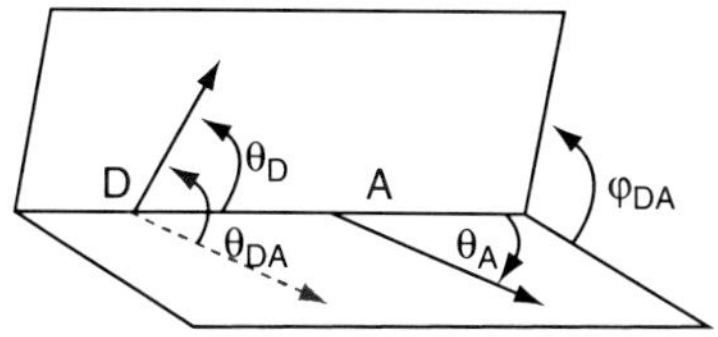

$\kappa^2 = \sin\theta_D \sin\theta_A \cos\varphi - 2\cos\theta_D \cos\theta_A$

Figure 13.11. Expression of κ^2 as a function of the respective transition dipole orientation.

$$k_{ET} = 1/\tau_F^\circ (R_0/r)^6$$

In this expression, τ_F° is the lifetime of the donor in absence of transfer (or $1/k_F$). R_0 is the Forster radius which corresponds to the distance r at which $k_{ET} = k_F$, or when half of the energy is transferred from the donor to the acceptor. R_0 can be obtained from experimental data, and depends of the overlap integral between the emission spectrum of the donor and the absorption spectrum of the acceptor, the refractive index n of the medium in the wavelength range of the overlap, the fluorescence quantum yield of the donor in the absence of transfer, and the orientation factor of the respective transition dipoles in D and A (see Figure 13.11). The mathematical expression of R_0 is then

$$(R_0)^6 = (9{,}000(\ln 10)\kappa^2 \Phi_D^\circ / 128\pi^5 N n^4) \int_0^\infty I_D(\lambda)\varepsilon_A(\lambda)\lambda^4 \, d\lambda$$

in which $I_D(\lambda)$ is the normalized fluorescence spectrum so that $\int_0^\infty I_D(\lambda) d\lambda = 1$, and $\varepsilon_A(\lambda)$ is the molar extinction of the acceptor.

For example, in the case of a zinc porphyrin energy donor D and a free base energy acceptor A, the overlap integral may be obtained by calculating the overlap between the emission bands of the zinc porphyrin (usually from *ca.* 550 to 700 nm), and the absorption of the lowest energy Q bands of the free base. For a series of spacers maintaining D and A at a given distance r in the same medium, the efficiency of the transfer will depend on the value of κ, the orientation factor. Although it is common to take an arbitrary value of 2/3 for energy dyads with flexible linkers, Figure 13.11 depicts the general expression of κ^2 as a function of the angles between the transition moments[13].

The Förster mechanism can be observed for Singlet–singlet long-range energy transfer, and also for energy transfer between $^1D^*–^3A^*$, or $^3D^*–^1A$ to give, respectively, $^1D–^3A^*$, and $^1D–^1A^*$ at some specific overlap conditions of the absorption spectrum of the acceptor with the emission spectrum of the donor.

The Dexter Mechanism This energy-transfer mechanism is possible in systems in which a spatial overlap, in terms of electronic structure, is present in the D–A dyad[17]. Only in this case is electronic motion similar to super exchange possible, as depicted in Figure 13.12.

In that case, the rate of energy transfer is expressed as

$$k_{ET} = (2\pi/h) K \exp(-2r/L) J'$$

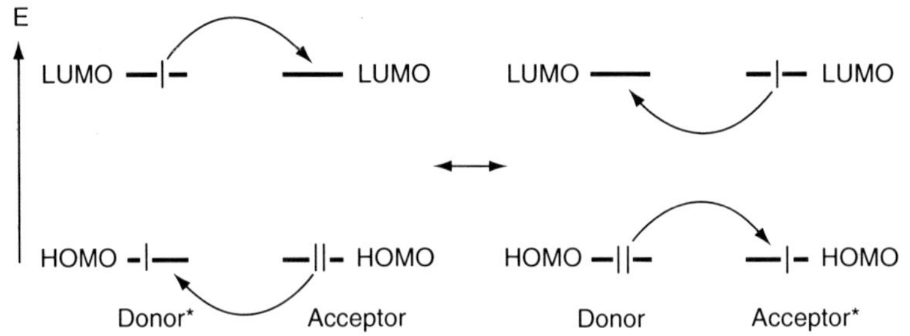

Figure 13.12. Electronic interaction between a donor and an acceptor due to orbital overlap (Dexter mechanism).

in which L is the effective van der Waals radius (4.8 Å for porphyrins), K is a constant, and J' is the overlap integral which can be expressed as

$$J' = \int_0^\infty I_\mathrm{D}(\lambda)\varepsilon_\mathrm{A}(\lambda)\mathrm{d}\lambda \text{ normalized so that } \int_0^\infty I_\mathrm{D}(\lambda)d\lambda = \int_0^\infty \varepsilon_\mathrm{A}(\lambda)\mathrm{d}\lambda = 1$$

Triplet–triplet energy transfers in which a $^3\mathrm{D}*-{}^1\mathrm{A}$ pair gives $^1\mathrm{D}$–$^3\mathrm{A}*$ can only occur via the Dexter mechanism because orbital overlap is required between the donor and the acceptor.

2.1.3. Nature of the Excitons

At this point, it is necessary to distinguish between two types of excitons: Frenkel excitons and Wannier–Mott excitons, because the nature of the exciton is closely related to their propagation mechanism.

As seen above, energy transfer via the Förster mechanism is due to resonant transitions, which can be seen as the emission of a virtual photon by the energy donor that is subsequently absorbed by the acceptor. In this case, the excited electron never leaves the energy donor and the virtual photon is transmitted via electrostatic coupling of the donor's transition dipole moment with that of the acceptor. Thus, the exciton electron–hole can be seen as a tight pair and is referred to as a Frenkel exciton. When the Dexter mechanism is involved in the energy transfer it requires an electronic coupling between the donor and acceptor. In such processes, the electron–hole pair is loose, and is referred to as a Wannier–Mott exciton. The preference for a Dexter or Förster energy transfer mechanism is strongly influenced by the nature of the bridge in covalent systems. If electron mobility is considered, then loose electron–hole pairs have a better chance of also playing a role in electron conduction and electron transfer.

Osuka, Anderson, and Warman have investigated the polarizability of two different covalently bridged porphyrin oligomers by flash photolysis time-resolved microwave conductivity measurements[18]. When the bridge forbids electronic interactions, only electrostatic interactions are present and the excitation generates Frenkel-type excitons. Upon photoexcitation, only a slight increase of polarizability is observed (<20 $Å^3$). When electronic communication is permitted by conjugation through the bridge, as demonstrated by strong bathochromic shifts of the Q band, the high polarizability of the singlet excited state is significant of the existence of Wannier–Mott-type excitons. It should be noted that in both cases, the T_1 state is localized on a single chromophore, despite the fact that triplet states are more likely to be involved in Dexter-type energy transfers. Albinsson has investigated the triplet–triplet energy transfer in various covalently linked zinc and free base dyads separated by a distance of 25 Å[19]. The use of saturated bridges affords electronically independent donor–acceptor pairs and no triplet energy transfer is observed. When unsaturated bridges are present, the triplet–triplet energy transfer is strongly dependent on the viscosity of the medium, which influences the degree of conjugation through the bridge. The dihedral angle between the aromatic bridges and the chromophore planes is also crucial, since coplanarity is required for electronic communication between the donor and the acceptor through the bridge.

2.2. Linking Chromophores: General Considerations

For energy transfer, the influence of the bridge on the mechanism of energy migration has been outlined above. Obviously, in the case of electron transfer in covalently linked species, the nature of the bridge will also be important because charge transfer can occur through space or through bonds. In a covalent molecular approach, the chromophores are usually linked to a molecular bridge. Still, several self-assembly approaches tend to use connections between chromophores and materials used in classical combined metal oxide semiconductor (CMOS) technology. It is necessary to succinctly present how excited states of porphyrin derivatives interact with the bridge structure or with a CMOS support.

An excellent definition of a molecular wire has been provided by Ratner and Wasielewski[20]. Conceptually, a molecular wire is a molecular bridge that can move charges over many chemical bonds. This implies a vectorial energy gradient that can generally be achieved by adjustment of the redox potentials of both the electron donor and acceptor at each end of the wire. Figure 13.13, based on the review article by Paddon–Row[14] on super exchange mediated charge transfer, considers only the interaction of a porphyrin excited state with a conducting or insulating covalent potential bridge. In situation (a), charge transfer will occur through space, and will involve an important distance dependence. In situation (b), charge transport can occur over larger distances, and its efficiency is related to the conducting properties of the bridge.

In porphyrin oligomers, two different cases should be distinguished. The first situation consists in several porphyrins linked by nonconducting bridges capable of achieving stepwise electron transfer according to situation (a), Figure 13.13. In this case, as mentioned above, a redox gradient will be required for a directed electron transfer. The second case considers molecular wires made of porphyrins. These will behave as unsaturated organic molecules and their efficiency will certainly depend on the degree of electronic coupling between the chromophores. In this case, electrical conduction can be measured by trapping the wires between nanoelectrodes in order to apply the required redox gradient[21]. Finally, an increasing number of articles describe the use of heterojunctions

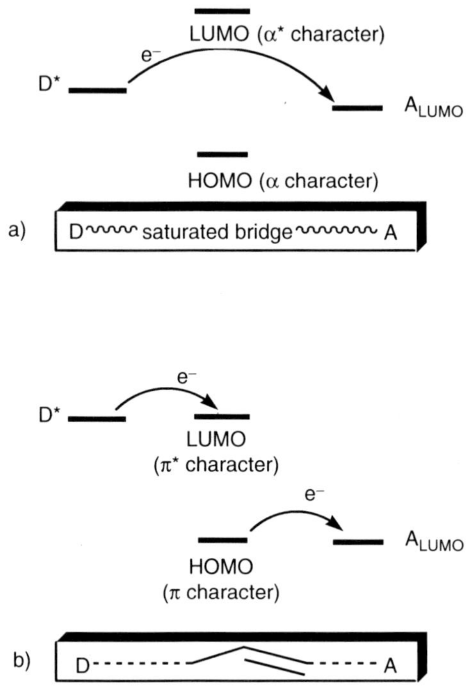

Figure 13.13. Energetic positioning of photoexcited state D* and accepter LUMO in electron transfer across saturated or conjugated covalent bridges.

between photoexcitable chromophores and semiconductors for applications in photovoltaics. In that case, considerations similar to those for unsaturated bridges can be made. As an example, Figure 13.14 shows the positioning of the free base of *meso*-tetrakis(4-carboxyphenyl)porphyrin H_2TPPC vs. the valence and conduction band of anatase TiO_2[22].

2.3. Energy and Electron Transfer in Photosynthesis

The general scheme of the photosynthetic process is represented in Figure 13.15. The photoinduced charge separation is initiated by collection of energy along two antennae systems, LH_1 (Light Harvesting) and LH_2, by three sets of bacteriochlorophylls. These are named B_{800}, B_{850}, and B_{880} after their respective absorption maximum. This energetic gradient illustrates the necessary spectral overlap between the energy donor and acceptor in energy conduction. Once localized on the LH_1, the photonic energy is trapped in the reaction center which will generate, through charge separation, electrochemical and proton gradients on both sides of the membrane, allowing the production of ATP. Charge compensation on the special pair occurs via a rather slow electron donation through a series of cytochromes *b* and c_1.

The initial boost for research in multiporphyrinic species was the resolution of the structure of the photosynthetic reaction center of *Rhodopseudomonas viridis*[23]. Indeed, the structure represented in Figure 13.16 provides extremely

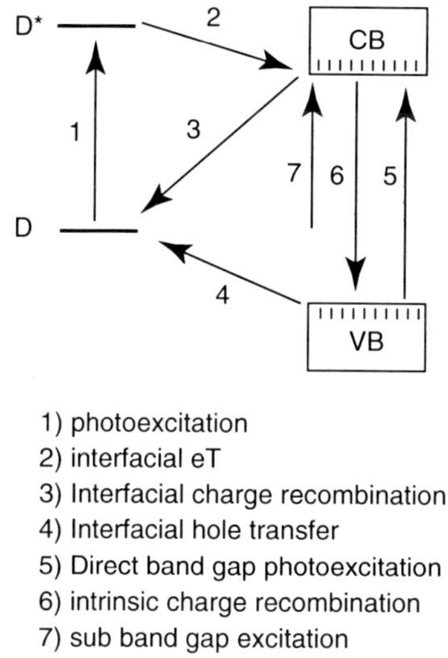

Figure 13.14. Energetic positioning of electron donor photoexcited state D* and valence/conduction bands (VB/CB) in anatase TiO_2.

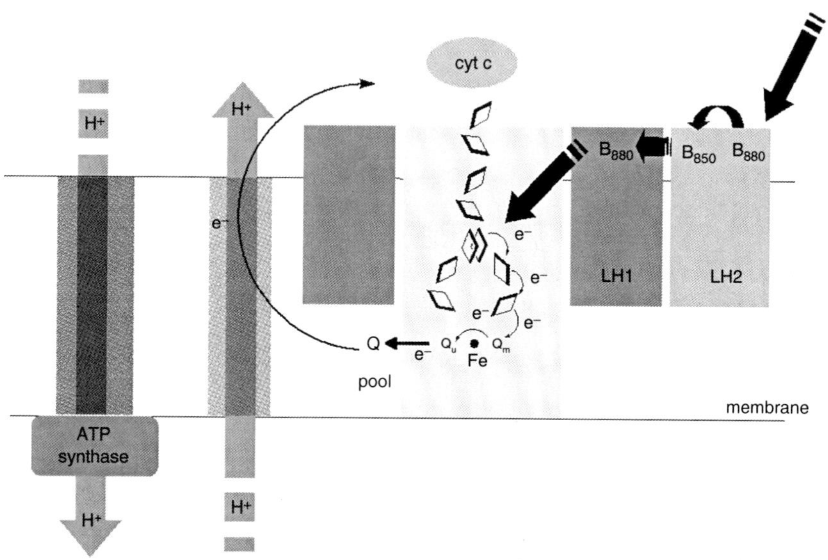

Figure 13.15. The bacterial photosynthetic process.

useful information on the topographical arrangement of the key chromophores involved in the charge separation process of the bacterial photosynthesis. Illustrating and anticipating the influence of this pioneering work in the field of synthetic energy and electron transport, this contribution of Deisenhofer, Huber, and Michel was readily rewarded by the Nobel prize of chemistry in 1988. Since the middle of the 1980s, numerous controlled arrangements of bis-porphyrinic species have

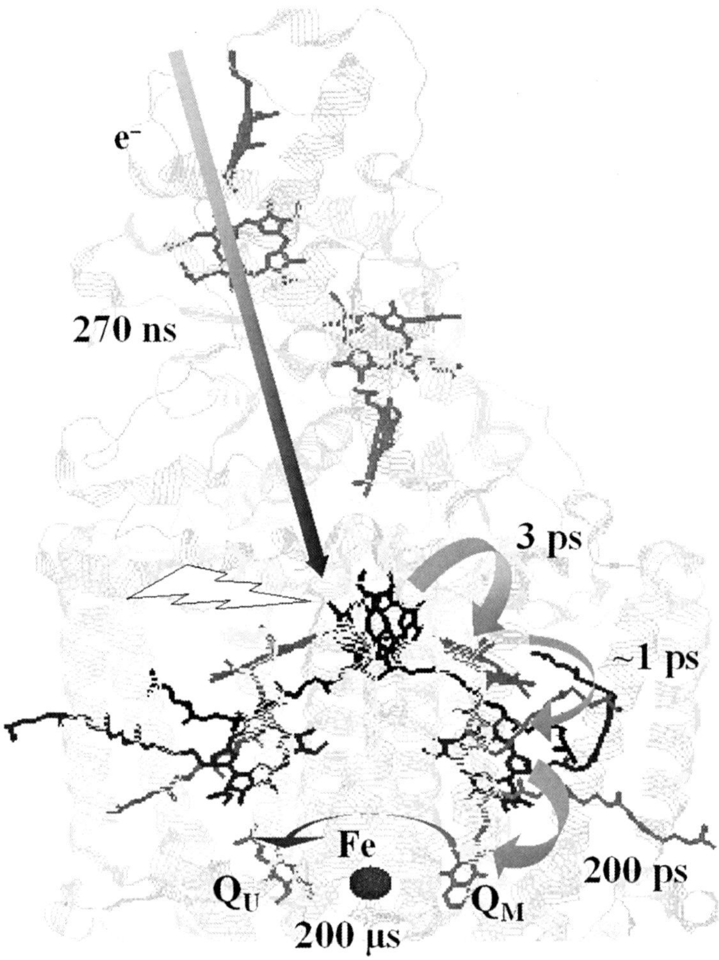

Figure 13.16. Charge transfer rates within the reaction center of *Rhodopseudomonas viridis*. Stick representation for tetrapyrrolic macrocycles only (2.3 Å res. generated with RasMol from PDF file 1PRC, ref.[23b]).

been designed to reproduce the interchromophore dihedral angles and distances as they are observed in nature.

Key rates of the photoinduced electron transfer process in the reaction center are noted in Figure 13.16. The excited special pair of bacteriochlorophylls (Figure 13.16) transfers an electron to an intermediate bacteriochlorophyll in *ca.* 3 ps. In *ca.* 1 ps, a second transfer takes place to the bacteriopheophytin, which is the last tetrapyrrolic chromophore involved in the electron transport. Electron transfer will then continue via a melaquinone Q_M (200 ps), a nonhemic iron, and a ubiquinone Q_U (200 μs). Charge compensation occurs in *ca.* 270 ns from the set of cytochromes. The left side of the reaction center does not participate in the electron transport, but its major role is probably to preclude back electron transfer

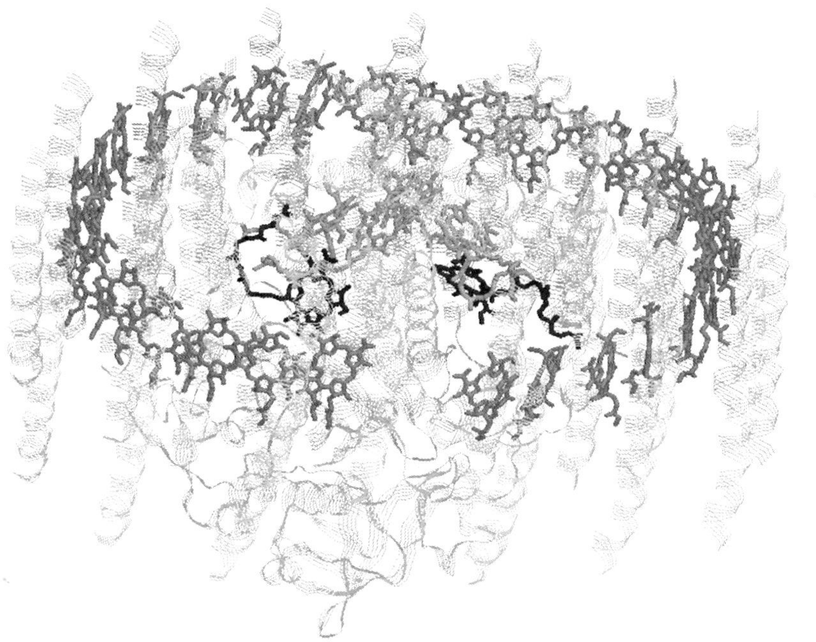

Figure 13.17. Circular arrangement of B_{880} chromophores in the LH_1 antennae system of *Rhodopseudomonas palustris*. Generated with RasMol from PDM file 1 PYH (ref.[24]).

from the Q_U to the special pair. None of the donor or acceptor moieties involved in this process are covalently linked to each other. It is important to note that the protein structure is responsible for the topographic adjustments made in order to meet physical requirements for efficient charge separation. The success of the process also lies in the extreme slowness of the back electron transfer through the only available pathway, that is, the unused left side of the reaction center.

Today, a keyword search of the PDB databank with the term "photosynthesis" will generate 190 deposited data. The more restrictive term "photosynthetic reaction center" will provide 59, and the keywords "light harvesting" will afford 33 hits. Very recently, Cogdell and Isaacs have reported the reaction center-LH_1 complex of *Rhodopseudomonas palustris* at 4.8 Å resolution[24].

The analysis of the structure provides an excellent summary of both photosynthesis and the role played by each tetrapyrrolic macrocycle in the successive processes of light harvesting and charge separation. In Figure 13.17, the reaction center (similar to that represented in Figure 13.16) is now located within a large ring shaped multichromophoric B_{880} assembly LH_1 in charge of energy collection. This circular arrangement of 15 pairs of bacteriochlorophylls collects the light energy from photons initially absorbed by the several LH_2 antennae, for which similar arrangements containing fewer B_{800} and B_{850} chromophores have also been reported[25]. In order to achieve this photonic energy collection, it is fascinating that all of the requirements for an optimized and efficient energy transfer detailed in the previous paragraphs are again fulfilled through the noncovalent assembly of pigments. Indeed, the bacteriochlorophylls B_{880} are arranged by coordination to helical α and β apoproteins.

2.4. Design of Synthetic Systems: The Molecular Electronics Challenge

By observation and better understanding of these sophisticated biological processes, the design of multiporphyrinic species capable of performing energy and electron transfer has been both stimulated and facilitated, leading to one of the major areas of research in molecular electronics in terms of activity and publications. In terms of applicability, the synthetic systems face two challenges.

The first challenge is the performance of the synthetic systems. The second aspect is the cost/performance ratio, which limits most of the results obtained so far to academic interest. At present, no synthetic system based on multiporphyrinic assemblies will reasonably compete with mass production of electronic devices via CMOS technology. For example, the estimated cost per bit of packaged DRAM is actually around 10 μcent (in US dollars)[26].

Completion of this second part of the challenge is conditioned by the conception and design of tailor-made, addressable molecular devices or molecular electronics with defined functions in terms of information propagation, storage, and processing. In this respect chemists have an important role to play, especially in designing efficient synthetic approaches for these devices. Molecular electronics is not limited to porphyrin derivatives, and the genesis of this general research area has been extremely well summarized in a recent review by Gorman[27]. In particular, this review emphasizes the role that porphyrin derivatives might play in the future development of molecular electronics.

Two complementary assembly strategies have been developed. The synthesis of covalently linked porphyrinic species offers the main advantage of very precisely controlling the geometrical features, in terms of distances, and dihedral angles, of the chromophores' respective arrangements. Thus, the covalent approach is extremely useful to obtain information on the parameters controlling the photophysics and the photochemistry of the assemblies. Twenty years ago, this approach was preferred to random aggregation or self-assembly involving weakly associated species. Nowadays, covalently linked species obtained via polymerization reactions complete the set of synthetic methods available. The explosion of supramolecular chemistry since the beginning of the 1980s has refined the tools available not only for the building of controlled self-assemblies, but also for the purification, characterization, and study of these species. Controlled arrangements similar to those obtained by covalent bonding can be achieved via self-assembly. Thus, this approach to multiporphyrinic species has a bright future because it usually involves a minimum number of synthetic steps to reach a programmed building block, and applications may be foreseen if the "simple" building block can be obtained on large scale. The multiporphyrinic species examined hereafter will be introduced according to the synthetic approach, covalent or noncovalent, used to generate them.

3. Scope and Limitations

Considering the enormous amount of literature concerning the use of porphyrin derivatives as photosensitizers in photoinduced processes, the title of this chapter has been deliberately restricted to multiporphyrinic assemblies. Thus, only species comprising more than one porphyrin for which energy or electron transfer data are available will be considered hereafter. The synthetic methods will not be discussed, despite their importance in terms of synthetic availability of potential new materials (see above). Again, concerning this aspect of multiporphyrinic species, several comprehensive literature reviews are available. Multiporphyrinic species can be assembled via coordination of external metals[28], axial coordination of the central metal of the porphyrin[29] especially in the case of pyridyl-porphyrin derivatives[30], weak interactions such as H-bonding[31], or by covalent bonding[32].

It is necessary to point out that numerous reports concerning the association of porphyrins and fullerene derivatives as primary electron acceptors have been published in the last decade. The interest in fullerene as an electron acceptor lies in the fact that its first one-electron reduction induces only small structural changes because the negative charge is delocalized over the entire carbon sphere. As a consequence, the internal reorganization energy term (involved in the definition of the electron transfer rate according to Marcus theory) is small. Charge separated species generated using fullerenes as an electron acceptor exhibit longer lifetimes because the back electron transfer occurs in the inverse Marcus region. For the reader interested in artificial photosynthetic applications of dyads, triads, and tetrads combining one porphyrin with nonporphyrinic electron donors or acceptors exhibiting stunning charge transfer properties and long-lived charge-separated states, several recent review articles are available[33–36].

4. Covalent Species

Two main sections will be devoted to well-defined porphyrin oligomers. The first deals with porphyrin dimers, trimers, and higher oligomers connected by various types of spacers. The second section will focus on similar compounds associated with nonporphyrinic species. It should be noted that assemblies of porphyrins connected to surfaces will be treated separately at the end of the chapter.

The properties of covalent dimers are primarily controlled by two distinct factors. First, the choice of the chromophores, depending on either the nature of the central metal core or the peripheral substitution of the porphyrin, will define the nature of the photoinduced event that will be observed (eT or ET). Second, the nature of the bridge that can be saturated or unsaturated. The bridge can force a perpendicular or orthogonal arrangement of the chromophores and will greatly influence the degree of communication between the chromophores. These parameters, which will be emphasized while analyzing porphyrin dimers, are also of importance for higher oligomers whose properties will be examined in light of the corresponding dimer's behavior when appropriate.

4.1. Covalent Dimers

4.1.1. Choice of the Chromophore

It is interesting to note that the use of porphyrinic structures in Nature usually deals with processes involving nonfluorescent species such as iron porphyrins in dioxygen storage, transport, or activation. When photoinduced events are performed, porphyrin structures are rather ignored by Nature, and partially reduced species, such as bacteriochlorin (BC) or bacteriopheophytin, are involved instead. Still, most photochemical and photophysical academic studies or applications inspired by natural photosynthetic processes extensively use porphyrins as chromophores. This deliberate choice may have important consequences.

When two chromophores are physically linked, a first analysis is usually made on the basis of their absorption properties. In the background introduction, the importance and the influence of the respective orientation of transition moment dipoles in the ground state absorption have been pointed out. Based on absorption, and especially the shape and positioning of the Q bands, the degree of electronic coupling can be estimated. As demonstrated by Yamaguchi, time-dependent density functional theory (TDDFT) studies have proven useful in this regard[37]. In the presence of identical bridges, *m*- or *p*-phenylene, the Q band couplings in free base (FBBC)–zinc bacteriochlorin (ZnBC) dimers **1** and **2** (Figure 13.18), and free base porphyrin (FBP)–zinc porphyrin (ZnP) dimers **3** and **4** have been examined based on both the pattern of the set of frontier orbitals and calculation of the angle dependent contribution of the bridge's π orbitals.

The oscillator strengths of low-lying excited states have been calculated and point out a significant difference between the bacteriochlorin and porphyrin monomers. In BC derivatives, degeneracy in HOMO and HOMO−1 orbitals is lifted compared to porphyrin analogs. In addition, the partial reduction of two opposite pyrroles also clearly differentiates the transition moments by affecting the HOMO and LUMO+1 in both the FBBC and the ZnBC. This has a strong effect on the oscillator strengths due to transitions from HOMO−1 and HOMO to the LUMO and LUMO+1 orbitals, for which self-compensation observed

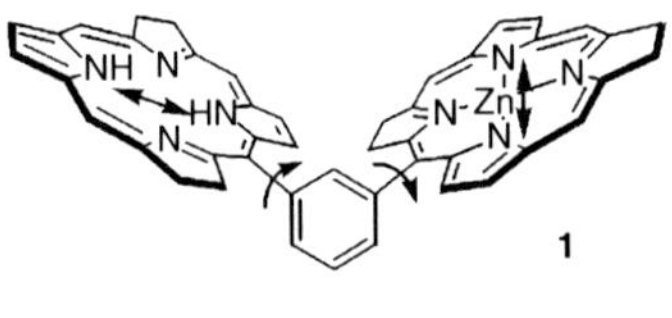

1

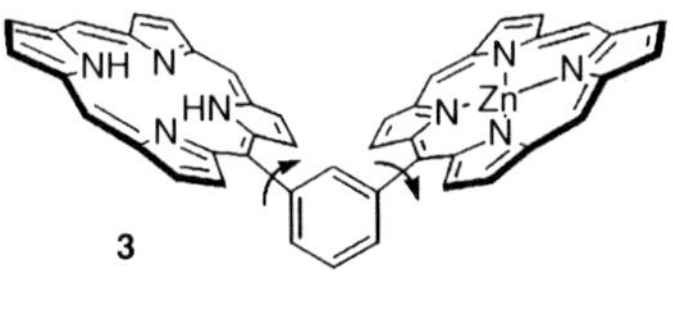

3

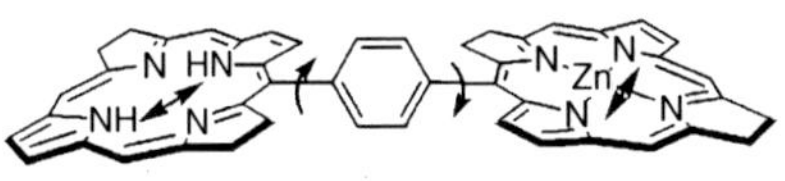

2

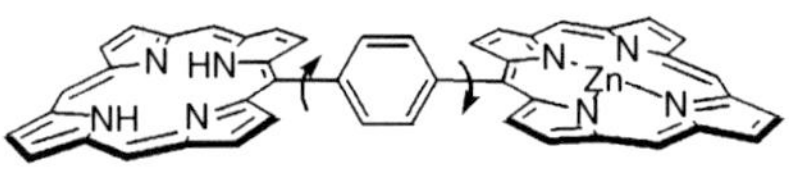

4

Figure 13.18. Phenylene bridged FBCC–ZnBC dimers **1**, and **2**, and porphyrin analogs **3**, and **4**. Note that in **1** and **2** the rotation around the *meso* carbon phenylene carbon bond will lead to situations in which the Q_x dipole moments (arrows) will be parallel in **1**, and orthogonal in **2**.

in porphyrin monomers does not occur in the corresponding bacteriochlorins. Indeed, in BC derivatives, no amplitude corresponding to a contribution of the sp^3 carbons is observed, and the remaining aromatic π system is more localized with shorter bond lengths. The red shift of the Q_x band in FBBC compared to that in FBP is explained by a narrower HOMO–LUMO gap in FBBC monomer. In dimers **1** and **2**, the coupling of transition moments induces both an unchanged set of Q bands plus a set of significantly red shifted Q' bands (0.5–0.7 eV). The latter shift is assigned to Q_x transition mixing, which in the case of **3** and **4**, produces only a small red shift (0.2–0.3 eV)[38].

In bacteriochlorin dimers **1** and **2**, a through space overlap of the unreduced pyrroles' LUMOs explains the enhanced mixing of monomer-like Q_x transitions. This overlap is in addition to π–π interactions through the phenylene bridge that are also observed for **3** and **4**. This is important regarding the use of porphyrin chromophores to model the properties of natural systems, because it implies that BC dimers in the photosynthetic apparatus are able to perform Q_x excitation energy transfer via the LUMO. This is not possible in model porphyrin dimers.

4.1.2. Electronic Coupling in Covalently Linked Dimers

In covalent dimers, the bridging element between the two chromophores may restrict the respective orientations of the two chromophores, and thus, influences the degree of their electronic coupling. In the absence of a bridge, the degree of electronic coupling between two porphyrins linked directly at the *meso* position can still be controlled by the peripheral substitution. Indeed, the peripheral substitution can affect the nature of HOMO frontier orbital.

As depicted in Figure 13.19, the tetraaryl porphyrin's HOMO has an A_{2u} character with significant electron density at the *meso* carbon, whereas alkyl-substitution of the eight β pyrrolic positions in a 5,15-diaryl porphyrin confers an A_{1u} character to the HOMO, with no electron density on the *meso* carbon. Since the *meso* carbons are implicated in both covalent bonding to moieties and in possible through bond electronic couplings, their involvement in the description of the HOMO is crucial. This is nicely illustrated in a paper by Osuka and Kim[39] who have investigated the energy transfer in *meso* linked hybrid Zn–H_2 bis-porphyrins, and compare the behavior of their directly linked species with various phenylene and Lindsey-type phenyl-ethynyl linkers. Table 13.1 reproduces

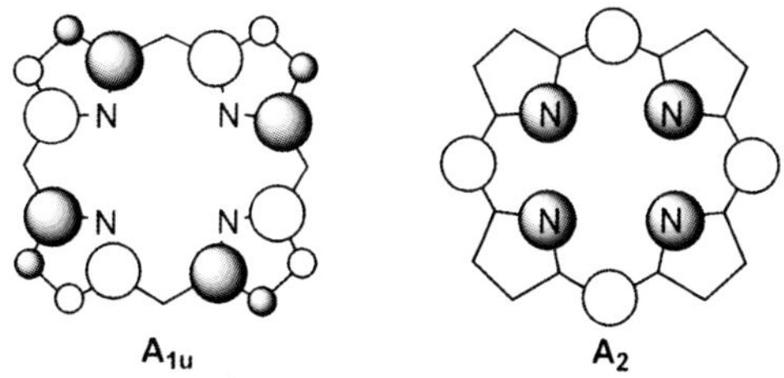

Figure 13.19. 5,15-Diaryl octa-alkyl porphyrin HOMO has an A_{1u} character. The tetraaryl porphyrin HOMO has an A_{2u} character.

Table 13.1. Energy Transfer Data for Hybrid Zn–H_2 Dimers Depicted in Figure 13.20.

Array	Chromophore								HOMO A_{2u}		HOMO A_{1u}		References
	R	R'	R"	Ar	Ar'	X	R (Å)	κ^2	$k_{ET}(10^{-12}s)$	$k_F(10^{-12}s)$	$k_{ET}(10^{-12}s)$	$k_F(10^{-12}s)$	
5	R^1	R^1	R^1	Ar^1	Ar^1	A	26.6	1.125	156	3200	—	—	[40]
6: X = A	26.6	R^4	R^1	—	Ar^1			1.125	—	—	1750	1450	[40]
7	R^1	R^1	R^1	Ar^2	Ar^2	B	19.6	1.125	24	745	—	—	[41]
8: X = B	19.6	R^4	R^2	—	Ar^1			1.125	—	—	417	233	[41]
9: X = B	19.6	R^1	R^1	Ar^4	Ar^4			1.125	—	—	240	1326	[41]
10	R^1	R^1	R^1	Ar^1	Ar^1	C	17.0	1.1	11	220	—	—	[39]
11: X = C	17.0	R^2	R^3	—	—			1.1	—	—	44	100	[39]
12	12.7	R^1	R^1	Ar^1	Ar^1	D	12.7	1.125	3	37	—	—	[39] [42]
13: X = D	12.7	R^2	R^3	—	—			1.125	—	—	10	17	[39] [42]
14: X = D	12.7	R^1	R^1	Ar^4	Ar^4			1.125	—	—	10	79	[39] [42]
15	R^1	R^1	R^1	Ar^5	Ar^5	E	11.0	0.531	3	33	—	—	[39]
16: X = E	11.0	R^2	R^3	—	—			0.531	—	—	10	15	[39]
17	R^1	R^1	R^1	Ar^3	—	—	—	—	0.55	—	—	—	[39]
18 X = -	—	R^1	R^1	Ar^4	—			—	—	—	0.55	—	[39]

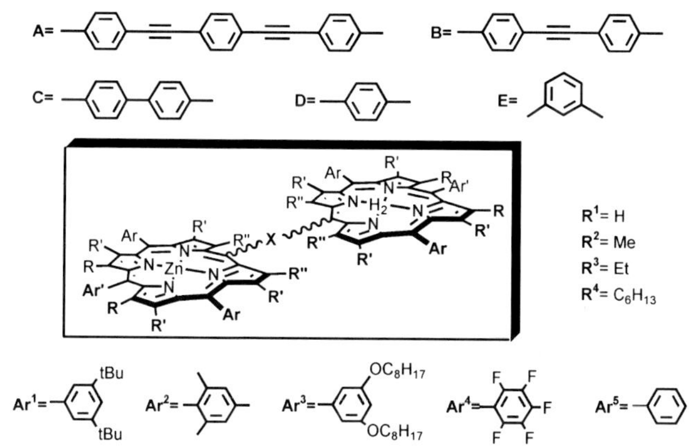

Figure 13.20. General structure of the hybrid dimers Zn–H_2 studied by Osuka and Kim in ref.[39–42].

experimental and calculated data concerning ET processes in the species depicted in Figure 13.20[39–42].

These data can be analyzed by comparing the calculated Förster energy transfer rate k_F to the experimental energy transfer rate k_{ET}. From this comparison, the contribution of Coulombic interactions to this process can be estimated. When discrepancies are observed, involvement of an electronic coupling contribution is claimed, and the larger the difference between k_F and k_{ET}, the stronger the electronic coupling.

Larger differences between k_{ET} and k_F are observed for long π-conjugated bridges. An extended conjugation results in low lying LUMO of the bridges, which enhances the electronic coupling and its contribution to the ET process (arrays **5–10** in Table 13.1). It is also clear that A_{2u} HOMO involving electron density on the connecting *meso* carbons significantly enhances the electronic coupling of the chromophores (arrays **5, 7, 10, 12,** and **16** in Table 13.1). In both series of HOMO character, it should be noted that the nature of both the bridge and the β substituent adjacent to the connecting *meso* carbon have a significant influence on the ET transfer rate. This is due to an enhanced perpendicular orientation between the *meso*-phenyl portion and the plane of the chromophore when the R$''$ substituents are methyl (R^2) and ethyl (R^3) groups (arrays **8, 11, 13, 16** in Table 13.1). This decreases the possibility of electronic coupling between π orbitals of the chromophores through the perpendicular π orbitals of the bridge. Smaller spacers (arrays **11–14**) tend to homogenize the behavior of A_{2u} and A_{1u} HOMO type compounds because through bond and Coulombic interactions are both increased.

Finally, direct linking of the chromophores affords species in which both chromophores are perpendicular, and thus, electronic interaction between the chromophores is minimum. In this case Coulombic interactions are responsible for a very fast ET between chromophores exhibiting independent excited states.

The case of directly *meso* linked chromophores deserves more attention due to their involvement in the construction of large porphyrin arrays that will be discussed later on. Exhaustive studies of directly linked Zn–Zn homodimers have been reported by the same authors, particularly in order to determine the influence of the dihedral angle between the two porphyrin rings on their photophysical behavior in both the ground and excited states[10b,c]. For this purpose, several strapped porphyrin dimers depicted in Figure 13.21 have been prepared. The size of the strap controls the dihedral angle between the chromophores and their electronic coupling. In both absorption and fluorescence spectroscopy, compounds **19** ($n = 10$) and **20** ($n = 10$) behave the same as the nonstrapped *meso–meso* linked dimers depicted in Figure 13.21. From the evolution of the B and Q absorption bands, it is obvious that as the length of the strap decreases, the electronic coupling increases with persistent red shifts and the progressive splitting of absorption bands.

Calculations of eight frontier molecular orbitals for D_{2h} dimers ($\theta = 90°$), and six frontier orbitals for D_2 dimers were carried out from the four frontier orbitals of the monomers. A simplified description of the frontier orbitals' evolution is reproduced in Figure 13.22, along with a transition dipole moment representation in the orthogonal and oblique bis-porphyrins. Raman resonance spectroscopy of the excited states shows that some transitions involve charge transfer between the two subunits, and that the contribution of charge transfer increases with the degree of coplanarity of the dimers. This is consistent with previous electric dipole measurements in the excited states.

Based on these observations, it is clear that the geometric restrictions introduced by the bridge in covalent dimers are very important. As seen in the beginning of this section and in previous literature coverage[1], phenyl and ethynyl linkers have been widely used to connect porphyrin homo- and heterodimers. We can differentiate bridging units in which an ethyne is directly connected to the *meso* position because this type of link does not affect the free rotation of the chromophore along the axis of the bridge. In this case, electronic coupling is efficient and communication between the two chromophores occurs preferentially via through-bond interactions, as shown by the study of the asymmetric ethyne bridged zinc bis porphyrins **21** and **22** represented in Figure 13.23.

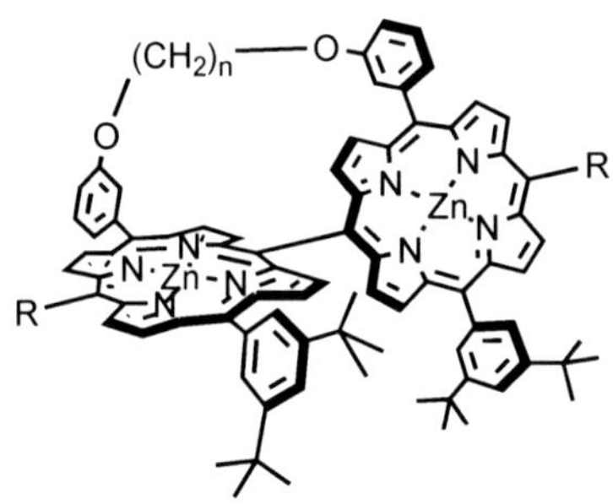

19 : R= Ph, n=1, 2, 3, 4, 5, 6, 8, 10

20 : R= H, n=1, 2, 3, 4, 5, 6, 8, 10

Figure 13.21. Osuka's strapped dimers.

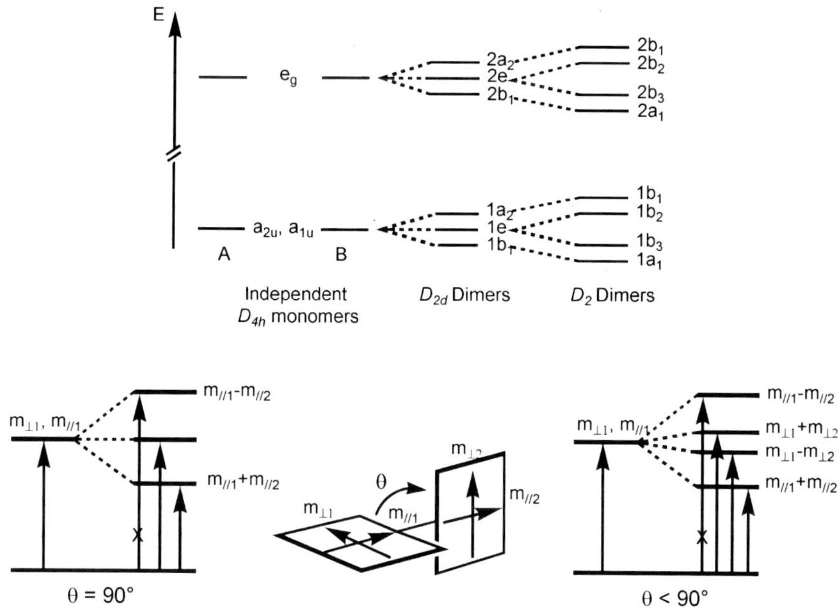

Figure 13.22. Frontier orbitals (top) and dipole moments (bottom) in orthogonal and oblique porphyrin dimers.

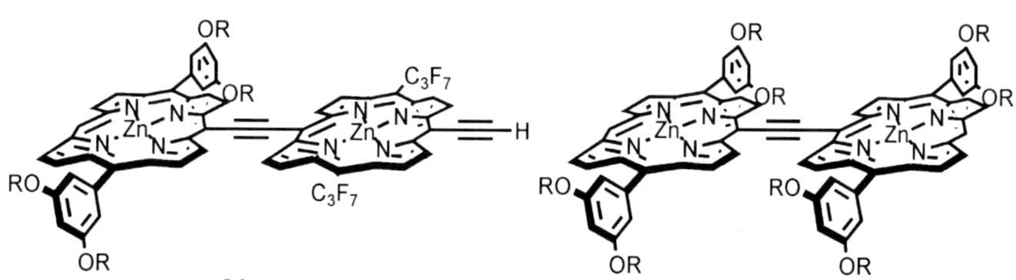

21 Donor D Acceptor A

22 D-D reference

Figure 13.23. *Meso–meso* ethynyl bridges result in efficient electronic coupling of the two chromphores.

Due to their conformational flexibility, compounds **21** and **22** both exhibit broad absorptions arising from conformers with large torsional angles for the high energy portion of the spectrum, and from conformers with maximum conjugation for the low energy components of the Q_x bands. Due to the electron withdrawing nature of the peripheral substituents in the acceptor A, the ground state of **21** displays a dipole moment of *ca.* 2 D, which is consistent with a larger Stokes shift in **21** (1,148 cm^{-1}) than in the symmetric apolar dimer **22** (324 cm^{-1}). This is explained by the larger energy consumed by outer sphere solvent reorganization in the case of the electronically asymmetric **21**, in which the acceptor's HOMO is of the a_{1u} type because the electron withdrawing *meso* substituents stabilize the

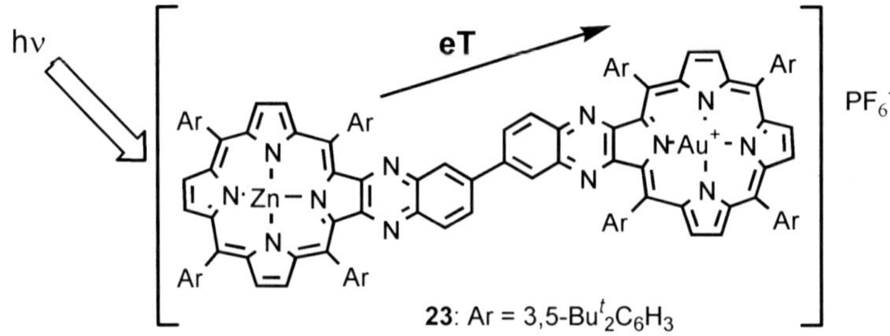

23: Ar = 3,5-$Bu^t_2C_6H_3$

Figure 13.24. $[ZnPQ\text{-}AuPQ]^+PF_6$ dyad.

a_{2u} orbital. Time resolved near infra-red (NIR) transient absorption spectroscopy experiments on both **21** and **22** were carried out for $S_1 \rightarrow S_n$ and $T_1 \rightarrow T_n$ transitions. Hole burning experiments using excitation on both sides of the Q_x band have been performed. Excitation on the blue side of the Q_x band involves conformers with large dihedral angles between the planes of the chromophores, while pumping in the red side of the Q_x generates S_n excited states of more conjugated species. Dynamic studies show that the respective x-polarized Q transitions ($S_0 \rightarrow S_1$) and ($S_1 \rightarrow S_n$) are parallel and aligned with the ethyne bond axis. Excited state spectral changes from **22** to **21** are explained by rotational dynamics around the ethyne bond, and a supplementary relaxation process in the case of the polar **21**, for which interaction with the THF solvent reduces the $S_1 \rightarrow S_0$ transition oscillator strength[43].

Of course, the influence of the bridge on the degree of electronic coupling appears to be an excellent way to promote electron transfer from a donor to an acceptor. Fukuzumi, Kadish, and Crossley have reported the use of covalently linked chromophores in which electronic coupling is enhanced by a bridge that allows a coplanar orientation of the donor and the acceptor[44]. In addition to the choice of a biphenyl type link, in which each phenyl is fused with the tetrapyrrolic rings, the subtlety of the hybrid dimer **23** depicted in Figure 13.24 is the use of a nonneutral dyad.

The link provides a dihedral angle-dependent degree of electronic coupling similar to that observed in the directly linked *meso-meso* dimers described previously. Thus, communication between the chromophore does not need to be established. By using a nonneutral acceptor, the redox potential of the gold porphyrin can be tuned with the help of solvent polarity. Indeed, the cationic gold porphyrin is easier to reduce in toluene ($E^0_{red} = -0.30$ V vs. SCE) than in benzonitrile (PhCN) ($E^0_{red} = -0.40$ V vs. SCE) due to weaker solvation in nonpolar solvent. The oxidation potential of the zinc donor remains virtually identical in both solvents. As a result, the fluorescence of the zinc porphyrin in dyad **23** is quenched 20-fold compared to that of the isolated zinc porphyrin reference. Not only is the electron transfer slower, but the charge separated (CS) state also lives longer than in neutral dyads in apolar solvents. In fact, neutral dyads in which electron transfer generates a charge separated state from a neutral species, and thus induces the presence of a strong dipole in the CS state, are poorly stabilized in apolar solvents. Consequently, the CS state is usually higher in energy than the triplet state that can be generated during the charge recombination. In the dyad in Figure 13.24, the electron transfer corresponds to the translation of the dipole already present in the ground state. In other words, a charge shift occurs instead of a charge separation. This correlates the prediction from Marcus theory which states that the smaller the reorganization energy of electron transfer, the slower the back electron transfer in the inverted region. The energy of the CS state in toluene (1.11 eV) is smaller than the 1.21 eV value in PhCN.

Experimental observation of $ZnPQ^{\bullet +}$ and AuPQ by transient absorption spectroscopy and EPR after 1.4 μs is consistent with the interpretation of the slow back electron transfer (BET), for which the slowest rate ($k_{BET} = 10\,\mu s$) is observed in cyclohexane (9.1 μs in toluene). No CS state was observed in PhCN, probably due to a larger reorganization energy during the electron transfer. The importance of the reorganization energy during the eT processes will be emphasized in multiporphyrinic species associated with C_{60} acceptors later in this chapter. When unsaturated bridges are used, the donor and acceptor porphyrins rotate freely. Thus, only the contribution of fully conjugated conformers enhances the electronic interaction between the two porphyrins, even over large distances.

A series of 12 porphyrins depicted in Figure 13.25 has been prepared and studied by Odobel and collaborators in order to analyze the linkage dependence of the electronic coupling, and its influence on the efficiency of energy transfer between a zinc porphyrin energy donor and a free base acceptor. Ground state absorption and MO calculations are, respectively, consistent with strong electronic coupling between the two chromophores, and the contribution of cumulenic forms of the bridge implies coplanarity of the porphyrin macrocycles[45]. In the four dyads highlighted in Figure 13.25, the fluorescence of the zinc porphyrin energy donor is quenched by more than 95%. It is remarkable that the most efficient energy transfer (**27ZH**, $k_{ET} = 12 \times 10^{10}\,s^{-1}$) is observed over the largest distance (28.6 Å center to center) for the tetrathiophene bridge in which ethynyl connections to the porphyrins authorizes a stericly unrestricted conjugation of the

Table 13.2. Absorption and Fluorescence Data for Dimers **24–27**

Dimer	λ_{max} (nm)	τ_f (ns)	Φf	$k_{ET}(s^{-1})$
24ZZ	624	1.6	0.10	
24HH	662	3.9	0.86	
24ZH	664	0.017	0.86	6×10^{10}
25ZZ	632	1.3	0.085	
25HH	667	3.5	0.78	
25ZH	666	0.32	0.78	3×10^{10}
26ZZ	676	1.3	0.17	
26HH	717	3.4	0.22	
26ZH	713	0.033	0.21	3×10^{10}
27ZZ	657	1.4	0.18	
27HH	690	6.2	0.34	
27ZH	688	0.008	0.31	6×10^{10}

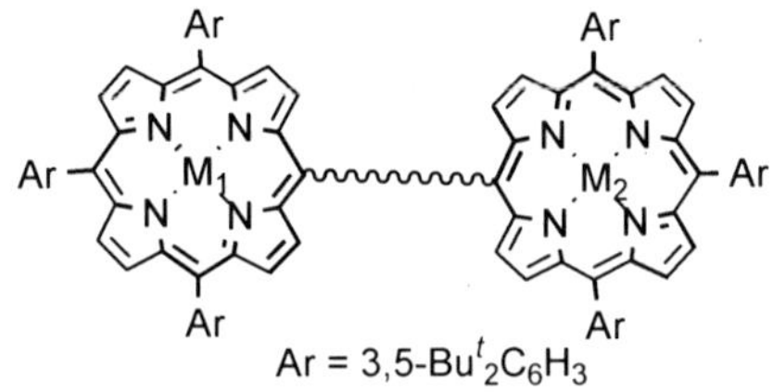

M_1	M_2	Bridge	Name
H_2	H_2		**24HH**
Zn	H_2	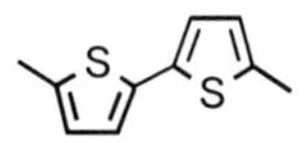	**24ZH**
Zn	Zn		**24ZZ**
H_2	H_2		**25HH**
Zn	H_2	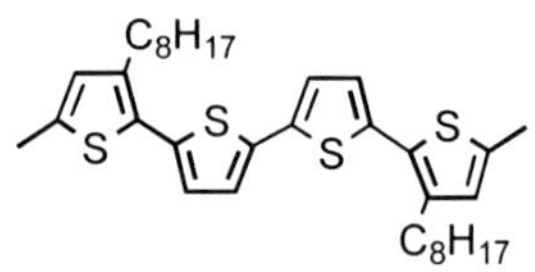	**25ZH**
Zn	Zn		**25ZZ**
H_2	H_2		**26HH**
Zn	H_2	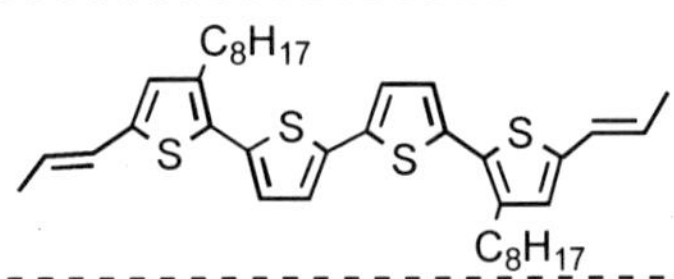	**26ZH**
Zn	Zn		**26ZZ**
H_2	H_2		**27HH**
Zn	H_2	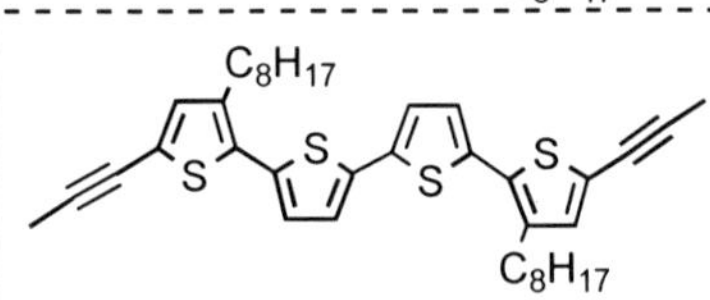	**27ZH**
Zn	Zn		**27ZZ**

Figure 13.25. Thiophene linked homo- and heterodimers.

dyad components. In these dyads, Förster mechanism is obviously ruled out in the energy-transfer process.

The case of dyads in which the chromophores are linked through *meso* aryl substituents has been investigated by several groups, and detailed studies prior to 2000 can be accessed through references[1,46]. In these systems, even though the *meso* aryl linker cannot be fully conjugated with the chromophore for steric reasons, the presence of π and π^* orbitals between the components of the dyad plays a role in ET processes for zinc free base hybrid dimers. Following their previous studies on such dimers, Albinsson and collaborators have performed several in depth analysis of both eT and ET in dimers **28–30** depicted in Figure 13.26. The influence of the bridge on the efficiency of hole transfer, which occurs when excitation is performed on the gold porphyrin (M is Au(III)), has been investigated. The quenching of ^{3}(AuP)* in **29-O,B,N** is highly efficient when a fully conjugated phenyl or naphthyl spacer is used. Quenching is inefficient for a bicyclo[2.2.2]octane bridge that does not provide an accessible π–π^* orbital as an electronic pathway for chromophore interaction. Even though a stepwise

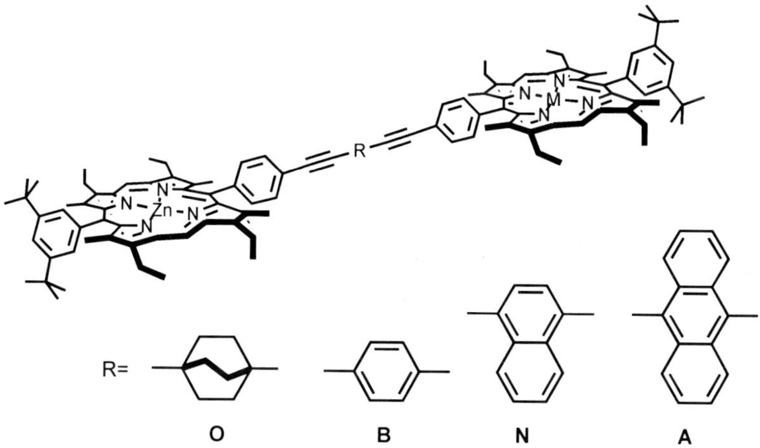

28-O: R = O, M = H_2 **29-O**: R = O, M = $Au(III)^+BF_4^-$ **30-O**: R = O, M = Fe(III)Cl
28-B: R = B, M = H_2 **29-B**: R = B, M = $Au(III)^+BF_4^-$ **30-B**: R = B, M = Fe(III)Cl
28-N: R = N, M = H_2 **29-N**: R = N, M = $Au(III)^+BF_4^-$ **30-N**: R = N, M = Fe(III)Cl
28-A: R = A, M = H_2 **29-A**: R = A, M = $Au(III)^+BF_4^-$ **30-A**: R = A, M = Fe(III)Cl

Figure 13.26. Albinsson's dyads M_1P–R–M_2P for electron, energy, and hole transfer.

mechanism with a primary hole transfer from the bridge to the triplet hole donor $^3(AuP)^*$ is conceptually permitted, it is ruled out due to the similar values of the room temperature lifetimes of this species in both the dyad and in a reference R–AuP^+ compound. Thus, no quenching occurs unless a ZnP hole acceptor is connected. The hole transfer is explained by a super exchange mechanism. Again, the hole transfer induced by triplet state formation on the electron acceptor (or hole donor) $^3(AuP)^*$ corresponds to a charge shift and not literally to a charge separation. The corresponding "CS" states display lifetimes of 30 and 50 ns for **29-N**, and **29-B**, respectively[47].

The same organic architecture has been studied using different metals for M. The introduction of Fe(III)Cl generates a nonfluorescent FeP species, associated with a photoactive ZnP species (dimers **30-O,B,N,A**). Subsequent excitation of the ZnP moiety generates a singlet S_1 state which, in the ZnP–R reference compound, is deactivated via fluorescence (S_1 emission) and phosphorescence (T_1 emission via intersystem crossing "isc"). Previous results show that in dyads **28-O,B,N,A** the nature of the bridge affects the rate of singlet energy transfer without significantly influencing the rate of isc. In dyads **30-O,B,N,A**, the Marcus theory predicts a favorable electron transfer from the S_1 state of the ZnP donor to the FeClP acceptor. In this series the S_1 quenching is independent of the solvent polarity, which influences λ_{out} (reorganization energy) in normal region eTs. Thus, the contribution of eT to the S_1 deactivation is minimum, and no $ZnP^{\bullet+}$ can be detected. A weak contribution of ET from the S_1 state of the ZnP donor to the FeClP acceptor is estimated by comparison of the dyads **30-O,B,N,A** with their **28-O,B,N,A** analogs, and from the knowledge of the ET process in ZnP–R–H_2P energy dyads previously reported[46]. Thus, instead of deactivation of the

ZnP donor S_1 state via eT or ET in dimers **30-O,B,N,A**, enhancement of the isc $S_1 \rightarrow T_1$ conversion is observed for R=B, N, and A. As no eT or ET was observed, the presence of paramagnetic Fe(III) near the ZnP S_1 state seems to be a remote promoter of isc. It was previously mentioned that the nature of the metal complexed in the tetrapyrrolic chromophore could enhance the isc mechanism, but no remote example of this enhancement had been observed so far. The crucial information pointed out by this study is that fluorescence quenching is not the only evidence for eT or ET processes[48]. Regarding the behavior of the ZnP–R–FeClP species (**30-O,B,N,A**), it is worth mentioning that R=A confers properties of an eT triad to the dimer, with around 20% efficiency. In polar solvents, excitation of the ZnP donor generates a CS state with a lifetime of at least 5 ns via a stepwise eT from the donor to the bridge A, followed by a second eT from the anthracene to the iron porphyrin, which subsequently releases the coordinated chloride anion. Given that the photogenerated iron(II) porphyrin can bind up to two axial bases, pyridine can be used to tune the efficiency of the eT in the triad[49]. We will see later that axial base coordination can actually act as a switch for dyad properties.

The mediating effect of the conjugated bridges has been previously observed by the same authors for energy dyads involving **28-B** and **28-N**. Although full conjugation of the bridge with the chromophores is not possible due to the methyl substituents at the β pyrrolic positions adjacent to the connecting *meso* carbon, it has been demonstrated that unsaturated bridges were required for the observation of singlet and triplet ET between the ZnP donor and the H_2P acceptor. Triplet–triplet ET has been further investigated, because, despite the mediating effect of the bridge, the triplet states of $^3R^*$ are too high in energy to serve as a relay. In a first series of papers, Kyrychenko and Albinsson[50,51] have pointed out particular features of the $^3ZnP^*$ or $^3H_2P^*$, and the corresponding dyads, using time resolved spectroscopy and DFT calculations. First, the triplet state of the ZnP* donor is composed of two states at different energies, corresponding to two different distortions of the porphyrin skeleton. The planar, high energy T_A state can be rapidly deactivated (100 ns at room temperature) via conversion into a saddled, lower energy T_B state. In the saddled T_B state the dihedral angle between the porphyrin and the phenyl spacer is significantly decreased. Consequently, the triplets' lifetimes for the ZnP donor are temperature and, most importantly, viscosity dependent. High viscosity media such as polystyrene films provide millisecond lifetimes and slow down both the T_A to T_B conversion and the triplet ET due to conformational restrictions lowering the degree of planarity[19].

Finally, studies of eT using M=Au(III) in the same architectures (**29-O,B,N,A**) has been pursued[52] and falls into the charge shift category rather than charge separation processes. Similar findings were also observed later by Crossley[44]. When R=B, or N, the eT proceeds directly from ZnP to AuP^+. In polar solvents, however, with R=A a stepwise eT is observed, as is the case for a FeClP acceptor[49]. In ZnP/AuP^+ electron transfer dyad systems, their very distinctive behavior depends on the way they are connected. This is emphasized by comparison of the works of Crossley[44] and Albinsson[52]. Both authors use similar donor/acceptor components with either a direct connection between possibly coplanar chromophores[44], or a roughly perpendicular mediating bridge with a gradually accessible π–π* MO pathway for electronic coupling[52]. With a direct connection and strong electronic coupling, eT is efficient, and low solvent polarity tunes the free energy associated with the eT and enhances the lifetime of the CS state[44]. In the case of weakly coupled chromophores ($V = 5$–$20\,cm^{-1}$), high solvent polarity is required, particularly to promote a stepwise eT to the anthracene bridge. The first reduction potential of the Au(III)P^+ species in the various solvents used is not discussed[51].

It was previously observed that the binding of an axial base can stabilize a photogenerated neutral Fe(II)P[49]. Such coordination can also be used as a switching tool for the photoactivity of a ZnP electron or energy donor. Otsuki and collaborators have reported a switch of this type, using a phenylazopyridine (PAP) axial base[53]. In the dyad **31** depicted in Figure 13.27, the two chromophores display ground state absorption that is the sum of the independent species and that are weakly coupled. Efficient ET (99% at $4.2 \times 10^{10}\,s^{-1}$) is observed upon excitation of the ZnP donor in the absence of PAP (association constant $k_{assoc} = 1.2 \times 10^4\,M^{-1}$, CH_2Cl_2, 298 K) or in the presence of a PAP competitor such as dimethyl aminopyridine (DMAP, $k_{assoc} = 2.9 \times 10^5\,M^{-1}$). In the presence of PAP, 97% eT to the diazo quencher is observed, precluding ET from ZnP to the H_2P. In addition to the fact that axial coordination brings the oxidative quencher

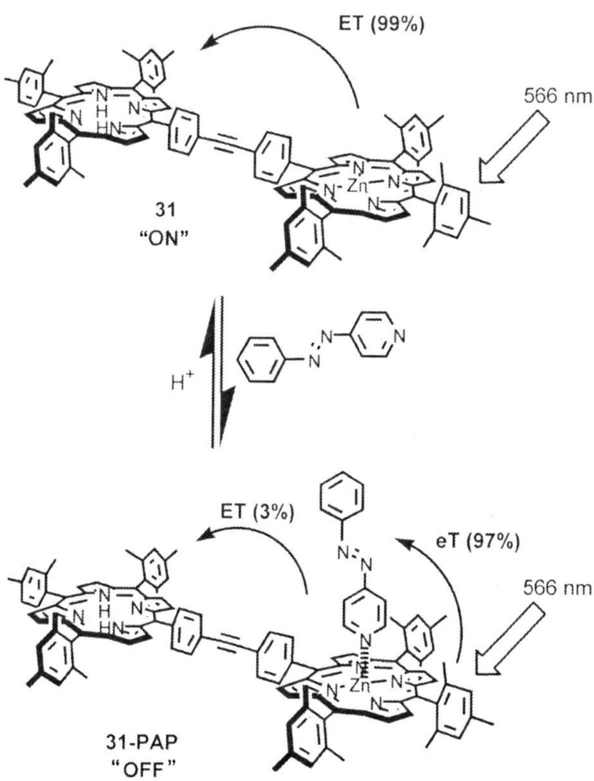

Figure 13.27. A molecular switch based on axial binding.

in the vicinity of the ZnP, it is important to note that the electronic enrichment of the ZnP by axial base coordination also enhances the reducing character of the pentacoordinated ZnP donor by approximately 0.1 V. The idea of tunable eT or ET is one obvious evolution that can be foreseen in the development of molecular devices. Applications may be developed not only for information storage, but also in the emerging field of "data treatment" with the help of molecular logic gates. This aspect will be treated in the next section, and the degree of innovation in these systems resides in the architecture and the topology of the devices[54].

From the examples discussed above, it is rather clear that the increase of electronic coupling through the bridge is not the only way to influence the efficiency of electron or energy transfer between chromophores. Bridge-induced topographic restrictions may enhance ET processes by affording properly oriented conformers for the chromophores. Lindsey, Bocian, and Holten[55] have reported a series of imine linked energy dyads **32–35** (Figure 13.28) in which H-bonding within the linker is associated with the 60° dihedral angle between porphyrins and their *meso*-aryl groups.

These noncovalent interactions introduce geometric restrictions that most probably affect the κ values. Ground state absorption properties show that the chromophores are weakly coupled when compared to reference monomers comprised of the bridge-substituted donor or acceptor. Highly efficient ET rates, ranging from 70 to 13 ps^{-1}, have been observed for ET from the zinc donor to the free base acceptor. Almost no competition with hole transfer, due to excitation of the free base, was observed, except for dyad **35** in Figure 13.28. Excited state lifetimes for the individual chromophores and ET quantum yields are given in Table 13.3.

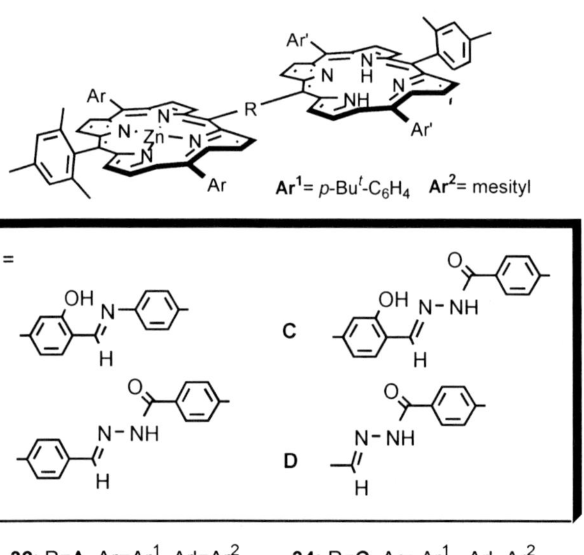

32: R=**A**, Ar=Ar^1, Ar'=Ar^2 **34:** R=**C**, Ar= Ar^1 , Ar'=Ar^2

33: R=**B**, Ar=Ar'=Ar^2 **35:** R= **D**, Ar=Ar^1, Ar'=Ar^2

Figure 13.28. Schematic representatin of dyads **32–36** from ref.[55].

Table 13.3. Room temperature data for dyads **32–34**

		Excited state lifetime (ps)		Fluorescence yield[a]					
Dyad	Solvent	Zn*	H_2*	Zn*	H_2*	$(k_{ET})^{-1}$ (ps)	Φ_{ET}	Φ_{eT}	Φ_{HT}
32	C_7H_8	13 ± 2	12,500	—	0.13 (0.13)	13	0.99	≤ 0.01	0.06
	PhCN		10,400	—	0.12 (0.11)		> 0.9	< 0.1	0.23
33	C_7H_8	65 ± 10	12,300	—	0.11 (0.12)	67	0.97	≤ 0.03	0.08
	PhCN		11,000	—	0.11 (0.11)		> 0.9	< 0.1	0.19
34	C_7H_8	68 ± 4	12,800	—	0.13 (0.16)	70	0.97	≤ 0.03	0.04
	PhCN		10,400	—	0.10 (0.09)		> 0.9	< 0.1	0.23
35	C_7H_8	5.4 ± 0.7	11,100	—	0.13 (0.13)	8.3	0.65	0.35	0.18
	PhCN	3.0 ± 0.2	4,200	—	0.10 (0.10)	6.0	0.50	0.50	0.61

[a]Referenced to *meso*-tetraphenylporphyrin $\Phi_f = 0.11$, excitation at 517 nm (free base excitation predominant); values in parentheses for excitation at 565 nm (zinc donor predominant). ΦeT is the electron transfer quantum yield from Zn* to the free base electron acceptor. Φ_{HT} is the hole transfer quantum yield (upper limit) from H_2* to the Zn hole acceptor.

It is worth noting that for the dyads **32–34**, due to H bonding within the bridge in an aprotic nonpolar solvent such as toluene, the average angle between the chromophores should be quite small, due to the planarity of the bridging element. In a polar, aprotic solvent like benzonitrile, some contribution from conjugation may be found in the same dyads, with a similar orienting effect on the chromophores. In dyad **35**, the situation is different because similar effects will maintain the bridge coplanar with the zinc donor and at *ca.* 60° with the free base. Thus, electronic delocalization on the bridge can occur from the zinc donor, and eT competes with ET, with the former contributing to 35% of the fluorescence quenching of dyad **35** in toluene, and 50% in benzonitrile. Two other dyads are reported in the same paper, but they are assembled via coordination around a metallic template and will thus be examined in the appropriate section (noncovalent). Globally, the efficiency of the ET process in these imine linked dyads are at least as efficient as in diphenylethyne linked dyads previously studied by these authors.

Even though the last example points out that other linkers can be used, linear unsaturated bridges are very attractive in order to deal with structurally well-defined species. The study of porphyrin dimers and either eT or ET dyads should now be considered as a source of knowledge for the design of functional devices.

The collection of knowledge from studies of dimeric species encouraged many groups to increase the size of the devices prepared, and to use the dimer's behavior to explain or predict the properties of oligomers. A typical example is provided by Osuka and Kim who have reported photophysical studies on rod-like species (**36**) depicted in Figure 13.29[56]. When excited to the S_2 state by irradiation at 400 nm, the excitonic splitting of the Soret band in the arrays affects the deactivation of the S_2 states. As explained in the introduction, the longer the arrays, the more split the Soret bands. In the longer arrays for which split bands are shifted to the red, as far as spectral overlapping with S_1 (Q bands), a ladder-type deactivation channel allows fast $S_2 \rightarrow S_1$ internal conversion (< 1 ps), which is considerably faster than the S_2 deactivation in the corresponding monomer. A minimum length of 6–8 porphyrin units is estimated for the existence of the deactivation channel, in other words the radiative coherent length.

In order to observe energy transfer more specifically, an appropriate energy acceptor must be introduced at one end of the arrays **37a–f**. In the array

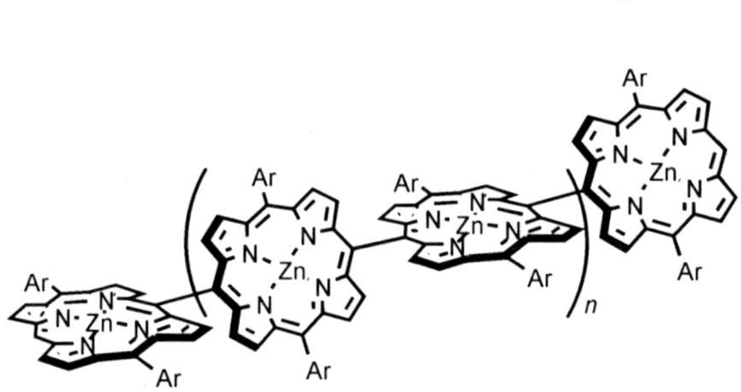

Ar=3,5-di(octyloxyphenyl) **36**: n= from 0 up to 63

Figure 13.29. Osuka's orthogonal *meso–meso* coupled oligomers.

represented in Figure 13.30, the energy acceptor is a 5,15-bis(phenylethynyl) zinc porphyrin that has a red shifted absorption spectrum due to extended conjugation of the ethynyl bridge with the lateral phenyl groups. The observed end-to-end energy transfer rate to the acceptor part of these arrays displays electronic coupling among the *meso–meso* coupled porphyrins that cannot be explained by Förster theory, except in the case of dimer **37a**. As shown in Table 13.4, calculated energy transfer rates are best fitted by statistical distribution of the excitation energy over coherent groups of chromophores consisting of four *meso-meso* coupled porphyrin units. Incoherence is found when larger chromophore groups are considered. While energy transfer is nearly quantitative for small arrays (**37a–c**), longer arrays display less efficient ET[57].

The *meso–meso* coupled arrays can be converted into the porphyrin tapes **38a–g** depicted in Figure 13.31, in which completion of the electronic coupling via coplanar cumulenic forms induces an increased red shift of the absorption maxima to the IR region (optical HOMO–LUMO gap for **38g** is 1,500 cm^{-1}).

These systems display a very fast and effective nonradiative deactivation because of the small energetic gap between electronic states S_0, S_1, and S_2, and

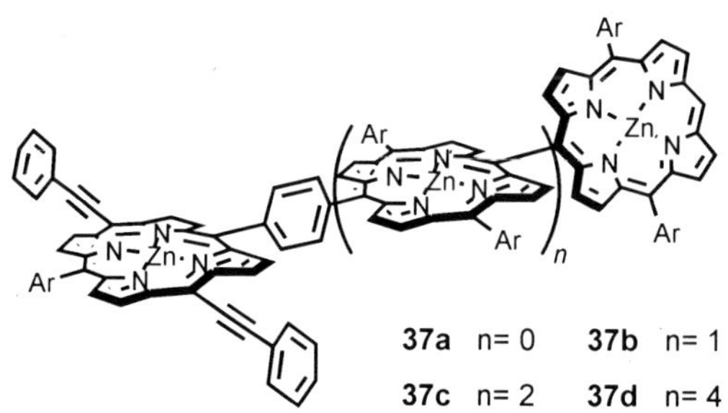

37a n= 0 **37b** n= 1
37c n= 2 **37d** n= 4
Ar=3,5-di(octyloxyphenyl) **37e** n= 10 **37f** n= 22

Figure 13.30. Osuka's arrays with an energy acceptor.

Table 13.4. ET rate constants in arrays **37a–f**

Array	R (Å)[a]	$k_{obs}^{-1\,b}$ (ps)	$k_{calc1}^{-1\,c}$ (ps)	$k_{calc2}^{-1\,d}$ (ps)	$k_{calc3}^{-1\,e}$ (ps)
37a	12.7	2.5 ± 0.1			
37b	16.9	3.3 ± 0.2	6.9	5.0	3.4
37c	21.0	5.5 ± 0.5	20	7.5	5.9
37d	33.6	21 ± 2	160	15	23
37e	58.6	63 ± 5	2,300	30	70
37f	109	108 ± 7	49,000	60	163

[a]Longest Zn–Zn distance.
[b]Transient absorption measurement.
[c]Calculation from Förster theory (ET from end to end).
[d]Calculated considering excitation distribution over N individual chromophores.
[e]Calculated considering coherent coupling over four porphyrin units.

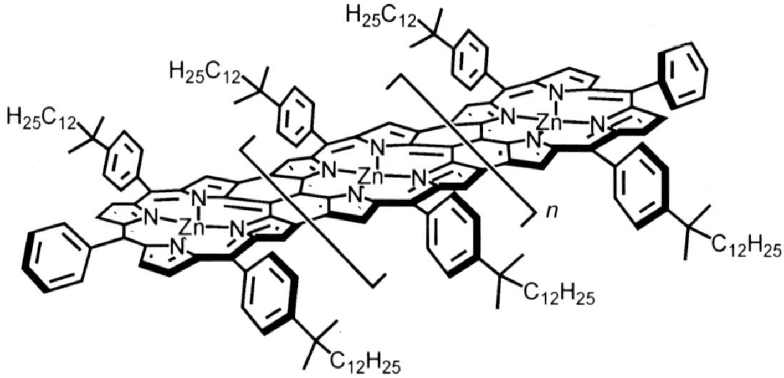

38a: n=0, **38b**: n=1, **38c**: n=2, **38d**: n=3, **38e**: n=4, **38f**: n=6, **38g**: n=10

Figure 13.31. Fused porphyrin tapes.

their coupling with vibrational motions that are enhanced by the size of the tapes. Again, this resembles the ladder type deactivation channel mentioned above. For example, the lowest excited state of dimer **38a** decays to the ground state within 4.5 ps compared to a decay time constant of 1.9 ns for the *meso–meso* coupled analog **32b**. Furthermore, the longer the tape, the faster the decay[58].

One problem that is still consistently recurrent in the study of porphyrin assemblies stems from very similar absorption and excitation properties of the species involved in the charge separation or the energy transfer processes. Very often, it is difficult to selectively generate the excited state of a zinc porphyrin donor without partial excitation of the free base porphyrin acceptor, even though no electronic coupling can be detected in ground state absorption spectra. This also stands for excitation of the electron acceptor or hole donor free base with a zinc electron donor or hole acceptor. It is thus useful to connect the multiporphyrin assembly to a nonporphyrinic electron/energy donor or acceptor, or both, that can be selectively excited in order to initiate the photoinduced process. This class of assemblies will now be examined.

4.2. Covalent Multiporphyrin Arrays Combined with Nonporphyrinic Acceptors or Donors

As we will see, the choice of electron donors associated with porphyrins has not evolved much over the years. In previous review articles, electron donors such as carotenes, aryl amines, and ferrocene derivatives have been examined. Concerning the class of energy donors, over the past decade, boron dipyrylmethanes (BODIPY) are probably the most widely used series of compounds. A very detailed study of their association and their excitation energy transfer to porphyrins has been carried out by Lindsey[59]. It should be noted that condensed aromatics such as anthracene, and especially perylene derivatives, may be good substitutes for BODIPYs.

The energy acceptors used in most energy conducting assemblies are usually free base porphyrins for the simple reason that, in such processes, the difficulty resides in the selective excitation of the donor moiety. Selectively detecting the output emission of the final acceptor is not a problem in itself. For a long time, the class of electron acceptors was restricted to quinones and aromatic imide-type compounds. However, the burst of fullerene derivatization methods in the early 1990s, together with the specific properties of C_{60} mentioned in the scope and limitation section, have changed the face of the electron acceptor world.

4.2.1. Energy Transfer in Linear Species

While most of the species combining several zinc porphyrin energy donors and one free base acceptor have been designed to mimic the photosynthetic antennae systems, linear arrays are conceptually different and aim at signal propagation and processing. Holten, Bocian, and Lindsey[54] have carefully rationalized the design of multiporphyrin arrays in this regard. The functional characteristics of the arrays **39–42** depicted in Figure 13.32 are based on the specific behavior of magnesium porphyrins which can exist as stable radical cations and can be incorporated into arrays via simple synthetic schemes. It is also worth mentioning that the HOMOs of the zinc/free base motif utilized in these arrays are of a_{2u} type, which favors the bridge mediating effect by localizing electron density on the *meso* carbon atoms. When a neutral magnesium porphyrin is used, excitation of the ZnP produces the highest S_1 level that can be deactivated by two ET processes, to MgP or H_2P, via a dominant through bond (Dexter) mechanism. The MgP* produced is then deactivated via a second ET to the H_2P acceptor, without mediation of the ZnP. When the MgP is oxidized to $MgP^{\bullet+}$, deactivation of the ZnP* no longer occurs via ET to H_2P, and only eT/ET to the $MgP^{\bullet+}$ is observed. Similarly, eT/ET deactivation of the H_2P^* is observed. As shown by the rates indicated in Figure 13.32, the fastest processes (eT and ET) are observed for *meso* linked porphyrin components with *para* connections in the phenylethyne bridges (**39**). Distinction between eT and ET quenching is not established and both mechanisms are possible based on theoretical and thermodynamic data[60a]. The same group of authors later reported that in triads and tetrads consisting of zinc porphyrin rods terminated by a magnesium porphyrin, the latter is the energy acceptor in the light-harvesting process[60b]. These systems allow tuning of the energy transfer process from the ZnP* to the H_2P unit, as already demonstrated in previous studies cited in refs.[54, 60].

To illustrate the need for energy donors that can be specifically excited, a BODIPY donor has been incorporated in these arrays (**43–44**), and in nonlinear analogs[61]. The structure and eT/ET transfer rates for these arrays with a BODIPY primary energy donor are given in Figure 13.33. The disadvantage of the BODIPY units lies in the existence of two energetically accessible, excited state conformers of the BODIPY[59]. As a result, two lifetimes are measured for the excited state conformers, which complicates data interpretation[62].

As mentioned above, BODIPY units tend to be replaced by other primary energy donors in order to facilitate the study of excited state deactivation. Lindsey and coworkers have undertaken comparative studies of the linear arrays incorporating perylene derivatives **45–47** represented in Figure 13.34. Table 13.5 shows

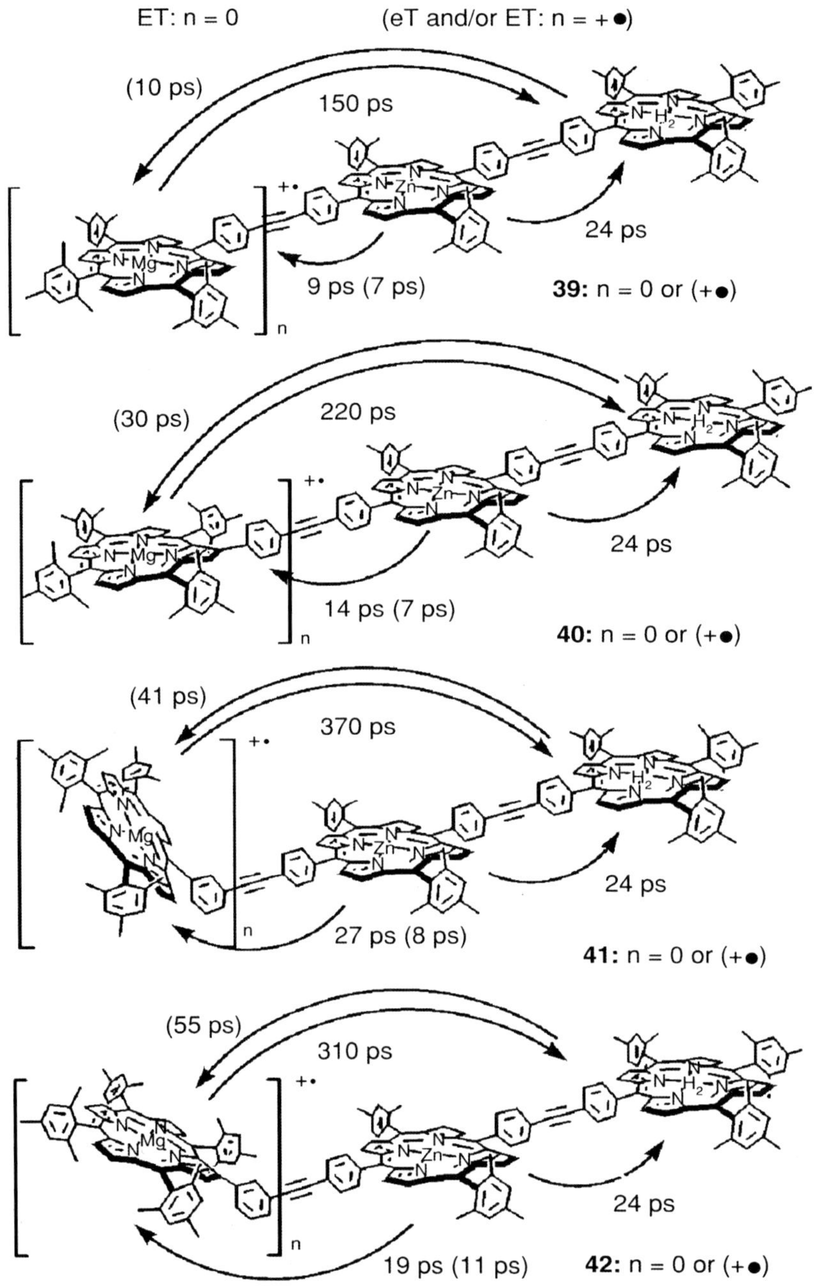

Figure 13.32. Porphyrin triads with MgP, ZnP, and H_2P components. The linking position of the MgP has been systematically varied using a *p*-phenyl spacer at the *meso* position in **39**, and at a beta position in **40**. In **41** and **42**, *m*-phenyl spacers are attached at *meso* and beta carbons, respectively.

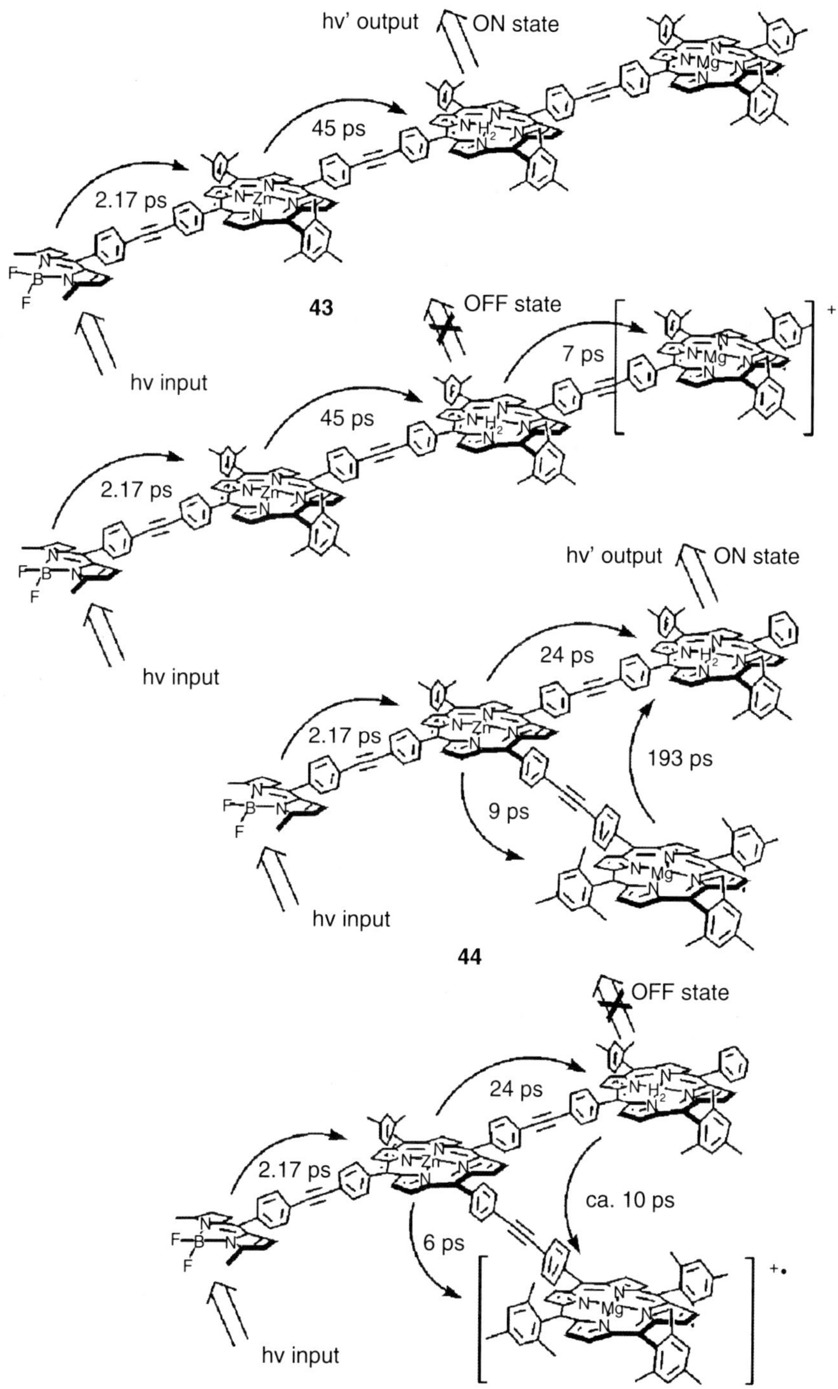

Figure 13.33. Linear and branched arrays with BODIPYs as primary energy donors. The "ON/OFF" states are a function of the oxidation state of the MgP.

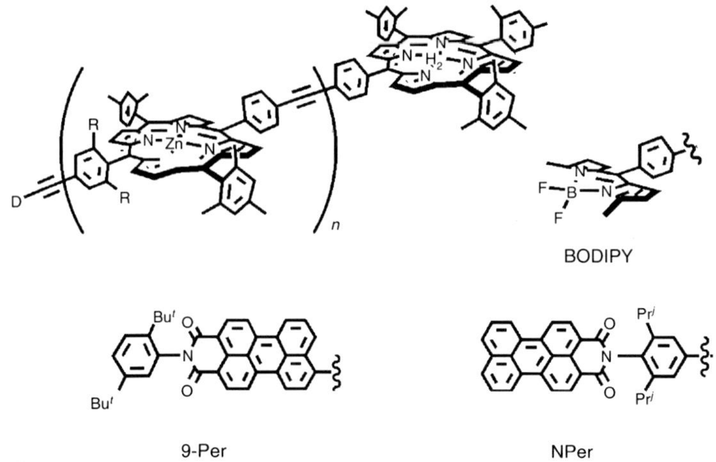

45: n=3, D= BODIPY, R=CH_3
46: n=1, D=9-Per, R=H
47a: n=1, D= NPer, R=CH_3
47b: n=3, D= NPer, R=CH_3

Figure 13.34. Photonic wires with BODIPY and perylene primary energy donors.

Table 13.5. Energy Transfer Efficiencies in Photonic Wires **45–47**

	Excited state lifetime (ps)		
Wire	D*	Zn*	Φtrans
45	25 ± 5 (1.7 ± 0.4)	160 ± 30	81 ± 9
46	≤ 0.4	26 ± 3	99
47a	2.5 ± 0.8	28 + 3	99
47b	5.3 ± 0.8	170 ± 30	86 ± 7

that transfer efficiencies along these wires are excellent and that perylene donors are valuable substitutes for BODIPYs. These examples show that when energy conduction is targeted, selective input of energy can be successfully achieved by introduction of a primary energy donor.

4.2.2. Photoinduced Charge Separation in Linear Arrays

When electron transfer processes are targeted, one of two different options can be chosen. The first deals with photoinduced charge separation. This requires the connection of the photoactive unit, the porphyrin or multiporphyrin architecture, to a nonporphyrinic electron donor and/or acceptor. The second option concerns electron conduction in multiporphyrin arrays. The donor and acceptor in this case can be electrode materials to which the array is connected on both sides.

For discrete, well characterized species, C_{60} derivatives are almost exclusively utilized as electron acceptors due to their small reorganization energies (0.48 and 0.6 eV)[63 and references cited]. A simple comparison between porphyrin-quinone and porphyrin-C_{60} dyads, on the basis of similar physical properties (oxidation/reduction potentials, electronic coupling) and thus similar values of $-\Delta G^{\circ}_{eT}$, shows that charge separation is accelerated *ca.* six times and charge recombination slowed down *ca.* 25 times in the case of a C_{60} type acceptor[28,29]. As a consequence, fullerene acceptors have been added to existing arrays. Previous results summarized in refs.[28,29] clearly show that the topography of the assemblies, and thus the geometric properties of the linkage between the porphyrins and the C_{60}, have an influence on the efficiency of the electron transfer due to variations in the electronic coupling. Here, we will consider only examples in which the localization of the C_{60} acceptor with respect to the porphyrin electron donor is unambiguous, and is either equatorial or apical to the tetrapyrrolic macrocycle. In Figure 13.35, four examples illustrate the incorporation of C_{60} acceptors in a linear arrangement with the porphyrin chromophore.

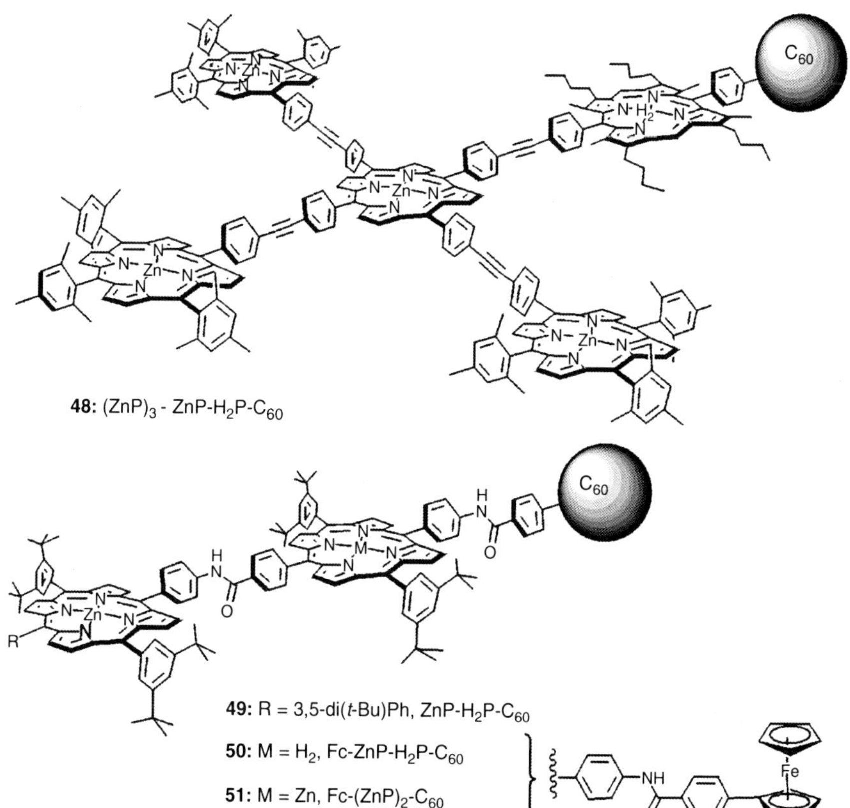

48: $(ZnP)_3$ - ZnP-H_2P-C_{60}

49: R = 3,5-di(*t*-Bu)Ph, ZnP-H_2P-C_{60}

50: M = H_2, Fc-ZnP-H_2P-C_{60}

51: M = Zn, Fc-$(ZnP)_2$-C_{60}

Figure 13.35. Fullerene acts as a final electron acceptor in these arrays. The presence of C_{60} results in increased lifetimes of the charge separated states compared to those arrays without C_{60}.

The combination of four energy collecting Zn porphyrins connected to the energy acceptor free base porphyrin and the final C_{60} electron acceptor (compound **48**) actually mimics the first stages of photosynthetic events. Although this $(ZnP)_3$–ZnP–H_2P–C_{60} hexad (**48**) was reported prior to the period covered in this chapter, it illustrates the advantages that can be taken from fullerene. The $(ZnP)_3$–ZnP antenna shows an energy transfer from three peripheral zinc porphyrins to the central unit within 50 ps. Energy is further quenched by the free base energy acceptor to yield the electron donor singlet excited state at the rate of $4.2 \times 10^9\ s^{-1}$. After electron transfer to the C_{60} acceptor, a charge separated state is efficiently generated (70%) that lives for 1.3 ns[64]. The system has been improved by replacing the octaalkyl free base porphyrin by a tetraaryl free base porphyrin in order to change the nature of the HOMO to a_{2u}. In this case, the efficiency reaches 98% and the lifetime of the CS state is much longer, 380 ps in THF[65].

Based on the use of a similar spacer, the linear ZnP–H_2P–C_{60} triad **49** (Figure 13.36) reproduces a similar series of events in which energy collected by the zinc porphyrin is transferred to the free base energy acceptor. A charge separated state ZnP–($H_2P^{\bullet+}$)–($C_{60}^{\bullet-}$) is produced in 40%, before giving the final ($ZnP^{\bullet+}$)–H_2P–($C_{60}^{\bullet-}$) charge separated state in 21 μs after a hole transfer of the charge shift type[66]. Differences in the efficiency of systems **48** and **49** may be partly explained, first, by the number of Zn porphyrins involved in the collection of energy, and second, by differences of through bond electronic coupling (discussed in previous sections) that may favor the charge recombination process in the hexad.

Further sophistication of these systems involves, of course, the introduction of a nonporphyrinic electron donor capable of stabilizing the photoinduced charge deficiency of the zinc porphyrin. This was achieved using ferrocene in the tetrad **50** that produced one of the very first, long lived charge separated states ($Fc^{\bullet+}$)–ZnP–H_2P–($C_{60}^{\bullet-}$), with a lifetime of 380 ms in benzonitrile[67]. Further implementation of the device has been recently described and consists in the replacement of the free base porphyrin H_2P by a second zinc porphyrin to generate the Fc–$(ZnP)_2$–C_{60} tetrad **51**[68]. This results in an increase of the excited state energy from 1.84 to 2.04 eV, and a decrease of the oxidation potential of the electron donor by 0.3 V. Thus a larger energy gap is observed for the CS state, which in turn slows down the charge recombination process. The CS state lifetime of 1.3 s at room temperature in dimethylformamide is among the longest lifetimes ever reported[68]. Here it should be mentioned that although ferrocene seems appealing for its electron donating properties, model compounds have shown that the charge recombination in $Fc^{\bullet+}/C_{60}^{\bullet-}$ species are in the normal region. This is not the case for carotene electron donors. Indeed, due to the energy level of the carotene's first triplet excited state that intercalates between the photogenerated radical pair and the ground state, deactivation occurs in the inverted region, at much slower rates[67,69]. Nevertheless, internal charge recombination involving $C_{60}^{\bullet-}$ and $Fc^{\bullet+}$ can be slowed down if the distance between the radicals in the radical pair is large enough. In this case, the rigidity of the porphyrin assembly plays a crucial role, as demonstrated by the longer lifetime of the photoinduced CS state generated

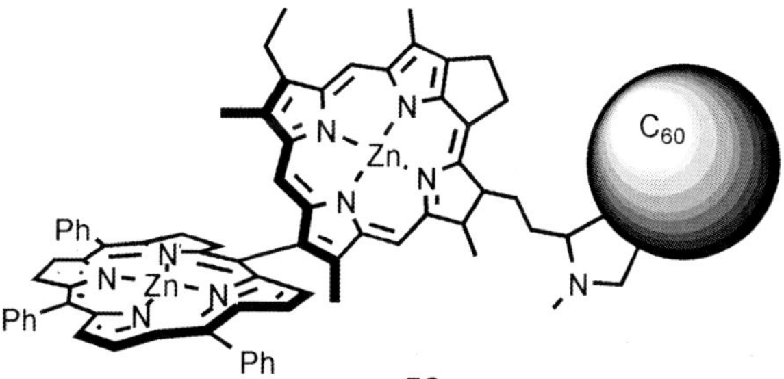

52

Figure 13.36. A porphyrin-chlorin-C_{60} triad.

with an analogue of tetrad **51** (not depicted) in which the flexible amide bond is replaced by direct *meso–meso* coupling of the two zinc porphyrins. This forces a stepwise eT process for charge recombination, and provides a lifetime of 19 μs in benzonitrile[70].

The concept of an energy cascade prior to charge separation can also be performed in triads comprising a zinc porphyrin (E donor) and a zinc chlorin (E acceptor, e donor) connected to a C_{60} electron acceptor (compound **52**). In this case, energy transfer between the E donor and the chlorin E acceptor is favored by the connection of the a_{2u} type zinc porphyrin *meso* carbon to the β-pyrrolic carbon of the a_{1u} type zinc chlorin. Despite the favorable coupling, only charge separation from the excited zinc chlorin to the fullerene acceptor is observed. No subsequent hole transfer type of charge shift is observed between the radical cation of the zinc chlorin and the zinc porphyrin. The charge separated radical pair formed upon excitation of the zinc porphyrin lives for 12 ps due to rather fast recombination processes involving folded conformers of the triad, since the linkage between the zinc chlorin and the C_{60} is rather flexible[71].

In the case of short flexible connections between C_{60} and the neighboring chromophore, the general conformation trend for the carbon sphere is to sit on the top of the aromatic tetrapyrrolic structure, for obvious π–π or lipophilic interactions. Multiporphyrinic architectures exhibiting rigid, linear arrangements can thus provide a way of quasicomplete topographic control, leaving only a small degree of freedom to the photoactive species at the end of the arrays. Diederich and coworkers have adapted Osuka's *meso–meso* coupled oligomer approach to the synthesis of linear, *meso–meso* porphyrin arrays terminated on both ends by C_{60} moieties (Figure 13.37).

Fluorescence quenching studies carried out on the first series of porphyrin dimer **53a**, trimer **53b**, and tetramer **53c** show that fluorescence of the porphyrin arrays is quenched approximately 100 times compared to arrays not bearing fullerenes[72]. Due to the properties of the arrays, namely strong excitonic coupling and red shifted, split Soret bands, no conclusions are drawn about the mechanism (ET or eT) of the quenching.

Further studies on a series of conformationally isomeric fused dimers, **54a**, and **54b**, show that in this case, the quenching is due to photoinduced electron transfer from the zinc porphyrins to the C_{60} acceptors, generating a lifetime of

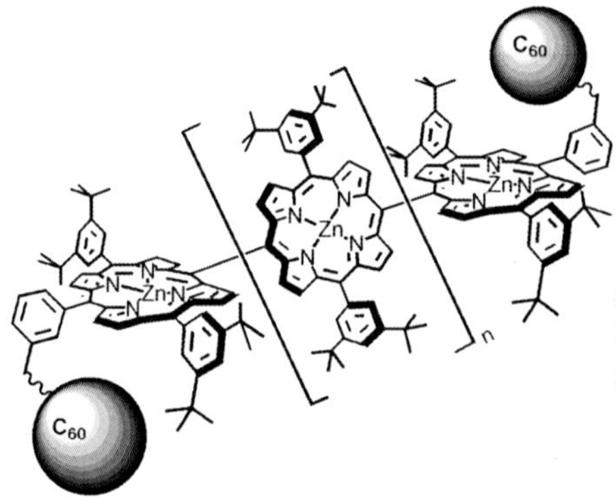

53a: n = 0
53b: n = 1
53c: n = 2

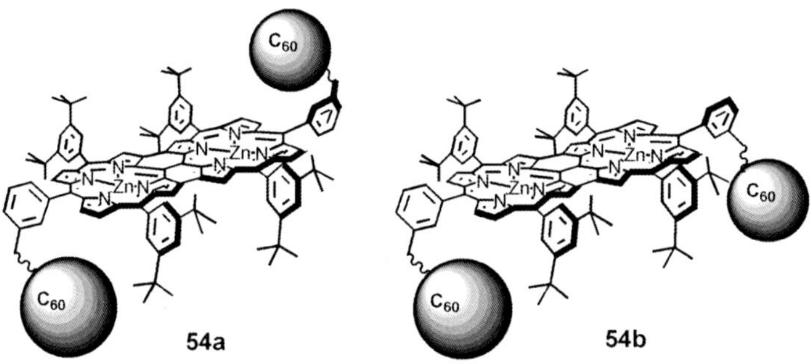

54a **54b**

Figure 13.37. Osuka-type *meso–meso* coupled ZnP arrays (top) and fused ring arrays (bottom) bearing fullerene electron acceptors.

630 ps for the CS state of **54a**[73]. Up to 15 possible redox processes, including zinc porphyrin oxidations, and reductions of both the zinc porphyrins and the C_{60}s, are possible in the fused conformers **54a–b**. Differences in the redox potentials are observed, depending on the side of the porphyrins with which the fullerenes interact. Interestingly, excitation of the C_{60} acceptor, by irradiation at 330 nm, induces ET from the carbon sphere to the porphyrins, without further charge separation. A typical porphyrin-based fluorescence is observed in this case.

As mentioned above, the electron donors and acceptors can be electrodes, given that manufacturing and connection are possible. Towards this goal, linear porphyrin arrays of the Osuka-type have been incorporated between nanoelectrodes in order to investigate electrical conduction in these species[74]. The orthogonal *meso–meso* coupled species **36** ($n = 23$ in Figure 13.29) have been used and show typical diode I–V curves, while the small band gap in the fused, tape-like oligomer **38f** (Figure 13.31) induces a higher electrical conductivity of the device. Both phenomena are temperature dependent, but as expected, the less flexible tape arrangement exhibits less temperature dependence.

Introduction of accessory pigments can also be performed in longer species, as demonstrated by Lindsey's approach to perylene-porphyrin oligomers[75]. In these soluble structures, a central core of phenylethynylphenyl bridged porphyrin oligomers is surrounded by a coating of perylene energy collectors, four per porphyrin.

This section and the previous show that linear arrangements of covalently linked porphyrins have clearly captured the attention of chemists regarding electron transfers and charge separation along redox gradients. As opposed to the linear approach, branched and dendritic structures are directly inspired from the architecture of photosynthetic antennae systems, and are therefore mostly oriented towards energy collection and funneling.

4.2.3. Energy Collection in Nonlinear Structures

A quick overview of branched structures has been recently published by Aida[76]. Examples of energy collecting devices that appeared in the time range covered in this chapter are examined in this section. The design of the devices reviewed hereafter is conceptually different from the approach leading to the species examined in the "energy transfer in linear species" section. Previously, the architectural features of the arrays were totally oriented towards the controlled processing of a photogenerated signal, whereas the aim is now to collect light energy with maximum efficiency by multiplying the number of energy transfers from primary energy donors to a limited number, and eventually one unique, energy acceptor.

Taking advantage of their efficient synthetic approach leading to *meso–meso* linked dimers, Osuka has recently reported a dodecameric porphyrin wheel (**56**, Figure 13.38) directly inspired by the LH1 and LH2 structures[77]. The circular structure, depicted in Figure 13.38, with a diameter of *ca.* 4 nm, is one of the closest covalent models (in terms of number of pigments) of natural systems. For the circular structure **56** and its open chain precursor **55**, several decay components are observed that can be explained by comparison with previously examined dimers (*vide supra*). The small time constant (*ca.* 0.24 ps) is assigned to incoherent exciton hopping within each of the *meso–meso* coupled dimers in the cyclic structure. The larger time constant (*ca.* 3.6 ps) is assigned to energy transfer between gable units separated by the phenylene linker, which is consistent with model compounds.

It should be noted that, based on similar inspiration, other wheel type structures have appeared in the literature but photophysical studies have not yet been reported, and thus will not be covered in this chapter. However, Kobuke has developed a very promising approach based on self-assembly that is worth mentionning[78]. It concerns the preparation of another dodecaporphyrin structure **57**, represented in Figure 13.39. The coordination driven self-assembled wheel bears olefinic R groups that can undergo metathesis, leading to the formation of a cyclic, ring-shaped, covalent structure containing 10 or 12 porphyrins (not depicted)[79].

The simplest approach to branched or denritic multiporphyrin structures is the combination of two arrays connected at one point. This has been successfully attempted by Osuka to produce the so-called "windmill" porphyrinic arrays **58–60** represented in Figure 13.40. The properties of these light harvesting arrays are adjusted taking into account previous results that show that the excitation energy of the S_1 state in the central *meso–meso* coupled species is smaller (2.07 eV) than that of the weakly coupled peripheral porphyrins (2.13 eV). To increase this

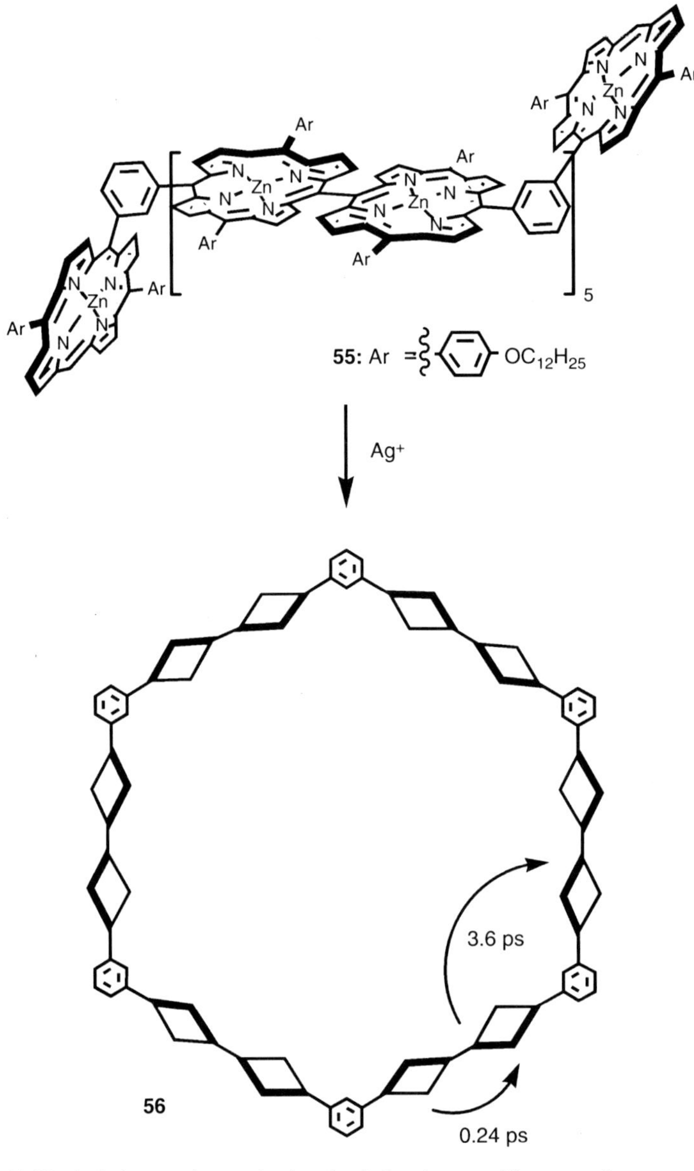

Figure 13.38. A dodecameric porphyrin wheel closely resembles natural antennae systems in terms of the number of chromophores.

difference and to allow energy transfer to the central *meso–meso* coupled chromophores, phenylethynyl axial substitution has been performed in the windmill array **58**. Fast and efficient energy transfer is then observed for this species, with 99.6% efficiency and $k_{ET1} = 3.3 \times 10^{11}\ s^{-1}$ for S_1–S_1 energy transfer from peripheral to central chromophores. Additionally, a S_2–S_2 energy transfer is observed due to the overlap of the emission spectrum of the peripheral components

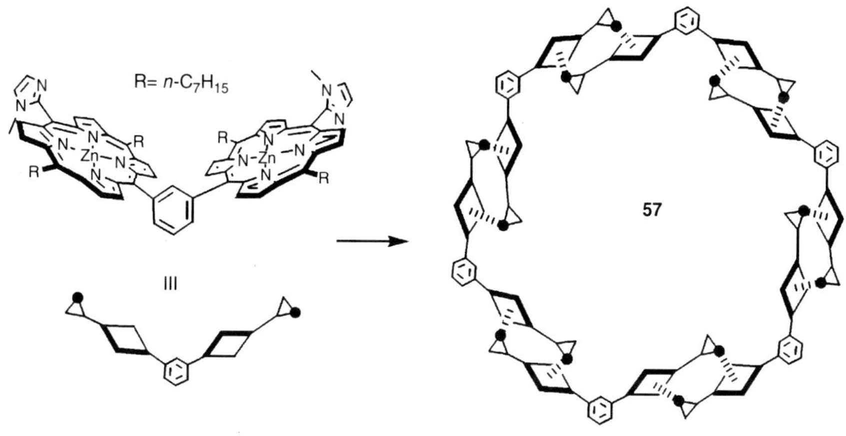

Figure 13.39. A noncovalent assembly of six porphyrin dimers that form a dodecameric porphyrin wheel.

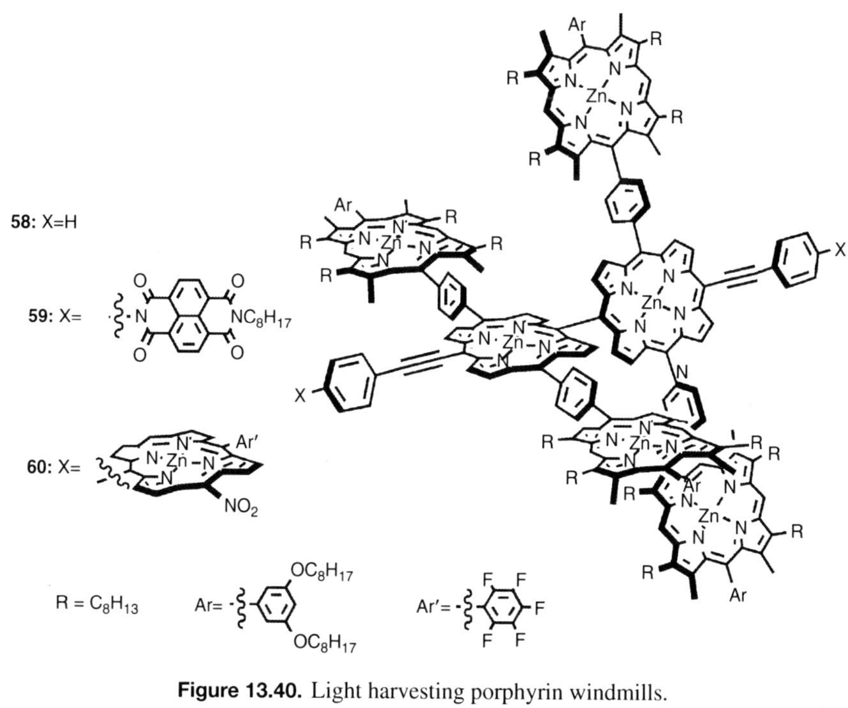

Figure 13.40. Light harvesting porphyrin windmills.

with the exciton-split Soret band of the central *meso–meso* coupled porphyrins. In that case, $k_{ET2} = 8 \times 10^{12}\ s^{-1}$.

For windmills **59** and **60**, electron acceptor species have been added in axial positions. Thus, electron transfer can occur from the photoexcited, central *meso–meso* coupled porphyrins to the electron acceptor diimide in **59**, or to the electron deficient pentafluorophenyl substituted nitroporphyrin in **60**. From **59** to **60**, an increase in the rate of charge separation, from 25 ps^{-1} to 10 ps^{-1}, and a decrease of the charge recombination rate, from 10 ps^{-1} to 55 ps^{-1}, are observed[80].

Dendritic approaches have also been developed by Aida, who has shown that in terms of energy collection, the properties of the chromophore arrays are strongly morphology dependent[81]. Only the structures of the larger species are represented in Figure 13.41, but smaller dendrimers and reference compounds have also been prepared.

Based on the fluorescence lifetimes of the ZnP units in dendrimers **61–64** comprised solely of porphyrins, the cone shaped wedge shows a dramatic decrease in the energy transfer efficiency from **62c** (86%) to **62b** (66%), and finally to **62a** (19%). Such a trend is not observed for the star shaped dendrimers of the **61a–c** series, with Φ_{ET} values of 87%, 80%, and 71% for **61c**, **61b**, and **61a**, respectively. Replacement of the ether linkage by an ester linkage (compounds **63–64**) does not have a major influence on the behavior of the species. Upon excitation at 544 nm, all species emitted almost exclusively from the H_2P moiety in the case of star shaped dendrimers. Time resolved measurements, with B band excitation at 415 nm and monitoring of the fluorescence decays at 585 nm, showed that the average lifetime of the ZnP* states in **61a** species was considerably decreased

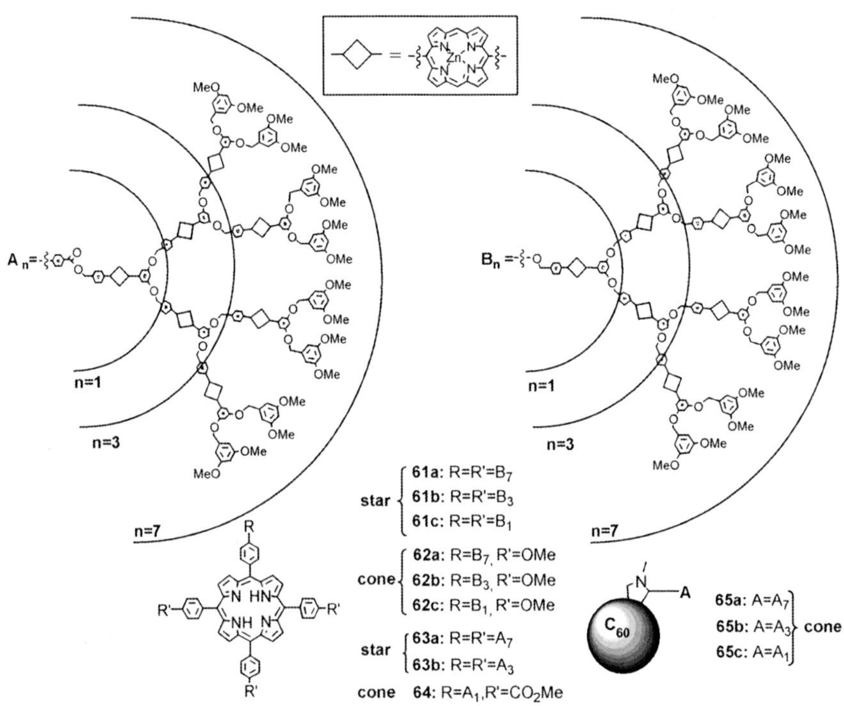

Figure 13.41. Star and cone shaped dendrimers for light harvesting and photoinduced charge separation.

(680 ps) compared to either an isolated ZnP moiety (~2,300 ns), or the analogous cone shaped dendrimer **62a** (1,899 ps).

The natural evolution of this work has been the attachment of an electron acceptor to the light harvesting species in order to convert the collected energy into a charge separated state. As seen above, fullerene is well adapted to be the terminal electron acceptor in a cone shaped dendron. The fullerene species **65a–c**, represented at the bottom right of Figure 13.41, have been studied, and showed 84% fluorescence quenching for the dyad **65c**, 60% for **65b**, and 62% for **65a**. The charge separation rate constants evolve similarly, with rates of $1.55 \times 10^9\ s^{-1}$ for **65c**, $0.40 \times 10^9\ s^{-1}$ for **65b**, and $0.43 \times 10^9\ s^{-1}$ for **65a**[82]. Interestingly, the charge recombination rate constants slow down [from $\sim 2.9 \times 10^6\ s^{-1}$ for **65c**, to $\sim 1.5 \times 10^6\ s^{-1}$ for **65a**] as the size of the dendrimer increases. This suggests, in agreement with electrochemical data, that the primary event in the process is the generation of a $ZnP^{\bullet +}-C_{60}^{\bullet -}$ pair followed by a fast migration of the hole among the peripheral ZnP moieties.

The recent dendrimer examples will close on a sophisticated architecture **66** consisting of 21 porphyrins, reported by Lindsey, Bocian, and Holten. In this array one free base energy acceptor is surrounded by four sets of five zinc porphyrins that act as energy collectors[83]. In contrast with the previous dendrimers, the compounds described in Figure 13.42 comprise weakly coupled chromophores. Due to the number of chromophores involved, the effects of the coupling are easily detected by the red shift and splitting of the Soret band. On the contrary, Q bands remain unchanged and the first oxidation and first reduction potentials of the zinc porphyrins confirm that electronic coupling is weak between individual chromophores. In **66**, the central free base porphyrin is the focal point of energy collected by the peripheral zinc porphyrins.

Excited state energy transfer in **66** has been compared with other species in which the central free base is surrounded by only 1, 2, or 8 zinc porphyrin collectors. As in the previous example, a decrease is observed in the efficiency, respectively 98, 96, 96 and 92% for 1, 2, 8, and 20 zinc porphyrins, together with a slight decrease of the energy travel time (respectively 45, 90, 105, and 220 ps). In addition, ground hole storage should be possible in these large zinc porphyrin assemblies, since a minimum of 21 electrons can be removed from **66** to generate a highly charged π-cationic multiradical[83].

This overview of the emerging trends in the design of covalently linked multiporphyrin devices obviously leads to the conclusion that the covalent approach is the most appropriate way to control and finely tune the coupling interactions between the chromophores involved in a photophysical or photochemical process. As the size grows, the overall efficiency of the synthesis usually drops consistently. Nevertheless, convergent strategies based on the use of building blocks, such as the dendrimer approach, have been developed recently. In this regard, the idea of building blocks that are easily accessible and ready to be assembled in an orderly fashion is not only applicable to convergent covalent synthesis, but also to the self-assembly of photo or electroactive components, which will be detailed in the next sections.

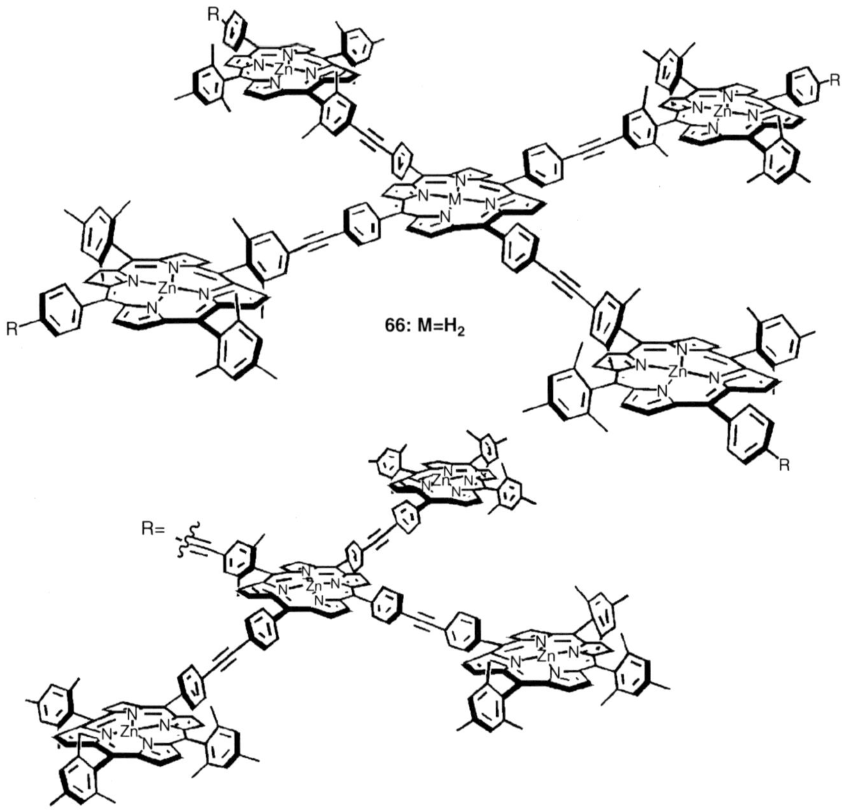

Figure 13.42. Weakly coupled large dendritic porphyrin arrays.

5. Noncovalent Species

5.1. Hydrophobic Interactions

5.1.1. Concave Hosts

One of the first types of noncovalent interactions that can be developed between porphyrins, independently of their free base or metallated character, is hydrophobic interactions. For decades, it has been well established that porphyrins have a strong tendency to aggregate in polar media, and more precisely to form J-aggregates. One of the crucial problems remaining, though, is the observation of well defined aggregates, particularly regarding the number of species involved, and the geometry/stoichiometry of the assembly[84]. Only a few examples have been reported in which J-aggregate formation leads to well defined species. These examples will be detailed later. However, it is necessary to mention that association of hydrophobic cavities with porphyrins has been previously used to enhance the interaction between porphyrin excited states and fluorescence quenchers like benzoquinone. The same principle can be applied to

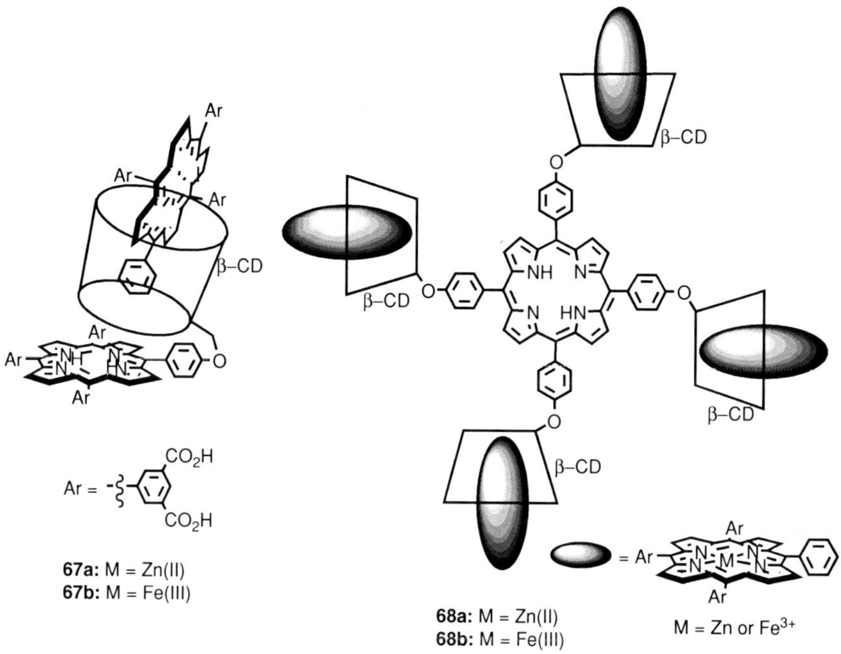

Figure 13.43. The hydrophobic pockets of βCDs provide binding cavities for porphyrin guests.

force the interaction between isolated porphyrins and porphyrins bearing concave hydrophobic substituents. This has been nicely illustrated by the group of Kano, who has reported scaffolds **67–68** represented in Figure 13.43[85].

The simplicity of the preparation of the scaffolds renders this approach extremely attractive, as it takes advantage of porphyrin structures that stem from similar building blocks and permethylated β-cyclodextrin (βCD). The association between the zinc porphyrin and the βCD is strong in aqueous solvents ($1.5 \times 10^7\,M^{-1}$ in phosphate buffer at pH $= 7.0$) and the construction of the assemblies consists in mixing the components at the right stoichiometry. Efficient fluorescence quenching is observed for heteroporphyrin assemblies, with effects similar to those observed in dendrimers, and a decrease from 93% to 85% in, respectively, the first "generation" dyad **67a** and the pentamer **68a** (Figure 13.43).

The zinc porphyrin guest shows a monoexponential decay and an expected lifetime of the excited state of 1.75 ns. In the pentamer **67a**, out of three decay components, two values 128 ps (34%) and 494 ps (22%), are ascribed to energy transfer processes arising from two different conformers of the self-assembled structure due to the presence of four βCDs substituents (binding sites) on the central free base energy acceptor. Indeed, in the model dyad (**67a**), in which a free base porphyrin bears one binding site for a zinc porphyrin energy donor, only one decay rate of 122 ps is attributed to ET process. When iron porphyrins are used as guests, fluorescence quenching of the central free base porphyrin is efficiently quenched in the pentamer **68b**, but weakly quenched in dyad **67b**.

Similar types of interactions have been used by Kuroda to assemble, in a dimer fashion, two free base porphyrins around two zinc porphyrin guests, as shown in array **69** in Figure 13.44[86].

Curiously, although the four βCDs substituents around the free base porphyrin could host four zinc porphyrins guests, only two of the latter are incorporated. The resulting structure (**69**) displays fluorescent quenching of the zinc porphyrins and enhancement of the free base emission corresponding to 81% ET. In a second array, a biphenyl bearing only two βCDs acts as a host for one zinc porphyrin and one free base porphyrin guest, forming a ternary complex. Whereas a solution of the ZnP and H_2P shows no energy transfer even in the presence of βCD, the ternary complex displays a 26% energy transfer from the ZnP to the H_2P.

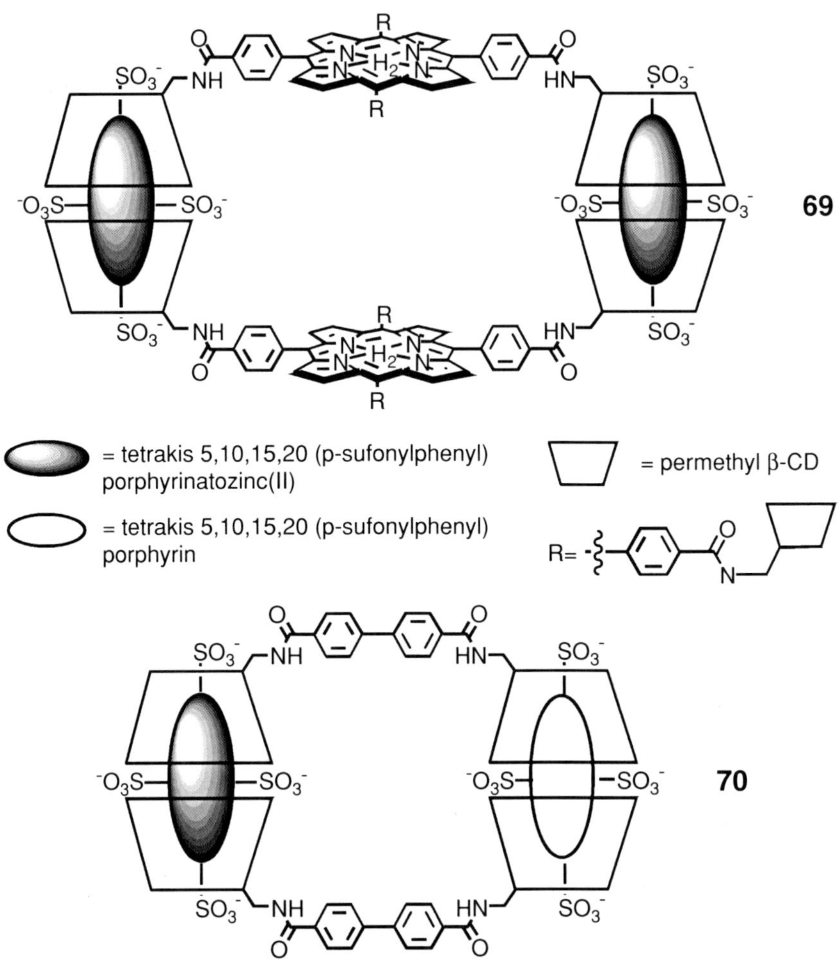

Figure 13.44. βCD hosts encapsulate sulfonated porphyrin guests, forming arrays with a host:ZnP:H_2P ratio of 2:2 for **69** and 2:1:1 for **70**.

5.1.2. J-Aggregates

Hydrophobic interactions can be developed between porphyrins themselves, and the use of a peripheral hydrophobic surface is not a necessary condition. This specific interaction, which originally led to ill-defined J-aggregates, has been refined and controlled rather recently. This type of porphyrin assembly has been now characterized geometrically.

The two examples selected and detailed below concern solid state structures that are obtained from controlled aggregation in solution. The first aggregate was obtained in a crystalline form during the search for π-functional coordination zeolites[87]. Its synthesis is reproducible to give, in high yield, a solid in which tetracationic (ethyl-pyridinium) free base porphyrins are stacked in a columnar J-aggregate. Most probably initiated by lipophilic interactions developed in a polar, anhydrous medium, the architecture is reinforced by the formation of CH-π bonds between adjacent chromophores at van der Waals distances, and CH-halogen bonds with the help of the counter anion $(CdI_4)^{2-}$. The tetragonal columnar stacks (**71a**, Figure 13.45), define areas of space in which polyaniline can be encapsulated (**71c**), which leads to inclusion compounds performing electron conduction (conductivity: 10^{-3} Scm^{-1}) 10^6 times more efficiently than the porphyrin stacks alone (10^{-9} Scm^{-1}). When lithium ions are present, they can be exchanged with the protons in the core of the free base porphyrins in the stacks. The conduction by polyaniline is then coupled with lithium ion conduction inside the porphyrin channels (**71d**)[88].

Instead of using peripheral quaternized pyridine substituents to solubilize the porphyrins in polar media, diprotonated porphyrins represent the second option to achieve the same goal. Okada and Segawa have used diprotonated free base

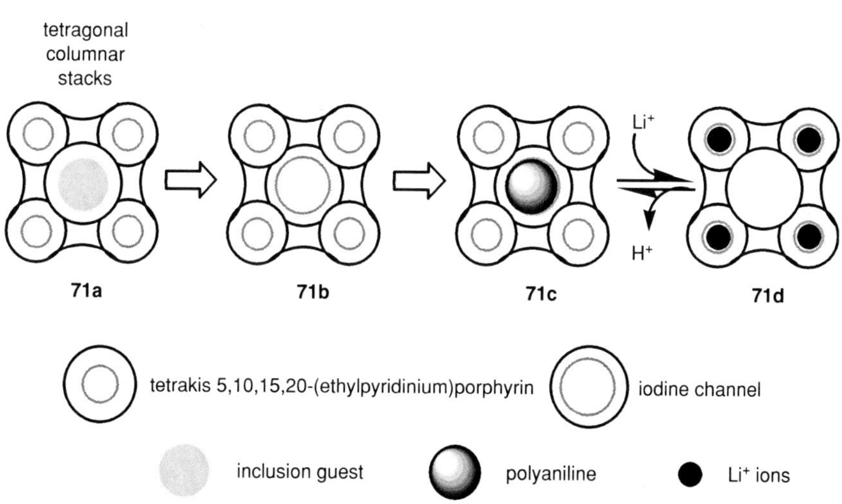

Figure 13.45. Electrical conductivity can be developed in J-aggregates of **71b** by polyaniline doping (**71c**) or by insertion of lithium ions into the porphyrin cores (**71d**).

meso-substituent (α, β, γ, δ)	β-substituent (1,2,3,4,5,6,7,8)	free base Soret (nm)	free base Q (nm)	acid dication Soret (nm)	acid dication Q (nm)	J-aggregate Soret (nm)	J-aggregate Q (nm)
H	H	392	614	402	547	n.d.	n.d.
H	$-CH_2CH_3$	397	619	406	593	n.d.	n.d.
Ph	H	417	646	438	652	475	725
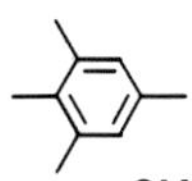	H	419	646	435	631	n.d.	n.d.
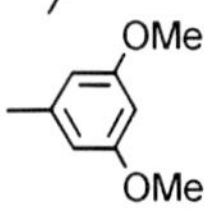	H	420	645	451	653	n.d.	n.d.
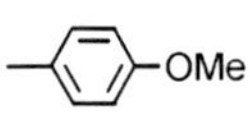	H	421	650	451	685	486	753
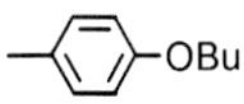	H	422	651	452	687	481	761
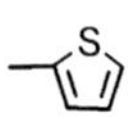	H	420	652	447	678	481	756
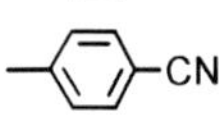	H	420	645	443	650	478	676
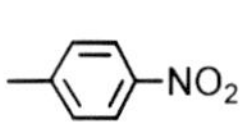	H	423	645	447	650	480	658

Figure 13.46. Absorption data for a series of porphyrins as their free-bases, dictations, and J-aggregates on glass substrates (n.d.: not determined).

porphyrins to produce J-aggregates in which a systematic modulation of the photophysical properties has been realized via peripheral substitution[88]. Figure 13.46 shows the variation of the substituents on the chromophore, and the effects observed on the electronic coupling within the J-aggregates (if formed) of this same chromophore.

It should be noted that the aggregates have been isolated at a liquid–liquid interface and could be deposited on glass substrates for characterization with AFM techniques. For this reason this example has been selected as a well characterized species.

This study shows that, while the S_0–S_2 transition is less affected by the peripheral substitution, the lower exciton corresponding to the S_0–S_1 transition affords Q bands whose position is substitution dependent, and electron withdrawing peripheral substituents have less effect on the S_0–S_1 transition[89].

5.2. Axial Bond Formation

The tendency of porphyrins to form aggregates, and the structural control of the assembly formed can be directed by the help of axial binding in the case of metalloporphyrins. In this section we will first examine examples in which the axial bond formed is a coordination bond for which dissociation is still possible. The formation of reversible bonds in self-assembled processes is extremely important because it permits the use of combinatorial type approaches, and leaves the possibility of correcting errors that may have occurred during the assembly process. Such synthetic design features usually afford thermodynamically stable scaffolds instead of a more or less random assembling of the chromophores.

Inspired by the circular structure of natural antennae systems, Kobuke and Ozeki have reported an interesting self-assembled heptaporphyrin species **82** (Figure 13.47) in which six energy collecting zinc porphyrins are arranged around a central free base energy acceptor[90]. The basic structure consists of the free base porphyrin covalently linked to three pendant zinc species. Three additional zinc porphyrins are associated with the pendant species through axial binding of imidazole to zinc. Energy transfer efficiency (11%) is lower than expected. An 80% decrease in the fluorescence of the zinc porphyrin collectors and a threefold increase of the emission from the free base are observed. However, a 16-fold increase in emission would be expected for quantitative ET. The species in Figure 13.47 has been successfully incorporated in unilamellar vesicles, and shows similar ET properties in confined medium.

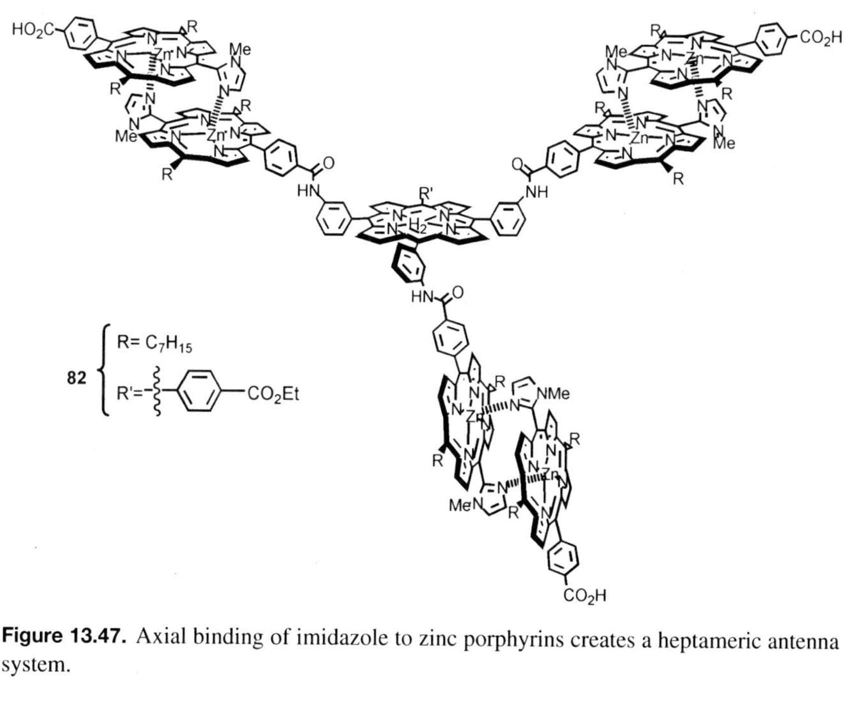

Figure 13.47. Axial binding of imidazole to zinc porphyrins creates a heptameric antenna system.

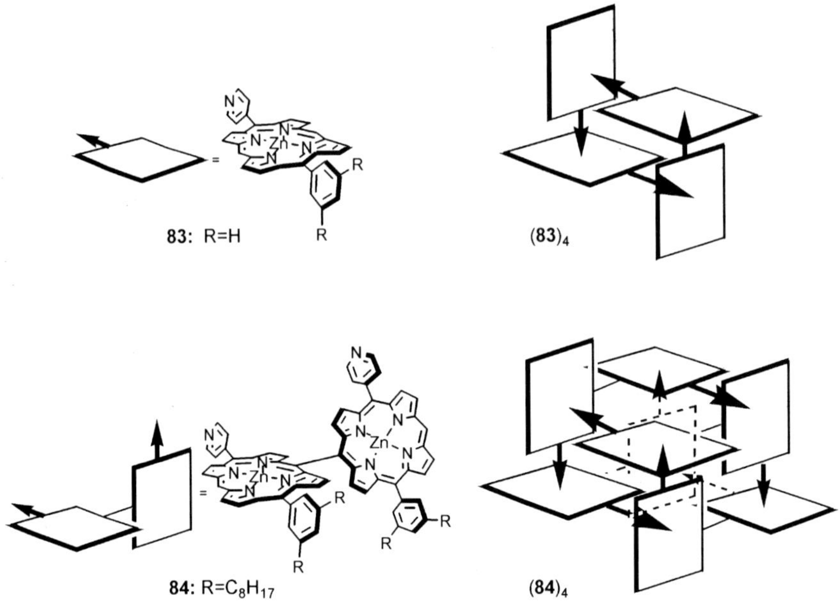

Figure 13.48. Pyridine-binding to zinc porphyrins leads to assemblies with four and eight porphyrins.

More rigid structures can be produced by axial base binding to zinc porphyrins, as demonstrated by the results of Osuka and Kim who studied the octamer box structure $(\mathbf{84})_4$ represented in Figure 13.48[91]. The box structure is actually a tetramer of the *meso–meso* linked dimer **84**. The rigidity of the ensemble is due to the direct *meso–meso* link between two porphyrin units. Comparison of the 4-pyridyl substituted dimer **84** with a model compound **83**, bearing regular aryl groups, shows that the presence of the pyridyl moiety does not affect the averaged lifetime (~1,500 ps with two decay components associated with conformational heterogeneity) or fundamental energetic properties of the dimer itself. The octamer's properties have been compared with those of the tetramer $(\mathbf{83})_4$ (Figure 13.48) that was reported earlier by Schaafsma and van Grondelle[92]. Photophysical data do show some similarities, but Osuka's porphyrin box $(\mathbf{84})_4$ shows a very fast anisotropy rise time of *ca.* 12 ps, which is consistent with an enhanced ET rate due to the excitonic dipole–dipole interactions in *meso–meso* linked dimers.

Zinc is not the only metal employed in the assembly of porphyrins via axial coordination. Among the metals that possess interesting photophysical behavior, gallium(III) offers the possibility of forming hexacoordinate complexes. In addition, the Ga(III) porphyrin is a cationic species which allows an additional electrostatic contribution to the driving force of the axial binding in the presence of anionic axial bases. This is illustrated by the staircase structure **85** prepared by Kobuke and represented in Figure 13.49[93].

These structures, which extend up to approximately 240 units per linear coordination polymer, have been characterized by tapping mode AFM imaging on

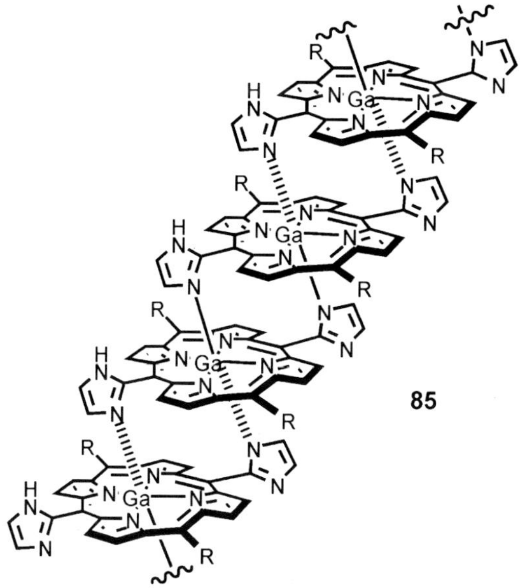

Figure 13.49. A staircase architecture based on axial binding to Ga(III).

Highly Oriented Pyrolitic Graphite (HOPG) surface. The slipped cofacial dimer arrangement confers UV–visible properties close to those of J-aggregates with red-shifted absorptions. The conductivity of multilayered films prepared with this material changes from $6.6 \times 10^{-10}\,\mathrm{Scm}^{-1}$ in the dark, to $5.5 \times 10^{-9}\,\mathrm{Scm}^{-1}$ (eightfold) upon irradiation at 532 nm. The weak, but significant current can be explained by boundaries between domains of the staircase structure. However, photocurrent intensities did not change during continuous light irradiation for 5 min.

As we have just seen, electrostatic interactions can reinforce the formation of axial base complexes in metalloporphyrins, but other thermodynamically favored processes, such as chelation around a bidentate base like pyrazine, can be used. This approach has led Kuroda to the preparation of self-assembled dendrimer-type structures **86–89** represented in Figure 13.50[94]. In these structures, the zinc porphyrin dimer formed by multihydrogen bonding of tetracarboxylate derivatives of TPP is considered as a chelate.

UV–visible titrations confirmed that in the experimental conditions of the photophysical experiments, for which concentrations of free base are in the low micromolar range, ~92% of the zinc dimers were chelated around pyrazines in the assembled structures of **86**, **87**, and **88**. Upon excitation at 564 nm, fluorescence occurred primarily from the central free base energy acceptor, due to energy transfer according to the Förster mechanism. Array **86** shows a 77-fold increase of the free base emission compared to the individual free base component (where R = PY2) with the same number of energy collecting zinc dimers located at a longer distance, array **87** shows only a 20-fold increase of the free base emission.

This is comparable with array **88** in which eight zinc dimers are also arranged around the free base, for a 12% fluorescence increase. In this latter

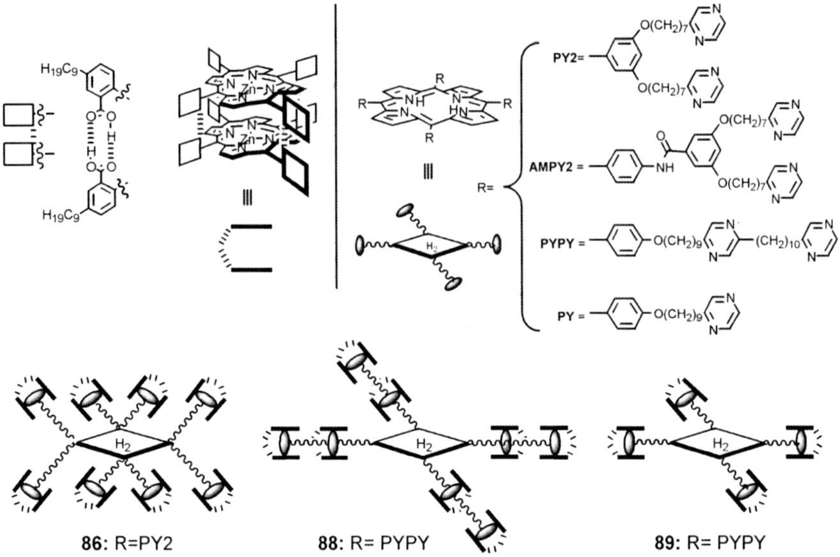

86: R=PY2 **88**: R= PYPY **89**: R= PYPY

87: R=AMPY2

Figure 13.50. Dendrimer-like antenna structure formed by H-bonded zinc porphyrin dimers bound to pendant pyrazines on a free base porphyrin.

assembly, a competing energy transfer between inner and outer pigments is invoked to explain this rather low efficiency of the energy collection.

Chelation of a free base porphyrin guest by a bis-zinc porphyrin species is again the driving force for assembly **90** reported by Sauvage and depicted in Figure 13.51[95]. A high association constant of $6 \times 10^8\ M^{-1}$ is established in toluene and energy transfer from the S_1 state of the zinc porphyrin donors to the free base acceptor is very efficient (98%), via a Förster mechanism. Interestingly, although the process is exergonic ($\Delta G^\circ = -0.12\,eV$), the energy transfer rate displays a temperature dependence, decreasing from $2 \times 10^{10}\ s^{-1}$ at 298 K to $2.3 \times 10^9\ s^{-1}$ at 77 K. A tentative explanation for this unusual behavior is the inhibition of molecular motions at 77 K, which may be responsible for a favorable orientation of transition moment dipoles at room temperature. Indeed, a too well organized structure may impose lower values of κ^2 that usually intervenes in the Förster equation as a statistical value that stems from a random chromophore orientation.

A similar design led Hunter to the porphyrin pentamer **91** in which a free base bearing four articulated pyridyl arms is chelated twice by bis-zinc porphyrin tweezers[96]. The structure depicted in Figure 13.52 shows a high association constant of $2.0 \pm 0.5 \times 10^6\ M^{-1}$ (vs. $10^3\ M^{-1}$ for usual ZnP-Py coordination). Upon excitation at 562 nm, isolated species have rather usual fluorescence decay times of 1,330 ps for the zinc porphyrin tweezers, and 9,500 ps for the free base tetradentate. After complete assembly, an energy transfer is observed at the rate of $2 \times 10^9\ s^{-1}$, with a 73% quantum yield.

Simultaneous axial coordination on zinc porphyrins has also been realized by Zenkevich and Rempel[97] in the continuation of previous work [see ref. [1].

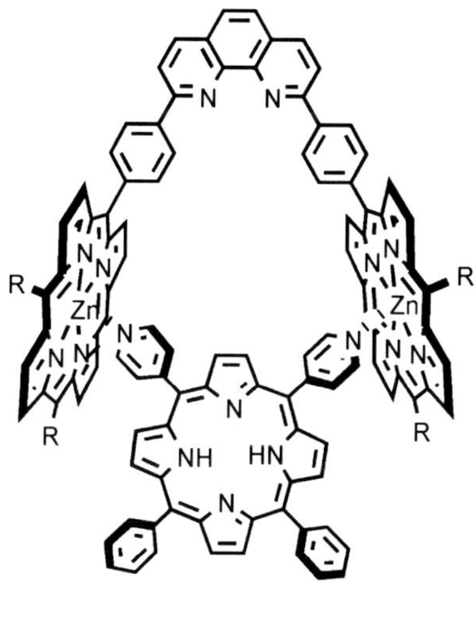

90: R = 3,5-(di-*t*-Bu)Ph

Figure 13.51. A bis(pyridyl) porphyrin bridges two zinc porphyrins and acts an an energy acceptor.

For the numerous structures schematically represented in Figure 13.53 the association constants in toluene at room temperature are in the range of 10^6–10^7 M^{-1}. The properties of the systems obtained depend upon the elements associated in the assembly, since the photoexcitable zinc porphyrin dimer can be surrounded by either an energy acceptor EAn, that could be a free base porphyrin, chlorin, or tetrahydroporphyrin, or an electron acceptor eAn, that could be a quinone or pyrromellitimide. If no electron acceptor is present (R = H) the assembly behaves like similar dyads described in this chapter, with the free base acting as an energy acceptor (for $[Zn_2]^*$). In the absence of an axially coordinated free base and in the presence of only an electron acceptor (R = eAn), the fluorescence of the zinc dimer is quenched due to electron transfer to eAn, with rates ranging from $k_{eT} = 0.03 \times 10^{10}\,s^{-1}$(R = eA4), to $k_{eT} << 0.08 \times 10^{10}\,s^{-1}$(R = eA3), $k_{eT} = 0.66 \times 10^{10}\,s^{-1}$(R = eA1), $k_{eT} = 2.86 \times 10^{10}\,s^{-1}$(R = eA2), and $k_{eT} = 19.9 \times 10^{10}\,s^{-1}$ and $2.4 \times 10^{10}\,s^{-1}$ for the two conformers with R = eA5. When triads are formed, three deactivation processes, summarized in Figure 13.54, can compete for fluorescence quenching of the zinc dimer (Zn_2*) excited state.

Singlet energy transfer to the S_1 state of the energy acceptor can occur, generating (EAn*) states that behave as a hole donor for the zinc dimer (Zn_2), or as an electron donor for the eAn species, with mediation of the high energy $(EAn)^+$–$(Zn_2)^-$ bridge by a super exchange mechanism to produce a CS state $(EAn)^+$–Zn_2^-–(eAn). As a consequence, deactivation of the (Zn_2*) states is accelerated by the presence, of axially coordinated H_2 species.

Imidazole-zinc binding is not sufficient to produce stable assemblies, unless two simultaneous coordinations are employed as in early work from Kobuke[98].

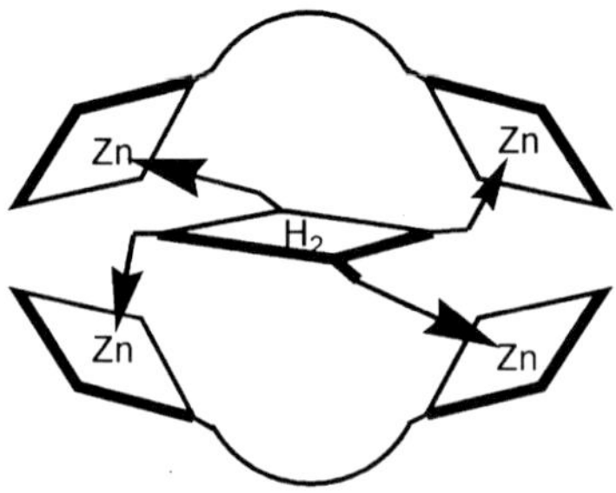

91

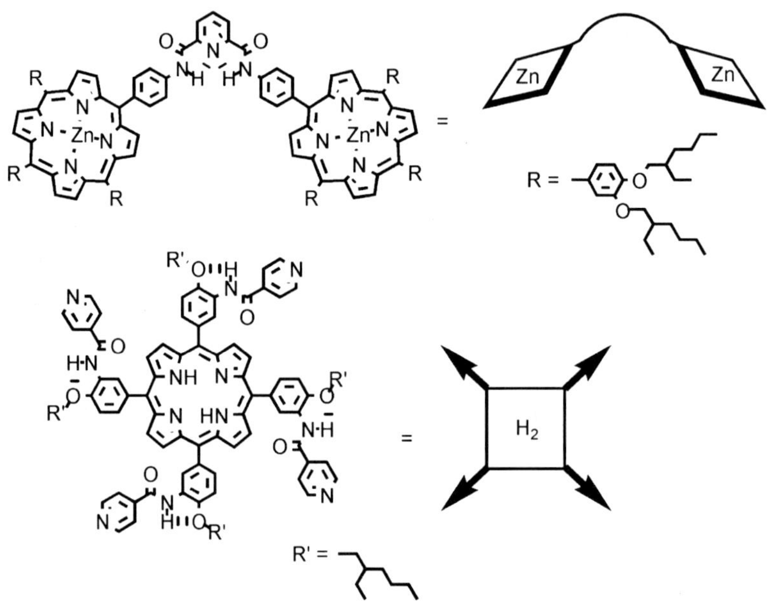

Figure 13.52. A rigid pentamer combining metal–ligand binding and H-bonds.

However, the combination of imidazole–zinc binding with H-bond and π–π interactions can afford association constants high enough to be used as an assembly tool between porphyrin moieties[99]. Based on K_{assoc} values in the range of 10^6–10^7 M^{-1} in CH_2Cl_2 for the phenanthroline-strapped zinc porphyrin **92** with even hindered imidazoles,[100] we have developed a series of self-assembled zinc/free base dyads **93–95** represented in Figure 13.55[101]. The high binding constants ensure that 1:1 D–A complexes are formed, except in dyad **93** where steric hindrance between the two porphyrins lowers the binding constant.

Efficient energy transfers of >99, 98, and 88% are respectively observed for **93**, **94**, and **95**. The phenyl acetylene spacer between the imidazole and the free base porphyrin of the acceptor in **94** seems to be the ideal compromise for efficient energy transfer and a high binding constant. This motif has thus been employed in a triad (**96**) in which a BODIPY has been added to the ZnP donor

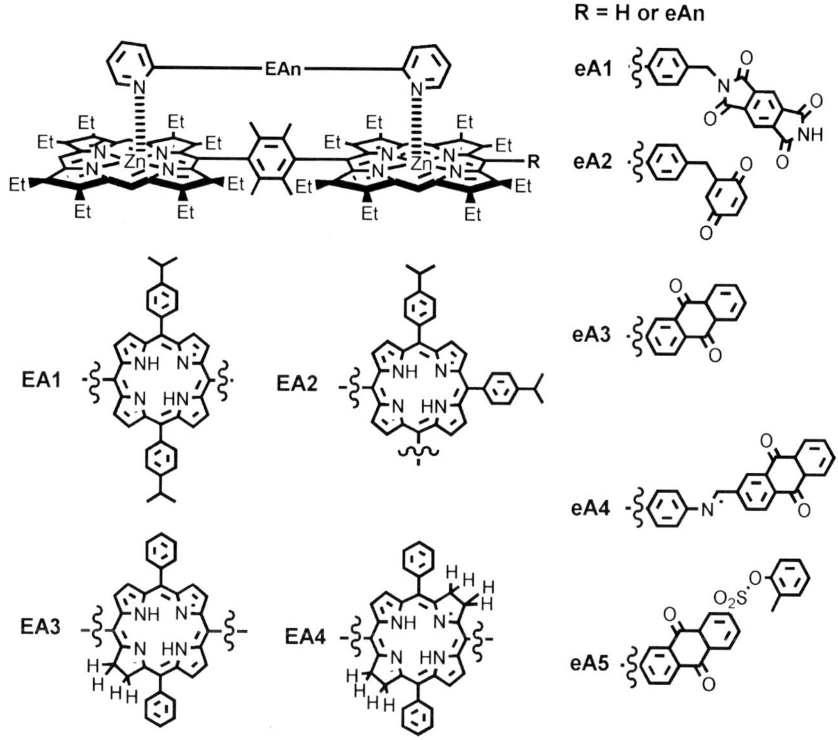

Figure 13.53. A series of ZnP dimers containing energy acceptors (EA) and/or electron acceptors (eA).

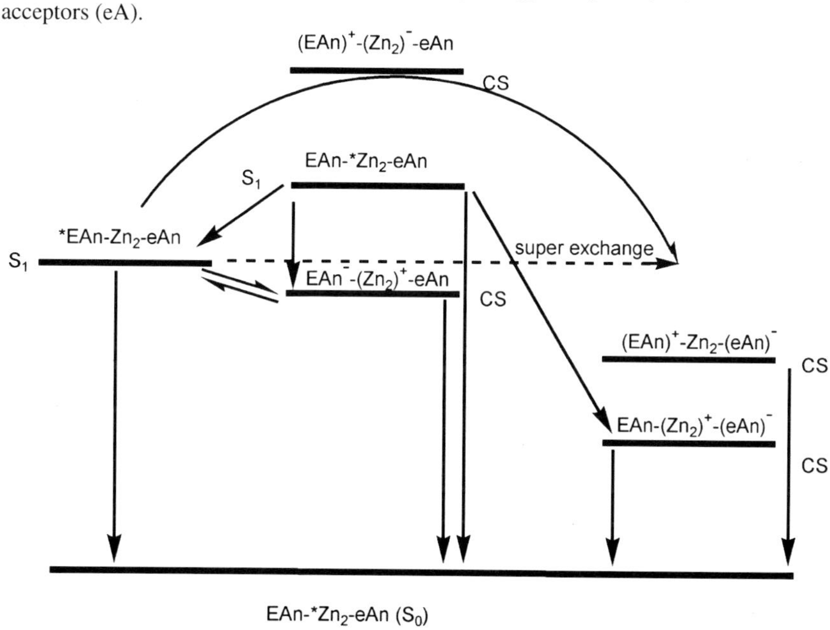

Figure 13.54. Deactivation processes for arrays shown in Figure 13.53.

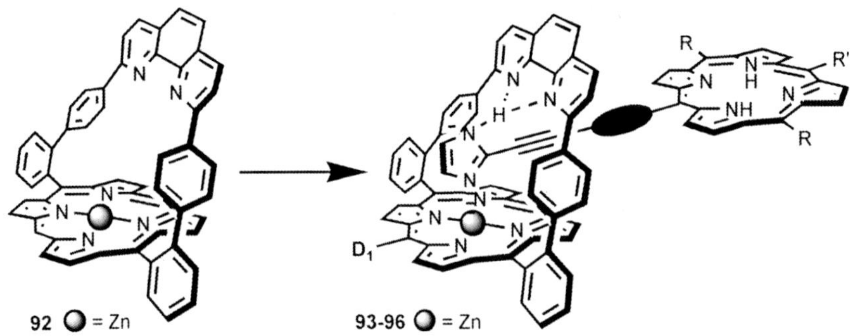

		R, R'	D_1
93	—	R = 3,5-di-*t*-BuPh, R' = *m*-xylyl	H
94	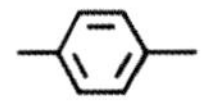	R = R' = 3,5-di-*t*-BuPh	H
95	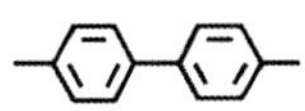	R = R'= phenyl	H
96	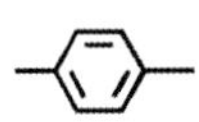	R = R' = 3,5-di-*t*-BuPh	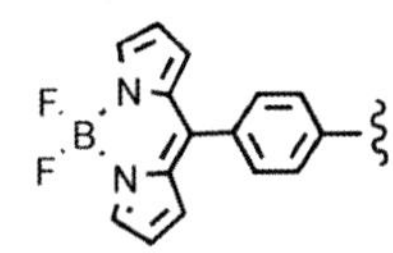

Figure 13.55. Strong imidazole binding within the phenanthroline-strap of the ZnP is used as an assembly tool for a series of dyads.

92. In this way selective excitation of the initial donor can be achieved, resulting in a cascade of energy transfers from the BODIPY to the ZnP (96% efficiency) and then to the final H_2P acceptor (92% efficiency)[102].

If only one coordination bond is developed, this might not be sufficient for a stable assembly in the case of Schiff bases and zinc porphyrins. Stronger coordination bonds such as alkoxy and aryloxy ligands with aluminum(III) have proved helpful in assembling porphyrins, as illustrated by the recent examples **97–98** reported by Maiya (Figure 13.56)[103].

One or two Al(III) cationic porphyrin(s) is (are) connected with a peripheral mono- or di-phenolic free base, and their photophysical properties are compared with previously described tin(IV), germanium(IV), and phosphorous(V) oxophilic porphyrins. Two processes can occur in these nonelectronically coupled species, the first being observed upon excitation of the aluminum porphyrin. When excited at 550 nm, the Al(II) species' fluorescence is quenched by the free base component mostly by energy transfer, but concomitantly by electron transfer in polar media. The second process occurs upon excitation of the free base species, and consists in a poorly efficient (low Q_{H_2}%) electron transfer from the free base to the Al(III) species. The same process is quite efficient in the case of P(V), Sn(IV), and Ge(IV) analogs with respective Q_{H_2} values of 93, 82, and 69%[103].

Table 13.6 collects the emission wavelength, fluorescence quantum yields, and quenching efficiencies for the monomers, dyad **H_2-97**, and triad **H_2-98** described in Figure 13.56.

In all of the previous examples in this section, at least one of the porphyrin species played both the role of electron or energy donor or acceptor in the device, and the role of the coordinating moiety. A judicious choice of both the metal in the core of the porphyrin and the heteroatom on the coordinating axial base allows the use of nonporphyrinic connectors that will be detailed in the next example.

The triad **99** depicted in Figure 13.57, uses a pyridine-carboxylate connector, with nitrogen binding to ruthenium(II) and oxygen binding to tin(IV). Due to the nonversatile affinities of the heteroatoms for the metalloporphyrins, the self-assembly is directed by N coordination to Ru(II), and O coordination to Sn(IV). Ruthenium porphyrins are non fluorescent as isolated species, and so is the triad **99**[104]. Upon excitation at 600 nm, where absorption is almost exclusively due to the tin species, the fluorescence of the central chromophore is strongly quenched ($\Phi_{em} = 0.007$, $\tau_{em} = 0.63$ ns) compared to an isolated tin porphyrin ($\Phi_{em} = 0.024$, $\tau_{em} = 1.48$ ns). Electron transfer from the tin(IV) porphyrin to the ruthenium(II) porphyrin is responsible for this quenching. For this eT, a free energy of 0.32 eV was estimated from the redox potential of the trimer and the energy of the S_1 excited state of the tin species (2.04 eV).

It should be noted that this approach, along with the previous examples in this section, usually generate neutral porphyrins arrays. Here for example, the double carboxylate binding to the dicationic tin(IV) porphyrin does not involve other counter ions. In these approaches that aim to efficiently produce long arrays of chromophores, it is important to work with neutral species, since it is known from coordination polymers involving charged species that counter ions may have non negligible structural impacts on the topography of the assemblies. This may be an advantage when flexible connectors are used between chromophores because the structuring influence of a counter ion will lead to a well defined arrangement of the respective components.

5.3. Exocyclic Coordination Complexes as Linkers

The attachment of metal binding sites at the periphery of porphyrins is a convenient way to link chromophores in a directed way. This approach has several advantages that will be illustrated in the examples detailed hereafter. The main advantage stems from the well defined geometries of transition metal coordination complexes. As a result, the chromophore assemblies produced by this approach are usually extremely well defined, not only in terms of topography, but also in terms of electronic coupling when the exocyclic coordination complex is coplanar or fused with the porphyrin ring, and in terms of super exchange coupling when the exocyclic coordination complex is not coupled with the porphyrin rings but displays some photoactivity as well.

Table 13.6. Photophysical Data for Isolated Monomers, Dyads, and Triads from Figure 13.56

	λ_{em} in (nm) (Φ_{em}, Q_{H_2} or $Q_{Al(III)}$ %)							
Compound	Toluene		CH_2Cl_2		CH_3CN		DMF	
	$\lambda_{exc} = 550$	$\lambda_{exc} = 650$	$\lambda_{exc} = 550$	$\lambda_{exc} = 650$	$\lambda_{exc} = 550$	$\lambda_{exc} = 650$	$\lambda_{exc} = 550$	$\lambda_{exc} = 650$
H_2P	–	721 (–, 0.4)	–	719 (–, 0.11)	–	717 (–, 0.09)	–	718 (–, 0.14)
AlP	597 647 (–, 0.049)	–	601 647 (–, 0.059)	–	606 660 (–, 0.11)	–	608 663 (–, 0.087)	–
H_2–97	599 657 723 (0.008, 84)	721 (0.13, 7)	601 656 721 (0.013, 78)	721 (0.10, 9)	607 655 720 (0.023, 79)	719 (0.07, 16)	608 660 722 (0.031, 65)	721 (0.12, 14)
H_2–98	595 663 727 (0.016, 67)	727 (0.11, 21)	599 658 725 (0.033, 45)	725 (0.07, 36)	606 665 726 (0.06, 45)	726 (0.05, 44)	609 675 732 (0.038, 56)	733 (0.09, 35)

Errors on Φ are 10%. λ_{exc} in nm.

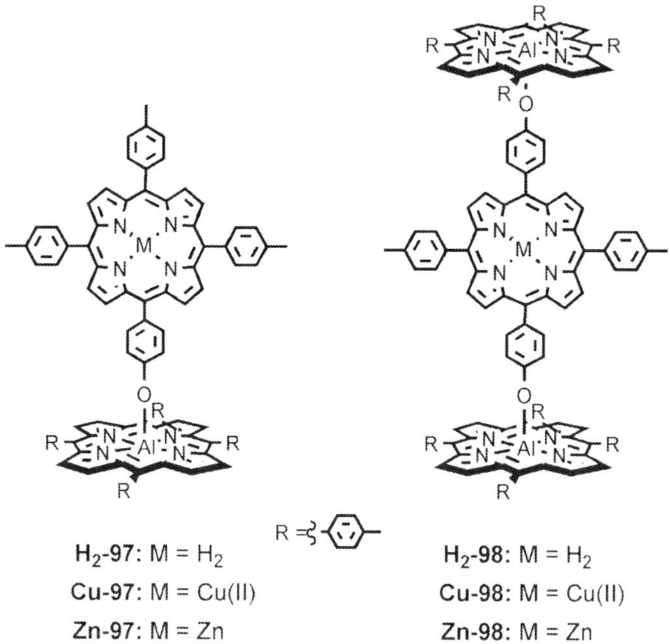

Figure 13.56. Oxygen ligands bind strongly to aluminum porphyrins, forming stable assemblies.

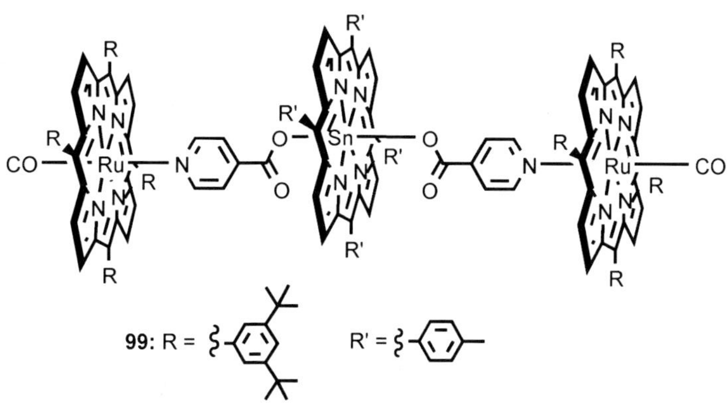

Figure 13.57. Directed self-assembly based on ligand affinity in a Ru(II)–Sn(IV)–Ru(III) traid.

The case of coplanar exocyclic coordination complexes is illustrated in assemblies **100–112** (Figure 13.58), reported by Callot and coworkers, who take advantage of their previous experience in porphyrin ring derivatization[105–107]. Obviously, from the numerous assemblies prepared by this methodology, the formation of exocyclic coordination complexes is extremely versatile regarding the combination of metals in both the porphyrin chromophores and the connecting complexes. The characteristic features of the assemblies are the drastic red shifts of the S_0–S_2 and S_0–S_1 transitions. The lower energy S_0–S_1 transitions are

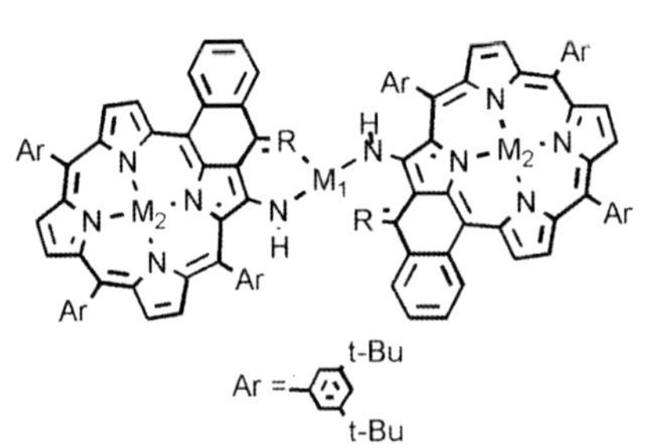

	R	M_1	M_2	Assembly
100:	O	Ni(II)	Ni(II)	Ni/Ni/Ni
101:	O	Cu(II)	Ni(II)	Ni/Cu/Ni
102:	O	Pd(II)	Ni(II)	Ni/Pd/Ni
103:	O	Pd(II)	Cu(II)	Cu/Pd/Cu
104:	O	Cu(II)	Cu(II)	Cu/Cu/Cu
105:	O	Pd(II)	Pd(II)	Pd/Pd/Pd
106:	S	Pd(II)	Zn(II)	Zn/Pd/Zn
107:	S	Pd(II)	Ni(II)	Ni/Pd/Ni

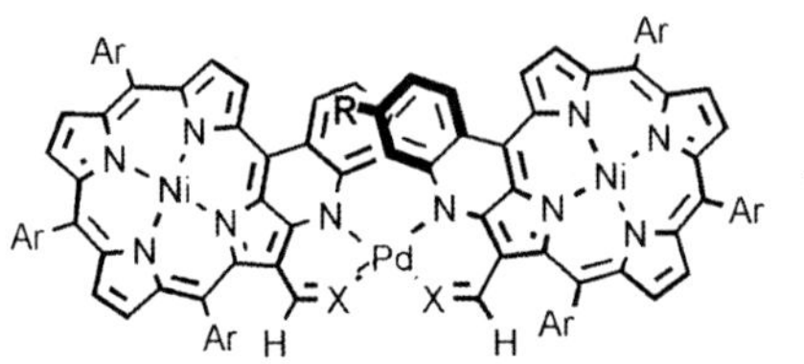

108: X = O, R = H, Ar = Ph
109: X = O, R = CH_3, Ar = tolyl
110: X = S, R = H, Ar = Ph

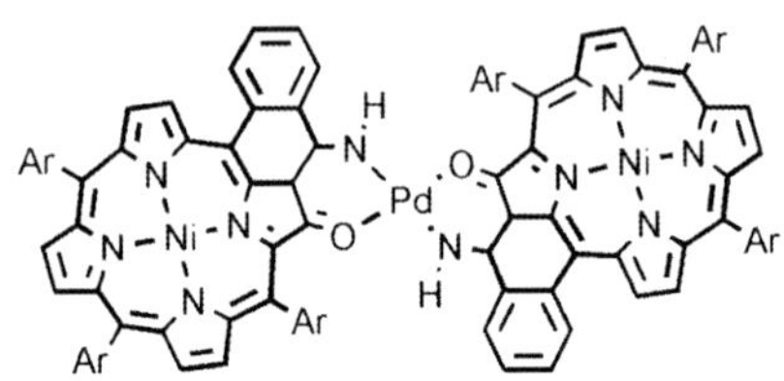

111: Ar = tolyl

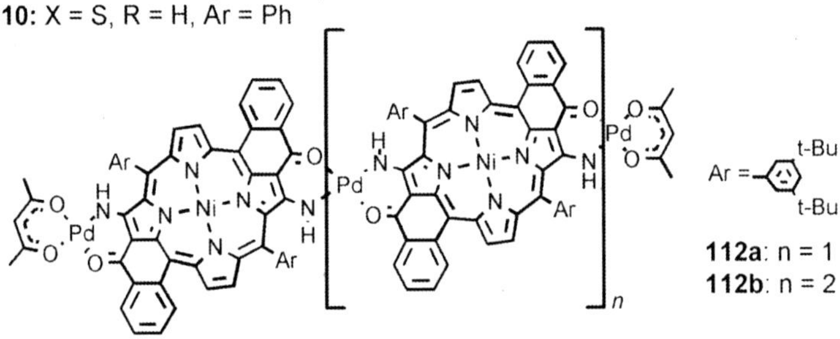

112a: n = 1
112b: n = 2

Figure 13.58. Dimers and trimers built by exocyclic metal complexation.

particularly affected by the primary coordination of the exocyclic transition metal. This can be seen in the visible part of the absorption spectra and correlated with variations in the first reduction and oxidation potentials by cyclic voltammetry. It is important to note that when two exocyclic binding sites are organized around the central porphyrin in a trimer, both metal complexes shift the lowest absorption band by ~ 50 nm. The presence of the terminal porphyrin rings in compound **112b** does slightly change the absorption, but the Soret band is affected the most. After assembly, a splitting of up to 190 mV is observed for the electrochemical peaks corresponding to oxidation and reduction of the macrocycles. This splitting denotes a strong porphyrin–porphyrin interaction via electronic coupling that is mediated by the connecting transition metal complex.

A similar approach can be developed using coordination complexes that are not photochemically inert. Lindsey has reported a neutral assembly that uses a zinc bis-dipyrrin (DIPY) complex as a connector. Due to the synthetic procedure employed, it is not yet possible to differentiate the chromophores in the trinuclear complexes represented in Figure 13.59. However, it is possible to show that the connector does not have a drastic impact on the absorption properties and the

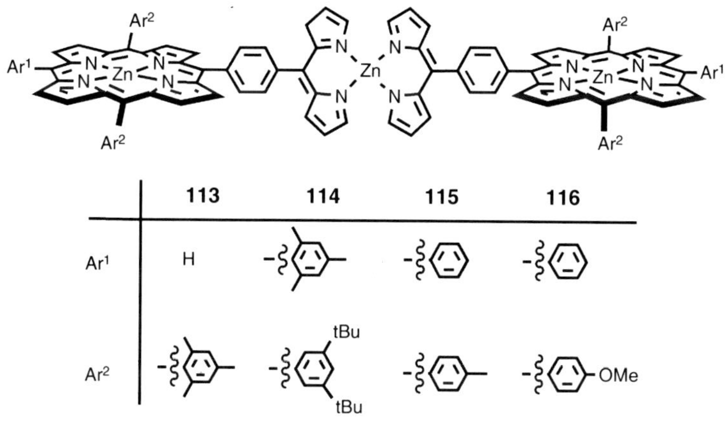

Figure 13.59. Dyads assembled via a dipyrrin metal complex.

Table 13.7. Room Temperature Measurements in Non Deaerated Solvents for **113** and **114**

Triad	Solvent	ZnP excited state decay Φ^{a}_{em}	τ (ps)	Φ_Q	$Zn(DIPY)_2$ excited state decay τ (ps)^{b}	Φ_{ET}
113	Toluene	0.022	2,600	< 0.1	1.4	0.97
	PhCN	(0.025)	~ 1,000	~ 0.6	0.8	~ 0.8
		0.010				
		(0.012)				
114	Toluene	0.043	2,400	< 0.1	1.4	0.97
	PhCN	(0.037)	~ 1,000	~ 0.6	0.8	~ 0.8
		0.030				
		(0.021)				

[a] ZnTPP ($\Phi_{em} = 0.03$) as reference for Φ_{em} values.
* ZnP decay with excitation at 400 nm. $Zn(DIPY)_2$ with excitation at 453 nm for **113**, and 487 nm for **114**.
[b] Typical isolated $Zn(DIPY)_2$ decay time of 9.3 ps.

ground state coupling of the two zinc porphyrins. This work also demonstrates that it is possible to take advantage of the photoexcitable zinc dipyrrin complex to induce ET processes in the assemblies. All measurements show that excitation of either the zinc dipyrrin complex or the zinc porphyrins generates emission from the zinc porphyrin chromophores due to an efficient ET from the dipyrrin complexes to the zinc porphyrins (Table 13.7)[108]. It should be noted that assemblies using Pd or Cu DIPY complexes are reported, but their photophysical properties are not described.

The final examples of metal templated assembly of porphyrins bearing an exocyclic coordination site were reported by Sauvage, who first described the preparation and study of triads **117a–b** built around a bis-terpyridyl complex of iridium(III) (Figure 13.60).

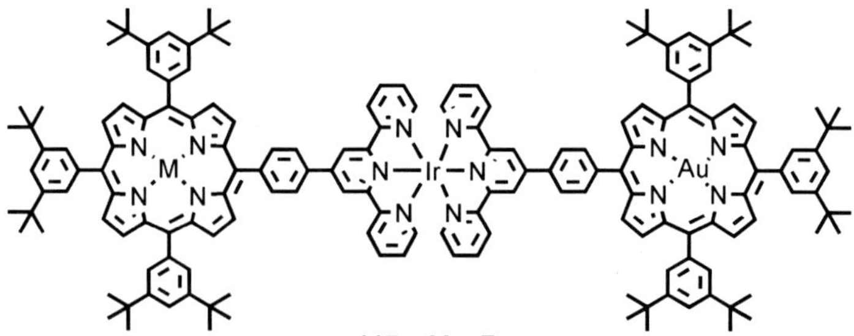

117a: M = Zn
117b: M = H_2

Figure 13.60. An iridium(II) bis-terpyridine complex used as an assembly tool.

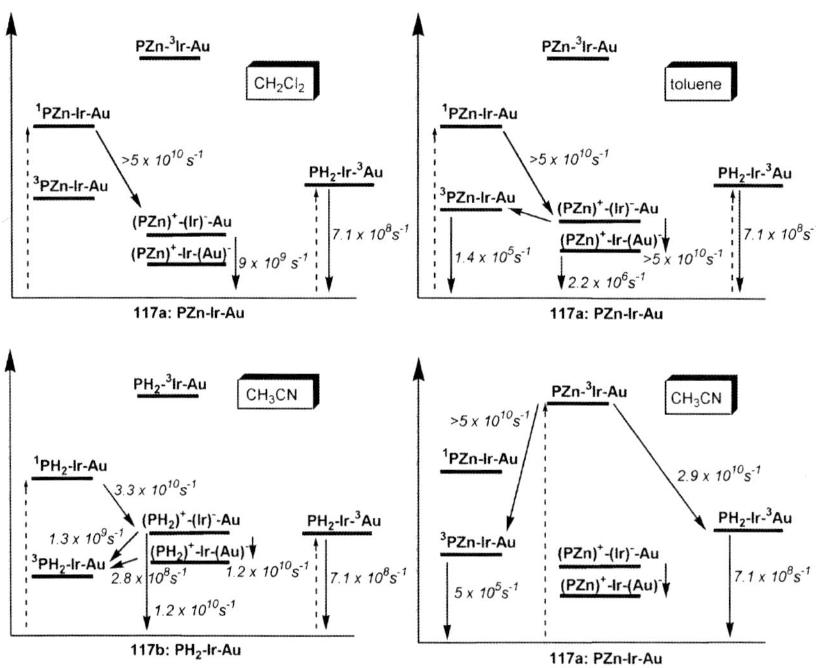

Figure 13.61. Energy diagrams for the energy transfer processes in triads **117a–b** in different solvents.

Let us first consider the situation when porphyrin chromophores are excited. In this case, only electron transfers are possible within the triads, and also within the model dyads (not represented) used as reference compounds. Upon excitation of the iridium chromophore in the UV, an ET process to both porphyrins occurs. In these systems, it is interesting to note that the nature of the solvent used for the photophysical studies has a strong impact on the photoinduced processes and on the lifetimes of the CS states generated. These data are summarized in Figure 13.61 in the form of energy diagrams.

The first two diagrams (Figure 13.61, top) show the evolution of the zinc porphyrin singlet excited state in the presence of two electron acceptors, the iridium complex and the gold porphyrin. In methylene chloride, the triad **117a** behaves like the model dyad comprising only the zinc donor and the iridium acceptor. The electron transfer in the triad **117a** stops at the iridium. The driving force for the second electron transfer to the gold porphyrin is too small (−0.05 eV) and the existence of a charge transfer ground state interaction between the zinc porphyrin and the iridium complex may be responsible for the fast deactivation of the ZnP^{+}–Ir^{-} state via an exciplex (low lying excited state with a charge transfer character).

In toluene, the stepwise electron transfer goes all the way to the charge separated species ZnP^{+}–Ir–AuP^{-}, and the photogenerated CS state has a 450 ns lifetime. It should be remembered, however, that the formal writing of "Ir^{-}" and "AuP^{-}" actually corresponds to a dicationic iridium complex and a neutral gold porphyrin. In the case of the H_2P–Ir–AuP triad **117b**, the nature of the solvent has less influence on the different processes, and excitation of the free base component leads to a full CS state H_2P^{+}–Ir–AuP^{-}, which has a very short lifetime of 3.5 ns. As the CR to the ground state is in the Marcus "inverted region", a longer lifetime could have been expected. However, the CR process is short-circuited by the presence of a low lying triplet state of the free base which also deactivates the corresponding model dyad[109]. Finally, irradiation of the iridium complex in acetonitrile (Figure 13.61, bottom right) generates a triplet excited state that can transfer energy to both the free base and the gold porphyrin[110].

Another way of templating the chromophore assembly, based on the copper(I) assisted synthesis of catenanes and rotaxanes developed by Dietrich-Buchecker and Sauvage in the 1980s, has led to the rotaxanes **118–120** represented in Figure 13.62. In this case, excitation of either the zinc or the gold porphyrins leads to photoinduced charge transfer to generate $[(ZnP)_2]^{+}$–$AuP^{\cdot}$.

For ZnP excitation, eT rate constants are quite fast (from 1.1×10^9 to $1.2 \times 10^{10}\ s^{-1}$), as are the hole transfer rates for AuP^{+} excitation (from $5 \times 10^8\ s^{-1}$ to

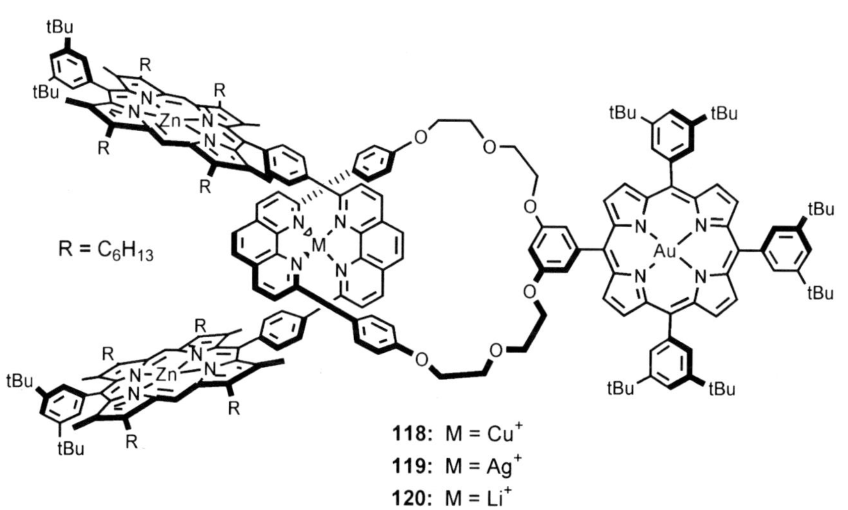

Figure 13.62. Rotaxane porphyrin triads built around a bis(phenanthroline) complex.

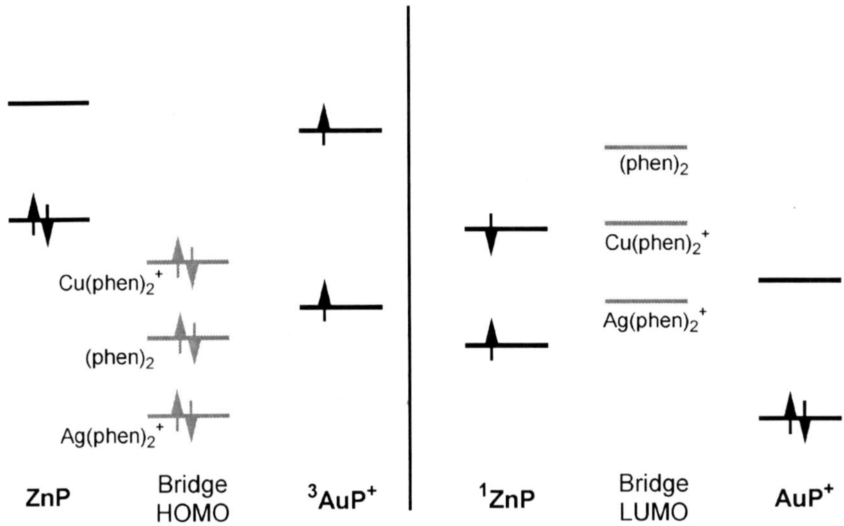

Figure 13.63. Energy diagrams of the frontier orbitals for porphyrin rotaxane traids **118–120**.

$3.9 \times 10^{10}\ s^{-1}$). The nature of the bridging metal complex is the governing factor in these systems, as the behaviors of the mechanically linked, copper(I)(phen)$_2$, or silver(I)(phen)$_2$ bridged complexes are significantly different as far as their mechanisms of eT are concerned. In fact, the presence or absence of a metal in the entwined bis-phenanthroline bridge affects the positioning of the frontier orbitals of the connecting bridge, as illustrated in the diagram in Figure 13.63.

In the case of hole transfer from the ^{3}AuP state, copper mediation is observed in the form of a hopping mechanism, even though no evidence for a Cu(II) intermediate was found. In the case of Ag(I), a super exchange-mediated mechanism is demonstrated and these systems offer an additional way to control and switch eT processes. In the case of electron transfer from ^{1}ZnP to AuP$^+$, the charge separated state lives for 10 to 40 ns, and its efficient generation (80% yield) competes with ^{1}ZnP deactivation through ^{3}ZnP, after a fast ($k_{ET} = 5.1 \times 10^9\ s^{-1}$) primary ET from ^{1}ZnP to Cu(I)(phen)$_2$, and a second ET process ($k_{ET} = 1.5 \times 10^9\ s^{-1}$) from *Cu(I)(phen)$_2$ to ZnP[111].

All of these examples show that the use of metal coordination at the periphery of the porphyrin chromophores can be extremely helpful in modular approaches targeting multiporphyrinic assemblies. As we will see later, the formation of coordination polymers has already found applications in the preparation of materials.

5.4. Membranes, Vesicles and Micelles: Templated Assembling

Natural systems use proteic environment and protein structures to spatially arrange chromophores. As we will see in this section, interactions of porphyrins with peptide fragments or organized monolayers can help to achieve such topographic control in artificial assemblies.

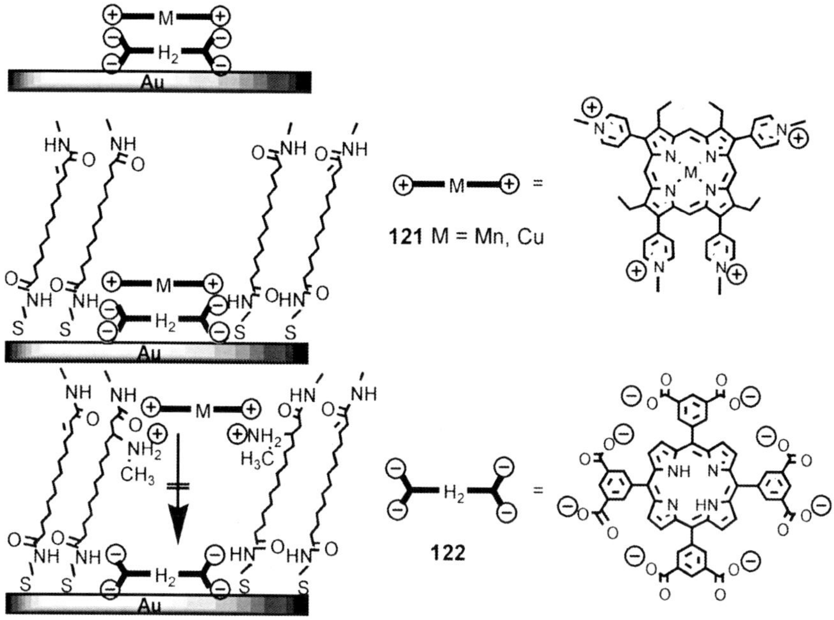

Figure 13.64. Porphyrin dimerization in organized monolayers on a gold surface. Top: electrostatic interactions between porphyrin substituents result in dimerization. Middle: dimerization within a monolayer. Bottom: dimerization is hindered by the presence of positively charged substituents near the top of the monolayer.

Fuhrhop and coworkers have used their expertise in monolayer construction to develop the concept of molecular landscaping on surfaces[112]. Using fluorescence quenching between noncovalently assembled porphyrins, they are able to control and monitor the formation of clefts in monolayers attached to gold surfaces, as depicted in Figure 13.64.

In these assemblies, the adsorption of the multianionic porphyrin prior to the formation of the monolayer, leaves a cleft that is accessible to molecules of electrolyte in solution. The electroactivity of the electrolyte that fills the cleft when it is accessible can be detected by electrochemistry. Addition of a cationic porphyrin containing a paramagnetic metal efficiently quenches the free base's first singlet excited state. This quenching is not efficient when ammonium residues are added at the entrance of the cleft. The positively charges residues prevent the fixation and motion of the cationic guest within the cleft. The addition of a second anionic porphyrin, that interacts strongly with the ammonium groups at the entrance of the cleft, acts as a lid on the top of the cleft and isolates the inside[113]. This work is interesting because it utilizes fluorescence quenching as a tool for detecting association between molecular components[114].

Controlled peptide architectures can be used to construct antennae-type structures. This option has been nicely illustrated by a dendrimer approach developed by Mihara[115]. In structure **123** depicted in Figure 13.65, two types of interactions are use to drive the assembly. The first is the interaction of the peptide with the flat, hydrophobic surface of the metalloporphyrin. The second

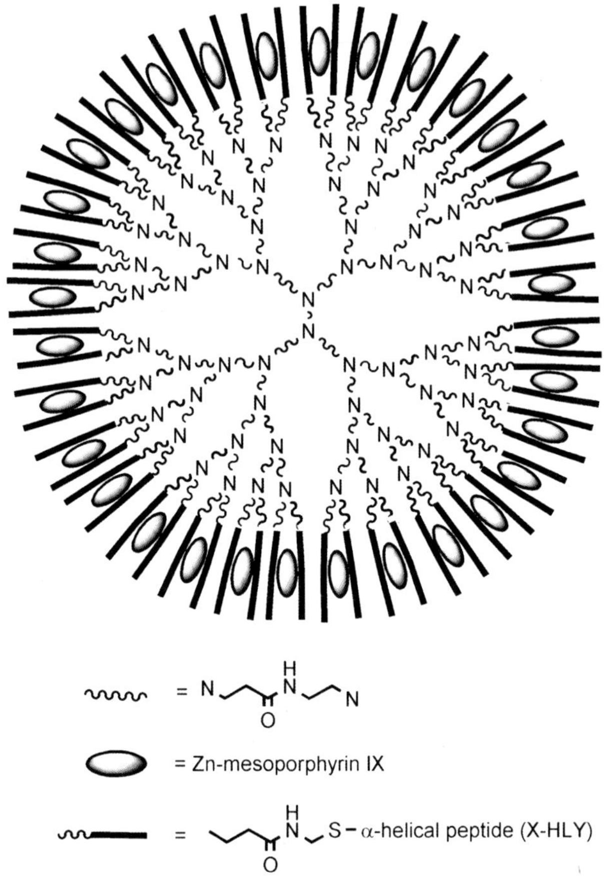

123	X-HLY	Ac-C-ALEE**Y**AKKHEE**Y**KKLA-G-**X**-NH$_2$		
	R-HLL	L	L	R
	E-HLL	L	L	E
	R-HLF	F	F	R

Figure 13.65. Pendant histidine residues at the periphery of a dendrimer bind zinc porphyrins.

is the axial coordination of the peptides' histidine residues to either zinc(II) or iron(III) porphyrins. These metalloporphyrins are rather strongly bound to these dendrimers, with $K_{\text{assoc}} = 1 - 2 \times 10^6\ \text{M}^{-1}$ for Fe(III), and 3–5 $\times 10^5\ \text{M}^{-1}$ for Zn(II). A maximum of 32 ZnPs are incorporated in the highest generation dendrimer ($n = 64$, n = number of helices), with a ratio of two helical peptides per porphyrin. When methylviologen (MV^{2+}) is added to a solution of **123**, photoinduced eT is observed from the singlet excited state of the zinc porphyrins to MV^{2+}. Due to interchromophore energy dispersion, the lifetime of the zinc porphyrins decreases from 1.8 ns for $n = 4$, to 1.2 ns for $n = 64$. Energy harvesting is more efficient and occurs at faster rates as the generation of dendrimer increases.

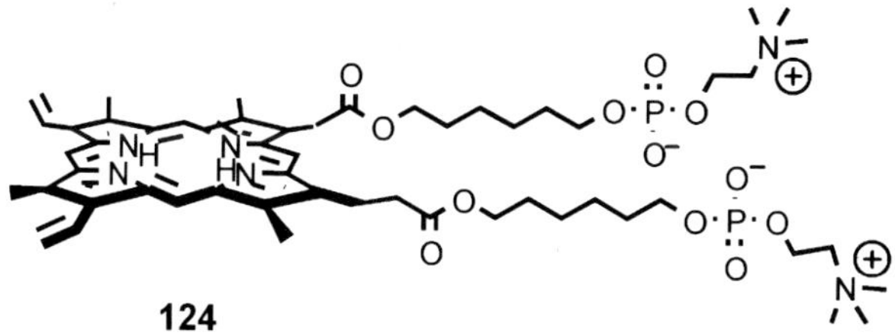

124

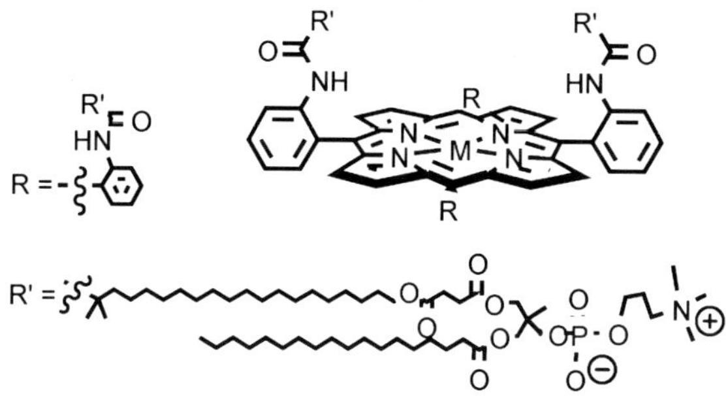

125(M): M = H_2, Zn, $Fe^{3+}Br^-$

Figure 13.66. Phosopholipid functionalized porphyrins that form vesicles with up to 23,000 porphyrins.

The rate of energy transfer between the chromophores varies from $0.8 \times 10^8\ s^{-1}$, for $n = 4$ and R-HLL, to $9.5 \times 10^8\ s^{-1}$, for $n = 64$ and R-HLL.

Tsuchida has developed a very elegant approach based on the use of amphiphilic porphyrins in which the lipophilic moiety of the amphiphile is the porphyrin and the polar head group is a classical phospholipid moiety[116]. The compounds **125(M)** represented in Figure 13.66 are able to form vesicles containing up to 23,000 porphyrins (vesicle diameter ~ 100–150 nm). This is without a doubt the largest structurally characterized multiporphyrinic assembly characterized to date.

It can be seen from UV–visible characterization of the species that the tetraaryl porphyrinic chromophores **125(M)** tend to form J-type aggregates (red-shifted absorptions). In these assemblies, the use of equimolar mixtures of **125(Zn)** and **125(H$_2$)** led to a very efficient (81%) ET process, via a postulated and reasonable Förster mechanism, with a rate constant of $k_{ET} = 3.1 \times 10^9\ s^{-1}$. When iron(III) porphyrins **125(Fe)** are used, free base protoporphyrin **124** bearing alkyphosphocholine side arms can be incorporated in the fluid part of the vesicles. In the presence of a sacrificial electron donor on the outside of the vesicle, a vectorial eT process occurs from the excited free base porphyrins to the iron(III) species, leading to photoreduction of the iron(III) to iron(II). Previous work had already shown that this could also be achieved between zinc porphyrins and iron porphyrins in vesicles[117].

Finally, an intermediate solution to the noncovalent assembly of porphyrins is the trapping of porphyrin assemblies within membrane models, as depicted in

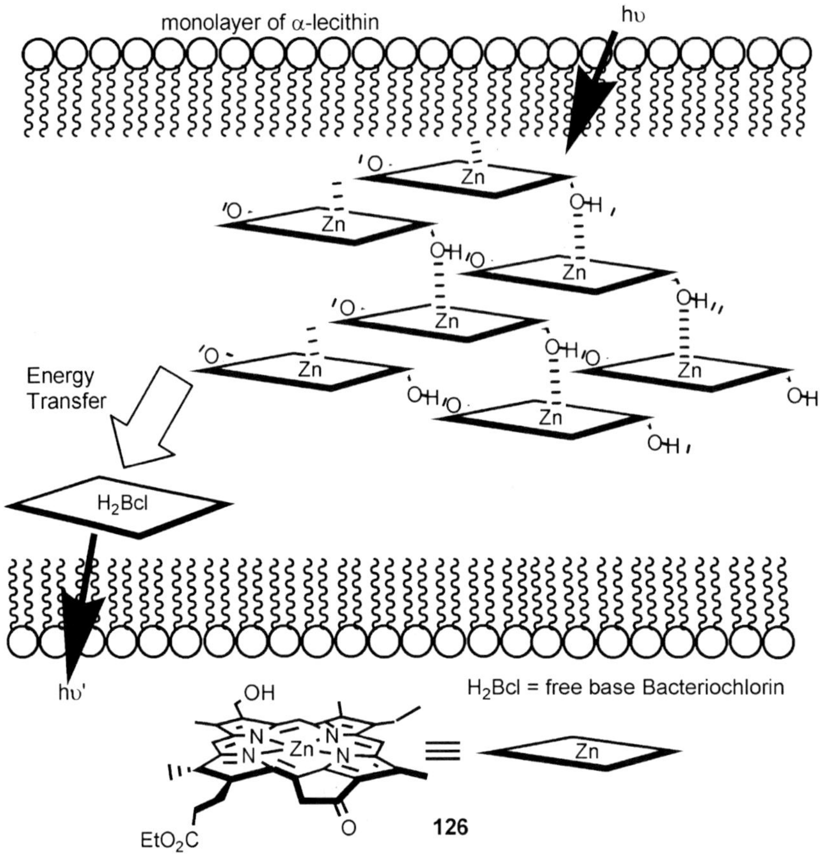

Figure 13.67. Zinc bacteriochlorin aggregates within a lipid bilayer are spectroscopically similar to bacteriochlorophyll aggregates.

Figure 13.67. As with most of the approaches in this section, the design of the assembly very closely resembles the features of LH_1 and LH_2 natural systems. Aggregates of zinc bacteriochlorins are trapped in a lecithin or Triton-X bilayer to generate chromophore assemblies with spectroscopic behavior that resembles that of the bacteriochlorophyll c aggregates in the antennae systems[118]. The aim of the work is indeed to mimic the function of chlorosomes. Chlorosomes are aggregates of BChl c, d, and e that are embedded in lipid monolayers.

The zinc(II) porphyrins **126** are held together and structured by H-bonds and by hydroxyl group coordination to zinc. Table 13.8 summarizes the properties of the aggregates as a function of the concentration of lecithin in the aqueous phase.

Aggregates secluded in the membrane start forming type II aggregates (tubular rod-like structures) at a concentration of around $10^{-3}\%$ in α-lecithin, which corresponds to a surfactant/ZnBChl ratio of 1.6. Incorporation of BChl a (free base BChl) shows that singlet energy transfer to the BChl a is efficient (60%),

Table 13.8. Spectral and Structural Characteristics of **126** as a Function of the α-lecithin Concentration in Aqueous Medium

[α-lecithin] [($\times 10^{-4}$% w/v]	0	4	10	20	40	100
Vis-absorption	Weak and broad			735 nm maximum		
CD	Weak		(+/−) type	(+/−) and (−/+) types		(−/+) type
Fluorescence	Weak		745 nm maximum			
Oligomeric structure assignment	labile		stable		Metastable	
	Type I aggregates		Chlorosome-type (type II) aggregates			

with properties comparable to those obtained when a covalent ZnChl–BChl a dyad is incorporated[119].

5.5. Noncovalent Multiporphyrin Assembling With Nonporphyrinic Electron Acceptors

Similar weak interactions can also be used to bring together multiporphyrin species and primary electron acceptors. A few examples involving more than one porphyrin have appeared in the last three years and are detailed hereafter.

In a series of articles, Johnston and Flamigni have detailed the properties of the porphyrin tweezers **127** in which the two porphyrins are maintained in a cofacial arrangement at a Zn–Zn distance of 22 Å[120–122]. These edifices appear in this section because their properties stem from interactions developed with a guest, and not from the cofacial character of the chromophore assemblies themselves. Indeed, the fundamental properties of the dimer **127** do not show drastic changes compared to independent monomers[120]. The well defined topography of the tweezers allows for strong binding of rigid guests, as observed for the electroactive diimides NIN and PIN represented in Figure 13.68[121,122]. The $N_{pyridyl}$–$N_{pyridyl}$ distance of *ca.* 15.5 Å in the guests NIN and PIN is well adapted to the zinc–zinc distance of the tweezers. An association constant on the order of 7×10^7 M^{-1} is observed for **127(NIN)**. Once the guests are tightly bound, excitation of the zinc porphyrins induces electron transfer to the diimide acceptor with a rate constant of 1.1×10^{10} s^{-1} in both cases. The CS states in both **127(NIN)** and **127(PIN)** have longer lifetimes (710 ps) than similar, covalently linked species[123]. Thus, in these systems, both forward and back electron transfers are slowed down compared to a global two- or threefold increase of the CS lifetime. The charge recombination rates for the **127(NIN)** and the **127(PIN)** complexes have been determined as 1.4×10^9 and 3.8×10^9 s^{-1}, respectively.

An interesting triad has been reported by Ito and D'Souza, who have combined a covalent free base porphyrin-fullerene dyad bearing a pyridine substituent,

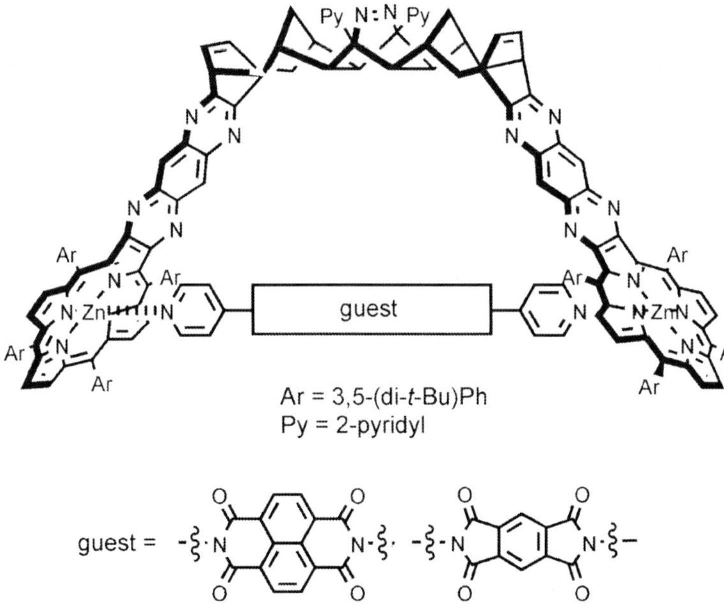

127(NIN) **127(PIN)**

Figure 13.68. In the presence of the NIN or PIN guest, electron transfer occurs from the ZnP to these acceptors.

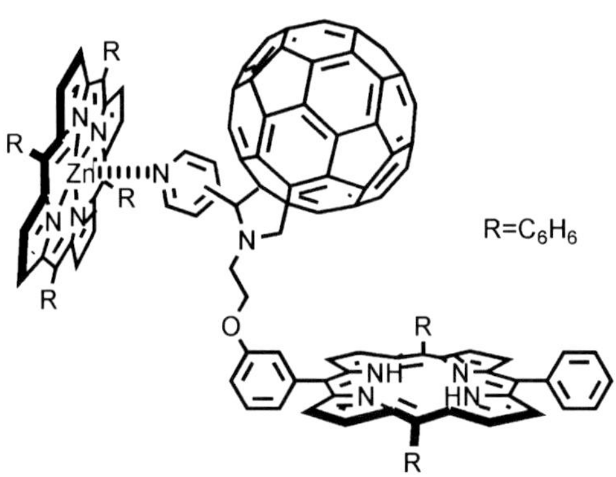

ZnTPP-128 = 4'-pyrridyl
ZnTPP-129 = 3'-pyrridyl

Figure 13.69. Axial binding of ZnTPP to pyridyl-C_{60}-H_2-P dyads results in triads with enhanced charge separation states.

with a ZnTPP moiety[124]. In the assemblies depicted in Figure 13.69, it could be expected that eT proceeds either from the ZnTPP or from the free base porphyrin.

Table 13.9 lists the photophysical data of the triads and the model dyads. Even though fluorescence properties of covalent assemblies of one porphyrin with

Table 13.9. Photophysical Data for **128, 129, 128-ZnTPP,** and **129-ZnTPP**

Compound	solvent	λ_{em} (nm)[a]	τ_{em} (ns) (%)[b]		$k_{ET}(s^{-1})$[c]	Φ_{CS}
ZnTPP	*o*-dichlorobenzene	600	2.10 (100)			
128	*o*-dichlorobenzene	720	0.44 (38)	1.46 (62)	2.2×10^9	0.97
128	benzonitrile	720	0.43 (63)	1.96 (37)	2.2×10^9	0.97
128 + ZnTPP (6:1)	*o*-dichlorobenzene	600	1.78 (100)		8.5×10^7	0.03
		720	0.47 (40)	1.49 (60)	2.1×10^9	0.97
128 + ZnTPP (1:1)	*o*-dichlorobenzene	600	1.99 (100)		2.6×10^7	0.01
		720	0.50 (42)	1.54 (58)	2.0×10^9	0.96
129	*o*-dichlorobenzene	720	0.71 (63)	1.92 (37)	1.3×10^9	0.95
129	benzonitrile	720	0.41 (66)	1.42 (34)	2.4×10^9	0.97
129 + ZnTPP (6:1)	*o*-dichlorobenzene	600	1.83 (100)		1.0×10^8	0.05
		720	0.71 (60)	1.84 (40)	1.3×10^9	0.95
129 + ZnTPP (1:1)	*o*-dichlorobenzene	600	1.99 (100)		2.6×10^7	0.02
		720	0.75 (52)	1.92 (48)	1.2×10^9	0.93

[a] 600 nm is the emission of ZnTPP and 720 nm is the emission of the free base in **128** or **129**.
[b] For the free base in **128**-ZnTPP and **129**- ZnTPP, a $\tau_{em} = 13.6$ ns was used.
[c] Calculated using $k_{CS} = (1/\tau_{em})_{sample}$ -$(1/\tau_{em})_{ref}$ using the shortest lifetime, since the longest lifetime is a delayed emission from species after charge recombination.

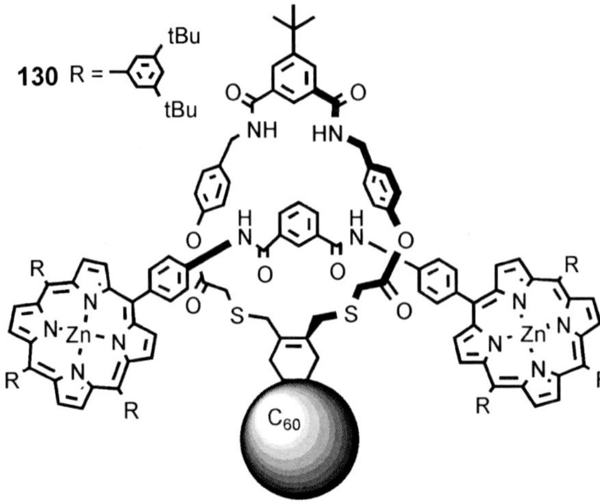

Figure 13.70. A rotaxane triad with ZnP stoppers and a fullerence electron acceptor.

C_{60} acceptors have not been covered in this chapter, the properties of the dyads **128** and **129** are comparable to those of similar species for which many examples have been reported[33,35]. However, the corresponding data have been incorporated in Table 13.9 to clearly show that, in dichlorobenzene, the triads display charge separation processes that are strongly enhanced in comparison to those of the dyads **128** and **129**. No evidence for eT occurring from the ZnTPP moiety was found by the authors. In addition to these data, a charge recombination rate of $5.0 \times 10^7\ s^{-1}$ was determined for the triads. In benzonitrile, an intermolecular electron transfer takes place between the triplet state of ZnTPP and the triplet state of C_{60}.

Inspired by previous rotaxane assemblies, Ito and Tanaka have reported the synthesis and the properties of triad **130** incorporating two zinc porphyrins and a mechanically attached C_{60} electron acceptor[125]. The behavior of the assembly depicted in Figure 13.70 is impressive in terms of lifetime of the CS state. The charge separation quantum yield is efficient (95%), and excitation of the zinc porphyrins leads to an eT process from their first singlet excited state of ZnP to the C_{60} acceptor, at the rate of $k_{eT} = 1.0 \times 10^{10}\ s^{-1}$, to produce a CS state with a lifetime of 180 ns. This lifetime is similar to that observed in covalently linked zinc porphyrin-C_{60} dyads[126].

Last but no least, inspired by the Sauvage-type rotaxanes, Schuster has recently reported the copper(I) bis-phenanthroline rotaxanes **131** and **132** (Figure 13.71)[127]. The nature of the chromophores and electron acceptor is similar to the rotaxane **130** reported by Ito,[125] but the copper(I) complex maintains some degree of organization. In addition, the presence of the copper(I) bis-phenanthroline complex adds a chromophore to the assembly that can be selectively excited at 460 nm to generate a metal to ligand charge transfer (MLCT) state (1.91 eV). In reference compounds without C_{60}, this MLCT state is quenched by the zinc porphyrins to produce a long lived ^{3}ZnP triplet state (1.53 eV) that can be detected

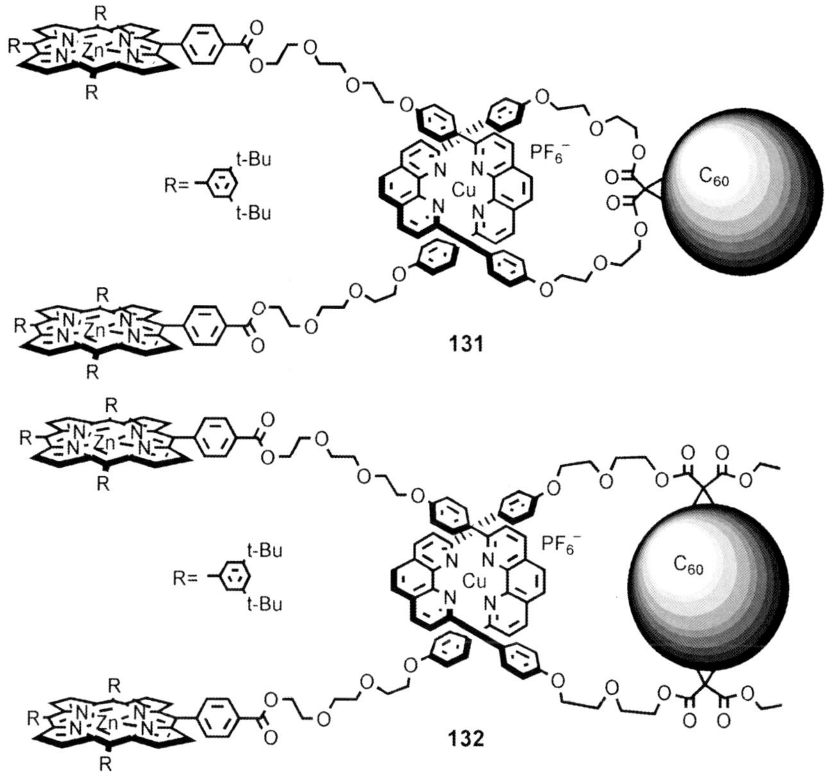

Figure 13.71. A $Cu(phen)_2$ complex assembles this bis-porphyrin-C_{60} rotaxane.

by transient absorption spectroscopy after excitation. In the same reference compound lacking the C_{60} electron acceptor, the fluorescence of the zinc porphyrins ^1S state is moderately quenched (75% compared to ZnTPP), with a decrease of the lifetime from 3.2 ns in a ZnTPP reference, to 1.0 ns. Thermodynamics rule out photoinduced eT between ^{1}ZnP and $[Cu(phen)_2]^+$ in the reference complex, and in **131** and **132**. ET is the proposed quenching pathway. In the fullerene-rotaxanes **131** and **132**, the behavior of the zinc moieties is similar to that observed in the absence of the C_{60} moiety. The luminescence of the MLCT state of the copper(I) complex is barely observed. This suggests that in the triads **131** and **132**, excitation energy reaches the copper complex to generate the MLCT state, whose decay is accelerated by the nearby fullerene via electron transfer. Transient absorption spectroscopy proves the presence of ZnP radical cations and of a C_{60} radical anion. This is in agreement with the formation of $(ZnP^{\bullet+})_2$–$[Cu(phen)_2]^+$–$(C_{60}^{\bullet-})$ that is probably in equilibrium with $(ZnP)_2$–$[Cu(phen)_2]^{2+}$–$(C_{60}^{\bullet-})$. The lifetime of the CS state is estimated from the decay of the transient species and reaches the exceptional values of 0.49 μs for **131**, and 1.17 μs for **132**. These values are among the longest lifetimes reported so far for this type of assembly. The location of the back electron transfer in the inverted Marcus region is again a key feature of the fullerene electron acceptor.

A global analysis of noncovalent approaches to multiporphyrinic assemblies is probably as difficult as for covalent approaches. From the examples detailed above, the properties of the assemblies are mostly dependent on the connecting tool used to control and perform the self-assembly process. The performance of the devices remains strongly dependent on the choice of the chromophores and the electron or energy acceptors involved. Device performance is controlled by the appropriate energetic positioning of the components' respective redox couples and excited state levels. On the other hand, the degree of electronic interaction is still dependent on the weak interactions that are used for the assembly. For long distance eT, the presence of intermediate transition metal complexes has a tendency to facilitate the super exchange. The rigidity and the control of the spatial arrangement of the chromophores within the assembly is also crucial, as demonstrated by the last two examples in the rotaxane approach. In Ito's rotaxane (**130**), direct interactions between the zinc porphyrins and the fullerene are still possible via π–π interactions, while the same interactions are precluded in Schuster's rotaxanes (**131–132**) due to the well-defined copper(I) coordination geometry. Finally, self-assembly has a bright future regarding the genesis of large assemblies, even though dendrimer-based covalent synthesis has produced large assemblies of porphyrins in the past. Self-assembled dimers are relatively young on the chemical time scale compared to covalently linked dimers, and just as basic concepts had to be established before leading to fully mastered covalent oligomers, the next decade will see the emergence of oligomeric self-assembled species.

6. Cofacial Arrangements

Cofacial porphyrin dimers were initially designed to mimic the cofacial arrangement of the special pair in the naturally occurring reaction center of photosynthesis. However, these dimers display very different optical properties due to the face-to-face arrangement of their chromophores. In terms of ground state absorption, for example, the special pair's slipped dimer arrangement displays red shifted absorptions typical of J-aggregates, while face-to-face dimers show strongly blue shifted transitions. The species reviewed hereafter have geometric characteristics that classify them as the Pacman type. In these architectures, for a given set of chromophores, the optical properties are usually governed by the distance separating the photoactive components.

Of course, sometimes the geometry of the spacer is not sufficient to fully control the behavior of the scaffold and the chromophores' flexibility can be responsible for unusual behavior. For example, Hupp has reported the properties of two collapsed dimers showing a compact structures due to ethynylpyridyl side arms used for binding rhenium complexes that act as spacers (Fig ure 13.72)[128].

In these two structures, **133** and **134**, the relative flexibility of the link with the spacer allows the chromophores to interact at a distance of 3.4 Å. This separation is smaller than the predicted van der Waals distance of 3.6 Å. The porphyrin rings are barely slipped (slip angle of 19°), and interact so strongly that even the

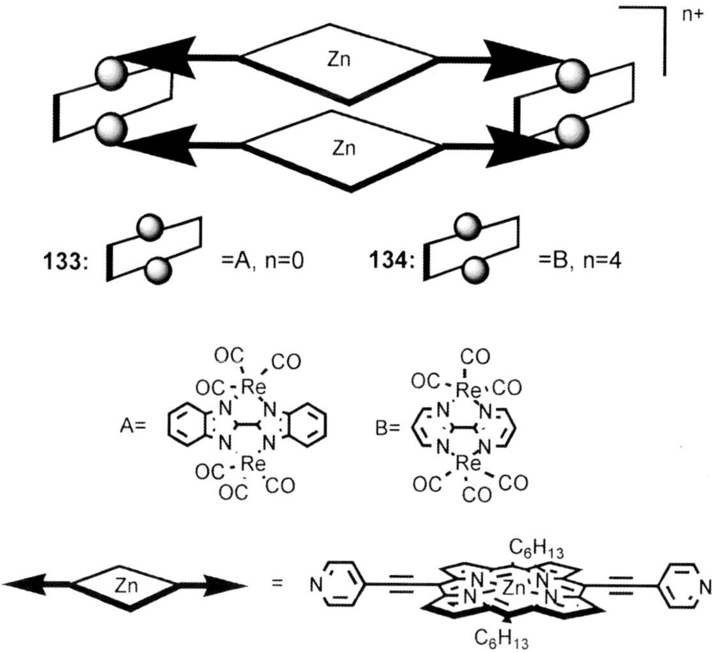

Figure 13.72. Collapsed porphyrin dimers.

bidentate binding of diazabicyclooctane between the two zinc porphyrins is weak ($\sim 5 \times 10^4\,M^{-1}$). Electroabsorption spectroscopy (Stark spectroscopy) shows a 5 Debye change from the ground state to the excited state dipole moment vector in the case of the B band, and a 2 D change for the Q bands, which are notably red-shifted in absorption. This unusual field effect behavior is explained by a polarization of the porphyrin rings, due to interchromophore interaction, that results in the alignment of the dipole moment (charge transfer vector) with the transition dipole moment (absorption vector), thus conferring a significant charge transfer character to the B and Q bands. Thus, the charge transfer behavior is analogous to the special pair of bacteriochlorophylls in the reaction center (RC), but the origin of this particular behavior is different in dimers **133**, and **134**. Indeed, the process in the natural system is an intermolecular charge transfer, while in **133** and **134** the charge transfer character is intramolecular.

In most cases, the nature of the spacer is the governing factor as far as optical properties are concerned. "Pacman" porphyrins, with rigid spacers such as anthracene and biphenylene, have received much attention due to the electrocatalytic properties of their cobalt complexes. Variations of both the dihedral angle and the distance between the porphyrin rings in these architectures have afforded examples of cofacial porphyrins linked by xanthene and dibenzofurane derivatives. Recent examples are detailed hereafter.

A systematic study, which nicely summarizes the influence of the spacer on singlet–singlet energy transfer in cofacial bis-porphyrins, has been reported by the groups of Guilard and Harvey for the free base, zinc, and gallium compounds

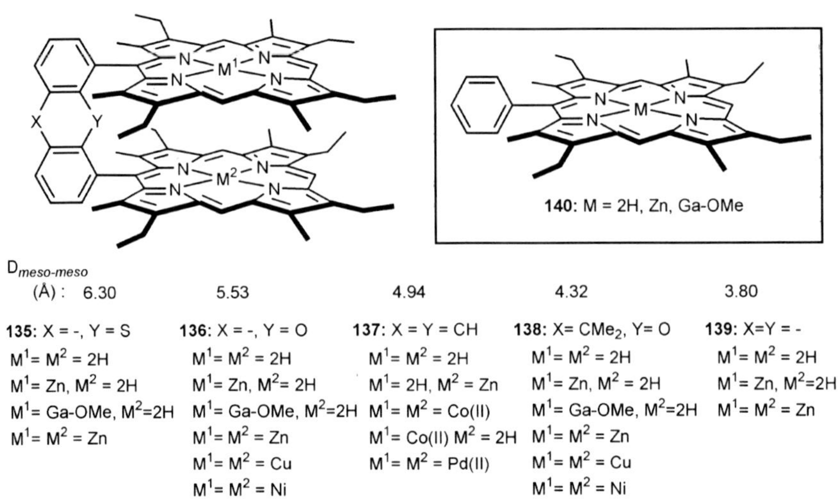

Figure 13.73. A series of cofacial dimers with varying interchromophore distances.

represented in Figure 13.73[129]. Table 13.10 clearly shows the blue shift of the Soret and Q bands of **135–139** compared to monomer **140**. The hypsochromic shifts are indicative of ground state chromophore interactions. Table 13.10 also includes data from Chang and Nocera, who have significantly developed the chemistry of xanthene and dibenzofurane derivatives[130].

From Table 13.11, the variation of the lifetimes and fluorescence quantum yields in the series of compounds shows the clear increase of homo-chromophore interactions in the excited states when the distance between the chromophores diminishes. The rate and efficiency of the energy transfer in hetero-dimers does not seem to be metal dependent. The distance dependence of the energy transfer rate has been analyzed using Förster and Dexter theories. Harvey and Guilard have established that in **135**-Zn–H_2 and **136**-Zn–H_2, energy transfer is dominated by a Förster mechanism, while in the case of hetero-dimers **137**, **138**, and **139**, it proceeds mainly via a Dexter mechanism. The critical distance at which the Dexter mechanism becomes inoperative is estimated between 5 and 6 Å[129,131]. By analogy with what has been discussed earlier in the case of linearly arranged covalent dimers, it should be noted that for compounds **135–139**, no electron density should be present on the *meso* carbons involved in the covalent connection to the spacer.

As mentioned above, interest in these structures dates back to the end of the 1970s and stems from the four electron reduction of dioxygen with Pacman-type porphyrins[132]. Thus, the study of oxygen interaction with cofacial dimers can provide valuable information for the adjustment of the tweezer's geometry in order to design efficient catalysts for electro- or photoinduced oxygen or dioxygen activation[130,133–135]. Only oxygen quenching of excited states fits the scope of this chapter, however, which leads us to the study of cobalt and palladium complexes of ligands **135–139**. The luminescence properties of the cobalt complexes, in the presence of dioxygen, have been determined using **137**-Co_2

Table 13.10. UV–visible Absorption Data for Compounds **135–140** in CH_2Cl_2

			λ_{max} (nm) ($\varepsilon \times 10^{-3}$ M^{-1} cm^{-1})				
Ligand	M^1	M^2	B band	Q bands			
140	H_2	—	402 (154)	502 (15)	532 (8)	578 (6)	626 (4)
—	Zn	—	410 (270)	540 (18)	—	576 (11)	—
—	Ga-OMe	—	408 (157)	538 (6)	—	576 (5)	—
135	H_2	H_2	398 (309.9)	502 (29.6)	536 (15.0)	570 (14.2)	622 (6.8)
—	Zn	H_2	402 (340.6)	502 (15.6)	534 (21.7)	570 (20.9)	624 (3.1)
—	Zn	Zn	402 (473.6)	536 (32)	—	572 (29)	—
—	Ga-OMe	H_2	402 (361.7)	502 (11.7)	536 (20.3)	574 (17.7)	622 (1.8)
136	H_2	H_2	396 (260)	502 (24)	536 (12.0)	572 (1.1)	624 (5.0)
—	Zn	H_2	400 (383.5)	502 (15.7)	534 (22.2)	570 (21.6)	622 (2.7)
—[a]	Zn	Zn	400 (512)	534 (30.6)	—	571 (29.6)	—
—	Ga-OMe	H_2	402 (252.7)	502 (11.1)	536 (14.8)	574 (13.9)	622 (2.8)
—[a]	Cu	Cu	397 (400)	528 (17.6)	—	565 (23.4)	—
—[a]	Ni	Ni	395 (411)	521 (20.6)	—	560 (39.4)	—
137	H_2	H_2	395 (190.5)	506 (14.1)	539 (5.1)	578 (6.0)	631 (3.3)
—	Zn	H_2	399 (196.6)	507 (7.0)	539 (10.5)	575 (11.6)	630 (1.3)
138	H_2	H_2	380 (200)	508 (12.0)	543 (5.4)	578 (6.0)	628 (3.3)
—	Zn	H_2	386 (268)	512 (9.5)	542 (11.7)	576 (12.0)	628 (2.0)
—[a]	Zn	Zn	389 (290)	541 (14.3)	—	576 (13.2)	—
—	Ga-OMe	H_2	388 (269.4)	510 (8.4)	542 (11.4)	580 (10.4)	628 (1.4)
—[a]	Cu	Cu	387 (265)	534 (16.7)	—	571 (19.8)	—
—[a]	Ni	Ni	388 (305)	526 (13.6)	—	564 (23)	—
139	H_2	H_2	379 (173.9)	511(6.3)	540 (2.0)	580 (34)	632 (1.8)
—	Zn	H_2	388 (200.0)	518 (4.1)	542 (5.2)	581 (6.8)	633 (0.8)

[a] From ref. [130].

and cobalt(II) octaethylporphyrin **141** as a reference[136]. The main conclusion that can be drawn concerns the poor fluorescence quenching properties of dioxygen. This stems from identical fluorescence quantum yields ($\Phi_{em} = 2$ % at 77 K) for **137**-Co–H_2 and **137**-Co-H_2-(O_2)(1-*t*-butyl-5-phenylimidazole), where Φ_{em} corresponds to emission of the free base moiety. This contrasts with the sensitivity of palladium bis-porphyrins reported earlier by the same authors[137]. Indeed, both fluorescence and phosphorescence of **137**-Pd_2 display a strong dependence on the presence of oxygen, with or without auxiliary imidazole binding, as demonstrated by the data listed in Table 13.12.

It is not too surprising that the quenching efficiency is much higher in the case of phosphorescence, since the triplet excited state is more likely to interact with dioxygen. When 1-*t*-butyl-5-phenylimidazole is added, it obviously offers a supplementary possibility for phosphorescence quenching. It should be noted that the τ_{fluo} and Φ_{fluo} values are small due to the heavy atom effect of the Pd(II) that facilitates the isc process[137].

These examples show that the introduction of a guest between the chromophores can clearly affect the fluorescence behavior of the species when it is

Table 13.11. Fluorescence data and energy transfer rates (when applicable) for Cofacial Dimers and Heterodimers

Ligand	M^1	M^2	τ_{em}(ns) (Φ_{em}%)[c]	λ_{max} (nm)	k_{ET}(ns^{-1})
135[a]	H_2	H_2	18.0 (8.87)	629, 697	—
—	Zn	H_2	0.19 (10.70)	583, 628, 697	4.7
—	Zn	Zn	1.95 (2.03)	581, 636	—
—	Ga-OMe	H_2	0.24 (7.88)	582, 629, 697	3.7
136[a]	H_2	H_2	18.5 (9.37)	628, 697	—
—	Zn	H_2	0.18 (9.89)	631, 697	5.0
—[b]	Zn	Zn	1.69 (2.94)	580, 635	—
—	Ga-OMe	H_2	0.09 (9.09)	628, 697	11
137[a]	H_2	H_2	13.7 (2.0)	634	
—	Zn	H_2	0.14 (10.70)	585, 630, 698	6.4
138[a]	H_2	H_2	14.1 (3.61)	635, 701	—
—	Zn	H_2	0.10 (1.90)	587, 641, 704	9.8
—[b]	Zn	Zn	1.73 (1.15)	584, 643	—
—	Ga-OMe	H_2	0.14 (6.66)	588, 639, 705	6.4
139[a]	H_2	H_2	11.7 (0.4)	641	—
—	Zn	H_2	0.05 (5.3)	623,690	20.8
—	Zn	Zn	0.63 (0.6)	643	—

[a] Ref. [131].
[b] From ref.[130].
[c] For **140-H$_2$** $\tau_{em} = 17.3$ ns, **140-Zn** $\tau_{em} = 1.70$ ns, **140-Ga** $\tau_{em} = 1.840$ ns.

Table 13.12. Fluorescence and Phosphorescence Data for Cofacial Dipalladium Species

	Φ_{fluo} (%) (τ_{fluo} ps)				Φ_{phos} (%)		
Compound	Argon	Air	Oxygen	τ_{phos} (Ar, ms)	Argon	Air	Oxygen
OEP-Pd	0.0373 (685)	0.0169 (554)	0.0108 (477)	0.27	10.24	0.042	0.00055
137-Pd_2	0.0335	0.0115	0.0044	0.23	0.34	$< 10^{-6}$	$< 10^{-6}$
137-Pd_2-$(Im)_2$	—	—	—	—	0.086	$< 10^{-6}$	$< 10^{-6}$

able to directly interact with the excited states. The introduction of a guest can also have an impact on the geometry of the tweezers when the spacer permits small changes. An example of such supramolecular tuning of excited states in cofacial chromophores has been reported by Nocera in the dibenzofuran spacer compound **136**-Zn_2 (Figure 13.74)[138]. They have shown that the insertion of 2-aminopyrimidine between the porphyrins (X-ray characterization) rigidifies the tweezer structure and significantly suppresses rotational motion around the C_{meso}-dibenzofuran bond. As a consequence, the triplet state, resulting from the transient

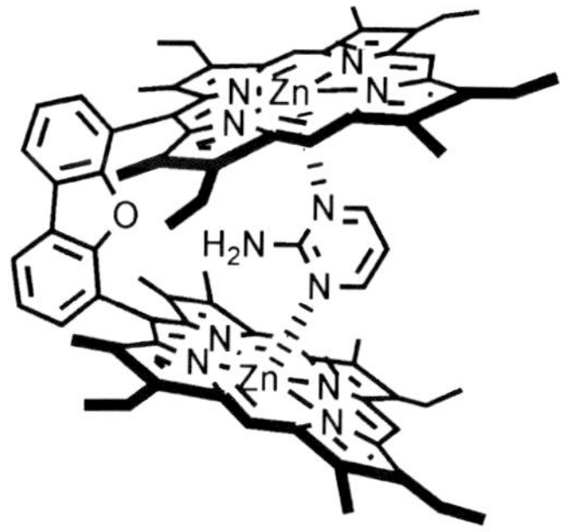

136-Zn_2(NH_2pyrimidine)

Figure 13.74. A host–guest tweezers complex.

Table 13.13. Transient Absorption Maxima and Decay Lifetime of bis-(zinc porphyrins).

	λ_{abs} (nm)	τ (μs)
136-Zn_2	443	21.4
136-Zn_2(NH_2-pyrimidine)	451	227
138-Zn_2	410	169

absorption of the host–guest complex, shows a drastic enhancement of its lifetime. The data collected in Table 13.13 show that, while the lifetime of the triplet excited state of **136**-Zn_2 is smaller than that of **138**-Zn_2, the lifetime of the host–guest complex's excited state is much larger. This effect correlates well with an increasing restriction of mobility from **136**-Zn_2 to **138**-Zn_2, and **136**-Zn_2(NH_2-pyrimidine), and the suppression of vibrational/rotational deactivation pathways for the triplet excited state.

Weak interactions may be responsible for the guest insertion within tweezers. Thus, strong binding due to a highly preorganized photoactive host allows the design of molecular switches for photoinduced electron transfer processes. The group of Yagi has developed the synthesis of diarylurea linked dimers depicted in Figure 13.75[139]. The affinities of these zinc porphyrin dimers for DABCO has long been known, and examples of viologen π–π interactions with porphyrin hosts were reported by Gunter in the mid 1990s[140]. The fluorescence of **142** and **143** is quenched by an eT process in the presence of hexylviologen **HV**, which is bound via π–π interactions in the tweezers, with association constants in the range of 10^6 M^{-1} for **142**. Upon displacement by DABCO, which is bound via more energetic coordination bonds, the photoinduced process is turned off, since no fluorescence quenching is observed with DABCO.

The presence of a guest may also induce the face-to-face arrangement of porphyrins, thus leading to sandwich-type structures. This design bears much resemblance to the self-assembly approach. The two next examples illustrate potential developments in this field.

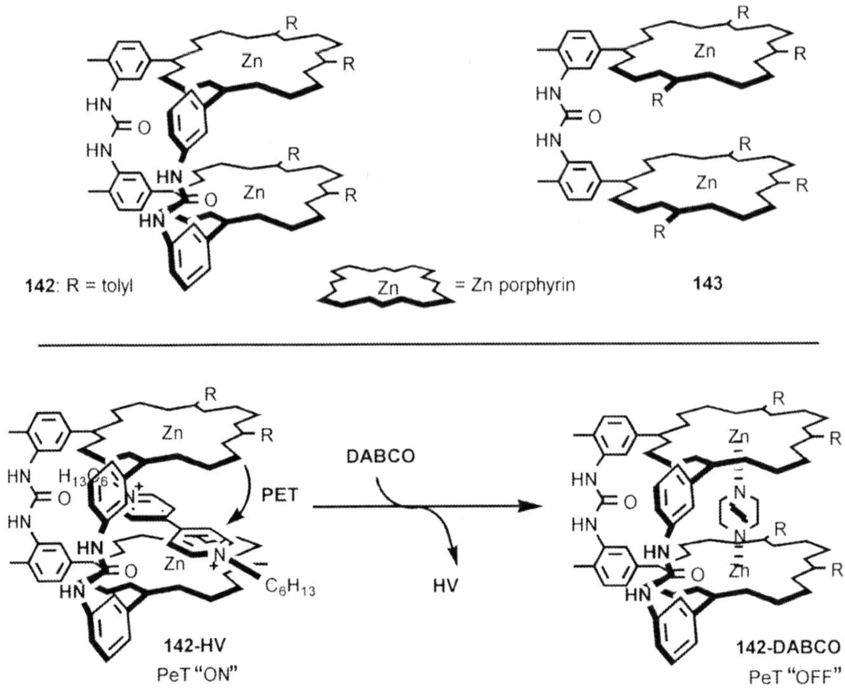

Figure 13.75. Supramolecular switches for photoinduced electron transfer (PeT).

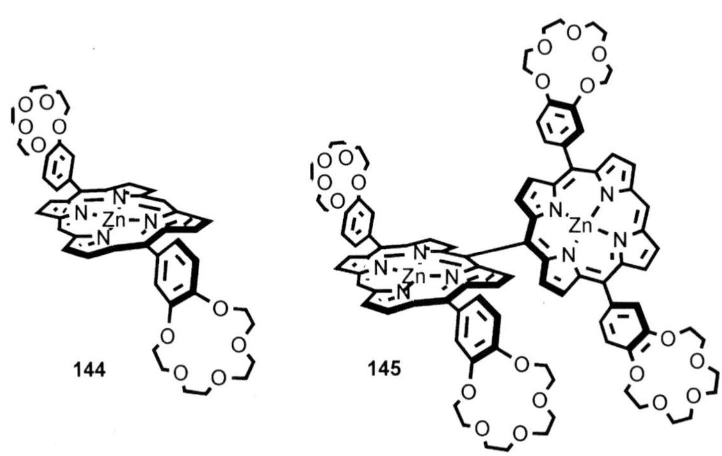

Figure 13.76. Crown ether appended porphyrins and *meso–meso* linked porphyin dimers.

Osuka and Shinmori have studied the aggregation of porphyrin **144** and the *meso–meso* coupled porphyrin dimer **145** bearing peripheral crown ether moieties[141]. For **144**, the addition of potassium ions in 2/1 $CHCl_3/CH_3CN$ mixtures induces a blue shift of the B band absorption, from 415 to 395 nm, and a fluorescence quantum yield decrease from 0.022 to 0.004 (reference ZnTPP = 0.03)

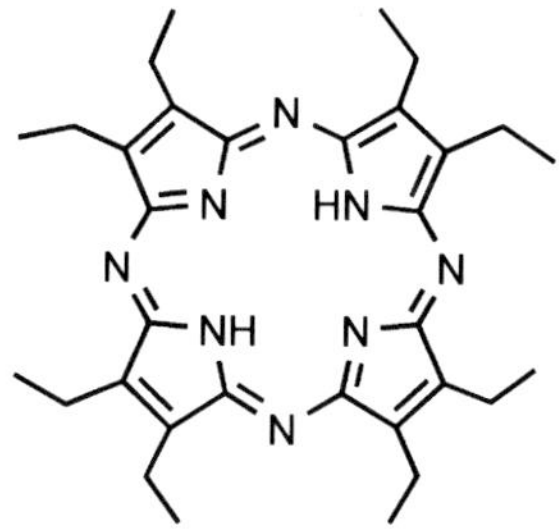

146

Figure 13.77. Tetraaza-octaethylporphyrin.

due to the formation of a compact face-to-face dimer. The compact nature of the dimer was confirmed by NMR measurements. Sodium has less influence because its ionic radius allows a better fit in the crown. On the other hand, potassium is too big and forces dimerization, as already observed for the stacking of crown ether appended phthalocyanines[142]. In the case of the cobalt analog of **144**, the formation of a stacked dimer in the presence of cations can be precluded by the binding of an axial base. The behavior of **145**, although not detailed in terms of fluorescence measurements, denotes a tendency to form linear assemblies through successive binding of cations, by self-assembled "pairs" of orthogonal dimers.

When the metals to be complexed in the porphyrin core are too big to fit in the plane, it is well established that sandwich-type complexes are formed in which the metal is bound by two porphyrins that are thus maintained at less than their van der Waals distance. The very compact arrangement drastically affects the redox behavior of such dimers, because the respective HOMOs of the porphyrin monomers are forced to overlap and combine to form porphyrin–porphyrin bonding and anti-bonding orbitals, raising the HOMO energy level. When lanthanide(III) cations are used to form the sandwich-type assembly, the resulting dimer is electron deficient and easily reduced. Combination of both types of complexes, one easier to reduce and one easier to oxidize, generates charge transfer complexes. Marchon and Collman[143] have reported the conductivity measurements of lanthanide(III) and zirconium(IV) sandwich-type complexes obtained with the octaethylporphyrin (OEP) and tetraazaoctaethylporphyrin **146**, represented in Figure 13.77. The interest of the OEP analog **146** is that it is easier to reduce than the OEP. Thus, its introduction in the structure of the Zr(IV) or the Ln(III) sandwich provides a way to tune the redox potential of the charge donor and/or the charge acceptor. Table 13.14 lists the charge transfer reaction percentages of **146** in different solvent systems.

The conductivities of these species (Table 13.15) are better than those of the individual components, but the low values observed are consistent with weak interactions between both sandwich complexes. These low values may also reflect the distortion of the porphyrin macrocycles in this type of assembly, and the steric hindrance of the ethyl substituents at the periphery of the macrocycles.

Table 13.14. Charge Transfer Efficiency in Mixed Sandwich Type Dimers

donor/acceptor	solvent[a]	% transfer
$Zr(OEP)_2/Gd(\mathbf{146})_2$	pentane	4 ± 2
	CH_2Cl_2	26 ± 3
	$CH_2Cl_2 + TBAPF_6$	79 ± 7
	CH_3CN	85 ± 7
$Zr(OEP)_2/Lu(\mathbf{146})_2$	CH_2Cl_2	6 ± 2
	$CH_2Cl_2 + TBAPF_6$	26 ± 1
$Zr(OEP)(\mathbf{146})/Gd(\mathbf{146})_2$	CH_2Cl_2	3 ± 1
	$CH_2Cl_2 + TBAPF_6$	9 ± 6
	CH_3CN	21 ± 3

[a]$TBAPF_6$: tetrabutylammonium hexafluorophosphate (0.1 M).

Table 13.15. Conductivity of the Sandwiched Charge Transfer Systems

donor/acceptor	conductivity ($S \times cm^{-1}$)	E_{act} (eV)
$Zr(OEP)_2/Gd(\mathbf{145})_2$	$5\text{-}8 \times 10^{-5}$	0.40
$Zr(OEP)_2/Lu(\mathbf{145})_2$	$0.6 - 1 \times 10^{-5}$	0.43
$Zr(OEP)(\mathbf{145})/Gd(\mathbf{145})_2$	1×10^{-7}	0.61
$Gd(\mathbf{145})_2$	$\leq 5 \times 10^{-9}$	—
$Zr(OEP)_2$	$\leq 5 \times 10^{-9}$	—

It should be noted that, in these systems, the type of conductivity observed involves a thermal activation barrier (E_{act}) determined from plots of $-\ln(\sigma)$ vs. $1/2kT$, as for thermally activated conductivity $\sigma = \sigma_0 \exp(-E_{act}/2kT)$, where k is the Boltzmann constant, and σ_0 a constant. This tends to show that the conductivity here should be seen as a charge hopping from one site to the next.

The use of lanthanide templates to form sandwich-type complexes has been long known in the case of phthalocyanines. Recent developments combining porphyrins and phthalocyanines deserve to be mentioned here, even though only one porphyrin is used in the scaffolds. Lindsey, Kuhr, and Bocian have previously reported a systematic study of the charge retention capacity (for information storage applications) of self-assembled zinc porphyrin monolayers on gold[144]. The method used for studying the monolayer properties was fluorescence imaging microscopy. This technique is based on the fact that the neutral zinc porphyrin monolayer is fluorescent, while the oxidized monolayer is not. As a follow-up, they demonstrated that the oxidized state of the monolayer is stable over long periods (> 100 s), and used this approach to optimize the link between the porphyrin species and the surface by studying the series of compounds represented

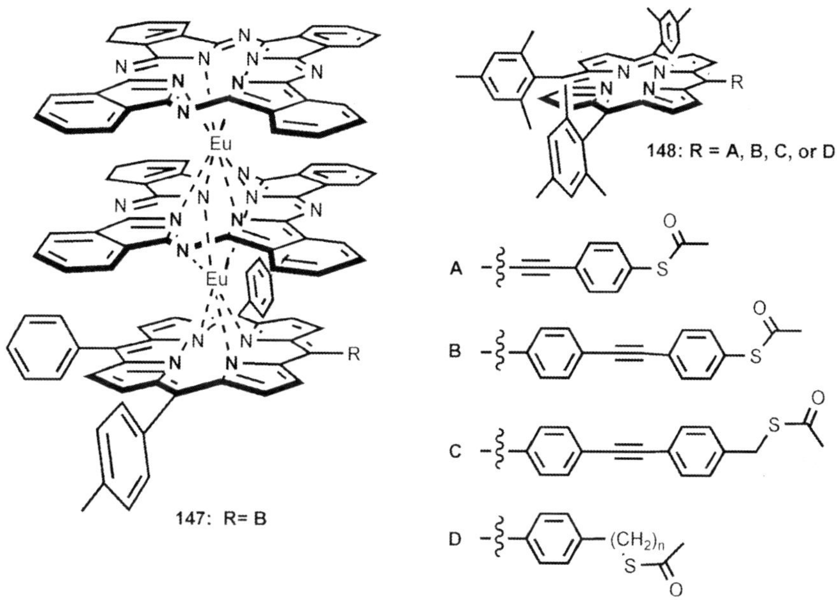

Figure 13.78. Components for charge storage in self-assembled monolayers (Au).

in Figure 13.78[145]. Thus triple-decker sandwich-type complexes improve the number of oxidations that can be performed on the molecular device, as **147** can undergo four distinct oxidation steps.

The facile cleavage of the S-acetyl group allows a reliable attachment to the gold surface in these compounds. When an appropriate potential is applied to the monolayers, oxidation is observed, and the rate of electron transfer, k^{ox}, varies from 10^4 to 10^5 s^{-1} for the electrochemical step. The half-life ($t_{1/2}$) of the charge retention that characterizes the charge dissipation rate has a variation trend that is parallel to k^{ox}. The rate decreases when the number of methylene groups increases, and also when conjugation of the linker is precluded (**148** with $R = B - D$). The rates for sandwich-type complexes are slower, both for k^{ox} and charge dissipation, reaching a high of $t_{1/2} = 170$ s for the tetracation of **147**.

These results led the same group of authors to the systematic investigation of triple-decker sandwiches for charge storage properties[146]. The series of europium complexes studied is represented in Figure 13.79. The tilt angle with respect to the surface normal direction is certainly a parameter that will slightly affect the properties of these charge storage devices. However, in a first analysis, it is clear that these devices perform well, in terms of charge storage capacity, as semiconductor-based trench capacitors that retain charges in the millisecond time range. The monolayers are stable and the read/write cycles can be repeated. Table 13.16 details the electron transfer rates for oxidation k^{ox} and the half-life $t_{1/2}$ of compounds **149–152**, **153**, and **154**, while **152** cannot be attached to the gold surface.

If molecular species are to be employed for the production of silicon-based devices, sooner or later, they must be interfaced with CMOS technology. In order

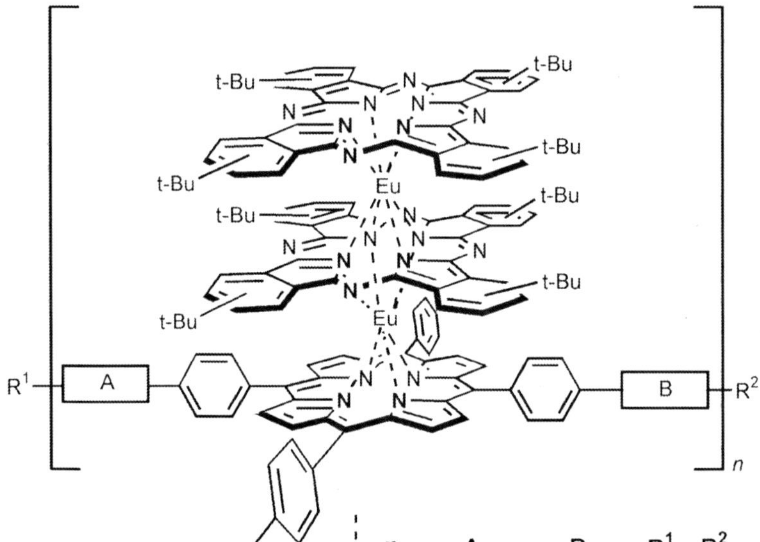

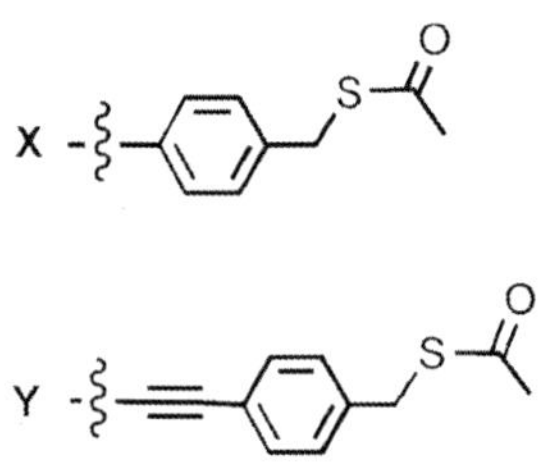

	n	A	B	R^1	R^2
149	1	-≡-	none	X	Y
150	3	-≡-	none	X	Y
151	2	-≡-	-≡-	X	X
152	1-5	-≡-	-≡-	H	H
153	1	-≡-	-≡-	Y	Y
154	2-4	-≡-	-≡-	Y	Y

Figure 13.79. Triple-deckers for self-assembled monolayers on gold surfaces.

to demonstrate that these molecular-based devices can resist the processing conditions of silicon devices, Lindsey and Bocian have examined the properties of the analog compound **155** (Figure 13.80), linked via an ether function to Si(100) surfaces[147] These compounds are stable under extreme conditions (1 h at 400°C), and can withstand up to 10^{12} read/write cycles. In conclusion, these compounds meet the operating requirements for CMOS interfacing, and thus, incorporation of porphyrin-based components into semiconductor operating devices can be seriously envisioned.

This leads us to the last section of this chapter that will consider the use of multiporphyrin scaffolds on electrode surfaces or semiconducting surfaces essentially for photovoltaic applications.

Table 13.16. eT Rates and Charge Retention Data for Self-assembled Monolayers of Triple-deckers

	Oxidation state + 1		Oxidation state + 2		Oxidation state + 3		Oxidation state + 4	
Compound	k^{ox} (10^3 s^{-1})	$t_{1/2}$ (s)	k^{ox} (10^3 s^{-1})	$t_{1/2}$ (s)	k^{ox} (10^3 s^{-1})	$t_{1/2}$ (s)	k^{ox} (10^3 s^{-1})	$t_{1/2}$ (s)
149	160	14	107	27	98	36	73	38
150	100	16	85	23	79	40	54	43
151	97	19	87	20	85	24	80	28
153	174	9	110	14	86	24	a	36
154	70	20	58	29	54	39	45	53

Approximate surface concentrations of **149**: 2.0×10^{-12} mol cm^{-2}, **150**: 1.0×10^{-11} mol cm^{-2}, **151**: 1.5×10^{-12} mol cm^{-2}, **153**:3.1×10^{-12} mol cm^{-2}, **154**: 5.4×10^{-12} mol cm^{-2}.

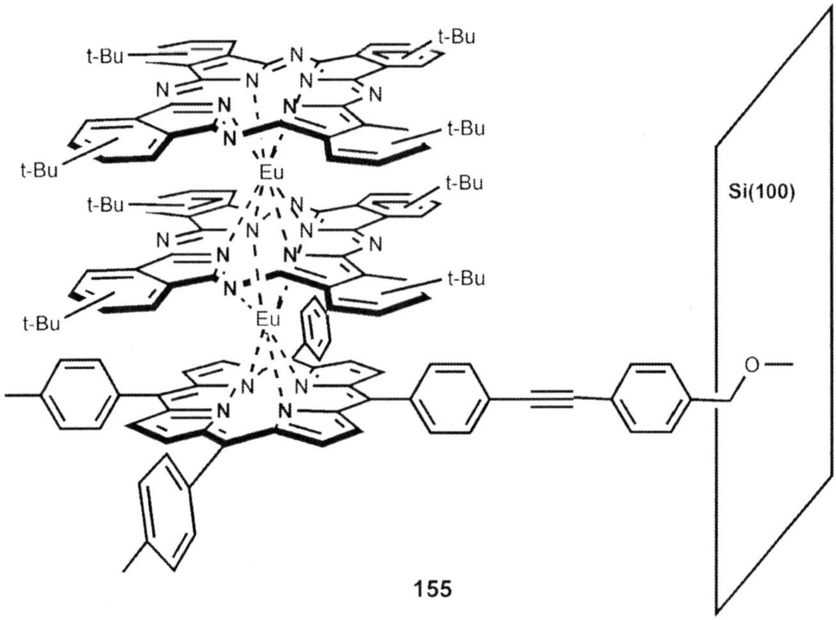

155

Figure 13.80. Highly stable triple decker compounds attached to a Si(100) surface.

7. Multiporphyrin Surface Assemblies

As mentioned in the introduction, the major focus in this overview is multiporphyrin architectures, and, just as processes involving only one porphyrinic chromophore have not been considered for solution studies, this section on modified surfaces will be limited to species and properties in which more than one tetrapyrrolic macrocycles are implicated. Of course, in order to develop photovoltaic materials, attachment of monoporphyrin species, alone or combined with acceptors and donors, is an extremely active field. Without going into detail, we feel that it is necessary to indicate a few leading references for readers interested in this type of approach. After having been an active area of surface modification, the anchoring of porphyrins alone on surfaces has slowed down in recent years, leading the way to the use of dyads and triads. Still, attachment of porphyrin monolayers to surfaces can proceed by the formation of a gold sulfur bond using thioalkane linkers, as reported by Fukuzumi and Imahori to fabricate stabilized gold clusters[148]. The same type of linkage can be obtained by cleavage of S-acetyl groups in order to attach imidazole derivatives on a gold nanoparticle. Further addition of zinc porphyrin derivatives produces a porphyrin covered gold surface via axial binding of the imidazole to the zinc(II)[149]. The spontaneous cleavage of disulfide bonds on gold has been recently used by Tamiaki to functionalize a gold surface with a chlorophyll analog[150]. Finally, in the case of films, classical deposition methods such as spin coating can also be used[22].

Considering their high performances in photoinduced charge separation processes, porphyrin-fullerene dyads are now a first choice candidate for modification of electrodes. It should be noted that an applied electric field can influence the rates of electron transfer in porphyrin-fullerene dyads, as demonstrated by Ohta and Fukuzumi with ITO electrodes spin-coated with dyad films[152]. Attachment of dyads can be performed on ITO or gold surfaces using, respectively, classical silicon oxide or thioalkane chemistry[152]. Surface structure characterization has been recently accomplished for the first time by the joint efforts of Yamazaki, Imahori, Kim, and Fukuzumi[153]. Deposition of porphyrin-fullerene dyads can also be performed on semiconducting tin oxide, SnO_2, in the form of nanocrystals, by simple soaking to generate photocurrent under subsequent irradiation[154]. Dyads can also be obtained from biomaterials, and electrostatic interactions with protein surface can be used as a driving force for a self-assembly type coating. ITO surfaces on which a polyanionic dendritic fullerene has been deposited form photoelectrodes upon deposition and electrostatic interactions with cytochrome *c*[155].

The modification of electrodes with molecular devices follows the same evolutionary trend as for the devices themselves, and logically yields interesting approaches involving the attachment of triads on surfaces. Typical electron donors such as ferrocene can be used, but again, the use of photoexcitable BODIPY energy donors allows the photosynthetic process to be closely modeled[156].

7.1. Covalent Attachment of Multiporphyrins

While the approaches summarized above limit the number of active chromophores in the vicinity of the surface, the attachment of multiporphyrins more closely resembles the natural process of photon collection. Two recent papers by Kobuke illustrate this approach quite well and take advantage of the axial imidazole binding to zinc porphyrins[157,158]. The phenomenon is better described by the more recent example in which a heterodimer is formed by self-assembly after the attachment of a zinc porphyrin to a gold surface, as depicted in Figure 13.81[157].

It should be noted that, for the assembly process, **157** and **158** exist as dimers in solution, except in pyridine solutions in which competition between the axial binding of imidazole and pyridine allows dissociation of the dimers. Photoinduced currents were measured by irradiating a $0.5\,cm^2$ modified electrode plate with a 150 Xe lamp (IR-cut-off filter) and applying a negative voltage (-200 mV vs. Ag/Ag^+). The plate was immersed in an aqueous solution of sodium sulfate containing MV^{2+} (5 mM) as charge carrier. Values of photocurrents are summarized in Table 13.17.

As we will see later in this section, oxygen may play a crucial role in the transfer of electrons to a surface, by providing photogenerated O_2^- species as electron carriers. However, under these conditions, O_2^- generation is doubtful. Yet, in a first analysis, it is reasonable to assign the better results of the "triad" **156(Zn)-157**, in terms of photocurrent generation, to the presence of the intermediate pyrolmellitimide electron acceptor that creates a better photoinduced charge separation.

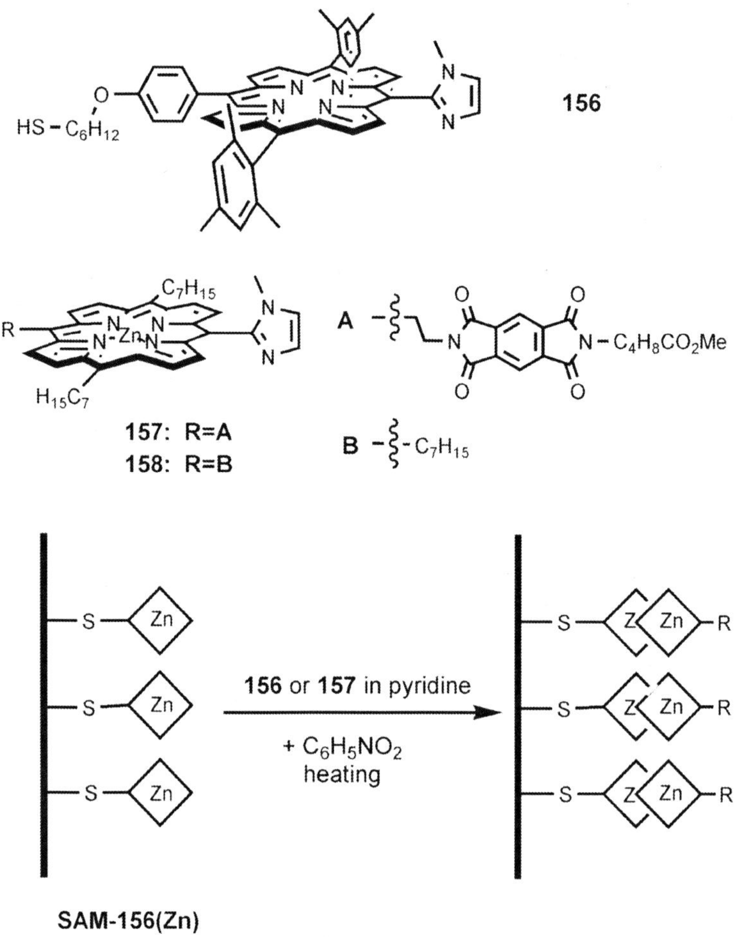

Figure 13.81. Covalent attachment of self-assembled heterodimers on gold.

Table 13.17. Photocurrent Generation in Covalently Attached, Self-assembled Heterodimers on Gold

Assembly	Under N_2	Under O_2
156(Zn)-**157**	145	230
156(Zn)-**158**	44	75

In a previous paper, a similar construction method has been applied to *meso–meso* linked dimers **159** represented in Figure 13.82[158]. The surface concentration of **156(Zn)** attached to the electrode is estimated to be 3.1×10^{-11} mol cm^{-2} from the intensity of its anodic peak in cyclic voltammetry. The low value is tentatively explained by the cleavage of the Au–S bond during the zinc metallation, which is carried out after the fixation of **156** onto the surface. Using this

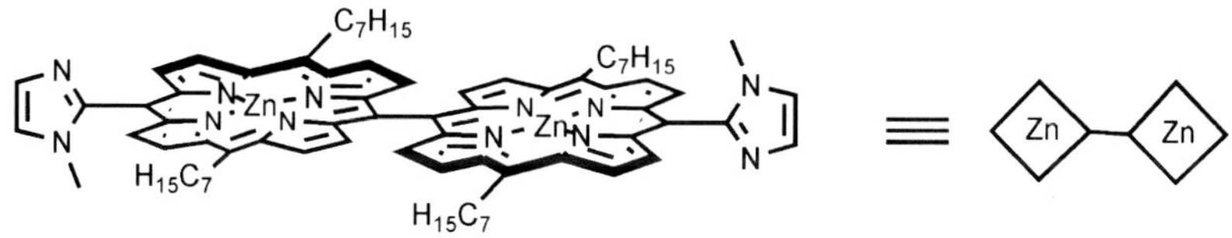

159

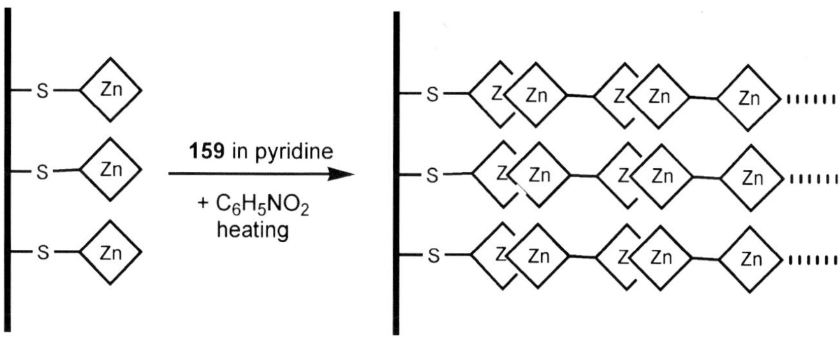

SAM-156(Zn)

Figure 13.82. Covalently linked, self-assembled light harvesting antennae on a gold electrode.

approach, the number of porphyrins attached to the electrode can be estimated from the number of repetitions of the deposition cycles, and from AFM imaging. Under the same conditions (150 W Xe lamp, $E_{Au} = -200$ mV, aqueous Na_2SO_4, 5 mM MV^{2+}), the photocurrent quantum yield efficiencies have been measured as 0.14% (430 nm), 0.11% (490 nm), 0.10% (580 nm), and 0.26% (650 nm), with intensities up to 10^3 nA after four deposition cycles. These values are comparable with those of systems which, as in this one, do not possess an additional electron acceptor like pyromellitimide.

7.2. Langmuir–Blodgett Multilayers of Mono- and Bisporphyrin Species

The incorporation of porphyrin derivatives in Langmuir Blodgett (LB) films or Langmuir Schäffer (LS) films have received much attention in the past years. LS films correspond to a horizontal lifting of the deposition surface, as opposed to the vertical dipping in LB films. The structural characterization of the multiporphyrinic architectures prepared by these methods is an important step. Recent studies have shown that the arrangement and orientation of the chromophores in LB films is imposed by several parameters. Although the analysis of the preparation methods for these types of structures is outside the scope of this chapter, nonexhaustive references concerning the formation of LB films are given here as an introduction.

Porphyrin dimers linked by barbituric acid derivatives can form well aligned species in which UV–visible absorbance and fluorescence properties are

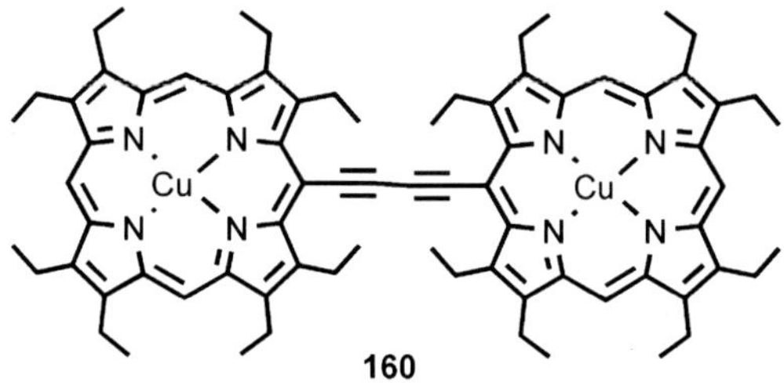

160

Figure 13.83. A Cu(II) OEP dimer as a NO gas sensor.

enhanced[159]. The use of prestructured chromophore assemblies, in the form of TPP derivatives copolymerized in a polyisobutyl methacrylate skeleton, provides control over the chromophore spacing in the films[160]. The structure of the films is generally studied by reflectance anisotropy spectroscopy (RAS) performed in the Soret band region. Large optical anisotropy due to chromophore ordering can be detected[161]. Using this technique, the influence of hydrophobic substitution on the behavior of porphyrin derivatives has been evaluated in LB films deposited on silver and quartz substrates[162]. The use of UV–visible, polarized UV–visible absorption, and fluorescence spectroscopy is also resourceful. The latter has been used alone to characterize mixed films containing tetracationic porphyrin derivatives, sodium dodecylsulfate and octadecylammonium derivatives, in which cooperative electrostatic interactions stabilize the films[163]. Finally, the latter techniques can be combined with low angle X-ray diffractometry for additional information[164].

The incorporation of chromophores in films as an organization and orientation matrix proceeds, from the direct inspiration of the protein environment in antennae systems. Any perturbation of the chromophore ordering should then be theoretically detectable using the above-mentioned techniques. This can lead to the extremely interesting developments of gas sensors based on LB porphyrin films. Two types of gas sensors are actually emerging from this approach using either the sensitivity of UV–visible absorption at 439 nm[165], or surface plasmon resonance[166] in porphyrin monomer LB films for detection of NO_2 in the 4–5 ppm range, or using changes in electrical conductivity in films of alkyne linked porphyrin dimers LB films to detect NO Figure 13.83. The relative variations of resistance in films incorporating **160** ranges between $\Delta R/R = 90\%$, for a 0.5 ppm NO concentration, and $\Delta R/R = (1.85 \times 10^5)\%$ for 1,000 ppm, independent of the film thickness[167].

The construction of dyad systems within a LB architecture provides a straightforward approach to study photoinduced electron transfer. Ito has reported the construction of successive polyimide LB layers incorporating a ZnTPP electron donor, an insulating film, and pyromellitimide as an electron acceptor[168]. The insulating film is equivalent to the spacer in covalent species in solution studies. It was shown that the efficiency of the eT process strongly depends of the thickness of the insulating layer.

Porphyrins alone can be incorporated in LB monolayer films to confer tunneling diode and switching diode capabilities to the assembly. A highly lipophilic

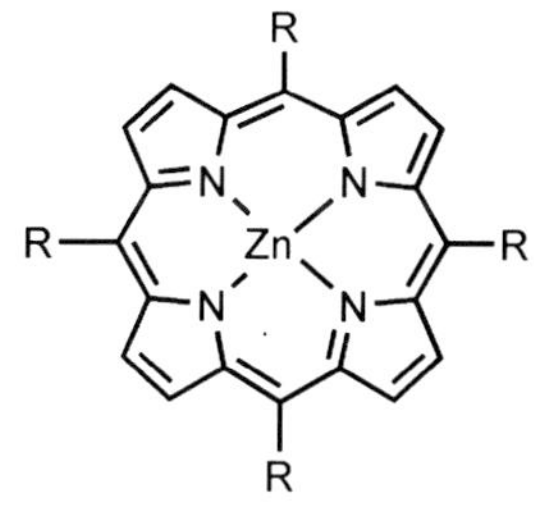

161: R=octadecyloxymethylphenyl

Figure 13.84. Lipophilic porphyrin for LB monolayer film.

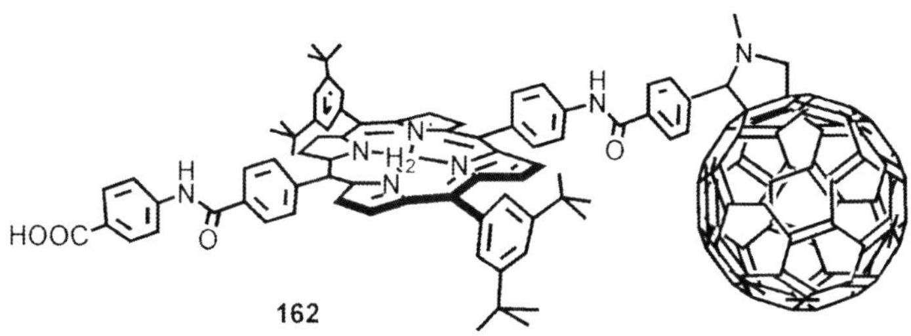

162

Figure 13.85. A dyad that adopts different conformations in LB films than in solution.

porphyrin **161** (Figure 13.84) has been used by Kim and coworkers to produce 4–5 nm thick LB films on a glass slide or Si. The Si wafers are topped by a bottom electrode (Al, Au, or ITO) covered with Al_2O_3, Al/Al_2O_3 bottom electrodes, and Al or Au top electrodes, providing 236 samples of which 60% showed typical rectification I–V curves[169].

When donor–acceptor systems are incorporated in LB films, it should be kept in mind that, in the case of flexible linkers, it is possible to favor one conformer over another due to the "immobilization" within the LB film matrix. In that case, the properties of the dyads in solution and in LB films can differ considerably (Figure 13.85). An example of this effect was reported by Tkachenko and Fukuzumi who incorporated the dyad **162** in LB films[170]. In addition to the fact that, in LB films (22 layers), the fluorescence decay profiles were wavelength dependent, the authors have also observed a third, fast component with a significant amplitude in LB films that is not observed in solution studies. Table 13.18 summarizes the emission decay lifetimes of the species in LB films.

The conclusion concerning the geometry adopted by **162** in LB films is that a folding of the amide bonds generates a conformer in which a distance of *ca.* 8 Å between the electron donor and acceptor is observed, instead of the 18 Å distance obtained from solution studies.

In a similar approach, Prato and Guldi have investigated LB and LS films of the dyad **163** (Figure 13.86)[171]. This dyad is based on components similar to but more rigid than those in assembly **162**. The LS multilayer deposition technique was used when the rigidity of the floating films did not permit ver-

Table 13.18. Emission Decay Lifetimes (τ) and Relative Amplitudes(a) for LB Films of Dyad **161**

Concentration mol%	τ_1 (ns)	a_1	τ_2 (ns)	a_2	τ_3 (ns)	a_3
Toluene					8.5	1
PhCN			1.6	0.9	9.4	0.1
LB films						
100	0.082	0.730	0.59	0.220	5.8	0.050
20	0.085	0.817	0.50	0.174	3.7	0.010
10	0.17	0.805	0.65	0.190	3.2	0.005
5	0.19	0.535	0.75	0.423	2.3	0.042
1	0.17	0.337	0.72	0.493	2.4	0.170

Decays measured at 655 nm.

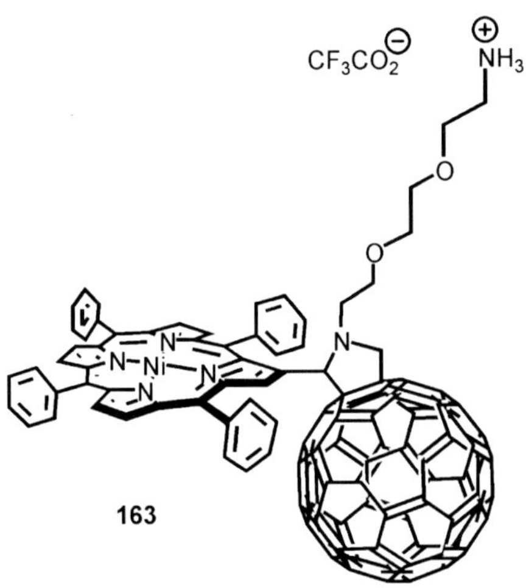

163

Figure 13.86. A rigid amphiphilic dyad.

tical dipping. When deposited on ITO electrodes, these films were studied for photocurrent generation. The IPCE efficiencies under various conditions for the dyad **163** are collected in Table 13.19. The IPCE value is the incident monochromatic photon conversion efficiency, which corresponds to the number of electrons generated in the circuit, divided by the number of incident photons which is expressed as:

$$\text{IPCE (\%)} = [1240 \times \text{photocurrent density}(\mu\text{A cm}^{-2})]/[\lambda\text{ (nm)} \times \text{photon flux}(\text{W m}^{-2})]$$

but can also be seen as the product of the light harvesting efficiency and the quantum yield for charge injection (Φ_{inj})

$$\text{IPCE} = (1 - 10^{-\varepsilon\Gamma})\Phi_{inj}$$

Table 13.19. IPCE Values for Films Incorporating **162**

Methods and conditions	IPCE values (%)
LS films	
transfer @ 2 mN m^{-1}	0.014
transfer @ 10 mN m^{-1}	0.018
transfer @ 15 mN m^{-1}	0.019
transfer @ 20 mN m^{-1}	0.021
LB films	
0.1 M NaH_2PO_4, 1 mM ascorbic acid N_2, no bias	0.01
0.1 M NaCl, 1 mM ascorbic acid, N_2	0.02
0.1 M NaH_2PO_4, 1 mM ascorbic acid, N_2, 200 mV bias	0.012

where Γ is the surface concentration of the dye, as defined in earlier work by Grätzel[172] or Kamat[173], upon photoconversion using ruthenium tris-bipyridine films on TiO_2 and SnO_2.

Photocurrent values increased with the degree of packing of dyad molecules. This increase corresponds to higher pressures for the formation of the solid films before LS transfer. LS films tend to perform more efficiently than films produced by LB transfers. Ascorbic acid was used in the electrolyte as a scavenger of superoxide radical anion O_2^-, which can form by eT from photogenerated C_{60}^-, and can mediate the electron transfer to the collecting electrode. Indeed, saturation of the electrolyte with oxygen, in the absence of ascorbic acid, led to a threefold intensification of the photocurrent in the absence of bias.

LB technique has been used by Isoda to produce molecular heterojunctions (MHJ) using porphyrins and flavin monolayers, as depicted in Figure 13.87a[174]. These heterojunctions were studied for transient photocurrent generation, and the experimental setup is described in Figure 13.87b, as a typical instrumentation example. In the absence of bias, the performance of the MHJs strongly depends on the number of layers involved, with a maximum value of 0.18 μA for MHJ with three porphyrin layers and four flavin layers.

The association of porphyrin and fullerene electron acceptors in LB films can also proceed via a stepwise, noncovalent formation of the assembly, as demonstrated by Shinkai and Ikeda (Figure 13.88). The host–guest interaction of C_{60} with homooxa-calix[3]arene is the noncovalent interaction that assists film formation[175]. The fullerene absorption in the host–guest complex is used to determine a surface concentration of 1.4×10^{-10} mol cm^{-2} in the C_{60} layer, which is then further covered with the polymer-sustained porphyrins **165** or **166**. Porphyrin surface concentrations are then controlled by UV–visible absorption and are found to be 7.6×10^{-11} and 5.5×10^{-11} for **165** and **166**, respectively. Photocurrent measurements performed under irradiation at 420 nm, with a 1.36 mW cm^{-2} for a coating of **165**, and at 430 nm, with a 1.53 mW cm^{-2} for a coating of **166**, in

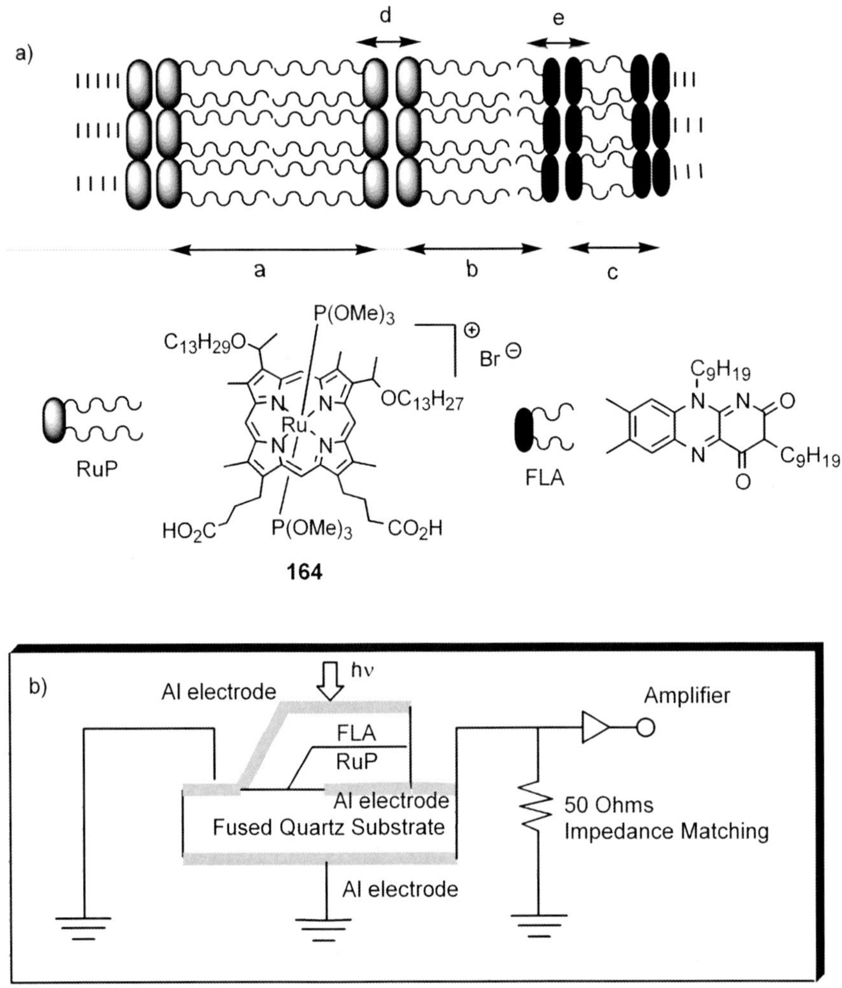

Figure 13.87. (a) Flavin-porphyrin heterojunctions, and (b) transient photocurrent measurement. Distances: (a) = 2.8 nm; (b) = 2.0 nm; (c) = 1.1 nm; (d) = 1.4 nm; (e) = 0.9 nm.

0.1 M Na_2SO_4 containing 50 mM of ascorbic acid that acts as a sacrificial electron donor, show a marked increase parallel to the increase of the bias voltage from −150 to +150 mV vs. Ag/AgCl. The maximum quantum yield efficiency is estimated at 21% for the most efficient C_{60} monolayer coated with **166**. This efficiency is comparable with the 25% efficiency measured earlier by Sakata in the case of a triad[176].

Engineering of LB films incorporating photoactive species and electron donors can even afford systems for the artificial production of hydrogen, if, instead of an electrode that collects photogenerated electrons, an enzymatic electron acceptor is incorporated in the "LB" membrane. Qian has reported the formation of films incorporating the water soluble tetrakis-5,10,15,20-methyl pyridinium

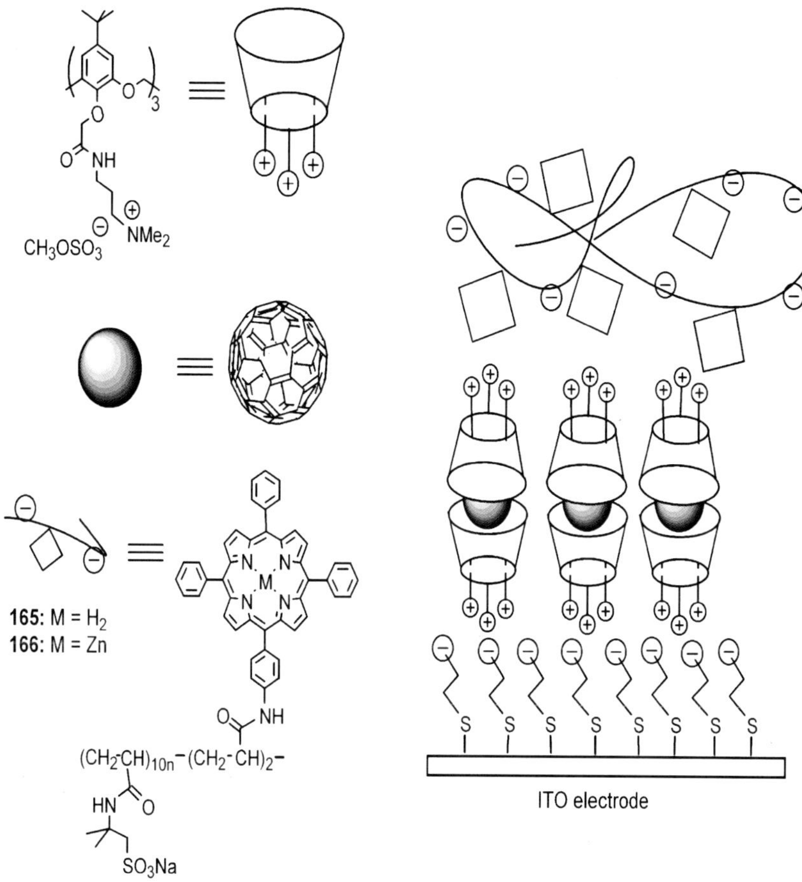

Figure 13.88. Self-assembled porphyrin-fullerene dyad in LB films.

zinc porphyrin photosensitizer, EDTA as an electron donor, and hydrogenase isolated from the bacterium *Thiocapsa roseopersicina*[177]. In the presence of various electron carriers of the violgen-type, photoinduced production of hydrogen reaches a maximum at pH 3.5.

7.3. Porphyrin Films

In this last section, recent results concerning the inclusion of multiporphyrins in films will be covered, and again, the preparation of the films will not be detailed. An attempt will be made to emphasize concepts that are attractive in terms of design of new materials. This, indeed, is probably the ultimate target of many of the approaches that have already been covered in the preceding pages and goes beyond the initial curiosity for phenomena performed by Nature. The doping of polymer materials with chromophores or multichromophore species, and the direct deposition of rather simple species, as well as the direct polymerization of basic porphyrin species, offer the tremendous advantage of simplicity.

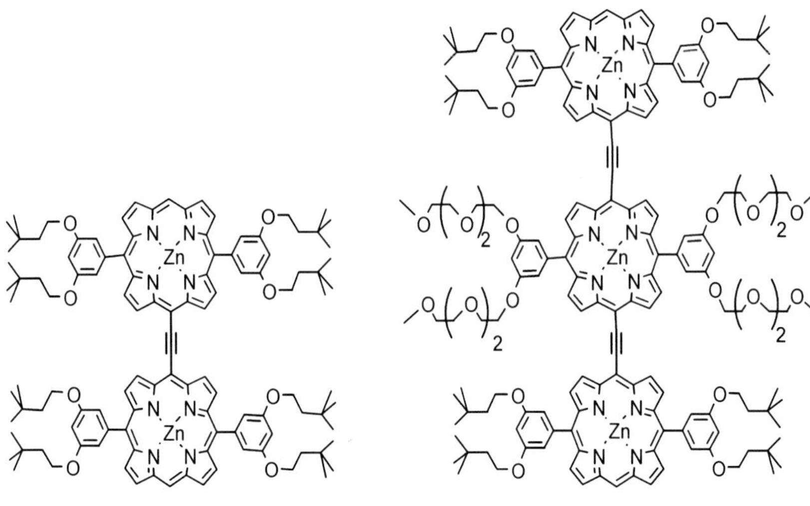

167 **168**

Figure 13.89. Porphyrin doping agents for light emitting polymers.

This is an extremely important parameter in the design of materials within an economically ruled environment.

In this regard, a very attractive approach to near-infrared (NIR) light emitting devices has been developed by Therien and Bazan, based on the properties of ethynyl-bridged porphyrin dimers and trimers. After proper functionalization with long alkyl chains, the dimer **167** and trimer **168**, depicted in Figure 13.89, have been incorporated as dopants into established light emitting polymers[178]. The blue emitting polymer polyvinylcarbazole (PVK) shows an overlap of its emission with the B band of **167**. Such an overlap is appropriate for Förster energy transfer from PVK to **167**. Similarly, the red emission of poly(2-methoxy-5-(2-ethylhexyloxy)-1,4-phenylenevinylene) (MEH-PPV) shows a reasonable overlap with the absorption spectrum of **168**. For electroluminescence studies, pure polymer hosts were employed, while for photoluminescence studies, these polymer species were combined with up to 40% of 2-tert-butylphenyl-5-biphenyl-1,3,4-oxatriazole (PBD) . This mixing improves electron transportation within the polymer matrix and allows a comparison with photoluminescence results.

The energy transfer from the polymer host to the porphyrin guests are quite efficient in some cases, with a 90% quenching efficiency in the case of a 5.0 wt % DD doping of a 60:40 PVK:PBD matrix, even though the spectral overlap in this case is the least favorable of all. The behavior of the doped host–guest polymer can be complicated by self-quenching processes when the dopant concentration is increased. Nevertheless, this approach yields efficiencies on the order of 0.1–0.3 photons per electrons in electroluminescence, for emissions at 720 nm for **167,** and 820 nm for **168**. These wavelengths are in a range suitable for applications in communications, optoelectronics, and sensing devices.

Liquid crystalline assemblies of porphyrins have been investigated for their photoconductive properties. A nice summary of findings has been provided by

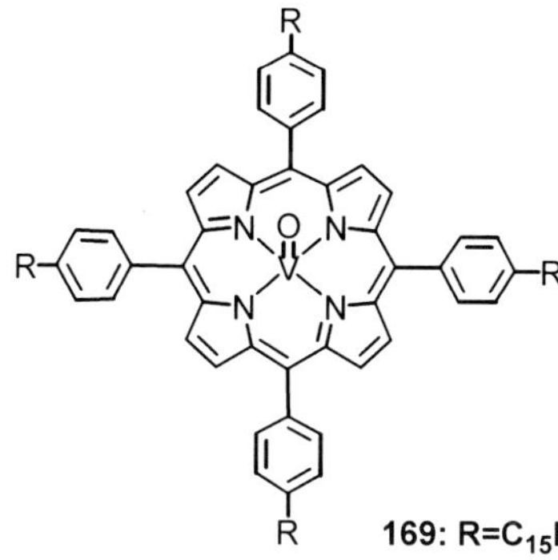
169: R=$C_{15}H_{31}$

Figure 13.90. A mesogenic porphyrin for photoconduction.

Eichhorn, who showed that charge carrier mobility of $0.28 \times 10^{-4}\,m^2\,V^{-1}\,s^{-1}$ in columnar phthalocyanine arrangements can challenge the charge carrier mobility of $0.1 \times 10^{-4}\,m^2\,V^{-1}\,s^{-1}$ observed in triphenylene columnar phases[179]. Note that the values for columnar porphyrin assemblies reached an averaged value of $0.06 \times 10^{-4}\,m^2\,V^{-1}\,s^{-1}$, independent of the central metal atom.

Regarding this statement, complementary information has been provided by the study of the photoconduction in the mesogenic porphyrin **169** (Figure 13.90) that shows two lamellar mesophases. Shimidzu and coworkers have shown that the photoconduction in these materials is strongly influenced by changes in the long range order and arrangements of the porphyrins due to phase transitions, and also by the presence of the metal in the central core[180]. Sandwich cells between ITO electrodes show current rectification I–V curves in the dark, with conductivities of $10^{-14}\,S\,cm^{-1}$ for the crystal and M_{LC} phase of **169**, and $10^{-13}\,S\,cm^{-1}$ for the M_L phase. This phenomenon is amplified with the increase of current intensities under illumination at 620 nm, and clearly differentiates **169** from its nonmetallated analog by the higher photocurrent generation efficiency in both the M_{LC} and M_L phases.

One clear demonstration of the influence of the metal on the photoconductivity in multiporphyrin assemblies has been established by Warman. First, porphyrin assemblies offer the possibility of using the triplet states for photocurrent generation. A good performance is expected for the generation of triplet states in photoconducting materials because of their longer lifetimes, which should allow migration over a longer distance. Using TiO_2-supported films of tetrakis-carboxyphenyl porphyrin **170** (Figure 13.91), or its metallated analogs, **171** and **172**, that are 70 nm thick (*ca.* 15 monolayers of tetrapyrrolic macrocycles), it has been shown that the fraction of incident photons absorbed was considerably affected by the nature of the metal, with an increase from 63% (**170**), to 68% (**171**), and 81% (**172**). The photocurrent generation efficiency measured by IPCE reaches a maximum of 12% for **172** under irradiation of 3×10^{12} photons cm^2. This efficiency is nearly six times higher than for **171**, and 20 times higher than for **170** for which photocurrent results almost exclusively from singlet excited states[181]. Obviously, the heavy-atom effect of the Pd(II) on the spin–orbital coupling favors the intersystem crossing that leads to triplet state formation.

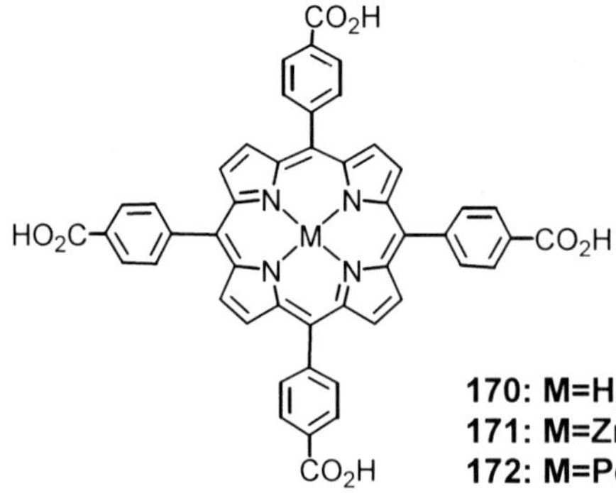

Figure 13.91. Tetrakis-carboxyphenyl porphyrin for TiO_2 supported films.

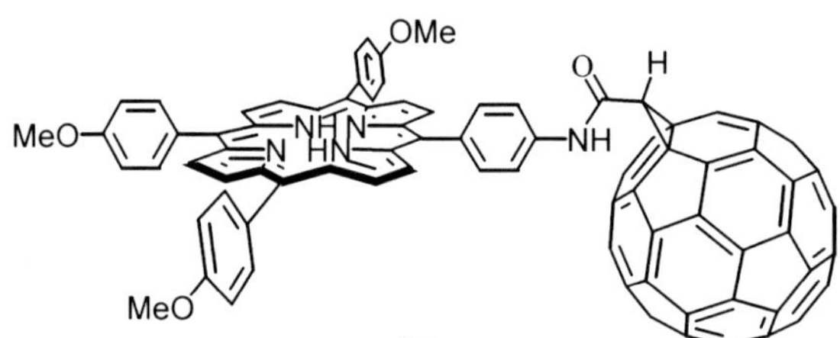

Figure 13.92. A dyad used in spin coating of SnO_2 supported films.

Naturally, film formation from fullerene-porphyrin dyads is attractive for photocurrent generation. The group of Sereno has reported the use of dyad **173** (Figure 13.92) in the spin-coating preparation of SnO_2-supported films[182]. Using this dyad, in which the distance separating the donor and the acceptor is estimated at 14 Å, photocurrent quantum yields of 20% are observed in the presence of hydroquinone as a sacrificial donor.

The covalent linking of the porphyrin with the fullerene moiety is not a prerequisite. Joint efforts from Fukuzumi, Imahori, and Kamat have shown that the same process can be efficient in nanoclusters incorporating C_{60} and 5,15-bis(di-tert-butylphenyl)porphyrin **170**, which are formed prior to electrophoretic deposition onto nanostructured SnO_2 films[183]. When a 0.2 V vs. SCE bias is applied, IPCE values of up to 17% are reached. In the absence of bias, the IPCE of 4.5% for the 1:1 **170**:C_{60} mixed clusters is significantly higher than that of C_{60} clusters (1.6%) or **170** clusters (0.06%) under the same conditions (430 nm irradiation).

Ultrathin films of cationic poly-L-lysine and anionic poly-L-glutamic acid, assembled by electrostatic forces, are isolating. However, when doped with hydrophilic hexacyanoferrate, they become conducting films. After sensitization of the film by repeated dipping in sulfonated ZnTPP **175** (Figure 13.93) solutions, the photoreduction of fullerene in a chlorobenzene solution has shown that photocurrents could be generated by these films, with IPCEs of 0.02%. This modest value is probably due to the heterogeneous nature of the film/solution junction[184].

In films incorporating porphyrins, the patterning is a parameter that must be controlled, mostly because the energy and electron transfer between porphyrin

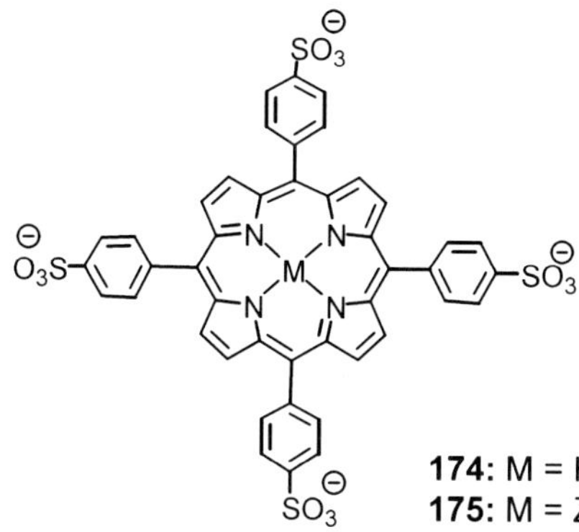

174: M = H_2
175: M = Zn

Figure 13.93. A tetrasulfonated ZnTPP for film sensitization.

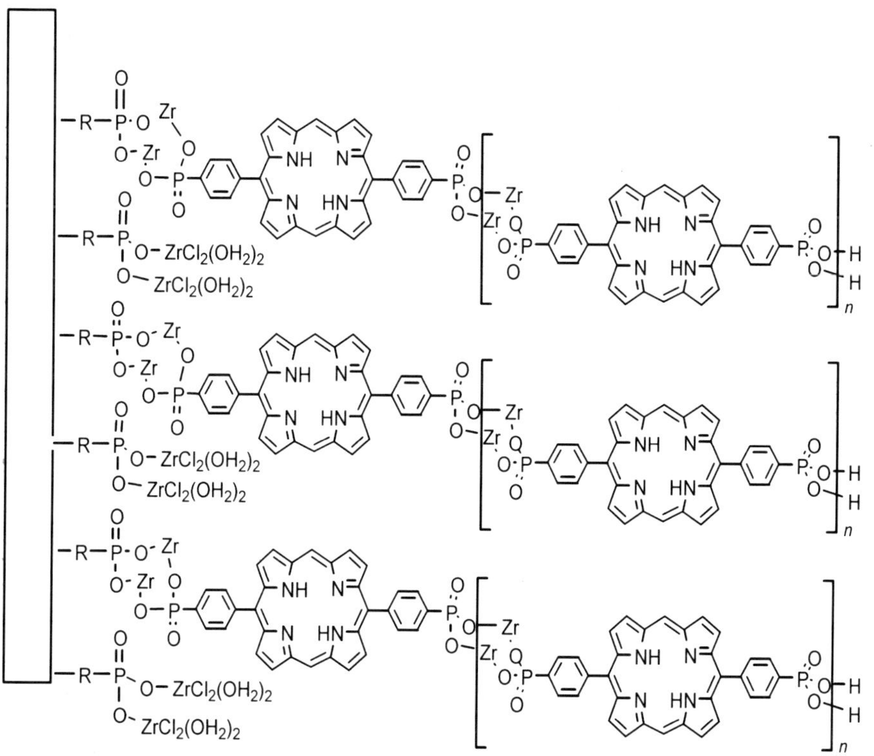

Figure 13.94. Assembly via zirconium coordination chemistry.

species depends on chromophore orientation. Chromophore orientation is a crucial parameter in solution that subsists in the film matrix. As we have seen in the case of self-assembled species in solution, exocyclic transition metal complexes are a powerful assembly tool for multiporphyrinic species. The same principle has been applied to the formation of films by the group of Hupp, using zirconium phosphate chemistry[185]. The constitution of the films is depicted in Figure 13.94.

Although energy or electron transfers have yet to be reported in these assemblies, this emerging approach deserves mentioning.

The photovoltaic behavior of patterned TiO_2 surfaces covered in a layer-by-layer method has been reported recently[186]. Composite films of oligoethylene glycol dicarboxylic acid (OEGDA), polyethyleneimine (PEI), and polyacryclic acid (PAA), with ionic conductivities of *ca.* 10^{-5} S cm^{-1}, have been dye-sensitized by impregnation with porphyrin **170** (Figure 13.91), porphyrin **174** (Figure 13.93), and the two porphyrins **176** and **177** represented in Figure 13.95. Using polymer on polymer stamping techniques, micropatterned surfaces with a pattern thickness of 4 μm have been produced. Table 13.20 lists the conversion efficiencies of photovoltaic devices constructed this way.

The chromophore J-aggregation at the TiO_2/electrolyte interface is controlled by the pH at the deposition stage. The performance of the devices is enhanced by stamp patterning rather than lithographic patterning. This approach opens the door to many potential applications such as the fabrication of flexible, thin film photovoltaic systems.

To close this section, just as the LB approach to porphyrinic assemblies can afford applications derived from the photoconducting properties such as the photogeneration of hydrogen (see above), the films formed from porphyrins or phthalocyanines offer potential catalytic applications. Bedioui and Zagal have reported a comparative study of the electro-oxidation of thiols for electrodes modified by the phthalocyanine or porphyrin derivatives represented in Figure 13.96. These compounds can be electropolymerized on glassy carbon electrodes by repeated cycling between -0.2 and 0.9 V vs. Ag/AgCl. The number of scans controls the film thickness (21 monolayers per scan for **178**, and 1.5 for **180**). The active sites, located on the surface of the films after deposition, catalyze the oxidation of the 2-mercaptoethanol in a potential range in which the electrode itself is inactive[187].

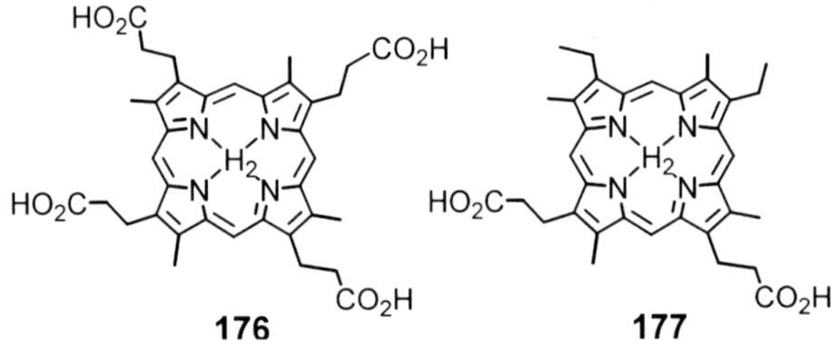

176 **177**

Figure 13.95. Porphyrin dyes for sensitizing composite films.

Table 13.20. Conversion Efficiencies of Photovoltaic Devices Consisting of **170**, Patterned TiO_2, and a $(LPEI_{5.0}/PAA_{0.5})/OEGDA_{3.0}$ Composite Electrolyte

Pattern size (μm)	None	10	5	3	1	0.32
Conversion efficiency (%)	0.04	0.06	0.09	0.15	0.66	0.55

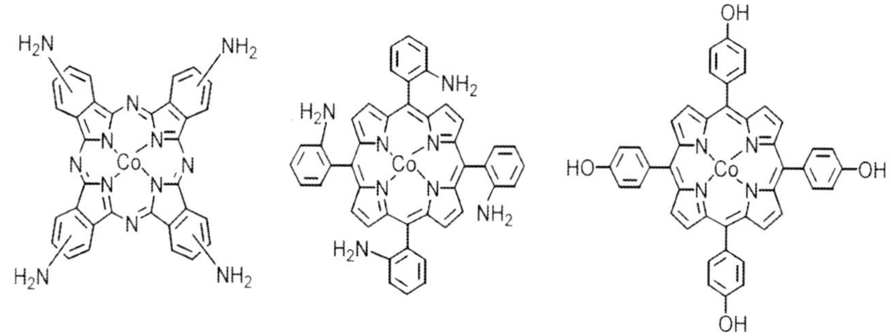

178 179 180

Figure 13.96. Compounds for electropolymerization.

8. Conclusions

If the evolution of Natural systems can be considered as complete, the design of the artificial devices that they inspire is still in its genesis stage. Thus, it is difficult to draw final conclusions on a topic which is continually evolving, and whose intellectual driving force is the imagination of chemists and physicists. It is crucial to realize, however, that despite the information obtained from academic research work targeting the comprehension of photoinduced events in Nature, only a few compounds studied will yield applicable devices due to the sophistication of the structures. The development of applications will always be ruled by laws of the economic jungle, and unfortunately not by the laws of chemistry and physics. Thus, beside the invaluable gain of knowledge afforded by academic studies using covalent synthesis, self-assembly approaches, based on design concepts afforded by covalent scaffolds, sounds extremely attractive for the preparation of molecular devices.

As we have seen throughout this chapter, several assemblies based on the fundamentals of covalent architectures, but that employ recent concepts of host–guest and supramolecular chemistry, now perform charge separation in bulk material. This opens new insights for the production of photovoltaic materials. In the introductory part of this chapter, the cost of a bit of packaged DRAM was given as an example, and we can now extend these economic considerations to the development of photovoltaic materials, as an additional example. Photovoltaic materials based on semiconductor technologies have long been available, and some performance levels already permit large scale production of solar cells. Why is application of this technology almost limited to the nonavailability of nuclear or combustion sources (e.g., solar panel power sources in satellites)? The semiconductor technology still has to prove that it is available at low cost. Porphyrin chemists should anticipate the production of low cost photovoltaic materials based on their, and Nature's, favorite chromophore. Thus, if one conclusion should be drawn from this contribution, it would be that the study of photoinduced events in fully mastered architectures has relatively little chance to afford an applicable outcome, but will certainly be the source of inspiration for the design of noncovalently organized materials that perform even less than individual molecules, but that are producible in bulk quantities.

References

1. Harvey, P. (2003). In K.M. Kadish, K.M. Smith, and R. Guilard (eds), *The Porphyrin Handbook*, Vol. 18. Academic Press, Amsterdam, pp. 63–250.
2. Gouterman, M. (1961). Spectra of porphyrins. *J. Mol. Spectrosc.* **6**, 138–163.
3. Fuchs, H., J. Zimmerman, and B. Röder (2003). Experimental determination of the orientation of excited state transition dipoles in tetrapyrroles with different molecular symmetries. *Opt. Commun.* **220**(1–3), 119–127 and references cited.
4. Klenina, I.B., I.V. Borvykh, A.Ya. Shkuropatov, P. Gast, and I.I. Proskuryakov (2003). Orientation of the Q_y optical transition moment of bacteriopheophytin in Rhodobacter sphaeroides reaction centers. *Chem. Phys.* **294**(3), 451–458.
5. a) Kasha, M., H.R. Rawls, M.A. El-Bayoumi, and M. Ashraf (1965). Exciton model in molecular spectroscopy. *Pure Appl. Chem.* **11**(3–4), 371-392; b) Kasha, M. (1963). Energy transfer mechanisms and the molecular exciton model for molecular aggregates. *Radiat. Res.* **20**, 55–70.
6. Hunter, C., J.K.M. Sanders, and A.J. Stone (1989). Exciton coupling in porphyrin dimers. *Chem. Phys.* **133**(3), 395–404.
7. Ribò, J.M., J.M. Bofill, J. Crusats, and R. Rubires (2001). Point-dipole approximation of the exciton coupling model versus type of bonding and of excitons in porphyrin supramolecular structures. *Chem. Eur. J.* **7**(13), 2733–2737.
8. Aratani, N., A. Osuka, Y.H. Kim, D.H. Jeong, and D. Kim (2000). Extremely long, discrete *meso – meso*-coupled porphyrin arrays. *Angew. Chem. Int. Ed.* **39**(8), 1458–1462.
9. Rubires, R., J.-A. Farrera, and J.M. Ribo (2001). Stirring effects on the spontaneous formation of chirality in the homoassociation of diprotonated *meso*-tetraphenylsulfonato porphyrins. *Chem. Eur. J.* **7**(2), 436–446.
10. a) Sendt, K., L.A. Johnston, W.A. Hough, M.J. Crossley, N.S. Hush, and J.R. Reimers (2002). Switchable electronic coupling in model oligoporphyrin molecular wires examined through the measurement and assignment of electronic absorption spectra. *J. Am. Chem. Soc.* **124**(31), 9299–9309; b) Jeong, D.H., S.M. Jang, I.-W. Hwang, D. Kim, N. Yoshida, and A. Osuka (2002). Investigation of interporphyrin charge resonance of dihedral angle controlled porphyrin dimers by resonance raman spectroscopy and MO approaches. *J. Phys. Chem. A.* **106**(46), 11054–11063; c) Yoshida, N., T. Ishizuka, A. Osuka, D.H. Jeong, H.S. Cho, D. Kim, Y. Matsuzaki, A. Nogami, and K. Tanaka (2003). Fine tuning of photophysical properties of *meso-meso*-linked Zn^{II}-diporphyrins by dihedral angle control. *Chem. Eur. J.* **9**(1), 58–75.
11. Zimmermann, J., U. Siggel, J.-H. Furhop, and B. Röder (2003). Excitonic coupling between B and Q transitions in a porphyrin aggregate. *J. Phys. Chem.* **107**(25), 6019–6021.
12. For reviews see: a) Pescitelli, G., S. Gabriel, Y. Wang, J. Fleischauer, R.W. Woody, and N. Berova (2003). Theoretical analysis of the porphyrin–porphyrin exciton interaction in circular dichroism spectra of dimeric tetraarylporphyrins. *J. Am. Chem. Soc.* **125**(25), 7613–7628; b) Huang, X., K. Nakanishi, and N. Berova (2000). Porphyrins and metalloporphyrins: Versatile circular dichroic reporter groups for structural studies. *Chirality* **12**(4), 237–255; c) Ogoshi, H. and T. Mizutani (1998). Multifunctional and chiral porphyrins: Model receptors for chiral recognition. *Acc. Chem. Res.* **31**(2), 81–89; d) Ogoshi, H., T. Mizutani, T. Hayashi, and Y. Kuroda (2006). In

K.M. Kadish, K.M. Smith, and R. Guilard (eds), *The Porphyrin Handbook*, Vol. 6. Academic Press, Amsterdam, pp. 279–340; e) Weiss, J. (2001). Zinc(II)–porphyrin receptors in multi-point molecular recognition: Recent progress. *J. Inclusion Phenom. Macrocyclic Chem.* **40**(1–2), 1–22.

13. Valeur, B. (2002). *Molecular Fluorescence: Principles and Applications*. Wiley-VCH, Weinheim.
14. Paddon-Row, M.N. (2003). Superexchange-mediated charge separation and charge recombination in covalently linked donor–bridge–acceptor systems. *Aust. J. Chem.* **56**(8), 729–748.
15. Marcus, R.A. and N. Sutin (1985). Electron transfers in chemistry and biology. *Biochim. Biophys. Acta* **811**(3), 265–322; b) Marcus, R.A. (1993). Electron transfer reactions in chemistry: Theory and experiment. *Angew. Chem. Int. Ed. Engl.* **32**(8), 1111–1121.
16. Förster, T. (1962). Electron spectra of coupled molecules. *Pure Appl. Chem.* **4**, 121–134.
17. Dexter, D.L. (1953). A theory of sensitized luminescence in solids. *J. Chem. Phys.* **21**, 836–850.
18. Piet, J.J., P.N. Taylor, B.R. Wegewijs, H.L. Anderson, A. Osuka, and J.M. Warman (2001). Photoexcitations of covalently bridged zinc porphyrin oligomers: Frenkel versus Wannier–Mott type excitons. *J. Phys. Chem. B* **105**(1), 97–104.
19. Andréasson, J., A. Kyrychenko, J. Mårtensson, and B. Albinsson (2002). Temperature and viscosity dependence of the triplet energy transfer process in porphyrin dimers. *Photochem. Photobiol. Sci.* **1**(2), 111–119.
20. Weiss, E.A., M.J. Ahrens, L.E. Sinks, A.V. Gusev, M.A. Ratner, and M.R. Wasielewski, (2004). Making a molecular wire: Charge and spin transport through *para*-phenylene oligomers. *J. Am. Chem. Soc.* **126**(17), 5577–5584.
21. Yoon, D.H., S.B. Lee, K.-H. Yoo, J. Kim, J.K. Lim, N. Aratani, A. Tsuda, A. Osuka, and D. Kim (2003). Electrical conduction through linear porphyrin arrays. *J. Am. Chem. Soc.* **125**(36), 11062–11064.
22. Kroeze, J.E., T.J. Savenije, and J.M. Warman (2002). Contactless determination of the efficiency of photo-induced charge separation in a porphyrin–TiO_2 bilayer. *J. Photophys. Photobiol. A: Chem.* **148**(1–3), 49–55.
23. a) Deisenhofer, J., O. Epp, K. Miki, R. Huber, and H. Michel (1984). X-ray structure analysis of a membrane protein complex. Electron density map at 3 Å resolution and a model of the chromophores of the photosynthetic reaction center from Rhopseudomonas viridis. *J. Mol. Biol.* **180**(2), 385–398; b) Deisenhofer, J., O. Epp, I. Sinning, and H. Michel (1995). Crystallographic refinement at 2.3 Å resolution and refined model of the photosynthetic reaction center from *Rhodopseudomonas viridis*. *J. Mol. Biol.* **246**(3), 429–457.
24. Roszak, A.W., T.D. Howard, J. Southall, A.T. Gardiner, C.J. Law, N.W. Isaacs, and R.J. Cogdell (2003). Crystal structure of the RC-LH1 core complex from *Rhodopseudomonas palustris*. *Science* **302**(5622), 1969–1972.
25. a) McDermott, G., S.M. Prince, A.A. Freer, A.M. Hawthornthwaite-Lawless, M.Z. Papiz, R.J. Cogdell, and N.W. Isaacs (1995). Crystal structure of an integral membrane light-harvesting complex from photosynthetic bacteria. *Nature* **374**(6522), 517–521; b) M.Z. Papiz, S.M. Prince, T. Howard, R.J. Cogdell, and N.W. Isaacs (2003). The structure and thermal motion of the B800–850 LH2 complex from Rps. acidophila at 2.0 (A) over-circle resolution and 100 K: New structural features and functionally relevant motions. *J. Mol. Biol.* **326**(5), 1523–1538.
26. R. Compañon, L. Molenkamp, and D.J. Paul (eds) (1999). Technology Roadmap for Nanoelectronics. European Commission IST programme; http://ftp.cordis.lu/pub/esprit/docs/melnarm.pdf

27. Lloyd, R.L. and C.B. Gorman (2002). The genesis of molecular electronics. *Angew. Chem. Int. Ed.* **41**(23), 4378–4400.
28. Iengo, E., E. Zangrando, and E. Alesio (2003). Discrete supramolecular assemblies of porphyrins mediated by coordination compounds. *Eur. J. Inorg. Chem.* (13), 2371–2384.
29. Wojaczynki, J. and L. Latos-Grażyňski (2001). Poly- and oligometalloporphyrins associated through coordination. *Coord. Chem. Rev.* **204**, 113–171.
30. Imamura, T. and K. Fukushima (2000). Self-assembly of metallopyridylporphyrin oligomers. *Coord. Chem. Rev.* **198**, 133–156.
31. Heitz, V., J.C. Chambron, and J.P. Sauvage (2000). In K.M. Kadish, K.M. Smith, and R. Guilard (eds), *The Porphyrin Handbook*, Vol. 6. Academic Press, Amsterdam, pp. 1–42.
32. Burrel, A.K., D.L. Officer, P.G. Plieger, and D.C.W. Reid (2001). Synthetic routes to multiporphyrin arrays. *Chem. Rev.* **101**(9), 2751–2796.
33. Guldi, D.M. (2002). Fullerene–porphyrin architectures: photosynthetic antenna and reaction center dyads. *Chem. Soc. Rev.* **31**(1), 22–36.
34. Imahori, H., Y. Mori, and Y. Matano (2003). Nanostructured artificial photosynthesis. *J. Photochem. Photobiol. C: Photochem. Rev.* **4**(1), 51–83.
35. Fukuzumi, S. and H. Imahori (2001). In V. Balzani (ed), *Electron Transfer in Chemistry*, Wiley-VCH, Weinheim, pp. 927–975.
36. Gust, D., T.A. Moore, and A.L. Moore (2001). In V. Balzani (ed), *Electron Transfer in Chemistry*, Vol. 3. Wiley-VCH, Weinheim, pp. 272–336.
37. Yamaguchi, Y., S. Yokoyama, and S. Mashiko (2002). Strong coupling of the single excitations in the *Q*-like bands of phenylene-linked free-base and zinc bacteriochlorin dimers: A time-dependent density functional theory study. *J. Chem. Phys.* **116**(15), 6541–6548.
38. Yamaguchi, Y., Y. Yokomichi, S. Yokoyama, and S. Mashiko (2001). Time-dependent density functional calculations of the *Q*-like bands of phenylene-linked free-base and zinc porphyrin dimers. *Int. J. Quantum Chem.* **84**(3), 338–347.
39. Cho, H.S., D.H. Jeong, M.-C. Yoon, Y.H. Kim, Y.-R. Kim, D. Kim, S.C. Jeoung, S.K. Kim, N. Aratani, H. Shinmori, and A. Osuka (2001). Excited-state energy transfer processes in phenylene- and biphenylene-linked and directly-linked zinc(II) and free base hybrid diporphyrins. *J. Phys. Chem. A* **105**(17), 4200–4210.
40. Osuka, A., M. Ikeda, H. Shiratori, Y. Nishimura, and I. Yamazaki (1999). Accelerated singlet energy transfer in bis(phenylethynyl)phenylene-bridged 5,10,15,20-tetraryl zinc-free base hybrid diporphyrins. *J. Chem. Soc. Perkin Trans. 2* (5), 1019–1026.
41. a) Osuka, A., N. Tanabe, S. Kawabata, I. Yamazaki, and Y. Nishimura (1995). Synthesis and intramolecular electron- and energy-transfer reactions of polyyne- or polyene-bridged diporphyrins. *J. Org. Chem.* **60**(22), 7177–7185; b) Yang, S.I., J. Seth, T. Balasubramanian, D. Kim, J.S. Lindsey, D. Holten, and D.F. Bocian (1999). Interplay of orbital tuning and linker location in controlling electronic communication in porphyrin arrays. *J. Am. Chem. Soc.* **121**(16), 4008–4018.
42. Yang, S.I., R.K. Lammi, J. Seth, J.A. Riggs, T. Arai, D. Kim, D.F. Bocian, D. Holten, and J.S. Lindsey (1998). Excited-state energy transfer and ground-state hole/electron hopping in *p*-phenylene-linked porphyrin dimers. *J. Phys. Chem. B* **102**(47), 9426–9436.
43. Rubtsov, I.V., K. Susumu, G.I. Rubtsov, and M.J. Therien (2003). Ultrafast singlet excited state polarization in electronically asymetric ethyne-bridged bis[(porphinato) zinc(II)] complexes. *J. Am. Chem. Soc.* **125**(9), 2687–2696.
44. Fukuzumi, S., K. Ohkubo, W.E.Z. Ou, J. Shao, K.M. Kadish, J.A. Hutchison, K.P. Ghiggino, P.J. Sintic, and M.J. Crossley (2003). Metal-centered photoinduced electron transfer reduction of a gold(III) porphyrin cation linked with a zinc porphyrin to

produce a long-lived charge-separated state in nonpolar solvents. *J. Am. Chem. Soc.* **125**(49), 14984–14985.
45. Odobel, F., S. Suresh, E. Blart, Y. Nicolas, J.-P. Quintard, P. Janvier, J.-Y. LeQuestel, B. Illien, D. Rondeau, P. Richomme, T. Häupl, S. Wallin, and L. Hammarström (2002). Synthesis of oligothiophene-bridged bisporphyrins and study of the linkage dependence of the electronic coupling. *Chem. Eur. J.* **8**(13), 3027–3046.
46. Andreasson, J., J. Kajanus, J. Mårtensson, and B. Albinsson (2000). Triplet energy transfer in porphyrin dimers: Comparison between π- and σ-chromophore bridged systems. *J. Am. Chem. Soc.* **122**(40), 9844–9845.
47. Andréasson, J., G. Kodis, T. Ljungdahl, A.L. Moore, T.A. Moore, D. Gust, J. Mårtensson, and B. Albinsson (2003). Photoinduced hole transfer from the triplet state in a porphyrin-based donor–bridge–acceptor system. *J. Phys. Chem. A* **107**(42), 8825–8833.
48. Kilså, K., J. Kajanus, S. Larsson, A.N. Macpherson, J. Mårtensson, and B. Albinsson (2001). Enhanced intersystem crossing in donor/acceptor systems based on zinc/iron or free-base/iron porphyrins. *Chem. Eur. J.* **7**(10), 2122–2133.
49. Kilså, K., A.N. Macpherson, T. Gillbro, J. Mårtensson, and B. Albinsson (2001). Control of electron transfer in supramolecular systems. *Spectrochim. Acta A* **57**(11), 2213–2227.
50. Kyrychenko, A. and B. Albinsson (2002). Conformer-dependent electronic coupling for long-range triplet energy transfer in donor-bridge-acceptor porphyrin dimers. *Chem. Phys. Lett.* **366**(3–4), 291–299.
51. Kyrychenko, A., J. Andréasson, J. Mårtensson, and B. Albinsson (2002). Sterically induced conformational relaxation and structure of *meso*-diaryloctaalkyl porphyrins in the excited triplet state: Experimental and DFT studies. *J. Phys. Chem. B* **106**(48), 12613–12622.
52. Kilså, K., J. Kajanus, A.N. Macpherson, J. Mårtensson, and B. Albinsson (2001). Bridge-dependent electron transfer in porphyrin-based donor–bridge–acceptor systems. *J. Am. Chem. Soc.* **123**(13), 3069–3080.
53. Otsuki, J., A. Yasuda, and T. Takido (2003). Non-covalent switch for intramolecular energy transfer. *Chem. Commun.* (5), 608–609.
54. For summary of logical architecture using eT and ET processes for processing signal see: Holten, D., D.F. Bocian, and J.S. Lindsey (2002). Probing electronic communication in covalently linked multiporphyrin arrays. A guide to the rational design of molecular photonic devices. *Acc. Chem. Res.* **35**(1), 57–69.
55. Sazanovich, I.V., A. Balakumar, K. Muthukumaran, E. Hindin, C. Kirmaier, J.R. Diers, J.S. Lindsey, D.F. Bocian, and D. Holten (2003). Excited-state energy transfer dynamics of self-assembled imine-linked porphyrin dyads. *Inorg. Chem.* **42**(21), 6616–6628.
56. Kim, Y.H., D.H. Jeong, D. Kim, S.C. Jeoung, H.S. Cho, S.K. Kim, N. Aratani, and A. Osuka (2001). Photophysical properties of long rodlike *meso–meso*-linked zinc(II) porphyrins investigated by time-resolved laser spectroscopic methods. *J. Am. Chem. Soc.* **123**(1), 76–86.
57. Aratani, N., H.S. Cho, T.K. Ahn, S. Cho, D. Kim, H. Sumi, and A. Osuka (2003). Efficient excitation energy transfer in long *meso–meso* linked Zn(II) porphyrin arrays bearing a 5,15-bisphenylethynylated Zn(II) porphyrin acceptor. *J. Am. Chem. Soc.* **125**(32), 9668–9681.
58. Cho, H.S., D.H. Jeong, S. Cho, D. Kim, Y. Matsuzaki, K. Tanaka, A. Tsuda, and A. Osuka (2002). Photophysical properties of porphyrin tapes. *J. Am. Chem. Soc.* **124**(49), 14642–14654.

59. Li, F., S.I. Yang, Y. Ciringh, J. Seth, C.H. Martin, III, D.L. Singh, D. Kim, R.R. Birge, D.F. Bocian, D. Holten, and J.S. Lindsey (1998). Design, synthesis, and photodynamics of light-harvesting arrays comprised of a porphyrin and one, two, or eight boron-dipyrrin accessory pigments. *J. Am. Chem. Soc.* **120**(39), 10001–10017.
60. (a) Lammi, R.K., A. Ambroise, R.W. Wagner, J.R. Diers, D.F. Bocian, D. Holten, and J.S. Lindsey (2001). Quenching of porphyrin excited states by adjacent or distant porphyrin cation radicals in molecular arrays. *Chem. Phys. Lett.* **341**(1–2), 35–44; (b) Loewe, R.S., R.K. Lammi, J.R. Diers, C. Kirmaier, D.F. Bocian, D. Holten, and J.S. Lindsey (2002). Design and synthesis of light-harvesting rods for intrinsic rectification of the migration of excited-state energy and ground-state holes. *J. Mater. Chem.* **12**(5), 1530–1552.
61. Lammi, R.K., R.W. Wagner, A. Ambroise, J.R. Diers, D.F. Bocian, D. Holten, and J.S. Lindsey (2001). Mechanisms of excited-state energy-transfer gating in linear versus branched multiporphyrin arrays. *J. Phys. Chem. B* **105**(22), 5341–5352.
62. Ambroise, A., C. Kirmaier, R.W. Wagner, R.S. Loewe, D.F. Bocian, D. Holten, and J.S. Lindsey (2002). Weakly coupled molecular photonic wires: Synthesis and excited state energy transfer dynamics. *J. Org. Chem.* **67**(11), 3811–3826.
63. Guldi, D.M. and K.-D. Asmus (1997). Electron transfer from C_{76} (C'_{2v}) and C_{78} (D_2) to radical cations of various arenes: Evidence for the Marcus inverted region. *J. Am. Chem. Soc.* **119**(24), 5744–5745.
64. (a) Kuciauskas, D., P.A. Liddell, S. Lin, T.E. Johnson, S.J. Weghorn, J.S. Lindsey, A.L. Moore, T.A. Moore, and D. Gust (1999). An artificial photosynthetic antenna-reaction center complex. *J. Am. Chem. Soc.* **121**, 8604–8614.
65. Kodis, G., P.A. Liddell, L. de la Garza, P.C. Clausen, J.S. Lindsey, A.L. Moore, T.A. Moore, and D. Gust (2002). Efficient energy transfer and electron transfer in an artificial photosynthetic antenna-reaction center complex. *J. Phys. Chem. A* **106**(10), 2036–2048.
66. Imahori, H., K. Tamaki, D.M. Guldi, C. Luo, M. Fujitsuka, O. Ito, Y. Sakata, and S. Fukuzumi (2001). Modulating charge separation and charge recombination dynamics in porphyrin–fullerene linked dyads and triads: Marcus-normal versus inverted region. *J. Am. Chem. Soc.* **123**(11), 2607–2617.
67. Imahori, H., D.M. Guldi, K. Tamaki, Y. Yoshida, C. Luo, Y. Sakata, and S. Fukuzumi (2001). Charge separation in a novel artificial photosynthetic reaction center lives 380 ms. *J. Am. Chem. Soc.* **123**(27), 6617–6628.
68. Guldi, D.M., H. Imahori, K. Tamaki, Y. Kashiwagi, H. Yamada, Y. Sakata, and S. Fukuzumi (2004). A molecular tetrad allowing efficient energy storage for 1.6 s at 163 K. *J. Phys. Chem. A* **108**(4), 541–548.
69. Kuciauskas, D., P.A. Liddell, S. Lin, S.G. Stone, A.L. Moore, T.A. Moore, D. Gust (2000). Photoinduced electron transfer in carotenoporphyrin–fullerene triads: Temperature and solvent effects. *J. Phys. Chem B* **104**(18), 4307–4321.
70. Imahori, H., K. Tamaki, Y. Araki, Y. Sekiguchi, O. Ito, Y. Sakata, and S. Fukuzumi (2002). Stepwise charge separation and charge recombination in ferrocene–*meso,meso*-linked porphyrin dimer–fullerene triad. *J. Am. Chem. Soc.* **124**(18), 5165–5174.
71. Ha, J.-H., H.S. Cho, D. Kim, J.-C. Lee, T.-Y Kim, and Y.K. Shim (2003). Time-resolved spectroscopic study on photoinduced electron-transfer processes in Zn(II)porphyrin–Zn(II)chlorin–fullerene triad. *Chem. Phys. Chem.* **4**(9), 951–958.
72. Bonifazi, D. and F. Diederich (2002). Strong intramolecular chromophore interactions in novel bis([60]fullerene)–oligoporphyrin nanoarrays. *Chem. Commun.* (18), 2178–2179.
73. Bonifazi, D., M. Scholl, F. Song, L. Echegoyen, G. Accorsi, N. Armaroli, and F. Diederich (2003). Exceptional redox and photophysical properties of a triply fused

diporphyrin–C_{60} conjugate: Novel scaffolds for multicharge storage in molecular scale electronics. *Angew. Chem. Int. Ed.* **42**(40), 4966–4970.

74. Yoon, D.H., S.B. Lee, K.H. Yoo, J. Kim, J.K. Lim, N. Aratani, A. Tsuda, A. Osuka, and D. Kim (2003). Electrical conduction through linear porphyrin arrays. *J. Am. Chem. Soc.* **125**(36), 11062–11064.
75. Loewe, R.S., K. Tomizaki, W.J. Youngblood, Z. Bo, and J.S. Lindsey (2002). Synthesis of perylene–porphyrin building blocks and rod-like oligomers for light-harvesting applications. *J. Mater. Chem.* **12**(5), 3438–3451.
76. Choi, M.-S., T. Yamazaki, I. Yamazaki, and T. Aida (2004). Bioinspired molecular design of light-harvesting multiporphyrin arrays. *Angew. Chem. Int. Ed.* **43**(2), 150–158.
77. Peng, X., N. Aratani, A. Takagi, T. Matsumoto, T. Kawai, I.-W. Hwang, T.K. Ahn, D. Kim, and A. Osuka (2004). A dodecameric porphyrin wheel. *J. Am. Chem. Soc.* **126**(14), 4468–4469.
78. Takahashi, R. and Y. Kobuke (2003). Hexameric macroring of gable-porphyrins as a light-harvesting antenna mimic. *J. Am. Chem. Soc.* **125**(9), 2372–2373.
79. Ikeda, C., A. Satake, and Y. Kobuke (2003). Proofs of macrocyclization of gable porphyrins as mimics of photosynthetic light-harvesting complexes. *Org. Lett.* **5**(26), 4935–4938.
80. Nakano, A., A. Osuka, T. Yamazaki, Y. Nishimura, S. Akimoto, I. Yamazaki, A. Itaya, M. Murakami, and H. Miyasaka (2001). Modified windmill porphyrin arrays: Coupled light-harvesting and charge separation, conformational relaxation in the S_1 state, S_2–S_2 energy transfer. *Chem. Eur. J.* **14**(14), 3134–3151.
81. a) Choi, M.-S., T. Aida, T. Yamazaki, and I. Yamazaki (2001). A large dendritic multiporphyrin array as a mimic of the bacterial light-harvesting antenna complex: Molecular design of an efficient energy funnel for visible photons. *Angew. Chem. Int. Ed.* **40**(17), 3194–3198; b) Choi, M.-S., T. Aida, T. Yamazaki, and I. Yamazaki (2002). Dendritic multiporphyrin arrays as light-harvesting antennae: Effects of generation number and morphology on intramolecular energy transfer. *Chem. Eur. J.* **8**(12), 2668–2678.
82. Choi, M.-S., T. Aida, H. Luo, Y. Araki, and O. Ito (2003). Fullerene-terminated dendritic multiporphyrin arrays: "Dendrimer effects" on photoinduced charge separation. *Angew Chem. Int. Ed.* **42**(34), 4060–4063.
83. Rosario Benites, M., T.E. Johnson, S. Weghorn, L. Yu, P.D. Rao, J.R. Diers, S.I. Yang, C. Kirmaier, D.F. Bocian, D. Holten, and J.S. Lindsey (2002). Synthesis and properties of weakly coupled dendrimeric multiporphyrin light-harvesting arrays and hole-storage reservoirs. *J. Mater. Chem.* **12**(1), 65–80.
84. See for example: Moschetto, G., R. Lauceri, F.G. Gulino, D. Sciotto, and R. Purrello (2002). Non-covalent synthesis in aqueous solution of discrete multi-porphyrin aggregates with programmable stoichiometry and sequence. *J. Am. Chem. Soc.* **124**(49), 14536–14537 and references cited therein.
85. Kano, K., R. Nishiyabu, T. Yamazaki, and I. Yamazaki (2003). Convenient scaffold for forming heteroporphyrin arrays in aqueous media. *J. Am. Chem. Soc.* **125**(35), 10625–10634.
86. Sasaki, K., H. Nakagawa, X. Zhang, S. Sakurai, K. Kano, and Y. Kuroda (2004). Construction of porphyrin–cyclodextrin self-assembly with molecular wedge. *Chem. Commun.* (6), 408–409.
87. Conductance (reciprocal of resistance) is expressed in Siemens or (Ohm^{-1}). In electronics, conductance is often expressed as Mho which is a backwards spelling of Ohm. Conductivity is expressed in Siemens per meter (in official SI) or centimeter.

88. Li, L.-L., C.-J. Yang, W.-H. Chen, and K.-J. Lin (2003). Towards the development of electrical conduction and lithium-ion transport in a tetragonal porphyrin wire. *Angew. Chem. Int. Ed.* **42**(13), 1505–1508.
89. Okada, S. and H. Segawa (2003). Substituent-control exciton in J-aggregates of protonated water-insoluble porphyrins. *J. Am. Chem. Soc.* **125**(9), 2792–2796.
90. Ozeki, H. and Y. Kobuke (2003). Incorporation of a photosynthetic supramolecular complex by using imidazolyl Zn porphyrin dimers in bilayer lipid membrane. *Tetrahedron Lett.* **44**(11), 2287–2291.
91. Hwang, I.-W., H.S. Cho, D.H. Jeong, D. Kim, A. Tsuda, T. Nakamura, and A. Osuka (2003). Photophysical properties of a three-dimensional zinc(II) porphyrin box. *J. Phys. Chem. B* **107**(37), 9977–9988.
92. Yatskou, M.M., R.B.M. Koehorst, A. van Hoek, H. Donker, and T.J. Schaafsma (2001). Spectroscopic properties of a self-assembled zinc porphyrin tetramer II. Time-resolved fluorescence spectroscopy. *J. Phys. Chem. A* **105**(51), 11432–11440.
93. Nagata, N., S.-I. Kugimiya, E.-I. Fujiwara, and Y. Kobuke (2003). Staircase-form assembly with 5,15-bis(imidazol-4-yl)porphyrinatogallium steps. *New. J. Chem.* **27**(4), 743–747.
94. Sugou, K., K. Sasaki, K. Kitajima, T. Iwaki, and Y. Kuroda (2002). Light-harvesting heptadecameric porphyrin assemblies. *J. Am. Chem. Soc.* **124**(7), 1182–1183.
95. Iengo, E., E. Zangrando, E. Alessio, J.-C. Chambron, V. Heitz, L. Flamigni, and J.-P. Sauvage (2003). A functionalized noncovalent macrocyclic multiporphyrin assembly from a dizinc(II) bis-porphyrin receptor and a free-base dipyridylporphyrin. *Chem. Eur. J.* **9**(23), 5879–5887.
96. Haycock, R.H., A. Yartzev, U. Michelsen, V. Sundström, and C.A. Hunter (2000). Self-assembly of pentameric porphyrin light-harvesting antennae complexes. *Angew. Chem. Int. Ed.* **39**(20), 3616–3619.
97. a) Zenkevich, E.I., C. von Borczyskowski, A.M. Shulga, S. Bachilo, U. Rempel, and A. Willert (2002). Self-assembled nanoscale photomimetic models: Structure and related dynamics. *Chem. Phys.* **275**(1–3), 185–209; b) Zenkevich, E.I., A.M. Shulga, C. von Borczyskowski (2002). Multistep photoinduced electron transfer in self-organised nanoscale porphyrin triads. *Physica E* **14**(1–2), 277–281.
98. a) Kobuke, Y. and H. Miyaji (1994). Supramolecular organization of imidazolylporphyrin to a slipped cofacial dimer. *J. Am. Chem. Soc.* **116**(9), 4111–4112; b) Nagata, N., S. Kugimiya, and Y. Kobuke (2000). Antenna functions of 5,15-bis(imidazol-4-yl)-10,20-bis(4-dodecyloxyphenyl)-porphyrin supramolecular assembly through imidazole–imidazole hydrogen bonding. *Chem. Commun.* (15), 1389–1390.
99. Paul, A., F. Melin, C. Hirtz, J. Wytko, P. Ochsenbein, M. Bonin, K. Schenk, P. Maltese, and J. Weiss (2003). Induced fit process in the selective distal binding of imidazoles in zinc(II) porphyrin receptors. *Inorg. Chem.* **42**(12), 3779–3787.
100. Paul, D., J.A. Wytko, M. Koepf, and J. Weiss (2002). Design and synthesis of a self-assembled photochemical dyad based on selective imidazole recognition. *Inorg. Chem.* **41**(14), 3699–3704.
101. Leray, I., B. Valeur, D. Paul, E. Regnier, M. Koepf, J.A. Wytko, C. Boudon, and J. Weiss (2005). Photodynamics of excitation energy transfer in self-assembled dyads. Evidence for back transfer. *Photochem. Photobio. Sci.* **4**(3), 280–286.
102. Trabolsi, A., M. Elhabiri, A.-M. Albrecht Gary, M. Koepf, D. Paul, J.A. Wytko, and J. Weiss (2005). Building blocks for self-assembled porphyrinic photonic wires. *Org. Lett.* **7**(7), 1279–1282.
103. a) Roa, T.A. and B.G. Maiya (1995). Axial-bonding-type hybrid porphyrin arrays: Design, synthesis and modulation of redox and photophysical properties. *J. Chem.*

Soc. Chem. Commun. (9), 939–940; b) Giribabu, L., T.A. Rao, and B.G. Maiya (1999). "Axial-bonding"-type hybrid porphyrin arrays: Synthesis, spectroscopy, electrochemistry, and singlet state properties. *Inorg. Chem.* **38**(22), 4971–4980.

104. Maiya, B.G., N. Bampos, A.A. Kumar, N. Feeder, and J.K.M. Sanders (2001). A supramolecular array assembled *via* the complementary binding properties of ruthenium(II) and tin(IV) porphyrins. *New J. Chem.* **25**(6), 797–800.
105. Richeter, S., C. Jeandon, C. Sauber, J.-P. Gisselbrecht, R. Ruppert, and H.J. Callot (2002). Preparation, mass spectrometry and electrochemical studies of metal connected porphyrin oligomers. *J. Porphyrins Phthalocyanines* **6**(6), 423–430.
106. Richeter, S., C. Jeandon, J.-P. Gisselbrecht, R. Ruppert, and H.J. Callot (2002). Syntheses and optical and electrochemical properties of porphyrin dimers linked by metal ions. *J. Am. Chem. Soc.* **124**(21), 6168–6179.
107. Richeter, S., C. Jeandon, J.-P. Gisselbrecht, R. Graff, R. Ruppert, and H.J. Callot (2004). Synthesis of new porphyrins with peripheral conjugated chelates and their use for the preparation of porphyrin dimers linked by metal ions. *Inorg. Chem.* **43**(1), 251–263.
108. Yu, L., K. Muthukumaran, I.V. Sazanovich, C. Kirmaier, E. Hindin, J.R. Diers, P.D. Boyle, D.F. Bocian, D. Holten, and J.S. Lindsey (2003). Excited-state energy-transfer dynamics in self-assembled triads composed of two porphyrins and an intervening bis(dipyrrinato)metal complex. *Inorg. Chem.* **42**(21), 6629–6647.
109. Dixon, I.M., J.-P. Collin, J.-P. Sauvage, and L. Flamigni (2001). Porphyrinic dyads and triads assembled around iridium(III) bis-terpyridine: Photoinduced electron transfer processes. *Inorg. Chem.* **40**(22), 5507–5517.
110. Flamigni, L., G. Marconi, I.M. Dixon, J.-P. Collin, and J.-P. Sauvage (2002). Switching of electron- to energy-transfer by selective excitation of different chromophores in arrays based on porphyrins and a polypyridyl iridium complex. *J. Phys. Chem. B* **106**(26), 6663–6671.
111. Andersson, M., M. Linke, J.-C. Chambron, J. Davidsson, V. Heitz, L. Hammarström, and J.-P. Sauvage (2002). Long-range electron transfer in porphyrin-containing [2]-rotaxanes: Tuning the rate by metal cation coordination. *J. Am. Chem. Soc.* **124**(16), 4347–4362.
112. For a review see: Li, G., W. Fudickar, M. Skupin, A. Klyszcz, C. Draeger, M. Lauer, and J.-H. Fuhrhop (2002). Rigid lipid membranes and nanometer clefts: Motifs for the creation of molecular landscapes. *Angew. Chem. Int. Ed.* **41**(11), 1828–1852.
113. Fudickar, W., J. Zimmermann, L. Ruhlmann, J. Schneider, B. Röder, U. Siggel, and J.-H. Fuhrhop (1999). Fluorescence quenching and size selective heterodimerization of a porphyrin adsorbed to gold and embedded in rigid membrane gaps. *J. Am. Chem. Soc.* **121**(41), 9539–9545.
114. Skupin, M., G. Li, W. Fudickar, J. Zimmermann, B. Röder, and J.-H. Fuhrhop (2001). Methylammonium groups at the solid walls of nanometer-sized, water-filled monolayer gaps as binding sites for a tetraanionic porphyrin. *J. Am. Chem. Soc.* **123**(15), 3454–3461.
115. Sakamoto, M., A. Ueno, and H. Mihara (2001). Multipeptide-metalloporphyrin assembly on a dendrimer template and photoinduced electron transfer based on the dendrimer structure. *Chem. Eur. J.* **7**(11), 2449–2458.
116. Komatsu, T., M. Moritake, and E. Tsuchida (2003). Molecular energy and electron transfer assemblies made of self-organized lipid-porphyrin bilayer vesicles. *Chem. Eur. J.* **9**(19), 4626–4633.
117. Yanagimoto, T., T. Komatsu, and E. Tsuchida (2000). Photoinduced electron transfer between lipid-porphyrinato-zinc(II) and -iron(III) complexes in a phospholipid vesicular membrane. *Inorg. Chim. Acta* **305**(1), 26–31.

118. Miyatake, T., H. Tamiaki, A.R. Holtzwarth, and K. Schaffner (1999). Self-assembly of synthetic zinc chlorins in aqueous microheterogeneous media to an artifical supramolecular light-harvesting device. *Helv. Chim. Acta* **82**(6), 797–810.
119. Miyatake, T., H. Tamiaki, A.R. Holtzwarth, and K. Schaffner (1999). Artificial light-harvesting antennae: Singlet excitation energy transfer from zinc chlorin aggregate to bacteriochlorin in homogeneous hexane solution. *Photochem. Photobiol.* **69**(4), 448–456.
120. Flamigni, L., G. Marconi, and M.R. Johnston (2001). Bis-porphyrinic clamp for photo- and electro-active guests: A spectroscopic and photophysical study. *Phys. Chem. Chem. Phys.* **3**(20), 4488–4494.
121. Flamigni, L. and M.R. Johnston (2001). Photoinduced electron transfer in a non-covalently linked donor-acceptor system: A bis-porphyrinic host and a naphthalene diimide guest. *New. J. Chem.* **25**(11), 1368–1370.
122. Flamigni, L., M.R. Johnston, and L. Giribabu (2002). Photoinduced electron transfer in bisporphyrin–diimide complexes. *Chem. Eur. J.* **8**(17), 3938–3947.
123. For examples see: a) Wiederrecht, G.P., M.P. Niemczyk, W.A. Svec, and M.R. Wasielewski (1996). Ultrafast photoinduced electron transfer in a chlorophyll-based triad: Vibrationally hot ion pair intermediates and dynamic solvent effects. *J. Am. Chem. Soc.* **118**(1), 81–88; and b) Shiratori, H., T. Ohno, K. Nozaki, I. Yamazaki, Y. Nishimura, and A. Osuka (1998). Coordination control of intramolecular electron transfer in boronate ester-bridged donor-acceptor molecules. *Chem. Commun.* (15), 1539–1540.
124. D'Souza, F., G.R. Deviprasad, M.E. Zandler, M.E. El-Khouly, M. Fujitsuka, and O. Ito (2002). Electronic interactions and photoinduced electron transfer in covalently linked porphyrin–C_{60}(pyridine) diads and supramolecular triads formed by self-assembling the diads and zinc porphyrin. *J. Phys. Chem. B* **106**(19), 4952–4962.
125. Watanabe, N., N. Kihara, Y. Furusho, T. Takata, Y. Araki, and O. Ito (2003). Photoinduced intrarotaxane electron transfer between zinc porphyrin and [60]fullerene in benzonitrile. *Angew. Chem. Int. Ed.* **42**(6), 681–683.
126. Imahori, H., M.E. El-Khouly, M. Fujitsuka, O. Ito, Y. Sakata, and S. Fukuzumi (2001). Solvent dependence of charge separation and charge recombination rates in porphyrin–fullerene dyad. *J. Phys. Chem. A* **105**(2), 325–332.
127. Li, K., D.I. Schuster, D.M. Guldi, M.A. Herranz, and L. Echegoyen (2004). Convergent synthesis and photophysics of [60]fullerene/porphyrin-based rotaxanes. *J. Am. Chem. Soc.* **126**(11), 3388–3389.
128. Benkstein, K.D., C.L. Stern, K.E. Splan, R.C. Johnson, K.A. Walters, F.M.W. Vanhelmont, and J.T. Hupp (2002). Collapsed molecular rectangles based on rhenium(I) coordination of ethynylpyridyl porphyrins—synthesis, structure, and bending-induced charge-transfer behaviour. *Eur. J. Inorg. Chem.* (11), 2818–2822.
129. Faure, S., C. Stern, R. Guilard, and P.D. Harvey (2004). Role of the spacer in the singlet–singlet energy transfer mechanism (Förster vs Dexter) in cofacial bisporphyrins. *J. Am. Chem. Soc.* **126**(4), 1253–1261.
130. Chang, C.J., E.A. Baker, B.J. Pistorio, Y. Deng, Z.-H. Loh, S.E. Miller, S.D. Carpenter, and D.G. Nocera (2002). Structural, spectroscopic, and reactivity comparison of xanthene- and dibenzofuran-bridged cofacial bisporphyrins. *Inorg. Chem.* **41**(12), 3102–3109.
131. Bolze, F., C.P. Gros, M. Drouin, E. Espinosa, P.D. Harvey, and R. Guilard (2002). Fine tuning of the photophysical properties of cofacial diporphyrins via the use of different spacers. *J. Organomet. Chem.* **643–644**, 89–97.
132. For an overview, see: Collman, J.P., P.S. Wagenknecht, and J.E. Hutchison (1994). Molecular catalysts for multielectron redox reactions of small molecules: The cofacial metallodiporphyrin approach. *Angew. Chem. Int. Ed. Engl.* **33**, 1537–1554.

133. Chang, C.J., E.A. Baker, B.J. Pistorio, Y.Q. Deng, Z.-H. Loh, S. E. Miller, S.D. Carpenter, and D.G. Nocera (2002). Structural, spectroscopic, and reactivity comparison of xanthene- and dibenzofuran-bridged cofacial bisporphyrins. *Inorg. Chem.* **41**(12), 3102–3109.
134. Pistorio, B.J., C.J. Chang, and D.G. Nocera (2002). A phototriggered molecular-spring for aerobic catalytic oxidation reactions. *J. Am. Chem. Soc.* **124**(27), 7884–7785.
135. Hodgkiss, J.M., C.J. Chang, B.J. Pistorio, and D.G. Nocera (2003). Transient adsorption studies of the pacman effect in spring-loaded diiron(III) μ-oxo bisporphyrins. *Inorg. Chem.* **42**(25), 8270–8277.
136. Bolze, F., M. Drouin, P.D. Harvey, C.P. Gros, E. Espinoza, and R. Guilard (2003). X-ray structures and luminescence properties of Co(II) and Co(III) complexes of cofacial diporphyrins. *J. Porphyrins Phthalocyanines* **7**(7), 474–483.
137. Bolze, F., C.P. Gros, P.D. Harvey, and R. Guilard (2001). Luminescence properties of a cofacial dipalladium porphyrin dimer under argon and in the presence of dioxygen. *J. Porphyrins Phthalocyanines* **5**(7), 569–574.
138. Chang, C.J., Z.-H. Loh, Y. Deng, and D.G. Nocera (2003). The pacman effect: A supramolecular strategy for controlling the excited-state dynamics of pillared cofacial bisporphyrins. *Inorg. Chem.* **42**(25), 8262–8269.
139. Yagi, S., M. Ezoe, I. Yonekura, T. Takagishi, and H. Nakazumi (2003). Diarylurea-linked zinc porphyrin dimer as a dual-mode artificial receptor: Supramolecular control of complexation-facilitated photoinduced electron transfer. *J. Am. Chem. Soc.* **125**(14), 4068–4069.
140. Gunter, M.J., D.C. Hockless, M.R. Johnston, B.W. Skelton, and A. H. White (1994). Self-assembling [2]-catenanes. *J. Am. Chem. Soc.* **116**(11), 4810–4823.
141. Shinmori, H., Y. Yasuda, and A. Osuka (2002). Control of *face-to-face* and extended aggregations of crown ether-appended metalloporphyrins. *Eur. J. Org. Chem.* (7), 1197–1205.
142. Engelkamp, H., S. Middelbeek, and R.J.M. Nolte (1999). Self-assembly of disk-shaped molecules to coiled-coil aggregates with tunable helicity. *Science* **284**(5415), 785–788.
143. Collman, J.P., J.L. Kendall, J.L. Chen, K.A. Collins, and J.-C. Marchon (2000). Formation of charge-transfer complexes from neutral bis(porphyrin) sandwiches. *Inorg. Chem.* **39**(8), 1661–1667.
144. Roth, K.M., J.S. Lindesy, D.F. Bocian, and W.G. Kuhr (2002). Characterization of charge storage in redox-active self-assembled monolayers. *Langmuir* **18**(10), 4030–4040.
145. Roth, K.M., D.T. Gryko, C. Clausen, J.Li, J.S. Lindsey, W.G. Kuhr, and D.F. Bocian (2002). Comparison of electron-transfer and charge-retention characteristics of porphyrin-containing self-assembled monolayers designed for molecular information storage. *J. Phys. Chem. B* **106**(34), 8639–8648.
146. Schweikart, K.H., V.L. Malinowskii, A.A. Yasseri, J. Li, A.B. Lysenko, D.F. Bocian, and J.S. Lindsey (2003). Synthesis and characterization of bis(*S*-acetylthio)-derivitized europium triple-decker monomers and oligomers. *Inorg. Chem.* **42**(23), 7431–7446.
147. Liu, Z., A.A. Yasseri, J.S. Lindsey, and D.F. Bocian (2003). Molecular memories that survive silicon device processing and real-world operation. *Science* **302**(5650), 1543–1545.
148. a) Imahori, H., M. Arimura, T. Hanada, Y. Nishimura, I. Yamazaki, Y. Sakata, and S. Fukuzumi (2001). Photoactive three-dimensional monolayers: Porphyrin-alkanethiolate-stabilized gold clusters. *J. Am. Chem. Soc.* **123**(2), 335–336; b)

Imahori, H. and S. Fukuzumi (2001). Porphyrin monolayer-modified gold clusters as photoactive materials. *Adv. Mater.* **13**(15), 1197–1199.

149. Fantuzzi, G., P. Pengo, R. Gomila, P. Ballester, C.A. Hunter, L. Pasquato, and P. Scrimin (2003). Multivalent recognition of bis- and tris-Zn-porphyrins by *N*-methylimidazole functionalized gold particles. *Chem. Commun.* (8), 1004–1005.
150. Saga, Y. and H. Tamiaki (2004). Facile synthesis of chlorophyll analog possessing a disulfide bond and formation of self-assembled monolayer on gold surface. *J. Photochem. Photobiol. B* **73**(1–2), 29–34.
151. Ohta, N., S. Mikami, Y. Iwaki, M. Tsushima, H. Imahori, K. Tamaki, Y. Sakata, and S. Fukuzumi (2003). Acceleration and deceleration of photoinduced electron transfer rates by an electric field in porphyrin–fullerene dyads. *Chem. Phys. Lett.* **368**(1–2), 230–235.
152. a) Yamada, H., H. Imahori, Y. Nishimura, I. Yamazaki, and S. Fukuzumi (2002). Enhancement of photocurrent generation by ITO electrodes modified chemically with self-assembled monolayers of porphyrin–fullerene dyads. *Adv. Mat.* **14**(12), 892–895; b) Yamada, H., H. Imahori, and S. Fukuzumi (2002). Photocurrent generation using gold modified with self-assembled monolayers of a fullerene–porphyrin dyad. *J. Mater. Chem.* **12**(7), 2034–2040.
153. Yamada, H., H. Imahori, Y. Nishimura, I. Yamazaki, T.K. Ahn, S.K. Kim, D. Kim, and S. Fukuzumi (2003). Photovoltaic properties of self-assembled monolayers of porphyrins and porphyrin–fullerene dyads on ITO and gold surfaces. *J. Am. Chem. Soc.* **125**(30), 9129–9139.
154. a) Fungo, F., L. Otero, C.D. Borsarelli, E.N. Durantini, J.J. Silber, and L. Sereno (2002). Photocurrent generation in thin SnO_2 nanocrystalline semiconductor film electrodes from photoinduced charge-separation state in porphyrin–C_{60} dyad. *J. Phys. Chem. B* **106**(16), 4070–4078; b) Fungo, F., L.A. Otero, L. Sereno, J.J. Silber, and E. Durantini (2000). Synthesis of porphyrin dyads with potential use in solar energy conversion. *J. Mater. Chem.* **10**(3), 645–650.
155. Zilbermann, I., A. Lin, M. Hatzimarinaki, A. Hirsch, and D.M. Guldi (2004). Electrostatically arranged cytochrome c-fullerene photoelectrodes. *Chem. Commun.* (1), 96–97.
156. Imahori, H., H. Norieda, H. Yamada, Y. Nishimura, I. Yamazaki, Y. Sakata, and S. Fukuzumi (2001). Light-harvesting and photocurrent generation by gold electrodes modified with mixed self-assembled monolayers of boron–dipyrrin and ferrocene–porphyrin–fullerene triad. *J. Am. Chem. Soc.* **123**(1), 100–110.
157. Nomoto, A., H. Mitsuoka, H. Oseki, and Y. Kobuke (2003). Porphyrin hetero-dimer as charge separating system for photocurrent generation. *Chem. Commun.* (9), 1074–1075.
158. Nomoto, A. and Y. Kobuke (2002). Photocurrent generation system incorporated with antenna function. *Chem. Commun.* (10), 1104–1105.
159. Ni, Y., R.R. Puthenkovilakom, and Q. Huo (2004). Synthesis and supramolecular self-assembly study of a novel porphyrin molecule in Langmuir and Langmuir–Blodgett films. *Langmuir* **20**(7), 2765–2771.
160. Ogi, T., H. Ohkita, S. Ito, and M. Yamamoto (2002). Preparation of porphyrin mono- and multi-layer films based on the polymer Langmuir–Blodgett method. *Thin Solid Films* **415**, 228–235.
161. Goletti, C., G. Bussetti, P. Chiaradia, R. Paolese, C. Di Natale, E. Mazzone, and A. D'Amico (2001). A reflectance anisotrophy spectroscopy investigation of porphyrin Langmuir–Blodgett films. *Phys. Stat. Sol.* **188**(4), 1339–1344.
162. a) Paolese, R., L. Valli, C. Goletti, C. Di Natale, A. Froiio, A. Macagnano, G. Bussetti, P. Chiaradia, and A. D'Amico (2002). Langmuir–Blodgett films of a modified

tetraphenylporphyrin. *Mater. Sci. Eng. C* **22**(2), 219–225; b) C. Goletti, R. Paolese, C. Di Natale, G. Bussetti, P. Chiaradia, A. Froiio, L. Valli, and A. D'Amico (2002). Optical anisotropy of porphyrin Langmuir–Blodgett films. *Surf. Sci.* **501**, 31–36.
163. Visue, M.I., A.M. Gonçalves da Silva, P. Antunes, and S.M.B. Costa (2002). Organization of cationic porphyrins in mixed Langmuir–Blodgett films. An absorption and steady-state fluorescence study. *Langmuir* **18**(15), 5772–5781.
164. Liu, H.-G., X.-S. Feng, L.-J. Zhang, G.-L. Ji, D.-J. Qian, Y.-I. Lee, and K.-Z. Yang (2003). Influences of hydrophilic and hydrophobic substituents on the organization of supramolecular assemblies of porphyrin derivatives formed at the air/water interface. *Mater. Sci. Eng. C* **23**(5), 585–592.
165. a) Pedrosa, J.M., C.M. Dooling, T.H. Richardson, R.K. Hyde, C.A. Hunter, M.T. Martin, and L. Camacho (2002). Characterization and fast optical response to NO_2 of porphyrin LB films. *Mater. Sci. Eng. C* **22**(2), 433–438; b) Richardson, T.H., C.M. Dooling, O. Worsfold, L.T. Jones, K. Kato, K. Shinbo, F. Kaneko, R. Treggoning, M.O. Vysotsky, and C.A. Hunter (2001). Taking advantage of optical and electrical porperties of organic molecules for gas sensing applications. *Thin Solid Films* **393**, 259–266; c) Richardson, T.H., C.M. Dooling, O. Worsfold, L.T. Jones, K. Kato, K. Shinbo, F. Kaneko, R. Tregonning, M.O. Vysotsky, and C.A. Hunter (2002). Gas sensing properties of porphyrin assemblies prepared using ultra-fast LB deposition. *Colloid Surf. A: Physicochem. Eng. Aspects* **198**, 843–587; d) Pedrosa, J.M., C.M. Dooling, T.H. Richardson, R.K. Hyde, C.A. Hunter, M.T. Martin, and L. Camacho (2002). The optical gas-sensing properties of an asymmetrically substituted porphyrin. *J. Mater. Chem.* **12**(12), 2659–2664.
166. Kato, K., C.M. Dooling, K. Shinbo, T.H. Richardson, F. Kaneko, R. T. Tregonning, M.O. Vysotsky, and C.A. Hunter (2002). Surface plasmon resonance properties and gas response in porphyrin Langmuir–Blodgett films. *Colloid Surf. A: Physicochem. Eng. Aspects* **198–200**, 811–816.
167. a) Arnold, D.P., A. Genga, D. Manno, G. Micocci, A. Serra, A. Tepore, and L. Valli (2002). LB multilayers of highly conjugated porphyrin dimers: Differentiation of properties and behaviour between the free base and the metallated derivatives. *Colloid Surf. A: Physicochem. Eng. Aspects* **198–200**, 897–904; b) Tepore, A., A. Serra, D.P. Arnold, D. Manno, G. Micocci, A. Genga, and L. Valli (2001). Study of gas sensing performances of Langmuir–Blodgett films containing an alkyne-linked conjugated-porphyrin dimer. *Langmuir* **17**(26), 8139–8144.
168. Okita, H., T. Ogi, R. Kinoshita, S. Ito, and M. Yamamoto (2002). Photoinduced electron transfer in nanostructures of ultrathin polyimide films containing porphyrin moieties. *Polymer* **43**(12), 3571–3577.
169. Koo, J.-R., H.-S. Lee, Y. Ha, Y.-H. Choi, and Y.K. Wim (2003). Electrical properties of porphyrin-based switching devices. *Thin Solid Films* **438–439**, 123–127.
170. Tkachenko, N.V., V. Vehmanen, J.-P. Nikkanen, H. Yamada, H. Imahori, S. Fukuzumi, and H. Lemmetyinen (2002). Porphyrin–fullerene dyad with a long linker: Formation of charge transfer conformer in Langmuir–Blodgett film. *Chem. Phys. Lett.* **366**(3–4), 245–252.
171. D.M., Guldi, I. Zilbermann, G.A. Anderson, K. Kordatos, M. Prato, R. Tafuro, and L. Valli (2004). Langmuir–Blodgett and layer-by-layer films of photoactive fullerene–porphyrin dyads. *J. Mater. Chem.* **14**(3), 303–309.
172. Vlachopoulos, N., P. Liska, J. Augustynski, and M. Grätzel (1988). Very efficient visible light energy harvesting and conversion by spectral sensitization of high surface area polycrystalline titanium dioxide films. *J. Am. Chem. Soc.* **110**(4), 1216–1220.
173. Bedja, I., S. Hotchandani, and P. Kamat (1994). Preparation and photoelectrochemical characterization of thin SnO_2 nanocrystalline semiconductor films

and their sensitization with bis(2,2′-bipyridine)2,2′-bipyridine-4,4′-dicarboxylic acid)ruthenium(II) complex. *J. Phys. Chem.* **98**(15), 4133–4140.

174. Isoda, S., Y. Hanazato, K. Akiyama, S. Nishikawa, and S. Ueyama (2003). Photoelectrical properties based on photo-induced electron transfer process in flavin-porphyrin hetero-type Langmuir–Blodgett films. *Thin Solid Films* **441**, 277–283.
175. Ikeda, A., T. Hatano, S. Shinkai, T. Akiyama, and S. Yamada (2001). Efficient photocurrent generation in novel self-assembled multilayers comprised of [60]fullerene-cationic homooxacalix[x]arene inclusion complex and anionic porphyrin polymer. *J. Am. Chem. Soc.* **123**(20), 4855–4856.
176. Imahori, H., H. Yamada, Y. Nishimura, I. Yamazaki, and Y. Sakata (2000). Vectorial multistep electron transfer at the gold electrodes modified with self-assembled monolayers of ferrocene–porphyrin–fullerene triads. *J. Phys. Chem. B* **104**(9), 2099–2108.
177. Qian, D.-J., S.-O. Wenk, C. Nakamura, T. Wakayama, N. Zorin, and J. Miyake (2002). Photoinduced hydrogen evolution by use of porphyrin, EDTA, viologens and hydrogenase in solutions and Langmuir–Blodgett films. *Int. J. Hydrogen Energy* **27**(11–12), 1481–1487.
178. Ostrowski, J.C., K. Susumu, M.R. Robinson, M.J. Therien, and G.C. Bazan (2003). Near-infrared electroluminescent light-emitting devices based on ethyne-bridged porphyrin fluorophores. *Adv. Mater.* **15**(15), 1296–1300.
179. Eichhorn, H. (2000). Mesomeric phthalocyanines, tetraazaporphyrins, porphyrins and triphenylenes as charge-transporting materials. *J. Porphyrins Phthalocyanines* **4**(1), 88–102.
180. Monobe, H., Y. Miyagawa, S. Mima, T. Sugino, K. Uchida, and Y. Shimidzu (2001). Photoconductive porperties of a mesogenic long-chain tetraphenylporphyrin oxovanadium(IV) complex. *Thin Solid Films* **393**, 217–224.
181. Kroeze, J.E., T.J. Savenije, and J.M. Warman (2002). Efficient charge separation in a smooth-TiO_2/palladium–porphyrin bilayer via long-distance triplet-state diffusion. *Adv. Mater.* **14**(23), 1760–1763.
182. Fungo, F., L. Otero, C.D. Borsarelli, E.N. Durantini, J.J. Silber, and L. Sereno (2002). Photocurrent generation in thin SnO_2 nanocrystalline semiconductor film electrodes from photoinduced charge-separation state in porphyrin–C_{60} dyad. *J. Phys. Chem. B* **106**(16), 4070–4078.
183. Hasobe, T., H. Imahori, S. Fukuzumi, and P.V. Kamat (2003). Light energy conversion using mixed molecular nanoclusters, porphyrin and C_{60} cluster films for efficient photocurrent generation. *J. Phys. Chem. B* **107**(44), 12105–12112.
184. Kakkassery, J.J., D.J. Fermin, and H.H. Girault (2002). Photocurrent responses at dye sensitised ultrathin polyelectrolyte multilayers supported on gold electrodes. *Chem. Commun.* (11), 1240–1241.
185. Massari, A.M., R.W. Gurney, M.D. Wightman, C.-H.K. Huang, S.T. Nguyen, and J.T. Hupp (2003). Ultrathin micropatterned porphyrin films assembled via zirconium phosphate chemistry. *Polyhedron* **22**(22), 3065–3072.
186. Tokuhisa, H. and P.T. Hammond (2003). Solid-state photovoltaic thin films using TiO_2, organic dyes, and layer-by-layer polyelectrolyte nanocomposites. *Adv. Funct. Mater.* **13**(11), 831–839.
187. Griveau, S., V. Albin, T. Pauporté, J.H. Zagal, and F. Bedioui (2002). Comparative study of electropolymerized cobalt porphyrin and phthalocyanine based films for the electrochemical activation of thiols. *J. Mater. Chem.* **12**(2), 225–232.

14

Vibrational Spectra and Surface-Enhanced Vibrational Spectra of Azamacrocycles

Marcelo M. Campos Vallette

1. Introduction

This chapter has been organized by considering several aspects. An introduction concerning the relevance of the electronic properties and applications of the azamacrocycles related to surface phenomena as well as the general aspects and characteristics of the vibrational techniques, instruments and surfaces normally used in the study of the adsorbate–surface interaction. The vibrational enhanced Raman and infrared surface spectroscopies, along with the reflection–absorption infrared spectroscopy to the study of the interaction of several azamacrocycles with different metal surfaces are discussed. The analysis of the most recent publications concerning data on bands assignment, normal coordinate analysis, surface-enhanced Raman and infrared spectroscopies, reflection–absorption infrared spectra and theoretical calculations on models of the adsorbate–substrate interaction is performed. Finally, new trends about modified metal surfaces for surface-enhanced vibrational studies of new macrocycles and different molecular systems are commented.

2. Surface Phenomena and Applications of Azamacrocycles

Electro- and photoelectro active materials such as phthalocyanine derivatives, which present remarkable chemical stability, have been intensively explored for their use in thin-film devices[1,2].

Molecular characterization of the film is a fundamental requirement to establish a correlation between the molecular structure, long range molecular

Marcelo M. Campos Vallette • Universidad de Chile, Faculty of Sciences, Department of Chemistry, P.O. Box 653, Santiago, Chile (facien05@uchile.cl)

N_4-Macrocyclic Metal Complexes, edited by José H. Zagal, Fethi Bedioui and Jean-Pol Dodelet. Springer Science+Business Media, Inc., New York, 2006.

organization and its physical properties. In order to search for electrooptic propertics very well characterized surfaces should be prepared: the molecules must be deposited as thin films with a unique molecular stacking and orientation[3].

Vibrational spectra of metalloporphyrins have been studied extensively because of their biological importance as prosthetic groups of a variety of heme proteins. Thus, many review articles have been published on this subject, and most of them discuss the vibrational spectra of metalloporphyrins together with those of heme proteins[4]. Nakamoto has reviewed several articles concerning these macrocycles and discussed the state of the art of the normal coordinate analysis, the assignment of the structure-sensitive vibrations, the metal-macrocycle ring coordination, the metal–metal bonded porphyrins and the π–π complex formation and dimerization[5].

Due to their extended π-electron-delocalized system, porphyrins and naphthalocyanines (H_2Nc) exhibit strong absorption bands in the longer wavelength region (700–900 nm) of semiconductor lasers[6]. This spectral characteristic and their photoconductive properties make these compounds suitable candidates for applications in solar energy conversion, laser electrophotography, electrocatalysis, chemical sensing, organic light emitting diodes (OLED) and nanoscale molecular devices[7–11].

Metallophthalocyanines, MPc, are known for their high thermal stability and very low solubility in any solvents. Transition metal-based phthalocyanines have attracted attention in the areas of fuel cells, gas sensors and biosensors[12,13]. Most of these applications require the use of phthalocyanines in the form of thin films. Further, it is desirable to have ordered, well-packed, and oriented molecular layers of phthalocyanines on various substrates. The metal phthalocyanines constitute an important class of compounds that provide models for theoretical and experimental studies involving electron-mediated processes; one of the main areas of interest is the electrocatalytic reduction of oxygen by metal phthalocyanines[14].

Transition metal complexes incorporating macrocyclic ligands are well-known to represent kinetically inert, thermodynamically stable systems. During the past three decades the field of macrocycles complexes has expanded greatly[15]. There is much interest in the metal complexes of macrocycles due to their diverse potential use, which includes radiometal uptake in cancer therapy[16]. Historically, azamacrocycles of varying ring size and degrees of insaturation were among the first azamacrocycles synthesized[17,18] and a large amount of literature has been gathered on the complexes of these tetra dentate macrocycles, see Figure 14.1. Most of the above mentioned applications of the macrocycles are strongly related to surface phenomena.

3. Surface Vibrational Spectroscopy

Two of the many enhanced optical phenomena in surface-enhanced spectroscopy are surface-enhanced Raman scattering (SERS) and surface-enhanced infrared absorption (SEIRA). These two phenomena and now analytical techniques can be described as a new branch of vibrational spectroscopy that deals with the spectra of molecules on specially fabricated nanostructures with the

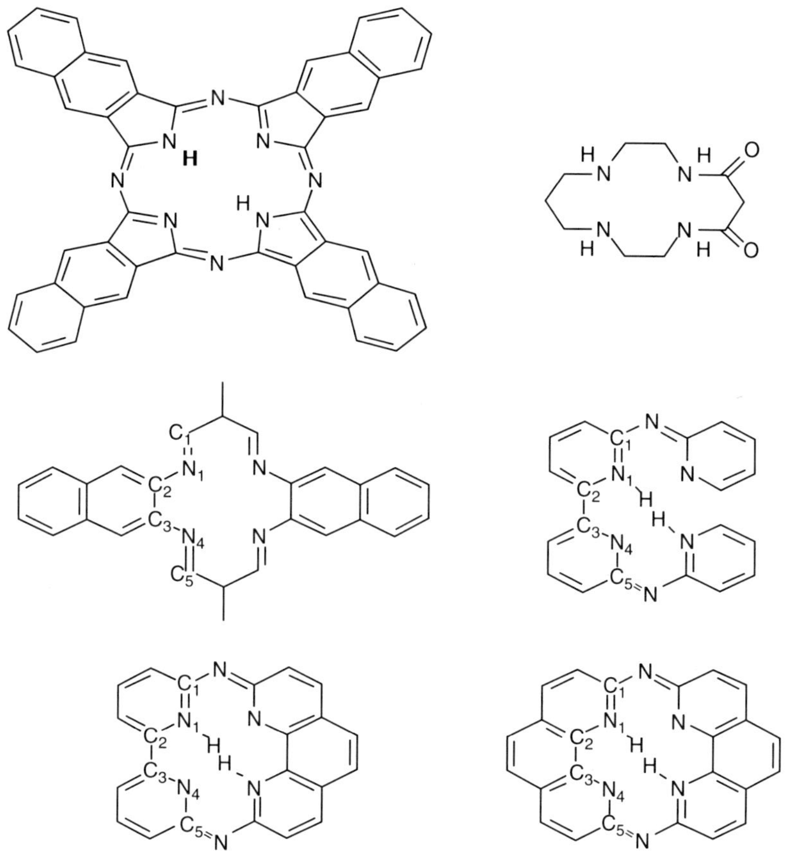

Figure 14.1. Azamacrocycles.

ability to enhance optical signals. This new branch of vibrational spectroscopy, now called surface-enhanced vibrational spectroscopy, (SEVS), is a powerful tool to obtain more information about both the surface phenomena, in particular the charge transfer processes between the adsorbate and the metal surface, and the influence that different surfaces have on the molecular structure, organization and orientation of a series of molecules as the N4-macrocycles.

SEVS is an active area with a broad range of analytical applications. In particular, SEIR and SERS have attracted much attention from the first observation of both phenomena[19,20]. Although these techniques are based on the enormous enhancement of the electromagnetic field occurring in the vicinity of metallic nanoparticles, mainly of Ag and Au, chemical contribution to the total enhancement in both effects, SEIR and SERS, has been also suggested[21,22].

SERS and SEIR have been the basis for the application of the vibrational spectroscopy to the analysis of molecular systems in the range of nanograms to femtograms. They currently are recognized as useful analytical techniques, because of their applicability in the surface chemistry to determine the identity and most probable orientation of molecules adsorbed onto a surface[23,24].

The electromagnetic enhancement of a signal in SERS strongly depends on different physical and physical chemistry characteristics of the surface. Recent investigations that involve silver and gold metals were oriented to obtain the most efficient amplifying surfaces[25].

3.1. Surface-Enhanced Raman Scattering (SERS)

According to the SERS selection rules, the spectral profile of the adsorbate is strongly dependent on the orientation of the main molecular axes with respect to the surface[26]. Thus, SERS intensities would also provide valuable information about the molecular orientation that the adsorbate adopts once adsorbed on the metal surface. For this reason, symmetry assignments are central to the discussion of molecular orientation of adsorbed species on the surface of island or colloidal metal particles. If the molecules, i.e. N4 macrocycles, are oriented face-on the metal surface, with the N atoms face to the metal atoms, the C_4 axis of the molecule and the normal to the surface are parallel; thus, the most symmetric vibrational modes, that derive their intensity from the *zz*-component of the polarisability derivative tensor α_{zz} will be the most intense at the surface plasmon resonance frequency and to the red of that frequency.

Most of SERS studies use the conventional Raman apparatus; the signal enhancement depends on the characteristics of the surface, the nature of the adsorbate, and on the incident wavelength. Colloids and electrodes surface preparation has been largely reported[27,28].

3.2. Surface-Enhanced Infrared Absorption (SEIRA)

The enhancement of vibrational intensities in infrared absorption spectroscopy SEIRA for molecules adsorbed or in close proximity to metal nanostructures, has been recently discussed[29]. In this chapter the differences with SERS are pointed out; the spectral interpretation of the role of the surface selection rules, along with a brief account of the SEIRA experimental conditions, are also included. SEIRA is explained as being the result of the enhanced optical fields at the surface of the particles when illuminated at the surface phonon resonance frequencies; this effect is analogous to plasmon resonance that is the basis for SERS. It has been suggested that electromagnetic and chemical contributions, as in SERS, are the responsible of the observed infrared enhancement[30]. The enhancement factors can be found in SEIRA in the range 10–100 at best. Even this small enhancement compared to SERS (10^{12}), SEIRA can be used to detect monolayers of films. It is less evident than in SERS to infer about the molecular orientation of the molecules on the surface. In general, it can be accepted that the intensity of the normal modes having a permanent dipole moment perpendicular to the incident radiation will be enhanced the most[31].

3.3. Infrared Reflection–Absorption Spectroscopy (IRRAS)

The most probable orientation of the molecules onto a smooth metal surface can be inferred from reflection–absorption infrared spectroscopic IRRAS measurements. The spectra are registered at different incident radiation angles in the range 30–80°. According to Perry *et al.*[32] the incident radiation angle of 90° corresponds to a beam parallel to the surface.

For a KBr pellet, a random orientation of crystallites exists in the solid, and the electric field of the IR radiation, perpendicular to the pellet, will excite both in plane and out of plane modes of the molecule. A solid film with a well defined molecular organization could produce IR spectra with different relative intensities for in plane and out of plane modes. Therefore, molecules arranged face-on to the smooth surface would strongly absorb the incident IR radiation normal to the surface through its plane molecular modes[2].

Thus, the infrared and SEIRA, Raman and SERS and IRRAS measurements allow us to determine the most probable orientation of the molecules onto different metal surfaces, to infer about the influence that these surfaces have onto the molecular structure and to conclude about the mechanism governing the spectral enhancement in IR and Raman.

3.4. Instrumentation

Reflection–absorption infrared spectra (RAIRS) are in general recorded with P-polarized light at different incident radiation angles in the range 30–80°, by using commercial variable angle accessory mounted on the IR optical bench.

Different laser lines for SERS studies in the IR–visible range are used, depending on the electronic characteristics of both the samples and the metal surfaces. To quantify the enhancement factors, the signal-to-noise ratios (S/N) must be calculated as the square roots of the number of scans both for the analyte film on glass and for the analyte film on the roughened metal substrate. The enhancement is then given as the ratio of the band heights corrected for the signal/noise ratios and for the incident laser powers.

Metal island films of controlled mass thickness are deposited in a commercial vacuum system evaporator using a glow discharge evaporation unit. The metal island films are in general fabricated onto preheated glass substrates. The film thickness is monitored using a commercial quartz crystal microbalance.

3.5. Surfaces

Smooth Au, Ag and Cu surfaces of high purity, generally manipulated under nitrogen atmosphere, are mostly used in RAIRS studies. Samples are dissolved in appropriate solvents and then dropped onto the smooth metal surface; the solvent is evaporated under vacuum. Samples can be also deposited by sublimation onto the surface with an adequate control of mass thickness.

Gold, copper and silver surfaces have the *d* band lower than the Fermi level which means that they enhance the vibrational signals. In the case of copper colloidal solutions or metal deposited as islands, the band corresponding to the surface plasmons is observed in the near infrared.[33,34]. The metal colloids are preferentially prepared by following the method reported by Lee and Meisel[35]. Reducing agents normally used are boronhydride, citrate or hydroxylamine[36], depending on the required physical and electronic characteristics of the surface.

Recent investigations that involve Ag and Au metals were oriented to obtain the most efficient amplifying surfaces[25]. Au and Cu electrodes have been surfaces largely used.

Metal island surfaces of controlled mass thickness for the SERS and SEIRA experiments are obtained by sublimation of the metal onto different plates according to general procedures described[37].

Germanium, an IR transparent substrate, is employed as substrate for the Au evaporation due to its higher IR transmission properties between 800 and 400 cm^{-1}; other supporting substrates commonly used are Si, CaF_2, BaF_2, KBr, ZnSe, sapphire, and MgO.

Aliquots of analyte solutions are added to colloidal solutions or deposited onto the metal island surfaces; thin layer spectra are obtained after diluting the sample on the surface using an appropriate solvent. All the spectra are scanned after complete solvent evaporation. Samples for SERS and SEIRA are also prepared by vacuum evaporation of controlled mass thickness.

Deposition rate of the metal is a crucial parameter for the shape and size of the islands; low rates give the best enhancement. Optimal film thickness for maximum enhancement depends on the deposition rate. Control of the substrate temperature during deposition permitted the growth of a film with a highly homogeneous distribution of particle shapes.

SEIRA and SERS are powerful techniques for structural characterization of ultra thin films and well-ordered monolayer on metal surfaces. Thin films at interfaces are prepared by different procedures and developed for various applications[29]. The fabrication and characterization of ultra thin films is a permanent area of research where some of the most interesting subjects are: (a) bilayers and monolayers at liquid–liquid interface, (b) adsorption monolayers and Langmuir (water-insoluble) monolayers at air–water interface, (c) adsorption films and self-assembled monolayers (SAMs) at liquid–solid interface and Langmuir–Blodgett films, cast (deposit) films and spin-coat films at air–solid interface. Studies about molecular organization of monolayers of porphyrins derivatives, of azamacrocycles and their metallic derivatives among the many SEIRA applications to films and interfaces, were published[23,38–40].

Metal films evaporated onto glass can be examined by using molecular spectroscopy and methods such as field emission scanning electron microscopy (FESEM), atomic force microscopy (AFM) and X-ray photoelectron spectroscopy (XPS). The characterization of organic surfaces has been reviewed by Perry and Somorjai[32].

Several techniques that provide information about composition and structure on the molecular level were discussed. For instance, secondary ion mass spectroscopy (SIMS), XPS which provide information about surface composition and the chemical environment and bonding of surface species, and ultraviolet photoelectron spectroscopy (UPS), which probes the density of electronic states in the valence band of materials. Also, the low energy electron diffraction (LEED) and high resolution energy electron loss spectroscopy (HREELS) are electron-scattering techniques that are uniquely suited to yield the structure of the surface monolayer. Raman spectroscopy (RS), Fourier-transform infrared spectroscopy (FTIR), sum frequency generation (SFG), are optical techniques used to study solid–gas interfaces at high pressures along with solid–solid and solid–liquid buried interfaces. Finally, scanning tunnelling microscopy (STM), AFM and surface force measurements (SFM) which provide atomic-scale resolution of surface structure, the forces between molecules at surfaces and, in some cases, measurements of hardness and friction properties of the surface region.

4. Infrared, Raman and Resonant Raman Spectra Analysis of Azamacrocycles

A complete IR and Raman bands assignment is necessary to interpret SEVS data. The following discussion on the bands assignment concerns mainly the macroring fragment of the present series of macrocycles. The spectral assignments reported for the macrocycles hexaazacyclophane, phthalocyanines, naphthalocyanines, azabipiridyl, bis(phenylhydrazine), cyclam and miscellaneous ligand and metal complexes, resulted from the analysis of several publications on macrocycles[2,41–55] and molecules whose structure is similar to the different fragments of the series such as phenantroline derivatives[56–59], and Schiff bases and their metal complexes[60,61]. In this sense, the bands assignment of related ligands and complexes was also reviewed[62,63]. Characteristic group frequencies must be considered[5,64,65].

4.1. Hexaazacyclophane and Its Cu (II) Complex

The mid-IR spectra of the ligand hexaazacyclophane, Figure 14.2, and its Cu(II) complex display a similar shape, which suggests that the structure of the corresponding macrocycles is not much different[66]. Large spectral differences were observed in the far-IR region where the metal-ligand vibrations are expected. Complex formation is mainly supported by the fact that the low energy bands below 450 cm^{-1} are found only in the spectrum of the complex Figure 14.3.

The asymmetric and broad band at 389 cm^{-1} in the Raman spectrum of the complex contains a metal dependent mode, see Figure 14.4. This band is more intense in the spectrum of the complex.

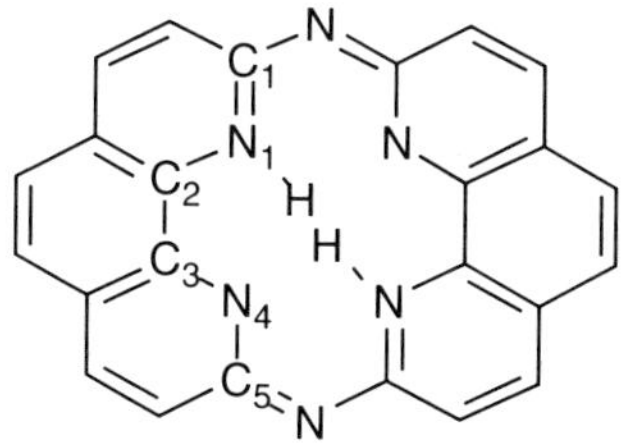

Figure 14.2. Hexaazacyclophane.

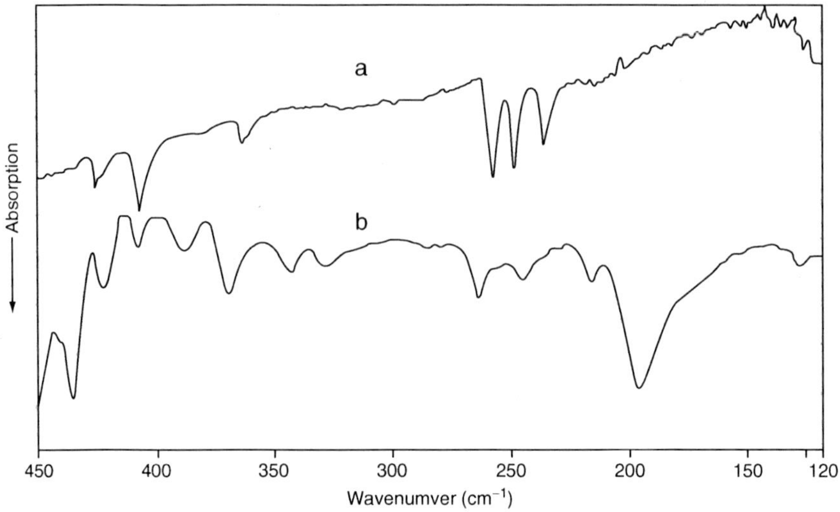

Figure 14.3. Far infrared spectra of (a) the ligand hexaazacyclophane and (b) its Cu(II) complex. Reprinted by permission of Elsevier B.V.

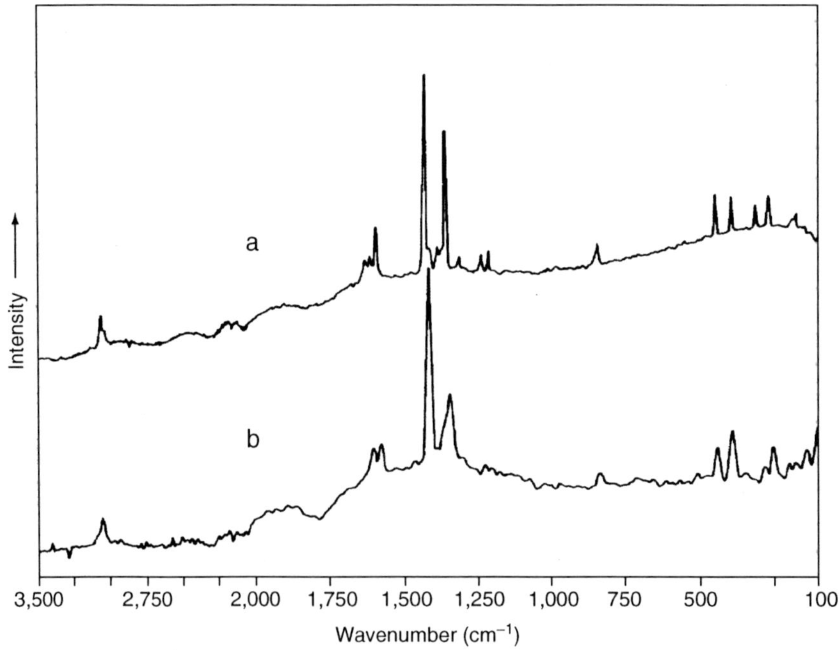

Figure 14.4. Raman spectrum of (a) hexaazacyclophane and (b) the Cu(II) complex in the region 100–3,500 cm^{-1}. Reprinted by permission of Elsevier B.V.

Gobernado-Mitre *et al.*[67] have proposed metal dependent frequencies at 341 and 270 cm^{-1} in the phthalocyanine Cu(II) complex; the band at 236 cm^{-1} was calculated by the same author as a stretching CuN mode (νCuN)[44]. Susi and Ard[54] and Gladkov and Solovyov[53] propose, from normal coordinate calculations performed for the copper porphin complex, that bands near 368, 234 and 206 cm^{-1} contain important contributions of the νCuN mode. From a vibrational study on the non-macrocycle phenantroline copper complexes it has been proposed the νCuN modes near 410 and 288 cm^{-1} [68]. In bipyridine complexes of Cu(II), the νCuN mode is proposed at 297 cm^{-1}[143]; a similar assignment is proposed in bipyrimidine complexes[69].

The νCC and νCN modes are highly coupled in molecules with extended π electronic system. The energy of the corresponding CN bonds near or directly involved in the metal coordination (CuN) are influenced by metal complexation. Then, the identification of the νCuN modes is of paramount importance to characterize the energy of the coordination site[70].

Some infrared bands in the 1,600–1,200 cm^{-1} region have participation of νCN and νCC modes[68]. Various bands of the coordination site are observed at the same frequency in the spectrum of the ligand and complex. Other bands shift to low or high energy regions by complexing. This effect is characteristic of an electronic redistribution in aromatic systems. A spectral shifting to lower energy by metal complexation is characteristic of Schiff base ligands[60,61,71]. The same sense of the shifting by complexation was observed for the bands at 1,502 and 1,383 cm^{-1} which were attributed to νCN modes of the phenanthroline ring[66].

Most of the remaining bands in the region are assigned to νCC modes. Bands slightly influenced by complexing, are due to CC bonds far from the coordination site. The band at 1,275 cm^{-1} is attributed to a νCN mode of the phen-N-phen bridge; the energy of this mode increases by complexing, suggesting that the corresponding bond does not participate in the complex formation. The macrocycle breathing mode, ascribed to a band near 640 cm^{-1}[150,51], displays an energy increase upon complexation; thus, an electronic change of all the macrocycle ring bonds is verified. Differences in the spectra of the ligand and complex in the regions of the macrocycle ring deformations, 480–400, 340 and 260–120 cm^{-1}, are evidences of an electronic redistribution upon complexation.

4.2. Phthalocyanines and Naphthalocyanines and Their Metal Complexes

The A_{2g} in plane vibrations of metallophthalocyanines (MPc), become Raman active, similar to metalloporphyrins, under resonance conditions. Several infrared and Raman spectra of MPc and MPc containing axial ligands have been reported by many investigators[44,72]. Hutchinson *et al.*[73] assigned the M–N stretching modes of different metal isotopes of MPc. The IR active νZnN (E_g) vibration in ZnPc was observed at 329 cm^{-1} and the observed ZnN distance is 1.980 Å[74].

Vibrational studies on naphthalocyanines (Nc) (see Figure 14.5) compounds are scarce. Kaplan *et al.* reported the IR energy of the most intense absorptions

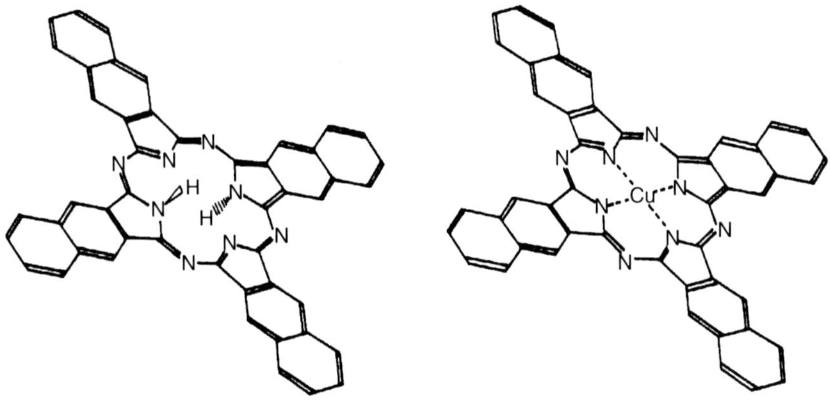

Figure 14.5. Naphthalocyanines. Reprinted by permission of Elsevier B.V.

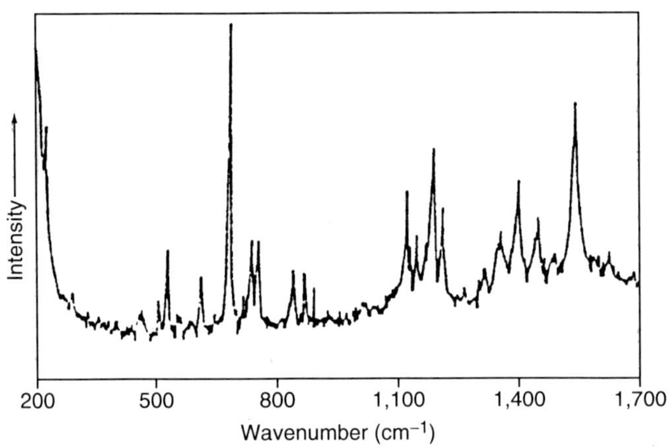

Figure 14.6. Resonance Raman Spectrum of CoNc in the regions 200–1,700 cm^{-1}.

of some Nc metal complexes[75]. Gobernado-Mitre *et al.*[76] assigned the medium-intensity Raman line at 223 cm^{-1} to a deformation with a large νCu–N contribution in CuNc.

A vibrational interpretation of the structure of the solid naphthalocyanines ligand (H_2Nc) and their Co(II), VO^{2+} and Pb(II) complexes was published[70]; the metal-macrocycle stretching mode was observed at 330, 308 and 273 cm^{-1}, respectively. The Co and VO ions slightly disturb the structure of the Nc moiety, constraining the whole system to adopt a deformed warped stereochemistry in the same way that in MPc macrocycles[77]. The collective vibration of the macrocycle (breathing) observed with a strong relative intensity at 680 cm^{-1} in the Raman spectra is one of the most characteristic bands of the Nc complexes, see Figure 14.6.

The spectral characteristics of this mode are not much different within the series Co(II) VO(II) and Pb(II), including the CuNc molecule (681 cm^{-1} and strong relative intensity)[76]. The macrocycle ring structure in that series is similar.

4.3. Vanadylphthalocyanine and Vanadylnaphthalocyanine

The bands assignment of phtalocyanines (H_2Pc) was performed in[50–52], whilst that from the naphtalocyanine (H_2NPc) spectra was performed by Clavijo *et al.*[70].

Spectral differences between the spectrum of vanadylnaphthalocyanine NPcVO and vanadylphthalocyanine PcVO[78], are mainly due to the different nature of the aromatic moieties. Infrared bands at 1,523, 1,476 and 1,461 cm^{-1} of PcVO belong to the pyrrole-benzene fragment, while the bands at 1,509 cm^{-1} of NpcVO and 1,499 cm^{-1} of PcVO are pyrrole stretch modes. Absorptions at 1,417 and 1,400 cm^{-1} of PcVO are νCN modes.

The band observed only in the spectrum of NPcVO at 1,342 cm^{-1} is associated to a naphthalocyanine νCC mode[78]. The νCC bands observed in both spectra in the region 1,330–1,260 cm^{-1} correspond to CC bonds far from the VO coordinating site. δCH deformations of the NPcVO and PcVO are observed in the region 1,200–1,000 cm^{-1}. The NPcVO band at 997 cm^{-1} corresponds to the νVO mode.

Macrocycle deformations are observed between 640 and 120 cm^{-1} in NPcVO and in the range 510–350 cm^{-1} in PcVO; the highest frequencies correspond to the ring deformations of benzene and pyrrole, while the lowest are macrocycle ring deformations[78]. Bands at 570 and 457 cm^{-1} in PcVO and the band at 537 cm^{-1} of NPcVO are characteristic of each macrocycle. Bands at 372, 308 and 235 cm^{-1} observed only in the vanadyl compounds correspond to VO–N vibrations. Other bands in this region are mainly due to macrocycle ring out of plane deformations.

4.4. Vanadyl naphthalocyanine and Vanadyl Porphine Phenyl Substituted

The infrared spectra of vanadyl-5,14,23,32-tetraphenyl-2,3-naphthalocyanine (VOTPNPc) and vanadyl-5,10,15,20-tetraphenyl-21H,31H-porphine (VOTPPORP) displayed in Figures 14.7 and 14.8 were reported by Carrasco *et al.*[78].

The bands at 1,509 and 1,365 cm^{-1} in VOTPNPc and at 1,421 and 1,337 cm^{-1} in VOTPPORP were ascribed to pyrrole fragment modes. The VO stretching modes were assigned to the bands at 944 and 919 cm^{-1} in VOTPNPc and at 968 and 922 cm^{-1} in VOTPPORP. The assignment of this mode is particularly important as the vanadyl group may be involved in an eventual substrate–adsorbate interaction. The macrocycle deformations at 527 and 507 cm^{-1} in the spectrum of VOTPPORP involve the VO group.

Some bands are observed only in one of the two tetraphenyl substituted complexes: this is the case for the band at 954 cm^{-1} which is assigned to a breathing mode of the VOTPNPc, the band at 802 cm^{-1} in VOTPPORP assigned to an out of plane CH mode of the pyrrole ring, the band at 791 cm^{-1} corresponding to the macrocycle deformation, and the ring porphine bands at 702, 662 and 575 cm^{-1}.

Important spectral changes are verified in the Raman spectra of solid VOTPNPc when passing from 514.5 to 633 nm laser lines[78] see Figure 14.9.

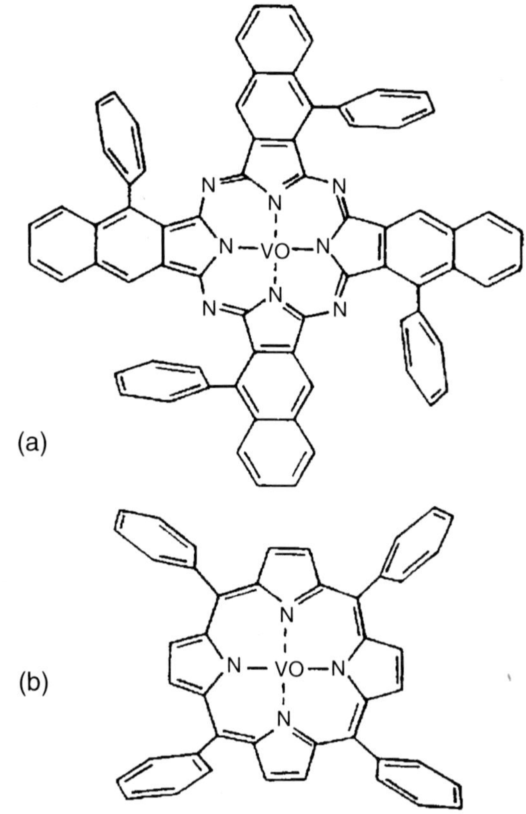

Figure 14.7. (a) Vanadyl-5,14,23,32-tetraphenyl-2,3-naphthalocyanine and (b) vanadyl-5,10,15,20-tetraphenyl-21H,31H-porphine. Reprinted by permission of Elsevier B.V.

The resonant effect, which concerns mainly vibrational modes belonging to the coordination site, is manifested by changes in the relative intensity of several bands. The band about 1,600 cm^{-1} decreases its intensity and the bands around 1,300, 900, 700 and at 340 and 298 cm^{-1} increase their intensities. Other bands at 1,367, 1,222, 1,006, 839 and 650 cm^{-1} maintain or slightly decrease their relative intensities. The intensity decreasing of the coupled νCC/νCN band at 1,600 cm^{-1} indicates that this mode is an asymmetric vibration with an important νCC component of the pendant group. The spectral behavior of the bands which increase their intensity indicates that the corresponding modes are symmetric vibrations.

The νVO modes are clearly observed by resonance effect near 900 cm^{-1}[78]. Some ρCH bands near 700 cm^{-1} are enhanced and other decrease in intensity which suggests that they could be assigned to different CH of distinct fragments of the macrocycle. The resonance effect can be explained by assuming a coupling of these modes to breathing motions as observed in azabipiridyl complexes[23]. The ring deformation near 320 cm^{-1} involves the molecular ring fragment close to the chromophore, the coordination site. Bands, which remain with about the

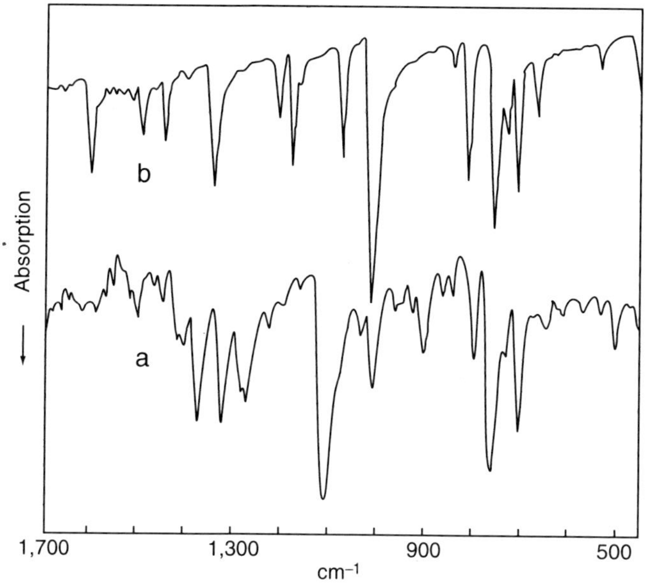

Figure 14.8. Infrared spectra of the solids (a) vanadyl-5,14,23,32-tetraphenyl-2-3-naphthalocyanine and (b) vanadyl-5,10,15,20-tetraphenyl-21H,31H-porphine, in the range 1,700–450 cm^{-1}. Reprinted by permission of Elsevier B.V.

same relative intensity when changing the laser lines, belong to modes vibrating far from the coordination site.

4.5. Azabipiridyl and its Ni(II), Cu(II) and Zn(II) Complexes

The metal Ni(II), Cu(II) and Zn(II) complex formation of azabipiridyl was followed by the analysis of their infrared spectra[79] and the assignment of the metal–ligand bands was based on several publications on related compounds (Figure 14.10).

The νNiN vibrations were assigned in 2,6-diacetylpyridine dioxime at 416, 341 and 276 cm^{-1}, and at 370 and 265 cm^{-1} in hexaamine complexes, at 410 and 334 cm^{-1} in triethylendiamine derivatives, around 240 cm^{-1} in pyridine derivatives and in imidazol complexes between 325 and 210 cm^{-1} [5]. Gobernado-Mitre *et al.*[76] attributed the νCuN mode in copper complex of naphthalocyanine to those bands observed at 341 and 221 cm^{-1}. In Cu(II) tri-azamacrocycles it has been proposed the bands at 383 and 314 cm^{-1} as due to νCuN[42]. This mode was observed in copper complexes of cyclam at 437 cm^{-1} [41] and in Cu(II) hexaazacyclophane at 390 and 280 cm^{-1} [66]. In phenantroline Cu(II) complexes the νCuN vibration was identified with the bands at 300 and 430 cm^{-1} [68]. In the case of the Zn complexes, some tetraamine derivatives display the νZnN band at 432cm^{-1}, and in triethylendiamine complexes were observed at 405 and 291 cm^{-1}; in imidazole complexes the νZnN mode was assigned to the bands between 325 and 210 cm^{-1} [5]. The νZnN mode of bis(phenylhydrazine)-1, 10-phenantroline Zn(II) was attributed to the bands at 376 and 267 cm^{-1} [55].

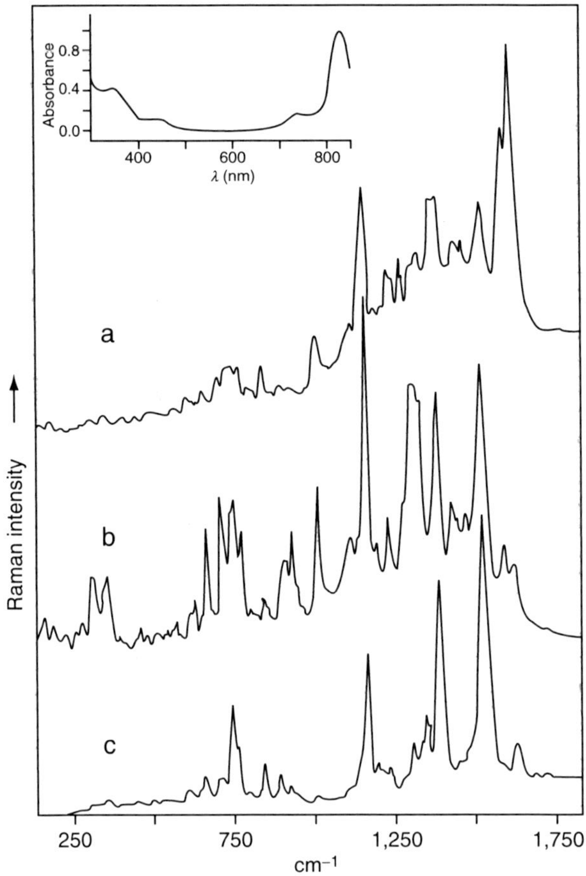

Figure 14.9. Raman spectra of the solid vanadyl-5,14,23,32-tetraphenyl-2,3-naphthalocyanine at: (a) 514.5, (b) 633 and (c) 1064 nm. The electronic spectrum is displayed in the upper left-hand side. Reprinted by permission of Elsevier B.V.

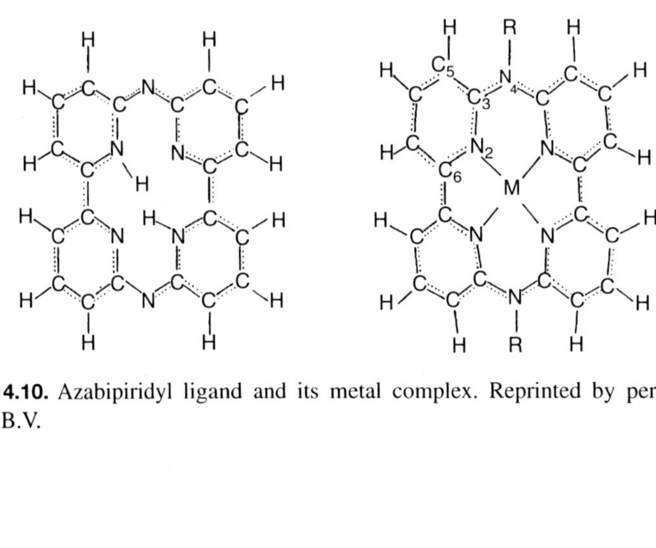

Figure 14.10. Azabipiridyl ligand and its metal complex. Reprinted by permission of Elsevier B.V.

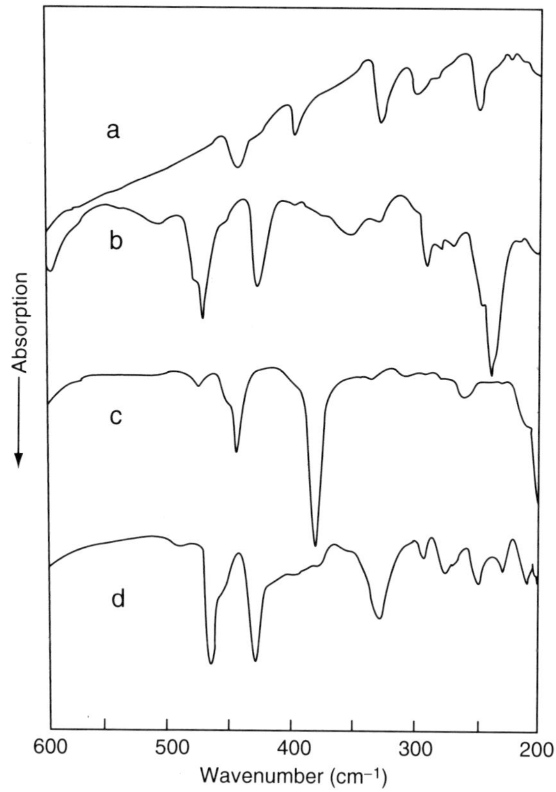

Figure 14.11. Infrared spectra of (a) azabipiridyl macrocycle ligand, and (b) the Ni, (c) Cu, and (d) Zn complexes in the range of 600–200 cm^{-1}. Reprinted by permission of Elsevier B.V.

In different phthalocyanine complexes of metal isotopes, the νM–N mode was observed at 376 and 317.8 cm^{-1} (Ni), 284 cm^{-1} (Cu) and at 240.7 cm^{-1} (Zn)[73]. On this basis, some bands below 500 cm^{-1} were assigned to the metal–ligand coordination[79], see Figure 14.11.

Bands at 426, 380 and 326 cm^{-1} were ascribed to νNi–N, νCu–N and νZn–N, respectively. Bands observed at 280 and 275 cm^{-1} in the spectrum of the Ni and Zn complexes are probably due to the same type modes.

An important spectral shift to lower energy when increasing the metal mass is observed. This is related to the relative stability of the complex formation, the size of the metal ions and basically to the electronic ion configuration. The most important difference between the spectra of the ligand and the complexes is observed in the case of the Ni complex[79]. The highest frequency for the Ni–N coordination indicates the highest relative stability in the series. The strong shift to lower energy is a consequence of different structural characteristics of the corresponding coordination sites. Thus, it has been proposed that the Ni ion should be accommodated into the coordination site plane while the Zn ion should be preferently situated out of this plane. This proposition was supported by the fact that the

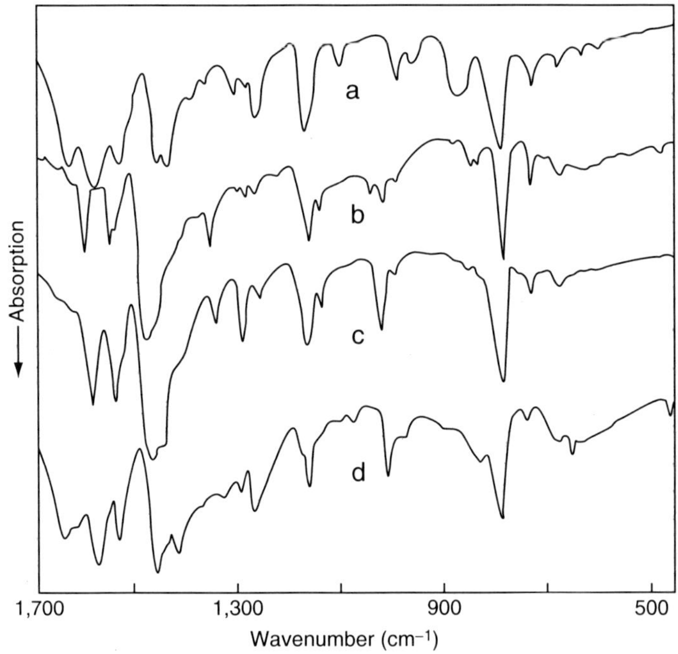

Figure 14.12. Infrared spectra of (a) azabipiridyl macrocycle ligand, and (b) the Ni(II), (c) Cu(II), and (d) Zn(II) complexes in the range of 1,700–450 cm^{-1}. Reprinted by permission of Elsevier B.V.

whole spectrum of the Zn complex is similar to that of the ligand, were the two hydrogen atoms in the macrocycle ring are decidedly out of the plane, see Figure 14.12. The spectral characteristics of the copper complex suggest that the copper cation could be situated slightly out of the coordination plane.

The resonance Raman (RR) spectra of the azabipiridyl macrocycle ligand and it Co(II), Ni(II) and Cu(II) complexes was studied[23]. The FT-Raman spectra at 1,064 nm of the macrocycles in the solid state show remarkable changes with regard to those obtained when exciting at 514 nm. The spectra of the ligand and Co complex are displayed in Figure 14.13. The selective intensity enhancement is a consequence of the RR effect of the vibrational modes of the chromophore[80]. In going from 1,064 to 514.5 nm a decreasing of the relative intensity of the band below 200 cm^{-1}, was observed. This band was assigned to an asymmetric deformation which is consistent with the decrease in intensity observed. The νM–N mode at about 390 cm^{-1} is observed with a low relative intensity; thus, this is an asymmetric mode. The other coupled modes in this region also display a lower relative intensity in the 514.5 nm spectra. The symmetric nature of the ligand macrocycle breathing near 680 cm^{-1} is concluded from the great intensity enhancement observed.

A RR enhancement was observed for at least two bands at about 1,180 cm^{-1}, assigned to δCH, coupled to a pyridinic breathing mode. The vibrational coupling explains the observed enhancement; thus, this mode should be basically

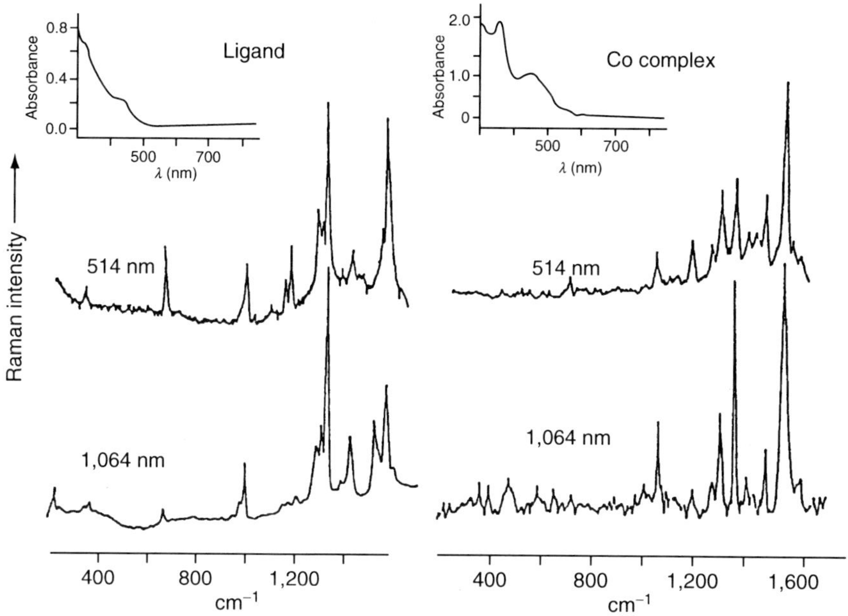

Figure 14.13. The Raman spectra of the ligand azabipiridyl and its Co(II) complex. In the left-up-side is the corresponding absorption spectrum. Reprinted by permission of Elsevier B.V.

symmetric. A great intensity decreasing of the strong band in the region 1,300–1,600 cm^{-1} at about 1,340 cm^{-1} is observed by resonance effect. The corresponding coupled νCC and νCN mode should be asymmetric. The band at about 1,560 cm^{-1} contains at least two components; the relative intensity of one of these components decreases, while the rest increases by changing the excitation laser line. Thus, the corresponding coupled νCC and νCN vibration should correspond to the asymmetric and symmetric modes, respectively. On changing the excitation wavelength from 514.5 to 1,064 nm, a relative decrease of the relative intensity of the bands at about 1,540, 1,464 and 1,130 cm^{-1} is observed. This effect suggests symmetric A_g modes. No evident influences of the excitation laser line variation on bands is consistent with internal modes far from the chromophore.

4.6. Bis(phenylhydrazine)-1,10-phenanthroline and its Co(II), Ni(II), Cu(II) and Zn(II) Complexes

The most relevant infrared spectral features due to the metal complexation of the bis(phenylhydrazine)-1,10-phenanthroline ligand (PHP), Figure 14.14, occur in the 450–100 cm^{-1} region[81]. In this region have been reported most of the coordination ring vibrations directly related to the metal.

The complex formation is verified in this spectral region by the appearance of several bands, see Figure 14.15. Bands about 430, 350 and 325 cm^{-1} only observed in the complexes are ascribed to metal sensitive vibrations; they are macrocycle ring deformations. The band at about 291 cm^{-1} in the spectrum of

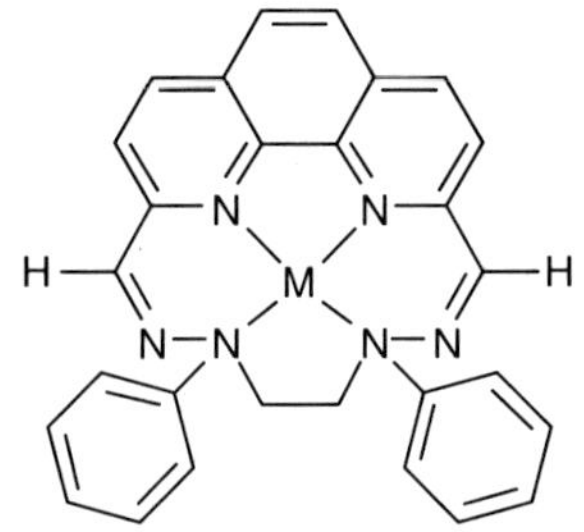

Figure 14.14. Bis(phenylhydrazine)-1,10-phenanthroline metal complex.

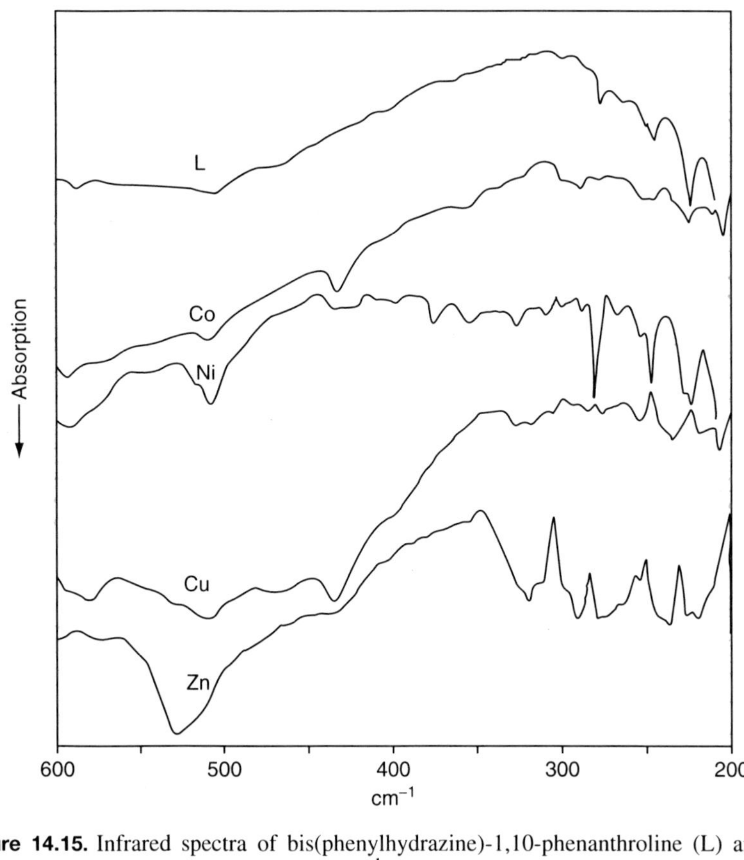

Figure 14.15. Infrared spectra of bis(phenylhydrazine)-1,10-phenanthroline (L) and its metal complexes in the region 600–200 cm^{-1}. Reprinted by permission of Elsevier B.V.

the Co complex was assigned to a νCoN mode following reported data on Schiff base complexes[60]. The band of the Ni complex at 435 cm^{-1} was ascribed to a νNiN mode following the assignment proposed in azabipiridyl and Schiff base Ni complexes[79,82]. Two bands were identified as νCuN modes in the spectrum of the Cu complex at 326 and 283 cm^{-1}. Data of azabipiridyl complexes[79], suggest that the Zn–N modes should be expected between 320 and 275 cm^{-1}. The observed

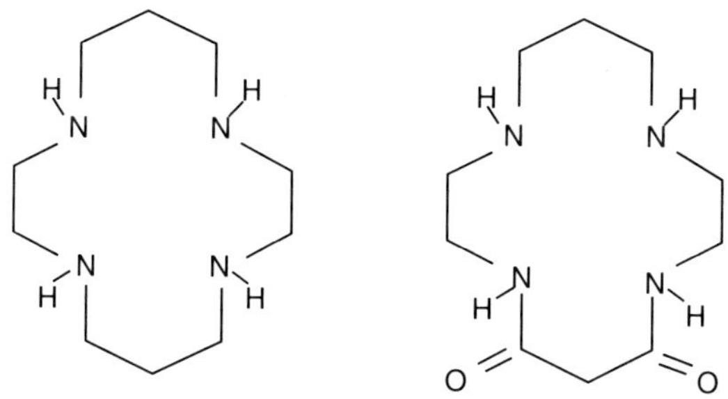

Figure 14.16. 1,4,8,11-tetraazacyclotetradecane, cyclam and 5,7-dioxo-1,4,8,11-tetraazacyclotetradecane, cyclamdione. Reprinted by permission of Elsevier B.V.

frequency shift to lower energy of the M–N bands at 291 and 283 cm^{-1} for the Co and Cu, is in agreement with the mass increasing.

Concerning the phenyl bands, in this region appears the *w* out-of-plane ring deformation. It is derived from the e_{2u} torsional mode of benzene (404 cm^{-1}) and exhibit a very weak intensity in the infrared spectra. This absorption has been estimated from combination bands at about 400 cm^{-1} for monosubstituted benzenes[83]. The *u* X-sensitive mode, which is related essentially to the Ph-hydrazine bending, appears at about 330 cm^{-1}. Finally, at about 300 cm^{-1}, it is expected the *x* X-sensitive mode, which is another out-of-plane ring deformation also with participation of the Ph-hydrazine bending. These bands are easily identified as giving rise to bands in both the infrared and Raman spectra, as is the case of the very weak band observed in all the spectra at about 300 cm^{-1}. Bands below 300 cm^{-1} are assigned to deformations of the coordination site.

4.7. Cyclam and Cyclamdione and Their Cu(II) Complexes

The IR spectra of the ligands 1,4,8,11-tetraazacyclotetradecane, cyclam and 5,7-dioxo-1,4,8,11-tetraazacyclotetradecane, cyclamdione (Figure 14.16), and their Cu(II) complexes, were exhaustively analyzed by Diaz *et al.*[41].

In the case of cyclam, the bands at 437 and 427 cm^{-1}, and 231 cm^{-1} observed only in the spectrum of complex were ascribed to νCuN modes and to a deformation involving the metal atom, respectively. The frequencies assigned to νCuN modes are in the range of those reported for νCrN in the complex Cr(III) 1,4,8,12-tetraazacyclopentadecane[84]. The energy of the νCuN modes is relatively higher than that observed in Cu (II) acyclic amines complexes, which is a consequence of the observed high stability of the macrocyclic metal complexes[15,85,86]. Other bands at 337, 210 and 188 cm^{-1}, observed only in the spectrum of the ligand, are macrocycle ring deformations involving the N atoms. Bands at 272 and 261 cm^{-1} in the ligand, shift to lower frequencies by complexation. This spectral behaviour is a consequence of a modified structure of the chelate ring. Intense bands below 170 cm^{-1} were ascribed to lattice vibrations, see Figure 14.17.

The νCO vibrations are observed in the ligand of cyclamdione at 1,672 and 1,562 cm^{-1} [41]. By complexing, these bands collapse in one asymmetric band at 1,588 cm^{-1} pointing out the equivalence of the two CO groups in the complex.

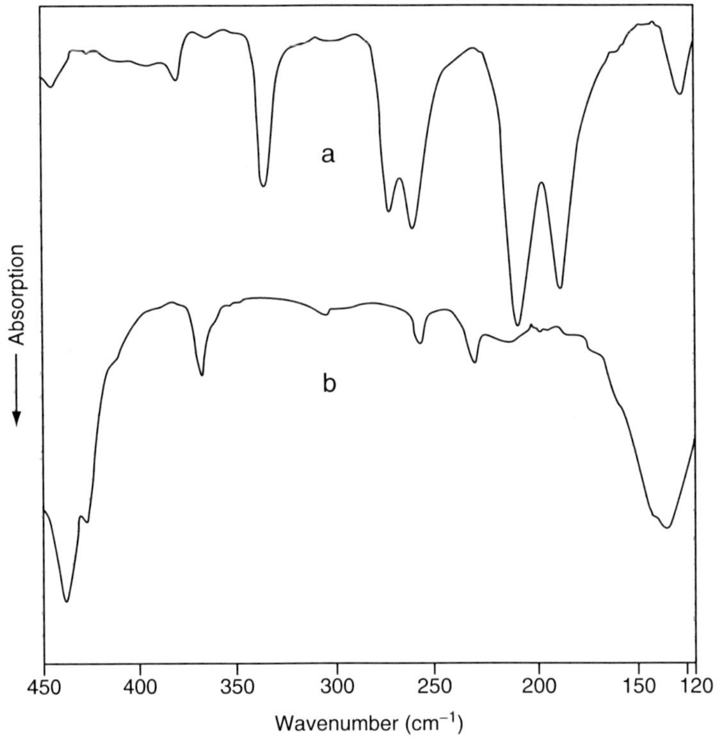

Figure 14.17. (a) Far-infrared spectrum of cyclam and (b) Cu(II) cyclam, in the region 450–120 cm^{-1}. Reprinted by permission of Elsevier B.V.

This situation represents an electronic redistribution on this fragment imposed by the loss of the amide hydrogens in the complex formation; this fragment adopts a less non-planar conformation. Bands in the region 650–500 cm^{-1} were assigned to CO vibrations. A CO mode has been assigned between 500 and 700 cm^{-1} in cyclohexanone[87] and, in the same range, although strongly coupled with ring deformations in the bis(glycinamido) Cu(II) monohydrate complex[88]. Between 450 and 250 cm^{-1} the spectrum of the complex displays more bands than in the ligand, see Figure 14.18. Common bands at 402, 373, 339, 273 and 285 cm^{-1} were assigned to macrocycle ring deformations without participation of the metal. Three bands at 430, 388 and 334 cm^{-1} are observed only in the spectrum of the complex and were assigned to metal-ligand vibrations. The band at 430 cm^{-1} is the most important participant of the νCuN_{amide} mode. The broad band at 305 cm^{-1} displayed only in the spectrum of the ligand is assigned to its coordination site. NCuN vibrations are expected in the region 250–160 cm^{-1}.

4.8. Dinaphthalenic Ni(II) and Cu(II) Azamacrocycle Complexes Methyl and Phenyl Substituted

The infrared and Raman spectra of dinaphthalenic Ni(II) and Cu(II) azamacrocycle complexes methyl and phenyl substituted, Figure 14.19, with different metal surfaces were analyzed[38].

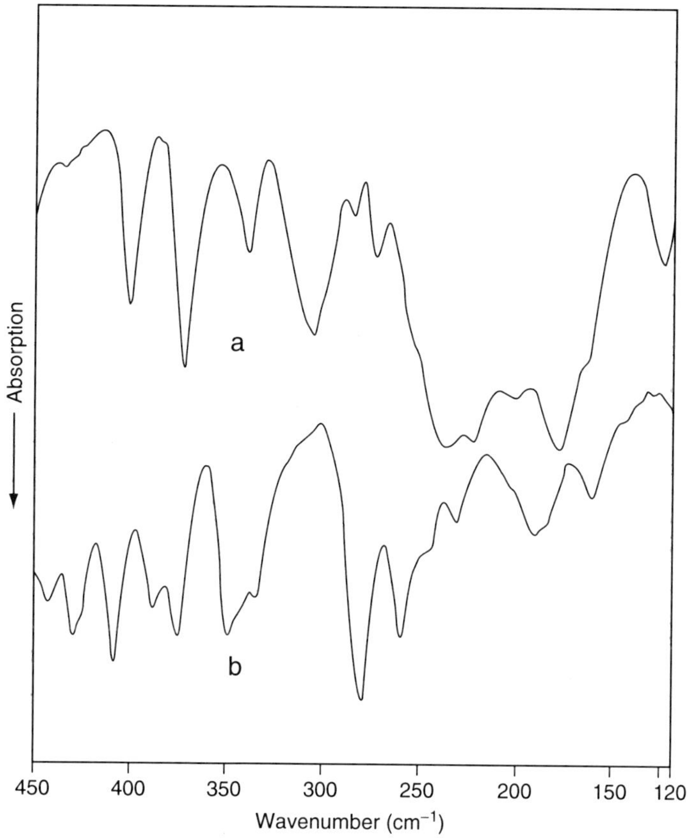

Figure 14.18. (a) Far-infrared spectrum of cyclamdione and (b) Cu(II) cyclamdione in the 450–120 cm^{-1} region. Reprinted by permission of Elsevier B.V.

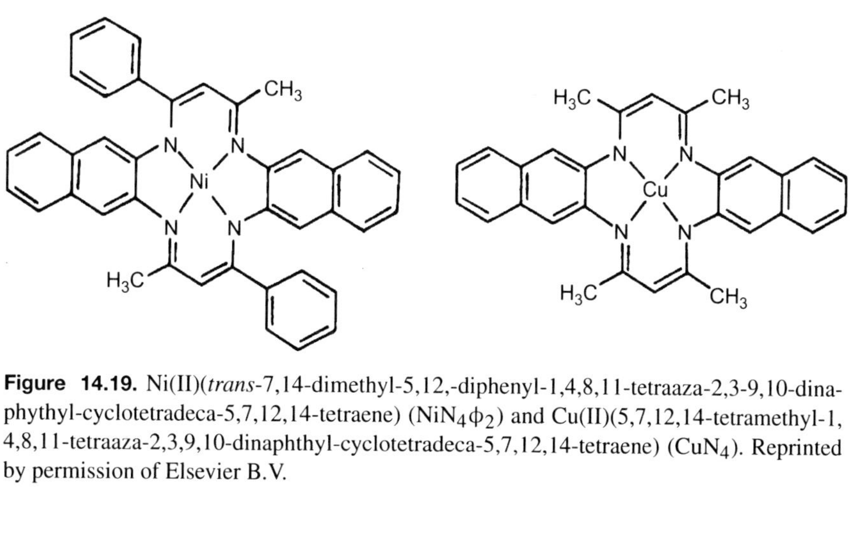

Figure 14.19. Ni(II)(*trans*-7,14-dimethyl-5,12,-diphenyl-1,4,8,11-tetraaza-2,3-9,10-dinaphythyl-cyclotetradeca-5,7,12,14-tetraene) ($NiN_4\phi_2$) and Cu(II)(5,7,12,14-tetramethyl-1,4,8,11-tetraaza-2,3,9,10-dinaphthyl-cyclotetradeca-5,7,12,14-tetraene) (CuN_4). Reprinted by permission of Elsevier B.V.

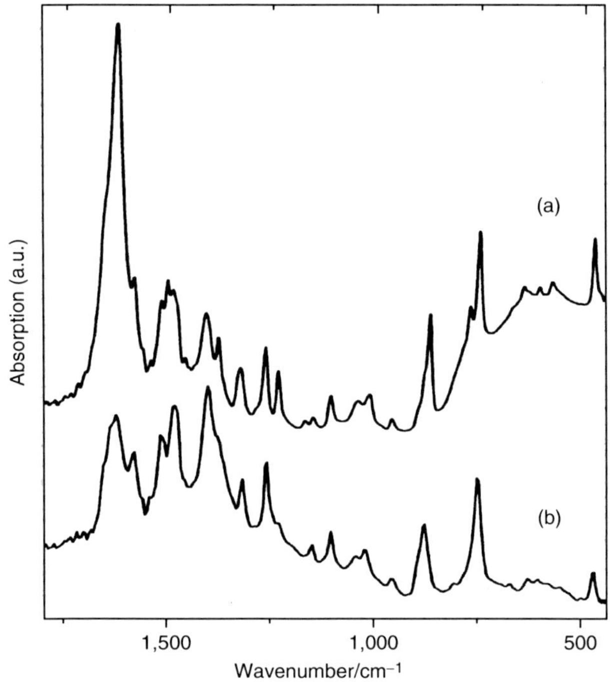

Figure 14.20. Infrared spectra of (a) $NiN_4\phi_2$; (b) CuN_4 in KBr in the 1,800–450 cm^{-1} range. Reprinted by permission of Elsevier B.V.

The IR spectra of the complexes in the middle region are shown in Figure 14.20. The spectral pattern of both complexes is different, but frequencies are very similar, suggesting that the structure of the macrocycles is not much different. The higher absorption intensity in the 1,450–1,650 cm^{-1} region in the case of the Ni complex, is explained by the presence of the phenyl groups, while the small frequency shifts are attributed to the different effect of the cation in each complex. δCH naphthyl modes observed at 1,014 cm^{-1} in Ni and at 1,023 cm^{-1} in the Cu complex correspond to C–H moieties closer to the metal coordination site; they display a large frequency variation due to the steric effect caused by the different structural characteristics of the methyl and phenyl substituents and the different metals in each complex.

The energy difference of bands at 644, 577 and 409 cm^{-1} in Ni and at 630, 552 and 420 cm^{-1} in the Cu complex, indicates that the corresponding vibrations belong to modes close related to the coordination site. That is not the case for bands at 606, 473 and 347 cm^{-1} in Ni and at 607, 475 and 342 cm^{-1} in the Cu complex, which correspond to ring modes far from the coordination site.

5. Normal Coordinate Calculations

The spectral assignment of vibrational bands in any molecular system, and in particular the assignment of the metal–ligand vibrations as well as the

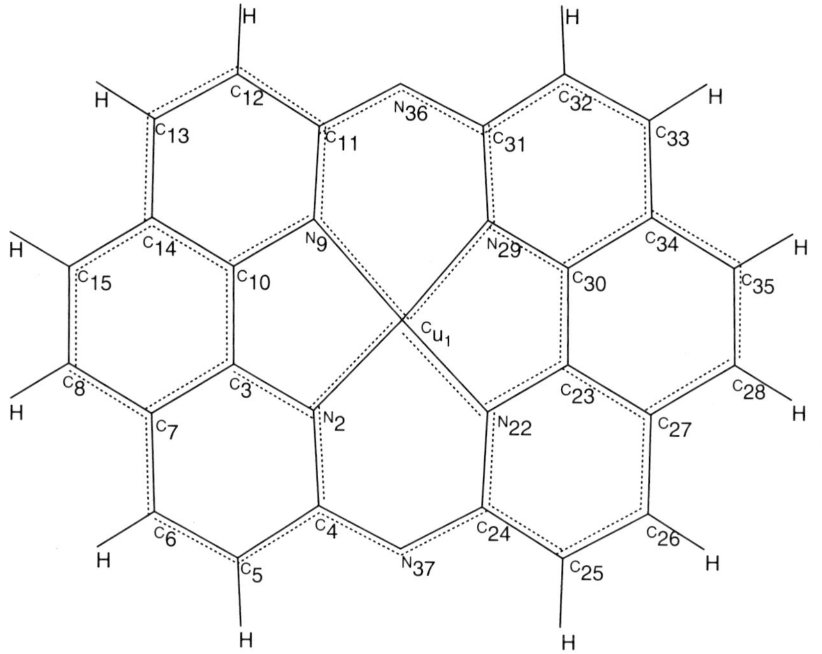

Figure 14.21. Molecular structure and atoms numbering of the hexaaza-cyclophane Cu(II) complex. Reprinted by permission of Elsevier B.V.

participation of other macrocycle ring modes in the metal complexation of the azamacrocycles must be completed by a normal coordinate analysis.

5.1. Hexaazacyclophane and Its Cu(II) Complex

A normal coordinate analysis (NCA) for hexazacyclophane Cu(II) complex was performed[66] by using the Wilson's force field geometrical matrix (FG) method[89]. An INDO/1 optimized molecular geometry was used to build the G matrix[90]. The CuN distance is 1.84 Å for the copper atom in the macrocycle plane; this bond length is relatively short in comparison to that reported for the non-macrocycle tris-(1,10-phenantroline)Cu(II) (2.1 Å)[91], the planar macrocycles copper porphin (2.031 Å)[53] and tetraazacyclotetradecane Cu(II) derivatives (2.08 Å)[92]. However, the CuN bond length of the present complex is not much different from that reported for metal phthalocyanine complexes (1.80 Å)[52]. A set of 113 internal force constants (F) was used in the calculation[66]. 77 non-diagonal terms (f), which represent the most relevant interactions, were also included. The valence force field was constructed using data given in the literature for the non-macrocycle bis-(1,10-phenantroline)Cu(II)[68], iron phthalocyanine[93], copper porphin[53,54] and benzene derivatives[94]. The analysis was performed by using a computer program developed for large molecules[95]. The force constants are listed in Table 14.1. Atoms numbering is given in Figure 14.21.

Table 14.1. Force Constants[a] for Hexaazacyclophane Cu(II) Complex

Diagonal	Non-diagonal
CuN = 1.0	CuN2, CuN9 = 0.25
N9C10[b] = 6.45	CuN9, N9C10 = 0.3
N9C11 = 6.6	N9C10, C10C3 = 1.28
N37C4 = 6.52	CuN2, CuN29 = 0.30
C3C10 = 7.0	CuN9, N9C11 = 0.22
C11C12 = 7.1	N2C4, C4N37 = 1.0
C10C14 = 7.2	C10N9, N9C11 = 1.03
C12C13 = 7.3	N9C11, C11C12 = 0.79
C14C15 = 7.1	C11C12, C12C13 = 0.52
C8C15 = 7.4	N9C10, C10C14 = 1.28
C13C14 = 7.0	C4N37, C4N37C24 = 0.54
CH = 5.1	C11C12, C12H19 = 0.14
N9CuN29 = 0.15	C15C8, C15C8H18 = 0.20
C11N36C31 = 1.45	N9C10, C10N9C11 = 0.54
N9C10C3 = 1.55	C24C25, C24C25C26 = 0.82
C11N9C10 = 1.40	C7C8, C7C8C15 = 0.82
CCC = 1.0	C10C3, C10C3N2 = 0.82
N9C11N36 = 1.5	C31N36, N36C11 = 1.3
C11C12H19 = 0.47	
CuN9C10 = 0.05	
CuN22C24 = 0.05	
H19C12C11C13 = 0.43	

Reprinted by permission of Elsevier B.V

[a]Units: stretching, mdyn $Å^{-1}$; angle bending, mdyn Å rad^{-2}.

[b]Atoms numbering is given in Figure 14.21.

The force constant values sequence for the CC bonds FCC has been proposed on the basis of INDO/1 bond orders. The same follows for the FCN values. No attempt to improve the difference between the experimental and calculated frequencies was performed[66].

The most relevant experimental and calculated frequencies along with the potential energy distribution (PED) are displayed in Table 14.2. Due to the extended π electronic system, the PED shows a high degree of coupling between most of the normal modes.

The strong coupling between the νCC, νCN and νCH modes prevent separating the contributions of the different modes. The averaged value 1.0 mdyn/Å for the CuN bond, obtained from data of copper porphyn complexes[53,54], reproduces satisfactorily the Raman lines at 389 and 278 cm^{-1} and the IR absorptions at 217 and 197 cm^{-1}. The assignment of the totally symmetric mode at 389 cm^{-1} is particularly important in Raman spectroscopy, since it may be used to probe the molecular orientation (for instance, in Langmuir–Blodgett semblies) in Raman polarizatuin studies. The PED suggests that the bands at 197 and 129 cm^{-1}

Table 14.2. Some Relevant Observed and Calculated Frequencies and Approximate Potential Energy Distribution (PED)[a] for the Normal Modes of Hexaazacyclophane Cu (II) Complex

Observed	Calculated	Approximate description
3045	3070	98νCH
1659	1676	23νC31C32 + 17δN36C31N29 + 7δC31N29C30 + 7νC31N36
1533	1527	7νC23C30 + 15δCCH + 5δN22C24N37
1492	1462	21δCCH + 6δCCC + 6νC3N2
1448	1433	36δCCH + 15νC11N9 + 8δC11N36C31
1412	1419	15νC33C32 + 12νC31N36 + 13δCCC
1379	1388	18νC22C23 + 13δCCC + 6νC24N22
1325	1335	19δCCH + 11νN2C4 + 9νC24N22 + 8δCCH
1275	1283	22νC30N29 + 24δCCC + 11νC23C30
1256	1238	81δ CCH
1222	1224	13δ C12C13H20 + 13δ C5C6H17 + 13δ C14C13H20
1182	1178	30δCCH + 6νN22C23 + 5νC30N29
1075	1090	22δCCH + 21δCCH + 19δCCH + 15δCCH
	936	29δCCH + 20δCCC
887	884	100ρCCCH
837	830	55δCCC + 7δC31N29C30
689	677	90δCCC
661	639	9δC31N29C30 + 16δCCC + 7δC11N9C10 + 6δC23N22C24+ 5δC3N2C4
516	522	35δCCC + 7δC30C23N22 + 4δN29C30C23
469	479	28δCCC + 7δN9C10C3 + 7δC10C3N2
436	449	30δCCC + 6δC11N36C31 + 6δC4N37C24 + 3δN29C30C23
390	389	7ν − CuN22 + 7ν − CuN9 + 7ν − CuN2 + 5ν − CuN29+ 11δCCC
329	328	7δCuN2C3 + 6δCuN9C10 + 6δCuN2C4 + 5δCuN9C11+ 18δCCC
280	232	23ν + CuN22 + 22ν − CuN29 + 13δNCuN
217	225	26ν + CuN9 + 23ν − CuN2 + 10δNCuN
197	206	40δCuNC + 10ν + CuN2 + 8ν − CuN29 + 8ν − CuN22 + 7ν+ CuN9
129	147	31δNCuN + 25δCuNC

[a]Contributions lower than 4% are not included. PED = $100L_{ik}^2 F_{ii}/\lambda_k$.

should be ascribed to macrocycle chelate ring deformations. The low force constant 1.0 mdyn/Å for the CuN bond was interpreted in terms of a low positive charge on the copper atom. Thus, an eventual redox reaction with gases such as CO_2 should involve the central metal ion.

5.2. Azabipiridyl Cu(II) Complex

The NCA for the azabipiridyl Cu(II) complex was performed[23], by utilising the GF matrix method and using the program Vibratz[96]. In absence of X-ray diffraction studies for the complex, the geometrical parameters needed to build the G matrix were taken from INDO/1 data. A valence force field with 14 diagonal type force constants, F, was used. Ten non-diagonal type terms representing the most relevant interactions were considered. Values for FCC and FCN and related angles were transferred from those used for similar aromatic macrocycles and bipyridine complexes[47,74,97–100], while the CuN stretching and bending force constants were estimated from current values given for Cu(II) macrocycles and polyamine complexes[59,66,68,101–103]. One of the most commonly employed strategies to reduce the number of off-diagonal elements is to neglect the stretch–stretch interaction constants that do not include a common atom. However, it has been pointed out that in the case of conjugated systems, this kind of non-diagonal force constant is important[104]. This concept has been convincingly supported by Neto *et al.*[105], in their studies of the vibrational spectra of polycyclic aromatic hydrocarbons. In addition, the interaction of the hydrogen rocking motion with the stretching coordinate that shares a common carbon atom and the stretch-rocking interaction with a bond in common was included in the calculation. The set of force constants from which the observed frequencies are best reproduced is listed in Table 14.3. It is expected in this type of conjugated systems that the normal modes have the instantaneous contribution of different internal coordinates.

In this sense, the NCA represents quite well this physical situation through out the pictorial vibrational vectors in each normal mode, according the sign of the normal modes form matrix L. The matrix point out that the A_g and B_{1g} species represent in plane vibrations. In Figure 14.22, some selected frequencies are shown.

5.3. Bis(phenylhydrazine)-1,10-phenanthroline and Its Cu(II) Complex

The normal coordinate analysis for bis(phenylhydrazine)-1,10-phenanthroline (PHP) and its Co(II), Ni(II), Cu(II) and Zn(II) complexes was performed by assuming a 35 atom planar model of C_{2v} symmetry, see Figure 14.23[81]. In the model the phenyl groups were deleted for the sake of simplicity and taking into account that the presence of these substituents does not mainly affect the vibrations of the rest of the molecule[106]. The model has $3N-6$ degrees of internal freedom which classify as $\Gamma_{vib} = 33A_1 + 17A_2 + 32B_1 + 17B_2$.

The secular equation was set up according to the Wilson's FG method. As in the case of the azabipiridyl complexes[23], in absence of X-ray diffraction studies for the PHP complexes, the geometrical parameters needed to build the G matrix were taken from those obtained through out the molecule built with the Hyperchem package and optimised by using the INDO/1 method[81]. The calculated parameters are similar to those currently observed for the ligands. Bonds and angles for the phenanthroline moiety are typical of condensed fully conjugated systems with sp^2hybridization (CC about 1.39 Å and CCC about 120°). In the case of the phenylhydrazine part of the macrocycle and the complexes, the geometrical parameters indicate clearly the double and single bonds C–C (1.44 Å), C = N (1.32 Å), N–N (1.45 Å), while the angles are slightly bigger than in systems with sp^2 hybridisation (132°). Concerning the coordination site, the calculated CuN bonds

Table 14.3. Valence Force Constants (mdyn/Å)

Diagonal force constants		
1	CuN	0.98
2	CN	6.5
3	CC	6.05
4	CH	5.0
5	NcuN	0.18
6	CNC	1.3
7	CCN	1.2
8	CCC	1.1
9	CuNC	0.16
10	HCC	0.51
11	CuNCN	0.25
12	CuNCC	0.15
13	CCCN	0.26
14	HCCC	0.27
Interactions		
2–3		0.43
1–1		0.12
2–2		0.25
3–3		0.77
2–6		0.32
3–7		0.25
3–8		0.12
3–10		0.15
CC/CC[a]	–	0.32 *m*-position
CC/CC[a]		0.28 *p*-position

[a]No atom in common.

distances were about 2.0 Å; the phenantroline and hydrazine NCuN angles are about 87°.

The potential constants were determined in a Modified General Valence Force Field (MGVFF)[81]. Valence force constants are more transferable between similar structures than are Urey–Bradley force constants, especially for aromatic molecules where interactions between non-adjacent internal coordinates have to be included[107]. In fact, as was pointed out by Scherer and Overend[104] in the case of highly conjugated systems, the stretch–stretch interactions concerning the CC bonds in position *meta* and *para* should be considered important. The values of the internal force constants for the PHP complexes were transferred from analogous ligands and metal complexes[68,93,102,103,108] such as polyamine, phthalocyanine, porphyne, 1,10-phenanthroline and porphyrinate metal complexes. The effects of the diagonal force constants (F) and those concerning the different types of interaction force constants (f) were examined systematically in trial and error calculations. The final force constants, which fit the observed frequencies, are listed in Table 14.4.

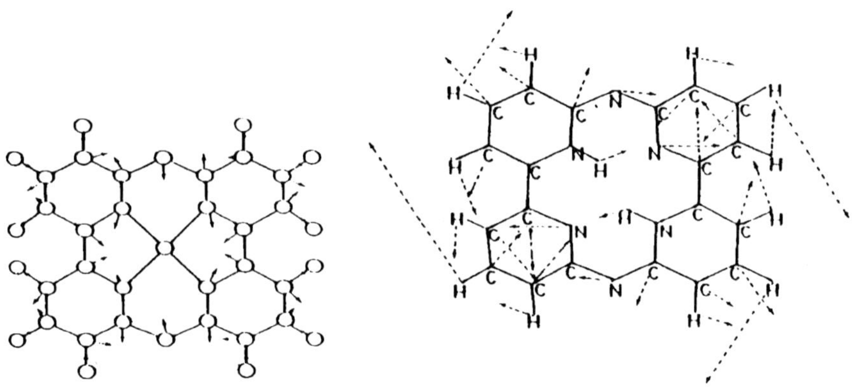

Figure 14.22. Pictorial vibrations of some selected NCA frequencies. Reprinted by permission of Elsevier B.V.

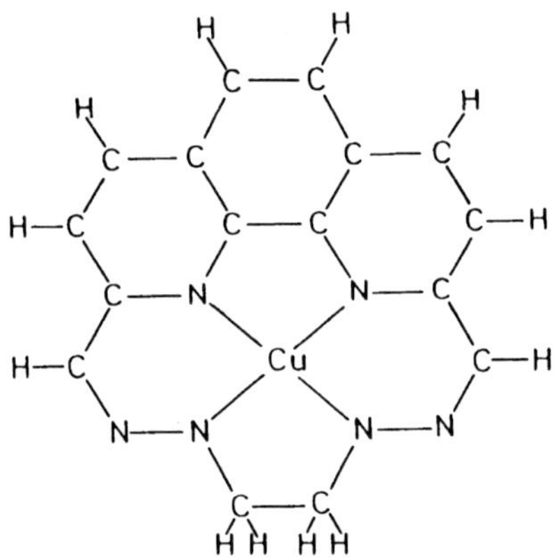

Figure 14.23. Molecular model for the NCA of Cu(II) bis(phenylhydrazine)-1, 10-phenanthroline. Reprinted by permission of Elsevier B.V.

5.4. Ironphthalocyanine

Melendres *et al.*[93] carried out a normal coordinate analysis on FePc: the Fe–N force constant was reported to be 1.0 mdyn/Å.

6. Surface-Enhanced Vibrational and RAIRS Studies

Several surface vibrational studies were oriented to determine the effect that different metal surfaces have on the structure and orientation of macrocycles displaying different ring macrocycle size and the electronic energy extension, and the number of the eventual metal coordinating nitrogen atoms[24,41,42,79,101]. New variables were introduced in recent studies, such as the addition of phenyl substituent (pendant groups) in the macrocycle complexes vanadyl-2,3-naphthalocyanine and vanadyl porphine[78], the addition of phenyl and methyl groups in

Table 14.4. Diagonal (F) and interaction (f) force constants (mdyn/Å), for Cu(II) bis(phenylhydrazine)-1,10-phenanthroline complex.

Diagonal force constants				Interaction force constants	
Bonds	F	Torsion angles	F	Bond/bond	f
CC	6.100	NCCN	0.082	CC/CC_{ortho}	0.750
CN	6.100	CCNN	0.120	CC/CC_{meta}	−0.320
NN	5.700	NCuNN	0.100	CC/CC_{para}	0.190
CH	5.100			CC/CN	0.100
CuN	1.000			CN/NN	0.100
				CC/NN	0.200
Angles		Bond–plane		Bond/angle	
HCH	0.520	HCCC	0.300	CC/NCN	0.090
CCH	0.540			CC/CNC	0.120
CCC	1.000			CC/CCC	0.082
CNC	0.950				
CCN	0.960				
NCN	1.100				
CuNN	0.200				

dinaphthalenic Ni(II) and Cu(II) azamacrocycles[38] and the asymmetry of the coordination site in bis(phenylhydrazine)-1,10-phenantroline (PHP) metal complexes[81].

6.1. Cyclam and Cyclamdione and Their Cu(II) Complexes

The reflection–absorption IR spectra RAIRS of the macrocycles cyclam and cyclamdione and their copper complexes onto a smooth copper surface were published[41]. It has been concluded that the orientation of the cyclam molecules is plane parallel to the surface with the interaction being directed by the nitrogen lone pair electrons. Figure 14.24 contains the IR and RAIR spectra of cyclamdione and its Cu(II) complex.

RAIRS of cyclamdione is not much different to that obtained in KBr, except for the νCO bands at 1,672 and 1,562 cm^{-1} which are shifted to 1,638 and 1,583 cm^{-1} by surface effect. The metal surface induces conformational restructuring in the CO molecular fragment. In the case of the complex, whose spectrum is very different to that of the ligand in KBr, it is observed only one νCO asymmetric band at 1,588 cm^{-1}, indicating a structural equivalence of both CO functional groups. The most probable conformation of the ligand displays the CO group towards the copper surface; the rest of the molecule is rather perpendicular to the surface, the interaction being through the CO groups. Spectral modifications mainly on the relative intensity of the νCO bands are imposed by the metal surface on the copper complex; the orientation of the whole molecule is perpendicular to the surface. The spectral behaviour of the νCN modes in the region 1,500–1,300 cm^{-1} suggests that this orientation is accompanied by an electronic redistribution within the ring fragment. The substrate–adsorbate interaction

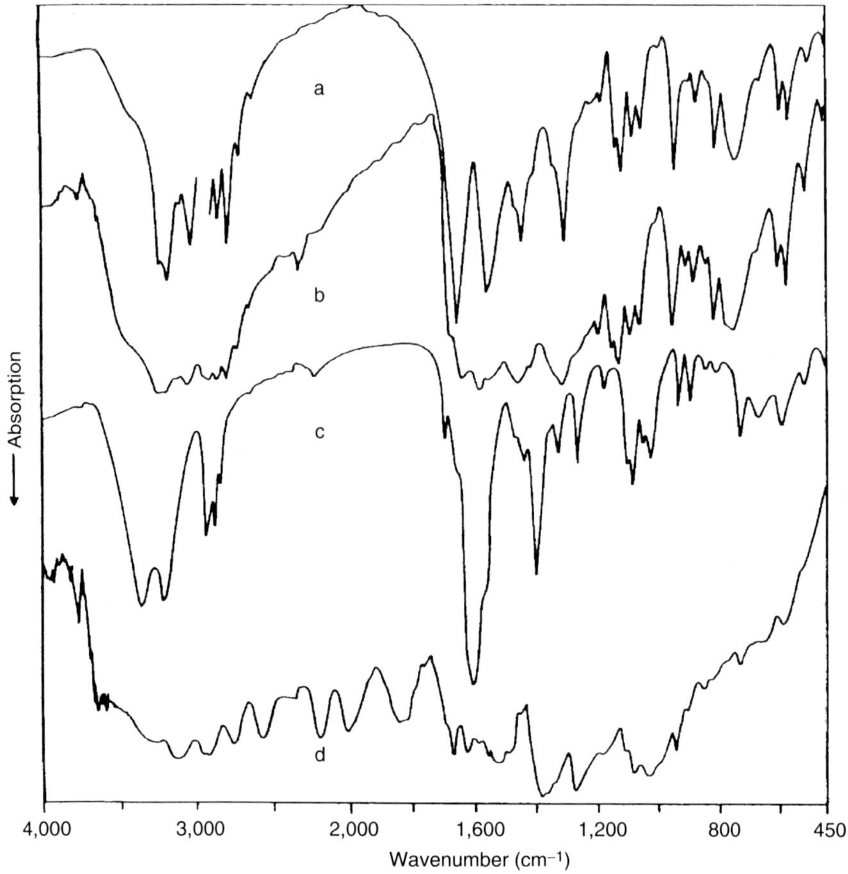

Figure 14.24. (a) IR spectrum of cyclamdione dispersed in KBr; (b) RAIRS of cyclamdione onto a smooth copper surface; (c) IR spectrum of Cu(II) cyclamsione dispersed in KBr and (d) RAIRS of Cu(II) cyclamdione onto a smooth copper surface, in the 4,000–450 cm^{-1} region. Reprinted by permission of Elsevier B.V.

of the cyclamdione compounds is of shorter range than of the cyclam systems; the CO groups are decidedly responsible of this characteristic.

In other molecular systems containing a nitro fragment, such as 2-nitrofluorene and 1-nitropyrene, the adsorbate–metal surface interaction is verified through the oxygen atoms of the nitro group[31].

6.2. Bis(phenylhydrazine)-1,10-phenanthroline and its Cu(II) Complex

A surface vibrational study of bis(phenylhydrazine)-1,10-phenanthroline (PHP) macrocycle complexes has been intended[81]. PHP macrocycles display a non-planar and asymmetric coordination site. No surface effect study is possible to perform on the ligand because it is easily decomposed by solvent dilution.

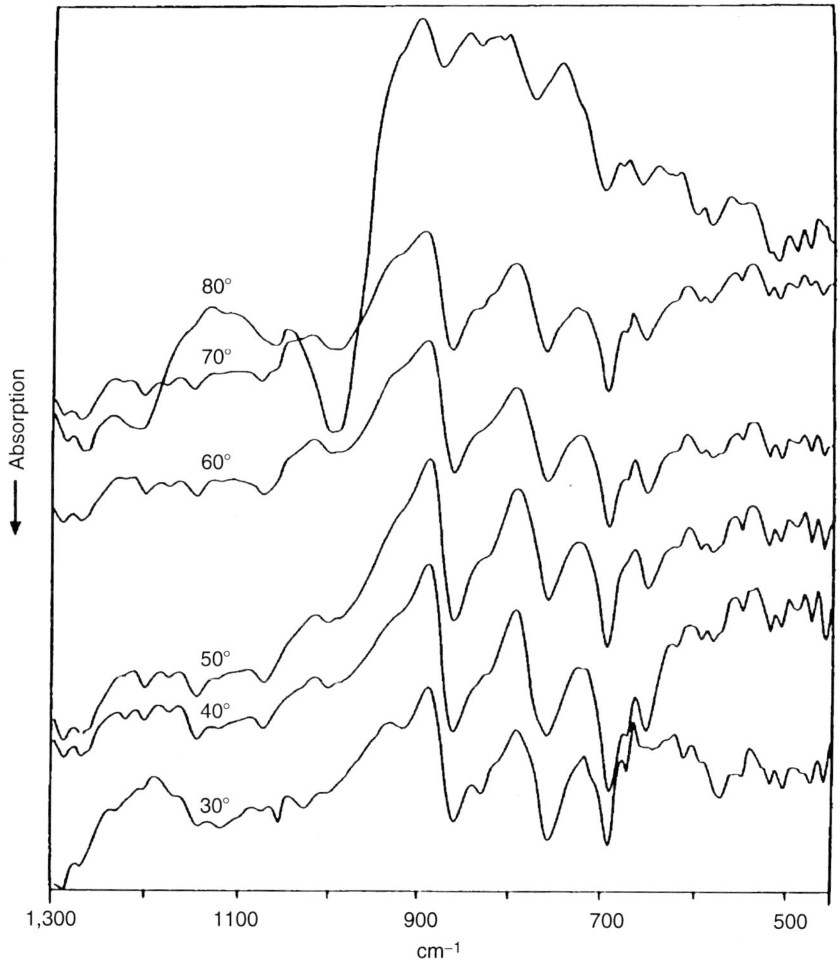

Figure 14.25. RAIRS of bis(phenylhydrazine)-1,10-phenanthroline Cu(II) complex at different incident radiation angles. Reprinted by permission of Elsevier B.V.

The infrared spectra of PHP metal complexes indicate that the presence of coordinated metals imposes structural modifications of the macrocycle ring and important electronic energy redistributions mainly around the coordination site. The metal coordination involves only the phenantroline and hydrazine nitrogen atoms. The reflection–absorption experiments are useful to discriminate normal modes of different fragments of the macrocycles. RAIRS data of the copper complex onto a smooth copper surface in Figure 14.25, indicate that the non-planar structure of the coordination site and the presence of phenyl groups disposed nearly perpendicular to the rest of the molecule sterically hindrance any interaction of the π electronic system with the smooth copper surface.

These structural characteristics also are responsible of the fact that the SERS spectrum is not observed. The same situation was found for the Co(II),

Ni(II) and Zn(II)[81] complexes for which the global molecular structure is not much different to the copper complex.

6.3. Azabipiridyl and Its Co(II), Ni(II) and Cu(II) Complexes

The FT-Raman spectra and SERS on silver colloid of the azabipiridyl ligand and its Co(II), Ni(II) and Cu(II) complexes was reported; a SEIRA study was also accomplished[23].

6.3.1. Raman Spectra and SERS of the Ligand and Its Complexes

The solid and SERS spectra of the ligand and Co complex are displayed in Figure 14.26, and the frequencies with the most probable assignments of the series of macrocycles are listed in Table 14.5.

Several bands which do not show spectral modifications by surface effect correspond to those parts of the molecule which are not directly involved in the adsorbate–substrate interaction. Other bands display frequency shifts to lower or higher energies; this behaviour is associated to bond energy modifications.

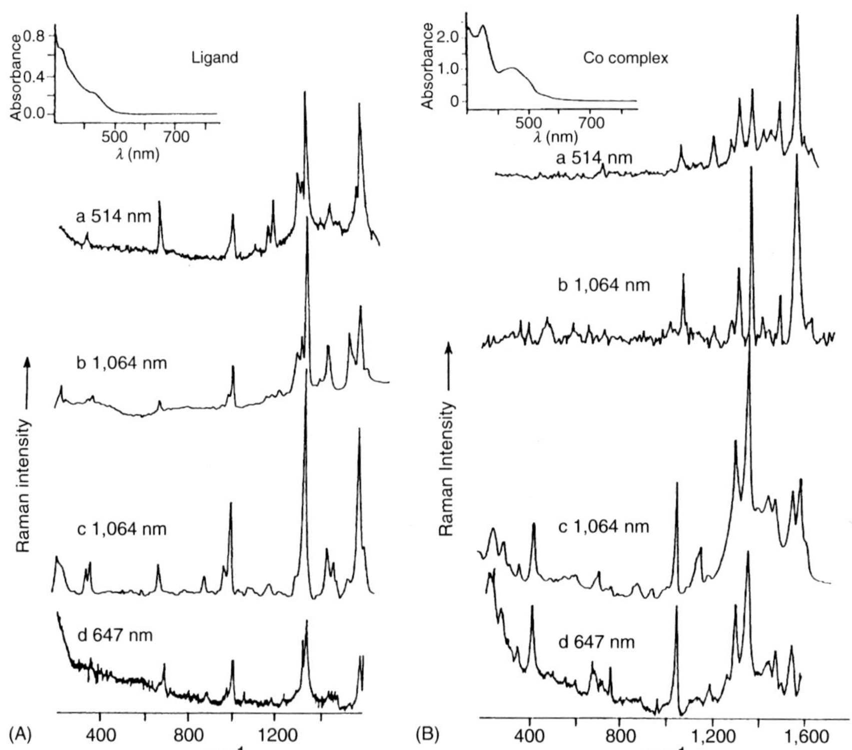

Figure 14.26. The solid (a,b) and silver colloid (c,d) Raman spectra of the ligand and Co complex. Reprinted by permission of Elsevier B.V.

Table 14.5. Mean Wavenumbers (cm^{-1}) of (a) Raman Spectrum and (b) SERS of Azabipiridyl (L) and Its Metal Complexes[a]

$L_{(b)}$	$L_{(a)}$	$Co_{(b)}$	$Co_{(a)}$	$Ni_{(b)}$	$Ni_{(a)}$	$Ni_{(b)}H$	$Ni_{(a)}H$	$Cu_{(b)}$	$Cu_{(a)}$	Assignment
1572 vs	1568 s	1585 s		1583 s		1584 s	1586 vs	1584 s		νCC + νCN
1524 w	1522 s	1551 s	1552 vs	1555 s	1549 vs	1556 m			1542 vs	
1462 m		1472 s	1472 m	1477 m	1471 w	1477 m	1485 m	1477 s	1464 m	
1432 m	1426 m	1441 s	1427 w	1445 m	1424	1451 m	1447 s	1448 m	1429	
	1390 w		1392 m		1393 w		1376 w		1384 w	
									1363	
1331 vs	1330 vs	1349 vs	1346 vs	1360 vs	1356 vs	1356 vs	1354 s	1356 vs	1340 vs	ν(C–C) + νCN
	1309 s			1338 s						
	1290 m	1294 vs	1289 s	1301 vs	1295 s	1298 vs	1300 s	1298 vs	1287 m	
			1257 w	1270 m	1262 w	1270 w		1269 w		δCH
						1239		1239 w	1250 w	
	1208 w					1205 w		1204 w		
1177 w	1178		1177 w	1187 w						
	1154	1143 m			1159				1160	Breathing +
						1129		1128		δCH
							1118			
1080			1070			1086		1080		
998 s	999 m	1036 s	1036 s	1035 vs	1032 m	1034 vs	1037 s	1035vs	1032 m	ρCH
				999 w		988 m		988 m		Breathing +
969 m	980 w									δNH
	952 vw									ρCH
	909 vw						913 vw			
879 w		880 dw		880 w		879 w		880 w		
	798 vw					788 w		790 vw		
		747 w								
						713 vw	709 vw	711 vw		

Table 14.5. *(Continued)*

$L_{(b)}$	$L_{(a)}$	$Co_{(b)}$	$Co_{(a)}$	$Ni_{(b)}$	$Ni_{(a)}$	$Ni_{(b)}H$	$Ni_{(a)}H$	$Cu_{(b)}$	$Cu_{(a)}$	Assignment
670 m	666 w	693 w	680 w	681 w	678 vw	670 vw		686 vw		Breathing +
		588 vw	608 vw			599 vw		597 vw		Macrocycles
										deformation
					570 vw	559 vw		560 vw		
			538 vw					502 vw		
		399 m	417 w	401 s	401 m	398 m	400 w	399 m	394 w	$\nu(M-N)$
350 m	353 w		335 w							Ring
331 w	334 vw	336 w	335 w	339 m	340 vw					
			298 w	298 vw	298 m	297 m	290 w	298 m		$\nu(M-N)$
				273 m	271 vw	281 m		281 m	271 m	
		265 m					250			Ring
	229 vw						224 w			
215 m		216 m		220 m		214 m		216 m		$\nu(Ag-N)$
194 m	205 w									Macrocycles
				160 w		153 vw				

[a] Abbreviations: s, strong; w, weak; m, medium; vw, very weak; d, doublet; χ, out of plane ring deformation.

Moreover, the intensity changes that are observed in the SERS spectra, are explained on the basis of the existence of a gradient of electromagnetic field on the surface, and the SERS selection rules which distinguish different molecular modes; this is a consequence of the orientation of the molecules on the colloid surface. Using the single sphere model and the approximate case for which the sphere shows a dipolar response to laser excitation, the field at the surface is greater in the perpendicular direction than in the tangential one[109,110].

In general, the Raman spectra of solids display a similar pattern in the series, with expected differences below 500 cm^{-1}, where the metal–ligand vibrations appear. An enhancement of the Raman lines falling in this region is observed in the SERS spectra (Figure 14.26), along with frequency shifts of several bands. Conclusions regarding the existence of an actual adsorbate–substrate interaction, as well as structural modifications of the macrocycle by the surface effect were derived from the analysis of this region[23].

The adsorbate–substrate interaction is verified by the appearance of the medium band at about 216 cm^{-1} in the series of complexes, ascribed to an Ag-macrocycle binding, involving mainly the N atoms of the coordinating site. In the SERS spectra, the M–N stretching bands at about 400, 298 and 280 cm^{-1}, are observed at about the same frequency for all the complexes. The higher differences observed in the SERS spectra respect to the Raman of the solids correspond to the cobalt and copper complexes. In the first case it is observed a shift from 417 to 399 cm^{-1} when it is adsorbed on the metal surface, while the copper system displays an exactly opposite behaviour, shifting from 394 to 399 cm^{-1}. This result points out that the structure of the copper complex in the colloid is less warped than in the solid[101,111], while the planar structure of the cobalt complex becomes slightly warped under the surface effect. The spectra of the solid and colloid are very similar in this region for the Ni complexes. These results suggest that the Ag surface constraints the complexes to adopt an equivalent structure when adsorbed on the metal surface. The fact that the M–N coordination is not altered during the interaction with the silver colloid is due to an interaction of the N coordinating atoms with the metal through the N π electrons. This suggests a planar parallel orientation of the adsorbate in relation to the surface.

At least one of the ring in and out of plane deformations is clearly observed in colloidal silver at about 680 cm^{-1} [23]. This frequency has been assigned to a macrocycle breathing in phthalocyanines[50,51,111] and naphthalocyanines;[76] most of assignments in aromatic systems propose this band to an out of plane CH or ring deformations[66,112].

From the resonance Raman spectra of phthalocyanine monolayers on different metals and a normal mode analysis, Palys *et al.*[74], conclude that the Raman band at 678 cm^{-1} is described by an in-plane macrocycle mode coupled to an out of plane CH deformation. Moreover they conclude that the phthalocyanine molecule is bonded via nitrogen atoms to a glassy carbon surface, and through the metal ion when the molecule interacts with a gold surface. On this basis it was proposed[23] that this band is due to an in and out of plane concerted π electronic system motion, corresponding to the breathing mode coupled to the out of plane CH vibration (ρCH). This assignment explains the presence of this band in the

SERS spectrum. Moreover, the shift to higher energy undergone by this metal ion dependent mode in the SERS confirms the macrocycle-Ag interaction. The longer up shift in the series corresponds to the cobalt complex.

The similar frequency of the ρCH mode observed at about 1,035 cm^{-1} for the complexes, and at 999 cm^{-1} for the ligand in the Raman of solids and in the SERS spectra, indicates that the adsorbate–substrate interaction does not involve the CH bonds. The shift to lower energy of the ligand band at 999 cm^{-1} respect to that of the complexes is related to both the different molecular structure and a lower π electronic energy around the carbon atom; this is further related to a different orientation of the ligand on the surface in relation to the complexes[23].

The band observed only in the spectra of the azabipiridyl ligand at 980 cm^{-1}in the solid and at 969 cm^{-1} in the SERS was ascribed to the NH deformation[23], in agreement with that proposed by Gobernado-Mitre *et al.*[76] in naphthalocyanine. The NH bonds are disposed out of the coordinating site plane. The frequency shift is due to a structural and/or conformational modification imposed by surface effect. This fact could be also the explanation of the spectral shifting to lower energy of the out of plane macrocycle ring deformation band (205 cm^{-1}) by colloidal metal effect. Most of the in plane CH deformations (δCH) bands observed in the 1,000–1,300 cm^{-1} spectral region display a very weak relative intensity and are not intensified in the SERS spectra. This result also supports a planar orientation of the macrocycle on the surface. Nevertheless, a band at 1,143 cm^{-1} is observed in the SERS of cobalt complex in this region. In naphthalocyanine[76] and phthalocyanine[50] it has been proposed a pyrrole breathing mode at about 1,140 cm^{-1}. If the enhanced band in the Co complex corresponds to the pyridinic ring breathing mode, its enhancement in the SERS spectrum can be explained on the basis of a great out of plane character of this vibration, which involves the motion of the π electrons in the ring.

Two intense bands observed in the 1,300–1,350 cm^{-1} spectral region of all the molecules are assigned to a mixture of νCC and νCN mode[23]. A shift to higher energy is observed for these bands in the SERS spectra. Likely, it is observed for these bands a slight trend to equalize their energies when adsorbed on the surface. This is consistent with an interaction of the adsorbate via the π electrons of N atoms which induces a structural change of the corresponding bonds involved in the adsorbate–substrate interaction. Moreover, these SERS bands display a similar relative intensity with respect to the same bands in the solid.

Two νCC bands at about 1,440 and 1,470 cm^{-1} are influenced by surface effect displaying the first a net shift to higher frequency; the relative intensity of one of these bands decreases by surface effect, Figure 14.26. These modifications indicate structural changes of bonds involved in the interaction. In the presence of the metal surface all the metallic complexes trends to make equal the energy of those bonds, which is interpreted as a π electronic redistribution in the adsorbed molecule. An evident relative intensity decrease of the most intense band in the 1,500–1,600 cm^{-1} is observed in the SERS of all complexes. In particular, the band at about 1,550 cm^{-1} undergoes a strong intensity decrease, being almost absent in the case of the cobalt complex, Figure 14.25. In the case of the ligand, the same situation is verified but it concerns the medium band at 1,522 cm^{-1}.

Bands in this region are ascribed to νCC and νCN aromatic vibrations. The observed relative intensity change observed in the SERS was ascribed (1) to the specific orientation that the molecules adopt on the surface, which is probably plane parallel to the surface, or (2) to a resonant effect which is a consequence of the charge-transfer mechanism contributing to the overall SERS effect. A general trend to equalise the energy of these bands by surface effect in the series of the complexes is also observed.[23]

Bands of the copper complex which are clearly intensified in the SERS spectra appear at 1,584, 1,448, 1,356, and 1,298 cm^{-1}, Table 14.5. Thus, those bands are attributed to B_{1g} modes, i.e. vibrational modes with a lower symmetry. The intensification of these less symmetric modes in the SERS spectrum indicates a contribution of the charge transfer mechanism producing the SERS effect via the Herzberg–Teller mechanism of resonance.[113] This assumption suggests that the 988 cm^{-1} band appearing in the SERS spectra could correspond to a B_{1g} asymmetric mode, and can be assigned to a coupled breathing-ρCH mode. All these changes occur in the opposite direction in the Raman of solids when going from non-resonant conditions (1,064 nm) to resonant ones (647 nm), thus confirming the above assignments and the possible importance of the Herzberg–Teller mechanism in the SERS enhancement.

The relative intensity enhancement of the out of plane bands suggests that the corresponding vibrations are mainly parallel to the normal of the incident radiation. Thus, the macrocycle complexes have a preferential face-on orientation on the surface.

In the SERS spectra of copper complex the bands appearing at 1,035, 1,298, and 1,584 cm^{-1} decrease their intensity by changing the excitation line from 1,064 to 647 nm, while the breathing mode at 686 cm^{-1} increases its relative intensity, which is not evident in the case of the other complexes. The coupled breathing-ρCH band at 988 cm^{-1} is not observed at 647 nm. This effect demonstrates the progressive importance of the symmetric modes (A_g) when the excitation is moved to lower wavelengths via a Franck–Condon mechanism.

6.3.2. SEIRA of the Ligand Azabipiridyl and Its Complexes

In Figure 14.27 are displayed the SEIRA spectra of the ligand adsorbed on a Au film.[23]

By dilution of the ligand sample on the gold surface, some of its bands display a shift to lower energy and some others to higher energy. Bands at 1,530, 1,431, 1,392 and 1,167 cm^{-1} change their intensity, while bands at 1,431 and 1,167 cm^{-1}, are resolved. Two bands at about 1,280 cm^{-1} trends to collapse by dilution. These modifications are explained by considering an adsorbate–substrate interaction. Changes observed on washing the sample are due to the molecules directly adsorbed on the surface. The adsorption clearly induces a new structure of the first layers of molecules adsorbed in the gold islands surface. In fact, the spectrum of the multilayer sample is very similar to that of the solid dispersed in KBr.[79]

The cobalt and the N-hexyl nickel substituted complexes structure in the surface is different from that of the compound at higher concentrations; this was

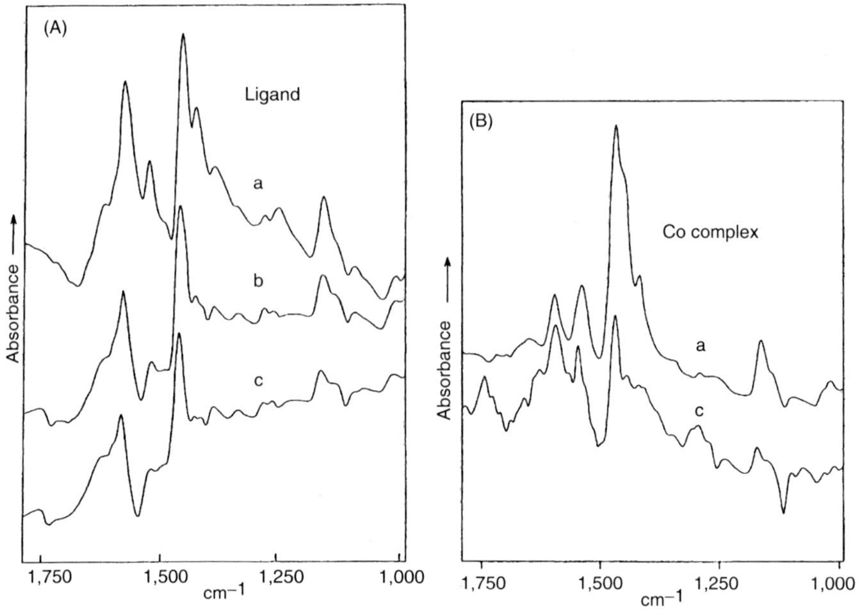

Figure 14.27. SEIRS of the ligand azabipirdyl and its Co complex: (a) multilayer, (b) first dilution and (c) thin film. Reprinted by permission of Elsevier B.V.

inferred from the important spectral changes observed in the whole frequency region, see Figure 14.27. Table 14.6 contains the frequencies and relative intensity of the macrocycles.

The multilayer spectrum of the Co complex is not different to that obtained in KBr pellets. The most relevant spectral changes in frequency and the relative intensity by dilution in the case of the copper complex, concern only the bands in the range 1,390–1,450 cm^{-1}, and the one near 1,170 cm^{-1}. The spectrum of the copper complex dispersed in KBr[79] is not much different to that of the multilayer sample on the gold surface. The surface effect is less important in this case than in the others three macrocycles.

The SEIRA study of the compounds deposited onto Au films reveals slight structural modifications upon dilution.

6.4. Naphthalocyanines

Little work dealing with vibrational studies of the macrocycle–surface interaction have been published for naphthalocyanine and its metal complexes. One deals with the ZnNc complex for which the orientation onto different substrates has been inferred from IR reflection–absorption spectroscopy and other measurements;[114] three types of molecular orientations were assigned for the deposited thin films depending on NaCl, graphite and glass at 250°C. The orientation of different metal complexes of naphthalocyanine was studied by X-ray

Table 14.6. Mean wave numbers (cm^{-1}) of SEIRS of azabipiridyl (L) and its metal complexes in the region, 1,100–1,700 cm^{-1} at different incrasing dilutions (D) in ethanol[a]

L			Co			NiH[b]		Cu	
D_1	D_2	D_3	D_1	D_2	D_3	D_1	D_2	D_1	D_2
1101 vw	1100 w						1105 w	1102 w	
		1141 vw					1139 m		
1167 ms	1168 mb	1170 m				1174 w		1170 mas	1171 w
			1167 mas	1173 was	1173 w				
						1195 w	1203 w		
1259 wb	1269 wd	1269 wd		1242 vw		1248 m	1243 m		
1287 w	1287	1287	1295 wt	1295 wd	1285 w			1289 mas	1288 w
1346 vw	1341 w	1339 w			1345 w				
1392 w	1392 w	1391 was				1379 mas	1398 w		1398
		1418 vw	1422 sh	1423 vw	1410 wm	1420 m	1418 m	1429 w	1416 tm
1431 m	1432 w	1432 vw		1447 sh	1444 w				
			1455 sh						
1459 s	1466 s	1467 s	1473 s	1471 s	1490 s	1479 s	1480 s	1471 s	1469 s
					1531 w			1533 mas	1532 m
1530 m	1525 w	1523 vw	1542 m	1550 ms	1550 m				
						1581 sb	1582		
1581 s	1584 s	1584 ms	1599 m	1596 s	1596 s		1588 mst	1588 ms	1588 m
				1628 sh	1628 m	1605 sh	1606		
1623 sh	1621 sh	1621 wb							
								1628 vw	
								1641 vw	
				1661 w	1662 m			1661 w	

[a] Abbreviations: s, strong; w, weak; m, medium; vw, very weak; sh, shoulder; d, double; t, triple; as, asymmetric; b, broad.
[b] *N*-hexyl Ni substituted complex.

diffraction, IR spectroscopy, transmission electron microscopy and electron diffraction:[115] metal-free H_2Nc and ZnNc took the face-to-face, eclipsed stacking, *P* orientation, in which their molecular planes came into a parallel contact to the substrate surface; AlNcCl and GaNcF, and VONc took the eclipsed, slipped stacking, *I* orientation, holding their molecular planes slightly inclined to the substrate surface.

Gobernado-Mitre *et al.*[76] studied the molecular organization of CuNc onto different substrates using IR and SERS; a tilted, close to face-on molecular orientation in the evaporated films on KBr and Ag was inferred from the transmission and reflection–absorption FT-IR spectroscopy.

An infrared and theoretical interpretation of the structure and orientation of H_2Nc and its Cu(II) complex deposited onto a smooth copper surface has been proposed.[101] In the case of H_2Nc no spectral differences among its spectrum dispersed in KBr and those of the macrocycle deposited onto the KBr monocrystal and onto the copper surface were observed (Figure 14.28).

This was interpreted in terms that there is no structural modifications caused by the interaction with the metal surface and that the ligand–metal interaction is not significant.

By changing the incident infrared radiation angle, the main spectral difference arises from the variation of the relative intensity of the CH wagging and ρ

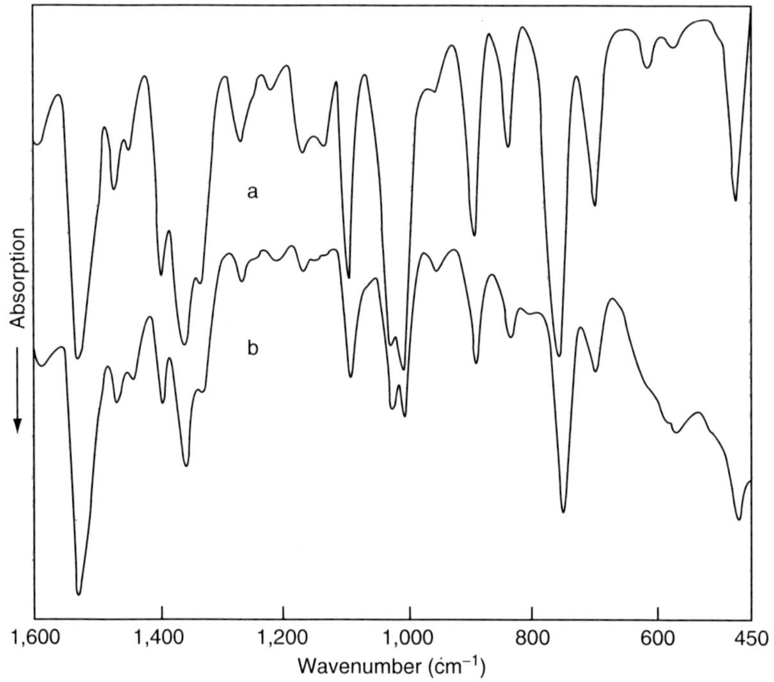

Figure 14.28. (a) IR spectrum of napthalocyanine dispersed in KBr, (b)RAIRS of naphthalocyanine onto a smooth Cu surface, in the region 1,600–450 cm^{-1}. Reprinted by permission of Elsevier B.V.

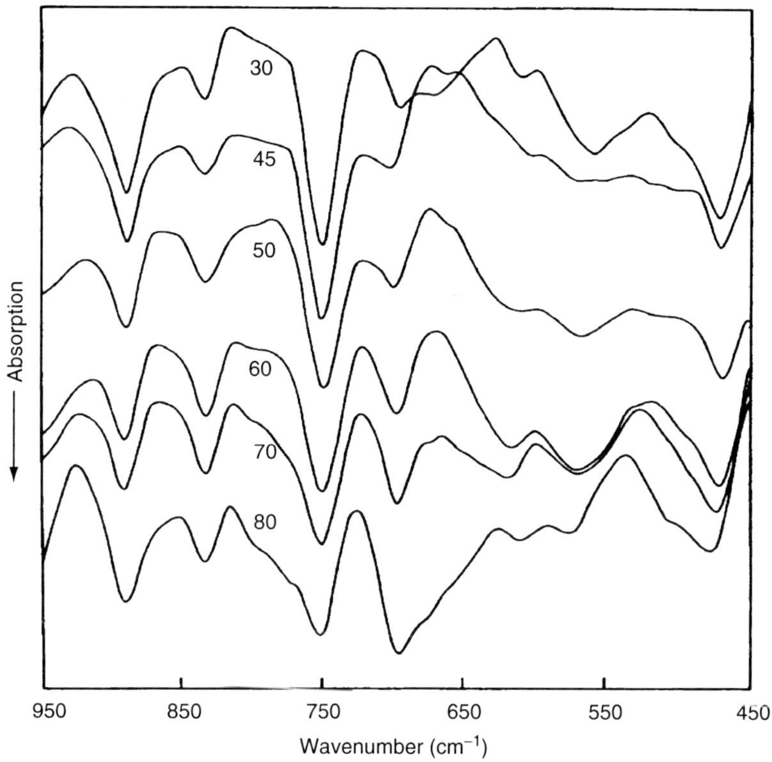

Figure 14.29. RAIRS of the naphthalocyanine onto a smooth copper surface. Relative intensity changes of the ω CH and ρ NH bands in the range 950–450 cm^{-1} with variation of the radiation angle. Reprinted by permission of Elsevier B.V.

NH bands at 756 and 697 cm^{-1}, respectively. Varying the angle between 80° and 30°, it is observed that the relative intensity of the CH wagging (ω CH) is maximum between 50° and 30° while the opposite is observed for the relative intensity of the ρNH band. Figure 14.29.

A similar spectral behaviour was observed by Gobernado-Mitre *et al.*[76] for the evaporated films of CuNc complex on silver and KBr, suggesting a nearly face-on or tilted molecular orientation onto those surfaces. The spectral modifications[101] suggest a preferential orientation of the molecules onto the surface: the molecules are oriented tilted to the plane and the CH bonds in one end of the molecule should be closer to the surface. This proposition is consistent with the unique and slight shifting to lower energy (7 cm^{-1}) of the ωCH mode originally observed at 756 cm^{-1}. A dramatic decreasing of the bands intensity by using the polarizer avoids to precise the preferential orientation angle of the macrocycle onto the surface.

The RAIR spectrum of CuNc onto the smooth copper surface displays significant differences respect to that obtained in KBr and to that of the complex deposited onto the KBr monocrystal[101] (Figure 14.30). These spectral differences

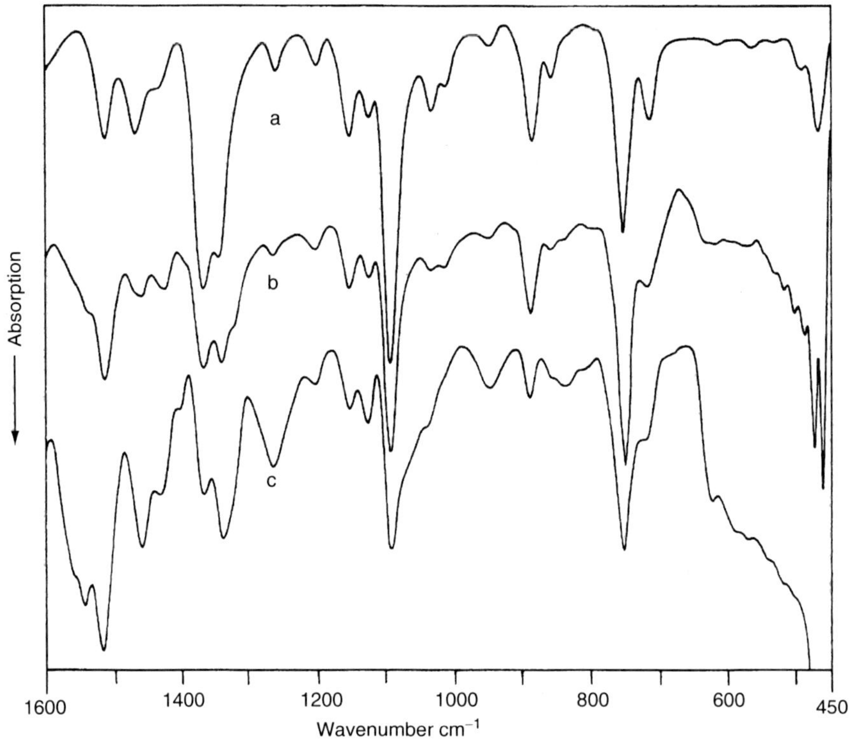

Figure 14.30. (a) IR spectrum of CuNc dispersed in KBr, (b) IR transmission spectrum of CuNc onto a KBr monocrystal, (c) RAIRS of CuNc onto a smooth copper surface for an incident radition angle of 45°, in the region 1,600–450 cm^{-1}. Reprinted by permission of Elsevier B.V.

are interpreted in terms of the influence of the surface on the structure of the complex rather than to a molecular interaction. The most relevant changes in the relative intensity are observed in bands at 1,260 and 1,337 cm^{-1}, assigned to in plane δCH and νCN modes, respectively. The general shifting to lower energy by surface effect and the variations in the relative intensity, suggest electronic energy redistribution around the coordination site, accompanied with an internal structural modification. The spectral shift to higher energy by interaction of the absorption at 1,260 cm^{-1} is that expected for a deformation mode[71,116]. The intensity variation of the in plane mode at 1,260 cm^{-1} and the out of plane vibrations at 750 and 887 cm^{-1} is rather aleatory when varying the incident radiation angle; this make no possible to infer about the molecular orientation of the complex onto the surface. Gobernado-Mitre[76] for the same complex deposited on to a–silver substrate proposed a face-on or a tilted orientation onto the surface. Moreover, it has been observed in general a broadening of the bands, which was interpreted as a crystalline effect due to an interaction of identically oriented molecules. The electronic density on the copper ion, resulting from complexation, moves to the copper surface during the interaction. The substrate–adsorbate interaction constraints the macrocycle complex to adopt a deformed warped structure.

The spectral variations, in particular the frequency shifting of CuNc interacting with the copper surface[101], were more significant respect to that observed by Gobernado–Mitre[76] for the interaction of the same complex with a silver substrate. The nature of the surface plays an important role in this type of adsorbate–substrate interaction.

The adsorption of vanadyl-5,14, 23, 32-tetraphenyl-2, 3-naphthalocyanine VOTPNPc and vanadyl-5,10,15, 20-tetraphenyl-21H,31H-porphine VOTPPORP macrocycles onto metallic surfaces has been studied by Carrasco *et al.*[78]. A weak adsorbate–substrate interaction and no significant structural modifications imposed by surface effect were inferred from the IR spectra of the compounds deposited onto a KBr monocrystal and onto a smooth copper surface.

The relative intensity ratio between selected bands measured in IRRAS suggested that in the case of the naphthalocyanine complex the compound is oriented with the naphthalocyanine moiety plane parallel to the copper surface; the VO and the phenyl groups in both complexes are oriented perpendicular to the macrocycles rings, with the oxygen atom opposed to the metallic surface. Therefore, it exist a preferential molecular orientation of the molecules deposited onto the copper surface.

The SERS spectral data obtained on colloidal Ag as well as Ag island films indicate a weak macrocycle interaction and small structural modifications of the naphthalocyanine complex on the surface. An energy transfer mechanism contribution to the observed enhancement was proposed[78]. The whole spectral data point that the naphthalocyanine complex is oriented with the naphthalocyanine plane face-on to the surface. In both vanadylnaphthalocyanine and vanadylporphine tetraphenyl substituted complexes the vanadyl group is perpendicular to the coordination site and opposed to the surface. In both complexes the phenyl substituents, oriented perpendicular to the macrocycle plane, are responsible of the weak adsorbate–substrate interaction. Several different experimental conditions such as scanning speed, laser power and excitation lines wavelengths, concentration and metal surfaces were intended to obtain SERS of the porphine complex without success; this was interpreted in terms that the complex has no interaction with the surface. Thus, it was concluded that the extension of the whole π-system plays a significant role in the mechanism involving the spectral enhancement by surface effect.

The advantage of systems with Ag nanoparticles and their assemblies for SERS spectral investigation, detection and determination of porphyrins species has been discussed[117]. SERRS spectral detection limits of the testing porphyrin species, including porphyrin aggregates, are 10^2–10^3 lower than detection limits by RRS. No negative effects on the detection limits of porphyrins were detected by using spacers as was the case for systems with nanoparticles modified by anionic organosulfur spacers, in particular thiopheneacetate-modified Ag particles prepared by laser ablation.

TEM images, UV–visible and SERS spectra of 2-dimensional assemblies of tetrapyridylporphine derivatized Au nanoparticles (Au colloid-H_2TpyP films) were reported[118]. The films were prepared by spontaneous 2-dimensional reassembling of multilayer interfacial films on H_2O surface and a subsequent deposition, forming hexagonally packed arrays of Au nanoparticles with some lattice defects. UV–visible spectra indicate stabilization of the array by porphyrin–porphyrin hydrophobic interactions.

Qu and Fredericks[119] found that the SERS behaviour of tetrakis (3-*N*-methylpyridyl) porphyrin chloride and Sn(IV) tetrakis (4-*N*-methylpyridyl) porphyrin chloride varies with different surface concentrations. Experiments were performed on electrochemically prepared Ag surfaces. In the same sense Praus *et al.*[120] studied, by means of the Raman correlation spectroscopy, the influence that the Ag colloid aggregation has on the SERRS spectra of 5,10,15,20-tetrakis (1-methyl-4-pyridyl) porphyrin.

6.5. Phthalocyanines

The smooth copper surface effect on the structure and orientation of phthalocyanine H_2Pc and vanadyl phthalocyanine VOPc onto the surface was also studied by IRRAS[78]. In H_2Pc and VOPc important frequency changes between 1,400 and 450 cm^{-1} were identified by surface effect suggesting that a compound-surface interaction is verified. The frequency changes are more significant in the case of the vanadyl complex, which indicates a more intense interaction. The orientation of H_2Pc in the surface was inferred from a study of the relative intensities variation with different incident infrared radiation angles. Results suggested that there is no a preferential orientation of the H_2Pc molecule on the surface; this effect is due to the out of plane position of the two NH bonds in the macrocycle ring. In the case of the VOPc complex the orientation is plane parallel to the surface, which is due to the inexistence of the out of plane NH bonds as in the case of the ligand H_2Pc. As the interaction exists and the νVO frequency (1,002 cm^{-1}) remains nearly constant by surface effect (999 cm^{-1}), it has been proposed[78], that this interaction does not involve the VO group. Thus, the molecule deposited face-on the surface displays the VO group perpendicular and opposed to the surface.

Faulques *et al.*[121]reported the SERS spectra of thin layers phthalocyanine (Pc) complexes deposited on Ag and Au supports. Molecular systems under study were MgPc and Lu*bis*Pc (Pc'LuPc) in different oxidation states. The enhancement of specific bands depends on the surface and the oxidation states.

Palys *et al.*[74] studied the resonance Raman spectra of monolayers of transition metal phthalocyanines; the spectra reveal specific interaction with the support. The mechanism in the case of ZnPc monolayers was elucidated by using Raman spectra and semiempirical MNDO calculations. The α-atoms dominate the interaction with the support and the usually assumed D_{4h} geometry does not represent the true energy minimum but the mean of two mesomeric forms.

Aroca *et al.*[122] characterized the lanthanide series of bisphthalocyanine complexes of $LnPc^t_2$ (free radical green material) and the corresponding blue material ($LnHPc^t_2$) by using Raman and SERS, IR and visible spectroscopy; interconversion of the green to blue material was observed by heating Langmuir–Blodgett (LB) assemblies and KBr pellets of the green material. Oxidation–reduction process was inferred to occur in the organic ring; the oxidation state of the central metal atom is not affected. Well-characterized LB multilayer assemblies of $PrPc_2$ and the tetra-*tert*-butyl derivative $PrPc^t_2$ were exposed to

$NO_x(NO_2/N_2O_4)$ gas and the effect of gas adsorption was monitored using visible and infrared spectroscopy;[123] the high sensitivity surface-enhanced resonance-Raman spectroscopy (SERRS) allowed to observe the reversible chemical absorption of NO_2 on a monomolecular LB layer of $PrPc_2$ and $PrPc^t_2$.

Sheng-Gao *et al.*[124] prepared novel asymmetrically substituted metal-free phthalocyanine, nitro-*tert*-butylphthalocyanine, which resulted suitable for fabrication as a thin film using the LB technique.

The molecular film formation of 2,9,16,23-tetraamino metal {metal = Co(II), Cu(II) and Fe(III)} phthalocyanines on gold and silver has been described; electrochemical methods and Raman spectroscopy were used to characterize these films[125]. These compounds absorb onto gold and silver electrodes with the nitrogen atoms anchoring onto the metal surface as it is revealed by Raman characteristic bands around 236 cm^{-1}; these bands result from the interaction between the metal substrate and the nitrogen of the $-NH_2$ group on the phthalocyanine molecules. The surface coverage is fairly high and close to a monolayer on both metal substrates.

The study of adsorbed molecular films of phthalocyanines has been also performed by other authors. Cook *et al.*[126], reported the preparation of phthalocyanines with one or two trichlorosilyl alkyl chains and subsequently formed self-adsorbed films on glass and silicon. The preparation and film formation of the thiol and disulfide-derivatized phthalocyanines on gold surfaces have been accomplished by the same group[127]. The IRRAS and evanescent wave-excited fluorescence emission studies suggested that the orientation of the phthalocyanine ring of the molecule depends on the length of the spacer between the ring and the gold surface. When the chain length is of the order of C11 hydrocarbon chain, the macrocycles assume a densely packed orientation with the phthalocyanine ring arranged perpendicular to the gold surface. A C3 alkyl chain induces the formation of a less closely packed monolayer with the phthalocyanine rings arranged parallel to the gold surface. Huc and coworkers[128], reported that ruthenium phthalocyanine can be grafted onto gold substrates by axial ligation using isonitriles as spacers between the naphthalocyanines and the metal. Li *et al.*[129] reported the formation of self-assembled monolayers of thiol-tethered, octasubstituted, and axially ligated silicon phthalocyanines.

Ultra thin Langmuir–Blodgett (LB) films of rhodium phthalocyanine (RhPc) were recently fabricated[130]. The LB film was characterized by using both UV–visible absorption spectra and Raman scattering. The Raman spectroscopy was carried out using 633 and 780 nm laser lines. LB films were deposited onto Ag nanoparticles to achieve the surface-enhanced pre-resonance Raman scattering (pre-SERRS) and SERS for both laser lines, respectively, which allowed the characterization of the RhPc ultra thin films. The morphology of the LB RhPc neat film was extracted from micro-Raman imaging.

Langmuir and Langmuir–Blodgett (LB) films of neat titanyl(IV) phthalocyanine (TiOPc) and mixed (TiOPC)-arachidic acid (AA) films were studied by Del Cano *et al.*[131]. The surface pressure–area isotherms (T–A) were recorded, and an unusual expansion of the surface area per molecule was detected for mixed TiOPc-AA Langmuir monolayers. The expansion is attributed to the TiOPc molecular orientation in the neat films being different from that in the mixed films with arachidic acid. The molecular orientation in LB films was monitored using RAIRS. It was found that the TiOPc molecules are tilted with respect to the substrate in the neat film, changing to a preferential face-on orientation in the TiOPc-AA mixed films. Complementary information about molecular organization and film structure was obtained using UV–visible absorption and micro-Raman imaging.

The degree of mixing and surface coverage in LB films of Yb bisphthalocyanine $YbPc_2$ mixed with stearic acid (sa) was probed using AFM and micro-Raman imaging[132]. The results show that a 5-layer mixed LB film with 75%/25% $YbPc_2$/sa displays smaller and more homogeneous distribution of aggregates compared with a 5-layer mixed film at 25%/75% $YbPc_2$/sa. Resonance Raman, SERS and SERRS were obtained using the 633 and 780 nm laser lines. The enhanced and unenhanced resonance Herzberg–Teller spectra of neat $YbPc_2$ LB films and mixed LB films excited with the 633 nm laser line are identical.

The vibrational spectra and SERRS of palladiumphthalocyanine (PdPc) evaporated thin solid films were reported by Gaffo *et al.*[133]. They have also performed SERRS, SERRS mapping of the film surface by using micro-Raman spectroscopy with the 633 nm laser line. SERRS of PdPc was obtained by evaporing an overlayer of Ag nanoparticles onto the PdPc film on glass. The SERS enhancement factor was estimated as $\sim 10^4$ with respect to PdPc evaporated films on glass. The molecular organization of the PdPc evaporated films was probed using transmission and RAIR spectra. A random molecular distribution found in PdPc evaporated films resulted to be independent of the temperature.

The solvent effect on the SERS activity of copper phthalocyanine on colloidal silver was performed by Mukherjee *et al.*[134]. The quality of the spectra was improved by optimizing the solvent for the substrate under study. The improvement is explained in terms of co-adsorption and replacement kinetics of the solvent molecules at the silver surface.

The surface-enhanced spectrum of mixed thin solid films of metallophthalocyanines (MPc; M: Cu, Co) and bis(n-propylimido) perylene materials fabricated by vacuum coevaporation onto Ag island films was scanned at different laser lines by Aroca *et al.*[135]. The SERS and SERRS imaging indicate that the dyes of the mixed films were physically adsorbed onto the Ag islands. The SERRS imaging is a useful tool to obtain complementary information about the quality of the films, homogeneity and phase separation.

An examination of the literature reveals that the formation of spontaneously adsorbed metal phthalocyanine molecular films on metal substrates and their applications have not been explored in detail. Surface-enhanced vibrational studies are rare.

6.6. Porphyrins

Surface-enhanced Raman spectroscopy has been used to investigate the physical and chemical processes of the porphyrins on surfaces in many scientific publications. Surface-reactions such as metallization, demetalization, aggregation, electrochemical redox and N-protonation of the porphyrins on colloids, electrodes or films have been revealed by SERS[136–138]. Although most SERS experiments have been carried out on noble-metals, enhanced Raman signals were also obtained from molecules adsorbed on transition metals, semiconductors and on Ag_2O[139]. Small Ag_n^+ clusters on Ag_2O are active sites of SERS. SERS spectra of metallo-tetraphenylporphyrins (MTPP; M: Ag, Cu, Pd, Mg) adsorbed on Ag_2O and hydroxyl-modified silver colloids have been obtained by Zhang *et al.*[140]. The spectra were found to be different from the ordinary resonance Raman spectra, which was interpreted in terms that the adsorbates undergo surface-reactions on Ag_2O colloids. Similar phenomena were observed for the compounds adsorbed on hydroxyl-modified Ag sol. New Raman bands were identified; one of them at 290 cm^{-1} probably belongs to the adsorbate–substrate νAg–N mode. The charge-transfer mechanism could be the major origin of the Raman enhancement observed.

A porphyrins-incorporated self-assembled monolayer (SAM) on gold, was prepared by formation of a 4-pyridinethiol SAM on gold followed by axial ligation of metalloporphyrins to it[141]. The porphyrins-SAM was characterized by SERS and electrochemical techniques; the macroplane of the porphyrin molecule assumes a nearly flat orientation with respect to the gold surface.

A SAM of 2,3,7,8,12,13,17,18-octaethylporphinatozinc formed on a Au surface premodified with a SAM of pyridinethiol was studied by SERS, SEIRA and UV–visible spectroscopies[40]. The SERS spectrum of the porphyrin monolayer on the pyridine–modified Au surface shows bands only from the underlying pyridine group. The frequency shift and change on the relative intensity of bands due to the pyridine group in the SERS spectra imply indirectly the binding of the metalloporphyrins to the pyridine group in the SAM. This result is supported by the SEIRA spectra of the porphyrin–pyridine SAM, in which bands arising from both the porphyrin moiety and the pyridyl group appear clearly. The UV–visible spectra suggest the formation of the porphyrin–pyridine SAM and a weak porphyrin-porphyrin interaction in the SAM.

The effects of surfactants on the SERS spectra of molecules adsorbed on solid substrates can be used to improve the qualities of SERS spectra and to reveal the distance-enhancement relationship of SERS. Liu *et al.*[142] studied the effect of cetyltrimethyl ammonium bromide (CTAB) on the SERS of both the free-base 5,10,15,20-tetrakis(4-*N*-methylpyridyl) porphyrin (H_2TMPyP) and the Ag-chelate (AgTMPyP) adsorbed on Ag colloid. It was found from the SERS spectra that the adsorbed H_2TMPyP molecules undergo silver-incorporation to form AgTMPyP which can change back to H_2TMPyP when added CTAB. The similar demetalation reaction was also observed in AgTMPyP/Ag sol/CTAB systems. The SERS signals were found to be obviously enhanced when CTAB was added. The implications of the findings in relation to the change of micro-environment on Ag particles by the addition of CTAB were discussed.

SERS of 5,10,15,20-tetrakis(1-decylpyridium-4-pyridyl)-21H,23H-porphyrintetrabromide adsorbed on silver hydrosols were compared with the FTIR and resonance Raman spectrum (RRS) in the bulk and in solution by Chowdhury *et al.*[143]. Comparative analysis of the RR and the FTIR spectra indicate that the molecule, in its free state, has D_{2h} symmetry. The SERS spectra, obtained on adsorption of this molecule on borohydride-reduced silver sol, indicate the formation of silver porphyrin. With the change in the adsorbate concentration, the SERS shows that the molecule changes its orientation on the colloidal silver surface. The appearance of a longer wavelength band in the electronic absorption spectra of the sol has been attributed to the coagulation of colloidal silver particles in the sol. The long wavelength band was found to be red-shifted with the decrease in adsorbate concentration. The excitation profile study indicates that the resonance of the Raman excitation radiation with the original sol band is more important than that with the new aggregation band for the SERS activity.

This has been interpreted as a large contribution of the electromagnetic effect to the surface enhancement.

Prochazka *et al.*[144, 145] demonstrated the suitability of surface-enhanced resonance Raman scattering (SERRS) spectroscopy to monitor silver coordination of free base 5,10,15,20-tetrakis(1-methyl-4-pyridyl) porphyrin adsorbed on the Ag colloidal surface. The approach was based on factor analysis (FA) which allows the analysis of the time development of SERRS spectra in terms of varying content of particular porphyrin forms. Their SERRS spectra can be isolated even if none of them is present singularly in any SERRS spectrum. Depending on the chemical properties of the molecular systems, SERRS spectra of two different porphyrin metalated forms have been obtained: Ag^+ and Ag^0 metalated. The latter form was found to be evoked by pre-treatment of the laser-ablated Ag colloids by the thiosulphate ions. The metalation kinetics obtained show a substantially slower metalation by Ag^0 than that by Ag^+ cations.

SERRS and voltammetry techniques were used to elucidate the mechanism of oxygen reduction on a silver electrode with iron(III) tetra-4-*N*-methylpyridylporphyrin[146]. The results indicate that after the oxidation-reduction cycle at pH 10 and pH 4 the iron porphyrin is adsorbed on the Ag surface as a high-spin, five-coordinated μ-oxo-bridged dimer. SERRS spectra show that the first electron transfer forms a mixed-valence Fe(II)-O-Fe(III) μ- oxo-bridged dimer.

It has been described the formation of porphyrin- and sapphyrin-containing self-assembled monolayers on electrochemically prepared Au surfaces by a multistep approach[147]. After electrochemical preparation the Au substrate was characterized by SERS and cyclic voltammetry. SERS characterization was also used to study the basic architecture and properties of the porphyrin- and sapphyrin modified SAMs and their interactions with fluoranthene, 1,10-phenantroline and adenine. In general, it can be proposed that macrocycle oligopyrrole functionalized SAMs could be useful in the analysis and detection of polyaromatic and heterocyclic compounds.

SERRS spectra of Fe-protoporphyrin IX adsorbed on Ag colloidal nanoparticles immobilized onto a polymer-coated glass slide were obtained at very low concentrations[148]. The spectra exhibit drastic temporal fluctuations on a time scale of seconds in both line frequency and intensity. This result is consistent with an approach to the single molecule limit. Sequences of spectra were analyzed in terms of an underlying continuum and of Raman peaks superimposed on this continuum. The statistical analysis of the spectrum intensity suggested that the main contribution to the intensity fluctuations arise from the continuum. It was found

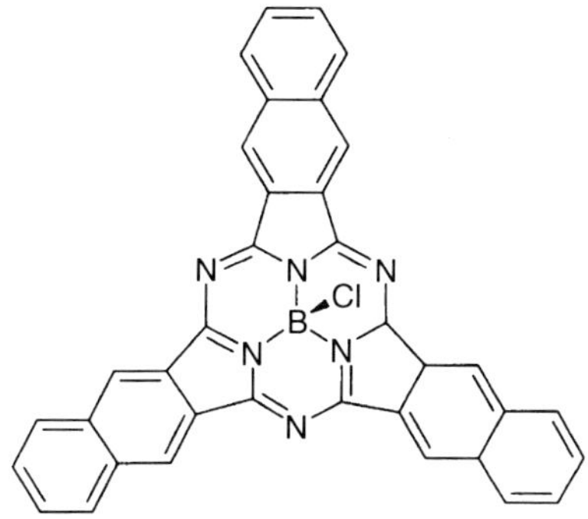

Figure 14.31. Sub-2,3-boronnaphthalocyanine chlorine. Reprinted by permissionn of Elsevier B.V.

a high correlation between the total integrated intensity and the intensity detected at different vibrational modes of the compound. The ratio between the intensity detected in correspondence of different vibrational modes of the molecule shows a temporal variability, which reflects the intrinsic dynamics of the molecules. The set of results was interpreted as a desorption–adsorption mechanism of the molecules at the Ag surface.

6.7. Miscellaneous Macrocycles

The subphthalocyanine (SubPc) macrocycles have been received a great deal of attention because of their interesting optical properties and their utility in the synthesis of asymmetrical non-planar phthalocyanines[149,150]. SubPc compounds consist of three coupled isoindole units with a boron atom at the centre. One component of this series is the sub-2,3-boronnaphthalocyanine chlorine (subBClNPc) compound, Figure 14.31. This compound has been considered in this chapter since it represents a new challenge in the optic of the surface vibrational studies because of the structure of this compound is conic, as in the case of the sub-2,3-boronphthalocyanine chlorine macrocycle[151], instead of the planar structure of the naphtalocyanine complexes already mentioned[70].

The adsorption of subBClPc compounds on metal surfaces gives rise to conformational changes at contacting interfaces of these macrocycles, which is matter of crucial importance for the design of technological functional devices based on these molecules like organic light emitting diodes and nanoscale molecular devices[11,152].

The spectral results and calculations performed by Saavedra *et al.*[153] indicate that the interaction of the macrocycle sub-2,3-boronnaphthalocyanine chlorine (subBClNPc) with different metallic surfaces is rather weak. The interaction with the surfaces takes place through the chlorine atom. This situation is mainly due to the cone-shape structure of subBClNPc, in contrast to other slightly warped naphthalocyanine complexes or planar azabipiridyl complexes, where the interaction with metal surfaces occurs through the whole π electronic system[41,42,76]. Such an orientation leads to slight differences in the band intensities on varying the incidence angle in RAIRS experiments (Figure 14.32).

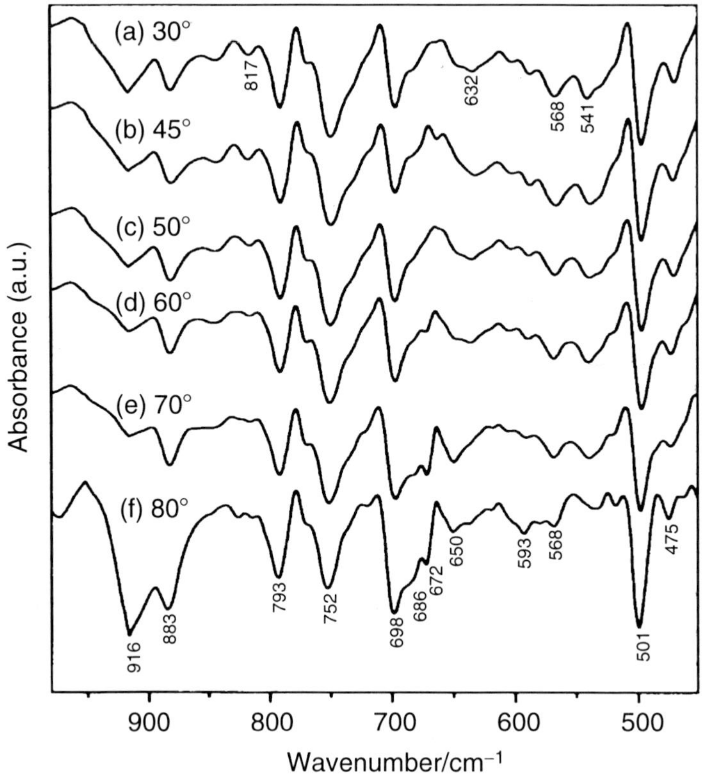

Figure 14.32. RAIRS of sub-2,3-boronnaphthalocyannie chlorine at different radiation angles. Reprinted by permission of Elsevier B.V.

The structure of subBClNPc is slightly modified by surface effect; this was concluded from the frequency changes observed in the RAIRS and SERS spectra. The extent of the subBClNPc surface interaction is rather limited in comparison to the other related NPc macrocycles. The relative intensities of the SERS spectra, Figure 14.33, were interpreted by using the two main mechanisms governing the SERS effect; that is electromagnetic and chemical.

The RAIRS, SEIRA and SERS spectra of the Ni(II) (*trans*-7,14-dimethyl-5,12,-diphenyl-1,4,8,11-tetraaza-2,3-9,10-dinaphthyl-cyclotetradeca-5,7,12,14-tetraene) ($NiN_4\phi_2$) and Cu(II) (5,7,12,14-tetramethyl-1,4,8,11-tetraaza-2,3-9,10-dinaphthyl-cyclotetradeca-5,7,12,14-tetraene) (CuN_4) in Figure 14.19, were reported[38]. For the RAIRS analysis, complexes were deposited onto a smooth copper surface and onto a KBr monocrystal. Both spectra were compared with the spectrum of the complex dispersed in KBr. No substantial frequency differences between them were observed, pointing out that the surface does not markedly influence the corresponding complex structure.

The RAIRS spectrum scanned at 70° of the $NiN_4\phi_2$ complex adsorbed on a smooth copper surface is displayed in Figure 14.34[38]. The RAIRS spectra were analysed in the 1,400–700 cm^{-1} region since bands corresponding to in-plane

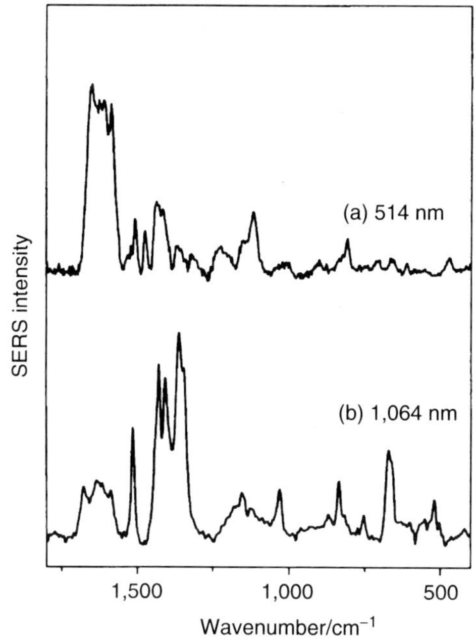

Figure 14.33. SERS of sub-2,3-boronnaphthalocyanine chlorine. Reprinted by permission of Elsevier B.V.

and out-of-plane CH deformations, which are very important for understanding the adsorbate orientation, usually appear in this region. This spectrum is compared to that of the same complex in KBr (Figure 14.34b). By taking the band at 1,265 cm^{-1} as a reference, an intensity decrease of the absorption bands appearing at 1,233 cm^{-1} and those falling in the 1,200–1,000 cm^{-1} region is observed. Since they correspond to δCH motions of the naphthalene fragments of the macrocycle, it has been deduced a parallel orientation of these groups with respect to the metal surface.

Below 1,000 cm^{-1} intense bands are observed at 887, 847 and 750 cm^{-1} in the RAIRS spectrum; they belong to the ρCH motions of the naphthyl moiety. These bands are intensified due to their perpendicular orientation supposing a parallel orientation of these residues with respect to the surface. In contrast the bands at 870 and 773 cm^{-1}, are markedly decreased in the RAIRS. These are assigned to phenyl ρCH motions. Their intensity decrease in the RAIRS suggested a perpendicular position of phenyl rings in the macrocycle complex.

The RAIRS spectra registered for the Cu complex follow a similar behaviour as compared to the Ni complex[38]. The bands appearing in the 1,200–1,000 cm^{-1} region are relatively weakened, thus meaning that the naphthyl moiety adopts a parallel orientation on the surface. However, bands below 1,000 cm^{-1} display a rather random intensity variation at different radiation angles. This result suggested that these modes are highly coupled to other vibrational modes, thus avoiding infer about the molecular organization onto the surface.

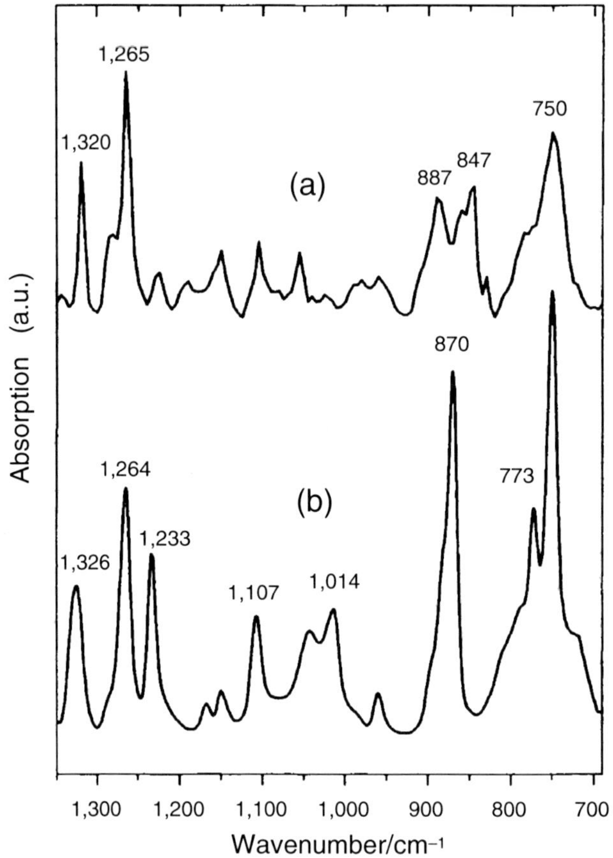

Figure 14.34. (a) RAIRS of $NiN_4\phi_2$ adsorbed on smooth copper surface at 70° of the incident radiation; (b) IR spectrum of the same complex in KBr. The spectra are baseline corrected for a better comparison. Reprinted by permission of Elsevier B.V.

The RAIRS analysis and differences observed in the relative intensities of complexes deposited onto KBr and Cu surfaces indicate that the organization of each macrocycle is different on both surfaces. Moreover, the structure of each complex is not markedly influenced by surface effect.

The SERS spectrum of the Ni complex on Ag colloid in Figure 14.35b displays a marked enhancement of the out-of-plane deformation naphthyl moieties bands at 471, 602 and 640 cm^{-1}.

The corresponding in-plane vibrations at 1,273, 1,380, 1,480 and 1,514 cm^{-1} undergo a relative intensity decrease. All the above results suggest that the coordination plane of the complex is face-on oriented on the metal. On the other hand, the observation of an enhancement in bands at 1,176 and 1,237 cm^{-1}, attributed to in-plane δCH phenyl modes, and at 1,453, 1,548, 1,590 and 1,630 cm^{-1} attributable to the νCC of the phenyl substituents, or to modes in which this group is involved, supports the idea of a perpendicular orientation of these groups with

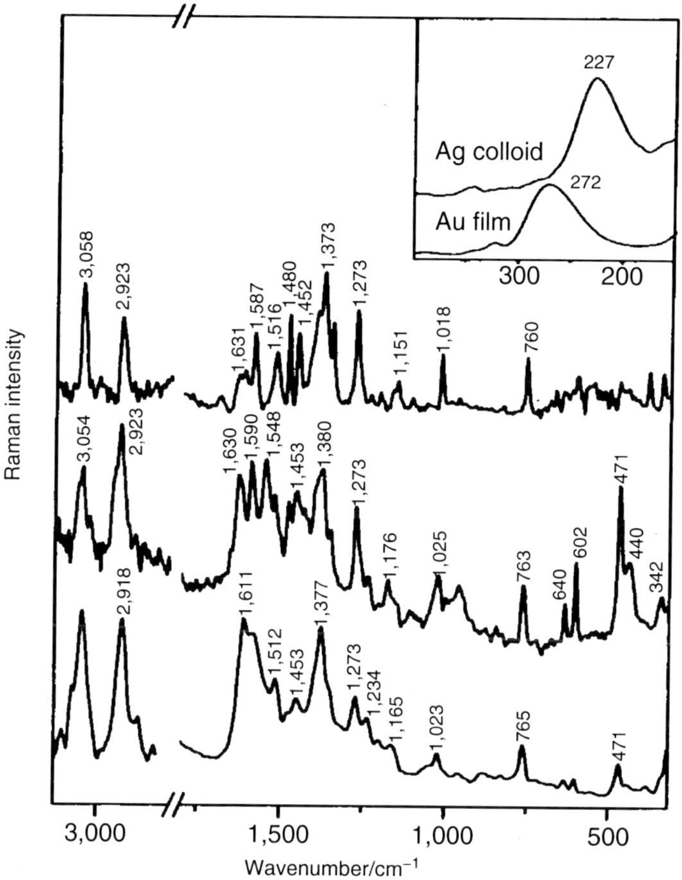

Figure 14.35. (a) Raman spectrum of $NiN_4\phi_2$ in the solid; (b) SERS in colloidal silver; (c) SERS on Au film. The inset figure corresponds to the (b) and (c) ones amplified in the 100–400 cm^{-1} region. Reprinted by permission of Elsevier B.V.

respect to the metal. This orientation is connected to the existence of a relatively intense band at 3,054 cm^{-1} due to ν CH modes of the phenyl group. The frequency shifts observed for some bands indicate the existence of an interaction between the complex and the metal surface. This is corroborated by the presence of an intense band at 227 cm^{-1}, assigned to νAg–N motions.

The SERS spectrum registered on Au film (Figure 14.35c) is different to that obtained on Ag colloid. The out-of-plane bands of the naphthyl group in the 400–700 cm^{-1} region are weaker, while the naphthyl in-plane δCH bands attributed to this group at 1,023 and 1,165 cm^{-1}, and the naphthyl νCC modes appearing at 1,377, 1,512 and 1,611 cm^{-1}, increase their intensities markedly. This increase is also extensive to the νCH modes at 3,054 cm^{-1}. All these results indicate that the organization of the Ni complex on Au changes with respect to that found on Ag colloids. On Au film it was not possible to conclude about a parallel or perpendicular orientation of the complex on the metal.

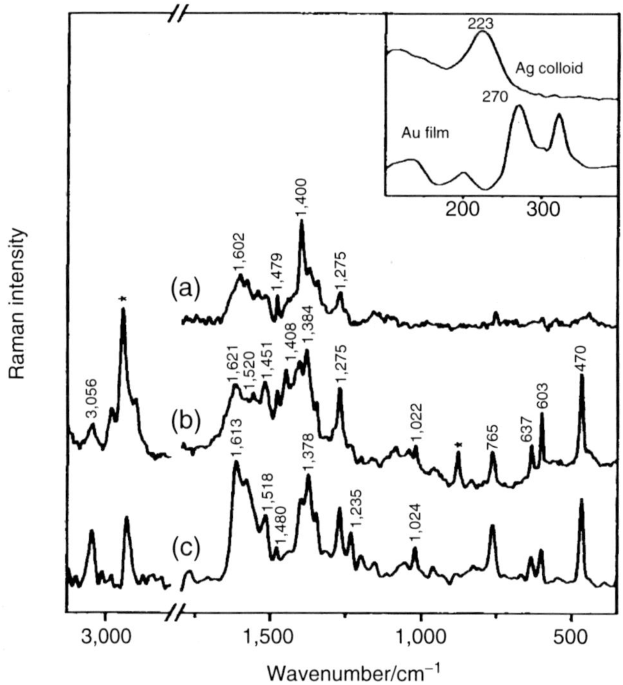

Figure 14.36. (a)Raman spectrum of CuN_4 in the soild; (b) SERS in colloidal silver; (c) SERS on Au film. Asterisks represent bands due to ethanol. The inset figure corresponds to the (b) and (c) ones amplified in the 100–400 cm^{-1} region. Reprinted by permission of Esevier B.V.

The SERS spectrum of CuN_4 on Ag colloid (Figure 14.36b) indicates that the complex adopts a face-on orientation on the metal as deduced also for the Ni complex.

This is concluded from the intensification of the naphthyl out-of-plane bands at 471, 602 and 637 cm^{-1}. The close position of these bands in relation to the Ni complex indicates a similar structure of the macrocycle when adsorbed on Ag, irrespective of the nature of the coordination metal. This result is also similar to that obtained for azabipiridyl macrocycle complexes[79]. The orientation also accounts for the relative intensity decrease of the νCH band at $3,056\,cm^{-1}$. In fact this band in the Cu complex corresponds to the νCH modes of the naphthyl moieties which are oriented parallel to the metal surface. The interaction of the macrocycle with the Ag surface is also deduced from the frequency shifts of some bands mainly those corresponding to the pyrrole moieties, which are the fragment through which the macrocycle-surface interaction takes place. The presence of the band at 224 cm^{-1} corroborates this interaction.

On Au film (Figure 14.36c) a marked intensification of the in-plane modes of the 1,500–1,650 cm^{-1} region is also observed along with a relative increase of the band at 3,056 cm^{-1}. However, the presence of prominent bands at 470, 603 and 637 cm^{-1}, suggests the existence of face-on oriented macrocycles. The simultaneous observation of both out-of-plane and in-plane intense modes on the Au film is probably an effect of the existence of a random orientation of the molecules on this metal.

SERS data indicate for both complexes an adsorbate–substrate interaction between the complex and the colloidal Ag surface. From the SERS experiments it has been also concluded that the complexes are oriented face-on to the surface being the number of Cu macrocycles oriented face-on to the surface larger than that of the Ni macrocycles. The organization of the complexes on Au film is different to that found on Ag colloids and rather random. Complexes are better organized on the Ag surface. The phenyl groups avoid a better organization of the Ni complex on the surface.

SEIR spectra of the complexes deposited on a Au film are displayed in Figure 14.37 together to those of the complexes deposited on neat CaF_2. In the Ni complex (Figure 14.37a–c) the most important feature is the remarkable intensity decrease of the band at 1,611 cm^{-1} which undergoes a strong up shift to 1,629 cm^{-1}.

The rest of bands remain more or less at the same frequency and intensity with the exception of that at 1,482 cm^{-1}, which shifts downward. In the case of the Cu complex a marked decrease of bands at 1,404, 1,515, 1,581 and 1,624 cm^{-1} is also noted, the last band appearing at a higher frequency (1,638 cm^{-1}). The other bands remain at more or less the same position and relative height. The intensity decrease of bands at 1,228/1,229, 1,107/1,105 and 860 cm^{-1} observed in the SEIRA spectra of both complexes is related to the parallel orientation of naphthyl groups in relation to the metal surface.

It is interesting the effect of the dilution with ethanol on the above SEIRA sample. This effect is more important in the case of the Ni complex; a relative decrease is observed for the 1,408 cm^{-1} band, which downward shifts to 1,404 cm^{-1}, whilst the band at 1,629 cm^{-1} shift to 1,638 cm^{-1}, and that at 1,583 cm^{-1} almost disappears. In contrast, the SEIR of Cu complex does not undergo a significant change except for the band at 1,580 cm^{-1}.

The different spectral behaviour of both complexes before and after washing the sample with ethanol reflects the fact that the Ni complex is less organized on the Au in relation to the Cu one, as it was deduced from the SERS spectra. The dilution lead to the removing of those molecules that are physisorbed onto the metal, i.e. that are weakly linked to the surface, remaining attached only the chemically adsorbed ones. The lower organization of the Ni complex is attributed to the presence of a phenyl group in the ligand. In the case of the Cu complex the amount of chemisorbed adsorbates is significantly higher, as concluded from the similar SEIRA spectra before and after washing the sample. A good agreement between SERS and SEIRA results is observed.

The profile of the SEIR spectra obtained for both complexes after washing is very similar, thus indicating a similar structure of both complexes on the Au film. Figure 14.37c,f). The small changes observed between the IR spectra of solid films and the SEIRA are indicative of a low organization degree in the Au film, since the molecules seem to be randomly oriented as in the case of the solid film. Nevertheless, the interaction of the metal with the macrocycles is demonstrated by

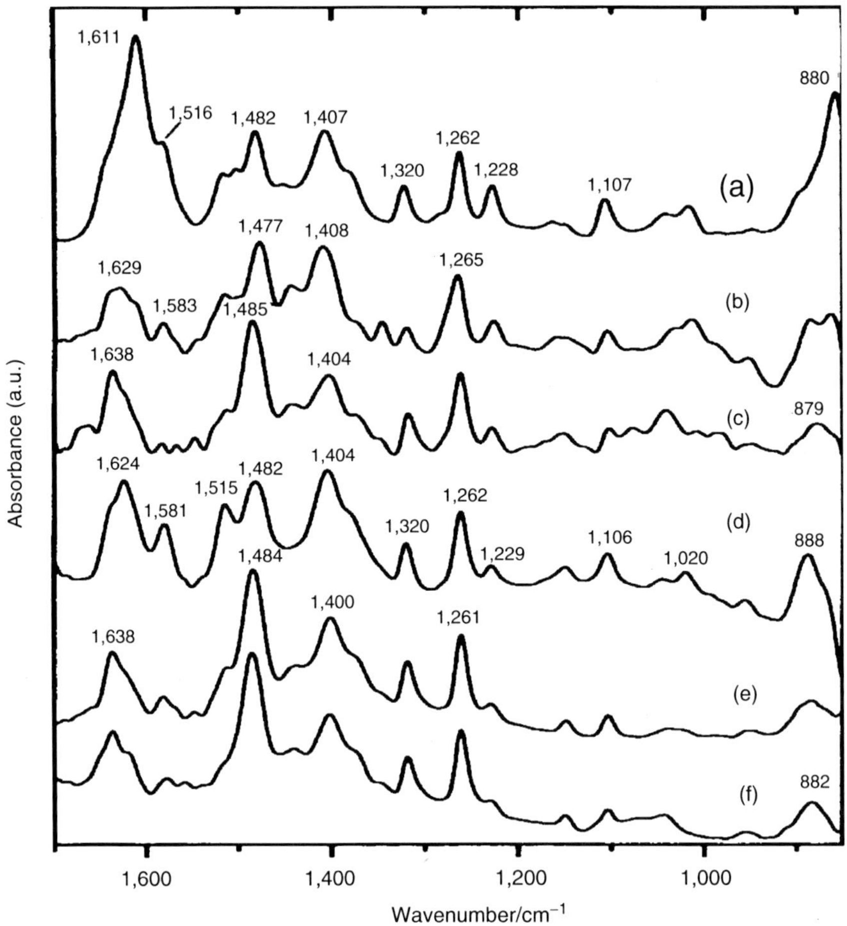

Figure 14.37. (a) Infrared spectrum of $NiN_4\phi_2$ deposited on CaF_2 and (c) SEIRA spectra of $NiN_4\phi_2$on Au film before and after washing with ethanol, respectively. (d) IR spectrum of CuN_4 on Au film before and after washing with ethanol, respectively. Reprinted by permission of Elsevier B.V.

the shifts that undergo the bands appearing above 1,400 cm^{-1}. This is consistent with an interaction of the macrocycles with the metal via the tetraaza moiety, directly linked to the metal cation.

7. Adsorbate–Substrate Interaction Vibrations

One of the most relevant evidences of a relatively significant interaction between the macrocycle complex and the metal surface is the appearance of a vibrational band. This kind of metal–ligand bands is better observed in the Raman spectrum below 500 cm^{-1}. Bands corresponding to the adsorbate–substrate interaction were observed in azabipiridyl, phthalocyanines, naphthalocyanines and miscellaneous complexes interacting with different metal surfaces, mainly colloidal silver.

7.1. Azabipiridyl Metal Complexes

In general, the Raman spectra of azabipiridyl and its Co(II), Ni(II) and Cu(II) complexes solids display a similar pattern in the series, with expected differences below 500 cm^{-1}, where the metal–ligand vibrations appear[79]. An enhancement of the Raman lines falling in this region is observed in the SERS spectra (Figure 14.27), along with frequency shifts of several bands. Conclusions regarding the existence of an adsorbate–substrate interaction, as well as structural modifications of the macrocycle by the surface effect were derived from the analysis of this region. The adsorbate–substrate interaction was verified by the appearance of the medium band at about 216 cm^{-1} in the series of complexes, ascribed to an Ag-macrocycle binding, involving mainly the N atoms of the coordinating site; this assignment was also proposed in amantadine[154] and in phthalimide adsorbed on colloidal silver[112]. This frequency is about the same in the series except in the case of the Ni complex where it appears at 221 cm^{-1}, suggesting that the Ni macrocycle–Ag interaction is the strongest in the series. The weaker interaction at 214 cm^{-1} occurs in the case of the substituted Ni complex; this situation is due to an steric hindrance caused by the hexyl groups, or to a π electronic redistribution imposed by the substituents, which is different in the unsubstituted Ni complex. This band is not clearly observed in the case of the macrocycle ligand.

7.2. Phthalocyanine Metal Complexes

The band observed in the SERS spectra of aluminium-oxo phthalocyanine films deposited on a Ag island film at 248 cm^{-1} was proposed as due to the Pc–Al bond[2]. The IR, Raman and SERS spectroscopic study indicate a random orientation of the molecule onto the surface.

The 2,9,16,23-tetraamino metal {metal = Co(II), Cu(II) and Fe(III)} phthalocyanines absorbed onto gold and silver electrodes[125] display Raman characteristic bands around 236 cm^{-1}, corresponding to metal-N vibrations, the nitrogen atom belonging to the NH_2 group on the phthalocyanine molecules.

7.3. Naphthalocyanine Metal Complexes

From the Raman spectra and SERS on silver colloid of vanadyl-5,14,23,32-tetraphenyl-2,3-naphthalocyanine VOTPNPc it was deduced a short distance complex-surface interaction;[78] this was verified by the appearance of a band at about 216 cm^{-1}, previously assigned to a Ag–N binding;[154] this band is not observed in the free molecular system.

The chemical interaction between the adsorbate and the substrate in the case of the sub-2,3-boronnaphthalocyanine chlorine adsorbed on metal surfaces was verified by a chemical Ag–Cl binding as deduced from the RAIRS and SERS data and theoretical calculations;[153] it was related to the appearance of the band at 244 cm^{-1} on the Ag film (Figure 14.33a) and at 239 cm^{-1} on the Ag colloid (Figure 14.33b)[155].

7.4. Dinaphthalenic Ni(II) and Cu(II) Azamacrocycle Complexes Methyl and Phenyl Substituted

The interaction of the macrocycle Cu(II) (5,7,12,14-tetramethyl-1,4,8,11-tetraaza-2,3-9,10-dinaphthyl-cyclotetradeca-5,7,12,14-tetraene) (CuN_4), with the Ag colloidal surface was deduced from different spectral facts; in particular, from the frequency shifts of some bands mainly those corresponding to the pyrrole moieties, which are the fragment through the macrocycle–surface interaction takes place[38]. The presence of the SERS band at 223 cm^{-1} assigned to a νAg–N mode corroborates this interaction (Figure 14.36). The interaction between the relative complex Ni(II) (*trans*-7,14-dimethyl-5,12-diphenyl-1,4,8,11-tetraaza-2,3-9,10-dinaphthyl-cyclotetradeca-5,7,12,14-tetraene) ($NiN_4\phi_2$) and the Ag colloidal surface was deduced from the appearance of the intense SERS band at 227 cm^{-1} (Figure 14.35).

7.5. Molecular Model and Theoretical Data of the Adsorbate-Substrate Interaction

With respect to the metal–adsorbate interaction it is of interest, for comparison with the experimental results, to develop theoretical calculations based on molecular models that involve the adsorbate molecules and a metal cluster. The INDO/1 semiempirical method has been proved to be very useful in the structure optimization and in the evaluation of the bond characteristics throughout the Wiberg index[65,101]. The interaction of azamacrocycle complexes with simulated copper surfaces studied by the same method was performed[79,100].

Density functional theory (DFT) is adequate to simulate such adsorbate–substrate systems[156,157]. This method is a non-empirical approach, alternative to Hartree-Fock-bases theory; it presently finds wider applications to chemical problems due to the possibility of including a significant part of electron correlation energy at a relatively low computational cost. Calculations on the structure and vibrational wavenumbers of the investigated compound are performed using the GAUSSIAN 98 program package[158], and the Becke's three-parameter hybrid method using the Lee–Yang–Parr correlation functional (B3LYP)[159]. Computed harmonic frequencies typically overestimate vibrational fundamentals due to the basis set truncation and neglect of electron correlation and mechanical anharmonicity. To compensate for these shortcomings, various scaling strategies exist for bringing the computed frequencies into a greater coincidence with the observed wavenumbers. Works by Scott and Radom[160] and Wong[161] showed that DFT consistently predicts harmonic vibrational frequencies in better agreement with observed fundamentals than conventional *ab initio* methods at a much lower cost. Comparison of observed fundamental with computed harmonics using the 6-31G(d) basis set gave scaling factors of 0.9833 for S-VWN, 0.9945 for BLYP and 0.9614 for B3LYP, whereas scale factors for conventional *ab initio* methods were 0.9427 and 0.9537 for MP2 and QCISD, respectively[160,161]. The 6-31G* Pople split-valence polarization basis set is generally used in the geometry optimisation and normal modes calculations.

Theoretical calculations about the energetic of the adsorbate–substrate interaction for approximate molecular models of the azamacrocycle series mainly containing free and metal coordinated azabipiridyl, naphthalocyanine and miscellaneous systems are now reviewed.

7.6. Azabipiridyl Metal Complexes

A theoretical study using the INDO/1 semiempirical method and a molecular model for the interacting adsorbate–substrate system (azabipiridyl metal complex-smooth copper surface) was performed[79]. This work pointed to determine the stability of the complex formation in the series Ni(II), Cu(II) and Zn(II) and to understand the energetic process of the complex–surface interaction. The structure of the macrocycle ligand and the corresponding metal complexes were optimized by means of the INDO/1 method. The hydrogen atoms of the ligand were constrained to adopt an out-of-plane position in agreement with which is expected for an averaged sp^3/sp^2 hybridization of the nitrogen atoms in the coordination site. In the case of the Ni and Cu complexes the energy minima were found for the ions located in the plane of the coordination site; the most stable geometry for the Zn complex was found for the cation located out of the coordination site in 0.5 Å. The metal surface was proposed to be a cluster composed by a minimum of five copper atoms representing a *001* face. Four of these atoms make a square and the fifth atom is located in the centre of the square at 2.55 Å from the corners; this distribution agrees quite well with experimental results[162], and it is not different to the metal surface characteristics of the model proposed by Lamoen *et al.*[163] for the computation of the structure, electronic and transport properties of PdP and perylene monolayers on the Au(*111*) surface. Padilla *et al.*[164], studied, at ab initio level, the adsorption of alkali-metals on a Cu(*111*) surface; they used small-cluster models (Cu_7) to represent the adsorption sites of the copper surface. The nitrogen atoms of the macrocycles were superposed onto the *001* face as it is indicated in Figure 14.38.

Theoretical data supported quite well the vibrational IR and RAIRS data and allowed an interpretation of the energetic process of the complex–surface interaction. The variation observed in the Wiberg index values for the metal–ligand coordination bond suggests an important π electronic redistribution around the cation by surface effect Table 14.7.

The bond character of the Ni–surface interaction is higher than in the Cu(II) and Zn(II) systems. The Wiberg index values indicate that the metal–surface interaction makes weak the metal ligand coordination. By comparing the metal–N bond character in the interacting systems with that obtained for the isolated complexes, it was deduced that the π electronic cloud moves from the complexes to the copper surface. The complex–surface distances are 1.9, 2.0 and 2.5 Å for the Ni, Cu and Zn complexes, respectively. The complex–surface interaction is favoured in the series Ni > Cu > Zn. This interaction is favoured when the stability of the complex increases. Moreover, the interaction between the macrocycles and the metal surface is energetically favoured by increasing both the electronic π system around the coordination site and the number of the N atoms in the macrocycle ring. The structure of the azabipiridyl ligand is uninfluenced by surface interaction; its molecular orientation onto the surface is random. This situation is due to the position of the macrocycle ring hydrogen atoms, which are oriented out

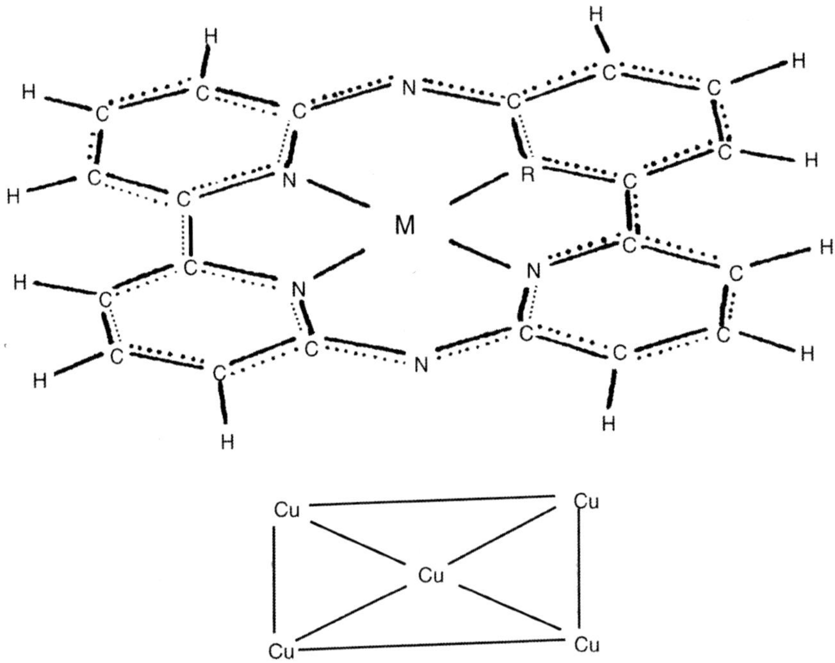

Figure 14.38. Diagram of the complex–substrate interaction model, M: Ni, Cu, Zn.

Table 14.7. Selected Wiberg index values for different interactions

Interaction type	Wiberg index
Ni–surface	0.451
Cu–surface	0.264
Zn–surface	0.070
Ni–N (onto the surface)	0.271
Ni–N (isolated)	0.315
Cu–N (onto the surface)	0.292
Cu–N (isolated)	0.358
Zn–N (onto the surface)	0.353
Zn–N (isolated)	0.366

of the plane of the coordination site. Ni and Cu complexes are preferently oriented plane parallel to the surface. The molecular structure of the Ni and Cu complexes is altered by the interaction. In the case of the Zn complex, where the ion is out of the plane of the coordination site, the interaction with the surface is negligible. As a consequence, the structure of the macrocycle is not modified by the surface effect.

In a rough approximation the macrocycle–metal colloid interaction model was represented by the macrocycle itself and a cluster composed by 12 Ag atoms

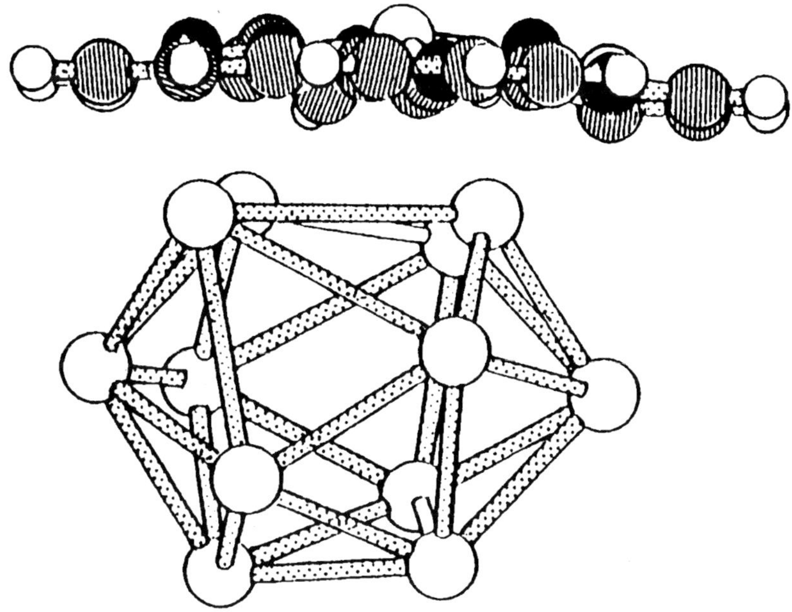

Figure 14.39. Diagram of the macrocycle–colloid interaction model. Reprinted by permission of Elsevier B.V.

Table 14.8. INDO/1 values for the Ag–N and Ag–metal mean distances (A°) and the interaction energy δ E (KJ mol), for the azabipiridyl metal complexes imteracting with the collodial silver Ag_{12}

System	Ag–N	Ag–metal	ΔE
Ag_{12}–Co	2.67	2.98	– 218.91
Ag_{12}–Ni	2.53	2.90	– 250.71
Ag_{12}–Cu	2.61	2.95	– 240.54

which mimics an icosahedral-type structure. The macrocycle structure evolves to a minimum of energy during the calculation. (Figure 14.39).

The most relevant data of the adsorbate-substrate interaction, listed in Table 14.8, indicate that the Ni complex–colloidal Ag interaction is the most stable in the series. The Wiberg index values of the free complexes and the adsorbate–substrate adduct in Table 14.9, point to an electronic energy redistribution by interaction effect, being the most influenced bonds those directly involved in the metal coordination site.

7.7. Naphthalocyanine and Its Cu(II) Complex

The molecular model proposed in the theoretical study of the azabipiridyl–surface interaction[79], was inspired in a similar idea proposed by Aroca *et al.*[165] to perform the normal coordinate analysis of maleimide chemisorbed on colloidal

Table 14.9. INDO/1 Wiberg index values for selected bonds of the free complexes and the adsorbate–substrate adducts.(See Figure 14.10)

Bond	Wiberg index					
	Co		Ni		Cu	
	Free	Adduct	Free	Adduct	Free	Adduct
M–N2	0.418	0.430	0.300	0.305	0.357	0.430
N2–C3	0.931	0.908	0.944	0.915	0.934	0.905
N2–C6	0.927	0.841	0.920	0.851	0.911	0.831
C3–N4	0.900	0.912	0.901	0.925	0.886	0.911
C3–C5	1.020	1.007	1.038	1.013	1.030	1.003
Ag–N[a]	0.017	0.060	0.050			

[a] Averaged.

Table 14.10. Distance from the complex and the lignand to the central atom of the copper surface and the energy of each interaction

Interacting model	d (Cu–X[a]) Å		ΔEg kJ mol^{-1}[b]
Complex/surface	2.15	−	913.0
Ligand/surface (plane parallel)	2.00	+	680.9
Ligand/surface (tilted)	4.00		0.0

[a] In the complex X corresponds to the copper ion, and in the ligand represents the geometrical center of the coordination site.
[b] Referred to an arbitrary zero energy of the ligand.

silver. This model was also used in the calculation of the naphthalocyanine (Npc) and CuNPc systems interacting with a copper surface[101] (Figure 14.40).

The geometry and energy parameters for the more stable ligand-surface and complex-surface interactions are listed in Table14.10. These values indicate that the complex-surface system is most stable than the ligand-surface system in about 913 kJ mol^{-1}. The complex-surface distance 2.25 Å and the energy of its interaction represent a strong van der Waals-type interaction for this plane parallel orientation of the complex onto the surface.

The displacement of the copper atom 0.3 Å out of the coordination site plane stabilizes the interaction just slightly in about 16 kJ mol^{-1}. The distance 4.0 Å from the geometrical centre of the ligand to the centre of the metal cluster is imposed by the out of plane position of the hydrogen atoms located in the coordination site, constraining the ligand to adopt a tilted orientation of about 27° respect to the surface. The structure of the ligand is uninfluenced by the metal surface interaction; the extended Hückel Wiberg index of the CN bonds of the adduct remain nearly constant respect to the isolated ligand. In the case of the isolated complex equivalent indolic and external CN bonds are observed.

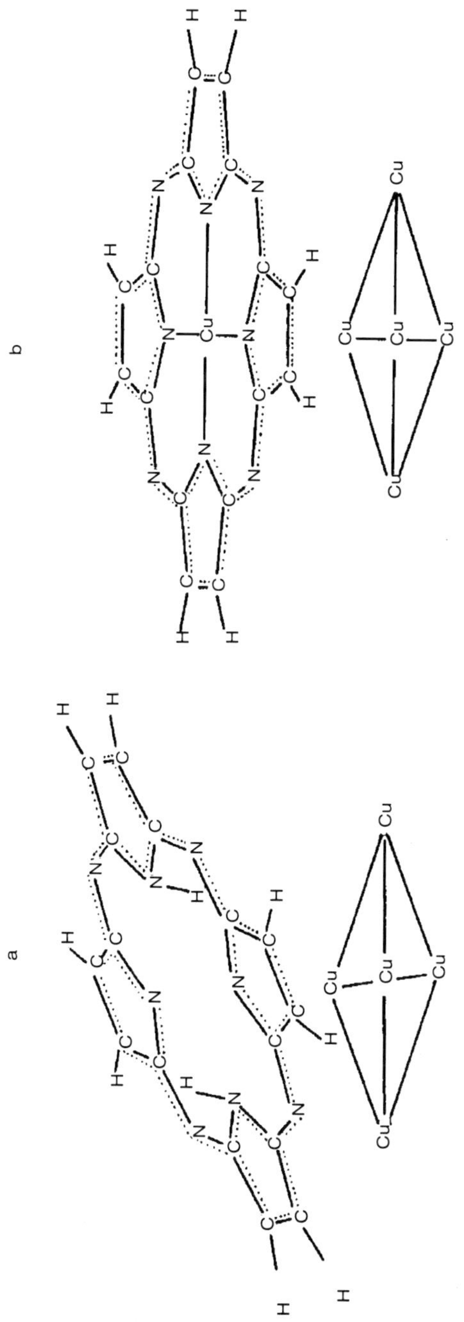

Figure 14.40. Diagram of the adsorbate–substrate interaction model (a) ligand, (b) complex. Reprinted by permission of Elsevier B.V.

The Wiberg index values for the coordination site of the complex during the interaction decrease in about 10%; for the external bonds the Wiberg index values remain nearly constant. Bonds far from the coordination site are uninfluenced by the interaction. During the interaction the net charge of the coordination nitrogen atoms become more positive and the copper ion of the complex and the copper atoms of the cluster modify their net charges. The copper ion becomes more positive in 0.04e while the central copper in the cluster gains 0.18e of negative charge; the other 0.14e come from other atoms of the cluster. The above results suggest an electronic transfer from the complex to the surface through the copper ion during the interaction, and explain the stabilization of the complex-substrate adduct. The modifications of the electronic energy of the complex should be accompanied by a structural change.

The number of the N atoms in the azamacrocycle ring has an important influence in the nature of the interaction with the surface; the bigger the number of the N atoms the more stable the complex-surface interaction is. The naphthalocyanine Cu(II) complex (NpcCu) displays 8 N atoms in the macrocycle ring and the total π electronic cloud is bigger than in the azabipiridyl Cu(II) complex. Although the NpcCu–Cu surface distance is longer (0.15 Å) than in the case of the azabipiridyl copper complex[79], and that the naphthalocyanine complex was simulated to be a porphyrine type system[101], the NpcCu-Cu interaction is stabilized in about 86.6 kJ mol^{-1}.

7.8. Dinaphthalenic Ni(II) and Cu(II) Azamacrocycle Complexes Methyl and Phenyl Substituted

The interaction of dinaphthalenic Ni(II) and Cu(II) substituted complexes with a smooth copper surface was performed by using the semiempirical method INDO/1 and simplified molecular models in which substituents were absent[38]. The structure of the model compounds is in Figure 14.41.

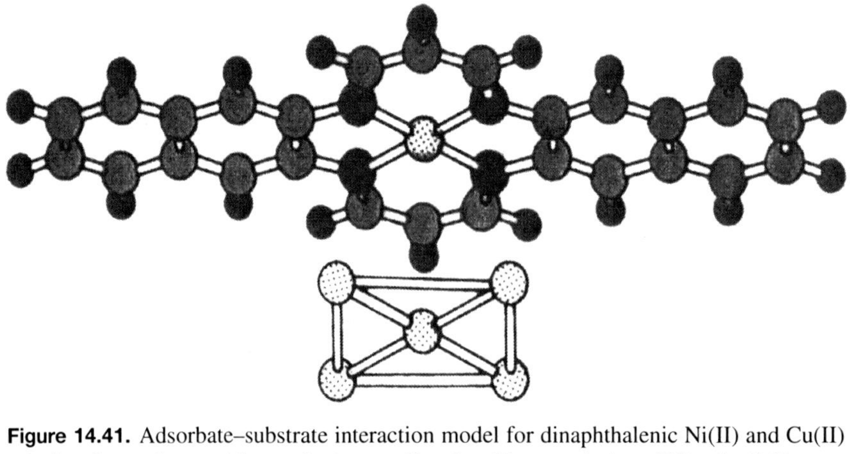

Figure 14.41. Adsorbate–substrate interaction model for dinaphthalenic Ni(II) and Cu(II) substitued complexes without substituents. Reprinted by permission of Elsevier B.V.

Table 14.11. Most relevant Wiberg index values for different interactions

Cu–$Cu_{surface}$	0.252
Ni–$Cu_{surface}$	0.400
Cu–N on the surface	0.435
Cu–N free	0.387
Ni–N on the surface	0.286
Ni–N free	0.325

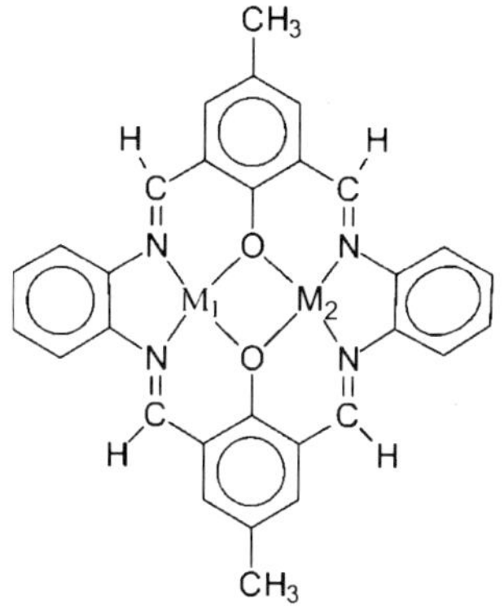

Figure 14.42. Robson-type azamacrocycle complex.

The surface model was identical to that used in the case of a similar study carried out for azabipiridyl[79], naphthalocyanine (NPc) and CuNPc macrocycles interacting with a copper surface[101].

The most relevant Wiberg index values for the isolated and the macrocycle-cluster interaction systems are displayed in Table14.11.

The stabilisation energy for the macrocycle-surface interaction model in Figure 14.41 is −2,078 and −2,401 kJ mol^{-1} for the Cu and Ni complexes, respectively; the M–Cu distance is equal to 2.0 Å for both systems. These data and the corresponding Wiberg index values for the macrocycle ring indicate that the adsorbate–substrate interaction is more significant in the case of the Ni complex. The Wiberg index values are not different for the rest of the macrocycle-copper surface systems in relation to the free macrocycles, which means that a π electronic redistribution is produced by the interaction effect, mainly involving the coordination site of the adduct; the rest of the macrocycle is predicted to be structurally unaltered[38].

8. New Trends

Robson type azamacrocyclic complexes in Figure 14.42 are of interest as new materials and as models of biomolecules. The availability of a wide variety of planar ligands with different cavities offers the opportunity to study in detail magnetic phenomena, metal ion recognition, encapsulations, transport and separation effects, catalytic process, and new molecular devices. Paredes *et al.*[166] reported the study of different Cu(II) complexes with the macrocyclic and hemicyclic ligands, derived from condensation reaction of 2-hydroxy-5-methyl-1, 3-benzenedicarbaldehyde (BDC) with 1,2-diamine-benzene (OPDA). These systems show a great delocalization and therefore a high planarity of the macrocycle. These complexes have been characterized by different spectroscopic techniques, elemental analysis, magnetic susceptibility and electrochemical studies. Less delocalized systems can be obtained if OPDA is replaced by 1,3-diamino-2-propanol (DAP)[167]. Depending on experimental conditions it is possible to obtain four different dinuclear Cu(II) complexes.

Today colloidal metal nanoparticles are starting to be mostly prepared by using hydroxylamine hydrochloride as reducing agent[36]. These nanoparticles display a high uniformity in size and shape. In addition, this kind of colloid could have advantageous electrochemical conditions on the surface for the detection of non-polar molecules, since the number of negative charges, mainly due to the residual chloride ions, should be much lower than in the case of citrate and borohydride colloids. These additional components interfere the SERS measurements[168]. The spectral enhancement, if it is verified, depends on both the surface and the analyte physical and electronic characteristics; moreover, an appropriate surface is specific for each analyte. This means that new surfaces should be built for each new adsorbate. Several studies point to employ assembly molecules to improve the adsorbate–substrate interaction. Recently, functionalized calixarene molecules have been employed to detect pollutants as polycyclic aromatic hydrocarbons PAHs[169]. The enhanced vibrational signals of PAHs were not obtained by using traditional surfaces due to their nearly absent polarity. The size of the internal cavity of calixarenes plays an important role when trying with different sizes of the adsorbates. Also, aliphatic acids of different chain length have been used as assembly molecules; the spectral enhancement is highly dependent on their chain lengths. Other self-assembled films as thiols, disulfides, silanes and nitrogen containing compounds on gold, silver and platinum substrates have been received wide attention.[170,171]. The potential use of these molecular films for the study of electron-transfer kinetics, wetting, corrosion, sensing, and other applications have been demonstrated[172,173].

Acknowledgements

This chapter has been written during the execution of the project 1040640 from Fondecyt, Conicyt Chile.

References

1. Debe, M.K. (1987). Optical probes of organic thin films: Photons-in and photons-out. *Prog. Surf. Sci.* **24**, 1.
2. Zeng, Z.Q., R. Aroca, A.M. Hor, and R.O. Loutfy (1989). Vibrational characterization of aluminum-oxo phthalocyanine films. IR, Raman and surface-enhanced Raman spectroscopic studies. *J. Raman Spectrosc.* **20**, 467.

3. Dann, A.J., H. Hoshi, and Y. Mayurama (1990). The structure and properties of phthalocyanine films grown by the molecular beam epitaxy technique. I. Preparation and characterization. *J. Appl. Phys.* **67**, 1371.
4. Kitagawa, T. and Y. Ozaki (1987). Infrared and Raman spectra of metalloporphyrins. *Struct. Bonding (Berlin)* **64**, 71.
5. Nakamoto, K. (1997). *Infrared and Raman Spectra of Inorganic and Coordination Compounds*, Part B, 5th Ed, Wiley, New York.
6. Fabian, J., H. Nakazumi, and M. Matzuoka (1992). Near-infrared absorbing dyes. *Chem. Rev.* **92**, 1197.
7. Minh, L.Q., T. Chot, N.N. Dinh, N.N. Xuan, N.T. Binh, and D.M. Phuoc (1987). A new organic photoconductor in the near infrared region of diode lasers. *Phys. Status Solidi (a)* **101**, K143.
8. Nalwa, H.S., T. Saito, A. Kakuta, and T. Iwayanagi (1993). Third-order nonlinear optical properties of polymorphs of oxotitanium phthalocyanine. *J. Phys. Chem.* **97**, 10515.
9. Chau, L.K., C. Arbour, G.E. Collins, K.W. Nebesny, P.A. Lee, C.D. England, N.R. Armstrong, and B.A. Parkinson (1993). Phthalocyanine aggregates on metal dichalcogenide surfaces: Dye sensitization on tin disulfide semiconductor electrodes by ordered and disordered chloroindium phthalocyanine thin films. *J. Phys. Chem.* **97**, 2690.
10. Rodriguez, M.L., R. Aroca, and J.A. DeSaja (1992). Electrochromic properties of Langmuir–Blodgett films of bisphthalocyanine complexes of rare earth elements. *Chem. Matter* **4**, 1017.
11. Berner, S., M. Brunner, L. Ramoino, H. Suzuki, and H.J. Güntherodt (2002). Time evolution analysis of a 2D solid–gas equilibrium: A model system for molecular adsorption and diffusion. *Chem. Phys. Lett.* **348**, 175.
12. Leznoff, C.C. and A.B.P. Lever (eds), *Phthalocyanines: Properties and Applications*, VCH, New York, Vol. 1 (1989), Vol. 2 (1993), Vol. 3 (1993), Vol. 4 (1996).
13. McKewn, N.B. (1998). *Phthalocyanine Materials*. Cambridge University Press, UK.
14. Tse, Y.H., P. Janda, H. Lam, J. Zhang, W.J. Pietro, and A.B.P. Lever (1997). Monomeric and polymeric tetra-aminophthalocyanatocobalt(II) modified electrodes: Electrocatalytic reduction of oxygen. *J. Porphyrins Phthalocyanines* **1**, 3.
15. Lindoy, L.F. (1989). *The Chemistry of Macrocycle Ligand Complexes*. Cambridge University Press, Cambridge.
16. Broan, C.J., J.P.L. Cox, A.S. Craig, R. Kataky, D. Parker, A. Harrison, A.M. Randall, and G. Ferguson (1991). Structure and solution stability of indium and gallium complexes of 1,4,7-triazacyclononanetriacetate and of Yttrium complexes of 1,4,7,10-tetraazacyclododecanetatraacetate and related ligands: kinetically stable complexes for use in imaging and radioinmunotheraphy. X ray molecular structure of the indium and gallium complexes of 1,4,7-triazacyclononane-1,4,7-triacetc acid. *J. Chem. Soc. Perkin Trans.* **2**, 87.
17. Melson, G.A. (ed), (1979). *Coordination Chemistry of Macrocycle Compounds*. Plenum Press, New York, NY.
18. Izatt, R.M. and J.J. Christensen (1978). *Synthetic Multidentate Macrocyclic Compounds*. Academic Press, New York, NY.
19. Fleischmann, M., P.J. Hendra, and A.J. McQuillan (1974). Raman spectra of pyridine adsorbed at a silver electrode. *Chem. Phys. Lett.* **26**, 163.
20. Harstein, A., J.R. Kirtley, and J.C. Tsang (1980). Enhancement of the infrared absorption from molecular monolayers with thin metal overlayers. *Phys. Rev. Lett.* **45**, 201.

21. Osawa, M. and M. Ikeda (1991). Surface-enhanced infrared absorption of *p*-nitrobenzoic acid deposited on silver island films. Contributions of electromagnetic and chemical mechanisms. *J. Phys. Chem.* **95**, 9914.
22. Burstein, E., Y.J. Chen, C.Y. Chen, S. Lundquist, and E. Tosatti (1979). Giant Raman scattering by adsorbed molecules on metal surfaces. *Solid State Commun.* **29**, 567.
23. Campos-Vallette, M.M., M. Saavedra, G. Díaz, R.E. Clavijo, Y. Martínez, F. Mendizabal, J. Costamagna, J.C. Canales, J.V. García-Ramos, and S. Sánchez-Cortés (2001). Surface-enhanced vibrational study of azabipiridyl and its Co(II), Ni(II) and Cu(II) complexes. *Vib. Spectrosc.* **27**, 15.
24. Johnson, E. and R. Aroca (1995) Surface enhanced infrared spectroscopy of monolayers. *J. Phys. Chem.* **99**, 9325.
25. Van Duyne, R.P., J.C. Hulteen, and D.A. Treichel (1993). Atomic force microscopy and surface-enhanced Raman spectroscopy. I. Ag island films and Ag film over polymer nanosphere surfaces supported on glass. *J. Chem. Phys.* **99**, 2101.
26. Moskovits, M. (1985). Surface-enhanced spectroscopy. *Rev. Mod. Phys.* **57**, 783.
27. Guzonas, D.A., G.F. Atkinsons, and D.E. Irish (1984). Surface-enhanced Raman scattering of the non-aqueous system silver/acetonitrile, KSCN. *Chem. Phys. Lett.*, **107**, 193.
28. Chang, R.K. (1987). Surface-enhanced Raman scattering at electrodes: A status report. *Ber. Bunsen Phys. Chem.* **91**, 296.
29. Aroca, R.F., D. Ross, and C. Domingo (2004). Surface enhanced infrared spectroscopy. *Appl. Spectrosc.* **58**, 324A.
30. Osawa, M. (1997). Dynamic processes in electrochemical reactions studied by surface-enhanced infrared absorption spectroscopy (SEIRAS). *Bull. Chem. Soc. Jpn* **70**, 2861.
31. Carrasco Flores, E., M.M. Campos-Vallette, P. Leyton, G. Diaz Fleming, R.E. Clavijo, J.V. García-Ramos, N. Inostroza, C. Domingo, S. Sánchez-Cortés, and R. Koch (2003). Study of the interaction of pollutant nitro polycyclic aromatic hydrocarbons with different metallic surfaces by surface-enhanced vibrational spectroscopy (SERS and SEIR). *J. Phys. Chem. A.* **107**, 9611.
32. Perry, S.S. and G.A. Somorjai (1994). Characterization of organic surfaces. *Anal. Chem.* **66**, 403.
33. Jennings, C.A., J. Kovac, and R. Aroca (1992). Near-infrared surface-enhanced Raman scattering from metal island films. *J. Phys. Chem.* **96**, 1340.
34. Jennings, C.A., J. Kovac, and R. Aroca (1993). FT-SERS of Langmuir–Blodgett monolayers on copper and gold island substrates. *Langmuir* **9**, 2151.
35. Lee, P.C. and D. Meisel (1982). Adsorption and surface-enhanced Raman of dyes on silver and gold sols. *J. Phys. Chem.* **86**, 3391.
36. Leopold, N. and B. Lendl (2003). A new method for fast preparation of highly surface-enhanced Raman scattering (SERS) active silver colloids at room temperature by reduction of silver nitrate with hydroxylamine hydrochloride. *J. Phys. Chem. B* **107**, 5723.
37. Sánchez-Cortés, S., C. Domingo, J.V. García-Ramos, and J.A. Aznarez (2001). Surface-enhanced vibrational study (SEIR and SERS) of dithiocarbamate pesticides on gold films. *Langmuir* **17**, 1157.
38. Olave, C.R., E.A.F. Carrasco, M.M. Campos-Vallette, M.S. Saavedra, G.F. Diaz, R.E. Clavijo, W. Figueroa, J.V. García-Ramos, S. Sánchez-Cortés, C. Domingo, J. Costamagna, and A. Ríos (2002). Vibrational study of the interaction of dinaphthalenic Ni(II) and Cu(II) azamacrocycle complexes methyl and phenyl substituted with different metal surfaces. *Vib. Spectrosc.* **28**, 287.

39. Z. Zhang, N. Yoshida, T. Imae, Q. Xue, M. Bai, J. Jiang, and Z. Lui (2001). A self-assembled monolayer of an alkanoic acid-derivatized porphyrin on gold surface: A structural investigation by surface plasmon resonance, ultraviolet–visible, and infrared spectroscopies. *J. Colloid Interf. Sci.* **243**, 382.
40. Zhang, Z., T. Imae, H. Sato, A. Watanabe, and Y. Ozaki (2001). Surface-enhanced Raman scattering and surface-enhanced infrared absorption spectroscopic studies of a metalloporphyrin monolayer film formed on pyridine self-assembled monolayer-modified gold. *Langmuir* **17**, 4564.
41. Diaz, G., R.E. Clavijo, M.M. Campos-Vallette, M. Saavedra, S. Diez, L. López, and R. Muñoz (1997). Specular reflectance infrared spectra of the macrocycles cyclam and cyclamdione and their Cu(II) complexes deposited onto a smooth copper surface. *Vib. Spectrosc.* **15**, 201.
42. Diaz Fleming, G., M.M. Campos Vallette, E. Clavijo, S. Diez, and M. Saavedra (1999). Vibrational study of copper(II) complexes with triaza and trithia macrocycles: [9]aneN3, [9]aneS3 and [6]aneS3. *Spectrochim. Acta A* **55**, 1827.
43. Kowal, A.T. and J. Skarzewski (1985). Far infrared study of iron(II) and copper(II) coordination compounds with 4,4′-didodecyloxy-2,2′-bipyridine and 4,4′-dioctadecyloxy-2,2′-bipyridine. *Spectrochim. Acta A* **41**, 563.
44. Aroca, R., Z.Q. Zeng, and J. Mink (1990). Vibrational assignment of totally symmetric modes in phthalocyanine molecules. *J. Phys. Chem. Solids*, **51**, 135.
45. Stichternath, A., R. Schweitzer-Stenner, W. Dreybrodt, R.S.W. Mak, X-Y. Li, L.D. Sparks, J.A. Shelnutt, C.J. Medforth, and K.M. Smith (1993). Macrocycle and substituent vibrational modes of nonplanar nickel(II) octaethyltetraphenylporphyrin from its resonance Raman, near-infrared-excited FT Raman, and FT-IR spectra and deuterium isotope shifts. *J. Phys. Chem.* **97**, 3701.
46. Aroca, R., A.H. Bolourchi, D. Battisti, and K.Najafi (1993). Gas adsorption and electrical properties of two Langmuir–Blodgett layers of cerium bisphthalocyanine. *Langmuir* **9**, 3138.
47. Abe, M., T. Kitagawa, and Y. Kyogoku (1978). Resonance Raman spectra of octaethylporphyrinato-Ni(II) and meso-deuterated and 15N substituted derivatives. II. A normal coordinate analysis. *J. Chem. Phys.* **69**, 4526.
48. Dodziuk, H., R.A. Kolinski, and B. Korybut-Daszkiewicz (1973). Infrared spectra and structure of 1,4,8,11-tetra-azacycotetradecanes. *Spectrochim. Acta* A **29**, 511.
49. Aroca, R., R.E. Clavijo, C.A. Jennings, G.J. Kovacs, J.M. Duff, and R.O. Loutfy (1989). Vibrational spectra of lutetium and ytterbium bis-phthalocyanine in thin solid films and SER(R)S on silver island films. *Spectrochim. Acta* A **45**, 957.
50. Jennings, C.A., R. Aroca, A.M. Hor, and R.O. Loutfy (1985). Raman spectra of solid films—IV. Pb and Sn phthalocyanine complexes. *Spectrochim. Acta* A **41**, 1095.
51. Aroca, R. D.P. Dilella, and R.O. Loutfy (1982). Raman spectra of solid films—I Metal-free phthalocyanine. *J. Phys. Chem. Solids* **43**, 707.
52. Kobayashi, T. (1970). The far infrared spectra of phthalocyanine and its metal derivatives. *Spectrochim. Acta* A **26**, 1313.
53. Gladkov, L.L. and K.N. Solovyov (1985). The normal coordinate analysis of porphin and its derivatives based on the solution of the inverse spectral problem for porphin and Cu porphin—II. A valence force field for in-plane vibrations of the Cu porphin molecule. *Spectrochim. Acta* A **41**, 1443.
54. Susi, H. and J.S. Ard (1977). A valence force field for nickel porphin and copper porphin. *Spectrochim. Acta* A **33**, 561.
55. Campos, A., J.R. Anacona, and M.M. Campos-Vallette (1999). Synthesis and IR study of a Zinc(II) complex containing a tetradentate macrocyclic Schiff base ligand: Antifungal properties. *Main Group Met. Chem.* **22**, 283.

56. Inskeep, R.G. (1962). Infrared spectra of metal complex ions below 600 cm-1. I. Spectra of the tris complexes of 1,10-phenantroline and 2,2′ bipyridine with the transition metals iron(II) through zinc(II). *J. Inorg. Nucl. Chem.* **24**, 763.
57. Baginski, M., P. Bruni, E. Maurelli, G. Natile, and G. Tosi (1994). Fourier transform infrared spectroscopy. Part V. Vibrational studies on phenantroline-perchloric acid complexes. *Vib. Spectrosc.* **7**, 231.
58. Bennett, A., G.A. Foulds, D.A. Thornton, and G.M. Watkins (1990). The infrared spectra of ethylenediamine complexes—II. Tris-, bis- and mono(ethylenediamine) complexes of metal(II) halides. *Spectrochim. Acta* A **46**, 13.
59. Diaz, G., S. Diez, L. Lopez, R. Muñoz, M.M. Campos-Vallette, V. Manríquez, and O. Wittke (1993). Spectra and structure of polyamine–copper(II) complexes. Infrared spectrum and normal coordinate calculations of mono(diethylenetriamine) copper(II) nitrate. *Vib. Spectrosc.* **6**, 37, and references cited therein.
60. Campos-Vallette, M.M., G. Diaz, J. Vargas, J. Costamagna M., and V.H. Poblete (1996). Vibrational study and structure of *N*-(substituted 5-bromo hydroxyphenylaldiminate) Co(II) complexes. *Spectrochim. Acta A* **52**, 13.
61. Faniran, J.A., K.S. Patel, and J.C. Bailar (1974). Infrared spectra of N, N'-bis (salicylidene)-1,1-(dimethyl)ethylene–diamine and its metal complexes. *J. Inorg. Nucl. Chem.* **36**, 1547.
62. Fabretti, A.C. and G. Peyronel (1978). Rhodanine complexes of zinc(II), cadmium(II), mercury(II) and mercury(I). *Spectrochim. Acta* A **34**, 667.
63. Michalska, D. and A.T. Kowal (1985). Vibrational spectra of Cu(II), Cu(I), Ni(II), Pd(II), Pt(II) and Hg(II) complexes with dithizone. *Spectrochim. Acta* A **41**, 1119.
64. Lin-Vien, D., N.B. Colthup, W.G. Fately, and J.G. Grasselli (1991). The Handbook of Infrared and Raman Characteristic Frequencies of Organic Molecules, 1st edn, Academic Press, Boston.
65. Nakamoto, K. (1963). Infrared Spectra of Inorganic and Coordination Compounds, 1st edn, Wiley-Interscience, London.
66. Campos-Vallette, M.M., R.E. Clavijo, F.E. Mendizabal, G. Díaz Fleming, J. Costamagna, J. Canales, and J. Vargas (1997). Vibrational study and semiempirical calculation for the hexaazacyclophane copper(II) complex. *Vib. Spectrosc.* **14**, 71.
67. Gobernado-Mitre, I., R. Aroca, and J.A. DeSaja (1994). Far infrared spectra of mono- and bisphthalocyanine derivatives. *Spectrochim. Acta* A **50**, 1243.
68. Campos-Vallette, M.M., R.E. Clavijo, F. Mendizabal, W. Zamudio, R. Baraona, and G.F. Diaz (1996). Infrared spectrum of the bis-(1,10-phenanthroline) Cu(I) and Cu(II) perchlorate complexes. *Vib. Spectrosc.* **12**, 37.
69. Battistuzzi, R. and G. Peyronel (1980). Copper (I) and copper (II) complexes of 4,6-dimethylpyrimidine-2(1H)-one. *Spectrochim. Acta A* **36**, 511.
70. Clavijo, R.E., M.M. Campos-Vallette, M.S. Saavedra, A. Alvarado, and G. Díaz Fleming (1997). Vibrational interpretation of the structure of Co(II), VO^{2+} and Pb(II) naphthalocyanine complexes. *Vib. Spectrosc.* **14**, 79.
71. Campos-Vallette, M.M., K. Figueroa, R.O. Latorre, G. Diaz Fleming, J. Costamagna, J.C. Canales, M. Rey-Lafon, and J. Derouault (1993). Vibrational study of *N*-phenyl-substituted hydroxynaphthylaldiminate copper(II) complexes. *Vib. Spectrosc.* **6**, 25.
72. Sievertsen, S., H. Schlehahn, and H. Homborg (1993). Darstellung, Eigenschaften und elektronische Raman-Spektren von Bis(chloro)phthalocyaninatoferrat(III), -ruthenat(III) und -osmat(III). *Z. Anorg. Allg. Chem.* **619**, 1064.
73. Hutchinson, B., B. Spencer, R. Thompson, and P. Neill (1987). Metal—nitrogen stretching assignments in some metallophthalocyanines and -oxo-$(Fephthalocyanine)_2$. *Spectrochim. Acta A* **43**, 631.

74. Palys, B.J., D.M.W. van den Ham, W. Briels, and D. Feil (1995). Resonance Raman spectra of phthalocyanine monolayers on different supports. A normal mode analysis of zinc phthalocyanine by means of the MNDO method. *J. Raman Spectrosc.* **26**, 63.
75. Kaplan, M.L., A.J. Lovinger, W.D. Reents, Jr., and P.H. Schmidt (1984). The preparation, spectral properties, and X-ray structural features of 2,3-naphthalocyanines. *Mol. Cryst. Liq. Cryst.* **112**, 345.
76. Gobernado-Mitre, I., R. Aroca, and J.A. DeSaja (1995). Vibrational spectra and molecular organization in thin solid films of 2,3-naphthalocyanines. *Chem. Mater.* **7**, 118.
77. Kasuga, K., M. Tsutsui, R.C. Petterson, K. Tatsumi, N.van Opdenbosch, G. Pepe, and E.F. Meyer, Jr. (1980). Structure of bis(phthalocyaninate) neodymium(III). *J. Am. Chem. Soc.* **102**, 4835.
78. Carrasco, E.A., M.M. Campos-Vallette, M. Saavedra, G. Díaz, R.E. Clavijo, J.V. García-Ramos, and S. Sánchez-Cortés (2001). Vanadyl naphthalocyanine and vanadyl porphine phenyl substituted macrocycles: SERS and thin film organisation studies. *Vib. Spectrosc.* **26**, 201.
79. Campos Vallette, M.M., R.E. Clavijo, J. Costamagna, J. Canales, G. Diaz, F. Mendizabal, J.M. Ramirez, and M. Saavedra (2000). Vibrational and theoretical study of azabipiridyl macrocycle and its Ni(II), Cu(II) and Zn(II) complexes deposited onto a smooth copper surface. *Vib. Spectrosc.* **23**, 39.
80. Hirakawa, A.Y. and M. Tsuboi (1975). Molecular geometry in an excited electronic state and a preresonance Raman effect. *Science* **188**, 359.
81. Diaz, G., M.M. Campos-Vallette, M. Saavedra, R.E. Clavijo, J.C. Canales, J. Costamagna, and J. Vargas (2002). Surface vibrational study of macrocycle complexes: Co(II), Ni(II), Cu(II) and Zn(II) bis(phenylhydrazine)-1,10-phenanthroline. *Vib. Spectrosc.* **28**, 223.
82. Campos-Vallette, M.M., K.A. Figueroa, R. Latorre, V. Manriquez, G. Diaz, J. Costamagna, and M. Otero (1992). *N*-(substituted phenylhydroxynaphthylaldiminate) nickel(II) complexes. *Vib. Spectrosc.* **4**, 77.
83. Dollish, F.R., W.G. Fateley, and F.F. Bentley (1974). Characteristic Raman Frequencies of Organic Compounds. Wiley, New York p. 163.
84. Chaudhuri, P. and K. Wieghardt (1987). The chemistry of 1,4,7-triazacyclononane and related tridentate macrocyclic compounds. *Prog. Inorg. Chem.* **35**, 329.
85. Hon, P.K., and C.E. Pfluger (1973). Crystal and molecular structure of tris (acetylacetonato) aluminum (III) and-cobalt (III). *J. Coord. Chem.* **3**, 67.
86. Hay, R.W. and M.T.H. Tarafder (1991). Chromium(III) Complexes of 1,4,8,12-Tetraazacyclopentadecane. *J. Chem. Soc. Dalton Trans.* 823.
87. Fuhrer, H., V.B. Kartha, P.J. Krueger, H.H. Mantsch, and R.N. Jones (1972). Normal modes and group frequencies. Conflict or compromise? In-depth vibrational analysis of cyclohexanone. *Chem. Rev.* **72**, 439.
88. Kieft, J.A. K. Nakamoto (1968). The normal coordinate analysis and infrared spectra of bis(glycinamido)copper(II) monohydrate. *Inorg. Chim. Acta* **2**, 225.
89. Wilson, E.B., J.C. Decius, P. Cross (1955). Molecular Vibrations, McGraw-Hill, New York.
90. Estiú, G.L., A.H. Jubert, J. Molina, J. Costamagna, J. Canales, and J. Vargas (1995). A theoretical investigation of the structure and electronic spectra of porphyrin homolog macrocycles. Hexaazacyclophane and its nickel and copper complexes. *Inorg. Chem.* **34**, 1212.
91. Anderson, O.P. (1973). Crystal and molecular structure of tris-(1,10-phenantroline) copper(II) perchlorate. *J. Chem. Soc. Dalton Trans.* 1237.

92. Chen, L.Q., L.K. Thompson, J.N. Bridson, J. Xu, S. Ni, and R. Guo (1993). Synthesis and characterization of 5,12-dimethyl-7,14-diphenyl-1,4,8,11-tetraazacyclotetradecane. *Can. J. Chem.* **71**, 1805.
93. Melendres, C.A. and V.A. Maroni (1984). Raman spectra and normal coordinate analysis of the planar vibrations of ion phthalocyanine. *J. Raman Spectrosc.* **15**, 319.
94. Draeger, J.A. (1985). The methylbenzenes—I. Vapor-phase vibrational fundamentals, internal rotations and a modified valence force field. *Spectrochim. Acta* A **41**, 607.
95. Orza, J.M., R. Escribano, and R. Navarro (1985). Out of plane vibracional assignments and potential function of pyrrole and deuterated derivatives. *J. Chem. Soc. Faraday Trans.* **81**, 653.
96. Dowty, E. Personal communication.
97. Piffat, C., D. Melamed, and T.G. Spiro (1993). Ruffling effect on porphyrin vibrations: Normal mode analysis for nickel octaethyltetraphenylporphine from resonance Raman and Infrared spectra of isotopomers. *J. Phys. Chem.* **97**, 7441.
98. Li, X., R.S. Czernuzewicz, J.R. Kincaid, P. Stein, and T.G. Spiro (1990). Consistent p orphyrin force field. 2. Nickel octaethylporphyrin skeletal and substituents mode assignment from ^{15}N, meso-d_4, and methylene-d_{16} Raman and infrared isotope shifts. *J. Phys. Chem.* **94**, 47.
99. Mallick, P.K., G.D. Danzer, D.P. Strommen, and J.R. Kincaid (1988). Vibrational spectra and normal coordinate analysis of tris(bipyridine) ruthenium(II). *J. Phys. Chem.* **92**, 5628.
100. Szafran, M. and J. Koput (1996). Calculation of the vibrational spectra of pyridine betaine. *J. Mol. Struct.* **381**, 157.
101. M. Saavedra, F. Mendizabal, M.M. Campos-Vallette, R.E. Clavijo, and G. Diaz (1998). Infrared and theoretical interpretation of the structure of naphthalocyanine and its copper(II) complex deposited onto a smooth copper surface. *Vib. Spectrosc.* **18**, 25.
102. Diaz, G., S. Diez, L. Lopez, R. Muñoz, H. Pessoa, and M.M. Campos-Vallette (1992). Vibrational study of polyamine copper(II) complexes. Infrared spectra and normal coordinate analysis of mono(diethylenetriamine)copper(II) complex ions. *Vib. Spectrosc.* **3**, 315.
103. Diaz, G. R.E. Shepherd (1987). Infrared spectra and normal coordinate analysis of bis(ethylenediamine) and bis(trimethylenediamine) complexes with copper(II). *Spectrochim. Acta* A **43**, 1141.
104. Scherer, J.R. and J. Overend (1961). Application of the Urey–Bradleyforce field to the in-plane vibrations of benzene. *Spectrochim. Acta* **17**, 719.
105. Neto, N., M. Scrocco, and S. Califano (1966). A simplified valence force field of aromatic hydrocarbons—I Normal co-ordinate calculations for C_6D_6, C_6D_6, $C_{10}H_8$, $C_{10}D_8$, $C_{14}H_{10}$ and $C_{14}D_{10}$. *Spectrochim. Acta* A **22**, 1981.
106. Diaz, G.F., S. Diez, M.M. Campos-Vallette, and J.I. Zink (1979). Vibrational analysis of metal diimine dithiolate mixed ligand complexes. *Spectrochim. Acta* A **55**, 979.
107. Califano, S. (1976). Vibrational States, Wiley, New York.
108. Sarma, Y.A. (1987). In plane vibrations of copper porphins. *Spectrochim. Acta* A **43**, 1337.
109. Moskovits, M., and J.S. Suh (1984). Surface selection rules for SERS: Calculation and application to the SERS of phthalocyanine on Ag. *J. Phys. Chem.* **88**, 5526.
110. Menendez, J.R., A. Obuchowska, and R. Aroca (1996). Infrared spectra and surface enhanced Raman scattering of naphthalimide on colloidal silver. *Spectrochim. Acta* A **52**, 329.

111. Aroca, R., C. Jennings, R.O. Loutfy, and A.M. Hor (1987). Raman spectra of thin solid films—VI. Davydov splitting. *Spectrochim. Acta* A **43**, 725.
112. Aroca, R., and R.E. Clavijo (1988). Molecular orientation of phthalimide adsorbed on colloidal silver. *Spectrochim Acta* A **44**, 171.
113. Corio, P., J.C. Rubim, and R. Aroca (1998). Contribution of the Herzberg–Teller mechanism to the surface-enhanced Raman scattering of iron phthalocyanine adsorbed on a silver electrode. *Langmuir* **14**, 4162.
114. Yanagi, H., T. Kouzeki, M. Ashida, T. Noguchi, A. Manivannan, K. Hashimoto, and A. Fujishima (1992). Molecular orientation of vacuum-deposited thin films of zinc-naphthalocyanine. J. *Appl. Phys.* **71**, 5146.
115. Yanagi, H., T. Kouzeki, and M. Ashida (1993). Epitaxial growth of naphthalocyanine thin films vacuum deposited on alkali halides. J. *Appl. Phys.* **73**, 3812.
116. Barraza, R.B., M.M. Campos-Vallette, K. Figueroa, V. Manriquez, and V. Vargas (1990). Hydrogen bond effect on the molecular vibrations of indole. *Spectrochim. Acta* A **46**, 1375.
117. Vlckova, B., P. Smejkal, M. Michl, M. Prochazka, P. Mojzes, F. Lednicky, and J. Pfleger (2000). SERRS spectroscopy of porphyrin and metalloporphyrin species in systems with Ag nanoparticles and their assemblies. *J. Inorg. Biochem.* **79**, 295.
118. Sloufova-Srnova, I., and B. Veckova (2002). Two-dimensional assembling of Au nanoparticles mediated by tetrapyridylporphine molecules. *Nano Letters* **2**, 121.
119. Qu, J., and P.M. Fredericks (2000). Investigation of the SERS behaviour of porphyrins with different surface concentrations on electrochemically prepared Ag-surface. *Spectrochim. Acta A* **56**, 1637.
120. Praus, P., M. Prochazka, J. Stepanek, and J. Bok (2001). Influence of Ag colloid aggregation on porphyrin SERRS spectra monitored via Raman correlation spectroscopy. *J. Mol. Struct.* **565–566**, 129.
121. Faulques, E., M. Baibarac, L. Galmiche, M. L'Her, A. Pondaven, and I. Baltog (2000). SERS spectra of mono and bisphthalocyanine complexes deposited on Ag and Au supports. *Spectrosc. Lett.* **33**, 625.
122. Battisti, D., L. Tomilova, and R. Aroca (1992). Spectroscopic characterization of rare earth octa-*tert*-butylbisphthalocyanine complexes. *Chem. Mater.* **4**, 1323.
123. Souto, J., L. Tomilova, R. Aroca and J.A. DeSaja (1992). Spectroscopic studies of Langmuir–Blodgett monolayers of praseodymium bis-phthalocyanines. *Langmuir* **8**, 942.
124. Sheng-Gao, L., L. Yun-Qi, Yu Xu, Z. Dao-Ben, Y. An-Chi, and Z. Xing-Sheng (1998). Synthesis, Langmuir–Blodgett film, and second-order nonlinear optical property of a novel asymmetrically substituted metal-free phthalocyanine. *Langmuir* **14**, 690.
125. Somashekarappa, M.P., J. Keshavayya, and S. Sampth (2002). Self-assembled molecular films of tetraamino metal (Co, Cu, Fe) phthalocyanines on gold and silver. Electrochemical and spectroscopic characterization. *Pure Appl. Chem.* **74**, 1609.
126. Cook, M.J., R. Hersans, J. McMurdo, and D.A. Russell (1996). Self-assembled monolayers of phthalocyanine derivatives on glass and silicon. *J. Mater. Chem.* **6**, 149.
127. Revell, D.J., I. Chambrier, M.J. Cook, and D.A. Russell (2000). Formation and spectroscopic characterization of self-assembled phthalocyanine monolayers. *J. Mater. Chem.* **10**, 31.
128. Huc, V., J.P. Bourgoin, C. Bureau, F. Valin, G. Zalczer, and S. Palacin (1999). Self-assembled mono- and multilayers on gold from 1,4-diisocyanobenzene and ruthenium phthalocyanine. *J. Phys. Chem. B* **103**, 10489.

129. Li, Z., M. Lieberman, and W. Hill (2001). XPS and SERS study of silicon phthalocyanine monolayers: umbrella vs octopus design strategies for formation of oriented SAMs. *Langmuir* **17**, 4887.
130. Gaffo, L., C.J.L. Constantino, W.C. Moreira, R.F. Aroca, and O.N. Oliveira (2004). Surface-enhanced Raman scattering and micro-Raman imaging of Langmuir–Blodgett films of rhodium phthalocyanine. *Spectrochim. Acta A* **60**, 321.
131. Del Cano, T., R. Aroca, J.A. DeSaja, and M.L. Rodriguez-Mendez (2003). Langmuir–Blodgett mixed films of titanyl(IV) pthalocyanine and arachidic acid. Molecular orientation and film structure. *Langmuir* **19**, 3747.
132. Gaffo, L., C.J.L. Constantino, W.C. Moreira, R.F. Aroca, and O.N. oliveira (2002). Atomic force microscopy and micro-Raman imaging of mixed Langmuir–Blodgett films of ytterbium bisphthalocyanine and stearic acid. *Langmuir* **18**, 3561.
133. Gaffo, L., C.J.L. Constantino, W.C. Moreira, R.F. Aroca, and O.N. Oliveira (2002). Vibrational spectra and surface-enhanced resonance Raman scattering of palladium phthalocyanine evaporated films. *J. Raman Spectrosc.* **33**, 833.
134. Mukherjee, K.M., and T.N. Misra (2000). Solvent effect on surface-enhanced Raman activity of copperphthalocyanine on colloidal silver. *Colloids Surf. B* **17**, 139.
135. Aroca, R., N. Pieczonka, and A.P. Kam (2001). SERS ans SERRS imaging of Pc mixed films. *J. Porphyrins Phthalocyanines* **5**, 25.
136. Woolley, P., B.J. Keely, and R.E. Hester (1996). Surface-enhanced resonance Raman spectra of water-insoluble tetraphenylporphyrin and chlorophyll a on silver hydrosols with a dioxane molecular spacer. *Chem. Phys. Lett.* **258**, 501.
137. Prochazka, M., P. Mojzes, B. Vlckova, and P.Y. Turpin (1997). Surface-enhanced resonance Raman scattering from copper(II) 5,10,15,20-tetrakis(1-methyl-4-pyridyl)porphyrin adsorbed on aggregated and nonaggregated silver colloids. *J. Phys. Chem. B* **101**, 3161.
138. Mou, C.H., D. Chen, X. Wang, B. Zhang, T. He, H. Xin, and F.C. Liu (1991). SERS and UV spectra of meso-tetrakis (4-sulfonatophenyl)-porphine adsorbed on Ag_2O colloids. *Chem. Phys. Lett.* **179**, 237.
139. Wang, X.Q., H. Wen, and T.J. He (1997). Enhancement mechanism of SERS from cyanine dyes adsorbed on Ag_2O colloids. *Spectrochim. Acta A* **53**, 2495.
140. Zhang, Y.H., D.M. Chen, T. He, and F.Ch. Liu (2001). Reaction of metallotetraphenylporphyrins on hydroxyl-modified silver colloid and Ag_2O colloid by surface-enhanced Raman scattering. *Spectrochim. Acta A* **57**, 2599 (2001); ibid 2001. Surface-enhanced Raman spectroscopy of metallotetraphenylporphyrins adsorbed on Ag_2O and hydroxyl-modified silver colloids. Dianhuaxue **7**, 59.
141. Zhang, Z.J., S.F. Hou, Z.H. Zhu, and Z.F. Liu (2000). Preparation and characterization of a porphyrin self-assembled monolayer with a controlled orientation on gold. *Langmuir* **16**, 537.
142. Liu, X., D.M. Chen, T.J. He, and F.C. Liu (2003). Effect of CTAB on the surface-enhanced Raman scattering of tetrakis (4-*N*-methylpyridyl) porphyrin. Chin. *J. Chem. Phys.* **16**, 176.
143. Chowdhury, J., M. Ghosh, P. Pal, and T.N. Misra (2003). Concentration-dependent surface-enhanced resonance Raman scattering of a porphyrin derivative adsorbed on colloidal silver. *J. Colloid Interf. Sci.* **263**, 318.
144. Prochazka, M., J. Stepanek, and J. Bok (2003). Ag^+ and Ag^o coordination of porphyrins adsorbed on laser ablated Ag colloids monitored by surface-enhanced resonance Raman scattering spectroscopy. *Main Group Met. Chem.* **26**, 317.
145. Prochazka, M., P.-Y. Turpin, J. Stepanek, and B. Vlckova (2002). SERRS of free base porphyrin in laser-ablated colloids: Evidence for three different spectral porphyrin forms. *J. Raman Spectrosc.* **33**, 758.

146. Rywkin, S., C.M. Hosten, J.R. Lombardi, and R.L. Birke (2002). Surface-enhanced resonance Raman scattering and voltammetry study of the electrocatalytic reduction of oxygen by the μ-oxo dimer of iron(III) tetra-4-N-methylpyridylporphyrin. *Langmuir* **18**, 5869.
147. Zaruba, K., P. Matejka, R. Volf, K. Volka, V. Kral, and J.L. Sessler (2002). Formation of porphyrin- and sapphyrin-containing monolayers on electrochemically prepared gold substrates: A FT Raman spectroscopic study. *Langmuir* **18**, 6896.
148. Bizzarri, A.R. and S. Cannistraro (2003). Temporal fluctuations in the SERRS spectra of single iron–protoporphyrin IX molecule. *Chem. Phys.* **290**, 297.
149. Geyer, M., F. Plenzig, J. Rauschnabel, M. Hanack, B. del Rey, A. Sastre, and T. Torres (1996). Subphthalocyanines: Preparation, reactivity and physical properties. *Synthesis* 1139.
150. Rojo, G., A. Hierro, M.A. Díaz-García, and F. Agulló-López (1997). Second harmonic generation from trinitro-substituted subphthalocyanines films: Evidence of noncentrosymmetric molecular organization. *Appl. Phys. Lett.* **70**, 1802.
151. Kietaibl, H. (1974). Die Kristall- und Molekülstruktur eines neuartigen phthalocyaninähnlichen Borkomplexes. *Monatsh. Chem.* **105**, 405.
152. Jung, T.A., R.R. Schlittler, and J.K. Gimzewski (1997). Conformational identification of individual adsorbed molecules with the STM. *Nature* **386**, 696.
153. Saavedra, M., M.M. Campos-Vallette, R.E. Clavijo, F. Mendizabal, G. Díaz, J.V. García-Ramos, and S. Sánchez-Cortés (2003). Vibrational study of sub-2,3-boronnaphthalocyanine chlorine adsorbed on metal surfaces. *Vib. Spectrosc.* **32**, 155.
154. Rivas, L., S. Sánchez-Cortés, J. Stanicova, J.V. García-Ramos, and P. Miskovsky (1999). FT-Raman, FTIR and surface-enhanced Raman spectroscopy of the antiviral and antiparkinsonian drug amantadine. *Vib. Spectrosc.* **20**, 179.
155. Rivas, L. (1999). Tesis Doctoral, Espectroscopía Raman sobre superficies metálicas (SERS) de fármacos antivirales: optimización de soportes coloidales metálicos de plata y oro. Universidad Autónoma de Madrid.
156. Parr, R.G., and W. Yang (1989). Density functional theory of atoms and molecules, Oxford University Press, New York, p. 225.
157. Chong, D.P. (1995). Recent advances in density functional methods, World Scientific, Singapore, vol. 1.
158. Frisch, M.J., G.W. Trucks, H.B. Schlegel, G.E. Scuseria, M.A. Robb, J.R. Cheeseman, V.G. Zakrzewski, J.A. Montgomery Jr., R.E. Stratmann, J.C. Burant, S. Dapprich, J.M. Millam, A.D. Daniels, K.N. Kudin, M.C. Strain, O. Farkas, J. Tomasi, V. Barone, M. Cossi, R. Cammi, B. Mennucci, C. Pomelli, C. Adamo, S. Clifford, J. Ochterski, G.A. Petersson, P.Y. Ayala, Q. Cui, K. Morokuma, D.K. Malick, A.D. Rabuck, K. Raghavachari, J.B. Foresman, J. Cioslowski, J.V. Ortiz, A.G. Baboul, B.B. Stefanov, G. Liu, A. Liashenko, P. Piskorz, I. Komaromi, R. Gomperts, R.L. Martin, D.J. Fox, T. Keith, M.A. Al-Laham, C.Y. Peng, A. Nanayakkara, M. Challacombe, P.M.W. Gill, B. Johnson, W. Chen, M.W. Wong, J.L. Andres, C. Gonzalez, M. Head-Gordon, E.S. Replogle, and J.A. Pople (1998). Gaussian 98, Revision A.9, Gaussian, Inc., Pittsburgh, PA.
159. Becke, A.D. (1988). Density-functional exchange-energy approximation with correct asymptotic behaviour. *Phys. Rev. A* **38**, 3098.
160. Scott, A.P., and L. Radom (1996). Harmonic vibrtional frequencies: An evaluation of Hartree-Fock, Møller-Plesset, quadratic configuration interaction, density functional theory, and semiempirical scale factors. *J. Phys. Chem.* **100**, 16502.
161. Wong, M.W. (1996). Vibrational frequency prediction using density functional theory. *Chem. Phys. Lett.* **256**, 391.
162. Ashcroft, N.W., and N.D. Mermin (1976). Solid State Physics, Holt, Rinehart and Winston, New York.

163. Lamoen, D., P. Ballone, and M. Parrinello (1996). Electronic structure, screening, and charging effects at a metal/organic tunneling junction: A first-principles study. *Phys. Rev. B* **54**, 5097.
164. Padilla, L., A. Toro-Labbe, and J. Maruani (1997). Theoretical investigation of the adsorption of alkali metals on a Cu(111) surface. *Surf. Sci.* **385**, 24.
165. Aroca, R., M. Scraba, and J. Mink (1991). SERS and normal coordinate analysis of maleimide chemisorbed on colloidal silver. *Spectrochim. Acta* A **47**, 263.
166. Paredes, V., S. Cortés, F. Rojas, D. Venegas-Yazigi, and E. Spodine (2004). Frontiers in materials research (A CIAM–CIMAT–CONICYT Workshop) Viña del Mar, Chile, Study of mononuclear and binuclear Cu(II) complexes derived from redox active macrocycles and their related ligands.
167. Spodine, E., Y. Moreno, M.T. Garland, O. Peña, and R. Baggio (2000). Molecular structure and magnetic properties of $[Gd(LH_4)(NO_3)_2(H_2O)NO_3(H_2O)_2$, $[Sm(LH_4)(NO_3)_2(H_2O)]NO_3(H_2O)1.5(CH_3OH)0.5$ and $[Cu_2(LH_2)(H_2O)_2]$ $(NO_3)_2$ complexes (LH4: Schiff base ligand derived from 4-methyl-2,6-diformylphenol and 1,3-diaminopropanol). *Inorg. Chim. Acta* **309**, 57.
168. Sánchez-Cortés, S. and J.V. García-Ramos (1998). Anomalous Raman bands appearing in surface-enhanced Raman spectra. *J. Raman Spectrosc.* **29**, 365.
169. Leyton, P., M. Campos-Vallette, R.E. Clavijo, C. Saitz, S. Sánchez Cortés, C. Domingo, and J.V. García Ramos (2004). Surface-enhanced Raman spectra of polycyclic aromatic hydrocarbons. International Bunsen Discussion Meeting. Raman and IR spectroscopy in Biology and Medicine, Jena Germany, February 2004.
170. Yan, L., W.T.S. Huck, X.M. Zhao, and G.M. Withesides (1999). Patterning thin films of poly(ethyleneimine) on a reactive SAM using microcontact printing. *Langmuir* **15**, 1208
171. Imahori, H., H. Norieda, Y. Nishimura, I. Yamazaki, K. Higuchi, N. Nato, T. Motohiro, H. Yamada, K. Tamaki, M. Arimura, and Y. Sakata (2000). Chain length effect on the structure and photoelectrochemical properties of self-assembled monolayers of porphyrins on gold electrodes. *J. Phys. Chem. B* **104**, 1253.
172. Cook, M.J. (1999). Phthalocyanine thin films. *Pure Appl. Chem.* **71**, 2145.
173. Semal, S., C. Bauthier, M. Voue, J.J.V. Eynde, R. Gouttebaron, and J.D. Coninc (2000). Spontaneous spreading of liquid droplets on mixed alkanethiol monolayers: Dynamics of wetting and wetting transition. *J. Phys. Chem. B* **104**, 6225.

Index

A

B

C

G

H

I

J

K

O

P

Q

R

S

T

U

V

W

X

Z